IFMBE Proceedings

Volume 55

The International Federation for Medical and Biological Engineering, IFMBE, is a federation of national and transnational organizations representing internationally the interests of medical and biological engineering and sciences. The IFMBE is a non-profit organization fostering the creation, dissemination and application of medical and biological engineering knowledge and the management of technology for improved health and quality of life. Its activities include participation in the formulation of public policy and the dissemination of information through publications and forums. Within the field of medical, clinical, and biological engineering, IFMBE's aims are to encourage research and the application of knowledge, and to disseminate information and promote collaboration. The objectives of the IFMBE are scientific, technological, literary, and educational.

The IFMBE is a WHO accredited NGO covering the full range of biomedical and clinical engineering, healthcare, healthcare technology and management. It is representing through its 60 member societies some 120.000 professionals involved in the various issues of improved health and health care delivery.

More information about this series at http://www.springer.com/series/7403

Victor Sontea · Ion Tiginyanu
Editors

3rd International Conference on Nanotechnologies and Biomedical Engineering

ICNBME-2015, September 23-26, 2015,
Chisinau, Republic of Moldova

Springer

Editors
Victor Sontea
Technical University of Moldova
Chisinau
Republic of Moldova

Ion Tiginyanu
Academy of Sciences of Moldova
Chisinau
Republic of Moldova

ISSN 1680-0737 ISSN 1433-9277 (electronic)
IFMBE Proceedings
ISBN 978-981-287-735-2 ISBN 978-981-287-736-9 (eBook)
DOI 10.1007/978-981-287-736-9

Library of Congress Control Number: 2015951970

Springer Singapore Heidelberg New York Dordrecht London

The IFMBE Proceedings is an Official Publication of the International Federation for Medical and Biological Engineering (IFMBE)

Printed on acid-free paper

Springer Science+Business Media Singapore Pte Ltd. is part of Springer Science+Business Media (www.springer.com)

Preface

This volume presents the Proceedings of the 3rd International Conference on Nanotechnologies and Biomedical Engineering (ICNBME) which was held on September 23–26, 2015 in Chisinau, Republic of Moldova. ICNBME-2015 continued the series of international conferences in the field of nanotechnologies and biomedical engineering with the main goal focused at bringing together scientists and engineers dealing with fundamental and applied research for reporting on the latest theoretical developments and applications in the fields involved.

The conference and related symposia covered a wide range of subjects of primary importance for research and development such as nanotechnologies and nanomaterials; bio-nanotechnologies and biomaterials; biosensors and biomedical instrumentation; biomedical signal and image processing; clinical engineering, health informatics and cellular and tissue engineering; biomedical engineering education; nuclear and radiation safety and security. Besides, the Proceedings include the papers presented at the special Symposium "Science and society: the use of light" organized in connection with the celebration of the International Year of Light and Light-based Technologies, following a global initiative adopted by the United Nations to raise awareness of how optical technologies promote sustainable development and provide solutions to worldwide challenges in areas such as energy, education, communications, health, and sustainability.

The papers included in the Proceedings reflect the results of multidisciplinary research undertaken by about one hundred of groups worldwide. Special attention is paid to the development of novel nanotechnologies and nanomaterials, in particular of bio-nanotechnologies and bio-nanomaterials. New bio-compatible materials are proposed for use in regenerative medicine, cellular and tissue engineering. Interesting data on novel chemical and bio-sensors are reported which are based on nanostructured metal oxides and hybrid nanocomposite materials. Considerable progress has been achieved at the intersection of nanotechnologies, information technologies and biomedicine as, for example, in health informatics, biomedical signal and image processing. New theoretical and experimental results are highlighted in such fields as superconductivity, novel magnetic materials, metamaterials, optoelectronic and photonic materials, photovoltaic structures, quantum dots, one- and two-dimensional nanomaterials, multifunctional hybrid materials like core-shell structures etc. The Proceedings reflects the state-of-the-art in controlling the properties of several classes of nanocomposite materials for important future applications in various fields. It is worth to note that the Proceedings include also a number of review papers reflecting the fascinating history and recent achievements in the development of novel solid-state structures, nanoelectronic and optoelectronic devices like quantum rings and room temperature polariton laser.

We hope that the papers included in the ICNBME-2015 Proceedings will be of interest for established researchers working in multidisciplinary fields of science and technology, young scientists, students and broad community wishing to get up-to-date information on progress in the fast-developing areas of nanotechnology and biomedical engineering.

Acad. Prof. Ion Tiginyanu
Prof. Victor Sontea
Editors

3rd International Conference on Nanotechnologies and Biomedical Engineering ICNBME-2015, September 23-26, 2015, Chisinau, Republic of Moldova

Organizers:

Technical University of Moldova
Academy of Sciences of Moldova
State Medical and Pharmaceutical University "Nicolae Testemiţanu"
Moldavian Society of Biomedical Engineering

Co-organized by:

Commission of the European Communities - Joint Project within the TEMPUS IV program "Biomedical Engineering Education Tempus Initiative in Eastern Neighbouring Area".

Sponsored by:

- International Federation for Medical and Biological Engineering
- Swiss Agency for Development and Cooperation
- Supreme Council for Science and Technological Development of the Academy of Sciences of Moldova
- State Agency on Intellectual Property of the Republic of Moldova

Committees

Conference Chairs

Ion Tiginyanu	Academy of Sciences of Moldova, Republic of Moldova
Victor Sontea	Technical University of Moldova, Republic of Moldova

International Advisory Committee

Rainer Adelung	University of Kiel, Germany
Victor Borisenko	Belarusian State University of Informatics and Radioelectronics, Belarus
Cepisca Costin	University Politehnica of Bucharest, Romania
Lee Chow	University of Central Florida, USA
Radu Ciupa	Technical University of Cluj, Romania
Hariton Costin	Grigore T. Popa University of Medicine and Pharmacy, Romania
Yury Dekhtyar	Riga Technical University, Latvia
Sergei Dik	Belarusian State University of Informatics and Radioelectronics, Belarus
Helmut Föll	Kiel University, Germany
Vladimir Fomin	Institute for Integrative Nanosciences, Germany
Sergey Gaponenko	National Academy of Sciences, Belarus
Nicolas Grandjean	Ecole Polytechnique Federale de Lausanne, Switzerland
Stanislav Groppa	Nicolae Testemiţanu State Medical and Pharmaceutical University, Republic of Moldova
Hans Hartnagel	Technical University Darmstadt, Germany
Vladimir Hotineanu	Nicolae Testemiţanu State Medical and Pharmaceutical University, Republic of Moldova
Manjon Javier Francisco Herrera	Polytechnical University of Valencia, Spain
Nicolae Jula	Military Technical Academy, Romania
Valeriu Kantser	Dumitru Ghitu Institute of Electronic Engineering and Nanotechnologies, Republic of Moldova
Lorenz Kienle	Institute for Materials Science, Kiel, Germany
Ratko Magjarević	University of Zagreb, Croatia
Hidenori Mimura	Shizuoka University, Japan
Radu Negoescu	Public Health Institute of Bucharest, Romania
Kornelius Nielsch	University of Hamburg, Germany
Nicolas Pallikarakis	University of Patras, Greece
Thierry Pauporte	Ecole Nationale Supérieure de Chimie de Paris, France
Abdel Salam Hamdy Makhlouf	University of Texas - Pan American, USA
Oliver G.Schmidt	Leibniz Institute for Solid State and Materials Research Dresden, Germany
Horia-Nicolai Teodorescu	Gheorghe Asachi Technical University of Iasi, Romania
Alexander Ushenko	Chernivtsi National University, Ukraine
Bruno Viana	Institut de Recherche de Chimie Paris, France
Adriana Velazquez Berumen	World Health Organization, Switzerland
Ionel Valentin Vlad	Romanian Academy of Science, Romania
Dan Zaharia	Grigore T. Popa University of Medicine and Pharmacy, Romania

International Program Committee

Organizing Committee

Serghey Railean	Technical University of Moldova, Republic of Moldova
Monaico Eduard	Technical University of Moldova, Republic of Moldova
Iavorschi Anatolie	Technical University of Moldova, Republic of Moldova
Ion Pocaznoi	Technical University of Moldova, Republic of Moldova
Alexandr Sereacov	Technical University of Moldova, Republic of Moldova

Reviewers List

Anastassia Rodina-Theocharaki
Anatoli Sidorenco
Anatolie Casian
Artur Buzdugan
Calin Corciova
Constantin Gaindric
Dan Zaharia
Dmitrii Nedeoglo
Dumitru Tsiulyanu
Eduard Monaico
Emil Rusu
Hariton Costin
Ion Tiginyanu
Konstantin Yashin
Liliana Verstiuc
Nellu Ciobanu
Oleg Lupan
Pavel Topala
Radu-George Ciorap
Vasile Tronciu
Veaceslav Ursaki
Victor Sontea
Victor Vovc
Viorel Nacu
Vladimir M. Fomin
Zhivko Bliznakov

Table of Contents

Nanotechnologies and Nanomaterials

Bio-nanotechnologies and Biomaterials

Biomedical Instrumentation and Biosensors

Biomedical Signal and Image Processing

Clinical Engineering, Health Informatics and Cellular and Tissue Engineering

Biomedical Engineering Education

Nuclear and Radiation Safety and Security

Part I
Nanotechnologies and Nanomaterials

Role of Charge-Transfer Complexes in Regulation of Processes Associated with Redistribution Electron Density in Biocomposite Systems

O.V. Yaltychenko and E.Yu. Kanarovskii

Institute of Applied Physics, Academy of Sciences of Moldova, Kishinev, Republic of Moldova

Abstract— **In this paper a simple and computationally convenient model for description of the influence of external periodic electric field on the electron transfer processes in the biocomposites is represented. The outlined theoretical approach allows to obtain the system of differential equations describing the electron and vibrational dynamics in the three-center charge-transfer complex integrated with an organic matrix As a result, by numerical simulation of the obtained system of differential equations two interesting sets of the electron density distribution regimes have been found. From the analysis and comparison these sets of regimes follows that this model gives a rich material for the biotechnological applications because help to reveal the optimal ways of control for the physical characteristics of organic nanocomposites and the kinetic processes in such systems.**

Keywords— **charge-transfer complex, matrix, biocomposite, periodic electric field, electron and vibrational dynamics.**

I. INTRODUCTION

At the present time the supra-molecular protein systems and synthetic biopolymers coordinated by the various nanoclusters have been widely used as the components of organic nano- and optoelectronics [1-4]. In such biocomposite materials the fundamental set of properties connects with a nonlinear dynamics of the charge and excitation transfer and there is a constant need in their detail characterization [1, 2].

In the last decades, the investigations developing the mathematical methods for description of regulation processes in the nonlinear dynamical systems with the aim of their biomedical applications are actively carried out. The main task of the theory of regulation in the nonlinear dynamical systems is a study of the process in time. For all processes occurring in the biological systems, an oscillatory behavior of the basic parameters is inherent, that leads a biosystem to a steady state. Therefore, it is important at the regulation to implement the tracking of oscillatory processes and maintaining of certain steady state.

For the biocomposites as the organic systems having a weak ordering the most interesting problems are related to a study of the trapping of charge or excitation on certain states and possibility of conformational changes connected with the specific localization regimes [2, 3]. The structural complexity of such biocomposites demands the computationally convenient models able to give a correct description of the localized and delocalized states, the electron and vibrational excitations, the protein conformation changes, the processes of structural relaxation and electron localization and their mutual influences. Such theoretical models allow picking those ways or processes that are the most suitable to experimentalists for the observation and effective control of the electron density redistributions and conformational rearrangements in the studied biocomposites. So, from this viewpoint, the study of opportunities for the regulation of catalytic and biological activity of such systems [2] through a control of the excitation and charge transfer processes is especially interesting.

The theoretical approach outlined in this paper allows to give an enough complete description of the kinetic processes in biocomposites with the account of most essential factors (both external and internal) influencing on the studied process. The main factors, which are taken into account in this theoretical model, are an electron-vibrational coupling, an influence of the surrounding matrix, an action of the external low-frequency electric field, and an existence of the long-lived coherence proved experimentally [3, 4]. At performing of the numerical simulation for certain biocomposite system the specific properties both the organic matrix and charge-transfer complex integrated with it can be easily taken into account by appropriate choice of the model parameters.

The aim of this paper to perform a comparison of two characteristic sets of the electron density distribution regimes obtained from the numerical calculations for the system of differential equations, which follows from this theoretical model.

II. THEORETICAL MODEL

A full mathematical formulation of the theoretical model is given in the previous paper of authors [5]. Here, it is represented a brief statement of the base features of this theoretical model. Hereafter, the final system of differential equations is given with the explanation of its outcome and identification of model parameters. In this model a matrix is taken into account as a nondissipative and a periodic electric field – as a low-frequency. While here, the dissipation in the matrix is not taken into account, but it is not difficult to include the dissipation in this theoretical model using different approaches [6, 7].

V. Sontea and I. Tiginyanu (eds.), *3rd International Conference on Nanotechnologies and Biomedical Engineering*,
IFMBE Proceedings 55,
DOI: 10.1007/978-981-287-736-9_1

The charge-transfer complex is considered as the three-center nanocluster (TNC) of bridge type, i.e. 1st and 3rd TNC's centers are tunnel-coupled exceptionally through intermediate (2nd) TNC's center. The 1st and 3rd TNC's centers are assumed as equivalent: they have an equal energy value of the electron levels, which is taken as the origin of count. At that TNC's centers are considered together with theirs nearest ligand environment. The intracluster vibrational modes accounted in the model are obtained after what a set of local vibrational modes was transformed into a set of generalized vibrational modes and the full-symmetric vibrational mode was eliminated from a consideration.

In the base of model the model Holstein's Hamiltonian was used. This Hamiltonian was supplemented by the terms responsible for the interactions of a TNC with an external periodic electric field and with a vibrational system of a nondissipative matrix. Eventually, the canonical Hamilton's equations are used for obtaining of the system of differential equations relatively the indicated variational parameters.

The final system of differential equations describes the electron and vibrational dynamics in the organic nanocomposites (such as the synthetic biopolymers and supramolecular protein systems) coordinated by the charge-transfer complexes and has the following form:

$$\frac{d\pi_1}{dt} = -\omega_1^2\beta_1 - g_1(|\alpha_1|^2 - |\alpha_3|^2) - \gamma_{1,i}Q_i, \quad \frac{d\beta_1}{dt} = \pi_1,$$

$$\frac{d\pi_2}{dt} = -\omega_2^2\beta_2 - g_2(|\alpha_1|^2 + |\alpha_3|^2) - \gamma_{2,i}Q_i, \quad \frac{d\beta_2}{dt} = \pi_2,$$

$$i\frac{d\alpha_1}{dt} = t_0\alpha_2 + g_1\beta_1\alpha_1 + g_2\beta_2\alpha_1 + dE_0\cos(\Omega t)\alpha_1, \tag{1}$$

$$i\frac{d\alpha_2}{dt} = t_0(\alpha_1 + \alpha_3) + \Delta\alpha_2,$$

$$i\frac{d\alpha_3}{dt} = t_0\alpha_2 - g_1\beta_1\alpha_3 + g_2\beta_2\alpha_3 - dE_0\cos(\Omega t)\alpha_3$$

In the model used the following parameters: t_0, g_m, Δ, and dE_0 which are, respectively, the parameter of electron tunneling between the TNC's centers, the constant of the electron-vibrational interaction for m-th (m = 1, 2) intracluster vibrational mode q_m and the detuning energy, which characterizes the nonequivalence of the intermediate TNC's center, the intensity of interaction between TNC and the electric field with a frequency Ω and amplitude E_0. The important feature of model is the use of the time-dependent wave function, which has a view of the superposition of electron-vibrational states in the coherent package form. This wave function is expressed via the time-dependent variational parameters: $\alpha_j(t)$, $\beta_m(t)$, $\pi_m(t)$, which are convenient for further numerical simulation. A physical sense of these parameters is following: $|\alpha_j(t)|^2$ – the electron population on j-th center (j = 1, 2, 3), $\beta_m(t)$ and $\pi_m(t)$ – the quantum-mechanical average values of amplitudes of the coordinate and momentum for m-th intracluster mode.

The system (1) include 4 equations for the values $\beta_m(t)$, $\pi_m(t)$, that describe the behavior of the TNC's vibrational subsystem, and 3 equations for the values $\alpha_j(t)$, that describe the behavior of the TNC's electronic subsystem. Note that in (1) the equations for the intracluster vibrational subsystem include the parameters, which characterize the TNC's electron subsystem and matrix's vibrational mode Q_i with frequency ω_{0i}, but the equations for the TNC's electron subsystem contains the parameters of TNC's vibrational subsystem. Also, it is assumed that the equation of motion for the external mode of matrix Q_i is the usual harmonic oscillator equation having in its right part a term $-(\gamma_{1,i}\beta_1 + \gamma_{2,i}\beta_2)$, i.e. a role of driving force plays the TNC's vibrational subsystem. The implementation of the numerical decision of equations (1) can be completed without account of equation for the external mode Q_i, because the matrix's mode is eliminated if to assume that the mode Q_i more slower than the intracluster modes and to suppose the constants $\gamma_{m,i}$ as an equal (i.e. $\gamma_{m,i} = \gamma$). As a result, neglecting in the equation of motion for Q_i the second derivative and substituting the resultant expression for Q_i in (1) to have the complete elimination of the matrix's mode. Thus, this leads to the explicit link of the equations for the intracluster modes. As follows from (1), due to the electron-vibrational interaction, there is also the implicit connection of the intracluster modes. These cross-relations significantly affect on the energy exchange between the intracluster modes.

Note that, such consideration of the vibrational system of matrix corresponds to the simplest approximation – continuum, i.e. a matrix is considered as a quasi-homogeneous and is characterized by a single mode. The model used here allows easy to consider a more general case when a matrix contains other frequencies and various $\gamma_{m,i}$, i.e. in the equations contained Q_i in (1) should perform to perform the summation on the index i.

III. Results and Discussion

The numerical decision of (1) and (2) for many different sets of model parameters was completed and some of the most interesting characteristic regimes leading to the stable electron density distributions on the TNC's centers represented graphically on Fig. 1 – the cases a), b), c) for the regimes with a charge accumulation on 2nd TNC's center and on Fig. 2 – the cases a), b), c) for the regimes without a charge accumulation on 2nd TNC's center. Thus, these two set of regimes have the differences in the charge state of the intermediate (2nd) TNC's center.

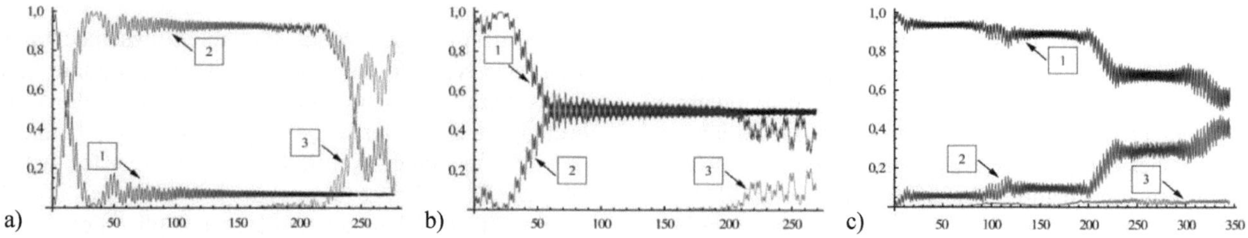

Fig. 1 The time dependence of the electron population $|a_j(t)|^2$ on j-th center of TNC (j = 1, 2, 3) – the regimes with a charge accumulation on 2nd center.

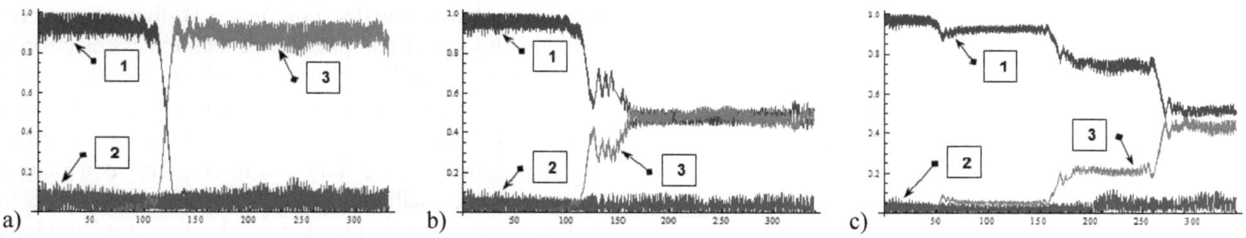

Fig. 2 The time dependence of the electron population $|a_j(t)|^2$ on j-th center of TNC (j = 1, 2, 3) – the regimes without a charge accumulation on 2nd center.

All frequencies in this model are taken in the units of some characteristic frequency ω_0 of the matrix's mode, i.e. it is taken that $\omega_{0i} = \omega_0$. The other model parameters are given in the energetic units normalized on a value $\hbar\omega_0$.

For all cases a), b), c) shown on Fig. 1 and Fig. 2: g_1=g_2=1 and the time is expressed in in units ω_0^{-1}.

For the cases a), b) on Fig. 1 the common parameters are $t_0 = 1/5$, $dE_0 = 4$ and $\Omega = 1/88$. The other model parameters for the cases shown on Fig. 1 are: in a) – $\Delta = 5.5$; $\beta_1(0) = 0.9$; $\pi_1(0) = 0.9$; $\omega_{1,2} = 2$; γ=0.235, in b) – $\Delta = 7$; $\beta_1(0) = 0$; $\pi_1(0) = 4$; $\omega_{1,2} = \sqrt{2}$, $\gamma = 0$, in c) – $t_0 = 1/4$; $dE_0 = 2.1$; $\Omega = 1/17$; $\Delta = 5.35$; $\beta_1(0) = 0.9$; $\pi_1(0) = 0.9$; $\omega_{1,2} = 2$; $\gamma = 0.13$. On the Fig. 1b the case without the influence of the surrounding matrix is shown (parameter $\gamma = 0$). Note also that taking $\gamma = 0.13$ from the regime on the Fig. 1a can to obtain regime, which is able to the regime with $\gamma = 0$ (Fig. 1b), only on the initial part of time diapason (before a full delocalization) the more deep oscillations in the populations of 1st and 2nd centers take place.

For all cases represented on Fig. 2 the same initial conditions for the intraclaster modes are chosen: $\beta_1(0) = 0$; $\pi_1(0) = 0.9$; $\beta_2(0) = 0$; $\pi_2(0) = 0$. Also, the same values are chosen for the following parameters: $dE_0 = 2.1$; $\Delta = 5.5$; $\omega_{1,2} = 2$. The other model parameters for the cases shown on Fig. 2 are: in a) – $t_0 = 1/1.5$; $\Omega = 1/77$; $\gamma = 0.235$, in b) – $t_0 = 1/1.8$; $\Omega = 1/77$; $\gamma = 0.235$, in c) – $t_0 = 1/2.4$; $\Omega = 1/34$; $\gamma = 0.13$.

If to perform a pairwise comparison of the regimes for the cases a), b), c) on Fig. 1 and Fig. 2, then not difficult to notice a similarity in the overall picture of the electron localization dynamics. So, an electron being initially at 1st TNC's center is localized almost completely on other TNC's center: in the case a) shown on Fig. 1 – on 2nd TNC's center, and in the case a) shown center on Fig. 2 – on 3rd TNC's center. The case b) shown on Fig. 1 and Fig. 2 correspond to electron delocalization mainly between two centers (Fig. 1b – between 1st and 2nd TNC's centers, and Fig. 2b – between 1st and 3rd TNC's centers). On Fig. 1c and Fig. 2c shown the case, when the electron delocalization, also as the case b), takes place between two centers (Fig. 1c – between 1st and 2nd TNC's centers, and Fig. 2c – between 1st and 3rd TNC's centers), but in a difference from the previous case an establishment of delocalization in the case c) occurs a slow stepwise. The common feature for the stepwise regimes (Fig. 1c and Fig. 2c) is that the frequencies of electric field are higher than in the cases a) and b) on Fig. 1 and Fig. 2, respectively.

From the obtained results follows that the external model parameters, such as the frequency and amplitude of periodic electric field, are the convenient regulation parameters for switching of the electron density distribution regimes in TNC. As it follows from the comparison of Fig. 1a and Fig. 1b, the electron density distribution in TNC is considerably determined by the interaction of intracluster modes with the external mode of surrounding matrix. The regimes of intracluster vibrational dynamics and their relationship with the regimes of the electron transfer are discussed more fully in [5].

It is revealed that in depending on the ratio of the internal and external parameters of the presented model can occur or accumulation (localization) of the electron density on any of TNC's centers according to the share of its population, and then two other centers actively exchange an electron (delocalization), or on all three centers a some certain distribution of electron density is established. At that the vibrational and electron subsystems of TNC remain during a time evolution in a form of the coherent electron-vibrational package distributed on the TNC's centers according to an extent of the electron localization on them.

The results represented on Fig. 1 allow to understand how the electron and vibrational dynamics in TNC can to influence on a biological activity of the biocomposite system in whole. So, the case b) on Fig. 1 correspond to the situation when the mutual influence of the protein matrix and TNC is absent because the interaction between them is equal to zero ($\gamma = 0$). In our model, it is understood as a situation before the coordination of TNC on the protein matrix, therefore a bioactivity of the protein matrix does not change in this case. After embedding of TNC in the matrix the situation changes substantially as can be seen on Fig. 1a ($\gamma \neq 0$). As a rule, the embedded nanocluster is coordinated in a vicinity to the active site of the supra-molecular protein system. The case a) corresponds to almost the maximum value of electron population on the intermediate TNC's center with a sharp activation 3rd TNC's center in the end of indicated time period. Such changing of the charge distribution in the embedded TNC can significantly affect on a bioactivity of the protein system. Note that, the interaction between the intracluster modes and matrix's mode ($\gamma \neq 0$) promotes to the accumulation of energy in the degrees of freedom of the protein matrix. In turn, this promotes to a significant rise of the probability of conformational rearrangement of the protein system and it can easily make the transition in the new conformation. But in the various conformations the protein system usually has the different level of biological activity. In this regard, we note that our model implies the possibility of realization of an interesting effect – let's call it the effect of the cascade changes of protein conformation. In general, for each new protein conformation is necessary to take at numerical simulation other values for the model parameters γ and ω_0. At the changed values γ and ω_0 the regime of electron density distribution in TNC and, respectively, the regime of intracluster and matrix's vibrational dynamics will vary and they can become again optimal for the next conformational transition in a protein system, and so on. The fitting of the optimal regime in TNC capable to induce the conformational changes in a protein system can be realized by varying of the frequency and amplitude of periodic electric field.

The electron localization dynamics presented on Fig. 2 deserves a separate discussion. This set of regimes demonstrates that the electron transfer from the 1st to the 3rd center through the 2nd center without a charge accumulation on the intermediate center, i.e. it occurs much faster without a delay on 2nd TNC's center.

The results of the modeling can be applied not only to the study of opportunities of regulation of bioactivity of organic nanocomposites. So, the second set of regimes (Fig. 2) has yet another useful interpretation, if to consider the TNC as a transition charge-transfer complex. At that should be regarded a molecular complex-agent with excess electron as 1st TNC's center, a molecular group of binding on the membrane as 2nd TNC's center (bridge), and the role of 3rd TNC's center plays a molecular group of membrane receptor. In this situation a periodic electric field can to simulate as a field associated with the oscillations in membrane potential. This interpretation gives an application of the proposed model to a simulation of charge transfer process into a membrane receptor. Note also that the regimes without a charge accumulation on the intermediate center, as it follows from a numerical simulation, are very sensitive to the choice of the initial conditions for the TNC's vibrational subsystem.

Thus, the analysis of the numerical results obtained within the given theoretical model allows to make the conclusions about the extent of electron localization in the TNC's centers, about the vibrational dynamics of intracluster modes, about the connections of electron and vibrational dynamics, and how they depend from a ratio of the internal and external model parameters.

IV. Conclusions

Thus, the offered model is quite acceptable to describe the picture of the biological activity of the organic nanocomposites such as the supra-molecular protein and synthetic biopolymer systems coordinated by the different type of TNCs and to reveal the methods for the regulation of their bioactivity. Indeed, the existence of the electron density distribution regimes in TNC coordinated in a vicinity of the protein's active site gives the possibilities to regulate the level of biological activity by adjusting the corresponding set of model parameters. So, if the electric field parameters (amplitude and frequency) are changed, then in TNC a new charge and vibrational regimes are established and theirs cumulative effect on the active site of protein system is modified (at that additionally may also be induced a conformational change of the protein system), so that its biological activity is altered too. Also, from the given model follows that in the organic nanocomposites can to take place the effect of the cascade changes of protein conformation.

It is necessary to draw attention to another important practical application of the theoretical model presented here, in the opto- and microelectronics. Due to the fact that the intracluster electron dynamics reflects the nanocluster's dipole moment changes, the simulation results can be used at a creation of the biopolymer composite films or plates with the controlled polarization properties (more details, see in [5]).

Thus, the theoretical approach represented here provides many opportunities for the effective modeling of different kinetic and optical processes in the complex organic nanocomposite systems.

ACKNOWLEDGMENT

This work was supported by the CSSDT of ASM, Institutional Project 15.817.02.03A.

CONFLICT OF INTEREST

The authors declare that they have no conflict of interest.

REFERENCES

1. Martinez-Duart J.M., Aqullo-Rueda F., Martin-Palma R.J. (2006) Nanotechnology for Microelectronics and Optoelectronics. Elsevier, Amsterdam.
2. Daniel M.C., Astruc D. (2004) Gold nanoparticles: assembly, supramolecular chemistry, quantum-size-related properties, and applications toward biology, catalysis, and nanotechnology. Chem. Rev. 104(1):293–346.
3. Chin A.W., Huelga S.F., Plenio M.B. (2012) Coherence and decoherence in biological systems: Principles of noise-assisted transport and the origin of long-lived coherences. Phil. Trans. R. Soc. A 370: 3638–3657.
4. Pachón Leonardo A., Brumer Paul (2011) Physical basis for long-lived electronic coherence in photosynthetic light-harvesting systems. J. Phys. Chem. Lett. 2(21):2728–2732.
5. Yaltychenko O.V., Kanarovskii E.Yu. (2014) Dynamics of Electron Localization in Low_Frequency Electric Field in Nanotrimer Considering the Influence of Nondissipative Matrix. Surf. Eng. Appl. Electrochem. 50(5):446–453.
6. Yaltychenko O.V., Kanarovskii E.Yu. (2012) Influence of stochastic processes on kinetics of two-electron transfer in dimer nanocluster embedded in organic matrix, SPIE Proc. vol. 8411, Advanced Topics in Optoelectronics, Microelectronics and Nanotechnologies VI, Constanta, Romania, 2012, pp. 84112U-1–84112U-5.
7. Silbey R. (2011) Description of Quantum Effects in the Condensed Phase. Procedia Chemistry 3(1):188–197.

Author: O.V. Yaltychenko
Institute: Institute of Applied Physics, ASM
Street: Academy, 5
City: Kishinev, MD-2028
Country: Republic of Moldova
Email: oialt@mail.ru

Influence of Resonant Optical Phonons on Intersubband Magnetoabsorption in Nanowires

E.P. Sinyavskii[1], E.Yu. Kanarovskii[1], and N.S. Kostyukevich[2]

[1] Institute of Applied Physics, Academy of Sciences of Moldova, Kishinev, Republic of Moldova
[2] T.G. Shevchenko State University of Pridnestrov'e, Tiraspol, Republic of Moldova

Abstract— **The investigation concerns the influence of the electron-phonon interaction on the form of bands of the intersubband absorption of a weak electromagnetic wave in quantum wires in a transverse magnetic field. The interaction of the electrons with the long-wavelength acoustic vibrations is taken into account using the relaxation time approximation; the interaction with the optical vibrations is considered in the resonance approximation (the energy of an optical phonon is the difference of the energies of the electronic states between which the transition occurs). In the paper the case of low temperatures is considered, when only the processes with an emission of a non-dispersion optical phonon are realized. It is shown that under these conditions the magnetoabsorption peak splits into two: the half-width of the long-wavelength peak is almost entirely determined by the interaction of electrons with the acoustic vibrations, and the half-width of the short-wavelength peak mainly depends on the interaction of electrons with the optical vibrations. The coefficients of the intersubband absorption of a weak electromagnetic wave for various polarizations were calculated. The influence of polarization of the electromagnetic wave on its absorption was analyzed.**

Keywords— **parabolic quantum wires, intersubband transitions, transverse magnetic field, half-width of optical absorption bands, polarization of light.**

I. INTRODUCTION

Calculation of the optical and kinetic characteristics of quantum wires can be significantly simplified using the model of the parabolic profile of the potential energy carriers in nanowires [1, 2]. This model has a reliable theoretical foundation [2] and is useful for the analysis of experimental data and comparing them with theoretical simulations [1-3].

While investigating the optical properties of a parabolic quantum wire (PQW) in an external magnetic field **H** directed along the *Z*-axis and perpendicular to the axis of the nanowire (*X*-axis), it is convenient to use the calibration of the vector potential in the form $\mathbf{A}(-By, 0,0)$. Therefore, the energy spectrum of electrons in the PQW in the transverse magnetic field is described by the following expression:

$$E_\alpha \equiv E_{mn}(k_x) = \frac{\hbar^2 k_x^2}{2m_c^*} + \hbar\omega\left(m + \frac{1}{2}\right) + \hbar\Omega_e\left(n + \frac{1}{2}\right), \quad (1)$$

$$m_e^* = m_e\left(\frac{\Omega_e}{\omega}\right)^2, \ \hbar\omega = \frac{\hbar}{R}\left[\frac{2\Delta E_c}{m_e}\right]^{1/2},$$

$$\Omega_e = [\omega^2 + \omega_c^2]^{1/2}, \ \omega_c = \frac{eB}{m_e c}$$

here: $\alpha = \{k_x, m, n\}$ is a set of quantum numbers describing the state of the band carriers, k_x is the electron wave vector with the effective mass m_e, ω_c is the cyclotron frequency, c is the speed of light, $\hbar\omega$ is the quantum confinement energy, associated with the height parabolic potential profile ΔE_c at the border of the PQW having radius R, and $\hbar\Omega_e$ is the energy quantization for the hybrid electron states. It should be noted that for the nanowires with a magnetic permeability $\mu = 1$, i.e. for the PQW made from nonmagnetic materials, the magnetic induction vector **B** coincides with the magnetic field vector **H**.

As it follows from (1), the energy spectrum of electrons (Fig. 1) is the additive composition of two sets of equidistant parabolas with the energetic distances $\hbar\omega$ and $\hbar\Omega_e$, respectively.

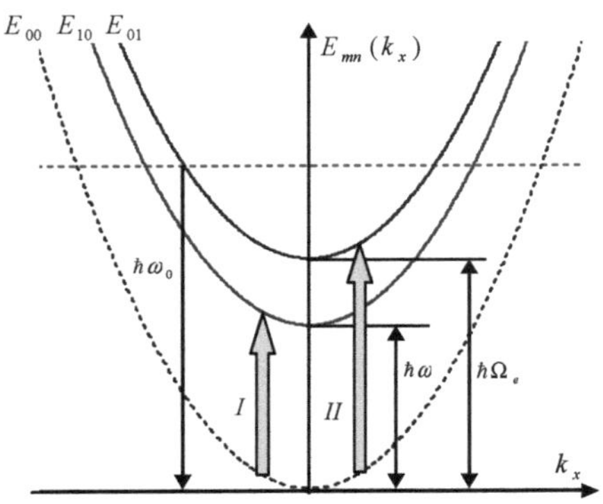

Fig. 1 Energy spectrum of electrons in a parabolic quantum wire in a transverse magnetic field, and studied optical transitions.

Note that according to (1) a curvature of the parabolas depends on the value of the magnetic field strength due to the dependence of the effective mass m_e^* on the magnitude **H**.

II. THEORETICAL APPROACH

In this paper, the absorption coefficient of weak electromagnetic wave (EMW) for the electron transitions between the quantized states of the conduction band (transitions I, II in Fig. 1) is calculated with the account of the interaction of electrons with two types of phonons – acoustic and optical. So that the interaction of band carriers with the long-

V. Sontea and I. Tiginyanu (eds.), *3rd International Conference on Nanotechnologies and Biomedical Engineering*, IFMBE Proceedings 55,
DOI: 10.1007/978-981-287-736-9_2

wavelength acoustic vibrations is considered in the relaxation time approximation [3] and their interaction with the optical phonons is considered in the resonance approximation. In this paper, the case of low temperatures is con-sidered when the processes with absorption of a dispersionless optical phonon can be neglected [4]. Also, note that the indirect optical intersubband transitions associated with the intraband scattering of the optical phonons are neglected.

The most interesting cases are when either the energy of size quantization $\hbar\omega$ or the energy of hybrid quantization $\hbar\Omega_e$ is equal to the energy of optical phonon $\hbar\omega_0$ (the resonance case). For example, if $\hbar\omega_0 = \hbar\omega$, then the first size-quantized state $E_{10}(k_x)$ (see Fig. 1) becomes doubly degenerate and splits into two due to the interaction of electrons with the optical vibrations. A similar situation is in the case of $\hbar\omega_0 = \hbar\Omega_e$. This is why the band of the intersubband light absorption for the considered electronic states (transition I, II in Fig. 1) splits into two optical absorption bands. In this approximation, using the method developed in [4], for the coefficient of the intersubband absorption of the weak EMW with the frequency Ω and polarization $\boldsymbol{\eta}$, the following expression is obtained:

$$K(\Omega) = \frac{8\pi e^2}{\hbar c V n_0 \Omega} \sum_{\alpha,\alpha_1} \left| \frac{\mathbf{P}_{\alpha\alpha_1}\boldsymbol{\eta}}{m_e} \right|^2 n_\alpha \times$$

$$\times \frac{\Gamma_\alpha + \Gamma_{\alpha_1} + \Gamma^{opt}_{\alpha\alpha_1}}{\left(\Gamma_\alpha + \Gamma_{\alpha_1} + \Gamma^{opt}_{\alpha\alpha_1}\right)^2 + \hbar^{-2}\left(E_{\alpha_1} - E_\alpha - \hbar\Omega\right)^2} \quad (2)$$

here: V is the volume of the nanostructure, n_0 is the index of light refraction in the given sample, $\mathbf{P}_{\alpha\alpha_1}$ is the matrix element of the momentum operator on the smoothed wave functions of electrons in the conduction band of the PQW, n_α is the equilibrium distribution function of electrons in the initial electron state α with energy E_α, $2\Gamma_\alpha$ is the quantum mechanical probability of elastic scattering of electrons per unit time on the long-wavelength acoustic vibrations of the PQW lattice in a magnetic field.

The calculations [1, 3] show that $\Gamma_\alpha \sim |k_x|^{-1}$: this dependence on the wave vector is typical for the one-dimensional quantum systems due to the peculiarities in the energy density of the states of band carriers. The value $\Gamma^{opt}_{\alpha\alpha_1}$ describes only the effect of optical phonons on the frequency dependence $K(\Omega)$ because, as it follows from the performed calculations, the contribution of the interaction of electrons with the acoustic oscillations in the value $\Gamma^{opt}_{\alpha\alpha_1}$ is negligible. At recording (2) it was assumed that in the state E_{α_1}, in which the optical transition of an electron occurs, there are no band carriers and therefore $1 - n_{\alpha_1} \cong 1$. This corresponds to the case of low temperatures T, i.e. $k_0 T \ll \hbar\omega$ (k_0 is Boltzmann's constant), hence at the studied optical transitions the electrons in the initial state are near the bottom of the lowest subband. For the simplicity, also it is assumed that the electron gas in the PQW is non-degenerate. Due to the foregoing assumption in the case of $\hbar\omega_0 = \hbar\Omega_e$, an electron realizes transition only at the emission of an optical phonon from the first level ($m_1 = 0$, $n_1 = 1$) to the lowest level ($m = 0$, $n = 0$). Thus, for the value $\Gamma^{opt}_{\alpha\alpha_1}$, it is easy to obtain:

$$\Gamma^{opt}_1 = \Gamma_0 \left(\frac{\hbar\omega}{\Delta}\right)^{1/2}, \quad (3)$$

$$\Gamma_0 = \frac{e^2 c_0 \omega_0 (1+\delta^2)^{1/2}}{2\hbar\left[1+(1+\delta^2)^{1/4}\right]} \left(\frac{m_e}{2\hbar\omega}\right)^{1/2},$$

$$c_0 = \frac{1}{n_0^2} - \frac{1}{\varepsilon}, \; \delta = \frac{\omega_c}{\omega}, \; \Delta = \frac{\hbar^2 k_x^2}{2m_e^*} + \hbar\Omega - \hbar\omega_0$$

here: ε is the static dielectric constant.

Note that the $\Gamma^{opt}_{\alpha\alpha_1}$ is different from zero only when $\Delta > 0$. So, let us consider the optical transition II in Fig. 1 (at the transition I the situation is analogical): the electron transfers into the hybrid electron state $E_{01}(k_x)$, but the emission of an optical phonon from this state, if $E_{01}(k_x) < \hbar\omega_0$, cannot take place since otherwise the energy conservation law is violated.

A direct consequence from (3) is the following: the higher the value of $\mathbf{H}$, the higher the value of $\Gamma^{opt}_{\alpha\alpha_1}$. Therefore, the increase of the carrier localization in the PQW due to the increase of the magnetic field strength leads to the enhancement of the electron-phonon interaction.

In the case when the optical phonon energy is equal to the energy of size quantization $\hbar\omega_0 = \hbar\omega$, for the value $\Gamma^{opt}_{\alpha\alpha_1}$, the following is obtained:

$$\widetilde{\Gamma^{opt}_1} = \Gamma_0 \left(\frac{\hbar\Omega_e}{\Delta}\right)^{1/2} = \frac{1}{\sqrt{\delta_0}} \Gamma^{opt}_1, \quad (4)$$

$$\delta_0 = \frac{\omega}{\Omega_e} = (1+\delta^2)^{-1/2},$$

$$\delta_0 < 1$$

It should be noted that according to (2) the coefficient of the intersubband light absorption is essentially dependent on the polarization of the incident EMW.

The calculations of the matrix elements $\mathbf{P}_{\alpha\alpha_1}$ for different polarizations of the incident EMW give the following expressions:

$$\left|\mathrm{P}^X_{\alpha\beta}\right|^2 = \frac{\hbar\omega\delta^2 m_c}{2} \delta_{k_x k_x'} \delta_{mm'} \{(n+1)\delta_{nn'-1} + n\delta_{nn'+1}\},$$

$$\left|\mathrm{P}^Y_{\alpha\beta}\right|^2 = \frac{\hbar\Omega_c m_c}{2} \delta_{k_x k_x'} \delta_{mm'} \{(n+1)\delta_{nn'-1} + n\delta_{nn'+1}\}, \quad (5)$$

$$\left|\mathrm{P}^Z_{\alpha\beta}\right|^2 = \frac{\hbar\omega m_c}{2} \delta_{k_x k_x'} \delta_{nn'} \{(m+1)\delta_{mm'-1} + m\delta_{mm'+1}\}$$

As it follows from (5) for the *X*- and *Y*-polarization of the EMW, the direct optical intersubband transitions can take place only between the hybrid electron states, but for the *Z*-polarization of the EMW the direct optical intersubband transitions occur only between the size-quantized states. Also, according to (5), in the absence of a magnetic field ($\delta = 0$), the absorption of the linearly-polarized EMW with the *X*-polarization is forbidden.

Finally, in the case of $\hbar\omega_0 = \hbar\Omega_e$ the intersubband absorption of the linearly-polarized EMW with the *X*-polarization is determined by the electron transitions from the lowest state $E_{00}(k_x)$ to the excited hybrid state of the conduction band $E_{01}(k_x)$ (transition II, Fig. 1). Thus, the intersubband absorption coefficient $K^{(X)}(\Omega)$ is expressed as:

$$K^{(X)}(\Omega) = K_0\delta^2(1+\delta^2)^{-1/4} \times$$

$$\times \frac{\omega}{\Omega}\int_0^\infty e^{-\tau}\frac{1+b_0\left(\frac{\tau}{\Delta_0}\right)^{1/2}}{\left(1+b_0\left(\frac{\tau}{\Delta_0}\right)^{1/2}\right)^2+\left(\frac{\Omega_e-\Omega}{\omega_f}\right)^2\tau_0\tau}d\tau, \quad (6)$$

$$K_0 = \frac{4e^2k_0T}{Sn_0c\hbar\omega_f}\left[\frac{(1+\delta^2)^{3/2}}{2m_e\hbar\omega_f}\right]^{1/2}e^{\frac{\xi}{k_0T}},$$

$$\omega_f = \left[\frac{\hbar\left(\frac{7}{4}\gamma_0\right)^2}{2m_e^*}\right]^{1/3}, \quad \gamma_0 = \frac{E_1^2k_0Tm_e^2\omega}{2\pi\hbar^4\rho v_0^2}\left(\frac{\Omega_e}{\omega}\right)^{\frac{5}{2}},$$

$$b_0 = \frac{\Gamma_0}{\omega_f}(\tau_0)^{\frac{1}{2}}\left(\frac{\omega}{\omega_f}\right)^{\frac{1}{2}}, \quad \Delta_0 = \tau_0\tau - \left(\frac{\Omega_e-\Omega}{\omega_f}\right), \quad \tau_0 = \frac{k_0T}{\hbar\omega_f}$$

here: ξ is the chemical potential in the studied quantum system (the value of ξ is counted from the bottom of the PQW conduction band in the magnetic field), S is the cross-section of the PQW, E_1 is the constant of a deformation potential, v_0 is the speed of sound in the nanostructure having the density ρ.

Calculation of the intersubband absorption coefficient for the linearly-polarized EMW with the *Y*-polarization (at $\hbar\omega_0 = \hbar\Omega_e$) is carried out in a similar manner, giving the expression $K^{(Y)}(\Omega)$, which has a quite similar view as for the expression $K^{(X)}(\Omega)$:

$$K^{(Y)}(\Omega) = K_0(1+\delta^2)^{1/4} \times$$

$$\times \frac{\omega}{\Omega}\int_0^\infty e^{-\tau}\frac{1+b_0\left(\frac{\tau}{\Delta_0}\right)^{1/2}}{\left(1+b_0\left(\frac{\tau}{\Delta_0}\right)^{1/2}\right)^2+\left(\frac{\Omega_e-\Omega}{\omega_f}\right)^2\tau_0\tau}d\tau \quad (7)$$

As it follows from (6) and (7), the expression $K^{(Y)}(\Omega)$ differs from the expression $K^{(X)}(\Omega)$ only by a factor, so that $K^{(Y)}(\Omega) = \delta^{-2}(1+\delta^2)^{1/2} \cdot K^{(X)}(\Omega)$.

The intersubband absorption of the linearly-polarized EMW with the *Z*-polarization (i.e., light propagates along the axis of the nanowire) is determined by the electron transitions in the conduction band from the lowest state $E_{00}(k_x)$ to the excited size-quantized state $E_{10}(k_x)$ (transition I, Fig. 1). Thus, the expression for the intersubband absorption coefficient $K^{(Z)}(\Omega)$ in the case of $\hbar\omega_0 = \hbar\omega$ has the following form:

$$K^{(Z)}(\Omega) = K_0(1+\delta^2)^{-1/4} \times$$

$$\times \frac{\omega}{\Omega}\int_0^\infty e^{-\tau}\frac{1+b_1\left(\frac{\tau}{\Delta_1}\right)^{1/2}}{\left(1+b_1\left(\frac{\tau}{\Delta_1}\right)^{1/2}\right)^2+\left(\frac{\omega-\Omega}{\omega_f}\right)^2\tau_0\tau}d\tau, \quad (8)$$

$$b_1 = (1+\delta^2)^{1/4}b_0, \quad \Delta_1 = \tau_0\tau - \left(\frac{\omega-\Omega}{\omega_f}\right)$$

Note that the structure of the expression $K^{(Z)}(\Omega)$ is also similar to that of the expression $K^{(X)}(\Omega)$. As it follows from the comparison of formulas (6) and (8), they can be converted into each other. So, (8) is obtained from (6) taking into account (3) and (4), then the value b_0 is replaced by b_1 and the value Δ_0 is replaced by Δ_1. Thus, the expression $K^{(Z)}(\Omega)$ is obtained from the expression $K^{(X)}(\Omega)$ if the substitution $\hbar\Omega_e \rightarrow \hbar\omega$ in integral in (6) is performed. Also, the expression $K^{(X)}(\Omega)$ unlike the expression $K^{(Z)}(\Omega)$ contains in a front of the integral the additional factor δ^2, which appears since, according to (5), the value $\left|P^X_{\alpha\alpha_1}\right|^2$ for the *X*-polarized EMW and the value $\left|P^Z_{\alpha\alpha_1'}\right|^2$ for the *Z*-polarized EMW differ from each other by this factor.

III. Results and Discussion

For an implementation of the numerical calculations and subsequent estimates, the following parameters, typical for semiconductor quantum wires, are considered: $m_e = 0.06m_0$, $E_1 = 10\ eV$, $\Delta E_c = 0.255\ eV$, $c_0 = 0.014$, $\rho = 5.4\ g/cm^3$, $v_0 = 3\cdot10^5\ cm/s$. In this case $\hbar\omega = 8.05\cdot R_0^{-1}\ eV$ (here, R_0 is the numerical value of the PQW radius R given in Å) and, therefore, the energy of the limiting optical phonon $\hbar\omega_0 = 0.03\ eV$ is equal to the energy of size quantization $\hbar\omega$ if $R \approx 270$ Å.

Note that when the magnetic field **H** is increased, the energy of the hybrid quantization $\hbar\Omega_e$ increases respectively, and the resonance case occurs at larger values of the PQW radius. (For example, if $\delta = 1$, then $R \approx 380$ Å). Also, note that at smaller values $\hbar\omega_0$ the resonance value of the nanowire radius increases.

Curves 1 and 2 in each of the cases considered below (see Figs. 2, 3, and 4) are normalized on the identical value K_0. For this, the reduced chemical potential $\xi = \frac{\xi}{k_0T}$ is chosen so that it has the same value in each of these cases.

Note that the pre-exponential factor in K_0 is not changed at varying $\mathbf{H}$ and T, since the value $(\hbar\omega_f)^{3/2} \sim T \cdot (1+\delta^2)^{3/4}$.

Fig. 2 (in arbitrary units) shows the influence of temperature on the frequency dependence of the intersubband absorption coefficient for the *X*-polarized EMW in the case of $\hbar\omega_0 = \hbar\Omega_e$ (transition II, Fig. 1) at the fixed magnitudes of $\mathbf{H}$ and R, respectively: $\delta = 1$ and $R = 380$ Å. Curves 1 and 2 in Fig. 2 were calculated at $T = 40\,°K$ and $T = 60\,°K$. For curve 1 – $\hbar\omega_f = 3.14 \cdot 10^{-4}\,eV$, $b_0 = 72.89$, $\tau_0 = 10.97$, and for curve 2 – $\hbar\omega_f = 4.12 \cdot 10^{-4}\,eV$, $b_0 = 52.00$, $\tau_0 = 12.56$. Here, the appropriate graphs only for the value $K^{(X)}(\Omega)$ are provided, since the influence of temperature on the value $K^{(Y)}(\Omega)$ is completely identical.

Fig. 3 (in arbitrary units) shows the influence of the magnetic field on the frequency dependence of the intersubband absorption coefficient for the *Z*-polarized EMW in the case of $\hbar\omega_0 = \hbar\omega$ (transition I, Fig. 1). Curves 1 ($\delta = 0$) and 2 ($\delta = 1$) in Fig. 3 were obtained at $R = 270$ Å, $T = 40\,°K$. This corresponds to $b_1 = 71.54$, $\tau_0 = 12.35$, $\hbar\omega_f = 2.79 \cdot 10^{-4}\,eV$ for curve 1 and $b_1 = 54.96$, $\tau_0 = 8.74$, $\hbar\omega_f = 3.95 \cdot 10^{-4}\,eV$ for curve 2.

Fig. 4 (in arbitrary units) shows the influence of temperature on the value $K^{(Z)}(\Omega)$. Curves 1 ($T = 40\,°K$) and 2 ($T = 60\,°K$) in Fig. 4 were obtained at $R = 270$ Å, $\delta = 1$. This corresponds to $b_1 = 54.96$, $\tau_0 = 8.74$, $\hbar\omega_f = 3.95 \cdot 10^{-4}\,eV$ for curve 1 and $b_1 = 39.20$, $\tau_0 = 10.00$, $\hbar\omega_f = 5.17 \cdot 10^{-4}\,eV$ for curve 2.

As it follows directly from Figs. 2, 3, and 4, the magnetoabsorption band of the linearly-polarized EMW splits into two peaks. Without an account of the resonant optical phonons, the band of the intersubband light absorption is centered at a frequency of $\Omega = \omega$ (or $\Omega = \Omega_e$) which corresponds to the transition I (or II) in Fig. 1. The rise of magnetic field strength $\mathbf{H}$ results in the broadening of both peaks and their shift away from each other (see Fig. 3), so that the maximum of the 1st peak (long-wavelength) moves into the low-frequency domain of the spectrum, and the maximum of the 2nd peak (short-wavelength) moves into the high-frequency domain of the spectrum. It should be noted that with the rise of temperature (see Figs. 2, 4) the half-width of the long-wavelength peak increases, but the half-width of the short-wavelength peak is weakly dependent on temperature. Also, with the rise of temperature both peaks are shifted into the low-frequency domain of the spectrum (this trend continues up to $T = 100\,°K$).

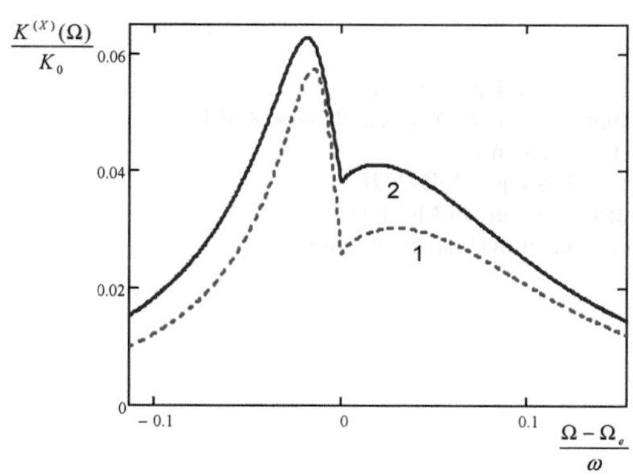

Fig. 2 Frequency dependence of intersubband absorption of linearly-polarized EMW with the *X*-polarization (arbitrary units) at different temperatures. Curve 1 – at $T = 40\,°K$ and curve 2 – at $T = 60\,°K$.

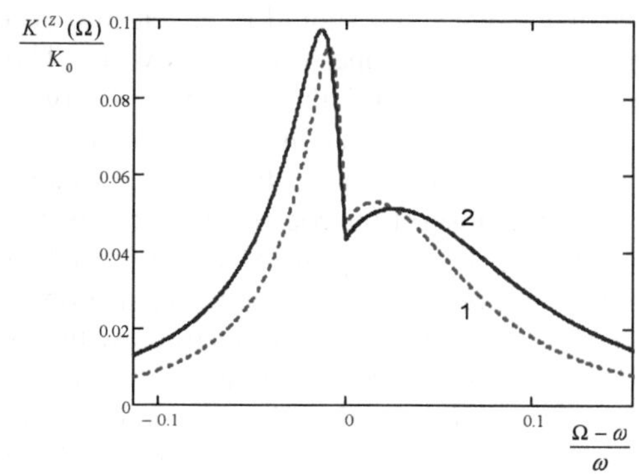

Fig. 3 Frequency dependence of intersubband absorption of linearly-polarized EMW with the *Z*-polarization (arbitrary units) at different values of magnetic field strength. Curve 1 – at $\delta = 0$ and curve 2 – at $\delta = 1$.

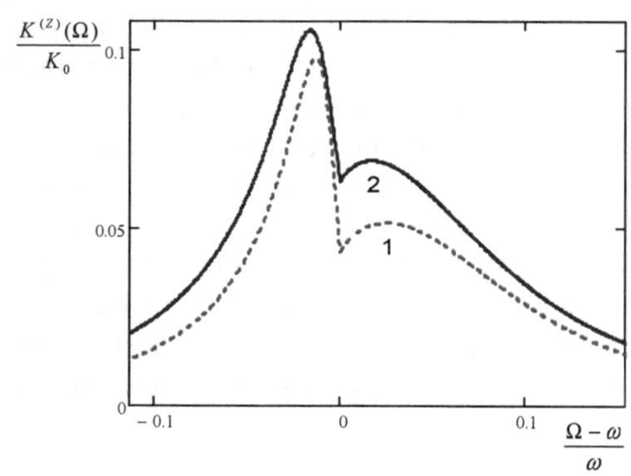

Fig. 4 Frequency dependence of intersubband absorption of linearly-polarized EMW with the *Z*-polarization (arbitrary units) at different temperatures. Curve 1 – at $T = 40\,°K$ and curve 2 – at $T = 60\,°K$.

From the obtained results it follows that the half-width of the 1st absorption peak is almost completely determined by the interaction of electrons with the acoustic phonons, and the half-width of the 2nd absorption peak mainly depends on the interaction of electrons with the optical phonons.

In the case of a circularly-polarized EMW, the intersubband absorption coefficient is equal to a half of the sum of corresponding components. So, for the EMW with the *XY*-polarization (light propagates in the direction of the magnetic field vector **H**): $K^{(XY)}(\Omega) = \frac{1}{2}[K^{(X)}(\Omega) + K^{(Y)}(\Omega)]$. Hence, the frequency dependence of the intersubband absorption coefficient for the circularly-polarized EMW has the features similar to those shown in Figs. 2, 3, and 4.

IV. CONCLUSIONS

The results presented in this paper show that the interaction of band carriers with the resonance optical phonons is an important factor, which should be taken into account in the study of the optical properties of nanowires. The resonance optical phonons essentially change the energetic spectrum of the carriers in the nanowires and that, as a result, leads to the emergence of the features in the frequency dependences of the absorption coefficients of a weak EMW. So, it was shown that the peak of the intersubband light absorption splits into two due to the resonance optical phonons. Herewith, the interaction of electrons with the resonance optical vibrations gives the main contribution into the half-width of the short-wavelength peak, and the interaction of electrons with the long-wavelength acoustical vibrations gives the main contribution into the half-width of the long-wavelength peak.

Thus, the offered theoretical approach is quite acceptable to describe the intersubband absorption of the light with different polarizations in the semiconductor nanowires and to investigate the influence of the external parameters (magnetic field, temperature) and the internal parameters of nanowires on their optical absorption spectra.

Finally, basing on the fact that the quantum wires along with other size-quantized systems (quantum dots and films), as the basic elements, are widely used in optoelectronics and nanoelectronics, it should be noted that the performed research is actual for biotechnological applications. Since, the results obtained here can be claimed at designing of the various biosensors and at creating of the new nanostructured biocomposites whose properties are controlled by the external fields. Also, the given results are interesting for the empowerment of diagnostic methods of organic materials, such for example, as a method of organic mass spectrometry [5], in which the quantum wires have found an effective implementation.

ACKNOWLEDGMENT

This work was supported by the STCU (Grant 5929) and the CSSDT of the Academy of Sciences of Moldova (Institutional Project 15.817.02.03A). The authors would also like to express their gratitude to Dr. Olga Iliasenco for technical assistance in preparing this manuscript in English.

CONFLICT OF INTEREST

The authors declare that they have no conflict of interest.

REFERENCES

1. Синявский Э.П., Костюкевич Н.С. (2013) Особенности поглощения слабой электромагнитной волны в нанопроволоках. Оптика и Спектроскопия 114(2):225–231.
2. Beenakker C.W.J., Van Houten H. (1991) Solid state Physics. Semiconductor Heterostructures and Nanostructures vol. 44, (Ed. by Ehrenreich H. and Turnbull D.). Academic Press, New York, p. 83.
3. Синявский Э.П., Хамидуллин Р.А. (2002) Особенности электропроводности параболической квантовой ямы в магнитном поле. ФТП 36(8):989–992.
4. Зенченко В.П., Синявский Э.П. (1975) К теории поглощения света в собственных полупроводниках в однородном магнитном поле. ФТТ 17(11):3138–3140.
5. Bedin S.A., Rybalko O.G., Polyakov N.B., et al. (2010) Metal micro- and nanowires fabricated by matrix synthesis and their application in mass spectrometry. Inorganic Materials: Applied Research 1(4): 359–364.

Author: E.Yu. Kanarovskii
Institute: Institute of Applied Physics, ASM
Street: Academy, 5
City: Kishinev, MD-2028
Country: Republic of Moldova
Email: kanarovskii@gmail.com

Peculiarity of High-Field Galvanomagnetic Effects in Bicrystals of Bi and Its Alloys with Sb

F.M. Muntyanu[1,2], A. Gilewski[2], V. Chistol[4], and K. Rogacki[2,3]

[1] Institute of Electronic Engineering and Industrial Technologies, Academy of Sciences of Moldova, Chisinau, Moldova
[2] International Laboratory of High Magnetic Fields and Low Temperatures, Wroclaw, Poland
[3] Institute of Low Temperatures and Structural Research, Polish Academy of Sciences, Wroclaw, Poland
[4] Technical University of Moldova, Chisinau, Moldova

Abstract— **We present the results of investigation of high-field (up to 40 T) galvanomagnetic effects in bicrystals of semimetalic Bi and 3D topological insulator Bi1−x Sbx (0.07 < x <0.22) with nano-width crystallite interfaces (~100 nm). At B>2T directed along the interface plane in the quantum oscillations spectrum of Bi and Bi-Sb bicrystals, two new harmonics have been detected. Their periods of oscillation characterize the much larger cross-sectional areas of Fermi surface of charge carriers than it is in single crystalline specimens. In small disorientation angle Bi bicrystals of an inclination type, a number of Hall quasi-plateaus were observed, which vanish by reversing the magnetic field. It has been also found that in bicrystals of Bi-Sb with small crystallite disorientation angle, the semiconductor-semimetal transition is induced in crystallites and interface layers at different values of magnetic field.**

Keywords— **Bismuth, Bismuth-antimony alloys, semiconductor – semimetal transition, superconductivity, quantum oscillations, ferromagnetism, 3D topological insulator.**

I. INTRODUCTION

Rhombohedral Bi is a semimetal with a small (~40meV) energy overlap between the conduction and valence bands. The highly anisotropic Fermi-surface (FS) consist of one hole ellipsoid of revolution (extended along trigonal C_3 axis) at the T point and three very elongated electron ellipsoids (tilted by a ~ 6,2° out of the plane defined by the bisectrix C_1 and binary axis C_2) at the L points of the Brillouin zone. Numerous studies carried out lately confirm that the electrons have a Dirac- like dispersion, while the energy-momentum dispersion of holes is parabolic. Recently, a number of impressive results in high magnetic fields directed along highest-symmetry C_3 axis were recorded in Bi: three peaks in the Nernst signal far beyond quantum limit (9 T) [1,2], jumps in magnetisation [3], Hall plateaus [1,3-5], etc.

The $Bi_{1-x}Sb_x$ $(0 < x < 1)$ alloy system has emphasized its importance [7-9] after discovery (at x=0.1) of the first 3D Z_2 topological insulator (3DTI) and the opportunity to extended study of the other two significant modern topics - three-dimensional Dirac particle [9] and charge quantum Hall fractionalization. Substituting Bi with Sb is expected to change both the spin-orbit interaction and critical energies of the band electronic structure especially in L points of the bulk Brillouin zone. So, for the x=0.04 the existing in Bi gap between L_a and L_s pockets closes and the 3D Dirac point is realized. If x grows again the gap in L points re-opens with inverted symmetry ordering. At x≈0.07 bismuth specific bands overlapping (L-T_{45}) disappears and the alloys becomes an inverted-band semiconductors up x=0.2, dominated by spin-orbit coupled Dirac particle at L. The surface band structure of semiconducting Bi_{1-x} Sb_x alloys has been studied for x=0.1, using incident-photon-energy-modulated angle-resolved photoemission spectroscopy. It was found that it consists [9] of single electron pocket surrounded Γ point and six elliptical hole pockets centered along the six line connecting Γ and M points of the surface Brillouin zone. The Fermi surface includes an odd number of Dirac points. The band structure of interfaces for Bi and Bi_{1-x} Sb_x alloys was investigated in [10]. It was proved that Fermi surface consisting of interfaces is similar to those of bulk specimens, but the shape, elongation, and volume of isoenergetic surfaces are essentially distorted.

II. EXPERIMENTAL PROCEDURE

The high-quality bicrystals with the Dingle temperature of charge carriers Td ≤ 1K were obtained by the zone recrystallization method using the double seed technique. The composition of the samples was controlled by a SEM equipped with Oxford and PV 9800 energy-dispersive X-ray (EDX) analyzers. The quantum oscillations of magnetoresistance and the longitudinal Hall effect were registered in stationary (up to 18) and pulse magnetic fields (up to 40T), directed along the CI plane near the C_3 axis. The measurements were performed in the International Laboratory of High Magnetic Fields and Low Temperatures (Wroclaw, Poland). The bicrystals of Bi and $Bi_{1-x}Sb_x$ (x ≤ 0.2) alloys, consisting of two single crystalline blocks (3D TIs) and the nano-width crystallite interfaces (CIs), have one of the external surfaces (i.e. interfaces with air) replaced with the CI composed of a solitary central part (the thickness about 60 nm) and two similar adjacent layers (~20 nm) on both sides of it. Two groups of bismuth and

V. Sontea and I. Tiginyanu (eds.), *3rd International Conference on Nanotechnologies and Biomedical Engineering*,
IFMBE Proceedings 55,
DOI: 10.1007/978-981-287-736-9_3

bismuth-antimony bicrystals were studied: bicrystals with small crystallite disorientation angle (SDA, $\theta_1 < 9^o$) and bicrystals with large crystallite disorientation angle (LDA, $\theta_1 > 12^o$). At low temperatures the CIs of Bi bicrystals are superconducting while interfaces of $Bi_{1-x}Sb_x$ bicrystals exhibit ferromagnetic and simultaneously superconducting properties. Note, that the single crystalline bulks (crystallites) do not show superconductivity or ferromagnetism. The magnetoresistance and Hall effects of bicrystls of Bi and $Bi_{1-x}Sb_x$ ($x \leq 0.2$) alloys were studied after transition of CI in the normal state (converted by magnetic field and/or a current).

III. Results and Discussion

Fig.1a shows that the spectrum of longitudinal Hall quantum oscillations of small disorientation angle (SDA) Bi bicrystals of inclination type is complex and contains frequencies from the Fermi surface of crystallites and interfaces. Above 2T, the oscillation peaks take an unusual configuration and their position is essentially shifted from of the oscillation maxima of Hall resistivity of elemental Bi as well as of magnetoresistanse in both bicrystals and single crystalline samples [11]. Between the peaks, along with minima in magnetoresistance, a number of Hall quasi-plateaus (~3T, ~6T, ~15T) were found that are indicator of bulk electron fractionalization. In addition to the known frequencies of quantum oscillations of Bi (see Fig.1b) two new harmonics are observed in our bicrystals since 2 - 2.5T (first frequency) and 10T (second frequency), respectively. Their periods of oscillation characterize larger than in single crystalline Bi cross-sectional areas of the FS and are related to the central and adjacent layers of interfaces. At least one of the harmonics defines extremal cross-sectional areas of Fermi surface which is almost 70 times higher than that of FS of crystallites. It should be noted, that the Hall plateaus and changing of the shapes of oscillation peaks are clearly pronounced only in SDA bicrystals of inclination type, while the new harmonics occur in all investigated Bi bicrystals (both SDA and LDA). In comparison with SDA bicrystals of inclination type, the LDA bicrystals have an increased disorder strength and high dislocation density; therefore, the Hall plateaus are not found and the localized states in the spectrum of Landau levels do not appear. According to our data, the FS of the majority carriers of LDA bicrystals considerably exceeds in volume [8] the holes isoenergetic surface of single crystalline Bi. Thus, the presence of non-interacting carriers that contributes to fractionalization seems to be problematic. Another interesting feature of the development of Hall plateaus (including the one at 15T) is they vanish by magnetic field reverse, therby indicating that in SDA bicrystals the flow of Dirac fermions along the CI plane is sensitive to the field orientation and as well as that the localization process runs only in a specific direction of the magnetic field.

The single crystalline $Bi_{1-x}Sb_x$ ($0.07 < x < 0.15$) alloys provide a remarkable opportunity to obtain ultra quantum limit (UQL) in lower magnetic fields than in Bi, where the electrons (or holes) occupy the lowest ($j = 0$) Landau level of the carrier pockets. Beyond UQL the band edge displacement $\Delta\varepsilon$ takes place (see inset of Figure 1b), depending on the ratio of spin $\Delta\varepsilon_s$ and orbital $\Delta\varepsilon_o$ level splitting: $\Delta\varepsilon = ½h\omega(1 - \Delta\varepsilon_s/\Delta\varepsilon_o)$. Consequently, the various electronic phase transitions occur in magnetic fields, including the semiconductor-semimetal transition.

Figure 1c shows the field dependency of the magnetoresistance $\Delta\rho/\rho$ in a single crystal and twisting SDA bicrystals of $Bi_{1-x}Sb_x$ ($0.07 < x < 0.15$) alloys. For the stationary magnetic field oriented along the CI plane, regular Shubnikov oscillations are observed up to 2 T. In higher fields charge carriers are in the UQL and at B ~8.5 T (see the maxima in Figure 1c) the semiconductor-semimetal transition takes place both in the single crystal and in the SDA bicrystals. Note, that for the SDA bicrystals, the first maximum in the $\Delta\rho/\rho$ versus field dependency is observed at the same field as for the single crystal, which means that the semiconductor-semimetal transition is induced in the bulk part of the bicrystals.

For fields higher than 11 T, an additional maximum is observed in the bicrystals only. This maximum appears at different fields, for the bicrystals with different disorientation angles Θ_1 and Θ_2. We believe that the second maximum reflects the semiconductor-semimetal transition in the adjacent layers of CI of the SDA bicrystals. The transitions appear at quite different values of the magnetic field, so the cyclotron mass of the relevant charge carriers and the ratio of spin and orbital level splitting $\Delta\varepsilon_s/\Delta\varepsilon_o$ in crystallites and in the CI region of the SDA bicrystals should differ considerably. This implies a significant increase of the spin-orbit interaction and the existence of gapless electronic states at the bicrystls interface. In fields higher than 25 T, third maximum appears which can be attributed to the semiconductor-semimetal transition in the central layer of CI.

The $\Delta\rho/\rho$ features related to the semiconductor-semimetal transition have not been observed for the LDA ($\Theta_1 > 12^o$) $Bi_{1-x}Sb_x$ bicrystals. For these bicrystals, however, a number of harmonics of the quantum oscillations of the resistance $\rho_{ii}(B)$ and the longitudinal Hall-effect resistance $\rho_{ij}(B_i)$ has been observed, similarly as revealed for Bi bicrystals. These oscillations are shown in Figure 1d and have been attributed to the Fermi surface properties of the CI areas.

The oscillation spectra of the LDA $Bi_{1-x}Sb_x$ bicrystals we analyse graphically based on B_n^{-1} position of the oscillation peaks versus the Landau level index, n, where B_n is the field

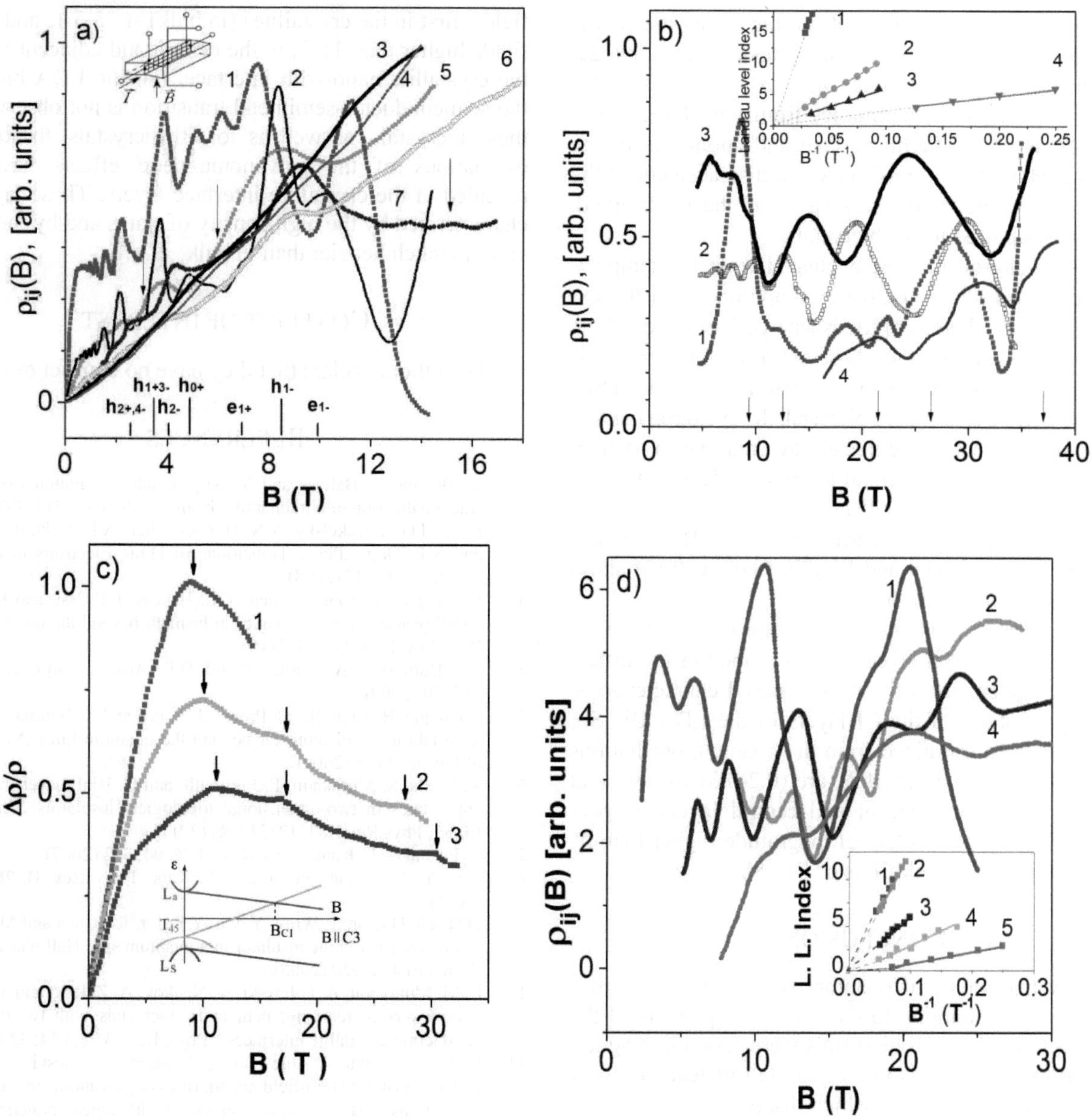

Fig. 1 The quantum oscillations of magnetoresistance ρii (B) and longitudinal Hall effect ρij (Bi) in Bi (a,b) and Bi 1-x Sb x (c, d) bicrystals of inclination (a) and twisting (b, c, d) types. (a) (1) ρij (Bi +), 4,2K, θ1 = 4o; (2) ρij (Bi), 2,1K , θ1 = 5o ; (3) ρij (Bi), 4,2K, θ1 = 4o; (4) ρii (B), 2,1K , θ1 = 5o ; (5) ρij (Bi), 4,2K, θ1 = 7o; (6) ρij (Bi), 4,2K, θ1 = 8o. The marks denote the position of oscillation peaks in Bi single crystals. (b) (1) ρij (Bi), 1,5K, θ1 = 5o , θ2 =5,5o ; (2) ρij (Bi), 4,2K, θ1 = 62o , θ2 =2o ; (3) ρij (Bi), 1,6K, θ1 = 3o , θ2 =5o ;(4) ρii (B), 4,2K, θ1 = 29o , θ2 =11o . Inset (b): The B-1 position of the Hall peaks versus their Landau level index: (1, 2) for the oscillation curve 4; (3, 4) for the oscillation curve 2. (c) (1) single crystal, x = 0.08; (2) and (3) bicrystals: (2) x = 0.08, Θ1 = 4o, Θ2 = 2o; (3) x = 0.09, Θ1 = 12o, Θ2 = 2o. Arrows point the fields at which the semiconductor-semimetal transitions appear. Inset: Band evolution in high magnetic fields (UQL) directed along the trigonal axis C3 in bulk (crystallites) Bi1-xSbx (0.07 < x < 0.15). BC1 is the critical magnetic field for the semiconductor-semimetal transition. (d) Twisting LDA bicrystals of Bi1-xSbx (x = 0.08, 0.12, 0.15) and Bi0.93Sb0.07Te at 4.2 K: (1) ρii (B), x = 0.08, Θ1 = 15o, Θ2 = 3o; (2) ρij(Bi), x = 0.12, Θ1 = 12o, Θ2 = 2o; (3) ρii(B), Bi0.93Sb0.07Te, Θ1 = 19o, Θ2 = 2o; (4) ρii(B), x = 0.15, Θ1 = 15o, Θ2 = 3o. The magnetic field is directed along CI plane (near C3 axis of crystalline bloks). Inset: Landau level index, n, versus the Bn-1 position of the oscillation

at which the Fermi level falls between two Landau levels. The relation between B_n and n is described by the formula: $1/B_n = ne/(hN_s)$, where: e is the elementary charge, h is the Planck's constant, and N_s is the surface density of states. The B_n-n relations obtained for three LDA bicrystals are shown in the inset of Figure 1d. Two oscillation frequencies related to the Fermi surface of the central and adjacent layers of the crystallite interfaces have been observed. The results show that the interface electronic states are of the Schrödinger type, since for these states, n takes integer values and for Dirac electrons, $n_D = n+½$. Note that for these $Bi_{1-x}Sb_x$ bicrystals, harmonics from very large cross-sectional areas of the Fermi surface and anomalies of the longitudinal Hall effect (quasi-plateaus, displacement of the oscillation peaks position, *etc.*) have not been observed, differently as reported for Bi bicrystals.

Carriers' cyclotron masses, evaluated from the temperature dependencies of the oscillation amplitudes and the extreme cross-sectional areas of the isoenergetic surfaces of the CI layers exceed several times those observed for crystallites. For example, in magnetic field oriented along the interface plane in the LDA $Bi_{0,93}Sb_{0,07}Te$ bicrystal, the cyclotron mass of carriers m_c/m_e (where m_e is the free electron mass) in crystallites (bulk) and CI adjacent and central layers is 0.05, 0.25 and 0.5, respectively. The quantum oscillations in LDA $Bi_{1-x}Sb_x$ bicrystals have similar spectrum characteristics as in Bi bicrystals, although Bi is a semimetal and $Bi_{1-x}Sb_x$ ($0.07<x<0.15$) alloys are narrow-gap semiconductors.

This is a strong argument for quantum origin of the effects observed for the CI layers, which properties are different and independent form the bulk material characteristics. The density of states of the CI layers of the LDA $Bi_{1-x}Sb_x$ bicrystals has been estimated from the quantum oscillations as described in. The values of N_s are (0.2-0.3) 10^{19} m^{-2} and (1.5-2.5) 10^{19} m^{-2}, for adjacent and central layers, respectively, which is several orders of magnitude higher than N_s obtained for similarly thick films of $Bi_{1-x}Sb_x$ alloys.

IV. Conclusion

Two new harmonics of quantum oscillation of longitudinal Hall effect belonging to the FS of components of the superconducting CIs of Bi bicrystals were found in magnetic fields up to 40T. We also detected Hall plateaus in SDA bicrystals of inclination type, which disappeared at magnetic field reversal. These results indicate that in bicrystals there are spin-oriented carriers which are located in spectrum of Landau levels at particular orientation of the magnetic field. We have specified quantum characteristic features of the twisting bicrystals of a 3D topological insulator $Bi_{1-x}Sb_x$ ($0.07 < x < 0.15$). For the SDA bicrystals, the semiconductor-semimetal transition is induced in high magnetic fields, first in the crystallites (in bulk) at ~8.5 T, and then, at fields higher than 11 T, in the central and adjacent layers of the crystallite nano-with interface. For the LDA bicrystals, the semiconductor-semimetal transition is not observed. For these bicrystals, as well as for Bi bicrystals, the quantum oscillations of the galvanomagnetic effects have been revealed in the crystallite interface layers. These layers are characterized by the high density of states and by the charge carriers much heavier than in bulk.

Conflict of Interest

The authors declare that they have no conflict of interest.

References

1. K. Behnia, L. Balicas and Y. Kopelevich, ''Signatures of electron fractionalization in ultraquantum bismuth", Science 317, 1729 (2007)
2. L. Li, J.G. Checkelsky, Y.S. Hor, C. Uher, A.F. Hebard, R.J. Cava and N.P. Ong, „Phase Transitions of Dirac Electrons in Bismuth", Science 321, 547 (2008)
3. B. Fauque, H. Yang, I. Sheikin, L. Balicas, J.-P. Issi, and K. Behnia, ''Hall plateaus at magic angles in bismuth beyond the quantum limit", Phys. Rev. B79, 245124 (2009)
4. F.J. Burnell, B.A. Bernevig and D.P. Arovas, Phys. Rev. B **79**, 155310 (2009)
5. B.Fauque, B. Vignolle, C. Proust, J.-P. Issi and K. Behnia, "Electronic instability in bismuth far beyond the quantum limit",New Journal of Physics **11**, 5 (2009)
6. V. Wada, S. Murakami, F. Freimuth, and G. Bihlimayer," Localized edge states in two-dimensional topological insulators: Ultrathin Bi films", Phys.Rev.B83, 121310(R) (2011).
7. L. Fu and C. L. Kane, Phys. Rev. B **76**, 045302 (2007)
8. J. C. Y. Teo, Liang Fu, and C. L. Kane, Phys. Rev. B. **78**, 045426 (2008)
9. D.Hsieh, D.Qian, L.Wray, Y.Xia, Y.S.Hor, R.J.Cava and M.Z.Hasan, ''A topological Dirac insulator in a quantum spin Hall phase", Nature (London) **452**, 970 (2008)
10. F. M. Muntyanu, A. Gilewski, K. Nenkov, A. Zaleski and V. Chistol, "Fermi-surface rearrangement in Bi bicrystals with twisting superconducting crystallite interfaces", Phys. Rev. B **76,** 014532 (2007)
11. F. M. Muntyanu, A. Gilewski, K. Nenkov, A. Zaleski, T. Palewski and V. Chistol, "High-field peculiarities of galvanomagnetic quantum oscillations in Bi bicrystals with nano-width superconducting crystallite interfaces", Solid State Comm. **158** 9 (2013).

Author: Vitalie Chistol
Institute: Technical University of Moldova
Street: Ştefan cel Mare, 168
City: Chisinau
Country: R. Moldova
Email: chistol_vitalie@yahoo.com

Effective Transfer of UV Energy to Red Luminescence in the Nanocomposites Polymer/Eu Coordination Compounds

V.I. Verlan[1], M.S. Iovu[1], I. Culeac[1], O. Bordian[1,2], V.E. Zubareva[2], and Iu. Nistor[1]

[1] Institute of Applied Physics, Academy of Sciences of Moldova, Chisinau, Republic of Moldova
[2] Institute of Chemistry of the Academy of Sciences of Moldova, Chisinau, Republic of Moldova

Abstract— **Nanocomposites (NCs) thin films based on oligomer poly-N-epoxypropylcarbazole (PEPC) and organic coordinated compounds (OCC) have been obtained by chemical methods. Absorption threshold for samples with different concentrations was determined as 3.34 - 3.4 eV. Photoluminescence spectra reveal a number of emission bands centered at 580, 590, 612-615, 651, and 700 nm, which can be associated to internal 4f - 4f transitions of Eu3+ ion 5D0 → 7Fi (i = 0,1,2,3 and 4). By comparing the photoluminescence spectra of the organic compound and of the PEPC/Eu(o-MBA)3Phen nanocomposite one can observe the amplification of the photoluminescence intensity in NC.**

Keywords— **Nanocomposites, rare earth, energy transfer, polymer.**

I. Introduction

The elements of rare earth and compounds on their base possess the unique properties for practical applications: as active materials for the electroluminescent devices (EL), lasers and luminescent biosensing applications with efficiency luminescence and temperature insensitivity, ferromagnetic materials, and high-temperature superconducting [1-3]. Its functional properties are determined basically by *4f* electrons layer which is shielded of outer electronic layers $5s^2 5p^6$.

For development of various light sources for the visible domain various compounds with Eu^{3+}, Gd^{3+} or and Tb^{3+} ions are most commonly used. The enhancing of their luminescence properties depends on the type of ligands used for surrounding of the rare earth ion. The research efforts are oriented to achieving of surrounding Eu^{3+} ion with ligands in order to improve luminescence efficiency, and to obtain a final stable compound for different practical applications. Advantage of application of OOC with rare earth ions with β-diketonates and ligands in nanocomposites is based on the so-called "antenna" effect. The intermolecular energy transfer to the Eu^{3+} ion takes place via the levels of ligands and their surroundings. However, their emission yields are yet low, because their coordination ligands are attributed to the nonradiative transition via relaxation of high-vibrational frequency *O–H* and *N-H* bonds of OOC structure [4,5].

It is necessary to find new coordination compounds of RE with a high thermo-stabilities and high emission efficiency for optoelectronic devices. In the previous paper we reported preliminary investigations of the NCs made of copolymer butylmetacrylate and styrene (1:1) with $Eu(TTA)_3H_2O$, $Eu(TTA)_3Phen$, $Eu(DBM)_3Phen$ [6], as well as made of polyvinylpirrolidone with $Eu(TTA)_2(Ph_3PO)_2NO_3$.

This paper is focused on investigation of organic compound with Eu^{3+} ion surrounded by ligands of o-MBA and of phenanthroline in the matrix of *PEPC* with the final goal to investigate the luminescence mechanism and energy transfer to Eu^{3+} ion.

II. Preparation Technology and Experimental Set-up

Ortho-methylbenzoic acid *o-MBA* is used as negative ionic ligand and phenanthroline (Phen) as neutral ligand for preparation of nanocristals $Eu(o\text{-}MBA)_3Phen$. The chemical structure of the constituent components are illustrated in (Fig. 1).

Synthesis of rare-earth complex $Eu(o\text{-}MBA)_3Phen$: a warmed ethanol solution (96%) containing 0.4 g (3 mmol) of o-methylbenzoic acid and 0.2 g (1 mmol) of 1,10 – phenanthroline was adjusted to pH 6.0 – 7.0 with 1M *NaOH* solution. Further, europium chloride (1 mmol) dissolved in 5 ml of water was added dropwise to the organic mixture and stirred. Light pink precipitate was formed immediately. The precipitate was filtered, washed with small portions of ethanol, dried thoroughly in air. The synthesis yield was 0.56 g (37.6%). For $C_{36}H_{29}O_6N_2$, it was calculated, %: *Eu* - 20.61, *C* - 58.63, *H* - 3.97, *N* - 3.80 . Found, %: *Eu* - 20.88, *C* - 59.23, *H* - 4.17, *N* – 3.75.

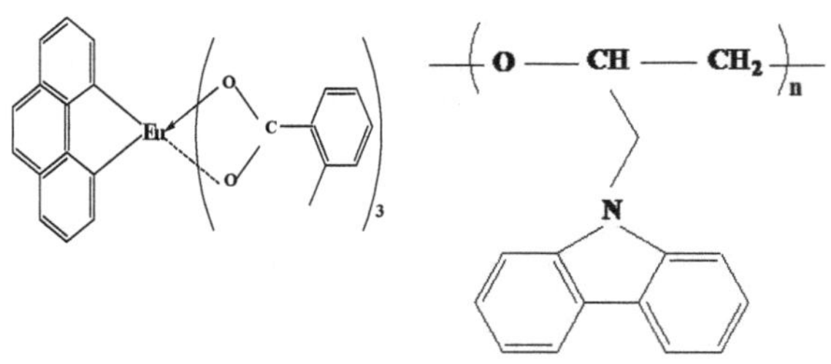

a b

Fig. 1 Illustration of chemical structure of the constituent components: $Eu(o\text{-}MBA)_3Phen$ compound (a) and poly-N-epoxypropylcarbazole (b).

V. Sontea and I. Tiginyanu (eds.), *3rd International Conference on Nanotechnologies and Biomedical Engineering*,
IFMBE Proceedings 55,
DOI: 10.1007/978-981-287-736-9_4

All reagents had analytical grade and were used without further purification.

The OCC-Europium complexes were successfully incorporated into organic matrixes of oligomer *PEPC* via sol–gel process. NCs *PEPC/Eu(o-MBA)$_3$Phen* were obtained from chemical solutions at different molar ratios of OCC into polymer matrix. The technology of NCs was similar of those described in [6]. Spin-coated thin films were obtained on optical glass and quartz substrates.

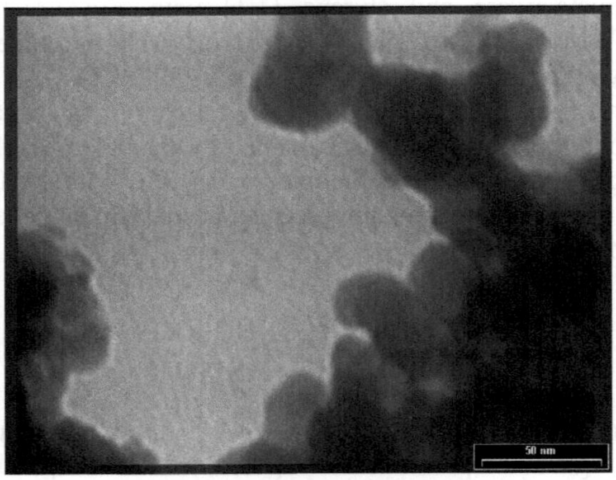

Fig. 2 TEM image of NC showing the distribution of NCs particles in PEPC.

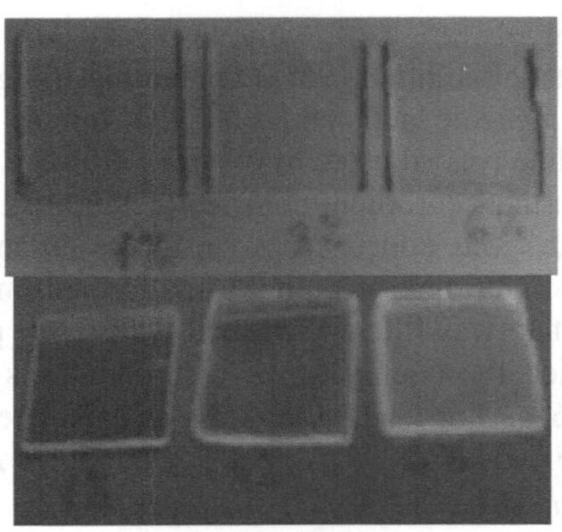

Fig. 3 Photos of thin films *PEPC/Eu(o-MBA)3Phen* nanocomposites on quarz substrates without (a) and during their lighting with ultraviolet light (b).

PEPC and polycristals *Eu(o-MBA)$_3$Phen* were dissolved in toluene (ρ(*PEPC*) = 161.9 mg/mL and ρ(*Eu(o-MBA)$_3$Phen*) = 9.3 mg/ml) by ultrasonic dispersing at room temperature. Solution of *Eu(o-MBA)$_3$Phen* of different concentrations (5.74%, 2.87%, 1.43%) was introduced in the polymer solution and the final composition was mixed using an ultrasonic bath. Deposition of nanocomposites on quartz substrates was performed applying the spin-coating method (Fig. 2). The samples were dried at room temperature 80-90 °C for 2 hours.

The dried layers with thickness around 1 μm (Fig. 2) are transparent in visible range of spectrum. Microscopic investigations of the morphology of the NC films (Fig. 3) show that the dimensions of OCC particles, incorporated in the polymer matrix are less than 100 nm.

III. Experimental Results

Fig. 4 shows the absorption spectra of nanocomposite *PEPC/Eu(o-MBA)$_3$Phen* at different concentrations of *Eu(o-MBA)$_3$Phen* (5.7%, 2.7%, 1.3%) in 350 – 650 nm range. The absorption threshold at different concentrations of NC were approximately the same (3.34 – 3.4 eV). With increasing concentration of *Eu(o-MBA)$_3$Phen* in NC it shifts to infrared.

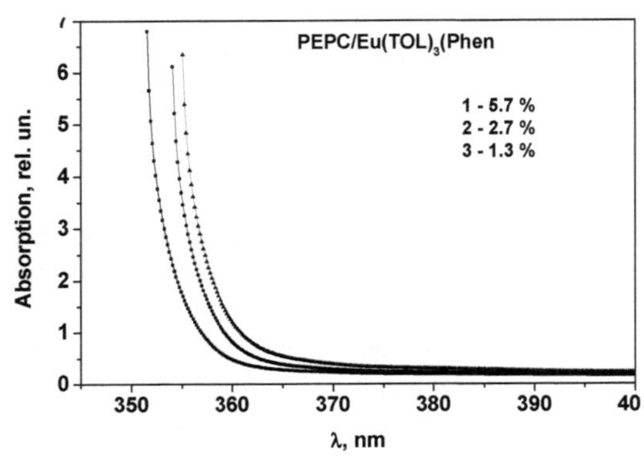

Fig. 4 Optical absorption spectra of NC with different concentrations of *Eu(o-MBA)$_3$Phen*.

Fig. 5 (a) illustrates the photoluminescence spectra of the *NC PEPC/Eu(o-MBA)$_3$Phen* under the excitation of the laser beam 405 nm. The luminescence bands are registered in the range 450-700 nm. When increasing the concentration of the Europium complex in the *NC* the PL intensity increases significantly.

The PL bands are positioned in a relatively wide range 570 – 700 nm (579, 590, (612, 620), 650, 687) nm (Fig. 5) and correspond to radiative transitions of internal energy levels of Eu^{3+} ion. The main emission band of the PL spectrum has a specific shape consisting of three maxima: first – at 612 nm and the third – at 620 nm, and their intensity is about 16 times higher than other PL bands, while its halfwidth is 15 nm. These bands can be assigned to optical transitions $^5D_0 \rightarrow {}^7F_i$ (i= 0, 1, 2, 3, 4).

IV. Discussion of Results

PL spectra characterize the efficiency of energy transfer from the LUMO energy levels of polymer and ligands of the compound to the energetic levels of Eu^{3+} ion. HOMO and LUMO levels of the complex *PEPC/Eu(o-MBA)$_3$Phen* are found to be situated below levels *S* and *T* of ligands, and this case is similar to the case of doping the semiconductor. The photoluminescence properties can be attributed to the internal transition of Eu^{3+} ion $^5D_0 \rightarrow {}^7F_i$ ($i = 0,1,2,3$ and 4), which are centered at 579, 590, (612, 620), 650, 687 nm.

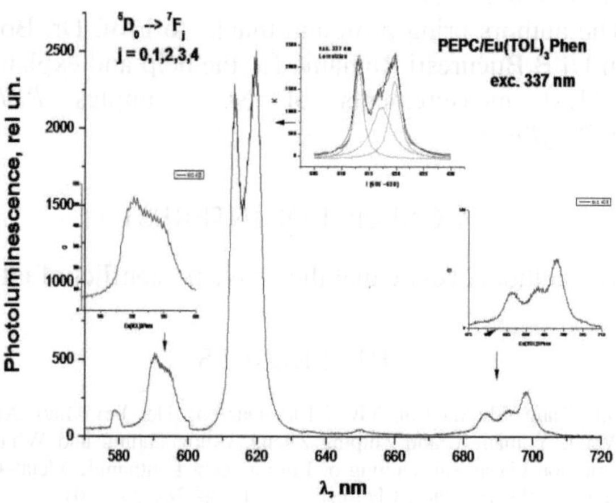

Fig. 5 Detailed photoluminescence spectrum of the NC PEPC/ Eu(o-MBA)3 Phen.

Optical absorption and luminescence properties of luminescent rare earth ions are affected by the local bonding environment and by the distribution of the rare-earth particles in the PEPC matrix. Important spectral changes reffering to the emission bands relative intensity are registered. The relative weakness of the photoluminescence band associated to the transition $^5D_0 \rightarrow {}^7F_0$ compared with $^5D_0 \rightarrow {}^7F_2$ transition reflects the asymmetric shell of the Eu^{3+} bonding environment. The PL spectra of the samples *PEPC/Eu(o-MBA)$_3$Phen* demonstrate that the effect of ligands is significant. These spectra show a significant enhancement of the fluorescence related to $^5D_0 \rightarrow {}^7F_2$ transition, as an indicative of low symmetry sites of *Eu(o-MBA)$_3$Phen*.

While variations in PL intensity of $^5D_0 \rightarrow {}^7F_0$ transition is almost unobservable in the absence of chelating agent (*o-MBA*) it becomes very significant upon chelation. The presence of the chelate organic ligands in the inner coordination shell reduces the nonradiative decay probability. PL increased intensity comes from the reduction in the symmetry on the europium species that makes the transition of the electrons allowed.

A rather weak and broad band at 579 nm is associated to the $^5D_0 \rightarrow {}^7F_0$ transition. Based on selection rules, the band at 590 nm is attributed to the $^5D_0 \rightarrow {}^7F_1$ parity-allowed magnetic dipole transition, which is independent of the surrounding symmetry. The $^5D_0 \rightarrow {}^7F_2$ transitions represent allowed electrical-dipole transitions. The most intensive band around (612, 620) nm belongs to the $^5D_0 \rightarrow {}^7F_2$ transition that is over sensitive to the symmetry of the crystal field, surrounding the Eu^{3+} ion. This PL band intensity will increase provided the symmetry is lower. In this respect, it is known, that red-to-orange emission intensities ratio $R = I(^5D_0 \rightarrow {}^7F_2)/I(^5D_0 \rightarrow {}^7F_i)$ is an asymmetry parameter for the Eu^{3+} sites and a measure of the extent of Eu^{3+} interaction with surrounding ligands, indicating on the enviro nment of the Eu^{3+} ion.

In Fig. 5b one can clearly observe two well developed emission lines centered at 652 and 702 nm that originate from $^5D_0 \rightarrow {}^7F_3$ and $^5D_0 \rightarrow {}^7F_4$ transitions, respectively. It is important to note that *Eu(o-MBA)$_3$Phen* nanocrystals itself similarly to NC show strong luminescence emission within these bands. Splitting of the lines of the transition $^5D_0 \rightarrow {}^7F_i$ ($i= 0 - 4$) into three components indicates on the absence of axial symmetry of the inner coordination sphere of the central ion Eu^{3+}. This also suggests a possible greater local symmetry of Europium ions (but not higher C_{2v}). Consequently, we can consider that the fine structure of the PL spectrum depends on the symmetry of the ligands molecules that determines the effectiveness of the energy transfer to Eu^{3+}.

We can observe that the absorption energy of the ligands and of the polymer is higher than emission energy of the Eu^{3+} resonance levels as well as of the energy of triplet level of polymer and ligands. This means that the probability of transition from *T* level to $^5D_0$ level is higher. Finally, the excitation energy received by Eu^{3+} ion from the ligands is determined by its absorption and by the rate of its nonradiative loss. Therefore it is necessary in future found ligands that would have strong absorption at energy near to absorption of the Eu^{3+} ion. In order to decrease the deactivation energy of nonradiative processes of the ligands containing *O-H*, *C-H*, and *N-H* groups it is necessary to replace the hydrogen atom in these groups by deuterium or an halogen (*F, Cl*) [5].

As we can see, the organic ligand plays a key role in the energy transfer processes. In addition to a negatively charged ligand, Europium ion also needs a zero-charged ligand, called "synergetic ligand", for structure stability. The synergetic ligand can reduce, for example, the rate of non-radiative decays and strongly enhance the fluorescence intensity of the complexes.

PEPC/Eu(o-MBA)$_3$Phen compound is a macromolecular system with the presence of donor groups in three-dimensional arrangement, flexible, which allows a cation to be not only complexed, but also encapsulated. The compound exhibits a good solubility in non polar organic solvents and high stability. The Europium ion has low non-radiative processes and thus provides high emission quantum yields. The coordination of Europium with the macrocyclic ligands o-MBA and Phenantroline is an interesting domain of investigation. The interest is determined by its potential application, particularly in supramolecular structures, yielding compounds which can be used in molecular light conversion devices, which absorb radiation in the UV region and emit in visible region. These devices may be used in medical applications such as radioimmunotherapy, positron emission tomography, as contrast enhancing agent in image resonance, luminescent markers, etc.

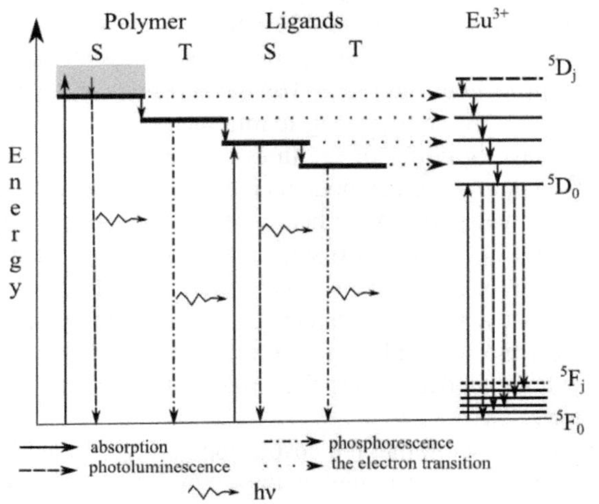

Fig. 6 The energy transitions in photoluminescence of NC *PEPC/ Eu(o-MBA)$_3$Phen*.

V. CONCLUSION

Photoluminescent NC material based on coordination compound *Eu(o-MBA)$_3$Phen* was synthesized and characterized by PL and UV-Vis spectroscopy. The preparation technology of compound *Eu(o-MBA)$_3$Phen* and NC *PEPC-Eu(o-MBA)$_3$Phen* is described. To protect nanocrystals *Eu(o-MBA)$_3$Phen* from the environment and ultraviolet irradiation it is necessary their to surround them with the oligomer PEPC. PEPC is an ideal environment, flexible and optically transparent, that surrounds the nanocrystals.

From the transmission spectra we have identified the absorption threshold at 3.34 – 3.40 eV. It slightly shifts to IR direction with increasing the concentration of *Eu (o-MBA)$_3$Phen* complex NC. Under UV excitation the nanocomposite *PEPC/Eu(o-MBA)$_3$Phen* exhibits an intensive red photoluminescence which is associated to internal optical transitions of Eu^{3+} ion $^5D_0 \rightarrow {}^7F_i$ (i= 0, 1, 2, 3, 4).

The nanocomposite *PEPC-Eu(o-MBA)$_3$Phen* is an perspective object of research and present scientific interest from fundamental and applied point of view.

ACKNOWLEDGMENTS

This work was supported by the Supreme Council for Science and Technological Development of the Academy of Sciences of Moldova (Project no. 11.817.05.03A and no. 11.836.05.04A).

The authors bring profound thanks to Prof. Dr. Bojin D. from UPB Bucuresti Romania for the help and explanations in TEM measurements of NCs samples *PEPC/Eu (o-MBA)$_3$Phen*.

CONFLICT OF INTEREST

The authors declare that they have no conflict of interest.

REFERENCES

1. Qun Tang, Shuxia Liu, Yiwei Liu, Danfeng He, Jun Miao, Xingquan Wang, Yujuan Ji, and Zhiping Zheng. Color Tuning and White Light Emission via in Situ Doping of Luminescent Lanthanide Metal–Organic Frameworks. Inorganic Chemistry, 53 (1), pp. 289-293, 2014.
2. Hui Xu, Jianzhe Wang, Ying Wei, Guohua Xie, Qin Xue, Zhaopeng DengWei Huang. A unique white electroluminescent one-dimensional europium(iii) coordination polymer. J. Mater. Chem. C 3, pp. 1893-1903, 2015.
6. A. Andriesh, S. Buzurniuc, V. Verlan, M.I. Caraman, S. Robu, N. Barbă, Fluorescent properties of nanocomposite organic luminophore compound – polymer. J. of Optoel. and Advan. Mater., 10, Vol. 10, No. 12, 2008, pp. 3353 - 3356 .
7. M.S.Iovu, A.M.Andriesh, S.A.Buzurniuc, V.I.Verlan, C.I.Turta, V.E.Zubareva, M.I.Caraman „New photoluminophore nanocomposite based on organic compound with Eu3+ ions and copolymer styrene-butylmethacrilate", J. of Non-Crystalline Solids 355(37-42), 1890-1892 (2009) http://dx.doi.org/10.1016/j.jnoncrysol.2009.04.047.
8. V.I.Verlan, M.S. Iovu, I. Culeac, Y. Nistor, C.I.Turta, V.E.Zubareva, S. Buzurniuc. Photoluminescence properties of PVP/Eu(TTA)2 (Phen3PO)2NO3 nanocomposites. Journal of Non-Crystalline Solids 2011, 357(3), p. 1004-1007. ISSN 0022-3093. Doi:10.1016/ j.jnoncrysol.2010.10.022.
9. M. S. Iovu, S. A. Buzurniuc, V. I. Verlan, I. P. Culeac, and Yu. H. Nistor. New composite materials for optoelectronic applications. Proc. SPIE Vol. 7297, 729703 (Jan. 6, 2009).

Author: Victor Verlan
Institute: Institute of Applied Physics, Academy of Sciences of Moldova
Street: 5 Academiei str.,
City: MD-2028 Chisinau
Country: Republic of Moldova
Email: vverlan@gmail.com

Transfer of Heat between Electrons and Phonons in Metallic Nanostructures

S. Cojocaru and D.V. Anghel

National Institute of Physics and Nuclear Engineering, Department of Theoretical Physics Magurele, Romania

Abstract— **It is demonstrated explicitly on an example relevant for nano- and micro-scale sensors, bolometric and microcaloric devices how the nanoscale confinement of electron and phonon subsystems affects their coupling and heat exchange. The temperature dependence of the transferred power, P(T), is considered for a layered metal-insulator structure. It is shown that below the phonon crossover temperature T* the heat transfer is strongly enhanced while its T-dependence becomes weaker than in the bulk material. The phonon modes responsible for this enhancement are identified and the analytic form of P(T) is derived, which suggests ways of controlling the heat flux. Quantitative accuracy is confirmed by comparing to the results of recent experiments.**

Keywords— **Enter up to five keywords and separate them by commas.**

I. INTRODUCTION

Coupling of electrons to oscillations of the ion lattice, $H_{e\text{-}p}$, in nanodevices is an area of active research for both practical and fundamental reasons [1]. For example, nanosensors serve various medicinal purposes or as gateways to building other nanoproducts, such as computer chips that work at the nanoscale, e.g. nanorobots. Electronic and thermal properties of all such devices are qualitatively different from their macroscopic analogs. The fundamental reason is that excitation spectra of electrons and phonons, as well as their coupling differ significantly from the bulk material and understanding this modification remains an open issue. Of particular interest is a non-equilibrium situation, commonly realized in nanoelectronic cooling or sensing devices [2] with a typical setup including a metallic element (e.g. a copper nanostrip) mounted on an insulating support element (e.g. a silicone-nitride thin film).

When a current is injected into a metallic strip, the main electron energy dissipation mechanism consists in emission and absorption of phonons. In a stationary regime each of the subsystems is characterized by its own temperature, T_e and T_p, and the heat flux is proportional to their difference. The key characteristics of the heat balance is the temperature dependent power *P(T)* transferred from the heated electron gas to the lattice. In earlier studies, [3], this process has been described theoretically for a bulk metal, resulting in the following expression:

$$P_{3D} = \Sigma_{3D} V_e \left(T_e^{\,5} - T_p^{\,5} \right), \tag{1}$$

where V_e is the volume of the sample and Σ_{3D} is determined by the deformation potential coupling which relates the mass density ρ fluctuations to variation of the Fermi energy E_F :

$$\Sigma_{3D} = 0.01 \cdot k_F^{\,4} \cdot k_B^{\,5} / (\hbar^3 c_l^4 \rho). \tag{2}$$

Here k_F is the length of the Fermi wavevector, k_B – the Boltzmann constant, c_l - longitudinal sound velocity. The temperature dependence (1) fits well the experimentally observed behavior, however the predicted values of Σ are an order of magnitude smaller than in real metals, $\sim 10^{9}\ WK^{-5}m^{-3}$. In [4] an experimentally more realistic situation has been considered, when a thin metallic film is deposited on a bulky insulating substrate. Although the interface introduces a thermal (Kapitza) resistance, this can be neglected at the low temperatures considered, when the dominating thermal phonon wavelength is much longer than the width of the metallic film. More importantly, the phonon spectrum itself is strongly modified: for a semi-infinite medium geometry there exist 5 types of eigen-modes, of which the Rayleigh surface waves become most important. As a result the authors have found a dependence $\sim T^{x}$, where $x > 5$ in Eq. (1), a behavior which was also observed in some experiments with a thick insulating support layer. Nevertheless, the large discrepancy for the magnitude of the heat flux, i.e. still a too small value of Σ, remains unexplained suggesting existence of an additional scattering mechanism.

On the other hand, a non-integer exponent $x < 5$ has been reported, see e.g. [5, 6], for thinner layered structures with an overall thickness *L* below *100 nm* (we have reserved *"d"* for the metal film). It can be seen that the latter enters the range of the thermal phonon wavelength at a crossover temperature,

$$T^{*} \simeq \pi v_s \hbar / \left(k_B L \right) \simeq 1K, \tag{3}$$

where the phonon sub-system becomes quasi two-dimensional. Respectively, for layers thinner than *10 nm* this temperature can reach some dozens of Kelvin. The phonon spectrum is then drastically modified by scattering and reflection from the boundaries as known from the theory of elastic plates [1]. Such nanoscopic systems have been investigated in the context of semiconductor heterostructures [7], e.g. Ge/SiGe, where due to confinement also the charge

V. Sontea and I. Tiginyanu (eds.), *3rd International Conference on Nanotechnologies and Biomedical Engineering*,
IFMBE Proceedings 55,
DOI: 10.1007/978-981-287-736-9_5

carriers show a quasi two-dimensional behavior: a quantum well (QW) state. It has been shown that $H_{e\text{-}p}$ essentially involves only the Lamb vibrational modes and of these just the flexural ones account for the electron scattering rate, τ^{-1}. This theory is based on the outstanding feature of the flexural acoustic Lamb waves of having quadratic (instead of standard linear) low energy dispersion. This ensures a high density of states, which, in turn, provides an enhanced coupling to electron excitations. But how does the scattering rate translate into the actual heat transfer, i.e. $P(T)$ behavior, remains unknown. This is an important question since excitations which actually count at low temperatures are those with long wavelength and the phonons with quadratic dispersion have a vanishing group velocity.

Moreover, for metallic nanostructures electron coupling to the Lamb modes has not been considered so far, i.e. both qualitative and quantitative understanding is missing. For instance, unlike semiconductors, the QW state is very rare in metals and has actually been observed only in ultrathin films. Because of the much shorter Fermi wavelength, i.e. $d\cdot k_F >> 1$, a more common situation in metals corresponds to the *3D* Fermi gas (FG). One of the consequences is that electrons can now be scattered also in the direction transverse to the propagation of Lamb waves, i.e. into the thickness of the layer. We will therefore mainly focus our discussion to the *3D* FG case.

II. Normalization and Electron – Phonon Coupling

The Hamiltonian describing our system is

$$H = \sum_{\mathbf{k}} E(\mathbf{k})\, c^{\dagger}_{\mathbf{k}} c_{\mathbf{k}} + \sum_{q} \hbar\omega(q)\, a^{\dagger}_{q} a_{q} + H_{e-p}, \quad (4)$$

where electron (c_k) and phonon (a_q) annihilation operators with respective wave-vectors, k and q, and dispersions $E(k)$ and $\omega(q)$ – solutions of the Schrodinger and continuum elasticity equations subject to the considered boundary conditions, see details in [7,9]. Since the transverse size of our system is comparable to thermal phonon wavelength, a single phonon distribution may be assumed for the whole sample viewed as a single elastic plate of volume $L\cdot A$ and mass M, where A is the surface of the plate. The electron wave function Ψ is normalized over the volume $V_{el} = d\cdot A$ and coupling to phonons is:

$$H_{e-p} = \frac{2}{3} E_F \int_{V_{el} = A\times d} d^3\mathbf{r}\; \Psi^{\dagger}(\mathbf{r})\, \Psi(\mathbf{r})\, \nabla \cdot \mathbf{u}(\mathbf{r}) \quad (5)$$

The displacement field $\mathbf{u}(\mathbf{r})$ has only (x,z) components and is expanded in eigenmodes, ξ, with in-plane (x,y) momentum $\mathbf{q}_{||}$

$$\mathbf{u}(\mathbf{r},t) = \sum_{\xi,\mathbf{q}_{||}} \sqrt{\frac{\hbar}{2M\omega_{\xi}(\mathbf{q}_{||})}} \left[a_{\mathbf{q}_{||}} + a^{+}_{-\mathbf{q}_{||}}\right] \mathbf{u}_{\xi}(\mathbf{q}_{||}, z) \exp\left(i\mathbf{q}_{||}\mathbf{r}_{||} - i\omega_{\xi} t\right). \quad (6)$$

Proper normalization is crucial for transport phenomena and in (6) it is fixed by orthonormality of the amplitudes $\mathbf{u}_{\xi}(q,z)$ on $z = [-L/2,\ L/2]$, $q_{||} = q$. With this choice the flexural eigenmodes are antisymmetric (a) upon reflection with respect to the $z = 0$ plane while dilatational modes are symmetric (s) (explicit expressions to be given elsewhere). The respective solutions of the Lamb eigenenergy equations

$$\left(\frac{4q^2 p_l q_t}{(q^2 - q_t^2)^2} = \frac{\tan(q_t L/2)}{\tanh(p_l L/2)}\right)_s, \left(\frac{4q^2 p_l p_t}{(q^2 + p_t^2)^2} = \frac{\tanh(p_l L/2)}{\tanh(p_t L/2)}\right)_a, \quad (7)$$

depend on the auxiliary parameters q_t and p_l determined by (7) and which also appear in the dispersion relation

$$\omega(q) = c_l \sqrt{q^2 - p_l^2} = c_t \sqrt{q^2 + q_t^2},$$

where $c_{t,l}$ are the longitudinal and transverse sound velocities with the ratio $\nu = (c_t / c_l)^2 < 0.5$. Fortunately, from the whole sophisticated set of solutions of (7) we can separate analytically the acoustic branches to account for the low temperature properties of the system:

$$\begin{aligned} \omega_s &\simeq 2qc_t\sqrt{1-\nu}, q_t^s \simeq q\sqrt{3-4\nu}, \quad p_l^s \simeq q(1-2\nu); \\ \omega_a &\simeq q^2 L c_t \sqrt{(1-\nu)/3}, \qquad (8) \\ p_t^a &\simeq -q + q^3 L^2 (1-\nu)/6, p_l^a \simeq -q + q^3 L^2 \nu (1-\nu)/6. \end{aligned}$$

The electron-phonon matrix element of the $H_{e\text{-}p}$ (4) describing emission and absorption of such a phonon with electron scattering from an initial state Ψ_n to a final state Ψ_m is given by

$$|g^{n,m}_{\xi,\mathbf{q}}|^2 = F_{\xi} \frac{8\hbar E_F^2}{9M\omega_{\xi}} S(n, m, \xi = s, a, p_l), \quad (9)$$

where F_{ξ} have absorbed the normalization factors:

$$\begin{aligned} F_s &= \left[q_t q\left(q^2 - p_l^2\right)\right]^2 \cos^2(Lq_t/2) / \sqrt{16q^6(3-4\nu)(1-\nu)^2}, \\ F_a &= 6\left[p_t q\left(q^2 - p_l^2\right)\right]^2 \sinh^2(Lp_t/2) / \left(\left(q^2 L\right)^3 (1-\nu)\right). \end{aligned} \quad (10)$$

The S - function in (8) is the overlap integral between the z-components of the electron wave functions modulated by the phonon displacement field:

$$S(n, m, \xi, p_l) = \left| \int_{L/2-d}^{L/2} \phi_n^*(z)\, \phi_m(z) \begin{Bmatrix} \cosh(zp_l), \xi = s \\ \sinh(zp_l), \xi = a \end{Bmatrix} dz \right|^2. \quad (11)$$

The wave function $\phi(z)$ for the *3D* Fermi gas is a plane wave

$$\phi(z) = \exp\left(ik_z\left(z + d - L/2\right)\right)/\sqrt{d}, \quad (12)$$

while the wave function for the quantum well electronic state can be written as

$$\phi_n(z) = \sqrt{\frac{2}{d}}\sin\left(\left(z + d - L/2\right)\pi n/d\right),$$

where n is the sub-band index. It is clear that in our final results the low temperature contribution will depend only on the long wavelength limit of the coupling constant (9). Then we can estimate the overlap integrals (11) in this limit as follows:

$$S(n, m, s, p_l) = \delta_{n,m}; \quad \delta_{k'_z, k_z},$$
$$S(n, m, a, p_l) = p_l^2\left|\int_{L/2-d}^{L/2}\phi_n^*(z)\,\phi_m(z)\,z dz\right|^2. \quad (13)$$

Eqs. (13) establish effective conservation laws for the electron - phonon scattering at low temperatures.

III. THE HEAT FLUX

The general expression for the heat power P, i.e. the energy transferred from hot electrons to phonons in a unit of time, is determined by the coupling in (9):

$$P = 2\sum_{\mathbf{k},\mathbf{q},\xi}\hbar\omega_\xi\left(\Gamma^{\mathrm{em}}_{\xi,n,m}(\mathbf{k}\to\mathbf{k}-\mathbf{q}_{||}) - \Gamma^{\mathrm{ab}}_{\xi,n,m}(\mathbf{k}\to\mathbf{k}+\mathbf{q}_{||})\right), \quad (14)$$

where Γ are the phonon emission and absorption rates,

$$\Gamma^{\mathrm{em}}_{\xi,n,m}(\mathbf{k}\to\mathbf{k}-\mathbf{q}_{||}) = \frac{2\pi}{\hbar}|g^{n,m}_{\xi,\mathbf{q}}|^2\left[n(\hbar\omega_\xi/k_BT_{ph}) + 1\right]$$
$$\times f(E_{\mathbf{k}}/k_BT_e)[1 - f(E_{\mathbf{k}-\mathbf{q}_{||}}/k_BT_e)]\,\delta(E_{\mathbf{k}-\mathbf{q}_{||}} - E_{\mathbf{k}} + \hbar\omega_\xi). \quad (15)$$

Functions n and f are the Bose and Fermi distribution functions, note that each subsystem is taken at its specific temperature. The absorption term $\Gamma^{ab}(k\to k+q)$ is obtained from $\Gamma^{em}(k\to k - q)$ by inversion q → - q and ω → - ω using the identity for the Bose distribution $n(-y) + 1 = -n(y)$. The power function P can then be presented in the form of a difference between terms separately dependent on T_e and T_p by using the following identity for the distribution functions $f(x)\ [1 - f(x-y)] = n(y)\ [f(x-y) - f(x)]$. Given that, $\hbar\omega \ll E_F$ this can be replaced at low temperatures by the expression $\hbar\omega n\left(\hbar\omega/k_BT_e\right)\delta\left(E_k - E_F\right)$, which effectively means that electron scattering occurs near the Fermi surface. The above arguments allow to single out the contributions of the two opposite processes to the power function (15): the rate at which electrons at T_e transfer their energy to phonons, ε $P = P_0(T_e) - P_0(T_p)$,phonons at T_p transfer their energy to electrons: where

$$P_0(T_e) = \sum_{\mathbf{k},\mathbf{q},\xi,n,m}\frac{32\pi\hbar^2E_F^2}{9M}F_\xi\omega_\xi n(\hbar\omega_\xi/k_BT_e)S(n,m,a,p_l)\,\delta(\epsilon_{\mathbf{k}} - E_F)$$
$$\times\,\delta(E_{\mathbf{k}-\mathbf{q}} - E_{\mathbf{k}} + \hbar\omega_\xi). \quad (16)$$

The first δ - function implies that $k = k_F$ and then the last δ - function only imposes restrictions on q and the scattering angle α ($\mathbf{k}\cdot \boldsymbol{q}_{||} = k\, q_{||} \cos\alpha = k\, q_x \sin\theta \cos\phi$) :

$$\delta\left(\frac{\hbar^2}{m_e}\left(k_Fq_x\sin\theta\cos\phi\right) - \frac{\hbar^2}{2m_e}\left(k_z^2 - \left(k_z'\right)^2 - q_x^2\right) - \hbar\omega(q_{||})\right) \quad (17)$$

here m_e is the electron mass and we have used spherical coordinates and the parabolic dispersion

$$E_{QW}\left(\mathbf{k}_{||}, n\right) = \varepsilon_n + \hbar^2\mathbf{k}_{||}^2/2m_e, \quad E_{FG}\left(\mathbf{k}\right) = \hbar^2\mathbf{k}^2/2m_e.$$

For the quantum well state one has to replace $\hbar^2k_z^2/2m,\ \ \hbar^2\left(k_z'\right)^2/2m$ in (17) by ε_n, ε_m respectively, $\varepsilon_n \sim n^2/(m_e\, d^2)$ corresponding to the quantized energy levels defining the QW. For the interaction with dilatational modes, due to (13), these two terms cancel each other in (17). We can then estimate that, since the sound velocity c is much smaller than the Fermi velocity, the remaining two terms in (17) are much smaller than the first term:

$$m\omega^s/\hbar k_F q \sim c_t/v_F \text{ and } q/k_F \ll 1.$$

This implies that the scattering by the dilatational modes is effectively orthogonal to the direction of electron momentum k: $\cos\alpha \approx 0$. Integration over the scattering angle removes the delta function (17) in (16), where the remaining delta-function allows to take also the integral over electron momentum. The summation over the two-dimensional phonon wave vector is replaced by integration in the usual way $\sum_q = (A/2\pi)\int dq$. As a result (16) takes the following form

$$P_0^s(T_e) = \frac{16m_e^2E_F^2}{9M\hbar^2}\frac{A^2d}{(2\pi)^2}\int_0^\infty dq \quad (18)$$
$$\times\int_0^\infty dq\,|dp_l|^2\,\omega_a|F_s|^2\left[q_tq(q_l^2 + q^2)\right]^2 n(\hbar\omega_s/k_BT_e)\cos^2(Lq_t/2).$$

With our previous results in Eqs. (8) and (10) we obtain for the contribution of the s-modes to the power density of the heat transfer

$$P_0^{FG,s}(T_e)/(dA) = (k_BT_e)^4\frac{0.009k_F^4\nu^2}{\rho L\hbar^2c_t^3(1-\nu)^{3/2}}. \quad (20)$$

Note that it depends neither on g or d parameters and the only effect of geometry is contained in the total thickness of the device, L. We will see below that the a-modes behave quite differently in this sense.

First, the flexural modes do not obey the momentum conservation law, as follows from (13). This complicates significantly the calculations of the power function (16), which will be given elsewhere in more detail. After taking analytically the five integrations (over the two angles in (18), the electron momentum $k_{||}$, the scattered momenum k'_z, and distance z in *(13)*) we obtain the following expression

$$P_0^{FG,a}(T_e) = \frac{32\hbar^2 E_F^2}{9M}\frac{(Ad)^2}{(2\pi)^4}\left(\frac{m}{\hbar^2}\right)^2 \tag{21}$$

$$\times \int_0^\infty dqI(q)\,|dp_l|^2\;\omega_a|F_a|^2\left[q_t q(q_t^2+q^2)\right]^2 n(\hbar\omega_a/k_BT_e)\sin^2(Lq_t/2).$$

Here $I(q) = \pi q\left(-\frac{(g-1)^2}{2}\ln(dq) + 0.807966*(g-1)^2 + \frac{1}{8}\right).$

The last expression is an approximation for the long-wave phonon limit assumed in the present theory. This leads to the following result

$$P_0^{FG,a}(T_e)/(Ad) = (k_BT_e)^4\frac{0.007\,5k_F^4\nu^2}{g^4\rho d\hbar^2 c_t^3(1-\nu)^{3/2}}$$

$$\times\left((g-1)^2\ln\left(\frac{4.4\hbar Lc_t\sqrt{1-\nu}}{d^2k_BT_e}\right)+\frac{1}{2}\right). \tag{22}$$

One can see that both contributions (21) and (22) have a weaker temperature dependence than in the bulk (1) and are roughly of the same order near the crossover temperature $T_e \approx T^* < 1K$, the dilatational modes gaining more weight at lower temperatures due to a slower T_e dependence. We also obtain a quantitative agreement for the values of parameters corresponding to the experiments in [6,7]: $\rho = \rho_{Cu}(d/L) + \rho_{SiN}(1-(d/L)) = 5107\ kg/m^3$, $\rho_{Cu} = 9\ g/cm^3$, $\rho_{SiN} = 3.29\ g/cm^3$, $k_F = 1.36\times10^{10}\ m^{-1}$, *see discussion in Ref.[4];* $c_t = 2325\ m/s$; $\nu = 0.27$.

IV. CONCLUSIONS

Quantitative agreement of the proposed description with experimental data on the sub-Kelvin heat transfer between electrons and phonons in layered metallic nanostructures is an important result of our work. It confirms that the scattering mechanism is correctly identified. Namely, the electronic state is the *3D* Fermi gas; it is coupled via deformation potential to the Lamb vibrational modes; the heat transport in this temperature range is dominated by dilatational and flexural acoustic modes. Density of the heat current grows roughly inverse to the thickness of the metal film d, however the strongest geometric effect is due to the thickness ratio between the insulating and metallic layers. A strong increase of heat transfer is produced in materials with larger Fermi momentum and lower sound velocity. The crossover from 3D to quasi-*2D* (Lamb) phonon at T^* spectrum was indeed observed in experiments and is associated with a strong increase of heat flux transferred between electrons and phonons due to a stronger coupling of the 3D Fermi gas of electrons to the 2D gas of confined phonons.

ACKNOWLEDGMENT

This work has been financially supported by CNCSIS-UEFISCDI (project IDEI 114/2011) and by ANCS (project PN-09370102).

CONFLICT OF INTEREST

The authors declare that they have no conflict of interest.

REFERENCES

1. Stroscio M A and Dutta M (2004) Phonons in Nanostructures. CUP, United Kingdom
2. Nguyen H Q, Meschke M Courtois H and Pekola J P (2014) Sub-50-mK Electronic Cooling with Large-Area Superconducting Tunnel Junctions, Phys. Rev. Appl. 2: 054001–054005
3. Wellstood F C, Urbina C, and Clarke J (1994) Hot-Electron Effects in Metals, Phys. Rev. B 49:5942-5955
4. Qu S., Cleland A N, and Geller M R (2005) Hot Electrons in Low-Dimensional Phonon Systems, Phys. Rev. B 72: 224301-1 - 224301-6
5. Karvonen J and Maasilta I (2007) Influence of Phonon Dimensionality on Electron Energy Relaxation, Phys. Rev. Lett. 99:145503-1 - 145503-4
6. Karvonen J and Maasilta I (2007) Observation of Phonon Dimensionality Effects on Electron Energy Relaxation, *J. Phys.: Conf. Series* 92: 012043-1 - 012043-6
7. Glavin B A, Pipa V I, Mitin V V and Stroscio M A (2002) Relaxation of a Two-Dimensional Electron Gas in Semiconductor Thin Films at Low Temperatures: Role of Acoustic Phonon Confinement, Phys. Rev. B 65:205315-1 - 205315-14
8. Muhonen J T, Meschke M and Pekola J P (2012) Micrometre Scale Refrigerator, Rep. Prog. Phys., 75: 046501-1 – 046501-23
9. Kühn K, Anghel D V, Galperin Y M, and Manninen M (2007) Interaction of Lamb Modes with Two-Level Systems in Amorphous Nanoscopic Membranes, Phys. Rev. B, 76: 165425-1 – 165425-8

Author: Sergiu Cojocaru
Institute: National Institute of Physics and Nuclear Engineering
Street: Reactorului, 30
City: Magurele
Country: Romania
Email: scojocaru@theory.nipne.ro

Theory of Catalytic Micro- and Nanoengines: From Self-propulsion Mechanisms to Remediation of Polluted Water

V.M. Fomin

Institute for Integrative Nanosciences (IIN), Leibniz Institute for Solid State and Materials Research (IFW), Dresden, Germany

Abstract— **A review is given of the theoretical ideas concerning the propulsion mechanisms of the catalytic micro- and nanoengines, which are prospective in robotics, biomedical engineering and environmental sustainability. Enhanced degradation of organic pollutants in water at small-scale environments underlies an important potential application of catalytic microengines.**

Keywords— **catalytic micro- and nanoengines, self-propulsion mechanisms, degradation of organic pollutants.**

I. INTRODUCTION

In the recent decade, considerable advances have been achieved in the field of self-propelled catalytic microengines, which exhibit self-propulsion in the presence of chemical fuels such as hydrogen peroxide [1, 2]. Several geometries have been designed and investigated [3, 4], including cylinders, cone-frustums, long spheroids, double truncated cones, double cones and spheres.

The propulsion of microengines is due to the decomposition of the hydrogen peroxide fuel into molecular oxygen and water. Recent review papers summarize the different mechanisms that have been proposed for these microengines [3, 5, 6]. In the particular case of tubular microjets, the propulsion force originates from the generation of oxygen bubbles that are expelled from the tube [2]. This mechanism makes microjets very interesting since, up to now, microtubular jets generate a higher propulsion force than other self-propelled micro- and nanoparticles, e.g., spherical Janus particles, nanorods and microhelices [7].

Catalytic micro- and nanoengines have shown numerous prospective applications [8], in particular, in robotics [9], biomedical engineering [10, 11] and environmental sustainability [12]. Recently, the effects of geometry on the driving force (see Refs. [13 to 15] and references therein) and the drag force (see Ref. [16] and references therein) for micro- and nanoengines of different shape have been intensively studied.

II. PROPULSION MECHANISMS OF THE CATALYTIC MICRO- AND NANOENGINES

There is a lack of an accurate theoretical model describing the origin of the self-propulsion. A few models of the self-propulsion have been advanced aimed at explanation of the recent experimental evidence. In Ref. [13], the propulsion of tubular microjets is explained by the *ejection* of a bubble from the tube. The average velocity of the microjet in Ref. [13] is found as a result of a three-step procedure: (1) equating the momentum of the drag forces acting on the subsystem, containing the separated bubble and the tube, to zero and assuming that (2) the separated bubble and tube stop at the same time (3) when they are separated by a distance equal to the tube diameter. This model is purely phenomenological because of the following reasons. (1) The whole system, for which momentum is conserved, consists of the bubble, the tube and the liquid. (2) After the separation of the bubble from the tube, the dynamics of the bubble are quite distinctive from those of the tube, because of specific drag forces leading to different stopping times. (3) Hence, the distance between the stopped bubble and the stopped tube is an *ad hoc* parameter.

In Ref. [14], the propulsion of catalytic tubular microjets is attributed to the *bubble growth* process implying a growth force governed by the Rayleigh-Plesset equation. The growth force, which arises from the growth of the bubble surface against the fluid, is assumed the only driving force acting on the tube. The instantaneous velocity of a microjet within this model consists of two kinds of jumps: low due to the growth force and high due to ejection of a bubble. Though such a behavior is reasonable, the model is not in a position to predict the overall ejection-governed microjet velocity. The explicit expression for the growth force contains an empirical (fitting) constant. The fitted reaction rate (as represented in Table 1 of Ref. [10]) varies within a factor of 5.

The experiments on catalytic microengines fabricated using rolled-up nanotech in 1% H_2O_2 and 1% SDS surfactant of Ref. [15] (see Fig. 1) lead to average microjet velocities of 100 $\mu m \cdot s^{-1}$ and ejection frequencies

V. Sontea and I. Tiginyanu (eds.), *3rd International Conference on Nanotechnologies and Biomedical Engineering*,
IFMBE Proceedings 55,
DOI: 10.1007/978-981-287-736-9_6

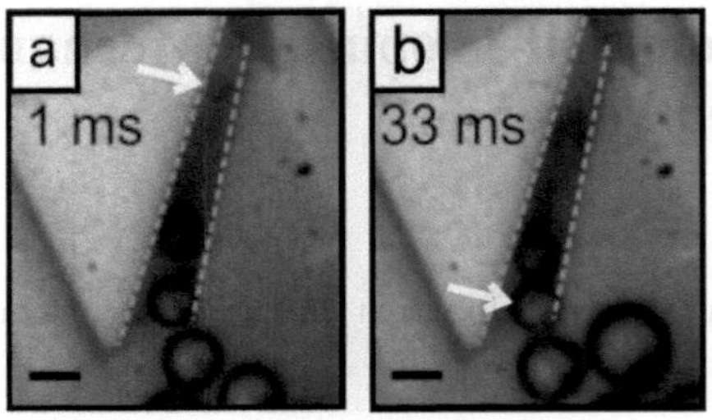

Fig. 1 Snapshots of the observed bubble propulsion taken at t=1 ms (a) and t=33 ms (b) from the nucleation. Dashed lines show the walls of a rolled-up tube. Scale bar is 10 μm. The inner surface of Pt catalyzes the decomposition of H_2O_2 and, consequently, the nucleation of an O_2 bubble, indicated with a white arrow. The bubble grows (a), moves to the larger opening of the tube and is released (b). (After Ref. [15])

for oxygen bubbles of 30 Hz. For the experimental conditions, the ejection frequencies and the microjet velocities are calculated using the bubble ejection model [13] and the bubble growth model [14]. From the comparison of the calculated and experimental values (see Table I in Ref. [15]), it is concluded that average ejection frequencies and microtube velocities, as calculated using the models of Refs. [13] and [14], are underestimated by 1 to 3 orders of magnitude and 2 orders of magnitude, respectively.

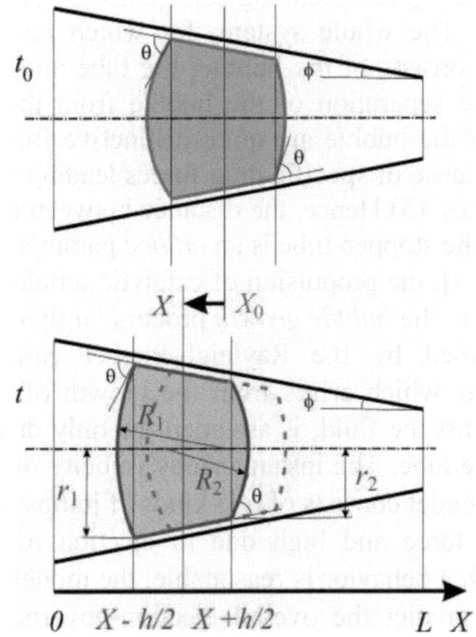

Fig. 2 Scheme of a bubble (blue at time t_0 and pink at a larger time $t > t_0$) confined to a conical tube. A heavy black arrow represents a shift of the central plane drawn at a half-height of the conical part of the bubble $X_0 \to X$. The conditions that the conical part of the bubble is completely touching the internal wall of tube are $X\text{-}h/2 \geq 0$ and $X+h/2 \leq L$. (After Ref. [15])

In Ref. [15], a *capillary-force-based* propulsion mechanism for the catalytic microjet engines is proposed and studied. A bubble, which is deformed due to its confinement to the microtube, consists of two spherical cups and a truncated cone, with a fixed total volume (see Fig. 2). Because of the presence of a fluid layer between the bubble and the tube wall, the contact angle θ in Fig. 2 can be much smaller than the equilibrium contact angle θ_{eq} between the liquid and the gaseous phases at the triple phase contact line.

The surface tension energy

$$E_{ST}(X) = \sigma S(X) \tag{1}$$

is determined by the surface tension σ (σ=33.8 mN·m^{-1} corresponds to the conditions T=25°C, 1% hydrogen peroxide, 1% SDS surfactant) and the total surface of the bubble:

$$S(X) = S_{cone} + S_{cap,1} + S_{cap,2} \tag{2}$$

The resulting 'capillary' force is defined as a derivative of the surface tension energy with respect to the coordinate

$$F_{cap}(X) = -\frac{dE_{ST}(X)}{dX}. \tag{3}$$

This thermodynamic force occurs due to *the broken symmetry of the bubble in a conical tube*: $r_2 < r_1$. It tends to propel the bubble towards the larger opening, thereby

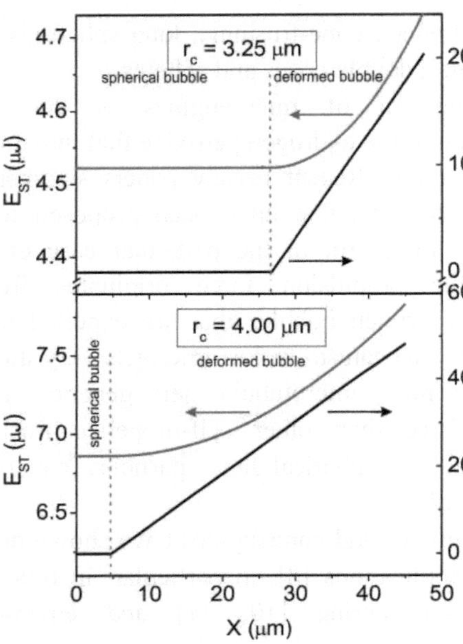

Fig. 3 Surface tension energy E_{ST} (red) and capillary force F_{cap} (black) driving a bubble calculated as a function of its position in an asymmetric microjet. Blue dashed lines separate the regions of deformed bubbles and spherical bubbles. Size of a spherical bubble is characterized by a radius r_c of a cross-section through the circumference of its contact with the wall. Calculations are performed for the maximal and minimal radii R_{max}=3.5 μm and R_{min}=3.0 μm, ϕ=0.57°, θ=24° and L=50 μm. (After Ref. [15])

decreasing its surface *S(X)* and, consequently, the surface tension energy of Eq. (2) (see Fig. 3). The calculated bubble velocity of the order of 1 mm/s is in a good agreement with the experiment represented in Fig. 1.

Due to the bubble motion in an asymmetric tube, there occurs a momentum transfer to the fluid. A jet force acting on the tube emerges in order to compensate this momentum transfer. The jet force, counterbalanced by the linear drag force in the low-Reynolds-number-regime (see Eq. (5-11.53) in Ref. [17]), enables tube velocities of the order of 100 μm/s [15]. The proposed mechanism provides a fundamental explanation for the development of driving forces acting on bubbles in tubular microjets, which is helpful for optimization of their geometry. Importantly, *other types of asymmetry* of a microengine, e.g. of the chemical composition of the fluid or the surface tension, lead to similar mechanisms of driving forces even in geometrically symmetric microtubes.

A unified solution for the drag force for various micro- and nanoengines of circular cross-section (cylinder, cone-frustum, long spheroid, double truncated cone, double cone and sphere) is derived in Ref. [16] in the low-Reynolds-number-regime. The effect of the geometric parameters of a micro- or nanoengine, such as the cone angle, the aspect ratio of the length to the larger radius on the drag coefficient is identified. It is revealed that the drag coefficient for a cone frustum diminishes with increasing the cone angle, e.g., from 2.15×10^4 to 1.75×10^4 when the cone angle rises from 0° to 3° at the ratio of the length to the larger radius equal to 10 [16]. As a result, the propulsion efficiency is enhanced by increasing the cone angle and/or decreasing the ratio of the length to the larger radius of a conical engine, when this ratio is larger than 6. Importantly, the conclusion based on the analysis of the drag force in Ref. [16], that for high-speed micro- or nanojets a conical shape is advantageous as compared to a cylindrical shape with the same length and radius, tends to confirm the key message of Ref. [15] derived from the behavior of the driving force.

III. SELF-PROPELLED MICROMOTORS FOR CLEANING POLLUTED WATER

The use of catalytically self-propelled microengines for degrading organic pollutants in water via the Fenton oxidation process is investigated in Ref. [12]. The tubular microengines are fabricated by the roll-up nanotech of Fe (100 nm)/Pt (1 nm) bilayers. Within their architecture, those microengines contain double functionality: (1) the inner Pt for the self-propulsion and (2) the outer Fe for the *in situ* generation of ferrous ions that boost the remediation of contaminated water. The degradation of organic pollutants occurs in the presence of hydrogen peroxide, which acts as a fuel to propel the microengines and as a reagent for the Fenton reaction [12]. As shown experimentally in Ref. [12], water solutions of dyes with microengines reveal an improved intermixing in liquids on the macroscopic scale (see Fig. 4). This results in the degradation of organic pollutants about 12 times faster in solutions with catalytically active microengines as compared to the Fenton oxidation process without them and provides considerable promise for the remediation of contaminated water at small-scale environments.

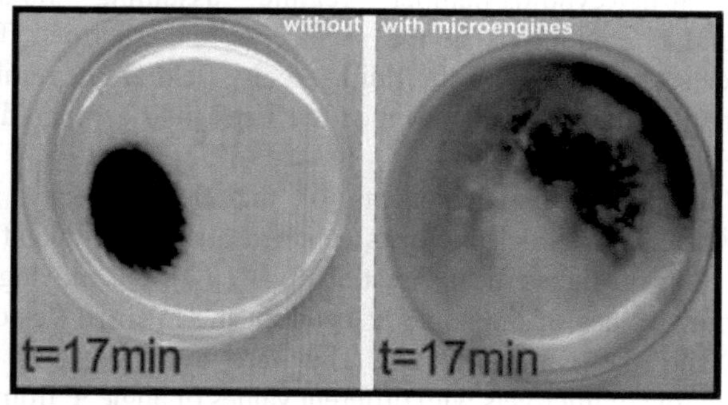

Fig. 4 Solutions without (left) and with (right) catalytic Fe/Pt microengines containing a sample with about 400 actively moving microtubes (fabricated of bilayers Fe 100 nm/Pt 1 nm). All beakers contain 3 mL of solution (6% H_2O_2, 1% SDS surfactant, and 10 μL of black ink as dye). The snapshots are taken at t=17 min after the beginning of diffusion. (After Ref. [12])

The enhanced diffusion provided by microengines is theoretically explained based on the stationary continuity equation for fluid dynamics (Ref. [17], Eq. (2-1.1)):

$$\nabla \cdot \mathbf{j}(\mathbf{r}) = 0, \tag{4}$$

where the mass flux density is a product of the local concentration and velocity:

$$\mathbf{j}(\mathbf{r}) = c(\mathbf{r})\mathbf{v}(\mathbf{r}). \tag{5}$$

The mixing along the vertical z-axis in a Petri dish with the bottom in the xy-plane is modelled by (1) assuming that the local concentration is a function of the z-coordinate alone and (2) averaging the z-component of the local velocity $v_z(\mathbf{r})$ over the available cross-sectional area in the Petri dish at a given value of the z-coordinate.

With the diameter of the Petri dish D=30 mm and the volume of the fluid V=1 mL, the height of the fluid in the Petri dish is L_{fluid}=1.41 mm. The integration of the continuity equation (4) over the coordinate z from a point z to L_{fluid} gives a relation

$$c(z)/c(L_{\text{fluid}})=v_z(L_{\text{fluid}})/v_z(z). \tag{6}$$

Within a standard approximation, when analyzing motion in a viscous fluid, elongated orthotropic bodies can be represented by prolate ellipsoids of revolution (Ref. [17], p. 222). In the model of Ref. [12], the axes of the ellipsoid are equal to diameter $2R$=120 μm and length L=500 μm of the microjet. The flow of a viscous fluid dragged by this ellipsoid moving with velocity v_{jet} (according to the experiment of Ref. [12], v_{jet}=538 μm/s) is described using the Navier-Stokes equation (see Ref. [17], Chapter 4). The solutions to the Navier-Stokes equation are obtained within the low-Reynolds-number regime assuming no-slip boundary conditions at the surface of the microjet. The effects of the walls on the fluid velocity field are negligible owing to the large diameter of the Petri dish compared to the sizes of the microjet.

The average z-components of the fluid velocity are estimated for model microjets centered at the half-height of the fluid in the vertical direction ($0.5\times L_{fluid}$) and moving at various values of the angle Ψ between the microjet velocity v_{jet} and the vertical z-axis. They are averaged over an ensemble of microjets with different values of angle Ψ from 0° to 90°. The corresponding profiles of the peroxide concentration calculated according to Eq. (4) are represented in Fig. 5 as a function of the z-coordinate.

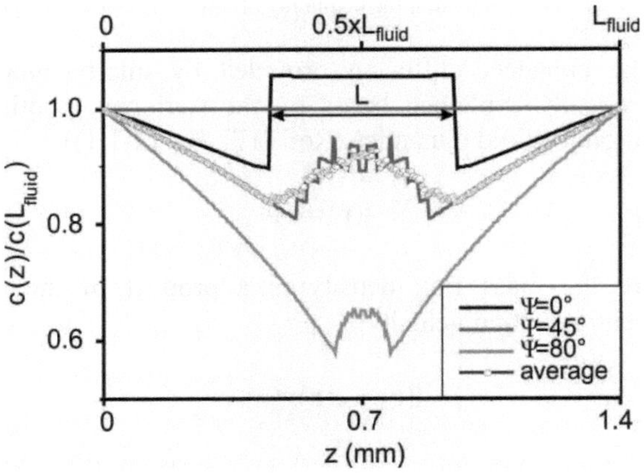

Fig. 5 Peroxide concentration related to its undisturbed value far away from the moving microjet calculated as a function of the vertical coordinate z for model microjets centered at the half-height of the fluid in the vertical direction ($0.5\times L_{fluid}$). The values of the angle Ψ between the microjet velocity v_{jet} and the vertical z-axis are given in the inset. The results averaged over an ensemble of microjets with different values of the angle Ψ are represented with open circles. The pink line represents the undisturbed value of the peroxide concentration. (After Ref. [12])

Within this model, mixing of the peroxide occurs due to the following physical reason: In order to keep the mass flux density along the vertical z-axis constant, the peroxide concentration must be an inverse function of the average z-component of the velocity [see Eq. (6)]. As shown in Fig. 5, the peroxide concentration shows minima just at the front and rear points of a moving microjet. Simultaneously, the peroxide concentration in the vicinity of the microjet is redistributed and decays far away from it. The highest modulation of the peroxide concentration ~20% is due to the vertical propulsion of microjets (Ψ=0°). Averaging over an ensemble of microjets with different values of the angle Ψ smears the profile of the peroxide concentration and provides the modulation of the order of 10%. Importantly, the region of rearrangement of the peroxide concentration moves in space together with propelling microjets. The drag of the solution due to the propulsion of microjets constitutes therefore an effective mechanism of mixing peroxide in the reaction vessel.

IV. CONCLUSIONS

The motion of bubbles inside conical microtubes provides a fundamental explanation for the development of driving forces acting on bubbles in tubular microjets with geometric asymmetry. The ability of the catalytically self-propelled microengines to improve intermixing in liquid solutions results in the enhanced removal of organic pollutants.

The overviewed above fundamental theoretical models have allowed for a new insight into propulsion mechanisms and motion of self-propelled catalytic micro- and nanoengines. A new playground for those models is provided by the research on novel microstructures with a high photocatalytic activity (e.g., in anatase crystalline structures [18]) as microengines moved by means of the UV-light for possible bioapplications, sensor design and ensuring environmental sustainability.

ACKNOWLEDGMENT

Fruitful collaborations with M. Hippler, V. Magdanz, S. Sanchez, L. Soler and O. G. Schmidt are gratefully acknowledged. This work has been partly supported by the bilateral BMBF (Germany) – ASM (Republic of Moldova) project TiNaTEng no. 01 DK13010.

CONFLICT OF INTEREST

The author declares that he has no conflict of interest.

REFERENCES

1. Mei Y, Huang G, Solovev AA, Ureña EB, Mönch I, Ding F, Reindl T, Fu RKY, Chu PK, Schmidt OG (2008) Versatile approach for integrative and functionalized tubes by strain engineering of nanomembranes on polymers. Adv Mater 20:4085–4090

2. Solovev AA, Mei YF, Urena EB, Huang GS, Schmidt OG (2009) Catalytic microtubular jet engines self-propelled by accumulated gas bubbles. Small 5:1688–1692
3. Sanchez S, Pumera M (2009), Nanorobots: The ultimate wireless self-propelled sensing and actuating. Chemistry-An Asian Journal 4:1402–1410
4. Pumera M (2011) Nanomaterials meet microfluidics. Chem Commun 47:5671–5680
5. Mirkovic T, Zacharia NS, Scholes GD, Ozin GA (2010) Fuel for thought: Chemically powered nanomotors out-swim nature's flagellated bacteria. ACS Nano 4:1782–1789
6. Wang J (2009), Can man-made nanomachines compete with nature biomotors? ACS Nano 3:4–9
7. Gao W, Sattayasamitsathit S, Wang J (2012) Catalytically propelled micro-/nanomotors: How fast can they move? Chemical Record 12:224–231
8. Solovev AA, Xi W, Gracias DH, Harazim SM, Deneke C, Sanchez S, Schmidt OG (2012) Self-propelled nanotools. ACS Nano 6:1751–1756
9. Ozin GA, Manners I, Fournier-Bidoz S, Arsenault A (2005) Dream nanomachines. Adv Mater 17:3011–3018
10. Wang J, Gao W (2012) Nano/Microscale motors: biomedical opportunities and challenges. ACS Nano 6:5745–5751
11. Balasubramanian S, Kagan D, Hu CMJ, Campuzano S, Lobo-Castanon MJ, Lim N, Kang DY, Zimmerman M, Zhang LF, Wang J (2011) Micromachine-enabled capture and isolation of cancer cells in complex media. Angew Chem Int Ed 50:4161–4164
12. Soler L, Magdanz V, Fomin VM, Sanchez S, Schmidt OG (2013) Self-Propelled Micromotors for Cleaning Polluted Water. ACS Nano 7:9611–9620
13. Li J, Huang G, Ye M, Li M, Liu R, Mei Y (2011) Dynamics of catalytic tubular microjet engines: Dependence on geometry and chemical environment. Nanoscale 3:5083–5089
14. Manjare MT, Yang B, Zhao Y (2013) Bubble propelled microjets: Model and experiment. J Phys Chem C 117:4657–4665
15. Fomin VM, Hippler M, Magdanz V, Soler L, Sanchez S, Schmidt OG (2014) Propulsion Mechanism of Catalytic Microjet Engines. IEEE Transactions on Robotics 30:40–48
16. Li L, Wang J, Li T, Song W, Zhang G (2015) A unified model of drag force for bubble-propelled catalytic micro/nano-motors with different geometries in low Reynolds number flows. Journ Appl Phys 117:1–6
17. Happel J, Brenner H (1983) Low Reynolds Number Hydrodynamics. Martinus Nijhoff Publishers, The Hague
18. Enachi M, Guix M, Braniste T, Postolache V, Ciobanu V, Ursaki V, Schmidt OG, Tiginyanu I (2015) Photocatalytic properties of TiO_2 nanotubes doped with Ag, Au and Pt or covered by Ag, Au and Pt nanodots. Surface Engineering and Applied Electrochemistry 51:3–8

Author: V. M. Fomin
Institute: Institute for Integrative Nanosciences (IIN), Leibniz Institute for Solid State and Materials Research (IFW) Dresden
Street: Helmholtzstraße 20
City: Dresden
Country: Germany
Email: v.fomin@ifw-dresden.de

Characterisation of Silicon Nanolayers Deposited by Plasma Enhanced Chemical Vapor Deposition on 3-D ZnO Templates for Hollow Silicon Microstructures

I. Hölken, S. Schröder, and R. Adelung

Functional Nanomaterials, Institute of Materials Science, Faculty of Engineering, University of Kiel, Kiel, Germany

Abstract— **Hollow inorganic microstructures have gained much interest in nowadays research fields as they offer unique properties (high specific area, thermal conductivity or density) when compared to their bulk equivalents [1]. On the other hand, for many applications, e.g. as anode material in lithium ion batteries [2], crystalline silicon microstructures (c-Si) would be of great interest. The development of thin hollow Si microparticles based on t- ZnO would therefore offer a new class of Si structures which are of special interest as they combine suitable physico-chemical properties with high porosity caused by their special morphology. One approach for the realization of these Si microstructures includes the usage of plasma enhanced chemical vapor deposition based on argon diluted silane source gases as it provides the opportunity to fabricate homogeneous nanolayers at relatively low temperatures. In order to obtain the desired morphology combined with an adequate deposition rate, in this work the process was optimized by parameter variations and subsequent investigations of the films by Raman spectroscopy, scanning electron microscopy and profilometry. First experiments for the deposition of Si onto t-ZnO were implemented and promising results were obtained.**

Keywords— **Aerographite, ZnO templates, Si nanolayers, plasma enhanced chemical vapor deposition, hollow microstructures.**

I. INTRODUCTION

The discovery of Aerographite in 2012 has gained much interest in ultra lightweight hollow microstructures due to their exceptional properties when compared to bulk materials. Mecklenburg et al. (2012) synthesized the unique material by using three dimensional flexible ZnO tetrapodal structures (t-ZnO) as template.

The variety of coverage for hollow microstructuress not only based on Aerographite includes drug delivery systems [1], lithium-ion batteries [2]. or lightning devices as to name only a few. Many of these applications are based on silicon and therefore the fabrication of 3-D Si microparticles would be of great potential [3]. With the help of 3-D ZnO microstructures as template material, the fabrication of these Si microparticles gets possible with relatively simple thin film deposition processes combined with adequate template removal techniques.

For deposition of silicon, chemical vapor deposition (CVD) offers the opportunity to deposit thin films on large scale at relatively low costs. Most often, plasma enhanced chemical vapor deposition (PECVD) in source gases like silane or disilane is used as this process features uniform film coverage and relatively low substrate temperatures compared to other CVD processes [4]. Additionally, it was shown elsewhere[5], that the deposition of highly crystalline Si films was successfully implemented by PECVD.

Aimed by Raman spectroscopy and scanning electron microscopy, it could be shown by this work that the deposition of highly crystalline Si films is possible with PECVD and that the coverage of 3-D ZnO templates could also be implemented successfully. The next step would include the complete dissolution of the template material in order to obtain hollow Si structures whose potential as e.g. anode material for Li-ion batteries should be proofed in the future.

II. EXPERIMENTAL

Various silicon thin films were deposited by a 13.56 MHz radio frequency SENTECH SI 500 PPD PECVD system on 20 mm * 20 mm amorphous quartz and crystalline Si {100} p-type substrates for 600 s with varying process parameters (TABLE I). Additionally, 5 mm * 5 mm partially photoresist covered Si bridges were added to the processes for subsequent profilometry measurements. All nanolayers were deposited from Ar diluted silane gas as in 1979 it was shown by Knights and Lujan [5] that nano- and microcrystalline grains appear in a-Si nanolayers deposited in argon diluted silane whereby the crystalline fraction can be increased.

Table 1 Variation of the process parameters for the deposition of Si.

Parameter	*Value range*
Substrate temperature	100 °C - 200 °C
Pressure	66.6 Pa - 106.6 Pa
Power	5 W - 100 W
Ar dilution	95 % - 99 %

Additionally, post-deposition experiments were conducted in order to influence the crystalline fraction of the nanolayers.

The influence of the parameters on the crystallinity and the deposition rate was evaluated for planar substrates and

V. Sontea and I. Tiginyanu (eds.), *3rd International Conference on Nanotechnologies and Biomedical Engineering*,
IFMBE Proceedings 55,
DOI: 10.1007/978-981-287-736-9_7

used as basis for the nanolayer deposition on different three dimensional tetrapodal ZnO templates (t-ZnO, Figure 1). Free-standing t-ZnO particles with a mean arm length of 20 µm were produced at the Faculty of Engineering from the University of Kiel by flame-transport synthesis [6]. Interconnected structures were obtained by slightly varied synthesis parameters, cobweb-ZnO (c-ZnO, Figure 2) interconnected networks with much smaller particle sizes were prepared in order to obtain branched Si structures.

Different preliminary approaches were conducted to obtain fully c-Si covered template structures based on the knowledge of Aerographite formation [6].

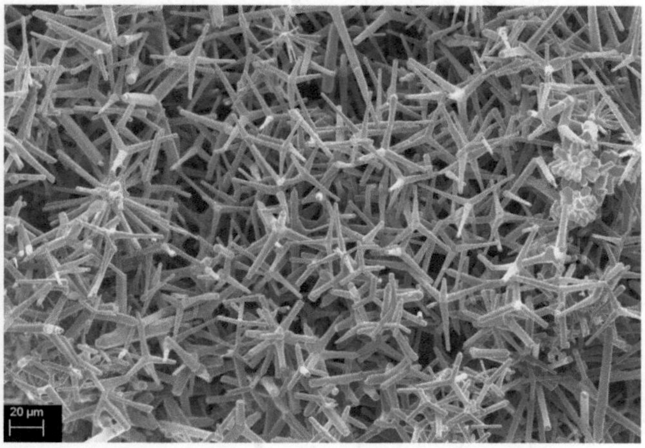

Fig. 1 SEM image of tetrapodal-ZnO obtained by flame transport synthesis.

Fig. 2 SEM image of interconnected cobweb-ZnO, obtained by deviating synthesis parameters.

For the evaluation of the deposition rate, an acetone-based lift-off process was performed on the silicon bridges and the remaining step was measured with a BRUKER DetaktXT profilometer at a scan speed of 40 µm/s and a scan range of 2000 µm. Arithmetic means were obtained from three measured datasets.

To determine the crystalline and amorphous fractions of the deposited Si films, Raman spectroscopy was performed with a WITec Alpha 300RA Raman setup equipped with an argon laser of 532 nm wavelength.

For the investigation of the nanolayer growth properties, scanning electron microscopy (SEM) investigations of the cross-sections were implemented with a ZEISS Ultra Plus microscope at an acceleration voltage of 5 kV.

III. RESULTS AND DISCUSSION

Information about the amorphous and crystalline phases of the samples was determined by Raman spectroscopy as they can be well distinguished by the transverse optical (TO) band of the spectrum. FIGURE 3 illustrates the TO bands of a-Si (broad band at 480 cm^{-1}) and c-Si (sharp peak at 520 cm^{-1}). The increased band width of a-Si is a result of the deviation from the c-Si bond angles caused by the disordered structure. The shift to lower wavenumbers can be explained by the increased bond length of a-Si [7,8].

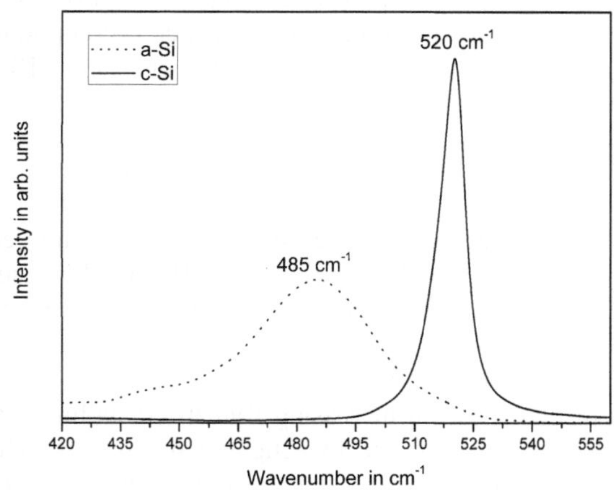

Fig. 3 Raman spectra of the transverse optical bands of amorphous-Si (broad peak at 485 cm^{-1}) and crystalline-Si (sharp peak at 520 cm^{-1}).

Figure 4 shows an exemplary spectrum in the range from 420 cm-1 to 560 cm-1 for a sample with amorphous and crystalline proportions. The TO mode of a-Si and c-Si are well distinguishable, the sharp peak at around 520 cm-1 can be attributes to c-Si, whereas the broad peak around 480 cm-1 belongs to a-Si [9,10].

Due to the overlapping Raman peaks, a deconvolution of the bands is performed to evaluate the amorphous and crystalline amounts and two Gaussian peaks are fitted at the respective band values 490 cm^{-1} and 516 cm^{-1}. For the crystalline fraction it holds:

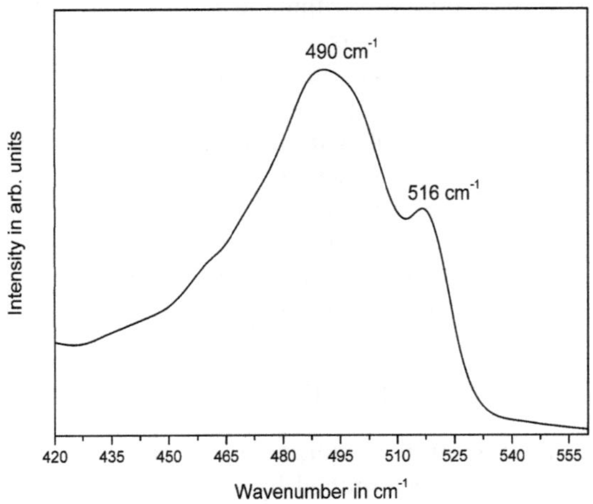

Fig. 4 Raman spectrum between 420 cm^{-1} and 560 cm^{-1} of amorphous-Si nanolayers with a remarkable crystalline fraction.

$$X_c = I_c / (I_c + yI_a) \quad (1)$$

where I_a and I_c are the intensity of the amorphous and crystalline contribution, y is a scattering factor which becomes $y \approx 1$ for grain sizes larger than a few nanometers. According to Amrani et al. (2012) [11], the fraction for $y \approx 1$ is defined as Raman crystallinity Φ_c which is compounded by the Raman intensities at 480 cm^{-1}, 510 cm^{-1} and 518 cm^{-1}: and can be expressed as [12]

$$\Phi_c = (I_{518} + I_{510}) / (I_{518} + I_{510} + I_{480}) \quad (2)$$

To investigate the influence of the process parameters pressure, power and Ar dilution of SiH_4 on the Raman crystallinity of the deposited nanolayers, the peak deconvolution was performed onto the measured data and the results are shown in Figure 5.

The results show an increasing Raman crystallinity with increasing numbers of the investigated parameters. These observations were already stated by Knights and Lujan (1979), who showed that crystalline grain formation inside the Si nanolayers is favored at high pressure, high power and low silane concentration. In order to obtain the deposition rates, the measured thicknesses were divided by the deposition time and the obtained results for varying process parameters are shown in Figure 6. The deposition rate decreases strongly with increasing pressure whereas it increases slightly with increasing temperature and strongly with power. This behavior is expected for hydrogen diluted silane as shown elsewhere [13,14]. In order to alter the crystallinity of the as-deposited nanolayers, post-deposition annealing was performed on four samples deposited at 200 °C at different temperatures (400 °C, 600 °C, 800 °C and 1000 °C) and the crystalline phases were characterized by Raman spectroscopy (Figure 7).

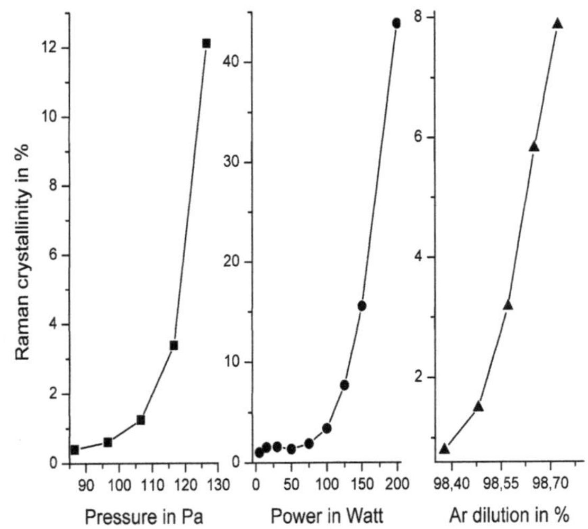

Fig. 5 Influence of the process parameters pressure, power and argon dilution on the Raman crystallinity of the deposited films.

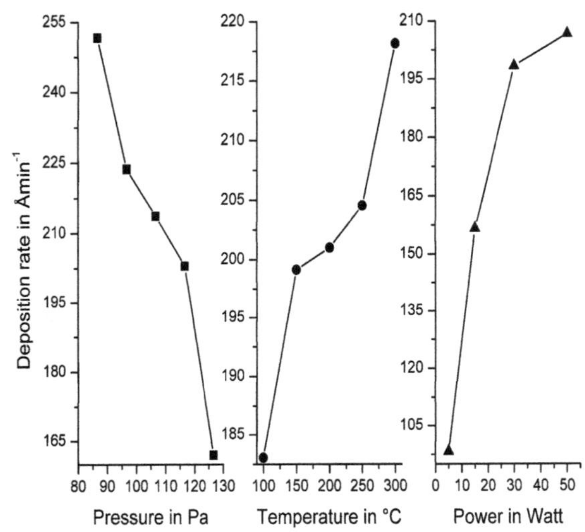

Fig. 6 Influence of the process parameters pressure, temperature and power on the deposition rate of the nanolayers.

Without annealing (200 °C curve), nearly no contribution to the crystalline phase is present as the broad peak at 484 cm^{-1} indicates an amorphous film. The crystalline amount starts to increase beyond 600 °C and reaches higher crystalline fractions with rising temperatures. The TO band of the amorphous phase decreases with increasing temperature and indicates a decreased disorder. At higher temperatures, the amorphous band vanishes completely as the amorphous phase is transformed nearly completely to the crystalline phase.

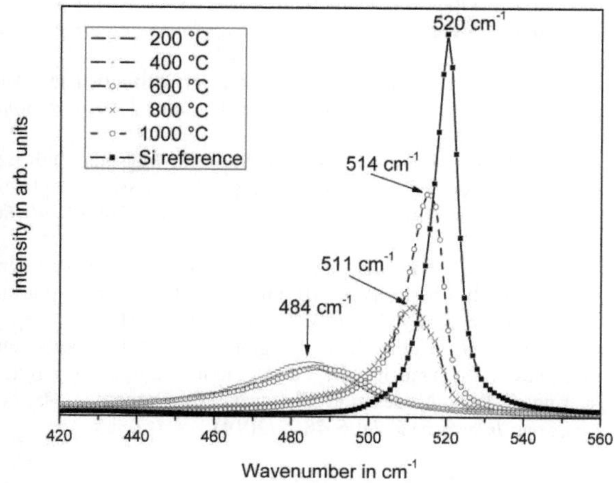

Fig. 7 Raman spectra of the annealed samples in the range between 420 cm^{-1} and 560 cm^{-1}, crystalline-Si is as well as a not annealed sample are shown as reference.

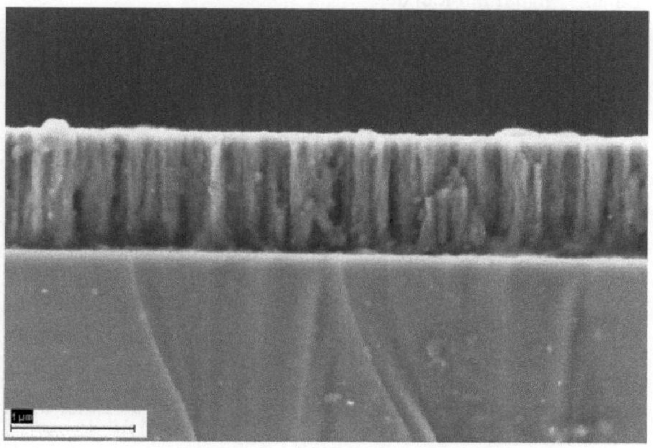

Fig. 8 SEM image of the fractured surface of a Si nanolayer deposited by PECVD on a c-Si{100} substrate.

SEM images of the sample cross-sections were taken in order to visualize the growth morphology of the nanolayers, an example is given in Figure 8.

A similar columnar growth morphology was obtained for all investigated samples. A related effect was already described by J. C. Knights et al. (1979) during the investigation of a-Si films deposited from Ar diluted silane. It could be shown, that no visible columnar structure was obtainable from pure silane whereas an increased Ar concentration combined with an increased plasma power led to columnar growth morphologies.

With the parameters optimized by Si nanolayer deposition and investigation, it was aimed to cover the 3 D ZnO template material for the future fabrication of hollow Si microstructures. To evaluate weather the Si deposition on the microstructures was successful, Raman spectroscopy was conducted and the results of pure t-ZnO and t-ZnO covered with Si were compared in Figure 7.

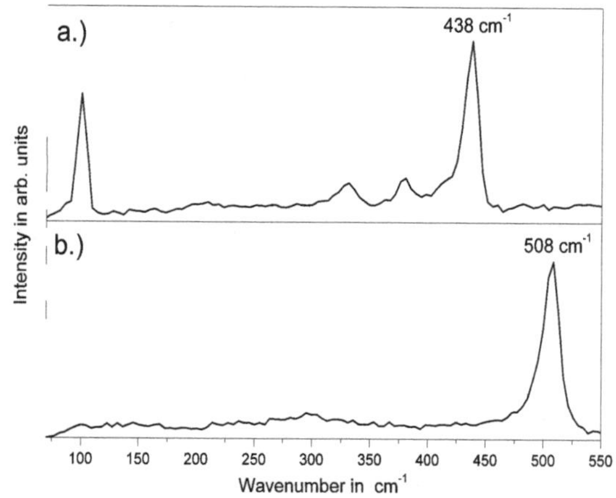

Fig. 9 Raman spectroscopy of a.) tetrapodal-ZnO and b.) tetrapodal-ZnO covered with Si.

The band at 438 cm^{-1} of the pure t-ZnO (Figure 9 a) can be assigned to oxygen atoms [15], the peak at around 99 cm^{-1} is associated with the Zn sublattice [15]. The Raman spectrum of the Si covered t-ZnO (Figure 9b) shows the characteristic Si peak with a shift of the TO peak to 508 cm^{-1} which indicates a decreased bond length maybe caused by small silicon grains embedded in the amorphous matrix.

IV. Conclusion

The characterization of silicon nanolayers deposited by plasma enhanced chemical vapor deposition at diverse process parameter variations was conducted with respect to crystallinity and deposition rate. At first, the deposition of a highly crystalline Si nanolayer with optimized process parameters was conducted. Subsequently, the developed process was transferred to prepare crystalline Si coated tetrapodal shaped ZnO templates and Raman measurements revealed the feasibility of this process. For the future, the complete removal of the template material should be implemented in order to obtain hollow three-dimensional microstructures.

Conflict of Interest

The authors declare that they have no conflict of interest.

References

1. Nomura, T., Morimoto, Y., Ishikawa, M., Tokumoto, H. & Konishi, S. Synthesis of hollow silica microparticles from bacterial templates. *Advanced Powder Technology* **21**, 218–222 (2010).
2. Wang, H. *et al.* Hollow porous silicon oxide nanobelts for high-performance lithium storage. *Journal of Power Sources* **274**, 951–956 (2015).

3. Chu, H. M., Nguyen, M. Van, Vu, H. N. & Hane, K. Fabrication of single-crystal silicon nanowires based on surface wet adhesion. *Materials Letters* **152**, 94–97 (2015).
4. Street, R. A. *Hydrogenated amorphous silicon.* (Cambridge Solid State Series: 2005).
5. Knights, J. & Lujan, R. Microstructure of plasma-deposited a-Si:H films. *Applied Physics Letters* **35**, 244–246 (1979).
6. Mecklenburg, M. *et al.* Aerographite: Ultra Lightweight, Flexible Nanowall, Carbon Microtube Material with Outstanding Mechanical Performance. *Advanced Materials* **24**, 3468–3490 (2012).
7. Liao, N. . *et al.* Raman study of a-Si:H films deposited by PECVD at various silane temperatures before glow-discharge. *Applied Physics A: Materials Science & Processing* **91**, 349–352 (2008).
8. Morell, G. *et al.* Raman study of the network disorder in sputtered and glow discharge a-Si:H films. *Journal of Applied Physics* **78**, 5120–5125 (1995).
9. Gogoi, P., Dixit, P. & Agarwal, P. Amorphous silicon films with high depositionrate prepared using argon and hydrogen diluted silane for stable solar cells. *Solar Energy Materials & Solar Cells* **91**, 1253–1257 (2007).
10. Dylla, T. *Electron Spin Resonance and Transient Photocurrent Measurements on Microcrystalline Silicon.* (2005).
11. Amrani, R. *et al.* Optical and structural properties of nc-Si:H prepared by argon diluted silane PECVD. *Journal of Non-Crystalline Solids* (2012).
12. Droz, C. *et al.* Relationship between Raman crystallinity and open-circuit voltage in micro-crystalline solar cells. *Solar Energy & Solar Cells* **81**, 61–71 (2004).
13. Fukawa, M., S.Suzuki, Guo, L., Kondo, M. & A. Matsuda High rate growth of microcrystalline silicon using high-pressure depletion method with VHF plasma. *Solar Energy Materials & Solar Cells* **66**, 217–223 (2001).
14. Niikura, C., Kondo, M. & Matsuda, A. Preparation of microcrystalline silicon films at ultra high-rate of 10 nm/s using high-density plasma. *Journal of Non-Crystalline Solids* **338-340**, 42–46 (2004).
15. Reimer, T. *et al.* Single Step Integration of ZnO Nano- and Microneedles in Si Trenches by Novel Flame transport Approach: Whispering Gallery Modes and Photocatalytic Properties. *Applied Materials and Interfaces* **6**, 7806–7815 (2014).

Author: Iris Hölken
Institute: Materials Science
Street: Kaiserstr. 2
City: Kiel
Country: Germany
Email: ih@tf.uni-kiel.de

Two-Dimensional Cavity Polaritons under the Influence of the Landau Quantization, Rashba Spin-Orbit Coupling and Zeeman Splitting

S.A. Moskalenko[1], I.V. Podlesny[1], E.V. Dumanov[1], M.A. Liberman[2], and I. Lelyakov[1]

[1] Institute of Applied, Chisinau, Republic of Moldova
[2] Nordita, KTH Royal Institute of Technology and Stockholm University, Stockholm, Sweden

Abstract— **The properties of the two-dimensional cavity polaritons subjected to the action of a strong perpendicular magnetic and electric fields, giving rise to the Landau quantization (LQ) of the 2D electrons and holes accompanied by the Rashba spin-orbit coupling, by the Zeeman splitting and by the nonparabolicity of the heavy-hole dispersion law were investigated. Our results are based on the exact solutions for the eigenfunctions and for the eigenvalues of the Pauli- type Hamilonian with third order chirality terms for heavy-holes and with first order chirality terms for electrons. They were obtained using the method proposed by Rashba [1]. We predict the drastic changements of the optical properties of the cavity polaritons. The main of them are related with the existence of a multitude of the polariton energy levels nearly situated on the energy scale, their origin being related with the LQ of the electrons and holes. Most of these levels have the nonmonotonous dependences on the magnetic field strength *B* with overlapping and intersections.**

Keywords— **polariton, Landau quantization, Rashba spin-orbit-coupling.**

I. INTRODUCTION

In the Refs. [2] the Hamiltonians describing in different approximations the electron-radiation interaction in the system of two-dimensional coplanar electrons and holes accumulated in the semiconductor quantum well and subjected to the action of a strong perpendicular magnetic field giving rise to the Landau quantization of their energy levels were obtained. In Ref. [3] the exciton-cyclotron resonance and the optical orientation phenomena arising under the influence of the circularly polarized laser radiation were studied but without taking into account the RSOC. The Hamiltonian derived in Ref. [3] in the e-h representation was transcribed in Ref. [4] with the purpose to described the magnetoexciton-photon interaction for the light propagating arbitrarily in the three-dimensional (3D) space as well as being confined in the microcavity. The dispersion law of the magnetoexciton-polariton in microcavity was deduced. The dependence of the Rabi frequency on the magnetic field strength as well as the selection rule for the numbers of the LQ levels of the e-h pair engaged in the dipole-active and in the quadrupole-active transitions were determined [3]. The influence of the RSOC on the optical properties of the magnetoexcitons and on the band-to-band quantum transitions arising due to the action of a supplementary electric field perpendicular to the plane of the QW was investigated in the Ref. [5]. In the mentioned Ref. [5] the effects related with the Zeeman splitting were not discussed. This lack is removed below.

II. THE HAMILTONIAN DESCRIBING THE 2D MAGNETOEXCITONS, CAVITY PHOTONS AND THEIR INTERACTION

The Hamiltonians describing the Coulomb electron-electron interaction as well as the electron-radiation interaction taking into account the Landau quantization, Rashba spin-orbit coupling and Zeeman splitting effects were deduced in Refs. [2]. Our further discussions are based on these results.

The zeroth order Hamiltonian describing free 2D magnetoexcitons, the cavity photons and their interaction has a quadratic form and consists of three parts:

$$H_2 = H_{mex}^0 + H_{ph}^0 + H_{mex-ph}. \tag{1}$$

In the Hamiltonian H_2 only the dipole-active magnetoexciton states F_1 and F_4 as well as the quadrupole-active states F_3 and F_5 were included as follows:

$$H_{mex}^0 = \sum_{n=1,3,4,5} E_{ex}\left(F_n, \vec{k}_{\|}\right)\psi_{ex}^{+}\left(F_n, \vec{k}_{\|}\right)\psi_{ex}\left(F_n, \vec{k}_{\|}\right). \tag{2}$$

The remained two states F_2 and F_6 were excluded because they are forbidden in both approximations.

The cavity photons have the wave vectors $\vec{k} = \vec{a}_3 k_z + \vec{k}_{\|}$ consisting from two parts. The longitudinal component is oriented along the axis of the resonator determined by the unit vector $\vec{a}_3$ perpendicular to the surface of the quantum well (QW) embedded inside the microcavity. It has a well definite values $k_z = \pm\pi/L_c$, where L_c is the cavity length. The transverse component $\vec{k}_{\|} = \vec{a}_1 k_x + \vec{a}_2 k_y$ is a 2D vector oriented in-plane of the QW and is determined by two in-plane unit vectors $\vec{a}_1$ and $\vec{a}_2$. Vectors of the light circular polarizations

V. Sontea and I. Tiginyanu (eds.), *3rd International Conference on Nanotechnologies and Biomedical Engineering*,
IFMBE Proceedings 55,
DOI: 10.1007/978-981-287-736-9_8

$\vec{\sigma}_{\vec{k}}^{\pm}$ can be constructed introducing two unit vectors $\vec{s}$ and $\vec{t}$ perpendicular to the light wave vector $\vec{k}$ and to each other.

The zeroth order Hamiltonian of the cavity photons with the wave vectors $\vec{k}$ and circular polarizations $\vec{\sigma}_{\vec{k}}^{\pm}$ described in Ref. [2] looks as

$$H_{ph}^{0}=\sum_{\vec{k}_{\parallel},k_z=\pm\pi/L_C}\hbar\omega_{\vec{k}}\left[\left(C_{\vec{k},+}\right)^{+}C_{\vec{k},+}+\left(C_{\vec{k},-}\right)^{+}C_{\vec{k},-}\right]. \quad (3)$$

The creation and annihilation operators of the photons with circular polarizations $\vec{\sigma}_{\vec{k}}^{\pm}$ and with the linear polarizations $\vec{s}$ and $\vec{t}$ are related as follows

$$\begin{gathered}C_{\vec{k},\pm}=\frac{1}{\sqrt{2}}\left(C_{\vec{k},s}\pm iC_{\vec{k},t}\right)\\ \left(C_{\vec{k},\pm}\right)^{+}=\frac{1}{\sqrt{2}}\left(C_{\vec{k},s}^{+}\mp iC_{\vec{k},t}^{+}\right)\\ \vec{s}C_{\vec{k},s}+\vec{t}C_{\vec{k},t}=C_{\vec{k},-}\vec{\sigma}_{\vec{k}}^{+}+C_{\vec{k},+}\vec{\sigma}_{\vec{k}}^{-}\\ \vec{s}C_{\vec{k},s}^{+}+\vec{t}C_{\vec{k},t}^{+}=\left(C_{\vec{k},-}\right)^{+}\vec{\sigma}_{\vec{k}}^{-}+\left(C_{\vec{k},+}\right)^{+}\vec{\sigma}_{\vec{k}}^{+}.\end{gathered} \quad (4)$$

The Hamiltonian describing the magnetoexciton-photon interaction including only the resonance terms and taking into account the two photon circular polarizations has the form:

$$\begin{aligned}H_{mex-ph}=&\sum_{\vec{k}_{\parallel},k_z=\pm\pi/L_c}\{\varphi\left(F_1,\vec{k}_{\parallel}\right)[C_{\vec{k},-}\left(\vec{\sigma}_{\vec{k}}^{+}\cdot\vec{\sigma}_{-1}^{*}\right)+\\ &+C_{\vec{k},+}\left(\vec{\sigma}_{\vec{k}}^{-}\cdot\vec{\sigma}_{-1}^{*}\right)]\psi_{ex}^{+}\left(F_1,F_{\parallel}\right)+\\ &+\varphi\left(F_4,\vec{k}_{\parallel}\right)\left[C_{\vec{k},-}\left(\vec{\sigma}_{\vec{k}}^{+}\cdot\vec{\sigma}_{1}^{*}\right)+C_{\vec{k},+}\left(\vec{\sigma}_{\vec{k}}^{-}\cdot\vec{\sigma}_{1}\right)\right]\psi_{ex}^{+}\left(F_4,\vec{k}_{\parallel}\right)+\\ &+\varphi\left(F_3,\vec{k}_{\parallel}\right)\left[C_{\vec{k},-}\left(\vec{\sigma}_{\vec{k}}^{+}\cdot\vec{\sigma}_{1}^{*}\right)+C_{\vec{k},+}\left(\vec{\sigma}_{\vec{k}}^{-}\cdot\vec{\sigma}_{1}\right)\right]\psi_{ex}^{+}\left(F_3,\vec{k}_{\parallel}\right)+\\ &+\varphi\left(F_5,\vec{k}_{\parallel}\right)\left[C_{\vec{k},-}\left(\vec{\sigma}_{\vec{k}}^{+}\cdot\vec{\sigma}_{-1}^{*}\right)+C_{\vec{k},+}\left(\vec{\sigma}_{\vec{k}}^{-}\cdot\vec{\sigma}_{-1}^{*}\right)\right]\psi_{ex}^{+}\left(F_5,\vec{k}_{\parallel}\right)+\\ &+H.c.\}.\end{aligned} \quad (5)$$

The coefficients $\varphi(F_n,\vec{k}_{\parallel})$ and their square moduli were determined in [2].

The equations of motion for four magnetoexciton annihilation operators $\psi_{ex}(F_n,\vec{k}_{\parallel})$ with $n=1,3,4,5$ as well as for the similar photon operators $C_{\vec{k},\pm}$ in the stationary conditions look as

$$\begin{gathered}i\hbar\frac{d}{dt}\psi_{ex}\left(F_n,\vec{k}_{\parallel}\right)=\left[\psi_{ex}\left(F_n,\vec{k}_{\parallel}\right),\hat{H}_2\right]=\hbar\omega\psi_{ex}\left(F_n,\vec{k}_{\parallel}\right)\\ i\hbar\frac{d}{dt}C_{\vec{k},\pm}=\left[C_{\vec{k},\pm},\hat{H}_2\right]=\hbar\omega C_{\vec{k},\pm}.\end{gathered} \quad (6)$$

The five order algebraic dispersion equations is

$$\begin{aligned}\left(\hbar\omega-\hbar\omega_{\vec{k}}\right)=&\frac{\left|\left(\vec{\sigma}_{\vec{k}}^{\mp}\cdot\vec{\sigma}_{-1}^{*}\right)\right|^2\left|\varphi\left(F_1,\vec{k}_{\parallel}\right)\right|^2}{\hbar\omega-E_{ex}\left(F_1,\vec{k}_{\parallel}\right)}+\frac{\left|\left(\vec{\sigma}_{\vec{k}}^{\mp}\cdot\vec{\sigma}_{1}^{*}\right)\right|^2\left|\varphi\left(F_4,\vec{k}_{\parallel}\right)\right|^2}{\hbar\omega-E_{ex}\left(F_4,\vec{k}_{\parallel}\right)}+\\ &+\frac{\left|\left(\vec{\sigma}_{\vec{k}}^{\mp}\cdot\vec{\sigma}_{1}^{*}\right)\right|^2\left|\varphi\left(F_3\vec{k}_{\parallel}\right)\right|^2}{\hbar\omega-E_{ex}\left(F_3,\vec{k}_{\parallel}\right)}+\frac{\left|\left(\vec{\sigma}_{\vec{k}}^{\mp}\cdot\vec{\sigma}_{-1}^{*}\right)\right|^2\left|\varphi\left(F_5,\vec{k}_{\parallel}\right)\right|^2}{\hbar\omega-E_{ex}\left(F_5,\vec{k}_{\parallel}\right)}.\end{aligned} \quad (7)$$

It is useful to introduce the dimensionless values and the detunings Δ_1:

$$\begin{gathered}\hbar\omega=\hbar\omega_c+E,\frac{\hbar\omega}{\hbar\omega_c}=1+\varepsilon,\varepsilon=\frac{E}{\hbar\omega_c},\\ \hbar\omega_c=\frac{E_{ex}(F_1,B,0)}{1-\delta_1},|\delta_1|<1,x=\frac{|\vec{k}_{\parallel}|L_c}{\pi}.\end{gathered} \quad (8)$$

The corresponding dispersion equation has the form:

$$\begin{aligned}\left(\varepsilon-\frac{x^2}{2}\right)=f_{osc}\cdot\{&\frac{|a_0^-|^2|d_0^-|^2\left(1-\frac{x^2}{2}+\frac{7}{16}x^4\right)}{\left(\varepsilon+\delta_1-\frac{n_c^2\hbar\omega_c}{M(F_1,B)c^2}\cdot\frac{x^2}{2}\right)}+\\ &+\frac{\frac{x^4}{16}}{\left(\varepsilon+1-\frac{E_{ex}(F_4,B,0)}{E_{ex}(F_1,B,0)}(1-\delta_1)-\frac{n_c^2\hbar\omega_c}{M(F_4,B)c^2}\cdot\frac{x^2}{2}\right)}+\\ &+\frac{|b_1^-|^2\left(\frac{\pi l_0}{L_c}\right)^2\frac{x^6}{32}}{\left(\varepsilon+1-\frac{E_{ex}(F_3,B,0)}{E_{ex}(F_1,B,0)}(1-\delta_1)-\frac{n_c^2\hbar\omega_c}{M(F_3,B)c^2}\cdot\frac{x^2}{2}\right)}+\\ &+\frac{|a_0^-|^2|d_0^-|^2\left(\frac{\pi l_0}{L_c}\right)^2\frac{x^2}{2}\left(1-\frac{x^2}{2}+\frac{7}{16}x^4\right)}{\left(\varepsilon+1-\frac{E_{ex}(F_5,B)}{E_{ex}(F_1,B)}(1-\delta_1)-\frac{n_c^2\hbar\omega_c}{M(F_5,B)c^2}\cdot\frac{x^2}{2}\right)}\}\end{aligned} \quad (9)$$

III. THE ENERGY SPECTRUM OF THE CAVITY MAGNETOEXCITON-POLARITONS

The energy spectrum of the cavity magnetoexciton-polaritons when the energy of the cavity mode is tuned to the magnetoexciton energy level $E_{ex}(F_1,B,0)$ was obtained solving the fifth order dispersion equation (9) and is depicted in the figure 1. They concern to the case of the cavity photons with circular polarizations $\vec{\sigma}_{\vec{k}_{\uparrow}}^{-}$, when the cavity mode energy $\hbar\omega_c$ was tuned to the magnetoexciton energy level $E_{ex}(F_1,B,0)$ in the way $\hbar\omega_c = E_{ex}(F_1,B,0)/(1-\delta_1)$ with dimensionless detuning $\delta_1 = -0.01$. Two values of the magnetic field strength B: 20 T and 40 T were considered. The energies $\hbar\omega_c$ in these cases are different. The obtained pictures essentially depend on the RSOC, arising under the influence of the external electric field applied perpendicularly to the surface of the QW being parallel to the external magnetic field. In the case of the 2D heavy holes the RSOC is characterized by the chirality terms of the third order in the wave vector k components, which inevitably lead to the collapse of the semiconductor energy gap. To avoid it in the heavy hole dispersion law the fourth order term of the type $|\vec{k}|^4$ was introduced being proportional to electron field strength with the parameter of the nonparabolicity C. Fig. 1 corresponds to the presence of RSOC with the electric field strength $E_z = 30$ kV/cm and the parameter C equal to 20. The four variants were calculated. They arised due to two values of the magnetic field strength 20 T and 40 T mentioned above, and due to two values of the heavy hole unknown g-factors $g_h = \pm 5$. The electron g factor was supposed to be equal $g_e = 1$. Looking at the figure 1 one can observe that the bare cavity photon branch is intersected by the bare magnetoexciton state F_1 branches. Only the interaction of the magnetoexciton with the cavity photons with the selected circular polarization $\vec{\sigma}_{\uparrow}^{-}$ gives rise to the well distinguishable polariton pictures. In another intersection points denoted as $(n-c)$ with n=3, 4, 5 the magnetoexciton-photon interactions are very small. The reconstructed polariton curves in the vicinities of these points are shown only qualitatively so as to help the eye. The real values of the splittings arising due to the anticrossings in these points are denoted as $\Delta(n-c)$. The coordinates of these intersection points $X(n-c)$ and $Y(n-c)$ together with the corresponding splittings are mentioned in the captions.

The forbidden bare magnetoexciton branches F_2 and F_6 are also drawn in the figures because their presence will influence on the kinetic and quantum statistic properties of the high density cavity magnetoexciton-polaritons. The description of another energy spectra in general outline is the same. But the positions of the different magnetoexciton energy branches on the energy scale essentially depend on the full set of the parameters $\{B, E_z, g_e, g_h, C\}$ which is demonstrated in the figure 1.

Our theoretical model taking into account the electron structure of the 2D magnetoexcitons arising under the influence of the perpendicular strong magnetic and electric fields is applicable for the GaAs-type QWs only in the range of magnetic field strength $B \geq 10$ T. A similar variant but without RSOC was described in the Ref. [6]. The experimental investigations of the cavity exciton-polaritons in a range of small and stronger magnetic fields up to 14 T were reported in the

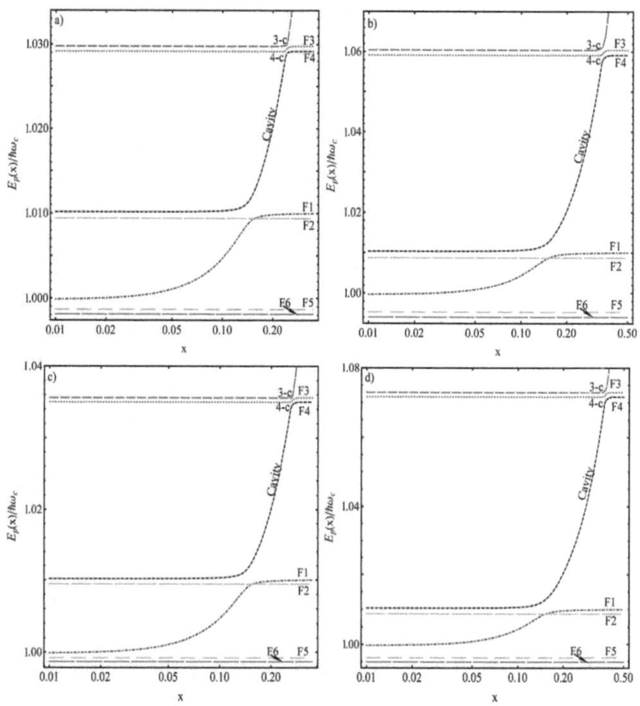

Fig. 1 Five cavity magnetoexciton-polariton energy branches in dependence on the dimensionless in-plane wave number x arising as result of the interaction of the cavity photons with circular polarizations $\vec{\sigma}_{\vec{k}_\uparrow}^{-}$ or $\vec{\sigma}_{\vec{k}_\downarrow}^{+}$ with four magnetoexciton energy branches F_1, F_4, F_3 and F_5 characterized by the circular polarizations $\vec{\sigma}_{\pm 1}$, two branches being dipole-active and two quadrupole active, when the cavity mode energy is tuned to the magnetoexciton energy level $E_{ex}(F_1,B,0)$. The influence of the external electric field and of the RSOC with the parameters $E_z = 30$ kV/cm and $C = 20$ is represented in four variants. They appear from the combinations of two values of the magnetic field strength B and of two values of the hh $g-$ factor g_h as follows: a) $B = 20$ T, $g_h = 5$; b) $B = 40$ T, $g_h = 5$; c) $B = 20$ T, $g_h = -5$; d) $B = 40$ T, $g_h = -5$. The electron $g-$ factor was selected $g_e = 1$. For the completeness two forbidden magnetoexciton energy branches F_2 and F_6 are added. The intersection points and the splittings of the polariton curves in these points:

a) $X(1-c) = 0{,}145$; $Y(1-c) = 1{,}01015$; $X_a(4-c) = 0{,}241$; $Y_a(4-c) = 1{,}02919$; $\Delta_a(4-c) = 11{,}23\cdot 10^{-6}$;
$X_a(3-c) = 0{,}245$; $Y_a(3-c) = 1{,}02980$; $\Delta_a(3-c) = 11{,}29\cdot 10^{-6}$;
b) $X_b(4-c) = 0{,}343$; $Y_b(4-c) = 1{,}05920$; $\Delta_b(4-c) = 15{,}74\cdot 10^{-6}$; $X_b(3-c) = 0{,}347$; $Y_b(3-c) = 1{,}06040$; $\Delta_b(3-c) = 16\cdot 10^{-6}$;
c) $X_c(4-c) = 0{,}264$; $Y_c(4-c) = 1{,}03492$; $\Delta_c(4-c) = 11{,}19\cdot 10^{-6}$; $X_c(3-c) = 0{,}268$; $Y_c(3-c) = 1{,}03561$; $\Delta_c(3-c) = 11{,}26\cdot 10^{-6}$;
d) $X_d(4-c) = 0{,}377$; $Y_d(4-c) = 1{,}07161$; $\Delta_d(4-c) = 15{,}60\cdot 10^{-6}$; $X_d(3-c) = 0{,}380$; $Y_d(3-c) = 1{,}07287$; $\Delta_d(3-c) = 15{,}91\cdot 10^{-6}$.

Ref. [7]. The more suitable theoretical model in this case was elaborated in Ref. [8], where the energy spectrum of the hydrogenic potential $V(r)=-e^2/r$ in two dimensions was studied for perpendicular magnetic field in two weak and strong regimes. Using a two – point Pada approximant a reliable interpolation between two limiting situations was proposed. This model was enlarged taking into account the dimension perpendicular to the QW. The existence of many theoretical models will stimulate the further experimental investigations in the regime of high magnetic fields $B \geq 10$ T. A short review of the comprehensive experimental findings found out in the Ref. [7] is needed to compare and to discuss them in two main models mentioned above. The experiments were performed using the 8nm-thick $In_{0.04}Ga_{0.96}As$ QW embedded into the resonator with the distributed Bragg reflectors.

The Wannier-Mott excitons have a resonance energy 1.484 eV, an exciton binding energy equal approx. 7 meV, and effective mass $0.046m_0$. The polariton linewidth was equal to 0.3 meV. The interest to these objects is due to the observation of the BEC phenomenon with the participation of the cavity polaritons accompanied by the superfluidity formation with zero viscosity. The effect of the magnetic field on the polaritons revealed in Ref. [7] is diverse already in the linear regime. The magnetic field induces the changes in the linewidth. The interpretation of these results on the base of the hydrogen interpolation model [8] takes into account the exciton energy shift including the Zeeman splitting, the modification of the exciton-photon coupling strength and the change of the Rabi splitting. The magnetic field is considered as a parameter changing the tuning of the exciton-photon resonances. The changes of the emission intensity and of the polariton linewidth were explained in Ref. [7] by the magnetically modified scattering processes with participation of the acoustical phonons. The increase of the exciton oscillator strength f_{osc} was directly determined in Ref. [7] measuring the Rabi splitting Ω of a polariton branches. Taking into account the relation $\Omega \sim \sqrt{N f_{osc}/L_{eff}}$, where N is the number of QWs embedded into the resonator and L_{eff} is the effective length, the f_{osc} was calculated. These results are represented in the fig. 6 of the Ref. [7]. One can see that the $\hbar\Omega$ value increases from 3.4 meV to 4.8 meV at $B=14$ T, whereas the f_{osc} increases by a factor of two.

In our model the dependences of the Rabi frequency Ω on the magnetic field strength is determined by the relation $\Omega \sim l_0^{-1} \sim \sqrt{B}$, where l_0 is the magnetic length. It leads to the dependence for the oscillator strength $f_{osc} \sim |\Omega|^2 \sim B$.

Another property of the cavity polariton spectra is the Zeeman splitting of the lower polariton branch. Two components of the bright exciton level are decoupled and couple independently with the photon of corresponding circular polarization. The observed experimentally in Ref. [7] the Zeeman splitting of the lower polariton branch reveals a decreasing in dependence on the magnetic field strength up to 14 T, instead of the expected increasing.

In the frame of our model the magnetoexciton states F_1-F_6 are represented by the non-degenerate branches with the being eventual dependence on the magnetic field strength with increasing splitting between them in the absence of the RSOC.

Only in the presence of the RSOC, as well as taking into account the nonparabolicity of the heavy-hole dispersion law, the intersections and overlapping between the magnetoexciton branches appear. The surprising behavior of the Zeeman splitting revealed in the Ref. [7] remains unclear.

IV. CONCLUSIONS

The properties of the 2D cavity polaritons subjected to the action of a strong perpendicular magnetic and electric fields were investigated. The dispersion equation describing the magnetoexciton-polariton energy spectrum includes the first four states F_1, F_3, F_4 and F_5 interacting with the cavity photons in two approximations. These four magnetoexciton degrees of freedom together with the branch of the cavity photons give rise to five order dispersion equation with five polariton-type renormalized energy branches.

The obtained picture was completed adding two magnetoexciton energy-branches F_2 and F_6 forbidden in the optical quantum transitions. The cavity polariton energy spectrum essentially depends on the detuning between the energy of the cavity mode $\hbar\omega_c$ and the energy of the magnetoexciton level. One variant of such detuning was studied. It concerns the detuning with the energy level $E_{ex}(F_1, B, 0)$ of the dipole-active state F_1 with circular polarization $\vec{\sigma}_{-1}$ interacting with the cavity photons with circular polarization $\vec{\sigma}^-_{\vec{k}_\uparrow}$ or $\vec{\sigma}^+_{\vec{k}_\downarrow}$ propagating inside the microcavity with wave vectors $\vec{k}_{\uparrow,\downarrow} = \pm\vec{a}_3 \pi/L_c + \vec{k}_{\parallel}$ correspondingly. The cavity mode energy was chosen to be equal $\hbar\omega_c(B) = E_{ex}(F_1, B, 0)/(1-\delta_1)$ with dimensionless detuning $\delta_1 = -0{,}01$. The value $\hbar\omega_c(B)$ changes in dependence on the magnetic field strength B and on other parameters of the theory simultaneously with the change of the magnetoexciton energy level $E_{ex}(F_1, B, 0)$.

Two different sign $\pm$ of the quantized component k_z of the wave vector $\vec{k}$ happen to be very important, giving rise to the gyrotropy effect. It consists in the dependence of the polariton energy spectrum picture on the direction of the photon propagation. In one case it is determined by the wave vector $\vec{k}_\uparrow$ with $k_z = \pi/L_c$ whereas in another case by the wave vector $\vec{k}_\downarrow$ with $k_z = -\pi/L_c$.

ACKNOWLEDGMENT

I.V.P. and E.V.D. thanks the Foundation for Young Scientists of the Academy of Sciences of Moldova for financial support (14.819.02.18F).

CONFLICT OF INTEREST

The authors declare that they have no conflict of interest.

REFERENCES

1. Rashba E I (1960) Properties of semiconductors with an extremum loop. 1. Cyclotron and combinational resonance in a magnetic field perpendicular to the plane of the loop. Sov Phys Solid State 2:1109-1123
2. Moskalenko S A et al. Effects of Rashba spin-orbit coupling, Zeeman splitting and gyrotropy in 2D cavity polaritons under the influence of the Landau quantization. Eur Phys J B (submitted)
3. Moskalenko S A et al. (2012) Two-dimensional magnetoexciton-polariton J of Nanophotonics 6:061806(1-16)
4. Moskalenko S A et al. (2009) Exciton-cyclotron resonance in two-dimensional structures in a strong perpen-dicular magnetic field and optical orientation conditions. Phys Rev B 79(12):125425
5. Hakioglu T et al. (2011) The influence of the Rashba spin-orbit coupling on the 2D magnetoexcitons. J Phys Cond Matt 23:345405(1-13)
6. Bardyszewski W et al. (2014) Nonlinear Zeeman splitting of magne-toexcitons.. Phys Rev B 90:075302
7. Pietka B et al. (2015) Magnetic field tuning of exciton-polaritons in a semiconductor microcavity. Phys Rev B 91:075309
8. MacDonald A H and Ritchie D S (1986) Hydrogenic energy levels in two dimensions at arbitrary magnetic fields. Phys Rev B 33:8336

Author: S.A. Moskalenko
Institute: Institute of Applied Physics
Street: Academiei, 5
City: Chisinau
Country: Moldova
Email: exciton@phys.asm.md

Detection in the Contacts with HTSC - InSb: Numerical Modeling of the Contact Area Role

Ia.I. Kerner

D.Ghitu Institute of Electronic Engineering and Industrial Technologies, Chisinau, Republic of Moldova

Abstract— **Diode detectors (DD) are widely used in electronic information and communication systems. The numerical modeling of the electrical properties in the contacts of the high temperature superconductor (HTSC) with semiconductor indium antimonite (InSb) had been made. There were analyzed the possibilities to create DD based on these contacts and working at liquid nitrogen temperature 77.4 K. The influence of the contact area on the DD parameters was analyzed.**

Also the numerical modeling of the electrical potential distribution and current passing in the contacts of normal metal or superconductor with semiconductor alloy bismuth-antimony (Bi-Sb) was made. There were analyzed possibilities to create DD based on these contacts and working at liquid helium temperature 4.2 K.

The comparison with existent literature data shows the proposed DD can be 10÷100 times better. Therefore these DD are perspective for cryogenic electronics and there is an actual problem to elaborate them.

Keywords— **Diode detectors, semiconductors, superconductors.**

I. INTRODUCTION

The diode detectors play an important role in radio technique and electronics. The use of high frequencies (above 1 GHz) stimulated the careful study of diodes with Schottky barrier. These diodes use the quick-acting metal-semiconductor contacts [1].

The further improvement of their parameters was achieved due to fall of the working temperature (T). This direction was named cryoelectronics [2], it allows to raise the nonlinearity of the current-voltage dependences (CVD) and current responsivity (CR). The thermal noise power decreases too. For example there were elaborated DD based on the contacts Pb-pGaAs [3, 4]. At the signal frequency f = 9 GHz and T= 4.2 K these diodes had CR ≈ 500 A/W and noise equivalent power NEP ≈ 5×10-15 W/$\sqrt{Hz}$. Also the deep cooling allows using the materials with little energy gap width but high mobility of electrons, such as solid solutions Bi-Sb [2, 5].

After the discovery of the high temperature superconductivity the possibilities to use HTSC in cryoelectronics were studied too. At the liquid nitrogen temperature 77 K and signal frequency f = 37.5 GHz the corresponding structures revealed the voltage responsivity VR=3000 V/W [6]. The further studies [7] allowed to create the structures with VR=5000 V/W and noise equivalent power NEP = 2×10-12 W/$\sqrt{Hz}$ at the signal frequency f=31 GHz and temperature T = 77 K. According to our results [8, 9] the diode detectors based on the contacts HTSC-InSb may have the next parameters: CR ≈ 40 A/W, VR ≈ 106 V/W and NEP ≈ 8×10-15 W/$\sqrt{Hz}$ at T = 77.4 K and f = 10 GHz.

At the same temperature and f = 30 GHz these DD may have CR ≈ 15 A/W, VR ≈ 3.5×105 V/W and NEP ≈ 2×10-14 W/$\sqrt{Hz}$.

On the other hand often there is an oxidation of semiconductor in HTSC-semiconductor contacts, because oxygen is an integral part of HTSC. Also cooling to the liquid nitrogen temperature 77.4 K may be insufficient to obtain the good DD parameters. In this situation, taking into account the rapid development of cryogenics, the study of DD based on the contacts traditional superconductor – semiconductor seems to be actual problem. Usually these DD work at liquid helium temperature (T = 4.2 K). In this article there are discussed both DD based on HTSC-semiconductor contacts and DD based on the contacts of superconductor (or normal metal) with semiconductor solid solution Bi-Sb.

II. RESULTS AND DISCUSSION

The calculated current responsivity, voltage responsivity and NEP dependences on the signal frequency are shown in figures (figs.) 1, 2 and 3 respectively.

In all figures the logarithmic scale for X-axes is used. An exponential form is often used for numbers of axes.

Figs. 1 - 3 show that current and voltage responsivities decrease and NEP increases at the frequencies more 3 GHz. At these frequencies the negative role of the barrier capacity is revealed and it begins to shunt the nonlinear contact resistance. On the other hand at high frequencies the contact capacity resistance becomes compared with ohm spreading resistance. The current redistribution occurs, it leads to reduction of the rectified current and DD parameters become worse.

The round flat contacts with contact area 100, 10 and 1 μ^2 were studied. Taking into account the little surface area these contacts may be considered as point contacts [1].

V. Sontea and I. Tiginyanu (eds.), *3rd International Conference on Nanotechnologies and Biomedical Engineering*,
IFMBE Proceedings 55,
DOI: 10.1007/978-981-287-736-9_9

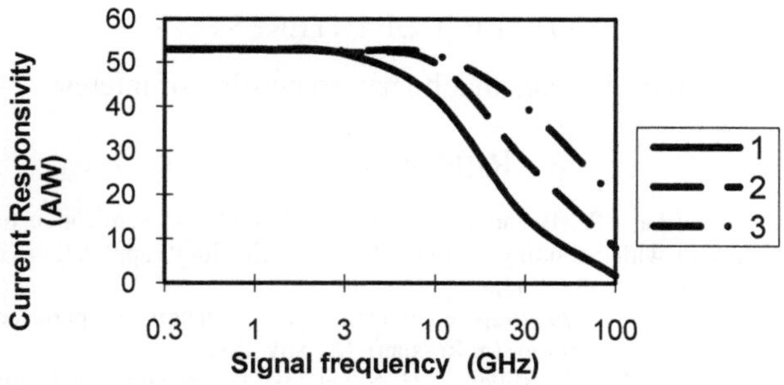

Fig. 1 The calculated current responsivity dependence on the signal frequency in the contacts with high temperature superconductor. The legend inscriptions 1, 2 and 3 correspond to the contact area 100, 10 and 1 μ2. T = 77.4 K.

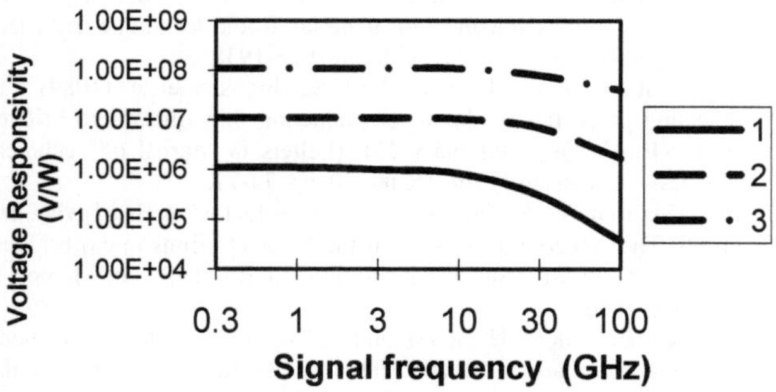

Fig. 2 The calculated voltage responsivity dependence on the signal frequency. The legend inscriptions and other data are similar to those in fig. 1. T = 77.4 K.

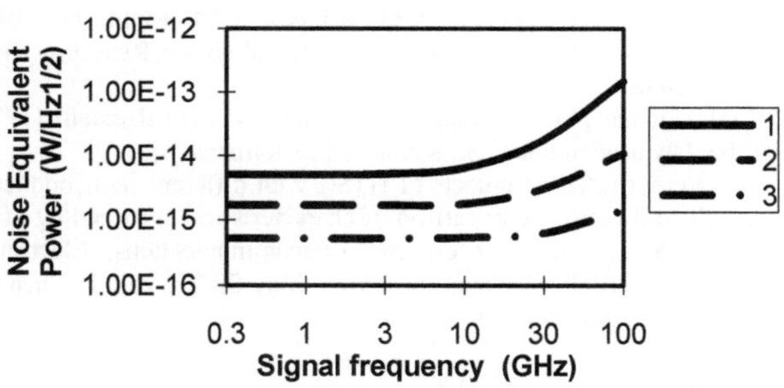

Fig. 3 The calculated noise equivalent power dependence on the signal frequency. The legend inscriptions and other data are similar to those in fig. 1. T = 77.4 K.

In this case the barrier capacity is proportional to S, where S is the contact area, and ohm spreading resistance is proportional to $S^{-1/2}$ [1]. In this situation, when the contact area decreases, the capacity resistance rises faster than the ohm spreading resistance. Therefore the redistribution of applied variable voltage occurs, the contact voltage grows and DD parameters become well (see curves 1, 2, 3 in fig. 1). When the contact area reduces the contact differential resistance rises and voltage responsivity rises too. On the other hand noise current falls and noise equivalent power falls too (see figs. 2 and 3).

Also contacts of semiconductor solid solution $Bi_{0.88}Sb_{0.12}$ with normal metal or superconductor were considered (for case S = 100 μ^2). The normal metal may be aluminum at T ≥ 1.2 K and silver or gold at lower temperatures. Niobium or niobium nitride (NbN) may be chosen as superconductors at liquid helium temperatures. Materials properties were taken from [10, 11]. Results of calculations are shown in figures 4 - 6. In all figures the logarithmic scale for X-axes is used.

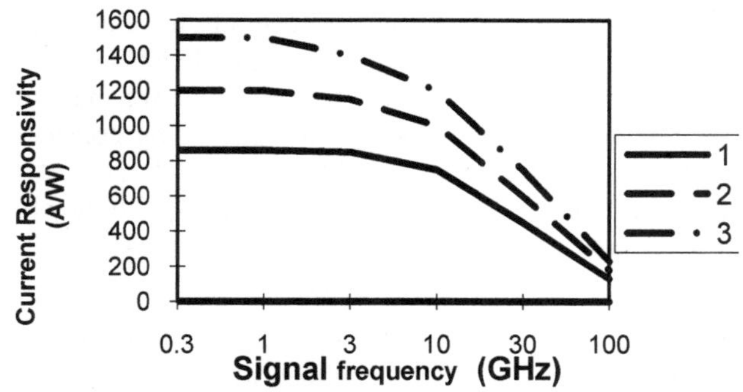

Fig. 4 The calculated current responsivity dependence on the signal frequency. The legend inscriptions 1, 2 and 3 correspond to the contacts of $Bi_{0.88}Sb_{0.12}$ with normal metal, niobium and niobium nitride accordingly. T = 4.2 K.

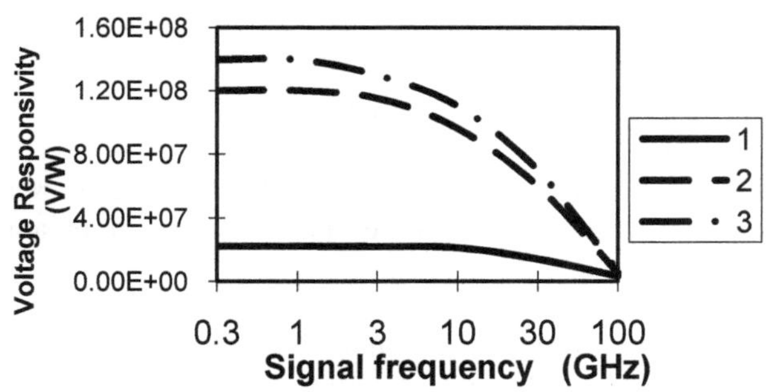

Fig. 5 The calculated voltage responsivity dependence on the signal frequency. The legend inscriptions are similar to those in fig. 4. T = 4.2 K.

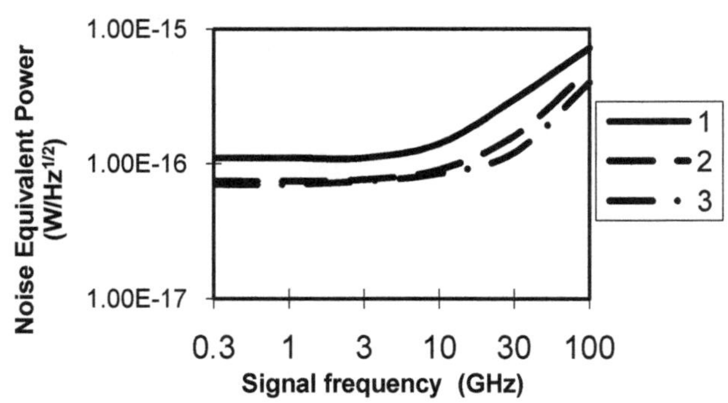

Fig. 6 The calculated noise equivalent power dependence on the signal frequency. The legend inscriptions are similar to those in fig. 4. T = 4.2 K.

Figs. 4 - 6 show that current and voltage responsivities decrease and NEP increases at the frequencies more 3 GHz. At these frequencies the negative role of the barrier capacity is revealed and it begins to shunt the nonlinear contact resistance. The current redistribution occurs, it leads to reduction of the rectified current and DD parameters become worse.

If metal is replaced by superconductor the DD parameters are improved essentially. This effect increases when superconductor energy gap is more (see curves 2 and 3 in figs. 4-6). This fact points out the necessity to select superconductor with wide energy gap (for example NbN).

Taking into account results [3, 4, 6-8, 12] we may conclude that contacts with Bi-Sb allow improving considerably DD parameters. On one hand they are much more effective than contacts HTSC-superconductor [6-8]. On the other hand they are better than contacts with GaAs [3, 4].

The main advantages of Bi-Sb are next:

(i) Little barriers heights due to narrow energy gap. This fact provides a big CVD nonlinearity and big current responsivity.

(ii) High mobility of electrons, which reduces ohm resistance and improves frequencies properties.

(iii) Little barrier capacity, due to little barriers heights and small effective masses of electrons, which also improves frequencies properties.

III. Conclusions

According to obtained results the temperature decrease from 77.4 K to 4.2 K allows to improve the diode parameters by two orders.

On the other hand the working temperature 77.4 K may be applied too, if the correspondent parameters are acceptable. In this case the reduction of the contact area may sufficiently improve the diode parameters.

The comparison with [3, 4, 6, 7] data shows that in the proposed DD current responsivity can be 2 times more and noise equivalent power can be 100 times less than the ones in existing DD (at the same temperature and signal frequency). Also they may have very high voltage responsivity.

This fact draws the conclusion the studied contacts are perspective to elaborate them.

Conflict of Interest

The author declares that he has no conflict of interest.

References

1. V.I. Striha, E.V. Buzaneva, I.A. Radzievsky (1974) Semiconductor devices with Schotttky barrier. Physics, technology, application (in Russian). M., Sov. Radio.
2. V.N. Alfeev (1979) (Superconductors, semiconductors and paraelectrics in cryoelectronics (in Russian), M., Sov. Radio.
3. M. Mc Call, M.F. Millea, A.H. Silver (1973) The superconductor-semiconductor Schottky barrier diode detectors. Applied Physics Letters, vol. 23, no. 5, pp. 263-264.
4. M. Mc Call, M.F. Millea, A.H. Silver et al. (1977) The super Schottky microwave mixer, IEEE Transactions on Magnetics. vol. MAG-13, no. 1, pp.221-227.
5. V.G. Alekseeva, T.M. Lifshits, E.G. Chirkova, A.Ia. Shul'man (1978) $Bi_{1-x}Sb_x$ is a new semiconductor material. Radio technique and Electronics (in Russian), vol. 23, no. 9, pp.1926-1938.
6. V.A. Kulikov, L.V. Matveets, A.Iu. Serebryakov et al. (1989) The detecting properties of the superconducting thin film micro bridges from $YBa_2Cu_3O_{7-x}$. Pis'ma v JTF (Letters in Journal of Technical Physics – in Russian)", vol. 15, no. 20, pp. 74-77.
7. L.S. Kuz`min, V. N. Ohrimenko, E. S. Soldatov, A. N. Tavhelidze (1990) The detecting properties of the YBaCuO films micro bridges. Sverhprovodimost` (Superconductivity– in Russian), vol. 3, no.11, pp. 2650-2660.
8. Ia.I. Kerner (2006) HTSC-Si and HTSC-InSb contacts for diode detectors: comparison of characteristics. Moldavian Journal of the Physical Sciences, vol. 5, no. 3-4, pp. 360-365.
9. Ia.I. Kerner (2008) Numerical modeling of detection in the contacts with bismuth- antimony alloy. 2-nd Int. Conf. "Telecommunications, Electronics and Informatics". Chisinau, Moldova, May 15-18, 2008. Proceedings, vol. 2. Chisinau: UTM, pp. 59-64.
10. D.V. Gitsu, I.M. Holban, V.G. Kantser, F.M. Munteanu (1983) Transport phenomena in bismuth and its alloys (in Russian), Kishinev, Stiinta.
11. Tables of the physical values. Reference book (in Russian) (1976). Edited by academician I.K. Kikoin, M., Atomizdat.
12. Ia.I. Kerner (2010) Contacts of HTSC with different semiconductors for diode detectors: comparison of characteristics", Proceeding of the 3-rd International Conference "Telecommunications, Electronics and Informatics", Chisinau, Moldova, May 20-23, vol. 1, Chisinau, UTM, 2010, pp. 268-273.

Author: Iakov Kerner
Institute: D.Ghitu Institute of Electronic Engineering and Nanotechnologies
Street: Academiei str. 3/3
City: Chisinau
Country: Republic of Moldova
Email: iacov@lises.asm.md

XRD and XPS of Cd_2SnO_4 Thin Films Obtained by Spray Pyrolysis

T. Potlog[1], V. Botnariuc[1], S. Raevschi[1], M. Dobromir[2], and D. Luca[2]

[1] Department of Physics and Engineering,Moldova State University, Chisinau, Moldova
[2] Department of Physics Al. I. Cuza University, Iasi, Romania

Abstarct— **Cd2SnO4 thin films with a 1:1 and 1:3 Sn/Cd weight ratio at different substrate temperatures were fabricated by spray pyrolysis. The structural and surface composition of these films by X-ray diffraction (XRD) and X-ray photoelectron spectroscopy (XPS) were investigated. XRD study reveals tetragonal and orthorhombic phases of SnO2 and cubic phase of Cd2SnO4. In all samples the grain size of the crystallites of the SnO2 phase is smaller than that of the Cd2SnO4 phase. The grain size of the crystallites of Cd2SnO4 cubic phase decreases with increasing the substrate temperatures. The full range XPS spectra of all Cd2SnO4 thin films contain the Sn3p, Sn3d, O1s, Cd3d and O2s characteristic peaks.**

Keywords— **Cd2SnO4 thin films, Cd/Sn ratio, XRD, XPS.**

I. INTRODUCTION

In the last few years, interest in transparent conducting oxides increases due to their applications in photovoltaic devices and modern optoelectronic. Among the various available TCO materials Cd_2SnO_4 thin films are more conductive and transparent, have lower surface roughness, are patternable, relatively less expensive than others and excptionally stable. CTO has been used as an *n*-type conducting window in thin film solar cells based on cadmium telluride [1]. The CdTe cells using Cd_2SnO_4 as the front contact material reached an efficiency of 16.5 % [2]. In this paper we report the preparation of dicadmium stannate oxide (CTO) thin films at lower deposition substrate temperature by spray pyrolysis technique and investigate the effect of Sn/Cd ratio on optical and structural properties.

II. EXPERIMENTAL

The deposition of Cd_2SnO_4 films is carried out by using simple, economic and home made spray pyrolysis technique. The parameters such as distance between the spray nozzle and the substrate (35cm), and spray angle (about 90°) were kept constant. The inner diameter of the spray nozzle for the carrier gas and of the capillary tube for the spray solution, were I and 0.5 mm, respectively. The carrier gas was oxygen (99.9%). The carrier gas flow rate was maintained at 6 l/min at a pressure of 3 x10^4 N·m^{-2}. The composition was $SnCl_4 \cdot 5H_20$ (purity 98.5%), $CdCl_2$ (purity 98.0%) and ethyl alcohol (purity 99.5%). A spray solution with Sn/Cd molecular weight ratios of 1:1 and 1:3 was used. The solution was stirred just before the spray deposition in order to obtain good reproducibility in the properties of the films. The resultant solution was sprayed on the preheated glass substrates at substrate temperatures of 350 °C and 430°C to prepare cadmium stannate thin films. Films were obtained for various spray time and spray intervals.

The temperature of the substrate was measured by a chromel-alumel thermocouple on the surface of the substrate, and monitored with BPT-3 temperature controllers that allow keeping a constant temperature with a precision of ± 0.5°C. A temperature drop in the surface of the hot substrate occurs during spraying, which depends on the quantity of the mist of spray solution arriving at the surface. A heterogeneous layer results from this temperature drop in the Cd_2SnO_4 films deposited. Therefore, the samples 187 and 189 were prepared by repeating the spray with a short duration time of 1 s, to obtain a homogeneous film. A temperature variation in the hot surface of ± 2.0% was observed. Optical transmittance spectrum in the wavelength range of 200–1100 nm was recorded using UV–VIS–spectrophotometer MultiSpec TEC 5 UV-VIS. X-ray diffraction system Shimadzu LabX XRD-6000 with a CuK_α X-ray source (λ = 1.5418 Å) was employed to record X-ray diffraction patterns of these films. X-ray photoelectron spectroscopy (XPS) survey and high-resolution measurements were used to derive the surface chemical composition of the samples using an ULVAC-PHI, 5000 Versa Probe spectrometer with monochromated AlK_α radiation (1486.7 eV).

The photoelectrons were collected for a take-off angle of 45°. The calibration of binding energy (BE) scale was performed with the C1*s* peak at 284.6 eV from the carbon contamination layer. The surface quantification has been carried-out following the standard procedure [3] by using the core level XPS narrow spectra of found elements. Peak deconvolution has been done using the PHI-MultiPak software and the elemental concentration values were derived from peak surface areas, taking into account the sensitivity factors of the analyzed elements [4].

III. REZULTS AND DISCUSION

Figure 1 shows the spectral transmittance of a set of the as-deposited Cd_2SnO_4 thin films with a Sn/Cd weight ratio of 1:1 and 1:3 at different substrate temperatures in the visible and near infrared regions. The variation in transparency is caused by variation in the film thickness. These

V. Sontea and I. Tiginyanu (eds.), *3rd International Conference on Nanotechnologies and Biomedical Engineering*,
IFMBE Proceedings 55,
DOI: 10.1007/978-981-287-736-9_10

films have the transmittance in the range of 70-88 % in the wavelength range of 180 nm – 1200 nm. The curves for all samples show that films are thicker. As shown in the figure, the film (190) obtained at 350°C with a Sn/Cd weight ratio of 1:1 have generally a higher spectral transmittance over all wavelengths measured in comparison with others and spectrum exhibits a higher transmittance in the UV region. The Cd_2SnO_4 film (187) with a Sn/Cd weight ratio of 1:3 has the same transmittance as the film (190) with a Sn/Cd weight ratio of 1:1 in the 600 -1000 nm, both obtained at the same substrate temperature. The other samples (186,188,189,191) have transmittance smaller than 80 %.

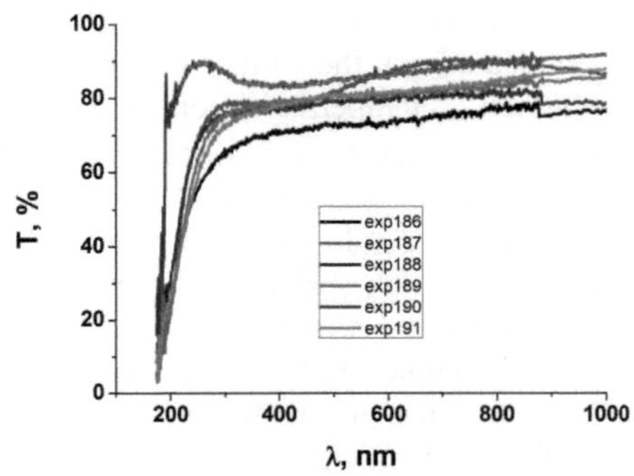

Fig. 1 Optical transmittance spectra of the Cd_2SnO_4 thin films.

All of the as-deposited Cd_2SnO_4 films exhibited a polycrystalline nature displaying diffraction peaks, corresponding to the (110), (200), (102) planes tetragonal phases and (111) plane orthorhombic phase for SnO_2. Also, was depicted the formation of Cd_2SnO_4 planes of cubic spinel phase when compared with the corresponding XRD of the PCPDFWIN data [5-8]. In the spinel structure, the O^{2-} anions are arranged in a cubic close packing array, which forms both octahedral and tetrahedral holes. Of the available octahedral holes, the Sn^{4+} cations are located in one half of these holes; the other half is occupied by Cd^{2+} cations. All of the available tetrahedral holes are occupied by the Cd^{2+}cations [8]. In particular, the low intensity of the Cd^{2+}pre-edge peak in the Cd_2SnO_4 samples indicates that Cd^{2+} is mostly located in highly centrosymmetric octahedral sites, i.e., high degree of inversion. Because in both normal and inverse spinel M^{2+}ions occupy octahedral sites, the proportion varying from 100% in normal spinels to 50% in inverse spinels. According to the XRD and XPS analysis in our samples dominate a partially inverse spinel structure. In Cd_2SnO_4, it has been suggested that reduced Cd^{2+} interstitials form n-type defects, contributing to conductivity [9]. Additionally, a small shift in the position of SnO_2 diffraction peaks was observed with the increasing of the substrate temperature, which is most likely due to a slight change in the lattice parameter associated with (1) oxygen vacancies, (2) Sn substitutes, and/or (3) stress [10-11].

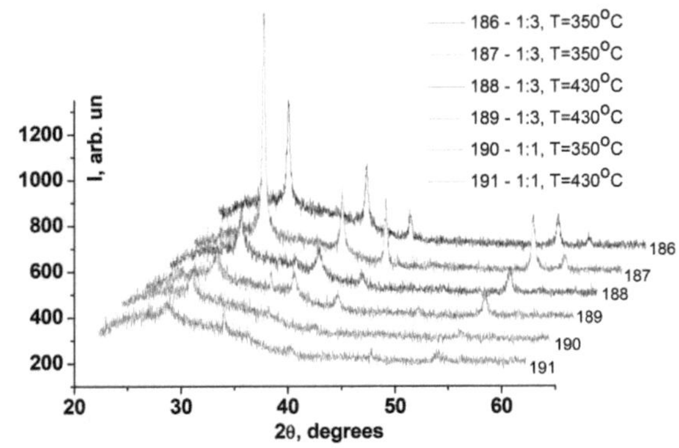

Fig. 2 X-ray diffraction pattern of the Cd_2SnO_4 thin films

The grain sizes from most intensive (110) , (111) planes of SnO_2 and (511) of Cd_2SnO_4 planes was calculated using Sherer's ecuation

$$D = \frac{0.9\lambda}{\beta\cos\theta} \quad (1)$$

where D is the grain size; 0.9 is a constant; λ is the wavelength; β is the FWHM for the crystal direction in the question; θ is the Bragg angle. The largest grains of Cd_2SnO_4 (511) plane formed were close to 50-55 nm for the layers obtained at Sn/Cd molecular weight ratio 1:3. At this Sn/Cd molecular weight ratio the grain size of the crystallites of Cd_2SnO_4 phase decreases with increasing the substrate temperatures. For the Sn/Cd molecular weight ratio 1:1 the grain size of the crystallites is smaller. In all samples the grain size of the crystallites of the SnO_2 phase is smaller than that of the Cd_2SnO_4 phase. The as-deposited films have the higher resistivity. According to the Agnihotri [12] if the cadmium stannate films are prepared in the normal air atmospheric conditions, the films can absorb excess amount of oxygen which can act as electron trap. Zhu [13] reported that the existence of adsorbed oxygen in films will trap a part of electrons, leading to the increase of resistivity. The full range XPS spectra of all SnO_2 thin films contain the characteristic peaks Sn3p, Sn3d, O1s, Cd3d and O2s. Fig. 3 shows XPS high resolution spectra of Cd3*d* of samples with Sn/Cd molecular weight ratios 1:1 and 1:3 obtained at the same substrate temperature. The XPS comparison of regions of Cd3*d* of 186 and 190 samples show difference only in signal intensity/ area of the peaks which increase with the increase of Cd concentration. As shown, $Cd3d_{5/2}$ and $3d_{3/2}$ peaks of the 186 and 190 samples are symmetric (404.9 ± 0.2) eV and (411.6 ± 0.2) eV and located at the binding energies which agree well with the 6.74 eV spin–orbits splitting between $Cd3d_{5/2}$ and $Cd3d_{3/2}$ states [14]. Fig. 4 shows the spectra in the regions of Cd3d with Sn/Cd molecular weight ratio 1:1 at substrate temperatures 350°C and 430°C. Evidently a shift has occurred in the films obtained at different substrate temperatures.

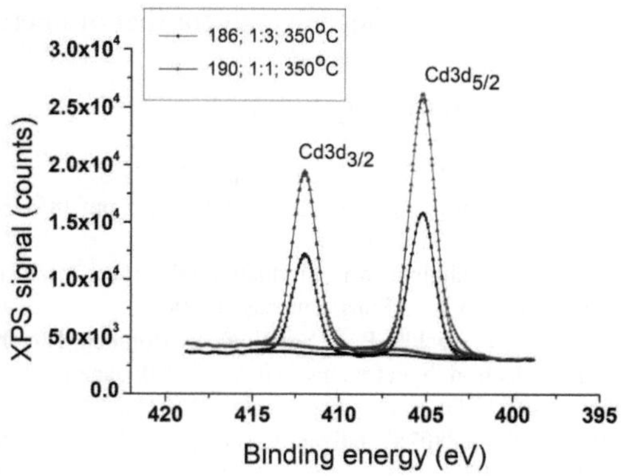

Fig. 3 XPS spectra of Cd3d region of as-deposited Cd_2SnO_4 films with varying Sn/Cd molecular weight ratio.

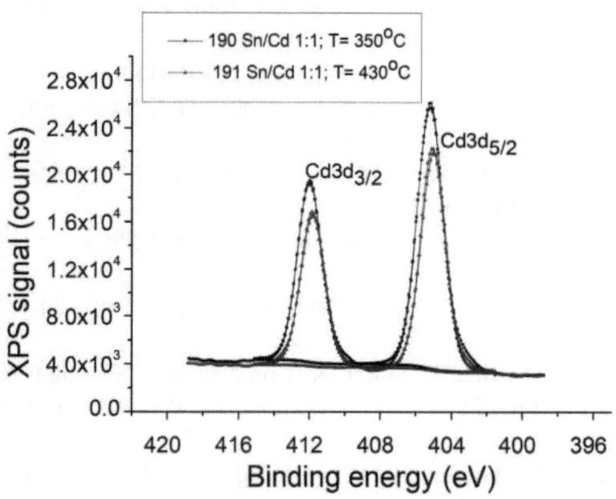

Fig. 4 XPS spectra of Cd3d region of as-deposited Cd_2SnO_4 films with Sn/Cd molecular weight ratio 1:1 at different substrate temperatures

The binding energies of Sn3d5/2 and Sn3d3/2 (Fig. 5, 6) located at 486.2 ± 0.2 eV and 494.5 ± 0.2 eV correspond to SnO_2 and agree well with the 8.4 eV spin-orbit energy splitting between Sn3d5/2 and Sn3d3/2 states, and are in good agreement with the published data [15]. Generally, the XPS high resolution spectra of oxygen are a convolution of four peaks for all samples regardless of the composition and substrate temperature. These peaks were marked as $O_{I\text{-st}}$, $O_{II\text{-nd}}$, $O_{III\text{-rd}}$ and O_{IV} peaks in order of increasing binding energies. The $O_{I\text{-st}}$ (529.0 ± 0.2) eV corresponds to the contributions from Cd-O. The $O_{II\text{-nd}}$ peak at 530.2 ± 0.2 eV can be attributed to basic peak of SnO_2 lattice oxygen (O_{lat}). A shift of this peak in the range of higher energies is observed for the samples with Sn/Cd molecular weight ratio 1:3.

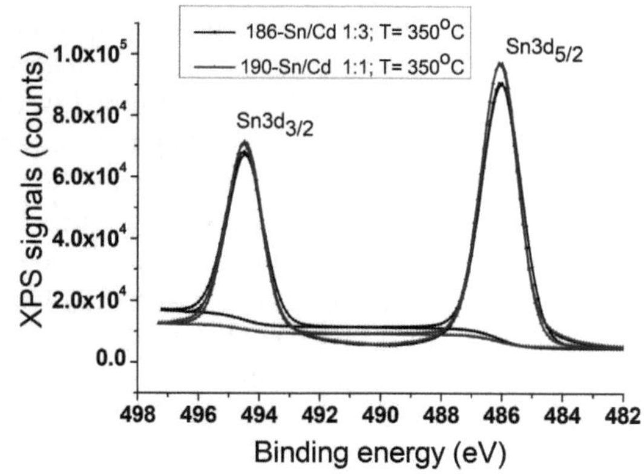

Fig. 5 XPS spectra of Sn3d region of as-deposited Cd_2SnO_4 films with varying Sn/Cd molecular weight

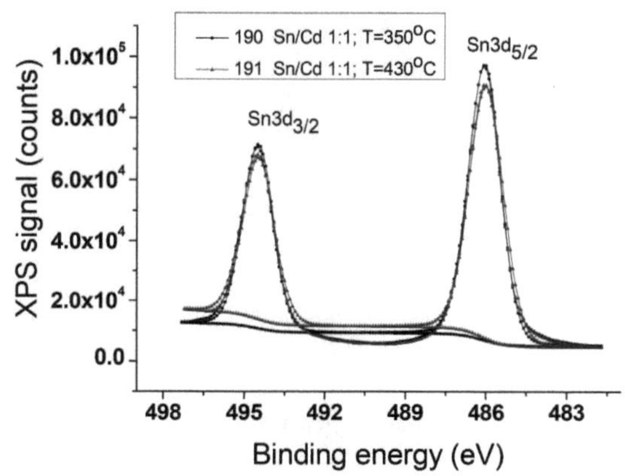

Fig. 6 XPS spectra of Sn3d region of as-deposited Cd_2SnO_4 films with Sn/Cd molecular weight ratio 1:1 at different substrate temperatures

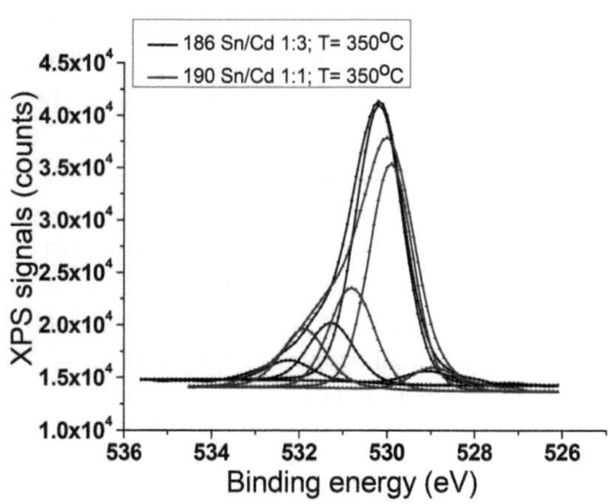

Fig. 7. The O1s XPS region of Cd_2SnO_4 thin films with Sn/Cd molecular weight ratios 1:1 and 1:3.

The small variation of the position of these peaks also is observed with the increasing of the substrate temperature. Authors [16-18] connect the $O_{III\text{-}rd}$ and O_{IV} peaks to hydroxide or hydroxyl groups OH and hydrate or adsorbed water. We suppose that part of species, corresponding to $O_{III\text{-}rd}$ and O_{IV} peaks can be connected also, with OH groups, which can have various forms with different bond energies. The OH groups can exist at least in two forms, such as "rooted" OH-groups, bounded with Sn atoms, and OH-groups, which can be formed during reaction of H_2O dissociation with participation of so-called "bridging" lattice oxygen.

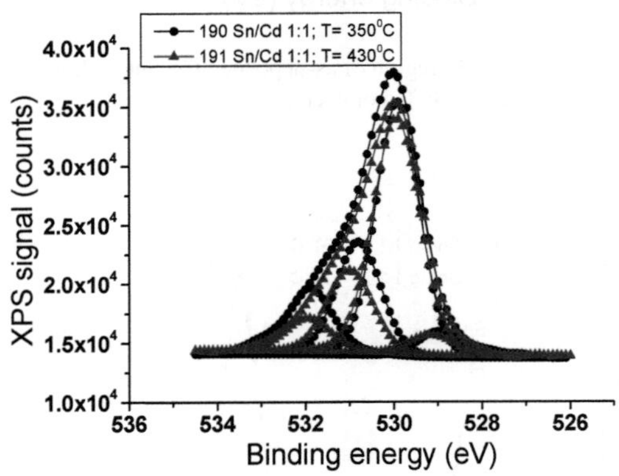

Fig. 8 The O1s XPS region of Cd_2SnO_4 thin films obtained at different substrate temperatures

IV. CONCLUSION

Cd_2SnO_4 thin films with varying Sn/Cd molecular weight ratio were deposited by spray pyrolysis method from the solutions of $CdCl_2$ and $SnCl_4$ at substrate temperatures of 350°C and 430°C. The as-deposited films were found to be polycrystalline starting with substrate temperature of 350°C. Structural differences between the samples with different Sn/Cd molecular weight ratio are relatively minor. The as-deposited films have the average optical transmittance of 70-85%. The XPS analyses have shown that from 530.5 eV to 531.1 eV, ionization characteristics of oxygen species are integrated in the material as OH^- or O_2^{2-} and therefore the films are higher resistive. Careful studies of post-heat treatments in H_2 atmosphere investigation could lead to improvement of the physical properties and conductivity for use in a solar cell made with CdTe.

ACKNOWLEDGEMENTS

This research was supported by the national project 15.817.02.39A

CONFLICT OF INTEREST

The authors declare that they have no conflict of interest.

REFERENCES

1. X. Wu. Solar Energy. V. 77, Iss. 6, 2004, pp. 803–814.
2. J. Britt and C. Ferekides . Appl. Phys. Lett. 62, Iss.22, pp. 2851-2852, 1993.
3. J. F. Watts, J. Wolstenholme, An Introduction to Surface Analysis by XPS and AES, John Wiley&Sons, Chichester, 2003.
4. J. F. Moulder, W. F. Stickle, P. E. Sobol, K. D. Bomben, Handbook of X-ray Photolectron Spectrosopy, ULVAC-PHI Japan, Physical Electronics, USA, 1995.
5. PCPDF WIN – PDF –880287. bolzan A.A. et.al. Sec.B: Structural Science, 53, 373 1997.
6. PCPDF WIN – PDF –801469.Bowden M.E, Cardile C.M, Powder Diffraction 5,36, 1990
7. M.E. Bowden and C.M. Cardile. Powder Diffraction 5,1, 1990.
8. PCPDF WIN – PDF –340928. Siegel L.J. Appl. Crystallor. 11, 284, 1978.
9. Y. Dou, R.G. Egdell, Phys. Rev., B 53 (23) (1996) 15405.
10. Z. Zhao, D. L. Morel, and C. S. Ferekides: Thin Solid Films 413 (2002) 203.
11. H. M. Ali, H. A. Mohamed, M. M. Wakkad, and M. F. Hasaneen: Thin Solid Films 515 (2007) 3024.
12. O.P. Agnihotri, B.K. Gupta, A.K. Sharma, J. Appl. Phys. 49 (1978) 4540.
13. M. Zhu, H. Huang, J. Gong, C. Sun, X. Jiang, J. Appl. Phys. 102 (2007) 043106.
14. J. Moulder, W. Stickle, P. Sobol, K. Bomben, in: J. Chastain (Ed.), Handbook of Xray Photoelectron Spectroscopy, Perkin-Elmer Corporation, Minnesota, 1992.
15. I. Hamberg, C. Granqvist, K.F. Berggren, B.E. Sernelius, L. Engstrfm, Phys. Rev., B 30 ,1984 3240.
16. J. Dupin, D. Gonbeau, P. Vinatier, A. Levasseur, Phys. Chem. Chem. Phys. 2 (2000) 1319
17. T. J. Chuang, C. R. Brundle and D. W. Rice, Surf. Sci., 1976, 59,413.
18. J. Haber, J. Stoch and L. Ungier, J. Electron. Spectrosc., 1976, 9, 459.

Author: Potlog Tamara
Institute: Moldova State University
Street: 60, A. Mateevici str.
City: Chisinau
Country: Republic of Moldova
Email: tpotlog@gmail.com

Superconductivity on the Background of the State of the Spin Density Wave in Anisotropic Systems

M.E. Palistrant, V.A. Ursu, and S.A. Palistrant

Institute of Applied Physics, Academy of Sciences of Moldova, Kishinev, Moldova

Abstract— **A theory of phase transitions in quasi-two-dimensional systems is developed in the case of doping. We take into account the presence of "nesting" on the Fermi surface and the structure of the lattice. A self-consistent system of equations for the superconducting order parameters Δ, magnetic M and chemical potential μ is obtained. For Δ=0, M ≠ 0 we have the magnetic state of the spin density wave (SDW). With the change of the density of charge carriers x phase transition - commensurate - incommensurate SDW state occurs. Against the background of this state (for Δ≠ 0 and M ≠ 0) superconductivity may appear, which is accompanied by magnetism. Numerical solutions for the thermodynamic quantities in magnetic and mixed phase are given.**

Keywords— **superconductivity, spin density wave, quasi-two-dimensional systems.**

I. INTRODUCTION

Over the past seven years a large number of high-temperature compounds (*Fe – pnictides* and *Fe–chalcogenides*) was received, whose superconducting transition temperature T_C can be up to 55K. These compounds are many band systems: on the Fermi surface overlap several electron and hole energy bands. In this connection arises an idea to describe properties of these materials on the bases of the many band model proposed many years ago by Moskalenko [1] and later by Sul et al. [2].In this theory it is necessary to consider several groups of charge carriers, taking into account the interaction both between the intraband and interband electrons. A very important conclusion is the assertion that in the multi-zone systems, high-temperature superconductivity can be achieved regardless of the sign of the constants of intraband V_{nn} and interbank $dV_{nm}(n \neq m)$interactions [3]. Therefore, this theory can be used to describe the properties of the system for the phonon and non-phonon mechanisms of superconductivity and to give a good agreement between the theoretical and experimental investigations. There are very much papers in this direction, but the history of research on the many band superconductivity can be found in [4]. Along with this, the above mentioned superconductors have a number of other features. In particular, they are layered quasi-two-dimensional and at the Fermi surface there is a „nesting", take place phase transitions when isovalent atoms of other valency are chemically introduced into system et al. A more detailed review of theoretical and experimental research development of properties of compounds based on FeAs in the past 7 years is given in papers [5]. The main conclusion is: superconductivity occurs against the backdrop of a spin density wave state after the suppression of magnetism by doping of the quasi-two-dimensional system by electrons or holes. Superconductivity is accompanied by magnetism.

In this paper we consider the quasi-two-dimensional system with an anisotropic energy spectrum. The deviation from half filling of the energy band is due to the introduction of impurities with variable density of charge carriers x.

II. MATERIAL AND METHODS

In our opinion, from theoretical point of view difficult decoding and additions of generally recorded two-band (multi-band) model [1] for the certain material from class of modern high-temperature compounds are performing.

Let us represent the Hamiltonian of the system in the mean-field approximation in the following form:

$$\widetilde{H} = \sum_{\vec{k}\vec{\sigma}} (\varepsilon(\vec{k}) - \mu) a^{+}_{\vec{k}\vec{\sigma}} a_{\vec{k}\vec{\sigma}} - \sum_{\vec{k}} [\Delta\, a^{+}_{\vec{k}\uparrow} a_{-\vec{k}\downarrow} + e.s.] - \sum_{\vec{k}\vec{\sigma}} [\vec{\sigma}\vec{M} a^{+}_{\vec{k}\vec{\sigma}} a_{\vec{k}-\vec{Q}\vec{\sigma}} + e.s.]. \quad (1)$$

Here Δ and M are the parameters of the superconducting and magnetic ordering, respectively, $\vec{Q}$- wave vector of the spin-density wave. Values Δ and M are given by expressions

$$\Delta = VT \sum_{\vec{k}} F^{\downarrow\uparrow}_{-\vec{k}\vec{k}}(\omega_n),$$

$$M = \frac{I}{2} \sum_{\vec{k}\omega_n\sigma} \sigma\, G^{\sigma\sigma}_{\vec{k}-\vec{Q}}\, n(\omega_n). \quad (2)$$

Here ω_n – is M at subara frequency. F и $G^{\sigma\sigma}$- Fourier - image of temperature Green's functions defined by relations

$$F^{\downarrow\uparrow}_{\vec{k},\vec{k}}(\tau - \tau') = -< T_\tau a^{+}_{-\vec{k}\downarrow}(\tau) a^{+}_{\vec{k}\uparrow}(\tau) >,$$

$$G^{\vec{\sigma}\vec{\sigma}}_{\vec{k}-\vec{Q}\vec{k}}(\tau - \tau') = -< T_\tau a_{\vec{k}-\vec{Q}\vec{\sigma}}(\tau) a^{+}_{\vec{k}\vec{\sigma}}(\tau') >. \quad (3)$$

Accounting of swapping processes is implemented by the summation over $\vec{k}$ on the entire Brillouin zone.

V. Sontea and I. Tiginyanu (eds.), *3rd International Conference on Nanotechnologies and Biomedical Engineering*, IFMBE Proceedings 55,
DOI: 10.1007/978-981-287-736-9_11

Because we are intended to study the properties of a system for a certain variable density of charge carriers x in addition to the system of equations for the Green's functions (3), we introduce the equation that determines the value$\mu(x)$.

This equation has the form

$$x = T\sum_{\vec{k}\sigma}\left[G^{\sigma\sigma}_{\vec{k}\vec{k}}(\omega_n) - G^{(-)\sigma\sigma}_{kk}(\omega_n)\right]. \quad (4)$$

Here $G_{\vec{k}\vec{k}}(\omega_n)$ corresponds to the above definition (3),but $\sigma^{(-)}_{kk}(\omega_n)$ corresponds to its value for replacement$\varepsilon_1$on(- ε_1)and $\varepsilon_2$on(- ε_2).

The system of equations (2) matches only account of "normal processes", which correspond to the electron - hole pairing in the first Brillouin zone. As a result of such accounting restrictions on the components of the vector$\vec{k}$are imposed.

After completing further summation on frequency $\omega_n = (2n+1)\pi T$, $(n = \pm 1, \pm 2 \ldots)$in the above system of equations in the standard way (see [7]), we obtain

$$\frac{1}{V} = -\sum_{\vec{k}}\left\{[-X_1 + \varepsilon_2^2 + \Delta^2 - M^2]\frac{th\,\beta\sqrt{X_1}/2}{2\sqrt{X_1}(X_1 - X_2)} - \right.$$
$$\left. -[-X_2 + \varepsilon_2^2 + \Delta^2 - M^2]\frac{th\,\frac{\beta\sqrt{X_2}}{2}}{2\sqrt{X_2}(X_1 - X_2)}\right\}, \quad (5)$$

$$\frac{1}{I} = -\frac{1}{4}\sum_{\vec{k}}\left(\frac{th\,\frac{\beta\sqrt{X_1}}{2}}{\sqrt{X_1}(X_1 - X_2)}[2M^2 - 2\Delta^2 - (\sqrt{X_1} + \varepsilon_1)(\sqrt{X_1} + \varepsilon_2) - (-\sqrt{X_1} - \varepsilon_1)(\sqrt{X_1} - \varepsilon_2) - \frac{th\,\frac{\beta\sqrt{X_2}}{2}}{\sqrt{X_1}(X_1 - X_2)}[2M^2 - 2\Delta^2 - -(\sqrt{X_2} + \varepsilon_1)(\sqrt{X_2} + \varepsilon_2)]\right),$$

$$x = -\frac{1}{X_1 - X_2}\sum_{\vec{k}}\left(\frac{th\,\frac{\beta\sqrt{X_1}}{2}}{\sqrt{X_1}}(X_1 + \varepsilon_2^2 + \Delta^2 - M^2) - \frac{th\,\frac{\beta\sqrt{X_2}}{2}}{\sqrt{X_2}}(X_2 + \varepsilon_2^2 + \Delta^2 - M^2)\right),$$

$$X_{1,2} = \frac{b \pm \sqrt{b^2 - 4c}}{2}, \qquad b = \varepsilon_1^2 + \varepsilon_2^2 + 2M^2 - 2\Delta^2,$$
$$c = \varepsilon_1^2\varepsilon_2^2 - 2M^2\varepsilon_1\varepsilon_2 + \Delta^2(\varepsilon_1^2 + \varepsilon_2^2) + (M^2 - \Delta^2)^2,$$
$$X_1 - X_2 = \sqrt{b^2 - 4c},$$
$$b^2 - 4c = (\varepsilon_1^2 - \varepsilon_2)^2 + 4M^2(\varepsilon_1^2 + \varepsilon_2)^2\,16\,M^2\Delta^2. \quad (6)$$

The method of performing of summation, taking into account flipping processes through the transition to the integration on energy in low-dimensional magnetic systems is developed in [7].

$$\frac{1}{V} = \frac{1}{4}\int_{-\tilde{W}}^{\tilde{W}} d\varepsilon N(\varepsilon)\sum_{\alpha,\beta}\Phi_1(\varepsilon, \mu_\beta^\alpha, M, \Delta),$$
$$\frac{1}{I} = \frac{1}{4}\int_{-\tilde{W}}^{\tilde{W}} d\,\varepsilon\,N(\varepsilon)\sum_{\alpha,\beta}\Phi_2(\varepsilon, \mu_\beta^\alpha, M, \Delta), \quad (7)$$
$$X = \int d\varepsilon\,N(\varepsilon)\sum_{\alpha,\beta}\Phi_3(\varepsilon, \mu_\beta^\alpha, M, \Delta).$$

Here

$$\Phi_1(\varepsilon, \mu_\beta^\alpha, M, \Delta) = \frac{M^2 + Y\,th\,\frac{\beta\sqrt{X_1}}{2}}{Y\sqrt{X_1}} - \frac{M^2 - Y\,th\,\frac{\beta\sqrt{X_2}}{2}}{Y\sqrt{X_2}},$$

$$\Phi_2(\varepsilon, \mu_\beta^\alpha, M, \Delta) = \frac{\Delta^2 + \mu_{\beta+Y}^{\alpha 2}\,th\,\frac{\beta\sqrt{X_1}}{2}}{Y\sqrt{X_1}} - \frac{\Delta^2 + (\mu_\beta^\alpha)^2 - Y\,th\,\frac{\beta\sqrt{X_2}}{2}}{Y\sqrt{X_2}}, \quad (8)$$

$$\Phi_3(\varepsilon, \mu_\beta^\alpha, M, \Delta) = \frac{\Delta^2 + \mu_\beta^{\alpha 2} + \varepsilon^2 + Y\,th\,\frac{\beta\sqrt{X_1}}{2}}{Y\sqrt{X_1}} - \frac{\Delta^2 + \mu_\beta^{\alpha 2} + \varepsilon^2 - Y\,th\,\frac{\beta\sqrt{X_2}}{2}}{Y\sqrt{X_2}},$$

$$X_{1,2} = \varepsilon^2 + (\mu_\beta^\alpha)^2 + M^2 + \Delta^2 \pm 2Y,$$
$$Y = \sqrt{(\mu_\beta^\alpha)^2(\varepsilon^2 + M^2) + M^2\Delta^2},$$

Where $\eta_{a,b}$ - parameters of an order of an incommensurable condition of SDW.

The system of equations (7) defines the order parameters M, Δ, as well as chemical potential μ for given values of the density of charge carriers and the temperature T. This system has a trivial solution Δ = M = 0, as well as non-trivial Δ≠ M ≠ 0.Below we sequentially consider these possibilities.

Let us first consider the state of the system in a magnetic state in the absence of superconductivity (Δ = 0, μ≠ 0). In this case on the basis of a system of equations for the transition temperature in the magnetic state (T_M), assuming second-order phase transition (M = 0), we obtain

$$\frac{1}{N_\varepsilon I} = \frac{1}{2}\sum_{\alpha\beta}\int_c^{\tilde{W}}\frac{N(\varepsilon)\,d\varepsilon}{c\quad\varepsilon}\left\{th\,\frac{\varepsilon + \mu_\beta^\alpha}{2T_M} - th\,\frac{\varepsilon - \mu_{\alpha\beta}}{2T_M}\right\}, (9)$$

$$x = 2N_0 \sum_{\alpha\beta} \int_c^{\tilde{W}} \frac{N(\varepsilon)}{c} \left\{ th\, \frac{\varepsilon + \mu_\beta^\alpha}{2T_M} - th\, \frac{(\varepsilon - \mu_{\alpha\beta})}{2T_M} \right\}, \quad (10)$$

$$\sum_{\alpha\beta} \alpha \int_c^{\tilde{W}} \frac{N(\varepsilon)}{N_C} d\varepsilon \left\{ \frac{1}{ch^2 \frac{\varepsilon + \mu_{\alpha\beta}}{2T_M}} - \frac{1}{ch^2 \frac{\varepsilon - \mu_{\alpha\beta}}{2T_M}} \right\} = 0. \quad (11)$$

For the system of equations (9) – (11) we should add an equation, which is obtained by replacing α→βandβ→α in (11). Expression (11) follows from imposed condition of maximum of value T_M, i. e.

$$\frac{\partial T_M}{\partial \eta_a} = \frac{\partial T_M}{\partial \eta_b} = 0.$$

III. RESULTS AND DISCUSSION

Let us further represent numerical solutions of the above equations. To determine the magnetic transition temperature T_M we solve a system of equations (9) - (11) by using the electronic density of states. It is convenient to begin consideration of this system of equations dependent of$\tilde{x}$, assuming $\eta_a = \eta_b = 0$. In this simplified case, we obtained that the ratio T_M / T_{M0} decreases with increasing $\tilde{x}$ up to the critical point (T_M, x_c), marked by a cross in Fig. 1. Near this point appears a solution satisfying instability of the state of SDW.

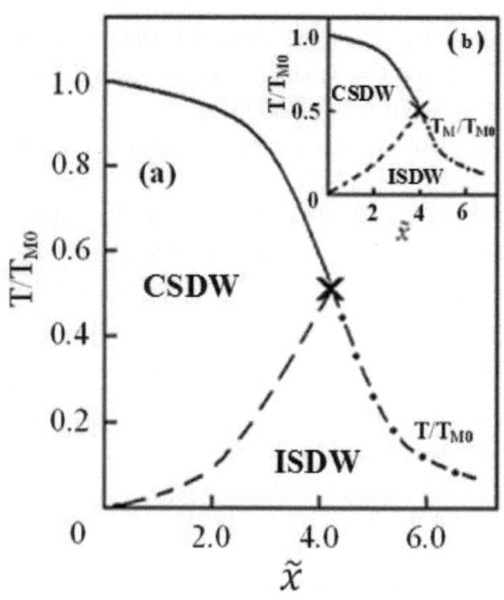

Fig.1 Phase diagram (T, $\tilde{x}$); CSDW and ISDW are the commensurate and incommensurate states of the spin density wave, respectively; (a) isotropic energy spectrum (W_2/W_1=1); (b) anisotropic energy spectrum (W_2/W_1=1.3). At the branching point (×), T_M/T_{M0} =0.49 (a) and 0.52 (b).

Solution of the system of equations (9) - (11) without simplifications mentioned above at point $\tilde{x}_c$ give values of the parameters η_a and $\eta_b \neq 0$, corresponding to the violation of "nesting" and shift for an electric circuit respectively to the Fermi surface $(q_x, q_y \neq 0)$.It should be noted that free energy of the system as a function of temperature has two minima. One of them disappears for $\eta_a = \frac{W_1 a a_x}{2}$ and $\eta_b = \frac{W_2 b a_y}{2} = 0$.Difference from zero of these parameters reduces the energy of the system and stabilizes the magnetic state by transition from commensurate state of SDW into incommensurate, transforming solutions for value T_M / T_{M0} in a stable state, which is determined by the incommensurate phase of SDW (dashed curve in Fig. 1).

Thus in Fig. 1, the solid line shows the dependence of the T_M / T_{M0} on $\tilde{x}$ in the process of phase transition into the commensurate SDW state, but dashed into the incommensurable. To construct the phase diagram $(T, \tilde{x})$, one should to examine the behavior of the SDW for $T < T_M$.

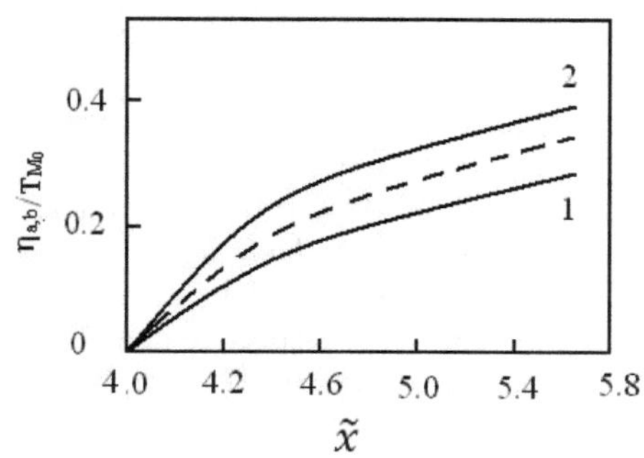

Fig. 2 Dependences of η_a / T_{M0} on $\tilde{x}$ for T/T_{M0} = 0.1 (curve 1), and 0.4 (curve 2), dashed curve corresponds to the isotropic case $\eta_a = \eta_b$.

Fig. 1 shows the phase diagram$(T, \tilde{x})$, as a result of analysis of solutions of these equations, the definition of dependency of parameters η_{ab} on the temperature and the concentrations of impurities (in detail see [7]).As example, the dependence of these parameters is given in Fig. 2. We have increase of parameters η_a, η_b with increasing of impurities concentration and their dependence on the anisotropy of the energy spectrum: in the isotropic case $\eta_a = \eta_b$ and anisotropic $\eta_a \neq \eta_b$ for $T = T_M$.

SDW (for low temperatures) by commensurate (for high temperatures).Comparison of two diagrams in Fig. 1 demonstrate the influence of the anisotropy of energy spectrum on the phase diagram $(T, \tilde{x})$.It follows that the anisotropy of the energy spectrum increase the area of ISDW (shifts dashed curve in the range of higher temperatures).It is chosen: (1a) isotropic case ($W_1 = W_2$); (1b) - anisotropic one W_2/W_1= 1,3.

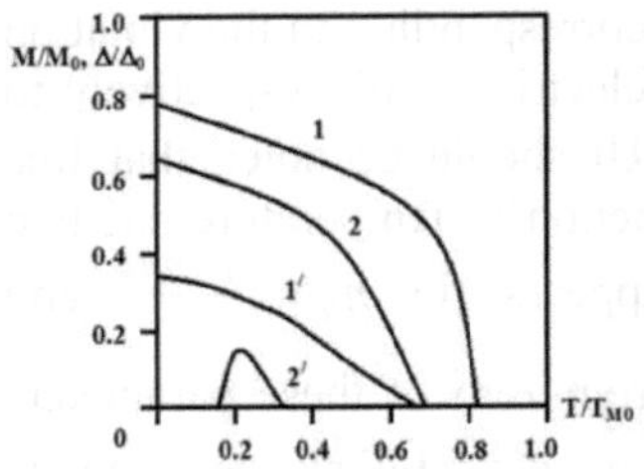

Fig. 3a Dependence of Δ and M on T/T_{M0} for $\tilde{x} = 4.0, t = 0.7$

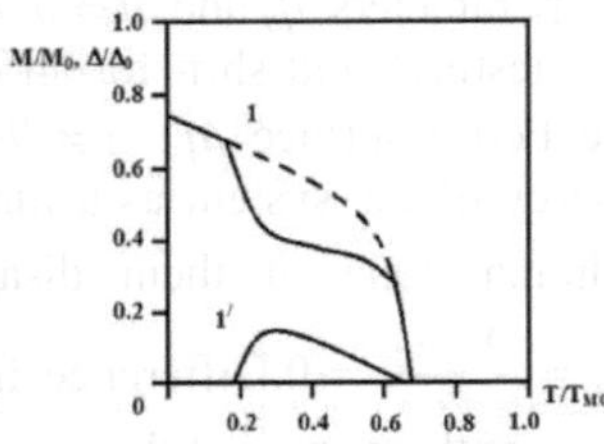

Fig. 3b Dependence of Δ and M on T/T_{M0} for $\tilde{x} = 4.2, t = 0.9$

Dependence of Δ and M as a result of solution of equation (5) is shown in Fig. 3a and 3b. It can be concluded that the coexistence of superconducting and magnetic phases is determined substantially by dependence on the parameters of the theory of $t=T_{C0}/T_{M0}$ and $\tilde{x}$ (Fig. 3a not shaded curves correspond $\tilde{x} = 4.0, t = 0.7$, and shaped correspond $\tilde{x} = 4.2, t = 0.5$and Fig. 3b $\tilde{x} = 4.2, t = 0.9$). There are various situations, namely, these phases coexist in a wide temperature range up to T=0 and at temperatures when superconductivity occurs at the point T_{c1} and suppressed at T_{C2} for given x. In this situation the superconductivity reduces the order parameter M, i.e. magnetism suppresses superconductivity.

Emergence of superconducting state corresponding to the shift of the energy gap with respect to the Fermi surface, showing the appearance of free carriers form Cooper pairs. It is possible when t is less than 1, which corresponds to the appearance of superconductivity at background of magnetic phase. For t more than 1 coexistence of considered phases is impossible.

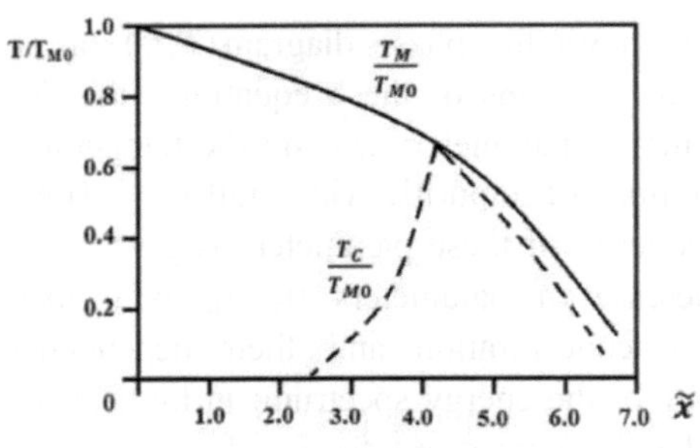

Fig. 4 Dependence of temperature of superconductingand magnetic transition on concentration of impurity for t = 0.7.

IV. CONCLUSIONS

This aim of this study was a theoretical description of the properties of new FeAs_based high_temperature superconducting materials at temperatures $T \leq T_M$, where T_M is the magnetic transition temperature. Our analysis was based on the Hubbard Hamiltonian in the mean field approximation, nesting on the Fermi surface, and rearrangement of the energy spectrum. We proceeded from the quasi_two_dimensional system approximating layered structures inherent in these materials, assuming that magnetism is associated with the spin density wave. The commensurability–incommensurability phase transition appearing, among other things, due to allowance for Umklapp processes in SDW_induced magnetism is a very interesting phenomenon.

CONFLICT OF INTEREST

The authors declare that they have no conflict of interest.

REFERENCES

1. В.А. Москаленко, Физикаметаллов и металловедиие, **8**, 503 (1959); Phys. Met. AndMetalog.**8**, 25 (1959).
2. H. Suhl, B.T. Matthias and L.R. Walker, Phys. Rev. Lett.**3**, 552 (1959).
3. M.E. Palistrant, F.G. Kochorbe, Physica C,**194**, 351, (1992).
4. M.E. Palistrant, Int. J. of Mod. Phys. B, **19**, 929, (2005).
5. A.A. Kordyuk. arXiv:15014154 (2015).
6. M.E. Палистрант, Теоретическая и Математическая Физика, **168**, № 3, 503, 2011.
7. M. E. Palistrant, V. A. Ursu, Journal of Experimental and Theoretical Physics, 2013, Vol. 116, No. 4, pp. 641–656.

Author address

Author: Palistrant Maria
Institute: Institute of Applied Physics
Street: Academie 5
City: Chisinau
Country: Moldova
Email: mepalistrant@yandex.ru

Peculiarities of Seebeck Effect in Strained Bismuth Nanowires

E. Condrea[1,2], A. Nicorici1, and A. Gilewski[2]

[1] Institute of Electronic Engineering and Nanotechnologies, Academy of Science of Moldova, Chisinau, Republic of Moldova
[2] International Laboratory of High Magnetic Fields and Low Temperatures, Wroclaw, Poland

Abstract— **The results on the effect of strain on the electrical resistance and thermopower of Pyrex-coated Bi nanowires were presented. The properties of Bi nanowires are examined in the light of a strain induced electronic topological transition. At low temperatures, the thermopower dependences on strain exhibit a non-monotonic behavior inherent in thinner wires, where the thermopower is dominated by the diffusion transport mechanism of holes. The hole-dominated transport can be transformed into electron-dominated transport through a smooth manipulation with the phonon spectrum and Fermi surface by applying a uniaxial strain. A fairly high value of the thermoelectric power factor was found above T = 80 K, where the dominant mechanism contributing to the thermopower is diffusive thermoelectric generation with electrons as the majority carrier.**

Keywords— **nanowires, thermopower, semimetal.**

I. INTRODUCTION

CURRENTLY Nanowires are widely used as building elements IN A VARIETY OF micro/nano devices. Interest in studying the transport properties of Bi nanowires (NWs) is supported by the theoretical prediction on enlarged thermoelectric efficiency of NWs, which results from the quantum size effect [1]. Although to date there has been no reliable experimental evidence showing an improved thermoelectric efficiency in Bi NWs, many published works indicate that the idea of studying pure Bi and Bi based crystals as a material with promising thermoelectric efficiency and high sensitivity to pressure is still attractive [2-7]. The prospect of the thermoelectric APPLICATION of NWs will require the ability to control THEIR PARAMETERS. In this context, Bi wires embedded in an individual glass capillary allow us to perform measurements of thermoelectric parameters in large ranges of temperature and elastic deformation. As is well known, the total thermopower (TEP), which is called also Seebeck coefficient (S), of bulk bismuth samples at liquid helium temperature is the sum of diffusion (S_d) and phonon drag term (S_g).

Systematic measurements on a set of bulk Bi samples with different sizes have shown the changes in magnitude of phonon drag thermopower with moderately varying the size of samples [8-11]. As concerns low-dimensional Bi samples, such as films, whiskers, and wires, there is a larger spread in the measurements of thermopower. Farther, a strong discrepancy was observed between experimental results for Bi wires with the same diameters obtained by different technological methods [12-16]. In different cases of positive S, indicating p-type transport for Bi NWs composites [13] and for individual polycrystalline NWs [2] or negative S of Bi/Al_2O_3 arrays [14], the authors suggest that an uncontrolled impurities could easily influence the value and sign of S.

Both the low-temperature TEP of Bi whiskers [15], which is of the order of a few $\mu V\ K^{-1}$, and the giant Seebeck effect observed in Bi nanocomposites [16], have a strange feature: the sign of S varies from sample to sample. These facts would imply that, in addition to classical or quantum size effects, other factors, such as structural defects and impurity must also be considered to understand anomalies found in the transport properties of small-diameter Bi wires. The data presented for bulk Bi crystals oriented along the bisectrix axis indicated an intrinsically negative phonon drag contribution that dominates TEP at low temperatures [9,10].

Our previous investigations of individual glass-coated Bi wires oriented along the bisectix axis have revealed a large positive thermopower dominated by the diffusion effect [17]. The diffusive origin of the TEP was attributed to structural imperfections, which strongly diminish the phonon-electron contribution to the total TEP. The suppression of the electron contribution to TEP was explained by a mechanism based on the model of selective carrier scattering by potential barriers, advanced by Ravich [18]. These structural imperfections with charge-selective scattering properties can be used to improve the efficiency of thermoelectric nanowires by removing some group of competing charge carriers from transport. Unfortunately, in most cases, structural defects can be naturally generated during the specific process of nanowire preparation. Some significant changes in the quality and parameters of Bi NWs can be obtained using different external impacts, such as thermal annealing, electric field effect, deformation, and magnetic field.

The final goal of the the present work is to understand the effect of size, thermal treatment, and uniaxial strain on the transport properties in order to improve the quality of Bi NWs. We report the results of the uniaxial strain effect on

V. Sontea and I. Tiginyanu (eds.), *3rd International Conference on Nanotechnologies and Biomedical Engineering*,
IFMBE Proceedings 55,
DOI: 10.1007/978-981-287-736-9_12

the thermopower and conductibility in glass-coated individual Bi wires. By combining thermal annealing, uniaxial deformation, and magnetic field, we have succeeded in conducting with thermoelectric parameters and in controlling with band structure by means of thermomagnetic measurements.

II. Results and Discussion

Galvanomagnetic size effect. The measurements of magnetoresistance and magnetothermopower were carried out for the set of wires with diameters of 0.1 – 3.0 µm in a magnetic field parallel to the axis of the wire. In figure 1 are presented the longitudinal MR and longitudinal MTEP data for 180 nm NW at temperatures of 4.2 K and 5.0 K, respectively. Longitudinal MR exhibits nonmonotonic field dependence consistent with previous reports on bismuth wires [21]. Longitudinal magnetoresistance achieves its maximum value in a magnetic field B_{max}, where the cyclotron radius of electrons r_c equals the wire radius, which is consistent with the theory of Chambers [22]. The cyclotron radius is given by: $r_c = h\, k_F /2\, e\, B_{max}$ where h is Planck constant, k_F is the Fermi wave vector for the appropriate carrier, heavy or light electrons or holes, and e is the electron charge.

The thermopower (curve S in figure 1) is positive in all magnetic fields. With increasing magnetic field MTEP increases and passes through two maxima at $B_1 \approx 1.7$ T and $B_2 \approx 3.6$ T. Unlike the MR curve that exhibited some weak (SdH) oscillations from electrons, there are no oscillations from electrons on MTEP. At the high magnetic field where only the holes are still in the quasi-classical range of magnetic fields, and one can observe weak oscillations of S resulting from the quantization of their spectrum.

As we shown (Ref. 17), the increase in the positive value of TEP takes place in the magnetic field region where condition: $(l_h\, d)^{1/2} > r > d/2$ is satisfied for holes (l_h is mean free path of holes, d is diameter of wire) and one of the more-pronounced peak at $B_2 \approx 3.6$ T on magnetothermopower curve is the artifact of size effect induced by the diffusion reflection of holes on the wire wells: $B_2 = 2cp_F/ed$, where p_F is the Fermi momentum of hole.

At magnetic field range of 12 - 15 T we observed the presence of some weak oscillations in the thermopower. According to the analysis of magnetoresistance curves, the quantum limit for all the electrons has already been reached at field B ≈ 2.7 T, therefore we can unambiguously claim that the peaks and wiggles at B > 3 T may not be related to quantum oscillations from electrons. Currently we should recognize that we can not explain the nature of the wiggles in the thermopower at magnetic field around 15 T.

The performed measurements of MTEP on the set of the NWs with diameters of 130-350 nm reveals the shift of the peak B_2, due to size effect of holes reflection, to lower magnetic fields with increasing wire diameter (figure 2). Decrease of the magnetic field value $B_2 = 3.8$ T for the 130-nm wire down to $B_2 = 2.7$ T for the wire with $d = 350$ nm is consistent with the values calculated by the formula $B_2 = 2cp_F/ed$.

The absence of well-defined size-effect features corresponding to electrons on $S(B)$ indicates that the reflections of the majority of these carriers at the wire boundary are specular [23].

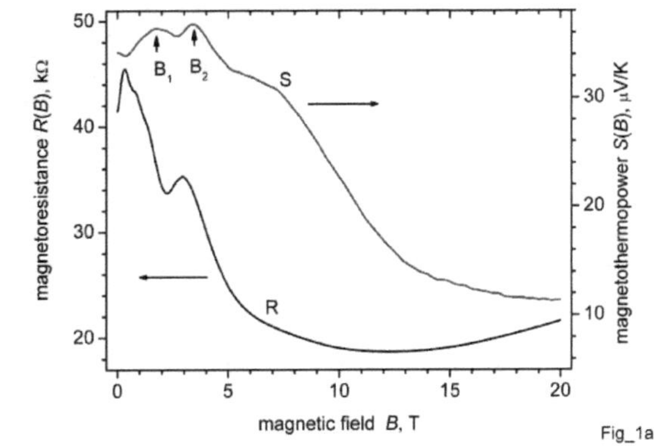

Fig.1 The magnetoresistance and magnetothermopower of Bi nanowire (d = 180 nm) at T = 4.2 and 5.0 K, respectively. The vertical arrow B indicates the magnetic field at which the cyclotron hole orbit equals the wire diameter.

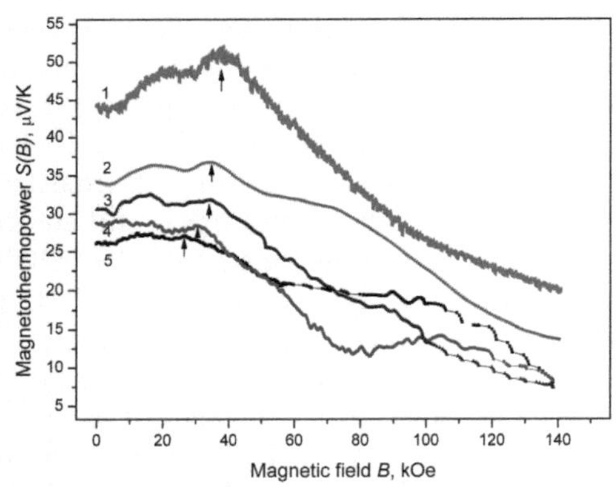

Fig. 2 MTEP for Bi NWs of various diameters: (1) 130 nm, (2) 180 nm, (3) 220 nm, (4) 300 nm, (5) 350 nm at T D 5 K. The vertical arrows indicate the position of B as the artifact of size effect induced by the diffusion reflection of holes on the wells.

Bi Wires Under Strain. Note that an increase in the contribution of electrons to TEP up to a sign change from positive to negative was observed in our earlier investigations on Bi wires under the uniaxial strain [25]. In the continuation of the investigations of the strained Bi NWs were measured the dependences of resistance and thermopower on strain ($0 \leq \varepsilon \leq 3$ %) and temperature ($4.2\ K \leq T \leq 300\ K$). In general, the shapes of the resistance dependences on strain $R(\varepsilon)$ resemble each other for the wires of all the investigated diameters.

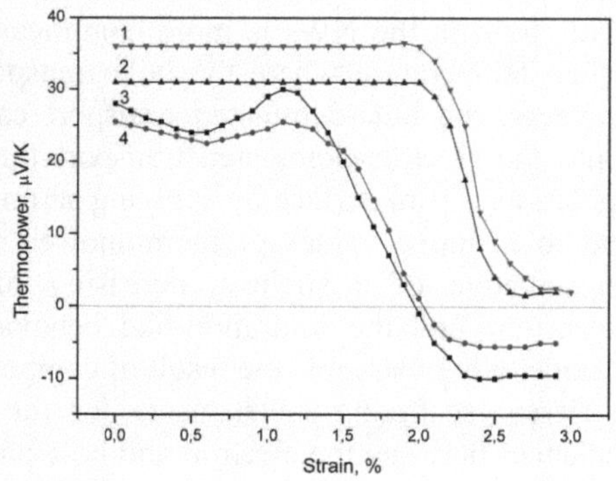

Fig. 3 Strain-dependence of TEP for Bi wires of different diameters at 4.2 K: 1 - 2.4 µm, 2 - 1.0 µm, 3 - 0.18 µm, 4 - 0.35 µm.

As concerns TEP, its strain dependences are quiet different for thinner and thicker wires in the investigated range of diameters (figure 3). Two types of behavior for two sets of samples were revealed: first type for NWs with diameters $0.1 \leq d \leq 0.35$ µm and second type for wires with $0.35 < d < 3.0$ µm. The application of strain upon the wires of the second type ($d > 0.35$ µm) does not produce any variation of the $S(\varepsilon)$ dependence up to a strong deformation (figure 3, curve 1,2). Only above $\varepsilon = 2.0$ %, the $S(\varepsilon)$ decreases and trends to zero. The second type of behavior is developed for wires when phonon drag contribution is dominant in the total S. In the case of first type ($d \leq 0.35$ µm) the $S(\varepsilon)$ dependences exhibit a non-monotonous behavior, which indicates a complex charge transport mechanism (figure 3, curve 3, 4). The initial positive TEP smoothly decreases, goes through fine minima and increases up to a peak around strain of 1.1 % with the following sharp decrease, a change of sign, and the nearly saturation behavior under a higher strain. Note that a non-monotonous behavior is inherent in the wires in which TEP is dominated by the diffusion transport mechanism. This complex behavior is representative of a competition among the different scattering mechanisms for the carriers and balances between the diffusive transport of electron or hole parts of the Fermi surface that changed during straining.

Electronic Topological Transition (ETT). Fourier analysis of SdH oscillations on MR dependences at different values of strain revealed an increase in the extreme cross-section of two equivalent electron ellipsoids, which is due to an electron transfer, resulting from the upward shift of the light electron ellipsoid relative to the heavy one up to its complete vanish under a strain higher than 0.6 %[26]. The emergence of a peak on S(ε) around $\varepsilon \approx 1.1$ % (figure 3) is indicative of a change in the Fermi surface topology under strain known as ETT, or Lifshitz transition.

Note that the presence of a peak on the $S(\varepsilon)$ dependence is characteristic only for the TEP dominated by diffusion mechanism. The absence of some peculiarities, due to ETT, on the strain dependences of TEP for the second type NWs (curve 1, 2 in figure 3), where phonon drag contribution is dominant, is in accordance with the prediction made by Abrikosov and Pantsulaya [27]. Analyzing of the non-monotonous behavior of the $S(\varepsilon)$ (curve 3, 4 in figure 3b) under a strain higher than the ETT point ($\varepsilon \geq 1{,}1$ %), we observed a sharp decrease in TEP and the sign change from positive to negative. Taking into consideration some predicted anomalies arising in the phonon spectrum for the materials undergoing an ETT of the Fermi surface [28], we can suppose that the increase in the negative contribution to TEP is due to the appearance of the phonon drag effect of electrons under strain. The phonon drag effect prevails in the TEP if the phonons acquire sufficient momentum to scatter the carriers across the Fermi surface changed after the ETT.

In order to draw more detailed information on how the diffusive and phonon drag term of TEP vary with strain, we measured the temperature dependences of thermopower $S(T)$ at various values of applied strain. Figure 4 shows the temperature dependences of TEP from the non-deformed state (curve a) up to maximal strain (curve g).

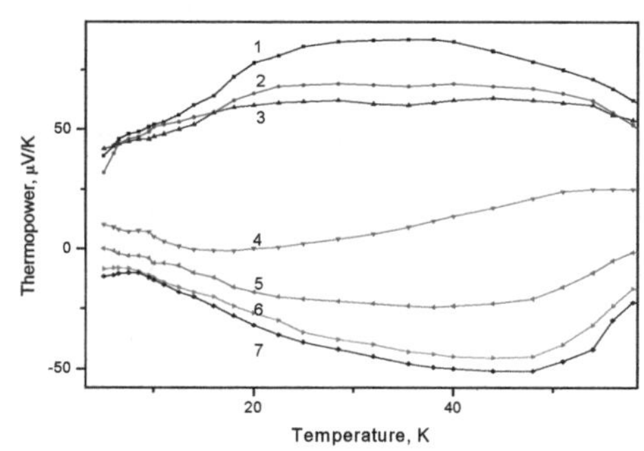

Fig. 4 Temperature dependence of thermopower of a 300 nm Bi wire at various strain: 0%, (b) 0.6%, (c) 1.1%, (d) 1.6%, (e) 1.9%, (f) 2.2%, (g) 2.7%.

A quasi-linear dependence $S(T)$ for thermopower in the non-deformed state persistent up to a strain of $\varepsilon \approx 1.1$ % is indicative of its diffusive origin (curves a-c, figure 4). A temperature dependence of thermopower measured after an ETT in the region of the abrupt drop of $S(\varepsilon)$ develops into a non-linear dependence. The observed non-linear dependence with a negative TCS ($\partial S/\partial T < 0$) indicates the dominance of the phonon drag effect in a narrow temperature range of 5 – 18 K (curve d, figure 4). In correlation with the predictions made in (Ref. 28), the phonon drag effect becomes dominant in the thermopower if the phonon momentum is sufficient for interacting of the phonons with the electrons after an ETT.

The increase in negative contribution under strain to the total TEP is indicative of the fact that a phonon drag term with a negative sign is added to the initial positive diffusion term. The predominance of the phonon drag effect of electrons implies that required condition $q \leq 2k_F$ (q is the phonon momentum, $2k_F$ is the maximum dimension of Fermi surface) for the phonons interacting with certain type of the charge carriers is satisfied for electrons after an ETT in the region of the abrupt drop of $S(T)$.

Some information on the Fermi momentum p_F of the electrons is provided by the set of curves of the LMR measurements under different strains. In accordance with the galvanic magnetic size effect theory theory, the extreme p_F of the electrons in the plane perpendicular to the magnetic field was determined in the non-deformed state and at a strain value of $\varepsilon \approx 1$ %: $p_F(\varepsilon = 0$ %$) = 1.1\times10^{-21}$ g cm s^{-1} and p_F ($\varepsilon \approx 1$ %) $= 1.25\times10^{-21}$ g cm s^{-1}. These rough estimates indicate to an increase in p_F of heavy electrons, which can contribute to the phonon drag effect.

III. CONCLUSIONS

Bismuth nanowires were prepared by the melt spinning method that makes it possible to easily obtain long elastic Pyrex-coated wires, protected from oxidation. One of the disadvantages of this method is the high rate of cooling in the process of solidification of the molten Bi core, which can produce some structural defects. Thus, specific imperfections inherent in thin NWs are defined by the technology. If a softened glass-coated NW is regarded in the solidification process as a composite, there is a critical diameter below which structural defects can be generated, due to the cylindrical geometry of the NWs composite, as was suggested [24]. For Bi wires, the critical diameter is around 350 nm, below which structural defects cause the appearance of internal electron-reflecting hole-transmitting barriers, which can lead to a localization-like behavior for the electrons and a metallic behavior for the holes.

Some improvements of the NWs quality were achieved by means of thermal annealing that has an effect on the lowering of potential barriers and the liberation of electrons for transport.

Note that an increase in the contribution of electrons to TEP up to a sign change from positive to negative was observed at low temperature for Bi wires under the strain. A fairly high value of thermoelectric power factor was found in the temperature range above T = 80 K, where the dominant mechanism contributing to the thermopower is diffusive thermoelectric generation with electrons as the majority carrier.

The effect of strain in the NWs is more complicated at the liquid helium temperature where the hole transport is dominant. However, the hole-dominated transport can be transformed into the electron-dominated transport through smooth manipulations Fermi surface by applying strain. The observed trend to negative values of thermopower under strain is representative of a slightly increasing of the mobility of electrons and the saturation-like behavior of negative TEP under a high strain is the result of competition among the different scattering mechanisms for the heat carriers and balances between the electron and hole phonon drag contribution.

The presented results on the effect of uniaxial pressure on the thermoelectric properties of Bi wires can be useful if they are considered in applying the "pressure tune" method for the synthesis of thermoelectric materials.

CONFLICT OF INTEREST

The authors declare that they have no conflict of interest.

REFERENCES

1. Lin Y-M, Sun X and M.S. Dresselhaus, Phys. Rev. B **62,** 4610 (2000)
2. Boukai A., K. Xu, J.R. Heath, Adv. Mater. **18,** 864 (2006)
3. Gurvitch. M.J., J. Low. Temp. Phys. **38,** 777 (1980)
4. Armitage N.P, R. Tediosi, F. Levy, E. Giannini, et all., Phys.Rev.Lett. **104,** 237401 (2010).
5. Cornelius T.W., M.E. Toimil Molares, R. Neumann, and S. Karin, J. Appl. Phys. **100,** 114307 (2006)
6. Chen F. and K.L.Stokes, R. Funahashi, Appl.Phys.Lett. **81,** 2379 (2002)
7. Ovsyannikov S. V., V. V. Shchennikov, G.V. Vorontsov, A.Y. Manakov, A. Y. Likhacheva, and V. A. Kulbachinsckii, J. Appl. Phys. **104,** 053713 (2008)
8. Issi J-P. and J H Manges, Phys. Rev. B **6,** 4429 (1972)
9. Uher C. and W. P. Pratt Jr., J. Phys. F: Met. Phys. **8,** 1979
10. (1978)
11. Boxus J. and J-P. Issi, J.Phys.C: Solid State Phys. **10,** L397 (1977)
12. Boxus J., C.Uher, J. Heremans, and J-P. Issi, Phys. Rev. B **23,** 449 (1981)
13. Bodiul P.P., D.V. Gitsu, V.A. Dolma, M.F. Miglei and G.G. Zegrea, Phys. Status Solidi (a) **53,** 87 (1979)
14. Heremans J. and C.M. Thrush, Phys. Rev. B **59** 12579 (1999)

15. Lin Y- M., O. Rabin, S.B. Cronin, J.Y. Ying and M.S. Dresselhaus, Appl. Phys. Lett. **81,** 2403 (2002)
16. 15. Yu.P. Gaidukov, Sov. Phys.-Usp. **27** 256 (1984), (1984 Usp. Fiz. Nauk **142** 571).
17. Heremans J.P., C.M. Thrush, D.T. Morelli and M.C. Wu, Phys. Rev. Lett. **88,** 216801 (2002)
18. Grozav A.D. and E. Condrea, J.Phys.: Condens.Matter **16,** 6507 (2004)
19. Ravich Y.I., In: CRC Handbook of Thermoelectrics, edited by D. M. Rowe (New York:CRC Press, 1995) ch 7.
20. Hagiwara M. and A. Inoue, Production Techniques of Alloy Wires by Rapid Solidification in Rapidly Solidified Alloys, edited by H. H. Liebermann (New York: Dekker, 1993) p.141.
21. Hansen O.P., I.F. Mikhail, M.Yu. Lavrenyuk, and N.Ya. Minina, J Low Temp.Phys. **95,** 481 (1994)
22. Brandt N.B., D.V. Gitsu, V.A. Dolma and Ya.G. Ponomarev, Sov. Phys. JETP **65,** 515 (1987)
23. Chambers R.G., Proc. R. Soc. London, Ser. A **202,** 378 (1950)
24. Tsoi V.S. and I.I. Razgonov, JETP Letters **23,** 92 (1976)
25. Gutkin M.Ya., I.A. Ovidko and A.G. Sheinerman, J.Phys.: Condens. Matter **12** 5391 (2000)
26. Azbel M.Ya, Phys. Rev. B **27,** 3852 (1983)
27. Condrea E., A.D. Grozav and N.I. Leporda, Proc. 6th European Workshop on Thermoelectrics of the European Thermoelectric Society 2001 (Freiburg, Germany, Sept. 2001) (Freiburg: Fraunhofer Institute of Physical Measurement Techniques IPM).
28. Abrikosov A A and Pantsulaya A V 1986 *Sov. Phys. Solid State* **28,** 1195

Author: Condrea Elena
Institute: Institute of Electronic Engineering and Nanotechnologies, Academy of Science of Moldova,
Street: 3/3 Academiei str.
City: Chisinau,
Country: Republic of Moldova
Email: condrea@nano.asm.md

X-Ray Photoelectronic Spectroscopy of GaN, AlGaN Layers, Grown on Silicon by the Chemical Transport Reactions Method

S. Raevschi[1], V. Botnariuc[1], L. Gorceac[1], T.Potlog[1], M. Dobromir[2], and D. Luca[2]

[1] Department of Physics, Moldova State University, Chisinau, Moldova
[2] Department of Physics, Al. I. Cuza University, Iasi, Romania

Abstarct— **Thin layers of GaN, and AlGaN were synthesized on Silicon substrates by chemical transport reactions method in the system (H_2-NH_3-HCl-Al-Ga).The composition of surface layers was studied by X-ray Photoelectron Spectroscopy Method. Besides Nitrogen, Aluminum, and Gallium, there were found also other elements - Carbon, and Oxygen in the atomic composition of layers. It was found a lower concentration of oxygen atoms in the composition of GaN layers deposited at relatively lower temperatures. It is assumed that the presence of Oxygen in the composition of layers is partly due to the decomposition of quartz glass, which the reactor is made from at high temperatures. It was found that atomic concentration of Gallium in AlGaN layers is much less than the Aluminum concentration, despite of the fact that the concentrations of Gallium and Aluminum precursors in the gaseous flow at the entrance in the synthesis zone were in equimolecular ratio. This demonstrates that the rate of chemical reactions of the ammonia with gaseous components of the Aluminum is significantly greater than with that of the Gallium. Therefore, the latter are removed by the gas stream from the growth zone.**

Keywords— **GaN, AlN, AlGaN, Si, NH_3, HCl, H_2, XPS.**

I. INTRODUCTION

Currently, Gallium Nitride layers for making luminescent diodes (LED) are grown on sapphire substrates. An alternative to sapphire as the substrate can be Silicon. However, it does not exist until now an appropriate technology of adaptation of GaN layers to Silicon substrates, and concrete ways of achieving this technology are not clear, despite the cost benefits of Silicon to sapphire (Silicon is ~ 10 times cheaper than sapphire). One can say that the implementation of projects of construction of GaN LEDs on Si is today in patent, initial stages. It should be noted that the structures GaN/Si have also other application areas: in the power electronics, at high frequencies and temperatures, in two-dimensional electronic gas field transistors, in the development of micrometer-sized sensors for detecting harmful gases (NO, NO_2) etc. The first works on deposition of GaN layers on Si began in 1970 and lasted for 20 years without significant success. Even now, obtaining of perfect GaN layers on Si faces many fundamental obstacles. Due to the high difference in the crystal lattice parameters (~17%) of GaN and Si, a high density of defects (dislocations due to non-coincidence) appears at the interface of layers in the initial process of deposition. In addition, the large difference between the thermal expansion coefficients of the GaN and Si (~ 116%) in the cooling process from the deposition temperature to room temperature, leads to mechanical stresses which cause cracking, and bending of structures. These factors lead to a concentration of the dislocations in the layers of the order of 10^{10} cm^{-2} and to an electron concentration of ~ 10^{20} cm^{-3}. A decisive step in overcoming the mentioned obstacles was made in the early 1990s, when for improving the crystallinity of GaN layers it was launched the concept of structures with intermediate (buffer) layers between the substrate and layer. This approach led to the increasing number of research results, and to real improvement in the quality of layers.

Currently, the main lines of research are directed towards reducing the concentration of defects and mechanical stresses in the structures (by applying the template structures, with AlN interlayers, with mono metal layers, buffer layers of different composition of AlGaN, interlayers with super-lattices etc.). It should be noted that the performed researches include the most directions of study of these structures, and a particular attention is drawn on their application in power and high frequency electronics. The most advanced results concerning the technology of synthesis of GaN /.../Si structures from organometallic precursors (MOCVD-Metalorganic Chemical Vapor Deposition) for developing LEDs are obtained using the method of chemical transport reactions. In this paper, the technological growth of III-N layers on Si were performed by chemical transport reactions method using the concept of structures with intermediate layers of AlN and AlGaN of different thicknesses, synthesized at different temperatures, etc. The AlN, AlGaN, and GaN layers on heterogeneous substrates were synthesized by HVPE method (Hydride Vapor Phase Epitaxy).The technology is relatively more simple to achieve, and it has a III-N layers growth productivity of ~ 10 times higher than other technologies (MOCVD, Molecular Beam Epitaxy, Vacuum Reactiv Evaporation, etc .), is suitable for producing quasi-massive GaN layers applicable as substrates. In the technology of

V. Sontea and I. Tiginyanu (eds.), *3rd International Conference on Nanotechnologies and Biomedical Engineering*,
IFMBE Proceedings 55,
DOI: 10.1007/978-981-287-736-9_13

fabrication of III-N compounds were used pieces of carbon and quartz reactors. In the synthesis of the corresponding layers by the HVPE method the carbon sheets were used to protect the reactor of interaction with aluminum, and the synthesized compounds. The synthesizing temperatures of the layers are quite high (1100°C). At these temperatures there is a partial decomposition of quartz with generation of oxygen in aggressive environments such as HCl and NH_3. Therefore the major aim of this investigation is to detect the influence of carbon and oxygen impurities on the properties of synthesized layers. Another aim of the research is to find the correlation between concentrations of Ga and Al precursors at the begining of the synthesis region with concentrations of these elements in AlGaN synthesized layers. The research results will be applied to the elaboration of the technology of fabrication of GaN layers on Si substrate. In this paper, we present the research results of surface composition of the AlGaN layers used as intermediate in the technology of producing GaN layers on Silicon. We will also present the surface component of GaN layers obtained on AlGaN/AlN/Si structures. The investigation of surface composition will be conducted by the spectroscopy method with electrons excited from the deep levels of atoms by a monochromatic X-ray beam.

II. EXPERIMENTAL

The GaN, AlN, and AlGaN layers were deposited on Silicon by chemical transport reactions method - HVPE. The pressure in the horizontally arranged quartz reactor, used in the synthesis, was 0.07 bar higher than atmospheric level. The hydrogen purified with a palladium filter was used as transfer gas. Ammonia, hydrogen chloride, Ga(5N) and Al(5N) were used as reagents. The Hydrogen Chloride flows in aluminum, and gallium pipes or the corrosion flows were dissolved in ~ 200 smlpm of H_2. The total consumption of hydrogen was 3.6 slpm and that of ammonia varied in the range of 400 to 1500 smlpm. The consumption of hydrogen chloride varied in the range of 2 to 15 smlpm at the layer deposition, and of ~ 50-100 smlpm at the corrosion of the reactor. The substrates were rotated during the deposition by a stream of hydrogen of 1 slpm with a rotational speed of 50 to 100 rpm. The thermal profile in the reactor was provided with a resistive heater. The temperature of the Al, and Ga sources was constant at a level of 850°C. The layers deposition was performed on Silicon wafers on crystallographic plane (111). Before placing into the reactor, the silicon was treated by standard method. The chemical composition of the surface layers was determined using the X-ray Photoelectron Spectroscopy (XPS). XPS spectra were measured using X-ray photoelectron spectrometer *PHI-5000 Versa Probe,* with monochromatic AlKα radiation (1486.7 eV). The photoelectrons were collected at an angle of 45°. The surface quantification was achieved by standard procedure and the high resolution XPS spectra were processed using the PHI MultiPak software. The peaks positions in the XPS spectra are calibrated with respect to the C 1s peak position whose binding energy (BE) is 284.6 eV. Identifying the different types of chemical bonds formed at the surface was achieved by deconvolution of high resolution spectra. Atomic concentrations of the elements were determined from the areas of peaks, taking into account the sensitivity factors of the elements analyzed.

III. RESULTS AND DISCUSION

The technological processes used to deposit the layers were carried out in several stages. At stage 1, it took one hour to replace the atmosphere of the reactor with the transporting gas (hydrogen).At stage 2, the temperature of the reactor was raised to deposit the first layer of aluminum nitride, AlN-1, at T1 ~ 1100°C.The second layer, AlN-2, as well as the AlGaN layer were deposited at the GaN layer growth temperature (T2).The temperatures and deposition time for the investigated layers are presented in Table 1. The cross-section scheme of the layers is shown in Figure 1. The mirror-like image of an AlN-1 layer deposited at ~ 1100°C is shown in Figure. 2. The XPS spectra were measured on 2 samples (153, 155) with AlGaN surface composition and 3 samples (156, 162,164) with GaN outer layer.

In Figure 3 the full XPS spectrum of an AlGaN layer is shown. It is seen from the figure that apart from gallium, aluminum and nitrogen the layer contains silicon (Si2s, Si2p), carbon (C1s) and oxygen (O1s).There are also in the spectrum two groups of Auger electrons (Ga LMM2, Ga LMM3,

Table 1 Temperature and deposition time of AlN, AlGaN, and GaN layers deposited on Silicon

N	AlN-1		AlN-2		AlGaN		GaN	
	T_1,°C	t,min	T_2,°C	t,min	T_2,°C	t,min	T_2,°C	t,min
153	1100	2	1010	2	1010	10	-	-
155	1106	2	1026	2	1026	10	-	-
156	1100	2	1005	2	1005	10	1005	30
162	1090	1.5	954	2	954	10	954	60
164	1106	1	954	1	954	10	954	60

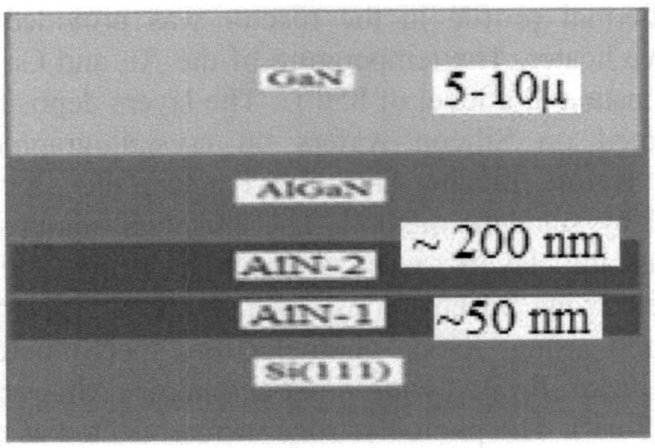

Fig. 1 Cross-section scheme of GaN/…/Si structures

Fig. 2 Surface of the AlN-1 layer(~50nm) grown on a silicon substrate of ~ 50 mm diameter.

Ga LMM4 and O-KLL, O-KLL). In GaN layers (156,162,164) spectra the silicon levels disappear, which shows that the diffusion of silicon from the substrate is blocked by the previous buffer layers. In the GaN layer, deposited for only 30 minutes (156), the aluminum concentration is 5 at.%, which is 5 times less than in the AlGaN previously grown layer. The aluminum concentration on the surface of layers grown for 6o minutes was lowered to the background level. The results of XPS measurements of samples investigated are shown in Table2. There is a high concentration of carbon and oxygen in all layers that might originate from or be generated in the reactor, because gases and metals used were of high purity (5N).The first element comes from the carbon foil used to protect quartz reactor against interaction with aluminum or nitride synthesized. The oxygen comes in part from the decomposition of the quartz glass of the reactor at high temperatures ($SiO_2 \rightarrow Si + O_2$). It should be noted that the oxygen concentration in 153,155 samples where GaN layer is missing is twice as high as in 156, 162,164 layers. In the thicker GaN layer grown at lower temperatures the oxygen concentration is lower, but still quite high that shows also the presence of other sources of oxygen. The value of the binding energy of 76.1 eV (155) of Al_{2p} correlated well with the value of 75.6 eV which is attributed to Al-O bond [1]. Please note that the oxygen concentration in this sample is in the range of the concentration of ~26% aluminum. The XPS spectra of gallium have two peaks that correspond to Ga2p1/2 and Ga2p3/2 states. The shape, as well as the values of their binding energy are in good aggreament with the reported [2,3]. In the spectrum of 156 sample besides the main binding energies located at 1118.2 eV and 1145 eV are present and one replica of a lower intensity shifted by ~1.2 eV to higher energies. These replicas were not found in available publications, but the provenience of them will be investigated in the future. The provenience of oxygen in synthesized III-N layers by various methods is currently intensively investigated.

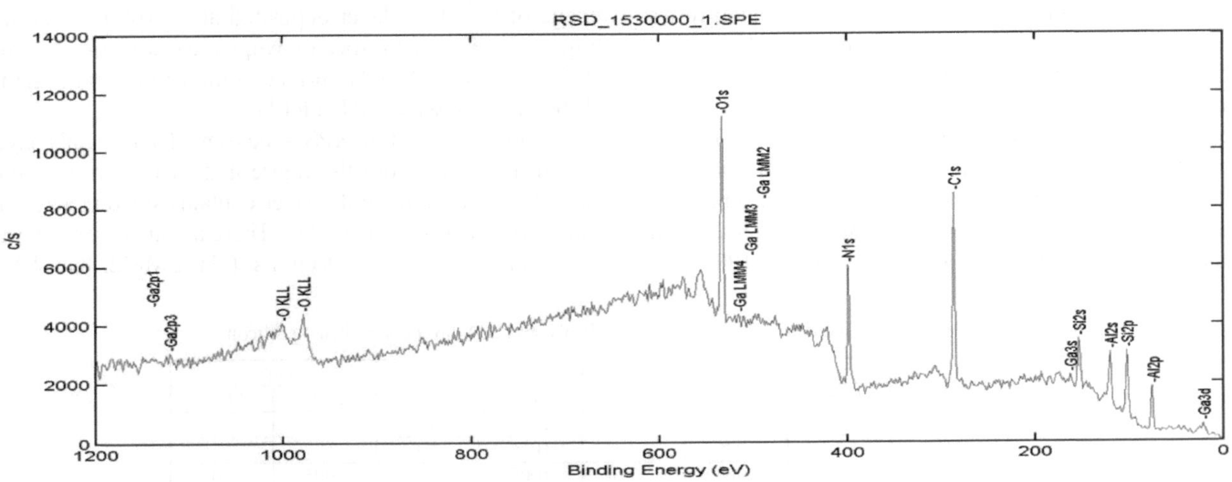

Fig.3 XPS survey scan spectra of the AlGaN/AlN-2/AlN-1/Si(III) layer (153) .

Table 2 Binding energy, eV, and Concentration, at. %

N	C1s eV; at. %	N1s eV; at. %	O1s eV; at. %	Al2p eV; at. %	Ga2p$_{3/2}$ eV;at. %	Ga2p$_{1/2}$ eV; at.%
153	285.9; 6.6	398.2; 45.6	532.9 ; 24.5	71.5 ; 21.5	1119.2 ; 1.8	1146.1 ; 1.8
155	287 ; 25	399.4 ; 22.7	533.9; 26	76.1; 25.6	1120.4 ; 0.7	1147.3 ; 0.7
156	285.8 ; 6.3	397.6; 61.6	532.1; 10.5	74.4 ; 5.1	1123.8 ; 16.5 1118.2 ; 16.5	1150.8 ; 16.5 1145.0 ; 16.5
162	285.4 ; 9.5	397.0 ; 61.6	531.6 ; 6.2	?; <0.1	1117.6 ; 22.6	1144.6 ; 22.6
164	285.6 ; 6.1	397.7 ; 61.5	532.2 ;11.9	?; 0.5	1118 ; 20	1144.6 ; 20
Average,E	285.94	397.98	532.34	74	1119.53	
Er [1,2]	288.23	403.78	538.25	76.753	1128.1	
(Er-E) /E ,%	0.8	1.45	1.07	3.72	0.76	

It was established that oxygen adheres to the surface of the layers forming a monolayer. Since the XPS investigations are carried out on the surface of the layer at depths that does not exceed 4 nm [4] it is evident necessary to remove this monolayer at the appreciation of the concentration in their own layers. The values of electron binding energies corresponding to the levels of detected elements are in satisfactory agreement, within one percent, with the values obtained by other researchers [5, 6]. Such deviations are caused by the structure and purity of investigated surfaces, the interaction forces between the structural elements, the possibilities of the research equipment and are characteristic also for other substances. The greater deviation of the binding energy of Al2p level can be attributed to the quality of corresponding layers. The surface composition analysis shows that the concentration of gallium in AlGaN layers is much less than the aluminum concentration (~ 12 times for the sample 153 and sample 155), in spite of the fact that gallium and aluminum precursors at the entry into the synthesis zone were in equimolecular ratio for the sample 155 and 2/1 for the structure 153 at a constant consumption of ammonia. This demonstrates that the rate of chemical reactions between the aluminum precursors and the ammonia is much greater than that of the gallium precursor. The difference between these rates increases with temperature.

IV. CONCLUSION

This study proves the fact that the substrate determines to a great extent the properties of the GaN layer. XPS analysis shows the presence of other elements at the surface of the layers - nitrogen, oxygen, carbon. It is difficult to determine in what form they are because these elements do not form solid compounds at mentioned temperatures. It should be noted that in samples 153, 155, where the GaN layer is less, the concentration of the oxygen is 2 times greater than in the samples 156, 162 and 164. A thick layer of GaN deposited at a lower temperature has a lower concentration of oxygen, but still quite high and which also shows the presence of other sources of oxygen. Electron binding energy values for all detected elements are of the order of one percent and are in a satisfactory agreement with the values obtained by other researchers [1-3,6]. A greater deviation of binding energy of Al2p may be attributed to the quality of the corresponding layer.

ACKNOWLEDGMENT

This research was partially supported by Institutional 15.817.02.34 A Project. Authors express profound gratitude professor Stefan Tiron for help given at work performance.

CONFLICT OF INTEREST

The authors declare that they have no conflict of interest.

REFERENCES

1. P. Motamedi, K. Cadien. XPS analysis of AlN thin films deposited by plasma enhanced atomic layer deposition.Applied Surface Science. 315(2014)104-109.
2. Yang Yingge, MA Honglei,HAO Xiaotao,Ma Jin, XUE Chengshan, ZHUANG Huizhao.Preparation and propertiesof GaN films on Si(111)substrates.Science in China(Series G).Vol.46, No.2,173-177.
3. G.N.Chaudhari, V.R.Chinchamalatpure, S.A.Ghosh.Structural and Electrical Characterization of GaN Thin Films on Si(100). American Journal of Analytical Chemistry, 2011,2,984-988.
4. Peter Y.Yu, Manuel Cardona.Fundamentals of Semiconductors.Physics and Materials Properties. Springer,2002.
5. D. Thomas. Bind. Energies of Electrons in Atoms from H (Z=1) to Lw (Z=103)., thomas@chembio.uoguelph.ca.
6. http://srdata.nist.gov. База данн. по полосам РФЭС

Author: Simion Raevschi
Institute: Department of Physics, Moldova State University
Street: Mateevici, 60
City: Chisinau
Country: Moldova
Email: raevskis@mail.ru

Birefraction in CdP_2 Photodiodes

I.G. Stamov[1], N.N. Syrbu[2], and L. Nemerenco[2]

[1] T.G. Shevchenko State University of Pridnestrovie, Tiraspol, Republic of Moldova
[2] Technical University of Moldova, Chisinau, Republic of Moldova

Abstract— **Characteristics of Me-CdP_2 photodiodes obtained by electro-chemical deposition of metal and thermo-chemical metal spraying in vacuum have been studied. The influence of birefringence and gyration on spectral characteristics of p-n photodiodes and Schottky diodes has been determined. The ability of controlling photodiodes' characteristics has been obtained using the features of gyration in CdP_2 crystals.**

Keywords— **Semiconductor compound, optical absorption and transmission spectra, CdP_2 crystals, birefringence, photodiodes.**

I. INTRODUCTION

CdP_2 is a wide band semiconductor material of A^2B^5 group, which possesses the anisotropy of optical properties with natural gyrotropy [1 – 3]. Polarized spectra are observed in the electronic transitions and vibrational modes, in photoluminescence, photoconductivity, absorption, and reflection [3 – 12]. Based on CdP_2 crystals there were developed devices for quantum electronics and nonlinear optics, which principle of operation is based on gyrotropic and nonlinear polarizability. The values of nonlinear polarization and gyrotropy of cadmium diphosphide are higher than for other crystals, and provides effective stabilization of the radiation field at the same time in space (cross-section) and in time [13]. Inertialess intensity saturation of the transmitted light through the sample in dependence on the intensity of incident light allows implementing instant feedback on the current radiation field and creating stabilizers of radiation field and laser pulse extenders. Deflectors' advantage is that they can be used in the visible and near-infrared spectrum [14, 15].

In cadmium diphosphide it was found a linear relationship between the rotating polarization of plane-polarized light and the magnetic field [16, 17] and the possibility of creating magneto-optical modulators, magneto-sensitive elements to measure the magnetic field, all based on cadmium diphosphide was shown.

Cadmium diphosphide has a high photosensitivity and intense luminescence [3 – 5]. The technology of producing *p-n* junctions and surface barrier diodes was elaborated based on CdP_2 crystals [2, 3, 5 – 9]. Photoresists built on CdP_2 have a small time constant of relaxation and are suitable for the registration of pulse radiation flows with a duration of $1 \cdot 10^{-9}$ s. Photoelectric emitters and photodiodes [18 – 20], electrical switches [21, 22] and Zener diodes [23] had been developed based on single CdP_2 crystals. CdP_2 crystals alter the optical activity with temperature, while maintaining the linear dependence of the rotational plane of polarization on the temperature (temperature sensors) [24].

II. EXPERIMENTAL METHODS

The process of growing single crystals of cadmium diphosphide from Cd and P precursors occurred in two stages. The cleaning the source of atomic materials and synthesis of CdP_2 chemical compounds was made in the first phase. Synthesis was carried out in quartz ampoules in a high pressure container (≈ 40 atmospheres). Up to 300-500 grams of the substance has been synthesized simultaneously. The synthesized compound was sublimed in a vacuum to obtain single crystals in the second phase. The maximum dimensions of CdP_2 ingots obtained from gas phase reached 15×15×30 mm. Single crystals had been cleaved perpendicularly to the C axis. Measurements were carried out on single crystals obtained in a form of prisms or plates, which have grown along the C axis, having mirror, non treated surfaces. The absorption spectra were measured at 2 K and 4.2K in a glass cryostat with liquid helium, using the DFS-24 spectrometer with a resolution of 5Å/mm. Optical transmission spectra were measured using crossed polarizers and the MDR-2 and JASCO-671 spectrometers. Low-temperature measurements were made in a LTS-32C330 Workhorse-type Optical cryostat.

III. RESULTS

When an author or the institution of the author has a relationship, financial or otherwise, with individuals or organizations that could influence the author's work inappropriately, a conflict of interest may exist. Examples of potential conflicts of interest may include but are not limited to academic, personal, or political relationships; employment; consultancies or honoraria; and finical connections such as stock ownership and funding. Although an author may not feel that there are conflicts, disclosure of relationships and interests that could be viewed by others as conflicts of interest affords a more transparent and prudent process. All authors must disclose any actual or potential conflict of interest. The Proceedings will publish such disclosures if judged to be important to readers.

V. Sontea and I. Tiginyanu (eds.), *3rd International Conference on Nanotechnologies and Biomedical Engineering*,
IFMBE Proceedings 55,
DOI: 10.1007/978-981-287-736-9_14

Table 1 Schottky diodes parameters obtained by electrochemical deposition of metals on the cleaved surface of CdP_2 crystals

Metal	φ_B, eV (LIV)	φ_B, eV (volt-capacity characteristic)	n, m^{-3}	η (300K)
In	0.68±0.02	0.66±0.01	3.6×10^{23}	1.08±0.02
Cu		0.85±0.01	2×10^{23}	1.07±0.02
Ni	0.90±0.02	0.90±0.01	2×10^{23}	1.03±0.01
Ni		0.83±0.01	7×10^{22}	
Ni		0.85±0.01	6×10^{22}	
Pt		1.20±0.02	5×10^{21}	1.1±0.02

Photodiode manufactured as Me-CdP_2 structure manufactured by using the electrochemical coating layer of metal on the surface of the etched CdP_2 crystal. Near-surface region of CdP_2 crystals has depleted phosphorus to a depth of no more than a micron. The p-domain is obtained by the diffusion of Hg or S for a number of structures. Nickel layer electrochemically deposited on the cleaved surface forms a "*close*" contact with the semiconductor. The volume of the crystal is an n-type conductivity semiconductor with a carrier density of $4.7\cdot10^{17}cm^{-3}$ order, due to an excess of phosphorus. Parameters of Schottky diodes were determined from current-voltage and capacitance-voltage characteristics and presented in Tables 1 and 2.

Table 2 The parameters of the Me - CdP_2 Schottky barriers obtained by metal spray in vacuum on the cleaved surface of CdP_2 crystals ($n=2\times10^{22}$, m^{-3})

Metal	Ga	In	Pt	Sb	Ag	Cu	Te	Au
ξ, eV	2.75	3.8	4.0	4.8	4.3	4.4	4.7	4.8
φ_{B}, eV	0.37	0.81	0.65	0.94	1.03	1.15	1.2	1.17

The photocurrent is conditioned by the emission from the metal into the semiconductor in the photon energy region $E<E_g$, in the absence of impurity absorption. According to the theory the frequency dependence of the photoemission current is determined by the parabolic law: $J_{ph}=0$ for $E < E_g$ and $J_{ph}=A_0^{\alpha}(E-E_g)^2/2K^2$ for $E > E_g$ [2, 6]. The dependence of photocurrent on the energy for Au-, In-, Cu-, Ni-CdP_2 barriers is presented in Ref. [10]. The barrier height depends on the metal and on the orientation of the crystal surface. As the voltage increases the displacement value φ_B decreases. Figure 1 shows the transmission spectra (T) of CdP_2 crystals with 470μm thickness, the interference of the transmission spectra of crystals with 290 μm thickness (T_*) and 187 μm thickness (T_{**}) placed between crossed polarizers and the spectral dependence $\Delta n=n_o-n_e$ determined from interference spectra. Main peak λ_0 and lateral satellites are observed in the spectra of crystals with 470μm thickness. Pronounced interference of transmission spectra (T_*) adjacent to the wavelength λ_0 is observed in crystals of 290μm thickness, in $\lambda > \lambda_0$ region. These spectra are due to the interference of ordinary (n_0) and extraordinary (n_e) lightwaves. The spectral dependence $\Delta n=n_o-n_e$, which is negative, is determined from these spectra. A feature of these spectra is that the distance between the peaks (β1, β2, β3,...) increases with wavelength decrease. This means that $\Delta n=n_o-n_e$ decreases as the light wavelength decreases (fig. 1, curve $\Delta n=n_o-n_e$). The distance between the peaks of the interference spectra decreases while moving to shorter wavelengths due to higher refractive index and the absence of an isotropic wavelength. Interference spectra and (α1, α2, α3, α4) maxima, which are gathering while decreasing the wavelength, are detected in the wavelength region $\lambda < \lambda_0$, in transmission spectra (T_{**}) of crystals with 187μm thickness.

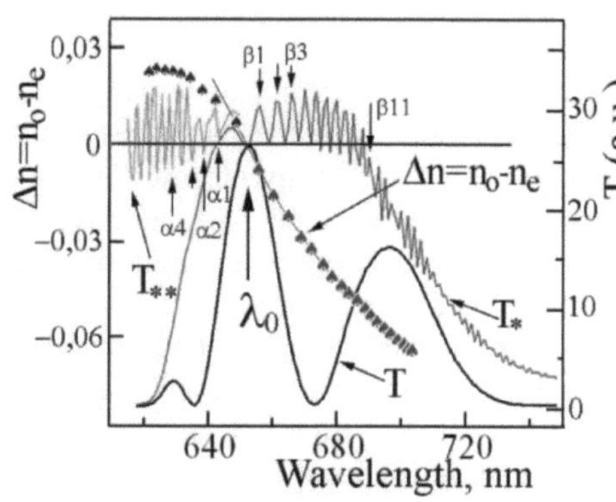

Fig. 1 Transmittance spectra (T) of CdP_2 crystals with 0.47 mm thickness, interference of transmittance spectra (T_*) of crystals with 0.29 mm thickness and 0.187 mm thickness (T_{**}) placed into crossed polarizers and the spectral dependence $\Delta n = n_o - n_e$ determined from interference spectra.

A series of maxima and minima are revealed at 10K in the transmission spectra T of CdP_2 crystals positioned between crossed polarizers at k||c (k – lightwave vector, c – crystal axis), which bifurcate at 7-10° change between the axis and the lightwave vector k (curve-T *), fig. 2. The difference value of the refractive index at 10K ($\Delta n=n^y-n^x$) is calculated from the interference spectra, where n^y - refractive index at E|| y polarization and n^x - refractive index at E||x polarization, fig. 2. The difference of the refractive index reaches, almost, the zero value at 550.7nm wavelength. This wavelength corresponds to λ_{01}, which is detected at 300 K and 574nm wavelength value.

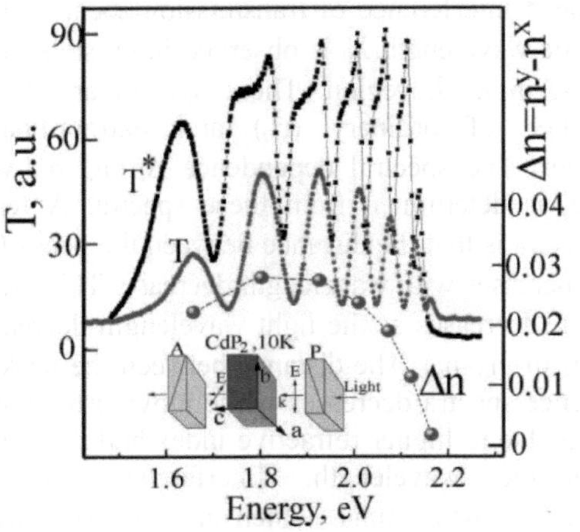

Fig. 2 Spectral dependence of transmittance of CdP_2 crystals placed inside crossed polarizers at k||c (k –lightwave vector , c – crystal's axis) and T^* - at incident angle of 7-10° between the axis and the lightwave vector k and calculated values $\Delta n = n^y - n^x$ (n^y - refractive index at E||y polarization and n^x - refractive index at E||x polarization).

Consequently, one can determine under what thicknesses, the maximum bandwidth with the narrowest band can be observed, considering the data gyrotropy. On the basis of gyrotropic crystals such as CdP_2, which have a narrow bandwidth, while placed between crossed polarizers, one can get narrow band sensitivity photodiodes. Figure 3 shows the structure of the two paired photodiodes obtained on the surface layer, connected by the ITO layer, while the C axis of layer 1 is perpendicular to C axis of layer 2, fig. 3. In the lower part of the figure it is shown the transmittance spectra T_a of CdP_2 crystals at E||a polarization and T_B at E||B polarization. The spectral dependence of photosensitivity J_{ph} is measured at E||a polarization of layer 1, and includes a central peak at λ_0 wavelength (654nm) and two side maxima from the short and long wavelength region. The spectral response has, almost, the same character at polarization change of on E||c of the layer 1, but EMF has the opposite sign. This characteristic, along with its peaks, is like in the case of light transmittance spectra along the C axis of the crystal, where there is a rotation of the polarization plane. The curve of the transmittance T spectral dependence of CdP_2 crystals placed between crossed polarizers are shown for comparison, i.e. when the light propagates along the C axis. The gyrotropy property and, naturally, the absorption (transmittance) coefficient and their difference at E||c and E⊥c affect partially the nature of the spectral dependence of J_{ph} near λ_0 (654nm). The wave vector K is perpendicular to the C axis in these measurements, however, the partial effect of gyrotropy occurs because light rays are focused on the surface of the photodetector, and thus, some of them are oriented at an angle to the C axis.

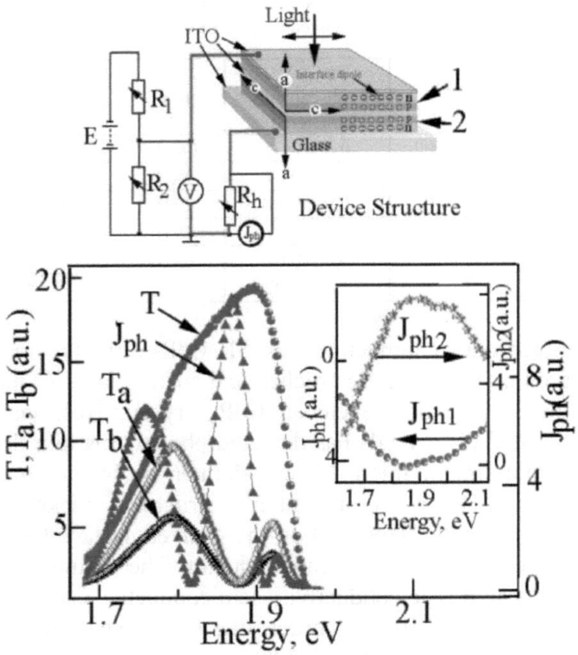

Fig. 3 Construction of two paired photodiodes obtained on the surface and connected by the ITO layer (C-axis of layer 1 is perpendicular to C axis of layer 2) (top); Transmittance spectra (bottom) of CdP_2 crystals at E||a polarization, T_B at E||B polarization, T is transmittance between crossed polarizers and the photo response J_{ph} of the coupled photodiode at unpolarized light.

The absorption coefficient is lower at E||c, E⊥B light polarization, that it is at E||B, E⊥c polarization in the shorter wavelengths' part of λ_0 (654nm). The lightwaves penetrate to the active region of photodiode 2 and give a conditionally positive sign of photo-voltage. The lightwaves are mostly absorbed in the active region of the diode 1 at E||B, E⊥c polarizations creating an inverse photo-voltage [2, 6, and 18]. The spectral characteristics of the device can be varied by changing the polarization of lightwaves incident on the surface of the coupled photodiode.

The transmittance spectra of doped crystals are higher at E||B (T_B) polarization than those at E||a (T_a) polarization, in the λ> 654nm (λ> λ_0) region, fig. 4, A. The lightwaves penetrate shallower at E||a polarization, i.e. create a photo-voltage J_{ph}^a in the active region 1 (conventionally, has a positive direction). The waves' polarization corresponds to E||B case if turning the polarization plane of the incident waves by 90°.

The transmittance coefficient T_B is higher at this polarization than in E||a case. The waves penetrate the spatial charge region of the diode 2 (fig. 4, B) creating a photo-EMF in this area. If the thickness of the crystal is corresponding and provides a plane rotation of polarized lightwaves up to 90°, the photo-voltage will fit to polarized waves at E||B. The J_{ph}^{b} photo-voltage has opposite direction to J_{ph}^{a} photocurrent (conventionally negative).

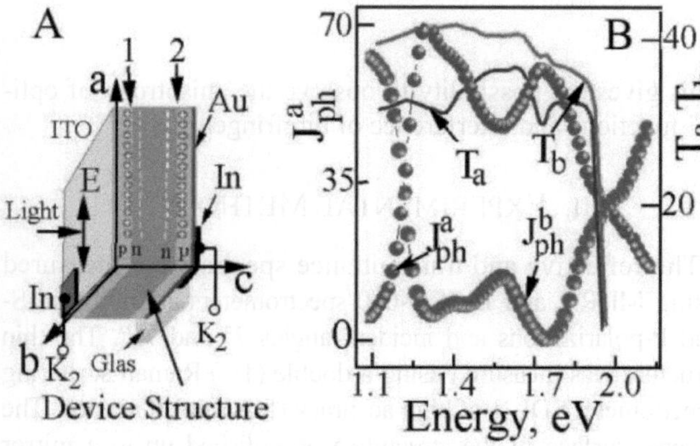

Fig. 4 A – orientation of the crystal's axes in the design of photodiodes with two pn-junctions; B – the photosensitivity of J_{ph}^{a} structures at E||a polarization and J_{ph}^{B} at E||B polarization in (EMF has opposite direction.

IV. CONCLUSIONS

The spectral dependences of the refractive indices n_o, n_e determined from the interference transmittance spectra between crossed polarizers and from reflection spectra using the Kramers-Kronig method intersect at λ_o wavelength in CdP_2 crystals. The crystals are isotropic at λ_o wavelength and there is a maximum transmittance between crossed polarizers and $\Delta n = n_o - n_e = 0$. In the long wavelength region of λ_0 the value of the refractive index $n^{k\|c}$ is higher than that at $n^{k\|y}$ polarization; the dispersion is positive. In the short wavelength region there is an inverse relationship, the dispersion is negative. The birefringence and gyration phenomenon affect the spectral characteristics of p-n photodiodes and Schottky diodes. The possibility of gaining control over the characteristics of photodiodes using the features of gyration in CdP_2 crystals was achieved.

CONFLICT OF INTEREST

The authors declare that they have no conflict of interest.

REFERENCES

1. Lazarev V.B., Sevcenco V.J., Grinberg J.H., Sobolev V.V. (1978) Semiconductors compounds $A^{II} B^{V}$. Nauca, Moscow
2. Syrbu N. N. (1983) Optoelectronic Properties of II–V Compounds. Kishinev
3. Syrbu N.N., Stamov I.G., Kamertsel A.I. (1992) Phys. and Tech. of Semiconductors 26(7): 1191
4. Bindemann R., Fisher H., Kreher K., Syrbu N.N. (1975) Phys. St. Sol. (b) 69(2): K79
5. Syrbu N.N., Stamov I.G. (1991) Semiconductors, 25(12): 2115
6. Syrbu N.N., Stamov I.G., Morozov V.I., Kiossev V.K., Peev L.G. (1980) Proc. I Int Symp. Phys. Chem. II-V Comp., Mogilany. Poland, pp. 237
7. Stamov I.G., Tkacenco D.V. (2008) Semiconductors, 42(9): 1079
8. Stamov I.G., Tkacenco D.V. (2006) Semiconductors, 40(10): 1196
9. Sobotta N., Neumann H., Riede V., Syrbu N. (1983) Solid State Communications 48(3): 297
10. Syrbu N.N., Livin V.E. (1991) Semiconductors 25(10): 1765
11. Syrbu N.N., Livin V.E. (1991) Semiconductors 25(2): 238
12. Syrbu N.N., Livin V.E. (1990) Semiconductors 24(11): 1911
13. Trukhan V.M., Sheleg A.U., Fekeshgazi I.F. (2004) Photoelectronics 13: 200
14. Zuev V.A., Fedotov V.G., Bickov A.G., Smolearenko E.M., Truhan V.M., Iakimovic V.N. (1988) Soviet Journal of Quantum Electronics 34: 57
15. Truhan V.M., Sheleg A.U., et. all. Patent Belorus №12388, 30.10.2009.
16. Borsc V.V., Lisica M.P., Vozoli P.E., Fecesgazi I.V. (1982) Soviet Journal of Quantum Electronics 22: 41
17. Sheleg A.U., Kutas A.A., Iakimovic V.N., Fedotov V.G., Karlikov D.N., Gorynia L.M.; Patent SU, №875959, 22.06.1981.
18. Syrbu N.N., Stamov I.G., Umanetc A.G.; Patent SU, №865040, 14.05.1981.
19. Syrbu N.N., Radautsan S.I., Stamov I.G., Kudinceva G.A., Kasmertsel A.Ju.; Patent SU, №950080, 07.04.1982.
20. Syrbu N.N., Stamov I.G., Umanetc A.G.; Patent SU, №816345, 21.11.1980.
21. Andreev A.A., Meleh B.T., Truhan V.M.; Patent SU, №621254, 27.04.1978.
22. Syrbu N.N., Stamov I.G., Umanetc A.G., Buga L.F.; Patent SU, №776440, 07.07.1980.
23. Stamov I.G., Syrbu N.N., Kiosev V.K.; Patent SU, № 824835, 22.12.1980.
24. Novikov V.P., Sheleg A.U., Truhan V.M., Fedotov V.G., Byckov A.G., Karlikov D.N., Gorynea L.M.; Patent SU, № 917004, 01.12.1981.

Author: prof. Nicolae Syrbu
Institute: Technical University of Moldova
Street: 168 Stefan cel Mare Avenue
City: Chisinau MD-2004
Country: Moldova
Email: sirbunn@yahoo.com

Birefractive Effects in Quantum Wells

A. Dorogan, V. Dorogan, N. Syrbu, and A. Tiron

Technical University of Moldova, 168 Stefan cel Mare Avenue, 2004 Chisinau, Republic of Moldova

Abstract— **Reflective spectra had been researched in $In_{0.3}Ga_{0.7}As$ quantum layers, modulated by the reflection and transmittance wavelength at P, P (S, S) and 45°, 45° (135°, 135°) light-wave polarizations at an incident angle near to the normal and Brewster ones. Isotropic wavelengths λ_0 - 1.137 μm (1.09 eV), λ_{02} - 1.11 μm (1.12 eV) and λ_{03} – 0.932 μm (1.09 eV) had been revealed. The refractive indexes *n* for P, P (S, S) and 45°, 45° (135°, 135°) were intersecting for these wavelengths and theirs difference $\Delta n = n_{PP} - n_{SS}$ ($\Delta n = n(45^o) - n(135^o)$) intersects the null axis. The isotropic wavelength (λ_0) was shifted towards the long wavelength region at Brewster angle in reference to the case of perpendicular incidence of light ($\varphi = 7^o$) on the QW surface.**

Keywords— **quantum wells, birefringence, isotropic point, optical functions, reflection and transmission spectra.**

I. INTRODUCTION

Heterostructures with $In_{0.3}Ga_{0.7}As/GaAs$ quantum wells possess optic properties determined mainly by a high effectiveness of interaction between a quasi bidimensional (2D) subsystem and light. Research of semiconductor nano- heterostructures is important, as from revealing new fundamental properties of excited states in QW and QD point of view, as realizing micro-optoelectronic devices of next generation [1 – 3]. A big assortment of optoelectronic devices was realized basing on QW structures, which are used in different industry branches [4 – 12]. Structures with periodical set of QW are like a single-axis crystal from optical properties point of view. In addition to the periodical set of QW, birefringence and the gyration phenomenon is present in such crystals.

The common point group of zinc blende Td is lowered down to D_{2d} symmetry for multiple GaAs/AlAs quantum wells grown along the [001] direction, which influences the optical properties: anisotropy, linear birefringence and optical activity [1 – 3] and, regularly, is accentuated for the light which propagates with q wavelength vector oriented along the QWs [7, 8].

The paper presents new information on the spectral dependencies of optical functions of $In_{0.3}Ga_{0.7}As$ QW at S- and P-polarizations at an incident angle near the normal (7°) and Brewster angle (76°). The reflective and transmittance coefficients in the region of resonance values of transitions change from 10 to 80% in the researched QW structures. This gives the possibility to observe the anisotropy of optical functions and interference of birefringence.

II. EXPERIMENTAL METHODS

The reflective and transmittance spectra were measured using MDR-2 and JASCO-670 spectrometers at 300K and S- and P-polarizations and incident angles 7° and 76°. The thin structure was measured using a double (1:2) Raman scattering spectrometer SDL-1 of high accuracy (5 Å/mm) if needed. The inverse surface of the structure was polished up to a mirror state for measuring the transmittance spectra. The incident and reflected light from the QW was leaded through the polarizers of island spar. The linear polarized radiation of the polarizer and analyzer was oriented parallel to the Z axis (PP indexes), perpendicularly to Z axis (SS indexes). It was used indication 45° in the case then the polarizer and analyzer were oriented at the angel of 45° to Z axis, if the polarizer and analyzer have the angle 135° with Z axis the index 135° used. The measurements were taken also with crossed polarizers, when the polarization orientation of light-waves of the polarizer and analyzer were corresponding to P, S or S, P, or 45°, 135° orientations.

III. RESULTS

Anisotropy of optical properties is revealed at MQW heterostructures obtained from cubic materials grown along [100] or [111] axis as well as at single axis crystals. The formation of MQW structure leads to material's symmetry lowering in quantum wells. Not only is the birefringence accentuated like in a single axis material of ***D_{2d}* group** but also the optical gyration in structures with QW. The off-diagonal component of the tensor of dielectric constant depends linearly on the wave vector, for the q parallel to [100] or [010] axis. This effect is not accentuated for the q oriented along the [110] axis due to the presence of (110) of the vertical reflection plane [1 – 3]. The presence of optical gyration leads to optical activity, i.e. to different refractive indexes for the right and left circularly polarized light. The experimental separation of optical gyration from birefringence is a problem. Birefringence is higher by value than the gyration. The existence of isotropic point, i.e. a certain frequency at which the linear birefringence disappears, restores the isotropy of linearly polarized light and,

V. Sontea and I. Tiginyanu (eds.), *3rd International Conference on Nanotechnologies and Biomedical Engineering*,
IFMBE Proceedings 55,
DOI: 10.1007/978-981-287-736-9_15

respectively, assures selection of both effects. The optical gyration is normally not revealed away from wavelength of the isotropic point. The experimental works presented in this paper show the research for revealing the isotropic wavelength in MQW structures. One can suppose that the wave vector is parallel to z axis at an incident light angle ≈ 7° (the axis of layers' growth).

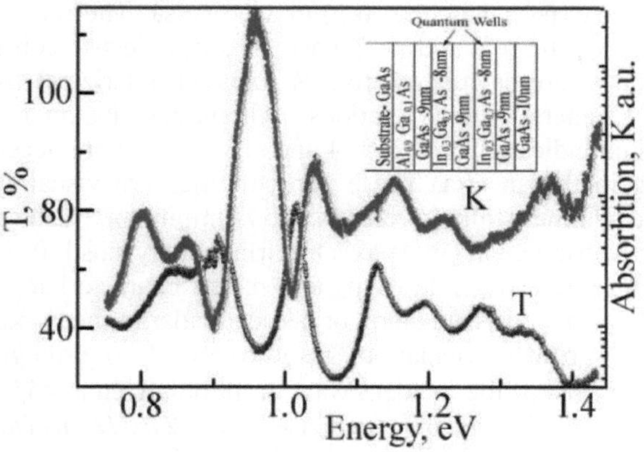

Fig. 1 Transmittance T and absorption K spectra of $In_{0.3}Ga_{0.7}As$ QW layers of 8 nm thickness, separated by GaAs barrier layers of 9nm thickness (layers' structure is shown in the insertion).

The light weakens practically at a few percent in the peak of intersubzone absorption in case of such geometry and at one time pass of light through the MQW structure, as it was mentioned in [2, 3]. A multipass geometry (Brewster angle) assures the sensibility increase of measurements mainly at one order. The change of absorption (fig. 1) and reflection (fig. 2) coefficients occurs strongly in case of fundamental transitions at the researched crystals consisted from two $In_{0.3}Ga_{0.7}As$ QW layers of 8nm thickness, separated by GaAs barrier layers of 9nm thickness. The transmittance coefficient in the region of 1 eV (region of quantic transitions hh1, lh1 – e1) changes from 10 up to 80% (fig. 1). The absorption coefficient changes in arbitrary units up to two orders.

Figure 2 shows reflective spectra consisted of two $In_{0.3}Ga_{0.7}As$ QW layers of 8nm thickness, separated by GaAs barrier layers of 9nm thickness. The reflective spectra of QW structure are measured in case of linearly polarized light. The structures had been placed between the analyzer and polarizer at a light incident angle of 7° and Brewster angle (fig. 1, A, B, C, D). The analyzer's and polarizer's angle was selected so that the wavelength polarization vector would be parallel to z axis, along the QW layers' growth orientation (fig. 1, indication P, P) and perpendicular to z axis (indication S, S). The reflective spectra had been also measured at 7° and Brewster angle, while orienting the polarizer and analyzer at 45°, 45° and 135° in reference to z axis (fig. 2, B and D).

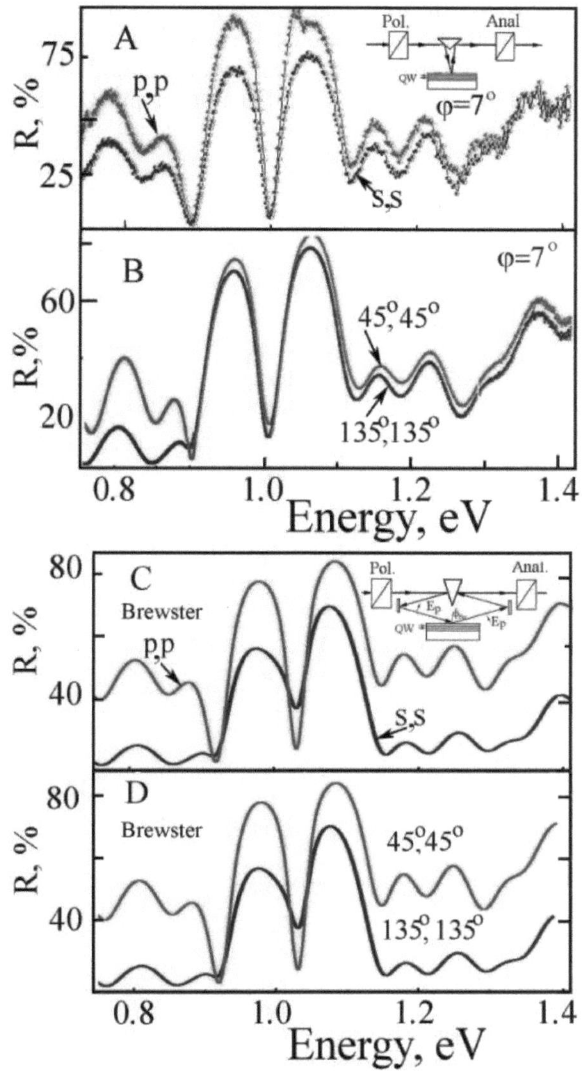

Fig. 2 The reflective spectra of $In_{0.3}Ga_{0.7}As/GaAs$ heterostructures at 7° (A, B) and Brewster (C, D) incident angle. The P, P indications represent the orientation of linearly polarized light of polarizer and analyzer parallel to z axis, S,S, indications – perpendicular to z axis. The indications: 45°, 45° and 135°, 135° concern that the polarization vector of lightwaves has the indicated angles in reference to z axis.

The reflective spectra with $In_{0.3}Ga_{0.7}As/GaAs$ surface layer structure and QW have practically the same character at 300K and incident light angle 7° for P, P and S, S polarization of lightwaves. Two intense maxima are observed at 0.9 and 1.0 eV at both polarizations, which are conditioned by electronic transitions hh1-e1 and lh1-e1. The magnitude of reflective coefficient at P, P polarization and 7° and 76° (Brewster angle) is higher than for S, S polarization. This difference is higher at Brewster angle (fig. 2, C). The difference between reflective coefficients with 45° and 135° lightwave orientation of polarization is lower at an incident light angle of 7° than at P, P and S, S polarizations. This is conditioned by the fact that the incident wave is linearly polarized, with the orientation of polarization vector under 45° and 135° in reference to z axis and the amplitudes of p- and s- component of incident light have, practically, the same value.

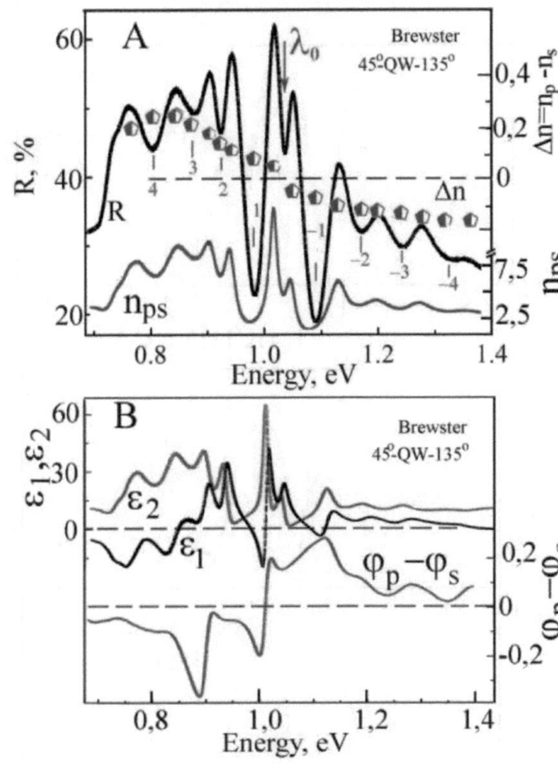

Fig. 3 Reflection spectra at Brewster angle of $In_{0.3}Ga_{0.7}As/GaAs$ QW heterostructures in case of crossed polarizers at 45° (polarizer) and 135° (analyzer) orientations. The refractive indexes n_{ps}, ε_1, ε_2 and the difference of phases φ_p-φ_s are obtained from Kramers-Kronig calculations, and Δn from interference spectra

The magnitudes of reflective coefficients at P,P and S,S polarizations, as for 45° and 135°, have a bigger difference, than for 7° angle (fig. 2). In this case, one can observe a significant difference of the magnitudes of reflective coefficients as for the lower energy region of electronic resonances, as for the resonance region, if comparing the reflective spectra measured at different light polarizations. The amplitude of reflective coefficient is higher on 25 – 30%, at Brewster angle and (P,P) light polarization along the growth of QWs, than it is for S,S polarization. Such difference of reflective spectra indicates that the refractive index of light depends on the relative orientation of the sample and polarization plane of light, i. e. the anisotropy of optical properties is present in the structure's plane.

The point group of zinc blende Td is lowered down to D_{2d} and, as a result, the linear birefringence is permitted only for the light, which passes along the orientation [110], while the optical gyration can also exist when the lightwave vector q is parallel [100] [2, 7].

Figure 3 shows the reflective spectra (R) at Brewster angle of QW layer placed between crossed polarizers oriented 45°-QW-135°. An analogic spectral dependence is also revealed in case of crossed polarizers, when the polarizer is oriented in P geometry (parallel to z) and the analyzer in S geometry (perpendicular to z) or vice versa. The refractive index n_{ps}, the real and imaginary part of dielectric constant ε_1 and ε_2 are calculated from R (crossed polarizers) using the Kramers-Kronig relations. Alternating maxima and minima indicated as 1, 2, 3, 4 and -1, -2, -3, -4 are accentuating in the spectra. These particularities are visually located symmetrically in reference to λ_0 minimum. Such spectra are observed in single axis birefringent crystals [10, 11].

The refractive indexes n_p and n_s are characterizing the light waves polarized along or perpendicular to the Z axis of layers' growth orientation, respectively. Two orthogonal components of the incident wave front polarized at $\boldsymbol{45^\circ}$ will be shifted by phase (dephased) at $\boldsymbol{2\pi d/\lambda\,[n_P(\omega) - n_S(\omega)]}$ value after passing the distance d through the sample. The intensity of detected light for crossed analysis will reflect in such a way the maxima (minima) at frequencies for which the phase differences is odd (even) and multiple to $\pm\pi$; the vice versa should, respectively, be revealed for the parallel analysis. The birefringence value can be derived from the energetic difference between consequent maxima (or minima). The following expression was obtained from interference relations:

$$2\pi d(n_p - n_s)\left(\frac{1}{\lambda_l} - \frac{1}{\lambda_{(l-1)}}\right) = \pm 2\pi, \qquad (1)$$

where l is the maximum number, and the difference of refractive indexes is taken as an average between consequent maxima (minima). The difference $\Delta n = n_p - n_s$, which intersects the null axis in the region 1.0 – 1.1 eV, i.e. in the region of hh1 - e1and lh1 - e1transitions, has been calculated from this expression (fig. 3).

The existence of isotropic point lower than the fundamental boundary of GaAs/AlAs quantum wells was researched in papers [7, 8]. The authors observed a big value of birefringence and revealed the change of scattering sign. The authors also note, that energies of light and heavy hole levels (zones) are changed in MQW relative to their position inside the crystal. The valence bands at q=0 in MQW are equivalent to the splitting conditioned by the single axis intense compression of the crystal. The optical selection rules for transitions from these states determine the sign of scattering component of the dielectric tensor near the beginning of fundamental boundary absorption.

The spectral dependencies of refractive index n, the values of real ε_1 and imaginary ε_2 part of dielectric constant (fig. 3) possess sharp changes in the region 0.9 – 1.1 eV for polarized light-waves, which are conditioned by the electronic transitions hh1 - e1 and lh1 - e1.

The functions ε_1 and ε_2 reach the value ≈ 300 at 1.1 eV for 45°, 135° lightwave polarization. This indicates that the maximum absorption is observed at the resonance value of lh1-e1transitions. The existence of electronic transitions with different interband gaps appears in the spectral dependence of the optical functions. The lightwaves polarized along the layers' growth of Z axis only excite the electronic transitions hh1-e1 due to the symmetry of the energy bands, i.e. are associated with the lowest energy gap. The lightwaves polarized perpendicularly to the Z axis are associated with both energy gaps hh1-e1 and lh1-e1, i.e. excite the electronic transitions from both levels - heavy and light holes. The polarization dependences of electronic transitions determine the polarization dependence of the optical spectra (reflective, transmittance) responsible for birefringence, reflected in the appearance of the difference between the real part of the dielectric function for the light polarized parallel and perpendicular to the z axis (fig. 3).

IV. CONCLUSIONS

The obtained experimental results indicate that there is no isotropic point (isotropic wavelength) in the transparency region. The spectral dependences of reflection R, transmission T, optical functions n, ε_1, ε_2 most strongly differ in the energy transitions region lh1-e1, both in magnitude and in energy position. The amount of maxima displacement of optical functions by energy is higher at Brewster angle than at $\varphi = 7^\circ$ in identical polarizations. Isotropic wavelengths λ_0 -1.137 μm (1.09 eV), λ_{01} - 1.11 μm (1.12 eV) and λ_{02} - 0.932 μm (1.09 eV) have been revealed in the region of electronic transitions from the heavy hh1-e1 and light lh1-e1 holes' levels. The interference of birefringence is observed in the neighborhood of isotropic wavelengths. The refractive indexes n for P, P and S, S and 45°, 45° (135°, 135°) polarizations intersect and their difference $\Delta n = n_{PP} - n_{SS}$ ($\Delta n = n(45^\circ) - n(135^\circ)$) intersects the null axis. The isotropic wavelength (isotropic point) is shifted to longer wavelengths relative to the normal incidence ($\varphi = 7°$) at Brewster incident angle of light.

CONFLICT OF INTEREST

"The authors declare that they have no conflict of interest".

REFERENCES

1. Fox Mark (2001) Optical Properties of Solids Oxford University Press
2. Ivchenko E.L. (2005) Optical spectroscopy of semiconductor nanostructures, Alpha Science International, Harrow, UK
3. Vorob'ev L.E., Ivchenko E.L., Firsov D.A., Shalygin V.A. (2001) Optical properties of nanostructures", Ed. By V.I. Il'in, A.Ya. Shik, Nauka St-Petersburg
4. Maximov M.V., Tsatsul'nikov A.F., Volovik B.V., Bedarev D.A., Egorov A.Yu., Zhukov A.E., Kovsh A.R., Bert N.A., Ustinov V.M., Kop'ev P.S., Alferov Zh.I., Ledentsov N.N., Bimberg D., Soshnikov I.P., Werner P.. Proc. ICPS24, Jerusalem, (World Scientific, 1998).
5. Ledentsov N.N., Grundmann M., Kirstaedter N., Schmidt O., Heitz R., Bohrer J., Bimberg D., Ustinov V.M., Shchukin V.A., Kop'ev P.S., Alferov Zh.I., Ruvimov S.S., Kosogov A.O., Werner P., Richter U., Gosele U., Heydenreich J. (1996) Sol. St. Electron., 40: 785-793
6. Kirstaedter N., Ledentsov N.N., Grundmann M., Bimberg D., Ustinov V.M., Ruvimov S.S., Maximov M.V., Kop'ev P.S., Alferov Zh.I., Richter U., Werner P., Gosele U., Heydenreich J. (1994) Electron. Lett., 30: 1416-1419
7. Fainstein A., Etchegoin P., Santos P.V., Cardona M., Tötemeyer K., Eberl K. (1994) Physical Review B 50: 11850-11856
8. Sirenco A.A., Etchegoin P., Fainstein A., Eberl K., Cardona M. (1999) Phys. Stat. Sol. (b) 215: 241-248
9. Maleev N.A., Zhukov A.E., Kovsh A.R., Mikhrin S.S., Ustinov V.M., Bedarev D.A., Volovik B.V., Krestnikov I.L., Kaiander I.N., Odnobludov V.A., Suvorova A.A., Tsatsul'nikov A.F., Shernyakov Yu.M., Ledentsov N.N., Kop'ev P.S., Alferov Zh.I., Bimberg D. (2000) Semiconductors 34: 612-618
10. Suslikov L.M., Slivka V.Ju., Lisita M.P. (1998) Solid state optical filters on gyrotropic crystals, Kiev, LTD 1-194
11. Syrbu N.N., Dorogan V., Dorogan A., Vieru T., Ursaki V.V., Zalamai V.V. (2012) Superlattices and Microstructures 52: 738-742
12. Syrbu N., Dorogan A., Ursaki V., Stamov I., Tiginyanu I.M. (2011) Optics Communications 284: 3552-3558

Author:	prof. Nicolae Syrbu
Institute:	Technical University of Moldova
Street:	168 Stefan cel Mare Avenue
City:	Chisinau MD-2004
Country:	Moldova
Email:	sirbunn@yahoo.com

Optical Properties of $ZnAl_2Se_4$ Crystals

A. Tiron[1], N. Syrbu[1], and V. Zalamai[2]

[1] Technical University of Moldova, Chisinau, Moldova
[2] Institute of Applied Physics, Academy of Sciences of Moldova, Chisinau, Moldova

Abstract— **Absorption, reflection and wavelength modulated reflection spectra were investigated in $ZnAl_2Se_4$ crystals. The energy positions of ground and excited states for three excitonic series (A, B and C) were determined. The main parameters of excitons and more precise values of energy intervals $V_1(\Gamma_7)$ - $C_1(\Gamma_6)$, $V_2(\Gamma_6)$ - $C_1(\Gamma_6)$ и $V_3(\Gamma_7)$ - $C_1(\Gamma_6)$ were estimated. Values of splitting due to crystal field and spin-orbital interaction were calculated. Effective masses of electrons (m^*_{C1}) and holes (m^*_{V1}, m^*_{V2}, m^*_{V3}) were estimated. Reflection spectra contours in excitonic region were calculated using dispersion equations.**

Keywords— **$A^{II}B_2^{III}C_4^{VI}$ compounds, excitons, wavelength modulated reflection and transmission spectra, effective masses of holes and electrons.**

I. INTRODUCTION

$ZnAl_2Se_4$ was attributed to the wide class of triple chalcogenide semiconductor compounds $A^{II}B_2^{III}C_4^{VI}$ and crystallized into the lattice with I4-S_4^2 space group. The crystals have a high photoconductivity and a bright visible luminescence. The band gap of these crystals at room temperature is equal to 3.5 eV [1 - 5]. Recent investigation of optoelectronic properties of $A^{II}B_2^{III}C_4^{VI}$ compounds doped by 3d transition and 4f rare-earth metals stimulate a large interest due to the influence of these impurities on properties of crystals [6 - 9]. The measurements of optical absorption in cobalt doped crystals of $CdGa_2S_4$:Co^{2+} and $ZnAl_2S_4$:Co^{2+} were carried out and lines with high absorption coefficient due to impurity of 3d transition metal were discovered. Electron transitions from Co^{2+} levels in structures of S_4 symmetry were revealed in corresponded absorption regions. Discovered absorption lines were considered on the base of possible splitting of Co^{2+} levels due to crystal field and spin-orbital interaction. Resealed properties show an investigation prospective of these crystals doped by transition and rare-earth metals. Optical properties of $ZnAl_2Se_4$ crystals were investigated near the edge absorption [1 - 4]. Optoelectronic properties and photoconductivity of thin epitaxial layers were analyzed [5].

II. EXPERIMENTAL METHODS

Crystals $ZnAl_2Se_4$ grown by the iodinless gas-transport method in ampoules had a plate form with mirror surfaces of 8×10 mm and thickness 20 -500 μm. Surfaces of samples were parallel to C axis. Low-temperature spectra of crystals deposed in closed helium optical cryogenic system LTS-22 C 330 were measured on the spectrometer MDR-2 with resolution 0.5 meV. Reflection and transmission spectra were also measured on spectrometers SPECORD-M40 and JASCO--670. All measuring systems are computerized.

III. RESULTS

A changing of absorption coefficient from 10^2 to 10^4 cm^{-1} was revealed in absorption spectra of $ZnAl_2Se_4$ crystals measured at 300 K in nonpolarized light in and around the minimum of interband gap (see Fig. 1). Reflection maxima A and B were revealed in absorption spectra at temperature decreasing to 30 K. Except of maxima A (3.512 eV) and B (3.491 eV) at temperature 10 K the new maxima C and D at energies 3.804 eV and 4.253 eV, respectively, were found out (Fig. 1). Maxima A, B and C have an excitonic nature. Maxima of photosensitivity at energies 3.5263 eV (A), 3.6147 eV (B) and 3.7177 eV (C) were revealed in the region of interband minimum in photoconductivity spectra of $ZnAl_2Se_4$ thin crystals measured at 10 K. These features were attributed to electron transitions in Brillouin zone center. Photoconductivity maxima at room temperature were discovered at energies 3.308 eV (A), 3.3959 eV (B) and 3.34994 eV (C) [10, 11].

Reflection spectra of these crystals were measured for an elucidation of discovered absorption maxima (A, B, C and D) nature, Fig. 2. Maxima A at energy 3.5813 eV (n^A = 1) and C at 3.9711 eV (n^C =1) were detected in reflection spectra of $ZnAl_2Se_4$ crystals at temperature 10 K and polarization E||c, k⊥c. Weak features at energies 3.6562 eV and 4.0399 eV were observed at the same polarization. These peaks were assigned to the excited states n = 2 of A and C excitons. An intensive reflection maximum at 3.6583 eV and a weak peak at 3.7126 eV were presented in polarization E⊥c, which are n = 1 and n = 2 states of excitonic series B. Reflection spectra have a traditional for excitons shape with maxima and minima. These extremes are due to presence of transversal and longitudinal excitons. Energy of longitudinal-transversal splitting is estimated based on these data. The longitudinal-transversal splitting (ω_{LT}) are equal to 6.45 meV and 7.5 meV for A and C excitons, respectively, and for B exciton this splitting corresponds to 5.42 meV.

V. Sontea and I. Tiginyanu (eds.), *3rd International Conference on Nanotechnologies and Biomedical Engineering*,
IFMBE Proceedings 55,
DOI: 10.1007/978-981-287-736-9_16

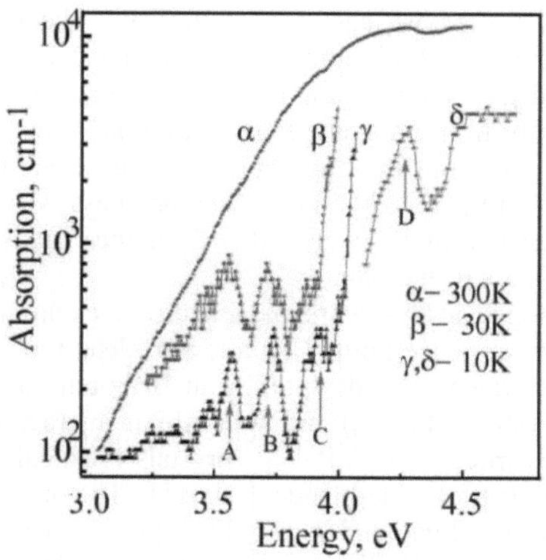

Fig. 1 Absorption spectra of ZnAl2Se4 crystal of 23 μm thicknesses measured at different temperatures (300 K - α, 30 K - β and 10 K - γ, δ) in nonpolarized light.

Table 1 The exciton and the free carrier mass band parameters deduced from the fits of Eq. (2) to the reflectivity spectra for $ZnAl_2Se_4$ crystal.

		A (eV), E‖c	B (eV), E⊥c	C (eV), E‖c	Δ_{cf} (meV)	Δ_{so} (meV)
Exciton state	n = 1	3.5813	3.6583	3.9711	93	344
	n = 2	3.6562	3.7126	4.0399		
	ω_{LT}	6.45	5.42	7.5		
R		0.101	0.0796	0.0918		
E_g (n = ∞)		3.6823	3.7379	4.0629	136	329
ε_b		6.3	9.9	6.6		
μ^*, m_0		0.6	0.55	0.55		
M, m_0		3.2	2.5	2.5		
m_C^*, m_0		0.8	0.8	0.8		
m_{V1}^*, m_0		2.4				
m_{V2}^*, m_0			1.7			
m_{V3}^*, m_0				1.7		

A, B and C excitons ground states contours of measured reflection spectra (Fig. 2, exp.) were calculated (Fig. 2, calc.) on the base of dispersion equations using one-oscillator model. These calculations of reflection spectra contours were made in the framework of a theory taking into account spatial dispersion and existence on the crystal border the "dead" layer (DL) of Thomas-Hopfild on the interior border of that the additional boundary conditions of Pekar is true. The permittivity of medium in nearby of excitonic resonance has the form [12, 13]:

$$\varepsilon(\omega,\vec{k}) = \varepsilon_b + \frac{2\varepsilon_b \omega_{LT}\omega_0}{\omega_0^2 - \omega^2 + \frac{\hbar^2 k^2}{2M}\omega_0 - i\gamma\omega} \quad (1)$$

where ε_b is a background dielectric constant depended on contributions of all interaction mechanisms except examined oscillator; ω_0 is a transversal frequency of exciton; $M = m^*_C + m^*_V$ is a translation mass of exciton: k is a wave-vector; $\omega_{LT} = \omega_L - \omega_0$ is a longitudinal-transversal splitting and ω_L is a longitudinal frequency of exciton. The reflectivity of crystal at normal incidence of light on the border crystal - DL - vacuum is determined by the next equation:

$$R = \left| \frac{\left(\frac{1-n_0}{1+n_0}\right) + \left(\frac{n_0 - n^*}{n_0 + n^*}\right) e^{i2kn_0L}}{1 + \left(\frac{1-n_0}{1+n_0}\right)\left(\frac{n_0 - n^*}{n_0 + n^*}\right) e^{i2kn_0L}} \right|^2 \quad (2)$$

where $n_0 = \sqrt{\varepsilon_b}$; $n^* = \frac{n_1 n_2 + \varepsilon_b}{n_1 + n_2}$; L is a depth of "dead" layer; k is wave-vector of exciton; n_1 and n_2 are refractive indices of transversal waves taking into account a damping constant γ. The magnitude of background dielectric constant in nearby of excitonic resonance was used in the calculations.

$$n^2_{1,2} = 0.5\left\{\left(\varepsilon_f + \frac{2Mc^2(\omega-\omega_0)}{\hbar\omega_0^2}\right) \pm \left[\left(\frac{2Mc^2(\omega-\omega_0)}{\hbar\omega_0^2} - \varepsilon_f\right)^2 + \frac{8Mc^2\varepsilon_f\omega_{LT}}{\hbar\omega_0^2}\right]^{1/2}\right\} \quad (3)$$

Figure 2 show measured and calculated contours of reflection spectra. Values of translation masses M according abovementioned calculations are equal to $3.2m_0$, $2.5m_0$ and $2.5m_0$ for A, B and C excitons, respectively.

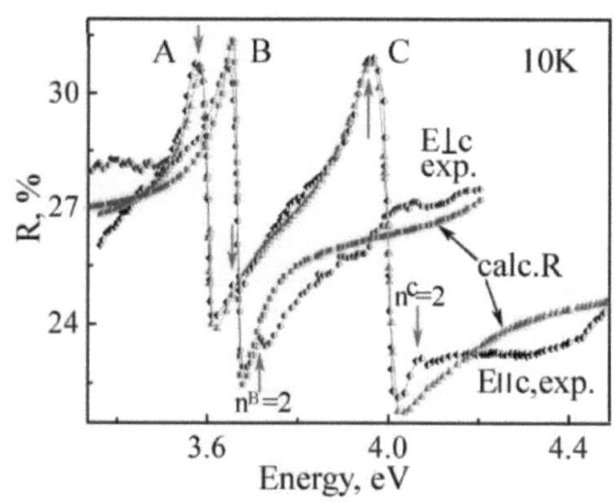

Fig. 2 Reflection spectra of $ZnAl_2Se_4$ crystals measured (exp.) at temperature 10 K in polarizations E‖c and E⊥c and calculated (calc.) contours of these spectra using dispersion equations.

Calculations of contour reflection spectrum for state n^A = 1 of A exciton give a good agreement between theory and experiment with next parameters ε_b = 6.3, ω_0 = 3.5813 eV, ω_{LT} = 6.45 meV, γ = 6.3 meV and M = $3.2m_0$. The value of longitudinal-transversal splitting equal to 6.45 meV was received from calculation for s - state of A exciton. We believe that these values are overstated due to the influence of high magnitude of damping constant (6.3 meV). The binding energy of A exciton in polarization E‖c determined from position of lines n^A = 1 and n^A = 2 is equal to 101 meV. The binding energy of C exciton in E||c polarization is a bit smaller and equals to 91.8 meV. The continuum energy (E_g) for both series is equal to 3.6823 eV and 4.0629 eV (see Table 1).

The minimum of interband gap is formed by direct electronic transitions in Brillouin zone center of $ZnAl_2Se_4$ crystals according theoretical calculations of the band structure [10, 11]. The lower conduction band in crystals of D_{2d} symmetry has a symmetry Γ_1 (s) [14, 15] without taking into account spin-orbital interaction. The lower conduction band has Γ_6 symmetry subject to spin-orbital interaction. Top valence bands V_1, V_2 and V_3 without taking into account spin-orbital interaction have symmetries $\Gamma_4(z)$, $\Gamma_5(x)$ and $\Gamma_5(y)$ and with taking into account this interaction the bands symmetries change to Γ_7, Γ_6 and Γ_7, respectively (fig. 3). The interaction of electrons of conduction band Γ_6 and holes of valence band Γ_7 are determined by the product of irreducible representation $\Gamma_1 \times \Gamma_6 \times \Gamma_7 = \Gamma_3 + \Gamma_4 + \Gamma_5$. The excitons Γ_4 allowed in E||c polarization, Γ_5 allowed in E⊥c polarization and Γ_3 forbidden in both polarizations are formed as a result of this interaction in long-wavelength region. The interaction of electrons of conduction band C_1 with Γ_6 symmetry and holes of valence band V_2 by Γ_6 cause an arising of three excitonic series Γ_1, Γ_2 and Γ_5. Γ_5 excitons are allowed and excitons Γ_1 and Γ_2 are forbidden according selection rules in polarization E⊥c [13].

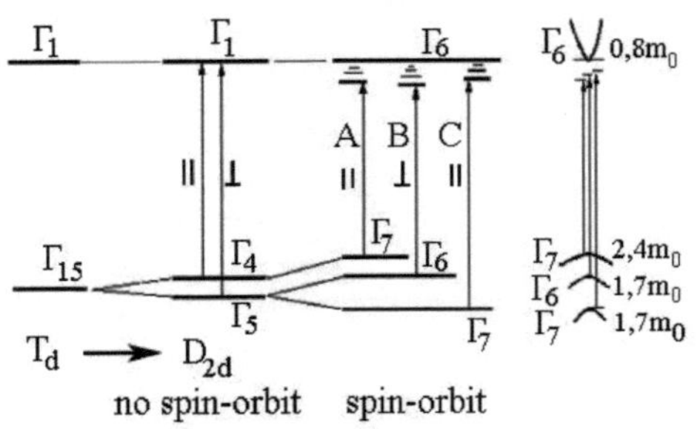

Fig. 3 Band structure at the Γ point showing the transition from zincblende (T_d) to chalcopyrite (D_{2d}) structure

Thus discovered maxima A and C are due to excitons of Γ_4 symmetry allowed in E||c polarization and are formed between valence band V_1 of Γ_7 symmetry and conduction band C_1 with Γ_6 symmetry. The maximum marked as B revealed in E⊥c polarization causes by excitons of Γ_5 symmetry and is formed by bands V_2 (symmetry Γ_6) and C_1 (symmetry Γ_6), fig. 3.

Based on received data (Table 1) and taking into account that $M = m^*_V + m^*_C$ and $1/\mu = 1/m^*_V + 1/m^*_C$, where m^*_C, $m^*_{V1\text{-}3}$ are effective masses of conduction and valence bands $\Gamma_7(V_1)$, $\Gamma_6(V_2)$ and $\Gamma_7(V_3)$ the effective masses of electrons in conduction band and holes in valence bands were determined. The effective mass of electrons m^*_C is equal $0.8m_0$ and effective mass of holes m^*_{V1} is equal $2.4m_0$ at $M = 3.2m_0$ and $\mu^* = 0.6m_0$. Parameters of Γ_5 excitons differ insignificantly from parameters of Γ_4 excitons. The effective masses of electrons m^*_{C1} and holes m^*_{V2} are equal to $0.8m_0$ and $2.4m_0$, respectively, for translation mass $M = 2.5m_0$, binding energy of B exciton R = 79.6 meV and reduced effective mass $\mu^* = 0.55m_0$.

Determined in this work parameters of excitonic series and its convergence limits (E_g) allow to determine the values of top valence bands splitting in Brillouin zone center due to crystal field (Δ_{cf}) and spin-orbital interaction (Δ_{so}). The band structure of $A^IB^{III}C_2^{VI}$ crystals is calculated as a closed analog of zinc-blende [10, 11]. The more strong changes of band gap and spin-orbital splitting were discovered than expected relative to their analogs. These effects were explained by the hybridization of p- and d- states, which determine top valence bands in Brillouin zone center.

Calculations of splitting values due to crystal field and spin-orbital interaction usually were carried out by energy positions of ground states (n = 1) of A, B and C excitons. In our table the calculated values of Δ_{cf} and Δ_{so} were calculated from energy positions of ground states of A, B and C excitons and threshold values of band gap energies (E_g $(n = \infty)$).

Band structure calculations of $ZnAl_2Se_4$ thiogallate show that in these crystals the extrema of valence and conduction bands determining the minimal band gap lay in Γ point of Brillouin zone (BZ) [10, 11]. Bands in the minimum of interband gap in $ZnAl_2Se_4$ crystals are formed by wave functions of Al and Se atoms. The top valence bands according Ref. [10, 11] are formed mainly by p- states of Se and negligible by p- states of Al atoms. The lower conduction band is formed almost in equal degree by p- states of aluminum atoms and s- states of selenium atoms. The energy bands were calculated in Z, Γ, X, P and N points of Brillouin zone. Almost the identical structure of energy bands is observed for $CdAl_2Se_4$ crystals. The results of calculations of band structure for $ZnGa_2Se_4$ crystals in Γ, T, H, C, N, P, I and T points give almost the same results. The minimum of interband gap is formed by direct transitions in Brillouin zone center. Minima of conduction bands are situated in Γ and N points and in vicinity of points T and I [10 – 12].

As mentioned above, the excitonic series A of Γ_4 symmetry is formed by holes of V_1 band with Γ_7 symmetry and electrons of C_1 band with Γ_6 symmetry. The excitonic series B (Γ_5 symmetry) is formed by holes of V_2 band with symmetry Γ_6 and electrons of C_1 band with Γ_6 symmetry. The short wavelength excitonic series C (Γ_4) is formed by holes of V_3 (Γ_7) and electrons C_1. The band symmetry was determined from theoretic-group calculations and polarizations measurements.

The theoretical calculations of energy band structure in nearby of all abovementioned points of Brillouin zone were carried out without taking into account spin-orbital interaction and crystal field. Valence bands in these points according approaches which used in theoretical calculation were degenerated. Lattice reducing from chalcopyrite to S_4^2 structure leads to bands spitting due to crystal field. Spin-orbital interaction also leads to bands splitting. Intensive maxima of reflection spectra in E||c and E⊥c polarizations due to excitonic states A, B and C were observed in $ZnAl_2Se_4$ crystals nearby the gap minimum localized in k = 0. Energy intervals between bands V_1, V_2 and V_3 were determined from energies differences of excitonic series A, B and C continuums.

IV. Conclusions

Energy positions of ground and first excited states of three excitonic series (A, B and C) were revealed in $ZnAl_2Se_4$ crystals by help of absorption, reflection and wavelength modulated reflection spectra investigations. Excitons with Γ_4(A, C) symmetry were formed in vicinity of energy intervals $V_1(\Gamma_7)$ - $C_1(\Gamma_6)$ and $V_3(\Gamma_7)$ - $C_1(\Gamma_6)$ and nearby interval $V_2(\Gamma_6)$ - $C_1(\Gamma_6)$ - excitons of Γ_5(B) symmetry. Values of splitting due to crystal field are less than ones due to spin-orbital interaction. Effective masses of electrons ($m^*_{C1} = 0.8m_0$) and holes ($m^*_{V1} = 2.4m_0$, $m^*_{V2} = 1.7m_0$ and $m^*_{V3} = 1.7m_0$) were estimated. Parameters of excitons Γ_4 and Γ_5 (damping constant, translation mass and longitudinal-transversal splitting) were calculated by the fitting of experimentally measured contours of reflection excitonic spectra. The optical constant were determined in interval 2 - 6 eV by Kramers-Kronig analysis of measured reflection spectra. Recognized electron transitions were interpreted on the base of the theoretical calculated band structure.

Conflict of Interest

The authors declare that they have no conflict of interest.

References

1. Georgobiani A.N., Radautsan S.I., Tiginyanu I.M. (1985) Wide-gap $A^{II}B_2^{III}C_4^{VI}$ semiconductors: Optical and photoelectric properties, and potential applications, Soviet Physics Semiconductors 19(2): 121-132
2. Radautsan S.I., Tiginyanu I.M. (1993) Defect Engineering in II-III_2-VI_4 and Related Compounds, Japanese Journal of Applied Physics 32: 5-9
3. Samanta L.K., Ghosh D.K., Ghosh P.S. (1989) Electronic and nonlinear-optical properties of some mixed thiogallates, selenogallates, and tellurogallates as candidates for laser device applications, Physical Review B, 39: 10261-10265
4. Park Tae-Young, Lim Ji-Young, Oh Seok-Kyun, Park Kwang-Ho, Song Ho-Jun, Kim Wha-Tek, Choe Sung-Hyu, Kim Chang-Dae, Yoon Chang-Sun (1998) Optical properties of $ZnAl_2Se_4$, $ZnAl_2Se_4:Co^{2+}$, and $ZnAl_2Se_4:Er^{3+}$ single crystals, Journal of Applied Physics 84: 1567-1571
5. You S.H., Hong K.J., Jeong T.S., Youn C.J. (2014) Growth and electrical/optical properties of the photoconductive $ZnAl_2Se_4$ layers grown by hot wall epitaxy method, Journal of Ceramic Processing Research. 15(1): 4-8
6. Hong K.J., Jeong T.S., Youn C.J., Hong K.J., Jeong T.S., Youn C.J. (2013) Optical properties of $ZnAl_2Se_4$, $ZnAl_2Se_4:Co^{2+}$, and $ZnAl_2Se_4:Er^{3+}$ single crystals Journal of the Korean Physical Society 62(4): 628-633
7. Koroleva L.I., Zashchirinskii D.M., Khapaeva T.M., Marenkin S.F., Fedorchenko I.V., Szymczak R., Krzumnska B., Dobrovol'skii V., Kilanskii L. (2009) Manganese-doped $ZnSiAs_2$ chalcopyrite: A new advanced material for spintronics, Physics of the Solid State 51: 303-308
8. Kim Mi-Yang, Baik Se-Jong, Kim Wha-Tek, Jin Moon-Seog, Kim Hyung-Gon, Choe Sung-Hyu, Yoon Chang-Sun (2003) Optical Properties of Undoped and Co^{2+} –, Ho^{3+} –, Er^{3+} and Tm^{3+} Doped $CaGa_2S_4$, $CaGa_2Se_4$, $CaIn_2S_4$, and $CaIn_2Se_4$ Single Crystals, Journal of the Korean Physical Society 43(1): 128-134
9. Angel S. (2011) Absorption spectra of cobalt-doped α-$ZnAl_2S_4$ spinel type single crystals, Optoelectronics and Advanced Materials - Rapid Communications 5(10):1081-1083
10. Jiang Xiaoshu, Lambrecht Walter R. L. (2004) Electronic band structure of ordered vacancy defect chalcopyrite compounds with formula II-III_2-VI_4, Physical Review B 69: 035201
11. Mishra S., Ganguli B. (2011) Electronic and structural properties of A Al_2Se_4 (A=Ag, Cu, Cd, Zn) chalcopyrite semiconductors, Journal of Solid State Chemistry 184(7):1614-1621
12. Gross E., Permogorov S., Travnikov V., Selkin A. (1972) Polariton emission from crystals, Solid State Communications 10(11):1071–1074
13. Syrbu N.N., Ursaki V.V. (2010) Exciton polariton dispersion in multinary compounds, in: Randy M. Bergin (Ed.), Exciton Quasiparticles: Theory, Dynamicsand Applications, Nova Science Publishersinc, USA

Author: dr. Victor Zalamai
Institute: Institute of Applied Physics
Street: Academy str., 5
City: Chisinau MD-2028
Country: Moldova
Email: zalamaiv@phys.asm.md

Nanomultilayer As_2S_3:Mn-Se Systems: Properties and Use as the Recording Media

O. Paiuk[1], A. Meshalkin[2], G. Triduh[2], A. Prisacar[2], E. Achimova[2], A. Stronski[1], V. Abashkin[2], O. Lytvyn[1], O. Senchenko[1], and A. Gubanova[3]

[1] V. Lashkaryov Institute of Semiconductor Physics NAS of Ukraine, Kyiv, Ukraine
[2] Institute of Applied Physics of the Academy of Sciences of Moldova, Chisinau, Moldova
[3] Kamianets-Podilsky National University, Kamianets-Podilsky, Ukraine

Abstract— **Properties of nanomultilayer As_2S_3:Mn systems and peculiarities of holographic gratings recording with their use are considered in this report. Optical constants were obtained from transmission data, spectral dependencies were analyzed within the frames of single oscillator model. Holographic gratings with spatial frequencies 1600 mm^{-1} were directly recorded using such nanomultilayer structures. Kinetic of gratings recording depended on polarization of recording beams. Such systems are perspective for the recording of planar optical elements.**

Keywords— **Nanomultilayers, optical properties, holographic diffraction gratings.**

I. INTRODUCTION

Chalcogenide glasses (ChG) attracted great attention due to diverse range of optical and electrical properties, and a number of applications, connected with light-induced changes of properties. Light-sensitivity effect of thin chalcogenide films was discovered fifty years ago [1] and perhaps photo-stimulated changes of their properties are the most interesting phenomena exhibited by these materials. They have complex nature, including various effects such as photodarkening (redshift in absorption edge), volume photoexpansion, photoinduced anisotropy, solubility change, etc. [2], Exposure of ChG by linearly polarized laser beam produces surface ripples, in which the rippling direction and periodicity seem to be governed, respectively, by the electric field and wavelength of light [3]. Selective etching after exposure makes possible to obtain surface relief and use of such media as high resolution inorganic resists [4-5]. With the use of thin layers of chalcogenide glasses and selective etching of exposed layers holographic diffraction gratings with high characteristics and other optical elements were obtained [6].

Properties of chalcogenide glasses can be changed doping by transitional metals [7]. X-ray diffraction measurements confirmed the amorphous structures of chalcogenide glasses doped by transition metals (including also Mn) and rare-earth elements. Radial distribution functions have not shown significant changes in distance for nearest neighborhood with the change of dopant (Mn) concentration. Differential scanning calorimetry (DSC) measurements have shown that T_g values were smaller for doped glasses as compared to undoped ones. The main feature of Raman spectra under the introduction of transitional and rare-earth impurities into chalcogenide glass matrix is change of relative concentration of the main and non-stoichiometric structural elements typical for initial glasses. Pure chalcogenide glasses are diamagnetics. Introduction of transitional and rare earth impurities changes the magnetic properties of investigated chalcogenide glasses.

Multilayer nanostructures on the base of chalcogenide structures and possibility of their use as registering media were considered in [8-9]. The properties and direct relief formation in chalcogenide nanomultilayer structures were considered in [10-15].

Physical phenomena used in fabrication of various micro-optical elements on a surface of ChG films, such as surface relief gratings (SRG), linear waveguides and microlenses are: photo-induced mass transport, which consists in lateral redistribution of amorphous material under illumination by near-bandgap light [10-15],. in addition, in some cases, illumination by linearly polarized light provides anisotropic changes, which are referred to as vectoral, which means that not only the photon energy value but also the direction of optical fields vector exerts some influence on the photostructural processes. Direct relief formation is interesting because in this case wet etching process which requires very strict control of many parameters (temperature, concentration, composition, etc.) is absent. In addition, often etchant is very toxic. So the development of methods for one step surface relief fabrication is considered as perspective for production of planar diffraction optical elements. This work is concerned with properties of nanomultilayer As_2S_3Mn 2 wt.%-Se systems and their use registering media for the fabrication of holographic diffraction gratings.

II. EXPERIMENTAL RESULTS AND DISCUSSION

The bulk samples of ChG were synthesized by melt-quenching technique from elemental components. ChG nanomultilayers were prepared by computer driven cyclic thermal vacuum deposition of two materials from two

V. Sontea and I. Tiginyanu (eds.), *3rd International Conference on Nanotechnologies and Biomedical Engineering*,
IFMBE Proceedings 55,
DOI: 10.1007/978-981-287-736-9_17

isolated boats on constantly rotated substrate at room temperature in one vacuum deposition cycle. Details of fabrication of chalcogenide glasses and nanomultilayer structures on their base are described in detail in [7, 14]. As_2S_3: Mn , Se layers and As_2S_3: Mn-Se nanomultilayer structures were prepared by thermal vacuum evaporation on glass substrate (90 As_2S_3: Mn nanolayers and 90 Se nanolayers for composite structure. Total thickness of Se layer was 935 nm ± 8 nm, As_2S_3: Mn 2 wt.%– 1031 nm ±3 nm, and of nanomultilyer composite As_2S_3: Mn 2 wt.%+ Se structure – 1966 nm. Cross section of sample is shown in Fig. 1. Overlapping part of samples (3 in Fig. 1) contains alternating nanolayers of As_2S_3Mn 2 wt.% and Se.

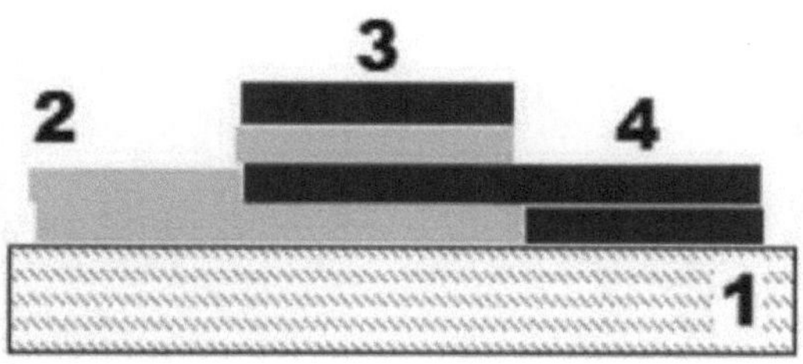

Fig. 1 Scheme of sample's structure: 1- substrate: 2- As_2S_3Mn 2 wt.% layer: 3- multilayer nanostructure As_2S_3Mn 2 wt.% layer-Se, 4- Se layer.

A. Optical Properties

The refractive index and its wavelength dependence other optical properties are among important parameters that determine the suitability of materials as optical media. Optical constants were obtained from transmission spectra using Swanepoel method. Transmission spectra (Fig. 2) were measured in 400-900 spectral range with the use of SPECORD M40 spectrometer.

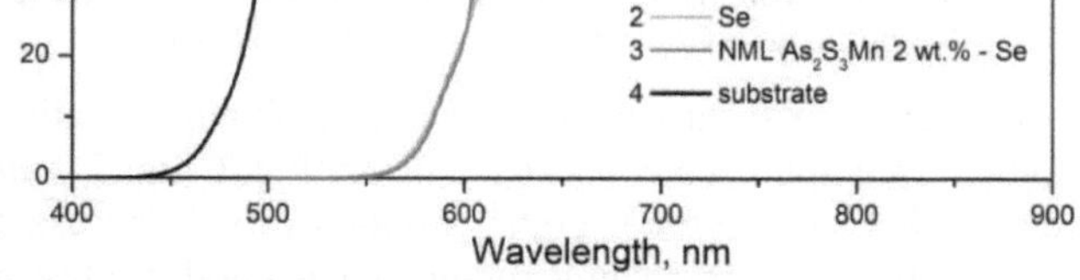

Fig. 2 Transmission of As_2S_3Mn 2 wt.% - 1 and Se -2 layers, 3- As_2S_3: Mn – Se nanomultilayer structure and 4- substrate.

Here it is necessary to note that transmission edge of Mn – Se nanomultilayer structure almost coincide with the transmission edge of Se layers. Spectral dependencies of refractive index for As_2S_3:Mn 2 wt.% and Se layers, and- As_2S_3: Mn 2 wt.% – Se nanomultilayer structure are presented in Fig. 3.

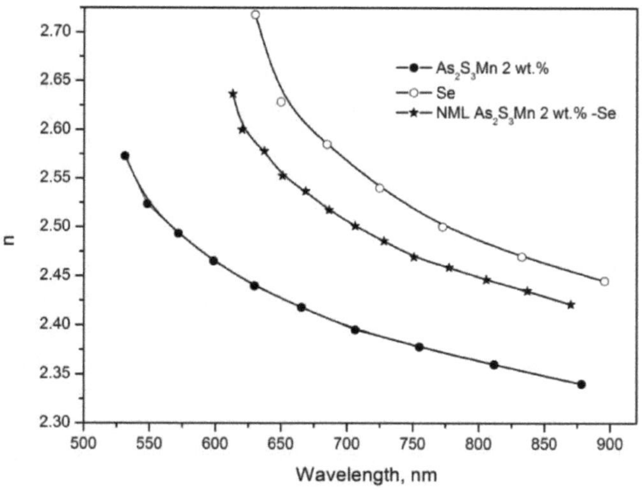

Fig. 3 Spectral dependencies of refractive index for As_2S_3Mn 2 wt.% and Se layers, and- As_2S_3: Mn 2 wt% – Se nanomultilayer structure.

Optical properties were analyzed within the frame of single oscillator model. According to this model refractive index n is related to energy of incident photon E by equation $n^2 - 1 = E_d E_0 /(E_0^2 - E^2)$, where E_0 is. single oscillator energy and E_d is dispersion energy. In this expression E_0 determines the position of the effective oscillator connected with an average energy gap and E_d is dispersion energy characterizing the strength of interband transitions. To obtain E_0 and E_d plots $(n^2 - 1)^{-1} = f(E^2)$ were obtained with the use of least squares fitting method to straight line. Parameters of single oscillator model for As_2S_3Mn 2 wt.% and Se layers, and- As_2S_3: Mn – Se nanomultilayer structure were obtained from such plots together with optical bandgap values obtained using Tauc plot $\alpha h\nu = \text{const}(h\nu - E_g)^2$, where $h\nu$ is the energy of light quantum, α is the absorption coefficient, are presented in Table 1. It can be seen from the table that values of optical absorption edge for Se layers (1.93 eV) and As_2S_3: Mn 2 wt.% – Se nanomultilayer structure (1.94eV) are close to each other.

Table 1 Parameters of single oscillator model for As_2S_3:Mn 2 wt.% and Se layers, and- As_2S_3: Mn 2 wt.% – Se nanomultilayer structure

Layer composition	n(0)	E_d, eV	E_0, eV	E_g, eV
Se	2.284	15.086	3.577	1.93
As_2S_3Mn 2 wt.%	2.248	18.696	4.611	2.32
As_2S_3Mn 2 wt.% - Se	2.286	16.714	3.954	1.94

B. Gratings Recording

An interferometric holographic recording was used to record grating on the NML As_2S_3–Se like described in [14]. The period of the grating $d=\lambda/(2\sin\alpha)$, where λ is the wavelength of laser beam, α is the angle between the incidence laser beams. CW DPSS single mode laser operated at wavelength λ=532nm (100mW) for different polarizations of recording beams. Gratings spatial frequency was ~1600 mm^{-1}. The holographic gratings were recorded by two symmetrical angled laser beams respecting to the sample surface normal. The intensity ratio 1:1 of recording beams was used in order to achieve maximum interference fringes contrast. The phase shift experiments were performed by angular turning the quarter wave plates in each optical path therefore we can change the state of polarization of both recording beams. The two interfering beams independently pass through phase turning quarter wave plates to provide a control over the polarization state of the writing beams. Interfering beams with S-S, L-R or (+45°)–(– 45°) polarization combinations were used for diffraction gratings recording. The experimental set up is sketched in Fig. 4.

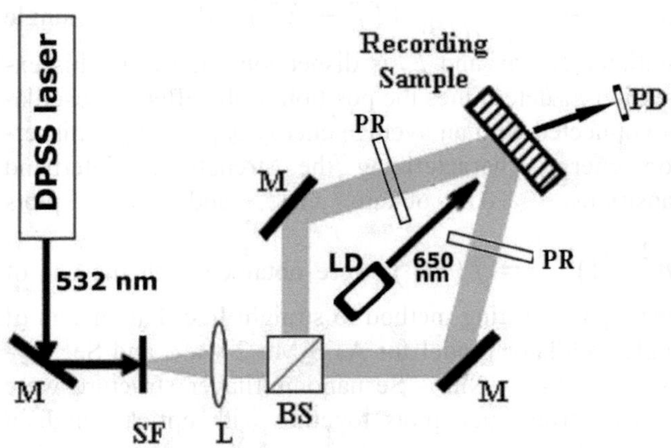

Fig. 4 Optical arrangement for holographic grating recording with real-time measurement of diffraction efficiency by photodetector. CW DPSS laser (λ=532 nm, power =100 mW), M – mirror; SF – spatial filter; L – collimating lens; BS – beam splitter; LD – laser diode (λ=650 nm) for monitoring recording process; PR- polarization retarder; PD – photodetector.

Diffraction efficiency η was controlled in real time by measuring the first order laser diode intensity in the transmission mode. Diffraction efficiency was determined as $\eta=I_{dif}/I$ 100%, where I_{dif}- diffractive beam, I-transmitted beam in 0 order.

It can be seen from Fig. 5 that kinetics of gratings recording and values of the obtained diffraction efficiency depend on polarizations of recording beams. From Fig. 5 it is seen that not only maximal shown diffraction efficiency η, but kinetics of grating recording process depends on state of polarization of recording beams. For S-S and L-R polarizations the recording process begins just after turn on laser illumination.

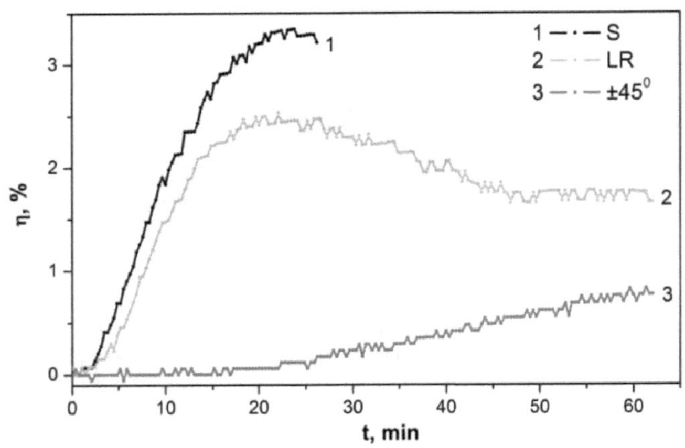

Fig. 5 Diffraction efficiency dependence on exposure time for different polarizations of recording beams. Recording media – As_2S_3: Mn 2 wt.% multilayer nanostructure.

It was found out that at some doze the diffraction efficiency reaches the saturation for polarizations pointed. For (+45°)–(–45°) polarizations we have observed initial induction period and the constant linear rise of η but with much smaller rate.

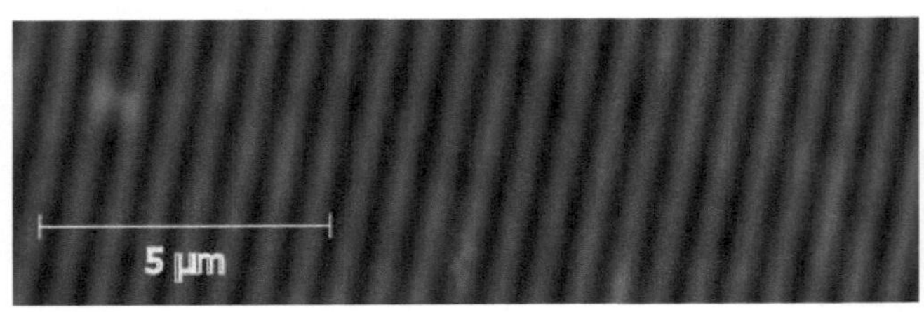

Fig. 6 AFM image of gratings surface obtained on As_2S_3: Mn multilayer nanostructure. Gratings spatial frequency – 1600mm^{-1}.

To study the surface and relief of the obtained gratings atomic force microscopy was used. From Fig. 6 and 7 it can be seen that surface-relief gratings obtained with the use of As_2S_3: Mn multilayer nanostructure have good quality of surface relief.

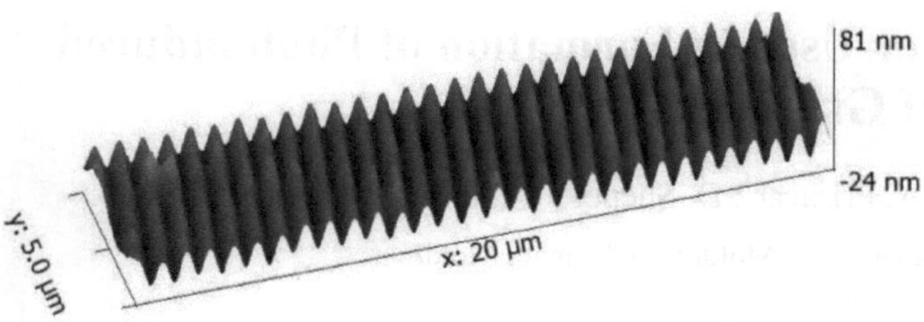

Fig. 7 3–D AFM image of gratings relief obtained on As_2S_3: Mn 2 wt.% nanomultilayer structure. Gratings spatial frequency – 1600mm^{-1}

III. CONCLUSIONS

We have demonstrated the holographic recording of 1600mm^{-1} gratings using As_2S_3: Mn –Se nanomultilayer structure. Direct one step grating surface relief formation in As_2S_3: Mn –Se nanomultilayer structure depends on polarizations of recording beams. Optical properties of As_2S_3: Mn - Se nanomultilayer structure and constituent films were studied and analyzed within the frame of single oscillator model. Parameters of model were obtained. Absorption edge of the As_2S_3: Mn –Se nanomultilayer structure is close to the absorption edge of Se multilayers obtained in the same deposition process. Nanomultilayerd compositionally modulated As_2S_3: Mn nanomultilayer structures are perspective for the recording of holographic optical elements.

ACKNOWLEDGMENT

The research was supported by the project FP–7 SECURE–R2I

CONFLICT OF INTEREST

The authors declare that they have no conflict of interest

REFERENCES

1. Kostishin M T, Mikhailovskaya E V et al. (1965) About the photographic sensitivity of the thin semiconductor layers. J Sci Appl Photogr Sinematogr 10 (6): 450–451
2. Stronski A V, Vlček M (2002) Photosensitive properties of chalcogenide vitreous semiconductors in diffractive and holographic technologies applications 4 (3): 699–704
3. Gotoh T., Hayakawa H. et al. (2000) Photoinduced anisotropy in the ion-conducting amorphous semiconductor Ag–As–S. J. Non-Cryst. Solids 266-269:944–947
4. Mizushima Y, Yoshikawa A (1982) Photoprocessing and lithographic applications. Technologies & Devices, Tokyo
5. Stronski A (1998) Production of metallic patterns with the help of highresolution inorganic resists. «Microelectronic Interconnections and Assembly»,NATO ASI Series, 3:High Technology, -Vol.54, G.Harman@P.Mach, Editors « Kluwer academic publishers », Netherlands, P. 263-293
6. Stronski A, Vlček M et al. (2000) Application of $As_{40}S_{60-x}Se_x$ layers for high-efficiency grating production. J.Non-Cryst.Solids 266-269:973–978
7. Stronski A, Paiuk O et al. (2015) Effect of doping by transitional elements on properties of chalcogenide glasses. Ceramics Int 41:7543–7548
8. Kökényesi S (2001) Light-stimulated structural transformations and optical recording in amorphous nano-layered structures. JOAM 3 (2):377–382
9. Kökényesi S (2006) Amorphous chalcogenide nano-multilayers: research and development. Ibid 8 (6):2093
10. Kaganovskii Yu, Beke D L, Kökényesi S (2010) Kinetics of photoinduced surface patterning in chalcogenide thin films Appl Phys Lett 97 (6):061906.
11. Trunov M L, Lytvyn P M, Dyachyns'ka O M (2010) Alternating matter motion in photoinduced mass transport driven and enhanced by light polarization in amorphous chalcogenide films Appl Phys Lett 97 (3):031905
12. Tanaka K, Terakado N et al. (2008) Photoinduced anisotropic deformations in covalent chalcogenide glasses. J Optoelectron Adv Mater 10 (1):124–130
13. Gertners U, Teteris J (2007) Direct holographic recording on amorphous chalcogenide films, ICANS22 Proc, Colorado, USA 2007 ThP13.7
14. Stronski A, Achimova E et al. (2015) Surface relief formation in $Ge_5As_{37}S_{58}$–Se nanomultilayers. J Non-Cryst Solids 409:43–48
15. Trunov M L, Lytvyn P M et al. (2011) Photoinduced mass-transport based holographic recording of surface relief gratings in amorphous selenium films. Appl Phys Lett 99 (5):051906

Author: Paiuk Oleksandr
Institute: V. Lashkaryov Institute of Semiconductor Physics
Street: 41 Nauki ave.
City: Kiev
Country: Ukraine
Email: paiuk@ua.fm

Properties of Carbazole-Based Azopolymer Used in Formation of Photoinduced Surface Relief Gratings

A. Meshalkin[1], S. Robu[1,2], E. Achimova[1], Yu. Boiarinov[1], A. Prisacar[1], D. Shepel[3], V. Abashkin[1], and G. Triduh[1]

[1] Institute of Applied Physics, Academy of Sciences of Moldova, Chisinau, Moldova
alexei@asm.md
[2] State University of Moldova, Chisinau, Moldova
[3] Institute of Chemistry, Academy of Sciences of Moldova, Chisinau, Moldova

Abstract— **New carbazole-based azopolymer was synthesized, characterized and applied for formation of photoinduced surface relief gratings. The surface modulation of spin-coated azopolymer film depending on the different polarization configurations is considered. It is shown that holographic surface relief gratings with relatively large amplitude could be obtained by direct one-step holographic recording without any subsequent processing steps. Right and left orthogonal circular polarization configuration is the best case, which allows to achieve surface relief grating with the largest amplitude up to 50% of original film thickness.**

Keywords— **azopolymer, carbazole, surface relief grating, polarization holographic recording.**

I. INTRODUCTION

Carbazole-containing polymer materials, photo-induced properties and application of this polymer as recording media were extensively studied by Andries and his group [1-3]. Carbazole-containing polymer materials attract wide interest due to their applications in optical data storage and information processing due to possessing of good photorefractive, photoconductive, and photochemical properties for hologram recording [4,5]. It was shown in [6] that holographic gratings could be obtained on carbazole materials that involves exposure of the thin film to an interference pattern. During the holographic recording two coupling gratings are formed simultaneously: 1 - phase grating, due to modulation of the refractive index and 2 - amplitude grating, resulting from the changes of absorption coefficient. The third type of holographic grating namely surface relief grating was obtained by a wet chemical process leads to film thickness changes. However, wet etching development techniques based on solvent dissolutions affect the fabrication process, leading often to a deformation of the fabricated structures and to increasing of background scattering.

Recently, new organic materials named azobenzene-containing polymers (azopolymers) have attracted much attention for the fabrication of surface structures [7-10]. Kim's group first showed a new possibility to direct formation of surface relief gratings (SRG) on azopolymer films without any subsequent processing steps thanks to mass transport effect under a modulated light irradiation [11]. It was reported that large surface modulations by a simple one-step process can be obtained on azo polymer films. Many theoretical models were proposed to explain the mechanism of SRG formation but the mechanism is still not fully explained [12-15]. The widely used model is the gradient force model [16]. According to this model, the resultant force can be achieved thanks to the light irradiation, which induces a trans-form ↔ cis-form photoisomerization effect in azobenzene molecules. In addition in [17] it was found that type of polymer matrix and molecular weight was a crucial parameter in SRG formation.

So the aim of our work was the synthesis and characterization of new carbazole-based azopolymer suitable for direct surface relief grating formation. The surface modulation of azopolymer film depending on the polarization of the writing beams was also considered.

II. EXPERIMENTAL

Azopolymer Synthesis

Carbazole-based polymer Epoxypropylcarbazole (EPC) and azodye Disperse Orange (DO) was utilized in this study. DO was purchased as a commercial product from Sigma-Aldrich Company [18]. DO molecule was chemically attached to EPC monomer by polycondensation scheme at a temperature of 120° C for 4 hours. The molar ratio between EPC and DO was 90/10. The copolymer was purified by precipitation in hexane and then in methanol. Scheme of synthesis and chemical structure of EPC:DO is presented in Fig. 1.

Fourier Transform Infrared Spectroscopy (FTIR) characterization was applied for copolymer structure characterizing and identifying organic molecules [19].

V. Sontea and I. Tiginyanu (eds.), *3rd International Conference on Nanotechnologies and Biomedical Engineering*,
IFMBE Proceedings 55,
DOI: 10.1007/978-981-287-736-9_18

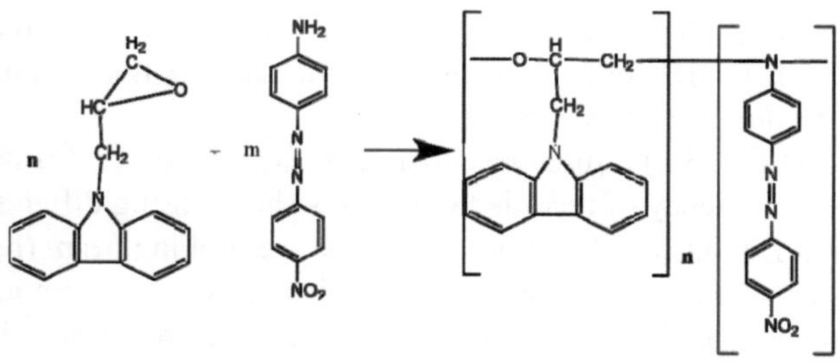

Fig. 1 Scheme of synthesis and chemical structure of EPC:DO.

With the purpose of firmly assessing the nature of the synthesized material, a detailed characterization by IR spectroscopy was applied. Before infrared analysis, samples were dried in exsiccator up to constant mass. Infrared spectra of the samples were recorded with a Perkin Elmer FTIR 100 spectrophotometer (4 cm^{-1} and 32 scans).

The introduction of azo group in synthesized copolymer EPC:DO was proved confirmed by the peak at 1576 cm^{-1} in the IR spectrum corresponding to the N = N stretching frequency. All constituents of copolymer (monomer EPC, azodye DO and copolymer EPC:DO) were the same characterized by FTIR spectroscopy with attribution of infrared absorption bands to main functional groups. The synthesized EPC:DO copolymer was soluble in common organic solvents namely toluene and formed good optical quality films by spin coating.

Formation of Polymer Films from Solution

A widely used technique to fabricate thin layers is spin-coating. In the first step, the material is poured or sprayed to the center of a glass disk. Next, the disk starts rotating in an axis normal to the surface of the disk. Thereby the material moves outward. In the stationary state, the viscous forces in the fluid dominate the thinning of the layer. Excess material may leave the disk at the edge as droplets. If the material contains a solvent, the solvent may evaporate. Spin-coating may be done at elevated temperatures. Various designs are in use.

In our study the thin polymer films were prepared from homogeneous polymer solution by spin coating procedure using programmable spin-coater "SGS Spincoat G3P-8"[20]. The thickness of the polymer film was varied by changing the concentration of polymer solution and the rotation speed of spin coating. In our case polymer thin films were prepared by spin coating of the 10 wt.% polymer solution in toluene onto glass substrate. Operation conditions for polymer solution deposited on 5 cm diameter optical glass substrate (BK7) was as follows: 1 cm^2 of liquid dispensed on the disk at rest, subsequently accelerated in about 3 s to 800 rpm and spun for 60 s. Obtained films were dried in a oven at 60°C for 6 hours.

Determination of Film Thickness

For the determination of film thickness in this work the modified interferometric PC based measurement based on MII-4 interference microscope was applied [21, 22].

A thickness characterization of the samples was achieved by a MII-4 interference microscope with CCD-camera recorded the micrographs. The interference pattern of light reflected from a flat reference surface and the investigated sample was recorded in PC. A magnification of 490 times was used. The area from which a data analysis is performed was 0.3 mm diameter circle. This enables a height resolution better than 100 nm

Spectral characterization.

UV-VIS transmission spectra were recorded on thin polymer films with Specord M40.

Holographic characterization.

Blue (473 nm) and green (532 nm) DPSS lasers were used for holographic recording. The laser beam was split into two identical secondary beams by a 50/50 non-polarized beam splitter. The two beams were then combined, by using two mirrors, and interfered on the azo-copolymer film. The polarizations of the two laser beams were controlled independently by using different wave plates (quarter-wave plate and half-wave plate). The incident angle was adjusted by tuning the two mirrors symmetrically, and grating period was 1 μm. The intensity of interference beams was from 160 up to 270 mW/cm^2 at the sample position. In order to monitor the dynamics of grating formation, a red laser beam (λ= 650 nm, power = 1 mW) was sent into the interference area, and the first-order diffraction intensity was measured as a function of time, indicating the formation of diffraction grating.

In this work, different gratings were experimentally realized and polarization dependence in the formation of SRG was investigated. We studied the formation of gratings with different polarization configurations. So we considered three polarization configurations:

1. S:S parallel linear polarizations;
2. +45°:-45° orthogonal linear polarizations with the polarization planes oriented at an angle of 45°;
3. RC:LC right and left orthogonal circular polarizations.

The resulted polarization distribution in the interference pattern is presented in Table 1.

For S-S polarization configuration, the polarizations are the same for both laser beams resulting in a maximal amplitude modulation (100%) of the intensity interference pattern and interference polarization keeps the same direction.

For +-45 polarization configurations the interference polarization varies periodically between linear and circular forms.

Table 1 The resulted intensity modulation and polarization distribution in the interference pattern for different polarization configurations.

Polarization of recording beams	Resulting electric field vector pattern (x-y plane)					Contrast of interf. pattern
x:	+Λ/2	+Λ/4	0	-Λ/4	-Λ/2	κ
S:S ↕ ↕	•	↕	↕	↕	•	1
+45°:-45° ↗ ↘	↔	○	↕	○	↔	0
RC:LC ○ ○	↔	↗	↕	↘	↔	0.01

For the case of RC:LC polarization configuration, the resultant polarization becomes a linear polarization with constant amplitude, but the polarization direction changes periodically in the xy-plane.

Surface Relief Measurements

After recording atomic-force microscopy (AFM) was applied for relief formation investigations. It was studied the variation of relief height with polarization configuration.

III. RESULTS

Good quality films were obtained after a drying period of 6 hrs. The thickness of spin-coated films determined from the fringe pattern in the interferogram was about 270 nm. UV–Visible transmittance spectrum of the spin-coated EPC:DO thin film is shown at Fig 2.

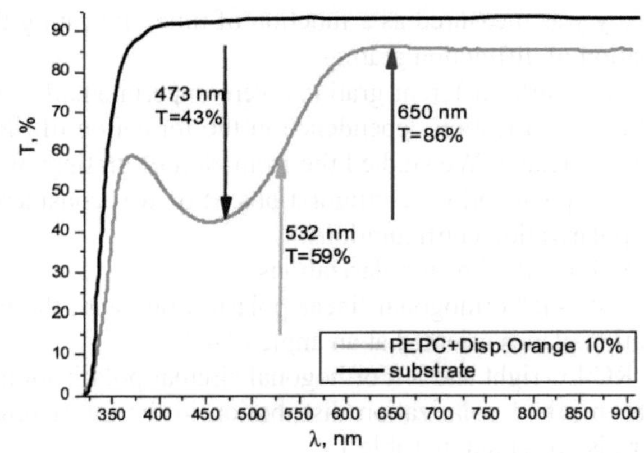

Fig. 2 Transmittance spectra of EPC:DO copolymer thin film and used substrate. The wavelength of used lasers (473 and 532 nm), and the wavelength of 650 nm of monitoring laser during holographic recording are shown insight.

The absorption band ranges from 400 nm to 580 nm with an absorption peak at 460 nm. Therefore a DPSS blue 473 nm and green 532 nm lasers are suitable for the holographic recording. The wavelength of used lasers, and the wavelength of 650 nm of monitoring laser during holographic recording are shown in Fig. 2.

At Fig. 3. the time dependence of the first-order diffraction efficiency of recorded gratings is shown using different polarization configurations. Note that the probing beam (red laser) applied to the film surface during the whole formation process has no influence on the sample structure because its wavelength is out of the absorption band of the DO molecules.

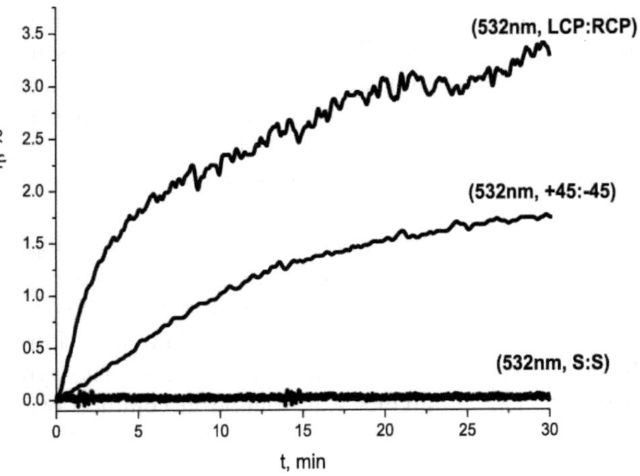

Fig. 3 Time evolution of the first-order diffraction intensity measured at 650 nm in transmission mode.

As it is seen the S:S polarization configuration, providing 100% of intensity modulation, doesn't lead to grating formation. But in the case of +-45° and RC:LC polarization configuration, which does not create intensity modulation, allows to obtain grating with high diffraction efficiency. So the formation of grating on azopolymer therefore depends strongly on the polarization distribution.

The surface relief structure of the gratings on polymer films were investigated by atomic force microscopy (AFM) after the holographic gratings were recorded. At the Fig. 4. the topography, three-dimensional view and profile of surface gratings recorded under +-45° polarization configuration are shown. AFM measurements showed very regularly spaced surface relief structures with a period of 1μm.

The depth of the surface relief patterns was about 80 nm which is approximately 30% of the original film thickness (270 nm).

At the Fig. 5. AFM images of grating recorded under RC:LC polarization configuration are shown.

In this case the depth of the surface relief patterns was about 130 nm which is approximately 48% of the original film thickness.

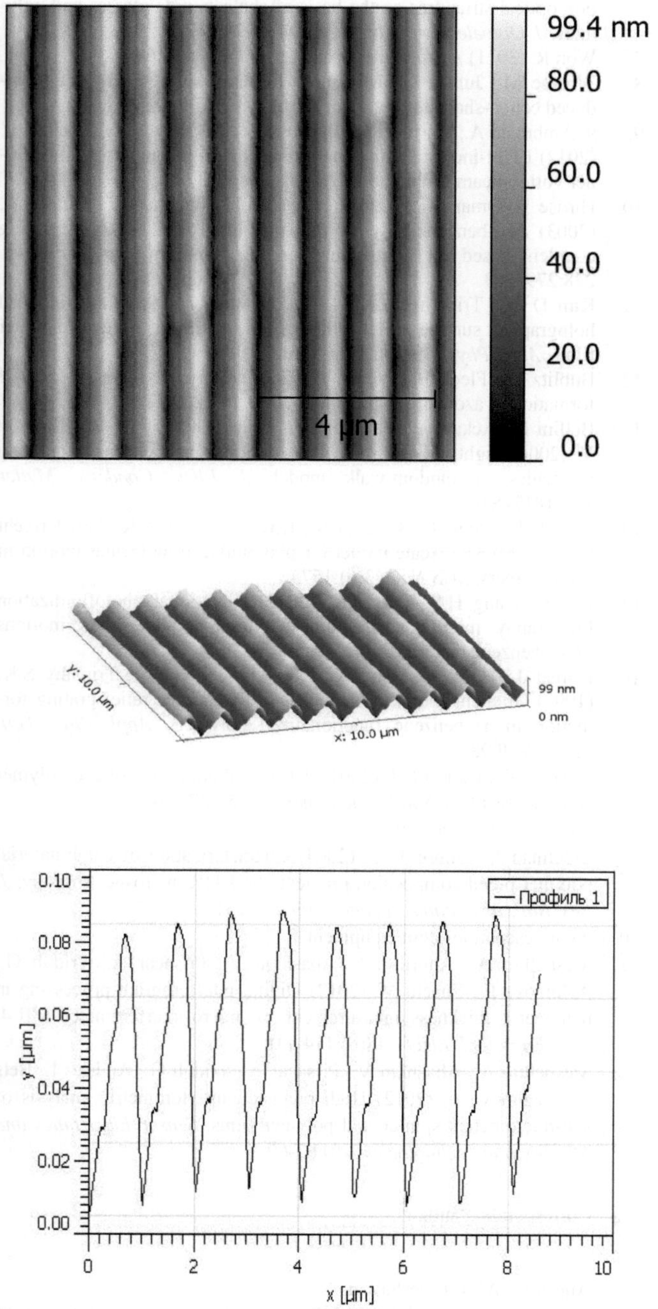

Fig. 4 Topography, three-dimensional view and profile of surface gratings recorded under +-45° polarization configuration.

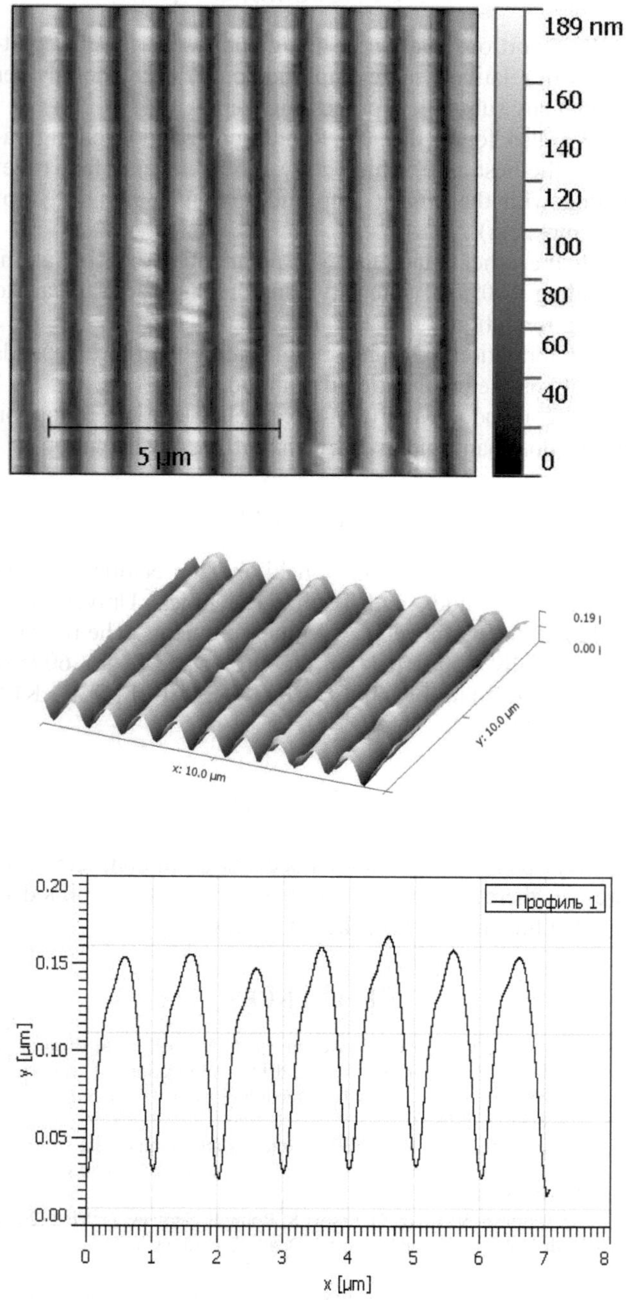

Fig. 5 Topography, three-dimensional view and profile of surface gratings recorded under RC:LC polarization configuration.

IV. Conclusions

We report observation of holographic surface relief gratings with relatively large amplitude on EPC:DO copolymer without any subsequent processing steps.

We also found that the RC:LC polarization configuration is the best case, which allows to achieve surface relief grating with the largest amplitude about 130nm (48% of film thickness).

On the other hand, the S:S polarization configuration, providing 100% of intensity modulation, doesn't lead to grating formation.

The formation of surface relief grating therefore depends strongly on the polarization distribution.

We conclude that the surface relief grating formation plays a dominant role in diffraction efficiency evolution.

Acknowledgments

The authors thank Mihail Enaki and the National Center for Materials Study and Testing (Technical University of Moldova) for performing AFM measurements. The research was partially supported by the project FP7 No 609534 SECURE-R2I and FP7-INCO-2013-9-ENER2I (CEI – KEP AUSTRIA).

Conflict of Interest

Every paper must contain a declaration of conflicts of interest. If there are no such conflicts write "The authors declare that they have no conflict of interest".

References

1. A. Andries, V. Abaskin, E. Achimova, A. Meshalkin, A. Prisacar, S. Sergheev, S. Robu, L. Vlad (2011) Application of carbazole-containing polymer materials as recording media. *Phys. Status Solidi A* 208(8):1837-1840.
2. A. Andriesh, M. Iovu (2009) Diffraction and luminescent structures based on chalcogenide glasses and polymers. *Phys. Status Solidi B* 246(8):1862-1865.
3. A. Andriesh, V. Bivol, O. Ersoy, S. Robu, S. Sergeev, L. Vlad (2004) Both Holographic and Electron-Beam Recording in New Carbazol-Containing Photoresists. *SPIE Proc.* 5581:531-545.
4. J. V. Grazulevicius, P. Strohriegl, J. Pielichowski, and K. Pielichowski (2003) Carbazole-containing polymers: synthesis, properties and applications. *Prog. Polym. Sci*, 28(9):1297–1353.
5. A. Andriesh, V. Bivol, A. Prisacar, S. Sergheev, A. Meshalkin, S. Robu, N. Barba, N. Sirbu (2005) Laser and electron induced structuring of thin films on the base of carbazolyl-containing polymers and polymer-ChGS compositions. *J. Optoelectron. Adv. M.*, 7(3):1169-1178.
6. A. Andriesh, S. Sergheev, G. Triduh, A. Meshalkin (2007) Diffraction optical structures on the basis of chalcogenide glasses and polymers. *J. Optoelectron. Adv. M.* 9(10):3007-3012.
7. Won R. (2011) Azopolymer option. *Nat. Photon.* 5:649.
8. Watabe M., Juman G., Miyamoto K., Omatsu T. (2014) Light induced conch-shaped relief in an azo-polymer film. *Sci. Rep.*, 4:4281.
9. 9. Ambrosio A., Marrucci L., Borbone F., Roviello A., Maddalena P. (2012) Light-induced spiral mass transport in azo-polymer films under vortex-beam illumination. *Nat. Commun.* 3:989.
10. Hirose T., Omatsu T., Kato R., Hoshino K., Watanabe T., Harada K. (2003) Azo-benzene polymer thin-film laser amplifier with grating couplers based on light-induced relief hologram. *Opt. Commun.* 228:279–283.
11. Kim D. Y., Tripathy S. K., Li L., Kumar J. (1995) Laser-induced holographic surface relief gratings on nonlinear optical polymer films. *Appl. Phys. Lett.* 66:1166–1168.
12. Bublitz D., Fleck B., Wenke L. (2001) A model for surface-relief formation in azobenzene polymers. *Appl. Phys. B* 72:931–936.
13. Bellini B., Ackermann J., Klein H., Grave Ch., Dumas Ph., Safarov V. (2006) Light-induced molecular motion of azobenzene-containing molecules: a random-walk model. *J. Phys. Condens. Matter* 18:S1817–S1835.
14. Juan M.L., Plain J., Bachelot R., Royer P., Gray S.K., Wiederrecht G.P. (2009) Multiscale model for photoinduced molecular motion in azo polymers. *ACS Nano* 3(6):1573.
15. Lee S., Kang H.S., Park J.K. (2012) Directional photofluidization lithography: micro/nanostructural evolution by photofluidic motions of azobenzene materials. *Adv. Mater.* 24(16):2069.
16. Kumar J., Li L., L. Jiang X., Kim D.Y., Lee T.S., Tripathy S.K. (1998) Gradient force: the mechanism for surface relief grating formation in azobenzene functionalized polymers. *Appl. Phys. Lett.* 72:2096–2098.
17. Yager K.G., Barrett C.J. (2001) All-optical patterning of azo polymer films. *Curr. Opin. Solid State Mater. Sci.* 5:487-494.
18. www.sigmaaldrich.com
19. Eidelman N., Simon C.G. (2004) Characterization of combinatorial polymer blend composition gradients by FTIR microspectroscopy. *J. Res. Natl. Inst. Stand. Technol.* 109:219-231.
20. www.scscoatings.com/equipment
21. Meshalkin A., Andriesh I., Abashkin V., Prisacar A., Triduh G., Achimova E., Enachi M. (2012) Digital interferogram processing in nanometer thickness measurement by micro-interferometer MII-4. *Surf. Eng. Appl. Elect+* 48(6):114-118.
22. Meshalkin A., Abashkin V., Prisacar A., Triduh G., Andries I., Bets L., Achimova E. (2012) High precision interferometric analysis of sub-micrometers spin-coated polymer films. *Sensor Electronics and Microsystem Technologies* 3(9):62-69.

Corresponding author:

Author: Alexei Meshalkin
Institute: Institute of Applied Physics of the Academy of Sciences of Moldova
Street: 5 Academiei str.
City: Chisinau
Country: Moldova
Email: alexei@asm.md

The Impact of Porosification upon Luminescence of HVPE Grown GaN and the Influence of the Porous Layer upon the Quality of the Overgrown GaN Film

T. Braniste[1], V. Popa[1], D. Martin[2], J.-F. Carlin[2], V. Ursaki[3], N. Grandjean[2], and I. Tiginyanu[1,3]

[1] National Center for Materials Study and Testing, Technical University of Moldova, Chisinau, Moldova
2 Laboratory of Advanced Semiconductors for Photonics and Electronics, ICMP, EPFL, Lausanne, Switzerland
[3] Institute of Electronic Engineering and Nanotechnologies "D. Ghitu" of the Academy of Sciences of Moldova, Chisinau, Moldova

Abstract— **In this paper, we show that the quality of the porous layers produced by photoelectrochemical (PEC) etching of HVPE GaN in oxalic acid is better than that of the initial material, and the quality of GaN films overgrown on porous layers can be improved by adjusting the conditions of PEC etching during the preparation of porous layers. The better quality of the porous layers produced in HVPE material is indicated by the photoluminescence (PL) analysis which shows a higher intensity of the luminescence related to the recombination of free excitons as well as the intensity of the blue luminescence in the porous layers.**

Keywords— **GaN, hotoelectrochemical etching, porous layer, luminescence.**

I. INTRODUCTION

The discovery of luminescent microporous Si [1,2] a quarter of century ago triggered considerable interest in porous semiconductors. Later on, other than Si porous semiconductors have been explored, the most noteworthy being III-V materials such as GaP, InP, GaAs [3,4], since these materials in porous form exhibit new properties with a large potential for applications. The most appropriate, versatile and cost-effective technique for producing porous materials is electrochemical pore etching. Electrochemical pore etching has been also applied to wider bandgap II-VI semiconductors such as p-ZnTe [5], p-$Cd_{1-x}Zn_xTe$ [6], n-CdSe [7], and n-ZnSe [8]. Electrochemical porosification of GaN is even more prospective from the point of view of practical applications, since GaN is a wider bandgap semiconductor. Being a direct bandgap semiconductor compound with E_g=3.4 eV at 300 K, GaN is widely used in industrial production of ultra-violet LEDs and lasers. GaN is a very hard, mechanically stable semiconductor material, transparent in the whole visible region, with high heat capacity, thermal conductivity, and strong piezoelectric effect. The last property is especially important from the point of view of preparation of nanocomposite materials on the basis of porous GaN templates. Previously, metal-semiconductor nanocomposite materials have been prepared by deposition of metal nanowires and nanotubes in InP [9], GaP [10], and ZnSe [11] porous templates. Taking into account the strong piezoelectric properties of porous GaN templates, they are prospective for the elaboration of cost effective magneto-electric sensors with small dimensions able to measure biomagnetic signals in the range of picoTesla on the basis of nanocomposite materials composed of a three-dimensional piezoelectric matrix (GaN) filled with one-dimensional (nanowires) from metals with magnetostrictive properties.

Most of experiments with photoelectrochemical (PEC) etching of GaN have been performed on MOCVD grown GaN layers, exhibiting either pyramidal morphologies with each pyramid consisting of many GaN nanowires merging at the top of the pyramid, or individual upstanding nanoneedles on a dense framework of nanowires being obtained [13]. Recently, porous GaN thin films with different surface morphologies were fabricated from GaN thin films using ultra-violet assisted electrochemical etching approach under various etching voltages, and luminescence and photoelectrochemical properties of these porous films were studied [14].

The goal of this work is to investigate the influence of porosification upon the photoluminescence properties of HVPE grown GaN and to explore the possibilities of improving the quality of GaN layers overgrown on porous substrates produced by PEC etching of MOCVD grown GaN layers.

II. EXPERIMENTAL

Specially non-doped bulk GaN substrates (~200 μm thick) were subjected to porosification in 0.3 M oxalic acid under anodic voltage of 15 V. During anodization the surface was irradiated with UV light from a 350 W Hg lamp focused to a spot of about 5 mm in diameter.

Apart from HVPE grown specimens, 2 μm-thick *n*-GaN layers grown by low-pressure MOCVD on (0001) *c*-plane sapphire substrates were used in our experiments. A buffer layer of about 25 nm-thick GaN was first grown at 510 °C. Subsequently a 2 μm-thick *n*-GaN layer was grown at 1100 °C. The concentration of free electrons was of the order of 10^{17} cm^{-3}, while the density of threading dislocations was in the range of 10^9–10^{10} cm^{-2}. PEC etching of MOCVD grown layers was carried out in a stirred 0.1 mol

V. Sontea and I. Tiginyanu (eds.), *3rd International Conference on Nanotechnologies and Biomedical Engineering*,
IFMBE Proceedings 55,
DOI: 10.1007/978-981-287-736-9_19

aqueous solution of KOH under in-situ ultraviolet (UV) illumination from a 350 W Hg lamp as described above. A bias of 15 V was applied to the sample during etching.

The morphology of samples was studied using a VEGA TESCAN TS 5130MM scanning electron microscope (SEM) and a NANOSTATION Atomic Force Microscope (AFM). The photoluminescence (PL) spectra were measured with a Horiba iHR320 spectrometer under excitation with radiation from a He-Cd cw laser with wavelength of 325 nm. The samples were mounted on the cold station of a closed cycle cryostat.

III. RESULTS AND DISCUSSIONS

SEM images of porous layers produced in HVPE grown GaN substrates after 15 minutes of PEC etching are presented in Figure 1. A 600 nm thick porous layer with a mean pore diameter of 70-80 nm is produced.

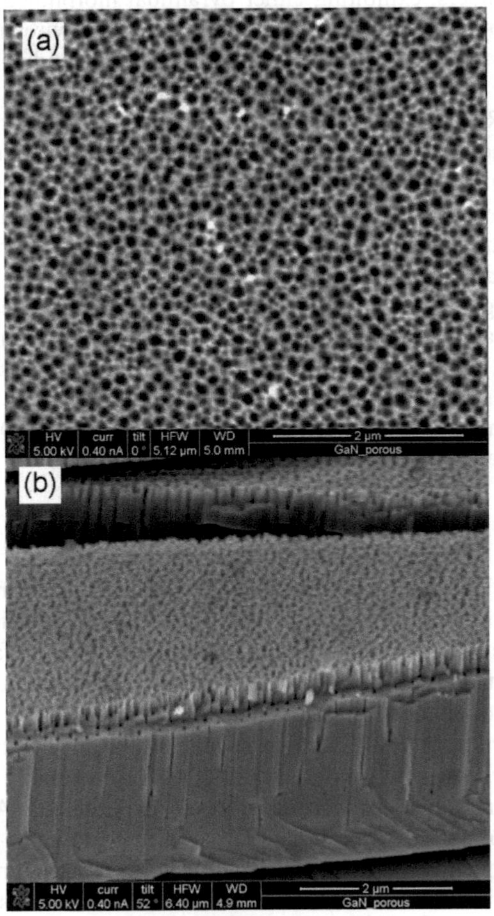

Fig. 1 Top view (a) and cross section view (b) SEM images of porous layers produced in HVPE GaN substrates.

The PL spectra of the bulk and porous HVPE layers measured at various temperatures are compared in Figure 2. In both cases the spectra are dominated by the near band edge emission at 3.471-3,477 eV. Apart from the near band edge emission, an ultraviolet PL band at 3.27 eV with LO replicas at 3.16 and 3.09 eV, a blue PL band at 2.9 eV, and a yellow band at 2.2 eV are observed in the spectra. The analysis of the PL band at 3.27 eV indicates that it is not elementary, but consists of several bands which are usually related to donor-acceptor pair (DAP) transitions [15]. Note that the narrow lines in the spectral range of 2.4 - 2.8 eV come from the laser plasma.

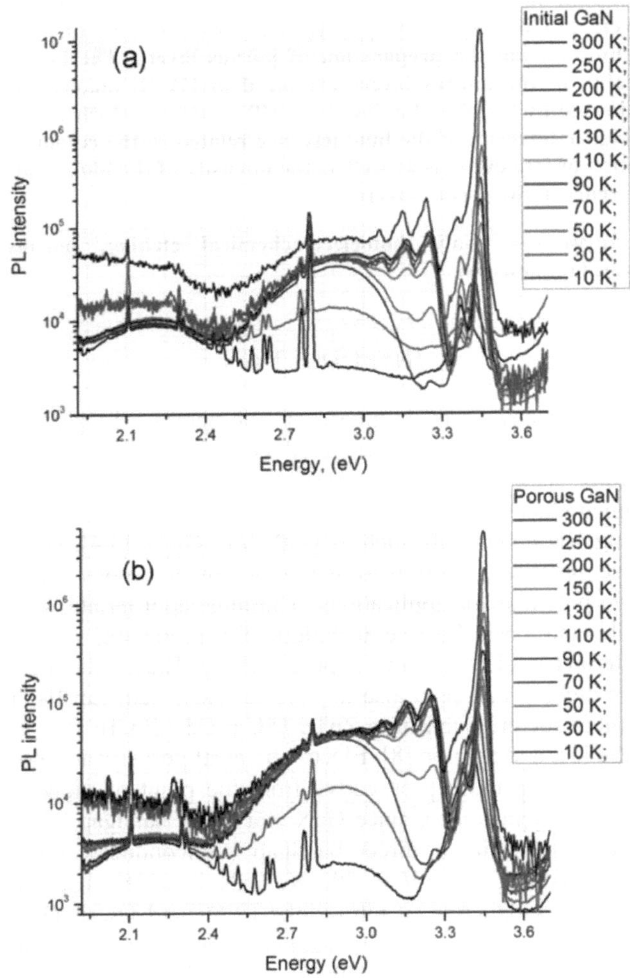

Fig. 2 PL spectra of the bulk (a) and porous (b) HVPE GaN layers measured at various temperatures.

The analysis of the near band edge emission is presented in Figure 3. The spectrum in this spectral range is dominated by the luminescence of excitons bound to neutral donors (D^0X) at 3.471 eV. The position of this line corresponds to that observed in totally unstressed GaN material [16]. On the high-energy side of the D^0X, one can see free exciton emission (XA) at 3.477 eV. On the low-energy side of the D^0X, a weaker luminescence originating from the recombination of acceptor-bound excitons (AX) is observed around 3.450 eV.

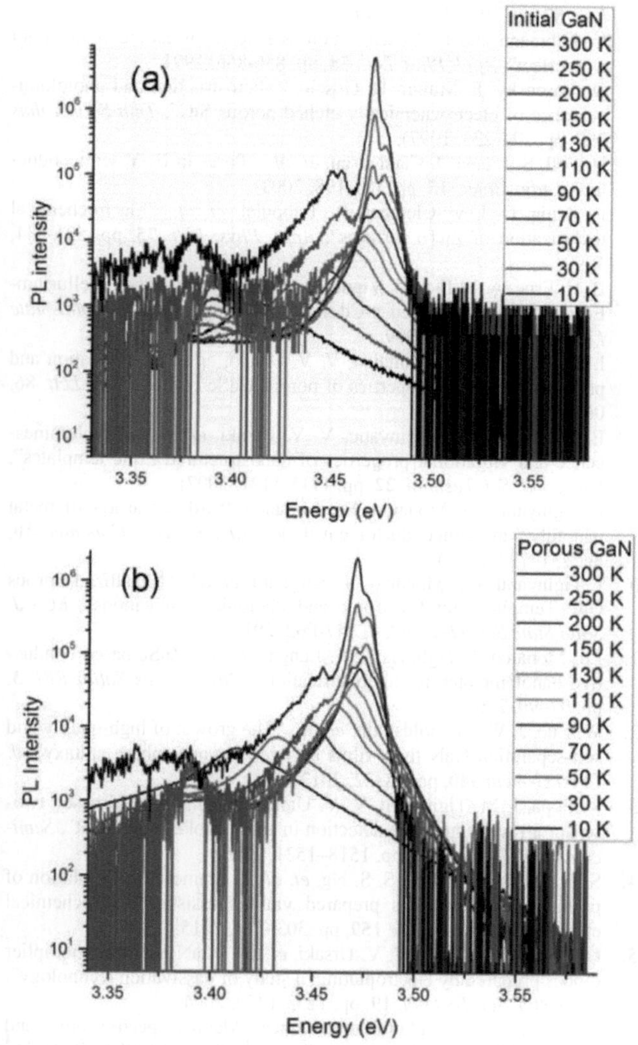

Fig. 3 PL spectra of the bulk (a) and porous (b) HVPE GaN layers measured at various temperatures in the near band edge spectral region.

An interesting observation is the different evolution of the near band edge PL in the bulk and porous layers with increasing temperature. While the intensity of this emission at low temperature (10 K) in the bulk layer is by a factor of 3 higher as compared to the porous layer, it becomes higher in the porous layer at temperatures higher that 70 K. Apart from that, this emission starts to be dominated by the free exciton luminescence at temperature higher thtan 70 K due to the quenching of (D^0X) with increasing temperature, since the binding energy of excitons bound to neutral donors is much lower as compared to the binding energy of free excitons (26 meV).

One can observe that at room temperature the intensity of the free exciton luminescence at 3.415 is by a factor of 5 higher in the porous layer as compared to the bulk material. That means that the quality of the porous material is higher as compared to the bulk one. This assumption is also supported by the analysis of the blue luminescence in Figure 2. One can see that the intensity of the blue luminescence at 2.9 eV is higher in the porous material as compared to the bulk one, while the intensity of the yellow luminescence is much lower in the porous layer. It is widely believed that the transitions between donors and acceptors induced by impurities or native point defects give rise to both yellow and blue luminescence. The available data do not allow us to establish the nature of defects responsible for the blue and yellow luminescence in our layers. However, it was found that the integrated intensities of the two bands may be closely related, and the edge dislocations play an important role in linking the blue and yellow luminescence in undoped *n*-type GaN films. It was established that the ratio of the blue to yellow luminescence intensity decreases with increasing edge dislocation density [17]. This suggests again that the dislocation density in our porous layer is lower as compared to the bulk material.

The angular dependence of PL was also measure at room temperature from the bulk and the porous GaN layers under the incident excitation radiation at an angle of 42°. The dependences are presented in Figure 4. One can observe that the emission from the bulk material is confined in a much narrow angle interval as compared to the porous layer. This observation is explained by the strong scattering of light inside the porous network indicating the strong photonic power of the GaN porous layer.

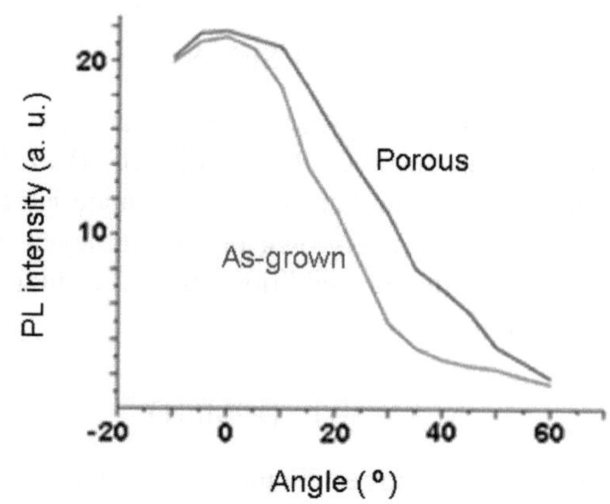

Fig. 4 Dependence of the PL intensity upon the emission angle.

By means of Scanning Electron Microscopy and Atomic Force Microscopy, we investigated also the influence of the porous layer morphology produced by PEC etching of MOCVD grown GaN layers, as described in the experimental section, upon the quality of a GaN film overgrown on the porous GaN layer. It was found that the quality of the film overgrown on the porous layer produced by etching during 5 minutes is better as compared to that overgrown on the initial layer, while the quality of the film overgrown on the layer produced by etching during 10 minutes is worse.

Figure 5 compares the near band edge PL spectrum of the initial GaN layer with those measured from porous layers produced by PEC etching during 5 and 10 minutes. This emission band is due to the excitons bound to neutral donors (D^0X) as described above. However, its position is shifted to higher photon energy by 10 meV as compared to non-stressed layers. That means than significant tensile strain is inherent to the initial MOCVD grown GaN layer. However, this stress is significantly relaxed in porous layers as indicated by the shift of the PL band to lower energies.

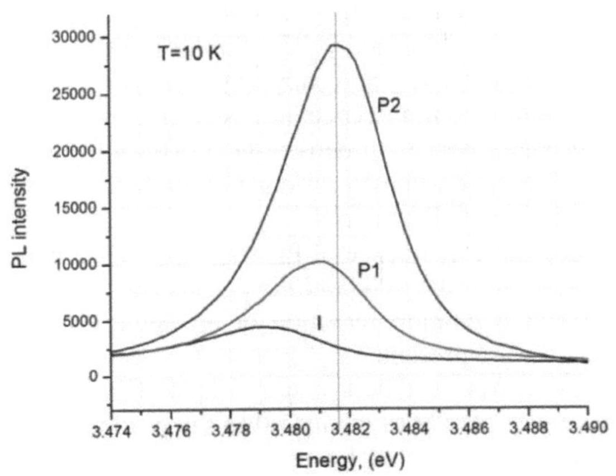

Fig. 5 Near band edge PL spectra of the initial GaN layer (I), and of porous layers produced by PEC etching during 5 minutes (P1), and 10 minutes (P2).

IV. CONCLUSIONS

The results of this study demonstrate that the quality of the porous layer produced by PEC etching of HVPE GaN in oxalic acid is better than that of the initial material, and the quality of GaN films overgrown on porous layers can be improved by adjusting the conditions of PEC etching during the preparation of porous layers.

ACKNOWLEDGMENT

This work was supported by the Swiss National Science Foundation under the SCOPES Programme (project no IZ73ZO_152273).

CONFLICT OF INTEREST

The authors declare that they have no conflict of interest.

REFERENCES

1. L. T. Canham L. T., "Silicon quantum wire array fabrication by electrochemical and chemical dissolution of wafers", *Appl. Phys. Lett.* **57**, pp. 1046-1048, 1990.
2. V. Lehmann and U. Gosele, "Porous silicon formation: A quantum wire effect", *Appl. Phys. Lett.* **58**, pp. 856-856, 1991.
3. O. Jessensky, F. Muller, U. Gosele, "Microstructure and photoluminescence of electrochemically etched porous SiC", *Thin Solid Films* **297**, pp. 224-228, 1997).
4. H. Foll, S. Langa, J. Carstensen, *et. al.*, "Pores in III-V semiconductors", *Adv. Mater.* **15**, pp. 183-198, 2003).
5. F. Zenia, C. Levy-Clement, R. Triboulet, *et. al.*, "Electrochemical texturization of ZnTe surfaces", *Appl. Phys. Lett.* **75**, pp. 531-534, 1999.
6. B. H. Erne, A. Million, J. Vigneron, *et. al.* "Porosity and Tellurium-Enrichment of Anodised p-$Cd_{0.91}Zn_{0.01}Te$", *Electrochem. Solid-State Lett.* **2**, pp. 619-621, 1999.
7. I. M. Tiginyanu, E. Monaico, V. V. Ursaki, *et. al.*, "Fabrication and photoluminescence properties of porous CdSe", *Appl. Phys. Lett.* 86, 063115, 2005.
8. E. Monaico, I. M. Tiginyanu, V. V. Ursaki, *et. al.*, "Photoluminescence and vibrational properties of nanostructured ZnSe templates", *Semicond. Sci. Technol.* **22**, pp. 1115-1121, 2007).
9. I. Tiginyanu, El. Monaico, Ed. Monaico, "Ordered arrays of metal nanotubes in semiconductor envelope", *Electrochem. Commun.* **10**, pp. 731–734, 2008).
10. I. Tiginyanu, Ed. Monaico, V. Sergentu, *et. al.*, "Metallized Porous GaP Templates for Electronic and Photonic Applications", *ECS J. Solid State Sci &Technol*, **4**, P57-P62, 2015.
11. Ed. Monaico, I. Tighineanu, S. Langa, *et. al.*, "ZnSe-based conductive nanotemplates for nanofabrication", *Phys. Status Solidi RRL* **3**, pp. 97–99, 2009.
12. W. Luo, J. Wu, J. Goldsmith, *et. al.*, "The growth of high-quality and self-separation GaN thick-films by hydride vapor phase epitaxy", *J. Cryst.Growth* **340**, pp. 18–22, 2012.
13. V. Popa, I. M. Tiginyanu, V. V. Ursaki, *et. al.*, "A GaN-based two-sensor array for methane detection in an ethanol environment", *Semicond. Sci. Technol.* **21**, pp. 1518–1521, 2006.
14. S. F. Cheah, S. C. Lee, S. S. Ng, *et. al.*, "Luminescence evolution of porous GaN thin films prepared via UV-assisted electrochemical etching, *J. Luminescence* **159**, pp. 303–311, 2015.
15. O. Cojocari, V. Popa, V. V. Ursaki, *et. al.*, "GaN Schottky multiplier diodes prepared by electroplating: a study of passivation technology", *Semicond. Sci. Technol.* **19**, pp. 1273-1279, 2004.
16. B. J. Scromme, "Photoluminescence, Magnetospectroscopy, and Resonant Electronic Raman Studies of Heteroepitaxial Gallium Nitride", *MRS Internet J. Nitride Semicond. Res.* **4**, 15, 1999.
17. D. G. Zhao, D. S. Jiang, J. J. Zhu, *et. al.*, "Role of edge dislocation and Si impurity in linking the blue luminescence and yellow luminescence in *n*-type GaN films", *Appl. Phys, Lett.* **95**, 041901, 2009.

Author: Veaceslav Ursaki
Institute: Institute of Electronic Engineering and Nanotechnologies
Street: 3/3 Academiei str.
City: Chisinau
Country: Moldova
Email: ursaki@yahoo.com

Chronic Toxicity of Silver Nanoparticles

I. Pavlovschi, V.Goncear, and C. Scutari

State University of Medecine and Pharmacy "Nicolae Testemiţanu", Chisinau, Republic of Moldova

Abstract— **The unique physical and chemical properties of silver nanoparticles make them excellent candidates for a variety of purposes in various fields, but the rhythm of using nanoscale materials is rapidly growing and possible ecotoxicological effects, are still unrated exactly. It requires determining the prospects for silver nanoparticles applications in medicine as one of the most attractive trading nanomaterials for applications. This paper attempts to provide a comprehensive understanding of the effects and the possible toxicological considerations of the silver nanoparticles both for animals and for human body.**

Keywords— **silver nanoparticles, chronic toxicity, nanomaterials.**

I. INTRODUCTION

During the last years, the explosion of emerging infectious diseases has become a significant burden on the global economy and public health. The transmission of infectious pathogens in the community caused epidemics such as grippe (A/H5N1), diarrhea (Escherichia coli), cholera (Vibrio cholera), and etc. all around the world. Complex treatments of the environments which contain infectious pathogens were proposed to prevent epidemics, using advanced disinfectant nanomaterials. Among such materials are silver nanoparticles (Ag-NPs) with unique properties of high antimicrobial activity, that has attracted a particular interest from scientists.

Silver's medical properties have been known for more than 2000 years. Currently, Ag-NPs are used as antibacterial / antifungal / antiviral / anti-inflammatory agents, in biotechnology and bioengineering, textile engineering, electronics, bio-sensing, food industry, paints, sunscreens, cosmetics, medical devices, in processing water, as well as consumer products [1, 6].

An enhanced interest presents the inclusion of Ag-NPs in a wide range of medical devices, bone matrix (bone cement), surgical instruments, surgical masks, wound dressings, antibacterial creams, cardiovascular implants, catheters, dental materials, in biodiagnostics.

The silver nanoparticles are particles with size from 1 nm to 100 nm. As particle's size decreases, surface to volume ratio of Ag-NPs increases dramatically, leading to significant changes in physical, chemical and biological properties.

The unique physical and chemical properties of Ag-NPs make them excellent candidates for a variety of purposes in the medical field, but the rhythm of using nanoscale materials is rapidly growing and their potential ecotoxicological effects are still not evaluated precisely. There are studies and reports that suggest that silver nanoparticles could cause adverse effects on humans and on the environment [2].

Although these studies tend to suggest that silver nanoparticles can induce toxicity in live beings, it must be taken into account that Ag-NPs toxicity studies were performed in conditions in vitro that are drastically different from those in vivo and at very high concentrations of the Ag-NPs.

II. MATERIALS AND METHODS

To achieve the objectives of the study was designed a preclinical study to determine the chronic toxicity of Ag-NPs colloidal solution.

The studies were performed in the Department of Pharmacology and Clinical Pharmacology of the State University of Medicine and Pharmacy "Nicolae Testemiţanu" and at the Vivarium.

For the realization of the experimental compartment were used for the determination of chronic toxicity 30 white mice, females, aged 2-2.5 months, divided by 10 per each group. The following doses were tested: I group: $LD50 = 30$ mg / kg, group $II = 10$ mg / kg, with daily intraperitoneal administration, $III =$ control group, with further supervision of the animals. The source of silver nanoparticles was commercial.

The occurred changes were submitted to observations. It was recorded the time of appearance of the phenomena of intoxication and death of animals. Visually was determined the clinical picture of intoxication. It was also determined the influence on erythropoiesis and on the peripheral blood [Fig.1].

The animals were kept in conditions close to the physiological optimum. Taking into consideration the importance of the circadian biological rhythms, seasonal etc., at the change of the content, structure and distribution of intracellular components of different tissues, experiments were performed under standard conditions [3, 4].

The researches were conducted according to "Methodical Recommendations on general toxicity study of pharmacological substances", approved by the Pharmacological

V. Sontea and I. Tiginyanu (eds.), *3rd International Conference on Nanotechnologies and Biomedical Engineering*,
IFMBE Proceedings 55
DOI: 10.1007/978-981-287-736-9_20

Committee of the HM of Russian State (2005) [5] and the institutional and national guide for the care and use of laboratory animals was followed.

Fig. 1 The evaluation of the pathological condition of the mice within the administration of Ag-NPs

III. RESULTS

The determination of the chronic toxicity aimed to establish the biological changes caused by the analyte in animals after repeated administration for 21 days.

The aim of the chronic toxicological experiments consisted in the characteristic of the harmful effect degree of the colloidal solution of silver nanoparticles in its long-term administration, identification of the most sensitive organs and systems and the study of the degree of reversibility of the damage caused by this.

The duration of administration of pharmacological substance in the chronic toxicity study depends on the expected duration of clinical application.

The aim of the chronic toxicological experiments consisted in the characteristic of the harmful effect degree of the colloidal solution of silver nanoparticles in its long-term administration, the identification of the most sensitive organs and systems and the study of the degree of reversibility of the damage caused by it.

The duration of administration of pharmacological substance in the chronic toxicity study depends on the expected duration of clinical application.

During the administration period (21 days), were highlighted deviations in the condition of the animals in varying degrees. Mice mostly were in active movement and were using actively food and water. The condition of the mucous membranes and skin was normal, without pathological changes. The urine color deviate from yellow to pink with blood threads.

The results showed that the mice did not presented significant changes in body weight reported to the Ag-NPs dose during the 21-day experiment.

Daily administration of the preparation (of the silver colloidal solution of nanoparticles) at a dose of 10 mg / kg induced no death of the animals. Mice lethality in dose of 30 mg / kg was 10% and occurred in the third week of administration of the substance [Table 1].

Table 1 Chronic toxicity. The lethality of the white mice (females) during the experiment at intraperitoneal administration of colloidal solution of silver nanoparticles

Groups	Dosage of the subst. mg/kg	The weeks of administration			Died	Lethality (%)
		1	2	3		
I	30	0	0	1	1	10
II	10	0	0	0	0	0
III	control	0	0	0	0	0

Ag-NPs influence on erythropoiesis and peripheral blood was tested three times: at onset of the study, at 7th day and 21st day of the experiment.

Erythropoiesis and peripheral blood were evaluated by the following indices: RBC content in a unit volume (10.12/L), total hemoglobin content in the blood (g/l) and hemtocritic index (% by volume of red cells in the blood).

The change of the number of red blood cells in peripheral blood is shown in Table 2. After analyzing the data presented in the table, continuous administration of the Ag-NPs solution induces some changes in this index. Thus, there is an increase in the number of erythrocytes in peripheral blood in the second week, that is maintained at the end of the experience.

But Ag-NPs did not essentially influence the content of erythrocytes in the peripheral blood.

Table 2 The changing of the content of the erythrocytes in the peripheral blood (10^{12}/L) after repeated administration of the solution of Ag-NPs

The administered dose	1st day of the study	On the 7th day of the study	On the 21st day of the trial
I group	8,53	9,87	10,01
II group	8,71	9,73	9,87
III group	8,65	8,58	8,61

Data obtained in the study of the hemoglobin content in the peripheral blood in the administration of the Ag-NPs solution are presented in Table 3. There was a growing tendency of the hemoglobin content toward the end of the experience - by 21st day. The increase in the hemoglobin content correlates with the number of red blood cells in the peripheral blood.

Table 3 The changing of the content of hemoglobin in the peripheral blood (g/L) in response to repeated administration of the Ag-NPs solution

The administered dose	1st day of the study	On the 7th day of the study	On the 21st day of the trial
I group	152	155	159
II group	145	157	162
III group	148	146	150

Data about the changing of the total content of leukocytes in peripheral blood are presented in Table 4. As shown, Ag-NPs did not essentially influence the total number of leukocytes, these manifesting only deviations, although in decreasing, however within the physiological limits, characteristic for the index.

Thus, all the oscillations of the total number of leukocytes have been to the amount of 7-15 × 10.9/l. Ag-NPs causes a slight decrease in white blood cell count to the end of the experience.

Table 4 The changing of the content of leukocytes in peripheral blood (109/L) after administration of the Ag-NPs solution

The administered dose	1st day of the study	On the 7th day of the study	On the 21st day of the trial
I group	10,02	8,42	7,04
II group	10,20	8,93	7,71
III group	9,95	**10,21**	10,04

According to the table data, the most notably changes suffer neutrophils and lymphocytes [Table 5]. The general law of the Ag-NPs is decreasing of the percentage content of neutrophils with the concomitant increase in the percentage of lymphocytes.

Table 5 The changing of the absolute content of neutrophils in the peripheral blood (%) after administration of the Ag-NPs solution

The administered dose	1st day of the study	On the 7th day of the study	On the 21st day of the trial
I group	4,19	3,50	1,92
II group	4,10	2,90	1,13
III group	4,00	4,21	4,12

Among the mononuclear leukocytes, essential changes suffer only lymphocytes [Table 6]. Ag-NPs administration causes an increase in the percentage of lymphocytes in the peripheral blood that is maintained throughout the experience.

Table 6 The changing of the absolute lymphocyte content in peripheral blood (%) after administration of the Ag-NPs solution

The administered dose	1st day of the study	On the 7th day of the study	On the 21st day of the trial
I group	75,7	77,9	78,8
II group	75,8	76,3	78,1
III group	75,6	75,8	75,7

The increase in the percentage of lymphocytes in the peripheral blood along with the increase of the absolute lymphocyte count would be a consequence of the regeneration stimulation and the mobilization in blood of the lymphocytes from the lymphoid tissue. This would suggest the fact that Ag-NPs have a stimulant role to the lymphocitopoesis and specific immunity.

IV. DISCUSSION

Although there are many mechanisms that would explain the activity of silver nanoparticles, they are not fully understood and therefore cannot be generalized acting in different organisms in different ways.

In case of chronic toxicity the obtained results suggest only a relative harmlessness of the Ag-NPs, requiring a further study in the field.

V. CONCLUSIONS

1. Daily administration of the preparation (of the colloidal solution of silver nanoparticles) in a dose of 10 mg / kg induced no death of the animals. The lethality of the mice at a dose of 30 mg / kg was 10% and this occurred in the third week of administration of the substance.

2. Ag-NPs has multiple biological properties: stimulates the synthesis and increase the total hemoglobin content in the blood; inhibits migration and reduces the content of neutrophils in the peripheral blood; stimulates lymphocitopoesis and increase the blood lymphocyte content in the peripheral.

CONFLICT OF INTEREST

The authors declare that they have no conflict of interest.

REFERENCES

1. Ahmed M, Al Salhi MS, Siddiqui MK et al. (2010) Silver nanoparticle applications and human health. Clin Chim Acta 14:1841-1848.
2. Quang HT, Van Quy N, Anh-Tuan L (2013) Silver nanoparticles: synthesis, properties, toxicology, applications and perspectives. Adv. Nat. Sci.: Nanosci. Nanotechnol 4:2043–6262.
3. Şteflea D (1984) Reflectări cronobiologice în medicină. Ediţie medical, Bucureşti.
4. Комаров Ф.И. (1989) Хронобиология и хрономедицина, Москва.
5. Методические рекомендации по изучению общетоксического действия фармакологических средств (2005), Москва, МЗРФ.
6. http://en.wikipedia.org/wiki/Silver_nanoparticles

Author: Ivan Pavlovschi
Institute: State University of Medecine and Pharmacy "Nicolae Testemiţanu"
Street: Ştefan cel Mare şi Sfânt street, 165
City: Chisinau
Country: Republic of Moldova
Email: ian_pavlovski@mail.ru

Effect of Spin Coating Technique on Mechanical Properties of Silicophosphate Thin Film Doped by Neodymium

D. Grabco[1], O. Shikimaka[1], M. Elisa[2], B. Sava[2], L. Boroica[3], E. Harea[1], C. Pyrtsac[1], A. Prisacaru[1], I. Feraru[2], Z. Barbos[1], and Ia. Vreme[1]

[1] Institute of Applied Physics, Chisinau, Moldova
[2] Institute of Optoelectronics Bucarest-Magurele, Romania
[3] National Institute for Laser, Plasma & Radiation Physics, Romania

Abstract— **This work is focused on the effect of spin coating technique on the mechanical properties of silicophosphate (SP) thin films belonging to system SiO2-P2O5-Nd2O3. The thin films have been obtained by spin coating technique for three rotation speeds: 2000, 3500 and 5000 rpm. A soda-lime-silicate (SLS) glass was used as a substrate. Strength characteristics (Young modulus, E, hardness, H, and plasticity index, H/E) of the SiO2-P2O5-Nd2O3 films were determined by using the dynamical nano/microindentation under Pmax=10, 30, 50, 100, 900 mN. As a result of the fulfilled studies, it was drown a conclusion: new materials with regulable mechanical properties (elastic modulus, hardness, plasticity, brittleness) can be fabricated for practical applications.**

Keywords— **Silicophosphate thin films, spin coating technique, mechanical properties, nanoindentation.**

I. INTRODUCTION

Silicophosphate thin films containing Nd3+ ions constitute active elements in multilayer planar waveguides for integrated optics [1]. Integrated lasers and amplifiers are mainly linked to rare earth doping level, which must be maximized, without quenching the fluorescence of the doped material. Neodymium doped SiO_2-TiO_2 planar waveguides have been obtained by sol-gel method, spin coating technique, and fluorescence lifetime has been measured [2].

The sol–gel route applied to prepare luminescent films possesses a number of advantages over conventional film formation techniques, i.e., low processing temperature, easy coating of large surfaces, possible formation of porous films and homogeneous multicomponent oxide films, simple and cheap experimental setups [3].

At the same time, mechanical properties of these materials are not always adequate to the technical requirements [4]. Therefore, it is necessary to study the mechanical parameter dependence on the manufacturing conditions, which can influence the elasticity, plasticity and brittle characteristics of fabricated coating/substrate structures. As such a manufacturing condition, the influence of rotation velocity in spin coating technique is considered in present paper.

II. EXPERIMENTAL

The synthesis and characterization of silicophosphates prepared by sol-gel method, using different SiO_2:P_2O_5 ratios as well as different precursors for these oxides have been the subject of recent studies [5-8].The preparation of sol-gel films started from tetraethoxysilane (TEOS) as precursor for SiO_2, triethylphosphate (TEP) as precursors for P_2O_5 and $NdCl_3$ as precursor for Nd_2O_3. The amount of reagents corresponds to the following molar ratios of the final oxide composition: P_2O_5/SiO_2 = 1/5; H_2O/TEOS = 1; $P_2O_5/NdCl_3$ = 5.4; SiO_2/ $NdCl_3$= 27; C_2H_5OH/TEOS = 1. The prepared sol had a single homogeneous and clear phase at pH about 4.5, which is efficient for the gelification of silicophosphate sols [9]. The thin films have been obtained by deposition of the sol onto a soda-lime-silicate (SLS) glass substrate by spin coating technique for three rotation speeds: 2000, 3500 and 5000 rpm. The films obtained by the deposition of $3\cdot10^{-3}$ L sol drops were dried in air at room temperature and then their mechanical properties were characterized. The thickness of the obtained films on the substrate surface varied within the range t=3÷5 μm.

Strength characteristics (Young modulus and dynamical nanohardness and microhardness) of the SiO_2-P_2O_5-Nd_2O_3 thin films were studied by using the Nanotester PMT-3-NI-02 equipped with a trihedral diamond Berkovich pyramid as an indenter. The determination of the hardness and Young modulus was made from the nano/microindentation curves using the Oliver-Pharr method [10]. Five maximum loads, P_{max}, were used for dynamical indentation: 10, 30, 50, 100 and 900 mN. The investigations were carried out on the samples aged during two years in ambient conditions after their preparation.

III. RESULTS

The AFM and light microscopy analysis revealed that Nd-doped silicophosphate films are relatively uniform and exhibit a smooth surface, these features depending on the spinning rate of the substrate. SEM investigations of the films for all three spinning speeds put in evidence low-sized

V. Sontea and I. Tiginyanu (eds.), *3rd International Conference on Nanotechnologies and Biomedical Engineering*,
IFMBE Proceedings 55,
DOI: 10.1007/978-981-287-736-9_21

units, about 0.1-0.2 μm, randomly distributed on the surface, and micro-units with round shapes, as well.

Each type of samples was subjected to mechanical tests by applying different dynamic nanoindentation maximum load (P_{max}). Five indentations for each P_{max} were made. The following mechanical parameters: the film nanohardness (H_f), the nano-microhardness of the film/substrate composite structure (H_{sc}) and Young's modulus (E_f, E_{sc}) were determined by analyzing the 'load-displacement' curves. (Tab.1).

From the table it follows, that (i) $E_f < E_{sc}$ and $H_f < H_{sc}$ for all rotation speeds; (ii) $H_{2000} < H_{3500}$, H_{5000} – for P_{max}= 10 mN; $H_{2000} > H_{3500}$, H_{5000} – for P_{max}=900 mN; such a tendency is also observed for E_{sc} values; E_f does not undergo significant changes with rate. In turn the analysis of 'load-displacement' curves provides information concerning the mechanism of plastic and brittle deformation of these complex structures.

The 'load-displacement' curves for the indentations made with two different loads (P_{max}) are presented in Fig.1. As it is seen from the shape of the curves, the deformation of studied materials develops according to common regularities for all rotational speeds (v=2000, 3500 and 5000 rpm) used for the preparation of CS-1, CS-2, CS-3 systems.

An inflection appears on all curves for P_{max}=10mN at the load $P_i \approx 1.5$-2.0 mN, when the indenter reaches a depth $h_i \approx 100$-150 nm. The inflection separates the curves into two parts, with different slopes φ. For example, φ'= 25°, whereas φ"= 40° for CS-1system (Fig. 1a) and φ'<φ" for all samples, CS-1, CS-2, CS-3, i.e. for all speeds (Fig. 1a,c,e). This fact indicates that at the initial stage of the loading process in the surface layers (to a depth of 100-150 nm) the material is easier deformed plastically. When the load increases above $P_i \approx 1.5$-2.0 mN and hence the indenter penetrates deeper into the film, the resistance of the samples increases, causing a change in the slope of the curve and an increase of the angle φ.

Table 1 Modification of Young modulus (*E*) and hardness (*H*) with the coating speed and the indentation load

Sample	v, rpm	E_f, GPa	E_{sc}, GPa	H_f, GPa	H_{sc}, GPa
		P_{max}, mN			
		10	900	10	900
CS-1	2000	52	74	3.0	6.3
CS-2	3500	54	61	4.2	5.8
CS-3	5000	52	62	4.1	6.1

As follows from Fig. 1a,c,e, the maximum depth of the indentations obtained at a load of 10 mN reaches a value of 0.3-0.4 μm, which is about 1/10 of the film thickness, and in accordance with the literature data [11], the hardness H_f is a hardness of the SiO_2-P_2O_5-Nd_2O_3 film itself, without the in-fluence of the substrate. Tab. 1 shows that H_f of the samples CS-2 and CS-3 is higher than H_f of the CS-1, i.e. the increase of film deposition speed increases its hardness.

The deformation of the samples with other maximum loads was also accompanied by an increase of the material resistance to the indenter penetration; however, this process took place more smoothly, without inflections or 'pop-in' effects on the *P-h* curves (Fig. 1b,d,f). This indicates that the indentation formation occurs homogeneously without

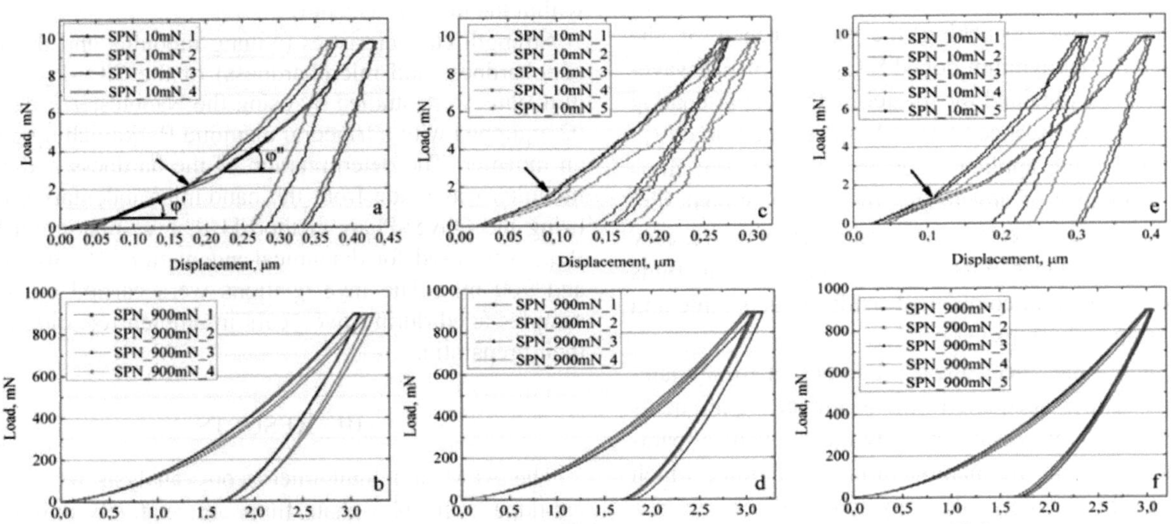

Fig. 1 Load–displacement curves for the SiO_2–P_2O_5–Nd_2O_3/SLG samples at two maximum loads, mN: (a,c,e) 10 and (b,d,f) 900 and three coating speeds, rpm: (a,b) – 2000, (c,d) – 3500, (e,f) – 5000. The inflections of curves at the loading stage are marked by arrows; φ' and φ" are the inflection slopes

brittle fracture. Note that the maximum depth of the indentations obtained by the highest load of 900 mN, reaches values of 3.0-3.5 μm, i.e., $\beta=h/t\approx1.0$. In this case, the measured hardness is a composite hardness, H_{sc}, which reflects the cumulative properties of the film and substrate. As can be seen from Tab. 1, in contrast to H_f, H_{sc} of the CS-1 sample was slightly higher than that of the CS-2 and CS-3. Note that the indentation depth at 900 mN reaches the boundary of the film/substrate interface. It is possible that in this case the degree of adhesion of the film to the sub-strate may have a certain influence on the hardness value: weak adhesion may be accompanied by some film delamination from the substrate, which in turn will increase the indentation depth, and as a result, will decrease the *H*.

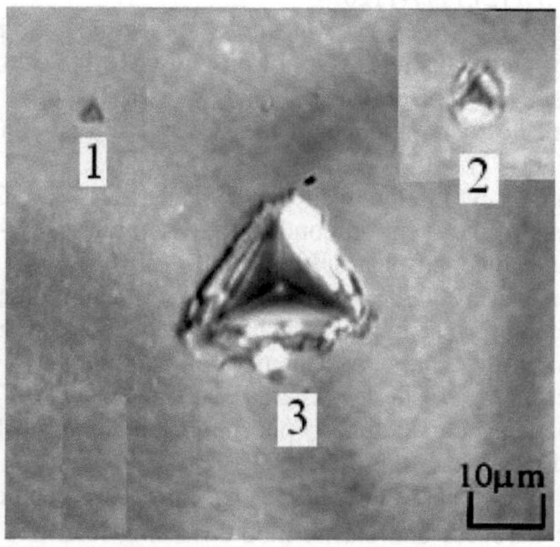

a

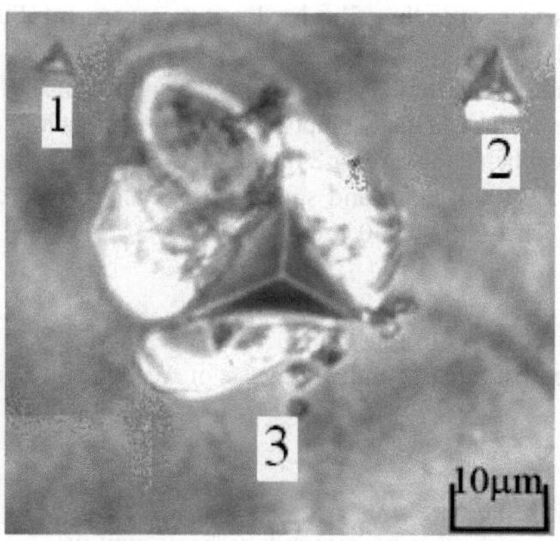

b

Fig. 2 The SiO_2-P_2O_5-Nd_2O_3/SCS coating structures obtained by spin coating technique at two rotation speeds: (a) 2000, (b) 3500 rpm. The shape of indentations made with different loads. *P*, mN: 1(a) – 10; 1(b) – 30; 2 – 100; 3 – 900

This assumption was confirmed by the analysis of the indentation microstructure. From Fig. 2 one can see that at low loads, 10 and 30 mN, the indentations and the surface relief in its neighborhood have a plastic aspect of deformation for both samples CS-1 and CS-2. But at higher loads, 50, 100, 900 mN, the destructions appear nearby the indentations caused by the delamination of the film from the substrate; at the same time the indentations keep a plastic aspect inside. The pronounced fragility of the films deposited at v=3500 and 5000 rpm influences on the values of hardness and elasticity of the samples that is confirmed by experimental data (Tab.1).

However, as it was mentioned above the 'load/displacement' curves (see Fig. 1b,d,f), do not show any pronounced 'pop-in' or 'pop-out' effects. This fact argues that the film delamination and the destruction of surface around the indentations took place not in the process of indentation, but 'post-factum', after complete removing of the load and had a relaxation character. It is possible that the destruction process begins at the loading stage by the formation of microcracks and local microdeliminations, which were not registered on the curve, but contributed to the decrease of the composite hardness value.

So, it was determined that at the loads P_{max}=50÷900 mN the Young modulus and hardness values of the films obtained under the velocity 3500 and 5000 rpm are somewhat lower than those ones of the film obtained under 2000 rpm. Microdestruction of indentations results in the increase of their size that leads to the reduced hardness.

As it is known, the value of Young's modulus characterizes the elastic properties of the material and hardness reflects the elastic-plastic ones.The obtained results showed that the SiO_2-P_2O_5-Nd_2O_3/SCS CS possess different elastic-plastic and brittle properties by changing the fabrication technique. In recent years along with the conventional parameters of the *E* and *H*, an other characteristic, the so-called 'plasticity index', is considered for the evaluation of material resistance to plastic deformation [12,13]. It is expressed as the ratio of hardness to its elastic modulus, *H/E*. We carried out a comparison of this parameter for the CS-1, CS-2 and CS-3 systems.

Table 2 The plasticity indexes, *H/E*, as a function of coating speed for composite systems: CS-1, CS-2, CS-3

Sample	v, rpm	H_f/E_f	H_{sc}/E_{sc}
CS-1	2000	0.058	0.085
CS-2	3500	0.078	0.095
CS-3	5000	0.079	0.098

Tab. 2 shows the values of the *H/E* parameters. As follows from the data presented, the plasticity index value increases with increase of the applied load: $H_f/E_f<H_{sc}/E_{sc}$. This means that the resistance of material to plastic deformation increases with increasing of indentation load applied to all CSs, due to the influence of harder substrate, and this is manifested as their common property.

On the other hand, the plasticity index increases with the increase of speed as well, which indicates the difference between the elastic-plastic properties of the structures obtained through the modification of technology. The composite created by using lower speed (v = 2000 rpm) appears to be more plastic. The increasing of speed (v = 3500; 5000 rpm) is accompanied by an increase of the *H/E* value, which indicates to higher resistance of the systems CS-2 and CS-3 to the action of the external concentrated load.

IV. CONCLUSIONS

Based on the presented results we can make the following conclusions. The obtained results have demonstrated that the modification of the manufacturing technology of the SiO_2-P_2O_5-Nd_2O_3/SCS composite by varying the rotation speed of film deposition allow obtaining materials with different mechanical properties. It was shown that the application of low loads (P_{max}=10 mN) gives information about the film behavior that is determined by mechanical properties of the film itself. Deformation with higher loads (P_{max}=900 mN) characterizes the film/substrate system as a whole. Increasing of the rotational speed results in the H_f and *H/E* growth, at the same time the fragility of the film increases to a certain extent. The value of the film Young's modulus, E_f, basically remains constant for all speeds and E_{sc} slightly decreases at high speeds. So, the mechanical parameters (*E*, *H*, *H/E*) of the composite structures are primarily dependent on the combination of elastic and plastic properties of their constituent components. Secondly, the mechanical properties can be controlled by the processing conditions thereby opening the opportunity of the manufacture of new composite materials of desired properties.

ACKNOWLEDGMENT

This work was supported by the Bilateral Collaborative Program between ASM (Moldova) and UEFISCDI-Romania - Project no. 10.820.05.21/RoF (2010-2012), Project no.13.820.05.20/RoF (2013-2014), and by the Research Project of the Institute of Applied Physics of the Academy of Sciences of Moldova (Project no. CSSDT 15.817.02.06A for 2015 - 2018).

CONFLICT OF INTEREST

The authors declare that they have no conflict of interest.

REFERENCES

1. Barbier D, Orignac X, Du X M et al. (1997) Improved composition for sol-gel rare-earth doped planar waveguides. J Sol-Gel Sci Tech 8:1013–1016
2. Orignac X, Barbier D, Du X M et al. (1996) Fabrication and characterization of sol-gel planar waveguides doped with rare-earth ions. Appl Phys Let 69(7):895 – 897
3. Yu M, Lin J, Fu J, Zhanga H J et al. (2003) Sol–gel synthesis and photoluminescent properties of $LaPO_4$:A (A ~ Eu^{3+}, Ce^{3+},Tb^{3+}) nanocrystalline thin films. J Mater Chem 13:1413–19
4. Galagan B I, Glushchenko I N, Denker B I et al. (2009) New high-strength neodymium laser glass on phosphate base. Quantum Electronics 39(12):1117-1120
5. Clayden N J, Aronne A, Esposito S et al. (2004) Solid state NMR study of phosphosilicate gels. J Non-Cryst Sol 345-346:601-605
6. Petrovskaya T S, Borilo L P, Kozik V V et al. (2008) Structure and properties of nanoproducts of the system P_2O_5 – SiO_2. Glass Ceram 65(11-12):410-414
7. Teterskii A V, Stefanovich S Yu, Turova N Y. (2006) Sol-gel synthesis of oxy-gen ion conductors based on apatite-structure silicates and silicophosphates. Inorg Mater 42(3):294-302
8. Shilova O A. (2005) Silicate Nanosized Films Prepared by the sol–gel method for use in planar technology for fabricating semiconductor gas sensors. Glass PhysChem 31(2):201–218
9. Zaharescu M, Vasilescu A, Badescu V et al. (1997) Hydrolysis-polycondensation in binary phosphorus alkoxides—TEOS system studied by GC-MS, J Sol-Gel Sci Tech 8:59-63
10. Oliver W C, Pharr G M. (1992) An improved technique for determining hardness and elastic modulus using load and displacement sensing indentation experiments. J Mat Res 7(6):1564-1580
11. Beegan D, Laugier M T. (2005) Application of composite hardness models to cooper thin film hardness measurement. Surf Coat Tech 199:32-37
12. Leyland A, Matthews A. (2000) On the significance of the H/E ratio in wear control: a nanocomposite coating approach to optimized tribological behaviour. Wear 246:1-11
13. Musil J. (2000) Hard and superhard nanocomposite coatings. Surf Coat Tech 125:322–330

Author: Daria GRABCO
Institute: Institute of Applied Physics of Academy of Science of Moldova
Street: 5, Academy str.
City: Chisinau
Country: Moldova
Email: grabco@phys.asm.md; grabco@mail.ru

Characterization of TiO_2 Nanoparticles and ZnO/TiO_2 Composite Obtained by Hydrothermal Method

E. Rusu[1], V. Ursaki[1], T. Gutul[1], P. Vlazan[2], and A. Siminel[3]

[1] Institute of Electronic Engineering and Nanotechnologies "D. Ghitu" of the Academy of Sciences of Moldova, Chisinau, Moldova
[2] National Institute for Research and Development in Electrochemistry and Condensed Matter, Timisoara, Romania
[3] Institute of Applied Physics of the Academy of Sciences of Moldova, Chisinau, Moldova

Abstract— **ZnO and TiO2 nanoparticles as well as ZnO/TiO2 core/shell structures have been prepared by hydrothermal method and characterized by means of scanning electron microscopy, XRD and EDX analysis, FTIR and luminescence spectroscopy. Porous ZnO/TiO2 nanocomposites with micron sized channels and walls consisting of nanoparticles with size of 30-80 nm for the ZnO component and 10-20 nm for the TiO2 component have been obtained. Wurtzite and anatase strucrures have been revealed for the ZnO and TiO2 constituents, respectively, from XRD analysis. The analysis of FTIR spectra indicates on a strong interaction of intrinsic atoms with species from organic solvents occurring at the surface of nanoparticles constituting the nanocomposite. A strong green emission coming from the TiO2 component has been observed in luminescence spectra.**

Keywords— **hydrothermal process, nanoparticles, core/shell structures, luminescence.**

I. INTRODUCTION

TiO_2 and ZnO are among the most prospective materials for nanostructure fabrication with various destination. Particularly, thick films of TiO_2, or ZnO nanoparticles that are sintered into a mesoporous network with a large internal surface area are widely used in dye-sensitized solar cells (DSSCs). Up to now, TiO_2 nanoparticles with sizes of 10–20 nm have been widely employed as the materials for fabricating mesoporous photoanodes in DSSCs. It was found that a combination of ZnO and TiO_2 in core-shell structure ensures high performances of DSSCs devices. Law et al. [1] prepared ZnO nanowires coated with thin shells of Al_2O_3 or TiO_2 and indicated significant improved efficiencies in power conversion of the DSSCs based on the core-shell architectures.

Another application of ZnO/TiO_2 core-shell structures is related to photocatalysis. The application of coupled semiconductors in the photocatalytic process is regarded as an alternative to enhance activity of photocatalyst by inhibiting the recombination of electron–hole pairs. ZnO is considered to be a suitable semiconductor to be coupled with TiO_2 due to its high photocatalytic activity, approximate band gap energy of 3.37 eV with a large binding energy and similar photocatalytic mechanism of reaction as compared to TiO_2.

The TiO_2/ZnO heterojunction significantly improved the performance in the photoelectrochemical water splitting cell for efficient H_2 production due to the beneficial synergetic effects of the heterostructured nanocomposite. One-dimensional core/shell structured TiO_2/ZnO heterojunction was superior to TiO_2 in the photoelectrochemical water splitting because of better charge separation and more favorable Fermi level. This combination is also beneficial for biological applications. It was recently shown that the cytotoxicity of nano-ZnO is significantly reduced by coating it with a TiO_2 layer in a sol-gel process [2]. In another biological demonstration, ZnO/TiO_2 core-shell quantum dots were used to label the tumor and normal tissue of human colorectal [3]. Nanostructured metal oxides such as ZnO and TiO_2 have also been extensively employed as self-cleaning coatings due to their large band gaps as well as their hydrophilic and photocatalytic properties [4].

As concerns the preparation methods, ZnO/TiO_2 core/shell nanostructures have been fabricated by various techniques including physical vapor deposition [5], radio-frequency sputtering [6], chemical vapor deposition [7] and microwave-assisted hydrothermal method [8].

A simple hydrothermal method was developed to coat thin TiO_2 nanosheets onto a ZnO nanorod array [4]. These ZnO/TiO_2 nanorod arrays demonstrated good hydrophilic properties, excellent photocatalytic properties for the degradation of dye molecules, and self-cleaning properties. Chen et al. [9] proposed a simpler spin coating method to obtain a shell layer of TiO_2 coated on ZnO arrays grown by a chemical bath deposition approach.

In this paper, we prepared ZnO/TiO_2 core-shell structures by a low temperature cost-effective hydrothermal method and investigated their morphological, structural, and optical properties.

II. MATERIAL SYNTHESIS

The synthesis of core-shell ZnO/TiO_2 structures was performed in two stages: ZnO nanoparticles are produced in the first stage of a hydrothermal process, while their covering with TiO_2 shells occurs during the second stage. ZnO nanopartcices are obtained with zinc nitrate hexahydrate

V. Sontea and I. Tiginyanu (eds.), *3rd International Conference on Nanotechnologies and Biomedical Engineering*,
IFMBE Proceedings 55,
DOI: 10.1007/978-981-287-736-9_22

($Zn(NO_3)_2.6H_2O$) and sodium hydroxide precursors. A mixture of 2.5mM of zinc nitrate hexahydrate dissolved in deionazed water and 1M NaOH was prepared, because the presence of alkaline solution is essential for the formation of ZnO nanoparticles. The mixture was vigorously stirred with a magnetic stirrer. The final pH value of the solution equals 9. The solution was subsequently transferred in an autoclave and treated during 5 ours at the temperature of 220 °C.

To produce TiO_2 nanoparticles in a hydrothermal process, 0.2g of polyvinyl alcohol is dissolved in 50 ml of ethanol and 2 mL of titanium isopropoxide is added to the solution, which is stirred during 1 hour. The mixture is subsequently introduced into an autoclave, and is treated at 220 °C during 2 hours. After that, the precipitate is washed and dried at 80 °C during 2 hours.

The SEM images of the ZnO and TiO_2 nanoparticles produce by hydrothermal method are shown in Fig. 1.

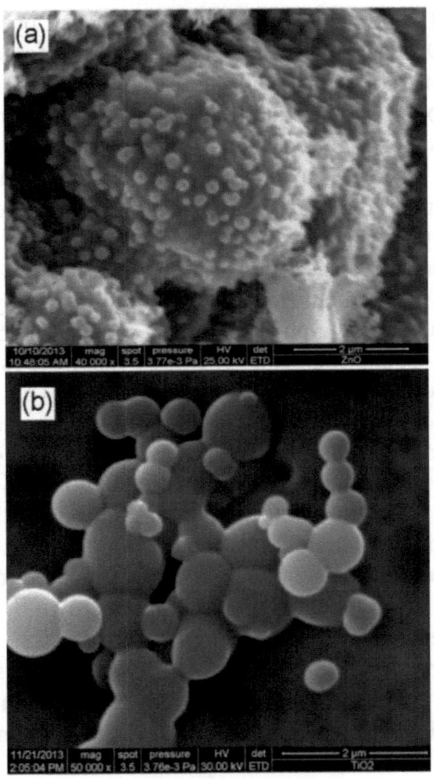

Fig. 1 SEM images of ZnO and TiO2 nanomaterilals produced by hydrothermal method.

One can see from Figure 1b that the obtained TiO_2 nanomaterial is composed of spherical nanoparticles with different sizes.

ZnO/TiO_2 core/shell structures have been prepared according to two technological routes, which are different from the point of view of both the precursors and the reaction media. The samples labelled as CS1 was obtained with a zinc nitrate hexahydrate ($Zn(NO_3)_2.6H_2O$) precursor with concentration of 3 mM dissolved in a solution of urea (3 mM). The precipitation was performed in a solution of 1M NaOH. 2 ml of titanium isopropoxide and 10 ml of ethylic alcohol was added to the obtained suspension under continuous stirring during 1 hour at 65 °C. The obtained precipitate was filtered out and washed several times in distilled water and ethylic alcohol. Subsequently, it was dried in an oven at 80 °C during 2 hours, and calcined at 500 °C during 4 hours.

The core/shell sample labelled as CS2 was prepared with a zinc acetate [$Zn(CH_3COO)_2x2H_2O$] precursor with concentration of 3mM dissolved in mixture of ethylic alcohol and distilled water (1:1) to which 0.2 moles of polyethylene glycol was added. A solution of 25% NH_4OH with pH=9 under continuous stirring at 60 °C during 1 hour was used for precipitation. Subsequently, 2 ml of titanium isopropoxide was added, and the stirring continued 1 hour more, after which the obtained suspension was introduced into a Teflon autoclave and subjected to thermal treatment at 220 °C during 5 hour.

III. CHARACTERIZATION

The morphology of the prepared ZnO/TiO_2 nanocomposite is illustrated in Fig. 2. One can see from Fig. 2 that the composite represents a porous structure with micron sized channels.

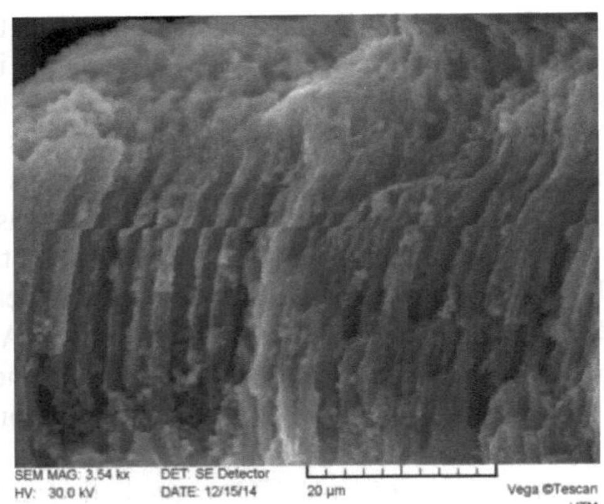

Fig. 2 SEM image of the CS1 nanocomposite.

Taking into account the results of XRD analysis, indicating that ZnO nanoparticles have sizes in the range of 30-80 nm, and the sizes of TiO_2 nanopraticles are within 10-20 nm, one can conclude the walls of the porous structure are composed of nanometric ZnO/TiO_2 core/shell particles.

Figure 3 presents XRD spectra of core/shell ZnO/TiO_2 nanomaterial calcined at 500 °C as compared to pure ZnO and TiO_2 nanoparticles. The reference patterns of anatase TiO_2 (00-001-0562 JCPDS card) and wurtzite ZnO (01-079-0205 JCPDS card) are also shown in the figure. One can see from this figure that the XRD reflexes from the TiO_2 powder correspond to the anatase structures with peaks at 2Θ equal to 25.5, 37.9, 48.1, 53.9, 55.3, 62.6, 68.8 and 70.5 degrees, corresponding to (101), (004), (200), (105), (211), (204), (116) and (200) reflexes, respectively. The main peaks detected in the XRD spectra of the ZnO sample at 32, 35, 37, 48, 57, 63, 67, 68, and 69 degrees correspond to the following lattice planes: (100), (002), (101), (102), (110), (103), (200), (112), (201), respectively. All the observed Bragg reflections are consistent with Wurtzite's $P6_3mc$ space group. The sizes of ZnO and TiO_2 nanoparticles deduced from the width of peaks according to the Scherrer formula lie in the range of 30-80 nm and 10-20 nm for the ZnO and TiO_2 nanoparticles, respectively.

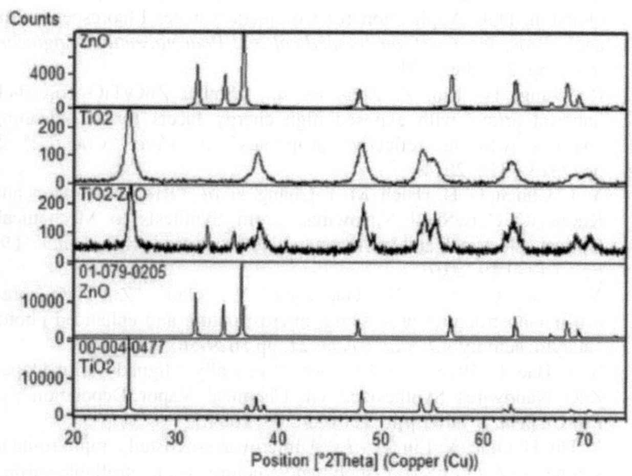

Fig. 3 XRD spectra of ZnO and TiO2 powders as well as the ZnO/TiO2 nanocomposite.

The XRD spectrum of the ZnO/TiO_2 nanoicomposite is dominated by reflexes from the TiO2 with anatase structure. However, peaks at 35.6° and 62.0° corresponding to the rutile structure are also observed in the spectrum.

The photoluminescence (PL) spectra of ZnO nanoparticles and the spectra of ZnO/TiO_2 nanocomposites CS1 and CS2 prepared according to the technological routes described above are presented in Fig. 4. The maximum of ZnO nanoparticles spectrum is in the near-bandedge UV region which is outside the spectral range presented in the figure, while the emission in the visible range is very weak.

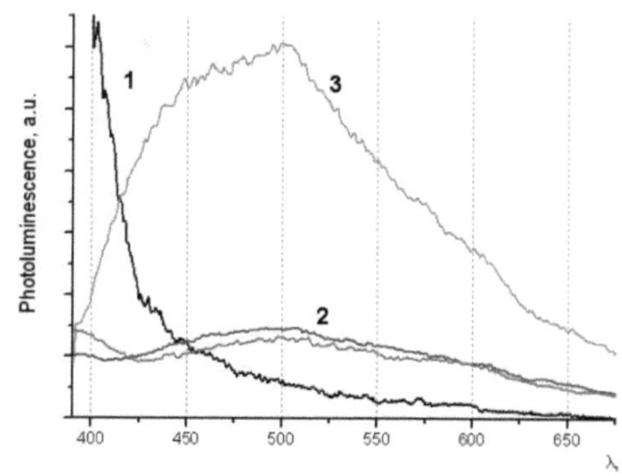

Fig. 4 PL spectra of ZnO nanoparticles (1), CS1 (2) and CS2 (3) core/shell nanocomposites measured at room temperature.

The ZnO/TiO_2 nanocomposites exhibit a broad green luminescence band with maximum around 500 nm, which most probably comes from the TiO_2 with anantase structure.

A green luminescence band is frequently observed in luminescence spectra of titania and is attributed to radiative recombination of selftrapped excitons localized on TiO_6 octahedra [10]. Selftrapped excitons are reported to be stable in anatase TiO_2, whose structure is based on a network of TiO_6 octahedral units. Excitons in such structures result from the interaction of conduction band electrons essentially located on Ti 3d states with holes essentially on oxygen 2p states [10]. One can see from figure 4 that the technological root used for the preparation of the sample CS2 leads to higher intensity of the green luminescence.

The FTIR analysis was performed to identify the eventual changes in the surface chemistry of the prepared samples (Figure 5). The metal oxides generally exhibit absorption bands under 1000 cm-1 generated by inter-atomic Me-O or Me-O-Me vibrations [11]. The vibration observed at $1067 cm^{-1}$ can be attributed to C=O stretching in carboxylic acids. The absorption band at 1252 cm^{-1} is related to C-O stretching vibrations in alcohols or ethers. The bands at 1395, 1401 and 1460 can be attributed to OH group bending vibrations in alcohols or carboxylic acids. The peaks at 1634, 1640 cm^{-1} correspond to C=O stretching and bending vibrations, respectively. The peaks at 1981 and 1987 cm^{-1} could be related to C=O stretching vibrations in carboxylic compounds. The absorption bands at 2160, 2304 and 2311 cm^{-1} can be attributed to stretching vibrations of CO_2 absorbed from atmosphere on the materials surface. The absorption bands in the range of 2800 – 3000 cm^{-1} can be attributed to C-H stretching vibrations in organic compounds, their exact position being dependent upon the hybridization state of the carbon. The broad intensive band situated in the interval of 3000 - 3800 cm^{-1} can be attributed to the OH group vibrations due to the adsorption of water on the metal oxide surface [12].

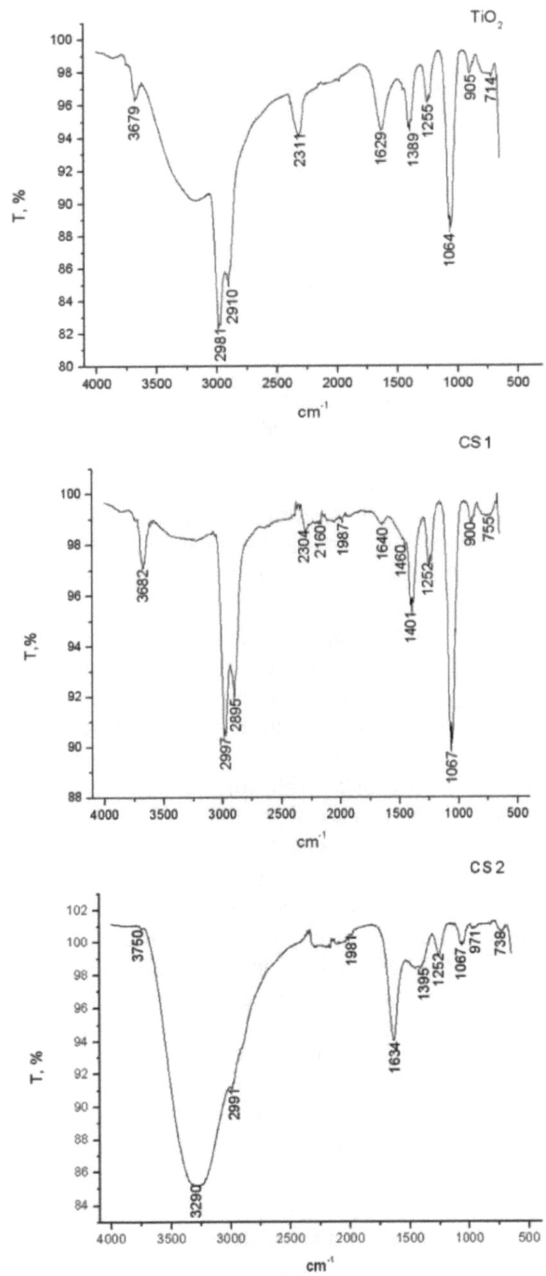

Fig. 5 FTIR spectra of TiO_2 nanoparticles, CS1, and CS2 core/shell nanocomposites.

The preliminary results of investigations suggest that these biometal nanoparticles manifest antibacterial and antifungal properties, as well as bio-stimulating effects on the biosynthesis of fungal proteases, which are prospective for the development of biotechnologies involving filamentous micro-organisms, particularly for the production of enzyme preparations.

IV. CONCLUSION

In this study we have developed hydrothermal technologies for the preparation of ZnO and TiO_2 nanoparticles as well as ZnO/TiO_2 core/shell structures with interesting morphological, structural, luminescent, and bio-stimulating properties for a range of solar cells, sensors, photocatalysts, and biological applications.

CONFLICT OF INTEREST

The authors declare that they have no conflict of interest.

REFERENCES

1. M. Law, L. E. Greene, A. Radenovic, *et. al.*, "ZnO-Al_2O_3 and ZnO-TiO_2 Core-Shell Nanowire Dye-Sensitized Solar Cells", *J. Phys. Chem. B* **110**, pp. 22652-22663, 2006.
2. I.-L. Hsiao and Y.-J. Huang, "Titanium Oxide Shell Coatings Decrease the Cytotoxicity of ZnO Nanoparticles", *Chem. Res. Toxicol.* **24**, pp. 303–313, 2011.
3. S. Fu, C. S. Lim, L. C. Kwek, T. C. Chia, "ZnO/TiO_2 Core-Shell Quantum Dots Application for Colorectal Cancer Fluorescence Image", *Proc. Int. Conf. on Biomedical and Pharmaceutical Engineering,* pp. 377-380, 2006.
4. R. Wang, H. Tan, Z. Zhai, *et. al.*, "Stable ZnO/TiO_2 core/shell nanorod arrays with exposed high energy facets for self-cleaning coatings with anti-reflective properties", *J. Mater. Chem. A* **2**, pp. 7313-7318, 2014.
5. Y. L. Chueh, C. H. Hsieh, M. T. Chang, *et. al.*, "RuO_2 Nanowires and RuO_2/TiO_2 Core/Shell Nanowires: From Synthesis to Mechanical, Optical, Electrical, and Photoconductive Properties", *Adv. Mater.* **19**, pp. 143-149, 2007.
6. X. Yan, C. Zou, X. Gao, and W. Gao, "ZnO/TiO_2 core–brush nanostructure: processing, microstructure and enhanced photocatalytic activity", *J. Mat. Chem.* **22**, pp.5629-5634, 2012.
7. S. Y. Bae, H. W. Seo, and J. Park, "Vertically Aligned Sulfur-Doped ZnO Nanowires Synthesized via Chemical Vapor Deposition", *J. Phys. Chem. B* **108**, pp. 5206-5210, 2004.
8. Y. Hu, H. Qian, Y. Liu, *et. al.*, "A microwave-assisted rapid route to synthesize ZnO/ZnS core-shell nanostructures via controllable surface sulfidation of ZnO nanorods", *Cryst. Eng. Comm.* **13**, pp. 3438-3443, 2011.
9. L.-C. Chen, S.-F. Tsai, J.-H. Chen, G.-W. Wang, "Preparation of Vertically Aligned ZnO/TiO_2 Core-Shell Composites for DSSC", *Int. J. Photoenergy*, **2013**, 417964, 2013.
10. H. Tang, H. Berger, P. E. Schmid, F. Levy, "Optical properties of anatase TiO_2", *Solid State Commun.* **92**, pp. 267- 271, 1994.
11. H. Kumar and R. Rani, "Structural and Optical Characterization of ZnO Nanoparticles Synthesized by Microemulsion Route", Int. Lett. Chemistry, Phys. and Astronomy **14**, pp. 26-36, 2013.
12. H. Becker, W. Berger, *et.al.*, "Chimie organic practica", Ed. Stiintifica si Enciclopedica, Bucuresti, pp. 114-115, 1982.

Author: Veaceslav Ursaki
Institute: Institute of Electronic Engineering and Nanotechnologies
Street: 3/3 Academiei str.
City: Chisinau
Country: Moldova
Email: ursaki@yahoo.com

Fabrication of Bismuth Telluride Wire Thermoelectric Devices

T.E. Huber[1], S. Johnson[1], K.A. Shirvani[1], Q. Barclif[1], T. Brower[1], A. Nikolaeva[2,3], and L. Konopko[2]

[1] Howard University, Washington, USA
[2] Ghitu Institute of Electronic Engineering and Nanotechnologies, ASM, Chisinau, Moldova
[3] International Laboratory of High Magnetic Fields and Low Temperatures, Wrocław, Poland

Abstract— **Bismuth telluride wires are interesting building blocks of thermoelectric devices. These are nanoscale heat to electric converters that have applications as uncooled detectors, generators and uncooled bolometers. Also, there is interest in bismuth telluride because it is an example of a topological insulator. The exploitation of the thermoelectric properties of devices based on wires of thermoelectric materials requires good electrical contacts between the wires and metal electrodes. The fabrication of the devices here, is based on contacting the wire ends to the device electrodes by depositing a platinum film using the focused ion beam method.**

Keywords— **Thermoelectrics, bismuth, telluride, wire.**

I. INTRODUCTION

The science and technology of bismuth telluride (BiTe) wires has advanced to the point that many methods for fabrication and for characterization are available. There have been several reports of bismuth telluride wire fabrication by laser ablation [1]. Reports have also confirmed the synthesis of individual bismuth telluride wires by cyclic electrodeposition/stripping coupled with electrochemical step edge decoration [2] where the wire diameter is determined by processing variables. We believe that there is a simpler approach to define the wire diameter by separating the fabrication steps of wire synthesis and isolation; wires can be grown in arrays employing the template strategy where materials consisting of arrays of channels of the chosen diameter, such as porous anodic alumina or other nanochannel array, are injected with the melt of the chosen composition. One method of injecting the channels is electrodeposition into porous anodic alumina templates [3-6]. The high pressure injection (HPI) method is successful in preparing thick, dense, large-area Bi wire array composites [7]. The HPI method benefits from the substantial body of knowledge regarding the preparation of bulk Bi2Te3-Sb2Te3-Bi2Se3 alloys of both n-type and p-type of optimal composition for thermoelectric operation because these alloys are prepared traditionally by mixing in the molten state prior to crystallization. The thermoelectric properties of the wires can be measured employing microfabricated devices. Here we discuss the fabrication of a device where the wire is contacted at the two wire ends. Unlike the four point devices (two contacts for current and two for voltage), the two point device is sensitive to the contact resistance between wire and electrodes. Also, the two-point device is a way to prepare devices for applications and is more practical than four-point devices. We use this device to evaluate the wire thermoelectric properties and the contact resistance. The plan of this paper is as follows. In section II we discuss BiTe wire arrays and in Section III we discuss the methods employed for isolating individual BiTe wires in a two-point device and the contact resistance that is observed.

II. BISMUTH TELLURIDE WIRE ARRAYS

In this section we discuss the fabrication and characterization of BiTe wire arrays. We have employed a template method, the high pressure injection (HPI) method, where the melt of the alloy of interest is injected in the template using high pressure. For this work, we used a commercial alumina membrane solid for microfiltration under the trade name Anopore (Whatam Laboratory Division). Anopore consists of an alumina plate 25 mm in diameter and about 55 μm thick, which supports an array of 200-nm diameter, parallel, largely non interconnected, cylindrical channels running perpendicular to the plate surface.

In order to inject bismuth telluride in the pores we place high purity, 99.999%, Bi_2Te_3 pellets (Sigma-Aldrich) in contact with Anopore in a high pressure reactor. The reactor is heated to a temperature above the melting point of the material to be injected; in the experiments presented here, this temperature is 650°C. The pressure is then gradually raised to 1 kbar, forcing the molten material into the matrix channels. After a few minutes, when the injection is complete and the melt solidifies inside the channels, the reactor is cooled and the pressure is then released. The wire array sample is then extracted from the capsule and cleaned of the surrounding excess material by standard mechanical polishing techniques.

The crystal orientation of the wire sample was determined by X-ray diffraction (XRD) experiments with the reflecting planes normal to the wire axis. The observed peak positions are very close to the peak positions in bulk BiTe, and correspond to planes that are perpendicular to the c-axis. This indicates not only that the rhombohedral structure of bulk BiTe is preserved in these fine wires but that the wires in the array are

V. Sontea and I. Tiginyanu (eds.), *3rd International Conference on Nanotechnologies and Biomedical Engineering*,
IFMBE Proceedings 55,
DOI: 10.1007/978-981-287-736-9_23

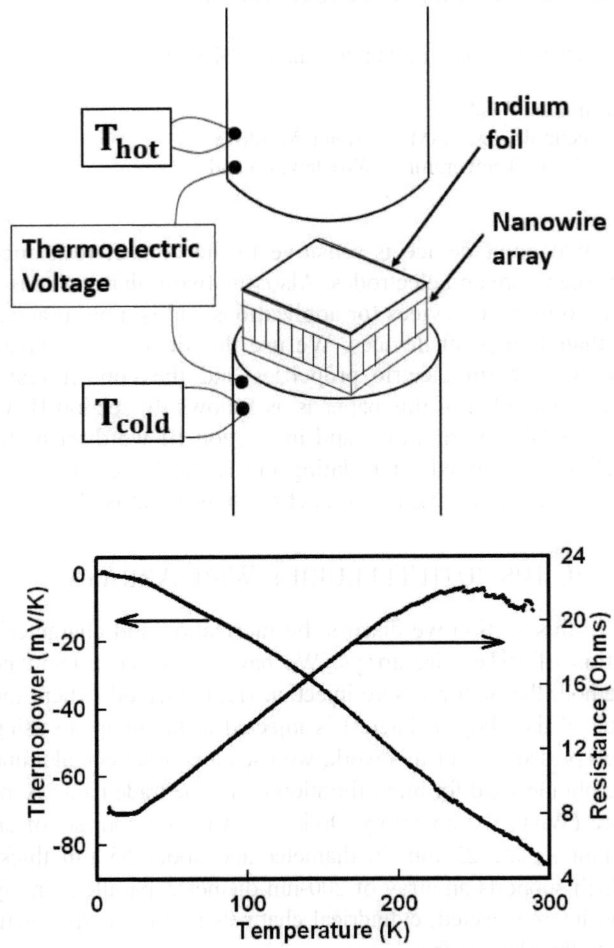

Fig. 1 Top: experimental equipment for wire array resistance and thermopower. Bottom: Temperature-dependent resistance R(T) and thermopower S(T) of a BiTe wire array. The number of wires contacted is unknown.

largely oriented with the c-axis perpendicular to the wire length. The resistance and thermopower of the arrays are measured by placing the sample, capped with indium foil, in a clamp designed to measure voltages and temperatures at either side of the clamp. Heaters establish a temperature difference between the clamp sides as required for thermopower measurements. This equipment is presented in the top of Fig.1. The wire's thermopower is negative (wires are type-n). The room temperature value is 90 μV/K .In comparison with the bulk that has $|S|$~200 μV/K, the thermopower is ~50% smaller in our case. Thermopower value is important for applications since the efficiency is ~ S^2. A recent paper by Hamdou *et al* [6] reports reaching a thermopower of -150 μV/K by using an annealing step after fabrication therefore we are not far off this target.

The individual wires in the present study are 200 nm; since the critical diameter for the semimetal to semiconductor transition SMSC transition in BiTe is roughly 2 nm [8], quantum confinement effects can be neglected and the Fermi surface of the carriers in the wire is roughly the same as in bulk BiTe of the same composition. However, the composition is not accurately known. Even though the material that we inject is pure Bi_2Te_3, the relative Bi/Te composition of the wires does not have to be stochiometric, which are 40 at% Bi and 60*at*% Te where "*at* " stands for atomic, due to Bi/Te differential surface tension. Composition control is a serious problem because the thermoelectric properties of BiTe are a strong function of the composition. (It only takes a 0.1% change in the Bi/Te ratio to change the material from type-p to type-n). The atomic composition of the arrays was studied with an S4500 Scanning electron microscope equipped with a Princeton Gamma-Tech EDX (Energy-dispersive X-ray) spectroscopy attachment. The wires are found to be stoichiometric within the precision of the measurement that is 1-3 %.

III. Isolating BiTe Wires

In this section we discuss the experimental methods employed for isolating a single wire of bismuth telluride.

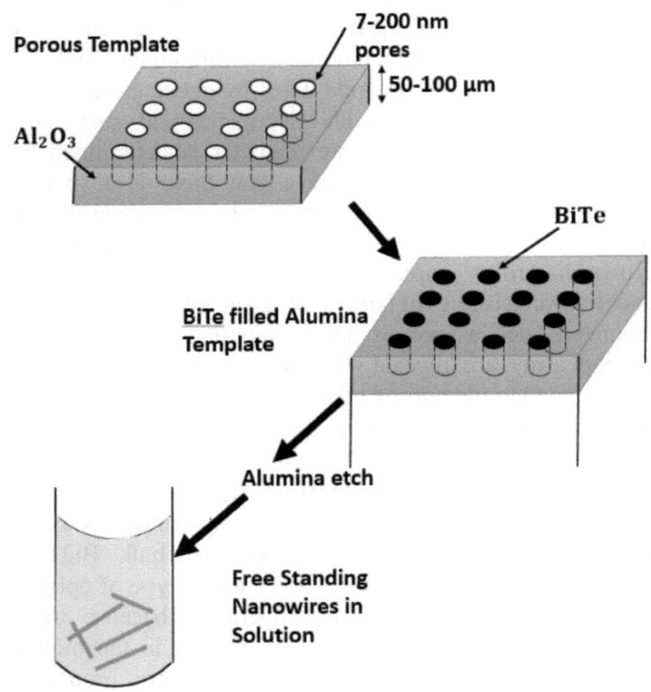

Fig. 2 Isolating individual BiTe wires via two steps; injection of bismuth telluride in aluminum template and etching the alumina using sodium hydroxide.

Removing the BiTe wires from the alumina template is difficult due to the chemical resilience of the alumina.

We employed a proven method [9] that consists of using a solution of 6N sodium hydroxide to dissolve the alumina while preserving the integrity of the BiTe wires.

The "solution" of wires is a good source of samples. To employ the solution, we deposited it with a pipette on the appropriate substrate: holey carbon for the SEM and TEM - experiments or patterned electrodes for the electrical measurements. The alcohol evaporates, leaving behind a sparse dispersion of wires that are loosely attached to the substrate. The individual BiTe wires were deposited on holey carbon for study with a 200 kV JEOL 2010F TEM with 0.25 nm resolution. This study will be discussed in a separate publication. Briefly, the selected area diffraction (SAD) pattern is consistent with the single crystal Bi_2Te_3 structure in PDF 15-863 and, therefore, consistent with the XRD study. Also consistent with the XRD study in Section II, we have found that the crystalline structure of the wire is oriented with the c-axis perpendicular to the wire length. SAD spots are dots in single crystalline material, as opposed to circular rings in polycrystalline materials or arc spots in textured structures. In fact we find that the wires are single crystalline along their whole length with few defects. A high resolution TEM image from the side wall of a Bi_2Te_3 wire is shown in Fig.3. The planes perpendicular to the <0001> direction are very clearly seen as horizontal lines (fringes) that are parallel to the side walls from this angle, therefore confirming that the c-axis is normal to the wire axis.

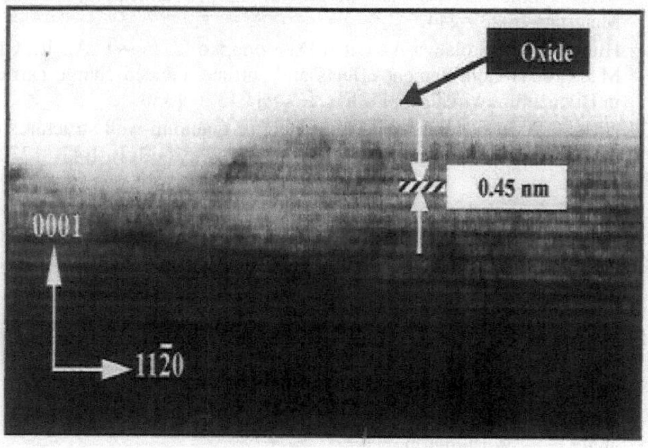

Fig. 3 High Resolution TEM image from the edge of a BiTe wire. The lattice fringes shown are (000L) planes where L=1,2,3,.. with an interplane distance of 0.5 nm for L=6. Fringes indicate that wires are single-crystal. The layer of oxide is clearly visible and is 1-2 nm thick. The image goes from clear to dark as we image the edge of wire.

The intensity of the image shows two periodicities, 0.5 and 1 nm, which correspond to the periodicity between planes perpendicular to the zone axis <0001>. The wires grown in the present study are single crystalline with few defects. The wire side walls are not smooth; they have sharp edges, but with ledges. This image shows an example of this morphology. A thin (1.5 nm) amorphous layer, presumably oxide, is observed along the side walls. The oxide layer is an important factor in achieving good electrical contacts to the wire. In the case of Bi wires [10], the presence of 7-nm thick oxide hampered the electrical measurements substantially.

In order to prepare a thermoelectric device we deposited a BiTe wire on patterned electrodes. The BiTe wire is electrically connected using a focused electron beam method. In this method, a conductive Pt composite joins wire and electrode. An example of the device that was assembled is shown in Fig.4. The resistance that is measured between the planar electrodes has contributions from the wire, from the Pt deposits and from the contacts. The current I as a function of applied voltage V was measured. The curves I=I(V) (I-V) were found to be non-linear, with a degree of non-linearity that increases for decreasing temperature. This temperature-dependent non-linear response is believed to be due to the contacts between the wires and the gold electrodes. A thin insulating layer would provide a tunnel barrier for the electrons passing through it. Similar phenomena is observed by Cronin in Bi [10].

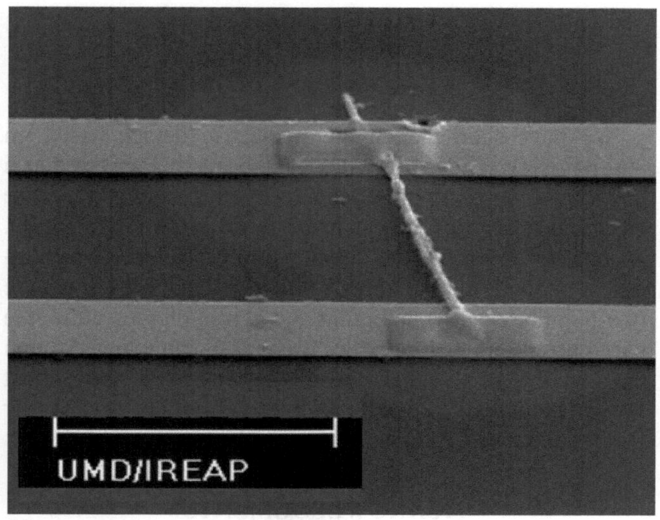

Fig. 4 Scanning electron microscope image of the two-point device. The gray bands are Au planar electrodes. The wire bridges two planar electrodes. The two light gray areas at the joint between planar electrodes and wire are Pt deposits. The wire is 200 nm in diameter, electrodes are 1 μm wide, and the distance between electrodes is 4 μm.

The room temperature contact resistance is 70 KOhms. We have observed that the characteristics of the I(V) curve at all temperatures depend upon the prior exposure of the wire to air and temperature.

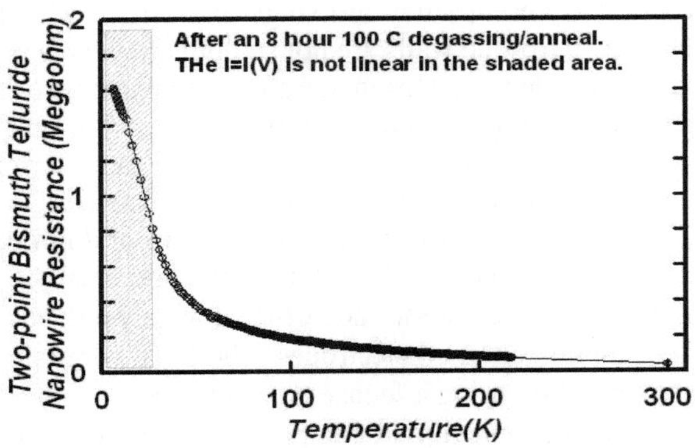

Fig. 5 R (T) of wires after an anneal/degassing.

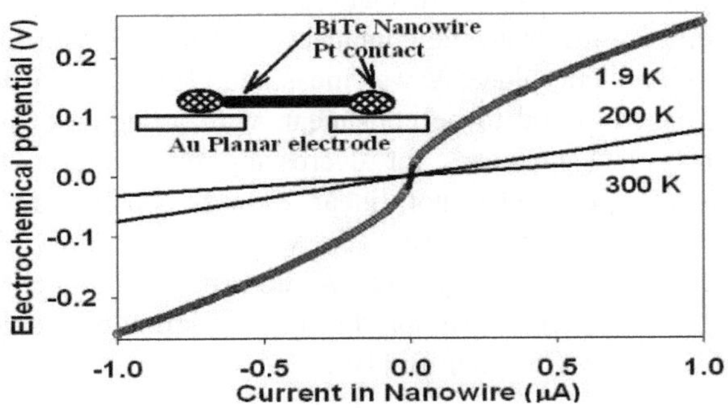

Fig. 6 I (V) of wires at various temperatures.

IV. CONCLUSIONS

We have fabricated a thermoelectric device based on 200-nm BiTe wires. The wires have been characterized as a part of the wire array. As shown in Fig.5 and 6, the wire electrical properties appear to be sensitive to air (oxygen). To test this property, we developed a procedure of degassing at modest temperatures (100 °C) conditions the wire two-point device. At that point and at room temperature, the wire current is not limited by the rate of tunneling through the Pt contacts at either end of the wire. We believe that further gains can be achieved by conditioning at a higher temperature and by degassing in a hydrogen atmosphere.

ACKNOWLEDGMENTS

We thank M. Ervin, ARL and J. Melngailis and J. Barry of the University of Maryland. This material is based upon work supported by the National Science Foundation through Partnership for Research and Education in Materials DMR-1205608 and Science and Technology Center DMR-1231319. We also acknowledge The Boeing Co, the Swiss National Science Foundation and the Science Technology Center in Ukraine # 5986. We also acknowledge Jeffrey H. Hunt and John H. Belk of the Boeing Company for their contribution to developing this technology.

CONFLICT OF INTEREST

The authors declare that they have no conflict of interest.

REFERENCES

1. Wei Q., Lieber C. M. (1999) Synthesis of single crystal bismuth-telluride and lead-telluride wires for new thermoelectric materials Mat. Res. Soc. Symp. Proc. Vol. 581, Boston, USA 2000, pp. 219-223.
2. Purkayastha A., Lupo F., Kim S., Borca-Tasciuc T., Ramanath. G. (2006) Low-Temperature, Template-free Synthesis of Single-crystal bismuth telluride nanorods, Advanced Materials 18:496-500.
3. Thomson M.A., Menke E. J., Martens C. C., Penner R. M. (2006) Shrinking wires by kinetically controlled oxidation, J. Phys. Chem. B 110:36-41.
4. Li. S., Liang Y., Qin J., Toprak M., Muhammed M. (2009) Template electrodeposition of ordered bismuth telluride wire arrays, J. Nanoscience Nanotechnology 2:1543-1547.
5. Trahey L., Becker C., Stacy A.M. (2007) Electrodeposited bismuth telluride wire arrays with uniform growth fronts, Nanoletters 7:2535-2539.
6. Hamdou B., Kimling J., Dorn A., Pippel E., Rostek R., Wolas P, Nielsch K. (2013) Thermoelectric characterization of bismuth telluride wires, synthesized via catalytic growth and post-annealing. Advanced Materials 25:239-244.
7. Huber T.E., Nikolaeva A., Gitsu D., Konopko L., Foss C.A., Jr., Graf M.J. (2004) Confinement effects and surface-induced charge carriers in Bi quantum wires. Appl. Phys. Lett. 84:1326-1330.
8. Hicks L.D. and Dresselhaus M.S. Effect of quantum-well structures on the thermoelectric figure of merit (1993) Phys. Rev. B47, 17727-17731.
9. Sapp S.A., Lakshimi B.B. and Martin C.R. (1999) Template Synthesis of bismuth telluride wires Adv. Mat. 11:402-405.
10. Cronin S.B., Electronic Properties of Bi Wires (2002) Ph.D. Thesis, MIT, Advisor M.S. Dresselhaus.

Author: Tito Hubor
Institute: Howard University
Street: 500 College St, NE
City: Washington
Country: USA
Email: titoehuber@gmail.com

Nanolayers with Advanced Properties for Superconducting Spintronics

R. Morari, E. Antropov, E. Zasavitsky, A. Prepelita, A. Socrovisciuc, E. Condrea, and A. Sidorenko

D.GHITU Institute of Electronic Engineering and Nanotechnologies ASM, Chisinau, MD2028 Moldova

Abstract— **The rapid development of semiconductor electronic industry requires revolutionary ideas for a jump to the next level of complexity. One of the possible level of 21st century microelectronics could be superconducting spin-sensitive electronics - spintronics. We are working for solutions of realization the superconducting spin-valve (spin-switch), the base component of spin-sensitive electronics with a high potential. The core of spin-valve comprises two separated ferromagnetic nano-layers and superconducting thin layer. In present work we report on main milestones of development of F/S/F - structures, giving description of the deposition process, characterization of grown samples, TC characterization of F/S/F sets patterns and discussion.**

Keywords— **Nanostructures, spintronics, advanced technology.**

I. NANOSTRUCTURES DEPOSITION AND CHARACTERIZATION

The core of spin switch comprises two ferromagnetic films (F) and superconductor (S). The F/S/F structure can be divided in two independent structures: F/S and S/F. For that reason we describe only S/F- type structures fabrication, others are deposited at the same fashion with certain differences which can be neglected. It is important to note that deposition process was developed and adjusted exactly on this configuration of samples.

A special advanced technological process of superconducting layers preparation for reliable fabrication of Superconductor/Feroomagnet (S/F) structures with the layer thickness of several nanometers is developed, using as a base magnetron sputtering system "Leybold Z400" [1]. The S and F layers were deposited by magnetron sputtering on commercial (111) silicon substrates at room temperature. The base pressure in the "Leybold Z400" vacuum system kept at about 2×10^{-6} mbar. Pure argon (99.999%, "Messer Griesheim") at a pressure of 8×10^{-3} mbar is used as sputter gas. A silicon buffer layer was deposited using RF magnetron, it produced a clean interface for the subsequently deposited niobium layer. To obtain flat and high-quality Nb layers with thickness in the range of 5-15 nm, the rotation of the Nb target around the symmetry axis of the vacuum chamber was applied. A dc-motor drive moved the operating magnetron along the silicone substrate of the 80×7 mm^2 size during the deposition process. Thus, the surface was homogeneously sprayed with the sputtered material. The constructed deposition mashine is shown in Fig.1. The effective growth rate of the Nb film in that case was 1.3 nm/sec (non-moving magnetron with sputtering target serves the deposition rate of 4 nm/sec).

Fig. 1 Magnetron sputtering deposition machine, constructed for preparation of the S/F nanostructures.

The next step of the procedure was deposition of a wedge-shaped ferromagnetic layer using the intrinsic spatial gradient of the deposition rate of the sputtering target. The $Cu_{41}Ni_{59}$ target (or Co target) was sputtered in RF mode to avoid inhomogeneous distribution of deposited materials in a grown ferromagnetic layer, resulting the same composition ($Cu_{41}Ni_{59}$) of the alloy in the film as in the bulk material of the target. The rate of growth was 3-4 nm/sec. To prevent a destructive influence by the atmospheric conditions, the last deposited layers were coated by a protective silicon cap of about 5 nm thickness.

II. TRILAYER STRUCTURES

Fig. 2 shows a sketch of the trilayer F/S/F sample – core structure for design of the superconducting spin-valve, Fig. 3 presents the Transition Electron Microscope image of it. The TEM demonstrate highest quality of the both interfaces between layers niobium/copper-nickel and copper-nickel/niobium. According to TEM image the thickness of the top $Cu_{41}Ni_{59}$, the niobium layer and the bottom copper nickel layer is 31.2 nm, 10.6 nm and 8.6 nm, respectively.

V. Sontea and I. Tiginyanu (eds.), *3rd International Conference on Nanotechnologies and Biomedical Engineering*,
IFMBE Proceedings 55,
DOI: 10.1007/978-981-287-736-9_24

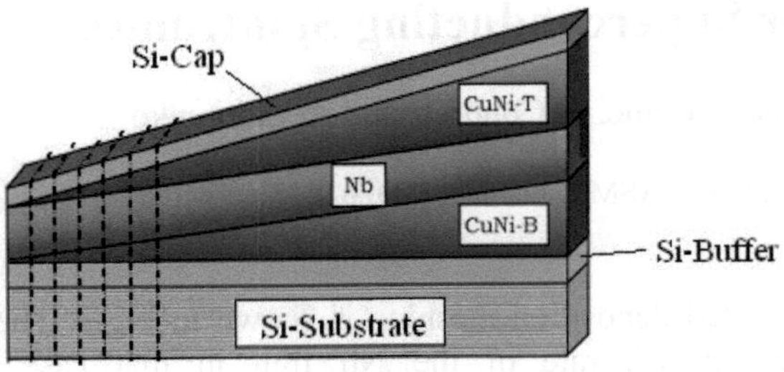

Fig. 2 Sketch of the layers stack in the deposited trilayer F/S/F-core structure of the spin-valve.

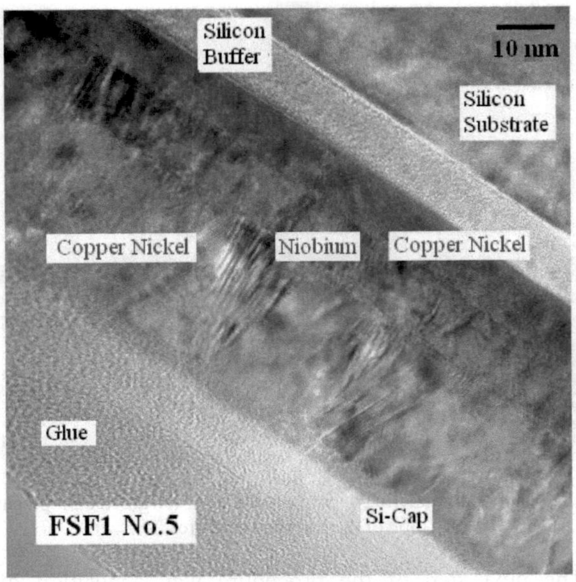

Fig. 3 Cross section TEM image of a F/S/F 3-layer structure (CuNi/Nb/CuNi)

Samples with a width of 2.5 mm were cut perpendicular to the wedge to obtain a set of strips (in our case F/S/F specimens) with varying $Cu_{41}Ni_{59}$ layer thickness d_F, for $T_c(d_F)$ measurements. Aluminum wires of 50 µm in diameter were bonded to the strips by ultrasonic bonder for resistance measurements.

To study the quality of interfaces between the layers we performed Auger electron spectroscopy (AES) measurements of specimens. A defocused Xe-ion beam erodes a crater into the film with inclination angles of the scarps of only a few degrees or below. An electron beam then scans the shallow crater. The emitted Auger electrons reveal the lateral distribution of elements. As a result, one reconstructs the elemental concentration as a function of the sample depth profile. The AES data for the $Nb/Cu_{1-x}Ni_x$ specimen are shown in Fig. 4.

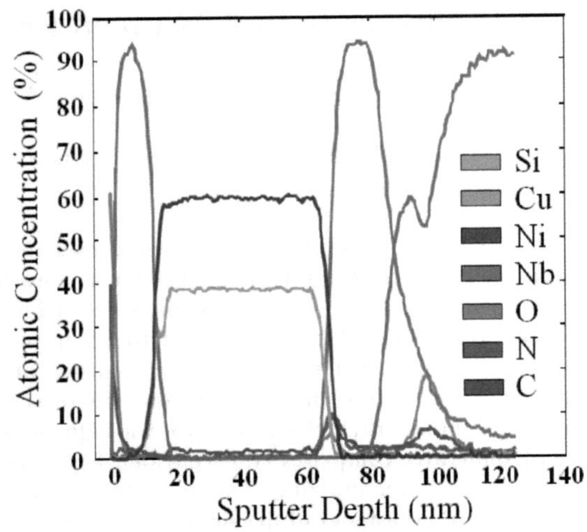

Fig. 4 Scanning Auger electron spectroscopy (AES) of a Si(substrate)/Si(buffer)/Nb/Cu1-xNix/Si(cap) sample (S/F specimen), dNb =7.5 nm and dCuNi=32.9 nm.

There are about 59 at.% Ni (in agreement with the RBS data) and 39.0 at.% Cu in the $Cu_{1-x}Ni_x$ film. There is a small concentration of O, C and N impurities at the $Nb/Cu_{1x}Ni_x$ interface as a result of physical absorption of gases from the residual atmosphere of the vacuum chamber. The $Cu_{1x}Ni_x$/Si (cap) interface is free of contaminations.

III. SUPERCONDUCTING PROPERTIES OF $CU_{41}NI_{59}$ /NB/$CU_{41}NI_{59}$ TRILAYERS

Fig. 5 demonstrates the dependence of the superconducting transition temperature for FSF samples on the thickness of the $Cu_{41}Ni_{59}$ layer. From the midpoint of the superconducting transitions the critical temperatures were evaluated and plotted as a function of the thickness of the $Cu_{41}Ni_{59}$ alloy layer in Fig.5. Depending on the thickness of the Nb layer as well deep critical temperature oscillations as also reentrant superconductivity could be realized. This behavior of critical temperature of two sample series, cut from a Nb wedge covered with $Cu_{41}Ni_{59}$ layers, is shown.

For specimens with $d_{Nb} \approx 15,5$ nm the transition temperature T_c reveals a non-monotonic behavior with a shallow minimum at about $d_{CuNi} \approx 12$ nm, it is just the qualitative behavior. For the series of specimens with $d_{Nb} \approx 12,8$ nm the transition temperature T_c decreases sharply for increasing ferromagnetic $Cu_{41}Ni_{59}$ layer thickness, until $d_{CuNi} \approx 17$ nm. Then, for $d_{CuNi} \approx 18$-49 nm, the superconducting transition temperature vanishes (at least T_c <40 mK, which is the lowest temperature measured).

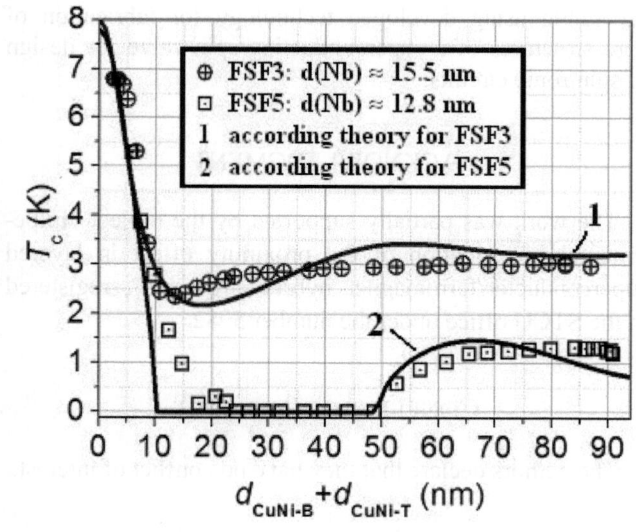

Fig. 5 Transition temperatures, T_c, of the investigated $Cu_{41}Ni_{59}/Nb/Cu_{41}Ni_{59}$ trilayer samples as a function of the thickness d_{CuNi-T} of the upper $Cu_{41}Ni_{59}$ alloy layer and the total thickness $d_{CuNi-B}+d_{CuNi-T}$ of the bottom and top layer, respectively. The thickness of the Nb layer is given in the figure. Here, sample series FSF3 and FSF5 result from a symmetric double wedge geometry, while series FSF1 and FSF2 result from a asymmetric single wedge geometry with one of the ferromagnetic layers kept constant.

For $d_{CuNi} > 50$ nm the transition into a superconducting state is observed again. Finally, T_c increases to a little bit above 1 K showing an outstanding reentrant superconductivity behavior with evidence for a second disappearance of the superconducting state at $d_{CuNi} > 90$ nm. Altogether, the $T_c(d_{CuNi})$ curves given in Fig. 5 represent all types of non-monotonic $T_c(d_{CuNi})$ behaviors predicted by the theory [2]. This phenomenon of the reentrant superconductivity in the S/F bilayer has been presented in our recent publications [3,4].

IV. FIVE-LAYER SPIN VALVE STRUCTURE

Finally, we introduced the inclusion of an antiferromagnetic (AF) layer (see Fig.6a) as a pinning layer against the rotation of the magnetization direction of one of the F-layers, while operating the superconducting spin valve.

The cross sectional TEM pictures in Fig.6(b) shows a perfect growth of the sample also in the case that the F/S/F core structure is grown on a cobalt oxide sublayer. The lattice constants of the antiferromagnetic cobalt oxide seem to indicate that cobalt monoxide has been grown.

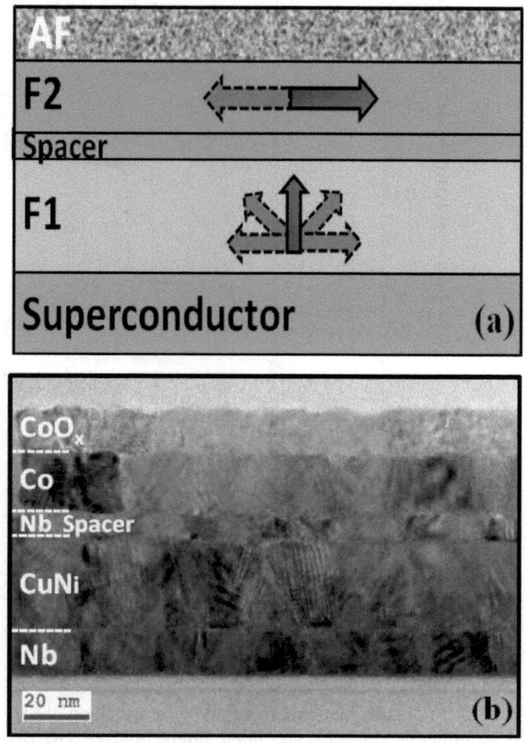

Fig. 6 a) Sketch of the layers stack in the deposited AF/F2/N/F1/S core structure of the spin-valve; b) Cross section TEM image of a AF/F2/N/F1/S structure (CoOx/Co/Nb-normal/CuNi/Nb-supercond./Si-subastrate). Arrows demonstrate the direction of magnetization of the F-layers at variation of the external magnetic field.

V. MAGNETIC PROPERTIES OF FIVE-LAYER STRUCTURE

An in-plane applied external magnetic field was used to control the magnetic configuration of the system from a parallel alignment, through a crossed one, towards an anti-parallel alignment of the F-layer magnetic moments (Fig. 7a). The intrinsic magnetization of $Cu_{41}Ni_{59}$ is perpendicular to the layer plane [6,7], Co has a natural magnetization parallel to the film plane [8].

The hysteresis loops of the samples #1, #16, cooled down at 10 kOe and recorded in the parallel to the film plane magnetic field along the in-plane easy axis of the Co film and the semi-easy axis of the $Cu_{41}Ni_{59}$ layer by a SC quantum interference device (SQUID) magnetometer are shown in Fig. 7a and in insert.

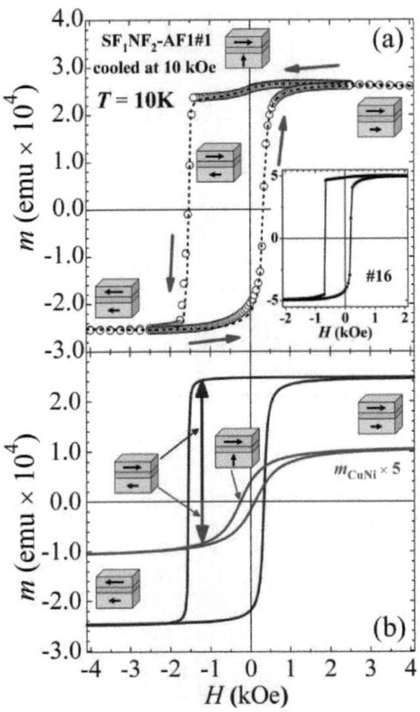

Fig. 7 (a) The magnetic moment, m, hysteresis loop of the Nb/Cu41Ni59/nc-Nb/Co/CoOx specimen, sample SF1NF2-AF1#1 (dCu-Ni ≈ 28 nm). The dashed line is a modeling according to Ref. [9]. Inset: Hysteresis loop of SF1NF2-AF1#16 (dCuNi ≈ 12 nm) adjacent to the sample used for the magnetoresistance, MR, measurements below. (b) Modeled components of the hysteresis loop: The blue line is the loop of the cobalt layer, and the red one the Cu41Ni59 (magnified by a factor of 5). Diamagnetic contribution of the Si substrate is subtracted in panel (b), however, appended to draw the model hysteresis loop (dashed line in (a)).

Upon sweeping the field ±4 kOe from the positive saturated configuration (PS) towards that negative (NS) starting at -1550 Oe, the sample passes through the state with crossed magnetic moments (CR) at approximately -250 Oe, towards the antiparallel alignment of the Co and $Cu_{41}Ni_{59}$ magnetic moments (APA) in the range from -250 to -1500 Oe. A similar sequence follows when sweeping the field in the reverse direction.

VI. CONCLUSIONS

It was developed advanced vacuum technology for reliable fabrication of multilayerd F/S/F and AF/F/S/F nanostructures. The range of controlled thicknesses of layers is accessible using developed technology for fabrication of core structures of a superconducting spin-valve for design of spintronic circuits.

ACKNOWLEDGMENT

The work was partially supported by the project "Experimental investigation of the proximity effect in layered superconductor/ferromagnet hybrid structures", registered in the STCU office under the number 5982.

CONFLICT OF INTEREST

The authors declare that they have no conflict of interest.

REFERENCES

1. A.S. Sidorenko, V.I. Zdravkov. R. Morari Device for preparation of superconducting layers, Patent of RM 175 (134) Y din 2010, Cl.Int. H01 L 21/00.
2. L.R. Tagirov, Physica C 307, 145 (1998).
3. V.I. Zdravkov, J. Kehrle, G. Obermeier, S. Gsell, M. Schreck, C. Müller, H.A. Krug von Nidda, J. Lindner, J. Moosburger-Will, E. Nold, R. Moari, V.V. Ryazanov, A.S. Sidorenko, S. Horn, R. Tidecks, and L.R. Tagirov, Phys. Rev. B 82, 054517 (2010).
4. Nanoscale Phenomena - Fundamentals and Applications. H. Hahn, A. Sidorenko, and I. Tiginyanu, Eds. Springer-Verlag, Berlin-Heidelberg, 2009, p.3-11.
5. Sidorenko Anatolie. Ventil Supraconductor de Spin. (Superconducting spin-ventil, Patent RM Nr.353 from 31.03.2011).
6. A. Ruotolo, *et al.*, J. Appl. Phys. **96**, 512 (2004).
7. I. S. Veshchunov, *et al.*, Pis'ma v ZhETF **88**, 873 (2008) [JETP Lett. **88**, 758 (2008)].
8. F. Radu and H. Zabel, in *Magnetic Heterostructures*, STMP v. **227** (edited by H. Zabel and S. D. Bader, Springer, Berlin-Heidelberg, 2008), Ch.3.
9. A. L. Geiler, *et al.*, J. Appl. Phys. **99**, 08B316 (2006).

Author: Anatolie Sidorenko
Institute: D.GHITU Institute of Electronic Engineering and Nanotechnologies ASM
Department: Criogenie
Street: Academiei 3/3
City: Chisinau
Country: Moldova
Email: anatoli.sidorenko@kit.edu

Application of the Strenghening Nanostructured Coatings Obtained at Electrodischarge Treatment by Tool Electrodes Manufacturing from Al-Sn Alloy

E. Yurchenko

T.G. Shevchenko Pridnestrovie State University, Tiraspol, Republic of Moldova

Abstract— **The results of the application of nanostructured coatings, obtained at electrodischarge treatment by tool-electrodes from Al-Sn alloy for restore and repair of machine parts are presented.**

Keywords—**Al-Sn alloy, electrospark alloying, nanostructuring, SnO2, wear resistance.**

I. INTRODUCTION

Currently, the use of parts made of aluminum alloys is a promising direction in the development of vehicles and machinery production, as their main advantages are: high resistance to corrosion and mechanical strength at low specific weight, weldability, cutability, good casting characteristics. The main disadvantage that is responsible for the underuse of aluminum alloys is their low wear-resistance when operating in the friction steams. Most of the vehicles and machinery that contain parts made of aluminum alloys (85-90%) lose operability not because of breakdowns, but as a result of the wear-and-tear of functional surfaces. Therefore, the increase of wear resistance, recovery and hardfacing of parts of aluminum alloys is a topical issue.

One of the promising methods for increasing the wear resistance, recovery and strenghening of surfaces of parts made of aluminum alloy, is the method of coating of the parts surfaces by means of electrodischarge treatment (electrospark doping, ESD). This method ensures high adhesion of the coating to the basis, possibility of local treatment of the surface, obtaining of the coating with desired functional properties, absence of high-temperature heating of the part while processing it, that excludes shrinkage and cracking, relative cheapness of the repairs due to low cost of equipment, low energy consumption, high ecological purity and simplicity of the technological process.

Earlier [1-3] it has been shown that under conditions of use of the ESD of aluminum surfaces by the tool electrode Al-Sn, the tin oxide nanowires are generated within the surface providing a high wear resistance of such coatings. The present work is devoted to the description of the fields of application for such coatings aimed at recovery and repair of parts made of aluminum alloys.

II. OBTAINING COATINGS USING ESD

A. Obtaining Tool Electrode

Rods having the diameters from $4 \cdot 10^{-3}$m to $8 \cdot 10^{-3}$m and length of $50 \cdot 10^{-3}$m casted from antifriction alloy AO20-1 (analogue AlSn20) have been used as tool electrodes. In conformity with the State Standard GOST 14113-78 the alloy contains [4] 17÷23% Sn; 0.7÷1.2% Cu; 0.02÷0.2% Ti; and the rest is aluminum.

Tool electrodes made of aluminum alloy AO20-1 were produced based on the following methodology:

The alloy with the predetermined chemical composition was melted in a graphite crucible, placed in a high-frequency inductor installation HFI-10U, and then poured into a specially produced metal mold (Fig. 1), where rods with the dimensions of Ø 8 X 50 X 10^{-3}m were produced. Tool electrodes of Ø 2 X 10^{-3}m and Ø 4 X 10^{-3}m were cast into a special mold using centrifugal machine with the vertical axis of rotation.

The technology of manufacturing tool electrodes for ESD involved the following operations:

- preparation of the burden materials;
- melting the alloy in an induction furnace;
- casting alloy into the metal mold or casting it into a mold of the centrifugal machine;
- cutting heads of casting, cleaning, turning radius of the electrode end.

Refractory elements (Ti and Cu) were introduced into the aluminum melt in the form of addition alloy (50% Al + 50% Cu) and (90% Al + 10% Ti), the remaining elements (Al and Sn) were introduced in their pure form. The calculation of the burden materials [5] was performed based on the average content of elements (wt.%): Sn – 20; Cu – 1; Ti – 0,1; Al – the rest. Electrodes of other aluminum alloys (Al-Pb, Al-Ni, Al-Cu, etc.) were produced using the same methodology.

Waste of elements during the melting was not taken into account, since the time of alloy melting did exceed 50 seconds. Crucible capacity is 100g.

The composition of the burden per 100g of melt AO20-1:
2 g of Al - Cu alloy containing 50% Al + 50% Cu;
1 g of Al - Ti alloy containing 90% Al+10%Ti;

V. Sontea and I. Tiginyanu (eds.), *3rd International Conference on Nanotechnologies and Biomedical Engineering*,
IFMBE Proceedings 55,
DOI: 10.1007/978-981-287-736-9_25

20 g of pure Sn grade O1;

77 g of pure Al grade A1.

Fluxing agent (50% NaCl + 50% KCl) was applied to the melt surface to prevent its oxidation. The metal mold was painted with refractory paint based on zinc oxide and sodium silicate (50% ZnO + 5% sodium silicate + 45% water) before pouring the electrodes. After being painted the metal mold was dried in an electric furnace SNOL 2.5 at 250 ° C, then the melt was immediately poured into it directly from the graphite crucible. Heads of casting were cut off the rods (tool electrodes) after solidification and cooling, the rods were cleaned from paint using sandpaper, and the end of the rods was turned with the radius at the top ≈ (2 ÷ 4) 10^{-3}m. (fig.1)

Fig. 1 Metal mold for casting tool electrodes made of aluminum alloy.

A series of experiments on determination of the elemental composition (EDX) were carried out during the production process with the view to study the composition of electrodes.

The original specimens of the tool electrodes that were not involved in ESD process have been studied. (Table 1).

Table 1 Chemical composition of electrodes

Composition (weight %)					Note
Al	Sn	Fe	O	Zn(Cu?)	
67.8	19.6	1.1	11.5	0.2	Original specimen of the electrode (top)
68.4	23.0	1.0	7.3	0.25	Original specimen of the electrode (bottom)

1. The content of Al and Sn in the original specimens of tool electrodes is within the limits of the State Standard GOST, i.e. Al ~ 68%, Sn ~ 20%.

2. The amount of oxygen on the surface of electrodes is ~ 9%, apparently in the form of oxides SnO_2, Al_2O_3, as the aluminium alloy surface is always covered with a continuous film of oxides.

3. A small content of Fe in the electrode is explained by diffuse transfer that is a result of the contact of specimens with the cast iron surface of the metal mold on pouring. A small segregation of Sn is observed in the electrode: the bottom part of the electrode with a diameter of 8 mm contains Sn in the concentration of ~ 23% and the top part - 19% , i.e. at the moment of casting the heavier tin (γ = 7.3 g / cm^3) goes downwards, and lightweighted aluminum (γ = 2.7 g /cm^3) goes upwards.

4. The presence of Zn in specimens may be associated with the use of paint based on oxide of Zn. Since no traces of Cu have beendetected in the specimens, it can be assumed that this is due to an inaccuracy of analysis.

A technological specimen has been developed for application of coating and subsequent study of its physical and mechanical properties (Fig. 2).

The specimens were prepared by milling of the alloy sheet D1 (GOST 4784) [6] with the following composition: Cu 3.8 ÷ 4.8%; Mg 0.4÷0.8%; Mn 0.4÷0.8%; the rest is aluminium.

Fig. 2 Technological specimen for study.

The apertures were designed for mounting on the mechanical ESD installation and on the friction testing machine. The working section of the specimen represented a plate $3 \cdot 10^{-3}$m thick and $5 \cdot 10^{-2}$ m long, the thickness of a specimen is $4 \cdot 10^{-3}$m. The ribs on both sides of the specimen are specially designed to prevent its buckling during manufacture and within the coating process by means of ESD. Another destination of ribs is to exclude the "edge effect" under ESD. When the tool electrode during the coating process reaches the edge of the specimen, there is a strong sparking and entrainment of material from the cathode-specimen.

Preparation of the specimen for coating involved the removal of scratches left after milling using sandpaper and the grinding machine for preparation of microsections. After grinding the specimens were labelled, weighed on the analytical balance VLR 200 up to the fourth sign, and then coated and re-weighed to determine the change in weight of the specimen.

B. Obtaining Coatings on the Aluminium Surface Using Al-Sn Electrode

The installation ALIER-31 served as the source of power for the ESD, in so doing the following modes were generally used 4÷6 (pulse width 125÷500 ms, current pulse amplitude 175A, pulse energy 0.39÷1.58 J), the amplitude of the electrode vibration – 0.5. The rods having the diameter of 8 mm produced from specially manufactured alloy Al-Sn, containing 20% of Sn and doped with copper and titanium (~ 1% Cu, 0.1% Ti) melted by a standard technique were used as electrodes. The plates measuring 3x5x55 mm from aluminium alloy D1 (GOST 4784) were used as specimens for ESD coating. The coating was applied in both manual and automatic modes.

C. Manual Mode of Coating

The following technology was used for the manual coating of aluminium alloy specimens:

— to ensure uniformity and equal conditions for the ESD process there were used technological specimens of only one size that have not undergone any special surface preparation (washing, degreasing, etc.);
— the speed of the tool electrode was chosen in such a way that the total processing time made up ~ 3 min/specimen;
— the electrode was held perpendicularly to the surface to be treated. The inclination of no more than 10÷15° was allowed;
— the working section of the electrode was in the form of a cone with a tip radius of $2\div4 \times 10^{-3}$m.;
— on wear-and-tear the form of the electrode tip was periodically restored by trimming;
— the speed of the electrode movement was maintained uniform. The coating was applied in stripes. The tool electrode was moved manually in circular motions with overlapping of previous layers for 0.25 of the stripe size;
— during the process of moving the electrode was not detached from the treated surface from the beginning of ESD treatment till the end of it. This was necessary because a strong sparking occurs on detaching the tool electrode from the specimen, it leads to the local erosion of the treated surface (in the form of a pit) and modifications of the parameters of weight gain (or entrainment) both of the technological specimen and tool electrode;
— during the manual coating only vibrating electrode with a constant vibration frequency (~100Hz) was used;
— all seven operation modes of installation ALIER-31 were used. After the process of ESD optimization there were selected modes 5 and 6, with the energy ratio of 0.8 and amplitude of vibration of 0.5;
— the coating was applied to a wide surface of the technological specimen.

After the coating the ribs were removed manually by filing;

It is difficult to maintain stable parameters upon manual coating, since the traverse speed of the tool electrode along the surface of the specimen, its die hold-down pressure to the specimen, the size of the applied layer overlapping are set up by the operator and can vary considerably (due to „human factor“). In such circumstances the important parameters of the technological process become poorly controlled.

D. Machine-Operated Mode of Coating

The surface for machine-operated mode of coating was prepared using the same methodology that was used for the manual mode of coating. Modernized turning lathe served as an experimental unit.

Upon preparation the specimens were fixed on a sliding carriage of the experimental unit. The tool electrode was fixed in the vibration exciter of the ALIER 31 installation. The necessary operating modes were set up on the control panel of the experimental unit, and namely: energy ratio, the amplitude of the vibration exciter, mode, the amplitude and frequency of the transverse vibration. Thereafter the unit was switched on and machine-operated coating was carried out due to movement of the sliding carriage back and forth. Immediately after the application of one or two layers of coating the specimens and electrodes were weighed again, therefore it was possible to determine the weight gain or entrainment of the specimen or the electrode material.

Two methods of coating were used: at constant number of applied layers and different traverse speed of the tool electrode relative to the specimen (method 1) and "at constant value of the transmitted charge" or at constant amount of energy introduced into the spark gap while maintaining a constant time of the specimen processing (method 2), as well at different speed.

While using method 1 the constant number of the applied layers was equal to four at different traverse speed of the tool electrode relative to the work piece. The traverse speed of the specimen varied from (0.3 ÷ 2) 10^{-3} m/s. There were carried out four passes of the tool electrode at a predetermined speed across the entire treated surface of the specimen.

While using method 2 the number of applied layers was increasing proportionally to the increase of the traverse speed of the tool electrode. Different number of layers was applied in case of processing by method 2, and namely: 4 layers - at a traverse speed of the tool electrode equal to 0.5 10^{-3} m/s and 16 layers at a traverse speed equal to 2 10^{-3} m/s. 8 and 12 layers were applied accordingly at a traverse speed equal to 1 10^{-3} m/s and 1.5 10^{-3} m/s.

Based on the results of specimens weighing, a specific speed of processing G (mg/s.cm^2) was determined, which could be both positive and negative value depending on the conditions of the process. In the latter case, the weight of the specimen after processing did not increase, but decrease. The overall time of processing, as well as the total area of the modified surface served as the basis for calculation of G.

The obtained surfaces were investigated by means of X-ray phase analysis (DRON UM1, FeKα cobalt radiation, Mn filter, θ/2θ method), as well as by means of scanning electronic microscopy (microscope TESCAN with the attachment for elemental analysis of the surface INCA Energy EDX (Oxford, UK)).

The study of the coating morphology revealed the presence of nanowires presumably from the tin oxide located in the aluminium matrix (Table 2, Figure 3), their hardness can reach 1200 kg/mm^2 by Vickers.

Table 2 Composition of nanowires in the coating

Element	Sn	Al	O	N
Weight %	61.5	19.1	11.5	4.9

Fig. 3 The morphology of the coating on a specimen from the alloy D1 carried out by tool electrode alloy Al-Sn.

III. Application of Tool Electrodes Made From Al-Sn for the Recovery of Matching Sites of Rolling Bearings

The property of high wear-resistance of the aluminium surfaces coatings can be widely used for the recovery of matching sites for bearings in bearing units of electric motors, generators and pumps, with housings made of aluminium-magnesium or duralumin alloys. The nomenclature of parts recovered via ESD is shown in Figure 4.

Fig. 4 Parts from aluminium alloys recovered via ESD by tool electrodes from alloy Al-Sn.

The service life of the majority of rolling bearings used in bearing units of electric motors and generators makes up on average 3000 hours and largely depends on bearing fittings in housing parts. The increase of the gap between the outer ring of the bearing and the matching site in the node appears as a result of fretting corrosion and worsens the distribution of load between the rolling elements, resulting in the reduction of durability of the bearing. The distribution of defects of electric motors and generators that require repairs is as follows: 50% - mechanical damage, the remaining 50% - defects of the electrical part. At the same time 30% of all electric motors and generators brought for repairs have the increased wear and tear of matching sites.

The causes of these defects are:

- outage of bearings or poor quality of their work;
- increased vibration due to the imbalance of the rotor or the driven mechanisms;
- bad quality centring of the engine with the driven unit;
- natural deterioration of the metal of bearings;

- insufficient reinforcement of the matching site;
- overload of mechanical part, shock loads from the drive;
- multiple pressing-out and press-fitting of bearings, resulting in impaired uniformity of the gap between the rotor and the stator, excessive vibration with corresponding negative consequences.

There are the following ways to restore the matching sites of bearings in housings from aluminium alloys:

- mounting of additional components - sleeves;
- argon - arc built-up welding;
- application of electrolytic coatings;
- application of polymer coatings based on anaerobic sealants.

Mounting of additional components - sleeves, is possible not in all cases due to the insufficient thickness of the housing wall for boring in the area of the bearing unit.

Due to insufficient thickness of the housing wall for built-up welding argon arc built-up welding can cause buckling and cracking of the housing surface. Furthermore, built-up welding does not provide for reinforcement of the worn surfaces;

Disadvantages of electrolytic coating are the complexity of the technological process, the need for expensive equipment, consequently increasing labour intensity and costs.

Application of polymer coatings based on anaerobic sealants is not always ensures the required service life of repaired electric motors, generators and pumps.

The material applied to the matching site should have a high module of elasticity and tensile strength in the operating range of temperature. The experiments have shown that such material, that meets the above requirements, is a coating obtained using ESD by electrodes from AO20-1 alloy. Due to high wear-resistance the coating counteracts the fretting - corrosion well. On the other hand, the wear and tear of matching sites of bearings usually does not exceed $(0.1 \div 0.3)\ 10^{-3}$ m, which allows recovering them by means of ESD using simple mobile equipment and relatively simple technology. Moreover, the process of recovery of matching sites is environmentally friendly, as no hazardous substances are used and there is no environmentally dangerous waste.

Owing to the high roughness of coatings for to ensure the dimensions and smoothness of the matching site, the coating was applied taking into account the subsequent mechanical processing by boring on a turning lathe to obtain the necessary coupling size and surface smoothness.

Thus, the recovery technology for matching sites of bearings in aluminium parts using ESD consists in the following:

- cleaning of matching sites from dirt;
- measuring the diameter of the aperture with an indicating hole gauge and determining the required thickness of the coating;
- application of ESD coating;
- mechanical processing of the aperture (boring or rolling-off) of the obtained coating.

IV. The Use of the Tool Electrodes From Al-Sn to Eliminate Leaks in Aluminium Tubes of Heat Exchange Units

ESD tool electrodes from Al-Sn alloy can be used to eliminate leaks in aluminium tubes of radiators and heat exchange units.

At the moment, there are the following methods of the radiators repairs:

- introduction of special crack sealing compositions into the coolant;
- plugging of defective tubes in various ways;
- sealing the tubes by means of compositions based on epoxy resins and anaerobic sealants;
- soldering of defective tubes by low-melting and refractory solders.

Practical application of the crack sealing compositions for tubes showed nondurability of this type of repair - after a short period of time the leak appears again due to frequent heating and cooling, and therefore thermal deformation of the tubes.

Plugging of leaky tubes reduces the area of heat transfer, worsens heat exchange and can lead to engine failure due to overheating.

Sealing the tubes by means of compositions based on epoxy resins and anaerobic sealants as well does not always lead to the desired results as the compositions often peel off and the working medium leaks occur again.

Soldering or welding of defective tubes by low-melting and refractory solders is the most reliable way to recover them, but encounters the following problems:

- poor weldability and solderability, as the refractory Al_2O_3 film with a high melting point (2500°C), preventing a reliable connection of welded or soldered metals is present on the surface of the tubes;
- high thermal conductivity of aluminium alloys, which hinders welding and soldering, as it requires a large supply of heat;
- high coefficient of linear expansion, which may lead to buckling of the parts
- thinness of the tube walls ($\delta=0{,}4\cdot10^{-3}$m), which often leads to burnout.
- low melting point (660°C).

Sophisticated technology is used for the preliminary removal of the oxide films from the surface of the defective tube: degreasing, etching, pickling in various environmentally harmful compounds. Moreover, it is necessary to apply fluxes, consisting of chlorides and fluorides which can cause corrosion of soldered joints; therefore it requires careful washing and neutralization of the salt solutions remaining on the surface of the soldered parts. In this regard, this technology is environmentally harmful.

Taking into account the drawbacks of existing technologies, there have been developed a new technology of flux-free soldering of automobile and tractor radiators. The place of the defect is cleaned of mud and using ESD the surface is modified by the electrode from AO20-1 alloy with subsequent application of the low-melting solder POS-40 using the ESD. ESD allows destroying the oxide film on the surface of the tubes and facilitating the subsequent tinning of the surface with a low-melting solder POS-40. After tinning the place of defect is soldered in a standard way and that does not present difficulties.

Application of the preliminary coating with the sublayer by means of the electrode from AO20-1 alloy and solder POS-40 on the defective surface with ESD allows to reduce the number of preparatory operations and improve the reliability of soldered joints and at the same time reduces repair costs of radiators with aluminium tubes.

For working off the technology of repair of aluminium radiator core there have been designed a stand that allows determining the place of a leak in the radiator by supplying in it compressed air under pressure.

Upon plugging of all apertures of the radiator using special plugs the compressor hose was connected to the pipe going to the expansion tank. The pressure of compressed air was regulated by the air reducer ranging from 0.03 to 0.15 MPa. When radiator was sunk into water, the air bubbles going up indicated the location of leakage. The technology was first worked off on aluminium tubes of a faulty radiator. Cuts and holes simulating defects were made on the tubes.

After that, the defect places were covered by means of ESD using the electrode from AO20-1 alloy by the coating, which destroyed Al_2O_3 oxide film on the surface of the tube and facilitated the preparation of the surface for soldering with the solder from tin - lead alloy POS40. Soldering was done using ordinary soldering iron with the capacity of 40 ÷ 60 W without flux. The objective of the soldering process was to level the coating on the surface of the tube so that the layer of solder around the hole is not less than 10×10^{-3} m wide. A layer of such width is necessary to increase the area of adhesion of the solder with the radiator tube and improve reliability and integrity of the soldered layer.

Testing of tubes on the stand under standard conditions (pressure 0,125MPa, lagged 10 minutes) showed that all 5 tubes withstood the test successfully. The technology was preserved when working directly with radiators. Subsequent tests of repaired radiators on the stand showed reliability of technology as there was no leak.

ACKNOWLEDGMENT

The work has been performed within the framework of the budget financing by Pridnestrovian State University after T.G. Shevchenko.

CONFLICT OF INTEREST

The authors declare that they have no conflict of interest.

REFERENCES

1. Yurchenko V.I., Yurchenko E.V., Fomichev V.M., Baranov S.A., Dikusar A.I. Obtaining of nanowires in conditions of electrodischarge treatment with Al-Sn alloy. *Surf. Eng. Appl. Electrochem.* 2009, 45 (4), 259 - 264.
2. Dikusar A.I.Obtaining Nanowires under Conditions of Electrodischarge Treatment In: *Nanowires – Implementations and Applications*, Abbas Hashim, Ed. INTECH. Org., 2011, pp. 357-374.
3. Agafii V.I., Petrenko V.I., Fomichev V.M., Yurchenko V.I., Yurchenko E.V., Dikusar A.I.
4. Deposition of Al-Sn Coatings on Aluminum Surface Using Electrospark Alloying and Their Wear Resistance under Dry Friction, *Surf. Eng. Appl. Electrochem.*, 2013, v.49, no3. pp.181 – 188.
5. Мотовилин Г.В, Масино М.А., Суворов О.М. Автомобильные материалы. Справочник. –М.: Транспорт, 1989. – С 147./ Motovilin G.V., Masino M.A., Suvorov O.M. Materials on Automobiles. Handbook. -M .: Transport, 1989. – p 147.
6. Иванов В.Н., Казенников С.А., Курчман Б.С., Лященко Н.Н. и др. под ред. Шкленника Я.И.,
7. Озерова В.А. Литье по выплавляемым моделям. - М.: Машиностроение 1984. - С. 246./ Ivanov V.N., Kazennikov S.A., Kurchman B.S., Lyashenko N.N. et al., editted by. Shklennik Ya.I., OzerovV.A. Casting by Dispensable Patterns. – M.: Engineering Industry, 1984. – p 246.
8. Арзамасов Б.Н., Макарова В.И., Мухин Г.Г., Рыжов Н.М. и др Материаловедение: Учебник для вузов. - М.: издательство МГТУ им. Н.Э.Баумана 2005. – С.363./ Arzamasov B.N., Makarova V.I., Mukhin G.G., Ryzhov N.M. et al., Materials Science: Textbook for universities. - M.: Publishing house of MSTU after N.E. Bauman 2005. – p 363.

Author: E. Yurchenko
Institute: T.G. Shevchenko Pridnestrovie State University
Street: str.25 Oktyabrya 128,
City: Tiraspol
Country: Republic of Moldova
Email: e.iurcenko@mail.ru

Structural, Optical and Electrical Properties of ZnO: Al Thin Films Synthesized by Sol-gel Method

A. Rogachev[1], A. Semchenko[1], V. Sidsky[1], V. Gaishun[1], D. Kovalenko[1], V. Gremenok[2], H. Zaretskaya[2], and L. Sudnik[3]

[1] Francisk Skorina Gomel State University, Gomel, Belarus
[2] Scientific-Practical Materials Research Centre of the National Academy of Sciences of Belarus, Minsk, Belarus
[3] State Scientific Institutions Powder Metallurgy Institute, Minsk, Belarus

Abstract— **The possibility of synthesis of active layers for solar cells by sol-gel method has been shown. The structural features of the surface (AFM), crystallization behavior (XRD) during the heating and properties of synthesized films are discussed. Achieved parameters suggest the possibility of using synthesized ZnO layers in solar cells. The perturbation imposed by Al atom incorporation leads to the atomic relaxation, which is computed and discussed in detail.**

Keywords— **sol-gel method, semiconductors, surface structure, crystallization, solar cells.**

I. INTRODUCTION

Zinc oxide ZnO is widely used in a number of optoelectronic devices such as solar cells [1-2]. ZnO films can be prepared by various methods. Of particular the opportunities to obtain ZnO films with controlled physical properties by sol-gel synthesis is very attractive. This method combines the simplicity of the process and the lower cost of used equipment and materials. In this paper, the structural and optical properties of ZnO:Al films formed by sol-gel method has been discussed

II. MATERIAL AND METHODS

The required amount of zinc acetate filled into absolute isopropyl alcohol and stirredin order to manufacture the ZnO/Al layer from the precursor (sol). Then sol was stirred for 30 minutes. Sol was kept at the room temperature (22±2)° C for 2-3 days. Monoetalamin was selected as the catalyst because reduced the exposure time to two days and ensure their stability during the month. After applying the sol onto the surface of glass, single crystal silicon etc., the samples were placed in the furnace and were heated stepwise at intervals of 20 °C to the temperature of 350 0C for 10 minutes. The process of applying and drying had been repeated until the desired thickness of the ZnO/Al layer was achieved.

X-ray phase analysis was performed by ARL X'tradiffractometer (Thermo Ficher Scientific, Switzerland), using Bragg-Brentano geometry and reflected Cu Kα1-and Kα1-radiation. To reduce reflections from the substrate small incidence angles was used. Angular 2θ-range was selected from 10° to 60°.

Diffraction peaks identification was performed using database JCPDS (with software Search Match). Diffractograms were processed using JANA 2006 software. The electrical resistivity ρ was measured at room temperature by four-probe method.

III. RESULTS AND DISCUSSION

A. Structural Properties of ZnO: Al Sol-gel Films

The ZnO:Al film synthesized by sol-gel method was investigated with AFM (Figure 1) wich confirmed the high

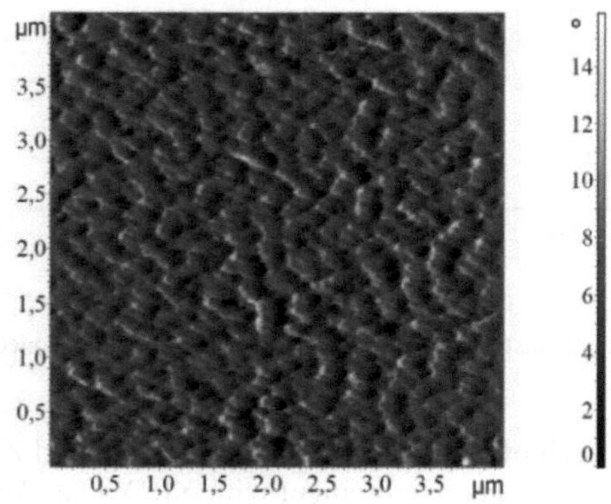

Fig. 1 AFM of ZnO:Al sol-gel films

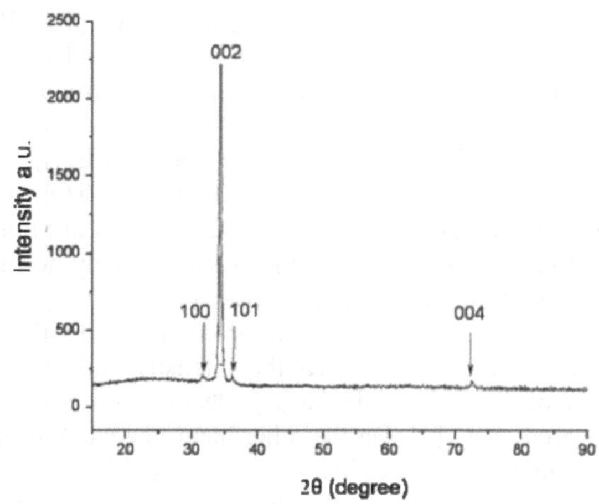

Fig. 2 XRD spectrum of ZnO:Al sol-gel films

V. Sontea and I. Tiginyanu (eds.), *3rd International Conference on Nanotechnologies and Biomedical Engineering*,
IFMBE Proceedings 55,
DOI: 10.1007/978-981-287-736-9_26

homogeneity of the films. According to the XRD ZnO:Al films are polycrystalline, single-phase and have a hexagonal wurtzite structure (Figure 2). The angular position of the peaks is consistent with the JCPDS-data for nominally pure zinc oxide (JCPDS, № 36-1451). The Al^{3+} ions do not violate the hexagonal wurtzite structure ZnO.

B. Optical Properties of ZnO: Al Sol-gel Filmsnsent

Figure 3 shows the transmission spectra as a function of wavelength for samples of zinc oxide obtained at different aluminum content (0.6-1.4 wt.%). Studies of the optical characteristics of the films ZnO: Al showed that zinc oxide films have a high transmittance T (λ) = 80 – 94%, indicating that this material is acceptable for use as a window layer and for tandem solar elements.

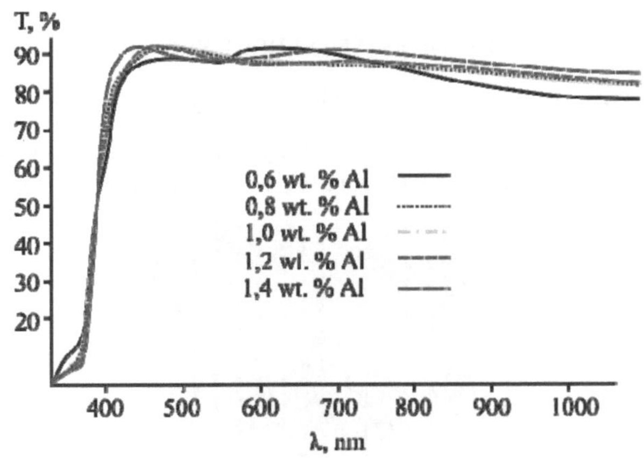

Fig. 3 Transmission of ZnO films depending on the concentration of aluminum

Research of Raman spectra (RS) of obtained ZnO:Al sol-gel films is in good agreement with XRD data. According to the theory of groups for ZnO, which is crystallized in hexagonal wurtzite structure with spatial symmetry C_{6V}, there are a set of phonon modes: $1A_{1\,(TO)} + 1A_{1(LO)} + 2B_1 + 1E_{1(TO)} + 1E_{1(LO)} + 2E_2$ [3]. Phonon modes are observed in ZnO at the following frequencies: $E_{2(low)}$ (low frequency) = 101 cm^{-1}, $E_{2(high)}$ (high-frequency) = 437 cm^{-1}, $A_{1(TO)}$ = 380 cm^{-1}, $A_{1(LO)}$ = 574 cm^{-1}, $E_{1(TO)}$ = 407 cm^{-1} and $E_{1(LO)}$ = 583 cm^{-1}, respectively (LO - longitudinal optical, TO - transverse optical modes). Among these $2B_1$ modes are not Raman activity [4]. In the fall of the beam is strictly normal to the surface of the ZnO layer are observed only $A_{1(LO)}$ and E_2 modes, while the other modes are prohibited under the rules of selection. It is known that the intensity and frequency position of $E_{2\,(high)}$ mode largely depends on the crystal lattice stress ZnO [4]. On the other hand, the study of spectral characteristics of oscillation $A_{1(LO)}$ provides the opportunity to get information not only on the chemical composition of the resultant ZnO: Al films, and about the presence of foreign phases and areas with locally ordered lattice. According to the data [3] all the bands observed in the Raman spectra of the resulting coatings (Fig. 4) can be attributed to scattering by phonons of zinc oxide. In the low-frequency Raman spectra of sol-gel ZnO:Al films (Series № 1-3) are observed mode $E_{2\,(low)}$ at 330 cm^{-1}, high-frequency mode $E_{2(high)}$ in the range of 427 - 439 cm^{-1} and mode $A_{1\,(LO)}$ in the range of 575-582 cm^{-1}. Mode at 330 cm^{-1} is a second-order oscillation mode phonon $E_{2\,(high)}$ and is due, mainly, the vibrations of atoms Zn [5]. Fixed high-$E_{2\,(high)}$ and $A_{1\,(LO)}$ mode are associated with the fluctuations of the oxygen atoms in a perfect crystal ZnO and local vibration of intrinsic lattice defects such as oxygen vacancies, intermediate cations [5].

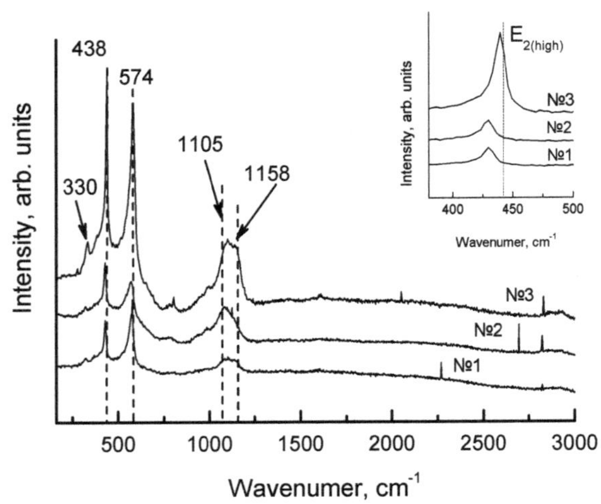

Fig. 4 Raman spectra of ZnO:Al films produced by sol-gel deposition of solutions based on: №1 -2-methoxyethanol - isopropylalcohol;№2 - dimethylformamide; №3 -isopropylalcohol.

The spectral position, intensity and half-width (FWHM) of $E_{2\,(high)}$ and $A_{1\,(LO)}$ oscillations are significantly different for the ZnO:Al coatings obtained from different starting compounds. Relatively to the standard ZnO, mode $E_{2(high)}$ and $A_{1(LO)}$ which are observed for specimens series № 1-2, are shifted to lower frequencies. A significant red shift of $E_{2\,(high)}$ mode (Δν≈ 8 cm-1) is associated with the disordering influence of the crystal lattice, a nonuniform distribution of Al impurities and indicates the presence of the coatings compressive stresses. With the improvement of the quality of the crystalline coating ZnO: Al modes $E_{2\,(high)}$ and $A_{1(LO)}$ are shifted to higher frequencies. Simultaneously, the intensity of these oscillations and their half-width (FWHM) are reduced (for mode $E_{2(high)}$ up to 8 cm^{-1} and for mode $A_{1\,(LO)}$ up to 10 cm^{-1}, respectively). Reflex from fluctuations $E_{2\,(high)}$ on the Raman spectra of the films ZnO: Al Series №3 is symmetrical (Figure 3). That fact is indicate the process of ordering the crystal lattice [6]. For the series of ZnO:Al films, obtained from isopropyl alcohol and monoetalamina (№3), oscillation position of $E_{2\ (high)}$ and $A_{1(LO)}$ is shifted to the standard ZnO. This means that the films of this series have no stress.

The fact that the mode $E_{2(high)}$, due to vibrations of oxygen atoms in the crystal lattice of ZnO, is subject of more obtaining conditions than mode $E_{2(low)}$, attributed to fluctuations sublattice Zn, suggests that defects in the oxygen sublattice are dominant. Increase of the intensity of mode $E_{2\ (high)}$ with the improving of crystal quality of the filmss indicates on the ordering of oxygen anion sublattice formed in layers with optimal choice of reactants and deposition conditions.

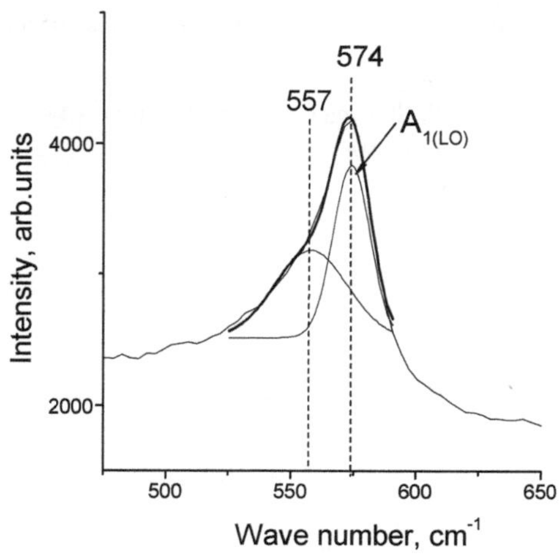

a)

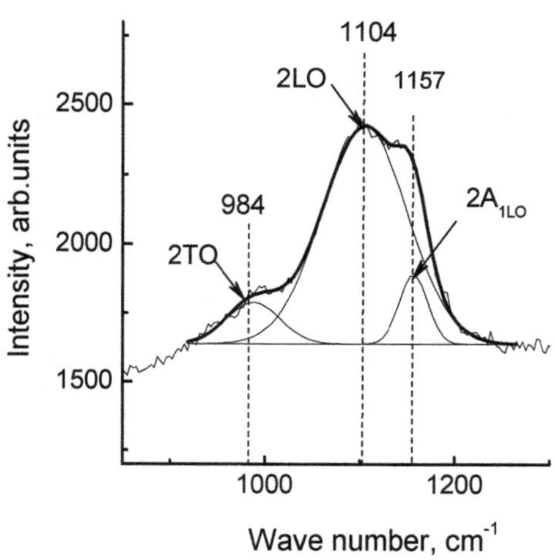

b)

Fig. 5.Raman spectra of ZnO: Alcoatingsproducedusing sol-gel precipitation from solutions based on isopropyl alcohol: a) the area of frequency-phonon mode A1 (LO); b) the frequency range of the phonon mode 2_{LO}

C. *Electrical Properties of Properties of ZnO: Al Sol-gel Filmss*

It should be noted that the mode $A_{1\ (LO)}$ of the resulting film at 574 cm^{-1} in the Raman spectra is asymmetric (Figure 4a). According to [7, 8] asymmetry of $A_{1(LO)}$ mode, observed in doped aluminum and gallium ZnO films, is attributed to local disordering of the crystal lattice, and created by oxygen vacancies. Therefore, we can conclude that the increase of the intensity fluctuations at 557sm^{-1} in the Raman spectra of the samples in the series №3 is due to the higher concentration of aluminum Al^{3+} in these films. Fig. 6 depicts the variation of resistivity (ρ), carrier concentration (n) and mobility (μ) of Al doped ZnO films with doping concentration. The resistivity first decrease with increasing current and a lowest resistivity (6.5 ohm cm) is obtained at a current 0,8*10^{-4} further increase in current the resistivity value started to increase significantly.

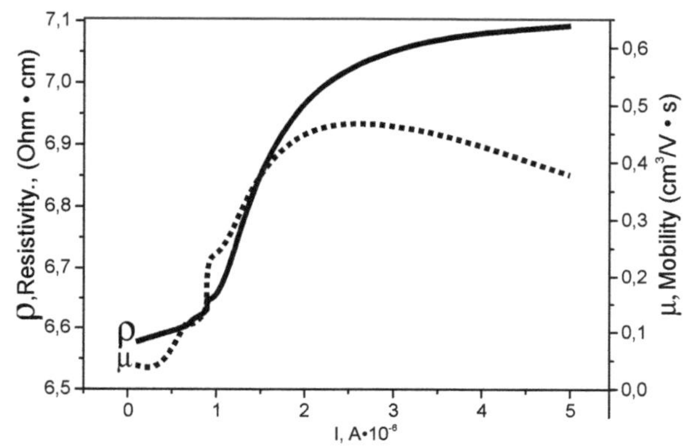

Fig. 6 The dependence of the resistivity of the charge mobility of the current ZnO: Al-film obtained by a sol-gel method

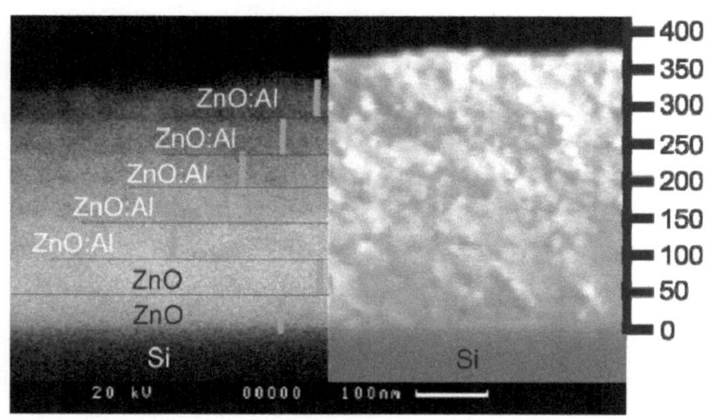

Fig. 7 SEM image ZnO:Al-Si film annealed at 550 ° C for 1 hour in an atmosphere of oxygen

Figure 7 depict the SEM micrographs of the surface of Al doped ZnO thin film with (0,4 at.%). The surface morphology of the deposited films is observed to be smooth and plane without crack. From the figure we find that the microstructure of the films consisted of many round shaped crystalline particles. The microstructure formed is found to uniform and compact interconnected grains. The average grain size of the ZnO thin films is around 20 nm.

IV. CONCLUSIONS

It has been shown that the use of sol-gel method allows to synthesize the functional nanostructured materials semiconductors (ZnO)) for electronic devices with required properties.

ACKNOWLEDGMENT

This research was partially conducted in the framework by the Belarusian Foundation for Basic Research, project T15MLD-033.

CONFLICT OF INTEREST

The authors declare that they have no conflict of interest.

REFERENCES

1. Sci J, Technol., Suwanboon, Songklanakarin et al. (2008) Fabrication and properties of nanocrystalline zinc oxide thin film prepared by sol-gel method, 30 (1), , pp. 65-69.
2. Mat. J, Today, L. Shmidt-Mende and J. L. MacManus–Driscoll. (2007) ZnO – nanostructures, defects, and devices, 10 (5), pp. 40-48.
3. Damen T., Porto S., Tell B., (1966) Phys. Rev. 570:142
4. Cuscó R, Alarcón-Lladó E, et al. (2007) Phys. Rev., 165202: B **75**
5. Serrano J, Manjon F. J et al. (2003) Phys. Rev. Lett., 055510:90
6. Arguello C.A,. Rousseau D.L, (1969) S.P.S. Porto, Phys. Rev. 1351:181
7. Lorite I., Villaseca L. et al. (2013)Thin Solid Films, 657:548.
8. Cuśco R, Alarćon-Lladó E, et al. (2007) Phys. Rev., 165202:B 75

Author: A.V. Rogachev
Institute: Francisk Skorina Gomel State University
Street: Sovetskaya 104
City: Gomel
Country: Belarus

Single Nanowire Nanosensors: A Case Study of the Effects of Metal Doping on ZnO

O. Lupan[1], L. Chow[2], Th. Pauporte[3], B. Viana[3], and R. Adelung[4]

[1] Technical University of Moldova/Department of Microelectronics and Biomedical Engineering, Chisinau, Republic of Moldova
[2] University of Central Florida/Department of Physics, Orlando, USA
[3] PSL Research University, Chimie ParisTech – CNRS, Institut de Recherche de Chimie Paris, UMR8247, Paris, France
[4] University of Kiel/ Institute for Materials Science/Functional Nanomaterials, Kiel, Germany

Abstract— **Individual nanowire/nanorod sensors of semiconducting oxides are of great interest for science and engineering. In current work, the effects of metal doping on ZnO nanowire properties are presented. It is found that metal-doping greatly improved the performance of ZnO nanowire nanosensors. Experimental results and possible mechanisms of the improved performance are discussed.**

Keywords— **Nanosensor, Nanowire, ZnO, UV Detector.**

I. INTRODUCTION

Sensors, micro-sensors and nanosensors are in great demand due to miniaturization and extensive development of bottom-up nanotechnologies and lately for various customer applications [1-4]. Recently, the investigations of numerous research groups around the world has focused on studies of semiconducting oxide nanostructures with different morphologies for fundamental science and engineering applications [1-4]. In this context, nanowires and nanorods, as well as, branched rods or tetrapods of semiconducting oxides are investigated thoroughly for their integration in various device structures and especially in single component ones [1-6]. Zinc oxide (ZnO) was selected as the main candidate for the nano-building block since it is an important *n*-type semiconductor material with a large band gap E_g = 3.37 eV at 300 K and can be prepared by simple synthesis technologies [1-2]. A single ZnO tetrapod or crossed-rods based device was developed in the first decade of the 21st century by a new bottom-up approach [1,2,4]. An individual micro- and nano-rod/wire sensors based on pure ZnO showed quite promising applications at that time [5,6]. Such nanodevices were able to operate as UV photodetectors and gas sensors even at room temperature, which was a breakthrough in this field [5]. However, further improvements during of the last few years were coming based on new requirements of higher response, better selectivity and lower power consumption of the nano-device structures [5]. In this work, we report on a single nanowire nanodevice made by employing metal maskless nanodeposition in the focused electron/ion beam (FIB/SEM) instrument. A case study of metal-doping effects on a single ZnO nanowire - nanosensor is discussed based on our experimental evidence.

II. EXPERIMENTAL

A. *Synthesis of Metal-Doped ZnO Nanowire Arrays*

Metal (Me, e.g. Ag, Al, etc)-doped ZnO nanowire arrays were deposited on Fluor-doped Tin oxide film supported on a glass with a resistance of 10 Ω/□ using an electrochemical deposition process [8-9]. The parameters of the deposition are an electrical potential of – 0.58 V vs. SCE (for Ag-doping) and of – 1.1 V vs. SCE (for Al-doping), a deposition temperature of 90 °C [8-9], and a growth time of 230 min. The potential was adjusted for each Me-dopant as reported in our previous works [7-10]. After ECD, the samples were subsequently rinsed with deionized DI water. Then, the samples were subjected to thermal annealing in air at 250 °C for 12 h to relax the ZnO lattice.

B. *Studies of Metal-doped ZnO Nanowire Arrays*

Scanning electron microscopy (SEM) images were acquired with a Ultra 55 Zeiss FEG (7 kV). The nanostructures were characterized with a high-resolution X-ray diffractometer (Siemens D5000) using the CuKα radiation (1.5406 Å). Secondary ion mass spectrometry (SIMS) was used for the analysis of the Ag or Al contents because of its depth resolution. SIMS measurements were carried out with a Physical Electronics ADEPT 1010 quadrupole analyzer with a 3 keV Cs^+ primary beam at 60° from the normal. The Raman spectrometer used in this study was a WITec system. A 532.2 nm line from Nd-YAG laser at an output power of 10 mW was used. The photoluminescence (PL) spectra were measured as described elsewhere [7-9].

C. *Integration of a Single Metal-Doped ZnO Nanowire on Chip*

The bottom-up approach consists on the assembly of atomic building blocks to construct up the desired nanoarchitecture. The advantages of such nanotechnology are the possibility to control the growth of nanostructure with near atomic precision, to synthesize it with specific chemical composition which may be not accessible or too expensive with conventional top-down technologies. Afterwards, such nanoarchitecture can be integrated in a desired device. In our case, the nanosensors were fabricated from a ZnO:Me nanowire grown by ECD, then by contacting a single NW

V. Sontea and I. Tiginyanu (eds.), *3rd International Conference on Nanotechnologies and Biomedical Engineering*,
IFMBE Proceedings 55,
DOI: 10.1007/978-981-287-736-9_27

on the chip substrate (quartz and SiO_2/Si with Au contacts as contact electrodes) by using metal maskless nanodeposition in the dual beam focused electron/ion beam (FIB/SEM) instrument Dualbeam Helios Nanolab (FEI) (5 kV, 0.17 nA) [3]. The single-nanowire contacts were imaged using FESEM [3]. Consistent experimental data from several single nanowire devices were obtained. Additional details on the connection of an individual tetrapod, a tripod, a single cross from rods and of a very thin ZnO NW (about 100 nm in diameter) using FIB or FIB/SEM are given in our recent works [1-6,10]. Also, it should be pointed out that a major difference in comparison to previous reports is that during the fabrication of a single Al-ZnO, Ag-ZnO NW device structure, a single NW was exposed to the electron beam for less than 1 min. Only the edges of NW were exposed to the beam about 2 -3 min. The H_2 sensing characteristics of a nanosensor structure were studied as described before [1-7]. By monitoring the output voltage across the NW-based sensor, the conductance was measured in air, under UV and in the test gas respectively. The gas sensing and photodetection performances of nanosensor structures toward UV radiation pulses at 365 nm were investigated as described below and in our previous works.[1-7] All measurements were performed in a quasi-steady state.

III. RESULTS AND DISCUSSION

A. Morphological and Structural Studies

Figure 1 represents a typical morphology of Al- and Ag-doped ZnO nanowire arrays on FTO substrate grown by the ECD process at 90°C. The Al- and Ag-doped ZnO NWs are fairly uniformly distributed on the substrate (Figure 1a-b). Magnified view (Figure 1c-d) clearly shows that Al- and Ag-doped ZnO NWs are quasy-1-D shape with a length of about 4 μm and diameters of 100–400 nm. Its length and thickness depends on the duration of ECD process.[7-10]

B. Chemical Composition and SIMS Results

The composition of the Me-, (Al- and Ag)-doped ZnO NWs product was analyzed by SEM/EDX as reported previously [7,10]. Initial [$Al_2(SO_4)_3$] and [$AgNO_3$] in the aqueous electrolyte was added in the range 1-20 μM. The [Ag(I)]/[Zn(II)] ratio in the starting aqueous electrolyte was 0.5% for 1 μM. Atomic Ag content determined by EDX was found as 1.8 %, defined as the [Ag]/[Ag+Zn] ratio for sample with 1 μM.[7] For the Al-doping it was lower as 0.4 %. SIMS results (not shown) indicated that the Al- or Ag-count rate closely follows the Zn count rate over most of the sample depth. This is an indication that Al- or Ag-dopant has been evenly incorporated into the ZnO nanowire arrays structure.

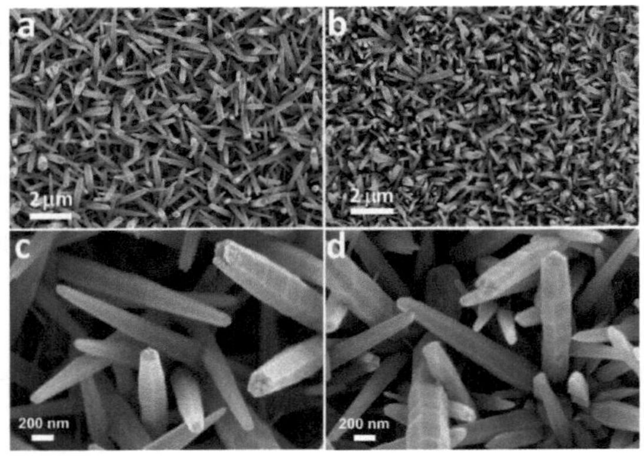

Fig. 1 SEM images of the nanowire arrays layers on FTO substrate: (a,c) Al-doped ZnO; and (b,d) Ag-doped ZnO at different magnifications

C. Micro-Raman Studies and Photoluminescence Studies

Micro-Raman spectroscopy is a sensitive technique to analyze crystallization, structural disorder and defects in nanomaterials. Micro-Raman study is an effective approach for nondestructive investigation of dopant incorporation.[7-11] It permits investigations on the local structure and vibrational modes in Ag- and Al-doped ZnO nanowires. Local vibrational modes may be related to a dopant and can be used as indicators for identifying the dopant's incorporation [11]. For example, additional/local vibrational modes (LVMs) were found, such as the mode at 411 cm^{-1} for Ag doped, at 531 cm^{-1} for Sb doped, at 631 cm^{-1} for Ga doped, at 638 cm^{-1} for Ni doped, and at 720 cm^{-1} for Fe doped ZnO thin films [11-12]. The origin of these modes is arguable, and Manjon et al. [12] attributed the additional or local vibrational modes to the breakdown of the translation symmetry of the lattice caused by defects or impurities [13]. Raman scattering of Ag- and Al-doped ZnO nanowires will help to understand the influence of different dopants on their microstructure. For instance, the origin of the dopant-related modes is still not clear and needs further work. We have measured micro-Raman spectra of the pure and Me-doped-ZnO nanowires to investigate how Ag and Al influence the Raman scattering of zinc oxide nanowires. The Raman measurement results (Figure 2) are consistent with the XRD spectra of the ZnO nanowires [7]. The intense high-frequency E_{2H} peak at 438 cm^{-1} (see Fig. 2), which is characteristic of the wurtzite structure [14] can be observed in all spectra. The Raman peaks at 382 and 410 cm^{-1} can be assigned to the $A_1(TO)$ mode and $E_1(TO)$ mode of ZnO, respectively. Two smaller peaks at about 334 cm^{-1} and 381 cm^{-1} are observed in Fig. 2 and correspond to E_{2H}-E_{2L} (multi phonon) and A_{1T} modes. It can be seen clearly difference in Raman spectrum of each sample and effect of doping Me in ZnO NW arrays. With increasing level of doping one observed a vanishing of Raman modes.

The photoluminescence (PL) spectra have been measured at room temperature and are presented in the normalized form in Figure 2b. The nanowire arrays are characterized by a strong near-band edge emission in the near-UV-violet wavelength region. There visible emission related to defects observed either for the pure ZnO, or of the Al- or Ag-doped nanowires at 20 μM and 2 μM, respectively. The NW arrays are of high structural quality. The PL emission is slightly shifted toward longer wavelength by several nm with the doping with Ag- and to shorter wavelength for Al-doping samples, which is consistent with previous works [15-17].

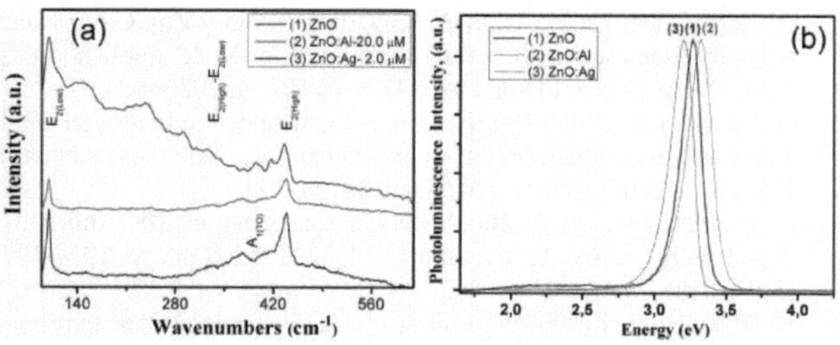

Fig. 2 (a) Micro-Raman spectra and (b) PL spectra of: (1) un-doped ZnO; (2) Al-doped ZnO, 20 μM; and (3) Ag-doped ZnO, 2 μM

D. Nanosensors Fabrication Procedure

Next, the in-situ nanodevice fabrication procedure is described based on our previous reports.[6,10,15,18] Figure 3a shows top-view SEM image of an Ag-doped ZnO nanowire arrays on FTO substrate. Detailed morphological studies of all concentrations of Ag-doped ZnO NW arrays on FTO and GaN can be found in previous work.[7] Figure 3b presents numerous NWs transferred from the initial substrate to an intermediate SiO_2/Si substrate for further dispersion to a lower density of NWs on the surface. The Al- or Ag-doped ZnO NWs can be released from the initial substrate by sonication in ethanol and then transferred to a SiO_2-coated Si substrate.[3,5,10,18] We also used a direct contact technique to transfer nano- and microwires from initial substrate to the clean Si sample by gently rubbing them few times. These procedures allow us to obtain a low density and uniformly distributed nanowires/rods on the second substrate for nanodevice fabrication (Figure 3(c-d)). More details on nanodevice fabrication and its characterizations are discussed in previous works.[1-6,10,17,18] Figure 3c shows an individually dispersed nanowires transferred as described above to an intermediate next SiO_2/Si substrate with gold pads for further integration in nanosensor structures. A single Ag-doped ZnO NW integrated on the chip is presented in Figure 3d. Similar procedures for bottom-up approach were developed for Al-doped ZnO NWs and other Me-doped ZnO NWs and after integration were studied for possible practical applications in near future.

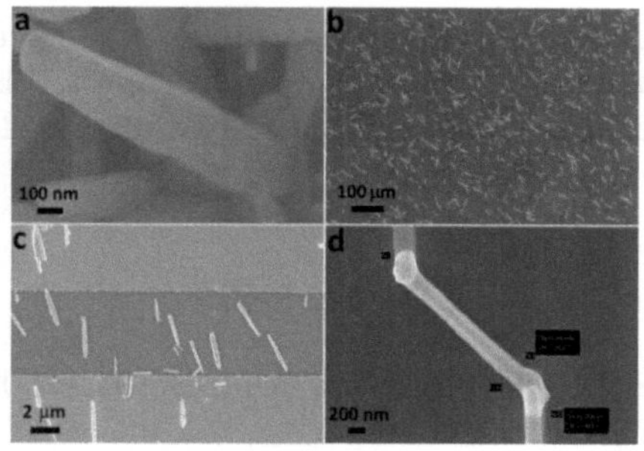

Fig. 3 (a) ZnO:Ag nanowires on initial FTO substrate; (b) ZnO:Ag NWs transferred from initial substrate to an intermediate one; (c) Ag-doped ZnO 2 μM NW transferred from an intermediate substrate to a sensor template between two gold pads; and (d) A single Ag-doped ZnO, 2 μM NW contacted by using FIB/SEM between two gold pads on insulating substrate

The results on bottom-up approach for Al-doped ZnO NW and its integration in nanosensor are not shown here, due to space limit, but will be presented in a forthcoming work with all details.

E. Nanosensors Performances

For investigation of sensing performances, a sensor was placed in a measuring apparatus.[1-5,10] Figure 4a shows that operating at room temperature, the single ZnO:Al nanowire based device shows good responses to UV and gas. It can be concluded that it is quite stable and moderately reversible at the rising and dropping UV light (Fig. 4a) or gas ambient (not shown). Although, it shows an initial quick response and gets rather equilibrated over time in the first 2 s, it is quite reversible in this experiment. Device was made from a single nanowire with a thickness of about 200 nm, total length of about 4 μm (Figure 3d) and it shows H_2 gas sensing response too. Figure 4b reveals the reversible switching of electrical current after 10 s illumination of UV light for ZnO:Ag specimen measured at room temperature. It can be seen that the device is working properly by using our experimental settings. Furthermore, the I_{UVON} and I_{OFF} currents are decreasing with temperature (not shown here). Results on pure ZnO NW and pure ZnO-tetrapod based sensor were reported in [4,19], respectively. According to

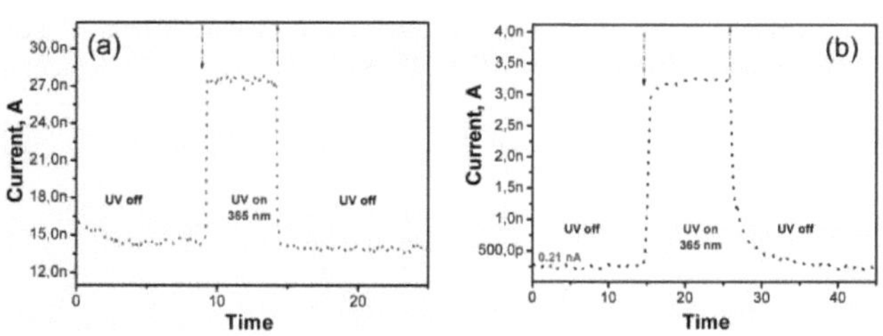

Fig. 4 Photoresponse of a single NW based nanodevice toward one pulse of UV light for sample: (a) Al-doped ZnO; and (b) Ag-doped ZnO

our detailed experimental investigations on another series of devices made from a doped NW at different levels leads to an enhancement in the response and an increased speed of response and will be in a forthcoming work.

IV. CONCLUSIONS

In this work, metal-doped ZnO nanowire arrays for gas nano-sensing investigations were synthesized by electrochemical method on Fluor-doped Tin oxide film supported on a glass. A successful integration of a single nanowire of Al-doped and Ag-doped ZnO on the chip was made possible by a new approach. A controlled change in response to UV light and reducing gases of Al- or Ag-doped ZnO based nanosensors was demonstrated. The metal-doped ZnO shows an enhanced response to hydrogen and to UV radiation. Such Al- or Ag-doped ZnO NWs can serve as nanomaterials for sensitive and ultra-rapid UV detectors and gas nanosensors in bottom-up nanotechnologies.

ACKNOWLEDGMENT

This work was partially supported by the STCU under the Grant no. 5989. OL acknowledge the AvH Foundation for the research fellowship for experienced researchers at the University of Kiel, Germany.

CONFLICT OF INTEREST

The authors declare that they have no conflict of interest.

REFERENCES

1. L. Chow et al (2010) FIB fabrication of ZnO nanotetrapod and cross sensor. Phys. Status Solidi B 247:1628-1632 DOI 10.1002/pssb.201090013
2. G. Chai et al (2009) Crossed zinc oxide nanorods for ultraviolet radiation detection. Sens. and Actuat. A 150:184-187 DOI 10.1016/j.sna.2008.12.020
3. O. Lupan et al (2014) Versatile Growth of Freestanding Orthorhombic α-Molybdenum Trioxide Nano- and Microstructures by Rapid Thermal Processing for Gas Nanosensors. J. of Phys. Chem. C 118, pp. 15068-15078 DOI 10.1021/jp5038415
4. O. Lupan et al (2009) A Single ZnO Tetrapod-based Sensor. Sens. and Actuat. B 141 (2), pp. 511-517 DOI 10.1016/j.snb.2009.07.011
5. O. Lupan et al (2010) Selective hydrogen gas nanosensor using individual ZnO nanowire with fast response at room temperature. Sens. and Actuat. B 144:56-66 DOI 10.1016/j.snb.2009.10.038
6. O. Lupan et al (2007) Fabrication of ZnO nanorod-based hydrogen gas nanosensor. Microelectronics J. 38:1211-1216 DOI 10.1016/j.mejo.2007.09.004
7. T. Pauporté et al (2015) Low Temperature Preparation of Ag-doped ZnO Nanowire Arrays, DFT Study and Application to Light Emitting Diode. ACS App. Mater. & Inter. DOI 10.1021/acsami.5b01496
8. O. Lupan et al (2011) Wavelength-Emission Tuning of ZnO Nanowire-Based Light-Emitting Diodes by Cu Doping: Experimental and Computational Insights. Adv. Funct. Mater. 21:3564-3572 DOI 10.1002/adfm.201100258
9. O. Lupan et al (2011) High Aspect Ratio Ternary $Zn_{1-x}Cd_xO$ Nanowires by Electrodeposition for Light-Emitting Diode Applications. J. Phys. Chem. C 115:14548-14558 DOI 10.1021/jp202608e
10. O. Lupan et al (2012) Highly sensitive and selective hydrogen single nanowire sensor operating at room temperature. Sens. and Actuat. B 173, pp. 772-780 DOI 10.1016/j.snb.2012.07.111
11. C. Bundesmann et al (2003) Raman scattering in ZnO thin films doped with Fe, Sb, Al, Ga, and Li. Appl. Phys. Lett. 83:1974-1976 DOI 10.1063/1.1609251
12. X. B. Wang et al (2006) Luminescence and Raman scattering properties of Ag-doped ZnO films. J. Phys. D: Appl. Phys. 39:4992-4996 DOI 10.1088/0022-3727/39/23/014
13. F. J. Manjón et al (2005) Silent Raman modes in zinc oxide and related nitrides. J. Appl. Phys. 97:053516(4) DOI 10.1063/1.1856222
14. C. A. Arguellor et al (1969) First-order Raman Effect in Wurtzite-Type Crystals. Phys. Rev. 181:1351-1363
15. L. Chow et al (2013) Synthesis and characterization of Cu-doped ZnO one-dimensional structures for miniaturized sensor applications with faster response. Sens. and Actuat. A 189, pp. 399-408 DOI 10.1016/j.sna.2012.09.006
16. S. T. Shishiyanu et al (2005) Photoluminescence of chemical bath deposited ZnO:Al films treated by rapid thermal annealing. Thin Solid Films 488:15-19 DOI 10.1016/j.tsf.2005.04.004
17. O. Lupan et al (2011) Comparative study of hydrothermal treatment and thermal annealing effects on the properties of electrodeposited micro-columnar ZnO thin films. Thin Solid Films 519:7738-7749 DOI 10.1016/j.tsf.2011.05.072
18. Lee Chow, Oleg Lupan, and Guangyu Chai. "Fabrication of ZnO Nanorod-based hydrogen gas nanosensor", **US Patent** No. 8,263,002 09/11/2012
19. O. Lupan et al (2010) Ultraviolet photoconductive sensor based on single ZnO nanowire. Phys. Status Solidi A 207:1735-1740 DOI: 10.1002/pssa.200983706.

Author: Lupan Oleg
Institute: Technical University of Moldova
Street: 168 Stefan cel Mare Blvd
City: Chisinau
Country: Moldova
Email: lupanoleg@yahoo.com

Anisotropic Thermoelectric Generator Made from Single Crystal Bi Microwire

L.A. Konopko[1,2], A.A. Nikolaeva[1,2], T.E. Huber[3], and A.K. Tsurkan[1]

[1] Ghitu Institute of Electronic Engineering and Nanotechnology, ASM, Chisinau, Moldova
[2] International Laboratory of High Magnetic Fields and Low Temperatures, Wroclaw, Poland
[3] Department of Chemistry, Howard University, DC 20059, Washington, USA

Abstract— **Currently, for thermoelectric conversion of heat most widely used is the approach based on the Seebeck and Peltier effects created at the interface of two materials with different values S of the Seebeck coefficient. Another type of thermoelectric converter is the anisotropic thermoelement (AT) using anisotropy of thermoelectric power. AT has some advantages: - the transverse thermopower, unlike a conventional thermocouple, is proportional to the temperature gradient (T1-T2)/h instead of the temperature difference T1-T2; - Voltage V is proportional to the length l of AT. To increase the AT output voltage we need either increase the length l of AT or decrease its thickness h. According to our experimental data, to obtain a thermoelectric voltage of 1 V at a transverse temperature gradient of 5 K, the microwire with a diameter of 2 μm and a length of 8 m must be used. In our experimental sample the long wire in glass coating was wound into a flat spiral.**

Keywords— **thermoelectric device, anisotropic thermoelement, bismuth, single-crystal microwire in glass coating, flat spiral.**

I. INTRODUCTION

There are many reasons we are looking for alternative energy sources. Using alternative energy helps to preserve many of the natural resources that are currently used as sources of energy. A source of heat for thermoelectric generators (the thermoelectric method of thermal energy conversion into electric energy) involves unconventional renewable sources of thermal energy: from solar energy to the heat of human body [1, 2]. Currently, for thermoelectric conversion of heat most widely used approach based on the Seebeck and Peltier effects created at the interface of two materials with different values *S* of the Seebeck coefficient. Such devices are commonly referred to as a thermocouple [3]. Along with its benefits, this thermocouple also has some disadvantages, for example, to obtain the required electrical voltage, they must be connected in series circuits, which leads to the design complexity and reduce reliability. Technological and operational difficulties at the joins between the n- and p-type thermocouple material motivate the search for alternative thermoelectric power converters. One of such converter, devoid of the above disadvantages, is the anisotropic thermoelement (AT) using anisotropy of thermoelectric power. The appearance of new more efficient materials for anisotropic thermoelements is reviving interest in the transverse thermoelectric effect. The principle of operation and the features ATs were extensively studied both in scientific and applied aspects [4-6]. A transverse AT, as a voltage source in measuring systems, has some advantages:

(a) The thermopower, unlike a conventional thermocouple, is proportional to the temperature gradient $(T_1\text{-}T_2)/h$ instead of the temperature difference $T_1\text{-}T_2$. Thus, decreasing the width *h*, it is possible to increase voltage at the same temperature difference.

(b) Voltage *V* is proportional to length *l*; thus, it is possible to increase voltage by increasing the length of the plate.

(c) To obtain voltage, we need no junctions that are required for increasing sensitivity. In the case of ATs, it is sufficient to increase the length of the crystal employed in order to enhance sensitivity.

In a homogeneous isotropic material difference in temperature creates an irrotational (potential) electric field. In materials with anisotropy of the thermopower, the temperature difference creates a transverse (perpendicular to the temperature gradient) the thermoelectric field [7-9]. Transverse thermoelectric field in contrast to the longitudinal, used in conventional thermocouples, may give rise to an electric current even in a uniform closed circuit [9, 10]. Therefore, even the homogeneous thermoelectrically anisotropic medium can serve as a generator of the thermoelectric power, i.e. thermoelement [11].

In the present paper we report measurements of transverse thermopower on a single crystal microwires of bismuth in a glass cover for the development of anisotropic thermoelectric generator

II. SAMPLE AND EXPERIMENT

We have studied the possibility of using a microwire of bismuth to design an anisotropic thermoelectric generator (ATG). Glass-coated single-crystal microwires of pure and Sn-doped bismuth were prepared by the Ulitovsky method (see schematic diagram at inset on Fig. 1); they were cylindrical single crystals with the (10$\underline{1}$1) orientation along the wire axis; the C_3 axis was inclined at an angle of 70° to the

V. Sontea and I. Tiginyanu (eds.), *3rd International Conference on Nanotechnologies and Biomedical Engineering*,
IFMBE Proceedings 55,
DOI: 10.1007/978-981-287-736-9_28

wire axis. The technique described in [12, 13] allows preparing single-crystal wires with diameters d from 50 µm to 40 nm. It is known that the size effect significantly changes the thermoelectric properties and leads to an increase in thermoelectric efficiency [13]. The developed technology allows obtaining a glass-coated single-crystal microwire of Bi and its alloys with Sn with a length up to a few meters and with a given diameter from 0.5 µm to 50 µm. Scanning electron microscope cross sections of the 1.0 µm Bi microwire in its glass coating is shown in Fig. 1.

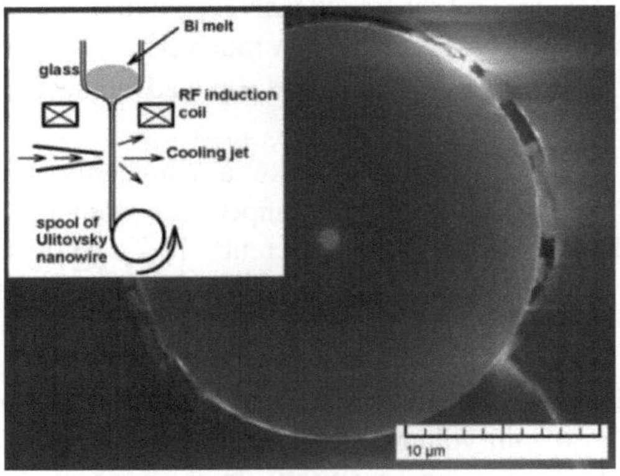

Fig. 1 Scanning electron microscope cross sections of the 1.0 µm Bi microwire in its glass coating. The inset illustrates the Ulitovsky method for synthesizing long, small-diameter wires in a glass fiber.

The transverse thermopower $S_{trans} = U/\Delta T$, where U is the voltage across the sample, ΔT is the transverse temperature gradient. To measure the transverse thermopower S_{trans} in microwire segments with a length of 10 cm, we made a special device consisting of two Al plates with different temperatures. A glass-coated microwire segment was placed between these plates in such a way as to keep good thermal contact between the glass cover of the microwire and the surface of the plates throughout the length of the microwire (Fig. 2). To obtain a uniform temperature gradient, resistive heaters were placed on the plates. The temperature gradient was measured by a differential copper-constantan thermocouple, two junctions of which were situated in the middle of the plates near the surfaces being in contact with the microwire. Displacement of plates relative to each other causes the rotation of the microwire, so we can record rotation diagrams of the transverse thermoelectric power on the direction of the temperature gradient. The presence of two heaters allows us to change the direction of the temperature gradient at a fixed microwire, and eliminates the inequality of the microwire's thermal contact with the surface of the plates during rotation of the microwire between the plates. Contacts to the microwires were made using liquid eutectic InGa, this method allowed us to eliminate twisting of the the microwire during its rotation. In order to ensure uniformity of rotation of the microwire, the top plate is replaced with a small step screw pair, the screw is rotated by the step motor. Step motor controlled by the computer and a rotation angle of the microwire can be identified by the number of steps. The top plate is moved around 1 mm for 900 steps of the motor that provides high accuracy and smooth rotation of the microwire. To determine the orientation of the crystallographic axes of the microwire setup allows to record the rotation diagrams of microwire transverse magnetoresistance. For this purpose, aside from the plates there are two neodymium permanent magnets, which create a transverse magnetic field ~ 0.2 T; at rotation of the microwire the changes of it resistance are recorded. Arrangement of the Bi microwire with crystallographic orientation $[10\bar{1}1]$ in the device for measuring transverse thermopower is shown at Fig. 3 (a), (b).

All measurements of the transverse thermopower of the microwires have been done at room temperature.

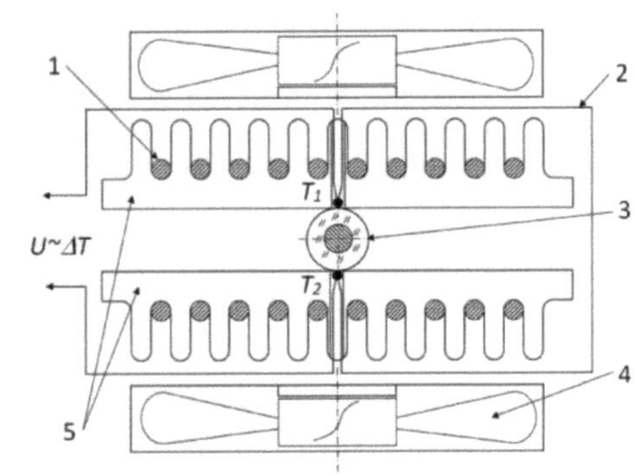

Fig. 2 Schematic drawing of the device for measuring transverse thermopower of the microwire in a glass coating. 1 - Resistive heaters; 2 - Differential thermocouple Cu-Constantan; 3 - Microwire in a glass coating; 4 - Fan; 5 - Radiators.

III. RESULTS AND DISCUSSION

We have measured the transverse thermopower of thin single-crystal bismuth microwires in glass coating with various diameters *d*. The results of measurements of sample of Bi microwire $D = 20$ µm, $d = 6.9$ µm are shown at Fig. 3 (c). Figure 3 also shows the location of the crystallographic axes relative to the transverse temperature gradient. From the rotation diagrams of transverse magnetoresistance is seen that the maximum of transverse thermopower occurs when the temperature gradient is directed along the C_3 axis. For subsequent use of the microwires in the anisotropic thermoelectric devices optimization of the parameters of the microwires based on Bi and Bi-0.05 at. % Sn was carried out. The highest values of the Seebeck coefficient were obtained in the microwire of Bi-0.05 at. % Sn ($D = 15$ µm, $d = 2.9$ µm), the specific thermopower $S_{special} = 250$ µV/

(K * cm) (see Fig. 4). According to our investigations we can conclude that the transverse thermopower increases with decreasing diameter of the microwires. Resistance of the microwire increases with decreasing its diameter, which adversely affects the efficiency thermoelectric devices. For example, to obtain a thermoelectric voltage of 1 V at a temperature gradient of 5 K, we need to use a microwire with a diameter of 2 μm and a length of 8 m. The resistance of the microwire will be R = 3 MOhm and the maximum current of that generator will be equal to $0.34*10^{-6}$ A.

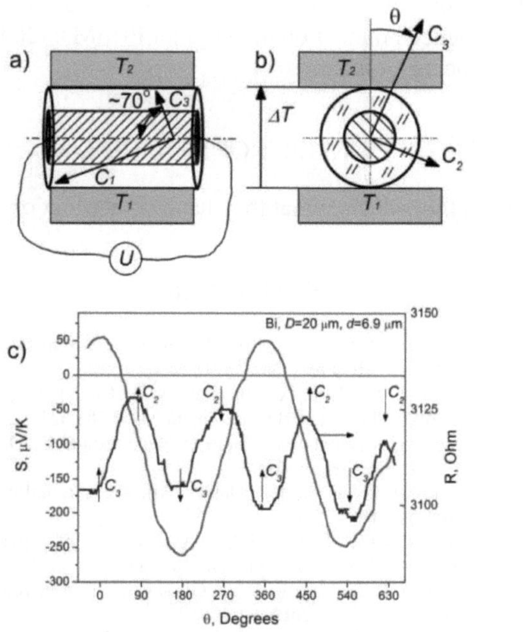

Fig. 3 (a), (b): Arrangement of the Bi microwire with crystallographic orientation (1011) in the device for measuring transverse thermopower; (c): The dependence of the transverse thermopower (left axis) and transverse magnetoresistance (right axis) for Bi microwire (D = 20 μm, d = 6.9 μm) on the rotation angle. T=300 K.

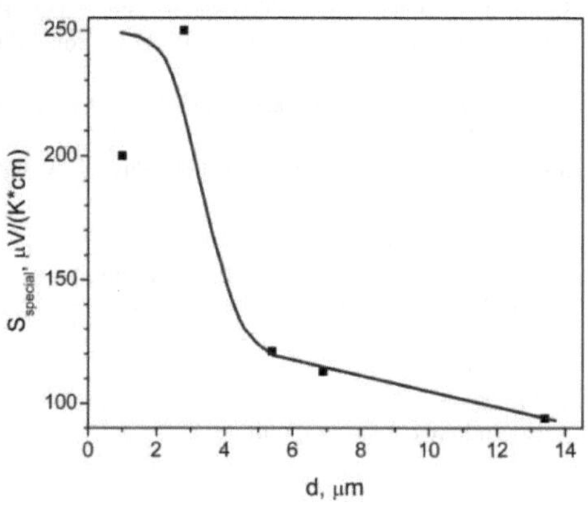

Fig. 4 The dependence of the specific transverse thermopower for Bi microwire on the microwire diameter.

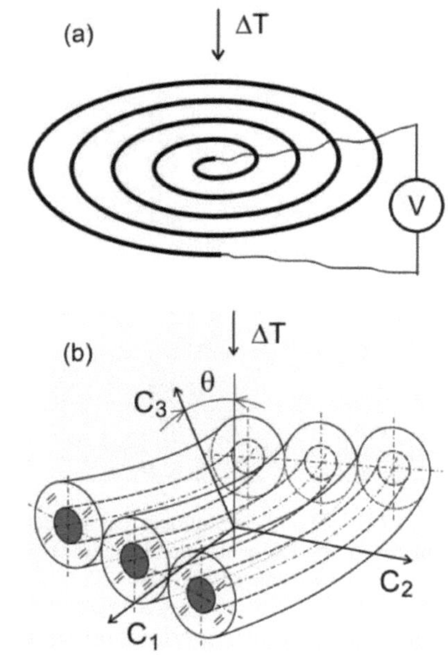

Fig. 5 Schematic of the anisotropic thermoelectric generator made of long Bi-Sn microwire coiled into a flat spiral; (b) Schematic drawing of a small piece of the flat spiral.

Today we focus on implementation of transverse thermoelectric effects on single-crystal semimetal microwires for localized electric generating and heat flux sensing. Our goal to develop anisotropic thermoelectric generator which will use human heat for generating voltages suitable for power supply of devices with low current consumption (for example, hearing aid). The voltage V that arises on the ATG outputs is equal to

$$V = S_{12}\Delta T\frac{a}{b} = (S_{\parallel} - S_{\perp})\sin\theta\cos\theta\,\Delta T\frac{a}{b} = (S_{\parallel} - S_{\perp})\sin 2\theta\,\Delta T\frac{a}{2b} \quad (1)$$

were $(S_{\parallel} - S_{\perp})$ - anisotropy of thermopower, θ - inclination angle of the crystallographic axes C_3, ΔT - transverse temperature gradient, a – length of the sample, b - thickness of the specimen. Maximum ATG output will be at inclination angle θ=45°. Thermoelectric efficiency or thermoelectric figure of merit, Z_a of anisotropic thermoelement is equal $Z_a \approx \sigma(S_{\parallel} - S_{\perp})^2/\kappa$, where σ is the electrical conductivity, and k is the thermal conductivity. It should be noted that the anisotropy of thermopower in Bi-0.05 at. % Sn microwires $\Delta S = (S_{\parallel} - S_{\perp})$ at room temperature is equal ~100 μV/K, which is 2 times higher than in pure bismuth microwire.

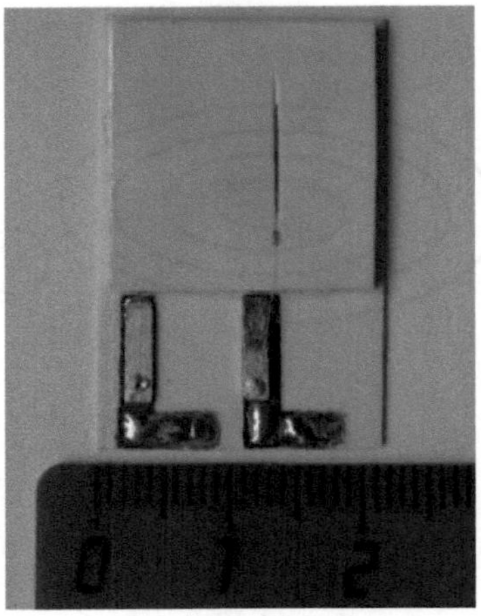

Fig. 6 Experimental sample of the anisotropic thermoelectric generator made of Bi-Sn microwire (l= 5 m) with outer diameter D= 45 μm and the core diameter d=11 μm.

We have manufactured an experimental sample of ATG by winding the microwire into a flat spiral, in which the direction of heat flow was perpendicular to the plane of the spiral. Schematic of the ATG and arrangement of main chystalographic axes of Bi microwire are shown in Fig. 5. During manufacturing of the flat spiral the direction of the C_3 axis was not controlled so when we tested the sample shown in Fig. 6 (we put the ATG in contact with a surface at 38 C (the temperature of the human hand) the output was only 3 mV (but theoretically, with the right crystalline orientation, the output should be ~1 V).

The developed ATG module can be used also in many other heat harvesting applications such as rapid nonselective IR detector and gradient heat flux sensor (GHFS) with a characteristic time above 10^{-3} s. GHFS is used for the direct measurement of heat fluxes to a flat plate in pulsed hypersonic gas flow.

IV. CONCLUSIONS

At room temperature we have investigated the transverse thermopower in thin single-crystal bismuth microwires. Bismuth microwire was placed between two polished aluminum plates, which were at different temperatures. At relative displacement of the plates the microwire was rotated around its axis. Transverse thermoelectric power varied by rotating microwires. The maximum thermoelectric power (S_{max} ~ 200 – 600 μV/K at the length of the plates L = 4.5 cm) occurred at the direction of the transverse temperature gradient along the C_3 axis. Therefore, to obtain a thermoelectric voltage of 1 V at a transverse temperature gradient of 5 K, the microwire with a diameter of 2 μm and a length of 8 m must be used. In our experimental sample of the anisotropic thermoelectric generator the long wire in glass coating was wound into a flat spiral. We have not achieved the desired results yet because the direction of the C_3 axis was not controlled.

ACKNOWLEDGMENT

This work was supported by STCU project #5986, National Science Foundation under grant PREM #1205608 and by the Boeing Company.

CONFLICT OF INTEREST

The authors declare that they have no conflict of interest.

REFERENCES

1. R. A. Messenger, J. Ventre (2010) Photovoltaic Systems Engineering. Third Edition. CRC Press, New York
2. J.-P. Fleurial, T. Olson, A. Borschevsky, T. Caillat, E. Kolawa, M. Ryan, and W. Philips (2001) U.S. Patent No. 6,288,321
3. Thermoelectrics Handbook: Macro to Nano, edited by D. M. Rowe (2006). CRC Press, New York
4. P.I. Baranskii, I.S. Buda, I.V. Dakhovskii, A.G. Samoilovich (1975) Phys. Stat. Sol. (b) 67:201
5. A.A. Snarskii, A.M. Palti, and A.A. Ascheulov (1997) Fizika i Tekhnika Poluprovodnikov 31:1281 (in Russian)
6. A.R. Regel', G.A. Ivanov, V.M. Grabov, L. I. Anatychuk, O. Ia. Luste (1968) USSR Patent No. 245859
7. A.G. Samoilovich and L.L. Korenblit (1953) Uspehi Fizicheskih Nauk 49:243 (in Russian)
8. L.D. Landau, L.P. Pitaevskii, E.M. Lifshitz (1984) Electrodynamics of Continuous Media. Butterworth-Heinemann, Oxford
9. W. Tomson (1882) Mathematical Physics Papers 1:248
10. C. Herring (1941) Phys. Rev. 59:889
11. A.G. Samoilovich, L.L. Korenblit (1964) UK Patent No. 1088764
12. N.B. Brand, D.V. Gitsu, A.A. Nikolaeva, and Ya.G. Ponomarev (1977) Zh. Exp. Teor. Fiz. 72:2332 [(1977) Sov. Phys. JETP 45:1226]
13. D. Gitsu, L. Konopko, A. Nikolaeva, and T. Huber (2005) Applied Physics Letters 86:10210
14. Y.M. Lin, X.Z. Sun, M.S.Dresselhaus (2000) Phys. Rev. B 62:4610

Author: L.A. Konopko
Institute: Ghitu Institute of Electronic Engineering and Nanotechnology, ASM
Street: Academiei Str. 3/3
City: Chisinau
Country: Moldova
Email: l.konopko@nano.asm.md

Removal of Barrier Oxide in the Anodized Aluminum Oxide Nanotemplates

N. Tsyntsaru[1,2]

[1] Institute of Applied Physics, Academy of Sciences of Moldova, Chisinau, MD-2028, Academiei 5, Moldova
[2] Vilnius University, Faculty of Chemistry, Vilnius, LT-03325, Naugarguko 24, Lithuania

Abstract— **Aluminum as other valve metals possesses an important feature: it can form on the surface a porous structure – aluminium anodized oxide (AAO), under proper anodization conditions (voltage, pH, temperature, and etc). In order to be used as a template obtained AAO should undertake the barrier oxide removing step, which will ensure the appropriate electrical contact for material deposition. Thus, in this study we show the simple and unique method for barrier oxide thinning/removing by anodizing solution itself, in comparison with conventional procedure used. The more precise control of barrier oxide removing/thinning was achieved in the same solution used as for anodisation, but at the temperature of 50 °C.**

Keywords— **aluminium anodized oxide, barrier oxide layer, zincate treatment, anodic solution treatment.**

I. INTRODUCTION

Aluminum and its alloys are important industrial materials due to their low weight and good corrosion resistance because of surface passivation. Aluminum possesses a unique feature, under proper conditions it can form on the surface a porous structure. This feature is known since 1960s and the as-prepared porous material is called anodized aluminum oxide (AAO). The size of the pores can be well controlled within micrometer and nanometer scales by careful selection of the processing conditions (temperature, voltage, type of electrolyte and electrolyte concentration, *etc.*). Due to the rapid progress in the nanotechnology, the fabrication of AAO attracts the attention of researchers as it provides convenient and cheap way to produce nanowires with diameters ranged from several tenths of nanometers.

The shape of pattern can be replicated using a templating method, wherein a material is printed, pressed or grown against the voids of a template. Often the template is then removed, leaving the inverse of its pattern; this pattern can be used as a template to achieve a replica of the original. Templating is a relatively fast, cheap and reproducible technique for the preparation of polymeric surfaces [1].

Various materials can be used as templates for nanostructures forming, such as lotus leaves, masters prepared by lithography to inorganic membranes. Woo Lee et al. [2] prepared AAO membranes and used them as templates to form nanopatterns by heating and pressure-driven nanoimprinting. Sun et al. Neto et al. [3] introduced the electrodeposition to prepare dense, aligned metal nanowires (nanocarpets) in the alumina membrane template, followed by dissolution of the membrane.

Masuda and Fukuda [4] proposed and successfully realized a two-step procedure to fabricate alumina membranes with regular pores throughout the entire membranes. Now this method became as standard procedure to prepare the highly ordered AAO structures.

The polishing of aluminum before anodizing is a one of the most important steps in the successful preparation of highly ordered AAO structures. This can be achieved by means of mechanical, chemical, or electrochemical polishing, and among which, the electrochemical polishing in a 60% $HClO_4$+C_2H_5OH (1:4, volume ratio) solution at 10 °C and 500 mA/cm^2 for 1 min is most commonly used to prepare extremely smooth Al surfaces before anodizing.

The anodization is commonly carried out in sulfuric, oxalic or phosphoric acids with voltages of 19–25 V, 40–60 V and 160–195 V, respectively [5]. In two-step anodization, during first anodization the non-uniformly ordered pore structure on the surface is formed. By removing this non-regular oxide layer, a periodic concave pattern is formed on the aluminum surface, and these hollows play role of centers for consequent act as seeds in a further anodization process. The second anodization is carried out at the same anodizing potential as that used for the first anodization, and well-ordered AAO surfaces are formed. In order to be used as a template the obtained AAO should undertake the barrier oxide removing step, which will ensure the reliable electrical contact required for material deposition.

This work is devoted to the fabrication of AAO nanotemplate by anodizing in sulfuric acid and to the investigation of the barrier oxide thinning/removing procedures by simple and unique method using anodizing solution itself.

II. MATERIALS AND METHODS

The first and second anodizing steps have been performed in a 20 wt% sulphuric acid solution using a two electrode electrochemical cell setup equipped with a magnetic stirrer rotating at 500 rpm. The temperature was kept at ~ 0 – 1°C; anodization potential 21 V. After each anodization step, the subsequent electropolishing and oxide removal steps, the influence of the condition on the each anodization step was controlled by means of SEM.

V. Sontea and I. Tiginyanu (eds.), *3rd International Conference on Nanotechnologies and Biomedical Engineering*,
IFMBE Proceedings 55,
DOI: 10.1007/978-981-287-736-9_29

Following the second anodization, the non-conductive barrier oxide dissolution, closing the pore bottoms and hindering the direct electrochemical growth of metallic nanostructures within the template, was performed using several procedures:

1. Treatment in the zincate solution;
2. Treatment in the acidic $CoCl_2$ solution;
3. Treatment in the anodization solution.

The zincate treatment was performed in the solution: 120 g\l NaOH + 20 g\l ZnO + 2 g\l $FeCl_3$. The treatment with acidic solution of cobalt chloride (0.1 M $CoCl_2$ + 0.001 M nitric acid) was performed at pH 2. The overview of the conditions applied during barrier layer dissolution is provided in Table 1.

Table 1 The description of samples

Sample number	Description	Barrier oxide layer, nm
1.	Zincate treatment, ultrasonic, 20 sec, 1 °C	31
2.	Zincate treatment, ultrasonic, 40 sec, 20 °C	Membrane dissolved
3.	Zincate treatment, ultrasonic, 20 sec, 20 °C	Membrane dissolved
4.	Zincate treatment, ultrasonic, 5 sec, 20 °C	22.5
5.	$CoCl_2$ treatment, pH 2 60 min	20
6.	Treatment by anodization solution, 120 min, 1 °C	28.7
7.	Treatment by anodization solution, 30 min, 30 °C	28.5
8.	Treatment by anodization solution, 60 min, 30 °C	15.7
9.	Treatment by anodization solution, 120 min, 30 °C	Membrane dissolved
10.	Treatment by anodization solution, 8 min, 50 °C	12.7
11.	Treatment by anodization solution, 15 min, 50 °C	Membrane dissolved

III. RESULTS AND DISSCUSIONS

The formation of porous alumina depends upon the current density or voltage employed for anodizing, the temperature, pH and composition of the electrolyte. These parameters determine the extent of chemical interaction between the anodic alumina and the electrolyte during anodizing. Such interaction is accelerated by the electric field [6].

After anodization, the films comprise a barrier layer next to the metal and an outer barrier layer, in which the pores, orientated normal to the film surface [7]. The topography of the AAO film as well as the formation rate of the oxide layer depends on the anodization condition which is closed related to the types of electrolyte, concentration, temperature and applied voltage.

Meantime, the most popular model for the self-ordering of pores in AAO is based on mechanical stress associated with expansion of aluminum during the oxide formation. This is the cause of a repulsive force between neighboring pores, which leads to the self-ordering process [8]. Mainly, the following chemical processes dominate in the anodization of the alumina membranes [9-12]:

(1) Al^{3+} ions form at the metal/oxide interface and distribute in the oxide layer near the oxide/metal interface:

$$Al \rightarrow Al^{3+} + 3e^- \tag{1}$$

(2) The electrolysis of water (a water-splitting reaction) occurs at the pore bottom near the electrolyte/oxide interface:

$$2H_2O \rightarrow 2O^{2-} + 4H^+ \tag{2}$$

(3) In the presence of electric field, the O^{2-}ions migrate within the barrier layer from the electrolyte/oxide interface to the oxide/metal interface, and react with the Al^{3+} ions there forming Al_2O_3:

$$2Al^{3+} + 3O^{2-} \rightarrow Al_2O_3 \tag{3}$$

(4) Then electric-field enhances oxide dissolution at the electrolyte/oxide interface:

$$Al_2O_3 + 6H^+ \rightarrow 2Al^{3+}_{(aq)} + H_2O \tag{4}$$

In the anodization process of the porous alumina membranes, a balance exists between the electric-field-enhanced oxide dissolution at the electrolyte/oxide interface and the formation of oxide at the oxide/metal interface. This balance is crucial to the formation of the porous alumina membranes, because it keeps the constant thickness of the barrier layer during whole duration of anodization process and hence allows steady-state pore propagation into the Al. Knowledge on the AAO formation including barrier oxide layer is important step for visualization of reaction occurring during barrier oxide layer dissolution.

Our investigations on zincate treatment (Table 1, Fig. 1) gave unsatisfactory results. Treatment at low temperatures (1°C) gives small growing of barrier layer instead of dissolution. Increasing the temperature promotes the rate of dissolution reaction resulting in complete dissolution of AAO, if the process is undertaken longer than 20 sec. Treatment at shorter times and 20 °C gives better results (Fig. 1), but still is difficult to achieve a good control over the dissolution process, even in the presence of ultrasonication.

Ultrasonication is a widely used process for equilibrating concentration differences and allows a controlled homogeneous dissolution of pore walls and bottoms, most probably

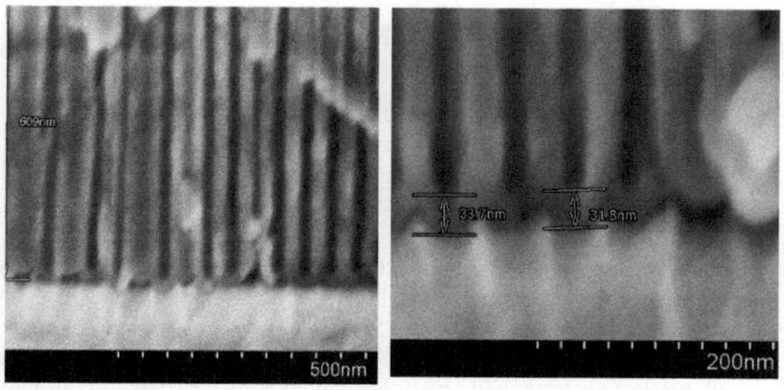

Zincate treatment, ultrasound, 20 sec, 1 °C

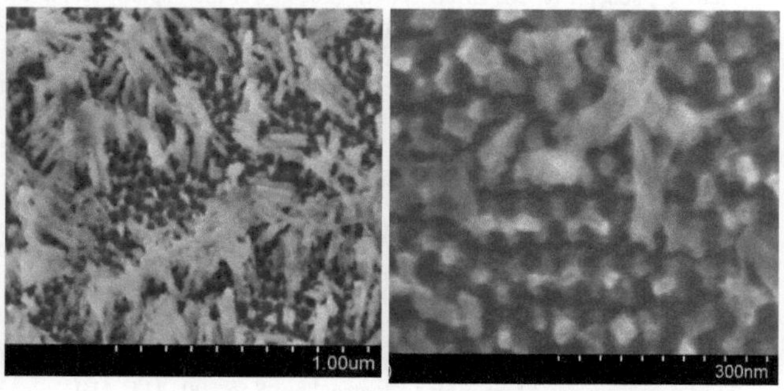

Zincate treatment, ultrasonic, 20 sec, 20 °C

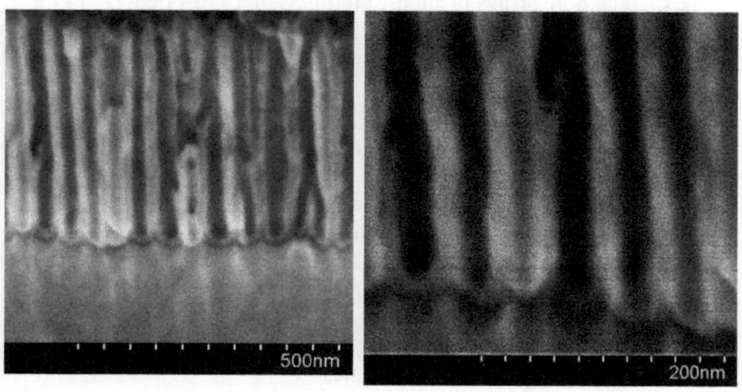

Zincate treatment, ultrasonic, 5 sec, 20 °C

Fig. 1 SEM images of the treated barrier oxide layers after zincate treatment.

by keeping an even concentration gradient throughout the entire pore structure.

The treatment performed in the acidic solution of the active metal as cobalt also did not result in some visible dissolution of the barrier oxide layer, but some tinning has been noticed (Fig. 2) in accordance with equation (4).

As acidic solution gives some tinning of the barrier oxide layer it was decided to explore the possibility to remove the barrier oxide layer in the same solutions used for aluminium anodization. Thus, several conditions were investigated varying the temperature and time (Fig. 2). The results show

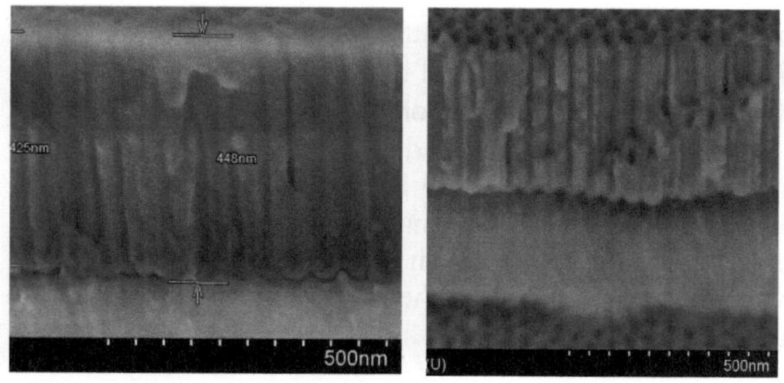

Treatment by anodization solution, 120 min, 1 °C

$CoCl_2$ treatment, pH 2 60 min

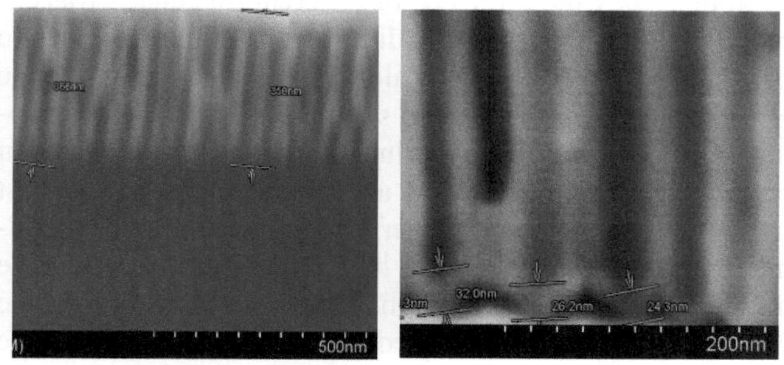

Treatment by anodization solution, 30 min, 30 °C

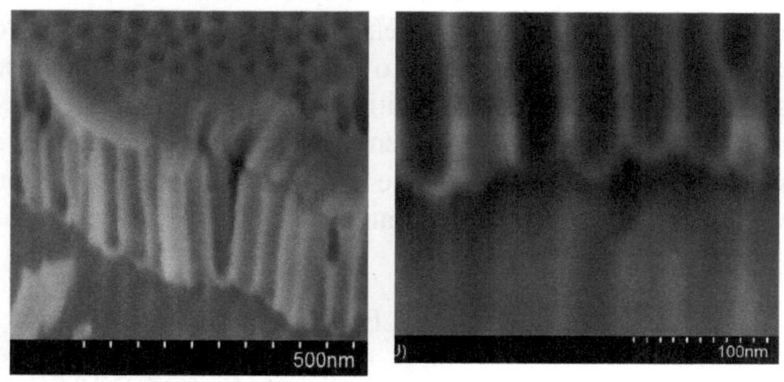

Treatment by anodization solution, 60 min, 30 °C

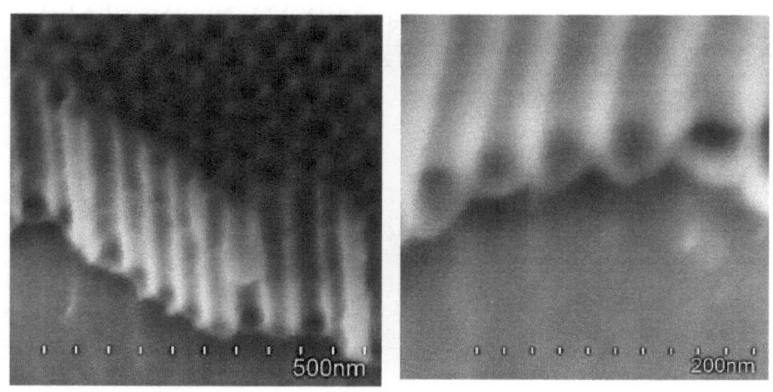

Treatment by anodization solution, 8 min, 50 °C

Fig. 2 SEM images of the treated barrier oxide layers after treatment by anodization solution.

that increasing in temperature promotes the reaction of dissolution, equation (4) and retards the rate of the reaction of alumina formation, equation (3).

It was found a reasonable dissolution rate of barrier oxide layer and practical removing it at the temperature of 50 °C. Nevertheless, some additional research needs to be performed in order to establish a good reproducible control over the dissolution rate of the non-conductive barrier oxide layer of AAO nanotemplate.

IV. CONCLUSIONS

In this study the influence of different treatments was explored in order to perform dissolution of non-conductive barrier oxide layer of anodic aluminum nanotemplate: in the zincate solution, in the acidic solution of $CoCl_2$, in the anodization solution. The experiments carried out reveal that treatment in the zincate and anodic solutions are the best options for barrier oxide layer removing/tinning, but treatment in the zincate solution is hardly controlled, because the dissolution rate is quite high: just several seconds. The treatment in the solution used for anodization can be the good option in view of performing the process under controllable conditions. Also, this simple method by treatment in the anodized solution itself provide a good option to conventional procedure, e.g. to the treatment by phosphoric acid. Nevertheless, some additional study on the efficiency of the ratio between treatment time and temperature are needed in order to determine optimal rate of dissolution with less damage of the template itself.

ACKNOWLEDGMENT

The authors acknowledge funding from FP7 Oil&Sugar project (295202). Also, partial funding was granted by the Research Council of Lithuania (MIP-031/2014) and Moldavian national projects (14.02.121A), (14.819.02.16F).

CONFLICT OF INTEREST

The authors declare that they have no conflict of interest.

REFERENCES

1. McHale G., Newton M. I., Shirtcliffe N. J. et al., (2011) Capillary origami: superhydrophobic ribbon surfaces and liquid marbles Beilstein J Nanotechnol 2: 145–151.
2. Lee W., Jin M. K., Yoo W. C., et al., (2004) Nanostructuring of a polymeric substrate with well-defined nanometer-scale topography and tailored surface wettability Langmuir 20: 7665-7669.
3. Neto C., Joseph K. R., Brant W. R., (2009) On the superhydrophobic properties of nickelnanocarpets Phys Chem Chem Phys 11: 9537-9544.
4. Masuda H., Fukuda K., (1995) Ordered metal nanohole arrays made by a two-step replication of honeycomb structures of anodic alumina Science 268: 1466–1468.
5. Lee W., Ji R., Gosele U., et al., (2006) Fast fabrication of long-range ordered porous alumina membranes by hard anodization Nat Mater 5: 741-747.
6. Garcia-Vergarai S. J., Iglesias-Rubianesi L., Blanco-Pinzon C. E. et al., (2006) Mechanical instability and pore generation in anodic alumina. The Royal Society A Proc. vol. 462, Math. Phys. & Eng. Sci., 2006, pp. 2345-2358.
7. Keller F., Hunter M. S., Robinson D. L., (1953) Structural features of oxide coatings on Aluminum J Electrochem Soc 100: 411–419.
8. Jessensky O., Mller F., Gösele U., (1998) Self-organized formation of hexagonal pore arrays in anodic alumina Appl Phys Lett 72: 1173.
9. Li F., Zhang L., Metzger R. M., (1998) On the growth of highly ordered pores in anodized aluminum oxide Chem Mater 10: 2470–2480.
10. Noh K., Brammer K. S., Seong T. Y. et al., (2011) Guided Nanostructures Using Anodized Aluminum Oxide Templates NANO 6: 541-555.
11. Thompson G. E., (1997) Porous anodic alumina: fabrication, characterization and applications Thin Solid Films 297: 192-201
12. Thompson G. E., Wood G. C., (1983) Treatise on Materials Science and Technology Academic Press New York 23: 205-329.

Author: Natalia Tsyntsaru
Institute: Institute of Applied Physics
Street: Academiei 5
City: Chisinau
Country: Moldova
Email: ashra_nt@yahoo.com

Cavity Field Suppression via Interference Effects

V. Ciornea, P. Bardetski, and M.A. Macovei

Institute of Applied Physics, Academy of Sciences of Moldova, Chisinau, Moldova

Abstract— **The purpose of the study is to show how interference effects are responsible for manipulating the output electromagnetic field of a microresonator in the good-cavity limit. The system of interest consists of a moderately strongly pumped two-level emitter embedded in a photonic crystal microcavity. This is due to the interference among the scattered light by a strongly pumped atom into the cavity mode and an incident cavity weaker laser field. The phase difference of the applied lasers can be a convenient tool to control the photon dynamics. The photon statistics shows super- or sub-Poissonian statistics.**

Keywords— **Microcavity, qubit, photon, phase, control.**

I. INTRODUCTION

Nowadays a lot of papers predict and propose various tools allowing a complete or partial control of the quantum mechanical interaction between light and matter, necessary mostly, for present and future quantum technologies. In particular, this control can be realized via light interference [1–5]. The control of the fluorescence processes is of particular interest for quantum computation processes [6] where, in addition, highly correlated photons are required [7, 8]. Typically, standard sources of incoherent light generate photons possessing a second-order correlation function $g^{(2)}(0)=2$ [9–11]. Very recently it has been shown that photons emitted by a single two-level system, ultra-strongly coupled with a thermal optical cavity, show photon correlation functions $g^{(2)}(0)>2$ or even smaller than unity exhibiting quantum light features, i.e., sub-Poissonian photon statistics [12]. The effect occurs when the atom-cavity coupling rate becomes comparable to the cavity resonance frequency. Therefore, various schemes generating highly correlated light are still important. Specifically, larger photon correlations are useful for a number of practical applications in many-body phenomena with strongly interacting photons [13] as well as in photonic quantum information processing or quantum networks [14–16].

Since a precise control over the system properties is highly required in such a network, here we investigate the feasibility of controlling a cavity-output-electromagnetic field in a system consisting of a moderately strongly pumped two-level emitter. If a second coherent driving is applied through one of the mirrors perpendicular to the first laser beam, then the output-cavity field can be almost completely inhibited in the good-cavity limit. Notice that the lasers are in resonance with the cavity mode frequency. We have found that the interference between the second weaker light beam and the light scattered by the two-level emitter into the cavity mode due to stronger pumping is responsible for the suppression effect. The destructive interference can be turned into a constructive one (or vice versa) via varying the phase difference of the applied lasers. Furthermore, the photon statistics exhibits interacting quantum behaviors.

II. QUANTUM DYNAMICS OF A PUMPED TWO-LEVEL ATOM INSIDE A DRIVEN PHOTONIC CRYSTAL MICROCAVITY

The Hamiltonian describing a two-level atomic system having the transition frequency ω_0 and interacting with a strong coherent source of frequency ω_1 while embedded in a pumped photonic crystal microcavity of frequency ω_c in a frame rotating at $\omega=\omega_1=\omega_2$ (see Fig. 1), is as follows:

$$H=\hbar\Delta S_z+\hbar\delta a^+a+\hbar g\left(a^+S^-+aS^+\right)$$
$$+\hbar\Omega\left(S^+e^{i\phi_1}+S^-e^{-i\phi_1}\right)+\hbar\varepsilon\left(a^+e^{i\phi_2}+ae^{-i\phi_2}\right), \quad (1)$$

where $\Delta=\omega_0-\omega$ and $\delta=\omega_c-\omega$. In the Hamiltonian (1) the components, in order of appearance, describe the atomic and the cavity free energies, the interaction of the two-level emitter with the microcavity mode, the atom's interaction with the first laser field with Ω being the corresponding Rabi frequency, and the interaction of the second driving field with the cavity mode with ε being proportional to the input laser field strength amplitude, respectively. The atomic bare-state operators $S^+=|e\rangle\langle g|$ and $S^-=|g\rangle\langle e|$ obey the commutation relations for $SU(2)$ algebra, i.e., $[S^+,S^-]=2S_z$ and $[S_z,S^\pm]=\pm S^\pm$. Here $S_z=|e\rangle\langle e|/2\ -|g\rangle\langle g|/2$ is the bare-state inversion operator. $|e\rangle$ and $|g\rangle$ are the excited and ground states of the atom, respectively, whereas a^+ and a are the creation and the annihilation operators of the electromagnetic field (EMF) in the resonator and satisfy the standard bosonic commutation relations, namely, $[a,a^+]=1$ and $[a,a]=[a^+,a^+]=0$. $\{\phi_1,\phi_2\}$ are the corresponding phases of the coherent driving sources.

V. Sontea and I. Tiginyanu (eds.), *3rd International Conference on Nanotechnologies and Biomedical Engineering*,
IFMBE Proceedings 55,
DOI: 10.1007/978-981-287-736-9_30

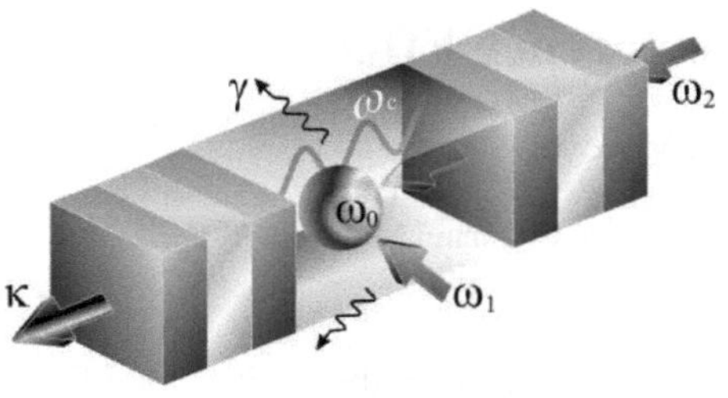

Fig. 1 Schematic of the model: A two-level emitter with transition frequency ω_0 embedded in a single mode (ω_c) microcavity is pumped with an intense laser field of frequency ω_1. A second coherent source of frequency ω_2 is driving the entire system through one of the mirrors. γ is the single-atom spontaneous decay rate, whereas κ describes the cavity photon leaking rate.

In the laser-qubit semiclassical dressed-state formalism [4] the Eq. (1) can be transformed in the following dressed-state Hamiltonian:

$$H = \hbar\bar{\Omega} + \hbar\delta a^{+}a + \hbar\varepsilon\left(a^{+} + a\right) + \hbar R_z\left(g_0^{*}a^{+} + g_0 a\right), \qquad (2)$$

with $\bar{\Omega} = \sqrt{\Omega^2 + (\Delta/2)^2}$, $g_0 = g\sin 2\theta\, e^{i\phi}/2$, $\phi = \phi_2 - \phi_1$, $\tan 2\theta = 2\Omega/\Delta$ and $\delta << \bar{\Omega}$. Here the new quasispin operator is $R_z = 2S_z\cos 2\theta + \left(S^{+} + S^{-}\right)\sin 2\theta$. The master equation describing a laser-dressed two-level atom inside a leaking pumped resonator and damped via the vacuum modes of the surrounding EMF reservoir is as follows [17]:

$$\frac{d}{dt}\rho(t) + \frac{i}{\hbar}[H,\rho] = -\kappa\left[a^{+}, a\rho\right] - \Gamma_0\left[R_z, R_z\rho\right],$$
$$-\Gamma_{+}\left[R^{+}, R^{-}\rho\right] - \Gamma_{-}\left[R^{-}, R^{+}\rho\right] + H.c., \qquad (3)$$

where

$$\Gamma_0 = \left(\gamma_0\sin^2 2\theta + \gamma_d\cos^2 2\theta\right)/4,$$
$$\Gamma_{+} = \gamma_{+}\cos^4\theta + (\gamma_d/4)\sin^2 2\theta,$$
$$\Gamma_{-} = \gamma_{-}\sin^4\theta + (\gamma_d/4)\sin^2 2\theta.$$

Here γ_0 and $\gamma_{\pm}$, respectively, are the single-atom spontaneous decay rates dependent on the density of modes g_k at the dressed state frequencies $\{\omega, \omega \pm 2\bar{\Omega}\}$, whereas γ_d signifies a pure dephasing rate, and κ describes the cavity photon leaking rate. Note that the master equation (3) was obtained either under the intense-field condition or under the far-off-detuned field, i.e. it is valid when $\bar{\Omega} >> \{\delta, g, \varepsilon, \Gamma_0, \Gamma_{\pm}\}$.

The equations of motion for the variables of interest can be easily obtained from the master equation (3). The steady state solution of Eq. (3) is used to find the mean-photon number in the microcavity mode [17], namely,

$$n_{mean} = \left\langle a^{+}a\right\rangle_s = \frac{\varepsilon}{\kappa^2}\{\varepsilon + |g_0|\left\langle R_z\right\rangle_s\cos\phi\}$$
$$+ \frac{|g_0|}{\kappa(\kappa + 2\Gamma_{+} + 2\Gamma_{-})}\{|g_0| + \varepsilon\left\langle R_z\right\rangle_s\cos\phi$$
$$-2(\Gamma_{+} - \Gamma_{-})(|g_0|\left\langle R_z\right\rangle_s + \varepsilon\cos\phi)/\kappa\}, \qquad (4)$$

and the degree of the second-order order coherence of microcavity photons defined as follows [9, 10]:

$$g^{(2)}(0) = \frac{\left\langle a^{+2}a^2\right\rangle_s}{\left\langle a^{+}a\right\rangle_s^2}.$$

In Eq. (4) $\left\langle R_z\right\rangle_s = -\frac{\Gamma_{+} - \Gamma_{-}}{\Gamma_{+} + \Gamma_{-}}$.

From the above expression (4), one can see that the mean-photon number due to weaker external pumping of the cavity mode is proportional to ε^2, whereas that due to stronger driving of the two-level qubit is proportional to $|g_0|^2$. There is also a cross contribution proportional to $\varepsilon|g_0|\cos\phi$. All these terms demonstrate interference effects among the contributions due to two pumping lasers and, hence, the minima nature in Fig. 2. In particular, from Eq. (4), for $\Gamma_{+} >> \Gamma_{-}$ we have that

$$\left\langle a^{+}a\right\rangle_s \approx \frac{\varepsilon^2}{\kappa^2} + \frac{|g_0|^2}{\kappa^2} - \frac{2\varepsilon|g_0|}{\kappa^2}\cos\phi, \qquad (5)$$

whereas for $\Gamma_{-} >> \Gamma_{+}$ we have

$$\left\langle a^{+}a\right\rangle_s \approx \frac{\varepsilon^2}{\kappa^2} + \frac{|g_0|^2}{\kappa^2} + \frac{2\varepsilon|g_0|}{\kappa^2}\cos\phi, \qquad (6)$$

respectively.

The nature of a destructive or constructive interference can be understood as follows: in the dressed-state picture both lasers are simultaneously in resonance with the dressed-state transitions $|+\rangle \leftrightarrow |+\rangle$ (where $|+\rangle = \sin\theta|g\rangle + \cos\theta|e\rangle$) and $|-\rangle \leftrightarrow |-\rangle$. (where $|-\rangle = \cos\theta|g\rangle - \sin\theta|e\rangle$) and, hence, different signs in front of the last term in Eqs. (5) and (6). Actually, if $\Gamma_{+} >> \Gamma_{-}$, then the atom is located on the lower dressed state $|-\rangle$, whereas it resides on the higher dressed state $|+\rangle$ when $\Gamma_{-} >> \Gamma_{+}$. However, the destructive interference can be turned into a constructive one (or vice versa) via varying the phase difference ϕ . (see Eqs. (5) and (6)). This allows a better control of the output- cavity field. The cavity-output-field suppression reported here occurs in the good-cavity limit, i.e., when $\gamma > \kappa$. and $g > \{\gamma, \kappa\}$.

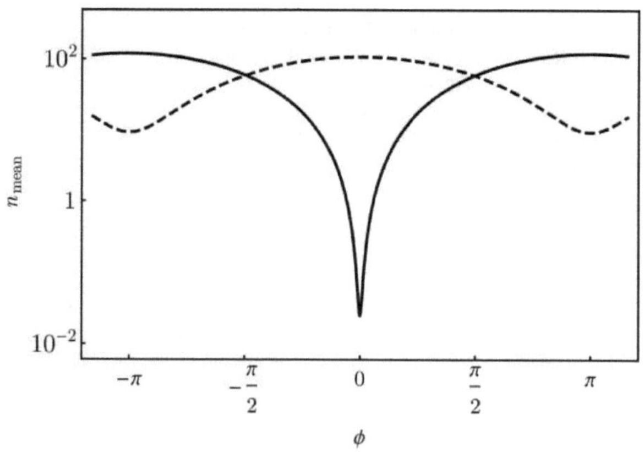

Fig. 2 Steady-state dependence of the microcavity mean-photon number n_{mean} vs variable ϕ . Other parameters are as follows: $\gamma_d / \gamma = 0.01$, $\kappa / \gamma = 0.1$, $g / \gamma = 2$, $\Delta = 0$ and $\varepsilon / \gamma = 0.54$. Solid line corresponds to $\gamma_- \to 0$ and dashed one to $\gamma_+ \to 0$.

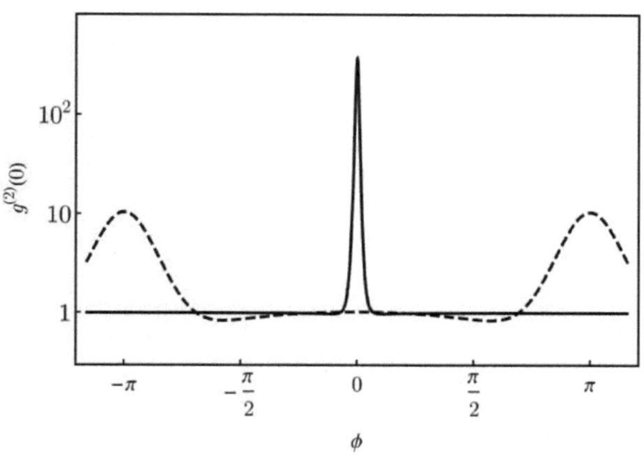

Fig. 3 Steady-state dependence of the second-order coherence $g^{(2)}(0)$ vs variable ϕ . Other parameters are as in Fig. 2.

Further, we focus on the steady-state expression for the second-order coherence function of microcavity photons (see Fig. 3). For $\Gamma_+ >> \Gamma_-$ and $\phi = 0$ we obtain larger photon correlations, i.e. $g^{(2)}(0) >> 2$, but this is due a small number of the microcavity mean-photon number. Larger photon correlations and relatively big number of photons are obtined for $\Gamma_- >> \Gamma_+$ and $\phi = \pm\pi$. Other interesting result occurs around $\phi = \pm\pi / 2$, where photon statistics exhibits quantum futures, i.e. sub-Poissonian photon statistics ($g^{(2)}(0) < 1$). Finally, in accordance with our analytical results, the quantum photon dynamics shows oscillatory behaviors as a function of ϕ.

III. Conclusions

To sum up, we have demonstrated a feasibility of a cavity-output-field control via interference effects. The destructive interference can be turned into a constructive one (or vice versa) via varying the phase difference of the applied lasers providing in this way a better control over the output electromagnetic field. Through variation of the phase difference of the applied lasers the second-order photon correlation function exhibits super- or sub-Poissonian photon statistics.

Acknowledgment

We acknowledge financial support from the Academy of Sciences of Moldova, Grant No. 15.817.02.09F.

Conflict of Interest

The authors declare that they have no conflict of interest.

References

1. Born M, Wolf E (1999) Principles of Optics. Cambridge University Press, Cambridge
2. Agarwal G S (1974).Quantum Statistical Theories of Spontaneous Emission and Their Relation to Other Approaches. Springer, Berlin
3. Allen L, Eberly J H (1975) Optical Resonance and Two-Level Atoms. Dover, New York
4. Ficek Z, Swain S (2005) Quantum Interference and Coherence: Theory and Experiments. Springer, Berlin
5. Kiffner M, Macovei M, Evers J et al (2010) Chapter 3–Vacuum-Induced Processes in Multilevel Atoms. Prog Opt 55:85–197 DOI 10.1016/B978-0-444-53705-8.00003-5
6. Nielsen M A, Chuang I L (2000) Quantum Computation and Quantum Information. Cambridge University Press, Cambridge
7. Ciornea V, Bardetski P, Macovei M A (2013) Enhanced photon correlations due to strong laser-atom-cavity coupling. Phys Rev A 88:023851 DOI 10.1103/PhysRevA.88.023851
8. Ciornea V, Bardetski P, Macovei M A (2013) Detuning-dependent dynamics of microcavity photon-statistics. Rom Rep Phys 65:1006-1011
9. Glauber R J (1963) The Quantum Theory of Optical Coherence. Phys Rev 130:2529 DOI 10.1103/PhysRev.130.2529
10. Glauber R J (2006) Nobel Lecture: One hundred years of light quanta. Rev Mod Phys 78:1267 DOI 10.1103/RevModPhys.78.1267
11. Scully M O, Zubairy M S, (1997) Quantum Optics. Cambridge University Press, Cambridge

12. Ridolfo A, Savasta S, Hartmann M J (2013) Nonclassical Radiation from Thermal Cavities in the Ultrastrong Coupling Regime. Phys Rev Lett 110:163601 DOI 10.1103/PhysRevLett.110.163601
13. Chang D. E., Gritsev V., Morigi G et al (2008) Crystallization of strongly interacting photons in a nonlinear optical fibre. Nat Phys 4:884 DOI 10.1038/nphys1074
14. Friedler I, Petrosyan D, Fleischhauer M et al (2005) Long-range interactions and entanglement of slow single-photon pulses. Phys Rev A 72:043803 DOI 10.1103/PhysRevA.72.043803
15. Gorshkov A V, Otterbach J, Fleischhauer M et al (2011) Photon-Photon Interactions via Rydberg Blockade. Phys Rev Lett 107:133602 DOI 10.1103/PhysRevLett.107.133602
16. Lvovsky A I, Sanders B C, Tittel W (2009) Optical quantum memory. Nat Photon 3:706-714 DOI 10.1038/nphoton.2009.231
17. Ciornea V, Macovei M A (2014) Cavity-output-field control via interference effects. Phys Rev A 90:043837 DOI 10.1103/PhysRevA.90.043837

Corresponding author:

Author: Viorel CIORNEA
Institute: Institute of Applied Physics
Street: Academiei 5
City: Chisinau MD-2028
Country: Republic of MOLDOVA
Email: viorel@phys.asm.md

Slow Relaxation of Magnetization in a Family of Linear $Mn^{III}M^{III}Mn^{III}$ (M = Fe, Ru, Os) Compounds

M. Revenco, M. Secu[1], S. Ostrovsky[2], O. Reu[2], A. Palii[2], O. Palamarciuc[1], K. Pedersen[3], R. Clerac[3], and S. Klokishner[2]

[1] State University of Moldova, Chişinău, Moldova
[2] Institute of Applied Physics, Academy of Sciences of Moldova, Chişinău, Moldova
[3] Centre de Recherche Paul Pascal, CNRS, Universite de Bordeaux, France

Abstract— **The experimental and theoretical study of a family of linear trimeric [NEt$_4$] {[Mn(L)]$_2$[M(CN)$_6$]} (M = Fe, Ru, Os) compounds is reported. All three complexes demonstrate slow relaxation of magnetization at low temperatures. The theoretical model for the explanation of the magnetic behavior (the magnetic susceptibility as a function of temperature and magnetization as a function of the applied magnetic field) includes the spin-orbital interaction acting within the ground $^2T_1$ multiplet of the Fe^{III}, Ru^{III} or Os^{III} ion, the axial crystal field that splits this multiplet into an orbital singlet and orbital doublet, the zero-field splitting for the Mn^{III} ions, the isotropic exchange interaction in the M-Mn pairs as well as the intercluster interaction taken within the mean field approximation. The proposed model provides a good agreement between the observed and calculated magnetic behavior. The low-lying energy levels form the energy barriers for magnetization reversal for all examined compounds. The calculated heights of these barriers are close to the observed ones.**

Keywords— **Single Molecule Magnet (SMM), Slow Magnetic Relaxation, Barrier for Magnetization Reversal.**

I. INTRODUCTION

Molecules that demonstrate magnetic bistability and slow magnetic relaxation, commonly referred to as Single Molecule Magnets (SMM), are of great interest due to their possible application in quantum and molecular electronics [1-3]. The majority of known SMMs contains transition metal ions with orbitally nondegenerate ground states. For these SMMs, the energy barrier for magnetization reversal appears as a result of the combination of a large ground-state spin of the cluster with a negative uniaxial zero-field splitting. Since the zero-field splitting parameter is usually small (a second-order effect with respect to the spin-orbit coupling), the relaxation of magnetization in this type of SMMs is very fast from the point of view of practical applications. In an attempt to increase the energy barrier for magnetization reversal, researches have turned to new types of SMMs [4-10] that contain ions with unquenched orbital angular momenta in the ground state. The anisotropy of transition metal ions with unquenched orbital angular momenta is a first-order effect with respect to the spin-orbit coupling and can be strong enough to result in a considerable barrier for magnetization reversal. In this contribution we report the experimental and theoretical study of a family of linear trimeric [NEt$_4$] {[Mn(L)]$_2$[M(CN)$_6$]} (M = Fe, Ru, Os) compounds containing ions with unquenched orbital angular momenta and demonstrating slow relaxation of magnetization at low temperatures.

II. MODEL

The examined complexes represent linear trimers with the central Fe^{III}, Ru^{III} or Os^{III} and terminal Mn^{III} ions. In all three complexes the central ions are surrounded by 6 carbon atoms and are in the low spin ground state ($^2T_1$ orbital triplet). The ground state of the Mn^{III} ion in the mixed ligand surrounding consisting of 3 oxygen and 3 nitrogen atoms is ^5E. The Hamiltonian for all three complexes can be written as:

$$\begin{aligned} \boldsymbol{H} = & -\kappa\lambda \boldsymbol{L}_1 \boldsymbol{S}_1 + \Delta\left(\boldsymbol{L}_{1Z}^2 - L_1[L_1+1]/3\right) - 2J_{ex}\boldsymbol{S}_1 \sum_{i=2,3} \boldsymbol{S}_i \\ & + D\sum_{i=2,3}\left(\boldsymbol{S}_{iZ}^2 - S_i[S_i+1]/3\right) - zJ\sum_{i=1,2,3} \boldsymbol{S}_i \langle S\rangle \\ & + \mu_B H\left(-\kappa \boldsymbol{L}_1 + g_e \boldsymbol{S}_1 + g\sum_{i=2,3}\boldsymbol{S}_i\right), \end{aligned} \tag{1}$$

where subscript 1 refers to the central ion, while subscripts 2 and 3 relate to the terminal Mn ions. In eq.(1) the first term represents the spin-orbit interaction within $^2T_1$ ground state of the central ion with λ being the spin-orbit coupling parameter equal to -486 cm^{-1}, -880 cm^{-1} and -3000 cm^{-1} for Fe^{III}, Ru^{III} and Os^{III}, respectively [11]. The second term of the Hamiltonian describes the axial cristal field and takes into account the distortion of the local surrounding of the central Fe^{III}, Ru^{III} or Os^{III} ions from the octahedral one. The third term in eq.(1) describes the exchange interaction between the cluster ions. Folowing the idea of Lines [12] this exchange is assumed to be isotropic, it operates with the real spins of the interacting ions (S_1=1/2, S_2=S_3=2). The fourth term in eq.(1) represents the zero-field splitting (ZFS) Hamiltonian for the terminal Mn^{III} ions with D being the axial ZFS parameter. The firth term in eq.(1) stands for the intercluster interaction with J being the interaction

V. Sontea and I. Tiginyanu (eds.), *3rd International Conference on Nanotechnologies and Biomedical Engineering*,
IFMBE Proceedings 55,
DOI: 10.1007/978-981-287-736-9_31

parameter between two nearest neighbors and z being the number of nearest neighbors. Finally, the last part of the Hamiltonian describes the Zeeman interaction. Since the central Fe^{III}, Ru^{III} and Os^{III} ions possess an unquenched orbital angular momentum, the Zeeman part of the Hamiltonian contains both the spin and orbital contributions.

III. RESULTS

The observed temperature behavior of the χT product for the MnRuMn complex is presented in Fig.1. The field dependence of magnetization measured at low temperatures is shown in Fig.2. With the temperature decrease the χT value increases, passes through the maximum at about 4K and then decreases. The magnetic behavior of two other complexes looks similar with the maxima of the χT product being slightly shifted to 5.5K and 3.5K for MnFeMn and MnOsMn, respectively.

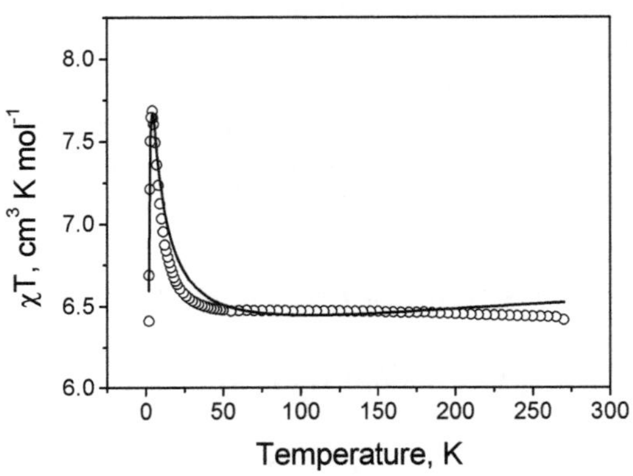

Fig. 1 Temperature dependence of χT for the [NEt$_4$] {[Mn(L)]$_2$[Ru(CN)$_6$]} compound. Symbols – experimental data, solid line – theoretical curve calculated with $\Delta = 60$ cm^{-1}, $J_{ex} = 5.3$ cm^{-1}, $g = 1.99$, $D = -3.6$ cm^{-1}, $zJ = -0.037$ cm^{-1} and $\kappa = 0.6$.

The magnetic behavior of all complexes under study was analyzed with the use of the Hamiltonian (1). The proposed model provides a good agreement between the observed and calculated magnetic behavior for all examined compounds. The result of the best fit procedure for the MnRuMn complex is shown in Figs.2, 3 as solid lines. The best fit parameters are parts of the Figure captions. For all three complexes the exchange interaction is found to be ferromagnetic with the corresponding exchange parameter being equal to 10.6 cm^{-1}, 5.3 cm^{-1} and 6.5 cm^{-1} for MnFeMn, MnRuMn and MnOsMn, respectively. The axial ZFS parameter for Mn^{III} ions varies in the range -2.9 ÷ -5.0 cm^{-1} that can be explained by the slight variation of the position of the nearest ligands (metal-ligand distances and corresponding angles) in the examined complexes. The axial crystal field parameter Δ for the central ion in all complexes is small (80 cm^{-1}, 60 cm^{-1} and 48 cm^{-1} for Fe, Ru and Os ions, respectively) that agrees well with the crystallographic data testifying the almost cubic symmetry of the local environment of the central ions.

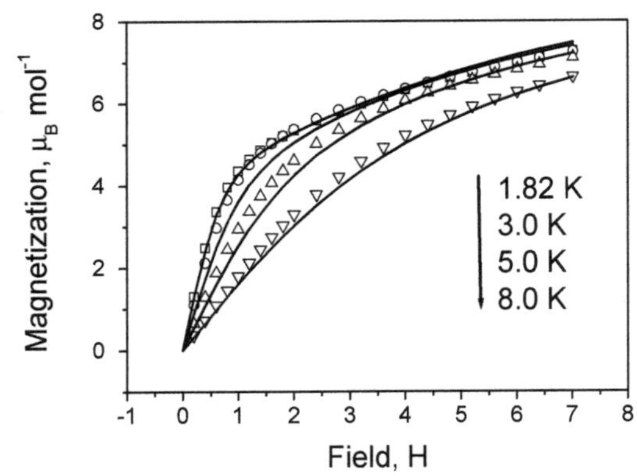

Fig. 2 Magnetization as a function of the external magnetic field for the [NEt$_4$] {[Mn(L)]$_2$[Ru(CN)$_6$]} compound. Symbols – experimental data, solid lines – theoretical curves calculated with $\Delta = 60$ cm^{-1}, $J_{ex} = 5.3$ cm^{-1}, $g = 1.99$, $D = -3.6$ cm^{-1}, $zJ = -0.037$ cm^{-1} and $\kappa = 0.6$.

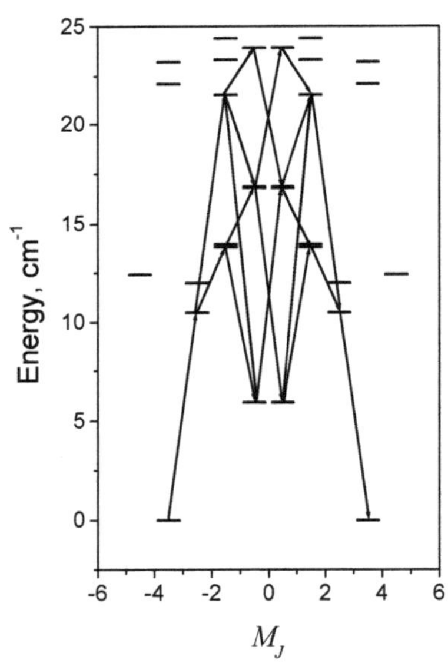

Fig. 3 Low-lying energy levels of MnRuMn cluster and the cascade responsible for magnetic relaxation in this complex.

For each cluster the pattern of low-lying energy levels has been rebuilt. As an example in Fig.3 the energy levels for the MnRuMn trimeric complex are presented. For all complexes under study the Kramers doublet with $|M_J|=7/2$ is the ground one. Levels with the smaller projections of the total angular momentum of the complex are higher in energy forming thus the barrier for magnetization reversal. The possible one-phonon transitions which depopulate the ground state of each cluster and are responsible for the magnetic relaxation were determined. In Fig. 3 the cascade responsible for relaxation in the MnRuMn complex is shown by arrows. It should be mentioned that in this scheme the allowed transitions which depopulate the state with $M_J = -7/2$ are only shown. The transitions which bring the system back to the state with $M_J = -7/2$ are not indicated in the figure for simplicity.

The analysis of one-phonon transitions populating and depopulating the ground cluster states gives the possibility to estimate the heights 16.1, 16.8 and 22.8 cm^{-1} of the energy barriers for magnetization reversal in the clusters MnFeMn, MnRuMn and MnOsMn, respectively. The calculated barrier values are close to the observed ones.

IV. SUMMARY

In this contribution the experimental and theoretical study of a family of linear trimeric $[NEt_4]\{[Mn(L)]_2[M(CN)_6]\}$ (M = Fe, Ru, Os) SMMs containing ions with unquenched orbital angular momenta was reported. Both the direct current magnetic behavior (the magnetic susceptibility vs. temperature and magnetization vs. magnetic field) and the existence of the barrier for magnetization reversal were explained for all complexes. The values of the key parameters were obtained. The calculated heights of the barriers for magnetization reversal are close to the experimentally observed values.

ACKNOWLEDGMENT

S.O., O.R., A.P. and S.K. thank the STCU (project 5929) and the Supreme Council for Science and Technological Development of the Republic of Moldova (project 15.817.02.06F) for financial support.

CONFLICT OF INTEREST

The authors declare that they have no conflict of interest.

REFERENCES

1. Gatteschi D, Sessoli R (2003) Angew Chem Int Ed 42: 268
2. Bogani L, Wernsdorfer W (2008) Nat Mater 7: 179
3. Leuenberger MN, Loss D (2001) Nature 410: 789
4. Choi HJ, Sokol JJ, Long JR (2004) Inorg Chem 43: 1606
5. Berlinguette CP, Vaughn D, Cañada-Vilalta C, Galán-Mascarós J-R, Dunbar KR (2003) Angew Chem Int Ed 42: 1523
6. Wang S, Zuo JL, Zhou HC, Choi HJ, Ke Y, Long JR (2004) Angew Chem Int Ed 43: 5940
7. Schelter EJ, Prosvirin AV, Dunbar KR (2004) J Am Chem Soc 126: 15004
8. Palii AV, Ostrovsky SM, Klokishner SI, Tsukerblat BS, Galán-Mascarós JR, Berlinguette CP, Dunbar KR (2004) J Am Chem Soc 126: 16860
9. Palii AV, Ostrovsky SM, Klokishner SI, Tsukerblat BS, Dunbar KR (2006) Chem Phys Chem 7: 871
10. Palii AV, Ostrovsky SM, Klokishner SI, Tsukerblat BS, Schelter EJ, Prosvirin AV, Dunbar KR (2007) Inorg Chim Acta 360: 3915
11. Abragam A, Bleaney B (1970) Electron Paramagnetic Resonance of Transition Ions. Clarendon Press, Oxford
12. Lines ME (1971) J Chem Phys 55: 2977

Corresponding author:	S.Klokishner
Institute:	Institute of Applied Physics, Academy of Sciences of Moldova
Street:	Academy str. 5
City:	Chişinău
Country:	Moldova
Email:	klokishner@yahoo.com

Slow Magnetic Relaxation in Dysprosium Based Single-Ion Magnets

K. Preuss[1], R. Clerac[2], O. Reu[3], S. Ostrovsky[3], A. Palii[3], and S. Klokishner[3]

[1] University of Guelph, Dept. of Chemistry, Guelph, Ontario N1G2W1, Canada
[2] Centre de Recherche Paul Pascal, CNRS, Universite de Bordeaux, France
[3] Institute of Applied Physics, Academy of Sciences of Moldova, Chisinău, Moldova

Abstract— **The experimental and theoretical study of two mononuclear dysprosium-based complexes is reported. Both complexes demonstrate slow relaxation of magnetization at low temperatures. The Hamiltonian for explanation of the magnetic behavior (the magnetic susceptibility as a function of temperature and magnetization as a function of the applied magnetic field) includes the crystal field of the nearest ligands of the dysprosium ion and the interaction with the external magnetic field. The crystal field parameters are evaluated in the framework of the exchange charge model, taking account for covalence effects. The proposed model provides a good agreement between the observed and calculated magnetic behavior. It is demonstrated that the low-lying energy levels for both studied complexes form the energy barriers for magnetization reversal that agrees with the observed single-ion magnetic behavior.**

Keywords— **Single Molecule Magnet (SMM), Single-Ion Magnet (SIM), Slow Magnetic Relaxation, Barrier for Magnetization Reversal.**

I. INTRODUCTION

Study of single-molecule magnets (SMMs) is of great interest due to their rich physical behavior. At lower temperatures crystals of these molecules demonstrate slow relaxation of the magnetization and magnetic hysteresis [1-3]. On the other hand, they show quantum phenomena like quantum tunneling of the magnetization, quantum coherence, or quantum interference, so they are proposed to be candidates for spin qubits in quantum computing [4-8].

Until now, the majority of the investigated SMMs contains polynuclear magnetic complexes composed of transition metal ions with orbitally nondegenerate ground states. In the past years, a new generation of SMMs has been discovered. These SMMs are based on mononuclear complexes of lanthanide ions and are known as single-ion magnets (SIMs) or mononuclear SMMs [9]. The energy barrier in this type of SMMs can be significant and can reach several hundreds cm^{-1} [10].

In this contribution we report the experimental and theoretical study of two mononuclear lanthanide-based complexes, namely, $Dy(hfac)_3$(boaDTDA) and $Dy(hfac)_3$(5-CNpyDTDA) that demonstrate slow relaxation of magnetization at low temperatures.

II. MODEL

The nearest local environment of the Dy^{3+} ion in the $Dy(hfac)_3$(boaDTDA) and $Dy(hfac)_3$(5-CNpyDTDA) compounds is similar and consists of two nitrogen and six oxygen atoms. The Hamiltonian of the crystal field acting in the space of the 4f orbitals of the Dy^{3+} ion can be written as:

$$H_{cf} = \sum_{l=2,4,6} \left\{ A_l^0 C_l^0 + \sum_m \left[A_l^m (C_l^{-m} + (-1)^m C_l^m) + B_l^m (C_l^{-m} - (-1)^m C_l^m) \right] \right\}, \tag{1}$$

where $C_p^m(\vartheta,\varphi) = [4\pi/(2p+1)]^{1/2} Y_{pm}(\vartheta,\varphi)$ are the tensor spherical operators with $Y_{pm}(\vartheta,\varphi)$ being the normalized spherical harmonics. The crystal field parameters A_p^m and B_p^m in Hamiltonian (1) are evaluated in the framework of the exchange charge model of the crystal field [11,12] and are:

$$A_p^{|m|} = A_p^{|m|(pc)} + A_p^{|m|(ec)}, \quad B_p^{|m|} = B_p^{|m|(pc)} + A_p^{|m|(ec)}, \tag{2}$$

where the parameters $A_p^{|m|(pc)}$ and $B_p^{|m|(pc)}$ describe the electrostatic interaction between 4*f* electrons of the Dy ion and the point charges of the surrounding ligands and can be written as:

$$A_p^{|m|(pc)} = (-1)^m \sum_\alpha \frac{Z_\alpha e^2 < r^p > (1-\sigma_p)}{2(R_\alpha)^{p+1}} \left[C_p^m(\vartheta_\alpha,\varphi_\alpha) + C_p^{m*}(\vartheta_\alpha,\varphi_\alpha) \right],$$

$$B_p^{|m|(pc)} = (-1)^m \sum_\alpha \frac{Z_\alpha e^2 < r^p > (1-\sigma_p)}{2(R_\alpha)^{p+1}} \left[C_p^m(\vartheta_\alpha,\varphi_\alpha) - C_p^{m*}(\vartheta_\alpha,\varphi_\alpha) \right]. \tag{3}$$

In eq.(3) $R_\alpha, \vartheta_\alpha$ and φ_α are the spherical coordinates of the α ligand with the effective charge $Z_\alpha e$. Parameters $\langle r^p \rangle$ are the radial integrals for the Dy^{3+} ion, and σ_p is the shielding factor. In the subsequent calculations the values $\langle r^2 \rangle = 0.849$ a.u., $\langle r^4 \rangle = 1.977$ a.u., $\langle r^6 \rangle = 10.44$ a.u. and $\sigma_2 = 0.527$, $\sigma_4 = -0.0199$, $\sigma_6 = -0.0316$ [13] are used.

The parameters $A_p^{|m|(ec)}$ and $B_p^{|m|(ec)}$ describe the contribution of exchange charges that appear due to the overlap of

V. Sontea and I. Tiginyanu (eds.), *3rd International Conference on Nanotechnologies and Biomedical Engineering*,
IFMBE Proceedings 55,
DOI: 10.1007/978-981-287-736-9_32

$4f$ orbitals of the Dy^{3+} ion with the ligand orbitals and are given as:

$$A_p^{|m|(ec)} = (-1)^m \frac{e^2(2p+1)}{7} \sum_\alpha \frac{S_p(R_\alpha)}{R_\alpha} \left[C_p^m(\vartheta_\alpha,\varphi_\alpha)+C_p^{m*}(\vartheta_\alpha,\varphi_\alpha)\right],$$

$$B_p^{|m|(ec)} = (-1)^m \frac{e^2(2p+1)}{7} \sum_\alpha \frac{S_p(R_\alpha)}{R_\alpha} \left[C_p^m(\vartheta_\alpha,\varphi_\alpha)-C_p^{m*}(\vartheta_\alpha,\varphi_\alpha)\right], \quad (4)$$

where the overlap integrals $S_p(R_N)$ and $S_p(R_O)$ for the nitrogen and oxygen ligands are expressed as:

$$S_p^L(R_\alpha)=G_L\left(S_{sL}^2(R_\alpha)+S_{\sigma L}^2(R_\alpha)+\gamma_p S_{\pi L}^2(R_\alpha)\right), \quad (5)$$

with $\gamma_2 = 3/2$, $\gamma_4 = 1/3$, $\gamma_6 = -3/2$. $S_{sL}(R_\alpha)=\langle 4f,m=0|2s\rangle$, $S_{\sigma L}(R_\alpha)=\langle 4f,m=0|2p,m=0\rangle$ and $S_{\pi L}(R_\alpha)=\langle 4f,m=1|2p,m=1\rangle$, ($L$ = O, N) are the overlap integrals of the $4f$ wave functions of the Dy^{3+} ion with $2s$ and $2p$ wave functions of the nitrogen and oxygen ligands. Numerical values of the overlap integrals were calculated with the use of the $4f$ and $2s$, $2p$ wave functions of Dy, O and N from [14,15]. The values G_O and G_N are the dimensionless phenomenological parameters of the model.

For the description of the magnetic characteristics of the Dy(hfac)$_3$(boaDTDA) and Dy(hfac)$_3$(5-CNpyDTDA) complexes the following Hamiltonian has been applied:

$$H = H_{cf} + g_J \mu_B \mathbf{H}\,\mathbf{J}, \quad (6)$$

where the first term represents the interaction of the Dy^{3+} ion with the crystal field of the nearest surrounding formed by two nitrogen and six oxygen ligands (eq.(1)), while the second term describes the interaction of the Dy^{3+} ion with the external magnetic field, μ_B is the Bohr magneton, $g_J = 4/3$ is the Lande factor for the ground $^6H_{15/2}$ term of the Dy^{3+} ion. Finally, $\mathbf{J}=\mathbf{L}+\mathbf{S}$ is the total angular momentum for the Dy^{3+} ion (L=5, S=5/2 and J=15/2). For both compounds under examinaion the obtained eigenvalues of the Hamiltonian have been utilized for the calculation of the main components $\chi_{\alpha\alpha}$ $(\alpha = x,y,z)$ of the magnetic susceptibility tensor, the magnetic susceptibility χ and magnetization M with the aid of the following formulae:

$$\chi_{\alpha\alpha} = N_A kT \frac{\partial^2}{\partial H_\alpha^2}\left[\ln Z(H_\alpha)\right]_{H_\alpha \to 0},$$

$$\chi = \frac{1}{3}(\chi_{XX}+\chi_{YY}+\chi_{ZZ}),$$

$$M = N_A kT \frac{1}{4\pi}\int_0^\pi \sin\theta d\theta \int_0^{2\pi} d\varphi \left.\frac{\partial \ln Z(H_\alpha)}{\partial H_\alpha}\right|_{H_\alpha \to H(\theta,\varphi)} \quad (7)$$

where

$$Z(H_\alpha)=\sum_i \exp(-E_i(H_\alpha)/kT) \quad (8)$$

is the partition functions and $E_i(H_\alpha)$ $(\alpha = x,y,z)$ are the energies of the complex in the applied magnetic field.

III. Results

For the Dy(hfac)$_3$(boaDTDA) complex the temperature dependence of the magnetic susceptibility for the powder sample and the main values of the components of the magnetic susceptibility tensor calculated with the set of the best fit parameters are shown in Fig.1.

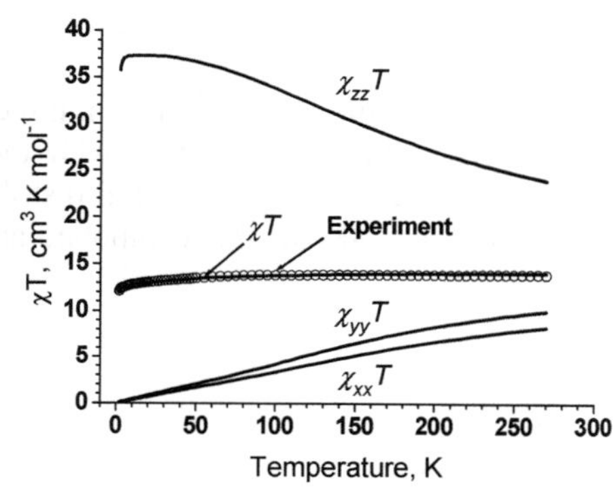

Fig.1 Temperature dependence of the magnetic susceptibility χT and its components $\chi_{\alpha\alpha}T$ for the Dy(hfac)$_3$(boaDTDA) complex calculated with Z_{N1}=1, Z_{N2}=0 and the set of the best fit parameters G_O=0, Z_O=0.4, G_N=9.5. Circles-experimental data [16], solid lines- calculated curves. Agreement criterion:

$$\delta_{\chi T} = \sqrt{\frac{1}{N}\sum_{i=1}^{N}\frac{((\chi T)_i^{cal}-(\chi T)_i^{\exp})^2}{((\chi T)_i^{\exp})^2}} = 0.8\%$$

It is seen that at low temperatures the components $\chi_{xx}T$ and $\chi_{yy}T$ coincide, while at higher temperatures they differ insignificantly. The latter testifies a small deviation of the complex symmetry from the axial one. The field dependence of magnetization for the Dy(hfac)$_3$(boaDTDA) complex is illustrated in Fig.2. It is seen that the suggested model also describes quite well this characteristics of the complex.

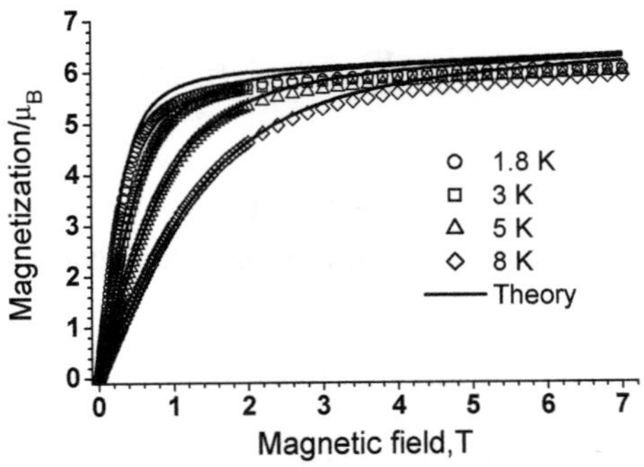

Fig. 2 Field dependence of magnetization for the Dy(hfac)$_3$(boaDTDA) complex in the temperature range 1.8 - 8 K. The theoretical curves are calculated with the same set of parameters as in Fig.1. Shapes-experimental data [16], solid lines - calculated curves. Agreement criterions:

$$\delta_M(T=1.8K)=5.7\%,\ \delta_M(T=3K)=3.3\%,$$
$$\delta_M(T=5K)=3.0\%,\ \delta_M(T=8K)=2.8\%.$$

The Stark structure of the ground $^6H_{15/2}$ multiplet for the Dy(hfac)$_3$(boaDTDA) complex calculated with the aid of the best fit parameters as well as the spatial distribution of the magnetization at T=1.8K for this compound are shown in Figure 3.

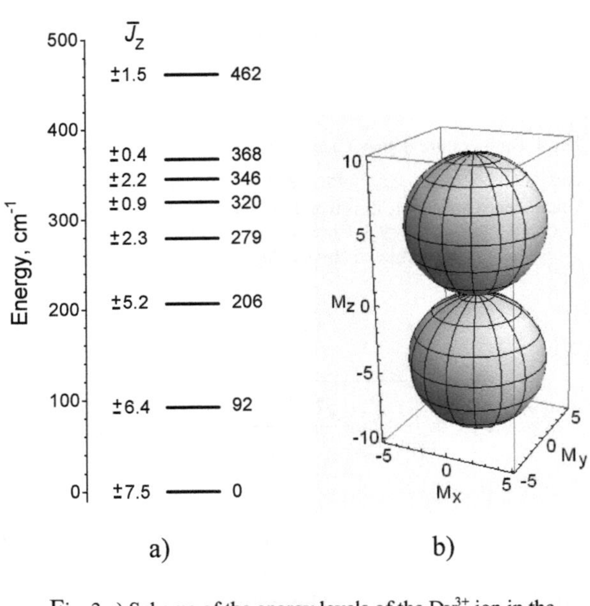

Fig. 3 a) Scheme of the energy levels of the Dy^{3+} ion in the Dy(hfac)$_3$(boaDTDA) compound calculated with the set of the best fit parameters G_O=0, Z_O=0.4, G_N=9.5, Z_{N1}=1, Z_{N2}=0. For each level the mean value of the operator J_Z is indicated. b) Spatial distribution of the magnetization at T=1.8K for the Dy(hfac)$_3$(boaDTDA) complex calculated with the set of the best fit parameters at the fixed magnitude H=1T of the external magnetic field.

The Stark structure of the Dy^{3+} ion obtained with the optimal set of the best fit parameters, which fit the magnetic susceptibility χT and magnetization of the Dy(hfac)$_3$(boaDTDA) complex principally evidences negative axial anisotropy and the presence of its SMM behavior. For all Stark levels the calculated mean value of the projection J_Z of the total angular momentum of the Dy-ion decreases with the increase of the level energy, thus the barrier for magnetization reversal is formed. The obtained barrier height together with the calculated temperature dependence of the χT components prove the SMM behavior of the Dy(hfac)$_3$(boaDTDA) complex.

The calculations of the magnetic susceptibility χT, magnetization as well as of the barrier for magnetization reversal have been also performed for the Dy(hfac)$_3$(5-CNpyDTDA) complex. In this case for the set of parameters G_O=7.6, Z_O=-0.47, G_N=29.3, Z_N =0 quite nice agreement between the observed magnetic characteristics and the calculated ones was obtained. The calculated Stark structure of the Dy^{3+} for the Dy(hfac)$_3$(5-CNpyDTDA) complex also proves the presence of negative axial anisotropy for this complex and its SMM behavior.The height of the barrier for reversal of magnetization in this complex is 106 cm^{-1}.

IV. Summary

In this contribution we have modeled the magnetic properties of the Dy(hfac)$_3$(boaDTDA) and Dy(hfac)$_3$(5-CNpyDTDA) complexes and, namely, the temperature dependence of the magnetic susceptibility and field dependence of magnetization. The performed calculations of the Stark structure of the ground $^6H_{15/2}$ multiplets of the Dy^{3+} ions allowed to estimate the heights of the barriers for the reversal of magnetization. The model reveals strong magnetic anisotropy for the Dy(hfac)$_3$(boaDTDA) and Dy(hfac)$_3$(5-CNpyDTDA) complexes with the easy axis of magnetization oriented along the Z-axis of the system. At the same time the developed model provides a satisfactory description of the magnetic susceptibility and magnetization of these complexes.

Acknowledgment

O.R., S.O., A.P. and S.K. thank the STCU (project 5929) and the Supreme Council for Science and Technological Development of the Republic of Moldova (project 15.817.02.06F) for financial support.

Conflict of Interest

The authors declare that they have no conflict of interest.

REFERENCES

1. Thomas L, Lionti F, Ballou R, Gatteschi D, Sessoli R, Barbara B (1996) Nature 383: 145–147
2. Friedman JR, Sarachik MP, Tejada J, Ziolo R (1996) Phys Rev Lett 76: 3830–3833
3. Wernsdorfer W, Ohm T, Sangregorio C, Sessoli R, Mailly D, Paulsen C (1999) Phys Rev Lett 82: 3903–3906
4. Gatteschi D, Sessoli R (2003) Angew Chem Int Ed 42: 268–297
5. Ardavan A, Blundell SJ (2009) J Mater Chem 19: 1754–1760
6. Troiani F, Affronte M (2011) Chem Soc Rev 40: 3119–3129
7. Stamp PCE, Gaita-Ariño A (2009) J Mater Chem 19: 1718–1730
8. Lehmann J, Gaita-Ariño A, Coronado E, Loss D, (2007) Nat Nanotechnol 2: 312–317
9. Sorace L, Benelli C, Gatteschi D (2011) Chem Soc Rev 40: 3092–3104
10. Ishikawa N, Sugita M, Ishikawa T, Koshihara SY, Kaizu Y (2003) J Am Chem Soc 125: 8694–8695
11. Malkin BZ (1987) In: Kaplyanskii AA, Macfarlane RM (Eds) Spectroscopy of Solids Containing Rare-Earth Ions. North-Holland, Amsterdam, p 13-50
12. Reu OS, Palii AV, Ostrovsky SM, Tregenna-Piggott PLW, Klokishner SI (2012) Inorg Chem 51: 10955-10965
13. Edvardson S, Klintenberg M (1998) J Alloys Comp 275-277: 230-233
14. Freeman AJ, Watson RE (1962) Phys Rev 127: 2058-2075
15. Clementi E, Mclean AD (1964) Phys Rev 133: A419-A423
16. Fatila EM, Rouzières M, Jennings MC, Lough AJ, Clérac R, Preuss KE (2013) J Am Chem Soc 135: 9596-9599

Corresponding author: S.Klokishner
Institute: Institute of Applied Physics, Academy of Sciences of Moldova
Street: Academy str. 5
City: Chişinău
Country: Moldova
Email: klokishner@yahoo.com

Electric Field Control of Magnetic and Polarizability Properties of Trimeric Mixed Valence Clusters

M. Roman, O. Reu, and S. Klokishner

Institute of Applied Physics, Academy of Sciences of Moldova, Chisinau, Moldova

Abstract— **We study the effects of the external *direct current (dc)* electric field in the magnetic and polarizability characteristics of trimeric*Cu*(II)–*Cu*(II)–*Cu*(III) mixed valence clusters. We calculate the temperature and field dependence of the magnitude of the mean dipole moment as well as of its components. We predict the anisotropy of the cluster polarizability.The magnetic moment at the low temperature limit is determined by the competition between the tunnel and Heisenberg interactions and the interaction with the external electric field.It is demonstrated that the external electric field gives an unique possibility of manipulation by the polarizability and magnetic characteristics of trimeric mixed-valence clusters.**

Keywords— **mixed valence clusters, *dc* electric field, polarization characteristics, magnetic moment**

I. INTRODUCTION

One of the hot topics in molecular magnetism refers to the search of molecule-based magnetic materials exhibiting multifunctionality. Especially the electric control of magnetism represents a subject of lively interest in modern physics [1]. The construction of molecule-based devices for which the magnetic state can be controlled electrically represents an important advance in molecular spintronics [2,3] and a basis for magnetic molecule-based quantum computing. The possibility to manipulate directly by the magnetic properties with the aid of the electric field can be provided by mixed valence (MV) clusters. In the present paper we examine the manifestations of the electric *dc* field in the magnetic and polarizability characteristics of trimeric MV clusters. The valence-delocalized trinuclear $d^9-d^9-d^8$ clusters $\lfloor Cu_3^{7+} \rfloor$ and $[Cu_3(\mu-S)_2]^{3+}$ [4,5] exemplify such systems.

II. THE MODEL

A crystal containing MV trinuclear copper clusters $d^9-d^9-d^8$ as structural units is examined. The "extra" hole in the *d*-shell migrates inside each cluster. The clusters are magnetically isolated from each other, the electrons of the unfilled metal shells do not form a band of noticeable width, and the crystal does not possess conductivity. Collective effects (magnetic and charge ordering) are not observed. In this regard the clusters can be considered as isolated ones. The full cluster spin acquires the values, $S=0,1$ since the spin of the Cu^{II} ((d^9)) ion is ½, while the spin of the Cu^{III}(d^8) ion is vanishing.

The full Hamiltonian of the cluster can be written in the following form:

$$H = H_0 - \mathbf{dE} \ , \tag{1}$$

where the first term H_0 is the Hamiltonian of an isolated cluster with eigenvalues determined by $\varepsilon(^{2S+1}A) = (-1)^S(-2p+J) - J$, $\varepsilon(^{2S+1}E) = (-1)^S(p+J) - J$, [6]where p and J are the transfer and Heisenberg exchange parameters, respectively ($A = A_2$ and A_1 for spins $S=0$ and 1, respectively, A_1, A_2 and E are the irreducible representations of the C_{3v} point group).The second term in Eq.(1) describes the interaction of the cluster with the external *dc* electric field of the intensity $\mathbf{E}$ and $\mathbf{d}$ is the operator of the cluster dipole moment.The field lies in the plane of the equilateral triangle formed by the cluster ions. The cluster is considered electrically neutral and its center is taken as the reference of the coordinate frame.

The system behavior in the external electric field is governed by two non-dimensional parameters $x = J/|p|$ and $y = d_0 E/|p|$. The parameter x characterizes the mutual arrangement of the isolated cluster levels. The second parameter y describes the competition between the intracluster delocalization and interaction of the cluster with the external electric field.

In the general case the energy levels of the cluster and corresponding wave functions in the *dc* electric field can be found by diagonalization of the matrix of Hamiltonian (1). For several values of the parameter x the field dependence of the energy levels of the $d^9-d^9-d^8$ cluster is presented in Fig. 1.

As it follows from Fig. 1 the external electric field leads to the following effects: (i) splitting of the doubly degenerate E-levels; (ii) lowering of the energy of the ground

V. Sontea and I. Tiginyanu (eds.), *3rd International Conference on Nanotechnologies and Biomedical Engineering*, IFMBE Proceedings 55,

DOI: 10.1007/978-981-287-736-9_33

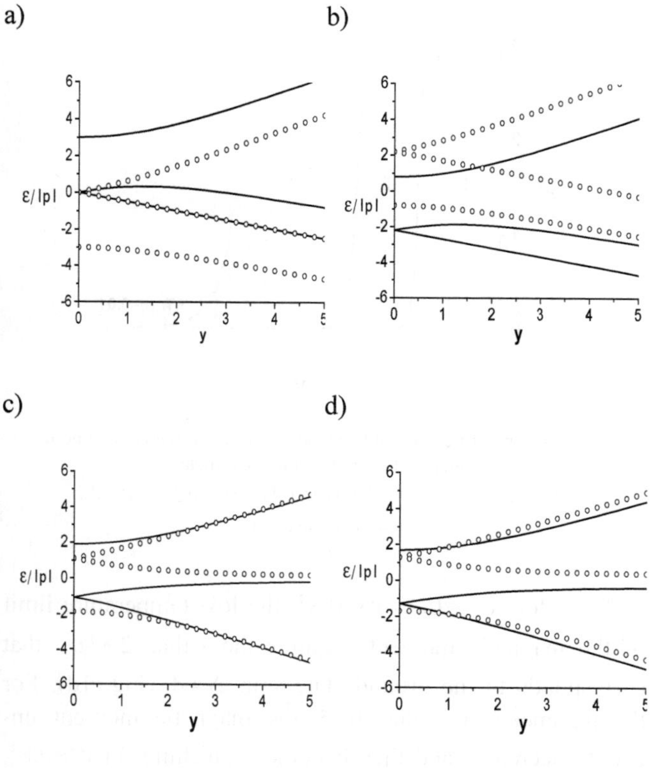

Fig. 1 Energy levels of the $d^9-d^9-d^8$ cluster as functions of the external *dc* electric field calculated with the following sets of parameters: a) $p<0, x=1.0, \varphi=\pi/6$; b) $p<0, x=-1.2, \varphi=\pi/6$; c) $p<0, x=-0.1, \varphi=0$. d) $p<0, x=-0.3, \varphi=0$. Solid lines– levels corresponding to $S=0$, circles – levels corresponding to $S=1$.

state with the increase of the field intensity; (iii) change of the spin of the ground state (Figs 1c, 1d); (iv) dependence of the level energies on the direction of the *dc* electric field.

III. ANISOTROPY OF POLARIZABILITY

The polarizability of the cluster i.e. its ability to acquire a dipole moment in the external *dc* electric field is caused by the localization of the "extra" hole in the $d^9-d^9-d^8$ cluster under action of this field. Further on we will deal with the magnitude $\bar{d}$ of the vector of the mean dipole moment and its components $\bar{d}_x$ and $\bar{d}_y$ which are calculated as

$$\bar{d}_x=\frac{Tr\left(Exp\left(-\frac{H}{kT}\right)d_x\right)}{Tr\left(Exp\left(-\frac{H}{kT}\right)\right)},\quad \bar{d}_y=\frac{Tr\left(Exp\left(-\frac{H}{kT}\right)d_y\right)}{Tr\left(Exp\left(-\frac{H}{kT}\right)\right)},\qquad (2)$$

where the Hamiltonian of the cluster in the external electric field is defined by Eq. (1).

Figure 2 illustrates the low temperature limit of the mean dipole moment as a function of the direction of the external *dc* electric field.

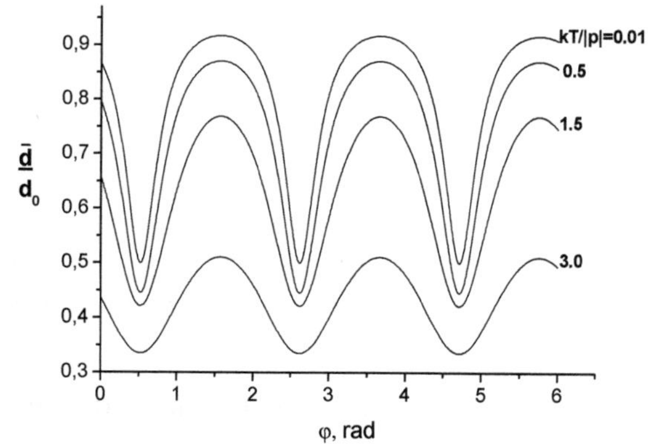

Fig. 2 Dependence of the mean dipole moment at low temperatureson the direction of the*dc* electric field for $p<0,\ x=-0.3,\ y=3$.

At $kT/|p|=0.01$ the mean dipole moment acquires the maximum value 0.92, when the electric field is oriented through the vertexes of the equilateral triangle formed by the cluster ions. Independent of temperature,the values of the mean dipole moment are minimal when the direction of the electric field coincides with the bisectors of the angles $\angle aOb$, $\angle bOc$ and $\angle aOc$.Also, from Fig. 2 it follows that with temperature increase the magnitude of the mean dipole moment decreases. The result obtained obviously shows the possibility of manipulation by the polarizability of the trinuclear cluster through the change of the direction of the external electric fieldas well as of temperature.

In Fig. 3 the field dependence of the low temperature limit of the mean dipole moment $\bar{d}/d_0$ is shown for $p<0$, $\varphi=0$ and different values of the parameter y that characterizes the strength of interaction of the cluster with the external electric field. For $x=-0.3$ and $0<y<1$ the low temperature limit of the dipole moment $\bar{d}/d_0$ smoothly increases from 0 to 0.30, then its value changes abruptly to 0.68.With further increase of y a gradual growth of $\bar{d}/d_0$ takes place.A similar dependence of $\bar{d}/d_0$ on y is observed for $x=-0.2$ and -0.1 with the difference that for these x values the jump occurs at higher intensities of the external electric field.

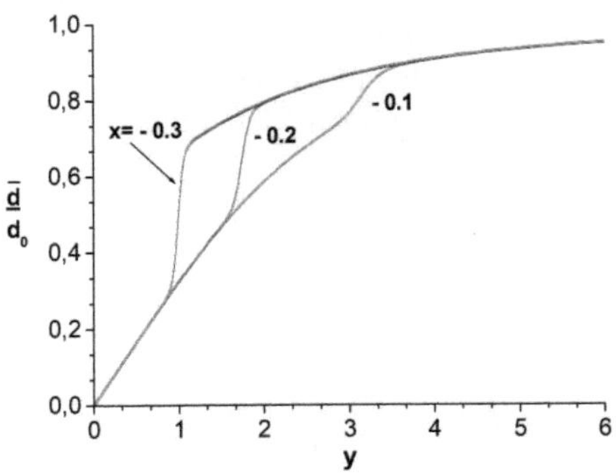

Fig. 3 Low temperature limit of the mean dipole moment $\bar{d}/d_0$ as a function of the external electric field for $p<0$, $\varphi=0$ and different values of parameter x

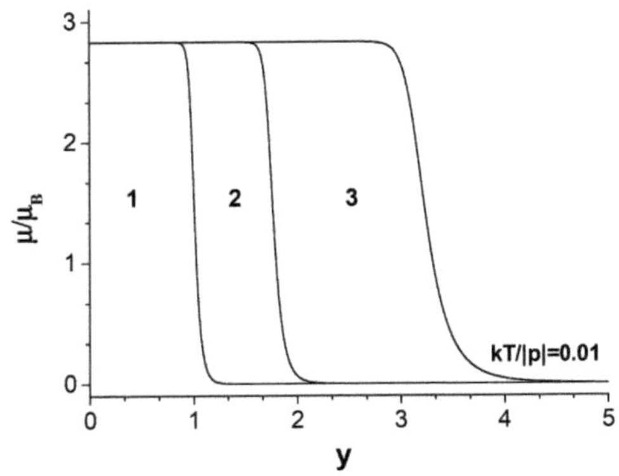

Fig. 4 Temperature and field dependence of the effective magnetic moment calculated withthe parameters: $p<0$, $\varphi=0$, $x=-0.3$ (curve 1), $x=-0.2$ (curve 2), $x=-0.1$ (curve 3).,

It is worth noting that for $x=-0.2$ and -0.3 the abrupt increase of $\bar{d}/d_0$ is accompanied by the change of the ground state spin (see, for instance, Fig1d) from $S=1$ to $S=0$ .Above all the ground becomes the level arising from the splitting of the ^1E – term for which as it was mentioned above the conditions of appearance of cluster polarizability are much softer [7].

IV. FIELD AND TEMPERATURE DEPENDENCE OF THE EFFECTIVE MAGNETIC MOMENT

The thermal average of the magnetic moment is determined by the following expression

$$\mu^2=\mu_B^2 g^2 \frac{\sum_{i,S}(2S+1)S(S+1)Exp\left(-\frac{\varepsilon_i(S)}{kT}\right)}{\sum_{i,S}(2S+1)Exp\left(-\frac{\varepsilon_i(S)}{kT}\right)}, \quad (3)$$

The temperature and field dependence of the magnetic moment is caused by the Boltzmann population of the levels $\varepsilon_i(S)$ in the external electric field.It is demonstrated that with the aid of the *dc* electric field, the valueof the magnetic moment can be significantly changed. For $p<0$, $\varphi=0$, $x=-0.3,-0.2,-0.1$ and $kT/|p|=0$ (Fig. 4.) the magnetic moment demonstrates a step-like dependence on the electric field intensity.

Thus, for $x=-0.3$, $y<0.85$ the low temperature limit of the magnetic moment acquires the value $2.83\mu_B$ that corresponds to the ground state spin $S=1$ (Fig. 1d). For the parameter y value 0.85 the magnetic moment unevenly decreases and then becomes vanishing for $y>0.85$ .Such a behavior of the magnetic moment can be simply explained by the change of the spin of the ground state from $S=1$ to $S=0$ at the point $y=0.85$ (Fig. 1d) as well as by the vanishing population of the first excited level at $kT/|p|=0$. The similar behavior of μ/μ_B is observed and for $x=-0.2$ and -0.1 with the difference that for these x values the decrease of μ/μ_B starts at higher intensities of y.In this case the effective magnetic moment also vanishes at strong electric fields.

Finally, we dwell on the question of the characteristic system parameters which allow to observe the predicted effects. The experiments on MV clusters reveal different relations between the parameters of Heisenberg exchange and electron transfer [8]. The transfer parameter takes on values from 0.1 to 10^3 cm^{-1}. The exchange parameter is a value of the order of 10–10^2 cm^{-1}. As to the characteristic energy of the applied electric field d_0E the values 160 -1600 cm^{-1} are acceptable [9].Therefore, the predicted effects of the electric field may be observed in systems with a sufficiently wide range of parameters p, J and d_0E.

V. CONCLUSIONS

In the present work we have investigated the manifestations of an external electric in the magnetic and electric properties of trimeric $d^9-d^9-d^8$ MV clusters. The low-temperature limit of the magnetic moment defined by the spin of the ground state can be changed by the external electric field. So the strong coupling of the MV cluster with the electric field provides an unique opportunity of governing by the cluster magnetic properties through the field. The most pronounced effects in the magnetic behavior of the cluster are attainable at low temperatures, when only the ground state is populated. Polarizability effects induced by the external electric field are also examined. In the delocalized $A-$ and $E-$ states the mean value of the cluster dipole moment vanishes. In the states with fixed valency clusters possess dipole moments. The electric field promotes a transition between these states. By playing with the magnitude and the direction of the applied field different types of the temperature dependence of the cluster mean dipole moment can be obtained. Besides the described effects the most important impact of the external electric field is the predicted anisotropy of the polarizability caused by this field. The examined MV clusters can be referred to type II multiferroics for which the magnetic (electric) moment can be changed by electric (magnetic) field.

ACKNOWLEDGEMENT

M.R., O.R. and S.K. thank the STCU (project 5929) and the Supreme Council for Science and Technological Development of the Republic of Moldova (project 15.817.02.06F) for financial support.

CONFLICT OF INTEREST

The authors declare that they have no conflict of interest.

REFERENCES

1. Vaz C A F (2012) J Phys: Condens Matter 24: 333201.
2. Sanvito S (2007) J Mater Chem. 17: 4455-4459
3. Bogani L, Wernsdorfer W (2008) Nature Materials 7: 179-186
4. Mezei G, McGrady JE, Rapits RG (2005) Inorg Chem 44: 7271-7273
5. Brown EC, York JT, Antholine WE, Ruiz E, Alvarez S, Tolman WB (2005) J Am Chem Soc 127: 13752-13753
6. Belinsky MI (2006) Inorg Chem 45: 9096-9106
7. Klokishner SI, Roman MA, Reu OS (2014) Eur J Inorg Chem 23: 3701-3709
8. Brown DB (1979) Mixed-Valence Compounds, D. ReidelPubl. Co., Dordrecht
9. Murga LF, Ondrechen MJ (1998) J Inorg Biochem 70: 245-252

Corresponding author: S.Klokishner
Institute: Institute of Applied Physics, Academy of Sciences of Moldova
Street: Academy str. 5
City: Chisinau
Country: Moldova
Email: klokishner@yahoo.com

Perspectives of Bulk and Nanosized II-VI Compounds for Light-Emission Application

I. Radevici[1,2], K. Sushkevich[2], G. Colibaba[2], H. Huhtinen[1], D. Nedeoglo[2], and P. Paturi[1]

[1] Wihuri Physical Laboratory, University of Turku, Turku, Finland
[2] Faculty of Physics and Engineering, Moldova State University, Chisinau, Moldova

Abstract— **Activation of II-VI semiconductors with transition metals (Cr) and rare-earths elements (Yb) results in interaction of the doping ions with the native defects (e.g., zinc vacancies). This interaction has the most significant influence on formation of emission properties of the wide bandgap semiconductors. Effects of defect-impurity interaction on emission spectra of bulk ZnSe:Cr and ZnSe:Yb crystals are shown. Possibility to use nanosized materials to control this interaction is discussed.**

Keywords— **II-VI semiconductors, transition metals, rare-earths elements, emission.**

I. INTRODUCTION

For several decades wide bandgap semiconductor materials from the II-VI group have attracted attention as promising candidates for the blue and green solid state lasers and diodes [1, 2]. Large scale application of these materials, however, been hampered by the low conductivity of either p-type or n-type [3]. Besides the fact that some progress in this direction was already achieved, e.g. well-conducting p-type ZnSe samples have been produced [4], it is still not known why both acceptor and donor doping of each II-VI semiconductor is so difficult.

A number of explanations for the problem have been suggested. The most commonly accepted one has been that native defects (e.g., cation self-interstitials) that act as donors are activated as a result of doping, resulting in compensation of acceptor impurities [2]. At the same time activation of II-VI semiconductor materials is absolutely necessary for controlling their emission properties. For example, doping of Zn and Cd chalcogenides with transient metals (TM), for example, Cr^{2+} and Fe^{2+}, results in apparition of a wide emission band in the middle infrared (IR) spectral region, which may be used for development of tunable solid state lasers [5].

In this work we will discuss influence of defect impurity interaction on emission properties of bulk ZnSe crystals doped with TM and rare-earths elements (REE). Usage of various growth and doping techniques [6, 7] allows to produce a wide range of point and complex defects that are most common for II-VI materials. Possibilities to control native-impurity defect interaction in the bulk materials will be analyzed and compared with the alternative to use nanosized materials to overcome the problem of near-equilibrium doping.

II. RESULTS AND DISCUSSION

In middle-IR spectral range of ZnSe:Cr emission spectra a wide structural photoluminescence (PL) band with a few maxima may be observed (Fig. 1). Total intensity of the band increases with increasing concentration of chromium doping impurity (Fig. 2), however, position of the main maximum, full width at half maximum (FWHM) as well as maximums' ratio strongly depend on the crystals doping method and temperature.

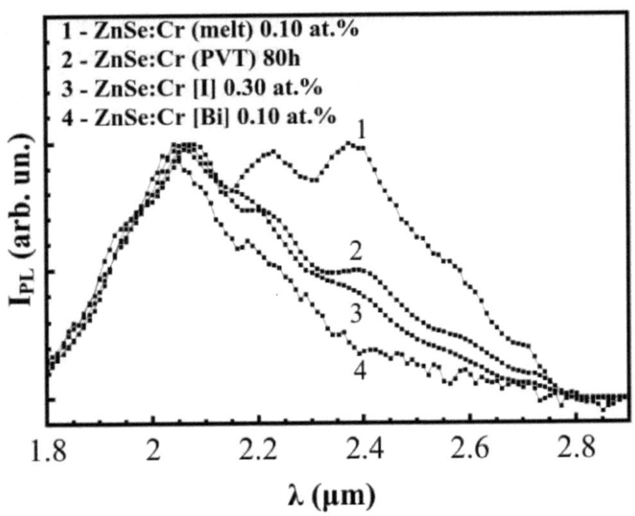

Fig. 1 Normalized PL spectra of ZnSe:Cr crystals in IR spectral range at T=6.5K

The large FWHM of the discussed band, presence of few maximums in its shape, as well as change in the maxima ratio depending on the chromium concentration or doping method, allows us to conclude that middle-IR band is complex, consisting of a number of elementary bands. This assumption does not contradict with literature about intrashell emission of the chromium ions but may explain

V. Sontea and I. Tiginyanu (eds.), *3rd International Conference on Nanotechnologies and Biomedical Engineering*,
IFMBE Proceedings 55,
DOI: 10.1007/978-981-287-736-9_34

differences in experimental data presented in literature and enumerated in the introduction of this article. Even more, secondary maxima mentioned above (Fig. 1) repeat with sufficient accuracy position of the middle-IR band given in other papers, e.g., 2.1 μm (0.59 eV) [8, 9], 2.2 μm (0.56 eV) [10], 2.4 μm (0. 51 eV) [5, 11].

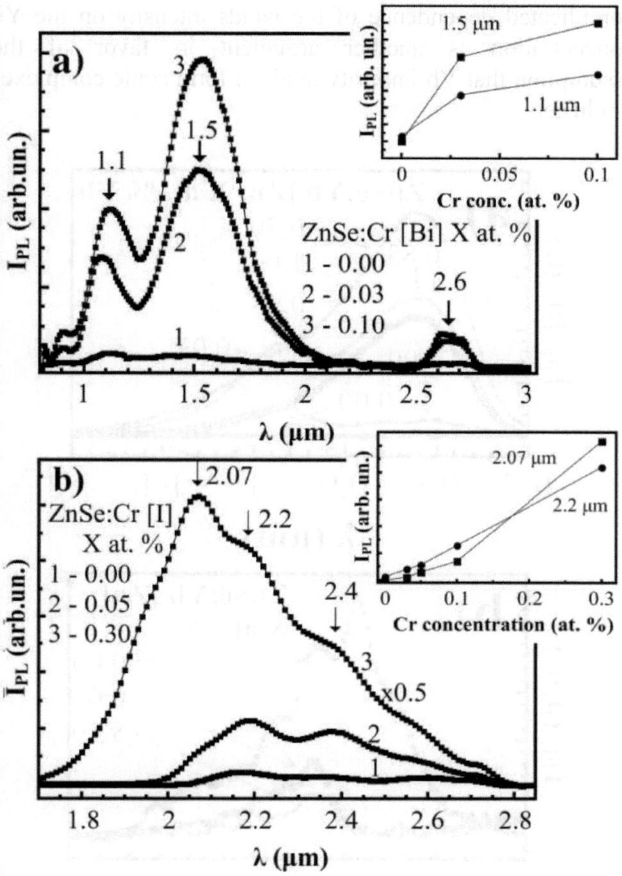

Fig. 2 PL spectra of ZnSe:Cr [Bi] (a) and ZnSe:Cr [I] (b) crystals in IR spectral range at T=6.5K. Inset: dependence of the emission intensity in the bands maximum on doping impurity concentration.

In case if band with the main maximum at room temperature located at ~2.07 μm is complex, we may try to associate its components with radiative transitions. The component located at 2.07 μm is most probably caused by intrashell transitions $^5E \rightarrow {}^5T_2$ within Cr^{2+} ion. This is confirmed by its intensity dependence on chromium concentration (Fig.2(b), inset); intensity increase with temperature, which may be the evidence of higher emission efficiency at room temperature;as well as the energy position of the emission being closest to theoretically calculated distance between ^5E and $^5T_2$ levels (2.04 μm) of Cr^{2+} ion [6]. Energy position of 2.2 μm component corresponds to energy position of [V_{Zn}, Al_{Zn}] complexes within the ZnSe band-gap [12] but 2.4 μm component may be caused by associative defects of shallow donor impurity (for example, iodine) and zinc vacancies [12]. It should be mentioned that assumptions about the defect-impurity composition of the crystals made during analysis of IR PL spectra are in the good agreement with the data referring to the visible PL spectra.

Annealing of ZnSe:Cr crystals in zinc vapors was performed to verify the assumption made about participation of complexes based on zinc vacancies in formation of the band in middle-IR spectral region (Fig. 3). It was established that annealing of ZnSe:Cr crystals in zinc vapors, which leads to decrease of zinc vacancy concentration in the crystals, results in significant decrease of the emission band in middle-IR spectral region. This result is in good agreement with the assumption about participation of zinc vacancy based centers in the formation of the band. At the same time, it should be mentioned that decrease of band intensity after annealing in zinc vapor cannot be caused by change of the chromium ion charge state, as can be seen from earlier EPR studies.

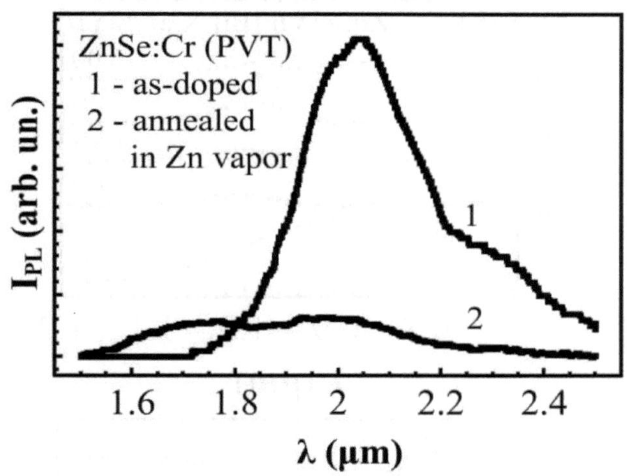

Fig. 3 Influence of the annealing in zinc vapors on PL spectra of the ZnSe:Cr (PVT) crystal doped by diffusion from the gas phase during 80h at T=85K.

Thus, experimental data show that the band in middle-IR spectral range of the ZnSe:Cr crystals emission spectra is complex and consists of both intrashell emission of Cr^{2+} ions, which are activators for middle-IR emission band, and emission caused by complexes based on zinc vacancies and background impurities. Obtained results show that emission of ZnSe:Cr crystals is determined not exclusively by the doping impurity but strongly depends on structural defects of thedoping crystals, their concentrations, and sets of background impurities.

In such a way, the problem of obtaining an active laser media on the basis of bulk TM doped II-VI semiconductors becomes much more complicated, because it requires simultaneous solution of 2 problems: doping with an TM activator and controlling concentration of the native defects.

The problem of defect-impurity interaction becomes even more complicated in the case of doping of II-VI semiconductor materials with the REE. Even in the early publications on the topic [13-15] it was showed that doping with f-element requires a co-dopant for charge compensation. Usage of a co-dopant will significantly change the environment, where the REE ion will incorporate. This leads to the drastically change of the emission spectra of the rare-earth ion depending on the compensating co-dopant used.

Usage a co-dopant has not only technological but also a practical reason, e.g. Er-Yb co-doping should improve Er up-conversion efficiency; Cr-Yb co-doping is believed to increase laser slope efficiency, and Bi-Yb co-doping is used to enhance the infrared emission of the material [7]. However, experimental data show that not only co-dopant participates in Yb activation.

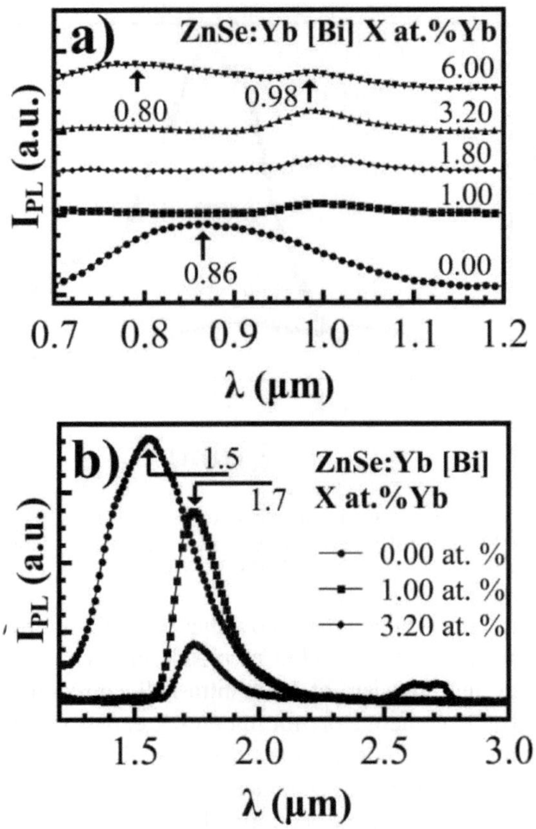

Fig. 4 PL spectra of ZnSe:Yb [Bi] crystals in near-IR (a) and mid-IR (b) regions at T = 6.5 K.

Thus, addition of Yb impurity to the Bi melt during annealing process leads to disappearance of the band with maximum at ~1.5 μm, attributed to the bismuth emission centers (Fig. 4). Instead of this band in spectra of ZnSe:Yb [Bi] crystals, a band with maximum at ~1.7 μm may be observed. Increase of the Yb concentration in the Bi melt leads to the decrease of the band intensity (Fig. 4(b)). The complicated dependence of the bands intensity on the Yb concentration is another argument in favor of the assumption that Yb impurity tends to form some complexes or clusters.

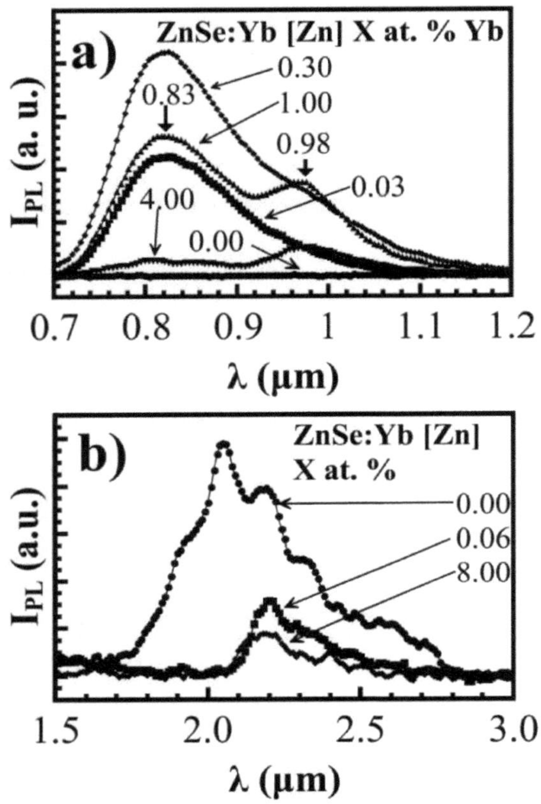

Fig. 5 PL spectra of ZnSe:Yb [Zn] crystals in near-IR (a) and mid-IR (b) regions at T = 6.5 K.

Structured bands in the near-IR spectral range (Figs. 4(a) and 5(a)) are characteristic for the Yb doped crystals [16]. Emission with the maximum at 0.98 μm is usually associated with intrashell transitions between the ground ($^2F_{7/2}$) and exited ($^2F_{5/2}$) states of the Yb^{3+} ions [18]. The presence of this characteristic band indicates the incorporation of the ytterbium impurity in the crystal lattice in the form of Yb^{3+} ions. Increase of the 0.98 μm maximum contribution to the spectra, with increase of the ytterbium concentration in the doping media, confirms the conclusion made. However, it should be mentioned that the intensity of the Yb intrashell emission is much higher in the crystals doped by annealing in the zinc melt. This leads to the conclusion of more significant influence of the Yb on emission properties of the crystals, when concentration of the selenium vacancies is higher.

III. CONCLUSIONS

Thus experimental data on emission properties of TM and REE doped bulk ZnSe crystals have shown that light-generation is determined not by the doping impurity exclusively but by interaction between native and impurity defects. Control over the concentration of the intrinsic defects is a big challenge for all II-VI semiconductor compounds and sometimes is crucial for practical applications. However, it is evident that nanostructuring will lead to significant increase in concentration of the surface defects comparative to the concentration of the native defects in volume, thus, allowing to control concentration of the intrinsic defects by changing the size of the particle. It allows to think that nanostrucuring will give possibility to change concentration of native and impurity defects independently and to control interaction between them. As it was shown, namely this interaction mainly determines the crystal properties. Thus, study of the physical properties of nanostructured II-VI semiconductor compounds doped with transition metals and rare-earth elements, for which control over interaction between native and impurity defects seems to be less complicated, may lead to development of tunable solid state lasers

ACKNOWLEDGMENT

This work is supported by BMUMID2011238 Erasmus Mundus Grant and RM Grant 14.02.044F. The Jenny and Antti Wihuri Foundation is acknowledged for financial support.

CONFLICT OF INTEREST

The authors declare that they have no conflict of interest.

REFERENCES

1. Lei G, Ge H, Sheng-Tao Z (2012) Defects energetics, electronic structure and optical properties of Cu-doping and Zn vacancy impurities in ZnSe. Acta Phys.-Cim. Sin. 28:2845 – 2851 DOI 10.3866/PKU.WHXB20121083
2. Kwak KW, Vanderbilt D, King-Smith RD (1995) First principle study of phosphorus and nitrogen impurities in ZnSe. Phys. Rev. B 52: 11912 – 11919 DOI 10.1103/PhysRevB.52.11912
3. Laks D B, Van de Walle C G (1993) Doping limits in ZnSe. Physica B 185:118-127 DOI 10.1016/0921-4526(93)90225-U
4. Nishizawa J, Suzuki R, Okuno Y (1985) p-type conductivity in ZnSe grown by temperature difference method under controlled vapor pressure. J. Appl. Phys 59: 2256 – 2258 DOI 10.1063/1.336372
5. DeLoach L D, Page R H, Wilke G D et al. (1996) Transition metal-doped zinc chalcogenides: spectroscopy and laser demonstration of a new class of gain media. IEEE J. Quant. Electr. 32:885 – 895 DOI 10.1109/3.502365
6. Radevici I, Sushkevich K, Colibaba G et al. (2013) Influence of chromium interaction with native and impurity defects on optical and luminescence properties of ZnSe:Cr crystals. J. Appl. Phys. 114: 203104 DOI 10.1063/1.4837596
7. Radevici I, Sushkevich K, Huhtinen H et al. (2015) Influence of the ytterbium doping technique on the luminescent properties of ZnSe single crystals. J. Lumin 158: 236 – 242 DOI 10.1016/j.jlumin.2014.10.002
8. Grebe G, Roussus G, Schulz H-J (1976) Infrared luminescence of ZnSe:Cr crystals. J. Lumin. 12/13: 701 – 705 DOI 10.1016/0022-2313(76)90164-2
9. Hommerich U, Jones I K, Nyein E E et al. (2006). Comparison of the optical properties of diffusion-doped polycrystalline Cr:ZnSe and Cr:CdTe windows. J. Cryst. Growth 287: 450 – 453 DOI 10.1016/j.jcrysgro.2005.11.064
10. Vivet N, Doualan J L, Morales M et al. (2010) Luminescence properties of Cr^{2+}:ZnSe: Comparison between a single crystal and a film deposited by RF magnetron co-sputtering. J.Lumin. 130: 1449 – 1454 DOI 10.1016/j.jlumin.2010.03.010
11. Godlewski M, Surma M, Ivanov Yu V et al. (2004) Mechanisms of radiative and nonradiative recombination in ZnSe:Cr and ZnSe:Fe. Low Temp. Phys. 30: 891 – 896 DOI 10.1063/1.1820019
12. Irmscher K, Prokesch M (2001) Spectroscopic evidence and control of compensating native defects in doped ZnSe. Mat. Sci. Eng. B 80: 168 – 172 DOI 10.1016/S0921-5107(00)00612-7
13. Watts R K, Holton W C (1968) Paramagnetic-resonance studies of rare-earth impurities in II-VI compounds. Phys. Rev. 173: 417 – 426 DOI 10.1103/PhysRev.173.417
14. Watts R K (1966) Host crystal effects on the electron paramagnetic resonance of Er^{3+} and Yb^{3+} in II - VI compounds. Sol. State. Comm. 4: 549 – 552 DOI 10.1016/0038-1098(66)90423-6
15. Kingsley J D, Aven M (1967) Paramagnetic resonance and fluorescence of Er^{3+} at cubic sites in ZnSe. Phys. Rev 155: 235-246 DOI 10.1103/PhysRev.155.235
16. Brown M R, Cox A F J, Shand W A (1971) Defect configuration for rare earth ions in zinc selenide. J. Phys. C: Solid St. Phys. 4: 2550 – 2556 DOI 10.1088/0022-3719/4/16/019
17. Szczurek I, Lozykowsky H (1976) Electroluminescence of ZnSe thin films doped with the rare-earth fluorides. J.Lumin 14: 389 – 402 DOI 10.1016/0022-2313(76)90008-9

Author: Ivan Radevici
Institute: University of Turku
Street: Vesilinnantie 5
City: Turku
Country: Finland
Email: ivarad@utu.fi

Spiropyran Based Smart Composites: Memorizing Polymer with Enhanced Molecular Switches

S. Shree[1], M. Schulz-Senft[2], X. Jin[1], Y.K. Mishra[1], A. Staubitz[2], and R. Adelung[1]

[1] University of Kiel, Functional Nanomaterials, Institute for Materials Science, Kiel, Germany
[2] University of Kiel, Otto-Diels-Institute for Organic Chemistry, Kiel, Germany

Abstract— **We demonstrate the fabrication and response of a smart material combined with spiropyran molecules. Spiropyrans belong to a class of compound that constitute molecular switch. They change color in response to different stimuli such as heat, UV light or mechanical stress. These molecules have been incorporated in polythiourethane (PTU) which is a shape memory polymer. The observations revealed that the molecular switches do not alter the physical properties of the PTU in general; however, both, the spiropyran molecules and smart polymer show changes individually when exposed to heat. The microscopic mechanisms of switching process in response to external stimuli have been discussed in detail.**

Keywords— **Spiropyran , Merocyanine), Polythiourathene, Shape memory polymer.**

I. INTRODUCTION

Polymer based functional composites, which exhibit the capability to respond to any external triggers, e.g., heat, UV light, stress, etc., are important material candidates for various applications. This kind of smart polymeric materials equipped with shape memory features, have found enormous applications in various fields ranging from aerospace to building industries [1]. Very recently, there has been significant progress in the direction of self-healing/reporting materials and these auto-responsive polymers are the most appropriate candidates in this context. The major drawbacks of the present heat/light sensitive smart polymer coatings are that they contain toxic metals and permanent deformations of the polymer decrease the life-time of the material.

The shape memorizing polymer matrix used in this paper is polythiourethane, a thermo-set polymer, which can be polymerized under solvent free conditions and is thermally stable. Hence it could have tremendous applications as a smart polymer. To make this shape memory polymer more efficient, spiropyran is added to the polymer matrix as a heat indicator.

Spiropyran has been used for its photo-responsive properties for several applications such as light modulated sensors [2] and for drug delivery [3].

These molecular switches have been used in our work and were incorporated into PTU. The usual ways of integrating spiropyran in a polymer matrix are either by polymerization of spiropyran based monomers or grafting of spiropyran on pre-formed polymer [4]. The downside of these methods of well defined covalent integration into a polymer is the accessible amount of material and its comparatively high cost, which might limit industrial usefulness. By adopting a simple strategy of dispersion of spiropyran molecules in the co-monomer, larger amounts can be obtained which is the main focus in present work.

II. EXPERIMENTS

A. Materials

Solvent free PTU, a two component system, was purchased from Fluid-& Prozesstechnik GmbH (Waltershausen, Germany). Preparation of the polymer was achieved by a poly-addition reaction between pentaerythritoltetrakis- (3 mercaptopropio-nate) (Tol) (2 mmol, 1 g) and 1.6-diisocyanatohexane (Iso) component (8.0 mmol, 1.4 g).

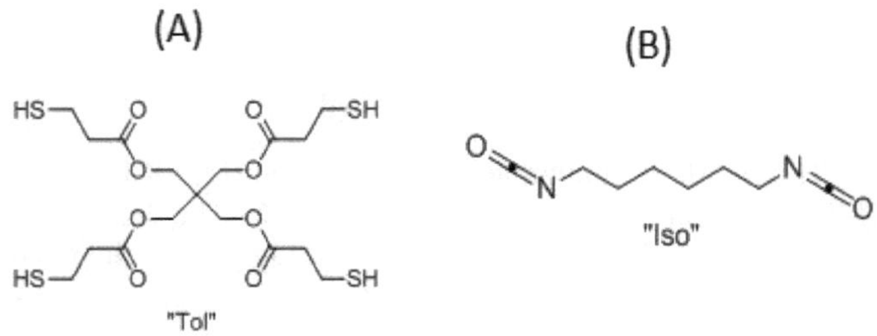

Fig. 1 Molecular structures of (A) the Tol component, and (B) the molecular structure of Iso the component.

B. Sample Preparation

The samples were prepared by dispersing 0.01 wt% spiropyran in the co-monomer Tol for 20 min at 2000 rpm in a Netzsch Mini-Master dispersion machine to obtain homogeneous distribution of spiropyran. To this mixture, an appropriate amount of Iso was added. After degassing the mixture in a desiccator, the polymer blend was set into the moulds and placed in an oven for 2 h at 75 °C for polymerization process.

C. Response of PTU to Heat

When heated, a pure PTU (without spiropyran) above its glass transition temperature (85 °C), softened. This allowed

V. Sontea and I. Tiginyanu (eds.), *3rd International Conference on Nanotechnologies and Biomedical Engineering*,
IFMBE Proceedings 55,
DOI: 10.1007/978-981-287-736-9_35

us to give the polymer any kind of shape which was maintained upon cooling. The deformed polymer returned to its original shape by an external stimulus such as heat.

D. Switching Mechanism of Spiropyran in Response to an External Stimulus

Spiropyran undergoes a reversible isomerization to the corresponding merocyanine form under the influence of an external stimulus such as heat. This causes changes in the molecular structure of thermochromic Spiropyran [5], resulting in the open ring structure called merocyanine as illustrated in Figure 2.

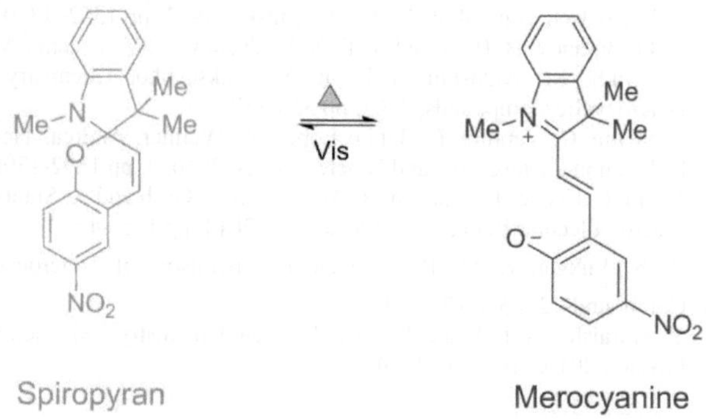

Fig. 2 Reversible isomerization between spiropyran and merocyanine in response to heat

III. RESULTS AND DISCUSSION

The PTU/spiropyran samples with a given flat shape (A) were heated to T = 100 °C slightly above their glass transition temperature which softened the samples. While cooling down from 100 °C, different shapes have been given to the polymer, such as (B) and (C) [Figure 3 (B, C)].

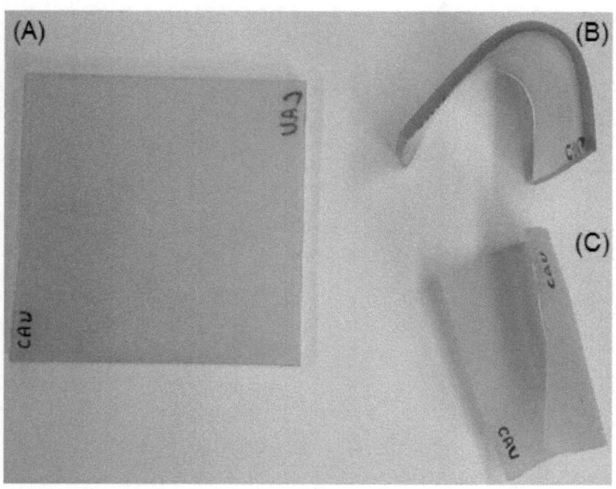

Fig. 3 Different shapes of the PTU/spiropyran under temperature controlled conditions. (A) Corresponds to the original shape of PTU/spiropyran, (B) and (C) are folded and twisted shape of PTU/spiropyran respectively

A folded PTU/spiropyran sample (Figure 3B) was clamped on one side and solely its free end was heated up to 100 °C gradually with a heating gun (Figure 4). Under heat treatment, a complete shape recovery of the folded sample could be achieved as demonstrated clearly in by Figure 4 (a-h).

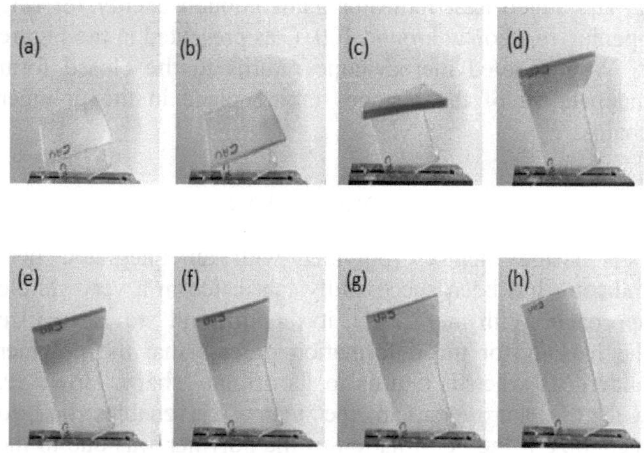

Fig. 4 PTU/spiropyran sample clamped on one end and only the folded end is heated with a heating gun. From (a) to (h) the sample was heated gradually up to 100 °C and the shape recovery behavior was clearly visible.

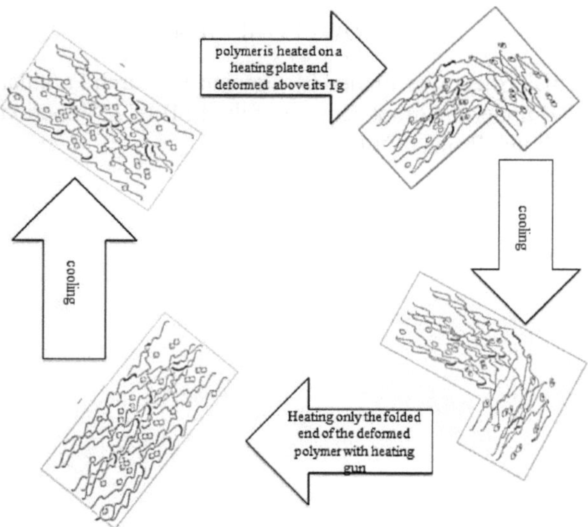

Fig. 5 Schematic representation of polymer chains incorporated with spiropyran undergoing structural changes independently

The PTU matrix consisted of a thermo-plastic and a thermo-set part. The thermo-plastic part of the polymer deforms upon heating. The thermo-set part of the polymer restrains deformation due to the cross-linking of SH-groups present in the Tol component with the NCO-groups in Iso component. When the temperature is above T_g, the polymer chains overcome inter-chain forces resulting in softening the polymer which is enough for deformation. Again when the polymer with temporary shape is heated, the chain relaxes which results in returning the polymer to its original shape. This has been described by schematic representation in the Figure 5.

The added heat indicator gains enough energy for ring opening reaction at around 100 °C as presented in the Figure 2. When cooled merocyanine returns to the closed form independent of the changes taking place in the polymer chains.

IV. CONCLUSION

A shape memory polymer with an integrated heat Indicator has been successfully fabricated in a very simple approach. With mere PTU, it is difficult to recognize the reasons for the deformation process that the polymer undergoes when it returns to its original shape. However, with the spiropyran in the matrix, it enables one to recognize if the deformation of the polymer was due to the heat applied by the change of its color. This PTU/spiropyran matrix was synthesized in an intelligent way where the physical properties of both the materials involved, were maintained independent of each other.

AKNOWLEDGEMENT

This project was funded by German Research Foundation (DFG) under the grant scheme of the SFB 677 "Function by Switching" Project C14.

CONFLICT OF INTEREST

The authors declare that they have no conflict of interest.

REFERENCES

1. D. Ratna, J. Karger-Kocsis (2008), Journal of Materials Science, 2008, pp 254-269
2. A. Radu, R. Byrne, N. Alhashimy, M. Fusaro, S.Scarmagnani, D. Diamond, Journal of Photochemistry and Photobiology A: Chemistry, 2009, pp 109-115
3. S. Son, E. Shin, B. Kim, Biomacromolecules, 2014, pp 628-634
4. R. Klajn, Chemical Society Reviews, 2014, 148
5. C. F. Koelsch, Journal of Organic Chemistry, 1951, pp 1362–1370
6. A. O. Bulanov, B. B. Safoklov, B. S. Luk'yanov, V. V. Tkachev,V. I. Minkin,S. M. Aldoshin, and Yu. S. Alekseenko, Chemistry of Heterocyclic Compounds, 2003, pp 350-357
7. R. Byrne, C. Ventura, F. Benito Lopez, A. Walther, Andreas Heise, D. Diamond, Biosensors and Bioelectronics, 2010, 4, pp 1392-1398
8. X. Jin, L. Heepe, J. Strueben, R. Adelung, S.N. Gorb and A. Staubitz, Macromolecular Rapid Communications, 2014, pp 1551-1570
9. B. S. Lukyanov, M. B. Lukyanova, Chemistry of Heterocyclic Compounds, 2005, pp 281-311
10. Y. Shiraishi, M. Itoh and T. Hirai, Physical Chemistry and Chemical Physics, 2010, pp 13737-13745

Author: Sindu Shree
Institute: University of Kiel, Functional Nanomaterials, Institute for Materials Science
Street: Kaiserstrasse 2
City: Kiel
Country: Germany
Email: sisi@tf.uni-kiel.de

Sensing Properties of Ultra-Thin TiO_2 Nanostructured Films Based Sensors

V. Postica[1], T. Reimer [2,3,*], E. Lazari[1], N. Ababii[1], S. Shishiyanu[1], S. Railean[1], V. Kaidas[3], S. Kaps[3], O. Lupan[1,3], W. Benecke[2], and R. Adelung[3]

[1] Technical University of Moldova/Department of Microelectronics and Biomedical Engineering, Chisinau, Republic of Moldova
[2] Fraunhofer Institute for Silicon Technologies (ISIT), Fraunhoferstr. 1, 25524 Itzehoe, Germany
[3] University of Kiel/Institute for Materials Science/Functional Nanomaterials, Kaiser Str. 2, D-24143, Kiel, Germany

Abstract— **In this work, ultra-thin TiO_2 nanostructured films, synthesized by atomic layer deposition method (ALD), were integrated in sensor structures. The effect of post-growth annealing and thickness of TiO_2 samples on UV and hydrogen gas sensing properties is investigated. An increase in current value of more than one order of magnitude (I_{UVON}/I_{UVOFF}~38) has been detected under exposure to UV light (365 nm) of sample with 45 nm thickness and annealed in furnace at 650 °C for 2 hours. Samples with 15 nm thickness and rapid thermal annealed at 450 °C for 3 min, have shown hydrogen gas response (~3.75).**

Keywords— **TiO_2, ALD, ultra-thin films, gas sensor, photodetector.**

I. Introduction

Titanium dioxide (TiO_2) is a wide band gap semiconductor (anatase 3.2 eV and rutile 3.0 eV), that is the most common compound of titanium and is an attractive material for a variety of applications in anticorrosion and self-cleaning coatings, photo-catalysts, optics, dye-sensitized solar cells, electronics, gas sensors etc. [1-5]. In the case of low temperature synthesis methods, usually TiO_2 nanostructures are amorphous or possess meta-stable anatase crystalline phase [6]. Due to the inert surface properties and high sensing abilities, a wide range of investigations for development of highly sensitive, selective, compact and reliable TiO_2 based sensors are being carried out [7]. Previous work [8] demonstrated improvement of response time and hydrogen sensitivity of a single TiO_2 nanotube integrated on the chip by post-anodization thermal annealing. Also, due to distinctive UV adsorption characteristics and such advantages like large device area, TiO_2 films are very suitable for UV detection [9]. Based on TiO_2 nanotube arrays, Zou et al [9] fabricated UV photodetectors with fast rise/decay time and high responsivity.

In this work, UV (365 nm) photo detection and hydrogen gas sensing performances of the sensor based on ultra-thin TiO_2 nanostructured films (15 nm and 45 nm thickness) deposited by ALD method is investigated. An increase in the current value of more than one order was observed for TiO_2 nanostructured films with 45 nm thickness under exposure to UV light. In the case of rapid thermal annealing (RTA) at 450 °C of samples with 15 nm thickness, the gas response to H_2 was observed. The UV and gas sensing mechanisms were proposed to explain obtained results.

II. Experimental

A. Synthesis of Ultra-thin TiO_2 Nanostructured Films

TiO_2 was deposited by atomic layer deposition (ALD) in a Picosun R-200 series atomic layer deposition reactor on 76 mm × 26 mm microscope glass slides as substrates (Thermo Scientific). The microscope slides were cleaned prior to TiO_2 deposition by wiping carefully with acetone and lint-free paper towel. They were then placed on 8" silicon dummy wafers as sample holders, with up to five glass slides per wafer, and the wafers were in turn placed in the stainless-steel sample holder of the ALD reactor in a stack of up to seven wafers.

The ALD chamber was maintained at 300 °C. After placing the samples, closing the lid and pumping to ~15 hPa, the ALD reactor chamber was allowed to re-heat to 300 °C, and upon reaching this temperature a stabilization time of 15 min was used to allow the substrates to reach thermal equilibrium.

TiO_2 was then deposited by alternating pulses of $TiCl_4$ and H_2O, with pulse lengths of 0.2 s for each, and purge times of 5 s for $TiCl_4$ and 7 s for H_2O. Lines were purged with the N_2 flow. The total deposition time for one cycle was therefore 12.4 s. Growth per cycle was ~0.045 nm. Thicknesses of deposited TiO_2 ranged from 15 nm to 45 nm, with 340 cycles for 15 nm and 1020 cycles for 45 nm. After deposition, the samples were taken out of the reaction chamber and allowed to cool naturally.

B. Characterization

A WITec Raman equipment was used for Raman scattering characterization of the ultra-thin TiO_2 nanostructured films. Excitation of samples was provided by a 532 nm line from a frequency doubled Nd:YAG laser at an output power of 10 mW.

A Keithley 2400 source meter (Keithley instruments, Germany) has been used for the electrical and photocurrent measurements by two-point probe method in the air at the

V. Sontea and I. Tiginyanu (eds.), *3rd International Conference on Nanotechnologies and Biomedical Engineering*,
IFMBE Proceedings 55,
DOI: 10.1007/978-981-287-736-9_36

room temperature. The detection of UV light is performed through using UV source (λ = 365 nm). The intensity of a UV light was measured using a photodiode, showing a range of value 10–15 mW·cm^{-2}. UV and hydrogen gas response were calculated by the ratio of current in the ON and OFF state (I_{ON}/I_{OFF}). For the gas sensing measurements the device was placed in the gas test system reported in previous work [10]. All sensing measurements were performed under a DC bias voltage of 30 V.

III. RESULTS

A. Micro-Raman Characterization

The micro-Raman spectra of 45 nm thickness TiO_2 nanostructured films thermal annealed (TA) at 450 ˚C for 2 hours in air is presented in Fig. 1 (a). TiO_2 is known to occur in three crystalline phase: anatase (tetragonal), rutile (tetragonal) and brookite (orthorhombic) [11]. All detected peaks are assigned to the anatase phase of TiO_2, that has a fundamental structural unit of [TiO_6] octahedron [12] and a primitive cell contains two TiO_2 chemical units [11]. Results are in agreement with our previous work [13]. Anatase belong to the $D_{4h}^{19} = I4_1/amd$ space group with Z = 4, and has six active Raman modes [12].

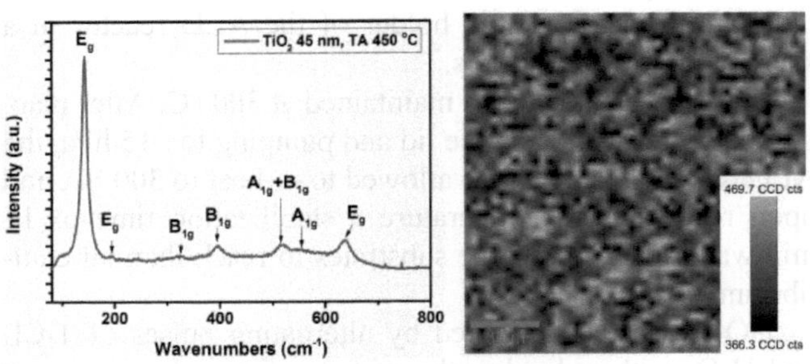

Fig. 1 (a) Micro-Raman spectra of TiO_2 treated TA at 450 ˚C for 2 h in air; (b) Micro-Raman mapping of the E_g phonon mode (143 cm^{-1}) of TiO_2 treated TA at 450 ˚C for 2 h.

Detected peaks can be assigned to the following phonon modes: E_g at 143 cm^{-1}, 196 cm^{-1} and 639 cm^{-1}, B_{1g} at 324 cm^{-1} and 395 cm^{-1}, A_{1g} at 555 cm^{-1} and A_{1g}+B_{1g} is the superposition of two peaks at 512 cm^{-1} and 516 cm^{-1} [13]. There were not observed essential differences in transition to rutile crystalline phase for samples treated TA at 650 ˚C for 2 hours and RTA at 450 ˚C for 3 min (not shown). Pal et al. [13] observed apparition of peaks that correspond to rutile crystalline phase at annealing temperatures higher than 750 ˚C.

The micro-Raman mapping technique was applied to investigate the spatial distribution of E_g phonon mode (143cm^{-1}) (integrated between 125 cm^{-1} to 160 cm^{-1}) for a certain local area of nanostructured film with thickness of 45 nm and treated TA at 450 °C for 2 hours. It can be clearly seen that the crystallinity of TiO_2 is relatively uniform.

B. UV Response

In the inset from Fig. 2 (a), is presented the sensor structure used in this work for UV and H_2 gas sensing investigations. Device preparation begins with deposition of TiO_2 nanostructured film of 15 nm or 45 nm thickness on the glass substrate, as was described above. After post-growth thermal annealing (TA) in electrical furnace or rapid thermal annealing (RTA) in air of samples, on top of TiO_2 layer was deposited Au contacts by sputtering in vacuum through a mask with meander configuration. The width of the active sensing layer between contacts is 1 mm. In Fig. 2 (a) is presented a typical current-voltage characteristic under UV illumination (365 nm) and in the dark for sensor based on the ultra-thin TiO_2 nanostructured film treated TA at 450 °C for 2 hours. Linear characteristics demonstrate formation of ohmic contacts at Au/TiO_2/Au structure. The sample set treated TA at 650 °C for 2 hours and RTA at 450 °C for 3 min demonstrated the same linear type of current-voltage characteristic (not shown). Good electrical contacts are an important factor for UV detection, especially for structures based on single nanostructure, and can essentially influence electrical transport in nanostructures of oxide metals [14]. In our case, ohmic type of contacts demonstrates that device signal will not be influenced by the Au/TiO_2 interface. Differences in values of device current under UV illumination and in the dark conditions, demonstrate a good ability of UV sensing for device based on the ultra-thin TiO_2 nanostructured films.

In Fig. 2 (c), (d) are presented data on photoresponse of devices based on TiO_2 nanostructured films with 45 nm thickness treated TA for 2 hours at: (a) 450 °C and (b) 650 °C. The four pulses of UV light (365 nm) with duration of 10 s and one pulse with a duration of 20 s was applied in order to demonstrate reproducibility of the device photocurrent variations. UV response and rise/decay time are very important parameters of photodetectors. The sample treated at 650 °C demonstrate higher UV response (S_{UV} ~ 38), in comparison with the sample treated at 450 °C (S_{UV} ~ 27). However, reproducibility of the photocurrent for the sample treated at 450 °C is better, due to almost identical rise and decay curves in first four cycles of UV illumination. It is clearly observable that rise and decay process of photocurrent pulses, for both types of samples, is occurred by an exponential function (Fig. 3 (c), (d)). To determine time constants for rising (τ_{r1}, τ_{r2}) and decay (τ_{d1}, τ_{d2}) of the photocurrent, bi-exponential fitting was used [15]:

$$I(t) = I_0 + A_1\left(1 - e^{-\frac{t}{\tau_{r1}}}\right) + A_2\left(1 - e^{-\frac{t}{\tau_{r2}}}\right) \quad (1)$$

$$I(t) = I_0 + A_3 e^{-\frac{t}{\tau_{d1}}} + A_4 e^{-\frac{t}{\tau_{d2}}} \quad (2)$$

where I_0 is dark current, A_1, A_2, A_3 and A_4 are positive constants. On the basis of curve fitting, for sample treated TA at 450 °C the values of time constants are $\tau_{r1} = 0.15$ s, $\tau_{r2} = 4.1$ s, $\tau_{d1} = 0.16$ s, $\tau_{d2} = 3.8$ s. Similar fitting were performed for device based on the TiO_2 nanostructured film treated TA at 650 °C, and resulted time constants are $\tau_{r1} = 0.17$ s, $\tau_{r2} = 0.6$ s, $\tau_{d1} = 0.08$ s, $\tau_{d2} = 1.6$ s.

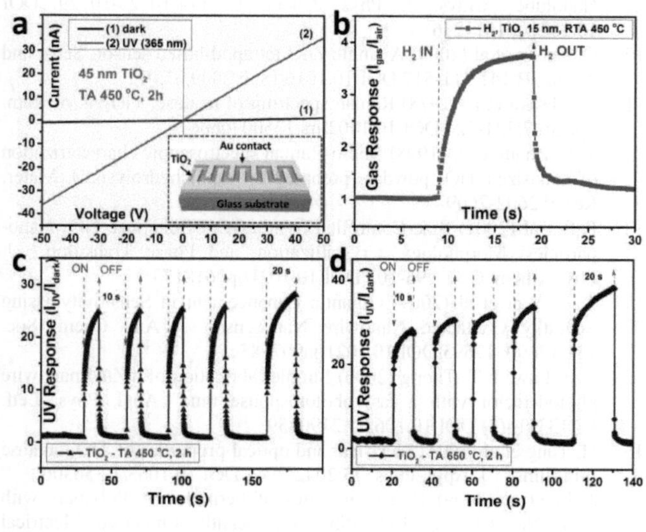

Fig. 2 (a) Current-voltage characteristic at room temperature of a device based on TiO_2 (45 nm) under UV light and in the dark. (b) Gas response to H_2 gas versus time of device based on TiO_2 (15 nm) treated RTA at 450 °C for 3 min. Photoresponse of a TiO_2 (45 nm) based device toward four pulses of UV light with duration of 10 s and one pulse with duration of 20 s for sample treated: (c) TA at 450 °C; (d) TA at 650 °C.

As can be seen, the sample treated at 650 °C shows fastest rise and decay time than sample treated at 450 °C, especially fastest τ_{r2} and τ_{d2} times. Such slow τ_{r2} and τ_{d2} times in case of treatment at 450 °C can be a cause of slower photo-desorption process of oxygen from the surface of TiO_2 nanostructured films.

Thinner nanostructured films (15 nm), demonstrated lower and instable UV response, thus samples are not of practical interest, and further will not be investigated. However, RTA treated nanostructured films (15 nm) at 450 °C for 3 min showed a response to H_2 gas ($S_{H2} \sim 3.75$) (Fig. 3 (b)), taking 4.1 s for 90 % of full response and 2.6 s for 10 % of recovery in ambient air.

C. UV Sensing Mechanism

It is generally accepted that at room temperature on surface of TiO_2 are adsorbed molecules of oxygen (O_2), and by trapping free electrons from conduction band, a depletion region with lower conductivity is created [9]:

$$O_{2(g)} + e^- \rightarrow O^-_{2(ad)} \quad (3)$$

Assuming that the carrier density in films is $n = 10^{19}$ cm^{-3} [16], the width of the depletion region (d) can be calculated according to:

$$d = \left(\frac{2\varepsilon_{TiO_2} \varepsilon_0 V_S}{qn} \right)^{1/2} \quad (4)$$

where q is the charge of an electron, ε_{TiO2} is the dielectric constant of the TiO_2, ε_0 is the vacuum permittivity and V_S is the height of interfacial potential as the result of oxygen species adsorption. Assuming that $\varepsilon_{TiO2} = 47$ [17] and ranging V_S from 0.1 to 1 V, the width of the depletion region can be estimated from 7.2 nm to 22.8 nm. These values are comparable to the thickness of TiO_2 nanostructured film, which can lead to a higher UV response [18]. Adsorption of UV light photons with a energy larger than band gap of TiO_2 by nanostructured film, will result in a transition of a electron from valence band (VB) to conduction band (CB), generating an electron-hole pair ($TiO_2 + h\upsilon \rightarrow e^- + h^+$). Photo-generated holes migrate to the surface along the potential slope produced by band bending, that results in photo-desorption of oxygen molecules and in an increase of the free carrier concentration, that decrease width of the depletion region [9,19]:

$$h^+ + O^-_{2(ad)} \rightarrow O_{2(g)} \quad (5)$$

Photo-generated electrons will contribute to increase in current of the ultra-thin nanostructured film. In the case of samples treated TA at 450 °C, the process of oxygen photo-desorption is slower than in the case of sample treated TA at 650 °C. Rise and decay time are dependent on the influence of in-diffusion, adsorption and ionization processes on the surface of metal oxides and photo-desorption of oxygen molecules [20]. Another important factor is the type of defects presented at the surface of TiO_2. Thus, there are needed more investigations on the UV sensing mechanism depending on ultra-thin TiO_2 nanostructured films phase type and thickness, that will be reported in further works.

D. Gas Sensing Mechanism

When molecules of hydrogen gas are adsorbed on surface of ultra-thin TiO_2 nanostructured films, they react with adsorbed oxygen species by releasing free electrons, which lead to an increase of conductivity and decrease of depletion region width [6]:

$$O_{2(g)} \rightarrow O_{2(ad)} + e^- \rightarrow O^-_{2(ad)} \quad (6)$$

$$2H_2 + O^-_{2(ad)} \rightarrow 2H_2O_{(g)} + e^- \quad (7)$$

The relatively high hydrogen response can be attributed to a comparable width of the depletion region and thickness of ultra-thin nanostructured TiO_2 film [18].

IV. CONCLUSIONS

In this work, ultra-thin TiO_2 nanostructured films with a thickness of 15 and 45 nm were deposited by the ALD method on a glass substrate for UV and H_2 sensing investigations. The micro-Raman studies were revealed anatase crystalline phase of samples treated at 450 °C and 650 °C. Due to a comparable width of depletion region with thickness of nanostructured films, was observed a hydrogen response of ~ 3.75 for sensors based on 15 nm TiO_2 samples treated RTA at 450 °C for 3 min. Samples with 45 nm thickness were demonstrated to show a UV response, with a S_{UV} ~ 27 and ~ 38, for TA treatment at 450 °C and 650 °C. Our results demonstrate that ultra-thin TiO_2 nanostructured films, by control of thickness and post-growth annealing treatment, are promising material for sensing applications.

ACKNOWLEDGMENT

This work was financially partly supported by the Project 45inst-15.817.02.29A. Partial support of the research by STCU under the Grant no. 5989 is acknowledged.

CONFLICT OF INTEREST

The authors declare that they have no conflict of interest.

REFERENCES

1. M. Reiners et al (2013) Growth and Crystallization of TiO_2 Thin Films by Atomic Layer Deposition Using a Novel Amido Guanidinate Titanium Source and Tetrakis-dimethylamido-titanium. Chem. Mater. 25:2934-2943 DOI 10.1021/cm303703r
2. D. V. Bavykin et al (2006) Protonated Titanates and TiO_2 Nanostructured Materials: Synthesis, Properties, and Applications. Adv. Mater. 18:2807-2824 DOI 10.1002/adma.200502696
3. Z. Li et al (2013) Hydrogen Sensing with Ni-Doped TiO_2 Nanotubes. Sensors 13(7):8393-8402 DOI 10.3390/s130708393
4. V. Galstyan et al (2013) TiO_2 Nanotubes: Recent Advances in Synthesis and Gas Sensing Properties. Sensors 13(11): 14813-14838 DOI 10.3390/s131114813
5. A. A. Haidry et al (2011) Hydrogen gas sensors based on nanocrystalline TiO_2 thin films. Cent. Eur. J. Phys. 9(5): 1351-1356 DOI 10.2478/s11534-011-0042-3
6. A. Hazra et al (2013) Studies on a resistive gas sensor based on sol-gel grown nanocrystalline p-TiO_2 thin film for fast hydrogen detection. Sens. and Actuat. B 183:87-95 DOI 10.1016/j.snb.2013.03.11
7. E. Sennik et al (2010) Synthesis of highly-ordered TiO_2 nanotubes for a hydrogen sensor. International Journal of Hydrogen Energy 35:4420-4427 DOI 10.1016/j.ijhydene.2010.01.100
8. M. Enachi et al (2015) Integration of individual TiO_2 nanotube on the chip: Nanodevice for hydrogen sensing. Phys. Status Solidi RRL 9:171-174 DOI 10.1002/pssr.201409562
9. J. Zou et al (2010) Ultraviolet Photodetectors Based on Anodic TiO_2 Nanotube Arrays. J. Phys. Chem. C 114:10725-10729 DOI 10.1021/jp1011236
10. O. Lupan et al (2009) A single ZnO tetrapod-based sensor. Sens. and Actuat. B 141:511-517 DOI 10.1016/j.snb.2009.07.011
11. T. Ohsaka et al (1978) Raman spectrum of anatase, TiO_2. J. of Ram. Spectr. 7:321-324 DOI 10.1002/jrs.1250070606
12. Y-H. Zhang et al (1998) Micro-Raman spectroscopic characterization of nanosized TiO_2 powders prepared by vapor hydrolysis. J. Mater. Res. 9:2602-2609
13. Pal et al (2007) Size-Controlled Synthesis of Spherical TiO_2 Nanoparticles: Morphology, Crystallization, and Phase Transition. J. Phys. Chem. C 111:96-102 DOI 10.1021/jp0618173
14. T.-Y. Wei et al (2009) Gigantic Enhancement in Sensitivity Using Schottky Contacted Nanowire Nanosensor. J. Am. Chem. Soc. 131:17690–17695 DOI 10.1021/ja907585c
15. J. B. Law, J. T. Thong (2006) Simple fabrication of a ZnO nanowire photodetector with a fast photoresponse time" Appl. Phys. Lett. 88:133114(3) DOI 10.1063/1.2190459
16. H. Tang et al (1994) Electrical and optical properties of TiO_2 anatase thin films. J. Appl. Phys. 75:2042-2047 DOI 10.1063/1.356306
17. J. Li et al (2009) Nanocomposites of Ferroelectric Polymers with TiO_2 Nanoparticles Exhibiting Significantly Enhanced Electrical Energy Density. Adv. Mater. 21:217–221 DOI 10.1002/adma.200801106
18. N. Yamazoe (1991) New approaches for improving semiconductor gas sensors. Sens. and Actuat. B 5:7-19
19. C. Soci et al (2007) ZnO Nanowire UV Photodetectors with High Internal Gain. Nano Lett. 7:1003-1009 DOI 10.1021/nl070111x
20. H.-R. Kim et al (2010) Ultra-fast responding and recovering C_2H_5OH sensor using SnO_2 hollow spheres prepared and activated by Ni templates. Chem. Commun. 46:5061–5063 DOI 10.1039/c0cc00213e

Author: Reimer Tim
Institute: University of Kiel
Street: Kaiser Str. 2
City: Kiel
Country: Germany
Email: tire@tf.uni-kiel.de

Nanotechnological Application Based on $CoFe_2O_4$ Nanoparticles and Electromagnetic Exposure on Agrotechnical Plant Growth

I. Bodale[1], M. Oprisan[2], C. Stan[3], F. Tufescu[4], M. Racuciu[5], D. Creanga[4], and M. Balasoiu[6,7]

[1] University of Agricultural Sciences and Veterinary Medicine, Iasi, Romania
[2] University Hospital „Sf. Spiridon" Iasi, Romania
[3] Politehnica University of Bucharest, Department of Physics, Bucharest, Romania
[4] "Alexandru Ioan Cuza" University, Physics Faculty, Iasi, Romania
[5] "L. Blaga" University, Faculty of Sciences, Sibiu, Romania
[6] Joint Institute for Nuclear Research Dubna, Russia, [7] IFIN-HH Bucharest, Romania

Abstract— **This study is focused on the bioeffects of nanoparticulate metallic matter and electromagnetic exposure on agrotechnical plant cultures, designed in the general context of soil, water and air pollution. Nanotechnological procedure was carried out to yield magnetic nanoparticles (MNPs) similar to those used for various biomedical purposes for human life improvement but finally leading to MNPs delivery in the waste waters and consequently loading soil with metal ions. $CoFe_2O_4$ MNPs were synthesized by chemical route and suspended in deionized water. Microstructural and magnetic properties were evidenced by investigations through standard solid state methods. Equal volumes of 80 µl/l MNP suspension equivalent to 10^{16} MNPs/ml were supplied to *Helianthus annuus* seedlings for 12 days. Identical sample array was exposed also to low power density microwaves for 1-2-4 hours daily. Plant response was observed using measurements of assimilatory pigment contents in green tissues of seedlings.**

Keywords— **Nanoparticles, cobalt ferrite, soil loading, plant metabolism.**

I. INTRODUCTION

Industrial era and especially the nanotechnology extensive development during last half century have raised environmental issues related to side effects of MNPs release in the industrial and domestic waste waters that are leaking into the soil and interfere with plant roots. The benefits of nano-technological applications in technical and life sciences seem to be balanced by the risk of local pollution of environment having further influences on human major food sources [1]. Nevertheless, electromagnetic fields in the domain of microwaves are more and more ubiquitous in our life being strongly involved in numerous telecommunication technologies and also in medicine – where, for example, experimental cancer therapy includes tumor attack with both magnetic nanoparticles and electromagnetic waves. Electromagnetic wave background suffered dramatic changes that are difficult to be controlled or estimated accurately, so that large culture terrains hosting agro-technical plants could be impaired because of electromagnetic pollution either in the presence or in the lack of magnetic contamination with residual MNPs.

Laboratory experiment designed to compare some bioeffects of the two threatening factors have evidenced significant influences on photosynthesis complex process in sunflower (*H. annuus L.*) young plantlets. $CoFe_2O_4$ MNPs were mostly chosen since heavy metals like iron and cobalt can be easily found in Earth soils. Iron and its anorganic derivatives are important compounds in the Earth crust and mantle but also in the human body cells, for example in hemoglobin and cytochromes. Important enzymes or co-enzymes contain iron or cobalt ions in the form of complex coordination structures; also $CoFe_2O_4$ MNPs are used in malign tissues treatment by hyperthermia [2].

Concentration of mixed iron oxide MNPs used in the present experiment was chosen due to the results of previous experiments where most remarkable effects of diminution of photosynthesis pigments were found in the same seedling species for 80 µl/l cobalt ferrite MNP suspension equivalent to 10^{16} cobalt ferrite MNPs/ml [3].

II. MATERIALS AND METHODS

A. Nanoparticles Synthesis

According to adapted Massart method [4] MNPs were synthesized by magnetic mixing of 100 ml solution containing 4.75g $CoCl_2 \times 6H_2O$ with equal volume of aqueous solution containing 10.67g $FeCl_3 \times 6H_2O$ at relatively high temperature (over 80 °C) in alkali reaction medium (12 g NaOH in 100 ml deionized hot water). Black brownish precipitate was decanted in magnetic field gradient and repeatedly washed with hot deionized water to remove residual reagent traces being then separated on filter paper to be further mixed with long chain molecule able to ensure MNP steric stabilization in water dispersion [5]. Sodium oleate, known as source of long chain oleate ions for hydrophilic dispersion media was mechanically stirred together with ferrophase sludge at the same high temperature for

V. Sontea and I. Tiginyanu (eds.), *3rd International Conference on Nanotechnologies and Biomedical Engineering*,
IFMBE Proceedings 55,
DOI: 10.1007/978-981-287-736-9_37

about 1.5 hours; its biocompatibility was also considered according to [6].

Crystalline features of raw MNPs and oleate surface modified MNPs were investigated with Shimadzu LabX XRD-6000 diffractometer (with Cu-Kα radiation of λ=1.54 Å) and magnetic properties with Vibrating Sample Magnetometer (VSM) MicroMag model 2900/3900 - at room temperature.

B. Biological Material

H. annuus seedlings grown from selected seeds provided by single plant – to diminish genetic variability – were cultivated in Petri dishes on watered porous paper in controlled conditions of temperature (24.0±0.5°C), humidity (90%) and illumination (dark/light cycle of 14 h:10 h).

Two sample series were arranged: first one containing seedlings exposed only to microwaves (MW) for 1-2-4 hours daily for 12 days after seed germination; to the second seedling series the porous paper support was daily impregnated with the 7 ml volume of MNP suspension. Low power density microwaves (about 1 mW/cm^2 in the plan of Petri dishes) were emitted by the horn antenna of laboratory assembled device based on IMPATT diode.

The influence on the contents of photosynthesis pigments of MNPs and that of MW and MNPs was estimated using SHIMADZU spectrophotometer to assay chlorophylls and carotenes in 85% acetone extract according to [7]. Three repetitions of every assay were done; statistical significance was ensured relatively to the threshold of 0.05.

III. RESULTS AND DISCUSSION

The investigation of MNPs physical properties are presented in Fig. 1.

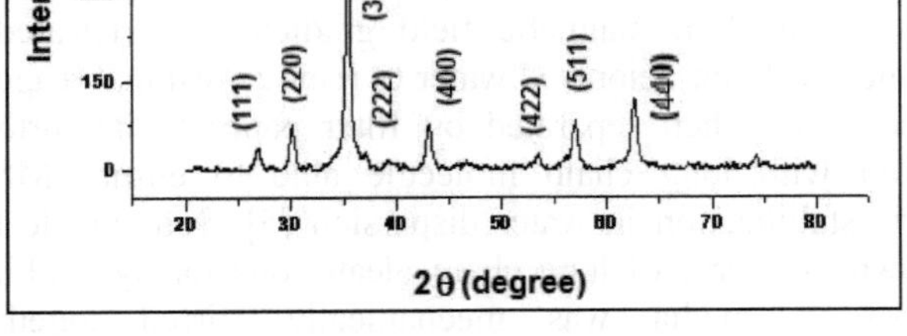

Fig. 1 Crystallinity investigation results

The X-ray diffraction revealed typical spinel crystal structure with all specific peaks at standard positions. Sodium oleate coating as double organic layer seems to generate additional peak (arrow in Fig. 1) corresponding to sodium ions array surrounding magnetic cores.

Magnetization curves shown in Fig. 2 indicate thin hysteresis loop with corresponding coercitive magnetic field of about 2000 Oe. This characteristics can be interpreted as evidence of overlapping of superparamagnetic smallest MNPs with larger ferrimagnetic ones as reported in [7].

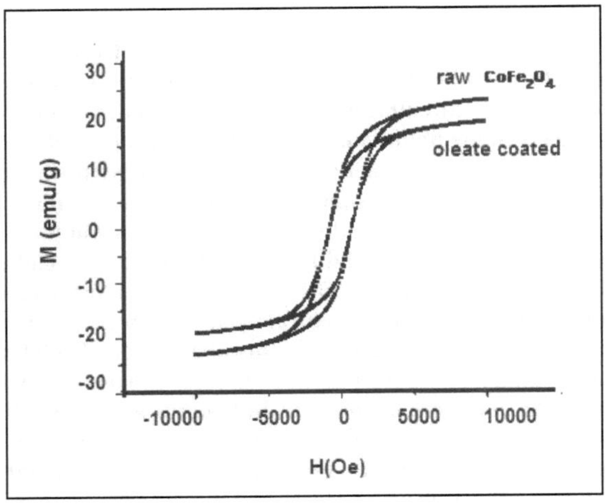

Fig. 2 Magnetic properties investigation results

Crystallite average size (eq.(1)) was found to have similar values in both raw ferrophase and coated ones - when estimated with Scherrer's formula:

$$D_{hkl} = \frac{0.94\lambda}{B\cos\theta} \qquad (1)$$

where λ is the X-ray wavelength, θ is the diffraction angle and B is the half width of (*hkl*) peak.

Surface modification with oleate ions appears to result in magnetic core diminution through non-magnetic layer yielding following oleate interaction with iron or/and cobalt ions from the external sheet of raw MNPs; maximal dimension of magnetic diameter D_M, could be estimated according to Langevin theory (eq. (2)):

$$D_M^3 = \frac{18k_BT}{\pi\mu_0 M_S m_S}\left(\frac{dM}{dH}\right)_{H\to 0} \qquad (2)$$

where k_B is Boltzmann constant, T - temperature, μ_0 - the free space magnetic susceptibility, M_S and m_S - the saturation magnetizations of bulk ferrite and of NPS analyzed sample, respectively.

Table 1 Physical characterization of $CoFe_2O_4$ MNPs

Physical parameter	Raw $CoFe_2O_4$	Oleate coated $CoFe_2O_4$
D_{XRD} (nm)	7.4	7.5
Ms (emu/g)	23.1	19.2
D_M (nm)	13.8	8.3
H_c (Oe)	198	203

The results presented in Table 1 are consistent with those reported in [8] with crystallite size of about 9 nm for $CoFe_2O_4$ MNPs.

Previous studies [3] found the range of the MNP concentration able to induce remarkable changes in *H. annuus* seedling pigments and demonstrated that the most affected pigment was chlorophyll A, which was diminished with about 28% for 80 μl/l. It was also reported the lack of influence of sodium oleate supplied in the amount equivalent to that from MNP/oleate coated suspension.

In the present investigation the concentration of MNP supply was kept constant at a level that induced the most visible effect (80 μl/l) while additional MW exposure was applied with various time durations.

Figures 3 and 4 show the experimental data for chlorophyll A and B contents in the two sample arrays.

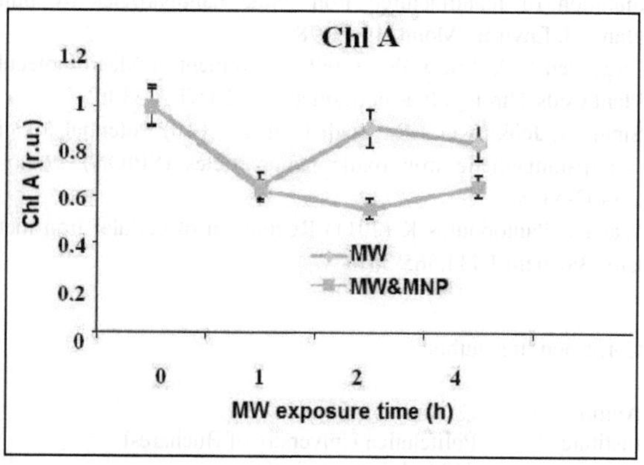

Fig. 3 Chlorophyll A content in relative units (r.u.)

Most significant MW effect is visible for shortest exposure time (1 h) both in the presence and the lack of MNPs – 40% diminished chlorophyll content in the irradiated seedlings compared to the control sample (with unitary value as reference level).

Nanotoxicity of MNPs is emphasized by 40% -35% diminished chlorophyll A level for exposure times of 2 and respectively 4 hours while MW exposure along resulted in only 15-20% decrease of chlorophyll A content ($p<0.05$). Quite similar behavior could be noticed from the results describing chlorophyll B situation (Fig.4) in MW exposed samples versus MNPs and MW exposed ones (standard deviation ranged between de 7.5 and 9%). Chlorophyll A content/chlorophyll B content exhibited slight non-significant difference only for the 4 h exposure to MW.

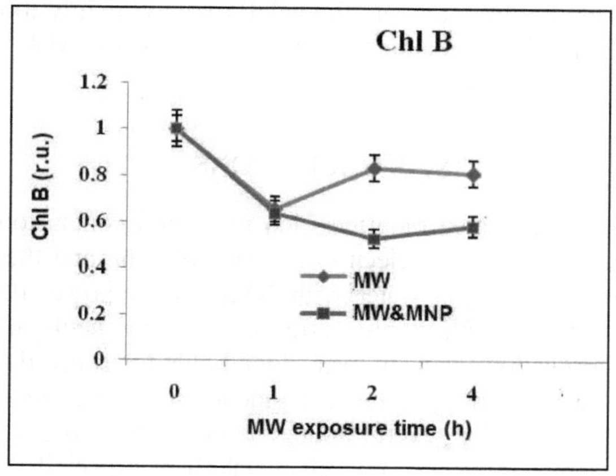

Fig. 4 Chlorophyll B content in relative units (r.u.)

Figure 5 present the experimental results of total carotene contents for the two samples.

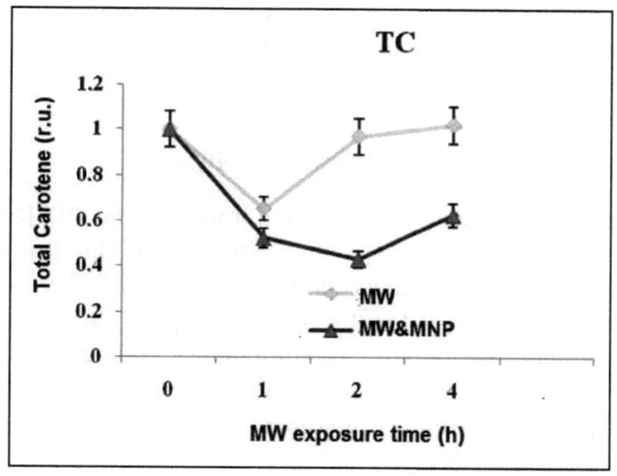

Fig. 5 Carotene content in relative units (r.u.)

As shown in Fig. 5, total carotene content was found diminished with around 50% for MNP supply and MW exposure, similar with chlorophyll changes.

According to [9-10] MNPs can be internalized in plant tissues when supplied at the root level i.e. in the culture medium. In [11] the authors assumed that plant tissues are able to receive MNPs through the plasmodesmata channels connecting neighbor cells with each other while internalized MNP fate seems to be decomposing through endocytosis and lysosymal digestion with release of ferric and ferrous ions [12-13]. Known for their catalytic role, iron ions and cobalt ones could trigger changes in the balance of reactive oxygen species by Fenton reactions with consequences on cell enzymatic equipment. In this way the cell metabolism could be impaired at the level of various biochemical parameters – like the levels of chlorophylls and carotenes, as evidenced in the present study. Additional electromagnetic exposure could potentiate the MNP toxicity mainly through non-thermal effects since low power density MW were chosen in this experiment.

IV. CONCLUSIONS

The investigated bioeffects led to several conclusions regarding environment electromagnetic pollution and the contamination soil and water with NMPs. Low power density microwaves - with thermal effects considered negligible for such irradiation intensity, were found able to reduce the biosynthesis of photosynthetic pigments in young seedlings exposed daily for several hours. Non-thermal effects could have subtle influences on cell metabolism especially at the level of complex metallo-proteins where iron or cobalt ions could experience electromagnetic interactions with MW fields and thus redox processes could be affected with consequences upon the levels of chlorophylls and carotenes. The overlapping of MW exposure with MNP supply emphasized significantly higher effects on photosynthesis pigments.

Further investigations are planned to assay enzymes involved in the oxidative stress.

ACKNOWLEDGMENT

Partially financially supported through the JINR-UPB Project No 34/23.01.2015 item 46 and by the European Social Fund in Romania, under the responsibility of the Managing Authority for the Sectored Operational Program for Human Resources Development 2007-2013 grant POSDRU/159/1.5/S/137750, project Doctoral and Postdoctoral programs support for increased competitiveness in Exact Sciences Research.

CONFLICT OF INTEREST

The authors declare that they have no conflict of interest.

REFERENCES

1. Buzea C, Pacheco Blandino I I, Robbie K (2007) Nanomaterials and nanoparticles: Sources and toxicity, Biointerphases 2:17–172
2. Mazario E, Menéndez N, Herrasti P et al (2013) Magnetic hyperthermia properties of electrosynthesized cobalt ferrite nanoparticles, J Phys Chem C 117:11405-11411
3. Ursache-Oprişan M, Focanici E, Creangă D et al (2011) Sunflower chlorophyll levels after magnetic nanoparticle supply, Afr J Biotechn 10:7092-7098
4. Massart R, Cabuil V (1987) Synthese en milieu alcalin de magnetite colloidale: contrôle du rendement et de la taille des particules, J Chim.Phys 84:967-973
5. Focanici E, Cirlescu A, Nica V et al (2009) Comparative study of magnetite and cobalt ferrite submicron particles, OAM-RC 3:326-329
6. Sun J, Zhou S, Hou P, et al (2007) Synthesis and characterization of biocompatible Fe_3O_4 nanoparticles. J. Biomed. Res. A, 80:333-341
7. Lichtenthaler H K, Wellburn A R (1983) Determinations of total carotenoids and chlorophylls a and b of leaf extracts in different solvents, Biochem Soc Trans 11:591 – 592
8. Kumar L, Kumar P, Narayan A et al (2013), Rietveld analysis of XRD patterns of different sizes of nanocrystalline cobalt ferrite, Int Nano Lett 3:2-12
9. Corredor E, Testillano P S, Coronado M J et al (2009) Nanoparticle penetration and transport in living pumpkin plants: in situ subcellular identification, BMC Plant Biol 9:45-56.
10. Zhu H, Han J, Xiao J Q et al (2008) Uptake, translocation, and accumulation of manufactured iron oxide nanoparticles by pumpkin plants, J. Environ. Monit. 10:71-78.
11. Jorgensen R A, Lucas W J (2006) Movement of Macromolecules in Plant Cells Through Plasmodesmata, Sci STKE. 323:tr2
12. Singh N, Jenkins G J S, Asadi R et al (2010) Potential toxicity of superparamagnetic iron oxide nanoparticles (SPION), Nano Rev. 1:3402-5358.
13. Wang J, Pantopoulos K (2011) Regulation of cellular iron metabolism, Biochem J 434:365-381

Corresponding author:

Author: C. Stan
Institute: Politehnica University of Bucharest
Street: 313 Spl. Independentei
City: Bucharest
Country: Bucharest
Email: cstan@physics.pub.ro

Application of Nano-Oxide Films on the Surfaces of Parts Made of Titanium Alloys in Order to Increase Their Corrosion Resistance

P. Topala[1], A. Ojegov[1], and P. Stoicev[2]

[1] Alecu Russo Balti State University, Balti, Republic of Moldova
[2] Technical University of Moldova, Chisinau, Republic of Moldova

Abstract— **The paper presents the results of nano-scale oxide films formation on sample surfaces made of titanium alloys that is material with a wide application in medicine and biomedicine. It was demonstrated that by the action of the plasma formed under ordinary conditions in the air media their surface is enriched with oxygen and nitrogen elements, the first substantially increases the corrosion resistance, and the other – the mechanical strength of the processed parts. The increase of corrosion resistance by at least 10 times in chemically active environments (30% aqueous solution of H_2SO_4 at the temperature of 80 °C) was observed for the processed titanium parts' surfaces.**

Keywords— **oxide films, processing, electro-erosion, surface, corrosion resistance.**

I. INTRODUCTION

Combating against corrosion is one of the oldest technical problems. It appeared once the man began to acquire metals and to use them in everyday life. Over the centuries the acquisition and the use of metals has expanded quite considerably. Although lately the use of technical polymer materials has begun, the metals and their alloys still remain primordial construction materials in machinery, equipment, devices, construction, medicine, etc.

Oxidation of metal surfaces has found applications in various fields of technology for corrosion protection [1, 2] and to change the electrical properties of surfaces in electrical and radio engineering [3, 4]. For this purpose, the method of electrolytic deposition of thermodynamically stable pellicles on some metal surfaces is applied, which can be obtained only at the so-called fed metals, which include aluminum, titanium, tantalum, niobium, etc. The spark-anodic electrolysis, also known as micro-plasma or micro-arc oxidation, has been developed [1]. Depositions formed by this method are a large range of structural, technological and operational requirements: micro-hardness, wear resistance, low coefficient of friction, and ability of electro-isolation, erosion resistance, and high adhesive properties. An important advantage of such depositions is that their properties are of complex operational character.

Low-alkaline and acidic electrolytes and metal salt solutions are used to realize this process. Deposition properties depend on the chemical composition of the alloy, concentration, composition and temperature of the electrolyte, electrical regimes of the process and its duration, etc. Coverings mainly consist of solid crystalline phases in the matrix of softer oxide phase of metals that form part of the alloy and the electrolyte. This structure provides a high micro-hardness, wear-proof and corrosion resistance, heat protection and electro-isolation properties of the covering. The application of this technology implies the use of special devices for electrolytic processes and the adoption of measures required to ensure safe working conditions for staff.

Recently, a number of works [2-4, 6] has been published where the results on the nano-oxidation by applying electrical discharges machining (EDM) of metal surfaces under ordinary conditions are presented.

The surface oxidation of titanium alloys with a wide applicability in medicine in order to produce the prosthesis and implants with different destination presents a particular interest for researchers and practitioners, which is why the experimental research results on the superficial oxidation of samples made of its alloys are developed in the present paper.

II. METHODOLOGY OF EXPERIMENTAL INVESTIGATIONS

The cylindrical tool-electrodes made of titanium alloy of technical purity BT-1-0 were designed for experimental investigations (Fig. 1).

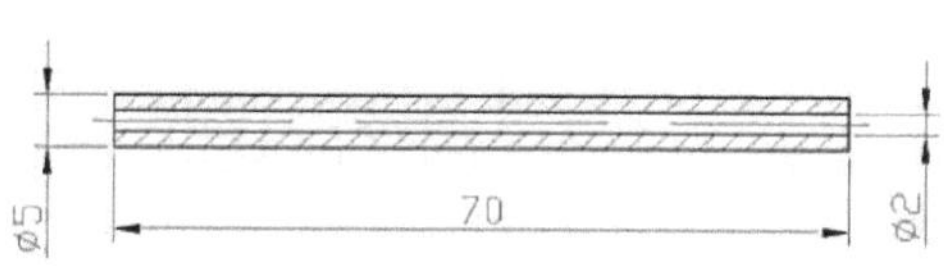

Fig. 1 Dimensions of the tool-electrode for dimensional EDM

Some of these tool-electrodes were subjected to the EDM processing of their lateral surface. The technological scheme for oxide films formation on the lateral surface is presented in Fig. 2. As generator of discharge impulses RC-type impulse generator designed in [5] was used.

V. Sontea and I. Tiginyanu (eds.), *3rd International Conference on Nanotechnologies and Biomedical Engineering*, IFMBE Proceedings 55,
DOI: 10.1007/978-981-287-736-9_38

The technology of oxide films formation is applied in the next consecutive order.

Analyzing the results obtained by the authors [5, 6] it has been established that to obtain oxide films on the parts' surfaces (for the case of "cold" spots plasma channel interaction with the machined surface) it is required that the energy density released on the machined surface must be less than the specific heat of the material melting and the latter can be evaluated by the relation [5, 6]:

$$Q < \frac{4W}{\pi d_c^2 \cdot S}, \qquad Q = q\rho, \tag{1}$$

where q and ρ are, respectively, the specific melting heat and the density of the workpiece material; W is the energy released in the gap, d_c is the diameter of the plasma channel and S is the gap size.

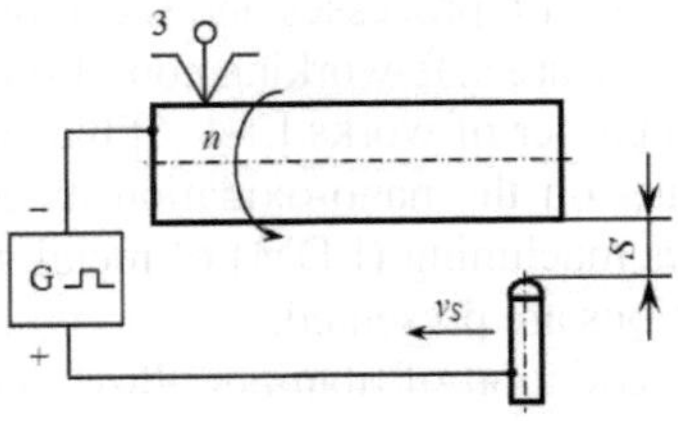

Fig. 2 Technological scheme for oxide films formation on the lateral surface of the tool-electrode by applying EDM [6]

Following the technological indications in the paper [6] on the formation of oxide films on cylindrical surfaces of titanium alloys the following input parameters and the energy processing regime of the power impulse generator have been set: the size of the gap S = 2 mm; the capacity of the condenser battery C = 600 μF; the charging voltage on the condenser battery U_C = 240 V; the impulse frequency f = 15 Hz; the power dissipated in the gap P_S = 24 kW; the frequency of rotation of the sample n = 120 rot/min; the feed speed v_s = 1 mm/s.

For qualitative and quantitative analysis of the properties of oxide films the SEM and EDX techniques were used. The investigations on the corrosion resistance were carried out according to the methodology described in [6].

III. RESULTS OF EXPERIMENTAL INVESTIGATIONS

The processing was performed under ordinary conditions in accordance with the method described above. SEM and EDX research (Fig. 3) of surface morphology and chemical composition allowed to establish that its roughness changes insignificantly, and in some cases decreases in relation to the original surface, and the enrichment of material with the elements from the plasma gas (air) occurs in the composition of the surface layer. In accordance with the results of EDX analyzes of the chemical composition of the surface layer the presence of oxygen and nitrogen was registered (Fig. 3). From these results we can see that the machined surface dissolve a considerable amount of oxygen (33.27% of the atomic oxygen) that can actually explain the properties that it possesses.

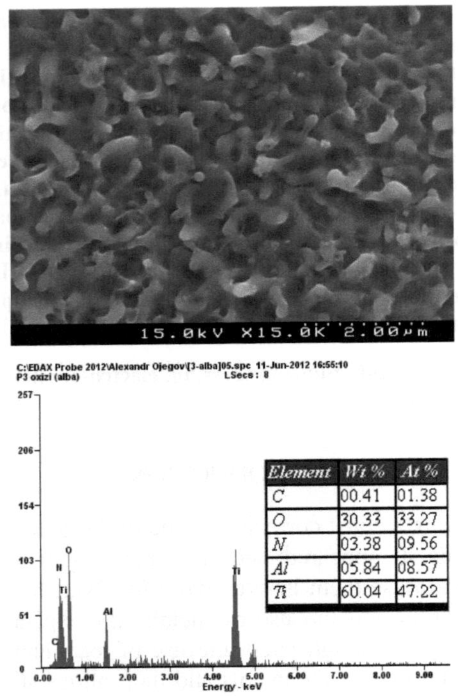

Fig. 3 SEM and EDX analysis of the machined surface made of titanium alloy BT 1-0

The asperities on the machined surface are in micro- and nano-metric limits. Ranging the electrical discharge energy we can obtain only nano-metric asperities on the modified surfaces. This increases considerably the active surface area, and, respectively, their behavior in aggressive media should manifest proportionately.

Further, the diffractogram shown in Fig. 4 shows that both oxygen and nitrogen form phases of TiO and TiN types, which explains the upper electrical surface resistance (1.6 MΩ/mm^2) [6].

In order to determine the corrosion resistance of samples' surfaces they were subjected to investigations on the determination of the corrosion speed in the chemically active medium (30% H_2SO_4 aqueous solution at 80 °C).

The comparative results of corrosion tests for unprocessed with plasma samples and those with films formed by air plasma application are presented in Table 1 and Table 2.

The material mass subjected to corrosion was determined by weighing with accuracy of 10^{-4} g, before and after the investigation, with the duration of anodic dissolution in a corrosive environment for one hour.

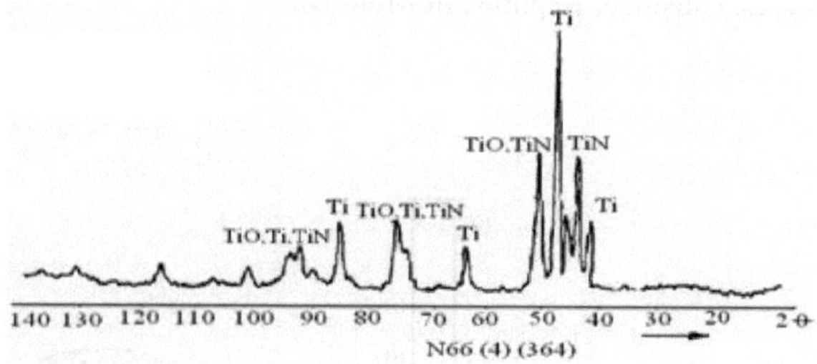

Fig. 4 BT 1-0 titanium surface sample diffractogram processed by electrical discharges

Table 1 The speed of anodic dissolution of samples made of BT 1-0 with raw surfaces

No	The initial mass of the sample m_i, g	The final mass of the sample m_f, g	Δm, g	The speed of corrosion K, $g/m^2 \cdot h$
1	7.4174	7.4150	-0.0024	95.50
2	7.5926	7.5902	-0.0024	95.50
3	7.6926	7.6903	-0.0023	91.52
4	8.1183	8.1160	-0.0023	91.52
5	7.5211	7.5186	-0.0025	99.48

Table 2 The speed of anodic dissolution of samples made of BT 1-0 with oxidized surfaces

No	The initial mass of the sample m_i, g	The final mass of the sample m_f, g	Δm, g	The speed of corrosion K, $g/m^2 \cdot h$
1	7.8245	7.8248	-0.0003	10.50
2	7.9362	7.9365	-0.0003	10.35
3	8.0214	8.0216	-0.0002	9.16
4	8.2216	8.2219	-0.0003	10.14
5	7.9837	7.9839	-0.0002	9.88

According to the test result it was shown that both mass corrosion and corrosion speed decrease at least by 10 times, that's why the proposed technology could be recommended for the implementation in the manufacture of bio-medical components.

IV. CONCLUSIONS

The SEM analysis shows the decrease of the roughness of the processed surface up to 0.1 μm, which can be applied in the formation of anti-wear surfaces.

Due to the formation of the oxide films in the amorphous state on the processed by EDM titanium surfaces its electrical surface resistance increases from 0.09 Ω/mm^2 up to 1.6 $M\Omega/mm^2$.

The speed of corrosion of the processed surfaces decreases at least by 10 times, that's why the proposed technology could be recommended for the implementation in the manufacture of bio-medical components, such as prosthesis and implants.

ACKNOWLEDGMENT

The authors wish to thank Professor Sobolev Nicolai and PhD Student Carlos Rosario for their assistance in SEM and EDX research execution, for logistical support and for fruitful discussions on the work.

CONFLICT OF INTEREST

The authors declare that they have no conflict of interest.

REFERENCES

1. Topala, P.; Slatineanu L.; Stoicev, P. Physical and chemical processes during the machining by means of the electro-erosive method. Nonconventional technologies review, 2010, No. 1, p. 50-54.
2. Topala, Pavel; Stoicev, Petru; Ojegov, Alexandr; Pinzaru, Natalia; Monaico, Eduard. Analysis of processes occurring on the tool and piece electrode surface during the formation of oxide pellicles by applying electrical discharges in impulse. ModTech International Conference – New face of TMCR. 20-22 May 2010. Slanic-Moldova, Romania, pp. 631-634. ISSN 2066-3919.
3. Pavel Topala, Alexandr Ojegov. Formation of oxide thin pellicles by means of electric discharges in pulse. Annals of the Oradea University. Fascicle of management and technological engineering, volume VII (XVII), 2008. CD-ROM Edition. Edition of the Oradea University, Romania. ISSN 1583-0691, CNCSIS „Clasa B+", pp. 1824-1829.
4. P.Topala, P. Stoicev, A. Ojegov, Experimental investigation on microoxidation of surfaces by means of applying electrical discharges in impulse under ordinary conditions. Creativity and Management. Proceedings of scientific papers. XIII Edition. Edition U.T.M., Chisinau, 2009, p. 172-175. ISBN 978-9975-45-115-4.
5. Topala P.; Stoicev P. Technologies of conductible material processing by applying electrical discharges in impulse. Chisinau, Edition TEHNICA-INFO, 2008, 265 p.
6. Ojegov A. Formation thin oxide pellicles on the metal surfaces of iron, copper, aluminum and titanium alloys, by applying electrical discharges in impulse. PhD thesis review. Chisinau, Edition "Tehnica UTM", 2014, 30 p.

Author: Pavel Topala
Institute: Alecu Russo Balti State University
Street: Pushkin 38
City: Balti
Country: Moldova
Email: pavel.topala@gmail.com

Entanglement among Photon and Phonon Degrees of Freedom

S. Cârlig

Institute of Applied Physics, Chişinău, Republic of Moldova
Institute of Electronic Engineering and Nanotechnologies, Chişinău, Republic of Moldova

Abstract— **We have investigated the quantum dynamics of a laser pumped quantum dot placed on a semiconductor beam suspended in an optical cavity. We demonstrate that optical and vibrational modes are correlated and we attest the generation of entanglement involving mechanical and optical degrees of freedom.**

Keywords— **quantum dot, correlation, phonon-photon entanglement.**

I. INTRODUCTION

Structures including nano-mechanical, atomic and optical subsystems are promising for practical applications in quantum informatics and generate great interest [1-11]. The multitude of interactions between implied subsystems put hybrid structures in the top of research, in order to obtain enhanced state preparation, transfer among disparate optical system [1-3], or cooling [4-6]. Another amazed aspect of quantum mechanics – entanglement – is an important source of applications in quantum information processing. In particular entanglement light due to nonlinear interaction was studied in [7] and aspects of entangling an optical cavity and a nano-mechanical resonator are presented in [8].

Here we analyze a scheme for entangling an optical cavity mode field and a nano-mechanical resonator via a laser pumped artificial atomic system. Flexions modify the energy of electronic states of quantum dot and thus entanglement between optical cavity and mechanical resonator vibrations can be generated.

The paper is organized as follow: The model Hamiltonian and analytical calculus are presented in section II. The master equation formalism is used in order to obtain equations of motion. In section III we analyze numerical results of mean phonon / photon number and continuous variable entanglement. Conclusions are giving in section IV.

II. ANALITICAL MODEL

We study a model presented in figure 1: a quantum dot is placed on nano-mechanical resonator suspended in an optical cavity. The quantum dot is pumped by a coherent laser source. Laser frequency is ω_L while cavity one is ω_C.

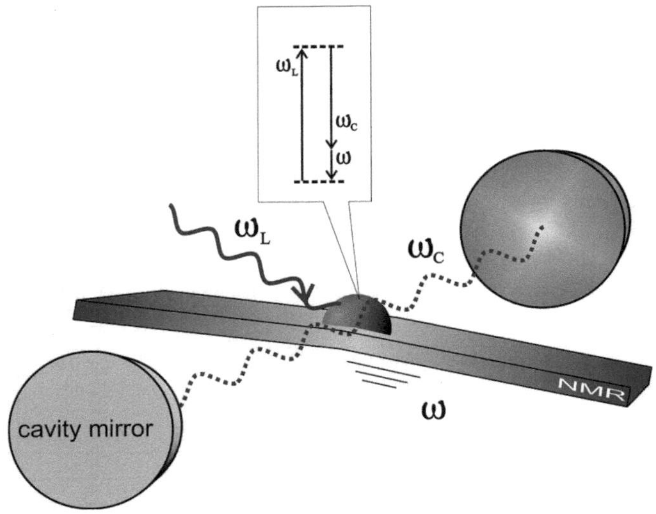

Fig. 1 Studied model. The two-level artificial atomic system placed on nanomechanical resonator suspended in an optical cavity.

The involved system will be described using master equation formalism [9]:

$$\dot{\rho} + \frac{i}{\hbar}[H, \rho] = -\gamma\{[S^+, S^-\rho] + [\rho S^+, S^-]\} - \gamma_c\{[S_z, S_z\rho] + [\rho S_z, S_z]\} - k_a\{[a^+, a\rho] + [\rho a^+, a]\} - k_b\{[b^+, b\rho] + [\rho b^+, b]\} \tag{1}$$

where Hamilton operator H is defined below by equation (2). γ. and γ_c are the single-qubit spontaneous decay and dephasing rates. k_a and . k_b are the photon and phonon resonator damping rate a^+, a and b^+, b are the generation and annihilation operators for photon and phonon subsystems. $S^{\pm}$, S_z are the qubit operators satisfying the standard commutation relations. The first two terms in right hand size in (1) describe qubit spontaneous decay and dephasing process, correspondingly, the last ones – cavity photon / phonon damping.

V. Sontea and I. Tiginyanu (eds.), *3rd International Conference on Nanotechnologies and Biomedical Engineering*,
IFMBE Proceedings 55,
DOI: 10.1007/978-981-287-736-9_39

The model Hamiltonian is formed by two parts. The first one consists from free energies of a cavity mode, phonon mode and quantum dot. The last part represents interaction energies between quantum dot and cavity, laser radiation and phonon mode respectively:

$$H = \hbar\omega_c a^+ a + \hbar\omega b^+ b + \hbar\omega_0 S_z + \\ + \hbar g\left(a^+ S^- + a S^+\right) + \hbar\Omega\left(S^+ e^{-i\omega_L t} + S^- e^{i\omega_L t}\right) + \\ + \hbar\lambda S_z\left(b^+ + b\right) \tag{2}$$

Here ω is the phonon mode frequency, ω_0 represents the difference between ground and excited quantum dot states. Ω is the corresponding Rabi frequency due to external laser pumping. g and λ are the interaction intensity among the two-level emitter and the involved optical and mechanical modes.

The system will be described in the dressed states representation and interaction picture, following the rotating wave and dipole approximation as well as Born-Markov ones. For a regime where the pumped two-level system is faster than the cavity photon and vibrational phonon subsystems, the quantum dot variables can be eliminated from the whole quantum dynamics, whereas $\Omega >> \gamma >> (k_a, k_b)$ and $\Omega >> (g, \lambda)$.

Thus for an operator Q describing phonon or photon subsystem we obtain the equations of motion [10]:

$$\left\langle \frac{dQ}{dt} \right\rangle + \frac{i}{2}(\Delta_1 - \omega)\left\langle\left[a^+ a + b^+ b, Q\right]\right\rangle = \\ = \left\langle [Q, a]\left(D_2^* b - A_1^* a^+\right) + \left(B_1^* a^+ - C_2^* b\right)[Q, a]\right\rangle + \\ + \left\langle\left[Q, a^+\right]\left(C_2 b^+ - B_1 a\right) + \left(A_1 a - D_2 b^+\right)\left[Q, a^+\right]\right\rangle + \\ + \left\langle [Q, b]\left(D_1^* a - A_2^* b^+\right) + \left(B_2^* b^+ - C_1^* a\right)[Q, b]\right\rangle + \\ + \left\langle\left[Q, b^+\right]\left(C_1 a^+ - B_2 b\right) + \left(A_2 b - D_1 a^+\right)\left[Q, b^+\right]\right\rangle, \tag{3}$$

where constants $(A, B, C, D)_{1,2}$ are giving [10] and $\Delta_1 = \omega_L - \omega_C$ is detuning of laser frequency from cavity one.

III. RESULTS

Thus equations of motion involving field correlators are [10]:

$$\frac{d}{dt}\left\langle a^+ a\right\rangle = \left\langle a^+ a\right\rangle\left(A_1 - B_1 + A_1^* - B_1^*\right) + \\ + \langle ab\rangle\left(C_2^* - D_2^*\right) + \left\langle a^+ b^+\right\rangle\left(C_2 - D_2\right) + A_1 + A_1^* \\ \frac{d}{dt}\left\langle b^+ b\right\rangle = \left\langle b^+ b\right\rangle\left(A_2 - B_2 + A_2^* - B_2^*\right) + \\ + \langle ab\rangle\left(C_1^* - D_1^*\right) + \left\langle a^+ b^+\right\rangle\left(C_1 - D_1\right) + A_2 + A_2^* \\ \frac{d}{dt}\langle ab\rangle = \langle ab\rangle\left(A_1 - B_1 + A_2 - B_2 + i\left(\Delta_1 - \omega\right)\right) + \\ + \left\langle a^+ a\right\rangle\left(C_1 - D_1\right) + \left\langle b^+ b\right\rangle\left(C_2 - D_2\right) + C_1 + C_2 \\ \frac{d}{dt}\left\langle a^+ b^+\right\rangle = \left\langle a^+ b^+\right\rangle\left(A_1^* - B_1^* + A_2^* - B_2^* - i\left(\Delta_1 - \omega\right)\right) + \\ + \left\langle a^+ a\right\rangle\left(C_1^* - D_1^*\right) + \left\langle b^+ b\right\rangle\left(C_2^* - D_2^*\right) + C_1^* + C_2^* \tag{4}$$

For steady states the left hand size of the equations became null and the roots of system $\langle a^+ a\rangle$, $\langle b^+ b\rangle$, $\langle a^+ b^+\rangle$, $\langle a\, b\rangle$ can be obtained. Therefore in Figure 2 are drawn the mean photon / phonon numbers in dependence of detuning between the laser and cavity frequencies Δ_1 normalized to qubit spontaneous decay rate γ. All other parameters are shown in the top of the graph. For $\Delta_1 \approx \omega$ one can observe a peak of $\langle a^+ a\rangle$, $\langle b^+ b\rangle$ where a laser photon absorption is following by the generation of a phonon and an optical cavity photon. The shift from resonance is treated in [10], and it is due to nonlinear effects.

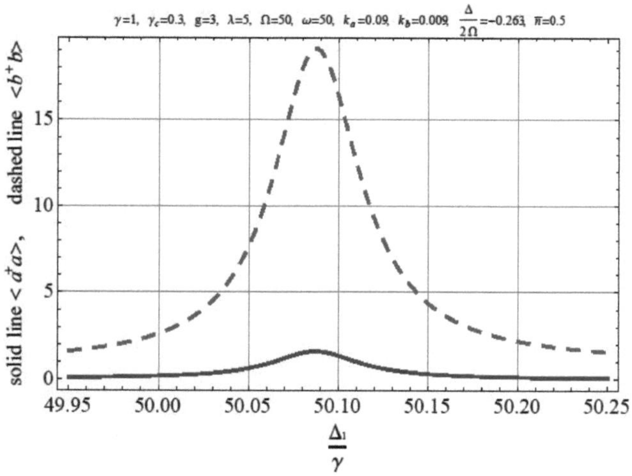

Fig. 2 (color online). Mean photon (solid line) and phonon number (dashed line) versus Δ_1/γ Here $\bar{n}$ is the mean phonon number corresponding to vibrational frequency ω. and environmental temperature T.

In order to describe entanglement between photons or phonons we will calculate continuous variable entanglement defines as [2, 7, 11]:

$$V = \sqrt{\langle a^+ a\rangle\langle b^+ b\rangle} - |\langle a^+ b^+\rangle| < 0 \qquad (5)$$

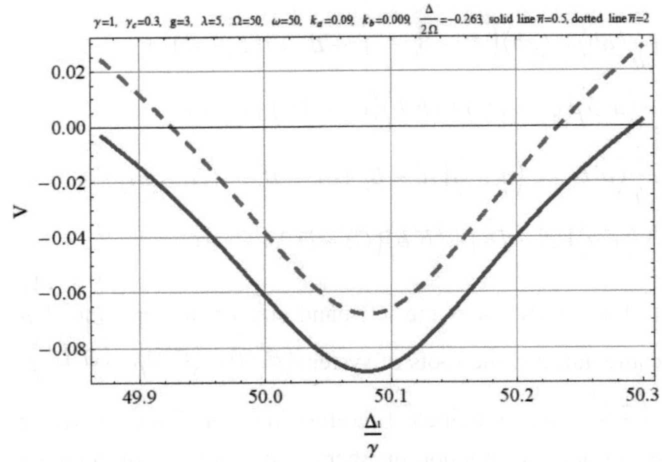

Fig. 3 (color online). Continuous variable entanglement versus Δ_1/γ. Solid line is for $\bar{n} = 0{,}5$ and dotted line – for $\bar{n} = 2$. $\bar{n}$ has the same meaning as in Fig.2.

In figure 3 it is drown the variable defined in eq. (5) in dependence of Δ_1/γ. This variable has values under zero around $\Delta_1 \approx \omega$ and their minimums are localized at the same value where the maximum of photon / phonon number occurs. Thus we attest a photon-phonon entanglement more evident at lower temperatures.

IV. CONCLUSIONS

We have studied an optical cavity inside of which a laser pumped quantum dot on nano-mechanical resonator is suspended. Two-level artificial atomic system acted by coherent radiation emits cavity photons and phonons. We have proposed a scheme to generate a steady state entanglement among cavity optical and vibrational modes of nano-mechanical resonator. This entanglement is more pronounced with decreasing of ambient temperature.

CONFLICT OF INTEREST

The author declare that they have no conflict of interest.

REFERENCES

1. C. Genes, A. Mari, D. Vitali, and P. Tombesi, Adv. At. Mol. Opt. Phys. 57, 33 (2009).
2. S. L. Braunstein and P. van Loock Rev. Mod. Phys., 77(2), 513, (2005)
3. M. Wallquist, K. Hammerer, P. Rabl, M. Lukin and P. Zoller, Phys. Scr. T137, 014001 (2009).
4. Y. Greenberg, Y. Pashkin, and E. Il'ichev, Phys. Usp. 55, 382 (2012).
5. A. Dantatn, B. Nair, G. Pupilo and C. Genes, Phys. Rev. A 90, 033820 (2014).
6. S. Carlig and M. A. Macovei, Phys. Rev. 90, 013817 (2014).
7. M. Macovei and Gao-xiang Li, Phys. Rev. A 76, 023810 (2007).
8. Xiao-Zhong Yuan, Phys. Rev. A 88, 052317 (2013).
9. H. J. Carmichael, An Open Systems Approach to Quantum Optics, (Springer, Berlin, 1993)
10. S. Carlig and M. A. Macovei, Phys. Rev. A 89, 053803 (2014).
11. M. A. Macovei, Phys. Rev. A 82, 063815 (2010).

Author: Cârlig Sergiu
Institute: Institute of Applied Physics
Street: Academiei 5
City: Chişinau
Country: Republic of Moldova
Email: carligsergiu@gmail.com

Activation Process of GaAs NEA Photocathode and Its Spectral Sensitivity

K. Mitsuno, T. Masuzawa, Y. Hatanaka, Y. Neo, and H. Mimura

Research Institute of Electronics, Shizuoka University, Hamamatsu, Japan

Abstract— **Negative electron affinity (NEA) Gallium Arsenide (GaAs) photocathodes with high photo-electric conversion quantum efficiency (QE) and high response speed are expected as an electron source for THz frequency vacuum devices. To achieve high QE and high response speed, further understanding of the electron emission mechanism on NEA GaAs photocathodes is necessary. This study focuses on the formation process of NEA photo-electric surface on a GaAs photocathode. The results suggest that NEA photo-electric surface allows the electron emission from both the Γ and L valleys in the conduction band. On the contrary, in the Cs or O excess state, vacuum level is between the conduction band level at Γ and L point, and electrons can emit only from L valley, and cannot be emit-ted from the Γ valley due to the barrier of vacuum level.**

Contribution of Γ and L valleys to the emission current was confirmed, of which balance depended on photo-electric surface conditions and excitation wavelength.

Keywords— **photocathode, GaAs, negative electron affinity, electron beam.**

I. INTRODUCTION

Recently, electromagnetic waves with a frequency of THz domain application to analysis of molecular structure and non-invasive medical diagnostics, because of the unique absorption spectrum observed at that frequency domain, is widely expected. We have studied about vacuum devices using electron beam for the high frequency generator. A high current density photocathode with high response speed is strongly demanded for such THz vacuum devices. One of the candidate for these devices is a photocathode excited by a high speed pulse lasers. Because, recently it is possible to get a many kind of high-speed pulse lasers. For these purposes, semiconductor photocathodes having low work function modified by cesium (Cs) and oxygen (O) are investigated. Electron affinity (EA) of GaAs photocathode covered with Cs and O is effectively observed as negative value [1-2]. These photocathode are called NEA photocathodes. NEA GaAs photocathodes with a high photo-electric conversion quantum efficiency (QE) when excited by a pulse laser could satisfy these requirements aiming to THz device applications. The NEA GaAs photocathode has been already put to practical use such as a photomultiplier tube [3-4], but adaptability to ultrafast modulation and high current density application requires further investigation of their operation mechanism. In this study we focused on the formation process of NEA GaAs photocathode photo-electric surface, to investigate the operation mechanism and characteristics of electron emission.

II. EXPERIMENTAL

In this study, two types of photoemission measurement were designed and conducted. In the first experiment, the formation process of the NEA surface was investigated by photoemission measurement during the surface activation. In the second experiment, photoemission was induced by lasers to measure the energy distribution of the electrons emitted from the GaAs photocathode.

Formation of an NEA surface, or surface activation, was conducted in a setup shown in Fig. 1. To form NEA surfaces, O and Cs were injected from (c) and (d) shown in Fig. 1. (c) is made of silver tube which is heated for introducing oxygen. (d) is a sleeve ribbon made by Cs composite used for the Cs injection. The EA of GaAs photo-electric surface was decreased by alternating exposure of Cs and O as in the reference Yo-yo method [1]. Photoemission spectroscopy measurement on such photo-electric surfaces suggests that the EA becomes extremely small, if not negative. The electrons emitted from the GaAs photo-electric surface due to the incident light were collected by a cylindrical collector. The shape of collector was designed to collect electrons without disturbing the incident light, and to prevent space charge effect by applying 50V between GaAs photo-electric surface and the collector. In this experiment, vacuum level and sample photo-electric surface condition greatly affects the electrons emission. Therefore it is necessary to remove harmful adsorbent such as oxide and carbon on the sample surface. For this purpose the sample has to be cleaned by etching about 2 minutes by H_2SO_4 : H_2O_2 (35%) : H_2O = 2 : 1 : 1 and after setup in the vacuum chamber, the sample surface was heated from backside to 650-680 °C by bombarding the sample from the back side with an electron gun [6]. The photoemission spectra were obtained in-situ, under a base pressure of 10^{-6} Pa order. In this experiments, p-type (100) GaAs wafers with an impurity Zn of 3 x 10^{18} cm^{-3} were used, and the wafer size was 7.5 mm square.

V. Sontea and I. Tiginyanu (eds.), *3rd International Conference on Nanotechnologies and Biomedical Engineering*,
IFMBE Proceedings 55,
DOI: 10.1007/978-981-287-736-9_40

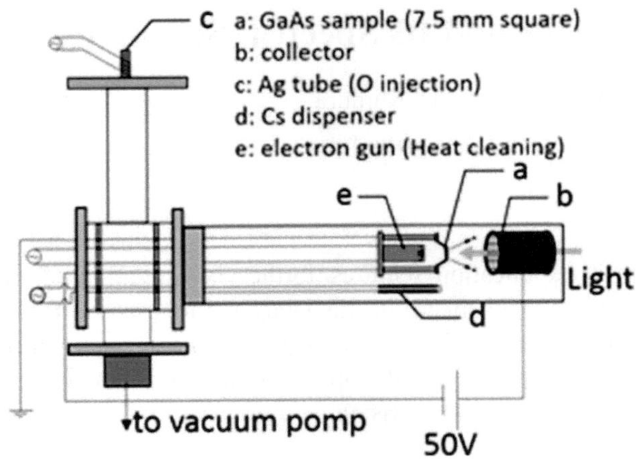

Fig. 1 Experimental setup for forming NEA GaAs photo-electric surface and measuring photoemission spectroscopy.

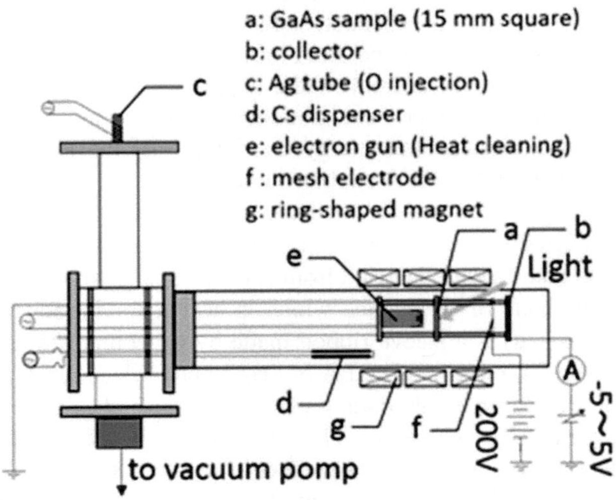

Fig. 2 Experimental setup for measuring electron energy distribution emitted from the NEA cathode.

Following the photoemission spectroscopy, energy distribution of emitted electrons was measured by a retardation method using experimental setup shown in Fig. 2. A vacuum setup construction is approximately same as in Fig. 1, but the arrangement and shape of the electrode are modified. A mesh electrode was placed between the emitter and the collector, and biased to 200 V to prevent space charge effect. The collector voltage was varied to change the retarding voltage from -5 to 5 V. A magnetic field of approximately 0.05 Tesla was applied parallel to the cathode-collector axial direction in order to suppress the diffusion of the emitted electrons. Electrons are thus assumed to land perpendicularly on the collector surface. The GaAs sample surfaces were heat-cleaned and evacuation condition by the vacuum pump was in the same manner as those of photoemission spectroscopy measurement, but the size of the sample was 15 mm square in this measurement. Semiconductor lasers of red (658 nm), green (532 nm) and blue (450 nm) were used to induce photoemission.

III. Results and Discussion

Photoemission spectroscopy was measured on three types photo-electric surface, Cs coating condition, O excessive additional condition and NEA condition. In the Yo-yo method, the emission current repeats the peak and bottom shown in Fig. 3. The Cs coating condition is firstly deposited Cs on the GaAs clean surface after heat cleaning. The O excessive condition is to be introduced O to the decreasing position indicating the bottom in the emission current. NEA condition is at the saturation point of the emission current. Figure 4 shows the measurement results of photoemission current on the three photoelectric surface type.

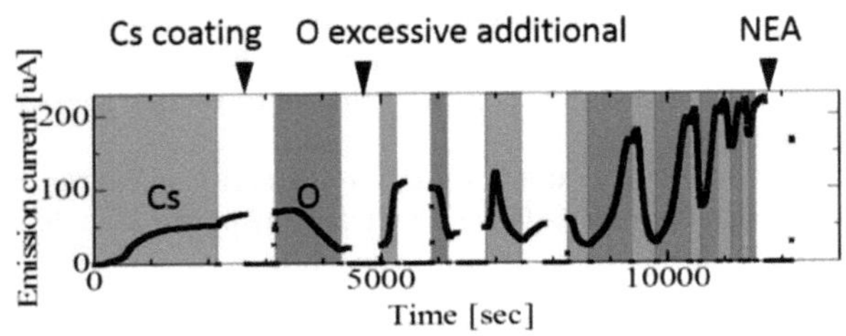

Fig. 3 Activation process by Cs (blue) and O (orange) alternating injections.

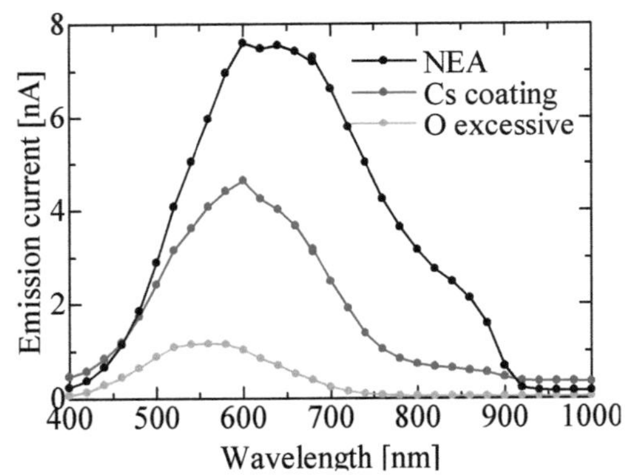

Fig. 4 Photoemission current spectra from different surface conditions.

The photoemission current is then converted to the quantum efficiency (QE) by Eqn. (1). Here, $N_{electron}$ is the number of electrons emitted, and N_{photon} is the number of incident photons. R shows the fraction of photons reflected from the surface, and T is the transmittance of the setup window, since the laser was attenuated by the absorption, scattering and reflection by experimental setup such as glass windows. This absorption rate was estimated to be 63.4 % of the initial power. The reflectance R of the GaAs surface is obtained from Eqn. (2).

$$QE = N_{electron} / \{N_{photon} \times (1-R) \times T\} \quad (1)$$

$$R = (n-1)^2/(n+1)^2 \quad (2)$$

Where n is refractive index having wavelength dependence. The number of the emitted electrons N_{electron} is estimated from photoemission current shown in Fig. 4, using Eqn. (3). The number of the incident photons N_{photon} is calculated from the optical power of the lasers using Eqn. (4).

$$N_{electron} = \frac{\text{Photoemission current[A]}}{\text{Elementary charge } (1.6 \times 10^{-19})} \quad (3)$$

$$N_{photon} = \frac{\text{Light power[W]} \times \lambda\text{[m]}}{6.626 \times 10^{-34} \times 2.998 \times 10^{8}} \quad (4)$$

The calculated QE was plotted against wavelengths of the incident light, to obtain a spectral sensitivity of the cathode, with incident light. Figure 5 shows spectral sensitivities of (QE) at the different photo-electric surface conditions. For the cathode with NEA, the QE was highest among three surface conditions, and showed constant change depending on the wavelength of the incident light up to the GaAs band edge, 1.43 eV. When the surface of GaAs was exposed to excessive amount of Cs or O, QE decreased rapidly with the energy lower than the energy level at L point, 1.74 eV, in addition to 1.43 eV. The kinks in 1.43 and 1.74 eV correspond to Γ and L points in the conduction band of GaAs, as shown in Fig. 6. It should be noted that the kink at 1.74 eV was always observed, in spite of difficulty to accurately control the vacuum level. It is considered that the excited electrons have the energy distribution on the verge emission. And its distribution is presumed to be due to electron energy in the Γ and L valleys. Therefore the NEA photo-electric surface allows the electron emission from both the Γ and L valleys in the conduction band, and in the Cs or O excess state, electrons can emit only from L valley in the higher energy than Γ valley. The spectral sensitivity suggests that vacuum level exists between Γ and L level. In that case, the vacuum level interfere the electrons emission from the Γ valley.

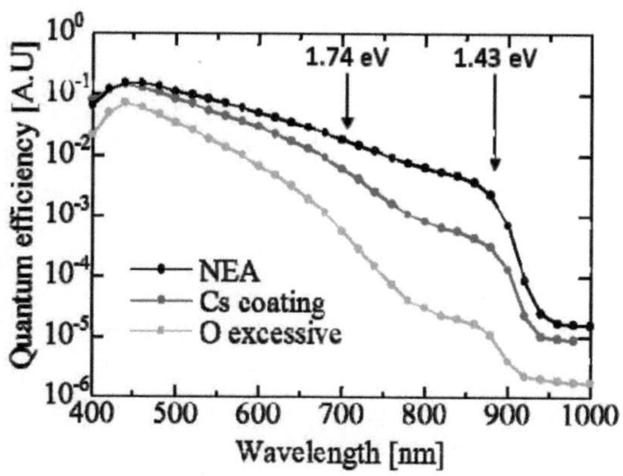

Fig. 5 Photo-electric conversion QE in the various surface conditions.

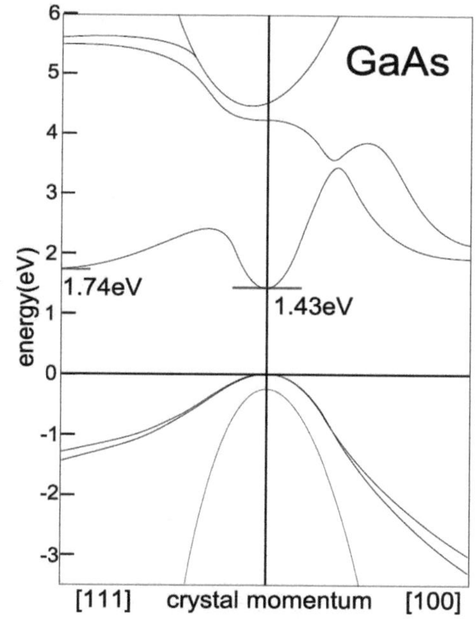

Fig. 6 Energy band structures of GaAs.

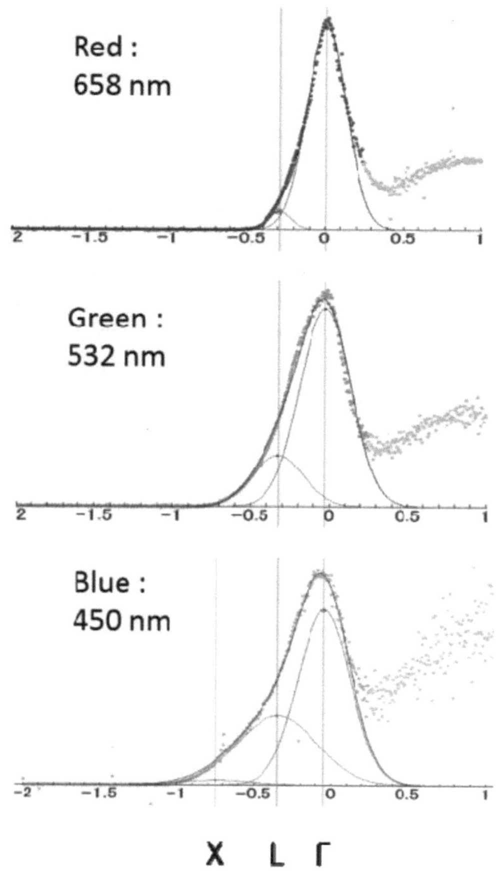

Fig. 7 Energy distributions of photo-emitted electrons by laser power.

Figure 7 shows the energy distribution of the emitted electrons by the retarding method using the lasers with three wavelengths. Peak fitting by Gaussian function suggests that the energy spectra of the photo-emitted electrons can be separated into three peaks corresponding to Γ, L and X valleys, which agrees with the results of spectral sensitivity measurement. The area ratio for Γ, L and X part depends on the energies of the incident light, as shown in Table 1. The more shorter laser wavelength was, the more high-energy peaks (L and X) became large. From these results, electrons in Γ, L and X are shown to involve in the photoemission from the NEA GaAs cathode.

Table 1 Laser wavelengths and distribution of Γ and L+X.

Laser color	Red	Green	Blue
Wavelength	658nm	532nm	450nm
eV	1.88	2.33	2,76
(L+X)/ Γ	5.2%	25.4%	62.7%

IV. SUMMARY

In this study, photoemission characteristics of GaAs photo-cathodes depending on the photon energy excited are investigated through the photo-electric spectroscopy and electron energy distribution measurements. It is suggested that Γ, L and X valleys in the GaAs conduction band contribute to electron emission for the NEA GaAs photocathode. With the NEA surface, the spectral sensitivity showed relatively high QE over the range of 450nm to 870nm. This is because the electrons in the conduction band bottom are emitted to vacuum. In Cs or O rich surfaces, spectral dependency of the photoemission showed a kink at the wavelength of 750nm, which corresponds to the L valley of GaAs conduction band.

The energy distributions of the emitted electrons were measured by the retarding potential method. The electrons excited by the short wavelength light are distributed to the higher energy level. Contribution of L and X valleys of the upper conduction band compared to Γ valley were observed as higher energy distributions. The present study suggests that short wavelength light is more suitable as excitation light than long wavelength one, even if the GaAs photo-electric surface condition influences to be degraded.

ACKNOWLEDGMENT

This work was partially supported by JSPS KAKENHI Grant Numbers 26289101. The authors would like to thank Dr. M. Niigaki, Hamamatsu Co. for useful discussions and Mr. Y. Momose, technical workshop, Shizuoka University for the preparation of glass tube setup of this experiment.

CONFLICT OF INTEREST

The authors declare that they have no conflict of interest.

REFERENCES

1. J. J Scheer et al., Solid State Comm. 3, 189, 1965.
2. W. E. Spicer, Appl. Phys. 12, 115, 1977.
3. Hamamatsu Co. Catalog 2014
4. Y. Hatanaka et al., IEEE Trans. ED28, 72, 1978.
5. Y. Hatanaka et al., IECE 62, B-2, 19
6. J. R. Chelikowsky et al. Phys. Rev, B14, 556, 1976

Author: Keigo Mitsuno
Institute: Research Institute of Electronics, Shizuoka University
Street:3-5-1 Johoku, Naka-ku
City: Hamamatsu 432-8011
Country: Japan
Email:mitsuno.keigo.15@shizuoka.ac.jp

Anticipated Synchronization of Passive Dispersive Reflector Semiconductor Laser

N. Ciobanu[1], S. Rusu[2], and V. Tronciu[2]

[1] Department of Human Physiology and Biophysics, State University of Medicine and Pharmacy "Nicolae Testemitanu", Chisinau, Republic of Moldova
[2] Department of Physics, Technical University of Moldova, Chisinau, Republic of Moldova

Abstract— **We discuss a model of anticipated synchronization of two distributed feedback lasers (DFB) with passive dispersive reflector. The influence of passive dispersive reflector parameters on the anticipated synchronization is discussed. The conditions for delayed, perfect and anticipated synchronizations are explored in detail.**

Keywords— **DFB laser, passive dispersive reflector, optical feedback, anticipated synchronization.**

I. INTRODUCTION

During the last decade the phenomenon of synchronization between various types of devices that imply chaotic systems play a significant role, where the controlling mechanism is essential. Such king of control is especially important for nonlinear systems, e.g., information security and transmission [1, 2], communication or complex systems like biological sciences [3].

The first model that describes anticipation synchronization of two chaotic dissipative systems with time-delayed feedback was proposed by Voss [4]. The idea of anticipated synchronization consists in the fact that for two or many identical systems connected in a "master-slave configuration" can be observed anticipated synchronization dynamics if the slave system is time-delayed from the rest components. The concept of anticipated synchronization represents an interesting phenomenon from theoretical [5] as well as experimental point of view [6] and was explored for various types of chaotic systems like semiconductor lasers with optical feedback [7, 8], quantum dot lasers [9], electronics circuits [10].

In this paper, we present a model of anticipation in the synchronization of two DFB lasers with passive dispersive reflector. For some parameters of the system, we observe a transition from delayed to anticipated synchronization regimes. A short description of the scheme setup and rate equations of the modes is presented in Section II. The Section III gives numerical results on the calculations and discussions. We finish with conclusions in Section IV.

II. MODEL SKETCH

We start this section with the description of our physical model. The master (*M*) and slave (*S*) DFB lasers with passive dispersive reflector section are connected in unidirectional configuration as shown in Figure 1. It is considered that *S* laser is connected to an external optical circuit with a delay time τ. The pumping mechanism of the lasers is realized via an electrical current notated by *I(t)*. The output powers for master and slave lasers are expressed by $P_{M(S)}$, while $n_{M(S)}$ correspond to the currier density.

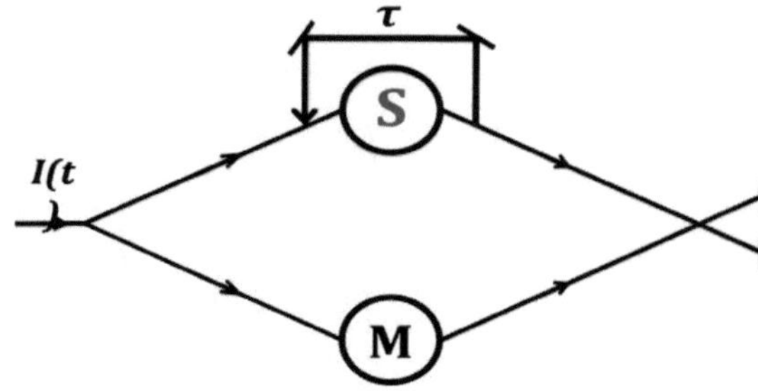

Fig. 1 Schema of master (M) and slave (S) DFB laser with passive dispersive reflector section connected in unidirectional configuration.

The equations of motion that describe the dynamics of such type of system are given by [11]

$$\begin{aligned}
\frac{dP_M}{dt} &= TG(n)P_M + I(t),\\
\frac{dn_M}{dt} &= J - n_M - (1+n_M)\Gamma(n_M)P_M,\\
\frac{dP_S}{dt} &= TG(n)P_S + I(t) + K(P_M(t) - P_S(t-\tau)),\\
\frac{dn_S}{dt} &= J - n_S - (1+n_S)\Gamma(n_S)P_S,
\end{aligned} \qquad (1)$$

where parameter *T* represents the ratio between the carrier and photon life times. The relative excess injection rate is notated by *J*, and its values correspond to $1 < J < 10$.

The functions $\Gamma(n_i)$ and $G(n_i)$ from the rate equations (1), where *i=M,S* characterize the passive dispersive reflectors of the lasers and have the form

$$\Gamma(n_i) = \Gamma_0 + \frac{AW^2}{4(n_i - n_0)^2 + W^2}, \quad G(n_i) = n_i + \alpha\Delta n \tanh\frac{n_i}{\Delta n}, \qquad (2)$$

where Γ_0, n_0, α and *W* coincide with those of [12].

V. Sontea and I. Tiginyanu (eds.), *3rd International Conference on Nanotechnologies and Biomedical Engineering*,
IFMBE Proceedings 55,
DOI: 10.1007/978-981-287-736-9_41

In what follows, we are interested in the dynamics of master and slave lasers, and how it is possible to obtain anticipated synchronization regime, for various parameter of the system.

iii. Numerical Results

In this section, we discuss the dynamics of master and slave lasers for various values of material and device parameters.

A. Pumping Parameter I(t)

Delayed synchronization: Figure 2 illustrates the evolution of laser power and currier density, where a delayed synchronization of slave laser can be observed. We mention that, for this fixed values of parameter I and others as in Figure 2, the time evolution of master laser is in advance comparing to slave laser.

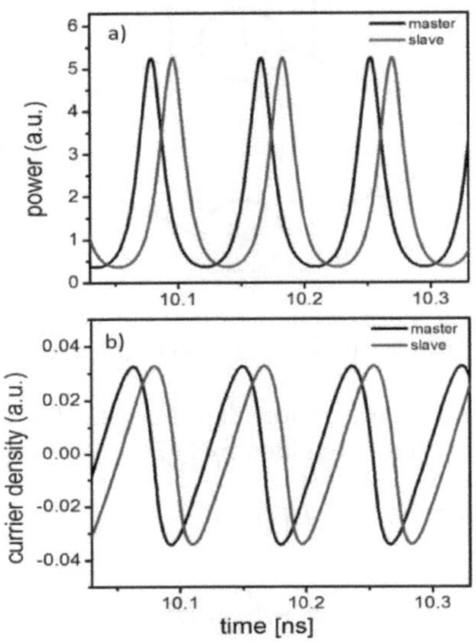

Fig. 2 Time dependence of the a) output power and b) currier density for master (black) and slave (red) lasers. The following parameters are used in simulations: I = 1, J = 2, T = 500, K = 1, $\alpha = 5$, $\Delta n = 0.1$, $\Gamma 0 = 1$, A = 0.5, W = 0.02, n0 = –0.01.

Perfect Synchronization: In the Fig. 3 we can distinguish an interesting regime, where the pulse traces curves of master and slave lasers overlap. Thus, changing the shape and value of injection current in the system, we can reach the same evolution for master and slave lasers. An identical situation is observed also for the currier density.

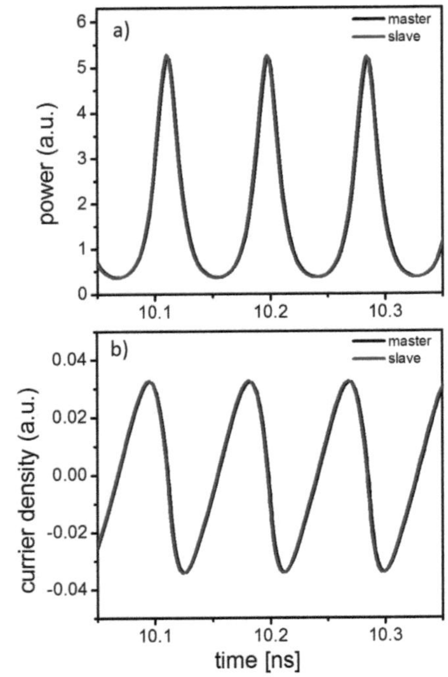

Fig. 3 Same as in Fig. 2 except $I(t) = 1.25\sin(0.74\, t)$.

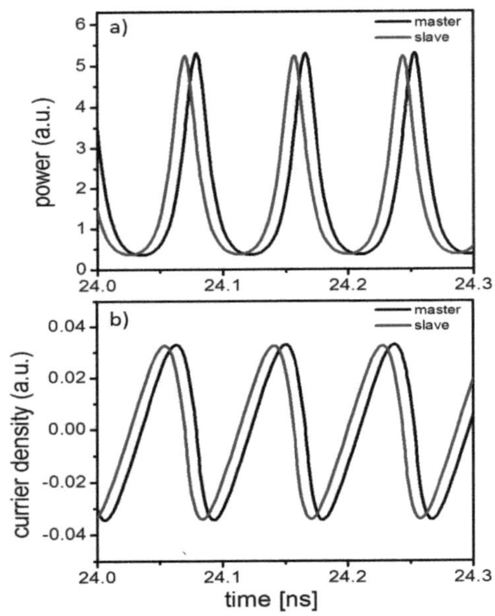

Fig. 4 Same as in Fig. 2 except $I(t) = 0.95\sin(0.58\, t)$.

Anticipated Synchronization: Figure 4 shows the process of anticipated synchronization of lasers. Here, we can observe anticipation in the pulse trace of slave laser due to the change of parameter *I*.

As follows from Fig. 2-4, the pumping parameters play an important role in the dynamics of our system and interesting anticipation regimes can be observed. However, it is interesting to explore the influence of passive dispersive reflector parameters, i.e., $\Gamma(n_i)$ and $G(n_i)$ functions, on the phenomena of anticipation synchronization.

B. Influence of Ratio Parameter between the Carrier and Photon Life Times T

Figure 5 shows the influence of the ration between currier and photon life time on the synchronization regime of master and slave lasers. One can observe the dynamics of master laser for output power as well as currier density, prevails in time. Here, we can distinguish a substantial delay between the master and slave lasers signals, comparing to case A.

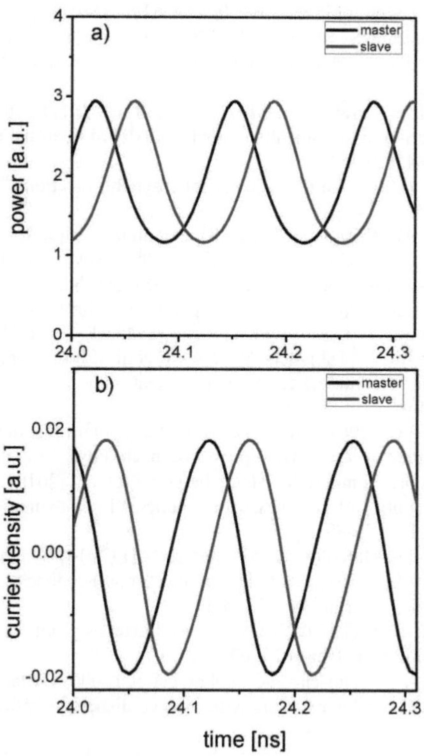

Fig. 5 Same as in Fig. 2 except A=0.1 and T=200

Changing the parameter T, we obtain anticipated synchronization shown in Figure 6. Thus, keeping all parameters of equations (2) constant, and increasing the ratio between the carrier and photon life times, we can switch to anticipation regime.

C. Influence of Amplitude of Reflected Signal (A)

Finally, we present the influence of reflected signal amplitude A on the anticipated synchronization. Figures 7 a) and b) shows delayed synchronization regimes in the dynamics of master and slave lasers, for A=35. On the other hand, for small values of amplitude of reflected signal, it is possible to get anticipation of synchronization (see Figure 7 c, d). As a result, we conclude that the change of amplitude of reflected signal can be experimentally tuned by the length of dispersive reflected section.

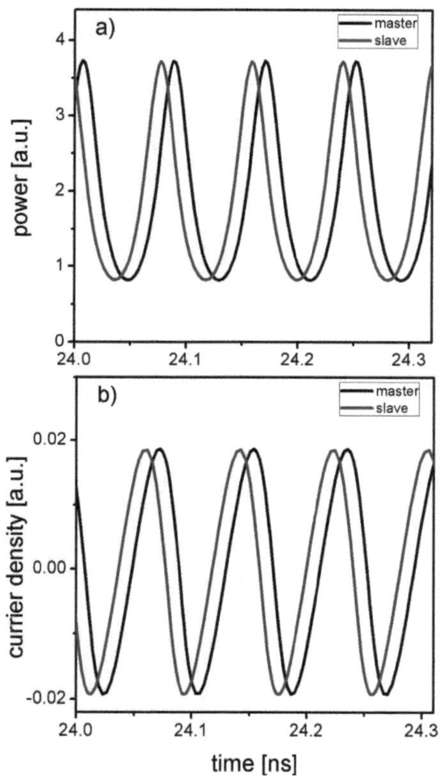

Fig. 6 Same as in Fig. 2 except A=0.1 and T=530

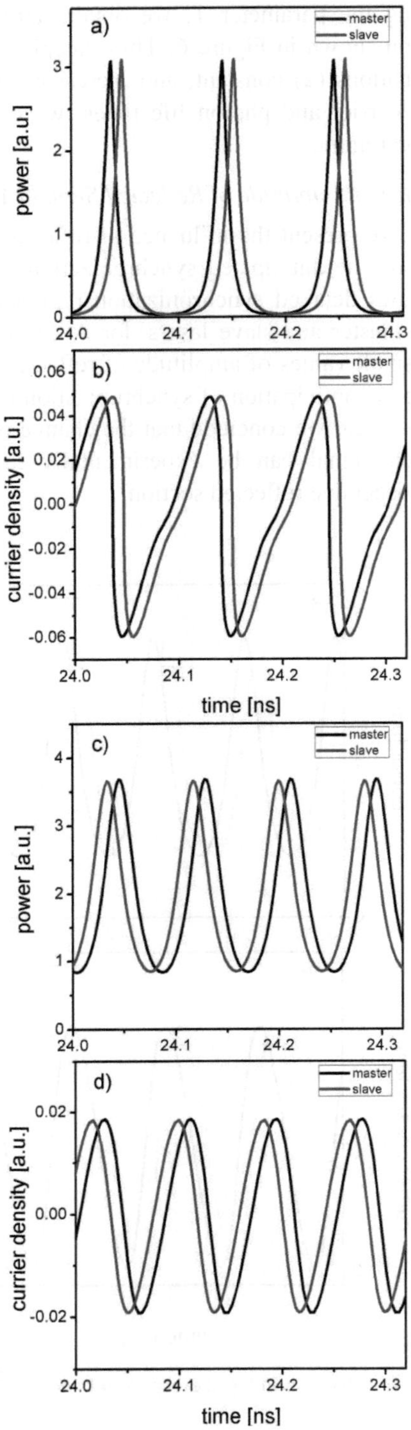

Fig. 7 Same as in Fig. 2 except a,b) A=35 and c,d) A=0.1.

IV. CONCLUSIONS

In this paper we present a model of anticipated synchronization for DFB lasers with passive dispersive reflector sections. The anticipation regime was explored for different values and shapes of the pumping parameter *I*, of ratio parameter between the carrier and photon life times T and the amplitude of reflected signal of the passive dispersive reflector sections of lasers. We demonstrated that all mentioned parameters of the system (I, T and A) are important in the anticipation synchronization regime.

ACKNOWLEDGMENT

This work was supported by the project STCU - 5993 and 14.02.116F.

CONFLICT OF INTEREST

The authors declare that they have no conflict of interest.

REFERENCES

1. Fischer I, Liu Y, Davis P (2000) Synchronization of chaotic semiconductor laser dynamics on sub-nanosecond time scales and its potential for chaos communition. Phys. Rev. A 62: 011801(R)
2. Sivaprakasam S, Shore K A (1999) Demonstration of optical synchronization of chaotic external-cavity laser diodes. Opt. Lett. 24: 466-468
3. Matias F S, Carelli P V, Mirasso C R et al. (2011) Anticipated synchronization in biologically plausible model of neuronal motifs. Phys. Rev. E 84: 021922
4. Voss H (2000) Anticipating chaotic synchronization. Phys. Rev. E 61: 5115
5. Pyragas K, Pyragiene T (2008) Coupling design for a long-term anticipating synchronization of chaos. Phys. Rev. E 78: 046217
6. Sivaprakasam S, Shahverdiev E M, Spencer P S et al. (2001) Experimental Demonstration of Anticipating Synchronization in Chaotic Semiconductor Lasers with Optical Feedback. Phys. Rev. Lett. 87: 154101; Liu Y, Takiguchi Y, Davis P et al. (2002) Experimental observation of complete chaos synchronization in semiconductor lasers. Appl. Phys. Lett. 80: 4306
7. Masoller C (2001) Anticipation in the synchronization of chaotic semiconductor lasers with optical feedback. Phys. Rev. Lett. 86: 2787
8. Dahms T, Flunkert V, Henneberger F et al. (2010) Noninvasive optical control of complex semiconductor laser dynamics. European Phys. J. 191: 71-89
9. Ghalib B A, Hafedh G A, Al Khursan A H (2014) Synchronization of Quantum Dot Lasers with an Optoelectronic Feedback Circuit. J. of Electronic Materials 44: 953-966
10. Weia H, Li L (2010) Estimating parameters by anticipating chaotic synchronization. Chaos 20: 023112
11. Tronciu V Z, Wunsche H-J, Sieber J et al. (2000) Dynamics of single mode semiconductor lasers with passive dispersive reflectors. Optics Comm. 182: 221-228
12. Tronciu V Z (2008) Excitability and coherence resonance of DFB lasers with passive dispersive reflector. Mold. J. Phys. Sci. 7: 218-223

Author: Nellu Ciobanu
Institute: Department of Human Physiology and Biophysics, State University of Medicine and Pharmacy "Nicolae Testemitanu"
Street: Testemitanue 27/2
City: Chisinau
Country: Moldova
Email: cnellu@gmail.com

Observation of Electron Spin Relaxation Time in Pnpn Structured GaAs

A. Morozumi[1], T. Ito[1], M. Ichida[2], and H. Ando[2]

[1] Research Institute of Electronics, Shizuoka University, Japan
[2] Department of Physics, Faculty of Science and Engineering, Konan University, Japan

Abstract— **To observe effect of electron-hole exchange interaction on electron spin relaxation time, we performed polarization- and time-resolved PL measurements by using pnpn structured GaAs. This structure is supposed spatial distance between excited electrons and holes are changed by excitation power density. Prolongation of spin relaxation time was observed when excitation power density was weak and distance between electron and hole was long. The change of spin relaxation time attributable to the reduction of electron-hole exchange interaction in pnpn structure.**

***Keywords*— Spin relaxation time, Electron-hole exchange interaction, Pnpn structure, Polarization resolved PL spectral measurement, Polarization- and time-resolved PL measurement.**

I. INTRODUCTION

An electron has internal degree of freedom called electron spin because it corresponds to the rotation of electron in classical mechanics. However, the behavior is different from the rotation in classical mechanics in actuality. A lot of studies about electron spin in a semiconductor have been performed, and they are mainly related to the electron spin relaxation phenomena. For example, band mixing of conduction and valence band (the Eliott-Yafet process) [1], electron spin splitting in noncetrosymmetric media (the Dyakonov-Perel' process) [2], and electron-hole exchange interaction (the Bir-Aronov-Pikus process) [3].

Recently, electron spin in semiconductor has been investigated for quantum information technologies and expected to be a candidate of quantum bit (qubit). However, electron spin relaxation time is too short in conventional semiconductors to control spin degree of freedom for the qubit.

In this study, we investigate effect of electron-hole exchange interaction on spin relaxation time in pnpn structure [4]. In pnpn structure, excited electrons and holes are spatially separated by optical means. We will discuss the electron-hole exchange interaction (which can be controlled by excitation power density) dependence of spin relaxation time.

First, we will describe structure and characteristics of the pnpn sample. Next, we will describe polarization-resolved PL spectral measurements [5] to identify carrier transition mechanisms in pnpn sample. Finally, we will describe polarization and time-resolved PL measurements to clarify relationship of spin relaxation time and exchange interaction.

II. PNPN STRUCTURE

A. *Structure of the Pnpn Sample*

A schematic illustration of pnpn structure and calculated result of band simulation in real space are shown in upper side and underside of Fig. 1, respectively. A sample we use in our experiments is GaAs based stacking structure consists of 4 layers of carbon (C) doped p-type GaAs with a doping density of 10^{18} cm^{-3} and 3 layers of silicon (Si) doped n-type GaAs with a doping density of 10^{17} cm^{-3}, widths of the p-layers and n-layers are 50 nm and 300 nm, respectively (pnpn structure). We designed the sample satisfying that depth of energy modulation at the depletion layer is the same as bandgap energy and almost all the regions in the sample are depletion layer, assuming that Fermi level of p-type and n-type semiconductor lies at upper edge of valence band and lower edge of conduction band, respectively and bandgap energy of GaAs is 1.5 eV.

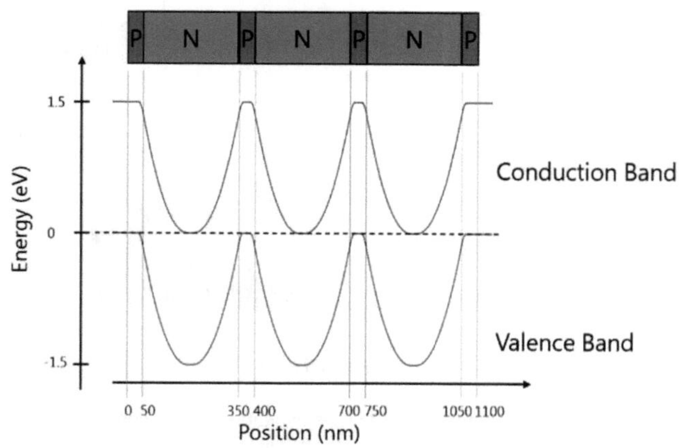

Fig. 1 Calculated band structure of pnpn structure in real space. Origins of horizontal and vertical axis mean sample surface and Fermi level, respectively. Curves above and under the Fermi level indicate conduction and valence band, respectively. Band gap energy of GaAs used in this calculation was 1.5 eV. Corresponding sample structure is illustrated in upper side.

B. *Characteristics of the Sample*

In pnpn structure, the distance between excited electrons and holes is determined by the magnitude of band modulation which changes with the density of ionized impurities. Because the ionized impurities are neutralized by the excited

V. Sontea and I. Tiginyanu (eds.), *3rd International Conference on Nanotechnologies and Biomedical Engineering*,
IFMBE Proceedings 55,
DOI: 10.1007/978-981-287-736-9_42

electrons and holes and the density of excited electrons and holes depend on excitation power density, the distance between excited electrons and holes can be controlled by excitation power density. In other words the exchange interaction is strong when the excitation power density is weak since electron-hole exchange interaction depends on the spatial distance between excited electrons and holes.

Circularly polarized laser pulses were irradiated on the surface of pnpn structure to generate spin polarized electrons. Excitation power density dependence of degree of polarization was measured by detecting circularly polarized PL components. Figure 2 shows PL spectra excited by circularly polarized light with excitation power densities of 0.5 W/cm^2 to 50.0 W/cm^2 by using a CCD camera and a spectrometer.

A few peaks are observed in each excitation power density conditions. We defined them as a peak (a), (b), and (c) from higher energy as shown in Fig. 2. The emission peak (a) is an emission of interband transition. The peak (b) is conduction band to acceptor level emission. The peak (c) is recombination emission of spatially separated electrons and holes by tunnel effect. We focused on recombination emission by tunnel effect to investigate relationship of electron-hole exchange interaction and spin relaxation time.

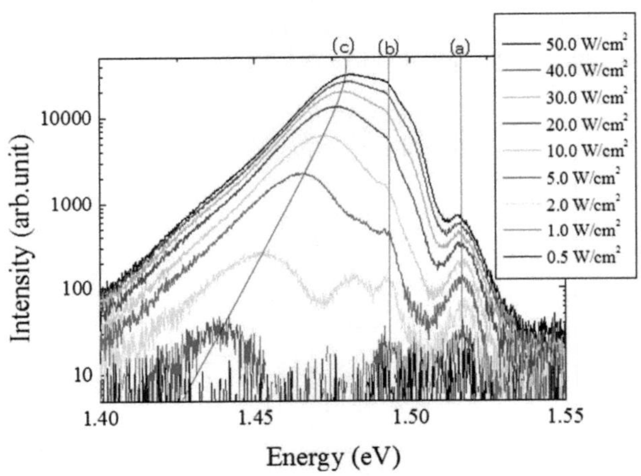

Fig. 2 PL spectra of pnpn structure with excitation power densities of 50.0 W/cm^2 to 0.5 W/cm^2

III. Polarization Resolved PL Spectral Measurements

A. Experiment

A sample used for polarization resolved PL spectral measurements was pnpn structure described in section 2. First, right-handed circularly polarized laser pulses from Ti-Sapphire laser with a wavelength of 754 nm were irradiated on the surface of pnpn structure. The pulse duration time was 2 ps and the repetition rate was 80 MHz. All measurements were performed at 35 K to clearly observe emission of tunnel effect by reducing emission of conduction band to acceptor level. Next, we measured spectra of right- and left-handed circularly polarized PL. To obtain spectra of degree of polarization, we calculated $(I^+ - I^-)/(I^+ + I^-)$ from the spectra of polarized PL components I^+ and I^-. Here, I^+ and I^- are right- and left-handed circularly polarized PL components, respectively. The spectra of degree of polarization reflect the degree of spin polarization in conduction band.

B. Result and Discussion

Figure 3 shows spectra of degree of polarization in pnpn structure with excitation power densities of 5.0 W/cm^2 to 100 W/cm^2 and the peaks are observed around the photon energy of 1.505 eV. As the excitation power density reduces, these peaks shift to a high photon energy side.

Figure 4 shows excitation power density dependence of the photon energy peaks of emission of tunnel effect and degree of polarization. The emission peaks of tunnel effect were determined from the spectra of sum of the measured right- and left-handed circularly polarized components. Because the peaks of degree of polarization shows the same trend as the emission peak of tunnel effect, the changes is supposed to be depending on the spatial distance between excited electrons and holes. Because the peaks of degree of polarization have good signal-noise ratios compared to that in other photon energy region, time evolution of the peaks of degree of polarization were measured by time-resolved PL measurements.

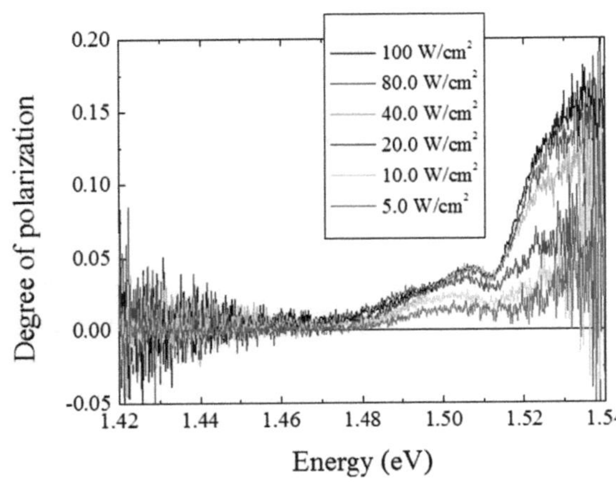

Fig. 3 Spectra of degree of polarization with excitation power densities of 100 W/cm2 to 5.0 W/cm2.

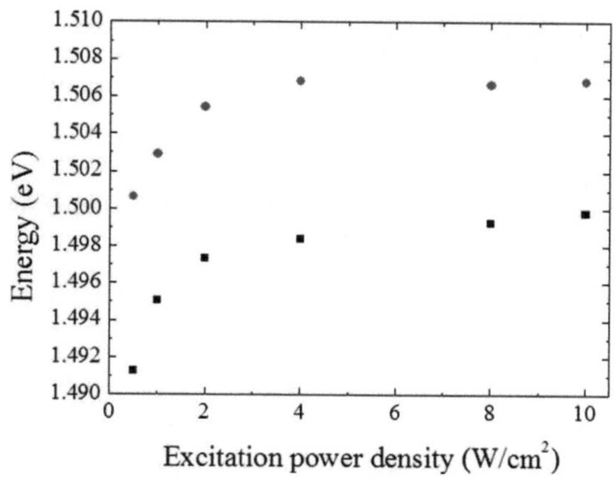

Fig. 4 Excitation power density dependence of the photon energy peaks of recombination emission by tunnel effect (■) and of degree of polarization (●).

IV. POLARIZATION AND TIME-RESOLVED PL MEASUREMENTS

A. Experiment

A sample used for time-resolved PL measurements is pnpn structure. First, we generated circularly polarized PL components in the same conditions as described in section 3.1. Next, photons at the peak photon energy of degree of polarization were selected by a spectrometer and were observed by a photomultiplier tube. Finally, the time-resolved measurements were performed by using Time to Amplitude Combater (TAC) system and Multi Channel Analyzer (MCA). The time resolution of the system is around 200 ps.

From the results of previous section, the photon energy to perform time-resolved measurements were the peaks of spectra of degree of polarization in Fig. 3. The peaks of spectra of degree of polarization with the excitation power density of 10.0 W/cm^2 and 80.0 W/cm^2 were 1.503 eV and 1.509 eV, respectively.

B. Result and Discussion

To confirm spin relaxation time dependence of the excitation power density, we compared time trace of degree of polarization with the excitation power densities of 10.0 W/cm^2 and 80.0 W/cm^2. Time trace of degree of polarization is obtained by calculating $(I^+ - I^-)/(I^+ + I^-)$ from the time trace of polarized PL components (I^+, I^-). Time trace of degree of polarization is represented by the formula $(A\exp[-2t/\tau_s])$. Here, τ_s is spin relaxation time, A is a constant. By a logarithmic scale, the reciprocal of spin relaxation time is expressed as the gradient. Figure 5 shows time trace of degree of polarization with the excitation power densities of 10.0 W/cm^2 and 80.0 W/cm^2 where the vertical axis is shown in logarithmic scale.

Although the time trace of degree of polarization in Fig 5 is noisy, we can confirm the gradient of spin polarization with the excitation power density of 80.0 W/cm^2 is larger than that of 10.0 W/cm^2 from Fig. 5. Because in pnpn structure, the electron-hole exchange interaction is weak when the excitation power density is small, the spin relaxation time is long when the electron-hole exchange interaction is weak.

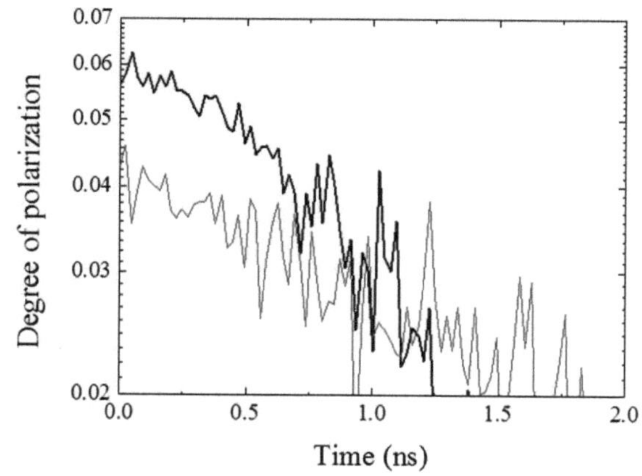

Fig. 5 Time trace of degree of polarization in single logarithmic scale. Thick black and thin red lines indicate results with the excitation power densities of 80.0 W/cm2 and 10.0 W/cm2, respectively.

V. CONCLUSION

To observe effect of electron-hole exchange interaction [3] on electron spin relaxation time, we performed PL spectra measurements and time-resolved measurements. The distance between excited electrons and holes can be controlled by excitation power density in pnpn structure. The exchange interaction depends on distance between the excited electrons and holes.

We performed PL spectra measurements to analyze origin of emission from the pnpn structure and determine the photon energy where time-resolved PL measurements performed. Because the peak of degree of polarization shifted in accordance with change of excitation power density, we consider the distance between excited electrons and holes was also changed with the excitation power density.

We performed time-resolved PL measurements at the peak of degree of polarization. From the results of time-resolved PL measurements, we obtained degree of polarization with respect to time with the excitation power densities of 10.0 W/cm^2 and 80.0 W/cm^2. We observed the spin relaxation time with an excitation power density of 80.0 W/cm^2 was longer than that of 10.0 W/cm^2. The spin relaxation time is long when the electron-hole exchange interaction is weak. The excitation power density dependence of spin relaxation time is seemed to be result from the electron-hole exchange interaction.

CONFLICT OF INTEREST

The authors declare that they have no conflict of interest.

REFERENCES

1. R. J. Elliott, Phys. Rev. **96,** 266 (1954).
2. M. I. D'yakonov and V. I. Perel', Sov. Phys. JETP **33,** 1053 (1971).
3. G. L. Bir, A. G. Aronov, and G. E. Pikus, Sov. Phys. JETP **42,** 705 (1976).
4. G. H. Döhler, H. Künzel, D. Olego, K. Ploog, P. Ruden, and H. J. Stolz, Phys. Rev. Lett. **47**, 864 (1981).
5. T. Ito, H. Sugata, S. Tanigaki, M. Ichida and H. Ando, 2012 SSDM, PS-9-9

Author: Atsushi Morozumi
Institute: Research Institute of Electronics, Shizuoka University
Street: 3-5-1 Johoku, Naka-ku
City: Hamamatsu
Country: Japan
Email: morosan8014@icloud.com

Evaluation of Spin Relaxation Time by Polarization- and Time-Resolved Pump and Probe Measurements

S. Fuma[1], T. Ito[1], H. Goto[2], M. Ichida[3], and H. Ando[3]

[1] Research Institute of Electronics, Shizuoka University, Japan
[2] NTT Basic Research Laboratories, NTT Corporation, Japan
[3] Department of Physics, Faculty of Science and Engineering, Konan University
fuma.sohichiro.15@shizuoka.ac.jp

Abstract— **Polarization- and time-resolved pump and probe (PP) measurements were performed to observe spin superposition in GaAs/AlGaAs multiple quantum wells (MQWs). We used a linearly polarized right for pump light under resonant excitation condition. Order of the relaxation times were very close to that of the spin superposition obtained by polarization- and time-photoluminescence (PL) measurements and dependence on quantum confinement energy also showed same trend as that of the PL measurement.**

Keywords— **pump and probe method, spin superposition, spin relaxation time, linearly polarized light.**

I. INTRODUCTION

Spin is an internal degrees of freedom which is different from the freedom of movement of the three-dimensional space. The possible state for spin of electron and hole is not only up spin state or down spin state, but also superposition state of up and down spin. It plays an important part of quantum mechanics. The behavior of the electron spins in quantum wells have been studied since the 1970s. The study aims to know about "spin relaxation time" [1-3], which is the time the polarity of the spins are maintained. In recent years, based on the knowledge obtained by such basic research, spin states are expected to be used in quantum information technology [4-10]. Creating and controlling the superposition of spin states are important skills for applying to quantum memory.

Temporal changing of spin polarized state in semiconductor can be measured by time-resolved measurement. There are two types of major method. They are polarization- and time-resolved photoluminescence (PL) [11] and pump and probe (PP) measurements. In PL measurement, we can obtain spin polarization of photo excited electrons and holes directly. But noise to signal ratio is not good for discussing spin-precession with high Larmor frequency. In PP measurement, we can obtain data which has good signal-to-noise ratio and time resolution. But since we measure change of absorption and refractive index in the samples, it is difficult to discuss what we are measuring. In this study, we have performed PP measurements to investigate creation and relaxation of the spin superposition. Results obtained by this way were compared with the results of PL measurements performed in previous studies [11].

II. CONDITIONS FOR OBSERVING SPIN SUPERPOSITION

Two conditions are necessary for observing the spin superposition states. One is linearly polarized excitation light which correspond to the superposition of right- and left-handed circularly polarized light, and the other is resonant excitation condition.

Beam emitted from the laser was polarized in the x-axis direction. Linearly polarized light is a superposition of right- (σ+) and left-handed (σ-) circularly polarized light. The laser pulse illuminated the samples to excite up- and down- spin superposition. The linearly polarized components which reflect the spin superposition were selectively measured using a half-wave plate and a polarizer.

We have to conduct resonant excitation to make spin superposition. Figure 1 schematically shows inter-band transition of electrons and holes when a non-resonant excitation is performed. The horizontal axis is the wave number, and the vertical axis is energy. Now, consider electrons of a superposition state of up and down spin is excited by linearly polarized light far from the band edge, as shown in Fig1 (a). Superposition state of hole-spins is also excited. The electrons and the holes move to the band edge immediately. Superposition of the hole-spin will collapse in the process, as shown in Fig (b). However the resonant excitation make such superposition of electrons and holes at band edge. So the superposition will not be collapsed. Results of PL measurement in previous studies indicate that result [11].

V. Sontea and I. Tiginyanu (eds.), *3rd International Conference on Nanotechnologies and Biomedical Engineering*,
IFMBE Proceedings 55,
DOI: 10.1007/978-981-287-736-9_43

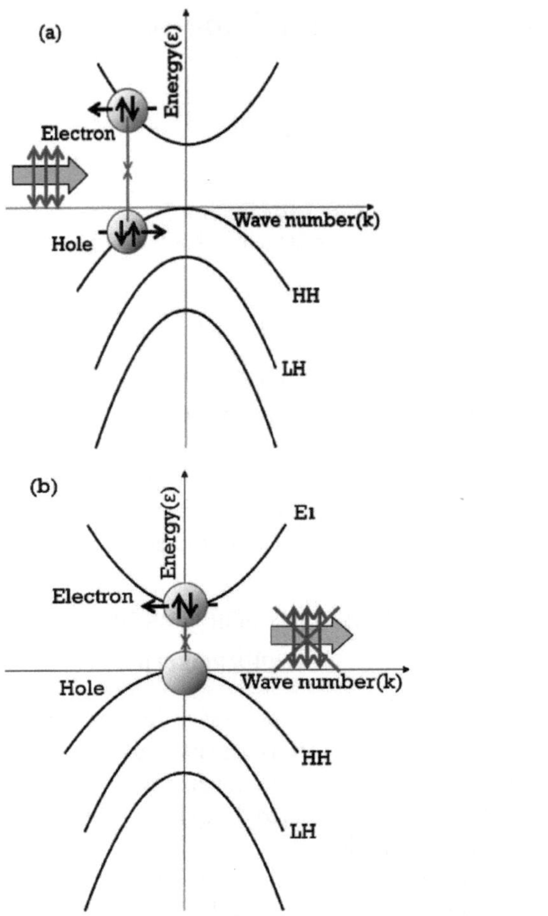

Fig. 1 Inter-band transition of electrons and holes under non-resonant excitation (a) and photoemission process through carrier recombination (b).

III. Pump-probe (PP) Measurements

A. Experimental

In this research we used GaAs/AlGaAs multiple quantum wells (MQWs). The effect of reflection, propagation and transmission is observed more clearly than when using a single quantum well. The MQWs have a layered structure that GaAs well layer is sandwiched by $Al_{0.35}Ga_{0.65}As$ barrier layers. We used a sample consists of 20 wells. Thickness of this well were 4, 8 and 12 nm. 6 nm thickness of GaAs are stacked on the surface to prevent oxidation. They were grown by MBE on a [001] GaAs substrate. Schematic representation of the layered structure is shown in Fig 2 and the experimental set-up is shown in Fig.3. In the experiment, MQW sample were illuminated by a Ti-doped sapphire laser with an excitation power of 20 W/cm^2 at 4.7 K. The pulse duration was 2 ps and the repetition rate was 80 MHz. In this research, we conducted linearly polarization resolved measurement by a PP method. The laser beam was

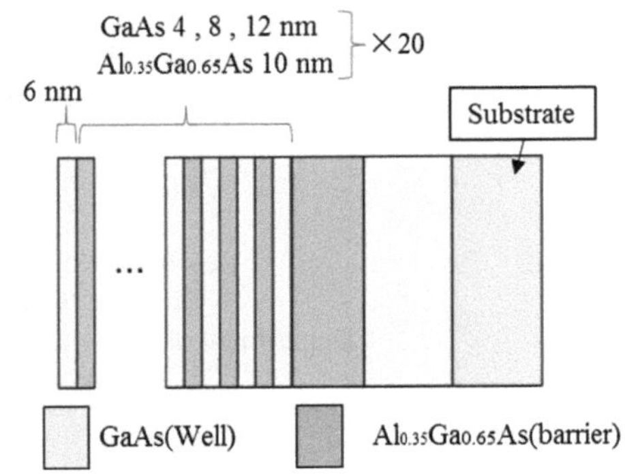

Fig. 2 Structure of GaAs/AlGaAs multiple MQWs samples.

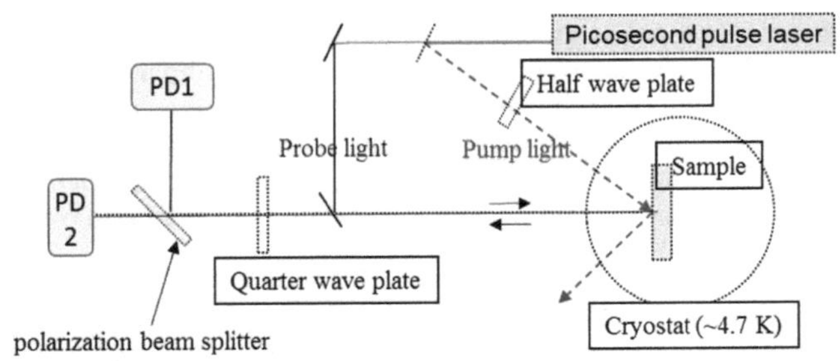

Fig. 3 The experimental set-up of pump and probe method. Reflected probe light was detected by photodiode (PD) 1 and PD 2.

divided into a pump and probe light. The probe light was focused on the sample surface where the pump light was illuminated, and the probe light was delayed from the pump light. When the spin in the sample is polarized by the pump light, the phase of reflected probe light will be changed compared to when the spin is not polarized. By measuring the phase change with respect to delay time between pump and probe pulses, we can know the time evolution of the spin states in the sample. In this experiment, differences of probe intensity between obtained from the photodiode (PD) 1 and PD 2 shows the change of the phase.

B. Results and Discussion

First, we examined wavelength dependence on output intensity. Figure 4 shows the probe intensity as a function of delay time between pump and probe beams for different-wave lengths. Each signal shows peak at zero delay time and the intensity decreased with increasing the delay time which correspond to spin relaxation time. The peak intensity show maximum when the excitation wavelength is 805.00 nm. This wavelength correspond to be heavy-hole absorption. Light-hole absorption wavelength is estimated to be 800.33 nm from Fig 5.

Next, we examined the probe power dependence on the relaxation time. Figure 6 shows time traces of spin polarization with different probe power. Slope of the line consist of

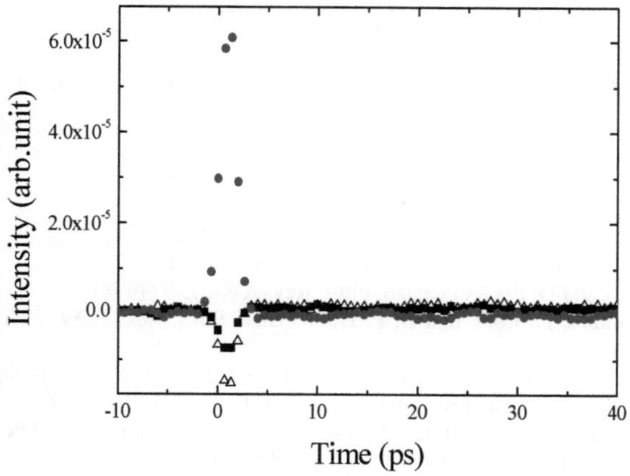

Fig. 4 Probe intensity as a function of delay time for different wavelength near the HH level. Measured wavelength was 805.40 nm (red circles), 805.20 nm (black squares) and 805.00nm (white triangles). Pump power was 60 W/cm2, probe power was 0.50 W/cm2. The well width was 12nm.

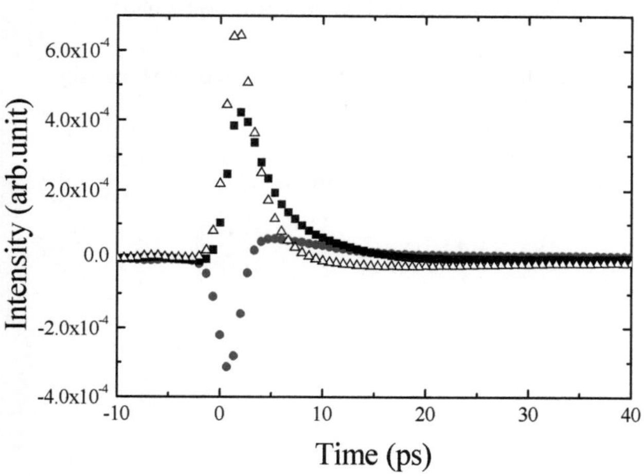

Fig. 5 Probe intensity as a function of delay time for different wavelength near the LH level. Measured wavelengths were 800.33 nm (red circles), 800.30 nm (black squares) and 800.15 nm (white triangles). Pump power was 60 W/cm^2, probe power was 0.50 W/cm^2. The well width was 12nm.

the plot corresponds to the relaxation time. Steeper slope means shorter relaxation time. We found that change of probe power did not have any influence on the relaxation time. We also examined the pump power dependence on the relaxation time. Figure7 shows time traces of spin polarization with different pump power. When the pump power was reduced, relaxation time became longer. Because, the electron and hole density was increased with increasing excitation power, interaction between electrons and holes increased in strong excitation power. Then electrons and holes interacted each other and lost polarization quickly.

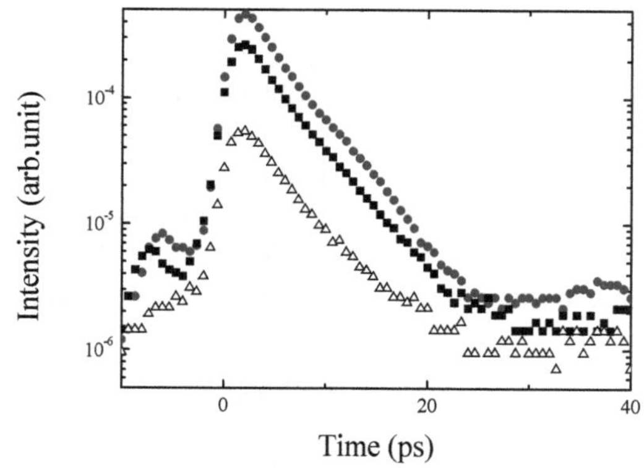

Fig 6. Probe intensity as a function of delay time for different probeop powers. Probe powers were 5.0 W/cm2 (red circles), 2.6 W/cm2 (black squares), 0.53 W/cm2 (open triangles). Pump power was 63 W/cm2. Wavelength was 805.20 nm.

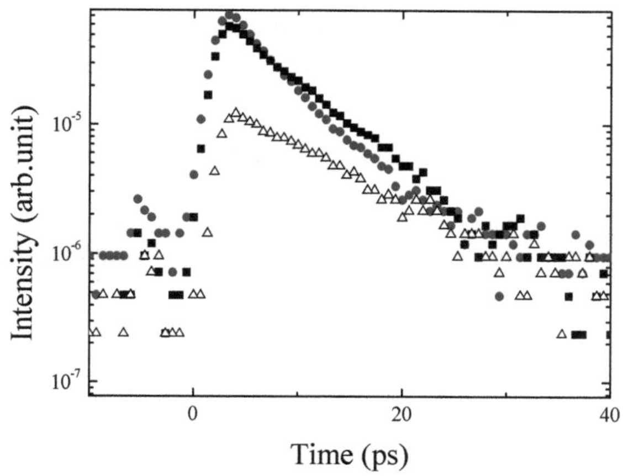

Fig 7. Probe intensity as a function of delay time for different pump powers. Pump powers were 30 W/cm2 (red circles), 6.1 W/cm2 (black squares), 1.6 W/cm2 (open triangles). Probe power was 0.50 W/cm2. Wavelength was 805.20 nm.

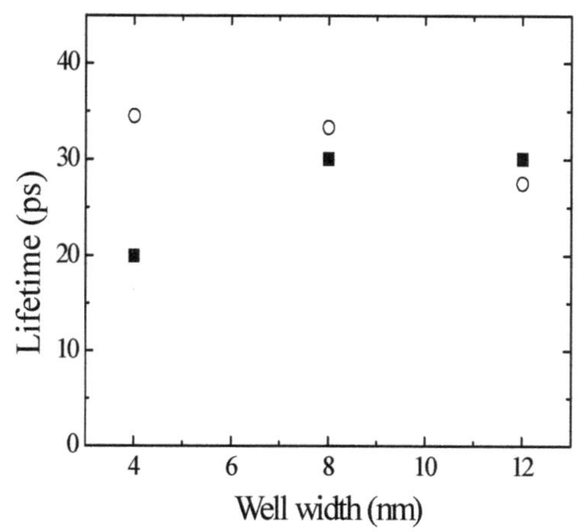

Fig. 8 Spin relaxation time dependence on the well width obtained by PL (open circles) and PP (solid squares) measurement.

Figure 8 shows dependence on the well width of spin relaxation time obtained by PL and PP measurement. The order of the spin relaxation time obtained by PP was almost the same as that by PL. The relaxation time obtained from PL measurement slightly increased with decreasing well width. Energy differences between light- and heavy-hole become large in smaller well width. So the Band-mixing effect become smaller and spin lifetime become longer [12].

IV. CONCLUSIONS

Hole-spin superposition is observed by polarization and time resolved pump-probe (PP) measurements. The results of the spin relaxation time obtained by PP measurements were almost same as those by PL measurement. Hole-spin relaxation time was increased with decreasing the excitation power. We will continue to investigate efficiencies of magnetic field and spin-spin interaction on the spin superposition.

ACKNOWLEDGMENT

A part of this work was carried out under the Cooperative Research. Project Program of the Research Institute of Electronics, Shizuoka University.

CONFLICT OF INTEREST

The authors declare that they have no conflict of interest.

REFERENCES

1. R. J. ELLiott, Phys. Rev. 96, 266 (1954)
2. M.I. Dyakonov and V.I. Perel. Phys. Lett. A 35 459 (1971)
3. G.L. Bir, A.G. Aronov, G.E. Pikus. JETP, 42, 705 (1976)
4. D. Loss and D. P. DiVincenzo, Phys. Rev. A 57, 120 (1998).
5. G. Burkard, D. Loss, and D. P. DiVincenzo, Phys. Rev. B 59, 2070 (1999).
6. A. Imamoglu, D. D. Awschalom, G. Burkard, D. P. DiVincenzo, D. Loss,M. Sherwin, and A. Small, Phys. Rev. Lett. 83, 4204 (1999)
7. T. D. Ladd, J. R. Goldman, F. Yamaguchi, Y. Yamamoto, E. Abe, and K. M. Itoh, Phys. Rev. Lett. 89, 017901 (2002).
8. S. M. Clark, K.-M. C. Fu, T. D. Ladd, and Y. Yamamoto, Phys. Rev. Lett. 99, 040501 (2007).
9. R. Vrijen and E. Yablonovitch, Physica E 10, 569 (2001).
10. H. Kosaka, H. Shigyou, Y. Mitsumori, Y. Rikitake, H. Imamura, T. Kutsuwa, K. Arai, and K. Edamatsu, Phys. Rev. Lett. 100, 096602 (2008).
11. T. Ito, H. Gotoh, M. Ichida, and H. Ando, Appl. Phys. Lett., 104, 252406 (2014).
12. R. Ferreira and G. Bastard, Phys. Rev. B 43, 9687 (1991).

Author: Soichiro Fuma
Institute: Research Institute of Electronics, Shizuoka University
Street: 3-5-1 Johoku, Nakaku
City: Hamamatsu
Country: Japan
Email: fuma.sohichiro.15@shizuoka.ac.jp

The Influence of External Cavity Optical Feedback on the Dynamics of Quantum Dots Lasers

A. Sanduta, S.S. Rusu, and V.Z. Tronciu

Department of Physics, Technical University of Moldova, Chisinau, Republic of Moldova

Abstract— **This paper reports the numerical results on the investigation of the dynamical behavior of a semiconductor laser with quantum dots active medium and a feedback from double cavity. Due to the influence of the external feedback, under the appropriate conditions, the system displays chaotic behavior appropriate for chaos-based communications. The optimal conditions and parameters for chaos generation are identified. It is found that the double cavity feedback requires lower feedback strengths for developing high complexity chaos when compared with a single cavity.**

Keywords— **quantum dots semiconductor lasers, optical feedback, pulse traces, bifurcations.**

I. INTRODUCTION

During recent years, the phenomena of self-organization and chaos have received considerable attention due to its fundamental and applied interests. The main aim of technological progress is the production of structures with stable properties and the possibility of their application in different areas. From the application point of view, chaos-based communications have become an option to improve privacy and security in date transmission, especially after the recent field demonstration of the metropolitan fiber networks of Athens [1]. In optical chaos-based communications, the chaotic waveform is generated by using semiconductor lasers with either all-optical or electro-optical feedback loops [2, 3]. In particular, synchronized chaotic waveforms have found applications in chaos based communication systems. Semiconductor lasers subject to the influence of optical feedback from a distant mirror have been investigated extensively for the past two decades and different dynamical behaviors have been characterized, including periodic and quasi-periodic pulsations, low frequency fluctuations and coherent collapse (for more details, see [4]). Optical feedback can considerably influence the dynamical behavior of a semiconductor laser Integrated lasers with ultra-short feedback cavities have also revealed similar characteristics if the feedback is properly amplified [5]. Due to the continuing technological progress, multi-section lasers with active medium quantum dots have reach stable operation. Arakawa [6] predicted that semiconductor lasers with active medium quantum dots have small temperature dependence performance than the existing semiconductor lasers, and that they will not degrade at the high temperature. Other advantages of quantum dots active medium includes the reduction of the threshold current and increase of the amplification coefficient [7]. In recent years, their dynamics has become the object of study and theoretical researches are necessary for the development and extension of the theory of nonlinear dynamics in semiconductor lasers with quantum dots active medium. Here we consider a configuration which includes feedback from an integrated double cavity.

II. MODEL AND EQUATIONS

In this paper, we focus on the investigation of the phenomenon of dynamical chaos of semiconductor lasers with quantum dots active medium shown in Fig. 1. The setup consists of a semiconductor laser operating under the influence of an external optical feedback from double cavity. The first mirror is located at distance l from the laser front facet. The distance between first and second mirrors is L. The phase φ in the air gap can be changed by a piezo-element. The optical feedback phase ψ in the second cavity can be controlled by injecting current into the passive section. We assume that the current injected into passive section is small enough to affect only the refractive index, so that the optical length of the resonator is changed in the sub-wavelength range. On the other hand, the feedback phase φ can be tuned, by the change in the delay time between the two mirrors.

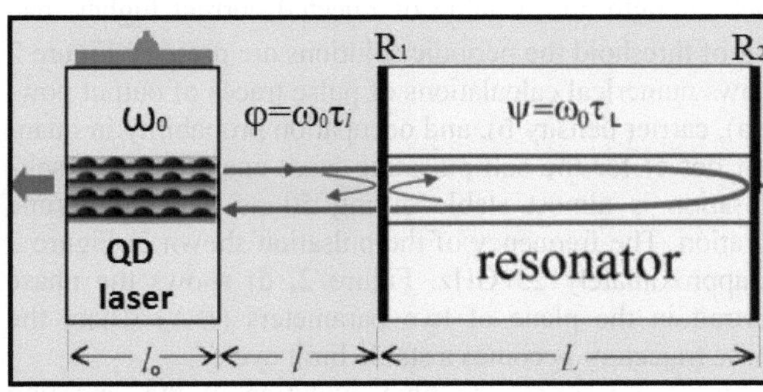

Fig. 1 Laser setup.

For modeling of the dynamics of quantum dots laser under the influence of double optical feedback we use the following equations [8, 9]

V. Sontea and I. Tiginyanu (eds.), *3rd International Conference on Nanotechnologies and Biomedical Engineering*,
IFMBE Proceedings 55,
DOI: 10.1007/978-981-287-736-9_44

$$\frac{dE}{d\tau}=\frac{1}{2}(1+i\alpha)\left[-\gamma_{np}+g(2\rho-1)\right]E+$$
$$+\Gamma_1 e^{-i\varphi}E(\tau-\tau_l)+\Gamma_2 e^{-i(\varphi+\psi)}E(\tau-(\tau_l+\tau_L)), \quad (1)$$

$$\frac{d\rho}{d\tau}=-\gamma_{ns}\rho-(2\rho-1)|E|^2+(CN^2+BN)(1-\rho), \quad (2)$$

$$\frac{dN}{d\tau}=J-N-2\left[(CN^2+BN)(1-\rho)\right], \quad (3)$$

where E is the complex amplitude of the electric field, N is the carrier density in the quantum well, and ρ is the occupation probability in the quantum dot. τ_l and τ_L are external cavity round trip times. g is the differential gain, N_d is the 2D density of dots, and J is pumping parameter. Γ_1 and Γ_2 represent the feedback levels. The parameters Γ and τ describe the feedback connection and the delay time, respectively. The constants B and C describe the transport of charge carriers through carrier-phonon interaction. α is the linewidth enhancement factor. φ is the feedback phase of air gap and ψ the phase of passive section.

III. NUMERICAL RESULTS AND DISSCUSIONS

Next we consider the case of optical feedback originated from double cavity. We have examined the dynamics of semiconductor lasers with active medium quantum dots using equations (1) - (3) and the following main parameters $\tau_1 = 0.2$, $\tau_2 = 0.5$, $\alpha = 2.0$, $\gamma_{np} = 500$, $C = 40$, $g = 1200$, $J = 20$ [10]. The parameter values are used for the calculated results that are shown in all figures of the paper.

A. Pulse Traces

In what follows we consider the pulse traces. In the case of a plane mirror for small values of the intensity of feedback strength and a value of injected current higher than that of threshold the periodic solutions are present. Figure 2 shows numerical calculations of pulse traces of output power a), carrier density b), and occupation probability in quantum dot c) for the self-pulsating laser operation. The self-pulsation is almost stable during 50 ns integration time duration. The frequency of the pulsation shown in Figure 2 is approximately 23 GHz. Figure 2, d) shows the phase portrait in the plane of two parameters (*P-N*) where the phase trajectory becomes a stable limit cycle.

Figure 3 presents the time evolution of the emerging power for the regular behavior, where $\psi = -\ \pi/2$ and the

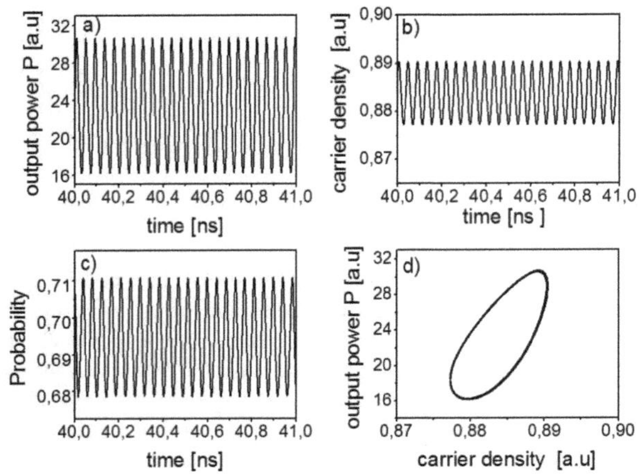

Fig. 2 Self-pulsations: Time evolution of a) output power P, b) density of carriers N and c) the occupation probability in quantum dots ρ. d) Phase portrait in the plane output power P - the density of carriers N. Other parameters: B = 0.01, Γ1 = 11, Γ2 = 20, γns = 1, φ = π/2, ψ = π.

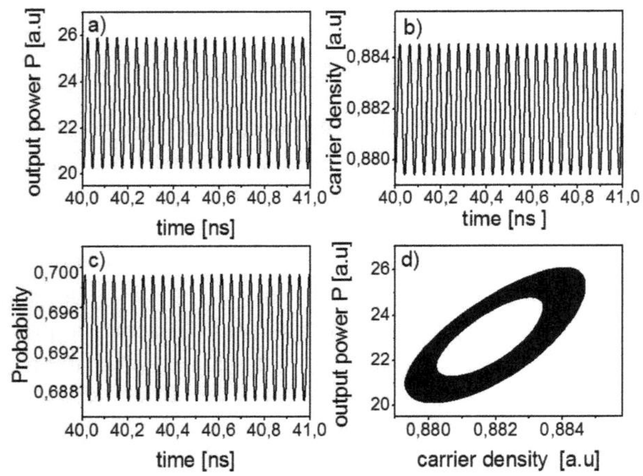

Fig. 3 Self-pulsations: Time evolution of a) output power P, b) density of carriers N and c) the occupation probability in quantum dots ρ. d) Phase portrait in the plane output power P - the density of carriers N. Other parameters: B = 0.01, Γ1 = 11, Γ2 = 20, γns = 1, φ = π/2, ψ = − π/2.

phase portrait is a slightly complicated limit cycle. Figure 4 illustrates time traces and phase portrait of a semiconductor laser under the influence of double feedback for quasi-periodic behavior for the parameters similar to those of Figure 2 except the phase φ which is fixed to $\varphi = -\ \pi/2$. One can observe the new harmonics, which appears in the system. When phase φ is changed to $\varphi = \pi$ (see Figure 5) the oscillations of the output power become more complicated and chaotic behavior appears and the phase portrait is a strange attractor.

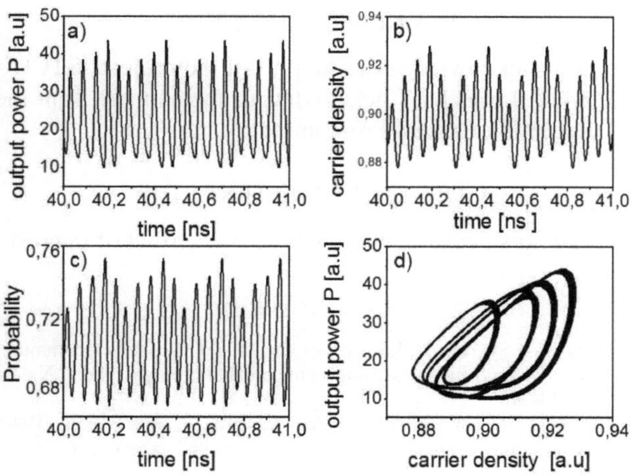

Fig. 4 Quasi-periodic behavior: Time evolution of a) output power P, b) density of carriers N and c) the occupation probability in quantum dots ρ. d) Phase portrait in the plane output power P - the density of carriers N. Other parameters: B = 0.01, Γ1 = 11, Γ2 = 20, γns = 1, φ = – π/2, ψ = π.

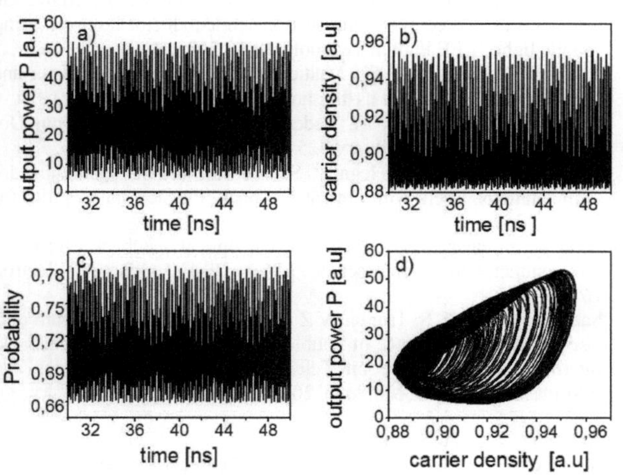

Fig. 5 Strange attractor: Time evolution of a) output power P, b) density of carriers N and c) the occupation probability in quantum dots ρ. d) Phase portrait in the plane output power P - the density of carriers N. Other parameters: B = 0.01, Γ1 = 11, Γ2 = 20, γns = 1, φ = π, ψ = π.

B. *Bifurcation Diagrams*

Next we examine the laser dynamics in terms of bifurcation diagrams. A typical calculation of bifurcation is shown in Figure 6, where the feedback strengths γ_1 are the bifurcation parameter. This figure shows the dependence of the maximum (black) and minimum (red) of the photon number on the feedback strength. When we increase the injected current, the continue wave operation is observed (see Figure 6). Then the laser begins to produce pulsations through a Hopf bifurcation marked by a circle in this figure.

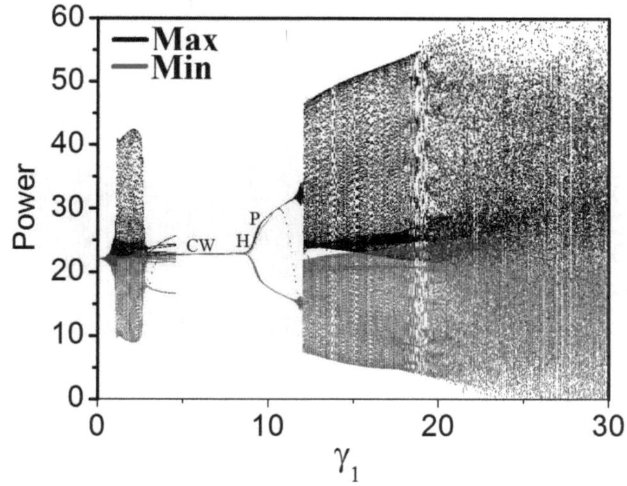

Fig. 6 Bifurcation diagram for feedback strength γ1 as bifurcation parameter. The H denote Hopf-bifurcations and P periodic solution. The parameters are B = 0.01, C = 40, g = 1200, J = 20, Γ2 = 20, γns = 1, γnp = 500, α = 2.

When we vary the parameter *g* we observe the chaotic behavior for lower values of *g* (see Figure 7). An increase of *g* lead to the appearance of continue wave operation in system.

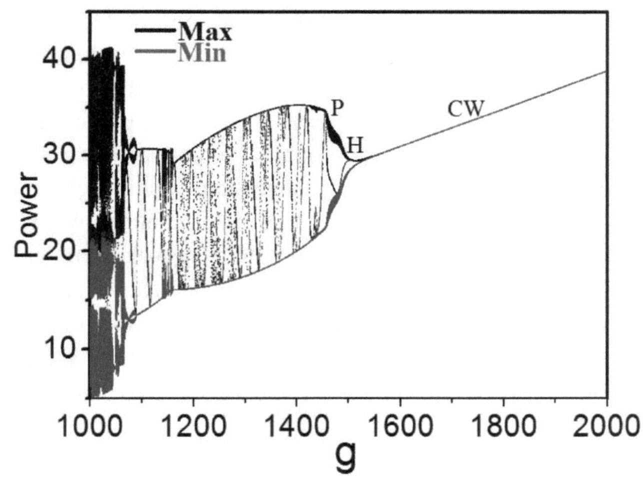

Fig. 7 Bifurcation diagram for g being a bifurcation parameter. Other parameters: B = 0.01, C = 40, J = 20, Γ1 = 11, Γ2 = 20, γns = 1, γnp = 500, α = 2.

Let consider an increase of α-factor. Figure 8 shows the bifurcation diagram when the α-factor is increased from zero to ten. For lower values of α-factor the continue wave operation is observed. For all values of α-factor higher than two only the chaotic behavior can be seen.

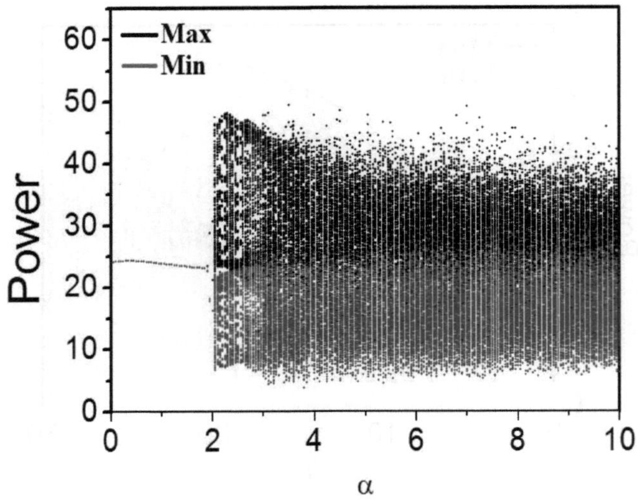

Fig. 8 Bifurcation diagram for bifurcation parameter α, and following parameters: B = 0.01, C = 40, J = 20, Γ1 = 11, Γ2 = 20, g = 1200, γns = 1, γnp = 500.

IV. CONCLUSIONS

In this paper we have studied the dynamics of a device composed by a semiconductor laser with active medium quantum dots subject to a double cavity optical feedback. Main advantages of proposed scheme include the existence of two feedback strengths, two feedback phases and two delay times that can be controlled separately. The results show the following features: under appropriate conditions the setup shown in Figure 1 is capable of generating a robust chaotic waveform within some regions. As a result continue waves, periodic and chaotic operations of output power were obtained. The bifurcation diagrams of semiconductor laser active medium under the influence of quantum dots optical feedback has been drown. Such devices are promising candidates for fast communications based on chaos.

ACKNOWLEDGMENT

Authors acknowledge the support of the projects STCU - 5993, 14.02.116F. VZT acknowledges the support from the CIM–Returning Experts Programme.

CONFLICT OF INTEREST

The authors declare that they have no conflict of interest.

REFERENCES

1. Argyris A, Syvridis D, Larger L et al. (2005) Chaos-based communications at high bit rates using commercial fibre-optic links. Nature 438:343-346
2. Mirasso C R, van Tartwijk G H M, Hernandez-Garcia E et al. (1999) Self-pulsating semiconductor lasers: Theory and experiment. IEEE J Quantum Electron 35:764–770
3. Annovazzi-Lodi V, Donati S, Scire A (1996) Synchronization of chaotic injected-laser systems and its application to optical cryptography IEEE J Quantum Electron 32:953-959
4. Krauskopf B, Lenstra D (ed) (2000) Fundamental issues of nonlinear laser dynamics. AIP Conf. Proc. 2000, pp. 548
5. Perez T, Radziunas M, Wünsche H-J et al. (2006) Synchronization properties of two coupled multisection semiconductor lasers emitting chaotic light. IEEE Photon Technol Lett 18:2135
6. Arakawa Y, Sakaki H (1982) Multidimensional quantum well laser and temperature dependence of its threshold current. Appl Phys Lett 40:939
7. Bimberg D, Grundmann M, Ledentsov N N (1999) Quantum Dot Heterostructures Britis Library 325
8. Huyet G, O'Brien D, Hegarty S P et al. (2004) Quantum dot semiconductor lasers with optical feedback. Physica status solidi (a) 201:345-352
9. Lang R, Kobayashi K (1980) External optical feedback effects on semiconductor injection laser properties. IEEE J Quantum Electron 16:347- 355
10. Sanduta A, Rusu S S, Tronciu V Z (2015) Dynamics of quantum dots lasers under the influence of double cavity external optical feedback. International Conference on Telecommunications, Electronics and Informatics, Chisinau, Moldova, 2015, pp. 62-65

Author: A. Sanduta
Institute: Department of Physics, Technical University of Moldova
Street: Bd. Stefan cel Mare 168,
City: Chisinau MD-2004
Country: Republic of Moldova
Email: sanduta.ana@gmail.com

PbTe Nanoparticles Obtaining and Studies of Their Electrical Properties

V. Nikorich[1], P. Ketrush[1], A. Nikorich[2], and A.Todosichiuc[2]

[1] Department of Physics and Engineering, Moldova State University, Chisinau, Moldova
[2] Electronics Engineering and nanotechnologies Institute D. Ghiţu of AS Republic of Moldova

Abstract— **Nano-and microparticles of PbTe: Ga (at 0.5%.) were obtained by mechanical milling. Samples of PbTe nanopowder of parallelepiped shape were made by pressing for measuring electrical parameters. The galvanomagnetic parameters were measured in the range of 77-350 K. Firstly crystalline samples were analyzed and their galvanomagnetic properties were compared with those of samples pressed from nanopowder . It was observed that the concentration of charge carriers in crystalline and pressed samples do not differ much from each other, but electrical conductivity temperature dependencies differ more essential. The same effect also affects the character of thermoelectric coefficient temperature dependence and causes its sign change.**

Keywords— **PbTe, Ga, nanopowder, galvanomagnetic properties, thermoelectric coefficient.**

I. INTRODUCTION

Recently nanotechnology and methods of nanomaterial obtaining advanced enormously, so there are a variety of methods of making nanoparticles. Depending on the work material, the quality and size of needed nanomaterial as well as the equipment and available chemical reagents, the optimal method for nanomaterial obtaining is chosen. Thus, a number of chemical and physical methods for PbTe nanocrystals manufacturing exist, for example, the hydrothermal method [1], the high temperature synthesis method from solution-phase [2], the electric explosion of [3], etc. AIV BVI compounds having high values of Bohr radius (PbS - 20 nm PbSe - 23 nm and PbTe - 46 nm) are more promising in this regard than Group AII-BVI materials or AIII BV (the latter having Bohr radii between 1 nm and 10 nm). The study of nanomaterials offer great opportunities in controlling and directing physical properties, which confirms the timeliness and importance of the research presented in the given paper.

II. EXPERIMENTAL REZULTS AND DISCUSSION

PbTe nano- and microparticles were obtained by mechanical milling. This method does not guarantee the particular nanomaterial quality (particle size varies within a wide range, their form cannot be controlled), but its simplicity gives the opportunity to get material in a quantity sufficient to realise several experiments. This provides more information about the properties of PbTe nanoparticles, which in turn allows to control their parameters. From a PbTe: Ga (0.5 at.%) crystal a few single crystal samples were prepared and subsequently a powder was prepared. For measuring the electrical parameters of obtained PbTe powder a sample of rectangular form was made from it by pressure. The powder is introduced into the press hole (hole size is 10×2 mm^2), and a pressure of 120 atm is operated on it in the hydraulic press, i.e. the hole in which the PbTe powder is containing is under the pressure of $1.6 \cdot 10^5$ atm. Thus, all of the samples made by pressing are the same as the material from which they are made and by dimensions. Fig. 1 shows the image of the compressed sample surface obtained by scanning electron microscope, the given powder particles size are clearly visible, ranging from 1 to 10 μm.

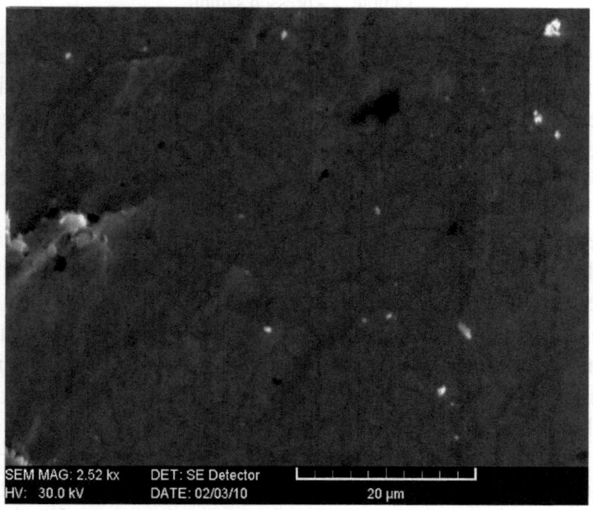

Fig. 1 SEM image of the pressed samples

To interpret the measurement results of nanomaterial samples at the first stage single crystal PbTe Ga (0.5% at.), samples were analyzed: then – then of the pressed one made of the same crystal. Measurement of electrical parameters was carried out in the temperature range 77-350 K.

All samples were of n-type conductivity. As a rule, the undoped PbTe crystals have electrical conductivity of p - type and very high value of hole concentration (higher than 10^{19} cm^{-3}), which is related to the native defects existence

V. Sontea and I. Tiginyanu (eds.), *3rd International Conference on Nanotechnologies and Biomedical Engineering*,
IFMBE Proceedings 55,
DOI: 10.1007/978-981-287-736-9_45

- Pb vacancies. Gallium creates donor impurities in PbTe, which causes the compensation of these vacancies and, depending on the concentration of introduced gallium the conductivity type of the crystals can be changed of from p- to n- type. The *lg n = f (1000/T)* dependence for crystalline samples (1) and pressed ones (2) is presented in Fig. 2.

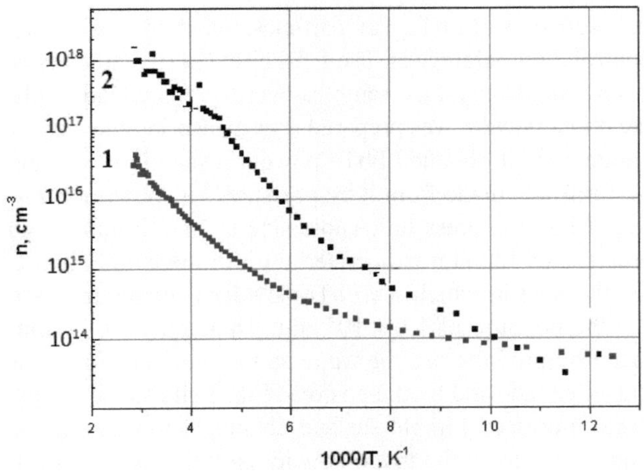

Fig. 2 Temperature dependence of the concentration for: 1 - crystalline sample, 2 - pressed sample

One can observe that for the crystals the electron concentration practically does not change with temperature increase up to T ≈ 130 K, which proves that in the band gap there are virtually no levels to participate in to the emission of free electrons. Subsequently, ranging from T ≈ 160 K the hole concentration increases more rapidly and up to 280 K the dependence is practically linear. Activation energy calculated from the curve 1 slope is 0.18 eV. By comparing the calculated activation energy value with the value of PbTe crystals band gap (E_g = 0.19 eV at 0 K), we conclude that the obtained activation energy corresponds to the intrinsic conductivity, so to the electrons transitions from the valence band into the conduction band.

For the pressed sample (curve 2) the concentration variation in the temperature range 80 -300 K is much higher, about 5 orders of magnitude compared to crystalline sample where the concentration varies by only 3 orders of magnitude. Up to T ≈120 K the curve 2 slope has a different character from those in the temperature range from 120 K to 300K where the concentration increase with temperature is faster. The slope of the two curves (1, 2) are close enough, however, the activation energy determined from this dependence for the pressed sample is E_g = 0.23 eV, so it is larger than that determined for the crystal. One can suppose that Eg increase by 40 meV for powder material is related to the appearance of surface states among crystallites and the PbO oxide formation on other hand. E_g determined from concentration temperature dependence is an average value of band gap.

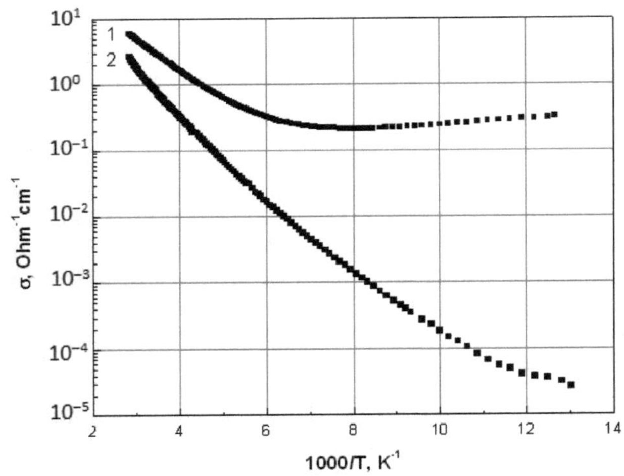

Fig. 3 Temperature dependence of the conductivity for: 1-cristalline sample, 2- pressed sample

A comparative temperature dependence of conductivity for crystalline and pressed samples is given in Fig. 3. One should note the essential difference between these two characteristics. At low temperatures (up to T ≈140 K) the crystal conductivity decreases slowly with temperature increase and beginning with T≈150K a sharp increase is observed, which indicates to the activation character of *σ(T)* dependence. From the curve 1 slope in the high temperature region the activation energy was estimated by taking into account the scattering coefficient ν.

The study of the scattering mechanism was based on the measurements of mobility μ temperature dependence, taking into account that $\mu = \sigma \cdot R_H$, where R_H is the Hall coefficient. For the crystal the mobility value reaches 26,000 cm^2/V s at 80K, then with increasing temperature first is observed a temperature region where mobility decreases very smoothly, and after T ≈ 150 K is decreasing sharply. Considering that mobility temperature dependence is given by $\mu \sim T^{\nu}$ law, the value of ν coefficient was determined, which determines the type of charge carriers scattering mechanism. . In the region of temperatures T> 150 K a value ν= -1.55 was obtained, so for crystalline sample the predominant scattering mechanism is on acoustic phonons. . As a result, the activation energy value equal to E_{g1} = 0.19 eV was obtained from the σ (t) dependence for the crystal , which is consistent with the band gap energy determined from the temperature dependence of concentration.

If in the whole temperature range the conductivity of crystalline sample varies only by two orders of magnitude and of the pressed sample conductivity vary by 5 orders of magnitude(Fig. 3, curve 2). If up to 90 K the conductivity growth feature is smoother, then further up to 350 K the conductivity growth has a very sharp increase. From the curve 2 slope the activation energy has been determined equal to 0.22 eV, which is consistent with the results described above.

The thermoelectric coefficient α (Seebeck coefficient) temperature dependencies are shown in Fig. 4. From these dependencies one should note that at low temperatures, up to ≈220 K T the *α(T)* growth for the crystal is quite sharp.

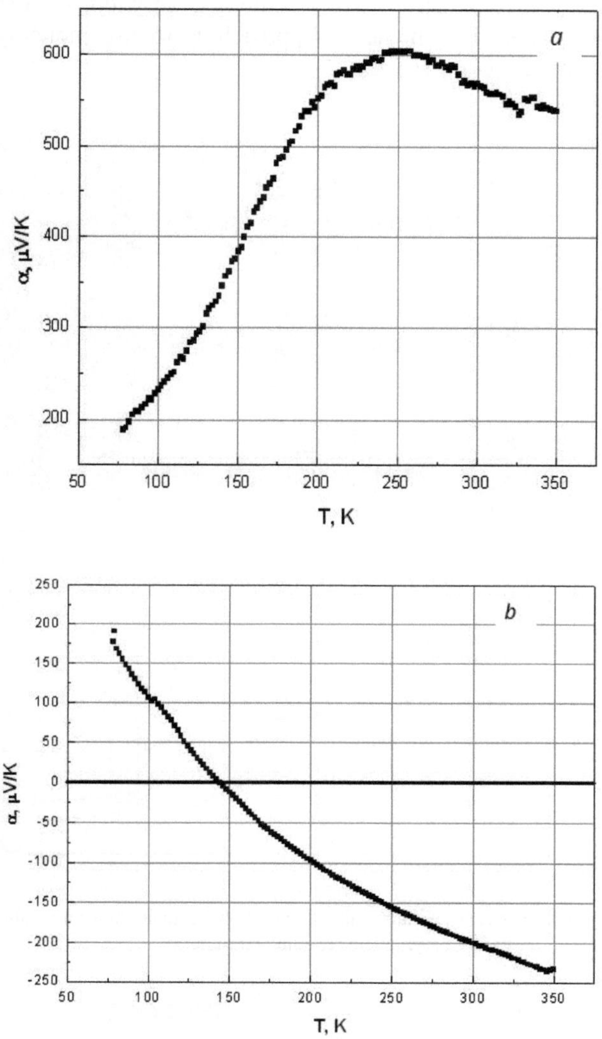

Fig. 4Temperature dependence of thermoelectric coefficient for: a) crystalline sample, b) pressed sample

Further, at T ≈ 270 K, the thermoelectric coefficient reaches a maximum, and with a subsequent temperature increase α value decreases. This is related to the appearance of the mixed conductivity region (which is confirmed also by the concentration temperature dependence) where, under the influence of the temperature gradient, the motion of the electrons as well as of the holes occur. This concomitant shift leads to mutual compensation of charge carriers flows and thermoelectric voltage potential difference is reduced, which also is observed in the graph. From the temperature dependence of the thermoelectric coefficient, the value of the electrons effective mass in the crystal was estimated. Thus, the effective mass of the electrons derived from calculations, is m_n= 0,056m_0, which is in agreement with the data from the literature [3]. One should note that for the pressed sample the behavior of α coefficient differs considerably from the one in crystal and thermoelectric coefficient value is much lower. Throughout the whole temperature range α coefficient value decreases, then at T ≈ 150K the thermoelectric coefficient becomes negative, which could be explained by the transition to mixed conductivity type.

III. CONCLUSIONS

From the presented data one can conclude that the charge carriers concentration in crystalline and pressed samples do not differ much from one another, but conductivity temperature dependencies differ more significantly. This is conditioned primarily by mobility decrease of the charge carriers in the pressed sample, which occurs at the boundaries between nanoparticles. The same effect also influences the feature of the of thermoelectric coefficient temperature dependence.

CONFLICT OF INTEREST

The authors declare that they have no conflict of interest.

REFERENCES

1. Guoan Tai, Wanlin Guo, Zhuhua Zhang.(2008) Hydrotermal synthesis and thermoelectric transport properties of uniform single-crystalline pearl-necklace-shaped PbTe nanowires. Cryst. Growth Des, vol. 8, pp. 2906-2911.
2. Gutsul T., Nicorici A. et al. (2008). High temperature solution-phase synthesis of SnTe and PbTe/SnTe nanoparticles passivated with organic shell. Proc. Int. Conf. on Materials Science and Condensed Matter Physics, Chisinau, Moldova, p. 182.
3. Akimov B.A., Dmitriev A.V, Khokhlov D.R et al (1993). Carrier transport and non-equilibrium phenomena in doped PbTe and related materials, Phys.stat.sol (a), vol. 137, p. 9-28.

Author: Valentina Nikorich
Institute: Moldova State University
Street: A. Mateevici, 60
City: Chisinau,
Country: Moldova
Email: vnicorici@yahoo.com

Plasmonic Effects for Enhanced Optical Mixing in View of THz Signal Generation

H. Hartnagel

Technical University of Darmstadt, Darmstadt, Germany

Abstract— **First various graphene realizations of electronic circuitry examples are described. These represent application examples for the new material graphene.**

Then compact THz sources are presented involving optical laser mixing, where resonant plasmonic structures at THz and optical frequencies are used to locally enhance the electromagnetic fields. Of additional particular interest is the usage here of graphene, which is optically transmitting and which is either a semimetal or which can be transformed into a semiconductor by reducing the width of its strips to several tens of nm, opening a band gap in the meV to tens of meV range. Continuous-wave THz photomixers were fabricated using Ag nanowires with diameters of 60 nm or 120 nm on low-temperature-grown (LTG) GaAs covered by graphene sheets and increased THz emission power was obtained due to these plasmonic effects.. In order to obtain acceptable power levels of such components, matrices of components in series and in parallel have to be implemented.

Keywords— **THz sources, plasmonic structures, THz photomixers.**

I. INTRODUCTORY REMARKS

Large scale graphene does not have a band gap and is a semimetal with charge carriers of zero mass at the Dirac point. A band gap can be formed by confinement in nano-ribbon or nano-constriction structures. For example, the induced bandgap by a 50 nm-wide nanoribbon is about 6 meV, a 30 nm constriction width results in about 14 meV. The charge carrier mass then increases, but it is still very small. The energy gap of graphene stripes is similar to that of Carbon NanoTubes (CNT), whose value is EG (eV)=0.8/d(/nm) (d = CNT diameter).

This presentation describes first the designs of a variety of semiconducting graphene structures where electrons (and holes) are heated appropriately by suitable stripe structures under the action of applied voltages. These electrons enter then a resonance region with two states. By dropping from the higher energy state to the lower one, energy of a THz photon is lost and the corresponding photon can be emitted. With an external resonator a laser can be realized, where the graphene nanostructures act as active medium, similar to the quantum cascade laser. Similarly, non-linear mixers are proposed for THz applications.

II. DESIGN OF OTHER TYPICAL GRAPHENE THZ SOURCES

To obtain a ballistic graphene resonator, we propose the following structure: narrow width – large width – narrow width (Fig. 1). The usual approach is to use monolayer graphene.

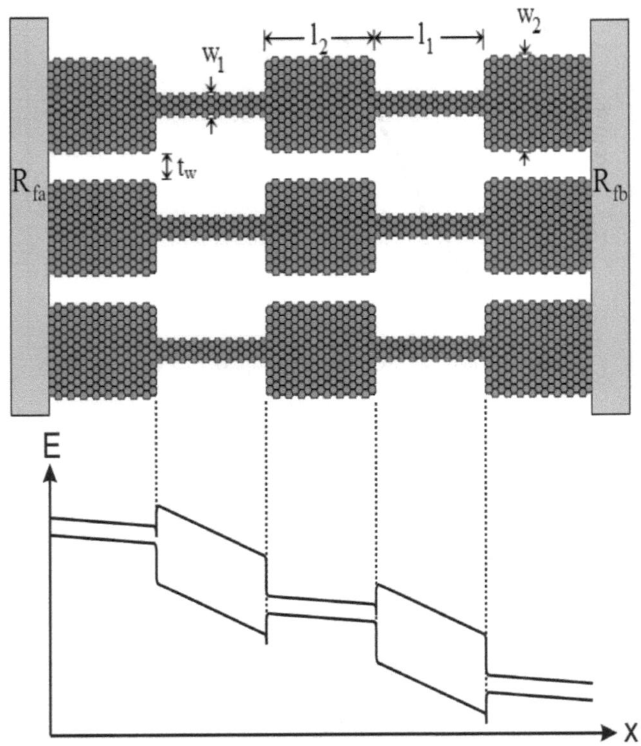

Fig. 1 Real space (top) and band diagram (bottom) under bias for a ballistic grapheme resonator structure.

III. FABRICATION TECHNOLOGY

The structure has to be fabricated by electron-beam lithography in the shape of a matrix of repetitions along x and y directions in order to increase the signal amplitude. In this way, a sheet of graphene structures with a large amount of THz ballistic resonances or laser transitions is possible and

V. Sontea and I. Tiginyanu (eds.), *3rd International Conference on Nanotechnologies and Biomedical Engineering*,
IFMBE Proceedings 55,
DOI: 10.1007/978-981-287-736-9_46

can be combined with an overlaying antenna structure for good emission efficiencies. Our design of such a matrix gives estimates on the signal power to be achieved. The external THz resonator can be an optical or an electrical cavity. A number of design solutions are to be presented.

IV. PLASMONICS AND GRAPHENE FOR THZ GENERATION BY OPTICAL MIXING

Compact THz sources involve optical laser mixing, including resonant plasmonic structures at THz and optical frequencies to locally enhance the electromagnetic fields. Of particular interest is the usage here of graphene, which is optically transmitting and which is either a semimetal or which can be transformed into a semiconductor by reducing the width of its strips to about 30 nm, opening a band gap in the meV to tens of meV range.

Continuous-wave THz photomixers are fabricated using 1D and 2D nanocontacts either on low-temperature-grown (LTG) GaAs or on nitrogen ion-implanted (N+i) GaAs and graphene sheets. The overlaying 1D and 2D nanocontacts were formed by silver nanowires with a diameter of 60 nm or 120 nm. They can handle currents of > 10 mA and > 30 mA, respectively, without electromigration, enabling reliably high photocurrents and both field enhancement at THz by plasmonic effects.

Resonant plasmonic structures can be made which locally enhance the electromagnetic field both for THz as well as for optics by the combination of semiconductor-graphene plasmons respectively by the semiconductor-metal-nanostructure plasmons. Plasmon polaritons are electromagnetic waves coupled to charge carriers at these interfaces.

The frequency ranges of THz and optics have numerous interactive involvements such as THz signal generation by optical mixing of two light frequencies, or of broad-band THz communication modulation. The opportunities of both THz and optical waves to be enhanced by interaction with plasmonics are well known.

Here then an experimental effort is undertaken correspondingly. To enhance optical penetration into the nonlinear semiconductor such as GaAs, Ag nanometric-sized wires are employed. The use of graphene as an optically transparent and electrically conducting layer is additionally of interest due to the existence of relevant THz plasmon polaritons.

The material low-temperature-grown (LTG)-GaAs is covered on one of the open wafer surfaces by a graphene sheet. In particular, this allows electrical conduction and optical transmission. On top of this Ag wires are deposited of 60 nm or 120 nm diameter. These permit the creation of THz antennas.

The generation of plasmonic resonances in graphene is possible at THz frequencies. This is due to the semi-metal properties of graphene.

Metal electrodes of very narrow conducting lines allow the use of optical-frequency light for THz generation in the LTG or N+i GaAs for optical illumination. It is necessary, not to have only a few of such wires spread over the semiconductor structure, but to space them in an appropriate regularity. This requires the separation of the wires from each other by careful consideration of the plasmonic distances outside the wire, particulary into the semiconductor. The existence of an intermediate thin graphene layer of one or even several atomic thicknesses does not greatly influence these effects. The generation of THz signals is highly successful and exeeds the results with similar structures without graphene and nanometric-sized wires.

V. CONCLUSION

Graphene possesses intrinsic plasmons that are tunable and adjustable and the combination of graphene with semiconductors promises a variety of exciting applications. Another plasmonic application is the uses of Ag wires of nanometric dimensions to enhance the optical penetration into the semiconductor.

The versatility of graphene means that graphene-based plasmonics may enable the manufacture of novel THz devices.

CONFLICT OF INTEREST

The authors declare that they have no conflict of interest.

Author:	Hans Hartnagel
Institute:	TU Darmstadt
Street:	Merckstr. 25
City:	D 64283 Darmstadt
Country:	Germany
Email:	hartnagel@mwe.tu-darmstadt.de

Features of Nanotemplates Manufacturing on the II-VI Compound Substrates

G.V. Colibaba[1,2], E.V. Monaico[3], E.P. Goncearenco[1], I. Inculet[1], and I.M. Tiginyanu[3]

[1] Moldova State University, A. Mateevici 60, MD-2009, Chisinau, Republic of Moldova
[2] Kazan Federal University, Kremlevskaya 18, 420008, Kazan, RF
[3] Technical University of Moldova, Stefan cel Mare 168, MD-2004, Chisinau, Republic of Moldova

Abstract— **Application of ZnSe, ZnS, ZnSSe, CdS, ZnCdS and ZnO single crystal substrates for the preparation of nanoporous matrices by electrochemical etching using various electrolytes is analyzed. We demonstrate prospects of using ZnSe and ZnCdS compounds for the fabrication of nanopore arrays with pore diameter down to 30 nm, as well as of ZnO substrates for the preparation of nanohills or nanopits arrays. The limitations for producing similar structures on the basis of ZnS and ZnSSe substrates are evidenced.**

Keywords— **II-VI semiconductor compounds, anodic etching, nanotemplates.**

I. INTRODUCTION

Wide band-gap II-VI semiconductors, such as ZnSe, ZnS and ZnO, are perspective materials to be used as the substrates for nanoporous matrices (NM) (nanotemplates). The nanotemplates have wide potential for fabrication of nanowires and nanotubes from various materials with defined values of diameter and length, as well as nanodots [1, 2]. These nanostructures are interesting for contact phenomena investigation between nanotube materials and nanoporous semiconductor matrix. These structures are promising for use in high-efficient solar cells with nanostructured p-n junctions, photonic elements, electronic sensors etc. [3, 4].

The easiest and most cost-effective method for NM manufacture is electrochemical etching (ECE). The dielectric nanotemplates from Al_2O_3 may be used only for manufacturing of nanowires and require special high-precision technology [5]. However, using ECE technology, it is necessary to manipulate with electrical parameters of initial material (substrate) which allows controlling the pore diameter [1, 2].

The present paper is devoted to review the manufacturing methods of nanoporous structures based on ZnSe, ZnS, CdS, ZnO single crystal substrates and ZnSSe, ZnCdS solid solutions with controlled electrical parameters.

II. EXPERIMENTAL TECHNIQUE

The growth of II-VI semiconductors and solid solutions on their basis was carried out from vapor phase in sealed quartz ampoules preliminary evacuated down to 10^{-4} torr. The ZnSe and ZnS crystals were grown by seeded physical vapor transport (PVT) method, in vacuum [6,7]. The ZnO, ZnSSe, CdS and ZnCdS crystals were grown by chemical vapor transport (CVT) method using HCl (H_2+HCl gas mixture for ZnO) as a transport agent [8,9]. High resistive as-grown ZnSe and ZnS crystals were doped by thermal treatment in Zn+Al melt [7]. To reduce the electrical resistivity, the ZnSSe:Cl and ZnCdS:Cl samples were annealed in Zn or Zn+Cd vapors, respectively [9]. Initial high conductivity of obtained CdS:Cl and ZnO:Cl samples is caused by shallow Cl donors and low compensation degree.

Electrical parameters (concentration of the charge carriers (n) and conductivity (σ)) were calculated from the Hall Effect measurements using a six-probe method with In contacts. Electrochemical etching was carried out at room temperature, in the dark, in various electrolytes. Anodization was performed in a potentiostatic regime [10]. Applied voltage between sample and Pt electrode was varied from +5 up to +160 V depending on the substrate conductivity. A Scanning Electron Microscope (SEM) Zeiss Sigma and Tescan Vega 5130 equipped with an Oxford Instruments INCA Energy EDX system were used to study the morphology of samples.

III. RESULTS AND DISCUSSION

A. ZnSe

Depending on the charge carrier concentration in the ZnSe:Al substrates (varied in the range of $2\cdot10^{15}$ – $2\cdot10^{18}$ cm^{-3}), using $K_2Cr_2O_7$:H_2SO_4:H_2O electrolyte with the ratio 5:100:10 allows to obtain nanopores with medium diameter from 400 down to 40 nm (for the most conductive material) (Fig. 1). The width of the nanopore walls correlates with the pore diameter, as a result, the pore diameter is about the same in all the samples [11]. The depth of porous layers up to 500 μm has been successfully achieved.

Preliminary results of nanostructuring of commercially available ZnSe single crystals grown either from the melt or by CVT method with the use of halogens have shown poor uniformity both electrical properties of this material and obtained nanoporous (Fig. 1(insert)). In the last case, probably, formation of halogen precipitates takes place; as

V. Sontea and I. Tiginyanu (eds.), *3rd International Conference on Nanotechnologies and Biomedical Engineering*,
IFMBE Proceedings 55,
DOI: 10.1007/978-981-287-736-9_47

essential shift of the edge PL band caused by exciton-impurity complexes to higher energy is observed in the case of ZnSe crystals doped by halogens. The use of electrolytes based on HCl, H_3PO_4 and HNO_3 does not give positive results. The influence of crystallographic orientation of the substrates on the anodic etching was not observed.

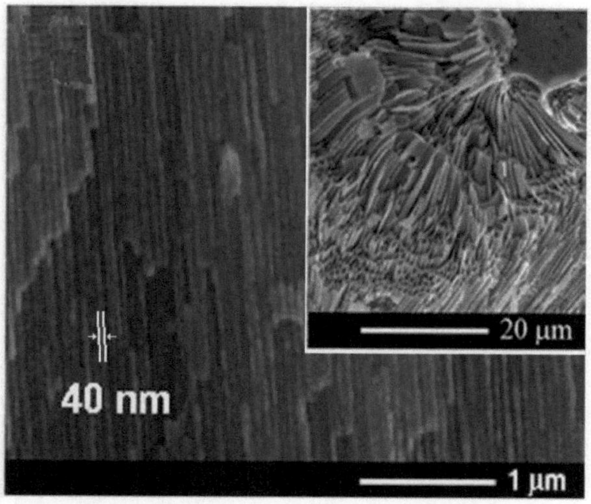

Fig. 1 SEM image of porous ZnSe:Al substrate obtained at applied voltage 5 V in cross-section; the insert illustrates porous ZnSe:I substrate.

B. ZnS and ZnSSe

The optimally doped ZnS:Al and ZnS_xSe_{1-x}:Cl (x = 0.3 – 0.5) samples have values of $n \sim 2{\cdot}10^{16}$ cm^{-3} and $\sigma \sim 0.3$ $(\Omega{\cdot}cm)^{-1}$ [7]. However, these materials are characterized by enormous surface resistance (~ $M\Omega{\cdot}mm^2$ for ZnS) due to the surface levels caused by the raised conduction band edge. Nanostructuring of these materials in various electrolytes (HCl, HCl:H_2SO_4, HCl:H_3PO_4) was unsuccessful because of extremely low values of current due to high surface resistance. As a result, ZnCdS solid solution crystals are of great interest, because they may have wider band-gap in comparison with ZnSe and good electrical parameters necessary for efficient electrochemical treatment.

C. CdS and ZnCdS

High conductivity of the CdS:Cl and $Zn_xCd_{1-x}S$:Cl (x = 0 – 0.5) samples (varied in the range of 20 – 100 $(\Omega{\cdot}cm)^{-1}$) and low surface resistance [9] gave the possibility to obtain uniform porous layers with thickness of 50 µm, the pore diameter down to 30 nm and the wall thickness of 20 nm, at the use of aqueous HCl electrolyte and relative low applied voltage ≤ 40 V [12] (Fig. 2).

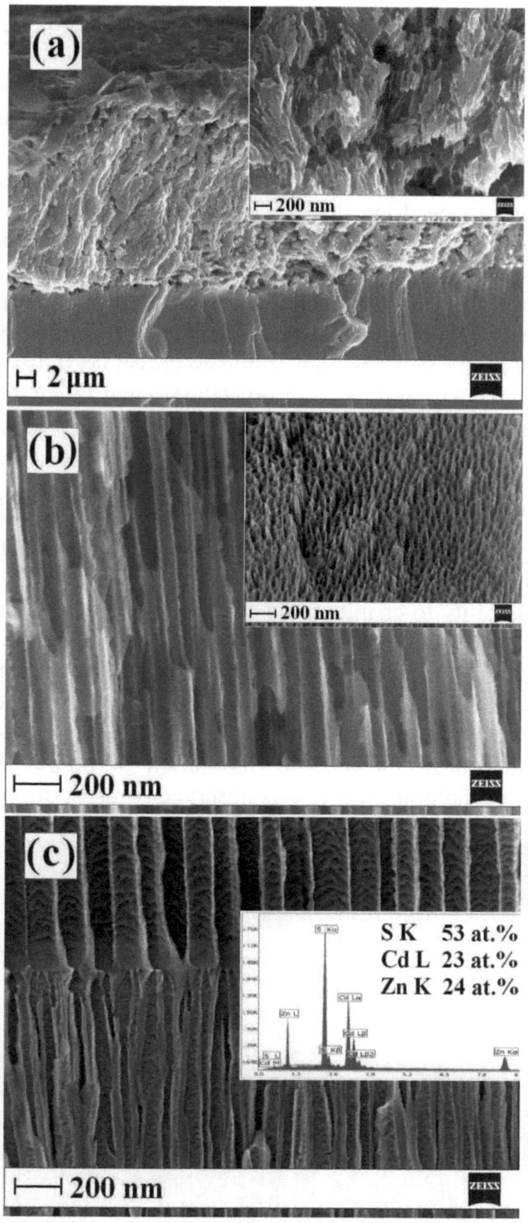

Fig. 2 SEM image of porous Zn0.6Cd0.4S layer anodized in 5 % HCl aqueous electrolyte at applied voltage 160 V (a) and 0.3 M HNO3 at 70 V (b) (the inserts illustrate magnified view (a) and the top view (b) of the porous layer). SEM image taken from Zn0.6Cd0.4S multilayer porous structure in cross-section obtained at switching of applied voltage from 40 to 30 V (c) (the insert illustrates EDX analysis of composition for porous Zn0.5Cd0.5S layer).

Porosification of more wide band-gap material $Zn_{0.6}Cd_{0.4}S$ ($E_g \sim 3.1$ eV) characterized by higher surface resistance of about 300 $\Omega{\cdot}mm^2$ can be realized only at extremely high voltage of 160 V, in aqueous HCl electrolyte. Porous layers fabricated at this high voltage are non-uniform in both the propagation directions of pores and the diameter values (Fig. 2(a)).

Anodization in HNO_3 aqueous electrolyte gives the possibility to reduce the anodization voltage in the case of $Zn_{0.6}Cd_{0.4}S$ crystals from 160 to 70 V; essentially more homogeneous porous layers have been successfully fabricated (see Fig. 2(b)). The use of HNO_3 aqueous electrolyte can be also useful for manufacturing multilayer porous structures by changing the applied potential during anodic etching, in one technological process (Fig. 2(c)). The compositions of porous layers coincide, within experimental error, with the compositions of the as-grown crystals (Fig. 2 (c), insert). The influence of crystallographic orientation of the substrates on the obtained porous layers was not observed.

D. ZnO

Variation of the loaded HCl pressure (0.1-1.75 atm at growth temperature) for growth of CVT ZnO crystals give the possibility to change the n and σ values in the ranges of $(4 - 12)\cdot 10^{17}$ cm^{-3} and 4 – 20 $(\Omega\cdot cm)^{-1}$ respectively. High conductivity of these materials is suitable for anodic etching. However, chemical instability of ZnO substrates in acid electrolytes limits both, the duration of ECE and the depth of porous layers; the dissolution of porous layers during technological process is observed (Fig. 3).

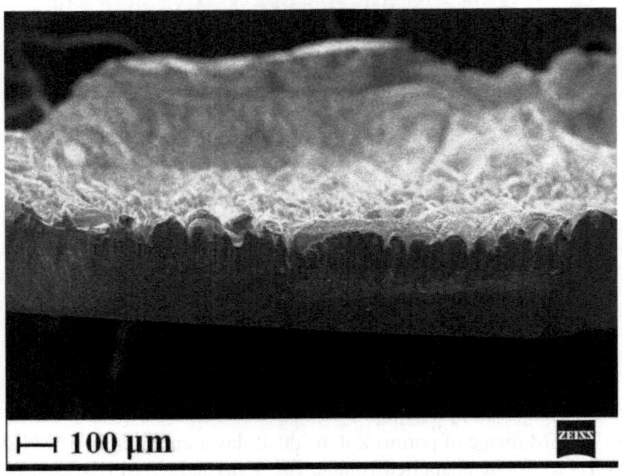

Fig. 3 Example of dissolution of porous ZnO:HCl samples anodized during 10 min in 5 % aqueous HCl electrolyte at 10 V.

The essential influence of crystallographic orientation of the polar substrates on the anodic etching and morphology of the nanostructured surface is observed. In the case of [0001] Zn and [000$\bar{1}$] O surface, hexagonal pits with depth of 1-5µm and hexagonal hills with diameter of 1-5µm can be obtained, respectively, using aqueous HCl electrolyte (Fig. 4(a,b)).

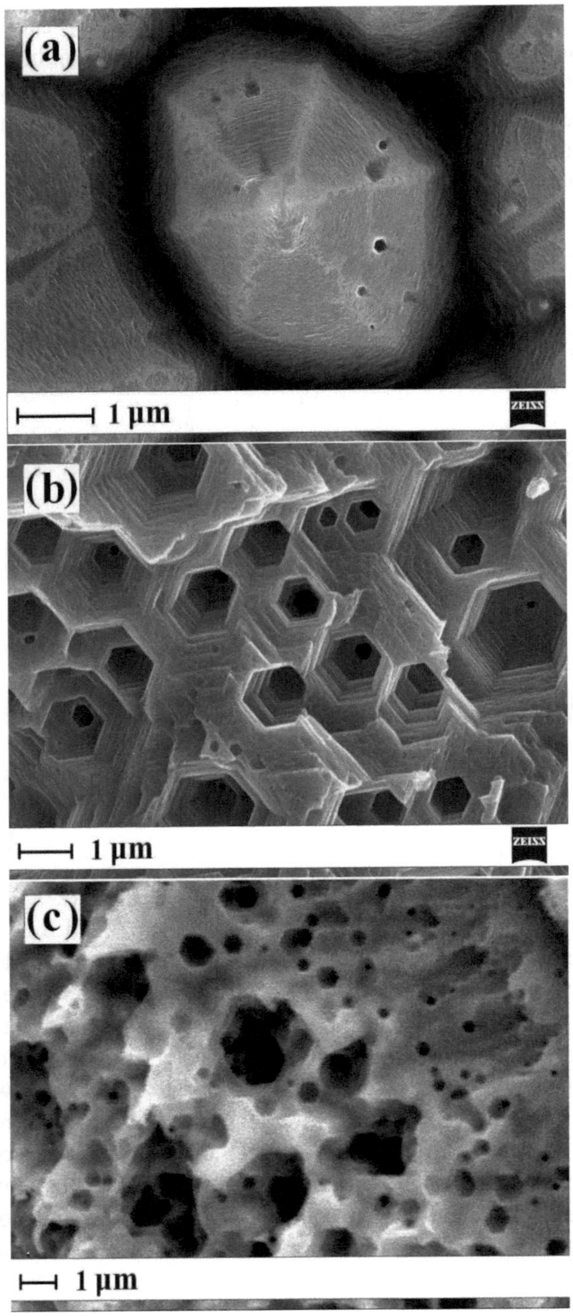

Fig. 4 SEM image of [000$\bar{1}$]O (a) and [0001]Zn (b) surface anodized during 2 min in 5 % aqueous HCl electrolyte at applied potential 5 V, and of [0001]Zn surface anodized at 40 V (c).

Low applied voltage (5-8V) and the best uniformity of the nanostructured surface are achieved on the surface cleaned by high temperature annealing at the optimal conditions. As-polished samples are characterized by high surface resistance; necessity of using higher applied voltage up to 40 V usually causes deterioration morphology of the nanostructured surface (Fig. 4(c)).

At the same time, uniformity of porous structures on the ZnO crystals is lower than that of other II-VI materials; is attributed to chemical instability of ZnO and creates a background for the further investigations and technology development.

IV. CONCLUSIONS

It was shown that porous layers with pore diameters from 400 down to 40 nm can be obtained on the basis of homogeneous ZnSe:Al substrates grown by PVT method. The use of ZnS and ZnSSe is limited by depletion surface layers of these materials. The ZnCdS single crystals grown by CVT method, with the band-gap energy larger than that of ZnSe, represent a promising material for nanotemplates with the pore diameter down to 30 nm. Nanostructuring of ZnO material is limited by chemical instability to acid solutions, nevertheless arrays of nano-micro hills and pits can be obtained. For the first time, the possibility of manufacturing multilayer porous structure on the ZnCdS substrates with controlled composition and morphology is demonstrated. The influence of crystallographic orientation of investigated II-VI compound substrates and their surface resistance on the ECE processes is discussed.

ACKNOWLEDGMENT

This work was supported by the Supreme Council for Science and Technological Development of the Academy of Sciences of Moldova under the project No. 14.819.02.14A, and partially by the Russian Government (agreement No.02.A03.21.0002) to support the Program of Competitive Growth of Kazan Federal University among World's Leaing Academic Centers. E.V. Monaico thanks to Alexander von Humboldt Foundation for support.

CONFLICT OF INTEREST

The authors declare that they have no conflict of interest.

REFERENCES

1. Monaico E. et al. (2009) ZnSe-based conductive nanotemplates for nanofabrication. Phys. Stat. Sol. RRL 3:97-99
2. Tiginyanu I., Monaico El., Monaico Ed. (2008) Ordered arrays of metal nanotubes in semiconductor envelope. Electrochemistry Communications 10:731-734
3. Jie J.S. et al. (2010) One-dimensional II–VI nanostructures: Synthesis, properties and optoelectronic applications. Nanotoday 5:313-336
4. Sergentu V.V. et al. (2007) Design of negative-refractive-index materials of the basis of rods with a gradient of the dielectric constant. Appl. Phys. Lett. 91:081103
5. Martin C.R. (1994) Science 266:1961
6. Colibaba G.V. et al. (2008) Growth technology of ZnSe single crystals having a low density of dislocation. Mold. J. Phys. Sci. 7:26-31
7. Colibaba G.V. et al. (2012) Growth of wide band-gap II-VI compound substrates with controlled electrical parameters. Mold. J. Phys. Sci. 11:361-365
8. Colibaba G.V. et al. (2012) Growth of ZnO crystals by CVT methods. Mold. J. Phys. Sci. 11:366–370
9. Colibaba G.V., Monaico E.V. et al. (2014) Growth of ZnCdS single crystals and prospects of their application as nanoporous structures. Semiconductor Science and Technology 29:125003
10. Tiginyanu I.M., Monaico E. et al. (2007) Electrochem. Solid State Lett. 10:D127
11. Tiginyanu I.M. et al. (2011) Quasi-Ordered Networks of Metal Nanotubes embedded in Semiconductor Matrices for Photonic Applications. Journal of Nanoelectronics and Optoelectronics 6:1–10
12. E. Monaico, G. Colibaba et al. (2014) Porosification of III–V and II–VI Semiconductor Compounds. Journal of Nanoelectronics and Optoelectronics 9:307-311

Author: MONAICO EDUARD
Institute: TECHNICAL UNIVERSITY OF MOLDOVA
Street: STEFAN CEL MARE 168, MD-2004
City: CHISINAU
Country: MOLDOVA
Email: m_eduard_y@yahoo.com

Excitonic Luminescence, X-ray Analysis and Local Band Structure of Chlorine Intercalated 2H- and 3R-MoS2 Polytypes

S. Anghel[1], Yu. Chumakov[1], A. Colev[1], V. Kravtsov[1], L. Kulyuk[1], C. Mamaliga[1], A. Mitioglu[1,3], K. Sushkevich[2], and G. Volodina[1]

[1] Institute of Applied Physics, Chisinau, Republic of Moldova
[2] State University of Moldova, Chisinau, Republic of Moldova
[3] Laboratoire National des Champs Magnétiques Intenses, Toulouse, France

Abstract— **Photoluminescence and X-ray diffraction are used to investigate 2H and 3R polytypes of MoS2 layered crystals intercalated with Cl2 molecules. The spectral structure of the observed bound excitons emission is individual for each polytype, that can be used for unambiguous identification of the 2H and 3R polytypes. The DFT band structure calculations have been done for three molecular layers of 2H-MoS2 and 3R-MoS2 polytypes without and with intercalation of Cl2 molecules. The spectroscopic experimental results confirmed theoretical calculations performed for both cases and it has been demonstrated that halogen intercalation completely changes the local band structure of 2H- and 3R-MoS2 polytypes.**

Keywords— **molybdenum disulphide, polytype, luminescence, bound excitons, halogen molecules.**

I. INTRODUCTION

In the past years, molybdenum disulphide (MoS_2) experienced a lot of investigations concerning optical, electrical, magnetic, mechanical and other properties, which is reflected in an essential growth of number of papers related to this topic [1]. Many useful properties of this material arise from the laminar structure, which allows one to obtain graphene-like MoS_2 layers [1-6]. Atoms within a single molecular MoS_2 layer are bound by strong ion-covalent forces, while the layers are bound by weak van der Waals interactions. The bulk material has an indirect band gap (1.2eV) [7], whereas a single layer film is a direct gap (1.9eV) semiconductor [2, 5]; the bandgap being an advantage over that of graphene and is a promising material for digital electronics and solar-energy conversion [6, 8]. Both good matching between the solar spectrum and the optical band gaps of molybdenum dichalcogenides and high resistance against photo corrosion provoke profound interest in their application in electrochemical and solid state solar cells. Moreover, the laminated structure of these materials open wider opportunities for fine tuning of their electronic properties by interlayers intercalation of different guest atoms or molecules [9], or by forming low-dimensional nanostructured materials [10, 11].

There are three crystal modifications of MoS_2. In 1T-MoS_2, molybdenum atoms are coordinated octahedrically by sulphur atoms [12], while in two polytypes 2H-MoS_2 and 3R-MoS_2, S atoms form the trigonal prismatic coordination around Mo atom. 2H-MoS_2 crystallizes in the $P6_3/mmc$ (D^4_{6h}) space group, while 3R-MoS_2 – in the $R3m$ (C^6_{3v}) space. Those two polytypes are built up by stacking of identical layers and differ only in their stacking sequence ABAB and ABCABC, respectively.

II. EXPERIMENTAL PROCEDURE

The synthetic 2H and 3R-MoS_2 single crystals were grown by means of the chemical vapour transport method (CVT), using Mo and S as starting materials. The $MoCl_5$ compound, which decomposes at high temperatures, was used as a source of Cl_2 molecules for the chemical vapour transport realized in a two-zone tube furnace. The two polytypes were obtained by choosing a different temperature gradient and a different concentration of the transport agent, which seems to have an important influence on the growth process.

The 2H- and 3R- polytypes were identified and structurally characterized by the single crystal X-ray method. The polymorphic purity of the as synthesized 2H- and 3R-polytypes (bulky samples) was confirmed by similarities between simulated and experimental X-ray powder diffraction patterns. The X-ray diffraction data were obtained at room temperature using an *Xcalibur E* diffractometer.

The photoluminescence (PL) spectra were registered using a standard lock-in technique. All spectra were corrected for the characteristics of the optical system. The excitation of the PL was provided by the second harmonic emission of a *cw*-operating YAG: Nd (λ = 532nm). The PL IR emission was detected by a cooled Hamamatsu photomultiplier tube (NIR-PMT Module). The samples were placed in a cryostat and the PL spectra, collected from the 001 plane, were registered in a broad temperature range (10 – 100 K).

V. Sontea and I. Tiginyanu (eds.), *3rd International Conference on Nanotechnologies and Biomedical Engineering*,
IFMBE Proceedings 55,
DOI: 10.1007/978-981-287-736-9_48

III. THEORETICAL CALCULATIONS

Since the X-ray single crystal study did not reveal any essential changes in the unit cell parameters in both studied polytypes, it was supposed that the concentration of Cl_2 molecules in crystals should be low and, as a consequence, that halogen molecules disturb the crystal lattice only locally. Due to the lack of structural information concerning of chlorine molecule position in the samples and to interpret the obtained experimental results and understand the effect of halogen molecules on the electronic structure of the material the DFT band structure calculations were carried out for the surfaces of 2H-MoS_2 and 3R-MoS_2 crystals. It is known also that in some layer-type crystals there is close similarity between the density of states of the bulk and a single-layer slab representing the surface parallel to layers [13]. The studied surfaces were modeled by three molecular layers of polytypes that are without and with intercalation of Cl_2 molecules (models I and II respectively). For model I, three layers and corresponding atomic coordinates were taken from the structure of bulk crystals. In model II we assume that Cl_2 molecules intercalated in the van der Waals gap only between two layers and move them apart while another interlayer gap remains unchanged. Possible positions of Cl_2 molecules intercalated between MoS_2 layers were chosen on the base of the bulk crystal structure analysis using PLATON tools [14]. The crystal packing of layers revealed two types of interstitial cavities in the interlayer of the van der Waals gap. These cavities are similar in both polytypes and formed by six or four surrounding sulphur atoms and have the shapes of a trigonal antiprism and a trigonal pyramid, respectively.

Although chlorine molecule has similar nearest surrounding, further environment in polytypes differs due to unlike mutual arrangement of MoS_2 layers. Even such small differences in halogen positions may affect the emission spectra of excitons bound to the halogen molecules.

Self-consistent ground-state calculations were performed with the ABINIT code [15] to obtain the detailed electronic structures. Electronic calculations were performed in the generalized gradient approximation with the Perdew–Burke–Ernzerhof exchange–correlation energy functional [16].

IV. RESULTS AND DISCUSSIONS

Halogen excitons related luminescence spectra at different temperatures of two different polytypes are displayed in Figs. 1a and 1b. The PL spectrum of 2H polytype involves two zero-phonon excitonic lines *B* and *C* (here the same notation is used as in [17] where the emission of the 2H polytype was in details investigated) followed by phonon replicas (B_1, B_2 and C_1, C_2). Line *A* that is very close to the *B*-line can be observed only at *T*~2K. As the temperature is increasing the intensities of *B* and *C* lines are redistributed in favour of line *C* (Fig. 1a). In contrast to the previous case, the halogen bounded exciton emission in the case of 3R-polytype is reported for the first time. There are some obvious differences between them, firstly the *B* and *C* lines are shifted to lower energies by 3.5meV and 6meV, respectively and, secondly, the temperature redistribution of the intensity between these lines is insignificant (Fig. 1b). For both polytypes the increase of temperature above 60K leads to a rapid quenching of luminescence. An analysis of the observed temperature behaviour of the excitonic spectral lines (for 2H- polytype) can be found in [17, 18 and references therein].

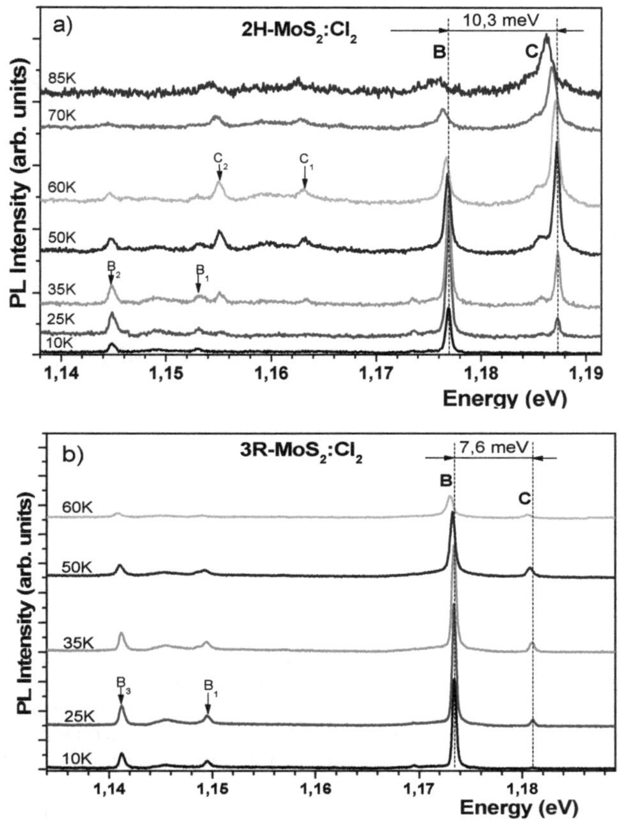

Fig. 1 Steady-state PL spectra at different temperatures of chlorine intercalated 2H-MoS2 (a) and 3R-MoS2 (b) polytypes.

The electronic structure of bulk and few-layered MoS_2 was thoroughly investigated both experimentally and theoretically in [2, 3, 19, 20], for example. An unusual electronic structure of few-layer MoS_2 and the resulting unique behaviour of the optical properties originate from the characters of *d*-electron orbitals that comprise MoS_2 conduction and valence bands. According to theoretical calculations, electronic states at different values of the wave vector possess electronic orbitals with different spatial distributions. At the Γ point the electronic states are strongly affected by the long-range interlayer Coulombic interactions. In our case, halogen molecules intercalated within MoS_2 layers should influence the interlayer interaction especially at the Γ point. It has to be taken into account that the intercalation of any molecules in the van der Waals gap leads to an enlargement of the adjacent layers, as was revealed in earlier publications [9, 21]. The interlayer interaction is more pronounced in the case of only few MoS_2 layers as was suggested in [20]. This assumption is confirmed by our theoretical calculations of the electronic structure of the two polytypes with and without halogen intercalation, presented in Fig. 2.

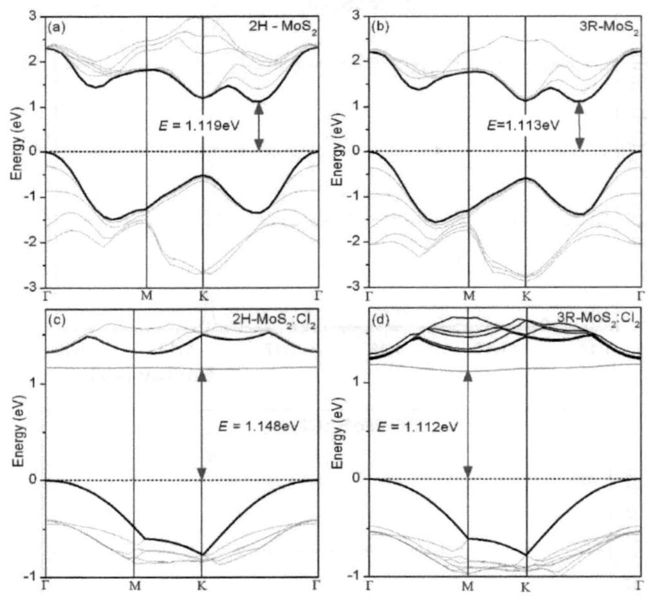

Fig. 2 Calculated band structures of two models containing three layers of MoS2. Band structures of 2H (a) and 3R (b) polytypes of MoS2 where layers are fixed at bulk positions. Local band structure of halogen intercalated 2H (c) and 3R (d) polytypes of MoS2. The orange line in (c) and (d) correspond to the halogen level within the band gap.

The model I has an indirect band gap structure, VBM being located at the Γ point, while the CBM – about halfway between Γ and K. The values of indirect gaps are equal to 1.119 and 1.113 eV for 2H and 3R polytypes (Fig. 2 a, b).

In the second model for two polytypes the chlorine intercalation creates energy levels in the band gap near the conduction band edge which consist of *2p* antibonding states of Cl atoms while the valence bands are comprise by *d*-electron orbitals of Mo atoms. Moreover, halogen intercalation completely changes the local band structure and energy states distribution in the Brillouin zone: both polytypes having now direct band gap transition at the Γ point with 1.323 and 1.246 eV values, for 2H and 3R polytypes, respectively (Fig. 2 c, d). The values of halogen levels within the band gap of 2H and 3R polytypes are equal to 1.148 and 1.112 eV, respectively. The intercalation of Cl_2 molecule has led to widening of the band gaps compared to the band structure of the studied polytypes in model I. The difference in the values of the given halogen levels is related to different positions of Cl_2 molecules between the layers where they have different spatial surrounding. For 2H-MoS_2, only one Cl atom in the molecule has a contact (4.32Å) with Mo atom that is less than above indicted threshold (4.5Å) when band gap does not depend on the inter-layers separation. For 3R polytype, the contacts of each of Cl atoms in the molecule with the metals are less than in 2H polytype and equal to 4.279 and 4.316Å, respectively. Since in Cl_2 molecule the electrons are shared (not transferred) in a molecule, there are no existing ions that are positive or negative charged, which leads to the creation of full outer shells of chlorine atoms. It means that the repulsive forces hinder the inter-layers coupling when Cl and Mo atoms are brought closer to each other. In turn, that leads to a greater splitting of anti-bonding states of Cl atoms in 3R-MoS_2, causing the lowering of the impurity levels in this polytype in comparison with 2H polytype. The band gap widening for both polytypes in model II in comparison with model I may be related to a long-range Coulombic interaction of chlorine molecules with adjacent layers. In order to understand the effect of halogen intercalation on the band structure transformation we calculated the energy bands for model II with increased distances but without Cl_2. The obtained values are equal to 1.319 and 1.248 eV for 2H and 3R polytypes respectively. These values are very similar to direct band gap transition at the Γ point in model II for 2H, 3R-MoS_2 which means that the intercalation of chlorine molecules have increased the distances between the adjacent layers which in turns have led to the local band gap widening.

Theoretical calculations have revealed that the difference in Cl_2 molecules positions in two polytypes should result in different luminescent spectra separated by a small amount of energy which indeed is observed experimentally. The temperature rise of the *C*-line in the luminescence spectra to the detriment of the *B*-line for 2H polytype occurs because the level responsible for the *C*-line has a radiative lifetime much shorter than the B level [18]. In the case of 3R polytype, despite a smaller (in comparison with 2H polytype) distance between the *B* and *C* excitonic levels (7.5meV instead of 10.3meV), the temperature increase of the *C*-line relative intensity is unimportant. This indicates that the radiative lifetimes of these two levels are not significantly different.

V. CONCLUSIONS

2H-MoS_2 and 3R-MoS_2 polytypes were identified and structurally characterized by the X-ray method to clarify its structure under intercalation of Cl_2 molecules. Luminescence related to the excitons bound on halogen molecules was investigated for the both polytypes. It was shown that there is an evident difference between the low temperature PL spectra of the investigated polytypes: in the case of the 3R polytype the main excitonic zero-phonon lines are shifted by 3.5meV to lower energies. The observed spectral shift and dissimilar behavior of the integral spectra as a function of temperature are explained as consequence of slightly different positions of halogen molecules in the interstitial space of the studied polytypes. To interpret the obtained experimental results the DFT band structure calculations were done for three molecular layers of 2H-MoS_2 and 3R-MoS_2 polytypes without (model I) and with (model II) intercalation of Cl_2 molecules. For model II, the structural features of polytypes were shown to affect the long-range interlayer interactions, which led to a difference in values of the band gaps in 2H-MoS_2 and 3R-MoS_2 due to slightly different positions of halogen molecules into the van der Waals gap. The DFT band structure calculations for model II confirmed the spectroscopic experiments. It was shown that the halogen intercalation completely changes the local band structure and energy states distribution in the Brillouin zone of both polytypes.

ACKNOWLEDGMENTS

This research was supported by HP-SEE project (the fast track access to HPC cluster resources available in the South-East Europe region), STCU project Nr. 5809 and Russian Science Foundation Grant № 14-12-01080.

CONFLICT OF INTEREST

The authors declare that they have no conflict of interest.

REFERENCES

1. Mingxiao Ye, Dustin Winslow, Dongyan Zhang, Ravindra Pandey, Yoke Khin Yap. (2015) Recent Advancement on the Optical Properties of Two-Dimensional Molybdenum Disulfide (MoS_2) Thin Films. Photonics, 2(1): 288-307.
2. Kin Fai Mak, Changgu Lee, James Hone, Jie Shan, Tony F. Heinz. (2010) Atomically Thin MoS_2: A New Direct-Gap Semiconductor. Phys. Rev. Lett., 105: 136805.
3. A. Splendiani, L. Sun, Y. Zhang, T. Li, J. Kim, C.-Y. Chim, G. Galli, F. Wang. (2010) Emerging Photoluminescence in Monolayer MoS_2. Nano Letters, 10: 1271-1275.
4. H. Li, Z. Yin, Q. He, H. Li, X. Huang, G. Lu, D.W.H. Fam, A.I.Y. Tok, Q. Zhang, H. Zhang. (2012) Fabrication of Single- and Multilayer MoS_2 Film-Based Field-Effect Transistors for Sensing NO at Room Temperature. Small, 8: 63-67.
5. Hiram J. Conley, Bin Wang, Jed I. Ziegler, Richard F. Haglund, Sokrates T. Pantelides and Kirill I. Bolotin. (2013) Bandgap Engineering of Strained Monolayer and Bilayer MoS_2. Nano Letters, 13: 3626-3630.
6. A. K. Geim and I. V. Grigorieva. (2013) Van der Waals heterostructures. Nature, 499: 419-425.
7. R. Coehoorn, C. Haas, R. A. de Groot. (1987) Electronic structure of $MoSe_2$, MoS_2, and WSe_2. II. The nature of the optical band gaps. Phys. Rev. *B*, 35: 6203-6206.
8. L. Britnell, R.M. Ribeiro, A. Eckmann, R. Jalil, B.D. Belle, A. Mishchenko, Y.-J. Kim, R.V. Gorbachev, T. Georgiou, S.V. Morozov, A.N. Grigorenko, A.K. Geim, C. Casiraghi, A.H.C. Neto, K.S. Novoselov. (2013) Strong Light-Matter Interactions in Heterostructures of Atomically Thin Films. Science, 340: 1311-1314.
9. E. Benavente, M.A. Santa Ana, F. Mendizábal, G. González. (2002) Intercalation chemistry of molybdenum disulfide. Coordination Chemistry Reviews, 224: 87-109.
10. R. Rosentsveig, A. Margolin, A. Gorodnev, R. Popovitz-Biro, Y. Feldman, L. Rapoport, Y. Novema, G. Naveh, R. Tenne. (2009) Synthesis of fullerene-like MoS_2 nanoparticles and their tribological behavior. J. Mater. Chem., 19: 4368-4374.
11. Reshef Tenne. (2014) Recent advances in the research of inorganic nanotubes and fullerene-like nanoparticles. Front. Phys., 9(3): 370-377.
12. F. Wypych, R. Schollhorn, 1T-MoS_2, a new metallic modification of molybdenum disulfide. (1992) J. Chem. Soc. Chem. Commun., 19: 1386-1388.
13. G. Jakub, G. Robert, W. Małgorzata, H. Jürgen, Relative stability of low-index V_2O_5 surfaces: a density functional investigation. (2009) J. Phys.: Condens. Matter., 21: 095008.
14. A. Spek. (2003) Single-crystal structure validation with the program PLATON. Journal of Applied Crystallography. 36: 7-13.
15. X. Gonze, J.M. Beuken, R. Caracas, F. Detraux, M. Fuchs, G.M. Rignanese, L. Sindic, M. Verstraete, G. Zerah, F. Jollet, M. Torrent, A. Roy, M. Mikami, P. Ghosez, J.Y. Raty, D.C. Allan. (2002) First-principles computation of material properties: the ABINIT software project. Computational Materials Science, 25: 478-492.
16. J.P. Perdew, K. Burke, M. Ernzerhof. (1996) Generalized Gradient Approximation Made Simple. *Phys. Rev. Lett.*, 77: 3865-3868.
17. L. Kulyuk, L. Charron, E. Fortin. Radiative properties of MoS_2 layered crystals. (2003) Phys. Rev. B, 68: 075314.
18. A. Colev, C. Gherman, V. Mirovitskii, L. Kulyuk, E. Fortin. (2009) J. Luminescence, 129: 1945-1947.
19. A. Ramasubramaniam, D. Naveh, E. Towe, Tunable band gaps in bilayer transition-metal dichalcogenides. (2011) Phys. Rev. B, 84: 205325.
20. J.K. Ellis, M.J. Lucero, G.E. Scuseria. The indirect to direct band gap transition in multilayered MoS_2 as predicted by screened hybrid density functional theory. (2011) Appl. Phys. Lett., 99: 261908.
21. D.M. D'Ambra, J.V. Marzik, R. Kershaw, J. Baglio, K. Dwight, A. Wold. (1985) Preparation and electronic properties of MoS_2 and WS_2 single crystals grown in the presence of cobalt. J. Solid State Chem., 57: 351-356.

Author: Leonid Kulyuk
Institute: Institute of Applied Physics
Street: 5 Academiei str.
City: Chisinau
Country: Republic of Moldova
Email: kulyuk@phys.asm.md

Exciton-polariton Laser

S.A. Moskalenko[1] and I. Tiginyanu[2]

[1] Institute of Applied Physics, Academy of Sciences of Moldova, Chisinau, Republic of Moldova
[2] Institute of Electronic Engineering and Nanotechnologies "D. Ghitu", Academy of Sciences of Moldova, and National Center of Materials Study and Testing, Technical University of Moldova, Chisinau, Republic of Moldova

Abstract— **In this paper, we present a review of the investigations realized in the last decades of the phenomenon of the Bose-Einstein condensation (BEC) in the system of two-dimensional cavity polaritons in semiconductor nanostructures. The conditions at which the excitons interacting with cavity photons form new type of quasiparticles named as polaritons are described. Since polaritons can form in a microcavity a weakly interacting Bose gas, similarly to the exciton gas in semiconductors, the microcavity exciton-polariton BEC emerged in the last decades as a new direction of the exciton BEC in solids, promising for practical applications. The high interest in BEC condensation of exciton-polaritons in semiconductor microcavities is related to ultra-low threshold lasing which has been demonstrated, in particular, for an electrically injected polariton laser based on bulk GaN microcavity diode working at room temperature.**

Keywords— **Excitons, polaritons, microcavity, quantum coherence, Bose-Einstein condensation, inversionless polariton lasing.**

I. INTRODUCTION

The rich information accumulated over the last decades by many groups of investigators on the way to achieve the Bose-Einstein condensation (BEC) of the cavity polaritons was published in a great amount of papers which cannot be cited here in a complete way. This information concerns not only to the fundamental aspects of the problem but also to the polaritonic devices. We will confine ourselves to the comprehensive review articles [1-7] and to some other initial publications [8-16]. Bhattacharya *et al* [11] succeeded recently to demonstrate room temperature electrically pumped inversionless polariton lasing from a bulk gallium nitride based microcavity diode. As was foreseen in the Ref. [17], this laser has two density thresholds for the injected electron and hole carriers forming in a semiconductor the Wannier-Mott type excitons, which in microcavity are forming the polaritons with the lower and upper polariton branches. The lasing obtained at lower threshold is due to the Bose-Einstein condensation (BEC) of the composite bosons, as polaritons are, similar with the BEC of atom-like structures described by the theory of the Bose gas elaborated by Bogoliubov in his microscopical theory of superfluidity [18]. In this paper, we review the investigations of the phenomenon of Bose-Einstein condensation in a system of two-dimensional cavity polaritons in semiconductor nanostructures realized in the last decades, leading to the development of exciton-polariton laser.

In the case of the inhomogeneous Bose gas in the traps, the atom like systems are described, in a coherent macroscopic state, by the Gross-Pitaevskii equation [19, 20].

Another higher density threshold corresponds to high density electrons and holes forming degenerate states characterized by the Fermi energy levels, when the particles from these surroundings take part in the coherent pairing of the Cooper type, similar with the electron Cooper pairs in the theory of the superconductivity elaborated by Bardeen, Cooper and Schrieffer (BCS) [21].

The Bose-Einstein condensation of these e-h pairs gives rise to new coherent macroscopic state and to the second threshold laser named as photon laser. The both types of Bose-Einstein condensations are described in the frame of the theory elaborated in the electron-hole language by Keldysh-Kozlov [22] in the low exciton concentration limit, and by Keldysh and Kopaev [23] in the higher electron-hole concentration limit leading to the formation of the excitonic dielectric. The two approaches form a unique Keldysh-Kozlov-Kapaev approach, as it was proposed in Ref. [24]. On this base both limiting cases were described from the unique point of view known as BEC-BCS crossover.

II. A SHORT REVIEW RELATED TO THE BEC IN A SYSTEM OF 2D CAVITY POLARITONS IN SEMICONDUCTOR NANOSTRUCTURES

First of all in Ref. [1] it was reminded that the exciton polaritons are composite bosons, containing not only the fermionic constituents as the electrons and holes are, but also the photonic components. The polaritons are the representatives of the light-natter interaction in the crystals. They appear as a result of the multiple, reciprocal reversible exciton-photon transformations, when the exciton-photon coupling is strong, whereas the radiative decay of the excitons and other decoderence processes do not succeed to take place before the polariton formation. In the case of QW excitons and cavity photons taking part in the transformations have the same in-plane wave vectors $\vec{k}_{\parallel}$ due to the translational

V. Sontea and I. Tiginyanu (eds.), *3rd International Conference on Nanotechnologies and Biomedical Engineering*,
IFMBE Proceedings 55,
DOI: 10.1007/978-981-287-736-9_49

symmetry in this direction. The coherent superposition states of excitons and photons in crystals were introduced by Hopfield [25]. Because the transformations are virtual the energies of the excitons and photons taking part in superposition must not be equal each other. The splitting of two polariton states gives rise to the upper (UP) and to the lower polariton (LP) branches. In the microcavity the photon branch has the cut-off frequency $\omega_c = \frac{\pi c}{n_c L_c}$, where L_c is the length of the resonator and n_c is its refractive index. The longitudinal component k_z of the light wave vector inside the resonator is quantized being equal to $\pm\pi/L_c$, what leads to the cavity frequency ω_c, when the in-plane wave vector $\vec{k}_{\parallel}$ equals to zero.

The existence of the cavity mode and of the bottom of the LP branch in a microcavity motivated the authors of Ref. [17] to study the possibility of BEC in these conditions. The possibility of the BEC at the bottom of the upper polariton branch in 3D crystals in the absence of the microcavity was suggested in Ref. [26].

As was mentioned in the Ref. [4], the confinement of the photon modes in the cavity leads to a small volume associated with each of them and so the exciton-photon coupling is strong. It happens to be stronger than the rate of the exciton-exciton scattering processes so that the cavity polariton formation becomes the primordial process in microcavity rather than the establishing of the exciton coherent macroscopic state in the form of a Bose-Einstein condensate. In the opposite scenario without microcavity the exciton condensate with small wave vector $\vec{k} \neq 0$ could transform into the coherent macroscopic beam of photons and their reciprocal transformations could lead to the formation of a condensate-photon mode [27] named as a giant polariton, as was proposed by Haken and Schenzle [28].

In our short review the quantum coherent macroscopic state is described on the base of the canonical transformation proposed by Bogoliubov in his theory of superconductivity. The Bogoliubov unitary operator is written in the electron Fermi operator representation, but it is similar with the Glauber-type [29] unitary transformation introduced in the theory of the coherent states. This method leads, according to Refs. [22, 23], to the Bogoliubov-type u-v transformations, in the case of electron and hole Fermi creation and annihilation operators, and to the Bogoliubov-type displacement transformation in his theory of the Bose gas. Some needed information concerning the polariton formation on the base of the coherent superposition states introduced by Hopfield [25] was included. The modifications of the standard theory of the Bose gas related to the finite life-time of the particles taking part in the condensation process are introduced. The main troubles are related to the question whether the quasiparticle lifetime will be sufficient to guarantee the effectuation of the relaxation processes and of the establishing of the thermodynamic quasiequilibrium. Unexpectedly, the very small effective cavity polariton mass and very small elastic polariton-polariton scattering time removed any doubts and permitted the realization of the desired phenomenon.

Due to the bosonic character of the cavity polaritons, they become a model system for demonstration of the dynamic character of the BEC in semiconductors [1-5]. There are some advantages related to the participation of the cavity polaritons in the BEC process. One of them was mentioned above and concerns the low effective polariton mass, four orders of magnitude lighter than the bare exciton mass. As a result the critical temperature needed to observe the polariton BEC is four orders of magnitude higher than for reaching the exciton BEC at the same particle density. The photon component of the polariton having an extent coherent phase gives rise to the second advantage of the polaritons in comparison with the excitons, which can be easily localized by the fluctuation of the crystal lattice potential [1].

As mentioned in the Refs. [1-5], the polariton system has a unique feature compared to the dilute atomic BEC and to the dense superfluid. It is related to a direct experimental accessibility to measure the quantum statistical properties of the condensate. The main decay channel of the cavity polaritons consists in the photon leakage from the cavity. During this process two conservation lows are fulfilled. The energy and the in-plane wave vector of each cavity polariton and of the leakage photon one-to-one coincide. It is a new experimental tool to study the nonequilibrium open system. The cavity polaritons are a unique system for exploring both the cavity quantum electro-dynamics as well as the many body physics, opening the possibility to realize experimentally different phase transitions [1, 4]. One of them is the BEC of the polariton gas. Another one takes into account the formation of the quantum vortices in the medium of the polaritons. While the spontaneous creations of one separate vortex is unlikely due to the necessity to spend an extensive amount of energy, nevertheless the creation of a pair vortex-antivortex can be easily realized even at small temperature due to the reciprocal compensation of the rotation velocities at big distances from the cores of the vortices. The binding and the unbinding of such pairs of vortices give rise to the Berezinskii [30] and Kosterlitz-Thouless [31] (BKT) phase transition.

The short lifetime of the Bose-Einstein condensed polaritons greater but not so far from their relaxation time leads to their description by the Gross-Pitaevskii equation with the loss and gain terms [1, 4]. As a result the dispersion law of the collective elementary excitations arising over a ground state deviates from the standard Bogoliubov form elaborated in his microscopic theory of superfluidity [18]. In the dispersion low some diffusive elements appear. They have the form of a flat section in the range of small wave vectors succeeded by a rising section approaching to the Bogoliubov spectrum. Note that up to now we have discussed the features of the Bose-Einstein condensed cavity polaritons arising due to their finite lifetime.

At the same time there are other fundamental properties related to the two-dimensional aspects of the phenomenon under consideration. As in the previous part of our short review, we will follow the Refs. [1, 4]. The two fold dimensionality of the described systems has an important role because, as established in Ref. [32], the BEC of the two-dimensional homogeneous Bose gas extended on an infinite surface area cannot occur at finite temperatures. It is possible only at the temperature $T=0$, which cannot be achieved experimentally.

A uniform two-dimensional system of bosons in the thermodynamic limit at finite temperatures has no phase transition to the highly ordered states, since the long wavelength thermal fluctuations destroy any long range order. In such a way the experimental realization of the BEC phenomena with the participation of the 2D cavity polaritons at finite temperatures faced an obstacle of principle.

Nevertheless, the ingenuity of the experimental physics permitted to avoid this obstacle and to reveal the basic properties of the nature hidden by such restrictions. The goal was to study the finite area 2D systems using for excitation the pumping light with finite spot area. It is to be noted, however, that the confinement of the Bose gas on a finite area gives rise to the uncertainty of the in-plane wave vector of the confined polaritons. The condensate wave vector also exhibits an uncertainty. It means that the condensate stretches to a finite wave vector, what was not foreseen by the standard Bogoliubov theory of superfluidity [18]. The spatial extent of the condensed polaritons leads also to the size-dependent critical density of the condensation. When the Bose-gas is confined by a spatially varying potential, the constant density of states of the uniform two-dimensional system is strongly modified. As a result the BEC phase at finite temperatures is recovered [1, 4]. Practically any experimental system has a finite size and a finite number of particles on the single-particles states. In a two-dimensional box system of size L the critical conditions for BEC can be fulfilled at $T \neq 0$. If the system is small, the finite size BEC phase transitions are expected to occur before the BKT transition [1-5]. As mentioned above, the tightly focused spot for optical excitations can provide localization of cavity polaritons. Due to the very small effective mass, the observation of their 3D quantization is possible for trap diameters on the order of a few micrometers. The lateral confinement of 2D polaritons creates opportunities for realization of the future quantum light emitters as a source of indistinguishable single photons. Polariton emission from photonic quantum boxes in planar microcavities of diameters 1-10 μm was probed. For the trap of 1-2 μm the quantization energy of the polariton equals a few meV [1, 4].

The possibilities to achieve the quasi-equilibrium states in the systems of 2D cavity polaritons are discussed in many of referred papers, underlying the role of the polariton-polariton and polariton-phonon stimulated scattering processes. The theory describing the nucleation and the buildup of the Bose condensate was elaborated by Stoof [33, 34], who argued that the solution of the problem does not lie in the kinetic stages of the phase transitions. In the proposed description there are an initial kinetic stage, the second stage named as a coherent stage and the last stage which is also a kinetic stage. During the first kinetic stage the initially highly nonequilibrium distribution created by the nonresonant pumping or by the electrically injected electron-hole pairs is transformed into the quasiequilibrium distribution. But this stage cannot lead to a macroscopic occupation of the single-particle ground state, since the production rate of the condensate fraction in thermodynamic limit $N \to \infty$, $V \to \infty$, is nonzero if the condensate already exists. The first kinetic stage is characterized by the elastic collision time τ_{el}. During this stage in the range of the critical temperature the number of elementary excitations in the system becomes so high that the instabilities in the system appear. Their evolutions give rise to the next stage of the phase transition named as a coherent stage. Its description proposed by Stoof in the language of the free energy density functional is represented in the figure 1 in dependence on the order parameter $< \psi(rt) >$.

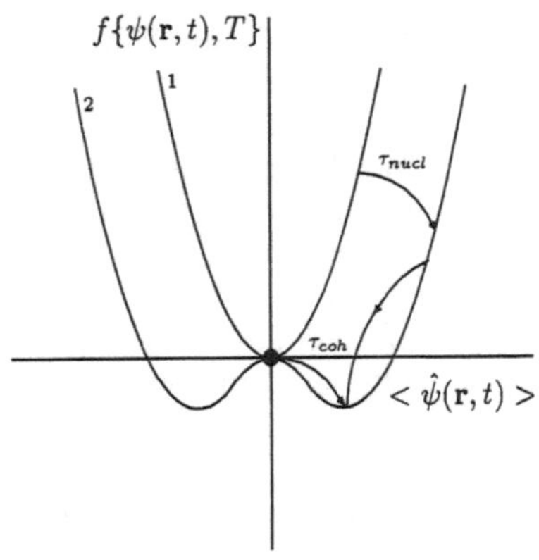

Fig. 1 The free energy density functional $f(\psi(r,t),T)$ in dependence on the order parameter $< \psi(rt) >$. This drawing is reproduced from the paper by H.T.C. Stoof published in Ref. [34].

There are two curves corresponding to two possible situations when the system conserves its gauge symmetry $U(1)$, or when this symmetry is broken. Two cases are denoted in figure 1 by the curves 1 and 2 correspondingly.

The developing of the instability in the system is denoted in the figure 1 by the nucleation time τ_{nucl}, which happened to not exceed and even to be smaller than the τ_{el}. This coherent stage gives rise to the appearance of the condensate embryo in a form of a seed which is not in a thermodynamic quasi-equilibrium with the noncondensed part of the weakly interacting Bose gas. This goal has to be achieved in the last stage of the phase transition, which is also a kinetic stage. Its duration τ_{rel} is much longer than that of the first kinetic stage and is limited by the polariton lifetime τ_{life}. Under the condition of $\tau_{\text{rel}} > \tau_{\text{life}}$, the complete quasi-thermalization cannot be achieved, what probably takes place in the case of cavity polaritons.

The number of applications of exciton polaritons, including their Bose-Einstein condensates, in semiconductor microcavities is expected to sharply increase in the coming years. Fortunately, the progress in materials science and technology provides excellent conditions for epochal breakthroughs. A good example, in this regard, is the recently demonstrated electrically pumped inversionless polariton lasing at room temperature from a bulk GaN-based microcavity diode [11]. Among many applications of gallium nitride, the industrial production of solid-state lasers and light-emitting diodes is the most amazing, especially taking into account the relatively high amount of dislocations inherent to GaN epilayers overgrown on foreign substrates such as sapphire, SiC, Si etc. We believe that the wonderful semiconductor compound GaN will contribute to the realization of new breakthroughs in lighting technologies, in particular through the development and implementation of ultra-low threshold solid-state lasers.

III. CONCLUSIONS

As a concluding remark of this short review based on previous literature data, one can mention the following. If the rate of the energy exchange between the cavity field and the quantum well (QW) excitons is faster than the decay and other decoherence rates both for cavity photons and excitons, then the excitons and the photons in the system are stored in a new type of quasiparticles named as polaritons. They are considered to form a weakly interacting Bose gas. The exciton density is well below the Mott transition density. The spontaneous establishing of the coherence and of the phase transitions is driven by the polariton-polariton scattering. It is a microscopic driving force of polariton dynamics. The correlation functions of the polariton condensates were measured experimentally, including the first and second orders temporal correlation functions and the first order spatial correlation function. The vortices arising in the frame of the polariton condensates were revealed [1-6].

The microcavity exciton-polariton BEC described above emerged in the last decades as a new direction of the exciton BEC in solids [22, 35-38] promising for practical applications. It is to be noted, however, that there are also other possibilities for the realization of the exciton BEC in solids, as recently mentioned, for example, in Ref. [39]. The field under consideration proves to be extremely dynamic with many opportunities for both fundamental and applied research.

ACKNOWLEDGMENT

This work was supported by the Supreme Council for Science and Technological Development, Academy of Sciences of Moldova.

CONFLICT OF INTEREST

The authors declare that they have no conflict of interest

REFERENCES

1. H. Deng, H. Haug, Y. Yamamoto. Exciton-polariton Bose-Einstein condensation. Rev. Mod. Phys. 82, 1489 (2010).
2. I. Carusotton and C. Ciuti, Quantum fluids of light. Rev. Mod Phys., 85, 299 (2013).
3. B. Deveand, Exciton-Polariton Bose-Einstein Condensates. Annual Rev. Condens. Matter. Phys. 6, 155 (2015).
4. J. Keeling, F. M. Marchetti, M. H. Szymanska and P.B. Littlewood. Collective coherence in planar semiconductor microcavities. Semicond. Sci. Technol. 22, R1-R9 (2007).
5. D. Bajoni, Polariton lasers. Hybrid light-matter lasers without inversion. J. Phys. D: Appl. Phys. 45, 313001 (2012).
6. T.C.H. Liew, I.A. Shelykh, G. Malpuech. Polaritonic devices. Physica E 43, 1543-1568 (2011).
7. I.A.Shelykh, A.V.Kavokin, Yu. G. Rubo et al. Polariton polarization-sensitive phenomena in planar semiconductor microcavities. Semicond. Sci. Technol. 25, 013001 (2010).
8. H. Deng, et al., Quantum degenerate exciton-polariton in thermal equilibrium. Phys. Rev. Lett. 97, 146402 (2006).
9. J. Kasprzak, et al. Bose-Einstein Condensation of exciton polaritons. Nature 443, 409 (2006).
10. A. V. Balatsky, Y. N.Joglekar and P. B. Littlewood. Dipolar superfluidity in electron-hole bilayer systems. Phys. Rev. Lett. 93, 266901 (2004).
11. P. Battacharya, T. Frost, S. Deshpande et al. Room temperature electrically injected polariton laser. Phys. Rev. Lett. 112, 236802 (2014).
12. C. Schneider et al. An electrically pumped polariton laser. Nature, 497, 348 (2013).
13. Md. Zunaid Baten et al. GaAs-based high temperature electrically pumped polariton laser. Appl. Phys. Lett. 104, 231119 (2014).
14. Cong Wei and Yong Sheng Zhao. Electrically pumped polariton lasers. J. Matter. Chem. C, 2, 2295 (2014).

15. J. Zuniga-Perez et al. Patterned silicon substrates: A comon platform for room temperature GaN and ZnO polariton lasers. Appl. Phys. Lett. 104, 241113 (2014).
16. Feng Li et al. Fabrication and characterization of a room-temperature ZnO polariton laser. Appl. Phys. Lett. 102, 191118 (2013).
17. A. I. Imamoglu, R. J. Ram, S. Pau and Y. Yamamoto. Nonequilibrium condensates and laser without inversion: exciton-polariton lasers. Phys. Rev. A 53, 4250 (1996).
18. N. N. Bogoliubov. Selected works. Gordon Breach Sci. Publ. N.Y. (1990-1991) v. 2.3.
19. L.P. Pitaevskii. Vortex filaments in the nonideal Bose gas. Zh. Eksp. Teor. Fiz. 44, 646 (1961).
20. E.P. Gross. Structure of a quantized vortex in boson systems. Nuovo Cimento. 20(3), 454 (1961).
21. J. Bardeen, L. N. Cooper, and J. R. Schrieffer. Microscopic Theory of Superconductivity. Phys. Rev. 106, 162 (1957).
22. L. V. Keldysh and A. N. Kozlov. Collective properties of excitons in semiconductors. Zh. Exper. Teoret Fiz. 54, 378 (1968).
23. L. V. Keldysh and Yu. V. Kopaev. Possible instability of the semimettalic state as regards to the Coulomb interaction. Fiz. Tverd. Tela (Leningrad) 6, 2791 (1964).
24. C. Comte, and P. Nozières. Exciton Bose condensation: the ground state of an electron-hole gas - I. Mean field description of a simplified model. J. Phys. France 43, 1069-1081 (1982).
25. J. J. Hopfield. Theory of the Contribution of Excitons to the Complex Dielectric Constant of Crystals. Phys. Rev. 112, 1555 (1958).
26. R.S. Knox. Theory of excitons, Academic Press - N.Y. and London, 1963.
27. S.A. Moskalenko et al. Bose-Einstein condensation of dipole-active excitons and photons. Sov. Phys. JETP 37, 902 (1973).
28. H. Haken and A. Schenzle. Giant polariton. Phys. Lett. A 41, 405 (1972).
29. R. J. Glauber. The Quantum Theory of Optical Coherence. Phys. Rev. 130, 2529 (1963).
30. V.L. Berezinskii. Breaking of the long range order in one-dimensional and two-dimensional systems with continuous group of symmetry. Zh. Eksp. Teor. Fiz. 59, 907 (1971).
31. J M Kosterlitz and D J Thouless. Ordering, metastability and phase transitions in two-dimensional systems. J. Phys. C: Solid State Phys. 6, 1181 (1973).
32. P. C. Hohenberg. Existence of Long-Range Order in One and Two Dimensions. Phys. Rev. 158, 383 (1967).
33. H.T.C. Stoof. Formation of the condensate in a dilute Bose gas. Phys. Rev. Letter. 66, 3148 (1991).
34. H.T.C. Stoof. Condensate formation in a Bose gas. In: Bose-Einstein condensation, eds: A.Griffin, D.W.Snoke and S.Stringari, Cambridge U. Press, Cambridge, p. 226 (1995).
35. S. A. Moskalenko. Reversible optico-hydrodynamic phenomena in a non ideal exciton gas. Sov. Phys. Solid State 4, 199-204 (1962).
36. J. M. Blatt, K. W. Boer and W. Brandt. Bose-Einstein condensation of excitons. Phys. Rev. 126, 1691 (1962).
37. S. A. Moskalenko and D. W. Snoke. Bose-Einstein condensation of excitons and biexcitons and coherent nonlinear optics with excitons, p. 428, Cambridge University Press. Cambridge (2000).
38. J. Kasprzak, M. Richard, S. Kundermann et al. Bose–Einstein condensation of exciton polaritons. Nature 443, 409 (2006).
39. D.W. Snoke and G.M. Kavoulakis. Bose-Einstein condensations of excitons in Cu_2O: progress over 30 years. Rep. Progr. Phys. 77, 1/6501 (2014).

Author: Acad. Sveatoslav Moskalenko
Institute: Institute of Applied Physics
Street: 5 Academiei str.
City: Chisinau
Country: Republic of Moldova
Email: exciton@phys.asm.md

Preparation of Fine Bentonite Suspensions in Cavitation Fields

P.G. Dumitras, M.K. Bologa, and T.D. Shemyakova

Institute of Applied Physics, Academy of Sciences of Moldova, Chisinau, Moldova

Abstract— **It was demonstrated that application of ultrasonic, hydrodynamic, and bifrequency cavitation for bentonite dispergation allows one to obtain suspensions of nanoscale particles. It was investigated how the dimensions of the particles depend on the amplitude of oscillations, duration of the treatment, and static pressure. The results of the electron microscopic examination of the suspensions were analysed.**

Keywords— **Bentonite, dispergation, ultrasonic, hydro-acoustic, bifrequency cavitation, wine, clarification.**

I. INTRODUCTION

In the wine-making industry bentonite suspensions are used, which is a natural ecologically friendly substance. Bentonite is usually imported in the Republic of Moldova from various countries. At the same time, the possibility exists to use the local bentonite [1] by virtue of preparation of activated finely dispersed bentonite suspensions at the level of nanotechnologies involving cavitation effects.

Clarification and stabilization of wines and juices are the main technological processes in the wine making. The essential properties of bentonite of swelling and absorbtion depend on the degree of dispergation of bentonite layered particles. The bentonite powder, especially Na bentonite, forms clusters in water or wine, and it is difficult to dispergate it using traditional methods. The clusters form due to the presence of the wet layer surrounding bentonite particles that swells, forms a gel and prevents penetration of water or wine inside the cluster. To prepare a uniform suspension one needs to dispergate fine bentonite in water or wine with intensive agitation of the mixture.

It is known that cavitation is a powerful tool for intensification of heat and mass transfer technological operations; it influences the structure of gas – liquid – solid systems. The changes of their phases occur under the action of nonlinear cavitation effects, which arise under the action of hydrodynamic flows behind the cavitators of different geometrical shape or in narrow channels. This induces high local temperatures and pressures and relatively high rated of their variation in comparison with the known technological processes. The cavitation actions lead to physical and chemical transformations in working media; this may be crucial in the development of technological processes and is especially promising for preparation of finely dispersed (nanodispersed) phases gas – liquid – solid. During the ultrasonic cavitation treatment of solid particles microjets and shock waves are formed directed to the surface [2]; this leads to the decreasing of the dimensions of particles.

The studies of the influence of ultrasonic treatment on the variation of particles' dimensions of such solid substances as bentonite (montmorillonite), talk, and vermiculite in aqueous suspensions have demonstrated that bentonite retains its crystalline structure; increasing of the duration of ultrasonic treatment increases the surface area of bentonite particles [3]. Application of the sonochemical method allowed ones to prepare bentonite in the form of nanorods with dimensions from 5 to 50 nm [4]. The vermiculite particles with micron and submicron dimensions and nanometric thickness (less than 10 nm) were prepared during the treatment in water and hydrogen peroxide [5]. It was shown that the dimensions of large particles in diluted suspensions diminish, and the percentage of the particles with the diameter less than 1.5 μm increases [6].

In this communication we provide the results of investigation of the process of preparation of bentonite suspensions with the initial dimensions of solid particles of 200 - 300 μm in ultrasonic, hydrodynamic, hydroacoustic, and mixed cavitation fields and of the possibility of their application for clarification of wines and juices.

II. EXPERIMENTAL

The experiments were performed using an ultrasonic cavitation installation (Fig. 1) at an excessive static pressure, hydroacoustic, and hydrodynamic installations with cylindrical cavitators (Fig. 2a), installation with a pulse rotor (Fig. 2b), and bifrequency installation. The technology was developed for application in the technological line for preparation of finely dispersed activated homogeneous bentonite suspensions in the pilot production conditions.

Application of ultrasonic cavitation for homogenization and dispergation is associated with high energy consumption, and the process becomes inefficient when large volumes of material should be treated. The main advantage of the ultrasonic cavitation method consists in the fact that the treatment duration considerably reduces; in some cases this method is unreplaceable. Application of hydrodynamic and hydroacoustic cavitation allows one to increase the process productivity; however, it demands to increase the pressure, temperature, and treatment duration

V. Sontea and I. Tiginyanu (eds.), *3rd International Conference on Nanotechnologies and Biomedical Engineering*,
IFMBE Proceedings 55,
DOI: 10.1007/978-981-287-736-9_50

for effective homogenization and dispergation. Independent on the method of generation of cavitation (ultrasonic, hydroacoustic, or hydrodynamic) it is desired to reduce the energy consumption, increase the quality of processing of multicomponent media, reduce or eliminate application of expensive and toxic chemical reagents.

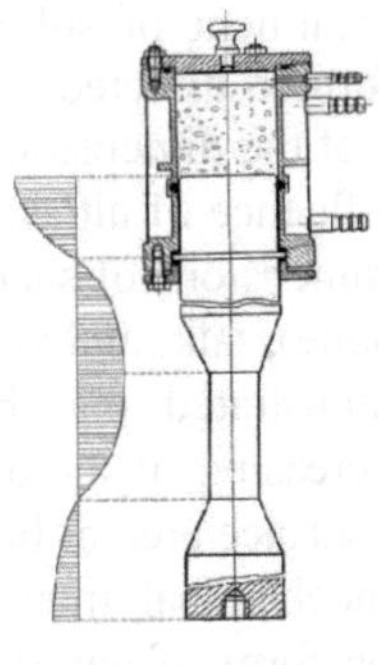

Fig. 1 Ultrasonic installation for fine dispergation

The results of the studies obtained earlier confirm that it is appropriate to use bifrequency cavitation. The combined action of oscillations in the range of various frequencies on the physico-chemical treatment processes considerably increases the efficiency of cavitation action [7 – 10]. While analyzing the action of two frequencies we have solved the modified Nolting-Nepairas equation using a decomposition of motion of the cavitation bubble into the 'slow' and 'quick' motions [10]. High frequency oscillations combined with the low frequency oscillations increase the ratio of the maximal dimension of the cavitation bubble to its minimal dimension. At the same time its implosion phase shifts; this leads to a substantial increasing of the erosion coefficient of the cavitation bubble. These effects defined the research directions related to the action of bifrequency cavitation on the physico-chemical processes that occur at the interfaces liquid – liquid, liquid – solid, liquid - liquid - gas, liquid - solid – gas.

The experimental treatment was chosen on the basis of the calculated and experimental data, which gave evidence that the acoustic and static pressures exert the maximal influence on the increasing of the efficiency of ultrasonic – hydroacoustic dispergation of bentonite suspensions [11]. The bentonite powder was loaded in a vessel with its volume of 1 dm^3, and then water was added. The concentration of the solid phase in bentonite varied from 5% to 20%. To perform the treatment, the resonance frequency and amplitude of oscillations of the ultrasonic installation were adjusted. The amplitude of oscillations was measured by virtue of an electrodynamic sensor [12]. The cavitation regime was registered using the waveguide signals according to the characterisitc images of the cavitation noise spectrum.

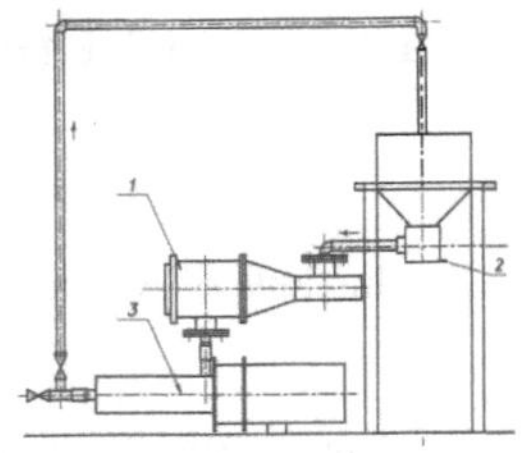

a)

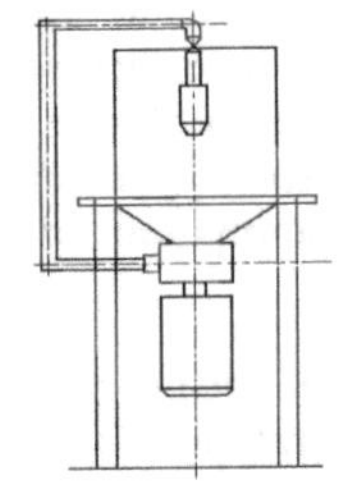

b)

Fig. 2 Scheme of the cavitation installation: (a) with hydrodynamic and hydroacoustisc apparatuses; (b) with pulse rotor

The process of nanodispergation of bentonite in water under the action of ultrasonic, hydroacoustic, and/or hydrodynamic cavitation was studied as a function of the duration of treatment, the amplitude of waveguide oscillations, the flow rate and pressure. The dimensions of particles and the degree of dispergation of the suspension were determined by electron microscopy.

III. RESULTS AND DISCUSSION

With the aim to determine the main regularities of cavitation dispergation of bentonite and propose a model of the process we studied their kinetic peculiarities and the relationship of the properties of the suspension of nanometric bentonite particles with the quality of the treated product (wine).

In the process of dispergation of solid bentonite particles using ultrasonic cavitation the duration of treatment exerts the most substantial influence (Fig. 3). Its increasing is accompanied by an intensive reducing of the dimensions of bentonite particles; when the duration of the treatment exceeds 12 min finely dispersed suspensions are obtained for all used amplitudes of oscillations.

The electron microscopy analysis has shown that the initial suspension consists of particles with different dimensions. The initial average dimensions of the particles amounted to 200 – 300 μm (Fig. 4a). The microphotographs of suspensions confirm the effectiveness of ultrasonic cavitation treatment, since the suspension after the treatment mainly contains submicronic particle with the same dimensions. The completely homogeneous fine bentonite

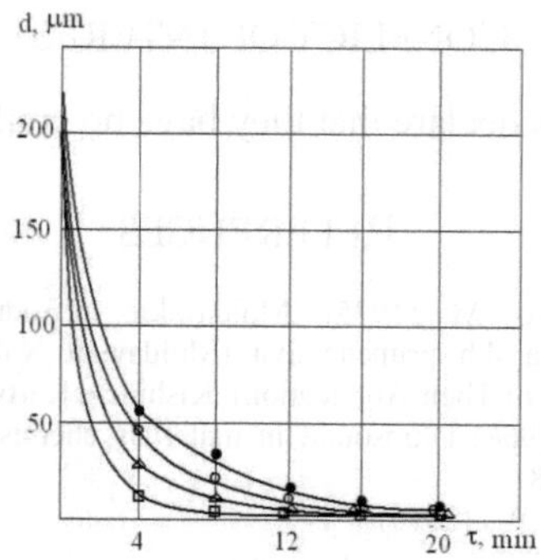

Fig. 3 Dimensions of bentonite particles (d, μm) vs the duration of ultrasonic cavitation treatment (μ, min). ξ μm: 1 – 6; 2 – 12; 3 – 20; 4 – 30

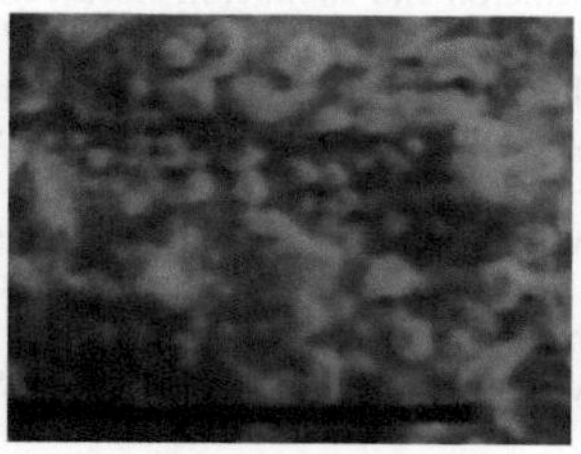

a)

b)

Fig. 4 Electron microphotographs of bentonite prepared using ultrasonic cavitation; a) initial dimensions (200 – 300) μm; b) ξ = 30 μm, τ = 20 min, d = 0.4 μm

suspension was obtained when the amplitude of oscillations was greater than 20 μm, and the duration of treatment exceeded 12 minutes (Fig. 4b).

Ultrasonic method does not allow one to treat large bentonite volumes. The method of hydrodynamic and hydroacoustic cavitation allows one to dispergate and homogenize large quantities of mixtures of liquid and solid particles. Bentonite dispergation was studied using hydroacoustic and hydrodynamic cavitation units at the cavitator frequences of 400-450 Hz, in the acoustic radiator at 10 – 12 kHz (Fig. 2a), and in the pulse rotor at frequencies of 2.5-3.0 kHz (Fig. 2 b).

The obtained results demonstrate (Fig. 5) that with the increasing of treatment duration the dimensions of bentonite particles reduce and reach the same value (35 – 40 μm with the hydrodynamic unit and 8 – 15 μm with the pulse rotor). After hydrodynamic treatment during more than 60 min the dimensions of bentonite particles do not change and remain the same (8 – 15 μm).

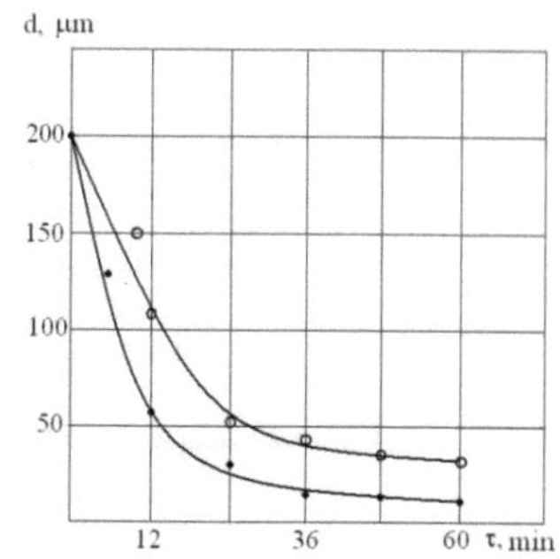

Fig. 5 Dimensions of bentonite particles vs the duration of cavitation treatment: o – hydrodynamic unit, f = 450 Hz; • – pulse rotor, f = 3600 Hz.

To increase the efficiency of dispergation and homogenization when various methods are used for treatment of large volumes it is appropriate to use a combined action of oscillations with different frequencies. The technological line shown in Fig. 6 was used for investigations. The action of bifrequency cavitation is efficient and optimal when the recirculation pressure of the suspension, duration of homogenization, and the phase ratio liquid – solid (concentration of dry substances in the suspension) are adjusted in a proper way. According to the obtained experimental data, for the efficient treatment the static pressure in the cavitation acoustic zone should be not less than 4 - 6 atm. Dispergation of bentonite under the action of bifrequency cavitation has shown that particles with the dimensions of (0.3 – 2 μm) prevail in the suspension (Fig. 7).

Fig. 6 Technological line for fine dispergation of bentonite

Bentonite suspensions dispergated in the field of bifrequency cavitation were used for clarification of wine according to the traditional process [13]. Bifrequency cavitation dispergation allows one to reduce by a factor of 6 the quantity of bentonite used for clarification.

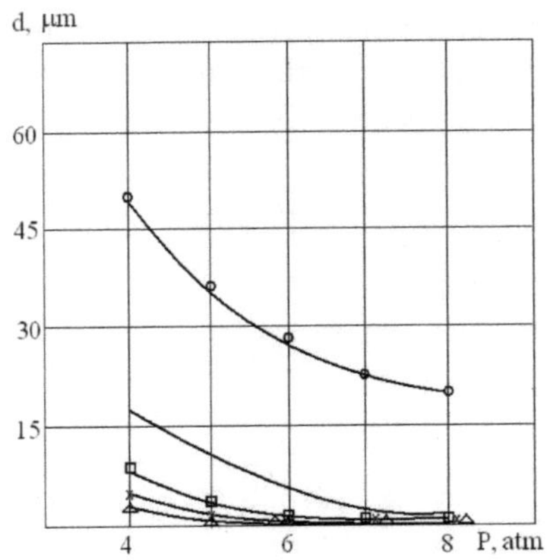

Fig. 7 Dimensions of bentonite particles (d, μm) vs the pressure in the recirculation line.o – τ=24, • - τ=12, x - τ=48, Δ - τ=60 min

Application of bentonite suspension prepared using the cavitation bifrequency technology to the maximal degree retains the gustative and nutritional qualitites of the end product (wine). The prepared suspention consists of water, nanometric gas bubbles, and solid bentonite particles. The contact area of the particles with the treated product is by 2 or 3 orders of magnitude greater than in usual suspensions; this allows one to considerably intensify the processes of adsorption and flocculation of substances in the treated product.

IV. CONCLUSIONS

The results of the studies of the process of preparation of bentonite suspensions in ultrasonic, hydroacoustic, hydrodynamic, and bifrequency cavitation fields are discussed. The initial dimensions of the solid particles amounted to 200 – 300 μm. The treatment allowed us to reduce the dimensions of the particles to submicronic, nanometric (0.3 – 2.1) μm. It was demonstrated that the prepared finely dispersed activated bentonite suspension reduces approximately by a factor of 6 the quantity of bentonite needed for clarification and deproteinization of wines.

CONFLICT OF INTEREST

The authors declare that they have no conflict of interest.

REFERENCES

1. Kerdivarenko M (1975) Moldavskie prirodnye adsorbenty i tekhnologiya ikh primenemiya (Moldavian Natural Sorbents and Technology of Their Application). Kishinev, Kartya Moldovenyaske
2. Peters D (1966) Ultrasound in materials chemistry. J Mater Chem 6:1605-1618
3. Mekhamer W (2010) The colloidal stability of raw bentonite deformed mechanically by ultrasound. J Saudi Chem Soc 14:301-306
4. Darvishi Z, Morsali A (2011) Synthesis and characterization of nano-bentonite by sonochemical method. Ultrason. Sonochem 18:238-242
5. Nguyen A, Reinert L, Lévêque J-M et al. (2013) Preparation and characterization of micron and submicron-sized vermiculite powders by ultrasonic irradiation. Appl Clay Science, 72:9-17
6. Lapides I, Yariv Sh (2004) The effect of ultrasound treatment on the particle-size of Wyoming bentonite in aqueous suspensions. J Mater Sci, 2004, 39:5209–5212
7. Margulis M, Khavskii N (1985) On the mechanism of combined action of acoustic oscilaation with two frequencies on the physico-chemical and chemical effects, Akusticheskaya kavitatsiya i primenenie ul'trazvuka v khimicheskoi tekhnologii. Slavskoe, Russia, pp. 93–94
8. Dmitrieva A (1985) Combined action of two frequencies of ultrasonic oscillations on the sonochemical reactions, Akusticheskaya kavitatsiya i primenenie ul'trazvuka v khimicheskoi tekhnologii. Slavskoe, Russia, p. 56
9. Agranat B, Khavskii N, Fel'dman A et al. (1987) On the problem of the mechanism of action of acoustic oscillations on the flotation process, 6th Vsesoyuzn. konf. po ul'trazv. metodam intensificatsii tekhnologicheskikh protsessov, Moscow, Russia, p. 83
10. Dubrovin M, Kitaigorodskii Yu, Khavskii N, (1987) Dynamics of pulsations of the cavitation bulb at bifrequency excitation, 6th Vsesoyuzn. konf. po ul'trazv.metodam intensificatsii tekhnologicheskikh protsessov, Moscow, Russia, p. 14
11. Agranat B, (1992) Ul'trazvukovaya tekhnologiya, Moscow, Metallurgiya
12. Herbertz J (1967) Non-contact electrodynamic measurement on metallic vibrators at 20 kHz. Ultrasonics, 5:239–240
13. Valuiko G, Zinchenko V, Mekhuzla N, (1999) Stabilizatsiya vinogradnykh vin, Simferopol', Tavrida

Author: Petru Dumitras
Institute: Institute of Applied Physics ASM
Street: Academiei 5
City: Chisinau
Country: Moldova
Email: pdumitras@yahoo.com

Influence of Fe Catalyst Morphology on the Growing of Carbon Nanotubes

A. Băra, C. Banciu, V. Marinescu, C. Morari, and D. Pătroi

National Institute for Research and Development in Electrical Engineering ICPE-CA/Advanced Materials Department, Bucharest, Romania

Abstract— **This study investigates the influence of Fe catalysts on the growing of carbon nanotubes (CNTs) by chemical vapour deposition (CVD) method on different types of substrates. The catalyst film was deposited on the silicon, ceramic or metallic substrates by magnetron sputtering before CNTs growing using C_2H_4 as carbon source. A detailed study about the influence of catalyst layer thickness and morphology and CVD process temperature on the growth of CNTs was performed, using different microscopy techniques.**

Keywords— **carbon nanotubes, chemical vapour deposition, catalyst film, magnetron sputtering.**

I. INTRODUCTION

Since their discovery in 1991 [1], carbon nanotubes (CNTs) have shown many remarkable mechanical and electronic properties: 100 times the tensile strength of steel, thermal conductivity better than all but the purest diamond, and electrical conductivity similar to copper [2–6]. Carbon nanotubes (CNTs) have been extensively investigated for the synthesis using arc discharge [7, 8], laser vaporization [9], pyrolysis [10], and plasma-enhanced chemical vapour deposition (PECVD) [11, 12], for its unique physical and chemical properties [13-15] and for applications to nanoscale devices [16-18]. Numerous potential applications for carbon nanotubes such as flat panel displays [19, 20], hydrogen storage [21, 22], chemical sensors [23], electrodes for secondary battery, fuel cells, memory devices etc. were proposed [24].

Since the properties of CNTs vary depending on their structure, synthesis method development was targeted towards obtaining CNTs with controlled diameter, length and number of walls. Among the various production methods, supported [25-27] and gas-phase [28-30], chemical vapour deposition (CVD) has been found to be efficient and selective for either single-walled (SWCNT) or multiwalled carbon nanotube growth (MWCNT) [31]. The CVD method has attracted much attention because of the advantage that the growth of CNTs can be achieved with high purity, high yield, and vertical alignment [24].

Here, we investigate the influence of Fe catalysts on the growing of carbon nanotubes (CNTs) by chemical vapour deposition (CVD) method on different types of substrates: silicon, ceramic or metallic. The catalyst film was deposited on the substrates by magnetron sputtering before growing CNTs using C_2H_4 as carbon source. A study about the influence of the thickness and morphology of the catalyst layer and the process temperature on the growth of CNTs was performed, using different microscopy techniques.

II. EXPERIMENTAL

A. *Obtaining Carbon Nanotubes*

CNTs were synthesized by CVD growth on the following substrates: silicon (100) type n, ceramic and stainless steel (1.4301) with an iron catalyst layer obtained through magnetron sputtering method. Prior to the CNT growth process, catalyst annealing was carried out at a temperature of 750°C for 15 minutes in an inert atmosphere inside the quartz tube of the CVD equipment (table 1). Within the CNT growth experiment by CVD method, C_2H_4 was used as gas source of carbon. General parameters of CVD process for nanotubes growth are presented in table 2.

Table 1 Catalysts annealing process parameters

Temperature [°C]	Time [minutes]	Pressure [torr]	H_2 flow [cm^3/min]
750	15	760	500

Table 2 CVD process parameters

Temperature [°C]	Time [min]	Pressure [torr]	Ar flow [l/min]	H_2 flow [cm^3/min]	C_2H_4 flow [cm^3/min]
750	300	760	1	400	300

B. *Characterization*

The obtained samples were characterized by different microscopy techniques (AFM, SEM and HRTEM).

AFM analysis was performed with a Scanning Probe Microscope model NTEGRA Aura produced by NT-MDT. AFM analysis was performed in non-contact mode, with a tip of Si_3N_4 and a resolution of 10 nm.

Morphological characterization of the CNTs by scanning electron microscopy (SEM) was performed by a FESEM/FIB/EDS Workstation Auriga produced by Carl

V. Sontea and I. Tiginyanu (eds.), *3rd International Conference on Nanotechnologies and Biomedical Engineering*,
IFMBE Proceedings 55,
DOI: 10.1007/978-981-287-736-9_51

Zeiss Germany, with an acceleration voltage of 5 kV, using the SESI detector.

Transmission electron microscopy studies were performed with a high resolution transmission electron microscope HRTEM model LIBRA 200FE-HR.

III. RESULTS AND DISCUSSION

Figures 1 and 2 display the AFM micrographs of the iron catalyst layer deposited on the silicon respectively ceramic substrate. The figures highlight the changes in size and shape of the catalyst grains after the annealing thermal treatment.

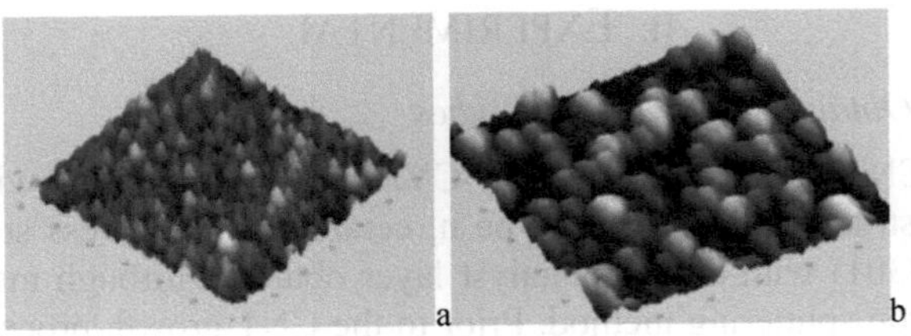

Fig. 1 AFM image of the magnetron sputtered Fe film on silicon substrate: a) before annealing; b) after annealing

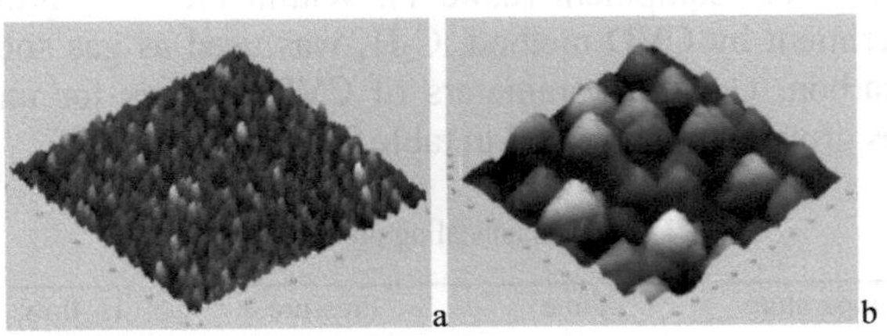

Fig. 2 AFM image of the magnetron sputtered Fe film on ceramic substrate: a) before annealing; b) after annealing

As it can be seen in the figures above, the annealing treatment led to an increase in the grain size of iron catalyst particles deposited on silicon and ceramic substrates.

Figure 3 shows the AFM micrograph of the CNTs grown by CVD at a temperature of 750°C for 300 minutes, on silicon substrate with iron catalyst without annealing.

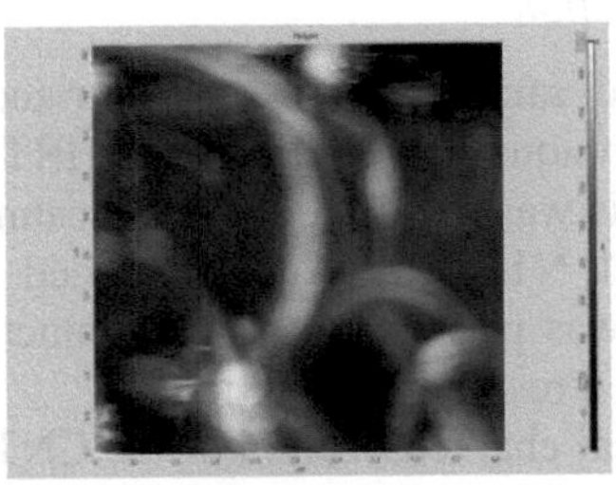

Fig. 3 AFM topography of CNTs grown by CVD at 750° for 300 minutes reaction time on silicon substrate with iron catalyst without annealing

The SEM images in the figures 4 - 6 correspond to the CNTs obtained in the same reaction conditions of synthesis: temperature 750°C, flow of Ar (1 l/min), H_2 flow (400 cm^3/min), C_2H_4 flow (300cm^3/min), pressure (70 torr), and reaction time (300 min) on the silicon, ceramic or metallic substrates.

Figure 3b displays CNTs obtained in the same reaction conditions after catalyst annealing at 700°C.

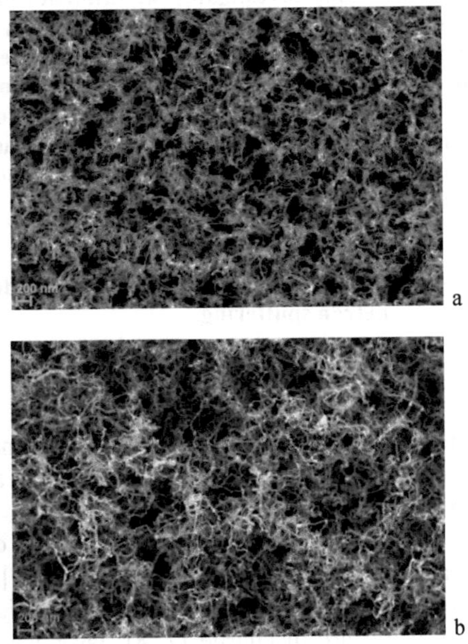

Fig. 4 SEM images of the CNTs grown by CVD at 750°C for 300 minutes reaction time on silicon substrate with iron catalyst: a) without annealing; b) with annealing

The SEM images presented above highlight the CNT growth when using iron catalyst deposited on silicon substrate. The average size of CNTs diameters obtained without catalyst annealing was between 8 nm and 15 nm. In the case of the CNT grown on the catalyst after annealing thermal treatment, the resulted CNTs show increased diameter in a wide range of values (from 9.3 to 25.7 nm).

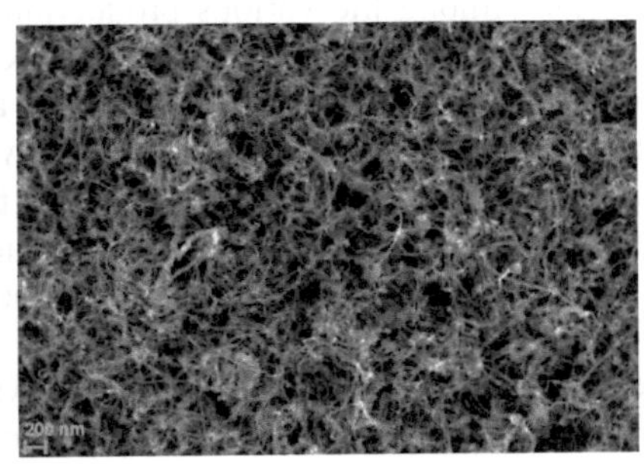

Fig. 5 SEM images of the CNTs grown by CVD at 750°C for 300 minutes reaction time on ceramic substrate with iron catalyst without annealing

In the case of the CNT growth on the iron catalyst deposited on ceramic substrate, the resulted CNTs show a diameter in the range 10.4-13.5 nm.

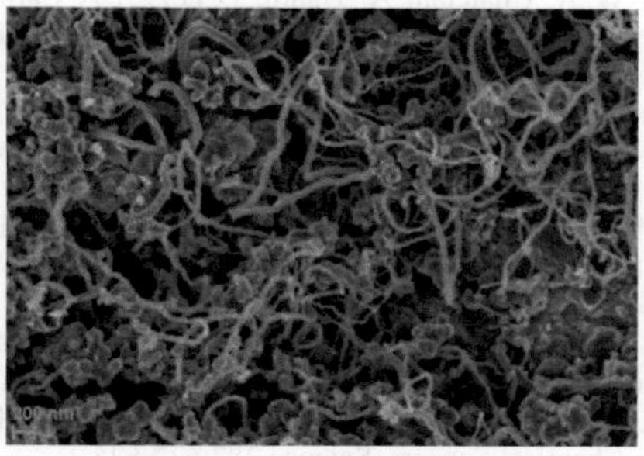

Fig. 6 SEM images of the CNTs grown by CVD at 750°C for 300 minutes reaction time on metallic substrate with iron catalyst

In the case of the CNT growth on the iron catalyst deposited on metallic substrate with annealing, the resulted CNTs show a diameter in the range 50-300 nm, probably due to the high roughness of the substrate.

Figures 7 and 8 display the HRTEM images of the CNTs grown by CVD at 750°C in 300 minutes on silicon and ceramic substrates.

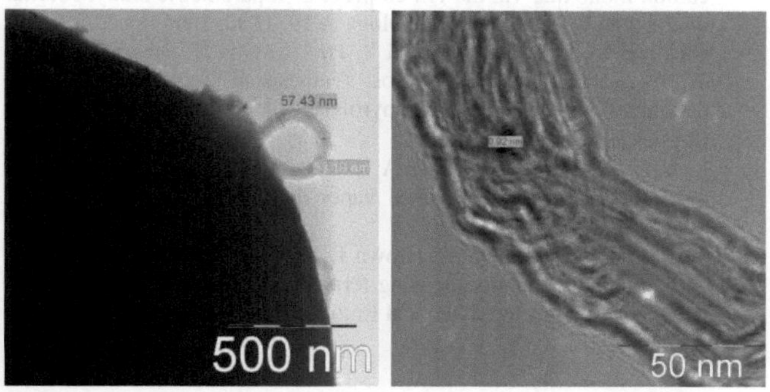

Fig. 7 HRTEM images of the CNTs grown by CVD at 750°C for 300 minutes reaction time on ceramic substrate with iron catalyst

The above HRTEM images show that the carbon nanotubes obtained by the CVD method on different substrates have inhomogeneous structure and they are multiwall carbon nanotubes (MWCNT). MWCNT grown by CVD method at 750°C on silicon substrate with iron catalyst have a bamboo-shaped structure.

In the case of CNTs grown on ceramic substrate it can be observed that the distance between two graphitic planes (002) is about 3.3 Å.

Fig. 8 HRTEM images of the CNTs grown by CVD at 750°C for 300 minutes reaction time on silicon substrate with iron catalyst

IV. CONCLUSIONS

CNTs were synthesized by thermal CVD method on different substrates using as catalyst iron obtained by magnetron sputtering. The morphology of the resulted CNTs was studied by different microscopy techniques. It was found that the CNTs diameter is influenced by the structure of the catalyst, depending on the substrate type and the annealing treatment.

The annealing treatment of the catalyst leads to the increasing of its grains. This fact influences the carbon nanotubes growth, leading to their increase in diameter.

HRTEM investigations show that by using magnetron sputtered Fe catalyst on different substrates (silicon, ceramic, metallic), the obtained carbon nanotubes are multiwall type, with diameters varying from few to tens nanometers, depending on the catalyst and the substrate. The carbon nanotubes grown by CVD method on silicon substrate with iron catalyst have a bamboo-shaped structure.

ACKNOWLEDGMENT

This work was supported by a grant of the Romanian National Authority for Scientific Research, in the frame of MNT ERA NET project CarPolCap ("Mini-supercapacitors technology, based on hybrid CNT/CNF - electroactive polymer networks“), project number 7-053/2012 and by the CORE Programme, project number 0935-0103/2009.

CONFLICT OF INTEREST

The authors declare that they have no conflict of interest.

REFERENCES

1. Iijima S (1991) Helical microtubules of graphitic carbon. Nature 354:56-58
2. Ebbesen T W (1997) Carbon Nanotubes: Preparation and Properties. CRC Press, Inc., Boca Raton, 1997, pp. 191–249
3. Ebbesen T W, Lezec H J, Hiura H, Bennett J M, Ghaemi H F, Thio T (1996) Electrical conductivity of individual carbon nanotubes. Nature 382:54-56
4. de Heer W A, Bacsa W S, Chatelain D A, Gergin T, Humphrey-Baker R, Forro L, Ugarte D (1995) Aligned Carbon Nanotube Films: Production and Optical and Electronic Properties. Science 268:845-847
5. Delaney P, Choi H J, Ihm J, Cohen M L (1998) Broken symmetry and pseudogaps in ropes of carbon nanotubes. Nature 391:466-468
6. Agboola A E, Pike R W, Hertwig T A, Lou H H, (2007) Conceptual design of carbon nanotube processes. Clean Technologies and Environmental Policy 9(4):289-311 DOI 10.1007/s10098-006-0083-2
7. Iijima S, Ichihashi T (1993) Single-shell carbon nanotubes of 1-nm diameter. Nature 363:603-605
8. Bethune D S, Kiang C H, deVries M S, Gorman G, Savoy R, Vazquez J, Beyers R (1993) Cobalt-catalysed growth of carbon nanotubes with single-atomic-layer walls. Nature 363:605-607
9. Thess A, Lee R, Nikolaev P et al. (1996) Crystalline Ropes of Metallic Carbon Nanotubes. Science 273:483-487
10. Terrones M, Grobert N, Olivares J et al. (1997) Controlled production of aligned-nanotube bundles. Nature 388:52-55
11. Ren Z F, Huang Z P, Xu J W et al. (1998) Synthesis of Large Arrays of Well-Aligned Carbon Nanotubes on Glass. Science 282:1105-1107
12. Sung S L, Tsai S H, Tseng C H, Chiang F K, Liu X W, Shih H C (1999) Well-aligned carbon nitride nanotubes synthesized in anodic alumina by electron cyclotron resonance chemical vapor deposition. Appl. Phys. Lett. 74:197-199
13. Dresselhaus M S, Dresselhaus G, Eklund P C (1996) Science of Fullerenes and Carbon Nanotubes, Academic Press, San Diego, 1996, Chapter 19
14. Treacy M M J, Ebbesen T W, Gibson J M (1996) Exceptionally high Young's modulus observed for individual carbon nanotubes. Nature 381: 678-680
15. Delaney P, Choi H J, Ihm J, Louie S G, Cohen M L, (1998) Broken Symmetry and Pseudogaps in Ropes of Carbon Nanotubes. Nature 391:466-468
16. Tans S J, Verschueren R M, Dekker C (1998) Room-temperature transistor based on a single carbon nanotube. Nature 393:49-52
17. Tans S J, Verschueren R M, Dekker C (1997) Individual single-wall carbon nanotubes as quantum wires. Nature 386:474-477
18. Choi Y C, Kim D W, Lee T J, Lee C J, Lee Y H (2001) Growth mechanism of vertically aligned carbon nanotubes on silicon substrates. Synthetic Metals 117:81-86
19. de Heer W A, Chatelain A, Ugarte D (1995) A Carbon Nanotube Field-Emission Electron Source. Science 270:1179-1180
20. Rinzler A, Hafner J H, Nikolaev P, Lou L, Kim S G, Tomanek D, Nordlander P, Colbert D T, Smalley R E (1995) Unraveling nanotubes: field emission from an atomic wire. Science 269:1550-1553
21. Dillon A C, Jones K M, Bekkedahl T A, Kiang C H, Bethune D S, Heben M J (1997) Storage of hydrogen in single-walled carbon nanotubes. Nature 386:377-379
22. Cao A, Zhu H, Zhang X et al. (2001) Hydrogen storage of dense-aligned carbon nanotubes. Chem. Phys. Lett. 342:510-514
23. Kong J, Franklin N R, Zhou C et al. (2000) Nanotube Molecular Wires as Chemical Sensors. Science 287:622-625
24. Zhang X, Cao A, Wei B et al. (2002) Rapid growth of well-aligned carbon nanotube arrays. Chemical Physics Letters 362:285–290
25. Cassell A M, Raymakers J A, Kong J, Dai H (1999) Large scale CVD synthesis of single-walled carbon nanotubes. Journal of Physical Chemistry B 103:6484–6492
26. Dai H, Rinzler A G, Nikolaev P, Thess A, Colbert D T, Smalley R (1996) Single-wall nanotubes by metal-catalysed disproportionation of carbon monoxide. Chemical Physics Letters 260: 471–475
27. Li Y, Zhang X, Shen L et al. (2004) Controlling the diameters in large-scale synthesis of single-walled carbon nanotubes by catalytic decomposition of CH_4. Chemical Physics Letters 398: 276–282
28. Bronikowski M J, Willis P A, Colbert D T, Smith K A, Smalley R E (2001) Gas-phase production of carbon single-walled nanotubes from carbon monoxide via the HiPco process: a parametric study. Journal of Vacuum Science and Technology A 19: 1800–1805
29. Cheng H M, Li F, Su G, Pan H Y, He L L, Sun X, Dresselhaus M S (1998). Large-scale and low-cost synthesis of single-walled carbon nanotubes by the catalytic pyrolysis of hydrocarbons. Applied Physics Letters 72:3282–3284
30. Li Y-L, Kinloch I A, Windle A H (2004) Direct spinning of carbon nanotube fibers from chemical vapor deposition synthesis. Science 304:276–278
31. Moisala A, Nasibulin A G, Brown D P et al. (2006) Single-walled carbon nanotube synthesis using ferrocene and iron pentacarbonyl in a laminar flowreactor. Chemical Engineering Science 61:4393 – 4402

Author: Adela Băra
Institute: National Institute for Research and Development în Electrical Engineering ICPE-CA
Street: 313 Splaiul Unirii
City: Bucharest
Country: Romania
Email: adela.bara@icpe-ca.ro

Investigation of the Generalized Anderson Impurity Model

V.A. Moskalenko[1,2], L.A. Dohotaru[3], D.F. Digor[1], and I.D. Cebotari[1]

[1] Institute of Applied Physics, Moldova Academy of Sciences, Chisinau, Moldova
[2] BLTP, Joint Institute for Nuclear Research, Dubna, Russia
[3] Technical University of Moldova, Chisinau, Moldova

Abstract— **The new canonical transformation was proposed for generalized Anderson impurity model and proved appearance of ordinary Anderson impurity model with the external magnetic field. The Dyson type equations for one-particle propagators of conduction and impurity electrons in the normal state of the system in the absence of the magnetic field have been established.**

Keywords— **Green's function, quantum dot, Dyson type equation, Anderson impurity model, strong electronic correlations.**

I. INTRODUCTION

The technological progress in nanoscale electronic circuits, such as quantum dots, stimulates the studies of quantum-impurity models. One of them is the Anderson single impurity model [1]. This model describes electronic correlations in small quantum dots.

Exist two models of quantum dots [2,3]: single two level quantum dot and double quantum dot where each dot harbors only a single level.

The original double dot system is transformed into generalized Anderson impurity model with local Zeeman magnetic field [2,3].

The fluctuations of the pseudo-spin electron degree of freedom are determined by two converting effects: the polarizing of the local magnetic field and the Kondo screening by the itinerant electrons.

The generalized Anderson Hamiltonian is [2,3]

$$\begin{aligned} H = &\sum_{\vec{k}\sigma} \varepsilon(\vec{k}) C^{+}_{\vec{k}\sigma} C_{\vec{k}\sigma} + \sum_{\sigma} (\varepsilon_{\sigma} - \sigma \frac{h}{2}\cos\theta_0) n_{\sigma} - \\ &- \frac{h}{2}\sin\theta_0 (d^{+}_{\uparrow} d_{\downarrow} + d^{+}_{\downarrow} d_{\uparrow}) + U n_{\uparrow} n_{\downarrow} + \\ &+ \sum_{\vec{k}\sigma} (V_{\vec{k}\sigma} C^{+}_{\vec{k}\sigma} d_{\sigma} + H.C.), \sigma = \pm 1, \end{aligned} \tag{1}$$

where $C_{\vec{k}\sigma}(C^{+}_{\vec{k}\sigma})$, and $d_{\sigma}(d^{+}_{\sigma})$ is lead and impurity electron annihilation(creation) operators correspondingly. The parameter U is the Coulomb interaction of the impurity electrons, h is the magnetic field.

We propose the new canonical transformation of localized electrons

$$d_{\uparrow} = \cos\theta a_{\uparrow} + \sin\theta a_{\downarrow}, d_{\downarrow} = -\sin\theta a_{\uparrow} + \cos\theta a_{\downarrow}, \tag{1}$$

The presence of strong Coulomb interaction invalidates the using of the Feynman diagram technique of statistical mechanics.

We elaborated [4-10] the new diagram technique for strong correlated electron systems based on the generalized Wick theorem and the occurrence of new elements of diagram named by us as irreducible Green's functions.

Using the new diagrammatic approach we derive the Dyson type equations for Green's functions of the localized and conduction electrons.

The system of equation can be closed by summing the infinite series of ladder diagrams containing irreducible Green's functions.

Our main approximation comes to the summation of the ladder diagrams which will be enough to obtain the leading contribution of the spin and charge fluctuations.

In the used approximation [9] only the simplest irreducible Green's functions of the localized electrons are considered and iterated many times.

We have analyzed in the such approach the simplest case of symmetric Anderson model [4], when the chemical potential is $\mu = U/2$.

II. CANONIC TRANSFORMATION

By using the canonical transformation (2) of electron impurity operator we obtain the following equations:

$$\begin{aligned} &d^{+}_{\uparrow} d_{\uparrow} = \cos^2\theta a^{+}_{\uparrow} a_{\uparrow} + \sin\theta\cos\theta (a^{+}_{\uparrow} a_{\downarrow} + a^{+}_{\downarrow} a_{\uparrow}) + \sin^2\theta a^{+}_{\downarrow} a_{\downarrow}, \\ &d^{+}_{\downarrow} d_{\downarrow} = \cos^2\theta a^{+}_{\downarrow} a_{\downarrow} - \sin\theta\cos\theta (a^{+}_{\uparrow} a_{\downarrow} + a^{+}_{\downarrow} a_{\uparrow}) + \\ &+ \sin^2\theta a^{+}_{\uparrow} a_{\uparrow}, d^{+}_{\uparrow} d_{\uparrow} + d^{+}_{\downarrow} d_{\downarrow} = a^{+}_{\uparrow} a_{\uparrow} + a^{+}_{\downarrow} a_{\downarrow}, \end{aligned}$$

$$\begin{aligned} &d^{+}_{\uparrow} d_{\uparrow} - d^{+}_{\downarrow} d_{\downarrow} = \cos 2\theta (a^{+}_{\uparrow} a_{\uparrow} - a^{+}_{\downarrow} a_{\downarrow}) + \sin 2\theta (a^{+}_{\uparrow} a_{\downarrow} + a^{+}_{\downarrow} a_{\uparrow}), \\ &d^{+}_{\uparrow} d_{\downarrow} + d^{+}_{\downarrow} d_{\uparrow} = \cos 2\theta (a^{+}_{\uparrow} a_{\downarrow} + a^{+}_{\downarrow} a_{\uparrow}) + \sin 2\theta (a^{+}_{\downarrow} a_{\downarrow} - a^{+}_{\uparrow} a_{\uparrow}), \\ &n_{\uparrow} n_{\downarrow} = d^{+}_{\uparrow} d_{\uparrow} d^{+}_{\downarrow} d_{\downarrow} = a^{+}_{\uparrow} a_{\uparrow} a^{+}_{\downarrow} a_{\downarrow} = m_{\uparrow} m_{\downarrow}. \end{aligned} \tag{3}$$

V. Sontea and I. Tiginyanu (eds.), *3rd International Conference on Nanotechnologies and Biomedical Engineering*,
IFMBE Proceedings 55,
DOI: 10.1007/978-981-287-736-9_52

These expressions permit us to obtain the equations

$$\sum_\sigma(\varepsilon_\sigma-\sigma\frac{h}{2}\cos\theta_0)n_\sigma-\frac{h}{2}\sin\theta_0(d_\uparrow^+d_\downarrow+d_\downarrow^+d_\uparrow)=$$
$$=\varepsilon_0\sum_\sigma(a_\sigma^+a_\sigma-\frac{h}{2}\cos\theta_0[\cos 2\theta\left(a_\uparrow^+a_\uparrow-a_\downarrow^+a_\downarrow\right)+\sin 2\theta\times$$
$$\times(a_\uparrow^+a_\downarrow+a_\downarrow^+a_\uparrow)]-\frac{h}{2}\sin\theta_0[\cos 2\theta(a_\uparrow^+a_\downarrow+a_\downarrow^+a_\uparrow)+$$
$$+\sin 2\theta\left(a_\downarrow^+a_\downarrow-a_\uparrow^+a_\uparrow\right)]=\varepsilon_0\sum_\sigma a_\sigma^+a_\sigma-\frac{h}{2}\cos(2\theta+\theta_0)\times \quad (4)$$
$$\times\left(a_\uparrow^+a_\uparrow-a_\downarrow^+a_\downarrow\right)-\frac{h}{2}\sin(2\theta+\theta_0)(a_\uparrow^+a_\downarrow+a_\downarrow^+a_\uparrow).$$

Now we choose the parameter θ from the condition

$$\sin(2\theta+\theta_0)=0, 2\theta+\theta_0=\pm n\pi, n=0,1, \quad (5)$$

and we obtain the simplest form for the dot Hamiltonian

$$H_d=(\varepsilon_0\pm\frac{h}{2})\sum_\sigma a_\sigma^+a_\sigma+Um_\uparrow^+m_\uparrow \quad (6)$$

as the form in the presence of the external field h and the Coulomb interaction U of impurity electrons.

The last term of (1) may be rewritten as

$$\sum_\sigma(b_\sigma^+a_\sigma+a_\sigma^+b_\sigma), \quad (7)$$

where

$$b_\sigma=\sum_{\vec{k}}(V_{\vec{k}\sigma}\cos\theta C_{\vec{k}\sigma}-V_{\vec{k},-\sigma}\sin\theta C_{-\vec{k},-\sigma}). \quad (8)$$

This equation determines the relation between the local and the wave vector $\vec{k}$ representation of the conduction electrons of the leads.

We have analyzed in the such approach the simplest case of symmetric Anderson model [4], when the chemical potential is $\mu=U/2$.

III. GREEN'S FUNCTIONS

In this Section, in the case of the absence of the magnetic field, we define the local operator for conduction electrons in the form

$$b_\sigma=\frac{1}{\sqrt{N}}\sum_{\vec{k}}V_{\vec{k}}C_{\vec{k}\sigma}. \quad (9)$$

The corresponding full conduction electron Green's function has the form

$$G_{\sigma\sigma'}(\tau-\tau')=-\left\langle Tb_\sigma(\tau)\bar{b}_{\sigma'}(\tau')U(\beta)\right\rangle_0^c, \quad (10)$$

where $U(\beta)$ is the evolution operator given by

$$U(\beta)=T\exp(-\int_0^\beta H_{\text{int}}(\tau)\,d\tau).$$

Because the matrix element of hybridization $V_{\vec{k}}$ is absorbed by the local operator b_σ it is convenient to introduce a new parameter λ, which will be associated to each vertex of the diagrams. In such a way the order of perturbation theory will be determined by λ and not by the matrix element $V_{\vec{k}}$ of hybridization which can be present even in zero order Green's function. In the last stage of the calculation λ will be put equal to one.

In zero order of perturbation theory the Fourier representation for the conduction electrons is determined by the formula

$$G_{\sigma\sigma'}^0(i\omega)=\delta_{\sigma\sigma'}G_\sigma^0(i\omega),\ G_\sigma^0(i\omega)=\frac{1}{N}\sum_{\vec{k}}\frac{|V_{\vec{k}}|^2}{i\omega-\varepsilon(\vec{k})}. \quad (11)$$

The presence in the definition of zero order Green's function G_σ^0 of the square of matrix element of hybridization is the consequence of our equation (9) and not of the perturbation.

The thermodynamic perturbation theory gives us in the normal state the results for one-particle Green's functions presented on the Fig.1 and Fig.2.

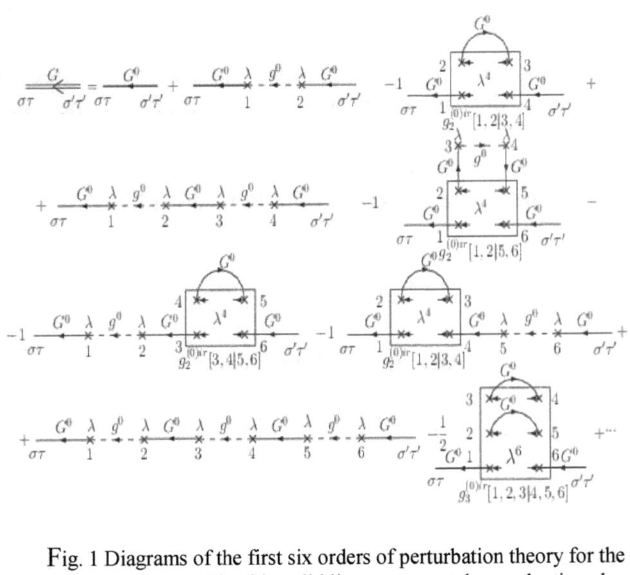

Fig. 1 Diagrams of the first six orders of perturbation theory for the conduction propagator. The thin solid lines represent the conduction electron propagator and dashed lines the impurity electron propagator of zero order. Double line represents the full propagator.

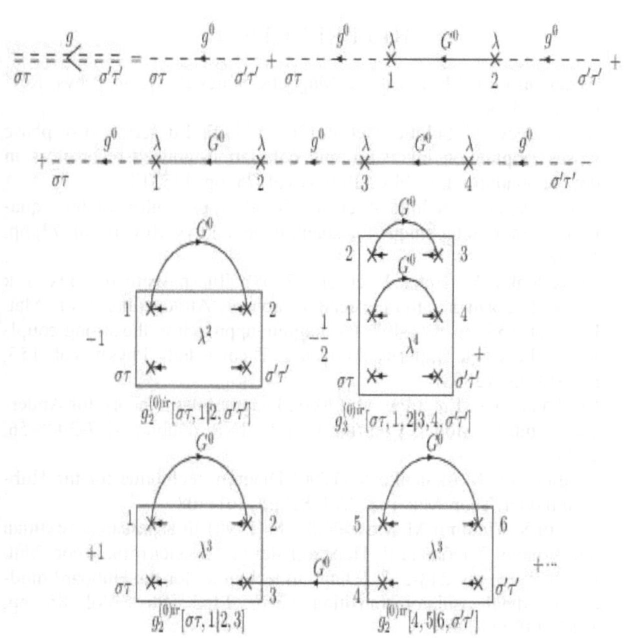

Fig. 2 Some diagrams for impurity electron propagator $g_{\sigma\sigma'}(\tau-\tau')$. The last three diagrams contain the correlation contributions. Two of them are strong connected and the last is week connected.

The double solid and dashed lines depict the renormalized propagators and the thin lines the bar propagators of conduction and impurity electrons. The lines connect the crosses which depict the impurity states. To crosses are attached two arrows one of which is ingoing and one outgoing. They depict the annihilation and creation of the electrons correspondingly. The crosses are the vertices of the diagrams and a λ multiplier is attached to each of them. The index n means (σ_n, τ_n). The summation on the index σ_n and the integration on the τ_n are intended. The rectangles with $2n$ indices and crosses depict the irreducible $g_n^{(0)ir}[1,...n\,|\,n+1,...,2n]$ Green's functions. The sign of diagrams is determined by the parity (even or odd) of the permutation of the Fermi operators necessary to obtain the diagram. Using Feynman's rules and the correspondence above, it is possible to establish the next equation for the diagrams shown in Fig.1:

$$G_{\sigma\sigma'}(\tau-\tau'\,|\,\lambda)=G^0_{\sigma\sigma'}(\tau-\tau')+\sum_{\sigma_1\sigma_2}\int_0^\beta d\tau_1\int_0^\beta d\tau_2\times \times G^0_{\sigma\sigma_1}(\tau-\tau_1)\lambda g_{\sigma_1\sigma_2}(\tau_1-\tau_2\,|\,\lambda)\lambda G^0_{\sigma_2\sigma'}(\tau_2-\tau'). \tag{12}$$

On the basis of the diagrams depicted on the Fig.2, instead it is possible to establish the following Dyson's type equation for $g_{\sigma\sigma'}$:

$$g_{\sigma\sigma'}(\tau-\tau'\,|\,\lambda)=\Lambda_{\sigma_1\sigma_2}(\tau-\tau'\,|\,\lambda)+\sum_{\sigma_1\sigma_2}\int_0^\beta d\tau_1\int_0^\beta d\tau_2\times \times\Lambda_{\sigma\sigma_1}(\tau-\tau_1)\lambda G_{\sigma_1\sigma_2}(\tau_1-\tau_2)\lambda g_{\sigma_2\sigma'}(\tau_2-\tau'), \tag{13}$$

where

$$\Lambda_{\sigma\sigma'}(\tau-\tau'\,|\,\lambda)=g^0_{\sigma_1\sigma_2}(\tau-\tau')+Z_{\sigma\sigma'}(\tau-\tau'\,|\,\lambda). \tag{14}$$

Here $Z_{\sigma\sigma'}$ is the new correlation function which contains an infinite sum of the irreducible Green's functions. As it was underlined above this function contains all spin, charge and pairing fluctuations and it is the main element of our diagram technique.

Diagram representation of the correlation function $\Lambda_{\sigma\sigma'}(\tau-\tau'\,|\,\lambda)$ is depicted on the Fig.3.

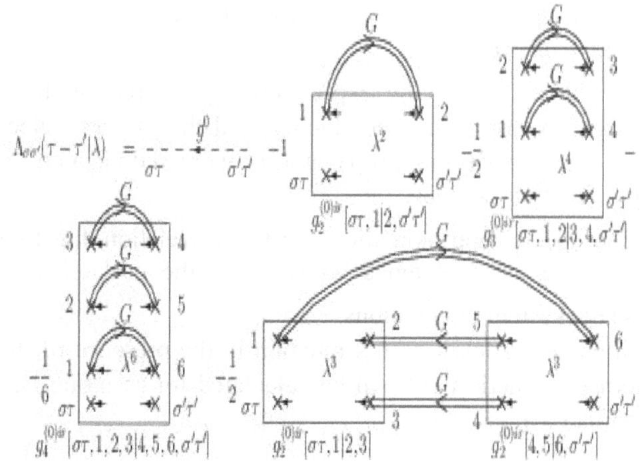

Fig. 3 Diagram representation of the correlation function $\Lambda_{\sigma\sigma'}$. Double solid lines depict the renormalized conduction electron propagators $G_{\sigma\sigma'}(\tau-\tau'\,|\,\lambda)$. The arguments of irreducible functions are supposed arranged in the clock wise direction.

The series expansion of the full propagator can gives us more detailed representation of this quantity.

By using the Fourier representation of Matsubara functions on the base of equations (12) and (13), we have:

$$G_\sigma(i\omega\,|\,\lambda)=\frac{G^0_\sigma(i\omega)}{1-\Lambda_\sigma(i\omega\,|\,\lambda)G^0_\sigma(i\omega)\lambda^2}, \tag{15}$$

$$g_\sigma(i\omega\,|\,\lambda)=\frac{\Lambda_\sigma(i\omega\,|\,\lambda)}{1-\Lambda_\sigma(i\omega\,|\,\lambda)G^0_\sigma(i\omega)\lambda^2}. \tag{16}$$

The equation (15) for conduction electron propagator is the Dyson one with mass operator determined by the correlation function of impurity electrons:

$$\Sigma_\sigma(i\omega|\lambda) = \lambda^2 \Lambda_\sigma(i\omega|\lambda). \qquad (17)$$

When λ is equal to one these quantities coincide.

The equation (16) for impurity electrons is of Dyson type and coincides, for $\lambda = 1$, with other equations obtained for strongly correlated electrons [4-10]. In equations (15), (16) and (17) the parameter λ can be taken equal to one and can be omitted.

IV. CONCLUSIONS

By using the new canonical transformation we transform the operators of generalized Anderson impurity model in new representation and obtain new operators structures which can be diagonalized by selecting the special value of the parameter θ of this canonical transformation. The result of such investigation is reduced to conclusion that $2\theta + \theta_0 = n\pi, n \in \mathbb{Z}$.

We have derived the expression for the renormalized propagators, for normal conduction and the impurity electrons. The impurity full propagator, is expressed via a correlation function $Z_\sigma(i\omega)$.

The special approximation for the correlation function $Z_\sigma(i\omega)$ has been obtained which gives the possibility to close the system of equations.

The one particle Green's function in the presence of the external magnetic field will be calculated for resulting Hamiltonian (6).

CONFLICT OF INTEREST

The authors declare that they have no conflict of interest.

REFERENCES

1. Anderson P (1961) Localized Magnetic States in Metals. Phys. Rev. B vol. 124, pp. 41
2. Kashcheyevs V, Schille A et al. (2007) Unified description of phase lapses, population inversion, and correlation-induced resonances in double quantum dots. Phys. Rev. B vol. 75, pp. 115313.
3. Kashcheyevs V, Schille A et al. (2006) Applicability of the equations-of-motion technique for quantum dots. Phys. Rev. B vol. 73, pp. 125338
4. Moskalenko V, Entel P et al. (2008) Диаграммный подход к сильной связи в однопримесной модели Андерсона. Teor. Mat. Fiz., vol. 155, pp. 474-497; [A diagram approach to the strong coupling in the single impurity Anderson. Theor. Math. Phys. vol. 155, pp. 914-935 (2008)]
5. Moskalenko V, Entel P et al. (2008) Diagrammatic theory for Anderson Impurity Model, Preprint BLTP, JINR, Dubna, E17-2008-56, 2008
6. Vladimir M, Moskalenko V (1990) Diagram technique for the Hubbard model. Teor. Mat. Fiz., Vol. 82, pp. 301-308
7. Vakaru S, Vladimir M, Moskalenko V (1990) Диаграммная техника для модели Хаббарда II. Переход металл–диэлектрик. Teor. Mat. Fiz., vol. 85, pp. 248-257; [Diagram technique for the Hubbard model. II. Metal-insulator transition Theor. Math. Phys.,Vol. 85, pp. 1185- 1192 (1990)]
8. Bogoliubov N, Moskalenko V (1991) К вопросу о существовании сверхпроводимости в модели Хаббарда Theor, Teor. Mat. Fiz., vol. 86, pp. 16-30; [On the existence of superconductivity in the Hubbard model. Math. Phys. vol. 86, pp. 10-19 (1991)]
9. Bogoliubov N, Moskalenko V (1992) Сверхпроводимость в модели Хаббарда с отклонением от половинного заполнения. Teor. Mat. Fiz., vol. 92, pp. 182-190; [Superconductivity in the hubbard model with deviation from half filling. Theor. Math. Phys. vol. 92, pp. 820-825 (1992)]
10. Moskalenko V, Entel P, Digor D (1999) Strong interaction of correlated electrons with phonons: A diagrammatic approach Phys. Rev. B vol. 59, pp. 619

Author:	L. Dohotaru
Institute:	Technical University of Moldova
Street:	Stefan cel Mare, 168
City:	Chisinau
Country:	Republic of Moldova
Email:	l_dohotaru@mail.utm.md

Thermoelectric Properties of $Bi_{1-x}Sb_x$ Alloys, Wires and Foils

A. Nikolaeva[1,2], L. Konopko[1,2], V. Shepelevich[4], V. Prokoshin[4], S. Gusakova[4], P. Bodiul[1,3], I. Popov[1], and R. Gritsko[1,3]

[1]Institute of Electronic Engineering and Nanotechnologies D.Gitsu, ASM, Chisinau, Moldova
[2] International Laboratory of High Magnetic Fields and Low Temperatures, Wroclaw, Poland
[3] Technical State University, Chisinau, Moldova
[4]Belarusian State University, Minsk, Belarus

Abstract— **In this study we focus on characterization of the transport properties of Bi-10at%Sb wires with different diameters and Bi-9at%Sb foils, prepared by the high speed crystallization. The nanowires samples were cylindrical single-crystals with (1011) orientation along the wire axis and diameters from 300 nm to 3mkm. The rapidly solidified Bi-9at%Sb foils have a microcrystalline structure and texture (1012) and thickness 20-40 mkm. Electrical resistivity and thermoelectric power were measured in the temperature range 4.2-300 K on various wire diameters d. Here we show that the quantum confinement effect in semiconducting $Bi_{0.9}Sb_{0.1}$ nanowires increases the energy gap from 8.2 meV to 17.3 meV. When the diameter of nanowires is increased, the temperature range of exponential growth of resistance shifts into higher temperature region. The dependence of thermoelectric figure of merit on diameter of wires, foils structure and Sn- doping foil was calculated in a broad temperature interval. The results obtained may provide a new way of enhancing the figure of merit in a wide temperature range.**

Keywords— **thermoelectricity, size effects, foil, semiconductor alloys, nanowires, Power factor.**

I. INTRODUCTION

Bismuth –antimony alloys have particular attention on the one hand due to features in a fundamental physics as the model materials for solid state physics, on the other hand the semiconductor $Bi_{1-x}Sb_x$ alloys is known one of the best thermoelectric material at low temperature [1-3]. The structure of their conduction and valence band dependences on different physical parameters such as alloys compositions, external presses, magnetic field [4-6].

The thermoelectric efficiency depends on the thermoelectric figure of merit of the thermoelectric material of which the thermoelectric device is made.

The figure of merit (ZT) of the materials is given by:

$$ZT = \frac{\sigma\alpha^2}{\chi}T \quad (1)$$

where σ is the electrical conductivity, α - the Seebeck coefficient, the total thermal conductivity $\chi = \chi_e + \chi_p$, (χ_e is the electronic thermal conductivity, χ_p is the lattice thermal conductivity), and T is the absolute temperature [1-3].

Searching for high ZT materials is essential in thermoelectric power generation and refrigeration.

One way to increase ZT is to maximize the Power factor $\alpha^2\sigma$ for a given χ through optimizing the electronic band structure of the material.

Previous theoretical studies of low dimensional systems for thermoelectric TE applications [7] have predicted an enhancement in ZT as the film thickness in two-dimensional (2D) systems [8] or the wire diameter in one-dimensional (1D) systems [9, 10] is decreased. The ZT benefits both from quantum effects related to carrier confinement and from pronounced phonon scattering at the boundaries.

In the present work we report the measurements resistance and thermopower data on $Bi_{0.91}Sb_{0.1}$ wires with different diameters, pure and heavily doped by Sn, $Bi_{0.091}Sb_{0.09}$ foils. The influence diameter, temperature and Sn doping on the power factor will be discussed.

II. SAMPLES AND EXPERIMENT

Thin single- crystal wires and foils of $Bi_{1-x}Sb_x$ semiconductor of various diameters in the range 300 nm- 3.0 mkm have been prepared by the high velocity crystallizations ($5*10^5$ K/s). Glass- coated single- crystal wires with diameters ranging from 300 nm to a few microns were prepared by the liquid phase casting [11-13]. X-Ray diffraction method indicate that wires all diameters are single crystals with (1011) orientation along the wire axis.

The alloy foils were prepared by crystallization of a thin layer of the melt on the inside polished surface of a rapidly rotating copper cylinder. The cooling rate of the liquid phase ≈ $5*10^5$ K/s [14].

The rapidly solidified alloys $(Bi_{91}\text{-}Sb_9)_{100-x}Sn_x$ have a microcrystalline structure and contain the dispersed thin particles. Their average size increases with increasing concentration of tin in the alloys and does not exceed 0.3 μm. The rapidly solidified foils have a microcrystalline structure and texture (10$\underline{1}$2).

To study the structure used a scanning electron microscope LEO 1455 VP with X-ray spectrum microanalyzer.

V. Sontea and I. Tiginyanu (eds.), *3rd International Conference on Nanotechnologies and Biomedical Engineering*,
IFMBE Proceedings 55,
DOI: 10.1007/978-981-287-736-9_53

The results showed that the composition of bulk samples and the composition of the foils corresponds to the composition of the alloy is initially given.

Figure 1 shows the microstructure and distribution of bismuth and antimony along a given scan line for bulk samples and foils.

To study the electrical and thermoelectric properties from the foil the samples were cut out up an down the lengths of a foil. The foil put on a special substrate (Fig. 2).

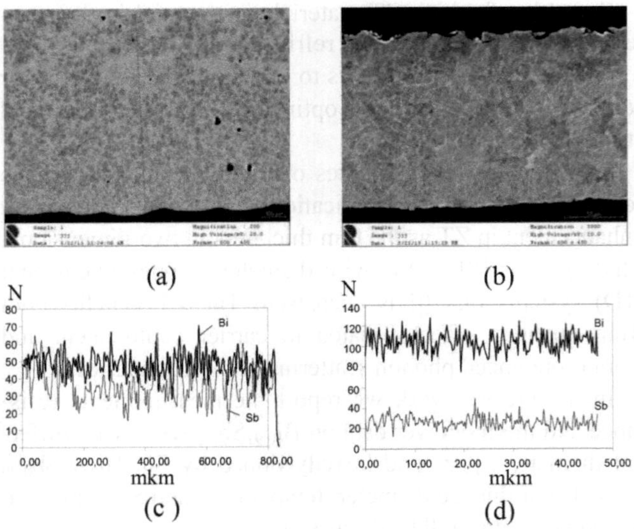

Fig. 1 Microstructure of bulk samples (a), the foil (b) and the distribution of bismuth and antimony in bulk samples (c) and foils (g).

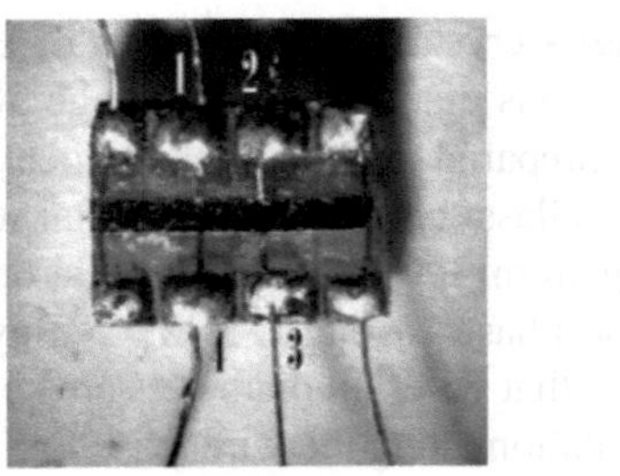

Fig. 2 Schematic representation of the experimental foil samples on the substrate with the potential (1-2), (3-4) and Hall contacts (2-3), (1-4).

Four-contact method was carried out using a copper wire with a silver paste and In solder (Fig. 2).

Type of solder does not play a significant role in the measurements. The studies were conducted in the temperature range 4.2 - 300 K and magnetic fields up to 0.4 T in transverse magnetic field (H⊥I).

Temperature dependences of kinetic coefficients in the temperature range of 77 – 300 K were measured on the experimental automated installation assembled on the basis of KAMAK. When measuring thermopower, temperature gradient 1 – 2° degrees was maintained on the sample. Copper-constantan thermocouples were used for control of temperature mode. Dependences of thermopower and resistance on temperature in the range of 4.2 – 80 K were studied on a special automated installation created on the basis of "Coolpower 4.2 GM" cryocooler by Leybold company. Temperature gradient when measuring α(T) varied by the linear law from 0.5 K liquid helium temperature to 2 K nitrogen temperature and higher. Temperature value below 80 K was controlled by Cu–CuFe thermocouple.

III. Results and Discussions

Fig. 3 shows temperature dependences of reduced resistance $R_T/R_{300}(T)$ for $Bi_{0.9}Sb_{0.1}$ wires of different diameters from 0.3 μm to 3 μm.

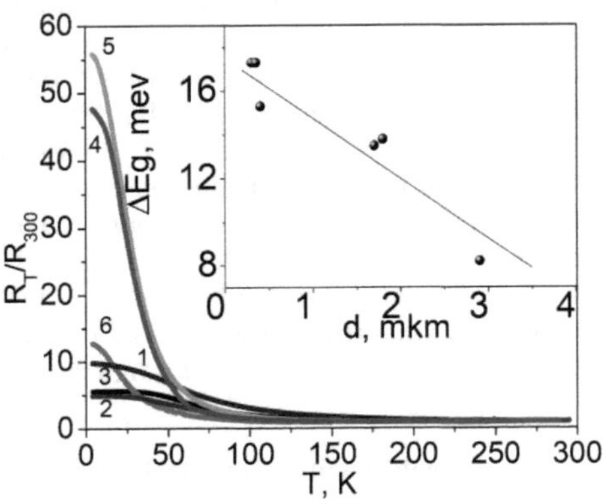

Fig. 3 Temperature dependences of reduced R_T/R_{300} resistance of single-crystal $Bi_{0.9}Sb_{0.1}$ wire: 1) d=0.3 μm; 2) d=0.35 μm; 3) d=0.4 μm; 4) d= 1.7 μm; 5) d=1.8 μm; 6) d=2.9 μm. The insert shows ΔEg vs the wire diameters.

The temperature dependences of electrical resistance of the semiconducting $Bi_{0.9}Sb_{0.1}$ wires is written as $\rho = \rho_0 * \exp(\frac{\Delta E_g}{k})$, where E_g is the conductivity activation energy, k- is the Boltsman constant, ρ_0 is temperature independent part of the resistivity. With decreasing diameter of microwires, the temperature range of exponential growth of resistance R(T) shifts into higher temperature region.

Figure 3 inset represents ΔE_g values calculated by the slope of exponential portion of $\rho(10^3/T)$ dependences for samples of $Bi_{0.9}Sb_{0.1}$ wires of different diameters. It can be seen that with increasing d the bend- gap ΔE_g decrease from 17.3 meV to 8.2 meV, that coincides with value $\Delta\varepsilon_g$=8.2 meV for the bulk $Bi_{0.9}Sb_{0.1}$ samples. Diameter dependence of energy gap ΔE_g (just increase of a gap with decreasing of the wire diameter) connected with appearance quantum size effect in nanowires. Energy shift due to the quantum size effect given by relation $\Delta\varepsilon \sim \frac{\pi^2\hbar^2}{m^* d^2}$, where

m^* is the average in-plane effective mass of the electrons and d is the wire diameter. This equation indicates that as the wire diameter decreases the energy shift increases. This leads to increase the energy gap ΔE_g with decrease diameter wires d, at it was observed in experiment.

Fig. 4 shows the temperature dependence of the resistance R (T) foil Bi-9at% Sb.

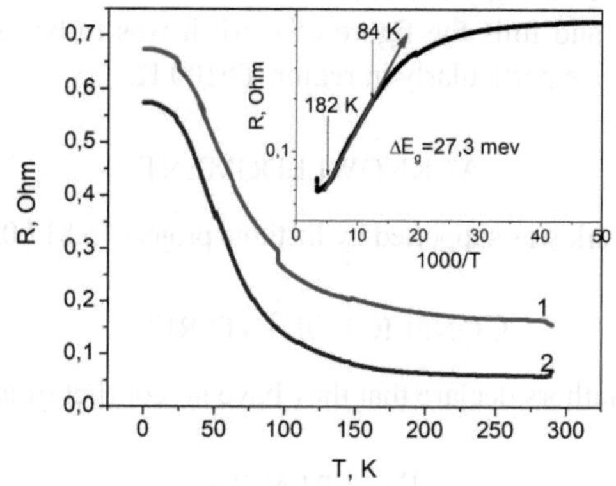

Fig. 4 The temperature dependence of the resistance R (T) Bi-foil 9% Sb, d = 45 mm. Inset: dependence logR (1000 / T). Curve 1, 2 samples cut out up an down the length of the foil.

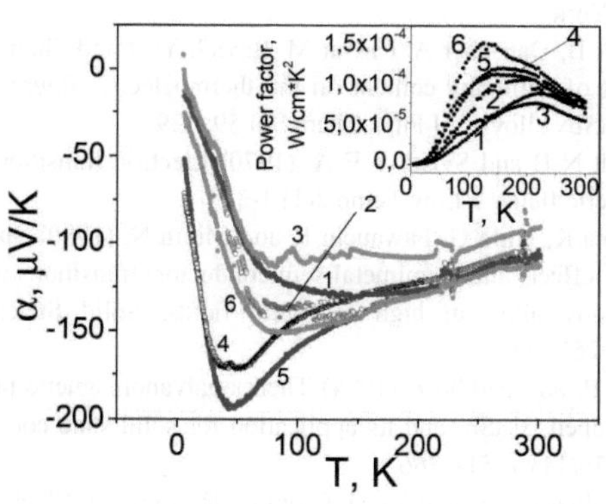

Fig. 5 Temperature dependence of the thermopower α(T) $Bi_{0,9}Sb_{0,1}$ wires: 1) d=0,3 μm; 2) d=0,35 μm; 3) d=0,4 μm; 4) d=1,7 μm; 5) d=1,8 μm; 6) d= 2,9 μm. Inset: temperature changes the Power factor ($\alpha^2\sigma$) for semiconductor $Bi_{0.9}Sb_{0.1}$ wires with diameter 0.3 – 3 μm.

Each curve has a linear section of the dependence logR (10^3/T) of which is defined by the gap ΔE_g, which was 26 meV. This difference in ΔE_g probably due to the imposition of the size effect and the scattering at the boundary of grains in objects of foil. The temperature dependence of the thermopower $Bi_{0.9}Sb_{0.1}$ wires and foils Bi-9at% Sb and Bi-9at% Sb + 0.2at% Sn are shown in Fig. 5, 6.

Over the entire temperature range from 4.2 K to 300 K the thermoelectric power has negative values in wires and foils pure $Bi_{0.9}Sb_{0.1}$, like in the bulk crystals of similar composition. The p- type doping (Sn) leads to sign of the change of the thermoelectric power α and in all temperature range α(T) has a positive value.

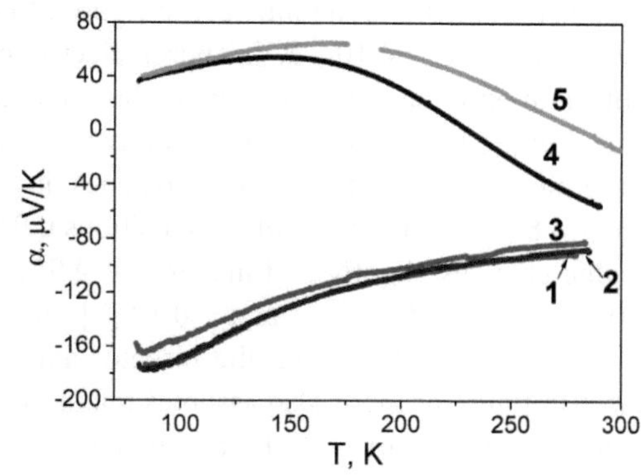

Fig. 6 Temperature dependence of the thermopower α(T) foils Bi-9% Sb (1, 2, 3), Bi-9% Sb + 0.2Sn (4), Bi-9% Sb + 0.4Sn (5) in the absence of a magnetic field (B = 0).

Figure 7 show dependences α(T) foils in transverse magnetic field (B⊥I), (B||C_3).

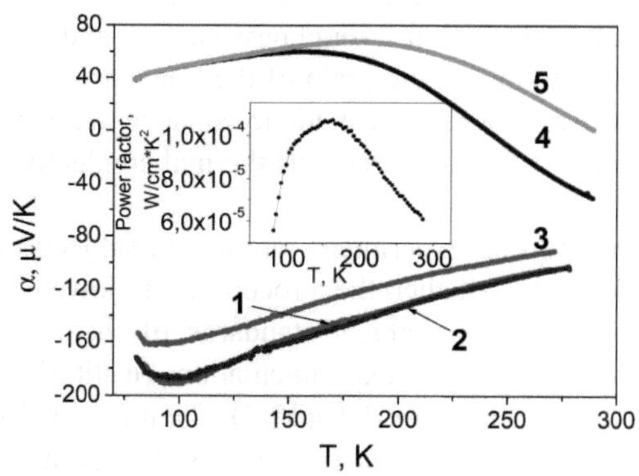

Fig. 7 Temperature dependence of the thermopower α(T) foils Bi-9% Sb (1, 2, 3), Bi-9% Sb + 0.2Sn (4), Bi-9% Sb + 0.4Sn (5) in transverve magnetoresistance (B = 0.4 T, B || C_3 (curves 1, 2), $B \perp C_3$ (curve 3)). Inset: temperature changes the Power factor ($\alpha^2\sigma$) for foil Bi-9at% Sb, d= 45μm.

In a magnetic field of 0.4 T, (B⊥I, B||C_3) value of the thermoelectric power is - 190 μV/K and weakly depends on the direction of the magnetic field. Thus investigated pure and alloyed by Sn semiconductor $Bi_{1-x}Sb_x$ foil can be used as n- and p- lags for thermoelectric converters of energy. Figures 5, 7 (inset) represents temperature changes the Power factor ($\alpha^2\sigma$) for semiconductor $Bi_{0.9}Sb_{0.1}$ wires with diameter 0.3 – 3 μm and foils. The largest value of Power factor $\alpha^2\sigma$=1.4*10^{-4} W/cm*K^2 corresponds to the wires with diameter 1.8 μm at temperature 150 K. In the

Bi-9at%Sb foil the maxima of the range Power factor are reached in the temperature 120-200 K.

Assuming the value of thermal conductivity χ in the wires is equal to the value of χ in the bulk samples of the same composition [16], the thermoelectric figure of merit Z = $(\alpha^2\sigma)/\chi$ will make $Z = 4{\cdot}10^{-3}$ K^{-1}. Thermoelectric properties of bismuth-antimony crystals along the trigonal axis were studied in [7, 17, 18]. On evidence derived from these papers, thermoelectric figure of merit reaches its largest value close to 80 K. In the range of compositions $0.05 \leq x \leq 0.12$ Z value changes slightly and makes $Z = 4.9*10^{-3}$ K^{-1} at $x = 0.05$ and $Z = 5.1*10^{-3}$ K^{-1} at $x = 0.12$ [7]. In the bulk $Bi_{0.93}Sb_{0.07}$ sample oriented along the bisector axis C_1, the thermoelectric figure of merit also assumes maximum value at the same temperature ($T \approx 80$ K) and makes $Z = 3*10^{-4}$ K^{-1} [18]. Thus, thermoelectric figure of merit of $Bi_{0.9}Sb_{0.1}$ wires approaches Z value in the bulk crystals along the C_3 axis and exceeds thermoelectric figure of merit along the bisector axis.

Although of the fact that value of the Power factor $(\alpha^2\sigma)$ are close to the value received on bulk $Bi_{1-x}Sb_x$ samples, it is necessary to expect the enhance the figure of merit thanks to the decrease of the heat conductivity in wires because the additional scattering on a roughness surface and in the foil because scattering on the grain of the boundary the investigated texture. Proposal are made to enhance the experimental value ZT at the measurements thermal conductivity even further.

When reporting experiments on human subjects, authors should indicate whether the procedures followed were in accordance with the ethical standards of the responsible committee on human experimentation (institutional and national) and with the Helsinki Declaration of 1975, as revised in 2000 and 2008. If doubt exists whether the research was conducted in accordance with the Helsinki Declaration, the authors must explain the rationale for their approach, and demonstrate that the institutional review body explicitly approved the doubtful aspects of the study. When reporting experiments on animals, authors should be asked to indicate whether the institutional and national guide for the care and use of laboratory animals was followed.

IV. CONCLUSIONS

Measurements of electrical resistivity thermoelectric power have been performed on single crystal $Bi_{0.9}Sb_{0.1}$ wires with different diameters and Bi-9% Sb, Bi-9% Sb + 0.2Sn, Bi-9% Sb + 0.4Sn foils in region 4.2-300 K. It is shown that the activation energy for conduction is found to be diameter dependent due to quantum size effect in the $Bi_{0.9}Sb_{0.1}$ wires. From the measured values of the electrical resistivity and absolute thermopower, the Power factor $(\alpha^2\sigma)$ was calculated. It is shown that with decreasing the wires diameter, the maximum of the Power factor is shifted in the high temperature region, that is important for applications. According strongly decrease the lattice thermal conductivity in wires and foils the figure of merit it was to be expected the enhance particularly in region T>100 K.

ACKNOWLEDGMENT

This work was supported by Institute project 15.817.02.09A.

CONFLICT OF INTEREST

The authors declare that they have no conflict of interest.

REFERENCES

1. Golin S Band Model for Bismuth-Antimony Alloys (1968) Phys Rev 176: 830
2. Rowe D M (1995) CRC Handbook of thermo-electric. CRC Press New York
3. Lenoir B, Dauscher A, Cassat M, Ravich Yu I and Sherrer H (1998) Effect of antimony content on the thermoelectric figure of merit of Bi1−xSbx alloys. J Phys Chem Sol 59: 129
4. Brandt N B and Svistova E A. (1970) Electron transitions in strong magnetic fields. J Low Temp 2(1):1-35
5. Hiruma K, Kido G, Fawauchi K and Miura N. (1980) Shubnikov de Haas effect and semimetal-semiconductor transition in bismuth-antimony alloy in high magnetic fields. Solid States Commun 33(2):257-260
6. Jandl P. and Birkholz, (1994) Thermogalvanomagnetic properties of Sn- doped $Bi_{95}Sb_5$ and its application for solid state cooling. J Appl Phys 76(11):7351-7366
7. Dresselhaus M S, Lin Y- M, Cronin S B, Rabin O, Black M R, Dresselhaus G, and Koga T (2001) Semiconductors and Semimetals: Recent Trends in Thermoelectric Materials Research III. T. M. Tritt Academic, San Diego chap. 1, pp. 1–121
8. Koga T, Sun X, Cronin S B, and Dresselhaus M S (1998) Appl Phys Lett 73:2950; *ibid.*. 75:2438 (1999). Koga T, Harman T C, Cronin S B, and Dresselhaus M S (1999) Phys Rev B 60:14286. Koga T, Rabin O, and Dresselhaus M S, (2000) Phys Rev B 62:16703
9. Sun X, Zhang Z, and Dresselhaus M S (1999) Theoretical modeling of thermoelectricity in Bi nanowires. Appl Phys Lett 74:4005
10. Lin Y- M, Sun X, and Dresselhaus M S (2000) Theoretical investigation of thermoelectric transport properties of cylindrical Bi nanowires. Phys Rev B 62:4610
11. Brandt N B, Gitsu D V, Nikolaeva A A and Ponomarev Ya G (1977) Investigation of Size Effects in thin Cylindrical Bismuth Single Crystals in a Magnetic Field. Sov Phys JETP 45(6):1226

12. Bodiul P P, Konopko L, Nicolaeva A A, Para G. Electric and magneto thermoelectric properties of $Bi_{1-x}Sb_x$ nanowires by semiconductor-semimetal transition induced by elastic elongation. The 21st International Conference on Thermoelectrics, Long Beach, California USA, August 25-29, 2002, Book of Abstracts, p.67.
13. Huber T E, Adeyeye A, Nikolaeva A, Konopko L, Johnos R C, and Graf M J. (2011) Surface state mobility and thermopower in semiconducting bismuth nanowires. Phys Rev B 83: 2354114
14. Shepelevich V G. (1986) Homogenizing influence on the kinetic properties of alloys of Bi – Sb. Inorganic Materials 22: 378 – 380
15. Ivanov G A, Kulikov V A, Naletov V L, Panarin A F, Regel A R (1972) Thermoelectric figure of merit of pure and doped bismuth- antimony alloys in a magnetic field. FTP 6(7): 1296 – 1299 [in Russian]
16. Kuznetsov V L, Vedernikov M V, Yandl P, Birkholz W (1994) Exploring the limits of thermoelectric refrigeration at liquid nitrogen temperatures Technical Physics Letters 20 :75 – 79
17. Red'ko N.A., Kagan V.D. (2008) Nonmonotonic magnetic- field dependence of transport coefficients for n- Bi-Sb semiconducting alloys Physics of the Solid State 50(3): 385 – 396

Author:	Prof. Nikolaeva
Institute:	IEEN D. Ghitu,
Street:	Academiei 3/3
City:	Chisinau,
Country:	Moldova,
Email:	A.Nikolaeva@nano.asm.md

Prospect Nanostructured Material for Thermoelectric Sensors of Infrared Radiations

A.I. Casian and I.I. Sanduleac

Faculty of Computers, Informatics and Microelectronics, Chisinau, Technical University of Moldova, Republic of Moldova

Abstract— **In the last years it is observed an increasing application of thermoelectric biosensors in medicine. Thermoelectric sensors of infrared (IR) radiations are widely used for detection of radiations in special cases, in the thermometers for contactless measurement of temperature including that of the human body ones, in the investigation of surface temperatures of the human body, in new thermal imaging systems, in the systems of night vision. The thermoelectric sensor is based on the thermocouple or the thermopile containing materials of *p*- and *n*-type. Earlier the organic crystals of tetrathiotetracene-iodide, TTT_2I_3, were proposed as effective material of *p*-type. In this paper the organic nanostructured crystals of $TTT(TCNQ)_2$ are proposed as material of *n*-type. The main parameter that determines the possibility of a given material to be used as sensitive element in the thermoelectric sensors is the thermoelectric power factor. The electrical conductivity, thermopower and thermoelectric power factor are modeled in function of the Fermi energy and their optimal values are calculated in order to achieve the maximum of the power factor.**

Keywords— **thermoelectric sensor; infrared radiation; nanostructured material; organic material; thermoelectric power factor.**

I. INTRODUCTION

Infrared (IR) radiation consists of electromagnetic waves in a large wavelength interval from 0.75 μm to 1000 μm, lying between visible light and microwave light. Advances in electronics in recent years have led to the creation of many different types of sensors of IR radiations. The thermoelectric sensors are distinguished by the ability for collection and detection radiation from a very large IR spectral interval. The IR radiation emitted by the human body is in a long wavelength interval with a maximum in the 10-μm wavelength band, and its detection requires some prudence because of its extremely low energy level compared with that of visible light. Nevertheless, in the last years it is observed an increasing application of thermoelectric biosensors in medicine [1-3]. Thermoelectric IR sensors are widely used for detection of radiations in special cases, in the thermometers for contactless measurement of temperature including that of the human body ones, in the investigation of surface temperature distribution of the human body, in the systems of night vision, in new thermal imaging systems.

Such sensors have advantages with respect to photo detectors of IR radiation as not cooled operation and broadband response over the infrared spectrum.

It is known that the main parameter that determines the possibility of a given material to be used as sensitive element in the thermoelectric sensors is the thermoelectric power factor $P = \sigma S^2$, where σ is the electrical conductivity and S is the thermopower (Seebeck coefficient). Values of P as higher possible are needed. But the requirement to increase in the same material simultaneously both σ and S is contradictory in ordinary materials. Usually the increase of σ leads to decrease of S and vice versa. New more complicated materials are needed in order to overcome this contradiction.

Very high values of the thermoelectric power factor $P \sim 25.5 \cdot 10^{-3}$ W/K^{2}m at 260K were obtained in Fe3O4/SiO2/*p*-Si(100) heterostructure [4]. In *n*-type and *p*-type PbTe/PbEuTe quantum wells values of $P \sim 6.2\text{-}6.6 \cdot 10^{-3}$W/m.K^2 and $\sim 1.6 \cdot 10^{-2}$ W/m.K^2 have been measured, respectively, at room temperature [5]. However, the technology for fabrication of such quantum structures is rather complicated and expensive.

Organic compounds attract more and more attention for biomedical sensors in the last years [6] as materials which are less expensive and their molecular structure can be easily modified to tune the desirable physical properties.

The thermoelectric sensor is based on the thermocouple or the thermopile the latter meaning a serial connection of a number of thermocouples. But in order to state a thermocouple it is necessary to have qualitative materials of both *p*- and *n*-type. In [7] the organic crystals of tetrathiotetracene-iodide, TTT_2I_3, were proposed as effective material of *p*-type.

The aim of this paper is to propose and investigate the organic nanostructured crystals of $TTT(TCNQ)_2$ as material of *n*-type. The electrical conductivity, thermopower and thermoelectric power factor are modeled in function of the Fermi energy and their optimal values are calculated in order to achieve the maximum of the power factor.

V. Sontea and I. Tiginyanu (eds.), *3rd International Conference on Nanotechnologies and Biomedical Engineering*,
IFMBE Proceedings 55,
DOI: 10.1007/978-981-287-736-9_54

II. HIGH-CONDUCTIVE COMPLEX TTT(TCNQ)$_2$

The structure of TTT(TCNQ)$_2$ complex is presented in [8]. The crystals are formed from segregated cation stacks or chains of TTT and anion chains of TCNQ directed along the c axis. The crystals are of n-type and the carries are electrons. They have needle like form with the length of 3 – 6 mm. The crystals were grown from solution and contained a certain amount of solvent molecules. Therefore, the room electrical conductivity is rather low and varies form 20 to 160 $\Omega^{-1}cm^{-1}$. However, the thermoelectric properties of TTT(TCNQ)$_2$ compounds have not been investigate in detail. The crystal perfection may be improved by applying for synthesis the gas phase method.

In the TCNQ chains the transfer energy for an electron to pass from one molecule to the nearest one along the chain w_1 is of the order of 0.125 eV. This means that in order to describe the charge transport along chains it is possible to use the band description. In the tight binding and nearest neighbors approximations the energy of a conduction electron with the 3D quasi-wave vector $\boldsymbol{k}$ and projections (k_x, k_y, k_z) measured form the bottom of the conduction band has the form

$$\varepsilon(\boldsymbol{k}) = 2w_1[1-\cos(k_x c)] + 2w_2[1-\cos(k_y a)] + \\ +2w_2[1-\cos(k_z b)] \qquad (1)$$

Here w_1, w_2 and $w_3$are the transfer energies of an electron from the given molecule to the nearest ones along lattice vectors $\boldsymbol{c}$, $\boldsymbol{a}$ and $\boldsymbol{b}$. The conducting chains are directed along $\boldsymbol{c}$ which is taken as the axis x. The axes y, z are directed along $\boldsymbol{a}$, $\boldsymbol{b}$. It is considered that w_1 is much bigger than w_2 and w_3, due to the quasi-one-dimensionality of the crystal. The analyze show that the interaction of electrons with the longitudinal acoustic phonons is the most important. The phonon dispersion law is

$$\omega_{\boldsymbol{q}}^2 = \omega_1^2 \sin^2(cq_x/2) + \omega_2^2 \sin(aq_y/2) + \omega_3^2 \sin(bq_z/2), \qquad (2)$$

whereω_1, ω_2 and ω_3 are the limit frequencies in the x, y and z directions and (q_x, q_y, q_z) are the projections of phonon quasi-wave vector $\boldsymbol{q}$. Due to the quasi-one-dimensionality ω_1 is much bigger than ω_2 and ω_3.

As in the case of TTT$_2$I$_3$ compounds, the crystal model takes into account two interactions of electrons with the longitudinal acoustic phonons. The first interaction is determined by the variation of electron transfer energies. The coupling constants of this interaction are proportional to the derivatives w_1', w_2' and w_3' with respect to intermolecular distances of the electron transfer energies. The second interaction is determined by the variation of polarization energy of molecules surrounding the conduction electron. The coupling constant of this interaction is proportional to the mean polarizabilityα_0 of TCNQ molecule.

The scattering of conduction electrons on impurities is also taken into account. The impurities are considered point like and neutral. The scattering rate is described by the dimensionless parameter D_0.The latter is proportional to the carrier concentration, and may be made very small, by increasing the crystal purity.

III. TRANSPORT PROPERTIES

Let consider that a weak electrical field and a weak temperature gradient are applied along the conductive TCNQ chins. At room temperature the scattering processes can be considered elastic. The transversal kinetic energy of the electron in the scattering processes can be neglected, because it is much smaller than the kinetic energy of the electron along the chains. Then the linearized kinetic equation takes the form of Boltzmann one. It can be solved analytically and the electrical conductivity along the chains σ_{xx} (the field is along x and the flows are along x), the thermopower S_{xx} and the thermoelectric power factor P_{xx} can been expressed through the transport integrals as follows

$$\sigma_{xx} = \sigma_0 R_0, S_{xx} = (k_0/e)(2w_1/k_0T)R_1/R_0,$$

$$P_{xx} = \sigma_{xx} S_{xx}^2, \qquad (3)$$

$$\sigma_0 = (2e^2 M v_{s1}^2 w_1^3 r)/(\pi^2 \hbar abc (k_0 T)^2 w_1'^2) \qquad (4)$$

Here e is the electron charge, M is the mass of TCNQ molecule, v_{s1} is the sound velocity along TCNQ chains, r is the number of molecular chains passing through the transversal section of the crystal elementary cell, k_0 is the Boltzmann constant. The transport integrals R_n have the form

$$R_n = \int_0^2 d\varepsilon \int_0^\pi d\eta \int_0^\pi d\zeta (\varepsilon-\varepsilon_F)^n \varepsilon(2-\varepsilon) n_{\varepsilon,\eta,\zeta}(1-n_{\varepsilon,\eta,\zeta}) \times \\ [\gamma_1^2(\varepsilon-\varepsilon_0)^2 + w_2'^2[2\varepsilon(2-\varepsilon)w_1'^2]^{-1}(1+\gamma_2^2+2\sin^2\eta + \\ +2\gamma_2\cos\eta) + w_3'^2[2\varepsilon(2-\varepsilon)w_1'^2]^{-1}(1+\gamma_3^2+2\sin^2\zeta + \\ +2\gamma_3\cos\zeta) + D_0]^{-1} \qquad (5)$$

$$\varepsilon_0 = (\gamma_1+1)/\gamma_1, \qquad (6)$$

$$\gamma_1 = \frac{2e^2\alpha_0}{c^5 w_1'}, \quad \gamma_2 = \frac{2e^2\alpha_0}{a^5 w_2'}, \quad \gamma 3 = \frac{2e^2\alpha_0}{b^5 w_3'} \qquad (7)$$

Here $n_{\varepsilon,\eta,\varsigma}$ is the Fermi distribution function in the variables $\varepsilon = 1 - \cos(k_x c)$, $\eta = k_y a$ and $\varsigma = k_z b$, $\varepsilon_F = E_F/2w_1$ is the dimensionless Fermi energy in the 1D approximation, $\varepsilon_0 = (\gamma_1 + 1)/\gamma_1$ is the dimensionless resonance energy in the relaxation time in units of $2w_1$. The parameters γ_1, γ_2 and γ_3 have the means of amplitudes ratios of second interaction to the first one along chains and in transversal directions.

It is seen form (5) that, if one puts w_2, w_3, w_2' and w_3' equal to zero, the integrals after η and ς are calculated directly and the results of 1D approximation are obtained. It is also seen that in this case the expression under the integral after ε has a maximum as a function of ε near ε_0, when $\varepsilon_0 < 2$ ($\gamma_1 > 1$), and this maximum must be very high, if D_0 or the scattering rate on impurities is very small. In the 3D model this maximum is limited not only by the scattering on impurities, but also by the electron scattering on adjacent molecular chains. Namely the last scattering begins to manifest itself and may become important in the case of very pure crystals, when the impurity scattering is strongly reduced. Unlike the *p*-type crystals, when the maximum of relaxation time is placed in the lower half of the energy ε, $\varepsilon_0 < 1$, in the case of *n*-type crystals the maximum is placed in the upper half of the energy ε, $\varepsilon_0 > 1$. Of cause, this situation will affect the behavior of the thermoelectric power factor.

IV. RESULTS AND DISCUSSIIONS

The numerical calculations of the electrical conductivity σ_{xx}, the thermopower S_{xx} and the thermoelectric power factor P_{xx} along the TCNQ chains have been performed after (3) – (7). The crystal parameters are: $M = 3.72 \cdot 10^5 m_e$ (m_e is the electron free mass), $c = 3.75$ Å, $b = 12.97$ Å and $a = 19.15$ Å, $w_1 = 0.125$ eV, $w_1' = 0.22$ eVÅ^{-1}, $v_{s1} = 2.8 \cdot 10^3$ m/s, $r = 2$. The mean polarizability of TCNQ molecule was estimated as$\alpha_0 = 9.5$Å^{-3} and this gives$\gamma_1 = 1.7$. The parameters $d_1 = w_2/w_1 = w_2'/w_1'$, and $d_2 = w_3/w_1 = w_3'/w_1'$ were introduced and were estimated as $d_1 = 0.015$ and $d_2 = 0.01$. Two sets of crystals are considered: with not very high degree of perfection, $D_0 = 0.3$, 0.1 and 0.05, and with higher degree of perfection, D_0 0.04, 0.03 and 0.02. To stoichiometric crystals ($\varepsilon_F = 0.35$, electronic concentration $1.1 \cdot 10^{21}$ cm^{-3}) with $D_0 = 0.3$ it corresponds $\sigma_{xx} = 0.48 \cdot 10^5$ Ω^{-1}m^{-1}.

In Fig.1 the dependences of σ_{xx} on dimensionless Fermi energy ε_F are presented. It is seen that with the increase of ε_F (or electron concentration) σ_{xx} grows considerable, especially in purer crystals, and the differences between the results of 3D and 1D models are negligible.

In the Fig. 2 the dependences of thermopower (Seebeck coefficient) along chains S_{xx} are presented. In stoichiometric crystals S_{xx} varies between -110 and -118 μV/K. With the further increase of ε_F the thermopower decreases by absolute value and changes the sign around $\varepsilon_F \sim 1.5$, after that the carriers become holes. The value of S_{xx} in the low minimum is -120 μV/K.

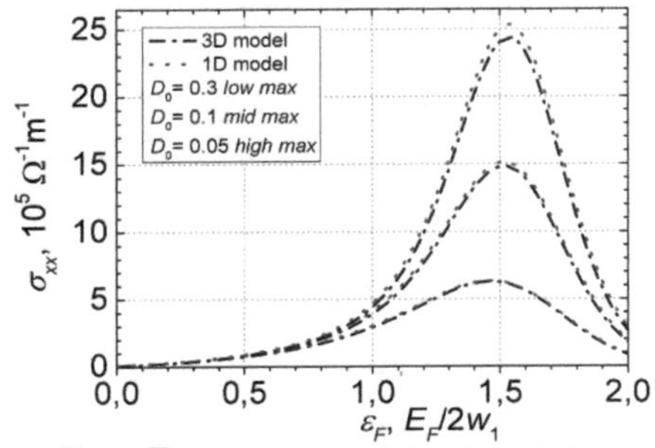

Fig. 1 Electrical conductivity along chains σ_{xx} as function of Fermi energy ε_F.

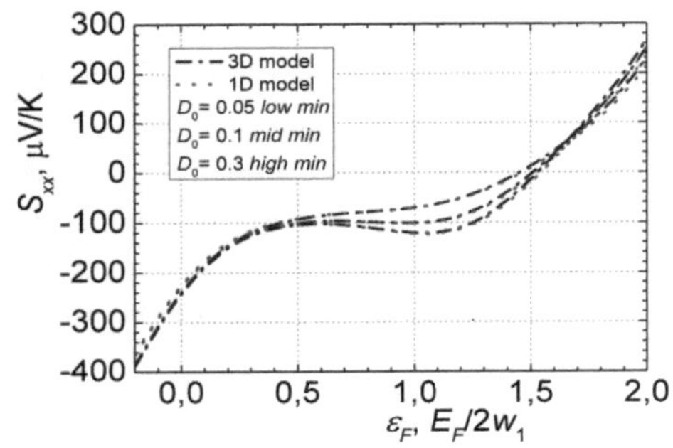

Fig. 2 Seebeck coefficient along chains S_{xx} as function of dimensionless Fermi energy ε_F.

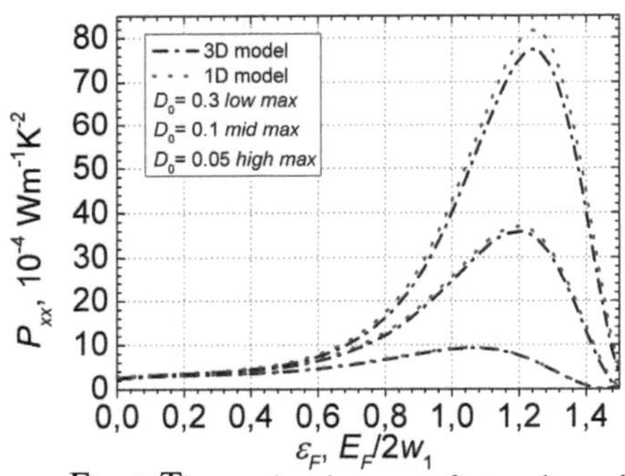

Fig. 3 Thermoelectric power factor along chains P_{xx} as function of Fermi energy ε_F.

The dependences of P_{xx} on dimensionless Fermi energy ε_F are presented in Fig. 3. In stoichiometric crystals P_{xx} is very small and almost does not depend of crystal perfection. But with the increase of electron concentration P_{xx} grows considerable and achieves in the most perfect crystal a maximum of $76 \cdot 10^{-4}$ Wm^{-1}K^{-2}, or almost two times higher than in Bi_2Te_3, widely used commercial thermoelectric material.

Let's consider the crystals from the second series with higher degree of perfection. In Fig. 4 the dependences of σ_{xx} on ε_F are presented. Although for stoichiometric crystals σ_{xx} are only slightly increased, the maximums have increased considerably, achieving $45 \cdot 10^5\ \Omega^{-1}\text{m}^{-1}$ in the most perfect crystals. Besides, the results of 3D model are significantly reduced in comparison with those of 1D model.

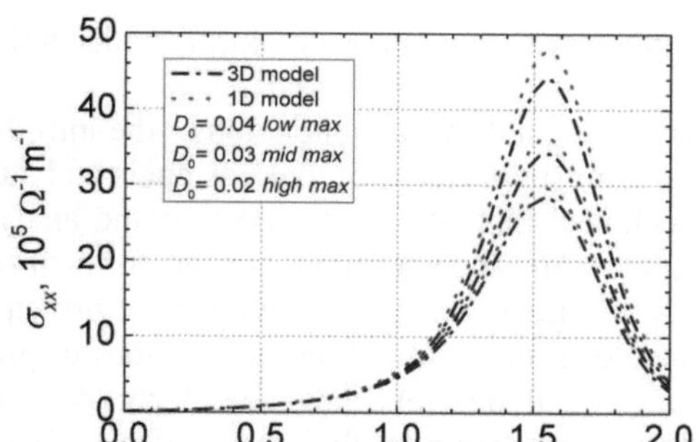

Fig. 4 Electrical conductivity σ_{xx} as function of Fermi energy ε_F in more perfect crystals.

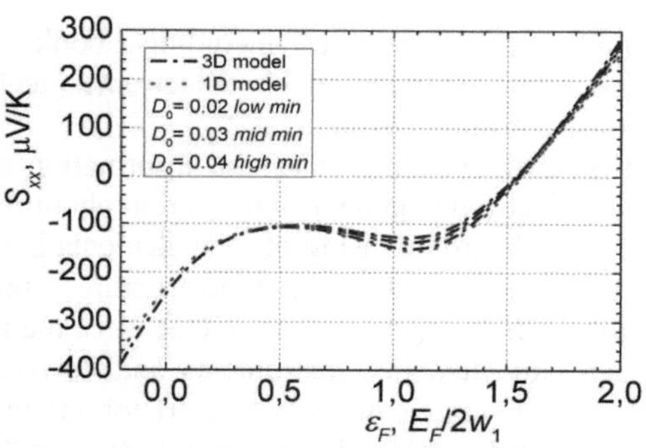

Fig. 5 The same as in Fig. 2 for more perfect crystals.

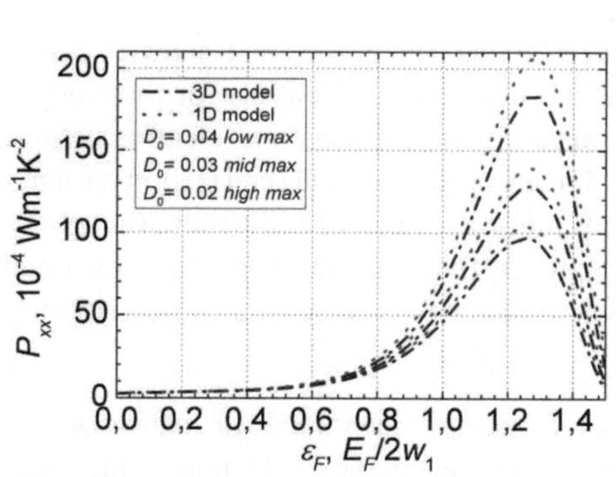

Fig. 6 The same as in Fig. 2 for more perfect crystals.

In Fig. 5 the dependences of S_{xx} on ε_F are presented for more perfect crystals. For stoichiometric crystals $S_{xx} \sim 119 - 120\ \mu$V/K, i.e. practically are the same as in previous case. The value of S_{xx} in the low minimum now is -152 μV/K.

The dependences of P_{xx} on dimensionless Fermi energy ε_F are presented in Fig. 6. The deviations between 3D and 1D models are more pronounced and the 3D model must be applied. But now the maximum achieves $184 \cdot 10^{-4}\ \text{Wm}^{-1}\text{K}^{-2}$, more than by 4.5 times higher than in Bi_2Te_3. This value is less than in TTT_2I_3 crystals of *p*-type, but is also good result. In order to achieve the maximum of P_{xx} it is needed to increase the electron concentration up to $2.5 \cdot 10^{21}\ \text{cm}^{-3}$, or by 2.3 times. It is possible by additional doping.

V. CONCLUSIONS

The organic nanostructured crystals of $TTT(TCNQ)_2$are proposed as material of *n*-type for thermoelectric sensors of infrared radiations. The electrical conductivity σ_{xx}, thermopower (Seebeck coefficient) S_{xx} and the thermoelectric power factor P_{xx} along conductive molecular chains are modeled as function of the dimensionless Fermi energy. It is found that improving the crystal perfection and increasing the electron concentration by 2.3 times, form $1.1 \cdot 10^{21}\ \text{cm}^{-3}$ up to $2.5 \cdot 10^{21}\ \text{cm}^{-3}$, it is expected to obtain an increase of P_{xx} up to $184 \cdot 10^{-4}\ \text{Wm}^{-1}\text{K}^{-2}$, or more than by 4.5 times higher than in Bi_2Te_3.

CONFLICT OF INTEREST

The authors declare that they have no conflict of interest.

REFERENCES

1. Ring E F J, Ammer K. "Infrared thermal imaging in medicine", *Physiol. Meas.* 33, R33–R46 (2012).
2. Kopparthy V L, TangutooruS M,Guilbeau E J. "Label Free Detection of L-Glutamate Using Microfluidic Based Thermal Biosensor", *Bioengineering*, 2, 2-14, 2015.
3. Dillner U, Kessler E, MeyerH.-G. "Figures of merit of thermoelectric and bolometric thermal radiation sensors", *J. Sens. Sens. Syst.*, 2, 85–94, 2013.
4. ViskadourakisZ, et al. "Very high thermoelectric power factor in a Fe3O4/SiO2/p-type Si(100) heterostructure", *Appl. Phys. Lett.* 101, 033505 (2012).
5. Harman T C, Taylor P J, et al. "Quantum dot superlattice thermoelectric materials and devices", *Science*, Vol. **297**, 2229 (2002).
6. Nambiar S,Yeow J T W. "Conductive polymer-based sensors for biomedical applications", *Biosens. Bioelectronics*, 26,1825, (2011).
7. Casian A, Balmus I, Sanduleac I, "Organic nanostructured material for thermoelectric sensors of infrared radiations". *2nd Int. Conf. on Nanotechn. and Biomed. Engineer.* April 18-20, 2013, Chisinau, Moldova, pp. 61-64.
8. BuravovL I, et al. "Structure and electromagnetic properties of a new high-conductive complex $(TTT)^+(TCNQ)_2^{-}$". JETP Lett., 20, 208, 1974.

Author: Anatolie Casian
Institute: Technical University of Moldova
Street: 168, Stefan cel Mare bd.
City: Chisinau
Country: Republic of Moldova
Email: acasian@mail.utm.md

The Generalization of Scientific and Educational Materials on Nanoelectronics

V.M. Spivak[1], A.G. Vlasiuk[1], and M.S. Tirsu[2]

[1] National University of Ukraine, "Kiev Polytechnic Institute", Kiev, Ukraine
[2] Institute of Power Engineering of ASM, Chisinau, Moldova

Abstract— **Summarizes the scientific and educational materials on nanoelectronics, as reflected in monographs and textbooks on the basis of copyright, the most significant results of research and compilation of the most important studies of leading companies and research institutions of the developed world.**

Keywords— **Nanoelectronics, nanotechnology, nanomaterials, quantum mechanical foundations, nanostructures.**

I. Introduction

In recent years the subject of nanotechnology and nanomaterials is consistently among the research priorities of the industrialized countries. Adopted and not without good evidence to believe that the development of nanotechnology and nanomaterials is a new industrial revolution. In the coming decades, the utilization of nanotechnology will determine the status of each developed country in the world.

An understanding of this has on the state level in Ukraine, Bulgaria and Moldova. For example, in general, the Ministry of Education and Science of Ukraine and the National Academy of Sciences on November 26, 2009 № 1066/609 «On the approval of the main scientific fields and the major problems of fundamental research for 2009-2013 in Ukraine .." separately entered chapters 1.4.5. "Nanophysics and nanotechnologies" and 1.6.5. "Nanostructured (nanodispersed, nanocrystalline) materials."

Therefore, generalization of scientific and teaching materials on nanotechnology and nanomaterials do together teachers and researchers: National Technical University of Ukraine "Kyiv Polytechnic Institute" and "Lviv Polytechnic"; Institute of Electronics and Energy Academy of Sciences, respectively, Bulgaria and Moldova. Together, developed and published 3 monographs [1-3] and 2 textbooks [4,5] based on its most significant, the results of research and generalization of studies of leading companies and research institutions of the developed world.

In [1-3], the authors summarized the research, published in more than 120 scientific papers and 45 patents, and leading scientific schools of the world in the field of nanomaterial knowledge synthesis of advanced materials nanodevices with ideas aimed at understanding the processes in materials nanodevices, development and production of new materials more appropriate for use in nanoscale dimensions, in order to increase the competitiveness of products through the use of nanostructuring of materials.

The purpose of publishing monographs - the introduction of the readers, mostly academics, professionals and students of higher technical institutions, perspective and analysis of the current state of nanoelectronics, in the form of a relatively short preliminary analysis of the general principles of modern nanoelectronics and properties, methods of preparation and study of modern and advanced technology of nanomaterials and, reception on their basis of new electronic nanoelements.

II. Basically Information

Monographs "Nanoelectronics" including 2 books.

Books no.1 "Introduction in nanoelectronics technologies" [1] consists from 3 chapters.

In the first chapter named "General information on micro- and nanoelectronic" general information about modern micro- and nanoelectronic, using of new elements in nanoelectronic, physical processes in nanostructures, physical interpretation of the ballistic motion of electrons, the idea of creating such nanoelements, instruments and devices such as electronic and magnetic memory, memristor and memristor systems, hybrid semiconductor circuits (such CMOL), the use of quantum dots in cellular automata are given.

The chapter 2 "Technological methods for obtaining of nanoelectronic elements" contains materials from main methods of analysis of surfaces and nanostructures namely: electron and ion analytical methods for the study of surfaces; depth analysis of fast ions and electrons; Methods and apparatus for imaging and characterization of nano-objects; methods for measuring the size of nanoparticles.

The chapter 3 "Technological methods for obtaining nanoelectronic elements" is dedicated to description of elaborations on:

- manufacturing technology of nano-sized thin film, deposition techniques of thin films by thermal evaporation in a vacuum, molecular beam epitaxy, atomic layer deposition, Langmuir - Blodgett, physical processes in ion sputtering, triode ion-plasma system, cutting to the dc magnetron systems for application thin films, thin film high-frequency sputtering.

V. Sontea and I. Tiginyanu (eds.), *3rd International Conference on Nanotechnologies and Biomedical Engineering*,
IFMBE Proceedings 55,
DOI: 10.1007/978-981-287-736-9_55

- ion and ion-plasma etching images into thin films, ion etching installations, particularities of the microstructure transport using ion etching;
- ion implantation of fast ions, the basic physical processes of penetration of the accelerated ions, radiation defects, evaluation of the surface resistance of the implanted layers, descriptions of equipment and practical application of ion implantation;
- ion implantation of fast ions, the basic physical processes of penetration of the accelerated ions, radiation defects, evaluation of the surface resistance of the implanted layers, descriptions of equipment and practical application of ion implantation;
- physical foundations of electron-beam lithography, brief information about the penetration and diffusion of fast electrons in thin films and solid samples, the experimental data path of the electrons penetrating the solid samples, the distribution of absorbed energy in the sample when it is irradiated with electrons;

Equipment and technology of electron and ion lithography, equipment for micro-lithography, resist exposure in electron and ion lithography, computer modeling of processes in lithography, the use of regression analysis to improve the quality of the lithographic process; lithography using a mechanical probe, STM-lithography using the resist by lithography local anodic oxidation of the metal surface, and the non-contact electrical microcontact lithography polymeric layers mechanical probe mode of an atomic force microscope, the deposition of materials on the basis of a scanning tunneling microscope, lithography using hybrid atomic force and scanning tunneling microscope.

The second book of monography "Nanostructured materials and functional devices" [2] is written by the authors in order to create educational and scientific material in the field of nanomaterials and nanoelectronics functional devices. You can consider this as a continuation of the introductory material of knowledge on nanotechnology and digital nanoelectronic components, which are presented in the book [1].

In a joint Ukrainian-Moldovan monograph [3], published in the Romanian language, we found the introduction, further development and a full complement of monographs [1, 2], the results of research of Moldovan specialists.

In monographs it is shown the difference of the nanostructured and conventional materials properties associated with the manifestation of size effects. Changes not only classics such parameters as: penetrating solvents, electrical and thermal conductivity, but the quantum-mechanical parameters: electron tunneling, the quantization of energy levels, the distribution of the density of states, the role of spins, etc.

As a result, non-transparent materials can become transparent, inert metals acquire catalytic properties, powders of stable metal becoming flammable, the melting point of nanopowders decreases, insulators can obtain conductive properties, and others. This defines the innovative potential of nanomaterials and their first practical application guarantee the correctness of predictions their importance.

There are two main directions of development of nanotechnologies and nanomaterials, described in the book. The first - the creation of new materials with unique properties due to their morphological features of these materials in the field of nano-sized (typically range in size from 1 to 100 nm). The second direction - finding opportunities for application of conventional (or improved conventional materials) to build nanodevices with nanoscale components.

The monographs are systematized and proved that integrated nanosystems, of course, can be realized with adequate interface and thereby get new or improved functions of the system. This applies to nanomaterials (nanophosphors for displays, nanoseparatoriv for electric batteries, nanocoatings for windows, solar panels, nano, and nanofiltration membranes, nanostructured and nanoporous materials, materials with high thermal conductivity and stability, nanosensors, additives to cosmetics and medicines, and many etc.).

Is shown to be noted that the progress in the sharing of nanocomponents and systems with information technology devices. Moreover, it is obvious reciprocal influence of these areas today and the prospects for future achievements. This success of computer production in the direction of reducing the size and power consumption, enhanced functions, replacing some miniature models of control information devices with sensitive sensor systems, the use of self-healing materials in components and the creation of dependable information systems, and last but not least - the future breakthroughs in the field of quantum information science and quantum computers. The book shows that will be created adaptive information systems that respond to changes in the environment and can make self diagnostics and work perfectly.

Research and generalizations made by the authors in the monographs [1-3] were used, arranged and adapted for higher education in the textbook "Fundamentals of nanoelectronics" in 2 books [4,5] by a team of teachers of the leading technical universities in Ukraine (Kiev and Lviv Polytechnic).

Tutorial [4] devoted to studying physical characteristics of nanomaterials, nanoelectronics structures, etc., which set out the quantum mechanical foundations of nanoelectronics, semiconductor and basic carbon structures and characteristic physical properties of these structures. Consider the behavior of the energy spectrum and the quantum-mechanical particles in potential wells of different shapes and dimensions, especially the tunneling particles through potential barriers of tunneling and resonant tunneling structures, patterns of change in the density of states of free electrons in the changes the dimension of the system, the electron dispersion law and application of strong external magnetic fields. The basic information about the two-dimensional crystals with two-dimensional structures of the leading channels, their composition: semiconductor superlattices, one-dimensional crystals, low-dimensional structures of carbon graphite, fullerenes, nanotubes. Highlight specific physical properties of two-, one-, nulvimirnih systems and superlattices caused by their dimensions and characteristics of spatial and energy structure.

Tutorial [5] consists of 3 parts containing 18 chapters devoted to:

- self-organizing structures of nanoscale silicon and composites;
- porous nanomaterials;
- research techniques of nanostructures and nanomaterials;
- the gas phase and molecular beam epitaxy layers of quantum organometallic compounds;
- nanolithography technology and methods for producing quantum wires, quantum dots and carbon nanostructures;
- quantum effects in submicron electronic components and devices to interference phenomena;
- photonic nanostructures and nanosystems touch;
- Molecular, nanobio- and magnetoelectronic, spintronics;
- nanoelectromechanical systems.

After each section the conclusions, test questions and assignments are presented as well as split ends problems for independent solution.

Authors sincerely hopes that this first attempt at the creation of monographs and textbooks on nanoelectronics will be successful and justified in the sense that the proposed research and educational material for nanoelectronics will be useful for specialists and training of students and teachers, which are designed to bring to the students of the latest achievements of world scientific and technical ideas in this important and dynamic field of human knowledge - nanoelectronics.

III. CONCLUSIONS

The monographs addresses questions about the achievements and the practical implementation of nanomaterials, nanodevices and elements. It is shown that these issues are in the area of intense competition of global corporations in the state economy - in the field of defense and security. At the same time, the existing experience of research and development points to the need to: develop a strategy of development of nanotechnologies and nanomaterials; create a platform for joint coordinated use of cash, scientific, technological and educational resources; close cooperation between the state and business.

The successful development of nanotechnology and nanomaterials requires both a favorable social atmosphere for nanoinnovations and highly Nanoscience and comprehensive training of personnel in these interdisciplinary fields of science and different fields of application of nanotechnologies and nanomaterials.

Complex knowledge needs and to create interfaces between the macro and nano (or micro) components to create bionanopristroiv or transfer of knowledge about the living or non-living nature in nanodevices.

As described monographs and textbooks shows that the development of nanoelectronics gives not only the technological systems for production, but also research methods, modeling, metrology and quality of nanomaterials, structures and devices. Measurements should be with atomic precision and temporal resolution in the near future can and must achieve tera- and atosekund.

In recent years, nanotechnology and nanomaterials attracted the attention not only of experts, but the whole of humanity, began a large-scale research and production direction in which is embedded a lot of money.

Therefore, national programs for scientific and technological development of Ukraine and Moldova, Bulgaria, in our opinion, should focus government and public attention on the effectiveness of investments and spent material resources, namely, the development of science, including nanoelectronics.

Collaborate on international monographs and textbooks scientists from different countries will strengthen friendly relations and mutually substantially developed scientific schools in nanoelectronics National Technical University (Kyiv and Lviv Polytechnic) and the Bulgarian Academy of Sciences Institute of Electronics and Energy Sciences of Moldova.

CONFLICT OF INTEREST

The authors declare that they have no conflict of interest.

REFERENCES

1. «Наноэлектроника » в 2 - х книгах: Книга 1. Введение в наноэлектронные технологии / Г. М. Младенов, В. М. Спивак, Е. Г. Колева, А. В. Богдан. - София - Киев : Аверс, 2010. - 332 с.
2. «Наноэлектроника » в 2 - х книгах: Книга 2 . Наноструктурированные материалы и функциональные устройства / Ю.И. Якименко, А.Н. Шмырева, Г.М. Младенов, В.М. Спивак, Е. Г. Колева, А. В. Богдан; под общей редакцией действительного члена НАН Украины Ю.И. Якименко. - Киев-София: Аверс, 2011. - 388 с.
3. Noтiuni generale despre micro- si nanoelectronica moderna. Andronati N.R., Spivak V.M., Tirsu M.S. et al. – Ch.: Tipografia Academiei de Stiinte a Moldovei, 2013.–340 p.
4. Основи наноелектроніки у 2-х книгах: Кн.1. Квантово-механічні засади, структури, фізичні властивості: Підручник / Д.М. Заячук, Ю.І. Якименко, В. М. Співак та ін. – К.; Кафедра, 2014. – 470 с.: іл..
5. Основи наноелектроніки у 2х книгах: Кн.2. Матеріали, технології і функціональні пристрої: Підручник / Ю.І. Якименко, Д.М. Заячук, , В. М. Співак, та ін. – К.; Кафедра, 2015. – 454 с.: іл..

Author: Victor Spivak
Institute: National Technical University of Ukraine “Kyiv Polytechnic Institute”
Street: 37, Prospect Peremohy
City: Kyiv
Country: Ukraine
Email: viktor_m53@mail.ru

Study of a New Colloidal Composite: Polymer-Magnetite Particles/Lyotropic Liquid Crystal

D. Mănăilă Maximean[1], O. Dănilă[1], B. Ştefănescu[1], R. Bena[1], C. Roşu[1], D. Donescu[2], and V. Eugeniu[3]

[1] Physics Department, Faculty of Applied Sciences, University „Politehnica" of Bucharest, Bucharest, Romania
[2] National Institute for Research & Development in Chemistry and Petrochemistry, Bucharest, Romania
[3] METAV-CD, Bucharest, Romania

Abstract— **Liquid crystals/magnetic particles composites have been shown to exhibit very interesting behaviors when subjected to different external fields, finding new applications in the field of functional nanomaterials. We present a new colloidal composite formed by combining polymer-magnetite nanoparticles and a lyotropic nematic liquid crystals. TEM investigations on the polystyrene magnetite show magnetite nanocrystals anchored in polymeric chains. The optical transmission of the ferronematic was investigated when the samples were subjected to different values of static magnetic field and the critical magnetic field corresponding to the Freedericksz transition was estimated.**

Keywords— **Ferronematics, magnetite nanomaterials, lyotropic liquid crystals, colloidal particles.**

I. INTRODUCTION

Liquid crystals(LC) are characterised by the presence of orientational order and the partial or total absence of positional order[1-3]. Lyotropic LC are formed by adding a solvent in a liquid, the mesomorphic state depending on concentration. Lyotropic LC are of hich interest in biology and medicine: biological lyotropic LC include biological membrane, myelin, DNA, viruses, blood in some extreme conditions etc.

Considering the shape of the molecules, liquid crystals can have calamitic or rod-like molecules, disk-shape or banana-shape molecules.

The nematic phase has the least order. In this phase, molecules do not present positional order, but have long distance orienatational order.

Nematic liquid crystals doped with ferrofluids have been called ferronematic liquid crystals[4-6]. The ferronematic are interesting composite systems containing magnetic particles and nematic liquid crystals with biomedical applications. To inhibit clumping, generally the particles are covered with a surfactant [7,8]. For a better interaction with the liquid crystal, in the synthesizing process of the particle, chemical groups compatible with the liquid crystal [9,10] are used and they will be found at the surface of the particle. A so-called "hairy" particle is thus formed.

In order to obtain a good compatibility between the colloidal particles and the liquid crystal, we prepared new polystyrene latexes containing magnetite with covalent bonds to the polymeric chains functionalized with alkoxysilanes containing double bonds. The particles have been characterized by TEM and magnetite nanocristallite embedded in polystyrene have been observed [11].

By dispersing the magnetite-polymer particles in a lyotropic liquid crystal based on sodium dodecyl sulfate, decanol and water, a new colloidal system was obtained. The lyotrop ferronematic mixture was stable over months. The system was filled in indium tin oxide (ITO) covered sandwich glass cells with homeotropic alignment.

The optical transmission of laser light (2.5 mW) was registered and discussed, when applying static (up to 3000Gauss) magnetic field. The critical magnetic field for Freedericksz transition has been estimated and found to be smaller than the corresponding value for the pure lyotropic nematic liquid crystal.

ii. EXPERIMENTAL PROCEDURES

A. Liquid Crystal/Magnetic Polymer Nanoparticles Sample Preparation

For a good compatibility between the colloidal particles and the liquid crystal, we prepared new hybrid-type polystyrene latexes containing magnetite (Fe_3O_4) [11] with covalent bonds to the polymeric chains functionalized with alkoxysilanes containing double bonds. The particles have been characterized by TEM and magnetite nanocristallite embedded in polystyrene have been observed.

Th electronic microscope used for the observation, HRTEM- High Resolution Transmission Electron Micoscope is type TECNAI F30 G2STWIN. The experiments were done at a 300kV acceleration voltage, the linear resolution of the microscope being of 1Å.

a)Sample preparation for TEM observation

For the observations of the magnetite or of the polystyrene-magnetite, solutions of concentration 1% in ethyl alcohol were prepared. The solution was homogenized by ultrasonication. A drop of this mixture was placed on the copper grid and covered with a carbon film.

V. Sontea and I. Tiginyanu (eds.), *3rd International Conference on Nanotechnologies and Biomedical Engineering*,
IFMBE Proceedings 55,
DOI: 10.1007/978-981-287-736-9_56

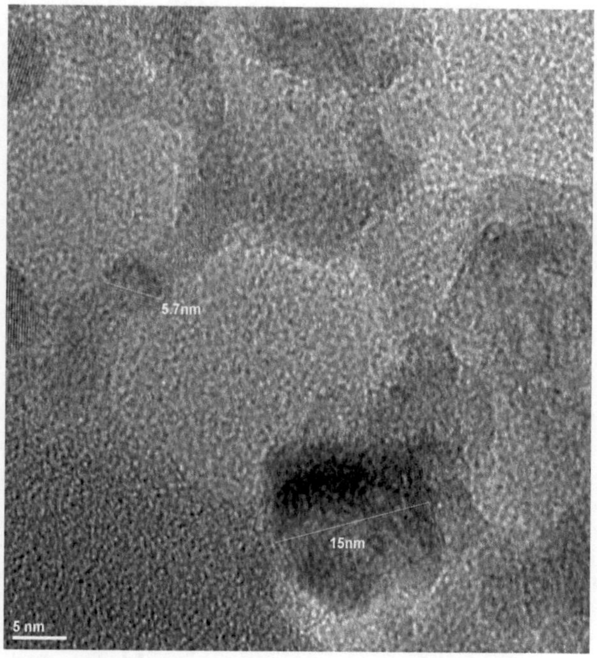

Fig. 1 HRTEM images collected for the polystyrene-magnetite composite

High-resolution image analysis - HRTEM (shown in Figure 1) executed for a polystyrene-magnetite probe (171) shows that the size of the magnetite crystals ranges between 5 and 15 nm; the magnetite crystals have different orientations and are anchored within the polymeric chains, the latter having a size of tens of nanometers.

b)Preparation of lyotropic LC and lyotropic-ferronematic samples

The polymer/magnetite (PM) nanoparticles have been mixed with a lyotropic liquid crystal a proportion of 0.1% and 0.2% b.w. by ultrasonication. The lyotropic liquid crystal (LLC) used is a ternary mixture composed of sodium dodecyl sulfate (SDS) 28.2%, decanol (DeOH) 5.7% and water 66.1%, exhibiting discotic nematic domains [12,13], noted N_D (see Figure 2). The lyotropic-ferronematic mixture (LC/PM) was stable over months.

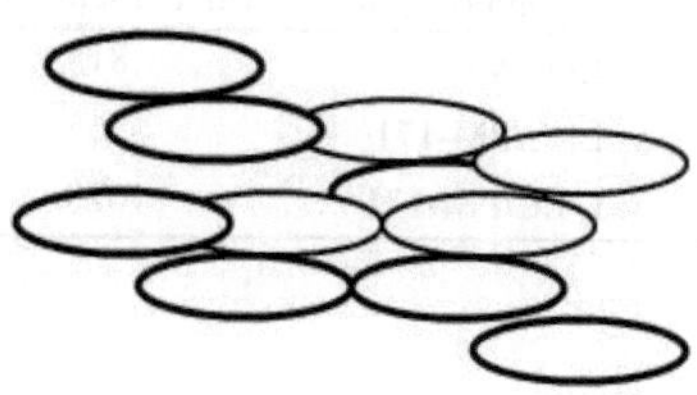

Fig. 2 Schematic presentation of LC in nematic discotic (ND) phase.

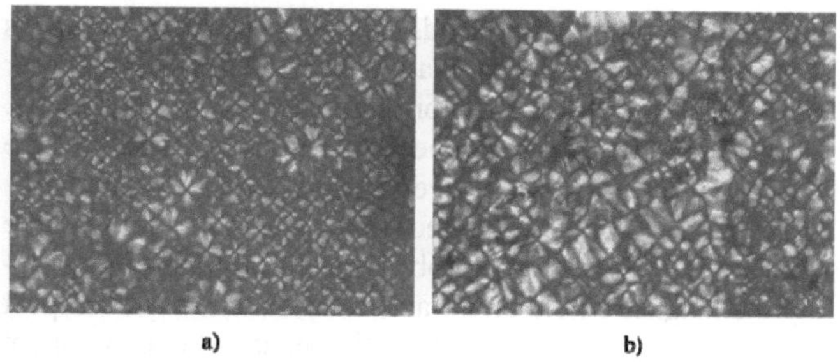

Fig 3 Polarization microscopy images of a) - liquid crystal and b) liquid crystal-composite samples

Different mixtures of lyotropic liquid crystal and copolymer-magnetite having the codename 171-Fe_3O_4/Pst in different concentrations were created (see Figure 3), homogenization being carried out through ultrasonication [14]. The composite system was filled in a cell of conventional sandwich type, consisting of two indium tin oxide (ITO) covered glass plates separated by Millar spacers of 130μm thickness for homeotropic alignment (H), that is with the director of the molecules perpendicular to the substrate.

Table 1 Nominal parameters for the liquid crystal/copolymer magnetite probes

Probe	Composition	Alignment	f[% of mass]	Cell thickness[μm]
H0	LLC N_D	H	-	130
H2	LLC/PM-171	H	0.1	130
H3	L LC/PM-171	H	0.2	130

The homeotropic-aligned cells were created by using DMOAP -metaxibenzen-butyl-anylin (N-dimethyl-N-octyl-3-aminopropyl trimetoxy-xylil) respecting the following ratios: 899 parts isopropanol, 100 parts distilled water, 1 part DMOAP. The solution was deposed onto the glass plates by spin coating, and after that polymerized for 15 minutes at a 100° C in a heating stage. The nominal parameters used for the samples are presented in Table 1.

B. Experimental Setup

The experimental setup [12] used to study the magnetic properties of the engineered samples composes of: A linearly polarized (500:1) laser having a c.w. optical power of 2.5 mW and an effective diameter of 0.52 mm, 1.52 mrad divergence, and a longitudinal modal distribution of 822MHz, two crossed polarizers having an 99.9% extinction ratio, a Si-based controlled-gain (0-70 dB), having a 400 - 1000 nm detection range, an electromagnet ranging in the 0.1-1 T, a high-precision multimeter and a current voltage sequencer in order to graph the electrical characteristics before the magnetic field was applied. The experimental setup is presented in Figure 4. Originally, the linearly polarized laser pulse is blocked by the two orthogonally crossed polarizers. Upon insertion of the probe between the polarizers, the liquid crystal anisotropy gives rise to a non-zero intensity laser pulse. For the homeotropic aligned samples, the transmission through the probe can be controlled by applying the magnetic field on a transversal direction with respect to the discotic liquid crystal alignment. The experiment setup is used to verify that the critical magnetic field at which the Freedericksz transition occurs satisfies $B < B_c^H$.

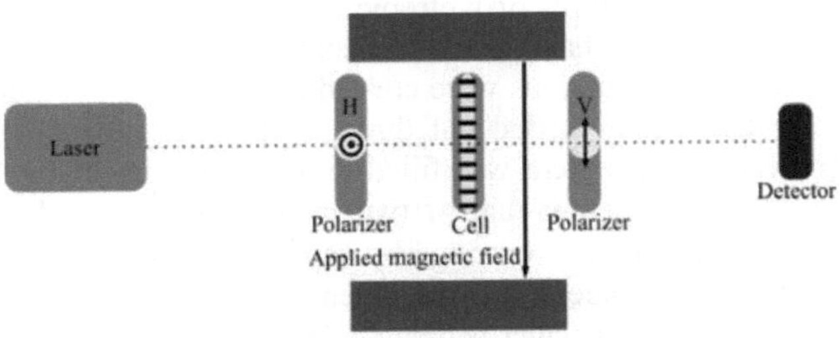

Fig. 4 Experimental setup for homeotropic alignment sample in magnetic field..

III. RESULTS AND DISCUSSIONS

The study of the static magnetic characteristics of the liquid crystals/composite material focused on the determination of the magnetic field at which the so-called Freedericksz transition appears. This transition manifests itself as a loss of reflectivity of the liquid crystal, and it has been theoretically hypothesized that magnetic nanoparticles tend to lower the Freedericksz transition threshold. Figure 5 presents the dependence of the transmitted light intensity as a function of the magnetic field for pure liquid crystal (fig.5a) and the doped cells (fig.5b), and highlights the critical transition field.

In the case of the pure liquid crystal cell - (B_c^{H0}), the Freedericksz transition threshold was measured at approximately 815 Gauss, whereas in the case of the doped cells, the threshold was lowered to approximately 765 Gauss - (B_c^{H2}) and 670 Gauss - (B_c^{H3}). At higher fields (over 1200 Gauss), however, parasitic oscillations (not shown) between these transitions appear, possibly due to convection phenomena and magnetic dipole resonance phenomena. The measured values are summarized in Table 2.

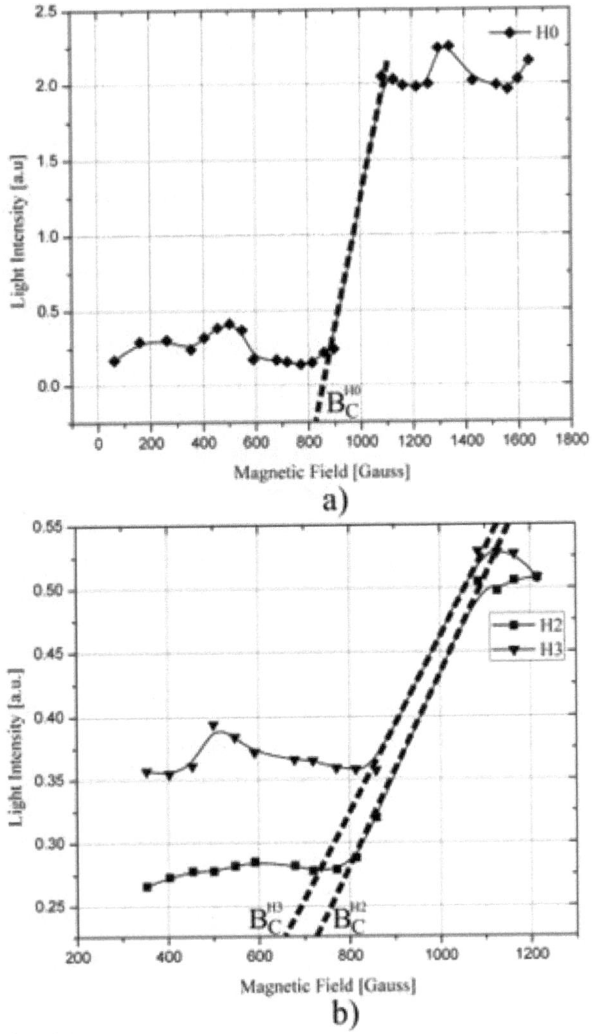

Fig. 5 Freedericksz transitions in the case of homeotropic alignment: a) - homeotropic pure liquid crystal; b) - homeotropic LC/PM-171 mixtures, H2 and H3.

Table 2: Determined critical field values for homeotropic alignment

Probe	Composition	Critical Field [Gauss]
H0	LLC N_D	815
H2	LLC/PM-171	765
H3	L LC/PM-171	670

IV. CONCLUSIONS

In order to obtain a good compatibility between the colloidal particles and liquid crystals, we prepared new polystyrene latexes containing magnetite with covalent bonds to the polymeric chains functionalized with alkoxysilanes containing double bonds. The particles have been characterized by TEM and magnetite nanocristallite embedded in polystyrene have been observed. By dispersing the magnetite-polymer particles in a lyotropic liquid crystal based on sodium dodecyl sulfate, decanol and water, a new colloidal system was obtained. The lyotropic ferronematic mixture was stable over months. The system was filled in ITO covered sandwich glass cells with homeotropic alignment. The optical transmission of laser light was registered when applying static magnetic field. The critical magnetic field for Freedericksz transition has been estimated and found to be lower in the case of composite samples.

Further experimental results and a theoretical model are under study and will be presented in the future.

CONFLICT OF INTEREST

The authors hereby declare that they have no conflict of interest.

REFERENCES

1. S. Chandrasekar, Liquid Crystals, (1992), 2nd ed. University Press, Cambridge.
2. P.G. De Gennes, J. Prost, (1993), The Physics of Liquid Crystals, 2nd ed., Claredon Press, Oxford.
3. P.G. De Gennes, (1994), *The Physics of Liquid Crystals*, Claredon Press, Oxford.
4. S.Odenbach (2009) Colloidal Magnetic Fluids: Basics, Development and Application of Ferrofluids. Volume 763 of Lecture Notes in Physics, Springer, ISBN: 3540853871, 9783540853879.
5. F. Brochard, P.G. de Gennes (1970) J. Phys. (Paris), **31**, pp. 691-708.
6. C. Y. Matuo, F. A. Tourinho, M. H. Souza, J. Depeyrot, and A. M. Figueiredo Neto, (2002), Lyotropic Ferronematic Liquid Crystals Based on New Ni, Cu and Zn Ionic Magnetic Fluids, Braz. J. Phys. vol.32 no.2b São Paulo.
7. H. Stark (2001) Physics of colloidal dispersions in nematic liquid crystal Physics Reports, 351, pp. 387-474.
8. D. Mănăilă-Maximean, G. Bossis, F. Giulieri, C. Metayer (2004) Mol. Cryst. Liq. Cryst., 417, pp. 227-233.
9. M. Ghiurea, R. Munteanu, D. Donescu, D. Mănăilă Maximean (2010) Journal of Optolectronics and Advanced Materials - Rapid Communications, vol. 4, pp 588-591.
10. C. Roşu, D. Mănăilă Maximean, D. Donescu, S. Frunză, A.R. Sterian (2010) Modern Physics Letters B, 24, No.1, p. 65-73.
11. D. Donescu, R. Somoghi, C. I. Spătaru, D. Mănăilă-Maximean, D. Mihaela Panaitescu, E. Vasile, C. L. Nistor (2013) Hybrid polymeric latexes containing magnetite, Colloid and Polymer Science, Volume 291, Issue 10, pp. 2345-2358.
12. P. O. Quist, B. Halle, I. Furo (1992) F. Chem. Phys. **96**, 3875.
13. P. Poulin, N. Francès, and O. Mondain-Monval, (1999), Suspension of spherical particles in nematic solutions of disks and rods, Phys. Rev. E 59, 4384.
14. D. Mănăilă Maximean, C. Roşu, E. Petrescu, C. Cîrtoaje, D. Donescu, E. Vasile, O. Dănilă (2011) Study of new polymer-magnetite particles/liquid crystal colloidal composite, 8th Liquid Matter Conference, 5-11 Sept, Vienna, Austria.
15. C. Cîrtoaje (2014) Ferrocholesteric - ferronematic transition in magnetic and laser field, University „Politehnica" of Bucharest Scientific Bulletin - Series A - Applied Mathematics and Physics, vol. 76, issue 1, pp. 187-196, ISSN 1223-7027 .
16. C. Cîrtoaje, C. Moţoc, E. Petrescu (2010) New magnetic method for determinig of elastic constants and rotational viscosity coeficient in nematic liquid cristals, University „Politehnica" Bucharest Scientific Bulletin- Series A- Applied Mathematics and Physics, vol. 72, Issue: 4, pp. 233–246, ISSN 1223-7027.
17. G. Iacobescu, A.L. Păun, C. Cârtoaje, (2008) Magnetically induced Freedericksz transition and relaxation phenomena in nematic liquid crystals doped with azo-dyes, Journal of Magnetism and Magnetic Materials, Vol. 320, Issue 17, Pages 2180–2184.

Author: Doina Manaila Maximean
Institute: University „Politehnica" of Bucharest
Department of Physics, Faculty of Applied Sciences
Street: Spl. Independenţei 313,
City: Bucharest, R-060082,
Country: Romania
Email: manaila@physics.pub.ro

Polaron Theory of the Emission Current in a Cathode-Adsorbed Nanofilm System at the Initial Stage of a High-Voltage Gas Discharge

Y.A. Barengolts[1] and S.I. Beril[2]

[1] Department of Mathematical Analysis, Shevchenko Dniester State University, Tiraspol, Moldova
[2] Department of Theoretical Physics, Shevchenko Dniester State University, Tiraspol, Moldova

Abstract— **A generalized formula has been derived for the current density of electron emission from a cathode coated with adsorbed non-metal nanofilm. Quantum contributions to the Nordheim function have been found which take into account the polaronic nature of the electron tunneling in this type of system.**

Keywords— **adsorbed non-metal nanolayers, gas discharge, electronic polyaron.**

I. INTRODUCTION

It was shown [1, 2] that in the field emission occurring in a strong electric field, the quantum nature of the interaction of an electron with the medium shows up, and the use of the classical formula for the image force potential in the calculation of the emission current density is incorrect.

The electronic polaron theory [3–5] provides a general expression for the field emission current density from which, in the limit of low and high temperatures, the general Fowler–Nordheim and Richardson–Schottky formulas follow, which remain valid for high fields ($E > 10^6$ V/cm).

The interaction of a band electron with the induced fast polarization of the medium (plasmons of valence electrons) transforms the electron into a quasiparticle, i.e. an electronic polaron. As the frequency of the plasma oscillations of valence electrons, , is of the same order as the inverse time of electron tunneling through a barrier, the effect of the fast polarization on the electron is dynamic, i.e., the state of the tunneling electron is self-consistent with the shape of the potential barrier.

The density of the emission current through the barrier formed at the interface between the metal and the adsorbed dielectric nanofilm can be described by the formula (see, e.g., Ref. 6)

$$j(E,T) = e\int_{-\infty}^{\infty} N(W,T)\,D(W,E)\,dW \,. \tag{1}$$

Here $j(E,T)$ is the emission current density, e is the elementary charge, E is the applied electric field strength, T is the absolute temperature, $N(W,T)$ is the electron energy distribution in the metal, and $D(W,E)$ is the transparency of the potential barrier at the interface between the cathode and the external medium.

Within the Sommerfeld model, the following relation is valid:

$$N(W,T) = \frac{4\pi mkT}{h^3}\ln\left(1+\exp\left(-\frac{W-W_F}{kT}\right)\right), \tag{2}$$

where W_F is the Fermi energy of the metal, h is Plank's constant, k is Boltzmann's constant, and m is the effective mass of an electron in the metal.

The transparency of the metal–nanofilm barrier, $D(W)$, is described by the relation

$$D(E,W) = \left[1+\exp\left(-\frac{4\pi i}{h}\int_{x_1}^{x_2} p(x)dx\right)\right]^{-1}, \tag{3}$$

where

$$p(x) = \sqrt{2m\left(W - W_p(x)\right)} \tag{4}$$

is the electron momentum depending on the coordinate ($W_p(x)$ is the total potential energy of the electron taking into account the polyaronic effect), x_1 and x_2 are the coordinates of the potential barrier boundaries at the Fermi level of the metal. We will use the system of reference where the coordinate is measured from the plane of the metal surface and the reference energy is taken to be the energy at the vacuum level [6].

II. ORIGINAL FORMULA FOR THE NORDHEIM FUNCTION

It was shown [5] that under the conditions of a high-voltage gas discharge operating in the presence of adsorbed non-metal nanofilm on the surface of a metal, the potential energy of the quantum image forces can be estimated by the extrapolation formula derived in the theory of electronic polarons:

$$W_{is} = -\frac{e^2}{16\pi\varepsilon_0\varepsilon(x + x_0/4)}. \tag{5}$$

V. Sontea and I. Tiginyanu (eds.), *3rd International Conference on Nanotechnologies and Biomedical Engineering*, IFMBE Proceedings 55,
DOI: 10.1007/978-981-287-736-9_57

Here ε_0 is the electric constant, x_0 is the interpolation parameter characterizing the size of the electronic polaron (tenths of a nanometer), and ε is the high-frequency dielectric constant of the adsorbed film material.

For the case where an electric field is present, the potential energy of an emitted electron, in view of (5), is given by the formula

$$W_p = -\frac{e^2}{16\pi\varepsilon_0\varepsilon\left(x+x_0/4\right)} - eEx\,. \qquad (6)$$

Then we have

$$p(x) = \sqrt{2m\left(W + \frac{e^2}{16\pi\varepsilon_0\varepsilon\left(x+x_0/4\right)} + eEx\right)}\,. \qquad (7)$$

The integration limits for the integral in (3) are found from the condition $p(x)=0$. Then, according to (7), we obtain

$$W + \frac{e^2}{16\pi\varepsilon_0\varepsilon\left(x+x_0/4\right)} + eEx = 0\,. \qquad (8)$$

To reduce the integral in (3), taking into account (8), to the form considered in Ref. 6, we use the change of variables

$$z = x + x_0/4\,. \qquad (9)$$

Then from (8) it follows that

$$W\left(1-\frac{eEx_0}{4W}\right) + \frac{e^2}{16\pi\varepsilon_0\varepsilon z} + eEz = 0\,. \qquad (10)$$

To find the integration limits, we introduce, as in Ref. 5, the parameter

$$\gamma = 1-\frac{eEx_0}{4W} \qquad (11)$$

and obtain from (8) a quadratic equation whose roots are

$$z_{1,2} = -\frac{\gamma W}{2eE} \pm \sqrt{\frac{\gamma^2W^2}{4e^2E^2} - \frac{e}{16\pi\varepsilon_0\varepsilon E}}\,. \qquad (12)$$

Introduce a new variable (different from the argument used in Ref. 6 by the presence of the parameters γ and ε):

$$y_a = \frac{e}{2\left|\gamma W\right|}\sqrt{\frac{eE}{\pi\varepsilon_0\varepsilon}}\,. \qquad (13)$$

Then, assuming that the energy W may take only negative values, we can write

$$z_{1,2} = -\frac{\gamma W}{2eE}\left(1 \pm \sqrt{1-y_a^2}\right). \qquad (14)$$

In view of the above notation, the integral in (3) becomes

$$I = -\frac{2\sqrt{2m}}{\hbar}i\int_{z_2}^{z_1}\left(\gamma W + \frac{e^2}{16\pi\varepsilon_0\varepsilon z} + eEz\right)^{\frac{1}{2}}dz\,. \qquad (15)$$

Introduce one more variable:

$$\rho_a = -\frac{2eE}{\gamma W}z\,. \qquad (16)$$

Then, for the integration limits in (15), we have

$$\rho_a = -\frac{2eE}{\gamma W}\left(-\frac{\gamma W}{2eE}\left(1\pm\sqrt{1-y_a^2}\right)\right) = 1\pm\sqrt{1-y_a^2}\,. \qquad (17)$$

Transform separately the integrand in (15) as

$$\gamma W + \frac{e^2}{16\pi\varepsilon_0\varepsilon z} + eEz = \frac{\gamma W}{2}\left(2 - \frac{e^3E}{4\pi\varepsilon_0\varepsilon\gamma^2W^2\rho_a} - \rho_a\right). \qquad (18)$$

Put the imaginary unit under the square root in (15):

$$I = \frac{\sqrt{m\left(\gamma W\right)^3}}{\hbar eE}\int_{1-\sqrt{1-y_a^2}}^{1+\sqrt{1-y_a^2}}\sqrt{\rho_a - 2 + \frac{y_a^2}{\rho_a}}\,d\rho_a\,. \qquad (19)$$

The resulting integral is similar to that considered in Ref. 6. Integration in (19) yields a tabulated function $\nu_a(a')$, which is different from the Nordheim function only by the argument a' instead of a:

$$\nu_a(a') = \sqrt{\frac{1+a'}{2}}\left(E\left[\sqrt{\frac{2a'}{1+a'}}\right] - (1-a')K\left[\sqrt{\frac{2a'}{1+a'}}\right]\right), \qquad (20)$$

where

$$a' = \sqrt{1-y_a^2}\,. \qquad (21)$$

It follows that

$$\nu_a(y_a)=\sqrt{\frac{1+\sqrt{1-y_a^2}}{2}}$$
$$\times\left(E\left[\sqrt{\frac{2\sqrt{1-y_a^2}}{1+\sqrt{1-y_a^2}}}\right]-\left(1-\sqrt{1-y_a^2}\right)K\left[\sqrt{\frac{2\sqrt{1-y_a^2}}{1+\sqrt{1-y_a^2}}}\right]\right).\quad(22)$$

For the case $y_a \le 1$ (where electrons tunnel from energy levels close to the Fermi level of the metal, we obtain from (22)

$$\nu_a(y_a)=\sqrt{1+y_a}\left(E\left[\sqrt{\frac{1-y_a}{1+y_a}}\right]-y_aK\left[\sqrt{\frac{1-y_a}{1+y_a}}\right]\right).\quad(23)$$

For $y_a>1$ we have

$$\nu_a(y_a)=\sqrt{2y_a}\left(E\left[\sqrt{\frac{y_a-1}{2y_a}}\right]-\frac{1+y_a}{2}K\left[\sqrt{\frac{y_a-1}{2y_a}}\right]\right).\quad(24)$$

The Nordheim function written in classical form is [6]

$$\nu(y)=\sqrt{\frac{1+\sqrt{1-y^2}}{2}}$$
$$\times\left(E\left[\sqrt{\frac{2\sqrt{1-y^2}}{1+\sqrt{1-y^2}}}\right]-\left(1-\sqrt{1-y^2}\right)K\left[\sqrt{\frac{2\sqrt{1-y^2}}{1+\sqrt{1-y^2}}}\right]\right),\quad(25)$$

where

$$y=\frac{e}{2|W|}\sqrt{\frac{eE}{\pi\varepsilon_0}}.\quad(26)$$

Comparing (13) and (47), we can write

$$y_a=\frac{y}{\gamma\sqrt{\varepsilon}}.\quad(27)$$

For an objective comparison, we put $\varepsilon=1$. In view of this condition, substitution of (27) in (22) yields

$$\nu_a(y)=\sqrt{\frac{\gamma+\sqrt{\gamma^2-y^2}}{2\gamma}}$$
$$\times\left(E\left[\sqrt{\frac{2\sqrt{\gamma^2-y^2}}{\gamma+\sqrt{\gamma^2-y^2}}}\right]-\left(\frac{\gamma-\sqrt{\gamma^2-y^2}}{\gamma}\right)K\left[\sqrt{\frac{2\sqrt{\gamma^2-y^2}}{\gamma+\sqrt{\gamma^2-y^2}}}\right]\right).\quad(28)$$

Figure 1 presents the Nordheim function $\nu(y)$ and the function $\nu_a(y_a)$ versus parameter y.

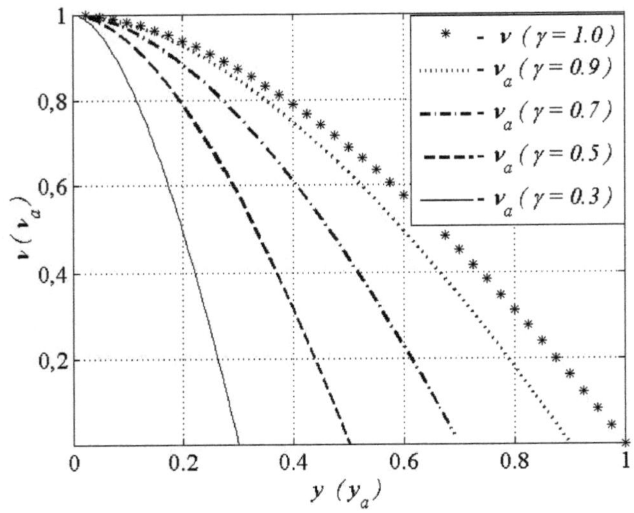

Fig. 1 Tabulated functions $\nu(y)$ and $\nu_a(y_a)$ vs parameter y for different values of factor γ.

As can be seen from Fig. 1, the adsorbed nanofilm substantially affects the tabulated function and, hence, the emission current density, as the function $\nu_a(y_a)$ enters into the exponent in the equation of the current-voltage characteristic. According to (11), this effect is due to the parameter x_0, which is the basic characteristic of the polaron accompanying a tunneling electron, and it is enhanced on increasing the applied electric field strength. This behavior of the function $\nu_a(y_a)$ can be considered abnormal, as in the field emission formula, the tabulated function enters into the numerator of the negative exponent.

This effect can be weakened if we take into account the actual value of the permittivity of the adsorbed nanofilm material. The obtained relations for the function $\nu_a(y_a)$ make it possible to calculate the emission current density by a formula similar to that proposed in Ref. 5, 6 (Hartree system):

$$j(E,T)=\frac{kT}{2\pi^2}\int_{-W_a}^{W_m}\frac{\ln\{1+\exp[-(W-W_F)/kT]\}dW}{1+\exp\left[\left(\frac{4}{3}\right)\sqrt{2}E_a^{-\frac{1}{4}}y_a^{-\frac{3}{2}}\nu_a(y_a)\right]}$$

$$+\frac{kT}{2\pi^2}\int_{W_m}^{\infty}\ln\{1+\exp[-(W-W_F)/kT]\}dW, \quad (29)$$

where W_m is the maximum potential barrier energy in the adsorbed dielectric nanofilm.

III. DISCUSSION

1. Formulas (23) and (24) actually determine the current density of field emission, thermionic emission, and thermal electron emission from a metal cathode coated with adsorbed non-metal nanofilm. Formula (29) is different from the classical formulas for emission current density given in Ref. 6 by the parameter x_0 taking into account the "polaron contribution" to the argument of the tabulated function $v_a(y_a)$. The elliptic integrals $E[k]$ and $K[k]$ are also functions of the electric field (see (13)).

2. Application of the polaron theory to the physical processes that occur in an emitter coated with adsorbed dielectric nanofilm eliminates the singularity of the potential energy of a tunneling electron at the emission boundary (at $x=0$). This is the case where the electric field penetrates into the metal to a depth of the order of the electronic polaron radius.

3. From the generalized formula for the current density (29) it follows that the polaron effects are negligible for fields $E<10^6$ V/cm and the Fowler–Nordheim and Richardson–Schottky formulas provide good agreement with experiment. For fields $E>10^6$ V/cm , at which the width of the potential barrier is comparable to the electronic polaron radius, the quantum contributions to the emission current density become significant. When dealing with the actual range of electric fields, at the initial stage of the study, one should use the classical formula (29) with corrections to the function $v_a(y_a)$.

CONFLICT OF INTEREST

Every paper must contain a declaration of conflicts of interest. If there are no such conflicts write "The authors declare that they have no conflict of interest".

REFERENCES

1. A. Harstein, Z. A. Weinberg, and D. J. Di Maria, "Experimental test of the quantum-mechanical image-force theory," *Phys. Rev. B,* vol. 25, pp. 7174–7182, June 1982.
2. A. Harstein and Z. A. Weinberg, "Unified theory of internal photoemission and photon-assisted tunneling," Phys. Rev. B, vol. 20, pp. 1335–1338, April 1979.
3. Y. Toyosawa "Theory of electronic polaron and ionization of trapped electron by an exciton," Phys. Rev. B, vol. 25, pp. 7174–7182, June 1982.
4. S.I. Beril, E.P. Pokatilov, E.P. Goryachkovskii, and N.N. Semenovskaya, "Polaron theory of the image potential considering the spatial dispersion," *Phys. Stat. Sol.(b)*, vol. 176, pp. 347 – 353, January 1993.
5. S.I. Beril and Y.A. Barengolts, "The Effect of Molecules Adsorbed to the Cathode Surface on the Characteristics of a High-Voltage Gas Discharge," *IEEE Transaction on Plasma Science*, vol. 42, pp. 3109 – 3112, October 2014.
6. E.I. Murphy and A.H. Good, "Thermionic emission, field emission, and ransition region," *Phys. Rev.,* vol.162, pp. 1464 – 1473, June 1956.

Author: Y.A. Barengolts
S.I. Beril
Institute: Shevchenko Dniester State University
Street: 25 October Street 128
City: Tiraspol
Country: Moldova
Email: barengolts@list.ru

Zero Frequency Spectrum of 3-D Metal Photonic Crystals Obtained by the 3-D Kronig–Penney Model

V.V. Sergentu[1] and V.V. Ursaki[2]

[1] Institute of Applied Physics of the Academy of Sciences of Moldova, Chisinau, Moldova
[2] Institute of Electronic Engineering and Nanotechnologies "D. Ghitu" of the Academy of Sciences of Moldova, Chisinau, Moldova

Abstract— **We demonstrate in this work the possibility of the existence of ultrashort (dark) pure electrical modes at extremely low frequencies in a system of metallic nanospheres inserted into a dielectric material. These modes resemble usual modes of photonic crystals, but they do not have counterpart for single nanospheres.**

Keywords— **metallic nanosphere, electromagnetic waves, ultrahort mode, dark mode, photonic crystal, plane wave, dispersion law.**

I. INTRODUCTION

Various systems of metallic nanoparticles, which can be used for the development of basically new materials, are produced by nanotechnologies [1,2]. It was shown [3] that dark modes of pure electrical nature (ultrashort modes of the transversal electrical field), which modify the Coulomb law in the system [4], emerge when a system of nanoparticles is embedded into a dielectric material.

The modified Coulomb law contributes to the formation of bound electron pairs from certain electronic states on the Fermi surface, if the wavelength of the dark mode is comparable with the particle radius. Stockman et al suggested for the first time that dark modes can exist in the volume of a disperse nanomaterial [5].

For the calculation of the dark mode spectrum, we used in this work the Kronig–Penney model for photonic crystals. The Kronig–Penney model was previously used for the description of properties of 1-D photonic crystals [6]. One can easily realized on the basis of elementary calculations that the emergence of zero frequency modes is not possible in 1-D photonic crystals. The carbon nanotubes represent a physical manifestation of the Kronig–Penney mathematical model for 2-D systems [7]. Therefore, models similar to the Kronig–Penney model have been used for the description of electromagnetic properties of carbon tubes systems arranged in a 2-D lattice taking into account the motion along the tube axis [8]. We have generalized this model for the case of a system of nanoparticles with cylindrical symmetry [4, 9,10]. Actually, the Kronig–Penney model for 2D photonic crystals represents a basis for the system investigated in this paper. The existence of dark modes is possible in a 2D photonic crystal, but only taking into consideration the motion in the direction perpendicular to the plane of the photonic crystal. The solution of this problem by means of usual methods from the theory of photonic crystals (the method of plane waves [11]) gave the same results [12].

We used in this paper the Kronig–Penney model for 3-D photonic crystals.

II. THE EQUATION FOR METALLIC NANOSPHERES

We will consider the propagation of electromagnetic waves in a system of simplest metallic nanospheres arranged in an infinite cubical lattice embedded into a dielectric material which interacts with the electromagnetic field for the case of extremely low frequencies $\omega \ll (2\pi c/a)$. The description is performed on the basis of the standard Maxwell equations system in vacuum [13]. It follows from the Maxwell equations that, for a nonmagnetic dielectric medium, the situation for the case of extremely low frequencies $\omega \ll 2\pi c/a$ is described by the following equations for the electric field $\vec{E}(\vec{r})\left(\vec{H}(\vec{r}) \equiv 0\right)$ and the potential $\Phi(\vec{r})$

$$\nabla\left(\varepsilon_0 \varepsilon_1 \vec{E}(\vec{r})\right) = \Omega(\vec{r}), \tag{1}$$

$$\vec{E}(\vec{r}) = -\nabla\left(\Phi(\vec{r})\right), \tag{2}$$

where $\varepsilon_0 \varepsilon_1$ is the dielectric permittivity of the medium surrounding the nanosphere, $\Omega(\vec{r})$ is the density of the free charge. For a nanosphere from nonmagnetic metal one can obtain the following equation instead of the equation (1)

$$div\left(grad(\Phi(\vec{r}))\right) - \Phi(\vec{r})/L_{sc}^2 = \Omega(\vec{r}), \tag{3}$$

where L_{sc} is the electrostatic screening length of the metal. Further we are looking for free modes, and assume that

$$\Omega(\vec{r}) \equiv 0. \tag{4}$$

We will show that the equations (1,2,3) have nontrivial periodically variable in the space solutions in the case of extremely low frequencies $\omega \ll (2\pi c/a)$, in addition to simple solutions of the type obtained in [14].

V. Sontea and I. Tiginyanu (eds.), *3rd International Conference on Nanotechnologies and Biomedical Engineering*,
IFMBE Proceedings 55,
DOI: 10.1007/978-981-287-736-9_58

III. A MODEL PROBLEM FOR A SIMPLEST-TYPE NANOSPHERE

Let us introduce the notion of simplest-type nanosphere, for which

$$\varepsilon_1 div(grad(\Phi(\vec{r}))) - \hat{u}(\theta,\varphi)\cdot\delta(r-R)\Phi(\vec{r}) = 0\,, \quad (5)$$

where R is the nanosphere radius, $\hat{u}$ is the operator acting only on angular variables φ and θ, so that it will be diagonal in the basis of spherical functions $Y_{lm}(\vec{n})$ with eigenvalues $\{\delta_{m,m'}\delta_{l,l'}u_l\}$ ($\vec{n}$ is the unity vector given by the angles (θ,φ) in the spherical coordinate system.

The method of obtaining a general solution for a single nanosphere of the type described above with radius R is standard. The center of the nanosphere is placed in the origin of the spherical coordinate system (r,θ,φ). The boundary conditions for (5) have the following form

$$\begin{aligned}&\Phi_{lm}(\mathrm{r})|_{R+0} - \Phi_{lm}(\mathrm{r})|_{R-0} = 0\\ &\frac{\partial\Phi_{lm}(\mathrm{r})}{\partial r}|_{R+0} - \frac{\partial\Phi_{lm}(\mathrm{r})}{\partial r}|_{R-0} = \frac{u_l}{\varepsilon_1}\Phi_{lm}(\vec{r})|_{R\pm0}\end{aligned}, \quad (6)$$

where ε_1 is the relative permittivity of the medium in which the nanospheres are embedded.

The solution of (6) is sought in the form of functions [15]

$$\Phi_{lm}(\vec{r}) = \begin{cases} r > R \rightarrow \left[r^l + D_l^{(ns)}/r^{l+1}\right] Y_{lm}(\varphi,\theta) \\ r < R \rightarrow C_{lm} r^l Y_{lm}(\varphi,\theta)\end{cases}, \quad (7)$$

where C_{lm} are certain constants obtained from the boundary conditions, J(Y) are Bessel functions, $\{D_{lm}^{(ns)}\}$ is a set of values fully describing the process of external electromagnetic wave scattering on a single nanosphere at ω→0 like similar values for a nanocylinder [16].

One can obtain from (6,7)

$$D_l^{(ns)} = -\frac{u_l R^{2(l+1)}}{u_l R + (2l+1)\varepsilon_1}\,. \quad (8)$$

We will use the standard method of plane in order to find solutions for a system of nanospheres arranged in a cubic lattice [11]. The solution of equations is sought in the form of Fourier series expansion on three-dimensional plane waves

$$\Phi(\vec{r}) = \sum_{\vec{q}_n} \tilde{\Phi}(\vec{k}+\vec{q}_n)\exp((\vec{k}+\vec{q}_n)\vec{r})\,, \quad (9)$$

where

$$\tilde{\Phi}(\vec{k}+\vec{q}_n) = = \frac{1}{V}\int_V d^3 r\,\Phi(\vec{r})\exp\left(-i(\vec{k}+\vec{q}_n)\vec{r}\right), \quad (10)$$

$\{\vec{q}_n\}$ is the vector of the reciprocal 3D cubic lattice, $\vec{k}$ is the vector of the 3D motion, which changes in the limits of the Brillouin zone, a is the lattice constant, and $V = a^3$ is the volume of the unit cell.

The term $\hat{u}(\theta,\varphi)\cdot\delta(r-R)$ is separately evidenced in the process of calculation of matrix elements in the left part of equation (4). The spherical harmonics (Y) of Bessel functions (J) of half-integer index are evidenced from 3D plane waves to take into account the scattering on a single nanosphere placed in the origin of coordinates [15].

Then, a system of linear equations is obtained from (3) taking into account (9)

$$\begin{aligned}&\left|\vec{k}+\vec{q}_n\right|^2 \tilde{\Phi}(\vec{k}+\vec{q}_n) = \\ &= -\frac{8\pi^3 R^2}{\varepsilon_1 a^3}\sum_{\vec{q}_{n'}}\sum_{l=0}^{\infty} u_l \frac{J_{l+0.5}\left(R\left|\vec{k}+\vec{q}_n\right|\right) J_{l+0.5}\left(R\left|\vec{k}+\vec{q}_{n'}\right|\right)}{R\sqrt{\left|\vec{k}+\vec{q}_n\right|\left|\vec{k}+\vec{q}_{n'}\right|}} \times \\ &\times \sum_{m=-l}^{l} Y^*_{lm}\left(\vec{n}_{\vec{k}+\vec{q}_n}\right) Y_{lm}\left(\vec{n}_{\vec{k}+\vec{q}_{n'}}\right)\tilde{\Phi}(\vec{k}+\vec{q}_{n'})\end{aligned} \quad (11)$$

where $\vec{n}_{\vec{k}} = \vec{k}/k$ is the unit vector.

There is a simple method to analytically solve the system of the type (11). The idea for solving the problem can be demonstrated on the instance of a system of linear equations of the following type:

$$\left\{ X_i = \sum_{i'}\sum_n V_n f_i^{(n)} f_{i'}^{(n)} X_{i'}\right.\,, \quad (12)$$

where V_n, f_i are constants, X_i are unknown values.

The method of solving such a system is elementary, and it is discussed, for instance, in [17]. The system (12) has non-trivial solutions at parameters values satisfying the analytical expression

$$\sum_{i,n} V_n f_i^{(n)\,2} = 1\,. \quad (13)$$

Let us assume that the nanosphere scatters all the spherical waves in the interval $L \ge m, m' \ge -L (L \rightarrow \infty)$. In this case the relation of the type (16) has the form of

$$\mathrm{Det}[Z] = 0\,, \quad (14)$$

where Z=(L-I), I is the unit matrix, L is a matrix for which

$$\begin{aligned}&L(lm;l'm') = -\frac{8\pi^3 R}{\varepsilon_1 a^3}\sum_{\vec{q}_n}\sqrt{u_l u_{l'}}\,Y^*_{lm}\left(\vec{n}_{\vec{k}+\vec{q}_n}\right)\times \\ &\times Y_{l'm'}\left(\vec{n}_{\vec{k}+\vec{q}_n}\right)\frac{J_{l+0.5}\left(R\left|\vec{k}+\vec{q}_n\right|\right) J_{l'+0.5}\left(R\left|\vec{k}+\vec{q}_n\right|\right)}{\left|\vec{k}+\vec{q}_n\right|^3}\end{aligned}. \quad (15)$$

One should notice that the equations (14,15) represent the dispersion law for spherical symmetry nanoparticles of different nature arranged in a periodic 3D lattice, for the case of extremely low frequencies $\omega \ll (2\pi c/a)$.

IV. ZERO FREQUENCY SPECTRUM OF 3D METAL PHOTONIC CRYSTALS

The boundary conditions for a single metal nanosphere have the form of [13]

$$\begin{aligned} \Phi_{lm}(\vec{r})|_{R+0} &= \Phi_{lm}(\vec{r})|_{R-0} \\ \varepsilon_1 \frac{\partial \Phi_{lm}(\vec{r})}{\partial r}|_{R+0} &= \frac{\partial \Phi_{lm}(\vec{r})}{\partial r}|_{R-0} \end{aligned}, \tag{16}$$

The solution of (16) has the form of

$$\Phi_{lm}(\vec{r}) = \begin{cases} r > R \rightarrow \left[r^l + D_l^{(met)} / r^{l+1}\right] \mathrm{Y}_{lm}(\varphi,\theta) \\ r < R \rightarrow C_{lm} (r/L_{sc})^{-1/2} I_{l+0.5}(r/L_{sc}) \mathrm{Y}_{lm}(\varphi,\theta) \end{cases}, \tag{17}$$

where

$$D_l^{(met)} = R^{(2l+1)} \times \times \frac{\varepsilon_1 l - R\, d\left(\ln((R/L_D)^{-0.5} I_{l+0.5}(R/L_D))\right)/dR}{\varepsilon_1 (l+1) + R\, d\left((R/L_D)^{-0.5} I_{l+0.5}(R/L_D)\right)/dR}. \tag{18}$$

In contrast to ideal nanospheres, for the dispersion law of a system of real nanospheres, the surface dielectric permittivity u_l should be replaced in (15) by another value u_l^{eff}, obtained from the relation

$$u_l^{eff} = -\frac{\varepsilon_1 l}{R} + \frac{d\left(\ln\left((R/L_{sc})^{-0.5} I_{l+0.5}(R/L_{sc})\right)\right)}{dR}. \tag{19}$$

The case of simple dielectric spheres is obtained at $L_{sc} \rightarrow \infty$

We used an approach employed previously in [9]. The expression (19) is deduced in a manner ensuring that the scattering efficiency $D_l^{(ns)}$ of a virtual nanosphere of the simplest type with an effective relative surface dielectric permittivity u_l^{eff} is exactly equal to the scattering efficiency D_l of a real nanosphere $(D_l \equiv D_l^{(nt)})$.

It remains to put (19) into (15) and to solve the obtained problem for a nanospheres system arranged in a regular cubic lattice. One should obtain as a result a surface family in the Brillouin space. However, the full solution of such a problem is possible only by numerical methods. Therefore, we assume that in the equation (14), written in the form of an equation system

$$ZX = 0, \tag{20}$$

the following inequality is fulfilled for the X vectors:

$$|L(lm;l'm')|_{lm;l'm' \neq 0,0} \ll |L(0,0)|.$$

Then, according to the perturbation method one can write a formal expression up to the second order of smallness inclusively

$$X(0,0) = -\frac{8\pi^3 R}{\varepsilon_1 a^3} \left[\begin{aligned} &\sum_{\vec{q}_n} \sqrt{u_0 u_0} \frac{J_{0.5}\left(R|\vec{k}+\vec{q}_n|\right) J_{0.5}\left(R|\vec{k}+\vec{q}_n|\right)}{|\vec{k}+\vec{q}_n|^3} X(0,0) + \\ &+ \sum_{l'm' \neq 0,0} \sum_{\vec{q}_n} \sqrt{u_0 u_{l'}} \mathrm{Y}_{l'm'}\left(\vec{n}_{\vec{k}+\vec{q}_n}\right) \frac{J_{0.5}\left(R|\vec{k}+\vec{q}_n|\right) J_{l'+0.5}\left(R|\vec{k}+\vec{q}_n|\right)}{|\vec{k}+\vec{q}_n|^3} X(l',m') \end{aligned} \right]. \tag{21}$$

$$X(l',m') = -\frac{8\pi^3 R}{\varepsilon_1 a^3} \sum_{\vec{q}_n} \sqrt{u_{l'} u_0} \mathrm{Y}^*_{l'm'}\left(\vec{n}_{\vec{k}+\vec{q}_n}\right) \frac{J_{l'+0.5}\left(R|\vec{k}+\vec{q}_n|\right) J_{0.5}\left(R|\vec{k}+\vec{q}_n|\right)}{|\vec{k}+\vec{q}_n|^3} X(0,0)$$

From which it follows that

$$1 = -\frac{8\pi^3 R}{\varepsilon_1 a^3} u_0 \left[\begin{aligned} &\sum_{\vec{q}_n} \frac{J_{0.5}\left(R|\vec{k}+\vec{q}_n|\right)^2}{|\vec{k}+\vec{q}_n|^3} - \\ &-\frac{8\pi^3 R}{\varepsilon_1} \sum_{l' \neq 0} u_{l'} \sum_{\vec{q},\vec{q}'_{nn}} P_{l'}\left(\vec{n}_{\vec{k}+\vec{q}_n} \vec{n}_{\vec{k}+\vec{q}'_{nnn}}\right) \frac{J_{0.5}\left(R|\vec{k}+\vec{q}_n|\right) J_{l'+0.5}\left(R|\vec{k}+\vec{q}_n|\right)}{|\vec{k}+\vec{q}_n|^3} \frac{J_{l'+0.5}\left(R|\vec{k}+\vec{q}'_n|\right) J_{0.5}\left(R|\vec{k}+\vec{q}'_n|\right)}{|\vec{k}+\vec{q}'_n|^3} \end{aligned} \right]. \tag{22}$$

where P is the Lagrange polynomial [15].

Further, we assume that

$$G_l = \frac{1}{a^3}\left|\sum_{\vec{q}_n \neq 0} \frac{J_{0.5}\left(R\left|\vec{k}+\vec{q}_n\right|\right) J_{l+0.5}\left(R\left|\vec{k}+\vec{q}_n\right|\right)}{\left|\vec{k}+\vec{q}_n\right|^3}\right| << . \quad (23)$$

$$<< \frac{1}{a^3}\left|\frac{J_{0.5}(Rk)J_{l+0.5}(Rk)}{k^3}\right|$$

Then one can obtain from (22)

$$1 = -u_0 \frac{8\pi^3 R}{\varepsilon_1 a^3} \frac{J_{0.5}(Rk)^2}{k^3}\left[1 - u_1 \frac{8\pi^3 R}{\varepsilon_1 a^3} \frac{J_{0.5}(Rk)^2}{k^3}\right]. \quad (24)$$

Assuming that $L_{sc} < R/\varepsilon_1$ when $u_0 \approx u_1 \approx 1/L_{sc}$, we obtain the value of the dark mode wave vector

$$k \approx \frac{2\pi}{R}\left(\frac{2\pi R}{a}\right)^{3/2}\left(\frac{R}{L_{sc}}\right)^{1/2}. \quad (25)$$

The parameter L_{sc} in metal nanoparticles can change in a wide diapason of values depending upon the considered electron state on the Fermi surface. Usually, L_{sc} is estimated from the formula [18]

$$L_{sc} \propto \sqrt{\varepsilon_o E_{sc} / e^2 n_0}\ , \quad (26)$$

where e is the electron charge, n_o is the concentration of free electrons, E_{sc} is a parameter with the units of energy. For a pure Fermi distribution of band electrons, E_{sc} is equal to Fermi energy. Therefore, for usual metals, taking into account only band electrons, we have L_{sc}>1nm. The low values of L_{sc}<0.1 nm can be explained by the influence of non-band electrons from atoms [3]. In conclusion of this section, one can mention the absence of the sought effect in this approximation for the case of a 3D photonic crystal, in contrast to the 2D photonic crystal in which the movement of the electron id allowed in the third direction.

V. CONCLUSIONS

In this study we have developed the Kronig–Penney model for 3D photonic crystals. The dispersion law was deduced for a system of nanoparticles with spherical symmetry arranged in a periodic 3D lattice. It was found that the conditions for the emergence of dark modes are less favourable in 3D photonic crystals from transparent materials as compared to 2D ones. The possibility of the existence of zero frequency dark modes in a 3D metal photonic crystal (a system of metal nanospheres) has been demonstrated.

CONFLICT OF INTEREST

The authors declare that they have no conflict of interest.

REFERENCES

1. C. N. R. Rao, G. U. Kulkarni, A. Govindaraj, *et. al.*, Metal nanoparticles, nanowires, and carbon nanotubes, *Pure Appl. Chem.* **72**, 21–33, 2000.
2. J. Ederth, Electrical Transport in Nanoparticle Thin Films of Gold and Indium Tin Oxide. Acta Universitatis Upsaliensis. Uppsala, 2003.
3. V. V. Sergentu, V. V. Ursaki, and I. M. Tiginyanu, Renormalization of the Coulomb Law in an Amorphous System of Metallic Nanospheres and Its Impact on the Electronic Subsystem, *J. Nanoelectron. Optoelectron.* **9**, 276-281, 2014.
4. V. Sergentu, V. Ursaki and L. Sirbu, Body zero frequency modes in photonic crystals of nanopores and nanocylinders, *Phys. Stat. Solidi B* **250**, 2215–2220, 2013.
5. M. I. Stockman, S. V. Faleev, and D. J. Bergman, Localization versus Delocalization of Surface Plasmons in Nanosystems: can one state have both characteristics? *Phys. Rev. Lett.* **87**, 167401 (2001).
6. F. Zolla, Foundations of Photonic Crystal Fibres, Imperial College Press, 2005.
7. G. Ya. Slepyan, S. A. Maksimenko, A. Lakhtakia, *et. al.*, Electrodynamics of carbon nanotubes: Dynamic conductivity, impedance boundary conditions, and surface wave propagation, *Phys. Rev.* **60**, 17136, 1999.
8. I. S. Nefedov, Electromagnetic waves propagating in a periodic array of parallel metallic carbon nanotubes, *Phys. Rev.* **82**, 155423, 2010.
9. L. Sirbu, V. Sergentu, V. Ursaki and I.M. Tiginyanu, Ultrashort electromagnetic nanodimensional modes spectrum in low frequency region from nanocylinder array system: as bio-application focusing membrane, *Proc. CAS Conf. 2012*, October 15-17, Sinaia, Romania, Vol.1, pp. 113-116, 2012.
10. L. Sirbu, V. Sergentu, R. Muller, *et. al.*, Ultrashort electromagnetic modes in the low frequency region in a nanocylinder array, in *Terahertz and Mid Infrared Radiation: Detection of Explosives and CBRN* (Using Terahertz), Springer, pp. 135-141, 2014.
11. J. D. Joannopoulos, R. D. Meade and J. N. Winn, Photonic Crystals: Modeling the Flow of Light, Princenton Univ. Press, Princenton, NJ, 1995.
12. V. Sergentu and M. Olednic, Tachyon-like spectrum of electromagnetic modes in the case of extremely low frequencies in a system of nanopores, *Proc. of the 2nd Int. Conf. on Nanotechnologies and Biomedical Engineering*, Chisinau, Republic of Moldova, April 18-20, pp. 363-365, 2013.
13. J. D. Jackson, Classical Electrodynamics, New York: Wiley, 1998.
14. R. Landauer, Electrical conductivity in inhomogeneous media. *AIP Conf. Proc.* **40**. American Institute of Physics. pp. 2–45. 1978.
15. L.A. Golovan, V.Yu. Timoshenko, P.K. Kashkarov, Optical properties of porous-system based nanocomposites, *Uspekhi Fizicheskikh Nauk*, **177**, 619 – 638, 2007.
16. X. Zhang, L.-M. Li, Z.-Q. Zhang, and C. T. Chan, Surface states in two-dimensionalmetallo-dielectric photonic crystal studied by multiple-scattering method, Phys. Rev. B 63, 125114, 2001.
17. A. I. Baz, Y. B. Zeldovich, and A. M. Perelomov, Scattering, Reactions and Decay in Nonrelativistic Quantum Mechanics, Israel Program for Scientific Translations, Jerusalem, 1969.
18. M. Dressel and G. Gruneru, Electrodynamics of Solids Optical Properties of Electrons in Matter, Cambridge: Cambridge University Press, 2002.

Author: Veaceslav Ursaki
Institute: Institute of Electronic Engineering and Nanotechnologies
Street: 3/3 Academiei str.
City: Chisinau
Country: Moldova
Email: ursaki@yahoo.com

Conditions for Plasma Obtaining in the Gaseous Media and Its Application in Nanotechnology

A. Hirbu

Alecu Russo Balti State University, Balti, Republic of Moldova

Abstract— **The conditions for electrical discharge plasma obtaining in the gaseous media, as well as its application in different fields of nanotechnologies are analyzed in this paper. Plasma obtaining under the investigated conditions avoids the use of auxiliary equipment for preventive ionization of the active media, which does not require synchronization between the base discharge and the moment when the ionization in the gap is maximal. As a result of the interaction of acquired plasma with different material surfaces the structures of nanometer order were obtained, and the plasma itself can be applied at the construction of quantum generators in non-aggressive gaseous media, as well as at the sterilization of objects in medicine.**

Keywords— **plasma, ionization, bio-medicine, nanostructures.**

I. INTRODUCTION

Plasma of the gaseous media presents a completely or partially ionized gas consisting of neutral atoms or molecules and charged particles (ions and electrons). A basic characteristic of the plasma is its quasi-neutrality, which means that the density of positively and negatively charged particles that form the plasma in a unit volume is approximately one and the same. The gas transforms in the plasma state when some of its components, atoms or molecules, in any case have lost or added one or several electrons, thus have transformed into positive ions or, in some cases, as a result of the electrons' "add" by neutral atoms, the negative ions can appear too. If there are no neutral particles and free electrons in the gas, then the plasma consists of positive ions and negative ions. If the neutral particles are not presented in the gas, then the plasma is called the completely ionized plasma. The free electrons and ions can condition the appearance of the electric current; therefore the plasma presents a conductor of electricity.

II. THEORETICAL PREMISES ON PLASMA OBTAINING IN THE GASEOUS MEDIA

One of the methods of plasma creating presents the substance heating. To receive the complete plasma ionization on the thermal way it is necessary the gaseous media to be heated up to tens or hundreds of thousands of degrees. In the gas at high temperatures, the number of particles which energy exceeds the ionization limit provides the possibility of the plasma creation in the partially ionized media. For this reason, to obtain the plasma in the laboratory conditions and in technique, the electrical discharges in the gas are used. In order to obtain the electric discharge in the gas it is necessary to apply a potential difference in the gap between two electrodes in which the gaseous medium subjected to investigation is included. The charged particles which move in the electric field are formed in this gap, and, as a result, the electric current appears. In order to sustain the current in the plasma it is necessary the negative electrode (the cathode) to emit electrons. The electron emission from the cathode may be carried out by various methods, such as: by heating the cathode up to the high temperature (thermo-emission) or by X-ray or γ radiation of the cathode, which is capable to tear out the electrons from the cathode surface. Such a type of discharge created by an external source is called the induced discharge [1, 2].

The induced discharges can be of sparks form, arc form and glow discharges. They differ fundamentally one from another by the way of formation (appearance) of electrons from the cathode or in the gap between electrodes.

The glow discharge is created in the tubes with a low pressure gaseous active medium at the ends of which the electrodes are placed to which a high voltage is applied. In this case the cathode emits the electrons (through the electronic auto-emission mechanism) that are torn out from the metal surface (cathode) under the action of the electric field. The electrical discharge plasma is comprised between the cathode and anode forming the dark zones (near the cathode and anode) and the glow zone called in physics the "positive" zone for which the electric field intensity is approximately constant value. The glow discharge in the molecular gas plasma is widely applied as the active medium for the construction of lasers at the oscillator-rotational transitions of the molecules. The ionization process in the electrical discharge plasma is accompanied by the current passing through the active gaseous medium and has the character of the avalanche with the ionizing effect. It follows that the appearance of free electrons in the gap is caused by the fact that at free pass of electrons accelerated by the electric field, they accumulate the sufficient energy to ionize neutral gas particles, which pro-

V. Sontea and I. Tiginyanu (eds.), *3rd International Conference on Nanotechnologies and Biomedical Engineering*,
IFMBE Proceedings 55,
DOI: 10.1007/978-981-287-736-9_59

duce the avalanche multiplication of electrons and ions able to maintain a steady current in the gap [1, 2].

Quasi-neutrality is the one of the most important features of plasma and consists in the fact that the negative charge of the electrons neutralizes the positive charge of the ions. In any type of action on the gaseous medium placed between the electrodes, the plasma tends to keep its quasi-neutrality. If there is a movement of a group of electrons in some place of medium between the electrodes (e.g. on account of the density fluctuation), where an excess of electrons (and in another place an electron deficit (by their concentration)) is attested, a strong electric field arises in plasma that annihilates the concentration of electric charge and immediately restores its quasi-neutrality.

Whether in a plasma layer of thickness Δx is created a volume of charge of the same density q, in this case, in accordance with the laws of electrostatics, an electric field $E=4\pi q\Delta x$ arises on this length. If Δn_e electrons are in a volume of 1 cm^3, more in number than those that neutralize the ions charge, then the volumetric density of the charge will be: $q=e\Delta n_e$, where $e=1.6\cdot10^{-19}$ C is the electron charge.

The electric field which arises as a result of the charge separation is equal to $E=1.8\cdot10^{-6}\Delta x$ V/cm. For air, $n=2.7\cdot10^{19}$ molecules/cm^3 are contained at the Earth's surface. If we suppose that, as a result of ionization, all the molecules become ionized, then the concentration of electrons in the plasma will be $n_e=2.7\cdot10^{19}$ electrons/cm^3. If we assume that the electron concentration has varied by 1% at the gap length of 1 cm, then the charge density will be $\Delta n_e=2.7\cdot10^{19}$ electrons/cm^3, and it will create an electric field of intensity of $E\approx10^{12}$ V/cm [1].

In order to create such a strong electric field it would require a very high energy, that's why we can affirm that the creation of high density plasma takes place at very short distances.

The space dimensions of the charge separation can be appreciated by a work that must be done for the charge separation at the distance d, that is performed by the forces occurring on the distance x out of the electric field intensity $E=4\pi n_e ex$. Taking into account the fact that $F=eE$, this work can be calculated by the relation [1]:

$$A=\int_0^d eEdx=2\pi n_e e^2 d^2.$$

This work cannot exceed the kinetic energy of the thermal movement of the particles in the plasma and for the one-dimensional case of the particles movement this will be equal to:

$$\overline{W_c}=\frac{1}{2}kT$$

where k is the Boltzmann constant, T is the absolute temperature, then $A\leq\frac{1}{2}kT$.

From this relationship the maximal dimension of the charge separation can be appreciated [1, 2]:

$$d=\left(\frac{\varepsilon_0 kT}{\mathrm{n_e e^2}}\right)^{1/2}$$

This measure is called the Debye length. For air plasma under ordinary condition $n_e=2.7\cdot10^{19}$ cm^{-3}, $T=273$ °K, $k=1.38\cdot10^{-23}$ J/K, $\varepsilon_0=8.85\cdot10^{-12}C^2/N\cdot m^2$:

$$d=2.2\cdot10^{-12}cm.$$

Length d (or Debye radius $r_0=\frac{d}{\sqrt{2}}$) represents a basic characteristic of the plasma; in particular, the electric field created by each of the charged particle from the plasma is screened by the field generated by the particle charge of the opposite sign. On the other hand, the measure d determines the depth of penetration of the external electric field in the plasma. Under ordinary conditions any volume of gas comprises a certain number of electrons and ions. On the average ~ 1000 of positive and negative ions contained in the air under ordinary condition that appears as a result of the interaction of the medium with ultraviolet and other types of radiation, including cosmic. Due to these conditions a certain number of electrons are torn from the electrode surface. The speed of the gas ionization and of the electrons emission from the electrode can be increased in the laboratory conditions by the additional application of radiation sources. When applying on the gap a potential difference, the current (which density j is proportional to applied electric field intensity E) will pass through the gaseous medium. For the values of the electric field within ~ 10÷20 V/cm the current intensity will constitute $j\approx10^{-9}$ A/cm^2 [3]. For these conditions the gaseous medium is in balance, therefore the recombination and appearance of the charged particles occur (electrons and ions mobility are almost equal). With the increase of electric field intensity E and of the current density j the balance is violated, which increases the recombination coefficient and decreases the number of charged particles, and, as a result, leads to the increase of the gap resistance and to the decrease of the current. In this case, the current density on the electrode surface will be [3, 4]:

$$j=S\cdot e\cdot\frac{dn}{dt},$$

where S is the gap size; e is the electron charge; $\frac{dn}{dt}$ is the overall speed of generation of charged particles in a unit of volume.

Therefore, the current density doesn't depend on the intensity of electric field, either on the mobility of the charged particles, and, in this case, the created situation leads to the phenomenon of current density saturation.

If the potential difference between the electrodes will continue to grow after the achievement of current density saturation, then it will be a moment when the current would increase again. The character of the current growth will depend on the gas pressure. With the increase of the electric field intensity E the electrons near the cathode are accelerated stronger and the energy that they acquire exceeds the ionization energy, which leads to the formation of a greater number of electrons and, respectively, the current will rise again. To explain the current increase Townsend introduced the measure α, called the first ionization coefficient, which determines the number of collisions with ionizing effect at one cm of the path in the direction of the electric field force lines.

The number of electrons that cross the surface, situated immediate near the cathode surface at a distance dx in the direction of the electric field force lines, will be [3, 4]:

$$dn = \alpha \cdot n \cdot dx.$$

By integration by x this relation, taking into account the fact that the distance between the cathode to the elementary surface which is situated at the distance dx, we obtain [3, 4]:

$$\int_{n_0}^{n} \frac{dn}{n} = \int_{0}^{x} \alpha \cdot dx.$$

Considering that α doesn't depend on x, in accordance with those described by the authors [3, 4], we obtain:

$$n = n_0 \exp(\alpha \cdot x),$$

where n_0 is the number of electrons that leave the cathode in a second.

Taking into account the fact that the distance between the electrodes (the gap size) is S, we will get for the current intensity the next relations [3, 4]:

$$i = n_0 \cdot e \cdot \exp(\alpha \cdot S) = i_0 \cdot \exp(\alpha \cdot S),$$

where $i_0 = n_0 \cdot e$ is the current of photoelectric nature near the cathode caused by the external radiation.

The difference i–i_0 presents the current caused by the positive ions in the immediate vicinity with the cathode surface.

The increase of dn, caused by the ionization of collision with electrons in the interval dx (n in a second), as well as those of photoionization along the same inteval dx, will be [3, 4]:

$$dn = \alpha \cdot n \cdot dx + n_i \cdot dx.$$

By the method of separation of variables from the above relation for the concentration of electrons we get:

$$\int_{0}^{n} \frac{dn}{\alpha \cdot n + n_i} = \int_{0}^{x} dx;$$

$$\alpha \cdot n + n_i = n_i \exp(\alpha \cdot x);$$

$$n = \frac{n_i}{\alpha}[\exp(\alpha \cdot x) - 1].$$

Or for the current [3, 4]:

$$i = \frac{n_i \cdot e}{\alpha}[\exp(\alpha \cdot S) - 1];$$

$$i_0 = \frac{n_i \cdot e}{\alpha};$$

$$i = i_0[\exp(\alpha \cdot S) - 1].$$

Out of the last relation it is observed that the total current in the gap depends on values of α and S. If the length of free path until the collision is λ, then the average energy that the electrons are accumulated between two consecutive collisions will be [3, 4]:

$$W = e \cdot E \cdot \lambda.$$

However, the length of free path is inversely proportional to the pressure of the active medium [3, 4]:

$$\lambda \sim \frac{1}{p},$$

The Townsend coefficient α is proportional to the electric field intensity E and the gas pressure p and can be expressed by the function:

$$\alpha = p \cdot f(e \cdot E \cdot \lambda),$$

or

$$\alpha = p \cdot F\left(\frac{e \cdot E}{p}\right),$$

where F is the function of the same type as f. Because e is a constant, then, according to the reports of the authors [3, 4], we get:

$$\frac{\alpha}{p} = \varphi\left(\frac{E}{p}\right).$$

III. Methodology of Experimental Investigations

The experimental installation which scheme is shown in Fig. 1 consists of the following elements [5, 6]: DC power supply operating at voltage U = 25 kV, discharger (Ecl.), condenser battery (C), ballast resistor R, special electrode 1 and the counter-electrode 2. The special electrode consisting of multiple wires with the clearly defined active resistance causes the division of the discharge current through all the formed channels. This facilitates the deployment of pulsed discharge in several channels simultaneously, leading to the formation of high current densities and thereby leading to the increase of the efficiency of the energy used for discharge and plasma homogenization in the gap.

The gap size is *S*=7 mm. The area of the cross-section of the special electrode is 17×7 mm, which were located a number of approx. 600 copper wires of 0.35 mm diameter isolated one from another. The measurement of current impulses (the duration, shape and amplitude value) [5, 6] and viewing them is performed through the oscilloscope with memory Osc (C8-13) shunted by the resistance R_s=0.003 Ω. Earthing of the scheme is done to protect the installation and the oscilloscope and provides the omitting of the influence of reactive elements on the discharge process.

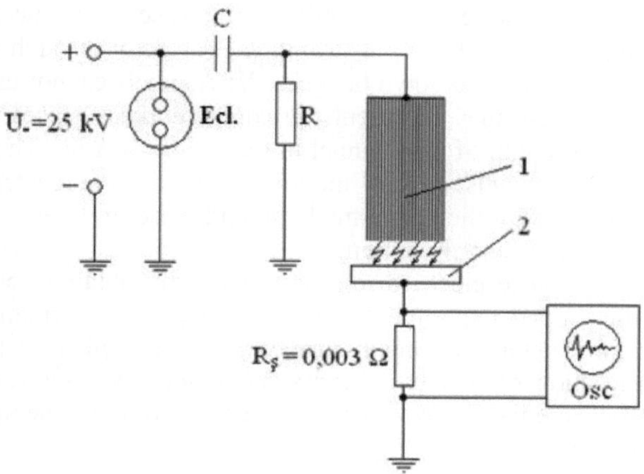

Fig. 1 Principal scheme of the experimental installation and measurement of the discharge impulse parameters [5, 6]: Ecl. – discharger; Osc - oscilloscope with memory, RS - shunt resistance; 1 - special electrode; 2 - counter-electrode

The shape and the duration of current impulses of the obtained discharges in the investigations are shown in Fig. 2. It is noted that the impulse shape directly depends on the value of the ballast resistance and capacity, which, with special electrode, form the generator of RLC type. In Fig. 2 the curve AB corresponds to auto-ionization (in this case special electrode functions as an electron gun), the curve BC corresponds to the base discharge (the plasma obtaining), the curve CD corresponds to the relaxation process (the gaseous media properties back to their original state), further, the process repeats from one discharge to another with the frequency required by the control electrical scheme.

The characteristic of electrical discharge impulses of the used experimental installation is shown in tab. 1. For the capacity of the condenser battery C=1/12 μF the current amplitude is 100 kA.

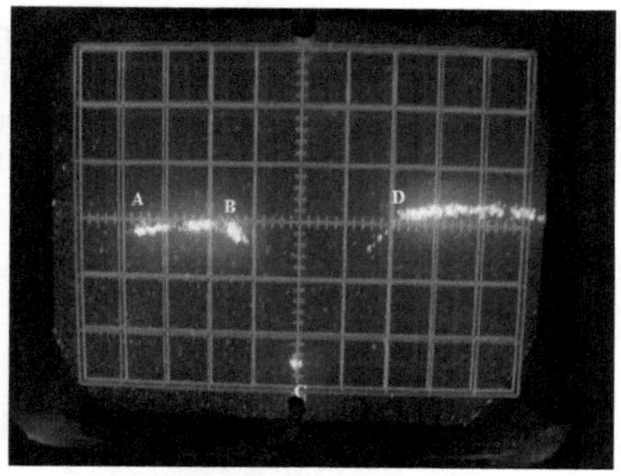

Fig. 2 General view of the current impulses for the value od the capacity C=1/12 μF, the ballast resistance R=8,2 MΩ [5, 6, 7]: the size of time divisions 0.1 μs/div, of voltage divisions 100 V/div

Table 1 The characteristic of electrical discharge impulses of the experimental installation [5, 6]

The capacity of the condenser battery C, μF	The energy accumulated on the condenser battery W_C, J	The impulse duration τ_{imp}, μs	The amplitude value of the impulse on the oscilloscope U_{osc}, V	The amplitude of the current impulse I_s, kA	The current intensity in the gap $\int_0^{\tau} i_s dt$, kA·s
1/12	26	0.15	300	100	50

The energy accumulated on the condenser battery (W_c= $\frac{CU^2}{2}$) is approximately equal with 26 J. Because the used scheme is unipolar and taking into account the losses that occur on the electrical circuit elements, about only 8 J transform in thermal energy.

IV. Results of Experimental Investigations

As a result of applying unipolar scheme of the experimental installation (Fig. 1) the stability of plasma formation was reached under laboratory conditions without active media preventive ionization. Referring to Fig. 2 the curve AB presents the intense ionization of the active media. Auto-ionization is due to the design of the special electrode 1 in Fig. 1 and to the increase of the electric field inhomogeneity in the space between the electrodes and consumes approx. 5-7% of total energy reserved in the condenser battery. The base discharge (the sector BC, fig. 2) occurs within 0.15 mS with the amplitude of the impulse discharge current I_s=100 kA. A part of the energy released in the gap transforms into the luminescent radiation of the intensity L= $10^8 \div 10^9$ W/cm^2. These types of discharges can be applied in the construction of quantum generators, which operate in non-aggressive active media. The general view of an impulse of electric discharge plasma obtained by the developed experimental installation is presented in Fig. 3.

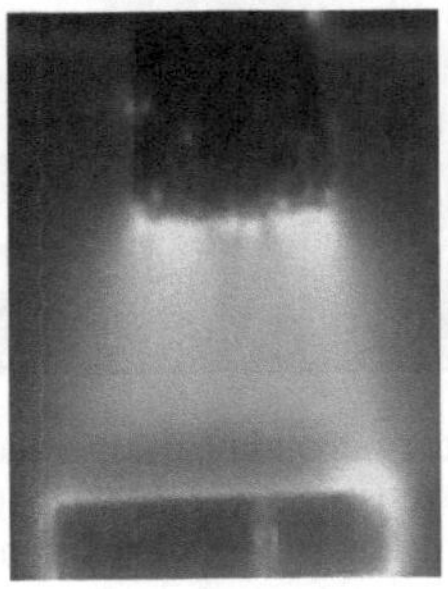

Fig. 3 The plasma of impulse electrical discharges obtained in the laboratory conditions

Out of this we can see that the amount of energy released on the electrode surfaces and in the gap is different.

The counter-electrode 2 from the scheme in Fig. 1 was made of different materials (steel 45, titanium alloy BT8, duralumin D16, bronze BrA5) of flat shape. As a result of the interaction of these electrodes with the obtained plasma, following the research of their surfaces with an electron microscope, it was detected the nanostructures formation on their surfaces which are distinguished by the shape and properties depending on the type of material of their execution. The morphology of the surface layer of samples processed by applying the obtained plasma using the experimental installation are presented in Fig. 4.

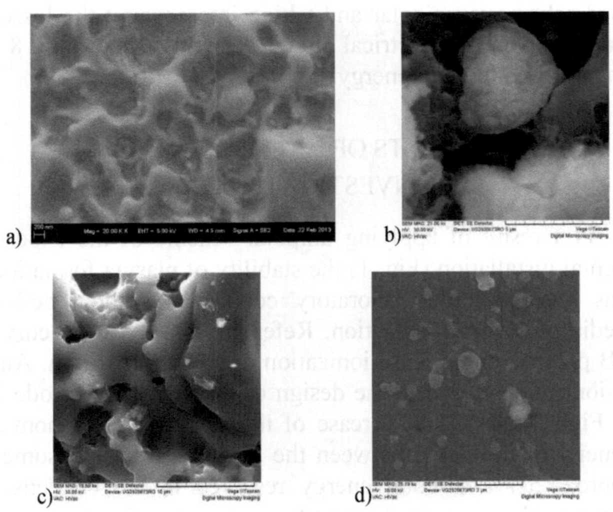

Fig. 4 The morphology of the surface layer of the counter-electrode 2 in Fig. 1, formed as a result of plasma interaction: a) steel 45; b) titanium alloy BT8; c) aluminum alloy D16; d) bronze BrA5

Due to the fact that the temperature of the plasma obtained in the developed experimental installation reaches values higher than 10^4 K with its duration of existence of hundreds of nanoseconds, it can be applied in technique at the formation of nanostructures on the metal surfaces, as well as at the destruction of micro-organisms in biomedicine.

V. CONCLUSIONS

Based on theoretical and experimental analysis on the plasma formation by the electrical discharges under laboratory conditions we can make the following conclusions:

The electrical discharges plasma in laboratory conditions was acquired without the use of special devices for the preventive ionization of the active media and does not require synchronization between the connecting these devices and base discharge.

The construction of the special electrode with separated from the discharge channel limits the increase of the current through a single channel distributing it uniform through all channels. The breakdown between the channels cannot take place because they represent the equipotential surfaces on the entire length of the channel in their cross section. These separate channels creates inhomogeneity of the currents passing through the active medium, and these, in their turn, lead to the plasma formation.

The impulse electrical discharges plasma obtained with the developed experimental installation permits formation of nanostructures on metal surfaces, can be applied at the construction of quantum generators in the non-aggressive gaseous media, as well as in order to sterilize medical instruments.

CONFLICT OF INTEREST

The authors declare that they have no conflict of interest.

REFERENCES

1. Жданов С.К. и др. Основы физических процессов в плазме и плазменных установках. М.: МИФИ, 2000, 184 с.
2. Чен Ф. Введение в физику плазмы. М.: Мир, 1987, 398 с.
3. Ховатсон А.М. Введение в теорию газового разряда: пер. с англ. – М.: Атомиздат, 1980, 182 с.
4. Велихов Е.П., Ковалев А.С., Рахимов А.Т. Физические явления в газоразрядной плазме. М.: Наука, 1987, 160 с.
5. Pavel Topala, Arefa Hirbu, Alexandr Ojegov. New directions in the practical application of electro erosion, Nonconventional Technologies Review – no. 1/2011, pg. 49-56.
6. Pavel Topala, Arefa Hirbu, Alexandr Ojegov. Acțiunea plasmei descărcărilor electrice asupra suprafețelor metalice, Fizica și Tehnologiile Moderne, vol. 10, nr. 1-2, 2012, pg. 13-21.

Author: Arefa Hirbu
Institute: Alecu Russo Balti State University
Street: Pushkin 38
City: Balti
Country: Moldova
Email: arefahirbu@yahoo.com

Part II
Bio-Nanotechnologies and Biomaterials

Quantum Information Processes in Protein Microtubules of Brain Neurons

N.A. Enaki[1], V.I. Koroli[1], S. Bazgan[1], A. Nistreanu[1], S. Palistrant[1], D. Bogoev[1], M. Turcan[1], T. Pislari[1], Y. Boshneaga[1], N. Lambropoulos[2], S. Patel[2], A. Khrennikov[3], M. Marinucci[4], S.C. Kwok[5], L. Pannese[6], M. Arniani[7], R. Torrenti[7], S. Maslobrod[8], V. Scherbakov[9], E. Kuznetsov[9], I. Moldovanu[10], O. Misic[10], S. Odobescu[10], A. Lupusor[10], A. Cernei[10], V. Vovc[11], O. Arnaut[11], N. Ciobanu[11], P. Tuzlucov[12], S. Kernbach[13], A. Sorli[14], and V. Anisimov[15]

[1] Quantum Optics and Physical Kinetics Laboratory, Institute of Applied Physics, Academy of Sciences of Moldova, Kishinev, Moldova
[2] Department of Informatics, London South Bank University, London, UK
[3] Department of Mathematics, International Center for Mathematical Modelling in Physics and Cognitive Sciences, Linnaeus University, Vaxjo, Sweden
[4] Forschungszentrum Informatik, FZI, Karlsruhe, Germany
[5] DPhil, Neuroimaging Laboratory, Santa Lucia Foundation, Rome, Italy
[6] Imaginary SRL, Milano, Italy
[7] Sigma Orionis, Buropolis 1, Valbonne Sophia Antipolis, France
[8] Laboratory of Applied Genetics, Institute of Genetics and Plant Physiology, Academy of Sciences of Moldova, Kishinev, Moldova
[9] Laboratory of Algebra and Topology, Institute of Mathematics and Computer Science, Academy of Sciences of Moldova, Kishinev, Moldova
[10] Laboratory of Functional Neorology, Institute of Neurology and Neurosurgery, Kishinev, Moldova
[11] Department of Biophysics, University of Medicine and Pharmacy "Nicolae Testemitanu", Kishinev, Moldova
[12] Department of Radiotherapy, Institute of Oncology, Kishinev, Moldova
[13] Research Center of Advanced Robotics and Environmental Science, Stuttgart, Germany
[14] Foundations of Physics Institute, Association for development of Consciousness and Health, Idrija, Slovenia
[15] Institute of Power Engineering, Academy of Sciences of Moldova, Kishinev, Moldova

Abstract— **We study biologically 'orchestrated' coherent quantum processes in collections of protein microtubules of brain neurons, which correlate with, and regulate, neuronal synaptic and membrane activity. In this situation the continuous Schrodinger evolution of each such process terminates in accordance with the specific Diosi–Penrose (DP) scheme of 'objective reduction' ('OR') of the quantum state. This orchestrated OR activity ('Orch OR') is taken to result in moments of conscious awareness and/or choice. We analyze Orch OR in light of advances and developments in quantum physics, computational neuroscience and quantum biology. Much attention is also devoted to the 'beat frequencies' of faster microtubule vibrations as a possible source of the observed electroencephalographic ('EEG') correlates of consciousness.**

Keywords— **quantum information processes, quantum consciousness, protein microtubules of brain neurons.**

I. INTRODUCTION

All living cells of plants, animals and humans continuously emit ultraweak biophotons (ultraweak electromagnetic waves) in the optical range of the spectrum, which is associated with their physiological states and can be measured using special equipment. Neural cells also continuously emit biophotons. The intensity of biophotons is in direct correlation with neural activity, cerebral energy metabolism, EEG activity, cerebral blood flow and oxidative processes [1, 2]. According to Van Wijk et al [3], there are significant correlations between the fluctuations in biophoton emission and fluctuations in the strength of electrical alpha wave production in the brain. Some unpublished observations suggest that the state of the biophoton field of a person may be connected to the state of the brain as measured by the EEG (e.g., degree of synchronization and coherence) [4]. Certain meditative states characterized by a high degree of coherence in the EEG may well be accompanied by a high degree of coherence in the biophoton field [4], although measurements correlating the coherence of the biophoton field and the EEG readings have not been made yet. In this situation it is very important to study the interaction of biomolecules with biophotons of brain neurons and to quantitatively investigate possible connections between the EEG and the biophoton production.

There is no doubt that EEG waves are deeply involved with the basic functioning of the brain. The EEG waves associated with two distant neurons are strongly correlated and this supports the view that EEG waves are related to the properties of the brain as a coherent quantum system.

In modern optogenetics much attention is devoted to the possibility to manipulate neurons with light [5-7]. In this situation spatially-precise neuronal control and monitor the activities of individual brain neurons are realized with the help of optogenetic actuators (channelrhodopsin, halorhodopsin, archaerhodopsin) and temporally-precise recordings can be made with the help of optogenetic sensors for calcium (Aequorin, Cameleon, GCaMP), chloride (Clomeleon) or membrane voltage (Mermaid).

V. Sontea and I. Tiginyanu (eds.), *3rd International Conference on Nanotechnologies and Biomedical Engineering*,
IFMBE Proceedings 55,
DOI: 10.1007/978-981-287-736-9_60

At present much attention is devoted to the nature of consciousness and mechanism by which it occurs in the brain [8]. In this situation consciousness depends on biologically 'orchestrated' coherent quantum processes in collections of microtubules within brain neurons [9]. It is very important to unify micro and macro levels: quantum-statistical properties of quantized fields insight MTs and EEG signals [10], because superradiant optical computing in networks of microtubules may provide a basis for biomolecular cognition and a substrate for consciousness [11], fMRI and SPECT (single-photon emission computed tomography) are also very important tools for investigation of cooperative processes in protein microtubules of brain neurons. There has been a strong recent interest in applying quantum theory in cognitive science [12]. It should be noted that special evaluation of altered states of consciousness (ASC), induction of ASC by different physical methods: electrical stimulation (direct and alternating current), magnetic stimulation, sound (binaural) stimulation compared with placebo in healthy volunteers will give more insight about nature of consciousness from quantum biophysical point of view [13-15].

II. Fluctuation Function of Protein Microtubules

At present much attention is paid to the coupling between fluctuation function $S_z(t)$ of protein MTs and EEG signal, influence of coherence in biophoton field (Fig. 1) on coherence in neuronal system and correlations between the fluctuations in biophoton emission and fluctuations in the strength of electrical alpha wave production in the brain [10]

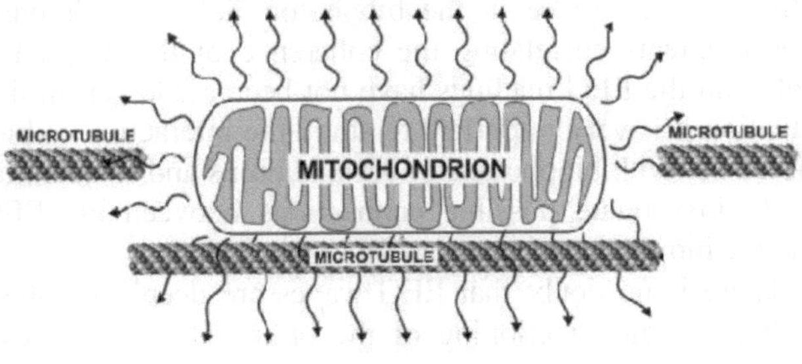

Fig. 1 Representation of biophotons produced by mitochondria and the interaction of biophotons with microtubules

Fig. 2 MTs are hollow cylinders composed of protein units called tubulin. The inner diameter of an MT is 17 nm and the outer diameter is 25 nm. The lengths of MTs vary widely from nanometers to micrometers. MTs have been considered to act as QED-cavities [16, 17]

It should be emphasized that MTs have been considered as optical cavities (Fig. 2) [16] with quantum properties [17], capable of supporting only a single mode [18] or perhaps a few widely spaced (in the frequency domain) modes. Rahnama et al [10] is successfully applied a fully quantum mechanical formalism of the Jaynes-Cummings model (JCM) [19-21]. It is noticed that MTs are biological hollow cylinders with a 17 nm inner diameter and a 25 nm outer diameter (Fig. 2) [22], composed of units called tubulin dimers, each of which has the dimensions $4\text{nm}\times 8\text{nm}\times 6\text{nm}$ [16]. Tubulin can be viewed as a typical two-state quantum mechanical system, where the dimers couple to conformational changes with 10^{-9} - 10^{-11} sec transitions due to electron transitions in hydrophobic pockets, corresponding to an angular frequency in the range $\omega_0 \sim O(10^{10}) - O(10^{12})Hz$ [16]. Using a first-order-approximation estimate of the quality factor for the MT cavities (i.e. Q_{MT}), it has been found that $Q_{MT} \sim O(10^8)$ [16]. High-quality cavities encountered in Rydberg atom experiments dissipate energy on time scales of $O(10^{-3}) - O(10^{-4})$ sec and have Q's which are comparable to Q_{MT} [16]. Frequencies of visible light are on the order of THz and Wang et al [23] have detected visible light in the brain as biophotons. Also, transition frequencies in tubulins are on the order of THz [16]. In order to get more physical insight about Quantum Brain Dynamics it is necessary to

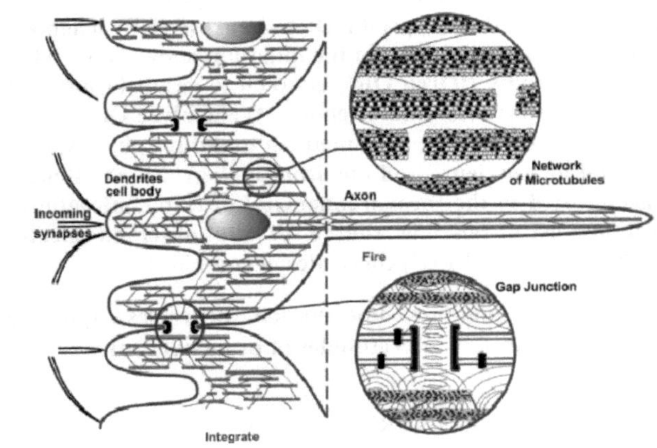

Fig. 3 An 'integrate-and-fire' brain neuron, and portions of other such neurons are shown schematically with internal microtubules. In dendrites and cell body/soma (left) involved in integration, microtubules are interrupted and of mixed polarity, interconnected by microtubule-associated proteins (MAPs) in recursive networks (upper circle, right). Dendritic–somatic integration (with contribution from microtubule processes) can trigger axonal firings to the next synapse. Microtubules in axons are unipolar and continuous. Gap junctions synchronize dendritic membranes, and may enable entanglement and collective integration among microtubules in adjacent neurons (lower circle right). In Orch OR, microtubule quantum computations occur during dendritic/somatic integration, and the selected results regulate axonal firings which control behavior.

carefully examine properties of a single protein microtubule [24] from quantum-mechanical point of view. This will give new possibilities for realization and manipulation of quantum information processes with the help of entanglement and non-local canals in the ensemble of microtubules of brain neurons (Fig. 3).

Let us investigate quantum brain dynamics and analyze fluctuation function of MTs in the case, in which MT has three energy states: ground state $|g\rangle$, first excited $|e_1\rangle$ and second excited $|e_2\rangle$ biomolecular states. In this situation it is assumed that at the initial moment t = 0 MT is situated in the first excited state $|e_1\rangle$ and quantized electromagnetic field in Holstein–Primakoff SU(1,1) coherent state [19, 20]

$$|\psi(0)\rangle_{biophotons} = (1-|\xi|^2)^k \sum_{m=0}^{\infty} \left\{\frac{\Gamma(m+2k)}{m!\Gamma(2k)}\right\}^{1/2} \xi^m |m,k\rangle \tag{1}$$

where $|m,k\rangle$ are the eigenvectors of the Casimir operator

$$C = K_0^2 - \frac{K_+K_- + K_-K_+}{2} = k(k-1)I \tag{2}$$

k is the so-called Bargmann index and $\xi = |\xi|\exp(i\varphi)$ $(0 \le |\xi| \le 1)$. In the present model k=½ and SU(1,1) CS can be written through oscillator number states

$$|\psi(0)\rangle_{biophotons} = (1-|\xi|^2)^{1/2} \sum_{n=0}^{\infty} \xi^n |n\rangle = \sum_{n=0}^{\infty} Q_n |n\rangle \tag{3}$$

In this case the initial state – vector of the coupled MT - field system takes the following form

$$|\psi(t=0)\rangle = |\xi\rangle \otimes |e_1\rangle = \sum_{n=0}^{\infty} Q_n |e_1, n\rangle \tag{4}$$

Exact analytical solution for state-vector of the MT - field system can be obtained by using Schrodinger equation:

$$|\psi(t)\rangle = \sum_{n=0}^{\infty} e^{-i\omega_0\{S_z+n\}t} Q_n \{\cos(\sqrt{n^2+\alpha^2(n+1)^2}\tau)|e_1,n\rangle - \frac{i\alpha(n+1)}{\sqrt{n^2+\alpha^2(n+1)^2}} \sin(\sqrt{n^2+\alpha^2(n+1)^2}\tau)|g,n+1\rangle - \frac{in}{\sqrt{n^2+\alpha^2(n+1)^2}} \sin(\sqrt{n^2+\alpha^2(n+1)^2}\tau)|e_2,n-1\rangle\} \tag{5}$$

where dimensionless time is $\tau = \lambda_1 t$, $\alpha = \lambda_2/\lambda_1$, λ_1 and λ_2 are MT - field coupling constants [10].

By using solution (5) one can obtain exact analytical solution for fluctuation function $S_z(t)$ in the following manner

$$S_z(t) = -(1-|\xi|^2) \sum_{n=0}^{\infty} (|\xi|^2)^n \times \sin^2(\sqrt{n^2+\alpha^2(n+1)^2}\tau) \frac{n^2-\alpha^2(n+1)^2}{n^2+\alpha^2(n+1)^2} \tag{6}$$

In figure 4 we show $S_z(t)$ as a function of time for cases $|\xi|$=0.6 and 0.9, α = 0.1 and 0.01. In this situation it is observed that the higher the initial mean biophoton number $n(t = 0) = |\xi|^2/(1-|\xi|^2)$, the higher $S_z(t)$. There are also some jumps in the fluctuation function, in which a strong energy exchange between the MTs and field subsystems takes place.

It should be noted that present JCM with intensity-dependent coupling [19, 20] in the best way describes quantum vibrations of MTs of brain neurons and gives more insight about faster microtubule vibrations as a possible source of the observed EEG correlates of consciousness from quantum biophysical point of view.

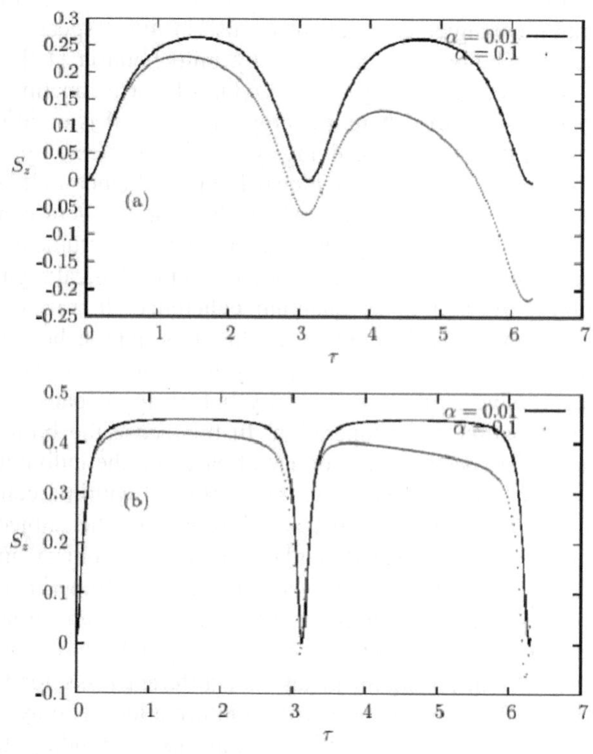

Fig. 4 Sz(t) as a function of the dimensionless time τ for cases α = 0.1 and 0.01: (a) |ξ|=0.6, (b) |ξ|=0.9.

III. Quantum Entanglement and Non-local Models of Quantum Information Processes in Microtubules of Brain Neurons

Quantum entanglement is, according to Erwin Schrodinger in 1935, the essence of quantum physics and inspires fundamental questions about the principles of nature. By testing the entanglement of particles we are able to ask fundamental questions about realism and locality in nature. Local realism imposes certain constraints in statistical correlations of measurements on multi-particle systems. Quantum mechanics, however, predicts that entangled systems have much stronger than classical correlations that are independent of the distance between the particles and are not explicable with classical physics.

In non-local models of quantum mechanics quantum system behaves as a whole no matter how far separated its components. In other words, correlations between electrons or photons, composing the whole quantum system exist at distances much larger than the dimensions of composed parts. It should be noted that experimental demonstration of quantum correlations over more than 10 km was realized by Tittel et al [25]. Moreover, quantum entanglement distribution over 100-kilometre free-space channels is successfully demonstrated by Yin et al [26]. It should be emphasized that quantum entanglement takes place also in living species, in which biophotons are generated by mitochondria [10]. In this situation each living cell is giving off, or resonating, a biophoton field of coherent energy. If each cell is emitting this field, then the whole living system is, in effect, a resonating field-a ubiquitous nonlocal field. And since biophotons are the entities through which the living system communicates, there is near-instantaneous intercommunication throughout, which is the basis for coherent biological organization - referred to as quantum coherence. In this case much attention is devoted to the timeless approaches and quantum vacuum [27, 28]. Capacity for evolution rests not on aggressive struggle and rivalry but on the capacity for communication and cooperation. In this sense the built-in capacity for species evolution is not based on the individual but rather living systems that are interlinked within a coherent whole: Living systems are thus neither the subjects alone, nor objects isolated, but both subjects and objects in a mutually communicating universe of meaning. Experiments with non-local signal transfer between seeds, plant organisms and bio-hybrid systems confirm this point [29-32]. Just as the cells in an organism take on different tasks for the whole, different populations enfold information not only for themselves, but for all other organisms, expanding the consciousness of the whole, while at the same time becoming more and more aware of this collective consciousness.

Because protein microtubules of brain neurons are quantum optical devices [9-11], [16, 17] it is instructive to study the generation of squeezing and entanglement in microtubules to obtain more insight about nature of consciousness from quantum biophysical point of view. In this situation non-local correlations takes place between different microtubules of brain neurons and therefore quantum information in the brain may be transmitted through non-local canals (https://www.youtube.com/watch?v=0fu_k8_tae0).

IV. Conclusions

In the present paper the possibility of quantum information processes in MTs of brain neurons is investigated. Much attention is devoted to the quantum-statistical properties of protein microtubules. It is shown that fluctuation function of MTs $S_z(t)$ has the tendency towards oscillations but exact periodicity of $S_z(t)$ oscillations is violated. It should be noted that exact periodicity of $S_z(t)$ oscillations takes place when one of the dipole moment matrix transition elements between neighbouring biomolecular levels tends to zero. In this situation fluctuation function determines number of generated biophotons in protein microtubules and is connected with oscillations of EEG signal. Moreover, entanglement and non-local correlations exist between biophotons and tubulin structures of different MTs of brain neurons, situated at distances much larger than their own dimensions.

Conflict of Interest

The authors declare that they have no conflict of interest.

References

1. Isojima Y, Isoshima T, Nagai K, Kikuchi K, Nakagawa H (1995) Ultraweak biochemiluminescence detected from rat hippocampal slices. NeuroReport 6: 658–660
2. Kobayashi M, Takeda M, Ito K, Kato H, Inaba H (1999) Two-dimensional photon counting imaging and spatiotemporal characterization of ultraweak photon emission from a rat's brain in vivo. J Neurosci Methods 93: 163–168
3. Van Wijk R, Bosman S, Ackerman J, Van Wijk E (2008) Correlation between fluctuations in human ultra-weak photon emission and EEG alpha rhythm. NeuroQuantology l6: 452–463
4. Bischof M (2005) Biophotons - The light in our cells. J. Optom. Photother. March: 1–5
5. Richner T et al (2014) Optogenetic micro-electrocorticography for modulating and localizing cerebral cortex activity. J. Neural Eng. 11: 016010–016021
6. Aravanis A M et al. (2007) An optical neural interface: in vivo control of rodent motor cortex with integrated fiberoptic and optogenetic technology. J. Neural Eng. 4: S143-S156

7. Wu F et al. (2013) An implantable neural probe with monolithically integrated dielectric waveguide and recording electrodes for optogenetics applications. J. Neural Eng. 10: 056012–056020
8. Manzelli P, Lambropoulos N, Kwok S C and Koroli V (2014) Quantum Creativity: the Possibility for Human Super Cognition. In the *Caos Management Journal*, Special Edition 92/2014-November, article #186. http://www.caosmanagement.it/186-quantum-creativity-the-possibility-for-human-super-cognition
9. Hameroff S and Penrose R (2014) Consciousness in the universe, a review of the 'Orch OR' theory. Physics of Life Reviews 11: 39-78
10. Rahnama M, Tuszynski J A, Bykkon I, Cifra M, Sardar P and Salari V (2011) Emission of Mitochondrial Biophotons and their Effect on Electrical Activity of Membrane via Microtubules. J. Integrative Neuroscience 10: 65-86
11. Jibu M, Hagan S, Hameroff S R, Pribram K H and Yasue K (1994) Quantum optical coherence in cytoskeletal microtubules: implications for brain function. Biosystems 32(3): 195-209
12. Khrennikov A, Basieva I, Dzhafarov E N and Busemeyer J R (2014) Quantum models for psychological measurements: an unsolved problem. PLOS ONE 9: e110909 - e110916
13. Moldovanu I and Vovc V (2013) A new conceptual dimension of functional neurology: altered states of consciousness. Bul. Acad. Sci. of Moldova. Medical Sciences 2(38): 8–15
14. Cojocaru N and Moldovanu I (2013) Altered states of consciousness induced by voluntary hyperventilation, Pilot study. Bul. Acad. Sci. of Moldova. Medical Sciences 2(38): 109-111
15. Moldovanu I, Vovc V and Cernei A (2015) Induction and phenomenological structure evaluation of altered states of consciousness through the use of various physical factors (transcranial electrical stimulation, binaural sound stimulation, their combination) compared with placebo in healthy volunteers (to be published)
16. Mavromatos N E, Mershin A and Nanopoulos D V (2002) QED-cavity model of microtubules implies dissipationless energy transfer and biological quantum teleportation. Inter. J. Mod. Physics B 16: 3623-3642
17. Mavromatos N E and Nanopoulos D V (1998) On Quantum Mechanical Aspects of Microtubules. Inter. J. Mod. Physics B 2: 517-542
18. Jibu M, Brinbram K H and Yasue K (1996) From conscious Experience to Memory Storage and Retrieval: The role of quantum brain dynamics and boson condensation of evanescent photons. Inter J Mod Physics B 10: 1735–1754
19. Enaki N A, Koroli V I and Bazgan S (2014) Nonclassical light in two-photon Jaynes - Cummings model of a three-level atom with different dipole transitions. Indian Journal of Physics 88: 25-29
20. Koroli V I (2013) Cooperative interaction of photons and phonons with radiation centers, Quantum Optics. LAP, LAMBERT Academic Publishing, Saarbrücken, Germany 128 pages
21. Jaynes E T and Cummings F W (1963) Comparison of quantum and semiclassical radiation theories with application to the beam maser. Proc IEEE 51: 89–109
22. Cifra M, Pokornэ J, Havelka D and Kučera O (2010) Electric field generated by axial longitudinal vibration modes of microtubule. BioSystems 100: 122 – 131
23. Wang Ch, Bykkon I, Dai J and Antal I (2011) Spontaneous and visible light-induced ultraweak photon emission from rat eyes. Brain Res 1369: 1–9
24. Sahu S, Ghosh S, Ghosh B, Aswani K, Hirata K, Fujita D and Bandyopadhyay A (2013) Atomic water channel controlling remarkable properties of a single brain microtubule: Correlating single protein to its supramolecular assembly. Biosensors and Bioelectronics 47: 141–148
25. Tittel W et al (1998) Experimental demonstration of quantum correlations over more than 10 km. Phys. Rev. A 57: 3229-3232
26. Yin J et al (2012) Quantum teleportation and entanglement distribution over 100-kilometre free-space channels. Nature 488 (7410): 185-193
27. Fiscaletti D and Sorli A (2014) Perspectives of the numerical order of material changes in timeless approaches in physics. Found. Phys. 45(2): 105-133
28 Fiscaletti D and Sorli A (2015) Space-time curvature of general relativity and energy density of a three-dimensional quantum vacuum. Annales UMCS, Physica 69(1): 53-78
29. Maslobrod S, Ganea A, Corlateanu L (2004) "Memory" of the System of Two Swelling Seeds of Maize. Maize Genetic Cooperation Newsletter. 78: 11–12
30. Маслоброд С Н (2012) Эффект дальней связи между прорастающими семенами, возникающий при их контакте в период набухания. Электронная обработка материалов 48(6): 99–113
31. Kernbach S (2015) Supernatural. Scientific approach (in Russian). Algorithm, Moscow
32. Kernbach S (ed). (2013) Handbook of collective robotics. Pan Stanford Publishing

Author:	Vlad Koroli
Institute:	Institute of Applied Physics, Academy of Sciences of Moldova
Street:	Academiei str. 5
City:	Kishinev MD 2028
Country:	Moldova
Email:	vl.koroli@gmail.com

Functional Ecofriendly Coatings for Marine Applications

I. Hölken, M. Hoppe, R. Adelung, and M. Baum

Functional Nanomaterials, Institute for Materials Science, Faculty of Engineering, University of Kiel, Kiel, Germany

Abstract— **The worldwide increasing number of off-shore wind turbines brings along the need for innovative polymer coatings which on the one hand can withstand the extreme forces acting on the rotor blades and on the other hand can provide adequate antifouling properties to counter biocorrosion of the basement. The development of environmentally friendly and commercially realizable coating systems is most important since the prohibition of tributyltin (TBT)-based antifouling coatings in 2008. In this study an alternative polymer/particle composite coating based on polythiurethan (PTU) with tetrapodal shaped ZnO (t-ZnO) as additive is investigated with respect to chemical, mechanical and antifouling properties. Overall, four different composites with filler amounts of 0 wt% t-ZnO, 1 wt% t-ZnO, 5 wt% t-ZnO, 10 wt% t-ZnO were tested. Preliminary antifouling experiments were carried out under artificial habitat conditions at an aquarium, imitating Pacific Sea water. For mechanical and chemical characterization tensile tests, adhesion tests and contact angle measurements were performed. An optimum of mechanical and antifouling properties were found for a filler amount of 5 wt% t-ZnO in PTU.**

Keywords— **polymer composite, polythiourethan, tetrapodal zink oxide, antifouling.**

I. INTRODUCTION

The increasing number of offshore wind parks driven by the demand for renewable energy sources brings along the necessity to adapt materials towards harsh and very unique environmental conditions. Starting with the rotor blades, where wear and resulting maintenance costs need to be reduced, down to the base, where biofouling (Figure 1) and biocorrosion are the major challenge, therefore efficient, long-lasting solutions are needed. But not only stationary maritime structures are affected. Also in shipping, where weight gain and drag induced by the accumulation of sessile organisms causes increased fuel consumption, an ecofriendly solution needs to be found. The presented solvent- and biocide-free polythiourethane-variant (PTU) combined with flexible ceramic particles mate high mechanical robustness with antifouling features.

II. MATERIALS AND METHODS

The polymer matrix was based on a two-component polythiourethane (PTU) thermoset (Fluid- & Prozesstechnik, Waltershausen, Germany) with the potential for spraying application. The polyaddition reaction of the components 1,6-Diisocyanatohexan (Iso) and Pentaerythritoltetrakis-(3 mercaptopropionat) (Tol) are shown in figure 2. The prepolymerisation is initiated via the four SH-groups of Tol and the reactive terminal Iso groups (NCO) directly after mixing and ends up in the formation of the characteristic urethane-linkage [1]. Due to their high reactivity and low price aromatic diisocyanates are typically used [2]. This kind of polymer is well suited as a coating with high impact resistance, non-yellowing features and UV stability used under natural conditions [1]. Therefore PTU is most suitable for highly mechanically stressed coatings.

Fig. 1 Biofouling on a floating device at Gammel Ålbo, Denmark.

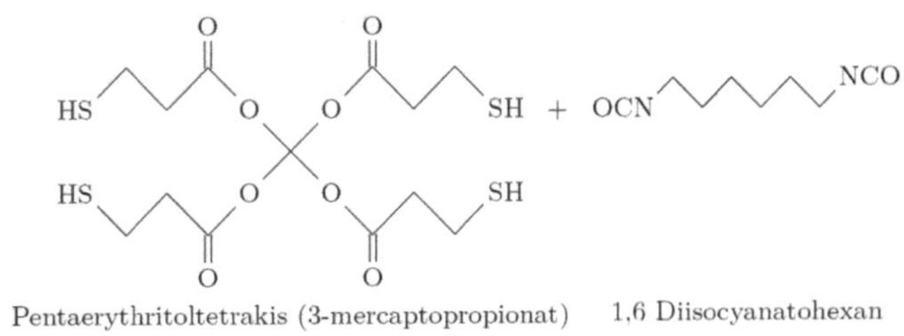

Fig. 2 Illustrated reaction of the single components 1,6-Diisocyanatohexan (Iso) and Pentaerythritoltetrakis- (3 mercaptopropionat) (Tol) synthesizing PTU.

As flexible ceramic particles tetrapodal shaped zinc oxide (t-ZnO) produced at the University of Kiel was used (Figure 3). Zinc oxide is a II-V semiconducting ceramic material with a direct bandgap at 3.37 eV [3]. The shape of

V. Sontea and I. Tiginyanu (eds.), *3rd International Conference on Nanotechnologies and Biomedical Engineering*,
IFMBE Proceedings 55,
DOI: 10.1007/978-981-287-736-9_61

ZnO can vary strongly depending on the manufacturing process [4,5]. After Moezzi et al. [6] manyfold shapes of zinc oxide particles like spheres, spring-like or needles are possible. The used tetrapodal ZnO showed a mean arm length of 20 µm (Figure 3) and is produced via the recently developed flame-transport synthesis [7].

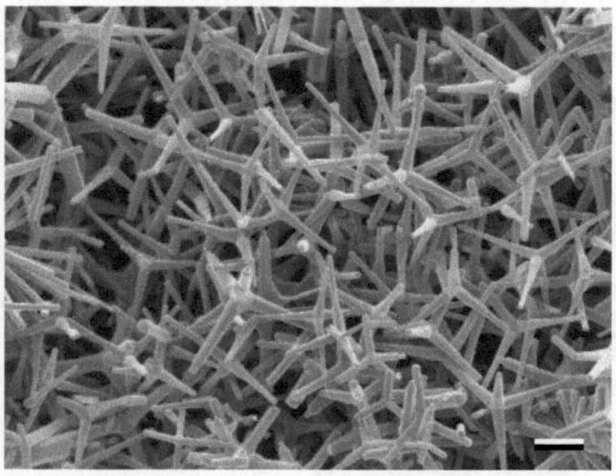

Fig. 3 Scanning electron microscopy (SEM)-micrographs of the inorganic tetrapodal-shaped ZnO filler particles, synthesized at the University of Kiel, Germany. Scale bar: 20 µm.

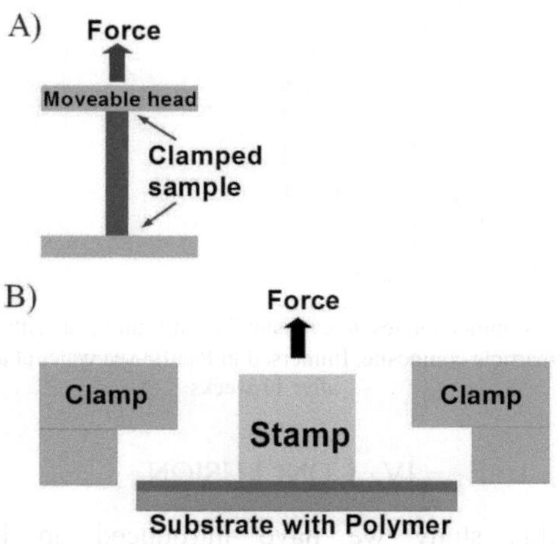

Fig. 4 Mechanical characterization of polymer/composites, by a) tensile test for the evaluation of fracture toughness and eleongation at fracture and b) Pull-off test for the evaluation of adherenace to the substrate.

The polymer/particle-composite was produced with varying filler amounts of t-ZnO (0 wt%, 1 wt%, 5 wt%, 10 wt%). The filler particles were mixed into the Tol-component. Afterwards the appropriate amount of the Iso-component was added and homogenized.

For the performance of the tensile tests a Zwick 1445 universal testing machine (Zwick GmbH & Co. KG, Ulm, Germany) equipped with a 10 kN load cell and an applied continuous strain rate of 1mm/min was used (Figure 4A). tThe investigated samples were dog-bone-shaped with a diameter of 1 mm, width of 5 mm and measuring length of 20 mm.

For the adherence test of the polymer/particle composite to AlMg3 substrates, a Zwick 1445 universal testing machine (Zwick GmbH & Co. KG, Ulm, Germany) was used. To evaluate the adhesion force, AlMg3-cylinders (diameter: 20 mm, length: 50 mm) were glued to AlMg3 plates (size: 80 mm by 80 mm by 3 mm) (Figure 4B) by the polymer/particle-composite with varying filler amounts. A perforce of 30 N was applied, the velocity was set to 1mm/min.

The contact angle measurements of water (density = 1.000 kg/m^3) were estimated aided by a high-speed optical contact angle measuring device OCAH 100 (DataPhysics Instruments GmbH). The measurements were performed by the free sessile drop method (drop volume: 2 µl).

The evaluation of the antifouling properties of the polymer/particle composite was done by immersion tests of three specimens with varying amounts of t-ZnO (0 wt%, 1 wt% and 5 wt%) into the Pacific-sea-tank of the Aquarium at Geomar in Kiel for 20 weeks. The degree of biofouling was recorded by photographs weekly.

III. RESULTS

Tensile tests of polymer/particle composite with varying filler amounts and two different particle shapes, spherical and tetrapodal, were performed (Figure 5). PTU without filling particles showed a maximum tensile strength of 57 MPa and a mean elongation at fracture of 15,5 %. The polymer/particle composite showed for filler amount of 1 wt% for tetrapodal shaped particles no effect on the maximum tensile strength, but for spherical particles a decrease to 55 MPa. For the elongation at fracture no influence of t-ZnO was found. The addition of s-ZnO lead to an increase to 23 %. The filler amount of 5 wt% influenced the average ultimate tensile strength and the elongation at fracture (t-ZnO: 59 MPa, respectively 12 %; s-ZnO: 54 MPa, respectively 18 %). No difference was shown for 10 wt% particle amount on the mean ultimate tensile strength but a difference regarding the elongation at fracture was shown for s-ZnO (18 %) and t-ZnO (11 %).

Adhering tests on different ZnO-structures showed for a filler amount of 1 wt% a strong increase in adhesion for tetrapodal shaped ZnO, where only a slight increase was observed for spherically shaped ZnO. Maximum adhesion was reached for the polymer/particle-composite with 5 wt% t-ZnO. Further increase in filler amounts of t-ZnO did not result in an increase in adhesion force. In general the addition of spherical ZnO did not improve the adhesion of PTU on the substrate (Figure 6).

The investigation of water contact angle showed with a contact angle of 80° the hydrophilic character of PTU without filler particles. Contact angle measurements on varying filler amounts showed a decrease in contact angle with increasing surface roughness (Figure 7).

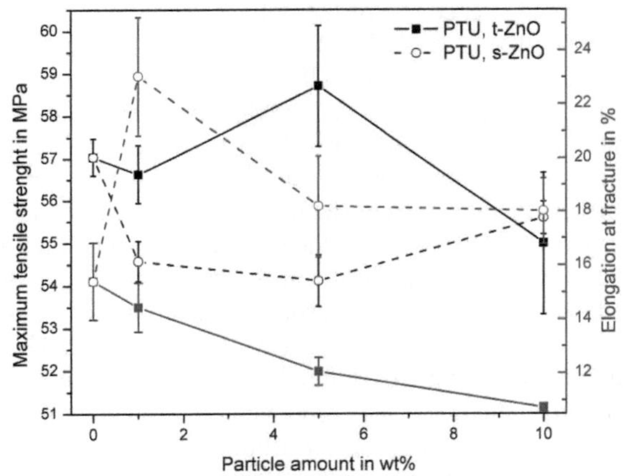

Fig. 5 Influence of different ZnO-structures on the fracture toughness and elongation at fracture.

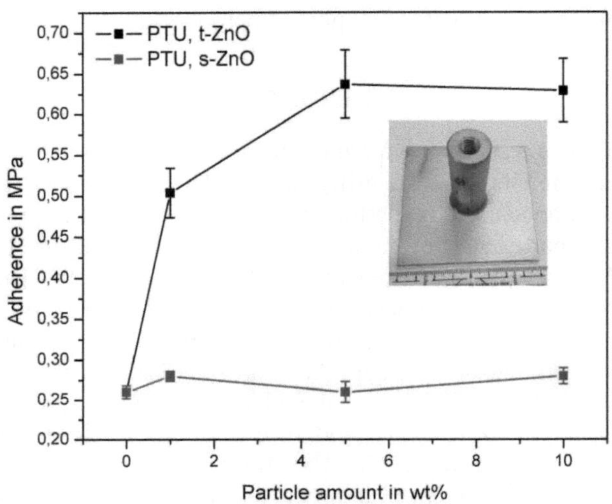

Fig. 6 Influence of different ZnO-structures on the adherence of PTU and reference polymer.

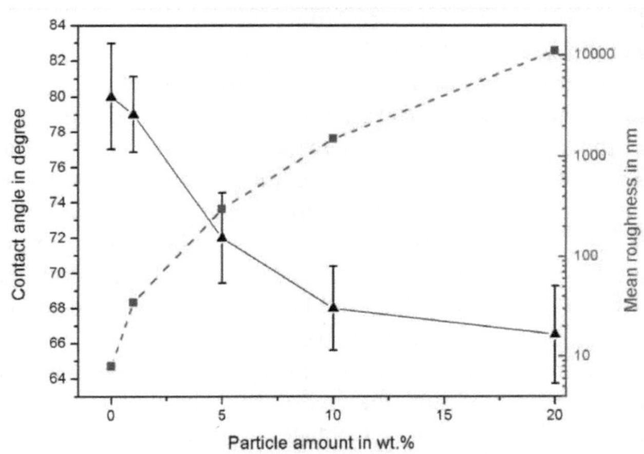

Fig. 7 Influence of t-ZnO on rougness and contact angle of PTU coated glass fiber reinforced material.

The immersion test at the pacific tank after 11 weeks showed strong differences in biofouling. The polymer/particle composite (Figure 8A) was almost completely overgrown, whereas the fouling on PTU with low filler amount of 1 wt% t-ZnO (Figure 8B) is strongly reduced. On the surface of the composite with a filler amount of 5 wt% t-ZnO only slight biofouling was observed (Figure 8C).

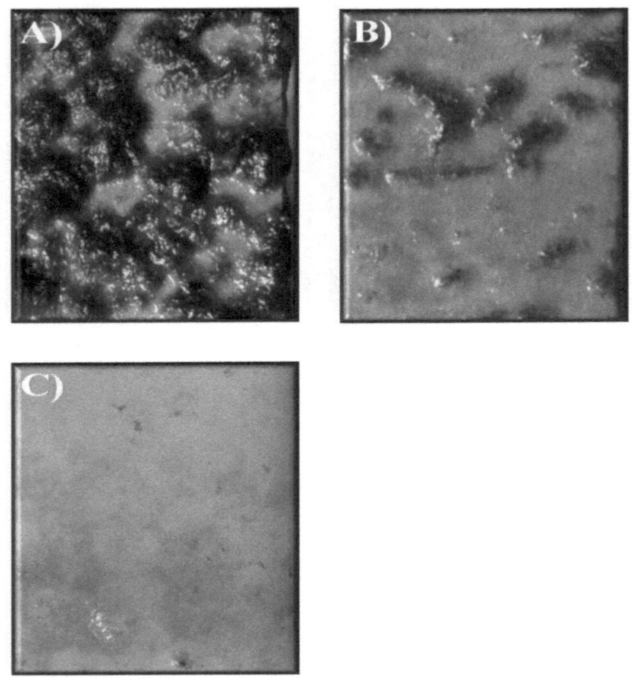

Fig. 8 Immersion test to evaluate the antifouling capacity of the polymer/particle composite. Immersed in Pacific-sea water at an aquarium after 11 weeks.

IV. CONCLUSION

In this study we have introduced an innovative environmentally friendly polyurethane-based composite coating which is assumed to offer long lifetimes due to its high overall mechanical strength. Additionally, preliminary immersion tests showed promising antifouling properties due to embedded flexible tetrapodal ZnO microparticles. Additionally, it provides the possibility for large-scale application due to its sprayability.

By tensile testing we have shown, that the polymer/particle-composites with an amount of 5 wt% t-ZnO has a reinforcing effect on the polymeric matrix which is not observable for higher or lower amounts of t-ZnO particles or for spherical shaped particles (s-ZnO).

The immersion tests were prepared with respect to the obtained materials characteristics. A filler amount of 5 wt%

tZnO was chosen as the highest amount to be tested due to its increasing viscosity with higher filler amounts, which prohibits the desired option of spray application.

The evaluation of water contact angle showed the hydrophilic character of this polymer/particle composite.

The lowest degree of biofouling of aquatic organisms exposed to artificial "Pacific" environmental conditions was found on PTU with a filler amount of 5 wt% t-ZnO.

These immersion tests have to be considered as preliminary and therefore further experiments have to be conducted to evaluate the antifouling properties of this polymer/particle composite under native environmental conditions.

ACKNOWLEDGEMENTS

The authors are thankful to the Aquarium Kiel at Geomar -Helmholtz Centre for Ocean Research Kiel for getting the opportunity to conduct the immersion test at their infrastructure. This work was funded within the projects DKL-WEA, „Nano-based coating technologies for erosion protection and de-icing of offshore-wind energy plants“ and Nano-WWZ, „Coating technologies for the application of nano-based antifouling coatings in the water transition zone of stationary maritime structures“ by the Federal ministry for Economic Affairs and Energy and the cluster of excellence "The Future Ocean" at the University of Kiel.

CONFLICT OF INTEREST

The authors declare that they have no conflict of interest.

REFERENCES

1. I. Clemitson, "Castable Polyurethane Elastomers", Taylor, Francis Group LLC, CRC Press, 2008.
2. G. W. Becker, D. Braun, "Kunststoff Handbuch Polyurethane", Carl Hanser Verlag Muenchen Wien, 1993.
3. T. Charinpanitkul, P. Nartpochananon, T. Satitpitakun, J. Wilcox, T. Seto, Y. Otani, "Facile synthesis of tetrapodal ZnO nanoparticles by modified French process and its photoluminescence", Journal of Industrial and Engineering Chemistry, 18, pp. 469–473, 2012.
4. J. Fang, A. Kelarakis, D. Wang, E. P. Giannelis, J. A. Finlay, M. E. Callow, J. A. Callow, "Fouling release nanostructured coatings based on PDMS-polyurea segmented copolymers", Polymer, 51, pp. 2636–2642, 2010.
5. X. Wang, D. Zhao, Y. Liu, J. Zhang, Y. Lu, X. Fan, "The photoluminescence properties of ZnO whiskers", *Journal of Crystal Growth*, **263**, pp. 316-319, 2004.
6. A. Moezzi, A. M. McDonagh, M. B. Cortie, "Zinc oxide particles: Synthesis, properties and applications", *Chemical Engineering Journal*, **1-22**, pp. 185-186, 2012.
7. M. Mecklenburg, A. Schuchardt, Y. K. Mishra, S. Kaps, R. Adelung, A. Lotnyk, L. Kienle, K. Schulte, "Aerographite: Ultra Lightweight, Flexible Nanowall, Carbon Microtube Material with Outstanding Mechanical Performance", *Advanced Materials*, **24**, pp. 3468-3490, 2012.

Corresponding author:	Martina Baum
Institute:	Functional Nanomaterials Institute for Materials Science
Street:	Kaiserstr. 2
City:	Kiel, 24143
Country:	Germany
Email:	marb@tf.uni-kiel.de

New Opportunities For Biomedicine

C. Yu. Zenkova, I.V. Soltys, and P.A. Ryabiy

Optics, Printing & Publishing Department, Chernivtsy National University, Chernivtsy, Ukraine

Abstract— **Theoretical and experimental approaches to diagnosing of internal spin and orbital optical flows and corresponding optical forces caused by these flows are offered. These approaches are based on the investigation of the motion of tested particles in the formed optical field. The dependence of the above-mentioned forces upon the size and optical properties of the particles is demonstrated. It is shown that the use of energy internal flows permits to create such optical tweezers which can control particles only by polarization of the incident radiation without changing the form of the beam and its intensity.**

Keywords— **spin flows, orbital flows, nanoobjects.**

I. Introduction

According to the classical concept, the picture of the energy current transported by the light flow with some restrictions is equivalent to the electromagnetic momentum. It is characterized by a spatial distribution of the averaged Poynting vector over time and is a natural instrument for describing light fields of an arbitrary structure [1]. Such an approach is most actual for the near field optics, for the description of mechanisms for controlling bioobjects of different properties and sizes as well as for the description of evanescent decaying waves, e.g. in plasma devices and for some new arrangements connected with micro-cavities, masking devices, super lenses and metamaterials, particular in cases where the Poynting vector controlled distribution is used. Alternative ways of creating optical manipulators and optically controlled systems are proposed. Optical flows are attractive for use to diagnose spatially inhomogeneous in energy optical fields where classical methods cannot be applied. The paper proposes to use the Rayleigh light scattering mechanism for studying and diagnosing optical fields. It is the mechanical action of inner energy fluxes on Rayleigh particles that permits to separate and define the effect of spin and orbital flows. Thus, the aim of this paper is to present an alternative way of controlling nanoparticles by optical flows for spatial displacement of bioorganisms, viruses, DNA structures.

II. General Concepts of Spin and Orbital Moments

Generally the macroscopic energy flow can be subdivided into "orbital" and "spin" parts, which show the peculiarities of orbital and spin degrees of freedom of the light field [1-3]. Modern investigations of orbital and spin energy flows are performed by studying the behavior peculiarities of the tested particles. The diagnostics of the spin flow can be performed by analyzing the effect of the optical field on the particle, namely, by investigating the peculiarities of transmitting a part of the inner angular momentum to the particle, followed by the absorption of this angular momentum and the rotation of the particle about its mass center. Such a rotational motion can be explained by the effect of the circularly polarized wave and the inner spin angular momentum. In some field configurations the influence of a circularly polarized field with a smooth wavefront upon the particle manifests itself as an orbital particle motion [4]. It is not only the transmission of the intrinsic "pure" angular momentum of the circularly polarized wave that may be responsible for the rise of orbital and, in some cases, transmission motion. The physical mechanism of the impact and interaction between the field and the particles is much wider and is complicated by the reverse influence of the particles on the field. The resulting field may be considered responsible for the complicated behavior of the tested particles. In this paper, we attempt to calculate the relationship between the force applied to the particle and the energy flow in the optical field before the particle is placed there ("the incident field" with vector amplitudes $\vec{E}$ and $\vec{H}$). The main idea is to determine the value of the electromagnetic field disturbed by the presence of particles, to calculate its momentum, and to compare the obtained result with the initial momentum of the "pure" incident field. Due to the presence of particles, the scattered field $\vec{E}_{sc}$, $\vec{H}_{sc}$ emerges, and it should be added to the incident field $\vec{E}$, $\vec{H}$ [3]. So the total field momentum density is represented as

$$\Delta\vec{p} = \frac{g}{c}\mathrm{Re}\left[\left(\vec{E}^* + \vec{E}_{sc}^*\right)\times\left(\vec{H} + \vec{H}_{sc}\right) - \vec{E}^*\times\vec{H}\right] = \frac{g}{c}\mathrm{Re}(\vec{E}_{sc}^*\times\vec{H}_{sc} + \vec{E}^*\times\vec{H}_{sc} + \vec{E}_{sc}^*\times\vec{H}) . \qquad (1)$$

The change of the field momentum results in the rise of the recoil force applied to the particle. The force can be defined by the field momentum flux through the spherical surface A_R with radius $R \to \infty$ surrounding the particle:

$$\vec{F} = -c\oint_{A_R}\Delta\vec{p}dA = -cR^2\oint_{A_R}\Delta\vec{p}d\Omega . \qquad (2)$$

V. Sontea and I. Tiginyanu (eds.), *3rd International Conference on Nanotechnologies and Biomedical Engineering*,
IFMBE Proceedings 55,
DOI: 10.1007/978-981-287-736-9_62

where dΩ indicates integration over the solid angle. This force depends on the particle position, and our aim is to find the correspondence between F and the incident field momentum at the point where the particle is placed. The calculation of the force affecting the particles is performed using the Mie theory approximation.

Thus, to simplify the modulation, the scattered field can be represented by two linearly or circularly polarized plane waves that meet at different angles, forming the resulting field. The formed field is described by its own intensity distribution or polarization, the distribution of nonzero spin, and orbital flows. The field created is different from the incident one, and this causes the change of the field momentum and the rise of the optical field that affects the particles. The force value depends on the size of the particles and their optical properties. These variations in the behavior of the resulting optical force, as well as the spin and orbital flows, can be used as a basis for the classification of these flows and their separate study and investigation. Bekshaev et al. [2-4] proposed a theoretical model that made it possible to separate, investigate, and calculate spin and orbital flows. According to their theory it is possible to define the optical field energy density and the components of the spin and orbital momentum densities using the mathematical model of the superposition of two plane waves. The results of the calculation show that the orbital momentum is absent in the x direction. In this case the whole flow and the generated force, which is calculated with allowance made for the changes of the scattered field momentum according to Eqs. (1) and (2), will be of spin nature. The x component of the optical force is determined only by the scattered field. Let us analyze the formation of a field with an inhomogeneous energy distribution. This field is produced by some symmetric superposition of circularly polarized plane waves. The angle between the direction of propagation of waves and the z axis is

$$\gamma_1 = -\gamma_2 = \gamma \qquad (3)$$

Both waves are identical, with a possible phase shift, i.e., $E_{x2} = E_{x1}e^{i\delta}$, $E_{y2} = E_{y1}e^{i\delta}$. We introduce the parameter $\Phi = ky\sin\gamma - \dfrac{\delta}{2}$ and we can calculate

$$w = 2g(|E_{x1}|^2 + |E_{y1}|^2)(1+\cos^2\gamma\cos 2\Phi) \qquad (4)$$

$$p_{sx} = -\frac{g}{c} i(E_{x1}E_{y1}^* - E_{y1}E_{x1}^*)\sin 2\gamma \sin 2\Phi \qquad (5)$$

$$p_{sy} = p_{oy} = p_{ox} = 0,$$

$$p_{sz} = \frac{2g}{c}(|E_{x1}|^2 + |E_{y1}|^2)\cos\gamma\sin^2\gamma\cos 2\Phi, \qquad (6)$$

$$p_{oz} = \frac{2g}{c}(|E_{x1}|^2 + |E_{y1}|^2)\cos\gamma(1+\cos^2\gamma\cos 2\Phi) \qquad (7)$$

The corresponding recalculation of the resulting momentum, with allowance made for the scattered field and the corresponding components of the optical field, according to Eq. (2), demonstrates that the force in the x direction is fully determined by the spin flow. It possesses a transverse momentum and is marked off along the line y = const. There are no inner orbital flows in the x and y directions of the resulting field of the given configuration. Correspondingly, a longitudinal momentum arises in the z direction, and it is identified with the effect of light pressure. The components of this momentum are the z components of the orbital and spin momentums. The inhomogeneous intensity distribution, as a rule, actualizes the y component of the optical force, which is identified with the gradient force. Thus, we can develop a model situation of linearly polarized fields, in which the influence of the longitudinal field component is enhanced. In this case transverse spin flows are absent and a longitudinal spin momentum in the z direction is observed. This model is realized when two mutually orthogonal fields that are linearly polarized in the incidence plane are superimposed, and the longitudinal component of the optical field becomes essential. Thus, according to the description presented, it is possible to speak about spin and orbital flows in the longitudinal direction, but it is impossible to separate and investigate the spin and orbital flows.

III. Mechanism of Spin and Orbital Moments

To analyze the spin flows by assessing the mechanical effect of the spin momentum on the tested particles, it is necessary to consider the experimental situation wherein the spin momentum manifests itself in a "pure" form.

Let us assume that the resulting field affecting the particle is formed by superposing circularly polarized waves of the same handedness [2]. Then we can expect the rise of the x-directed spin flow and the propagation of energy in the x direction. The spin flow will not be masked by the action of the orbital flow. In this case the result of the simulation essentially depends on the size of the particles, their properties, and parameter Φ, which is a function of the angle of incidence and the wave phase. In accordance with the particle size, the optical force component values F_x, F_y, F_z, which are calculated using Eq. (2), will be quite different (Fig. 1). The force values connected with the effect of orbital and spin flows differ even for particles of the same order, i.e., Rayleigh particles of about λ/100. Let us introduce the particle size parameter $\xi = ka$, where a is the particle radius and $k = 1.33\cdot 10^5$ cm^{-1} is the radiation wavenumber for a He–

Ne laser. Because this type of interaction causes the formation of a field with corresponding distributions of intensity, the Fy component presets the action of the gradient force (i.e., the linear behavior of the force at small particle sizes), and Fz determines the traditional effect of light pressure. The most interesting conclusions can be made by analyzing the F_x (F_{x+}, F_{x-}) component of the optical force. The direction and the size of the force x component are changed with the change in the circularly polarized beam helicity ($\sigma = \pm1$). We analyze the mechanical effect of the x- or y-polarized beams, $\sigma = 0$, and see that the F_x component disappears completely (Fig. 1).

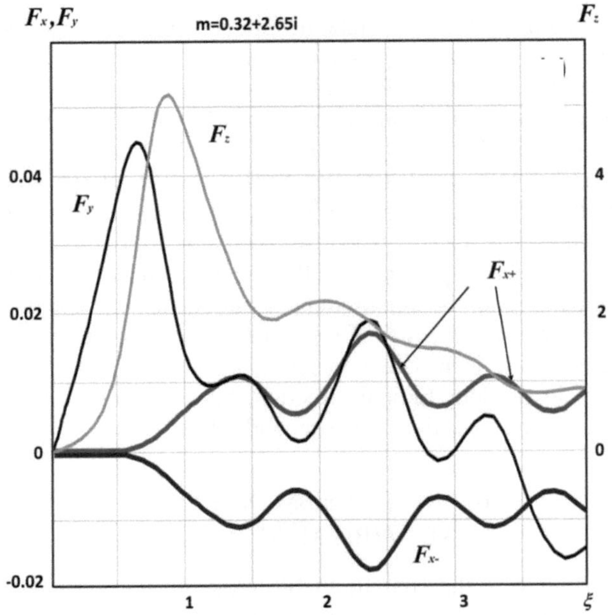

Fig. 1 Dependence of the force components on the particle size parameters in the inhomogeneous incident field with circular polarization. The particle refraction index is $m = 0{,}32+2{,}65i$

a) The data presented in this figure are obtained by normalizing all the calculated forces to the total momentum flow of the incident field through the particle cross section:

$$F_0 = 2g(|E_{x1}|^2 + |E_{y1}|^2)\pi a^2 \quad (8)$$

To differentiate the spin flows we analyze the behavior of the transverse spin flows [2], which arise in the dissipated field and are caused by the interaction of the field with hydrosol of gold particles ($m = 0{,}32+2{,}65i$) of different sizes and under different irradiation conditions (Fig. 2).

The data presented in this figure are obtained by normalizing by means all momentum components to the total momentum flow of the incident field through the particle cross section as well:

$$P_0 = \frac{2g}{\mu}(|E_{x1}|^2 + |E_{y1}|^2)(1+\cos 2\Phi)\cdot \pi a^2 \quad (9)$$

Depending on the polarization state of the irradiated wave the value and the direction of the generated transverse spin momentum can be quite different (Fig. 2, curves $P_{sx}^{+}(\pm)$). If the particles are irradiated by linearly polarized light, the rise in the transverse spin flow is not observed (Fig. 2, curves $P_{sx}^{xy}(\pm)$). Because Rayleigh scattering is characterized by symmetry, the redistribution of the scattered energy must be symmetrical in the forward and backward directions. Solid (dashed) lines describe the momentum flux into the forward (backward) directions (Fig. 2).

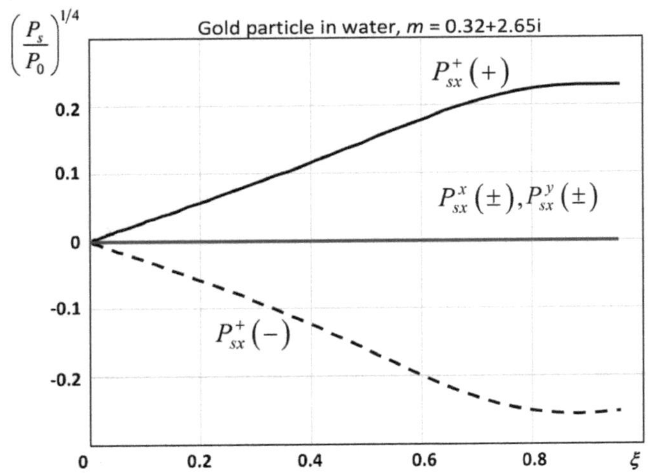

Fig. 2 Dependence of the normalized transverse momentum flow components of the scattered field on the particle size parameter

In this case the scattered radiation must not hold flows of the transverse spin momentums when they are irradiated by linearly polarized light. The irradiation of particles by circularly polarized light, independent of helicity, causes the generation of x-directed transverse spin flows in the scattered field in both the direct and backward directions. The behavior of such generated momentums essentially differs in both directions. The difference between the absolute values of the momentum presets the force that is identified with the intrinsic noncompensated effect of the scattered field. The equilibrium of this force with the recoil force exerted on the particle causes the origin of the translatory ponderomotive motion in the inhomogeneous circularly polarized field. Since the analysis of the momentum is performed in the x direction, according to the mathematical model, the optical force, which arises in that direction, can be considered as a mechanical action of the energy spin flow. To compare the effect of spin and orbital flows, it is necessary to analyze different model situations where corresponding momentums arise in a "pure" form.

Their local values must be equal and determine the identical regularity of ponderomotive effects. They are independent of other field parameters. As noted above, the x direction presets the prerequisites to generating spin flows, and the y direction presets orbital flows. To analyze the transverse spin momentums we consider the above described experimental situation of circularly polarized field interaction when the distribution of intensity is inhomogeneous. The y-directed gradient force arises. In this situation it is impossible to distinguish the orbital flow, because it is also axially y directed.

b) To separate the "pure" orbital flow we investigate a simple field configuration formed at the superposition of two plane circularly polarized waves, when $\gamma_1 = \gamma_2 = \gamma$ [3]. The model fields are chosen so that a plane wavefront is formed if $\delta = \pi/2$.

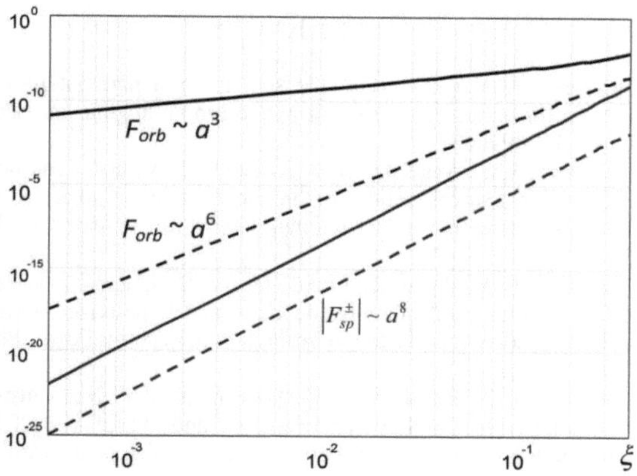

Fig. 3 Comparison of mechanical actions associated with spin and orbital internal energy flows for metallic and dielectric spherical particles suspended in water. Solid lines describe the behavior of metallic particles; dashed lines describe the behavior of dielectric particles.

The resulting circular polarized wave irradiates the tested particles, and an orbital flow is generated in the scattered field. In this field the configuration of the orbital flow corresponds to the x-directed spin flow described in the above model situation. As the results of our simulation show (Fig. 3), the generated spin and orbital flows essentially depend upon the size of the particles. Thus, $F_{orb} \approx a^3$ will do for metallic particles, which is explained by the predominant influence of the absorption. At the same time, $F_{orb} \approx a^6$ fits well for dielectric particles, where the physical mechanism for transmitting the light force pressure to the particles is determined by the properties of the particles themselves. Correspondingly, the spin-induced force $F_{sp}^{\pm} \approx a^8$ practically does not depend upon the type of particles.The force value progressively increases with the increase in the size of the particles, and they become practically equal when $\xi>2$. Thus, both spin and orbital contributions to the energy flow can cause the translatory motion of the tested particles, but the quantitative properties of the spin- and orbital-induced motion and their dependence on the size and properties of the particles essentially differ. Correspondingly, the spin and orbital flows differ experimentally if we use tested particles of specially chosen sizes and properties.

IV. Results of Experimental Modulation for Generated Spin Flows

The results of the modulation are experimentally confirmed in the schematic (Fig. 4), where the interaction of two circularly polarized beams determine the interference distribution around the focus [5].

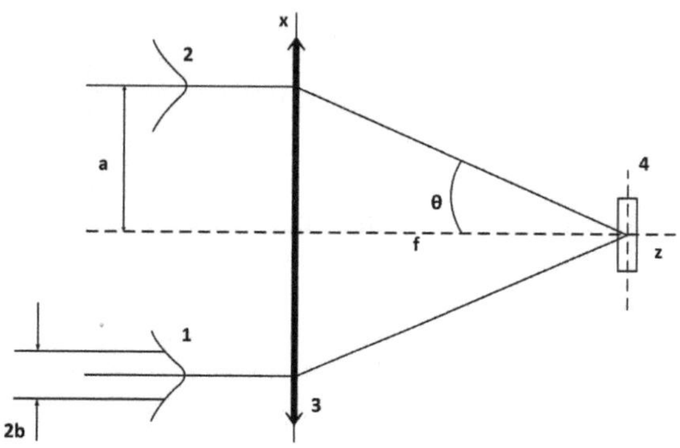

Fig. 4 Schematic of the experimental setup. 1, 2, input beams 2 (b = 0.7 mm) of semiconductor lasers (λ = 0.67 μm); 3, objective lens (f = 10 mm); 4, cell with tested particles suspended in water. a = 1.3 mm, θ = 7.4°, NA = 0.16

An inhomogeneous intensity distribution is formed. A circulation of energy takes place within each maximum, and it determines the rise of spin flows. A strict focusing of the circularly polarized beam causes the transformation of the spin flow into the orbital one [6]. In this case the mechanical effect of the generated flow on the particles doesn't allow us to separate the effects of the spin and orbital momentums. That is why, to avoid this ambiguity, the focusing strength should not be high (in accordance with the known data [7], the spin–orbital conversion is negligible and does not exceed 1% at NA ≈ 0.2, with the focusing angle θ ≈ 11°). Of course, this leads to a certain loss in the energy concentration. However, one can avoid essential reduction of the focal region of the spin momentum density if the decrease of intensity is compensated by an increase in the beam inhomogeneity. It is obvious that the circulatory flow of spin nature exists within each lobe, while the orbital momentum density is completely radial, attributed to the beam divergence. This radial field momentum can be used for probing particle confinement at a desirable off-center position [8], allowing us to observe the spin-induced orbital motion. Within an inhomogeneous optical field any tested particle is subjected to the gradient force [9,10] that pulls the particle toward the intensity maximum. The radial orbital momentum of a divergent beam pushes the particle away from the axis. As a result, both forces can compensate each other at certain off-axial points within the central lobe of the interference pattern. The spin momentum in the two-beam interference pattern is approximately 2.5 times higher than in a single Gaussian beam focused with the same NA objective. The interference technique of the focal pattern formation facilitates the avoidance of this undesired spin in orbital conversion and promotes the observation of the mechanical action of the "pure" spin flow without any contamination influence from the orbital one. To define the "pure" effect of spin flows in the experiment, a cell was used that contained an ensemble of latex microparticles (refractive index 1.48) suspended in water. The particles were chosen so that their shape was close to ellipsoidal with an approximate size of 1.5 μm× 1 μm, which made it possible to observe individual particles within a single lobe of the interference pattern formed in the focal region. Experimental observations of the motion of the trapped particles when both superposed beams are circularly polarized are represented in Fig. 5. The asymmetric particle spins around its own center of mass, which is explained by partial absorption of the incident circularly polarized light and its inherent angular momentum [6].

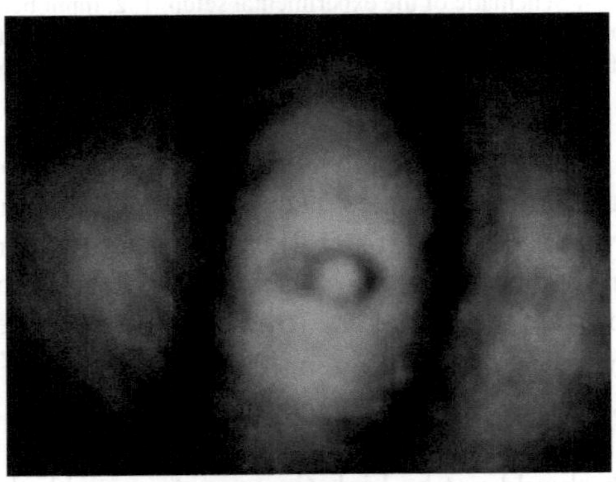

Fig. 5 Motion of a particle trapped within the central lobe of the interference pattern

A new observation is that simultaneously the particle's center of mass performs an orbital motion, which can only be associated with the azimuthal light pressure originating from the spin momentum circulation. This attribution is confirmed by the reverse rotational direction when the sign of the circular polarization is changed. When both beams are linearly polarized, the particle stops.

CONFLICT OF INTEREST

The authors declare that they have no conflict of interest.

CONCLUSION

A new approach to modulating optical flows for fields with scattered tested particles is proposed. In accordance with the structure and form of the incident beam wave front, spin or orbital flows arise in the scattered field. The essential dependence of the recoil force, determined by the spin and orbital flows on the size and properties of the particles, makes possible to separate, analyze and investigate the mechanical effect not only of orbital flows, but of spin flows as well. The use of optical flows (spin and orbital) is a new alternative method to control nanoobjects, contactless diagnostics and for direct shifting of nanostructures.

REFERENCES

1. Bekshaev A Y (2009) Oblique section of a paraxial light beam: criteria for azimuthal energy flow and orbital angular momentum. J Opt A 11: 094003.
2. Bekshaev A Y, Angelsky O V, Sviridova S V, and Zenkova C Yu (2011) Mechanical action of inhomogeneously polarized optical fields and detection of the internal energy flows. Adv Opt Technol 2011: 723901.
3. Bekshaev A Y, Angelsky O V, Hanson S G, and Zenkova C Yu (2012) Scattering of inhomogeneous circularly polarized optical field and mechanical manifestation of the internal energy flows. Phys Rev A 86: 023847.
4. Angelsky O V, Bekshaev A Y, Maksimyak P P, Maksimyak A P, Mokhun I I, Hanson S G, Zenkova C Yu, and Tyurin A V (2012) Circular motion of particles suspended in a Gaussian beam with circular polarization validates the spin part of the internal energy flow. Opt Express 20: 11351–11356.
5. Angelsky O V, Bekshaev A Y, Maksimyak P P, Maksimyak A P, Hanson S G, and Zenkova C Yu (2012) Orbital rotation without orbital angular momentum: mechanical action of the spin part of the internal energy flow in light beams. Opt Express 20: 3563–3571.
6. Bekshaev A Y (2010) A simple analytical model of the angular momentum transformation in strongly focused light beams. Central Eur J Phys 8: 947–960.
7. Nieminen T A, Stilgoe A B, Heckenberg N R, and Rubinsztein-Dunlop H (2008) Angular momentum of a strongly focused Gaussian beam. J Opt A 10: 115005.
8. Ashkin A (2006) Optical Trapping and Manipulation of Neutral Particles Using Lasers. World Scientific.
9. Zenkova C Y, Soltys I V, and Angelsky P O (2013) The use of motion peculiarities of particles of the Rayleigh light scattering mechanism for defining the coherence properties of optical fields. Opt Appl 2: 297–312.
10. Angelsky O V, Zenkova C Y, Gorsky M P, Soltys I V, and Angelsky P O (2013) The use of new approaches to estimating the coherence properties of mutually orthogonal beams. Open Opt J 7: 5–12,

Author: Zenkova Claudia Yu.
Institute: Chernivtsy National University
Street: 2, Kotsyubinsky Str.
City: Chernivtsy
Country: Ukraine
Email: zenkova@itf.cv.ua

New Perspective for Biomedical Productions: Application of Cast Amorphous Microwire for Electromagnetic Absorption

S.A. Baranov[1,2]

[1] Institute of Applied Physics, Academy of Sciences of Moldova, Chisinau, Republic of Moldova
[2] Departament de Genie Physique, Ecole Polytechnique de Montreal, Montreal, Canada

Abstract— **The technological use of high and ultra-high frequencies in medical engineering has led to the need of creating electromagnetic shielding's. The absorption properties of films made of a parallel array of cobalt based microwires have been studied by noise suppression experiments. The absorption S_{21} parameter values observed are of -2 dB at 40 GHz, the insertion losses are registered below -0.2 at 1 GHz, for two films placed in the top of a 50 μm width CPW signal line. The losses profile registers mainly eddy current losses as the origin of the observed absorption.**

Keywords— **microwires, noise suppression sheet, ferromagnetic resonance, eddy current losses, high frequency devices.**

I. Introduction

We present the results of a research regarding new applications of cast amorphous microwires in the realization of some new composite material and the possibility to use it in electromagnetic shielding and protection equipment manufacturing. The composite material and the technical problems raised by its manufacturing technology are described [1 – 14], and its shielding, rejection and polarization properties are evaluated.

The search for materials exhibiting outstanding properties in high frequency range is thus relevant in the development of those applications. In this article, we focus on a particular family of materials, glass-coated amorphous magnetic microwires, exhibiting exciting microwave behavior with enormous technological applications in particular fields: as elements for absorption of electromagnetic radiation from medical engineering .

Before entering into their microwave properties, let us briefly summarize some important aspects of these composite materials. Glass-coated magnetic microwires are characterized by a nucleus out of a magnetic alloy, structurally amorphous and metallic conductor, with diameter between around 2 and 20 μm, covered by a Pyrex-like coating 2 to 10 μm thick. They are fabricated by a particular rapid solidification technique, sometimes called quenching and drawing method, following the early Taylor technique later modified by Ulitovsky [2 – 9, 12 – 14] (TU). That coating, besides insulating the metallic nucleus from corrosion and electrical points of view, induces strong mechanical stresses in that nucleus [8, 9] that couple with magnetostriction to determine large magnetoelastic anisotropy, and consequently a unique magnetic behavior. Particularly, the strong magnetoelastic anisotropy in magnetostrictive Fe-rich alloys amorphous microwires gives rise to the existence magnetic bistability characterized. These microwires exhibit also outstanding high frequency properties as for example natural ferromagnetic resonance, *NFMR*, [2 – 7, 12 – 14] appearing typically within the frequency range 0,1 to 12 GHz depending on the magnetic alloy composition.

The above mentioned amorphous microwires can be thus employed as radio-absorption elements when embedded in suitable matrix [1 – 9, 12]. The equipment was used such that to ensure mechanical protection, to perform the shielding function in electromagnetically polluted environments. The microwave electromagnetic response has been analyzed for a composite consisting of dipoles of amorphous magnetic glass-coated microwires in a dielectric. This material for medical engineering can be employed for radio absorbing screening. The spontaneous *NFMR* phenomenon observed in glass-coated microwires has opened the possibility of developing novel materials with broad-band of radio absorbing materials.

The actual need to operate increasingly miniaturized devices at high speeds has led to noise and signal interference caused by radio frequency devices to nearby components. This fact has strongly promoted the development of thin film technology for noise suppression purposes [10]. As this research has evolved, the need to control the magnetic anisotropy in thin films came into attention since the near field absorption of films is related to the ferromagnetic resonance and eddy current losses. In terms of ferromagnetic resonance high anisotropy field and coherent rotations of spins at high frequencies are the most desirable characteristics for the optimum power absorption observed at the FMR frequency value. In terms of eddy currents losses a high resistivity of the magnetic film is required, since this allows a higher skin depth penetration value and therefore higher interaction with the HF field. In both cases a large loss generation at a high frequency range is required [11]. In general terms the ideal role of a noise suppressor is to prevent a raise in insertion losses in the pass-band, and to give as large attenuation as possible in the stop band frequency region.

V. Sontea and I. Tiginyanu (eds.), *3rd International Conference on Nanotechnologies and Biomedical Engineering*,
IFMBE Proceedings 55,
DOI: 10.1007/978-981-287-736-9_63

Several attempts have been successfully made, especially in the case of granular thin films for noises suppressors as a integrated type [11]. For the non-integrated type a new magnetic material is suggested as absorption element for a noise suppression experiment. Amorphous glass coated magnetic microwires present strong shape anisotropy due to their fabrication method, a controllable value of ferromagnetic resonance [2 – 9, 12 – 14] and a relatively high resistivity due to their amorphous state.

Depending on the microwires alloy composition, microstructure and geometry it has been shown that is possible to produce absorption characteristics within a required frequency band in the range of several megahertz as single elements and as composites in the range of several GHz [2 – 9, 12, 13]. In this contribution we suggest a film of made of a parallel array of microwires and test their absorption properties.

II. DATABASE AND METHODS

In the case of cobalt based amorphous microwires obtained by TU method, the soften glass capillarity is filled with a melted metal alloy that is in turn quenched by a stream of water and frozen into an amorphous alloy. Thus, strong quenching stresses are present in the metallic core of the wire that couple with magnetostriction to give rise to a magnetoelastic anisotropy that mainly determines the magnetization easy axis. In the case of Co based negative magnetostriction amorphous alloy microwires a radial easy axis is determined particularly near the surface [9] owing to the negative character of radial and axial stresses in that region. The surface under layer region defining the skin effect penetration depth is of particular interest for high frequency applications.

The microwires alloy composition to be tested for absorption experiments is $Co_{68.15}Fe_{4.35}Si_{12.5}B_{15}$, with a saturation magnetostriction constant of $\lambda_s = -10^{-7}$ [8, 9], 310 parallel aligned pieces of microwires (metallic radius $r_m<3$ μm,13 and length L= 5 mm) with a separation distance of 2 μm, where confined in 5 x 4 mm^2 area.

As can be observed in the micrograph on figure 1, the glass shell (< 4μm thickness), and the inner metallic radius are homogeneous within a considerable length, as should therefore be the interaction of the magnetic nucleus with the HF electromagnetic field. The resistivity value of a single wire is 1.78 μΩm somehow lower than those usually considered in films.

A first magnetic characterisation of the films of microwires has been performed measuring the axial hysteresis loops in a vibrating sample magnetometer at room temperature and 50Hz in signal frequency. Two regions can be distinguished in each hysteresis loop with a noticeable anhysteretic region at very low field which should be understood as the magnetization reversal of a given region in the core of the microwires where the easy axis is longitudinal.

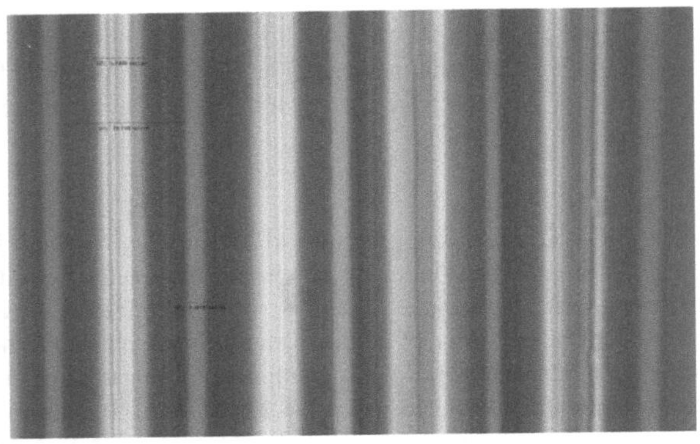

Fig. 1 Micrograph (100 time enlarged) of the parallel array (grid) of Co based microwires. The distance between wires is 2 μm. the inner diameter is 5μm and the thickness of the glass shell is 4 μm. Film dimensions (4mmx5mmx4μm)

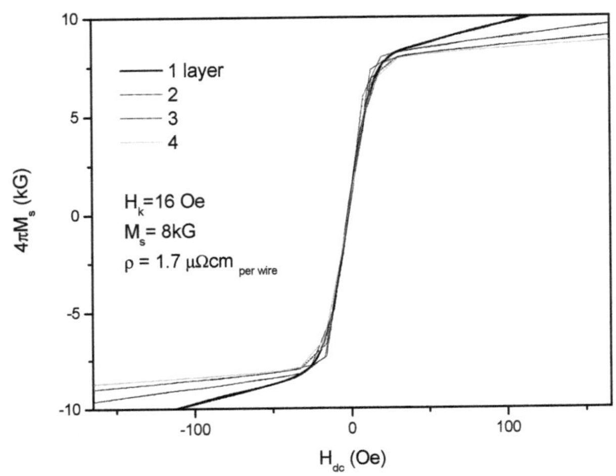

Fig. 2 Axial hysteresis loops for up to four aligned. An increase in anisotropy H_k from 16 to 32 is observed with the applied number of layers. The saturation magnetization value (which is not reach here) is 8 kG.

For moderate fields, the magnetization increases nearly linearly with applied field, until reaching saturation (not given here). This behaviour is ascribed to the nearly reversible magnetization reversal at a region closer to the surface of the wires with radial easy axis. As observed in figure 2 there is an increase of the effective anisotropy field, H_k, from 16 to 32 Oe, which could be ascribed to an effective magnetic interaction between microwires from different layers.

The passage of electromagnetic field through an absorbing screen can be characterized by the transmission parameters S_{21} (attenuation) and S_{11} (transmission). In this work only the absorption factor and the losses profile will be considered as we will focus on the properties of microwire layers in the first attempt as a non-integrated type noise suppressor.

The used coplanar transmission line (50 ohms characteristic impedance) has been fabricated with 50 μm width in the signal line and 3 μm in thickness on a 7059 corning glass (permittivity, ε_r = 5.84). The details of the micro fabrication process and design of the line can be found elsewhere [3, 13].

The electric performance of the line is measured with two ground signal ground (GSG) pin type wafer probes mechanically touching the left and right most ends of the transmission line using an Agilent E8363B network analyser from 0.1 to 40 GHz.

The microwires magnetic layer has been placed on the top of the coplanar line to suppress the noise of the signal line. Two microwires layers have been added, stacking one on top of the other, in the test of absorption. The distance between layers was kept at minimum using a micrometer XYZ station adding negligible pressure in the Y direction.

The measured absorption coefficient (S_{21}) on the coplanar transmission line is shown in figure 3. On adding subsequent layers the attenuation increased. As can be observed the absorption coefficient decreases only for the line -0.96 dB (labelled A in fig. 3) and from -0.96 dB to -2 dB for the cases from one and two microwire layers (labelled as B to C in fig. 3). The glass shell of the microwires is thicker than we would prefer because beside the induced stress on the outer magnetic structure in the experiment it also determines how close the magnetic nucleus can get to the signal line for absorption. The closer we can get to the line is therefore the better, and in this case the distance is fixed at 4μm.

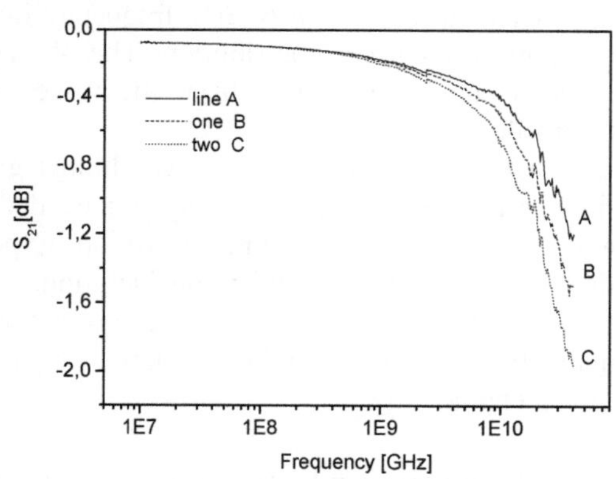

Fig. 3 Signal attenuation S_{21} on a coplanar transmission line with a frequency increment up to 40 GHz. The dimension of the signal line is 50μm The microwire film dimension are (4 mm x 5 mm x 4μm).

The ratio of the power absorbed by the microwire film to the input power (1W) was calculated using the expression P_{loss}/P_{in} = 1- ($|S_{11}|^2+|S_{21}|^2$) and can be observed in figure 4. The observed losses profile indicates that the absorption is due mainly to eddy current losses since the peak usually assigned to resonance losses is absent.

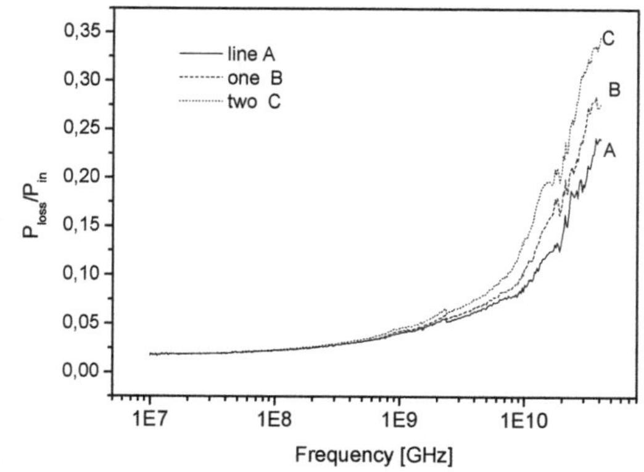

Fig. 4 Tangent loses profile from 0.1 to 40 GHz. A relatively low value of losses at high frequency is observed. This profile is assigned mainly to losses generated by eddy currents, for the peak usually observed in the case of FMR losses is absent.

III. RESULTS

Measurements of the permeability of the microwires films have been performed with the in lab 9 GHz permeabilometer.

The film sample is enclosed in a pick-up coil through all the measurements. An RF power is supplied form a network analyser to a side open TEM cell to apply the RF magnetic field to the hard axis direction of the sample; in the case of the microwires this direction is perpendicular to the circumferential direction of the outer shell. The permeability of a single microwire layer is shown in figure 5. As can be observed, a deviation from a typical behaviour of ferromagnetic resonance response is observed.

The resonance frequency of (0.8 – 1) GHz observed in figure 5 should be treated carefully since the value is in agreement with the theoretical prediction by Kittel [15], that assumes a magnetization component lying parallel to its axis.

When a cylinder without any magnetic anisotropy except for the shape anisotropy is submitted to an AC field along its axis,

$$\Omega/\gamma \sim 2\pi M + He,$$

where: He = Hy + Hk , Hy is outside field, Hk is inside field, M is magnetization (determined in [12]) and γ = 2.8 MHz/Oe is the gyromagnetic ratio.

The demagnetizing field for a precession by a magnetization lying parallel to the cylinder axis is $2\pi M$ (by Kittel [15]).

The effective magnetization M can be calculated by [12, 14]

$$M \approx 2M_0 \int_{\rho_0}^{1} d\rho \, \rho \cos\left[\theta(\rho)\right] exp\{ (\rho - 1) / \zeta \} \quad ,$$

where $\zeta \sim \delta / r_m < 1$, $4\pi M_0 \sim$ 0,8 T,

$$\theta(\rho) = c\ln(\rho/\rho_0), \qquad c = \pi/(2\ln(1/\rho_0)), \quad \rho \sim 0{,}01.$$

The skin penetration depth is defined as $\delta \sim [R/(\Omega\mu\mu_0)]^{1/2}$, where $\mu\mu_0$ is the effective magnetic permeability and R is the microwire electrical resistivity.

Therefore, for the cast amorphous microwire with negative magnetostriction the relative magnetization is, approximately,

$$M/M_0 \sim 0.05.$$

which, for the present sample, predicts $\Omega/2\pi$ to be ~ 1 GHz. [12, 14].

The widening of the imaginary part μ'' of the permeability profile is due to a distribution of time constants within the film due to the large number of absorption elements.

The measure profile agrees with the calculated values based on the LLG equation [15] up to 9 GHz that considers eddy current contribution.

The permeability profile for the cases of 1 to 4 layers is shown in figure 2 - 6. As can be observed a considerable increase in permeability value and in ferromagnetic resonance frequency value (from 0.85 to 0.95 GHz) is obtained.

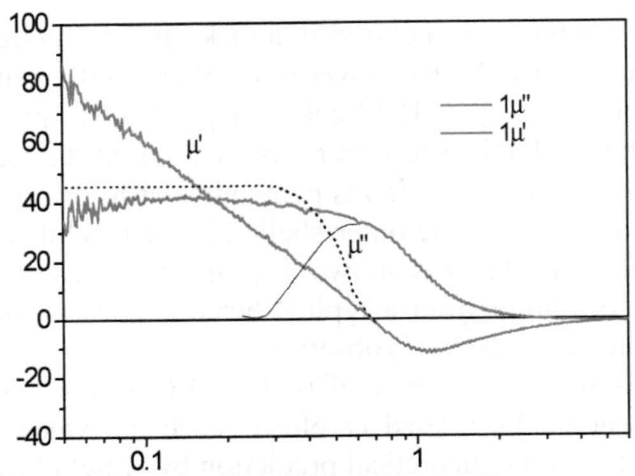

Fig. 5 Permeability profile of a single layer, with a resonance frequency value of 0.67 GHz. The widening observed in both real and imaginary part is due to eddy currents contribution. The microwires axis is parallel to the ac field.

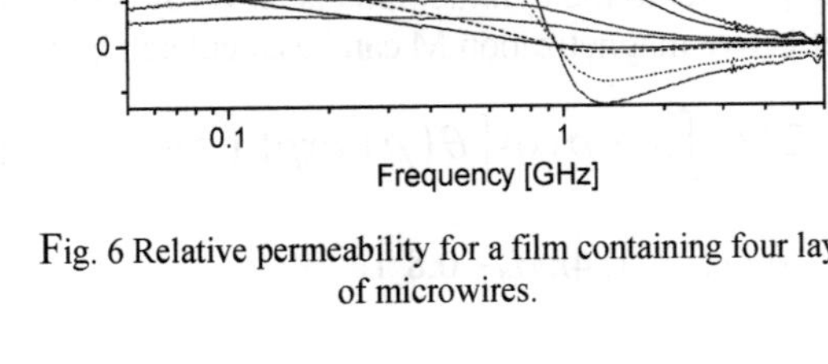

Fig. 6 Relative permeability for a film containing four layers of microwires.

Initial permeability at high frequencies has been measured using the in-lab 9 GHz permeameter. (The initial permeability profile of a single microwire layer, thus obtained, is shown in figure 5.)

A rapid decrease of the real part of the permeability μ', was observed. This can be attributed to eddy current caused by the high initial permeability.

This estimation is significant when compared to the microwire diameter of 2 μm.

However, a possible contribution due to wall displacement loss should not be excluded; domain walls could contribute significantly to initial permeability observed at lower frequencies, at higher frequencies domain wall movement is damped by microscopic eddy currents rapidly [9].

A multilayer configuration increases the value of field penetration depth δ as seen in Figure 5. This behavior is a direct consequence of a subsequent decrease on the permeability, and is consistent with the axial loops measured.

IV. CONCLUSIONS

The FMR frequency of an absorption sheet of 5 x 4mm2 made of 310 parallel aligned Co based microwires is 0.67 GHz. Films with up to two layers containing each a grid of microwires have shown absorption values, S_{21} of -2 dB, with insertion losses below 0.2 a 1 GHz at 4μm distance into a 50 μm width signal line and a magnetic permeability μ of 10^2.

The permeability profile denotes the presence of several time constants active in the 0.1 to 6 GHz frequency range and a dominating eddy current contribution. The observed absorption results can be explained from the losses and permeability profile.

We believe that a thicker metallic nucleus, thinner glass shell, and higher Hk and higher resistivity are needed in order to optimized the microwires film absorption. Experiments in this respect are currently under consideration.

The composite structure can be used with good results for electromagnetic shielding in a relatively large frequency range in medical engineering.

CONFLICT OF INTEREST

The authors declare that they have no conflict of interest.

REFERENCES

1. Faxiang Qin and Hua-Xin Peng, "Ferromagnetic microwires enabled multifunctional" Progress in Material Science, **58**, pp 183-259. 2013.
2. 2 S.A. Baranov, "Microwire-based composites for radio absorption screens" Surface Engineering and Applied Electrochemistry. **39** **(3)**, pp. 72-75, 2003.

3. S.A. Baranov, M. Yamaguchi, K.L. Garcia and M. Vazquez. "Application of amorphous microwires for electromagnetic shilding" Moldavian Journal of the Physical Sciences. **9 (1)**. pp 76 – 82. 2010.
4. S.A. Baranov, "Magnetic properties of an amorphous microconductor in the microwave range", Technical Physics **43 (1)**, pp 122 – 123. 1998.
5. S.A. Baranov, "Use of microconductor with natural ferromagnetic resonance for radio-absorbing materials", Technical Physics Letters, **24 (7)** , pp 549-550. 1998.
6. S.A. Baranov, "Generalized conductivity and optimum energy release ", Technical Physics **44 (7)**, pp 853-854. 1999.
7. S.A. Baranov, "Permeability of an amorphous microwire in the microwave band", J. Communications, Technology and Electronics, **48** pp 226 – 228. 2003.
8. S.A. Baranov, " Magnetic Models of Cast Amorphous Microwires", Surface Eng. and Appl. Electrochemistry. **47**(4), pp. 316-330. 2011.
9. S. Baranov " Magnetic Properties of Co-based amorphous microwires" J. Magn. and Magn. Mat. ***(JMMM)*** **266**, pp 278-281. 2003.
10. K. H Kim and M. Yamaguchi, "Ferromagnetic integrated devices – RF Integrated Noise Suppressors and Integrated Inductors" The physics of Metals and Metallography. **102**, Suppl. 1 pp S74-S79. 2006
11. Y. Shimada, M. Yamaguchi, S. Ohnuma, T. Itoh, W. D. Li, S. Ikeda, K.H. Kim and H. Nagura. "Granular thin films with high RF Permeability" IEEE Trans. On Magn. **39**, No 5. 2003.
12. S.A Baranov, M. Vazquez. "Thin microvire with negative magnetostriction in a zero magnetic field.", Technical Physics Letters. **30 (23),** pp. 989-990. 2004.
13. S. A. Baranov, M. Yamaguchi, K. L.Garcia, M. Vazquez. „Dimensional absorption high-frequency properties of the cast glass coated microwires." Surface Engineering and Applied Electrochemistry. **44 (6),** pp. 245-247. 2008.
14. S. A. Baranov, D. Laroze, P. Vargas, M. Vazquez. „On micromagnetic theory of thin cast amorphous microwires." Physica B **372** , pp. 320-323. 2006.
15. C. Kittel „On the theory of ferromagnetic resonance absorption." Physical Review **73 (2)** , pp. 155-161. 1948

Author:	Serghei Alexei Baranov.
Institute:	Institute of Applied Physics, Academy of Sciences of Moldova.
Street:	5 Academia str. MD-2028.
City:	Chisinau.
Country:	Republic of Moldova.
Email:	baranov@phys.asm.md

Assessment of the Antimicrobial Activity of Polymer Materials with Added Nanosilica Modified by Silver Compounds

M.D. Zheliba[1], I.I. Gerashchenko[2], L.V. Karabanova[3], E.F. Voronin[2], R.M. Chornopyshchuk[1], T.P. Osolodchenko[4], and M.I. Burkovskyi[1]

[1] Vinnytsia National Pirogov Memorial Medical University, Vinnytsia, Ukraine
[2] Chuiko Institute of Surface Chemistry of the National Academy of Sciences of Ukraine, Kyiv, Ukraine
[3] The Institute of Macromolecular Chemistry of the National Academy of Sciences of Ukraine, Kyiv, Ukraine
[4] Mechnikov Institute of Microbiology and Immunology of the National Academy of Medical Sciences of Ukraine, Kharkiv, Ukraine

Abstract— **The problem of prevention and treatment of the wound infection remains urgent. One of the ways of struggle against a similar complication is the use of the materials with antimicrobial properties. Among them there are polyurethane (PU) and interpenetrating polymer networks (IPNs) with different ratios of polyurethane (PU) and poly(2-hydroxyethyl methacrylate) (PHEMA) containing nanosized silica modified with an antimicrobial agent. The objective of the work was to study the antimicrobial properties of composite materials created on the basis of PU and IPNs, consisting of PU and PHEMA, and nanosilica modified with silver compounds, and to estimate their influence on the biofilm-forming ability of microorganisms. The study was performed using PU and IPNs with 83 % PU and 17 % PHEMA. 10 % of the nanosilica with silver salt immobilized as one "layer" on the surface was inserted into each of them. The antimicrobial activity of the known central venous catheter with the immobilized silver compound and created in this work composites were compared. The microbiological studies included determination of the antimicrobial activity of the investigated samples and determination of their influence on the ability of the microorganisms to form a biofilm. The obtained results of the microbiological studies prove that the proposed samples are equal to the existing analogues and according to some criteria even surpass them.**

Keywords— **nanocomposites, PU, PHEMA, nosocomial infections, antimicrobial activity, biofilms.**

I. INTRODUCTION

Despite the development and implementation of new effective surgical materials the problem of prevention and treatment of wound infection remains urgent. This problem has become especially acute at present when the number of infectious complications conditioned by the opportunistic infection increases [1]. The frequency of nosocomial infections does not decrease either. 5-20 % of patients get these infections at the pre-hospital stage already (in the developed countries these cases reaches 25–40 %) [2]. During the postoperative period the risk of infectious complications increases considerably: the frequency of wound infection is 14–16 %, infection of the urogenital system – 42–45%, lower respiratory tract – 10–19 %, blood-vascular system – 2–5 % [3]. One of the reasons for development of such complications is the use of routine medical means and devices being both the "reservoir" for living of microorganisms and the "bridge" for their direct ingress into the patient's body, especially under the influence of local and general contributory factors. Taking into account this fact, the use of materials performing both the main function and having antimicrobial properties can be prospective [4]. Indeed, antimicrobial agents applied to the surface of medical devices (catheters, drainages, etc.), especially those directly contacting the microorganism, either affect microorganisms or often influence their ability to propagate and form biofilms [5]. In the last century there was a great interest in the possibility to use xenogeneic materials with application of different acting substances thereon [6], but despite positive results of their use they enjoyed neither mass consumption nor further development. Synthetic materials were much more widespread. In 1964 for the first time Volkmann conducting an experiment on animals showed that silicone could be used as a drug carrier for low-molecular compounds [7]. Later it was established that among numerous materials polyurethane possessed the highest biocompatibility [8]. Obviously, that's why further studies were concentrated on the use of this material exactly as a matrix for application of active substances.

But the majority of the well-known materials coated with the antimicrobial component have a limited ability of even and prolonged release of active substances. That's why, we have proposed a new approach to formation of nanocomposite materials: by the method of gas-phase mechanosorption modification of nanosilica with silver compounds a filler was created with further possibility of its insertion into the polymers: polyurethane (PU) and

V. Sontea and I. Tiginyanu (eds.), *3rd International Conference on Nanotechnologies and Biomedical Engineering*,
IFMBE Proceedings 55,
DOI: 10.1007/978-981-287-736-9_64

interpenetrating polymer networks (IPNs) with different ratios of polyurethane (PU) and poly(2-hydroxyethyl methacrylate) (PHEMA). The prior studies confirmed the prolonged silver ion release from the proposed materials during the whole observation period (15 days).

The objective of the work was to study the antimicrobial properties of composite materials (nanocomposites) created on the basis of PU and IPNs, consisting of PU and PHEMA, and nanosilica modified with silver compounds, and to estimate their influence on the biofilm-forming ability of microorganisms.

II. MATERIALS AND METHODS

The study was performed using two types of polymer matrixes – PU and IPNs with 83 % of PU and 17 % of PHEMA. 10 % of the nanosilica with silver salt immobilized as one "layer" on the surface was inserted into each of them (Sample 1 and Sample 2, respectively). We studied for comparison the antimicrobial activity of the known central venous catheter with the immobilized silver compound, represented in the market (Control 1), and polymer matrixes without the filler (PU – Control 2; IPN with 83 % of PU and 17 % of PHEMA – Control 3). The weight of the studied fragments was 0,04 g. The materials were sterilized by soaking in 8 % solution of Lysoformin 3000 for 60 min and further washing with sterile 0,9 % NaCl solution [9].

The antimicrobial activity of the studied samples was determined by the agar well diffusion method on two layers of the solid nutrient medium poured into Petri dishes [10]. The test strains of *Staphylococcus aureus* ATCC 25923, *Escherichia coli* ATCC 25922, *Pseudomonas aeruginosa* ATCC 27853, *Bacillus subtilis* ATCC 6633, *Proteus vulgaris* ATCC 4636, *Candida albicans* ATCC 885/653 were used. The microbial load was 10^7 microbial bodies per 1 ml of the medium that was determined by the hardware method according to the McFarland standard. Muller-Hinton agar was used for the study. The measurements were performed on the 1st; 3rd; 5th; 7th and 14th days. The results of the microbiological study by the agar diffusion method were estimated according to the degree of the growth inhibition in the area of application of the researched composition. The following criteria were used therefore: absence of inhibition zones of microorganisms around the hole and inhibition of up to 10 mm pointed at absence of the microorganism sensitivity to the studied drug; the inhibition zone of the diameter of 10-15 mm pointed at low sensitivity of the culture to the studied concentration of the antibacterial substance; inhibition zone of the diameter of 15-25 mm was considered as the index of the microorganism sensitivity to the studied drug; inhibition zone of more than 25 mm proved high sensitivity of the microorganism to the studied drugs.

The influence of the materials on the ability of microorganisms to form a biofilm was studied according to the method of Yu. M. Romanova et al. [11]. For this purpose the clinical strains of *Staphylococcus aureus* 8 and *Pseudomonas aeruginosa* 24 were used. Their discharge and identification was determined through the study of their cultural, biochemical and morphological characteristics in accordance with normative documents. The obtained results of the optical density study were taken as the criteria of the ability of the investigated microorganism strains to form biofilms: $< 0{,}120$ c.u. – no ability to form biofilms, 0,120–0,240 c.u. – moderate ability, $> 0{,}240$ c.u. – high ability.

III. RESULTS

The results of the microbiological study showed that 1 day after beginning of the experiment the composite materials with immobilized silver based on the PU matrix (Sample 1), IPNs with PU/PHEMA=83/17 matrix (Sample 2) and polymer matrix without the filler of IPNs PU/PHEMA=83/17 (Control 3) manifested antimicrobial properties.

On the 3rd day of observation Sample 1 preserved the antimicrobial properties with respect to *Staphylococcus aureus*, *Pseudomonas aeruginosa*, *Candida albicans* and Sample 2 with respect to *Staphylococcus aureus*.

On the 5th day the diameter of the growth inhibition of all the studied samples did not exceed 10 mm that proved the absence of the microorganism sensitivity to them.

From the very beginning the microorganisms showed weak sensitivity to the samples Control 1 and 2 or were insensitive at all.

As a result of this study on influence of the investigated materials on the ability of the microorganisms to form a biofilm it was established that after 24 hours the strain of *Staphylococcus aureus* 8 in the presence of Samples 1 and 2 manifested a significantly weaker biofilm-formating ability as compared to the influence of the control samples. After 48 hours the ability to inhibit formation of a biofilm remained only in Sample 1 (composite based on the PU matrix, filled with nanosilica modified with a silver compound).

After 24 and 48 hours with respect to *Pseudomonas aeruginosa* 24 there was no significant difference between the ability of the studied materials to influence the ability of this microorganism to form a biofilm.

IV. Conclusion

The results of the microbiological studies prove that polymer materials based on the PU and PHEMA with added nanosilica modified by silver compounds (nanocomposites) are equal to the well-known analogues in antimicrobial properties and according to some criteria even surpass them.

It is reasonable to continue the study of the developed materials, including *in vivo* experiments on animals.

V. Funding

The work has been performed within the framework of the State Target Scientific and Technical Program "Nanotechnologies and Nanomaterials" (Project No. 6.22.7.21).

Conflict of Interest

Every paper must contain a declaration of conflicts of interest. If there are no such conflicts write "The authors declare that they have no conflict of interest".

References

1. V V. Dmitrieva, I.N. Petukhov, Postoperative infectious complications: diagnosis, prevention, treatment. Moscow: Prakticheskaya medicina, 2013.
2. R.P. Wenzel, G.M.L. Bearman, T.F. Brewer, J.-P. Butzler, Gide to infection control in the hospital, 4th edn., May, 2010.
3. Guideline for Prevention of Surgical Site Infection, 1999. Infect. Control. Hosp. Epidemiol, 1999, pp. 247-280.
4. E.V. Rymsha, L.K. Sorokoumova, "Characteristics of effective use of urethral catheters with antimicrobial coating," Biomedical and Biosocial Anthropology, No. 22, 2014, pp. 160-163.
5. V. M. Moroz, V.P. Kovalchuk, V.M. Kondratyuk, T.I. Antonets, "Susceptibility to antibiotics and antiseptics of microorganisms which colonize venous and urinary catheters," Biomedical and Biosocial Anthropology, No. 11, 2008, pp. 1-4.
6. A.A. Putilin, V.V. Melnikov, "Application xenogeneic saturated drug substances, in the treatment of purulent wounds," Vestnik Khirurgii, vol. 140, 1988, pp. 77-80.
7. J. Folkman, D.M. Long, "The use of silicone rubber as a carrier for prolonged drug therapy," Journal of Surgical Research, vol. 4, 1964, pp. 139-142.
8. T.E. Lipatova, L.N. Chuprina, V.A. Tsirkel, A.V. Rudenko, "Biocompatible polyurethanes containing in its composition antibiotics and proteolytic enzymes," Clinical Surgery, No. 1, 1985, pp. 85-88.
9. Methodical instructions for the use of "Lysoformin 3000" for disinfection, presterilization cleaning and sterilization, 2010.
10. Yu.L. Wolyanskyi, I.S. Gritsenko, V.P. Shyrokobokov, Study of specific activity of antimicrobial drugs: Method. recommendations. Kyiv, 2004.
11. Yu.M. Romanova, N.V. Alekseeva, T.A. Smirnova [et al.], "Biofilm formation by different strains of Salmonella typhimurium in artificial systems," Zh. Microbiol. (Moscow), No. 4, 2006, pp. 42-45.

Author: Mykola Zheliba.
Institute: Vinnytsia National Pirogov Memorial Medical University.
Street: Pirogov str., 56.
City: Vinnytsia.
Country: Ukraine.
Email: ro_man@mail.ru.

Antibacterial Properties of the Nanoparticles with the Zinc Sulfide Quantum Dots

A.M. Mieshkov, L.I. Grebenik, T.V. Ivahnuk, and L.F. Sukhodub

Sumy State University, Ministry of Education and Science of Ukraine Sumy,Ukraine

Abstract— **Synthesis of the nanoparticles based on ZnS quantum dots, which were incorporated into the natural alginate (Alg) polymer, has been conducted. The sample synthesis was using the method of chemical base deposition from an aqueous solution. Analysis of the morphological characteristics and chemical composition of the obtained ZnS-Alg nanocomposite has confirmed the creation of nanoparticles with the specified physicochemical properties. The ZnS-Alg composite sample has showed the high antibacterial activity towards *Peptostreptococcusanaerobius*, *Streptococcus pyogenes*, *Bacteroidesfragilis*, *Escherichia coli* and *Klebsiella pneumonia*.**

Keywords— **quantum dot, zinc sulfide, alginate, antibacterial, biopolymer, chamical base deposition, transmission electron microscopy.**

I. INTRODUCTION

Polymeric nanocomposite biomaterials attract the attention of researchers thanks to the possibility of their application in medicine, cosmetology, food industry, mechanical, optical and electronic equipment.

A growing number of studies in the creation of the biologically active composites in recent years is explained by the necessity to solve the actual issues caused by the development of civilization. In particular, the fact that bacteria are becoming resistant to antibiotic is an acute biological and medical problem nowadays. Modern nanotechnologies can help to solve this problem by creating antibacterial nanoparticles on the basis of metal compounds with well-known bactericidal properties (gold, silver, copper, zinc and other). The use of biopolymers as a matrix for the synthesis of metal nanoparticles allows to increase the bactericidal action and reduce the toxicity of nanomaterials towards the cells in the human body.

Such biopolymers as chitosan, collagen, sodium alginate are good matrixes for the synthesis of the nanocomposite biomaterials due to such their advantages over the synthetic polymer as availability, low cost, harmlessness to the environment, etc. [1-4]The interest to the synthesis of the nanomaterial composites, created on the basis of dispersion of metals and semiconductors nanoparticles which are incorporated into the polymer matrix, is growing. This modification of a biopolymer matrix allows to improve the basic properties of the polymer such as mechanical strength, stability and leads to the increase of the surface area.

The nanoparticles synthesized by the use of embedded quantum dots within porous materials are of great interest nowadays.

Numerous studies show that the creation of semiconducting sulphide nanomaterials inter alia on the basis of zinc sulfide (ZnS) considerably expands the possibilities of the application of quantum dots in bionanotechnology and biomolecularengineering.

The popularity of carbohydrates as matrixes for encapsulating quantum dots nowadays is explained by the possibility of their biodegradation in the cells of the living organisms which allows to synthesize nanomaterials with the high biological activity. Mono-and polysaccharides which are viewed as capping agents are used for the stabilization of the particles structure[5]. Besides, it has been found that the carbohydrate biopolymers, which constitute the nano-sized particles, avert their precipitation and agglomeration [6].

It is known that the biopolymer matrix chosen for the synthesis of biologically active spherical nanoparticles such as alginate gel, is a good carbohydratemacroporous encapsulating material [7].

Methods used for the synthesis of nano-sized structures on the basis of ZnS are, as a rule, technologically sophisticated and expensive. The synthesis of zinc sulphide nanoparticles by the use of magnetron sputtering, pulsed laser deposition, electron beam evaporation, electrochemical deposition, laser ablation, zol-gel method and so on is described in the literature[8-10].

In recent years the method of chemical vapor deposition from the aqueous solution is used for the synthesis of the multicomponent semiconductor nanoparticles [11-16].This method is simple, effective and therefore very popular for making stable and biocompatible composites [17]. The basis of the synthesis is the process of joining of the organic chemical active substances with the surface of nanocrystals.

Chemical base deposition method is regarded as a very promising approach to the obtaining of nanocrystals with the crystallites sizes ranging from 20 to 100 nm.

The aim of our research is to create nanoparticles with the antibacterial properties on the basis of ZnS quantum dots encapsulated in sodium alginate gel by chemical vapor deposition from the aqueous solution. Nowadays studies in this field are extremely insufficient

V. Sontea and I. Tiginyanu (eds.), *3rd International Conference on Nanotechnologies and Biomedical Engineering*,
IFMBE Proceedings 55,
DOI: 10.1007/978-981-287-736-9_65

II. DATABASE AND METHODS

Quantum dots (QD) of zinc sulfide were obtained by the chemical vapor deposition from the aqueous solution [18-20]. The mixture of zinc nitrate and thiourea was used as a precursor in an amount 1:1. The ammonia solution was added to the mixture for raising pH level to 8. The temperature of the chemical reactor for the synthesis was maintained at 90° C. The 3%-solution of sodium alginate was added to the mixture for the formation of the composite of zinc sulfide with alginate (ZnS-Alg).

Study of the mineral composition of the obtained nanomaterial was conducted by X-ray fluorescence analysis (XRF) using XRF-spectrometer «ElvaX Light SDD» (Ukraine). The samples for the elemental analysis of the ZnS-Alg composite were prepared by the standard method. The sample amount, sufficient to completely fill the cuvette bottom, was used for XRF.

The spectra were recorded in "light" and "hard" modes. Spectrum accumulation time was 120 sec. Spectrum analysis was carried out using the method which depends on fundamental principles.

Morphological studies of the ZnS-Alg composite structure were carried out by transmission electron microscopy (TEM). Samples for TEM were prepared by depositing isopropyl alcohol solution of ZnS-Alg on a copper grid covered with a thin carbon film with the subsequent drying in air at room temperature. TEM analyses were performed using "PEM-125k" (Ukraine).

The antibacterial properties of ZnS-Alg nanoparticles were studied using pure cultures of microorganisms which play an important role in the development of inflammatory wound complications. Such strains of the microorganisms as *Staphylococcus aureus*, *Staphylococcus epidermidis*, *Peptostreptococcusanaerobius*, *Streptococcus pyogenes*, *Bacteroidesfragilis*, *Escherichia coli*, *Klebsiellapneumonia* and *Pseudomonas aeruginosa* have been used in the experiment.

Previous studies have been conducted to confirm the high adhesion activity of all strains of test cultures, which were used in the next series of experiments to determine the antibacterial properties of nanoparticles in vitro. The antibacterial activity of the test sample was studied by applying slurry of test cultures (102 CFU/ml) to the surface of a dense nutrient medium with the addition of ZnS + Alg suspension. After incubation for 24 hours at 37 ± 0,1°C, the colonies were counted (CFU/ml), the number of which corresponds to the number of viable cells of test cultures. To control the experiment, the test microorganism strains were deposited in the same amount on the culture media supplemented with a suspension of ZnS and Alg. The experiment was performed ten times for each test culture.

III. RESULTS

TEM-analysis of the obtained ZnS-Alg nanoparticles has shown that the conditions created in the chemical synthesis reactor appeared to be optimal for the formation of the particles of the proper spherical form and similar size.Information about the phase structure and the average size of nanocrystals was obtained by electron diffraction TEM. Results of TEM-analysis illustrate the presence of homogeneous crystal structure of the composite material despite the presence of the organic polymer.The pictures at different magnification, which allow to determine the ZnS-Alg spheres' size are shown in the Figure 1. According to our estimates, it is 50-80 nm. Further nanocomposite's TEM-analysis at a greater magnification let us get convinced that the nanoparticles have the nucleus formed by the ZnS quantum dots and friable polymeric sodium alginate shell.

It is known that the fluorescence peak of nanodots is thin and symmetrical, which allows to distinguish its signal reliably. Spectrogramms of the ZnS-Alg nanocomposite allowed to prove the presence of zinc sulfide salt in the studied sample. Sodium alginate is responsible for the presence of Na ions fluorescence peak in the "light" spectrum. Rh, Ru, Ca, Fe densities in this case are caused by the noise of the device's X-ray tube.

Difference of the Zn fluorescence densities in "light" and "hard" spectrum should be considered as a specific feature of the XRF-analyzer's work in a "light" recording mode.

The formation of the ZnS nanoparticles covered with the natural biopolymer (Alg) allowed us to assume the changing of the zinc salts biological properties, in particular, towards microorganisms. High degree of bactericidal activity of the test sample was detected during the studying of the antibacterial activity of ZnS-Alg nanoparticles towards the tested strains of microorganisms. It was found that the ZnS-Alg sample had bactericidal activity to gram-positive and gram-negative test cultures such as *Peptostreptococcusanaerobius, Streptococcus pyogenes, Bacteroidesfragilis, Escherichia coli* and *Klebsiella pneumonia* which were used in the experiment. The growth of the bacterial colonies after being cultured for 24 hours on a nutrientmedium containing ZnS-Alg nanoparticles was not detected.

To estimate the degree of the antibacterial activity of the sample, a series of control experiments were conducted on the nutrient media which contained individual components of the composite. It has been shown that the number of colonies on the nutrient media varied after adding ZnS, but a significant difference in the quantitative content indicators in the initial suspension of test cultures and in the test cultures after incubation was not detected ($p > 0.05$). In the series of control experiments in nutrient media supplemented with Alg, the antibacterial activity of this biopolymer towards the test cultures was identified. However, it should be noted that the bactericidal activity of Alg was significantly lower than the activity of the sample containing ZnS-Alg nanoparticles(p ℜ 0.05).

The bactericidal activity is explained, as a rule, by the influence of metal nanoparticles on the permeability of the bacterial cell wall and by the activation of metabolic processes which cause the death of the microorganism. [21] It has been found that the bacterial wall thickness can limit efficiency of nanomaterials. In particular, gram-positive bacteria, which have thicker and more chemically complicated cell wall, are distinguished by the enhanced resistance to the action of antibacterial nanoparticles. [22].

Results of our study show that the cell wall thickness of the gram-positive bacteria cannot be the limiting factor of the bactericidal action of the obtained ZnS-Alg nanoparticles. The high antimicrobial activity of ZnS-Alg nanoparticles including with regard to gram-positive bacteria can be explained not only by the structural and physicochemical peculiarities of nanoparticles but also by the increasing effect of biomatrix. It has been shown in our

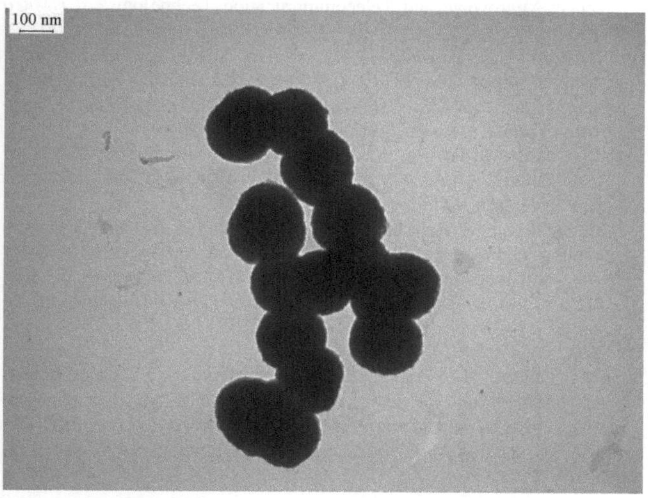

Fig. 1 TEM-images of the nanocomposite of the zinc sulfide with alginate at different magnifications.

study that Alg has also antibacterial action (though insignificant) towards the tested strains. From our point of view, incorporating of quantum dots of metal salt (Zn) known for its antimicrobial properties into the structure of the porous bactericidal biopolymer (Alg), allowed to create nanoparticles with high antibacterial activity.

The confirmed antibacterial activity of the nanoparticles with the zinc sulfide quantum dots towards some strains of microorganism will allow to direct our further activity to the investigation of the possible practical application of the indicated ZnS-Alg properties in medicine and/or in cosmetology. Moreover, we are considering the perspective of the application of the synthesized nanoparticles in creating biofilms, which are used for the production of the packing material, capable of protecting food products from contamination.

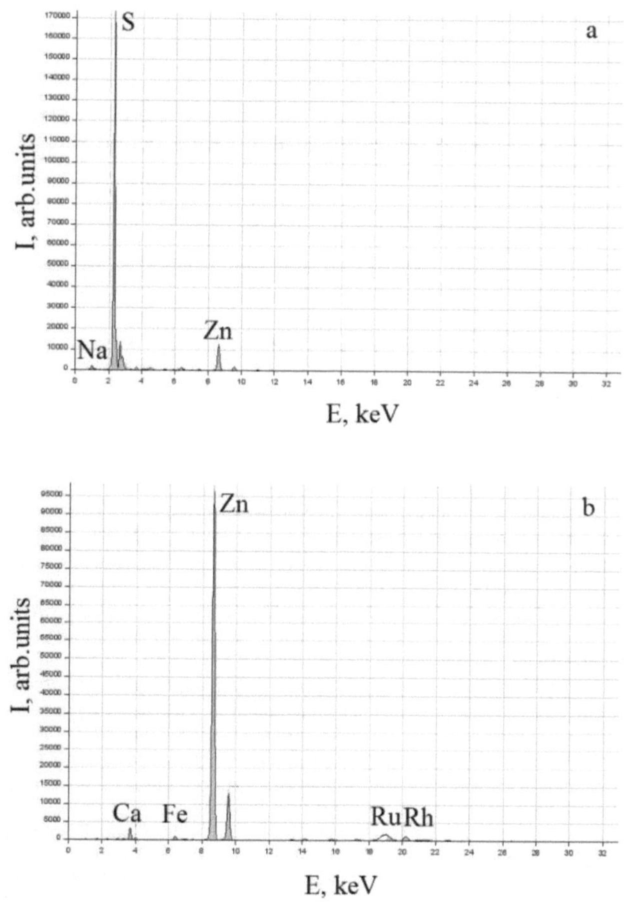

Fig. 2 XRF spectra of the nanocomposite material of zinc sulfide with alginate at a "light" (a) and "hard" (b) recording mode

IV. Conclusions

The nanocomposite material on the basis of the zinc sulfide quantum dots incorporated into the alginate was obtained by the chemical vapor deposition method from the aquatic solution. It was determined that the nanoparticles growth occurs by the means of the nucleus formation from the zinc sulfide molecules with the subsequent formation of the external polymeric shell from the sodium alginate.

The investigation of the element composition and morphological characteristics of the synthesized nanocomposite has confirmed the structure homogeneity of the obtained spherical nanoparticles based on the zinc sulfide salt.

The study results of the antibacterial properties of ZnS-Alg nanoparticles have convincingly proved the high bactericidal activity of the sample towards the gram-positiveand gram-negative microorganisms *Peptostreptococcusanaerobius*, *Streptococcus pyogenes*, *Bacteroidesfragilis*, *Escherichia coli* and *Klebsiella pneumonia.*

ACKNOWLEDGMENT

This work was supported by EU TEMPUS project "Biomedical Engineering Education Tempus Initiative in ENA".

CONFLICT OF INTEREST

The authors declare that they have no conflict of interest.

REFERENCES

1. H. Mittal, E. Fosso-Kankeu, S. Mishra, A.K. Mishra,"Synthesis, characterization, and swelling behavior evaluation of hydrogels based on gum ghatti and acrylamide for selective absorption of saline from different petroleum fraction–saline emulsions",*Int. J. Biol. Macromol. 62,370(2012).*
2. S. Ghorai, A. Sinhamahpatra, A. Sarkar, A.B. Panda, S. Pal,"Novel biodegradable nanocomposite based on XG-g-PAM/SiO 2: application of an efficient adsorbent for Pb 2+ ions from aqueous solution",*Bioresour. Technol.119, 181(2012).*
3. S.S. Banerjee, D.H. Chen, J. Hazard, "Fast removal of copper ions by gum arabic modified magnetic nanoadsorbent.",*Mater.147, 792(2007).*
4. Y.C. Chang, D.H. Chen, J. Colloid. "Preparation and adsorption properties of monodisperse chitosan-bound Fe_3O_4 magnetic nanoparticles for removal of Cu(II) ions.",*Interface Sci.283, 446(2005).*
5. P. Bansal., N. Jaggi. and S.K. Rohilla., "Green" Synthesis of CdS nanoparticles and effect of capping agent concentration on crystallite size, Res.J.Chem.Sci., 2, 69-71(2012)
6. YanpingXie, Yiping He, Peter L. Irwin, Tony Jin,3 and Xianming Shi. "Antibacterial Activity and Mechanism of Action of Zinc Oxide Nanoparticles against Campylobacter jejuni_APPLIED AND ENVIRONMENTAL MICROBIOLOGY", *Apr. 2011, p. 2325–233*
7. S.Suryajaya, A.V.Nabok, A.K. Hassan, A. Tsargorodskaya, "Detection of low molecular weight toxins using optical phase detection techniques, Sensors and Actuators B",*Jurnalfisikadanaplikasinya 7(1) 110107-1 (2011).*
8. G. Ferblantier., F .Mailly.,R. Al. Asmar.et.al. "Deposition of zinc oxide thin films for application in bulk acoustic wave resonator",*Sens. Actuat.A.(2005). Vol. 122. P. 184–188.*
9. E. Fortunato.,P. Barquinha.,A. Pimentel.et.al."Recent advances inZnO transparent thin film transistors",*Thin SolidFilms.(2005). Vol. 487. P. 205–211*
10. N. Mukherjeea., S. F. Ahmedb., K. K. Chattopadhyayc., Mon-dal A. "Role of solute and solvent on the deposition of ZnO thin films",*Electrochim. Acta.(2009). Vol. 54. P. 4015–4024*
11. [W. Vogel, P.H. Borse, N. Deshmukh, S.K. Kulkarni, Langmuir 16, "Structure and stability of monodisperse 1.4-nm ZnS particles stabilized by mercaptoethanol",*2032(2000).*
12. Raymond C. Rowe, Paul J. Sheskey, Marian E. Quinn, ed. (2009), "Adipic Acid", *Handbook of Pharmaceutical Excipients (Rowe, Handbook of Pharmaceutical Excipients) (Sixth ed.), Pharmaceutical Pr, pp. 11–12, ISBN 0-85369-792-2*
13. H.J. Lee, S.I. Lee, "Deposition and optical properties of nanocrystallineZnS thin films by a chemical method", *Current Applied Physics. 7, 193(2007).*
14. F. Long, W.M. Wang, Z. Cui, L.Z. Fan, Z. Zou, T. Jia "An improved method for chemical bath deposition of ZnS thin films", *Chemical Physics Letters462, 84087(2008).*
15. A.U. Ubale , D.K. Kulkarni, "Preparationandstudyofthicknessdependentelectricalcharacteristicsofzincsulfidethinfilms", *Bull. Mater.Sci.28, 43(2005).*
16. Q. Liua, M. Guobing, A. Jianping, "Structural and Optical Properties of ZnS Thin Films Deposited by Spray Pyrolysis Technique", *Applied Surface Science, 254, 5711 (2008).*
17. T. Ben Nasr, N. Kamoun, C. Guasch, "Physical properties of ZnS thin films prepared by chemical bath deposition", *Materials Chemistry and Physics96, 84 (2006).*
18. T.O. Berestok, D.I. Kurbatov, A.S.Opanasyuk, N.M. Opanasyuk, V.M.Kuznetsov, "Proceeding: 22st International Crimean Conference: Microwave and Telecommunication Technology", *(CriMiCo 2012).*
19. D. Barreca, E. Tondello, D. Lydon, T.R. Spalding and M. Fabrizio "Single-Source Chemical Vapor Deposition of Zinc Sulfide-Based Thin Films from Zinc bis(O-ethylxanthate)", *Chemical Vapor Deposition Volume 9, Issue 2, pages 93–98, March, 2003.*
20. P. Charoensirithavorn, S. Yoshikawa, "Controllable Synthesis of ZnONanorod Arrays via Simple Solution-Based Method", *MRS Online Proceedings Library / Volume 957 / 2006,*
21. C. Malarkodi, S. Rajeshkumar, K. Paulkumar, M. Vanaja, G. Gnanajobitha, and G. Annadurai,"Biosynthesis and Antimicrobial Activity of Semiconductor Nanoparticles against Oral Pathogens Hindawi",*Publishing Corporation Bioinorganic Chemistry and Applications Volume 2014, Article ID 347167, 10 pages http://dx.doi.org/10.1155/2014/34716*
22. M. Vanaja, G. Gnanajobitha, K. Paulkumar, S. Rajeshkumar, C.Malarkodi, and G. Annadurai, "Phytosynthesis of silver nanoparticles by Cissusquadrangularis: influence of physicochemical factors," *Journal of Nanostructure in Chemistry, vol. 3, no. 17,pp. 3–17, 2013*

Author: Mieshkov Artem
Institute: Sumy State University
Street: 2, Rymskogo-Korsakova st.
City: Sumy
Country: Ukraine
Email: guhosotsu@gmail.com

Influence of Dispersed Solutions of Copper, Silver, Bismuth and Zinc Oxide Nanoparticles on Growth and Catalase Activity of *Penicillium Funiculosum*

T. Sirbu[1], S.N.Maslobrod[2], Yu.A. Mirgorod[3], V.G. Borodina[3], N.A. Borsch[3], and L.S. Ageeva[3]

[1] Institute of Microbiology and Biotechnology of ASM, Chisinau, Moldova
[2] Institute of Genetics, Physiology and Plant Protection of ASM, Chisinau, Moldova
[3] Southwest State University, Kursk, Russia

Abstract— **It was shown that water dispersed solutions of silver at concentrations of 10-9 mol/l, 10-7 mol/l, 10-5 mol/l and bismuth at concentrations of 10-7 mol/l, 10-5 mol/l, 10-1 mol/l stimulate growth of biomass and catalase activity at *Penicillium funiculosum CNMN FD 11* strain. The maximum effect for the two parameters was obtained with the dispersed solution of silver at a concentration of 10-5 mol/l, bismuth at a concentration of 10-7 mg/l and zinc oxide at a concentration of 10-7 mg/l. The use of water dispersed solutions of copper at concentrations of 10-7 mg/l, 10-5 mg/l, 10-1 mg/l showed the inhibiting effect or the absence of physiological effects on growth and catalase biosynthetic processes of *Penicillium funiculosum.* The inhibiting effect was produced by the solution of copper at a concentration of 10-7 mg/l. The method can be used for optimising conditions for the submerse cultivation of non-pathogenic microorganisms.**

Keywords— **catalase, nanoparticles, bismuth, silver, copper.**

I. INTRODUCTION

Water dispersed solutions of metal nanoparticles exert bactericidal, fungicidal and virucidal effect on living organisms. The development of the nanotechnology allowed to create substances and materials of the new generation that are widely used in microbiology [1,2], medicine [3,4] and plant growing [5,6]. According to our data, water dispersed solutions at different concentration of copper and silver nano-particles substantially influence germination of seeds of grain and vegetable crops. Treatment of seeds with the nano-solutions boosts their resistance to fungal pathogens *Helminthosporium avenae* and *Alternaria alternata* which positively affects efficiency of plants in field conditions [7]. Multidirectional activity of these nanoparticles compromises induction of microbial defensive mechanisms, affects energy metabolism and genetic material of non-pathogenic microorganisms [8]. The effect of the nanostructured silver on wheat seeds was similar to the effect of the exometabolites of *Streptomyces* bacteria [9].

Nowadays, there is a growing concern about increasing resistance of some microorganisms to disinfecting agents. Therefore, research efforts focus on developing novel disinfectants of chemically stable properties and high biocidal effectiveness [10]. Pharmaceutical preparations containing nano-structured copper exert marked therapeutic effect on wound healing [11]. Preparations incorporating nanosilver demonstrate strong inherent anti-bacterial activity towards a wide range of harmful microorganisms, i.e. bacteria, fungi and viruses.

In the case of pathogenic microorganisms *E.coli, P.vulgaris, S.aureus, K.pneumonia, P.aeruginosa* treated with nanostructured silver, a direct relationship between survival of cells and concentration of water-dispersed solutions was revealed [1]. Thus, the data show that nano-solutions of silver with 1:80 and 1:40 concentrations have a bactericidal effect on germinating spores of *E.coli, P.vulgaris* and *S.aureus*, the solution with a 1:10 concentration has a bactericidal effect on *E.coli* and *P.vulgaris* and a bacteriostatic effect on *P.vulgaris*, without influencing the behavior of the other of microorganisms.

Water dispersed solutions of copper at different concentration and phase composition have an antibacterial effect and are active against Gram-positive and Gram-negative bacteria [2]. The revealed properties of nano-structured copper made accessible the use of copper nano-particles as antimicrobial agents at creation of antiseptic materials in medicine, industry and agriculture.

Nanotechnology proves to be a field of growing scientific interest as the properties of engineered nano-materials can be utilized in a broad spectrum of applications. An important and promising direction in plant growing could therefore be the use of nano-preparations based on silver, copper, bismuth and other metals as they aim to enhance resistance of biological objects to the adverse external factors, favoring suppression of pathogenic factors and adaptation of biological objects to environmental changes. Preparations incorporating nanosilver represent one of the prospective forms for use in livestock breeding in particular for the treatment of bacterial infections in the gastrointestinal tract providing a possibility to remove the need for antibiotics [12].

No less effective and may be even more prospective is the usage of nanobismuth complexes with amino acids. The preparation containing bismuth oxide was shown to exhibit

V. Sontea and I. Tiginyanu (eds.), *3rd International Conference on Nanotechnologies and Biomedical Engineering*,
IFMBE Proceedings 55,
DOI: 10.1007/978-981-287-736-9_66

immunomodulating properties in treatment of oncological patients. A distinguishing feature of the preparation is that it is less toxic and more efficient than other anticancer pharma- ceuticals [13].

Nanotechnology allowed to create substances and materials which demonstrate strong inherent anti-bacterial activity towards a wide range of harmful microorganisms. In addition, there is a possibility for the prospective use of nanosubstances for stimulation of the vital activity of non-pathogenic microorganisms. Non pathogenic soil micro-organisms, actinomycetes genus *Streptomyces* and fungal genus *Penicillium* play a special role for plant growing as their metabolic products stimulate plant development [14] and inhibit the growth of pathogenic fungi and bacteria [15].

II. DATABASE AND METHODS

The object of the investigation was *Penicillium funiculosum CNMN FD - 11* as a producer of intracellular catalase, which was taken from the National Collection of Nonpathogenic Microorganisms (NCNM). The experiment was performed using nanoparticles obtained from chemical method at Southwest State University. The particles had spherical forms, their medium sizes were the following: nano silver – 35 nm [17], nano copper – 2-3 nm [18], nano bismuth – 30-40 nm [19], nano zinc oxide – 20-30 nm [19]. The nanoparticles were introduced into the nutrient medium immediately before inoculation of the fungus. The concentration of the metal particles was as follows: nano silver – 10^{-9} mol/l, 10^{-7} mol/l, 10^{-5} mol/l; nano copper, nano bismuth, nano zinc oxide – 10^{-7} mol/l, 10^{-5} mol/l, 10^{-1} mol/l. Selection of these concentrations was justified by the pronounced stimulating and biocidal effect on vegetable and cereal seeds [7].

The microscopic fungi were grown in Erlenmeyer flask of volume 500 ml in optimal nutrient medium, on a shaker (180-200 rpm) by the temperature 28-30 ^{0} C during 6 days and nights [20]. We used as culture inoculums aqueous suspension of spores cultivated for 14 days on the wort-agar medium, the number of fungal cells in a sample was 5 x 10^{6}. Catalase activity was determined in the culture fluid by titration [21]. Culture fluid was separated from the biomass by filtration through paper filters; the purified product was washed with distilled water and dried in a drying oven at the temperature of 105 °C. The arithmetic mean of the data was calculated based on the set of results from three experiments.

III. RESULTS

A. Water Dispersed Solutions of Silver Nanoparticles

As it is seen from the figure 1, dispersed solutions Ag in different concentrations have different effect on the growth of the fungus. In all embodiments of the experiment, the amount of biomass produced in response to application of nanosilver exceeded by 16.0 % the similar indicator of the control as well as the activity of the intracellular catalase was higher than in the control by 4-12%.

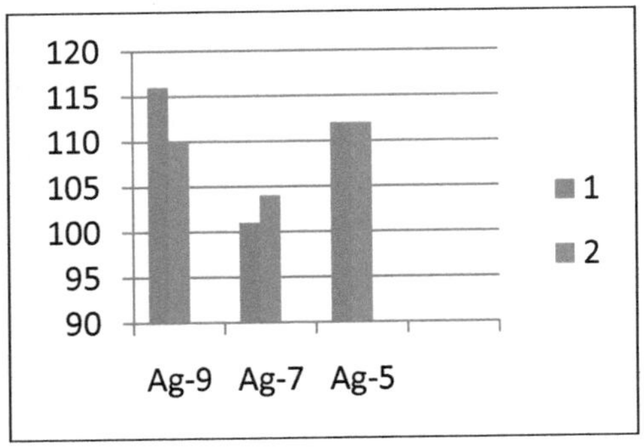

Fig. 1 Amount of biomass and catalase activity of *P.funiculosum CNMN FD 11* strain, expressed as a percentage of the control, in response to application of water dispersed nanosilver solutions in the different concentrations.

1– amount of biomass, 2– catalase activity; Ag-9, Ag-7, Ag-5 – concentrations of water dispersed nanosilver solutions, equal accordingly to: 10^{-9} mol/l, 10^{-7} mol/l, 10^{-5} mol/l

Maximum increase in amount of biomass (by 16.0 %) was observed at a concentra- tion of 10^{-9} mol/l, catalase activity (by 12.0 %) at a concentra- tion of 10^{-5} mol/l (see fig. 1). In summary, the most favorable concentration for the both parameters was 10^{-5} mol/l; in response to application of nanosilver solution at this concentration, the amount of biomass and catalase activity by 12 % exceeded those of control.

B. Water Dispersed Solutions of Copper Nanoparticles

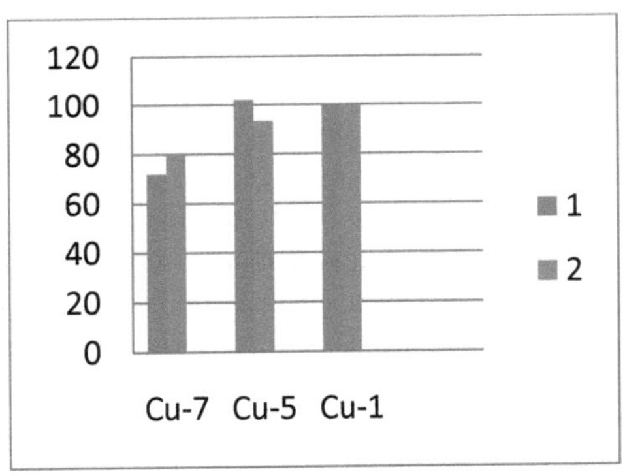

Fig. 2 Amount of biomass and catalase activity of *P.funiculosum CNMN FD 11* strain, expressed as a percentage of the control, in response to application of water dispersed nanocopper solutions in the different concentrations. 1– amount of biomass, 2– catalase activity; Cu-7, Cu-5, Cu-1 – concentrations of water dispersed nanocopper solutions, equal accordingly to: 10^{-7} mol/l, 10^{-5} mol/l, 10^{-1} mol/l

The addition of nanocopper solution at a concentration of 10^{-7} mol/l to the fungal culture medium resulted in a change of biomass growth and catalase activity, it was shown in loss of growth rate of the microfungal population by 28% and catalase synthesis by 20% (fig. 2). In the case of concentra-tions of 10^{-5} mol/l and 10^{-1} mol/l the inhibitory effect was not observed, the values of parameters were on the level of the control.

C. Water Dispersed Solutions of Bismuth Nanoparticles

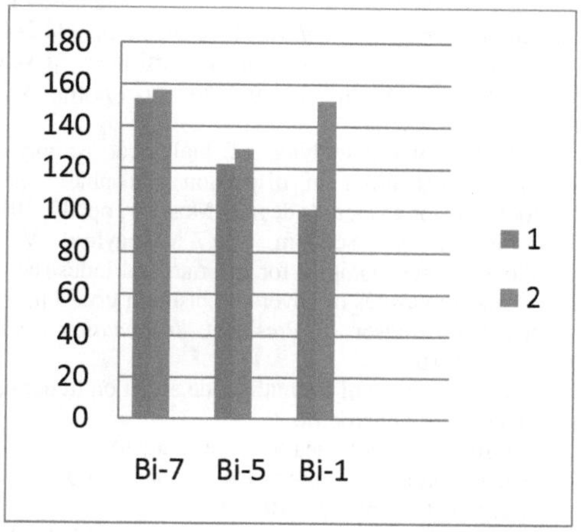

Fig. 3 Amount of biomass and catalase activity of *P.funiculosum CNMN FD 11* strain, expressed as a percentage of the control, in response to application of water dispersed nanobismuth solutions in the different concentrations. 1 – amount of biomass, 2 – catalase activity; Bi-7, Bi-5, Bi-1 – concentrations of water dispersed nanobismuth solutions, equal accordingly to: 10^{-7} mol/l, 10^{-5} mol/l, 10^{-1} mol/l

As indicated in figure 3, the nanobismuth factor produced stimulating effect on the *Penicillium funiculosum* strain, at all concentrations. In all embodiments of the experiment, the amount of biomass produced in response to application of nanobismuth exceeded by 22-53 % the similar indicator of the control as well as the activity of the intracellular catalase was higher than in the control by 29-57%. Maximum increase in amount of biomass (by 53 %) and catalase activity (by 57 %) was observed at substance concentration of 10^{-7} mol/l.

D. Water Dispersed Solutions of Zinc Oxide Nanoparticles

After treatment with zinc oxide, expressed stimulation that led to significant increase of biomass and catalase activity was registered. The best variant was the use of nanosolution of zinc oxide at a concentration of 10^{-7} mol/l: amount of biomass increased by 40 % and catalase activity by 24 %. A concentration of 10^{-5} produced 30% increase in the amount of biomass and 10% increase in catalase activity. Maximum growth rate of the fungal population (by 50%) was observed at application of water dispersed solutions of zinc oxide at a concentration of 10^{-1} mol/l (see fig. 4).

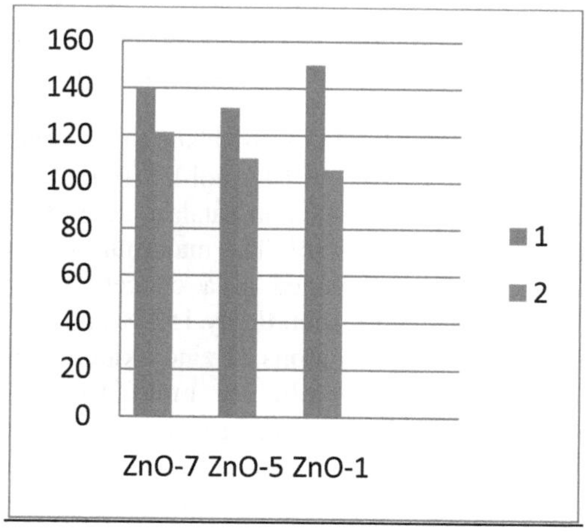

Fig. 4 Amount of biomass and catalase activity of *P.funiculosum CNMN FD 11* strain, expressed as a percentage of the control, in response to application of water dispersed nanosolutions of zinc oxide in the different concentrations 1 – amount of biomass, 2 – catalase activity; ZnO-7, ZnO-5, ZnO-1 – concentrations of water dispersed solutions of zinc oxide, equal accordingly to: 10^{-7} mol/l, 10^{-5} mol/l, 10^{-1} mol/l

Comparative experimental data about the influence of water dispersed solutions of nanoparticles of different substances on growth and catalase biosynthetic processes of *Penicillium funiculosum* strain show that some nanostructured substances (Ag, Bi, ZnO) have stimulatory effects on fungal development, and some (Cu) have inhibitory effects. In certain embodiments of the experiment the effect was achieved at lower water concentrations of nanoparticles and, therefore, the method is more economical and technologically feasible. The results obtained from the study showed the potential use of nanotechnology not only for prevention of harmful microorganisms growth but for stimulation metabolic activity of non-pathogenic fungi. In our studies, *Penicillium funiculosum* strain was selected as an active catalase producer among the microscopic fungi, yet we look with moderate optimism to a variety of the potential applications of nanotechnology for the development of next generation sequencing techniques in microbiology.

IV. CONCLUSION

1) Water dispersed solution of silver nanoparticles at a concentration of 10^{-5} mol/l causes increase in growth and catalase activity of *Penicillium funiculosum CNMN FD 11* strain by 12%, and at a concentration of 10^{-9} mol/l increases the biomass growth of the microscopic fungi by 11 %.

2) Water dispersed solution of copper nanoparticles at concentrations of 10^{-5} mol/l, 10^{-1} mol/l do not significantly influence the biomass growth and catalase activity of *Penicillium funiculosum CNMN FD 11* strain. A concentration of 10^{-7} mol/l is likely to cause a loss of growth rate of the microfungal population and catalase synthesis by 20-28%.

3) Water dispersed solution of bismuth nanoparticles at concentrations of 10^{-7} mol/l, 10^{-5} mol/l, 10^{-1} mol/l stimulate substantially the growth rate and catalase activity of fungus as compared with the control. The maximum effect for the two parameters was obtained at a concentration of 10^{-7} mol/l, and comprise correspondingly 153 % and 157 %.

4) Water dispersed solutions of zinc oxide nanoparticles at concentrations of 10^{-7} mol/l, 10^{-5} mol/l, 10^{-1} mol/l cause increase in reproduction and catalase activity of *Penicillium funiculosum CNMN FD 11* strain. A concentration of 10^{-7} mol/l causes the growth of biomass and catalase activity by 140% and 124 %, respectively. A concentration of 10^{-1} mol/l causes the growth of the parameters by 150% and 105%.

Conflict of Interest

The authors declare that they have no conflict of interest.

References

1. T.S. Dultseva, "Antibacterial activity of silver nanoparticles obtained in different ways" (in Russian), *Materials of the 69th international scientific conference*, Tomsk, pp. 297-298, 2010.
2. T. A. Shulgina, I. A. Norkin, D.M.Puchinyan, "Antibacterial influence of water dispersions of silver nanoparticles on gram-negative microorganisms (on the example of Escherichia Coli)" (in Russian), *Basic researches*, **7**, pp. 424-426, 2012.
3. A. I. Archakov, O. M. Ipatova, I. V. Medvedeva, Yu. D. Ivanova, A. I. Drozhzhin, "Nanobiotechnology and medicine" (in Russian), *Biomedical chemistry*, **6**, pp. 529-546, 2006.
4. M.Singh, S Singh, S Prasad, I.S. Gambhir, "Nanotechnology in medicine and Antibacterial Effect of Silver Nanoparticicles", *Digest Journal of Nanomaterials and Biostructures*, **3**, pp. 115-122, 2008.
5. V.A. Scriabin, "Nanostructures on the basis of bismuth and silver for storage system of grain and production of compound feeds" (in Russian), conf. Actual problems in the field of creation of innovative storage technologies of agricultural raw materials and foodstuff, Moscow, pp. 230-233, 2011.
6. A.P. Raykova, L.A. Panichkin, N. N. Raykova, "Studies of the effect of ultra-dispersed metal powders obtained in different ways on growth and development of plants" (in Russian), Materials of International scientific and practical conference: Nanotechnologies and information technologies of the XXI century, Moscow, pp. 108-111. 2006.
7. S.N. Maslobrod, Yu.A. Mirgorod, V.G. Borodina, N.A. Borsch, "Stimulation of Seed Viability by Means of Dispersed Solutions of Copper and Silver Nanoparticles", *Journal of Nano-Electronic Physics*, **5** (4), 04018 (3pp), 2013.
8. N. N. Glushchenko, O.A.Bogoslovskaya, I.P. Olkhovskaya, "Comparative toxicity of metal salts and metal nanoparticles and specific nature of their biological effect" (in Russian), Materials of International scientific and practical conference: Nanotechnologies and Information Technologies the XXI Century, Moscow, Russia, pp. 93-95, 2006.
9. S. N. Maslobrod, S. A. Burtseva, M. N. Birsa, G. A. Lupashku, "Influence of silver nanoparticles and exometabolites of streptomycetes on wheat seeds resistance to Helminthosporium avenae pathogenic fungus" (in Russian), *Scientific works, state agriculture university of Moldova*,Chisinau, 41, pp 338-340, 2014.
10. M. V. Sukovnitsky, "Mechanisms of resistance development to antibiotics at bacteria" (in Russian), *Biological products*, **2**, page 44, 2011.
11. A. A. Rakhmetova, "Studying of biological activity of copper nanoparticles that differ in dispersion and phase structure" (in Russian), thesis for a Doctor's degree, Moscow, pp. 30, 2011.
12. L. Machikhina, V. Scriabin, Yu. Mikhaylov, V. Boldyrev, "Innovative nanotechnologies for agrarian and industrial complex of Russia. Nanocomposites of silver and bismuth give a positive effect" (in Russian), Advisor of President. *Information and analytical edition*, (86), 2010.
13. M. Levasov, "Studying of bismuth oxide effect on tumor cells", 2012, http://mixail-lavasov.narod.ru.
14. S. A. Burtseva, "Biologically active agents of streptomycetes (biosynthesis, properties, prospects of application)", thesis for Doctor's degree, Chisinau, pp. 40, 2002.
15. S. N. Maslobrod, S. A. Burtseva, T.F.Sirbu, O. M. Postolaki, L. N. Namolovan, Yu.I. Gapii, "Assessment of the antagonism of micro-organisms of Moldovan soils to phyto-pathogens and influence of the microorganisms metabolites on seeds germination" (in Russian), international conference: Innovations transfer in agriculture activities in the context of climate change, Moldova, Chisinau, pp. 230-241, 2009.
16. O.S.Miroshnichenko, "Biogenesis, physiological role and properties of a catalase", *Biopolymers and cell*, 3(6), pp. 59-80, 1992.
17. Yu.A.Mirgorod, N.A. Borsch, V. G. Borodina, G.Yu.Yurkov, "Obtaining and characterization of cotton fabric modified by copper nanoparticles", Chemical industry, 89(6), pp. 310-316, 2012.
18. Yu. A. Mirgorod, V.G. Borodina, "Preparation and Bactericidal Properties of Silver Nanoparticles in Aqueus Tea Leaf Extract", Inorg. Mat., 49 (10), pp. 980-983, 2013.
19. Unpublished data.
20. T.F. Syrbu, "The searching of active catalase producers among the microscopic fungi", *jourlib Journal*, 2009
21. V. I. Bilay, "Bases of the general pathology", Kiev, high School, pp. 156-157, 1974.

Author: Tamara Sîrbu
Institute: Institute of Microbiology and Biotechnology of ASM
Street: 1, Academiei St.
City: Chisinau
Country: Moldova
Email: sirbutf@rambler.ru

Effect of Aqueous Dispersions with NPAg, NPCu, NPBi, and ZnNO, Millimeter-Wave Radiation, and Weak Magnetic Fields on the Germination of Triticale and Wheat Seeds under the Action of a Pathogenic Fungus and Low Temperatures

S.N. Maslobrod[1], Yy.A. Mirgorod[2], G.A. Lupashku[1], N.A. Borsch[2], V.G. Borodina[2], L.S. Ageeva[2], A. Shibaev[3], and I. Groisman[3]

[1] Institute of Genetics, Physiology and Plant Protection, ASM, Chisinau, Moldova
[2] Southwest State University, Kursk, Russia
[3] Institute of Electronic Engineering and Nanotechnologies, ASM, Chisinau, Moldova

Abstract— **It has been found that the exposure of triticale and wheat seeds to aqueous dispersions containing low concentrations of Ag, Cu, Bi, and ZnO nanoparticles significantly stimulates seed germination (up to 30%). Aqueous dispersions with low concentrations of nanoparticles exhibit a fungicidal effect on the seeds in seed treatment prior to the exposure of the seeds to pathogenic fungus *Helminthosporium avenae.* The most informative parameter was seed germination capacity. Potassium permanganate, which is a conventional factor of seed disinfection, has proved to be less effective. In comparative experiments on studying the effects of the nanofactor and weak electromagnetic fields on the seeds, it has been generally found that there are some similarities between the factors in the responses of both normal seeds and the seeds subsequently exposed to the action of the pathogenic fungus and low temperatures. In some versions, the effects of the nanofactor and then of the magnetic field were more pronounced.**

Keywords— **nanoparticles, aqueous dispersions, millimeter-wave radiation, pathogenic fungus, seeds.**

I. INTRODUCTION

In recent decades, crop growers have shown keen interest in the products of nanotechnology, i.e., metal nanopowders that have a pronounced bactericidal effect on plant objects [1, 2]. In addition, metal nanopowders function as long-acting microelements of the mineral nutrition of plants, which increase the adaptive capacity of the plant organism (productivity and environmental sustainability) [1, 2].

Metal nanopowders are characterized by an enormous surface area (on the order of a few hundred square meters per gram); therefore, they can be effectively used in microdoses [1–3]. Thus, it is possible to simultaneously provide the ecological safety of both the environment and the bioproduct. The toxicity of metal nanopowders is 10–40 times lower than the toxicity of salts of the same metals [2, 3].

According to our data [4], treatment of seeds of cereals and vegetables with aqueous dispersions containing low concentrations of silver and copper nanoparticles (10^{-7} mol/L NPAg, 10^{-7} mg/L NPCu) has a stimulation and fungicidal effect on the seeds and thus has a positive impact on the productivity of plants in the field environment. In this work, in addition to copper and silver nanoparticles, we have studied nanoparticles of bismuth (NPBi) and zinc oxide (NPZnO). These nanoparticles are used as stimulation, bactericidal, and fungicidal factors in medicine and plant industry [3, 5]. In comparative experiments, the seeds were additionally exposed to weak electromagnetic fields, i.e., millimeter-wave radiation (MWR) with a wavelength of 5.6 mm and a low-intensity magnetic field (MF). MWR is an effective factor in improving the viability of living objects [6], as was shown in [7]. MFs have a similar function [8]. This report presents data on the effect of the nanofactor and the above electromagnetic fields on the resistance of seeds to pathogenic fungus *Helminthosporium avenae* and a low temperature of +4°C.

II. DATABASE AND METHODS

The objects of research were seeds of winter triticale (Ingen 93) and winter wheat (Select and BT 43-02 and a Select x BT 43-02 hybrid). The seeds were subjected to the action of the following factors: 1) aqueous dispersions containing different concentrations of NPAg, NPCu, NPBi, and NPZnO with exposure times of 1h (for the NPCu aqueous dispersion) and 24h (for the NPAg, NPCu, NPBi, and NPZnO aqueous dispersions); 2) MWR with a wavelength of 5.6 mm, a power density of 7 mW/cm^2, and an exposure time of 8 min (according to [7], this mode has a stimulation effect on cereals); 3) an alternating MF with a magnetic induction of 1–10 Hz, an intensity of 40–50 mT, and an exposure time of 30 min; 4) a 1% solution of potassium permanganate $KMnO_4$ with an exposure time of 1h; 5) a culture filtrate of fungus *Helminthosporium avenae* with an exposure time of 8h; and 6) a temperature of +4°C

V. Sontea and I. Tiginyanu (eds.), *3rd International Conference on Nanotechnologies and Biomedical Engineering*,
IFMBE Proceedings 55,
DOI: 10.1007/978-981-287-736-9_67

for 12 h. After exposure to these factors, the seeds were germinated in Petri dishes on distilled water in a factorostatic room at a temperature of 22°C. Fifty seeds were placed in each dish; ten dishes were used in each version. The recorded parameters were the seed germination capacity ((SGC), the number of germinated seeds on the second day), the seed vigor (at the end of the experiment), the length of sprouts and roots of seven-day-old seedlings, and the number of right-handed seedlings ((NRHS), the first leaf of these seedlings twists clockwise [7]). The nanoparticles were prepared by chemical methods at the Southwestern State University. The nanoparticles had a spherical shape and the following diameters: NPAg, 35–60 nm [9]; NPCu, 2–3 nm [10]; NPBi, 30–40 nm [11]; and NPZnO, 20–30 nm [11]. The culture filtrates of fungus were prepared on a Czapek-Dox liquid medium at 22–24°C for 21 days [12].

III. RESULTS

A. Effect of Nanoparticle Concentration in Aqueous Dispersions

Aqueous Dispersions with NPCu

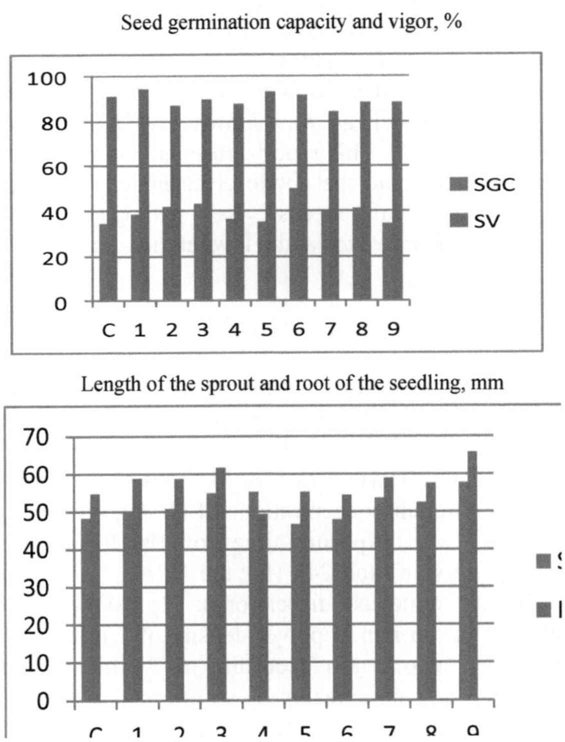

Fig.1 Parameters of the triticale seeds and seedlings (Ingen-93) under the action of dispersions containing different NPCu concentrations (in mg/L): *1*) 1×10^{-3}, *2*) 2×10^{-4}, *3*) 4×10^{-5}, *4*) 8×10^{-6}, *5*) 16×10^{-7}, *6*) 32×10^{-8}, *7*) 64×10^{-9}, *8*) 128×10^{-10}, and *9*) 256×10^{-11}, K is the control

Figure 1 shows that, after the exposure of the seeds to aqueous dispersions with different concentrations of NPCu, the parameters of the triticale seeds and seedlings vary nonlinearly. Thus, with respect to SGC, the versions of aqueous dispersions containing 2×10^{-4}, 4×10^{-5}, and 32×10^{-8} mg/L of NPCu had a stimulation effect. In the last-mentioned case, a significant exceedance of the control value—16%—was observed. The change in the SV in the versions was not considerable (up to 7%). With respect to the sprout and root lengths, substantial changes were observed in the last version of 256×10^{-11} mg/L.

Aqueous Dispersions with NPZnO

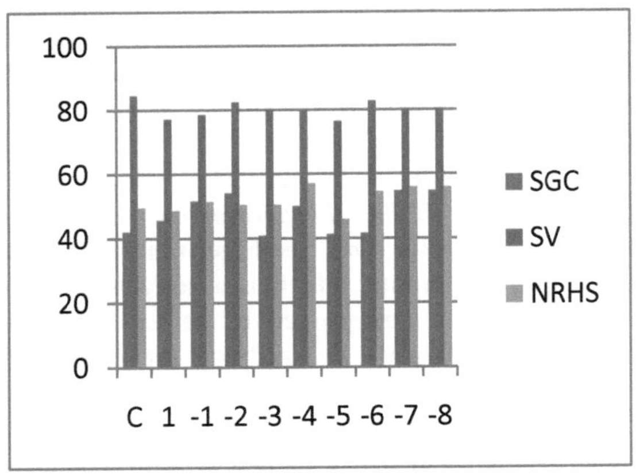

Fig. 2 SV, GC, and NRHS of the triticale seeds exposed to the action of aqueous dispersions containing different concentrations of NPZnO (%): (*1*) – (*-8*) – 10^{-1}–10^{-8} mg/L, respectively

Comparison of three parameters—SGC, SV, and NRHS—has revealed significant changes with respect to the first and third parameters (Fig. 2). The SV was 42.0 ± 2.75% in the control; 51.6 ± 2.48 at 10^{-1} mg/L; 54.0 ± 4.97 at 10^{-2} mg/L; 54.8 ± 4.46 at 10^{-7} mg/L; and 54.8 ± 2.05% at 10^{-8} mg/L. The greatest stimulation was 13% in the last two versions. The SV of the seeds, as in the first experiment, did not change significantly. The lowest concentrations of nanoparticles were the most effective. Thus, in the two experiments, the most informative parameter was the SV.

B. Estimation of the Fungicidal Effect of Aqueous Dispersions Containing Nanoparticles on the Seeds

Aqueous Dispersions with NPAg: Effect of the Term of Recording of the SGC Parameter

Taking into account the data of Fig. 1, in this experiment, the SGC parameter was used as a criterion for estimating the fungicidal effect of the nanofactor. The instructional guidelines report on different terms of recording of this parameter depending on the genotype and state of the seeds [14]. Therefore, the optimal period for recording this parameter is determined by conditions of the particular experiment. We used aqueous dispersions with three NPAg concentrations, one of which (10-7 mol/L), according to [4], has a stimulation effect.

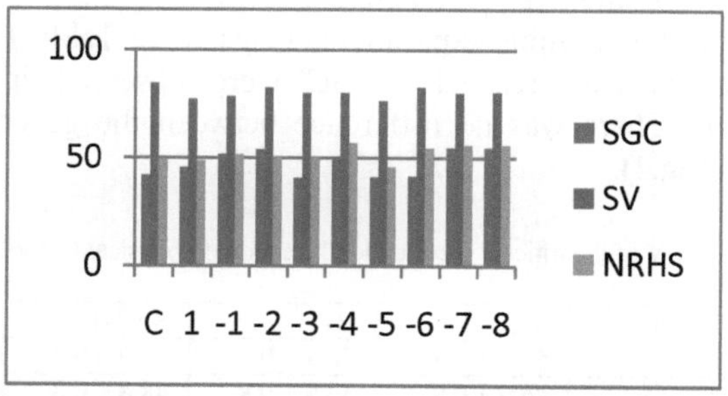

Fig. 3 Number of germinated triticale seeds under the separate and joint action of the NPAg aqueous dispersion, fungus *Helminthosporium avenae*, and $KMnO_4$, %. K is the control; F is the fungus; 1, 2, and 3 are the NPAg concentrations of 8×10^{-6}, 16×10^{-7}, and 32×10^{-8} mol/L; PP is $KMnO_4$; II, III, and IV day are the days after exposure of the seeds to the factors

It is evident from Fig. 3 that the highest change in this parameter of the triticale seeds fell on the second day after soaking the seeds. Here, the greatest stimulation is obtained in versions of aqueous dispersions with NPAg concentrations of 16×10^{-7} and 32×10^{-8} mol/L. The exceedance was 70 and 60% with respect to the control value and 14.8 and 13.4 times with respect to the *fungus* version, which showed the highest inhibitory effect. The *potassium permanganate* version gave a 20% lower value than the control value. The simultaneous action of the nanofactor and the fungus showed that the former contributed to a considerable increase in the parameter with respect to the latter, particularly in the case of the water dispersion with an NPAg concentration of 16×10^{-7} mg/L (fourfold), although it was lower than the control value. *Potassium permanganate* exceeded the *fungus* version as little as by 40%. Hence, with respect to the fungicidal effect on the triticale seeds, the nanofactor has proved to be significantly more effective than potassium permanganate, which is a conventional antifungal factor. On the third day of recording of SGC, the differences smoothed out. The stimulation versions gave as low as 30–60%, while the fungicidal effect of the nanofactor decreased to 8–12%. On the fourth day, the parameter values almost approached those of the control and the version with the fungus. Thus, to obtain the greatest stimulation and fungicidal effect from the nanofactor, the SGC parameter should be recorded on the second day after soaking of the triticale seeds.

C. Comparative Study of the Stimulation and Fungicidal Effect of the Nanofactor and Weak Electromagnetic Fields

The following parameters were used: the NRHS, because it correlates with the informative parameter SGC (Fig. 2), and SV. The triticale seeds were soaked in water dispersions containing NPBi and NPZnO and then subjected to the action of a MF and MWR. According to Table 1, all of the versions exceeded the control with respect to the absolute value of NRHS. A significant stimulation effect (26% versus the control) was obtained in the version with the water dispersion containing NPZnO (at a concentration of 10^{-3} mg/L); in the version with the water dispersion containing the same con-centration of NPBi, the stimulation effect was insignificant (7%). The parameters of the version of EMF, potassium per-manganate, and fungus were at the level of the control values. A significant fungicidal effect (approximately 26%) was obtained from the aqueous dispersion with NPBi (at a con-centration of 10^{-7} mg/L), MWR, and potassium perman-ganate. No differences were found with respect to the SV of the seeds in all the versions, as in the previous experiments (Figs. 1, 2). The experiment with a MF was conducted sepa-rately. It was found that it has a significant stimulation and fungicidal effect with respect to SGC (35.6 ± 0.98, 60.4 ± 0.40 and 42.8 ± 0.80% for control, MF, and MF + F, respectively); the effect with respect to the NRHS was less pronounced (44.9 ± 4.19, 53.1 ± 3.41, and 48.4 ± 3.41% for control, MF, and MF + F, respectively); no effect on the SV was observed.

Table 1 Parameters of triticale seeds and seedlings under the action of aqueous dispersions with NPBi and NPZnO, fungus, MWR, and potassium permanganate, %

No.	Version	NRHS	SV
1	C	48.9 ± 3.84	62.8 ± 4.96
2	Bi-3	52.5 ± 2.90	66.4 ± 2.48
3	Bi-7	50.8 ± 3.41	71.2 ± 4.54
4	ZnO-3	61.7 ± 2.62*	71.6 ± 6.66
5	ZnO-7	55.4 ± 3.90	72.0 ± 2.76
6	MWR	52.4 ± 3.40	64.8 ± 3.81
7	PP	52.9 ± 4.51	76.4 ± 1.99
8	F	45.1 ± 3.17	66.0 ± 5.47
9	Bi-3+F	57.5 ± 4.28	67.6 ± «.78
10	Bi-7+F	60.7 ± 5.47*	78.8 ± 4.84
11	ZnO-3+F	54.0 ± 3.61	60.4 ± 3.49
12	ZnO-7+F	52.7 ± 3.07	66.0 ± 6.80
13	MWR+F	60.2 ± 2.29*	62.0 ± 4.61
14	PP+F	63.7 ± 1.69*	70.4 ± 6.01

Note: C is the control; Bi-3, Bi-7, ZnO-3, and ZnO-7 are the concentrations of bismuth and zinc oxide nanoparticles of 10^{-3} and 10^{-7} mg/L, respectively; F is fungus *Helminthosporium avenae*; and PP is potassium permanganate.

D. Comparative Study of the Thermal Stability of Seeds Subjected to Treatment with the Nanofactor and Weak Electromagnetic Fields before Exposure of the Seeds to a Low Temperature

In this experiment, the response of the germinating triticale seeds to the action of a low temperature (+4°C for 8h) was studied. The external protective factors were water dispersion with NPAg (a concentration of 10^{-7} mol/L, which, according to [7], has a stimulation effect (Fig. 3)), MWR, and a weak MF (as in Table 1). The objects were winter wheat seeds of two cultivars (Select, BT 43-02) and a hybrid thereof.

None of the versions had a stimulation effect with respect to the SV (Table 2). The exposure of the seeds to external factors before subjecting them to the action of a low temperature made it possible to estimate the effect of these factors on the thermal stability of the objects. After this procedure, the seeds did not achieve the level of the control values and were strongly inhibited.

Table 2 SGC of the triticale seeds under the action of physical factors, %

Version	S	BT	S x BT
C	78.5 ± 4.1	86.5 ± 6.6	87.0 ± 6.6
LT	09.5 ± 4.3	23.5 ± 4.5	16.0 ± 4.2
NPAg	83.5 ± 4.7	82.0 ± 7.1	84.5 ± 4.3
NPAg+LT	18.0 ± 3.5*	39.5 ± 3.6*	33.0 ± 4.2*
MWR	80.5 ± 4.2	82.0 ± 6.0	79.0 ± 1.6
MWR+LT	12.0 ± 2.8	31.5 ± 4.4	18.5 ± 5.7
MF	81.0 ± 2.4	88.0 ± 2.9	80.0 ± 2.7
MF+LT	17.5 ± 3.4*	36.0 ± 4.9*	28.5 ± 5.6*

Note: S, BT, and S x BT are cultivars Select and BT 43-02 and a hybrid thereof, respectively; C is the control; LT is low temperature (4°C); NPAg is the aqueous dispersion with an NPAg concentration of 10^{-7} mol/L; and MF is a low-intensity magnetic field

However, in the versions with the action of a nanofactor and an MF, a significant exceedance of the parameter with respect to the "low temperature" version was recorded for the three genotypes (on average by 1.8 times); that is, the protective effect of these factors was observed.

With respect to the NRHS, the picture is fundamentally similar: the absence of stimulation from external factors and the presence of a protective effect of these factors on the temperature factor (Table 3).

Table 3 NRHS under the action of physical factors on the triticale seeds, %

Version	S	BT	S x BT
C	53.5 ± 1.7	57.6 ± 4.4	54.5 ± 5.0
T	42.6 ± 3.1	46.4 ± 2.1	47.3 ± 3.4
Ag	64.3 ± 5.1	52.3 ± 2.4	58.1 ± 2.4
Ag+T	68.6 ± 5.4*	57.7 ± 1.2*	55.4 ± 3.9
MWR	53.5 ± 1.2	55.2 ± 2.1	55.8 ± 3.6
MWR+T	54.8 ± 4.9	56.2 ± 3.6*	63.8 ± 0.5*
MF	54.5 ± 4.5	53.4 ± 4.8	63.8 ± 3.4
MF+T	54.4 ± 3.5*	56.8 ± 3.0*	49.7 ± 0.3

Note: S, BT, and S x BT are cultivars Select and BT 43-02 and a hybrid thereof, respectively; C is the control; T is temperature; and MF is a low-intensity magnetic field.

In this case, in addition to the nanofactor and the MF (for Select and BT 43-02), MWR also proved effective (for BT 43-02 and the hybrid). The exceedance of the parameter value with respect to the "low temperature" version was 1.3–1.6 times. The response of the genotypes to the impact was less intense with respect to the NHRS and less similar compared to the SGC parameter.

As in the previous experiments (Figs. 1, 2; Table 1), no differences with respect to GC were observed in the versions. There was no difference between the genotypes either (Fig.4).

Table 4 SV of the triticale seeds under the action of physical factors, %

Version	S	BT	S x BT
C	52.4 ± 0.8	56.4 ± 0.9	54.9 ± 0.8
T	46.9 ± 0.9	53.9 ± 0.8	48.8 ± 0.8
Ag	53.6 ± 0.8	57.8 ± 1.0	57.5 ± 0.8
Ag+T	46.7 ± 1.3	54.9 ± 0.9	48.6 ± 0.8
MWR	52.8 ± 0.8	59.9 ± 1.0	56.4 ± 0.8
MWR+T	44.3 ± 1.0	52.4 ± 0.9	52.1 ± 0.9
MF	53.8 ± 0.8	58.3 ± 0.9	58.1 ± 0.9
MF+T	46.5 ± 0.9	51.6 ± 0.9	49.1 ± 0.8

Note: S, BT, and S x BT are cultivars Select and BT 43-02 and a hybrid thereof, respectively; C is the control; T is temperature; and MF is a low-intensity magnetic field.

IV. CONCLUSION

1) In the case of soaking of winter triticale and winter wheat seeds in aqueous solutions containing NPAg, NPCu, NPBi, and NPZnO nanoparticles in different concentrations, a substantial stimulation of seed germination relative to the control value has been revealed in the versions of low concentrations of the nanofactor (10^{-9} mol/L for NPAg and 10^{-7} mg/L for the other nanoparticles). No fundamental qualitative and quantitative differences in the effects of the different types of the nanofactor have been identified.

2) The informativity of the parameters with respect to the effect can be arranged in the following descending order: "seed germination capacity vigor–the number of right-handed seedlings–the length of the sprout and root of the seedling– seed vigor." Quantitatively, this fact was evident from the exceedance of the parameter to 70% at the beginning of the series and the absence of any exceedance at the end of the series. With respect to seed germination capacity, the effect was most pronounced at the early stages of recording of this parameter (on the second day after the beginning of the experiment).

3) Aqueous dispersions with low concentrations of nano-particles have a fungicidal effect on the seeds. Soaking of the seeds in a culture filtrate of pathogenic fungus *Helminthosporium avenae* after seed treatment in the

stimulation versions of the nanofactor has led to a significant—severalfold—increase in the parameters with respect to the "fungus infestation" version. In this case, the most informative parameter was also seed germination capacity and the above order of the stimulation effect was preserved. Potassium permanganate, which is a conventional factor of seed disinfection, has proved to be less effective.

4) A similar conclusion can be drawn in regard to an increase in the resistance of the germinating seeds to a low temperature of +4°C.

5) In comparative experiments on studying the effects of the nanofactor and weak electromagnetic fields (millimeter-wave radiation of 5.6 mm and an alternating magnetic field with a frequency of 1–10 Hz and a magnetic induction 40–50 μT) on the seeds, it has been generally found that there are some similarities in the response of both normal seeds and the seeds subsequently exposed to the effects of the pathogenic fungus and low temperatures. However, in some versions, the effects of the nanofactor and then of the magnetic field were more pronounced.

6) The data of this report supplement our previous findings on this issue. The data can be used both for the development of optimum conditions of the preplant seed treatment with a nanofactor and physical fields and for the determination of the mechanism of action of low-intensity factors on the vitality of plants.

Conflict of Interest

Every paper must contain a declaration of conflicts of interest. If there are no such conflicts write “The authors declare that they have no conflict of interest”.

References

1. H. Zhu, J. Han, J.Q. Xiao, Y.Jin, «Uptake, translocation and accumulation of manufactured ironoxide nanoparticles by pumpkin plants», Journal Environment Monitoring, (10), pp.713–717, 2008.
2. А.П Райкова, Л.А. Паничкин, Н.Н. Райкова, «Исследование влияния ультрадисперсных порошков металлов, полученных различными способами, на рост и развитие растений», Материалы Между-народной научно-практической конференции «Нанотехнологии и информационные технологии – технологии XXI века». Москва, с. 108–111, 2006.
3. Н.Н. Глущенко, О.А. Богословская, И.П. Ольховская, «Сравнительная токсичность солей и наночастиц металлов и особенность их биологического действия», Материалы Международной научно-практической конференции «Нанотехнологии и информационные технологии – технологии XXI века». Москва, с. 93–95, 2006.
4. С.Н. Маслоброд, Ю.А.Миргород, В.Г.Бородина, Н.А.Борщ, «Влияние водных дисперсных систем с наночастицами серебра и меди на прорастание семян», Электронная обработка материалов, 50(4), с. 103- 112, 2014.
5. Скрябин, В.А. Наноструктуры на основе висмута и серебра для системы хранения зерна и производства комбикормов / В.А. Скрябин [и др.] // Материалы Всерос. науч.-практич. конф. «Актуальные проблемы в области создания инновационных технологий хранения сельскохозяйственного сырья и пищевых продуктов»/ Углич, ГНУ ВНИИМС. – М.: Россельхозакадемия, 2011. – С. 230–233.
6. О.В.Бецкий, Н.Н.Лебедева. «Применение низкоинтенсивных миллиметровых волн в биологии и медицине», Миллиметровые волны в биологии и медицине. (1), с.12-18, 2007.
7. S.N. Maslobrod, L.B. Korlatyanu, A.I. Ganya, «Influence of Millimetric Radiation on the Viability of Plants: Changing the Metabolism of Seeds at the factor s Influence on Dry Seeds. Surface Engineering and Applied Electrochemistry», 46(5), pp.477-488, 2010.
8. С.И.Аксенов, Т.Ю.Грунина, С.Н.Горячев, «Особенности влияния низкочастотного магнитного поля на набухание семян пшеницы на разных стадиях», Биофизика, . 46(6), с. 1127-1132, 2001.
9. Ю.А. Миргород, Н.А. Борщ, В.Г. Бородина, Г.Ю. Юрков, «Получение и характеризация хлопчатобумажной ткани, модифицированной наночастицами меди», Химическая промышленность, 89(6), с.310-316, 2012.
10. Yu.A.Mirgorod, V.G.Borodina, Preparation and Bactericidal Properties of Silver Nanoparticles in Aqueus Tea Leaf Extract, Inorg. Mat. , 49 (10), pp.980-983, 2013.
11. Unpublished works
12. Методы экспериментальной микологии, Киев, Наукова думка, 550 с.,1982.
13. International rules for seed testing. Москва, Колос, 310 p, 1984

Author: Serghei Maslobrod
Institute: Institute Of Genetics Physiology and Plant Protection of ASM
Street: 20, Padurii St.
City: Chisinau
Country: Moldova
Email: maslobrod37@mail.ru

Theoretical Treatment of Millimeter and Terahertz Radiation Action on Biological Media

N. Ciobanu[1,2], V. Vovc[1], A. Saulea[1], and V. Tronciu[3]

[1] Department of Human Physiology and Biophysics, State University of Medicine and Pharmacy "Nicolae Testemitanu", Chisinau, Republic of Moldova
[2] Institute of Applied Physics, Academy of Sciences of Moldova, Chisinau, Republic of Moldova
[3] Department of Physics, Technical University of Moldova, Chisinau, Republic of Moldova

Abstract— **In this paper we give a model that describes the nonlinear cooperative stationary phenomena at interaction of Bose-condensed phonons with millimeter or terahertz electromagnetic radiation in a biological media. The expressions that characterize the real and imaginary parts of the dielectric susceptibility and permeability, as well as the refraction and reflection indexes caused by condensed phonons were obtained.**

Keywords— **millimeter and terahertz waves, biological system, Bose-Einstein phonon; complex dielectric functions.**

i. INTRODUCTION

During the last years, a great attention is devoted to the action mechanism between electromagnetic radiation and biomedical systems. It was observed that from the total domain of electromagnetic radiations, millimeter (30 ÷ 300 GHz) frequencies (MM) and terahertz (TH) known as submillimeter (0.3 ÷ 10 THz) prevail [1-7]. Many apparatus and devices that emit MM and TH waves are used in the treatment of various diseases such cardiovascular disorders [8], gynecology [9], ulcers [10], traumatism and tumors [11], and the results can be considered significantly.

In order to understand the action mechanism, various theoretical models were proposed, however, the mechanism is still not clear. The first attempt to explain electromagnetic waves interaction with living systems was done by Fröhlich [12]. According to his idea, living organisms are able to generate electromagnetic waves in the MM diapason. Most important characteristics of MM waves are extremely high frequency, coherence, i.e., concrete frequency, polarization, phase and well fixed wave vector. As was demonstrated, the generation of MM waves in the bio-molecules and cells are conditioned by the metabolic energy state that can be transformed in collective vibrational modes or phonons. The phonons play a fundamental role in the activity of biological systems, and the resonance absorption of MM as well as TH waves can change drastically metabolic activity of the cells.

The Fröhlich model was rapidly extended to Bose-Einstein condensation (BEC), which assume that generated energy by the cell can be supplied to the phonon with a greater rate than a critical value, and the condensation mechanism to the lowest excitation of a single mode can happen. An equivalent type of condensation that occurs at low excitation vibrational mode may serve as mechanism action of energy storage as well as channel for various molecular or cells processes, e.g., syntheses, transmission of information into or between the cells, cell division etc. Based on BEC phonons generation in alive systems, other interesting theoretical models were proposed, like collective state excitation in one-dimensional molecular structures [13] or spiral protein molecules that can form solitons [14].

In this paper we give a theoretical study on the investigation of nonlinear cooperative phenomena at the interaction of Bose-condensed phonon with the millimeter electromagnetic radiation generated in a biological media. A short description of the model Hamiltonian at the interaction of MM radiation with coherent phonons in the square form for the low concentrations of Bose-condensed phonons is presented in Section II. The Section III is dedicated to the dielectric functions and their properties. We finish with conclusions in Section IV.

ii. MODEL HAMILTONIAN AND EQUATION OF MOTIONS

Our model is based on the Fröhlich idea that coherent phonons excited in biological objects turn into coherent internal photons and form the self-correlated interval millimeter electromagnetic field. As was demonstrated previously, the coherent state of the phonon in the biological objects can be similar to the coherent state of the exciton in the condensed media [15-17]. These coherent states of the elementary excitations can appear in different regimes, i.e., in regime of ultrashort pulses existing a time period shorter that the relaxation times, as well as in conditions of Bose-Einstein condensation during a time period longer that the relaxation times, but shorter that the quasiparticle life time. In both cases the corresponding elementary excitations are characterized by certain phase, wave vector $\vec{k} \equiv k$ and macroscopic value of the single-particle state, i.e., the number of excitations in condensate N_V and system volume V. Based

V. Sontea and I. Tiginyanu (eds.), *3rd International Conference on Nanotechnologies and Biomedical Engineering*,
IFMBE Proceedings 55,
DOI: 10.1007/978-981-287-736-9_68

on this assumption our system can be described by the Hamiltonian

$$H = \sum_k \hbar\Omega a_k^+ a_k + \frac{1}{8\pi}\sum_k \int\left(\varepsilon E_k^2 + \mu H_k^2\right)dv_k - \sum_k d_k\left(E_k^+ a_k^+ + E_k^- a_k\right) + \frac{1}{2V}\sum_{k_1,k_2,k_1',k_2'} g(k_1 - k_1')\delta_{kp}(k_1 + k_2, k_1' + k_2')a_{k_1}^+ a_{k_2}^+ a_{k_1'} a_{k_2'}. \quad (1)$$

Here, $\hbar\Omega$ is the dipole-active phonon energy. $E_k = E_k^+ + E_k^-$, $H_k = H_k^+ + H_k^-$ are the intensities of the electrical and magnetic fields (positive or negative frequency parts), respectively. $a_k^+(a_k)$ are the operators of creation (annihilation) of the dipole-active phonons that satisfy the commutation relation $[a_k, a_{k'}^+] = \delta_{kk'}$, $[a_k, a_k^+] = 0$. ε and μ are the dielectric and magnetic permeability of the biological medium. $g(k)$ is the phonon-phonon interaction strength constant, while d_k represents the dipole momentum of the transition into phonon state.

In the Hamiltonian (1), both the field and the Bose-condensed phonons are supposed to have the same wave vector $k \neq 0$, polarization and phase. The amplitudes are considered to be macroscopically large, i.e. proportional to the system effective volume V. We mention that the effects of spatial dispersion were neglected being insignificant in the actual spectrum range, and phonon mass $m \approx \infty$.

The Heisenberg motion equation for the dipole-active phonon operator a_k has the form:

$$i\frac{da}{dt} = \Omega a + \frac{g}{\hbar}\frac{|a|^2}{V}a + \frac{d}{\hbar}E^+ - i\gamma a, \quad (2)$$

where the last term $-i\gamma a$ was introduced phenomenologically and takes into account the Bose-condensed phonons leaving from condensed state. On the other hand, we assume that the attenuation term can be considered strictly within the framework of the quantum theory of attenuations and fluctuations using the Fokker-Plank master equation [18], where the polarization is given by

$$P^+ = -\frac{1}{V_0}\frac{\partial H}{\partial E^-}, \quad P^- = -\frac{1}{V_0}\frac{\partial H}{\partial E^+}, \quad (3)$$

where $V_0 = V/N_d$, and N_d is the atomic dipole number. According to the expressions (1) and (3) the polarization can be expressed:

$$i\frac{da}{dt} = \Omega a + \frac{g}{\hbar}\frac{|a|^2}{V}a + \frac{d}{\hbar}E^+ - i\gamma a, \quad (4)$$

$$P^+ = \frac{da}{V_0}, \quad P^- = \frac{da^+}{V_0}. \quad (5)$$

We assume that the Fröhlich MM field is spread along the axe x, which equation is equivalent to

$$\frac{\partial^2 E^+}{\partial x^2} - \frac{1}{c^2}\frac{\partial^2 E^+}{\partial t^2} = \frac{4\pi}{c^2}\frac{\partial^2 P^+}{\partial t^2},$$

where c is the light velocity in the biological media.

Equations (2) and (5) describe completely the spatial–temporal evolution of the Bose-condensed phonons in the biological media, taking into account self-interaction and the interaction with the Fröhlich field. We write the macroscopic amplitudes of the phonons and the field in the form of modulated plane waves with the frequency ω and the wave vector k

$$a(x,t) = \sqrt{V}\tilde{a}(x,t)e^{-i\omega t + ikx}, E^+(x,t) = \sqrt{V}\tilde{E}(x,t)e^{-i\omega t + ikx} \quad (6)$$

where $\tilde{a}(x,t)$ and $\tilde{E}(x,t)$ are the slowly-varying amplitudes satisfying the following conditions:

$$|\partial\tilde{a}/\partial t| << \omega|\tilde{a}|, |\partial\tilde{a}/\partial x| << k|\tilde{a}|,$$
$$|\partial\tilde{E}/\partial t| << \omega|\tilde{E}|, |\partial\tilde{E}/\partial x| << k|\tilde{E}|.$$

Taking into account the above assumption, the equation Heisenberg equation of motion (2) is transformed in

$$\frac{d\tilde{a}}{dt} = i\left[\Delta + i\gamma - \frac{g}{\hbar}n\right]\tilde{a} + i\frac{d}{\hbar}\tilde{E}. \quad (7)$$

In the relation (7) $\Delta = \omega - \Omega$ represents the resonance detuning between the Fröhlich field frequency and the phonon frequency, wile $n = |\tilde{a}|^2$ represents the concentration of the Bose-condensed phonons.

III. Steady State, Dielectric Functions and Their Properties

In this Section we discuss the steady states ($d\tilde{a}/dt = 0$) of the equation of motion (7). For the considered situation the dependence of amplitudes of phonons on photons is given by:

$$\tilde{a} = -\frac{d/\hbar}{\Delta - gn/\hbar + i\gamma}\tilde{E}. \quad (8)$$

The dielectric susceptibility of the biological media is given by the relation parameter $\chi = P^+/E^+$. According to relations (4) and (8) for the real and imaginary parts of dielectric susceptibility we obtain

$$\chi' = -\frac{d^2}{V_0\hbar}\frac{\Delta - gn/\hbar}{(\Delta - gn/\hbar)^2 + \gamma^2}, \quad \chi'' = \frac{d^2\gamma}{V_0\hbar}\frac{1}{(\Delta - gn/\hbar)^2 + \gamma^2}. \quad (9)$$

Thus the dielectric permeability, that takes into account the phonon level contribution, is described by the expression

$$\varepsilon = \varepsilon_\infty + 4\pi\chi = \varepsilon' + i\varepsilon'', \qquad (10)$$

where $\varepsilon', \varepsilon''$ are the real and imaginary parts of the dielectric permeability

$$\varepsilon' = \varepsilon_\infty - \frac{4\pi d^2}{V_o \hbar}\frac{\Delta - gn/\hbar}{(\Delta - gn/\hbar)^2 + \gamma^2},\ \varepsilon'' = \frac{4\pi d^2 \gamma}{V_o \hbar}\frac{1}{(\Delta - gn/\hbar)^2 + \gamma^2}.$$

Here, ε_∞ is the phonon dielectric permeability that takes into account the dielectric permeability of all excitations besides phonons of the biological media. The real and imaginary parts of the permeability determine the frequency of dispersion and depend significantly on Bose-condensed phonon density n.

In what follows it is convenient to switch to the following dimensionless quantities

$$N = \frac{g}{\gamma\hbar}n, f = \frac{\widetilde{E}}{E_s}, \frac{1}{E_S^2} = \left(\frac{d}{\hbar}\right)^2 \frac{g}{\hbar\gamma^3}, \Omega_0 = \frac{4\pi d^2}{\hbar V_0}, \delta = \frac{\Delta}{\gamma}. \qquad (11)$$

Taking account (11) from (10) we obtain

$$\varepsilon' = \varepsilon_\infty - \frac{(\Omega_O/\gamma)(\delta - N)}{[(\delta - N^2) + 1]},\ \varepsilon'' = \frac{\Omega_O/\gamma}{[(\delta - N)^2 + 1]}, \qquad (12)$$

$$N[(\delta - N)^2 + 1] = f^2.$$

For a plane wave, where the surfaces of the field components are planes, the dielectric permeability, perpendicular to the spreading direction, is connected with the refraction index $\overline{n}$ and with the absorption index κ by the expression

$$\varepsilon = \overline{n}^2 - \kappa^2 + 2i\overline{n}\kappa. \qquad (13)$$

Thus, we can write the refraction and absorption indexes

$$\overline{n} = \sqrt{\frac{\varepsilon' + \sqrt{\varepsilon'^2 + \varepsilon''^2}}{2}},\quad \kappa = \sqrt{\frac{-\varepsilon' + \sqrt{\varepsilon'^2 + \varepsilon''^2}}{2}}. \qquad (14)$$

It is well known that the reflection coefficient R is defined as the ratio of the time average energy flow reflected from surface to the incident one. For the case of perpendicular flow the reflection coefficient can be written [19]

$$R = \frac{(\overline{n} - 1)^2 + \kappa^2}{(\overline{n} + 1)^2 + \kappa^2}. \qquad (15)$$

In what follows, let consider that the electromagnetic field is uniformly distributed in the biological media $(\partial\widetilde{E}/\partial x = 0)$. From (5) we obtain the equation for the field temporal evolution in the slowly-varying amplitude approximation

$$\frac{d\widetilde{E}}{dt} = i\left[\frac{\omega^2 - c^2k^2}{2\omega} - \frac{4\pi d^2}{V_O\hbar}\right]\widetilde{E} + \frac{2\pi d\omega i}{V_O}\left[1 - \frac{2}{\omega}\left(\omega - \Omega - \frac{g}{\hbar}n\right) - \frac{2i\gamma}{\omega}\right]\widetilde{a}. \qquad (16)$$

In the stationary case and dispersion limit, when the attenuation of the dipole-active Bose-condensed phonons can be neglected, from (16) we obtain

$$k^2 = \frac{\omega^2}{c^2}\left[1 + \frac{\Omega_O}{\Omega + gn/\hbar - \omega}\right]. \qquad (17)$$

The equation (17) determines the dispersion law of the nonlinear polaritons, i.e. the energy values of the elementary excitations in biological medium at the Fröhlich radiation interaction with dipole-active Bose-condensed phonons can be realized at the same value of the wave vector. The position and the form of the polariton branches depend not only on the parameters of the phonons and of the electrical field, but also on the stationary concentration of the Bose-condensed phonons. The concentration depends on the Fröhlich field intensity according to the equation (12). Thus, the polariton appearance is caused by the intersection of the coherent photon dispersion curves with Bose-condensed phonon for small wave vectors. Close to the intersection point the energies and pulses of both excitations coincide and the bond between them becomes significant.

IV. NUMERICAL RESULTS

The dependences of real (left) and imaginary (right) parts of the dielectric permeability of the biological object on field intensity, caused by Bose-condensed phonons are shown in the Figure 1. For simplification, the following dimensionless variables were used: $\overline{k} = k/k_O$, $\overline{\omega} = \omega/(ck_O)$, $\overline{\omega}_O = \Omega_O/(ck_O)$, where $ck_O = \Omega + gn/\hbar$. All curves correspond to $\varepsilon_\infty = 5.2$, $\overline{\omega} = 5.0$ and various values of detuning. When the resonance detuning values are smaller than the critical δ_c, the real and imaginary parts of the dielectric permeability are single-valued functions of the field intensity. However, with increasing of detuning, the curves $\varepsilon'(f)$ and $\varepsilon''(f)$ become deformed, and some hysteresis dependences appear at $\delta > \delta_C$.

The dependences of the refraction $n(f)$ and absorption $\kappa(f)$ indexes on the field amplitude for different values of resonance detuning are given in Figures 2.

As can be observed, for high values of detuning, n, κ dependencies become more complicated and display hysteresis behavior caused by the nonlinear dependence of the Bose-condensed phonon density on the coherent MM (TH) electromagnet field amplitude, when the resonance detuning exceeds the critical value δ_C.

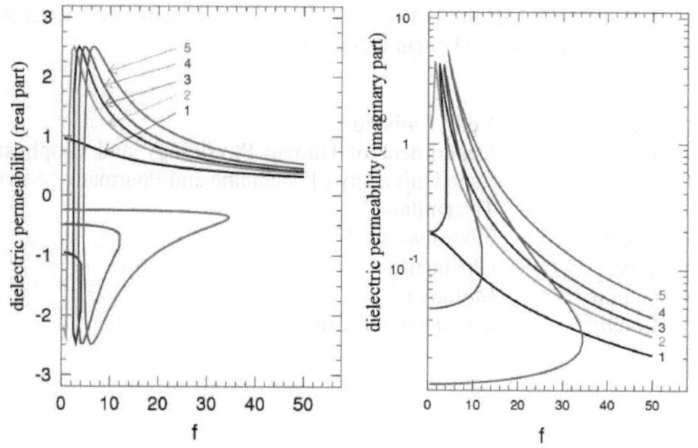

Fig. 1 The dependences of the real and imaginary parts of the dielectric permeability on the amplitude of the field for ε_{∞}=5.2, $\overline{\omega}$=5.0 and various values of the resonant detuning: 1) δ=-5.0, 2) δ=1.7, 3) δ=5.0, 4) δ=10.0 and 5) δ=20.0.

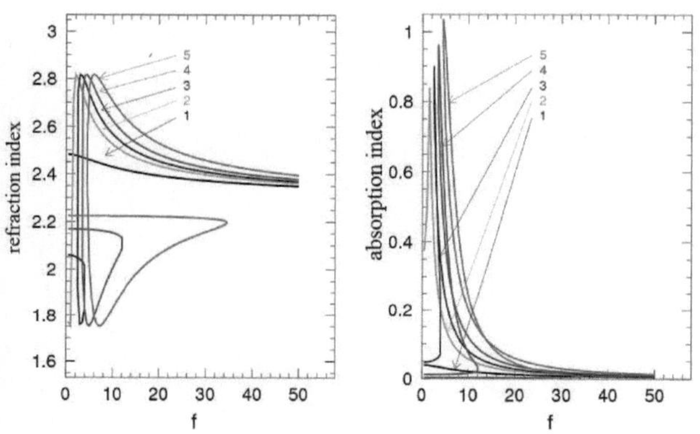

Fig. 2 The dependences of the refraction and absorption indexes on the field amplitude for the same parameters as in Figure 1.

V. CONCLUSIONS

In this article we developed a model that describes the interaction of Bose-condensed phonons with MM or TH electromagnetic radiation in a biological media. As was demonstrated the dielectric functions, i.e., absorption, reflection and refraction coefficients strictly depend on the coherent MM (TH) electromagnet field amplitude and resonance detuning. According to the dispersion law (17) in some conditions of the biological media the polaritons can occur, and the position as well the form of the polariton branches depend on the stationary concentration of the Bose-condensed phonons. In the case when the interaction of the coherent phonons with the thermostat is neglected, we demonstrate that equation of motions have soliton solutions. We believe that our model provides a good basis for future study and, in particular, gives some pointers for more detailed investigations of dynamics of Bose-condensed phonons with MM or TH electromagnetic radiaton in the biological media.

ACKNOWLEDGMENT

This work was supported by the project STCU - 5993 and 14.02.116F.

CONFLICT OF INTEREST

The authors declare that they have no conflict of interest.

REFERENCES

1. Deviatcov N D, Betskii O V (1994) Biological Aspects of Low Intensity Millimeter Wawes, Moscow
2. Zhadobov M, Saileau R, Viè V et al. (2006) Interactions between 60-GHz millimeter waves and artificial biological membranes: dependence on radiation parameters. IEEE Tras. MW Theory and Tec. 54: 2534-2542
3. Zhadoboy M, Nicolaz C N, Sauleau R et al. (2009) Evaluation of the Potential Biological Effects of the 60-GHz Millimeter Waves Upon Human Cells. IEEE Tans. on Antennas and Propagation 57: 2949-2956
4. Titushkin I A, Rao V S, Pickard W F et al. (2009) Altered calcium dynamics mediates P19-derived neuron-like cell responses to millimeter-wave radiation. Rad. Res. 172: 725-736
5. Ramundo-Orlando A (2010) Effects of Millimeter Waves Radiation on Cell Membrane - A Brief Review. J. Infrared Milli. Terahz Waves 31: 1400-1411
6. Davies A G, Linfield E H, Johnston M B (2002) The development of terahertz sources and their applications. Phys. Med. Biol. 47: 3679–3689
7. Globus T, Bykhovskaia M, Woolard D et al. (2003) Sub-millimetre wave absorption spectra of artificial RNA molecules. J. Phys. D: Appl. Phys. 36: 1314–1322
8. Black D R, Heynick L N (2003) Radiofrequency (RF) Effects on Blood Cells, Cardiac, Endocrine and Immunological Functions. Bioelectromagnetics Supplement 6: 187-195
9. Markovic M, Manderson L, Wray N et al. (2006) Complementary medicine use by Australian women with gynaecological cancer. Psycho-Oncology 15: 209-220
10. Li X H, Tang J T, Liao Y P et al. (2008) Millimeter wave in the treatment of acute radiation-induced cervical skin ulcers. J. Clin. Rehab. Tissue Eng. Res. 12: 663-666
11. Logani M K, Szabo I, Makar V et al. (2006) Effect of millimeter wave irradiation on tumor metastasis. Bioelectromagnetics Supplement 27: 258-264
12. Fröhlich H (1968) Bose condensation of strongly excited longitudinal electric modes. Phys. Lett. A 26: 402-403, 1968
13. Bolteraljer H, Tuszynski J A (1989) Friöhlich's Condensation in a Biological Membrane Viewed as a Davydov Soliton. Journal of Biological Physics 17: 41-50.
14. Davydov A S (1977) Solitons and energy transfer along protein molecules. J. Theor. Biol. 66: 379–387

15. Zaloj V A, Rotaru A, Barbaroshie A E et al. (1995) Nonlinear dynamics of the immune system interaction with the bilocal cancer tumor. Journal of Biological Physics 21: 155-176
16. Rotaru A, Shura C (1998) Two-photon ultrashort non-linear dynamics of coherent bosonic quasiparticles in condensed media. Zh. Eksp. Teor. Fiz. 86(6): 132-1139
17. Tronciu V, Rotaru A H (1998) Non-linear propagation of light in semiconductors in the presence of two photon exciton of biexcitons. Phys. Solid State 40(11): 1809-1812
18. Rotaru A H, Shibarshina G D (1985) The Lorenz attractor in a system of coherent excitons, photons and biexcitons. Phys. Lett. A 109: 292-294
19. Landau L D, Lifshits E M (1982) The electrodynamics of continuous media. Moscow: 624 p. (in Russian)

Author: Nellu Ciobanu
Institute: Department of Human Physiology and Biophysics, State University of Medicine and Pharmacy "Nicolae Testemitanu"
Street: *Testemitanue 27/2*
City: Chisinau
Country: Moldova
Email: cnellu@gmail.com

Fabrication MEMS Platform for Sensors Applications by Laser Micro Engraving

A.V. Ivanova, K.Y. Oblov, S.A. Soloviev, N.N. Samotaev, B.V. Gurkovskiy, and V.D. Mironov

Micro- and nanoelectronics department
National Research Nuclear University MEPhI (Moscow Engineering Physics Institute)
Moscow, Russia

Abstract— **The target of this work is the demonstration of advanced novel approaches able to provide rapid prototyping by using laser technology ceramic MEMS platforms for chemical sensor operating under harsh environmental conditions and, on the other hand, to assure microhotplate stable at high temperature, which can be used for the deposition of high working temperature gas sensing materials, for example, oxides of tin, gallium, zirconium and hafnium. As substrate ceramic material in work using alumina oxide.**

Keywords— **laser, MEMS, microhotplate, gas sensor.**

I. INTRODUCTION

Many research groups in the World investigate now the possibility of the fabrication of metal oxide gas sensors based on micromachining technology [1]. All of these technologies have really low production cost starting only from large scale production (more than ~ 1 million units per year). Moreover, some of these technologies (first, silicon based approaches) require expensive microelectronic technological equipment and clean rooms for efficient manufacturing; this equipment and tools give additional non-transparent contribution to the cost of production [2].

In our work, we tried to develop an ultra fast, low cost and technology effective process for the production of MEMS microhotplates to be used in gas sensors orientated to the laboratory application, when where it is sufficient to produce 10-100 samples with different layout of heater and membrane per day. Of course, this technology can be used as well for middle-scale mass production of gas sensors, up to one thousand units per day (under full load of equipment using in describing work).

Using the philosophy of rapid prototyping technologies, it is possible to form two approaches to this problem – add material or remove one for form the MEMS construction. As an example of technology with the materials addition (layer by layer) is LTCC (low temperature co-fired ceramic) technology. There is possible to avoid the process of etching of ceramic MEMS substrate using multi-layer ceramic system, for example stack of low LTCC sheets. These green tape sheets can be cut by laser or stamp and then the stack of these sheets can be co-fired to form rigid frame. The microhotplates based on LTCC were studied by several research groups from Technical University of Wroclaw (Poland) [3, 4], University of Bayreuth (Germany) [5], National Institute for R&D in Microtechnologies (Bucharest, Romania) [6], and others. LTCC ceramics is, obviously, very suitable material for the tooling and fabrication of all elements of microhotplate. It is possible to fabricate microhotplates suspended with ceramic beams in rigid ceramic frame, which can be much thicker than the microhotplate itself. This rigid frame provides easy manipulation with sensor during packaging. However, the microheaters made of LTCC ceramics are fabricated using commercial green tape. This green tape has minimum thickness of about 100 µm. Because of this thickness and because of the application of screen-printing for the fabrication of microhotplate elements, the microhotplates are characterized by relatively high heating power consumption equal to 600 mW at 400^0C, as it was reported in [4]. This power consumption is even higher that power consumption of traditional thick film gas sensors [7].

Lack of LTCC technology for rapid prototyping in the cost of the equipment set to work with the ceramic green sheet is several million Euros required clean room class ISO7 complex and expensive materials (ceramic tape and paste). Therefore, in our study, we chose an approach to the use of material removal by laser treatment monolithic ceramic aluminum oxide. The cost of a standard substrate 96% Al_2O_3 with 60x48 mm size and a thickness of 0.5 mm used in the work is a few Euros. Prevalence and cost of this type of substrates to order more than green ceramics for LTCC technology. Platinum paste to create metallization can be prepared by hand in the presence of small skills in the field of chemistry or used from the list of LTCC materials. It is also desirable, but not necessary to have equipment for screen printing, to apply the paste uniformly to form a layout.

Our main goal in this work was to fabricate MEMS microhotplate with thermal response time comparable with the level typical for ceramic MEMS [8-11] or screen-print technologies [12, 13]. Also additional requirement for material for microhotplate is stability up to 600^0C for using in harsh environmental conditions [14, 15] where low power consumption of MEMS platform is not critical. The aim of the present work is to develop a process allowing quick design

V. Sontea and I. Tiginyanu (eds.), *3rd International Conference on Nanotechnologies and Biomedical Engineering*,
IFMBE Proceedings 55,
DOI: 10.1007/978-981-287-736-9_69

and production of low-scale series of ceramic microhotplate for gas sensors. Microhotplate made by LTCC technology [9] with power consumption of 480mW at 330°C and characteristic heating time below 10 seconds were taken as a guideline for acceptable results.

II. EXPERIMENT

Laser micromachining using Minimarker installation with parameters given in Table 1 was taken as a basic technological tool. The base material for the microhotplate substrate was monolithic ceramics from aluminum oxide containing 96% alumina because of its cheapness (5 times cheaper than 99.9% alumina) and sufficient robust for sensor fabrication. As metallization base material, we took platinum containing composite with additives of organic binder components and borosilicate glass microparticles, which are used for the preparation of finely dispersed paste usable in screen printing process. Microhotplate layout was designed using AutoCAD program (Fig. 1a), and then loaded to the computer of laser engraving machine. According to the previously obtained relationship between the number of laser ablation passes (Fig. 2a), laser power and depth of engraving (Fig. 3a), the micromachining process was pre-set (Fig. 2b). The next steps of fabrication process were the thinning of the whole surface by engraving, cutting of the membrane obtained during the thinning process leading to the formation of beam suspended microhotplate, and dicing the substrate into chips. In the final microhotplate manufacturing step, platinum ink was deposited on its surface, and then its local laser firing was implemented. The residual, unfired, ink was removed by washing to form microhotplate. Or, alternatively, the ink was fired in oven, and the heater was formed by local laser ablation.

III. RESULTS AND DISCUSSION

Microhotplate obtained within technology tests was made during several hours; the top view of the microhotplate is presented in Fig. 1d. Microhotplate was packaged in TO-8 housing, and studied for the power consumption and heating characteristic time. The following results were obtained: power consumption 525mW at 350°C (Fig. 3b) and heating characteristic time less than 10 seconds. Subsequent studies showed that for more precise design, the simulation of microhotplate in the Ansys program, which gives more accurate heat and power parameters of the manufactured chip, should be carried out initially.

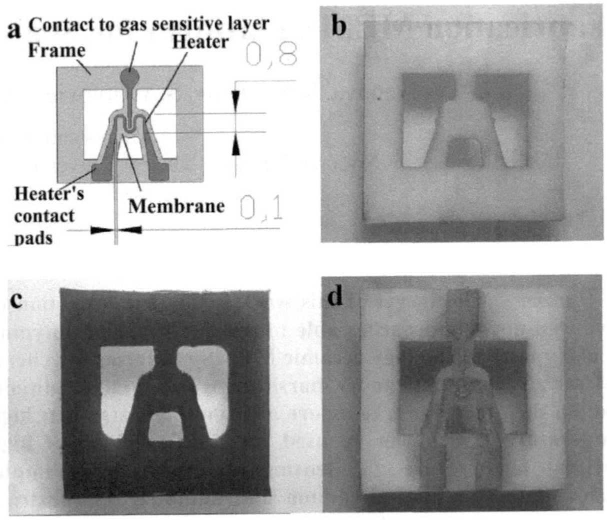

Fig. 1 Production stages of microhotplate (a) Design layout in AutoCAD software, dimension given in mm, chip size of microhotplate is 6×6 mm. (b) Engraving by laser Al_2O_3 substrate. (c) Deposition and annealing of Pt paste on top of chip microhotplate. (d) Engraving by laser Pt paste on the Al_2O_3 substrate.

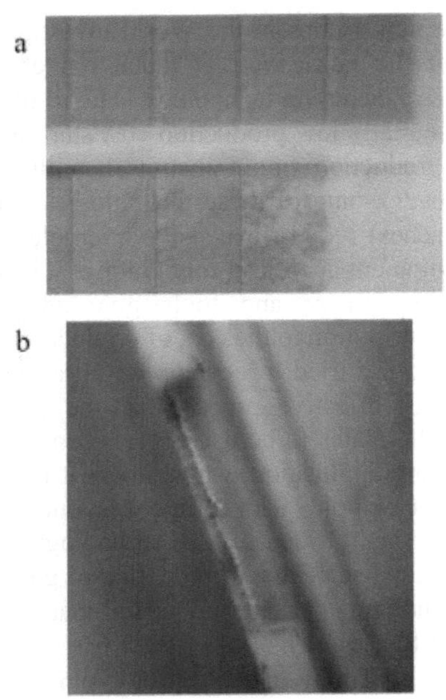

Fig. 2 (a) Test structure for investigation relationship between the number of laser ablation passes, laser power and depth of engraving (Each step is related to 5 laser ablation passes for 20W laser source power (average) and bottom structure for 16W respectively; (b) Photo of microhotplate's section with bottom side. Thickness of Al_2O_3 substrate is 500 μm. Thickness of engraving Al_2O_3 membrane is 110 μm. Roughness of engraving Al_2O_3 membrane measured by Veeco DEKTAK 150 Profilometer is 19 μm. Al_2O_3 membrane is 110 μm. Roughness of engraving Al_2O_3 membrane measured by Veeco DEKTAK 150 Profilometer is 19 μm.

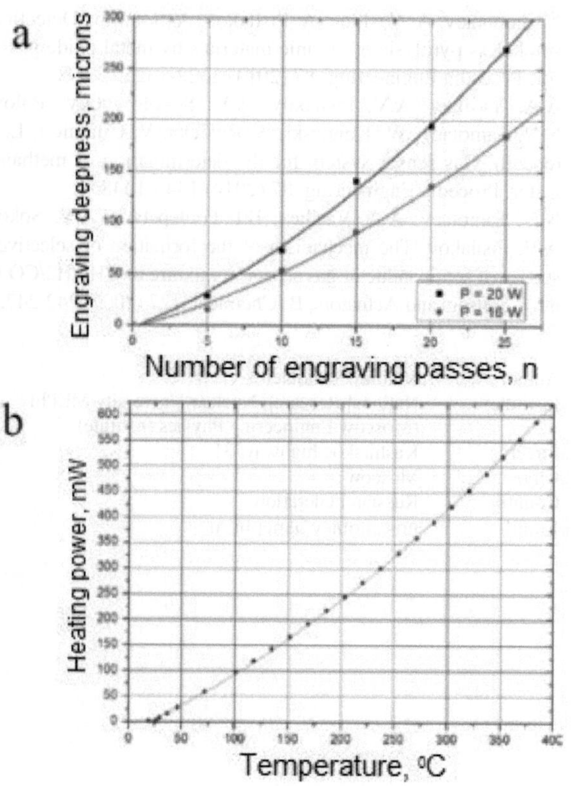

Fig. 3 (a) Dependence deepness from numbers of engraving passes measured by Veeco DEKTAK 150 Profilometer. (b) Thermal characteristics of microhotplate.

Table 1 Main Characteristics of the Laser Facility. Minimarker-2

Laser pulsed source type	ytterbium fiber
Wavelength, µm	1,064
Pulse frequency (regulated), kHz	from 20 to 100
Power (average),W	20
Pulse energy, mJ	1,0
Focused spot diameter, µm	50
Laser position system - two-axis scanner	G325DT
Maximum speed of the laser beam, mm/s	8700
line width with automatic filling, mm	From 0,05 to 3
Software and hardware solution, µm	2,5

IV. CONCLUSIONS

Using combination of cheap monolithic ceramic substrate, thick film Platinum paste, laser engraving and AutoCAD software for sensor layout design, microhotplates with good mechanical properties were fabricated. These chips can be fixed and bonded in TO-8 packaging; power consumption and response time characteristics are close to level typical for LTCC ceramic MEMS technologies. Production speed (laser engraving) of single MEMS structure was the order of 1 minute and 15 minutes for group annealing of Platinum paste in belt furnace. Our experiments show that describing technology in experience hands gives fantastic speed the design and manufacture of ready-to-use microhotplate in the order of 1 hour. The cost of materials (alumina substrate and Platinum paste) for the manufacturing single microhotplate size 6×6 mm is about 0.50 Euro.

The concept of the fabrication process also enables the integration of an array of sensor elements with different sensing layers and/or operating temperatures on to one single ceramic chip to be used for enhanced gas mixture analysis.

ACKNOWLEDGMENT

The research was performed within a frame of Russian Federation President Grant for support young scientists №14.Y30.15.7910-MK from 16.02.2015.

CONFLICT OF INTEREST

The authors declare that they have no conflict of interest.

REFERENCES

1. Simon, N. Barsan, M. Bauer, U.Weimar, Micromachined metal oxide gas sensors: opportunities to improve sensor performance, Sensors and Actuators B: Chemical. 73 (2001) 1-26.
2. D. Briand, J.Courbat, Micromachined semiconductor gas sensors. Semiconductor Gas Sensors. (2013) 220-260.
3. H.Teterycz, J.Kita, R Bauer, L.J.Golonka, B.W.Licznerski, K.Nitsch, K.Wi niewski. New design of an SnO_2 gas sensor on low temperature cofiring ceramics. Sensors and Actuators B: Chemical. 47 (1998) 100–103.
4. J. Kita, A.Dziedzic, L.J.Golonka, A. Bochenek. Properties of laser cut LTCC heaters. Microelectronics Reliability, 40 (2000) 1005-1010.
5. F. Rettig, R. Moos, Ceramic meso hot-plates for gas sensors, Sensors and Actuators B. 103 (2004) 91-97.
6. C.Moldovan, O.Nedelcu, P.Johander, I.Goenaga, D.Gomez, P.Petkov, U.Kaufmann, H.-J.Ritzhaupt-Kleissl, R.Dorey, K.Persson. Ceramic micro heater technology for gas sensors. Romanian J. of Information Science and Technology. 10 (2007) 43-52.
7. J.H. Kim, J.S. Sung, Yu.M. Son, A.A. Vasiliev, E.A. Koltypin, A.V. Eryshkin, D.Yu. Godovski, A.V. Pisliakov, A.V. Malyshev, S.S. Yakimov. Propane/butane semiconductor gas sensor with low power consumption. Sensors and Actuators, B. 44 (1997) 452-45.

8. E.E. Karpov, E.F. Karpov, A. Suchkov, S. Mironov, A. Baranov, V. Sleptsov, L. Calliari, Energy efficient planar catalytic sensor for methane measurement, Sensors and Actuators, A: Physical, 194 (2013) 176-180.
9. A.A. Vasiliev, R.G. Pavelko, S.Yu. Gogish-Klushin, D.Yu. Kharitonov, O.S. Gogish-Klushina, A.V. Sokolov, A.V. Pisliakov, N.N.Samotaev, Alumina MEMS platform for impulse semiconductor and IR optic gas sensors, Sensors and Actuators, B: Chemical. 132 (2008) 216-223.
10. A.A. Vasiliev, A.S. Lipilin, A.M. Mozalev, A.S. Lagutin, A.V. Pisliakov, N.P. Zaretskiy, N.N. Samotaev, A.V. Sokolov, Gas sensors based on MEMS structures made of ceramic ZrO2/Y 2O3 material, Proceedings of SPIE - The International Society for Optical Engineering. 8066 (2011) art. no. 80660N.
11. H. Teterycz, J. Kita, R Bauer, L.J. Golonka, B.W. Licznerski, K. Nitsch, K. Wi niewski, New design of an SnO2 gas sensor on low temperature cofiring ceramics, Sensors and Actuators B: Chemical. 47 Issues 1–3 (1998) 100–103.
12. N.N. Samotaev, B.I. Podlepetsky, A.A. Vasiliev, A.V. Pisliakov, A.V. Sokolov, Metal-oxide gas sensor high-selective to ammonia, Automation and Remote Control. 74 (2013) 308-312.
13. N. Samotaev, A. Vasiliev, A. Pisliakov, A. Sokolov, Detection of smokeless pyrolysis of organic materials by metal oxide gas sensor, Procedia Engineering, 87 (2014) 1322-1325.
14. A.A. Vasiliev, A.V. Pisliakov, A.V. Sokolov, O.V. Polovko, N.N. Samotaev, W. Kujawski, A. Rozicka, V. Guarnieri, L. Lorencelli, Gas sensor system for the determination of methane in water, Procedia Engineering. 87 (2014) 1445-1448.
15. N.N. Samotaev, A.A. Vasiliev, B.I. Podlepetsky, A.V. Sokolov, A.V. Pisliakov, The mechanism of the formation of selective response of semiconductor gas sensor in mixture of CH4/H2/CO with air, Sensors and Actuators, B: Chemical, 127 (2007) 242-247.

Author:	Nikolay Samotaev
Institute:	National Research Nuclear University MEPhI (Moscow Engineering Physics Institute)
Street:	Kashirskoe highway, 31
City:	Moscow
Country:	Russian Federation
Email:	nnsamotaev@mephi.ru

A Novel Bioactive Compound of Palladium(II) with Mercaptoethanol

D.B. Tagiyev[1], A.N. Azizova[1], S.R. Imamverdieva[2], and M.M. Asadov[1]

[1] Institute of Catalysis and Inorganic Chemistry, ANAS, Baku, Azerbaijan
[2] Institute of Control Systems, ANAS, Baku, Azerbaijan

Abstract— **A method for synthesis of the complex compound by reaction of palladium ion Pd(II) with mercaptoethanol in solution and equilibrium properties at complexation have been offered. The synthesis of the complex of Pd(II) with mercaptoethanol, and the identification and study of properties among various polar and nonpolar solvents are conducted. The models for calculating the stability constant of the complex formed in this system, with Pd(II) are considered. Processing of the experimental data using the selected multi-parameter model equation may provide new information on the behavior of the complex under various conditions.**

Keywords— **Palladium ions(II), mercaptoethanol, solution, equilibrium properties.**

I. INTRODUCTION

Recently, there has been a growing itnerest towards palladium complexes as modulators of pharmacological activity in pharmaceuticals. Such substances can increase the effect and reduce the overall toxicity of bioactive compounds [1-4].

In particular, in [3] pointed out that recently some of already developed pharmaceutical substances containing very small amounts of coordination dipyridyl (dipy) palladium compounds (~10^{-7} mol/l) with bridged L-cysteine molecules (CysH) - $[Pd_2(SCys)(SCysH)(dipy)_2](NO_3)_3\cdot 4.5H_2O$ are preclinically tested and pass through phases of clinical trials. Also polynuclear complexes with participation of Pd(II) cluster type are known. For example, one of them is heterometallic carbonyl-phosphine cluster [5]. In review [6] the results of the synthesis, structure and properties of homo- and heterometallic cluster containing palladium including structure type $Pd_n(CO)_x(PR_3)_y$ clusters with palladium nanoparticles are considered.

II. RESULTS AND ANALYSIS

It is known that -mercaptoethanol – $HS–CH_2CH_2OH$ has both S- and O-donor atoms, which leads to the fact that depending on the nature of the metal the ligand can be coordinated both mono- and bidentally. Thus, at first deprotonation of SH-group ligand has to occur and Cl-monodentate coordination is to be implemented.

If such a donor ligand reacts with the electron acceptor, such as, for example, palladium, wherein the coordination of ligand has to occur with sulfur atom of the ligand.

In this paper the problem of determining the conditions for formation of palladium (II) complex with polyfunctional ligand with sulfur-containing fragments is considered. The synthesis of Pd(II) complex with mercaptoethanol, and the identification and study of properties in media containing various polar and nonpolar solvents are carried out.

For experiments the reagents and solvents of industrial production, which were purified in accordance with standard techniques, were used. IR spectra of the substances were recorded in the spectral frequency range of 200-4000 cm^{-1} on FT-IR spectrometer IFS-113V in the form of suspensions in the vaseline and fluorinated oils between windows made from KBr and polyethylene. Spectral analysis was conducted according to the reference data.

Thermal analysis was performed on STA 449 F3. X-ray structure analysis was performed on Bruker SMART APEX II; IR spectroscopic analysis was performed on Nicolet S10 FT-IR.

Conclusion on the structure of the complex is based on the comparative study of IR spectra and X-ray crystallography.

Complex compound $[C_{24}H_{60}O_{12}S_{12}Pd_6]$ has been synthesized by the following method. $PdCl_2$ salt in NH_4OH solution at 50 °C and continuous stirring has been dissolved and the formed transparent solution was hot filtered. Then, according to the theoretical yield of the $[Pd(NH_3)_4]Cl_2$ (M) complex, the ligand of mercaptoethanol $HSCH_2CH_2OH$ (L) with M:L = 1:4 ratio was added to it. As a result of the reaction, the yellow precipitate is formed, which precipitates completely over 6 hours. Start a new paragraph by indenting it from the filtered, washed first with water, then with alcohol and ether; the precipitate was dried in vacuum to constant weight. From synthesized $C_8H_{26}O_4S_4N_2Pd_3Cl_2$ complex the aqueous suspension which with monoethanol amine forms an alkaline medium (pH=12) was prepared; at 80 °C the resulting precipitate dissolves and forms transparent solution. Ligand (mercaptoethanol) was added to that transparent solution until the light red solution forms. At decreasing temperature from 80 °C to 20 °C during 24 hours, red color crystals start to precipitate. After complete precipitation crystals were filtered, washed with alcohol and ether and then dried in vacuum over $CaCl_2$ to constant weight.

Obtained results of differential thermal analysis for $[Pd_6(\text{-}SCH_{22}CH_2OH)_{12}]\cdot 3H_2O$ complex allowed to establish

V. Sontea and I. Tiginyanu (eds.), *3rd International Conference on Nanotechnologies and Biomedical Engineering*,
IFMBE Proceedings 55,
DOI: 10.1007/978-981-287-736-9_70

temperature regime for $[Pd_6(-SCH_{22}CH_2OH)_{12}]\cdot 3H_2O$ formation, which is 600 °C. On the thermogram the three regions of mass change are showing.

The first region in the 20–325 °C temperature range characterized by a thermal effect and associated with removal of crystallization water molecules. The exothermicity of a process and relatively high initial decomposition temperature (294 °C) can be associated with intense oxidation of the ligand. The initial mass loss (27.13%) in the thermal analysis of the complex can be associated with the mercaptoethanol molecule loss. The next two regions in 325–510 °C and 510–850 °C temperature intervals are accompanied by one endo- and exo-effect respectively. With increasing temperature, decomposition of the complex takes place. Complete decomposition of the complex occurs at 455.9 °C. The final stage of decomposition is PdO formation, and at 897.7 °C decomposition of PdO to metallic palladium takes place.

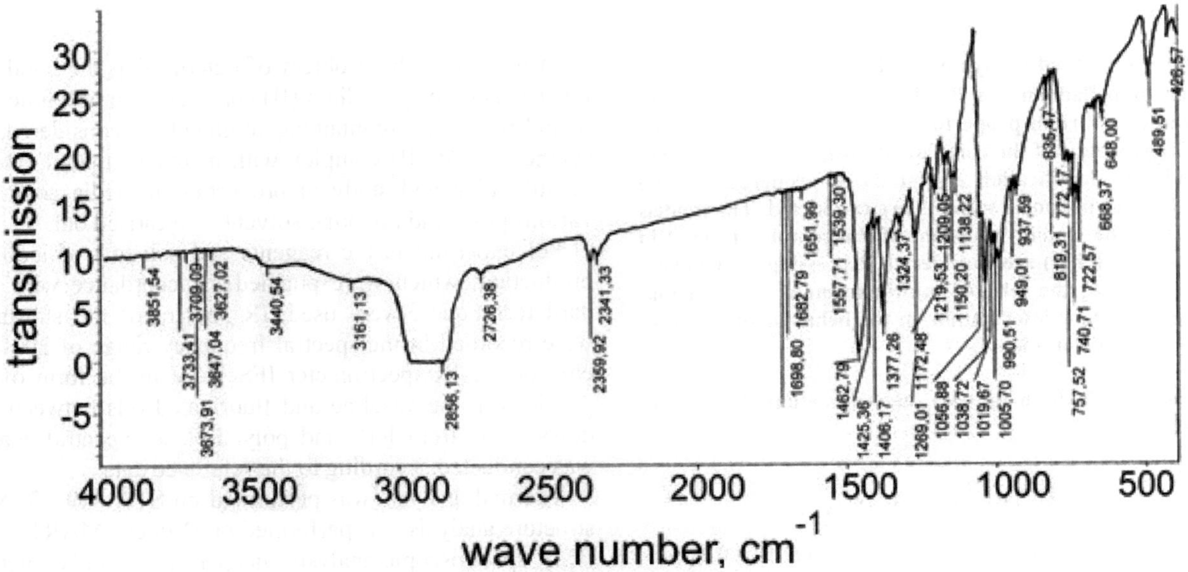

Fig. 1 IR spectrum of $[Pd_6(-SCH_{22}CH_2OH)_{12}]\cdot 3H_2O$.

Based on comparative study of IR spectra for ligand and complex the conclusion about a complex structure is made. In particular, the fact that monodentate coordination of -mercaptoethanol on thiolate sulfur atom occupying a bridging position, follows from IR spectra of the complex (Fig. 1); ν (Pd-S) = 284 and 280 cm^{-1}.

The characteristic IR band for -SH-group ligand - 2605 cm^{-1} in the IR spectrum of the complex is not observed. Two absorption bands for one valence bond indicates a bridging position for sulfur atoms of the ligand. The observed IR spectrum in the 3586–3620 cm^{-1} band of the ligand hydroxyl group in alcohol, indicates the neutral ligand. Observed in IR spectrum regions of 3586–3620 cm^{-1} - are bands of alcoholic hydroxyl group of the ligand indicate the neutrality of that ligand group.

XRD results confirm the IR spectroscopic data. The structure of the complex is a centrosymmetrical molecule - $[Pd_6(\mu_2\text{–}SCH_{22}CH_2OH)_{12}$ (Fig. 2), where each palladium atom is surrounded by four sulfur atoms of four (μ_2–$SCH_{22}CH_2OH$) - ligands on a flat, deformed square, in which PdS edges and planes form a ring.

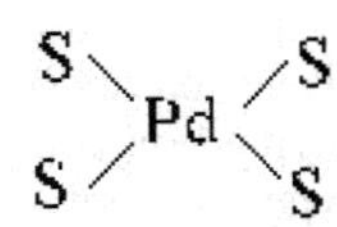

Fig. 2 Design of centrosymmetrical molecule, where each palladium atom is surrounded by four sulfur atoms.

In the fragments indicated in Fig. 2 bond angle S- is in the 81.229(18) and 98.976(18) range, and the bond length Pd-S - is in the 2.6094–2.3378 Å range. Six palladium atoms form a hexagon with 0.001 Å deviation from Pd6 plane. Pd-Pd distance is in the 3.0931 (2)-3.1098(2) Å range, and the range of vertex angles is 119.613(6)–120.659 (6) Å. The average distance between two opposing palladium atoms is 6.22 Å.

The two μ_2–SCH_2CH_2OH bridges arise from the combination of two - mercaptoethanol ligands with neighboring palladium atoms, wherein one of the μ_2–SCH_2CH_2OH atoms is above and one is below the Pd_6 plane, thus twelve μ_2–SCH_2CH_2OH radicals form double crowns. As a result, six Pd(II) atoms show virtual C_6 symmetry.

Plane with Pd(1), S(1) and Pd(3) creates a dihedral angle of 83.715(17) with a plane formed by Pd(1), S(6) and Pd(3). These two planes form a dihedral angle of 83.715(17) and 84.165(17) with planes of Pd_6 ring respectively.

In the crystal packing of $[Pd_6(\mu_2\text{–}SCH_2CH_2OH)_{12}]$ compound hydroxyl groups of - mercaptoethanol form strong intermolecular hydrogen bonding with one another, with the donor-acceptor distance from 2.676(3) to 3.454(2) Å, which leads to supramolecular network formation.

In order to establish regularities of complex formation reaction in solutions, the following solution for the problems is to be carried out: thermodynamic calculation of the energetics of intermolecular interactions in solutions containing bioactive compounds and ligands; detection of interaction mechanism for studied molecules.

For calculation of stepwise equilibrium complexation and constants in solutions based on experimental data, we have developed software for Windows operating system using Delphi environment [7].

III. Conclusions

Thus, in an aqueous medium the hexanuclear palladium complex $[Pd_6(\mu_2\text{–}SCH_{22}CH_2OH)_{12}]$ is synthesized and characterized. It was found that the complex is centrosymmetrical metallocrown molecules that bind with neighboring molecules due to hydrogen bonds and forming a supramolecular structure.

Conflict of Interest

The authors declare that they have no conflict of interest.

References

1. Eremin A.V., Antonov V.G., Panina N.S., Belyaev A.N., Simanova S.A. (2009) Coordination compounds of palladium and iron with oxygen-bridged ligands in oxidation reactions of thioamine acids. Russian Chemical Journal LIII. 1:135–139
2. Balazovsky M.B., Antonov V.G., Belyaev A.N., Eremin A.V. (2009) Low molecular weight modulators of pharmacological activity. Int. Patent. WO 2011/008132 A1. Corporation "IVA FARM". PCT,RU2010 / 000391. Application 13.07.2009
3. Stepanova M.A., Belyaev A.N. (2011) Bi- and oligonuclear palladium(II) complexes with thioamine acids - potential modulators of pharmacological activity of drugs. Contest of projects for young scientists. Chemistry Moscow, D. Mendeleev University of Chemical Technology 35–36
4. Stepanova M.A. (2011) Aminate coordination compounds of palladium(II) - potential modulators of pharmacological activity of drugs. Polzunovsky almanac 4/2:79–83
5. Kawano M., Bacon J.W., Campana C.F., Winger B.E., Dudek J.D., Sirchio S.A., Scruggs S.L., Geiser U. and Dahl L.F. (2001) High-Nuclearity Close-Packed Palladium-Nickel Carbonyl Phosphine Clusters: Heteropalladium [Pd16Ni4 (CO)22 (PPh3)4]2 and [Pd33Ni9(CO)41(PPh3)6]4 Containing Psuedo-Td ccp Pd16Ni4 and Psuedo-D3h hcp Pd33Ni9 Cores. Inorg Chem 40: 2554–2569
6. Mednikov E.G., Dahl L.F. (2010) Syntheses, structures and properties of primarily nanosized homo / heterometallic palladium CO / PR3 ligated clusters. Phil Trans R. Soc. A.:1301–1332
7. Tagiyev D.B., Asadov M.M., Azizov A.N., Imamverdiyev S.R. (2015) The method for determining the equilibrium constants of reactions of complex compounds. Patent of Azerbaijan. 2015 a 0007. Application. 30.03.2015

Author: M.M. Asadov
Institute: Institute of Catalysis and Inorganic Chemistry, Azerbaijan National Academy of Sciences
Street: H. Javid pr. 113
City: Baku
Country: Azerbaijan
Email: mirasadov@gmail.com

Copper-Containing Polyoxometalates: Syntheses and Anticancer Activity against the SH-SY5Y Human Neuroblastoma Cell Line

T. Gutul[1], A. Dimoglo[2], and T. Mironic[1]

[1] Institute of Electronic Engineering and Nanotechnology "D. Ghitsu", Chisinau, Republic of Moldova
[2] Gebze Technical University, Gebze- Kocaeli, Turkey

Abstract— **A new method for synthesising new copper-containing polyoxometalates based on selenium and tungsten has been developed. The structure of one of the complexes has been decoded; the cytotoxic effect of the complex on the human neuroblastoma SH-SY5Y has been studied. The cell lines have been cultured at different concentrations of polyoxometalates in a DME medium for 24 h; the percentage of cell viability has been evaluated by 3-(4,5 dimethyl thiazol-2yl)-2,5 diphenyltetrazolium bromide and neutral red. Polyoxometalates have shown the dose dependent inhibition of cell proliferation. These results suggest that the polyoxometalates exhibit strong anticancer and apoptosis activity against human neuroblastoma SH-SY5Y cell line.**

Keywords— **polyoxometalates, SH-SY5Y cell line, MTT, NRU.**

I. Introduction

Polyoxometalates (including heteropolyoxometalates and isopolyoxometalates) are oligomeric aggregates of transition-metal cations (so-called polyatoms, usually the d^0 species V(V), Nb(V), Ta(V), Mo(VI), and W(VI)) bridged by oxide anions [1]. Polyoxometalates (POMs) are remarkable for their molecular and electronic structural diversity and promising for catalysis [2]. POMs are discrete early transition metal oxides with a variety of compositions and structures and manifold properties which give rise to potential applications in different areas.

In addition to the design and synthesis of well-defined polynuclear clusters, POMs have also received considerable attention for biological applications, such as inhibitors of enzymes and bacteria. They exhibit interesting biological effects mainly related to their antiviral and antitumor properties.

In recent years, the application of POM in the medicinal field has made great progress [3]. $(NH_4)_{17}Na[NaSb_9W_{21}O_{86}]$ nH_2O (HPA-23) was the first POM applied to clinical trials as antiviral agent [X]. In the context of the antitumor studies of POM, Yamase and et al. [4] reported that $[NH_3Pri][Mo_7O_{24}]$ $3H_2O$ (PM-8) showed potent *in vivo* inhibitory effects on Meth A sarcoma and MM-46 adenocarcinoma. The authors concluded that polyoxomolybdates and polyoxotungstates exhibit high antitumor activity.

Here we report on the selenium-containing polyoxotungstates $Na_7\ [Cu_3Na_3(H_2O)_9(\alpha\text{-}SeW_9O_{33})_2]\ 21\ H_2O$ *(I)* and $\beta\text{-}Ala_2Na_5[Cu_3Na_3(H_2O)_9(\alpha\text{-}SeW_9O_{33})_2]\cdot XH_2O$ *(II)* which was isolated as a hydrated sodium salt. In this study, in order to obtain some valuable information to find an effective antitumor agent, we investigated the *in vitro* inhibitory effect of various selenium-containing POMs on the human neuroblastoma SH-SY5Y cell line.

II. Materials and Methods

The previously developed method for POM production was taken as the base for new synthesis. The advanced synthesis gave a higher yield of the reaction products and single crystals and thus provided successful PCA analysis.

Complexes (I) and (II) have been time synthesised by us [5-8]; later on, the structure of polyanion $[Se_2W_{18}Cu_3O_{66}(H_2O)]^{10}$ was crystallographically confirmed [9] (Krebs) for salt $K_9Na[Cu_3Se_2W_{18}O_{66}(H_2O)]\cdot 18H_2O$. In the present work, we describe the complexes synthesized by us on the basis of the new method and the physicochemical, structural and citotoxical properties of the complexes.

A. Synthesis of Complex $Na_7\ [Cu_3Na_3(H_2O)_9(\alpha\text{-}SeW_9O_{33})_2]$ 21 H_2O (I)

16.5 g (50 mmol) of $Na_2WO_4\cdot 2H_2O$ и 1.22 g (5.5 mmol) of Na_2SeO_3 were dissolved in 20 mL of boiling distilled water (at a mole ratio of Se:W = 1: 9). The solution was acidified with 6 M HCl till pH 5.8 under heating and admixed with 7.5 mL of a 1 M $Cu(NO_3)_2$ solution (a mole ratio of Se:Cu = 1:1.5). the resulting greenish-yellow solution was heated with a reflux condencer at 80°C for 1.5 h. A few hours later, precipitation of white and green crystals was observed. The green crystals were separated and recrystallized from a aqueous solution with a 85% final yield.

For complex (I) found (%): W 60,23; Cu 3,47; Se 3,20; Na 5,08; H_2O 8,05. Calc.: (%):W 61,35; Cu 3,55; Se 2,92; Na 4,26; H_2O 8,35.

The selected crystals were subjected to X-ray analysis.

V. Sontea and I. Tiginyanu (eds.), *3rd International Conference on Nanotechnologies and Biomedical Engineering*,
IFMBE Proceedings 55,
DOI: 10.1007/978-981-287-736-9_71

B. Synthesis of the Complex with β-Ala: β-Ala_2Na_5 [Cu_3 $Na_3(H_2O)_9$(α-$SeW_9O_{33})_2$]·$16H_2O$ (II):

1.08 g (0.2 mmol) of complex (I) was dissolved in 40 mL of distilled water; previously dissolved β-Ala (0.4 mmol) was introduced under pH 4.5 into this solution under heating (a mole ratio of the complex of 1:2). The solution changed its colour initially from green to dark-green and then to dark blue. A few hours later, precipitation of dark-blue crystals was observed. The crystals were selected and recrystallized from an aqueous solution. The yield was 35%.

For complex (II) found (%): W, 53.05; Cu, 3.10; Se, 2.24; H_2O, 3.80; Na 2.54. Calc.: (%):W, 54.45; Cu, 3.13; Se, 2.59; H_2O, 4.74; Na 3.03

C. Cell Lines

The SH-SY5Y human neuroblastoma cell lines were received from Biotechnology Institute of Yeditepe University (Istanbul, Turkey). Briefly, SH-SY5Y cells were cultured with 1:1 DMEM (Dulbecco's Modification Eagle Medium) and Ham's F12 Medium supplemented with 10% FCS, a penicillin-streptomycin (100 U/mL - 100 ug/mL) solution and 2 mM L-glutamine and humified, 5% CO_2, 37°C incubator.

D. Preparation of the Cells for Experiments

The SH-SY5Y cells were harvested by trypsinization after achieving adequate numbers for tests and suspended in culture media as 40 000 cell/mL counting by Neubauer slide. SH-SY5Y cells were seeded into each well of 96 well-plates for MTT (3-(3,4-dimethylthiazol-2-yl)-2,5-diphenyltetrazolium bromide), NRU (Neutral Red Up-take) tests. Stock solutions of polyoxomethalates were solved in dimethyl sulfoxide (DMSO, Sigma) at 250 mM concentration. MTT and NR solutions were protected from light and the entire experimental procedure was conducted avoiding the light. Then, polyoxometalate solutions were prepared in the culture media at 0.01-0.1- 1.0- 10 and 100 mM final concentrations for MTT, NR and CV assays. The same procedure was used for all tests; after 24 h of cultivation, the media were removed from plates. After 24, 48, 72, and 96 h after incubation of cells with POMs concentrations, the culture media were removed and dyes were applied [10].

E. MTT Assay

Viability/cell metabolism was tested by using MTT assay, which is a colorimetric test based on the formation of a complex between mitochondrial succinate dehydrogenase enzyme and tetrazolium salts. As a stock solution, MTT (3-(4,5-dimethylthiazol-2-yl)-2,5-diphenyl tetrazolium bromide) (Sigma) was dissolved as 5 mg/mL in phosphate buffer saline (PBS). It was filtered through a membrane filter with a pore size of 0.22 μm and stocked at -20°C. The MTT working solution was prepared from the stock solution with a dilution of 1:10 in a culture medium (final concentration of MTT was 500 μg/mL). At the end of incubation periods of 24, 48, 72, and 96 h, the MTT working solution was added to each well (100 mL/well) and the plates were incubated at 37°C for 4 h. After that, the media were removed and the cells were lysed with 100 mL DMSO. The cells were incubated under gentle shaking at room temperature for 10 min more. Absorbances at 540 nm were determined using a micro titer plate reader (Bio-Tek) [11].

F. Neutral Red Assay

This method was used for measuring the lysosomal activity of the surviving cells, since neutral red is a vital dye, which accumulates in the lysosomes of living and uninjured cells. (Xie SP). The stock solution of neutral red was prepared as 0.4% dye in distilled water and sterilized by using a membrane filter with a pore size of 0.22 μm. The working medium was prepared from the stock solution with a dilution of 1:100 in a culture medium giving a final concentration of 40 μg/mL. The fresh fixative/wash solution was prepared by using 1.3 mL formaldehyde (37%), 10 mL 10% calcium chloride, and 89 mL distilled water; the reading solution was prepared of 1.0 mL glacial acetic acid dissolved in 100 mL 50% ethanol.

At the end of incubation periods of the cells with POMs, a 100 mL/well NR working solution was added onto cells and the plates were incubated at 37°C for 4 h. the dye was removed from the plates by the end of 4 h, and the cells were rinsed once with a 200 mL/well fixative/wash solution. The plates were decanted and turned down on paper towels and drained for 24 h. One hundred mL/well of the reading solution was added onto the cells followed by incubation under gentle shaking at room temperature for 30 min. Absorbances were read at 540 nm using a microtiter plate reader.

G. Statistical Analysis

Statistical analyses were performed using Microsoft Excel 2002 and SPSS (version 11,5) Software. The results are expressed by mean-S.E.M. from four independent experiments. For all in vitro assays, the percentage of cell viability at each test concentration was calculated For statistical comparisons, quantitative data were analyzed by one-way analysis of variance (ANOVA) followed by Tukey-test. A value of less than 0.05 was regarded as significant.

III. RESULTS AND DISCUSSION

A. Antitumor Activity of Various Polyoxotungstates

The complex formation and bioactivity of POMs with varying amino acids can be found in a number of publications [8]. For the assessment of antitumor properties of (I) and (II), the influence of the POMs on the metabolism of the human neuroblastom SH-SY5Y cell line was investigated. This selection is caused by the high sensitivity of the cells to the components of the nutrient medium and their common use in the toxicity tests.

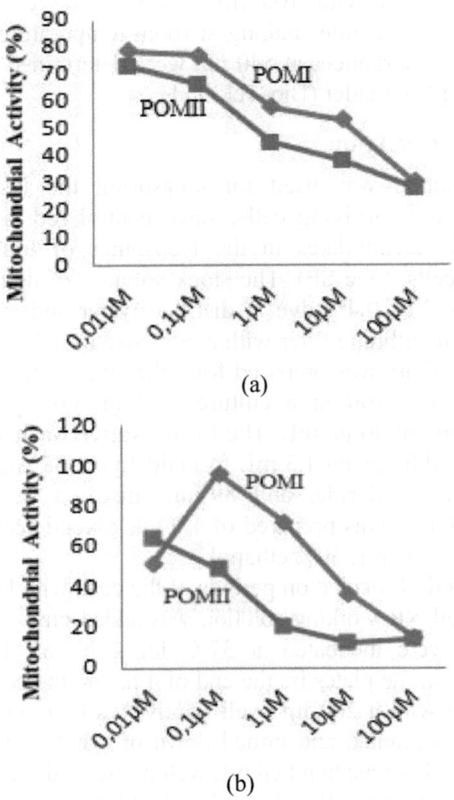

Fig. 1 Mitochondrial activity (MTT-test) of polyoxotungstates POMI and POMII in a concentration rage of 0.01-100 μM in the SH-SY5Y medium in the case of incubation for 24 (a) and 72 h (b).

Changes in the metabolic conditions of the cells were estimated according to the following indices:

1. A decrease of the total activity of mitochondrial dehydrogenases in the micro-tetrazolean test (MTT test – photometric approach).
2. A reduction in the absorption level (entocytose) for the neutral red dye (NRU), which correlates with lysosomal activity.

Figures 1 and 2 show the results of the mitochondrial activity relative the SH-SY5Y cell line for polyoxotungstates (I) and (II), while their lysosomal activity is shown in Figs. 3 and 4. The cells were incubated during the time intervals of 24-72 h with different concentrations of the complexes in a range of 0.01-100 μM.

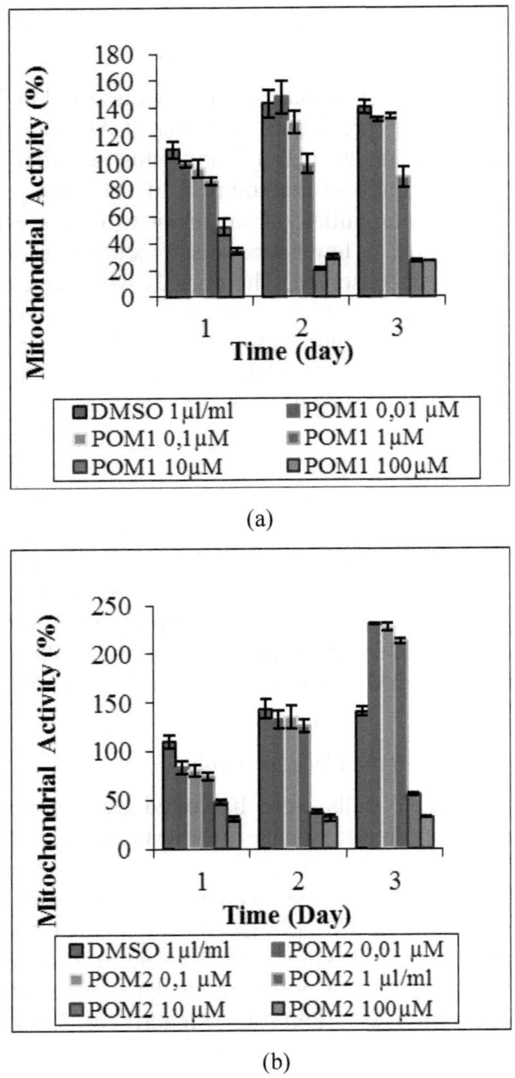

Fig. 2 Dependence of the mitochondrial activity (MTT-test) of polyoxotungstates POMI- (a) and POMII- (b) on incubation time in the SH-SY5Y medium at different concentrations (0.01-100 μM).

The results of the study have shown that both compounds exhibit a pronounced cytotoxic effect and cause the time- and dose-dependent response of the cells. For Cu salt (I), the numbers of survived cells are 52% after 24 h of incubation at salt concentrations in a range of 0.01-100 μM. For complex (II) with β-alanine, the number of survived cells decreases 64% of incubation at the same concentrations.

Antitumor activity increases with increasing incubation time and POM concentration. This fact suggests that the untitumor activity of heteropolyanion can be increased by imparting organic groups of amino acids onto the surface of POM that facilitates their penetration through the cell membranes. The conclusion is in agreement with the point of view in the works of Hill [3].

Figure 3 shows the lysosomal activity of POM(I) and POM(II) as a function of concentration and time of incubation of the cells.

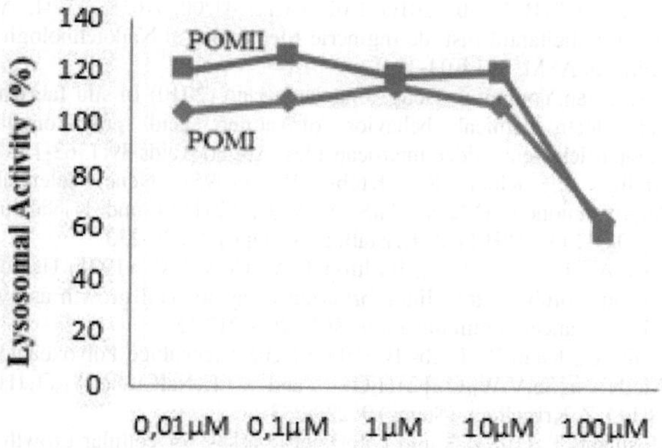

(a)

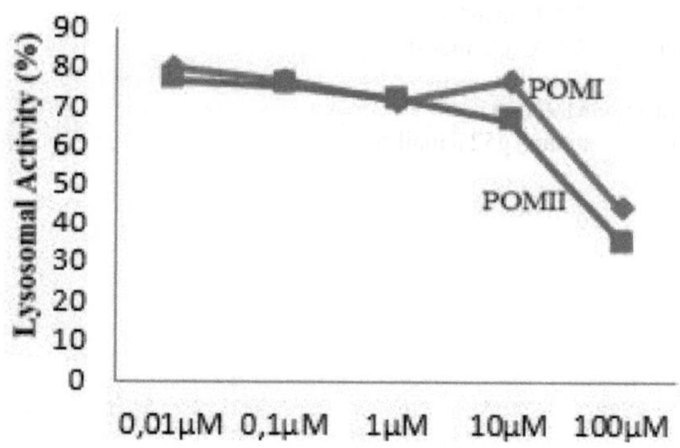

(b)

Fig. 3 Lysosomal activity (NRU-test) of polyoxotungstates POMI and POMII in a concentration rage of 0.01-100 µM in the SH-SY5Y medium in the case of incubation for 24 h (a) and 72 h (b).

The proportional decrease in the NRU dye, which is dependent on the POM concentration (see Figs. 3, 4), is also attributed to the lysosomatic activity of complexes (I) and (II). Curve *1* shows the dependence of the mitochondrial activity of the complexes on their concentrations; a peak of metabolic activity is observed at minimum concentrations. Thus, at a concentration of 0.1 µM, the values of mitochondrial activity for complexes (I) and (II) are 95 and 98%, respectively. This fact can be apparently attributed to the stress undergone by the cells under the action of chemical agents, as it has been noted earlier.

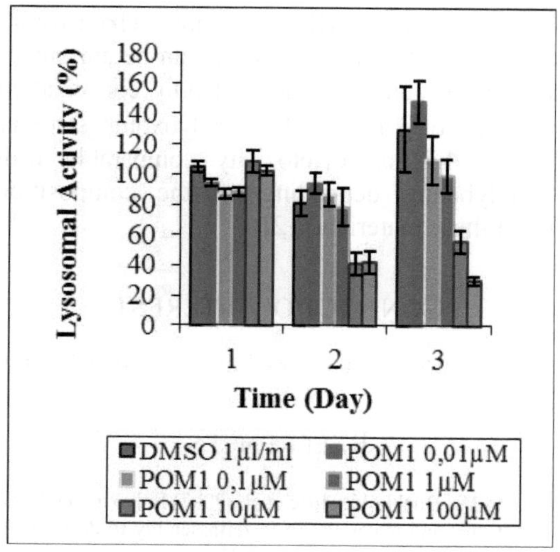

(a)

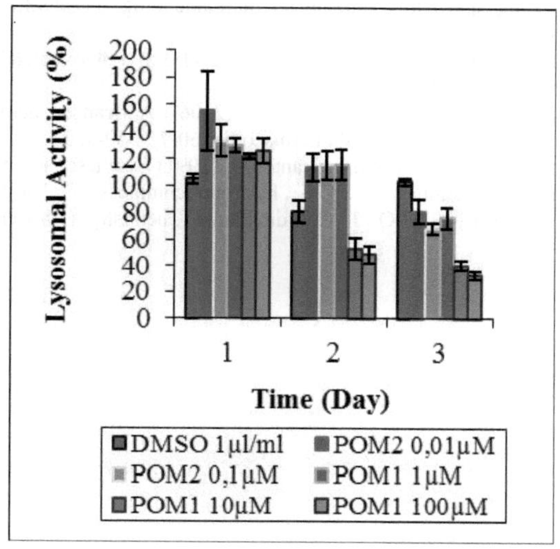

(b)

Fig. 4 Lysosomal activity (NRU-test) of polyoxotungstates POMI- (a) and POMII- (b) as a function of incubation time in the SH-SY5Y medium at different concentrations (0.01-100 µM).

IV. CONCLUSIONS

For the compounds under study, the IC_{50} index was calculated with the use of the dose-dependent action curve and non-linear regression analysis Hill-Modells (EDX 2.0 program). After 24-h incubation of the SH-SY5Y cells together with POMs with concentrations varying in a range of 0.01-100 µM, it was shown that the 50%-inhibiting concentration is 8µg/mL for complex (I) and 0.45 µg/mL for complex (II). A strong cytotoxic action demonstrated by complexes (I) and (II) is not characteristic of polyoxotungstates POMs. This result can be attributed to the fact that the polyanion contains not only copper atoms, but also amino acid of β-alanine. Thus, polyoxotungstates are also capable of exhibiting cytotoxicity comparable with that polyoxomolybdates depending on the composition and structure of these materials [12].

CONFLICT OF INTEREST

The authors declare that they have no conflict of interest.

REFERENCES

1. Liang Long De, Buckholder Eric et al. (2007) Polyoxometalate clusters, nanostruct ures and materials: from selfassembly to designer materials and devuces. Chem.Soc.Rev 36:105-121
2. Yamase T et al. (2005) Anti-tumor,-vital, and -bacteral activities of polyoxometalates for realizing an inorganic drug. J.Mater.Chem 15: 4773-4782
3. Rhule J.T., Hill C.L., Judd D.A. et al. (1998) Polyoxometalates in Medicine. Chem.Rev 96: 327-367
4. Yanagie H., Ogata A., Mitsui S. (2006) Anticancer activity of polyoxomolybdate. Biomed Pharmacother 60(7): 349-352
5. Gutsul T., Geru I., Bozan I., Nantoi O. (1991) Synthesis and physico-chemical studies of heteropolicomplecs $Na_{10}[Se_2W_{18}Cu_3O_{66}(H_2O)_2]\cdot 23\ H_2O$. J. Coordinational Chemistry (USSA) 17(5): 540-542
6. Gutsul T., Yurcenco E. (1992) Synthesis and structure of some Se(IV) and Te(IV) polyoxometalates. J. Coordinational Chemistry (USSA) 18(9): 944-947
7. Guţul T., Dimoglo A., Petrenko P., Korkmaz S., Mironic T. Utilizarea complecşilor polioxometalaţi cu atomi de seleniu centrali în calitate de compuşi cu activitate antitumorală. Republic of Moldova patent 4282. Int. Cl.: C01B 19/00, C01G 1/02, C01G 41/00, A61K 33/04, A61P 35/00. Solicitantul Inst. de Inginerie Electronică şi Nanotehnologii „D. Ghiţu" al AŞM. 2014-04-30.
8. Dawei Fan,Yongming Deng, Jingcheng Hao (2010) In situ fabrication and electrochemical behavior of amino acid polyoxometalate nanoparticles-embedded microcapsules. Amino Acids 39:1363-1367
9. Thülig C., Klein R., Krebs B. (1995) Neue selenhaltige Polyoxoanionen: $(Me_4N)_7Na[Se_2V_2W_{16}O_{60}]\cdot 21H_2O$ und $K_9Na[Cu_3Se_2W_{18}O_{66}(H_2O)]\cdot 18H_2O$, Z. Kristallogr. (Suppl.) 9: 229-233
10. Cory A. H., Owen T. C., Barltrop J. A., Cory J. G. (1991) Use of an aqueous soluble tetrazolium/formazan assay for cell growth assays in culture. Cancer communications 3(7): 207–212 15.
11. Thülig C., Klein R., Krebs B. (1995) Neue selenhaltige Polyoxoanionen: $(Me_4N)_7Na[Se_2V_2W_{16}O_{60}]\cdot 21H_2O$ und $K_9Na[Cu_3Se_2W_{18}O_{66}(H_2O)]$ $18H_2O$, Z. Kristallogr. (Suppl.) 9: 229-233
12. Mosmann T. (1983) Rapid colorimetric assay for cellular growth and survival: application to proliferation and cytotoxicity assays, J. Immunol. Methods 65: 55-57

Author: Tatiana Gutul
Institute: Institute of Electronic Engineering and Nanotechnology "D. Ghitsu"
Street: 3/3 Academiei str.
City: Chisinau
Country: Moldova
Email: tatiana.g52@mail.ru

Genotoxicity of Nanoparticulate Zinc Ferrite – Possible Application in Plant Biotechnology

G. Vochita[1], M. Oprisan[2], M. Racuciu[3], and D. Creanga[4]

[1] Institute of Biological Research, Iasi, Romania
[2] University Hospital "Sf. Spiridon", Iasi, Romania
[3] L. Blaga Univ., Faculty of Sciences, Sibiu, Romania
[4] Alexandru Ioan Cuza Univ., Physics Faculty, Iasi, Romania

Abstract— **Nanoparticulate matter, mainly nanoparticles containing metal ions could represent a promising biotechnological tool when preliminary testing could indicate the ability of inducing chromosomal changes in the plant tissues. Chemical route was applied as synthesis method of ZnFe2O4 powder with nanometric size, thus enabling plants to internalize the particles through endocytosis for further lysosomal digestion. Surface modification with long chain molecular shell was chosen to ensure nanoparticle stability in water dispersion. We present the results of some genetic effects induced by such prepared zinc ferrite nanoparticles as diluted suspension when supplied to agrotechnical plant species of large interest namely sunflower. Various and numerous chromosomal aberrations were identified that could be perpetuated in the form of genetic mutations through adequate biotechnology. Quantitative analysis emphasized remarkable influence of ZnFe2O4 nanoparticles on meristem tissues of freshly germinated seeds which suggested possible new biotechnological tool based on nanotoxicity application in genetics research.**

***Keywords*— zinc ferrite nanoparticle, oleate shell, vegetal tissue, cytogenetic analyses.**

I. INTRODUCTION

Nanotechnology development resulted in various applications of magnetic nanoparticles for biology and medicine [1]. Iron oxides are preferred for new medical methods of diagnosis and therapy due to their biocompatibility and magnetic properties. Being characterized by high opacity and heat resistance, zinc ferrites are used as pigments in applications requiring heat stability and corrosion protection [2].The relative toxicity of zinc derivatives do not exclude its utilization in cosmetics [3] and for some biomedical applications under controlled conditions like contrast agent in magnetic resonance imagistic [4]. Zinc ferrite nanoparticle synthesis by chemical coprecipitation seems to represent a highly available method characterized especially by its versatility, meaning that particle size and magnetic properties are dependent on the reaction conditions [5]. According to early reports [6] the growth of nanoparticles of mixed oxide of iron and other metal requires particularly higher temperature of the reaction medium comparing to the case of magnetite or maghemite that could be yielded even at room temperature. Other study [7] underlined also that temperature influences the formation of ferrites during the boiling of precursor salt solutions when dissolution-crystallization processes could take place. Recently zinc ferrite nanoparticles were obtained [8] at medium temperature (around 40 °C) with particle size of 5-10 nm but rather modest saturation magnetization of around 7 emu/g. Temperature raising could increase magnetic moment favoring an exchange of lattice sites of Fe^{3+} and Zn^{2+} ions [9], leading to the partial transformation of normal spinel antiferromagnetic zinc ferrite into a mixture of normal and inverse spinel structures with superparamagnetic features. The role of pH controlling during crystallite growth is also an important issue of coprecipitation method [5, 6] influencing intermediate hydroxides conversion into final oxides. Beside medical interest in zinc ferrite nanoparticles for increasing the contrast in magnetic resonance imagistic, there are some scientific reports focused on different bioeffects of zinc ions at different biological levels [10, 11]. Since zinc salt solutions were found to induce chromosomal changes in cereal plantlets [12] we carried out experimental investigation with nanoparticulate zinc ferrite aiming to identify chromosomal aberration - that could be further transmitted to next plant generation as genetic mutations; further selection of individuals with benefic phenotypic features could underlie new biotechnological approach in agrotechnical species cultivation.

II. MATERIALS AND METHODS

A. Nanotechnology of Zinc Ferrite Yielding and Characterization

First step of yielding technology was consistent with zinc ferrite nanoparticles preparation by chemical coprecipitation in alkali medium of zinc and ferric iron ions from corresponding metal salt as precursors in their respective stoichiometry ($Zn^{2+}/Fe^{3+}= 0.5$) i.e. 100 mL deionized water containing 10.8g $FeCl_3 \times 6H_2O$ and 100 mL deionized water containing 2.72g $ZnCl_2$. The temperature was raised over 80 °C during reagent stirring as well as during the dropwise addition of 12 g NaOH in 100 mL, (also hot solution). After one hour of continuous mixing the $ZnFe_2O_4$

V. Sontea and I. Tiginyanu (eds.), *3rd International Conference on Nanotechnologies and Biomedical Engineering*,
IFMBE Proceedings 55,
DOI: 10.1007/978-981-287-736-9_72

nanoparticles resulted with brownish color and magnetic responsivity. Ferrophase decantation in magnetic field gradient was followed by intensive washings with about 1 L of hot deionized water to remove impurities, mainly chloride ions (Scheme 1).

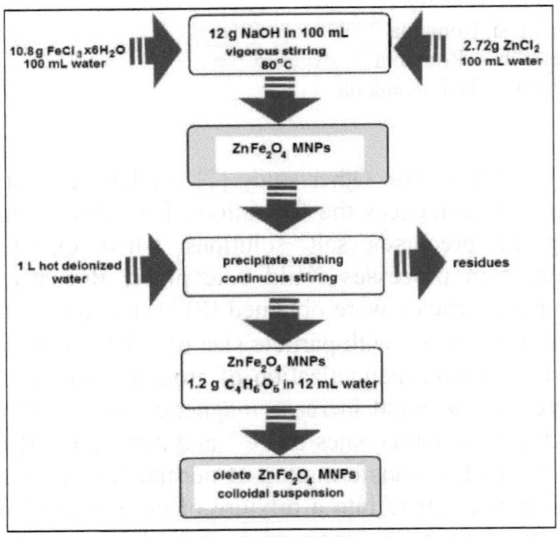

Scheme 1. Technology of nanoparticle yielding

During the second technological step the magnetic sludge separated on filter paper was carefully and continuously stirred with 1.2 g sodium oleate ($C_{33}H_{18}NaO_2$) in 12 mL distilled water at over 80 C° temperature that was ensured all over the time duration of ferrophase coating - for final colloidal suspension stability. Microstructural and magnetic investigation was accomplished by X-ray Diffraction (XRD), Vibrating Sample Magnetometry (VSM).

B. *Genotoxicity Analysis*

Biological material was consistent with seeds of sunflower germinated in darkness and 22 °C in INCUCELL room on moisten filter paper in Petri dishes. Three series of samples were arranged: one control series (5 Petri dishes with about 25 seeds each) supplied with deionized water; two test sample series differing by supplied liquid: one corresponding to 3.8 μg/mL $ZnFe_2O_4$/oleate nanosystems (contained in 40 μl/mL colloidal suspension) and the other corresponding to 7.6 μg/mL $ZnFe_2O_4$/oleate nanosystems (contained in 80 μl/mL colloidal suspension). When roots were no longer than 5-8 mm, tissue aliquots were harvested for cytogenetic analyses according to standard methods (squash method and rapid Feulgen dying technique) [13]. Nikon microscope with 40× objective was used to screen such prepared slides. Nikon Eclipse 600 camera with 100× objective with immersion was used to take images. Five slides for each experimental variant were analyzed; ten microscopic fields per slide were screened to count all cells exhibiting normal division (mitosis) and chromosomal aberrations.

III. RESULTS AND DISCUSSION

A. *Zinc Ferrite Oleate Coated Nanoparticles*

The dried colloidal suspension of zinc ferrite nanosystems was found to contain crystallites in spinel lattice either before or after surface modification with oleate ions (Fig. 1).

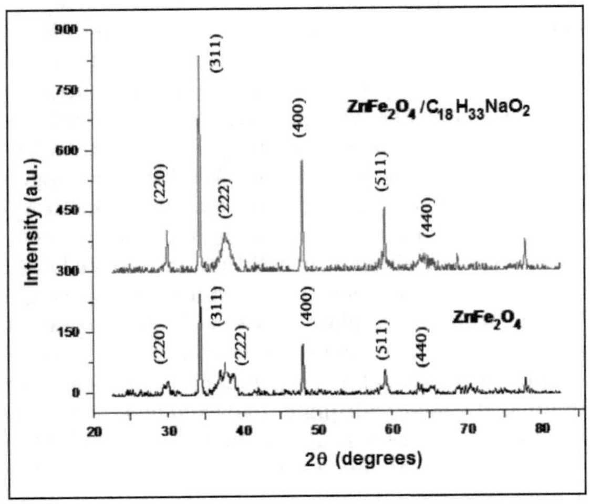

Fig. 1 X-ray diffraction analysis; results for zinc ferrite powder before and after surface modification with oleate

From Scherrer's formula the crystallite average diameter was calculated from most intense diffraction peak (3 1 1) resulting in D_{XRD}= 9.4 nm [14]. Other authors reported 9.6 nm crystallite size when similar experimental approach was applied to yield zinc ferrite nanoparticles [15]; in [8] crystallite size of 5-10 nm was reported for reaction temperature ranging between 20 and 80 ° C. Magnetic properties investigation at room temperature has evidenced that magnetization density over 30 emu/g was obtained for 10 kOe applied magnetic field and the sample magnetization still not saturated at this level (Fig. 2), which indicates the presence of superparamagnetic and single domain particles [8]. Other study [9] reported considerable higher saturation magnetization of zinc ferrite nanoparticles prepared also in alkaline reaction medium but applying ultrasound based physical method; the estimated magnetic diameter of about 10 nm resulted by applying Langevin's theory [16]. Thus, the zinc ferrite nanoparticles prepared as discussed above seem to be suitable for the testing of suggested new biotechnology since nanometric size permits their internalization in plant cells [17-19].

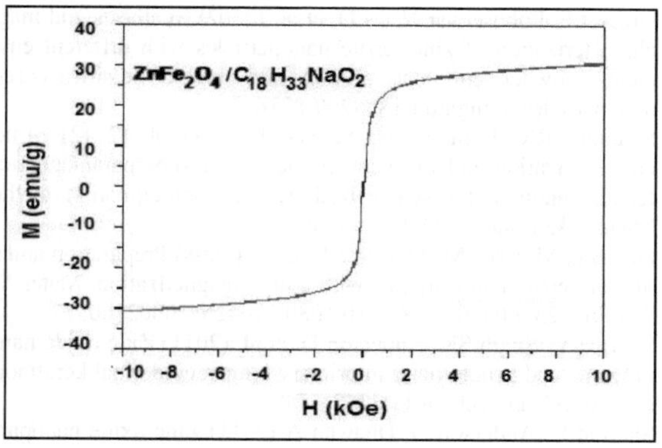

Fig. 2 Magnetization density versus applied magnetic field

B. Cytogenetic Analysis Results

The qualitative cytogenetic investigations led to the results shown in Fig. 3 where normally dividing cells (Fig. 3a) and aberrant divisions (aberrant mitoses) (Fig. 3 b, c, d) are presented.

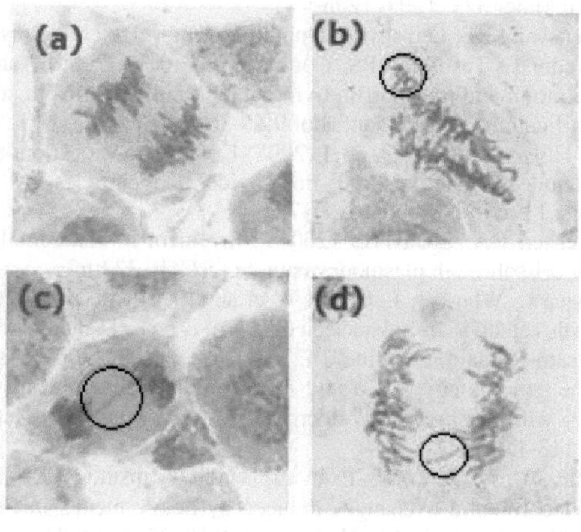

Fig. 3 Normal (a) and aberrant mitoses: (b) expelled chromosomes, (c) interchromatidal bridge, (d) laggard chromosome

Expelled chromosomes (Fig. 3 b) and laggard chromosomes (Fig. 3 d) are bearing part of genetic information that could be lost. Interchromatidal bridges (Fig. 3c) allow the two chromosome arrays to remain together in one single daughter cell – abnormal cell. The quantitative results are presented in Figs. 4-5. From Fig. 4 one can see that cell division rate (mitosis rate) was diminished by $ZnFe_2O_4$/oleate nanoparticles with more than 35% for 7.6 μg/mL ($p<0.05$) suggesting considerable influence on the mitotic activity of meristem tissue.

We mention that oleate control sample, corresponding to sodium oleate amount equivalent to higher concentration of colloidal nanoparticle suspension, shown no significant differences when compared to water control sample.

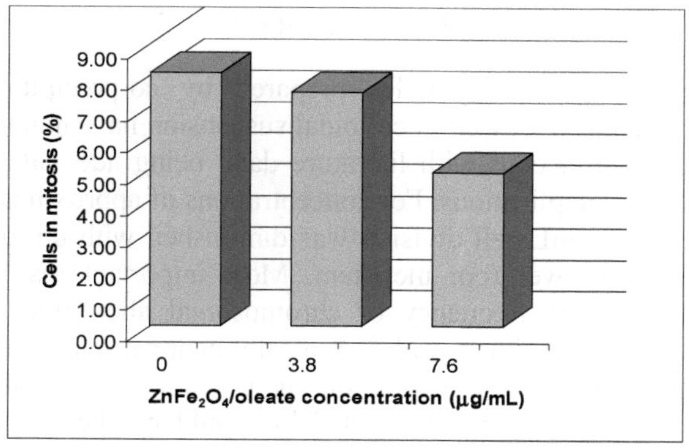

Fig. 4 Zinc ferrite nanoparticle influence on division rate

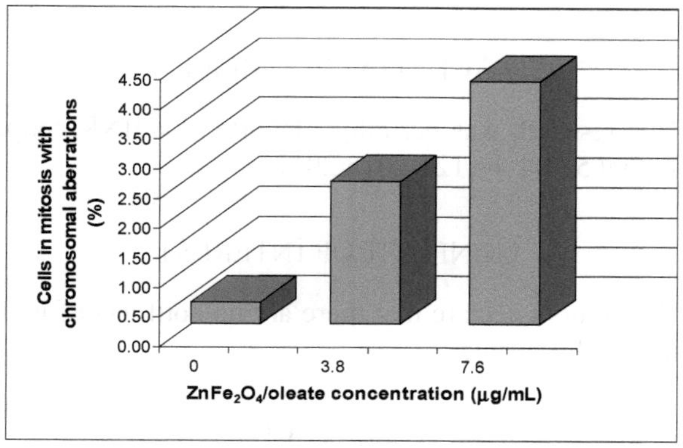

Fig. 5 Zinc ferrite nanoparticle genotoxicity

The frequency of chromosomal aberration sum (Fig. 5) was enhanced more than 5 times for 3.8 μg/mL $ZnFe_2O_4$/oleate nanparticles ($p<0.01$) and more than 8 times for 7.6 μg/mL ($p<0.01$. In control sample small number of chromosomal abnormalities could be present because of uncontrollable environmental factors like radiation background or unpredictable variability of biological material in spite of seed selection. Older scientific reports [20] have emphasized genotoxicity of zinc ions supplied from zinc salt solutions in anion root meristem based both on diminished mitotic index and enhancing of abnormal mitoses. Alternatively other papers mentioned relatively low zinc toxicity in bean plants compared to other heavy metals [21-22]. In the case of nanoparticulate zinc/iron oxide it is possible that specific properties of colloidal particles – mainly their large area to volume ratio confer supplementary potential of interference with plant tissues. The presence of iron ions could also influence plant ability of uptake some zinc ferrite nanoparticles due to the fact that iron is needed at certain levels for several essential metabolic processes (involving cytochromes and peroxidase enzymes); so that part of zinc ferrite nanotoxicity could be interpreted in this larger and more complex context of zinc ions administration to cultivated plants.

IV. CONCLUSIONS

$ZnFe_2O_4$/oleate particles prepared by co-precipitation method and stabilized in colloidal suspension have nanometric size in accord with literature data, being adequate for biological applications. For concentrations of approximately 4 and 8 µg/mL cell division was diminished with up to 35 % in sunflower root meristem. Most important result is related to the frequency of chromosomal aberrations induced by supplying $ZnFe_2O_4$/oleate nanoparticles which recommend them for possible new plant biotechnology based on their genetic effect. This could be the basis of nano-biotechnological protocol resulting in controlled yielding of new cultivars with selected beneficial biological parameters for improving cultivation of this plant species.

ACKNOWLEDGMENT

This research was partially supported by JINR cooperation grant 57/04-4-1121-2015/2017.

CONFLICT OF INTEREST

The authors declare that there are no conflicts of interest related to this article.

REFERENCES

1. Trana N, Webster TJ (2010) Magnetic nanoparticles: biomedical applications and challenges. J Mater Chem 20:8760-8767
2. Ayana YMA, El-Sawy SM, Salah SH (1997) Zinc-ferrite pigment for corrosion protection. Anti-Corros Meth Mater 44: 381–388. DOI: http://dx.doi.org/10.1108/00035599710367681
3. Landsiedel R, Ma-Hock L, Van Ravenzwaay B, et al. (2010) Gene toxicity studies on titanium dioxide and zinc oxide nanomaterials used for UV-protection in cosmetic formulations. Nanotoxicol, 4:364-381 DOI: 10.3109/17435390.2010.506694
4. Bárcena C, Sra, AK, Chaubey GS et al. (2008) Zinc ferrite nanoparticles as MRI contrast agents. Chem Commun 19:2224-2226 DOI: 10.1039/B801041B
5. Thirupathi G, Singh R (2012) Magnetic properties of zinc ferrite nanoparticles. IEEE Trans Magn 48: 3630-3633
6. Kamiyama T, Haneda K, Sato T, et al. (1992) Cation distribution in $ZnFe_2O_4$ fine particles studied by neutron powder diffraction. Solid State Commun 81: 563-566
7. Grasset F, Labhsetwar N, Li D, et al. (2002) Synthesis and magnetic characterization of zinc ferrite nanoparticles with different environments: powder, colloidal solution, and zinc ferrite-silica core-shell nanoparticles. Langmuir 18: 8209-8216
8. Shahraki RR, Ebrahim M, Seyyed ESA, et al. (2012) Structural characterization and magnetic properties of superparamagnetic zinc ferrite nanoparticles synthesized by the coprecipitation method. J Magn Magn Mater 324:3762–3765.
9. Drofenik M, Krist M, Makovec D, et al. (2008) Preparation and study of zinc ferrite nanoparticles with a high magnetization. Mater Manufact Proc 23: 603–606 DOI: 10.1080 / 10426910802160577
10. Sharma V, Singh SK, Anderson D, et al. (2011) Zinc oxide nanoparticle induced genotoxicity in primary human epidermal keratinocytes. J Nanosci Nanotechnol 11:3782-3788
11. Sharma V, Anderson D, Dhawan A (2011) Zinc oxide nanoparticles induce oxidative stress and genotoxicity in human liver cells (HepG2). J Biomed Nanotechnol 7:98-99
12. Truta E, Gherghel D, Bara I Cs, Vochita G (2013) Zinc induced genotoxic effects in root meristems of barley seedlings, Not Bot Horti Agrobo 41: 150-156
13. Singh RJ (2003) Plant Cytogenetics. Second Edition, CRC Press, Boca Raton, USA
14. Patterson A (1939) The Scherrer formula for X-ray particle size determination. Phys Rev 56: 978–982
15. Gomes JA, Azevedo GM, Depeyrot J et al. (2011) $ZnFe_2O_4$ nanoparticles for ferrofluids: A combined XANES and XRD study. J Magn Magn Mater 323: 1203–1206
16. Rosensweig RE, Ferrohydrodynamics, Cambridge Univ Press, 1985
17. Corredor E, Testillano PS, Coronado MJ, et al (2009) Nanoparticle penetration and transport in living pumpkin plants: in situ subcellular identification. BMC Plant Biol 9:45-56
18. Zhu H, Han J, Xiao JQ, et al (2008) Uptake, translocation, and accumulation of manufactured iron oxide nanoparticles by pumpkin plants. J Environ Monit 10: 71-78
19. Jorgensen RA, Lucas WJ (2006) Movement of macromolecules in plant cells through plasmodesmata. Sci STKE. 323:tr2
20. Dongua L, Whuseng J., Chunli W et al. (1996) Effects of Zn^{+2} on root growth, cell division and nucleoli of *Allium cepa* L. J Environ Sci 8:21-27
21. Marcato-Romain CE, Pinelli E, Pourrut B, et al. (2009) Assessment of the genotoxicity of Cu and Zn in raw and anaerobically digested slurry with the *Vicia faba* micronucleus test. Mutat Res-Gen Tox En 672:113-118
22. Gómez-Arroyo S, Cortés-Eslava J, Bedolla-Cansino RM, et al. (2001) Sister chromatid exchanges induced by heavy metals in *Vicia faba*. Biol Plant 44:591-594 **DOI**: 10.1023/A:1013703022584

Author: Creanga, D.
Institute: Univ. "Alexandru Ioan Cuza", Physics Faculty
Street: 11, Blvd. Carol I
City: Iasi
Country: Romania
Email: mdor@uaic.ro

Enhancement of Antioxidant and Antibacterial Activities by Immobilization of Natural Bactericide into Hybrid Supra-molecular Chitosan Bio-composite Gel

A. Gonta[1], T. Lupascu[1], I. Povar[1], N. Timbaliuc[1], T.E. Sukhanov[2], V. Petrova[2], and I.A. Skorik[2]

[1] Institute of Chemistry, Academy of Sciences of Moldova, Chisinau, Moldova
[2] Institute of Macromolecular Compounds, Russian Academy of Sciences, St. Petersburg, Russian Federation

Abstract— **Natural polyphenols are of significant morphological and physiological importance in plants. In particular, natural polyphenols have been reported to possess antioxidant, antibacterial and antifungal properties. However due to their low solubility in water, modification and improvement of physico-chemical properties of these compounds are of high interest among researchers. Therefore after depolymerization processes of these tannins, a product with increased water solubility have been obtained. Nevertheless these new products could undergo autooxidation reactions and shortening the action of biological activities. Following this, bio-matrix such as chitosan is considered to be the best choice for efficient entrapment of these modified polyphenols with prolongation of their biological activities.**

The difficulties in finding the right treatment for open trauma skin infections and efficient drugs inhibitors for bacteria growth are well known. For these reasons, in this work has been purposed immobilization of modified natural polyphenol bactericide into hybrid supra-molecular chitosan bio-composite gel with enhanced antioxidant and antibacterial properties for biomedical applications. For determination of antioxidant activities, ABTS and DPPH methods have been employed and it was observed that new bio-composite gel have an increased antioxidant activity with the inhibition of 91% of DPPH radicals in comparison with chitosan gel (55%) alone. It has to be mentioned that a novel bio-composite material has been prepared with enhanced bacteriostatic action against S.aureus (MIC of 0.02 mg/mL), E.coli (MIC of 0.62) and B.cereus (MIC of 0.01 mg/mL) bacteria pathogens frequently identified in open trauma skin infections.

Keywords— **antioxidant, antibacterial, bio-composite, chitosan, tannin.**

I. INTRODUCTION

Recent investigations have correlated the properties of some plants with their antioxidant capacity to defend the human body against oxidative stress [1]. On this basis, among the natural active sources, the grape seed extract represents an alternative, being a rich source of high active bio-flavones [2]. The grape extract is a complex mixture of oligomer proanthocyanidins with catechin as monomer unit, belonging to the class of the flavonoid polyphenols [3]. The biological properties of flavonoids have been extensively studied and reported [4]. These active compounds are known to exert anti-inflammatory, antibacterial and antiviral ones [5].

In order to improve physicochemical properties of tannins, such as the solubility in water, change of polarity, a product entitled Tanoxil was obtained after oxidation and depolymerization reactions of initial tannins and reported elsewhere [6]. This product represents a mixture of compounds such us catechin and quercetin monomers (see Fig. 1), quinic acid and also others components with carboxyl, aldehyde and ketone groups. During investigations of the antioxidant and antibacterial properties, Tanoxil has manifested amplified biological activities in all experiments. Although phenolic compounds are known for broad therapeutic activity, their use is limited due to chemical instability and low endogenous bioavailability.

To overcome these drawbacks, polymeric materials have been used for efficient immobilization or grafting the active molecules in order to protect phenolic compounds from degradation and act as protectors and prolongation of polyphenol biological activities.

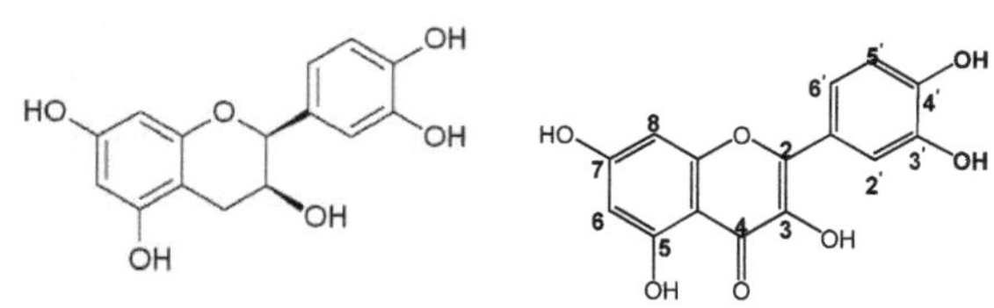

Fig. 1 Catechin structure – left side and quercetin - right side

Chitosan is a linear polymer saccharide (see Fig. 2), which contains free amino groups ($-NH_2$), used in a broad range of biomedical application due to its physical-chemical and biological properties [7].

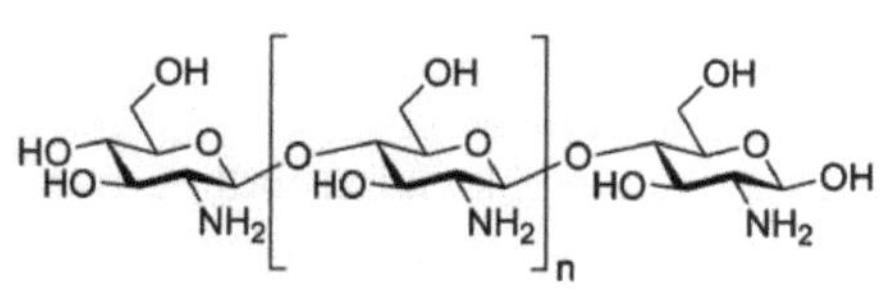

Fig. 2 Chitosan general structure

Moreover, chitosan is considered one of the first choices in the preparation of bio-composite gels, films intended for im-

V. Sontea and I. Tiginyanu (eds.), *3rd International Conference on Nanotechnologies and Biomedical Engineering*,
IFMBE Proceedings 55,
DOI: 10.1007/978-981-287-736-9_73

mobilization, entrapment and encapsulation of bioactive molecules, as it does not have high toxicity and its synthesis process is relatively easy to manage. Chitosan bio-matrix has been proposed as enhancer for the oral bioavailability of amino acids, essential oils, polyphenols and other compounds [8].

II. MATERIALS AND METHODS

A. Materials

Chitosan (M_w = 210 kDa, 77% deacetylated), Tanoxil (natural product was obtained in our laboratory), acetic acid (ACS reagent, purity ≥99.7%), catechin (purity ≥98.5%), ABTS (2,2′-Azino-bis(3-ethylbenzothiazoline-6-sulfonic acid, HPLC purity ≥98%), (2,2-diphenyl-1-picrylhydrazyl) DPPH (purity ≥98%), ammonium persulphate (APS, ≥99.8%) and Tween 80 have been purchased from Sigma-Aldrich (Germany), ethanol solution (purity ≥96%) from local chemicals provider, double distilled water was obtained in our laboratory. For homogenization of chitosan-Tanoxil solution Bandelin Sonopuls HD 2200 (20 kHz, amplitude setting from 10 to 100%). UV/Vis measurements have been carried out on the Jenway 6505 Spectrophotometer.

For microbiological investigation has been used two Gram-positive bacteria, S.aureus *ATCC 6538 P* and B.cereus *ATCC 11778,* and also one Gram-negative bacterium, E. coli *ATCC 25922*. Luria broth (LB) agar has been used as nutrient-rich media with bacteria inoculum.

B. Green Synthesis of Chitosan-Tanoxil Biomedical Gels

The preparation of bio-composite gel from *chitosan* (Cts) with *Tanoxil* (Tox) bactericide embedded into the bio-matrix, was carried out as follows: Tox (0.2 % , w/v) was dissolved in double distilled water (100 mL) until its complete solubility, then chitosan (1.5 g) was dissolved in Tox aqueous solution for swelling processes. Afterwards bulky solution was homogenized for 10 min in an ultrasonic homogenizer, Sonopuls HD 2200, with amplitude setting between 40-50%. Then, the solution is left for complete entrapment at room temperature in a dark place for 12 hours. Next, 0.5 mL of concentrated acetic acid was added to the solution and a gel-like solution was obtained.

C. Screening of the Antioxidant Activities in Vitro

The antioxidant activities *in vitro* were evaluated by ABTS assay and DPPH method.

$ABTS^{+\bullet}$ assay was employed according to the protocol by *Re Roberta et al.,* with minor modifications [9]. Briefly stock solution of ABTS+ (7 mM) was mixed with ammonium persulphate solution (2.45 mM). Afterwards, 40 µL of antioxidant was added to 4 mL of $ABTS^{+\bullet}$ ethanol solution (70%, w/v). The UV-Vis absorbance was measured at wavelength of 734 nm, exactly after 1 minute up to 6 minutes, which was compared with the reference sample (Abs = 0,700 a.u. ±0.02). The percentage of inhibition (PI, %) of $ABTS^{+\bullet}$ has been calculated as following: PI (%)=((A_0-A_1)/A_0)*100%, where A_0 is the UV/Vis absorbance value of reference sample solution (40 µL of H_2O) and A_1 is the UV/Vis absorbance value for sample with antioxidant addition, taken exactly after 1 minute.

Another method for determination of antioxidant scavenging activity is determination after inhibition with DPPH radical [10]. Briefly, 70µL of antioxidant was vortexed with 3 mL of DPPH (C_{DPPH}= 23.2mg/100mL of 70% ethanol solution), with measuring absorbance of sample at wavelength of 517nm, exactly after 3 minutes. The results are expressed as percentage of remaining DPPH radical (%) compared to the absorbance of reference sample (Absorbance = of 0,700 a.u., ±0.02), which has been calculated as: %, $DPPH^{\bullet}_{rem}$= $[DPPH^{\bullet}_{rem}]_{T=3}/[DPPH^{\bullet}_{init}]_{T=0}$*100%, where $DPPH^{\bullet}_{rem}$ denotes remaining concentration of DPPH radical after inhibition and $DPPH^{\bullet}_{init}$ is the initial concentration of DPPH radical, according to the values taken at initial time (T=0), before adding the antioxidant (as reference sample), and after 3 minutes (T=3) of the antioxidant addition.

D. Estimation of the Antimicrobial Activities

The antibacterial properties of chitosan gel, Cts-Tox gel and Tanoxil forming solutions have been tested towards inhibition of bacterial strains growth of three common identified bacteria in wound skin infections: two Gram-positive bacterium, S.aureus (3,2 x10^8 - 9,0 x10^8 UFC/ml), B.cereus (1,9 x10^7 - 1,6 x10^8 UFC/ml), and one Gram-negative bacterium, E. coli (1,2 x10^8 - 4,1 x10^8 UFC/ml). These microbiological investigations have been done in collaboration with the Moldovan State University of Pharmacy and Medicine "N. Testimitanu".

For antibacterial determination, the serial broth dilution method has been used, according to the Standards for Antimicrobial Disk Susceptibility Tests, developed by The Clinical and Laboratory Standards Institute (CLSI), Wayne, Pennsylvania, USA—January 2015, with minor modifications [11]. Briefly, 1 mL of Cts-Tox bio-gel has been added to 1 mL of Luria broth agar with 0.1 mL of bacteria inoculum and 1 mL of Tween 80, for decreasing the viscosity of Cts-Tox gel. Afterwards from this final solution, 1 mL has been taken for further twofold serial broth dilutions (up till 1:1024). Same experimental procedures have been carried out for the chitosan gel (1.5 %, w/v) and Tanoxil (0.2 %, w/v) solution.

III. RESULTS AND DISCUSSIONS

A. Antioxidant Activities

Polymeric matrix such as chitosan has been reported to exert antioxidant effects both *in vitro* and *in vivo* [12].

Inhibition of free radicals by the H-abstraction mechanism is possible due to the presence in the chitosan polysaccharide structure of two hydroxyl groups and one amino group, which can both react with free radicals according to the free radical theory [13].

Polyphenolic compounds may disrupt the oxidative cleavage mechanism initiated by reactive oxygen species. The hydroxyl groups in the 3' and 4' positions of the polyphenol catechol B-ring are the main structure characteristics for exerting antioxidant activity. Among these polyphenols are catechin, quercetin, kaempferol and others [14].

For the estimation of antioxidants a sensitive method, $ABTS^{+\bullet}$ assay, has been employed in this paper. According to the $ABTS^{+\bullet}$ method mechanism, the interaction of $ABTS^{+\bullet}$ with antioxidants will result in decolorization of the dark green color of the $ABTS^{+\bullet}$ ethanol solution. This decolorization has been monitored by UV/Vis absorbance measurements, expressed by the line charts in the Fig. 3.

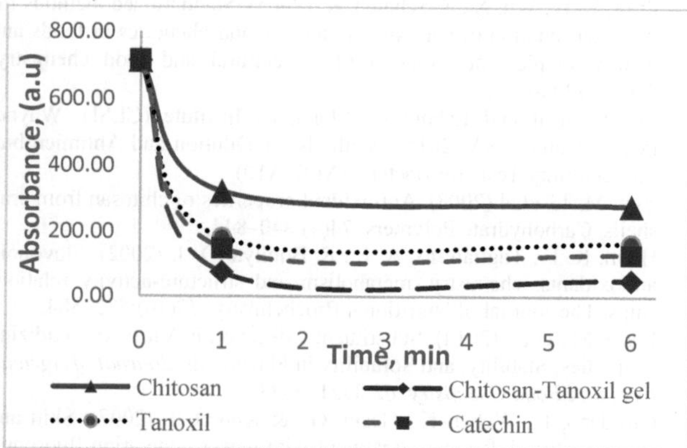

Fig. 3 Absorbance values after reduction of 4 mL ABTS+● with a) 40 µL of 1.5% (w/v) chitosan-acetic acid solution, b) 40 µL of 0.2% (w/v) Tanoxil solution, c) 40 µL of 0.2% (w/v) Catechin ethanol-aqueous (1:1) solution and d) 40 µL of Chitosan-Tanoxil gel

According to the results presented in Fig. 3, one can observe that the highest antioxidant activity is being exerted by the chitosan-Tanoxil gel solution, which could be explained by the synergy of the antioxidant mechanism of chitosan functionalized with Tox bactericide compared to chitosan or Tanoxil alone. Moreover, biopolymer with Tanoxil have a higher antioxidant activity than the natural polyphenol Catechin, which is considered to be one of the most potent natural antioxidant used in food, cosmetic and pharmaceutical applications.

The oxidation reactions initiated by DPPH radicals were efficiently inhibited by Cts-Tox gel due to an increase in its antioxidant activity by 50% as compared to chitosan alone, which is mainly based on the electron transfer reaction involving hydrogen – atom abstraction, in the first place, and electron-donor mechanism as a consequence.

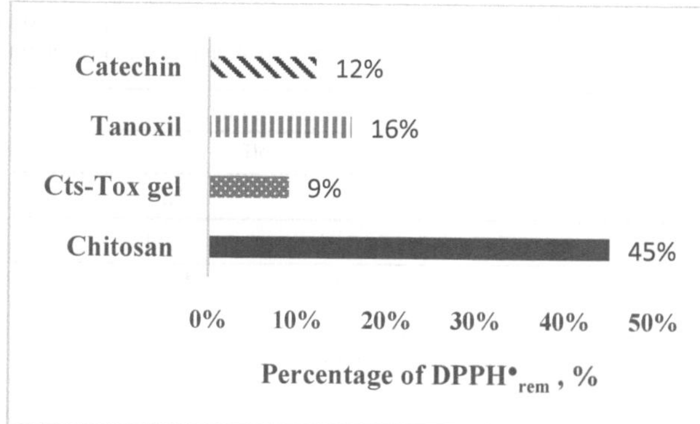

Fig. 4 Percentage of remaining DPPH● after addition of (A) 70µL of 0.2% Catechin ethanol-aqueous (1:1) solution, (B) 70µL of 0.2% Tanoxil solution (C) 70µL of chitosan gel and (D) 70µL of Chitosan-Tanoxil gel to 3mL of DPPH ethanol solution

Moreover, it has to be mentioned that antioxidant activity results by DPPH radical method are in concordance with ABTS cation radical assay data, with the following relation: Cts-Tox > Catechin > Tanoxil > Chitosan.

B. Antibacterial Properties

A common problem encountered in patients with wound skin infections is the contamination with Gram-positive bacteria, *Staphylococcus aureus*, which is the predominant bacteria found in patients with skin infections (in 46 % of skin infection cases) [15]. Nevertheless, Bacillus cereus infections in patients with open wound fractures have been also reported as a high potential in contamination of patients with open trauma cases. The Gram-negative bacteria E.coli could seriously affect in 8% of wound skin infections [16].

The main problem in finding the right treatment is the occurrence of bacterial-resistance to synthetic drugs, which in most of the cases should be administrated per-oral. Therefore, topical treatments are considered as a better approach for inhibition of bacterial growth in open trauma skin infections.

It should be mentioned that the addition of Tanoxil into chitosan gel leads to a significant increase of the antibacterial activity as compared to chitosan gel or Tanoxil solution alone. The results are presented in the Table 1. This phenomenon could be explained by the produced synergistic antibacterial effect when Tanoxil was entrapped into chitosan bio-matrix. The presence of both mechanisms, such as the interaction of the polycationic chitosan and the negatively charged surface of Gram-positive bacteria, as well as the formation of irreversible peptide bond between catechol B-ring of polyphenol and the peptidoglycan, improves antibacterial properties of Cts-Tox gel against S.aureus (MIC=0.02 mg/mL) and B.cereus (MIC=0.01 mg/mL) in comparison with Cts gel.

Table 1 Minimal inhibitory concentration (MIC) of chitosan gel, Tanoxil solution and Cts-Tox gel against S.aureus, E.coli and B.cereus bacteria

Tested material	MIC (mg/mL)		
	S.aureus	E.coli	B.cereus
Chitosan gel[a]	1.25	2.5	1.25
Tanoxil[b]	0.11	0.61	0.11
Cts-Tox gel	0.02	0.62	0.01

[a] chitosan (1.5 %, w/v) gel solution
[b] Tanoxil (0.2 %, w/v) aqueous solution

IV. CONCLUSIONS

In this work has been synthesized novel chitosan bio-composite gel with modified tannins as bactericide for the first time. The provided research results for antioxidant determination have highlighted the amplified antioxidant activity of prepared bio-gel in comparison with chitosan alone. Moreover the increased antibacterial properties of the bio-composite gel, especially against Gram-positive bacteria, S.aureus and B.cereus, could determine further in vitro and in vivo research for the use of such material as valuable bio-gel adhesive in open trauma skin infections.

ACKNOWLEDGMENT

Authors are grateful for the financial support accorded by the European Union's Seventh Framework Programme (FP7-PEOPLE-IRSES-2013) under the grant agreement Nr. PIRSES-GA-2013-612484.

CONFLICT OF INTEREST

The authors declare that there is no conflict of interest regarding the publication of this paper.

REFERENCES

1. García-Lafuente, A., et.al (2009). Flavonoids as anti-inflammatory agents: implications in cancer and cardiovascular disease. Inflammation Research, 58(9):537-552
2. Franco, D., et.al (2008). Polyphenols from plant materials: Extraction and antioxidant power. Electronic Journal of Environmental, Agricultural and Food Chemistry, 7(8):3210–3216
3. Vinson, J. A., Dabbagh, Y. A., Serry, M. M., & Jang, J. (1995). Plant flavonoids, especially tea flavonols, are powerful antioxidants using an in vitro oxidation model for heart disease. Journal of Agricultural and Food Chemistry, 43(11): 2800-2802.
4. Xia, E. Q., et.al (2010). Biological activities of polyphenols from grapes. International journal of molecular sciences, 11(2):622-646
5. Havsteen, B. (1983). Flavonoids, a class of natural products of high pharmacological potency. Biochemical pharmacology, 32(7):1141-1148
6. Gonta, A., Lupascu, T., Timbaliuc, N., & Meghea, A. (2014). Obtaining and characterizing modified tannins by physical-chemical methods. Open Chemistry, 12(7): 757-762
7. Croisier, F., & Jérôme, C. (2013). Chitosan-based biomaterials for tissue engineering. European Polymer Journal, 49(4):780–792. http://doi.org/10.1016/j.eurpolymj.2012.12.009
8. Kumirska, J., et.al (2011). Biomedical activity of chitin/chitosan based materials—influence of physicochemical properties apart from molecular weight and degree of N-acetylation. Polymers, 3(4):1875-1901
9. Re, R., Pellegrini, et.al (1999). Antioxidant activity applying an improved ABTS radical cation decolorization assay. Free radical biology and medicine, 26(9):1231-1237.
10. Prior, R. L., Wu, X., & Schaich, K. (2005). Standardized methods for the determination of antioxidant capacity and phenolics in foods and dietary supplements. Journal of agricultural and food chemistry, 53(10):4290-4302.
11. The Clinical and Laboratory Standards Institute (CLSI), Wayne, Pennsylvania, USA (2015). Methods for Dilution and Antimicrobial Susceptibility Tests for Bacteria (M07-A10);
12. Yen, M. T., et.al (2008). Antioxidant properties of chitosan from crab shells. Carbohydrate Polymers, 74(4):840–844.
13. Heim, K. E., Tagliaferro, A. R., & Bobilya, D. J. (2002). Flavonoid antioxidants: chemistry, metabolism and structure-activity relationships. The Journal of Nutritional Biochemistry, *13*(10):572–584.
14. Plaza, M., et.al. (2014). Substituent Effects on in Vitro Antioxidizing Properties, Stability, and Solubility in Flavonoids. *Journal of Agricultural and Food Chemistry*, *62*:3321–3333.
15. Giordano, P., Weber, K., Gesin, G., & Kubert, J. (2007). Skin and skin structure infections: treatment with newer generation fluoroquinolones. Therapeutics and Clinical Risk Management, 3(2), 309–317.
16. Dubouix, A., et.al (2005). Journal of infection, 50(1):22-30

Author: Alexandru Gonta
Institute: Institute of Chemistry of Academy of Sciences of Moldova
Street: Academiei 3, MD 2028,
City: Chisinau
Country: Republic of Moldova
Email: alexandru.gonta@chem.asm.md

Antimicrobial Reagents as Functional Finishing for Textiles Intended for Biomedical Applications. II. Metals and Metallic Compounds: Silver

F. Tanasa and M. Zanoaga

"Petru Poni" Institute of Macromolecular Chemistry, Romanian Academy, Iasi, Romania

Abstract— **Silver is known to exhibit strong toxicity towards a wide range of microorganisms and, therefore, it has been intensively and extensively used for bactericide purposes. This article reviews some of the most interesting studies concerning Ag biocide activity, the mechanism of action and the corresponding finishing techniques employed in order to obtain Ag-functionalized textiles for biomedical applications.**

Keywords— **silver, biocide effect, mechanism of action, functional textiles, biomedical applications.**

I. INTRODUCTION

Functional textiles intended for biomedical applications represent a rapidly growing research area, when it comes to textiles industry. Medical textiles, as a particular variety, provide a series of technical and functional properties according to specific demands in the field of medical care. They are commonly used as bandages, sutures, protective clothing, but more recently they are to be found as substrates to promote wound healing, scaffolds for tissue engineering able to assist cell growth, in stents and other intracorporeal devices [1].

The most challenging issue in the healthcare institutions is the microbial contamination of surfaces, medical textiles included, which can readily lead to infections. Therefore, irrespective of their specific application (intra- or extracorporeal devices, in direct contact with the wound or simply auxiliary materials), medical textiles need to possess bioactive properties, wether they are biocide or biostatic. New classes of antimicrobials with enhanced activity, ranging from synthetic reagents to naturally occurring biocids, were tested with promising results [1-6]. The use of metals and metallic compounds as bactericide agents is more recent, although some knowledge of silver antibacterial effect dates back to antiquity when people used to keep water, wine and vinegar in silver vessels to ensure freshness.

Many metals (heavy, transitional, noble) are toxic towards microorganisms at very low concentrations, either as atoms or in their corresponding compounds. They bind to specific intracellular proteins, inactivating them through the oligodynamic effect. This mechanism was discovered in 1893 and described as a toxic effect, to various degrees, of metal ions (Hg, Ag, Cu, Fe, Pb, Zn, Au, Al etc.) on viruses and living cells, even in relatively low concentrations [7]. Unlike viruses, bacteria are affected by the oligodynamic effect, which suggests that the biocide mechanism is interfering with the cell metabolism, since viruses do not carry out metabolic processes typical to living cells [8]. The mechanisms of action are not yet fully known and depend on the chemistry of the metal [9]. In some circumstances, the growth inhibition and cellular death are likely to be the result of a combination of different mechanisms. Experimental data suggest that these ions alter certain enzymes of the target cell by binding to their reactive groups, yielding in enzymes precipitation and inactivation.

Silver (Ag) is one of the oldest and best known metal with antimicrobial activity. However, systematic studies on various formulations of Ag and its mechanism of action are relatively recent and cover a wide range of applications, from healthcare to building, home appliances, food packaging and automotive industry. Along with the development of modern antibiotics, the interest in Ag as a broad spectrum antimicrobial decreased until recently.

This article reviews some of the most interesting studies concerning Ag biocide activity, the mechanism of action and some specific uses, as well as the corresponding finishing techniques employed in order to obtain Ag-functionalized textiles for biomedical applications.

II. SILVER FORMULATIONS

Silver was used in different forms throughout the time, but there are, basically, only two possibilities: Ag as atom (Ag^0) or ion (Ag^+). Ionic silver can be generated by the electrolysis of the metallic silver or dissolving the silver compounds (mainly salts) to yield in ionic solutions. It appears that the electrically generated silver ions have superior antimicrobial activity to the silver compounds [10] which can be explained considering their purity, the effect of the solvent, the hydration effect upon solubilization, the dynamic balance between hydration and salt binding, effect of pH, temperature etc. Various silver-based medical textiles absorb exudates and simultaneously release silver ions into the wound, acting on the microorganisms both inside the wound and within the dressing. The release of

V. Sontea and I. Tiginyanu (eds.), *3rd International Conference on Nanotechnologies and Biomedical Engineering*,
IFMBE Proceedings 55,
DOI: 10.1007/978-981-287-736-9_74

Ag^+ is favoured by the presence of bodily fluids and other exudates, which enable the silver compound to ionise and deliver the biologically active species [11].

Colloidal silver can be manufactured by a physical method (the electro-colloidal approach in which particles, consisting of 15 atoms or less, are electrically dispersed in deionized or distilled water) [12] or by synthesis (the reduction of soluble silver salts with reducing agents, carried out in water or organic solvents), when quasi-spherical particles with a narrow size distribution can be obtained [13-18]. It is known that silver accumulates in the body [16] and, upon arbitrarily intake, it can result in an accumulation of silver or silver sulfide particles in the skin, disorder known as *argyria* [16-18].

A modern approach refers to silver nanoparticles, but their preparation must be assisted by a stabilization stage using specific reagents able to prevent the growth and secondary aggregation of nanoparticles. Polymers such as poly(N-vinylpyrrolidone) (PVP) and substituted phosphane ligands were used as stabilizers with good results. The reproducible synthesis of silver nanoparticles proved to be more difficult than expected, due to the initially formed nuclei of metallic silver which can evolve into crystals having various morphologies and dimensions, depending on the reaction conditions (concentration, reducing agent, temperature, additives, impurities, mixing regime etc.) [18]. Silver nanoparticles are obtained by physical methods, too, performed typically in the gas phase [19] or plasma [20]. These nanoparticles are not surface-functionalized, and not redispersable in water, but are highly agglomerated. Laser ablation from metallic silver in liquids has been also used to prepare colloidal silver dispersions [21]. The biocide effect of Ag nanoparticles was significantly more pronounced on the gram-negative species, than on the gram-positive ones. This behaviour may be attributed to the nanoparticles stability in the cell and their effect on the bacterial proteins and bacterial growth cease [12].

III. MECHANISM OF BIOCIDE ACTION

Despite the well-known antimicrobial effect of silver, the mechanism of action is not yet fully understood, though is strongly connected to Ag chemistry.

It is known that, according to the Pearson's acid-base theory, Ag(I) is a soft acid and readily associates to soft bases, such as sulphhydryl (-SH) or thiol groups in proteins structure. Therefore, the biocide effect is directly proportional to Ag affinity towards –SH moieties. In Fig. 1, the most known mechanisms of the Ag biocide activity are presented.

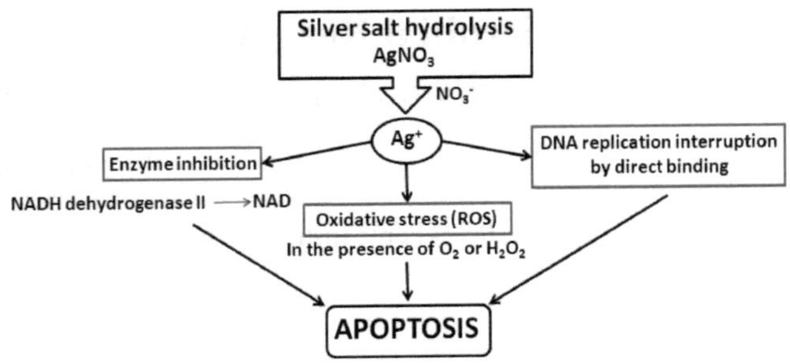

Fig. 1 Mechanisms of silver biocide action

One mechanistic hypothesis refers to the proteins denaturation, dysfunction and loss of the corresponding enzyme activity [22,23]. Another approach is the oxidative stress, in the presence of reactive oxygen species (ROS), along with the antioxidant depletion. It was proved that, in highly O_2-containing media, Ag catalyzes the complete destructive oxidation of bacteria. The same effect was noticed when Ag and H_2O_2 act synergistically, yielding in apoptosis due to irreversible damages of cellular proteins [24]. Apoptotic cells occur when their wall is permeabilized by Ag^+ that irreversible binds to cysteine residues in phosphomannose isomerase and interrupts the cell wall synthesis [25]. Genotoxicity was also considered, experimental data indicating that Ag^+ induced structural changes in DNA macromolecules by its interactions with –SH groups, causing condensed packing of DNA which lost their replication ability [12,24,26]. Despite its biocide effect, Ag has not yet been associated to increased mutation rates [9]. Bacterial signal transduction is another mechanism suggested. It was found to be affected by the phosphorylation of certain protein substrates, as decreased phosphorylation may indicate an inhibition of enzymes activity (*e. g.*, RNA polymerase sigma factors, UDP glucose dehydrogenases), with critical effect on gram-negative bacteria growth [12], although some positive results were also reported on exopolysaccharide and capsular polysaccheride biosynthesis and transport in both gram-negative and –positive bacteria [12,27].

IV. AG-FUNCTIONALIYED TEXTILES FOR BIOMEDICAL APPLICATIONS

Many metals have attracted attention as effective biocide agents for textiles (Cu, Zn, Co, Ni, Mn, Fe, etc.). Ag is by far the most widely used in functional textiles for biomedical applications [3-5,11,12,16,28-30], mainly as Ag^+ releasing dressings.

Ag-functionalized finishing of textiles is appropriate for both natural and synthetic fibers, but the protocol must be wisely selected depending on the nature of fibers, as it must afford the maximum efficiency in bioactive reagent upload.

Some of the most common commercially available textiles are presented in Table 1.

Natural fibers can be functionalized with silver only during the finishing stage. The main process indicators are the uptake and durability. Different approaches have been designed in order to optimize the process. Thus, a pre-treatment of the textile substrate is required to enhance the Ag salt adsorption. In the case of cotton fibers, the oldest and most widely used in healthcare, succinic acid anhydride was employed as ligand for Ag salts; the final product proved to be very effective as biocide dressing [4]. As for the protein fibers, such as wool, it is known they have polar and ionizable groups on the side-chain of constituent aminoacids. Ag^+ is most likely bonded to free carboxylic groups in aspartyl and glutamyl fragments which provide negatively charged binding sites over a wide range of pH conditions. Pre-treatments were also considered to obtain tightly binded Ag-wool fibers [31]: tannic acid was used to increase the number of binding sites, EDTA dianhydride was employed for its capacity to chelate Ag^+. Wool-Ag textiles showed low levels of metal release at acidic pH and prominent antimicrobial activity against bacteria, which recommend them as fabrics with enhanced barrier properties against microorganisms. Silk fibers (*Bombyx mori*) were also used in biomedical applications (tissue engineering, regenerative medicine) due to their biocompatibility, versatility and mechanical robustness, controllable degradability and positive impact on wound healing [32]. Two strategies were applied to obtain Ag-functionalized silk fibers/mats: the simple coating or direct inclusion of the Ag-containing reagent. The coated silk dressings enabled a slightly faster wound recovery than the drug loaded silk fibers, but their level of performance is higher as compared to existing wound dressings.

Synthetic fibers, given their structure and properties, are more readily functionalized, Ag being included into the polymer prior to the extrusion stage or nanofiber formation by electrospinning [4]. Upon textiles exploitation, Ag migrates onto the fibers surface and, in the presence of water (body moisture, exudates), ionic species is formed and gradually released over an extended period, thus providing biocide activity [33]. Besides direct incorporation, colloidal silver nanoparticles have been applied on synthetic and cellulose fabrics by padding [34], with good results in terms of finishing durability.

Commercial synthetic fibers, yarns and fabrics are largely produced for biomedical applications, as the finishing technology was specifically adapted. Ultrafine Ag particles were applied to polyester fabrics during the finishing stage (Ultra-Fresh® and Silpure®); a zirconium phosphate-based ceramic ion-exchange resin containing silver was added during the extrusion process of polyester or nylon fibers (AlphaSan®); silver ions are dispersed into zeolites which are subsequently incorporated into the polymer or coating in order to achieve the biocide effect by an ion-exchange mechanism (AgION® Technology).

Table 1 Commercial Ag-functionalized textiles

Product	Silver compound	Remarks
Acticoat™ Antimicrobial Barrier Dressings	Silver nanocrystals	Finishing agent
Allevyn™ AG Hydrocellular Antimicrobial Dressing	Colloidal Silver	
SilverSeal® Wound Contact Dressings	Ag_2O	
AgION®	Silver zeolite	Additive
AlphaSan®	Silver-based masterbatch	Additive
iSys AG	AgCl	Textiles finishing agent with an organic-inorganic binder
Ruco-Bac AGP	AgCl	Finishing agent
Nylcare	Ionic Ag	Ag^+-containing intellingent polyamide yarns
SeaCell®active	Ionic Ag	Cellulose fibers obtained from seaweeds, enriched with Ag^+
Silpure	Silver microparticles	Finishing agent

V. CONCLUSIONS

Antimicrobial Ag-functionalized textiles for biomedical applications provide benefits in hygiene, exudates and odor control and protection against microbial attack and proliferation, promoting wound healing, and securing sterilization and pro-active bacterial barriers.

A toxicological aspect still must be addressed: appart from the already known colloidal silver side effects (see argyria), the long-term effects of silver nanoparticles are still under assessment. Persistence in the human body of ingested formulations of silver was found to range between 30 weeks and 12 months, its effects are not life-threatening and has limited genotoxicity. Still, when it comes to nanoparticles, living organisms do not have appropriate mechanisms to deal with nanowaste, the interaction of nanoparticles with biological systems are not yet fully understood. This is mostly due to the fact that a nanoparticle shows many characteristics that go beyond its chemical composition. The greater nanoparticles surface area to volume ratio, the higher their chemical and biological activity, which yield in an increased production of reactive oxygen species (ROS), including free radicals. the environmental exposure to nanoparticles and its effects are already under study [18,35].

Given the constantly increasing use of metal-functionalized antimicrobial textiles in consumer and health-care sectors, the challenge to ensure no occurrence of unintended side-effects to human health and environment is becoming a race against time.

CONFLICT OF INTEREST

The authors declare that they have no conflict of interest.

REFERENCES

1. Sanders D (2007) Development in medical textiles. Tex Intel 70:97-130
2. Windler L, Height M, Nowack B (2013) Comparative evaluation of antimicrobials for textile applications. Environ Int 53:62-73
3. Ristić T, Zemljič LF, Novak M et al Antimicrobial efficiency of functionalized cellulose fibers as potential medical textiles, Science against microbial pathogens: communicating current research and technological advances, vol. 1 FORMATEX "Microbiology Series No. 3", Badajoz, Spain, 2011, pp. 36-51
4. Gao Y, Cranston R. (2008) Recent advances in antimicrobial treatments of textiles. Text Res J 78:60-72
5. Simoncic B, Tomsic B (2013) Structures of novel antimicrobial agents for textiles – A review. Tex Res J 80:1721-1737
6. Zanoaga M, Tanasa F (2014) Antimicrobial reagents as functional finishing for textiles intended for biomedical applications. I. Synthetic organic compounds. Chem J Mold 9:14-32
7. von Nägeli KW (1893) Über oligodynamische Erscheinungen in lebenden Zellen. Neue Denkschr. Allgemein. Schweiz. Gesellsch. Ges. Naturweiss. Bd XXXIII Abt 1.
8. Raoult D, Forterre P (2008) Redefining viruses: Lessons from mimivirus. Nat Rev Microbiol 6:315-319 DOI 10.1038/nrmicro
9. Lemire JA et al (2013) Antimicrobial activity of metals: mechanisms, molecular targets and applications. Nat Rev Microbiol 11:371-384
10. Jung WK et al (2008) Antibacterial activity and mechanism of action of the silver ion in Staphylococcus aureus and Escherichia coli. Appl Environ Microb 74:2171-2178
11. Aziz Z et al (2012) A systematic review of silver-containing dressings and topical silver agents (used with dressings) for burn wounds. Burns 38:307-318
12. Singh M, Singh S et al (2008) Nanotechnology in medicine and antibacterial effect of silver nanoparticles. Dig J Nanomater Bios 3:115-122
13. Silvert PY et al (1997) Preparation of colloidal silver dispersions by the polyol process. Part 2. Mechanism of particle formation. J Mater Chem 7:293-299
14. Silvert PY et al (1996) Preparation of colloidal silver dispersions by the polyol process. J Mater Chem 6:573-577
15. Sun Y, Xia Y (2002) Large scale synthesis of uniform silver nanowires through a soft, self-seeding polyol process. Adv Mater 14:833-837
16. Fung MC, Bowen DL (1996) Silver products for medical indications: risk-benefit assessment. J Toxicol Clin Toxicol 34: 119–126
17. Hadrup N, Lam HR (2014) Oral toxicity of silver ions, silver nanoparticles and colloidal silver – A review. Regul Toxicol Pharmacol 68:1-7
18. Chernousova S, Epple M (2013) Silver as antibacterial agent: ion, nanoparticle, and metal. Angew Reviews Angew Chem Int Ed 52:1636-1653
19. Backman U et al (2002) The effect of boundary conditions on gas-phase synthesized silver nanoparticles. J Nanopart Res 4:325-335
20. Loher S et al (2008) Micro-organism-triggered release of silver nanoparticles from biodegradable oxide carriers allows preparation of self-sterilizing polymer surfaces. Small 4:824-832
21. Stelzig SH et al (2011) Compatibilization of laser generated antibacterial Ag- and Cu-nanoparticles for perflorinated implant materials. Eur Polym J 47:662-667
22. Pandian SRK et al (2010) Mechanism of bactericidal activity of silver nitrate – a concentration dependent bifunctional molecule. Braz J Microbiol 41:805-809
23. Liau SY et al (1997) Interaction of silver nitrate with readily identifiable groups: relationship to the antibacterial action of silver ions. Lett Appl Microbiol 25:279-283
24. Feng QL et al (2000) A mechanistic study of the antibacterial effect of silver ions on Escherichia coli and Staphylococcus aureus. J Biomed Mater Res 52:662-668
25. Wells TN et al (1995) Mechanism of irreversible inactivation of phosphomannose isomerases by silver ions and flamazine. Biochemistry 34:7896-7903
26. Nover L et al (1983) Formation of cytoplasmic heat shock granules in tomato cell cultures and leaves. Mol Cell Biol 3:1648-1655
27. Mijakovic I et al (2005) Protein-tyrosine phosphorylation in Bacillus subtilis. J Mol Microbiol Biotechnol 9:189-197
28. Purwar R, Joshi M (2004) Recent developments in antimicrobial finishing of textiles–A review. AATCC Rev 4:22-26
29. Pavlidou V (2005) New multifunctional textiles: antimicrobial treatments. Proc. ITS-App., Prod. and Test. Internl. Workshop, Thessaloniki, Greece 2005
30. Hermans MH (2006) Silver-containing dressings and the need for evidence. Am J Nurs 106:60-68
31. Freddi G et al (2001) Binding of metal cations to chemically modified wool and antimicrobial properties of the wool–metal complexes. J Appl Polym Sci 82:3513-3519
32. Gil ES et al (2013) Functionalized silk biomaterials for wound healing. Adv Healthcare Mater 2:206-217
33. Williams JF et al (2005) Antimicrobial functions for synthetic fibers: recent developments. AATCC Rev 5:17-21
34. Lee HJ et al (2003) Antibacterial effect of nanosized silver colloidal solution on textile fabrics. J Mater Sci 38:2199–2204
35. Neal AL (2008) What can be inferred from bacterium–nanoparticle interactions about the potential consequences of environmental exposure to nanoparticles? Ecotoxicol 17:362-371

Author:	Fulga Tanasa
Institute: "	Petru Poni" Institute of Macromolecular Chemistry
Street:	41A Grigore Ghica Voda Alley
City:	Iasi - 700487
Country:	ROMANIA
Email:	ftanasa@icmpp.ro

Antimicrobial Reagents as Functional Finishing for Textiles Intended for Biomedical Applications. III. Other Metals and Metallic Compounds

M. Zanoaga and F. Tanasa

"Petru Poni" Institute of Macromolecular Chemistry, Romanian Academy, Iasi, Romania

Abstract— **Some metals and metallic compounds are known to exhibit high toxicity at very low concentrations towards a wide variety of microorganisms. This characteristic – the antimicrobial activity – has been used in medicine and health care for bactericide purposes. This paper summarizes some of the most interesting studies concerning metals and metallic compounds (oxides, salts, complexes etc.) with biocide activity, their formulation and mechanism of action, as well as specific finishing techniques employed in order to obtain metal or metallic compound-functionalized textiles for biomedical applications.**

Keywords— **metals, biocide effect, mechanism of action, functional textiles, biomedical applications.**

I. INTRODUCTION

Metals and metallic compounds have found significance and use into the medical field since the ancient times. Egyptians used copper (Cu) to sterilize water (3000 BC) and Chinese used gold (Au) in a variety of medicine (2500BC); Hippocrates (~400 BC) used Cu to treat leg ulcers associated with varicose veins and also used mercury (Hg); Paracelsus pioneered the use of metals in medicine using antimony (Sb), arsenic (As), magnesium (Mg) salt (1600s) – these are just a few examples.

Nowadays, metals and their salts are used in metal-based drugs, non-invasive radio-pharmaceuticals, magnetic resonance imaging (MRI), antimicrobial materials (textiles and beyond), mineral supplements [1]. Their specific mechanisms of action are investigated by the means of interdisciplinary studies (organic and inorganic chemistry, biochemistry, medicine, pharmacology) that aim to elucidate the complex interaction between metals and metallic compounds and biologic systems, their effect on biochemical processes (nutrition, energy production, DNA replication etc.) and toxicity (metals transport through media and storage in inert/active deposits), as well as to assess the capability of natural systems to deal with the corresponding residues [2-4].

Concerning the metal functionalized textiles used in biomedical applications, aside from their fundamental function as protective media (surgical masks, gowns and apparel), they proved able to respond to more specific requirements due to their corresponding specialized finishing. Thus, medical textiles can be found in sutures (surgical threads) and various implants (*e.g.*, vascular grafts), as well as gauzes, bandages, nappies, tampons etc.

Antimicrobial metal compounds, including metallic surfaces and coatings which are used, for example, on indwelling medical devices, chelates and nanomaterials, have found numerous applications in healthcare due to certain advances concerning the metal-cell biochemistry: it was proven that metals disrupt antibiotic-resistant biofilms [5], display synergistic biocide activity along with other biocides [6], selectively inhibit bacterial metabolic pathways [7] and destroy multidrug-resistant bacteria [8].

The mechanisms of inhibiting bacterial growth and killing pathogens are supposed to be independent of hyperosmotic shock [9]. Toxic doses of metals selectively disrupt processes involved in the bacterial cell growth and their action depends on the physical and chemical properties of both the metal atoms and the accessible donor ligands within intracellular biomolecules. Thus, metals take part in spatially localized reactions that inactivate enzymes, disrupt cell membrane or alter DNA replication. Data concerning the biophysical and biochemical mechanisms of metals toxicity towards bacteria, yeast or other microbial eukaryotes are scarce, probably due to the large metabolic diversity of microorganisms. Still, to elucidate the antimicrobial activity of metals becomes of high importance given the increasing amounts of metals and metallic compounds in the environment. On the other hand, a better understanding of metal biocide mechanisms will conttribute to the development of new generations of metal-based antimicrobials.

II. FORMULATIONS AND MECHANISM OF ACTION

Antibacterial mechanisms of metal toxicity depend on the nature of the metal and physiological mechanisms of bacteria and fungi, which can vary along with the corresponding biochemistry. An illustration of these mechanisms is given in Figure 1. It was proven that metals can induce proteins denaturation (Figure 1a), they can promote the production of intracellular reactive oxygen

V. Sontea and I. Tiginyanu (eds.), *3rd International Conference on Nanotechnologies and Biomedical Engineering*,
IFMBE Proceedings 55,
DOI: 10.1007/978-981-287-736-9_75

species (ROS) (Figure 1b), impair the membrane function (Figure 1c), interfere with metabolic processes (Figure 1d) and even alter the DNA replication (Figure 1e).

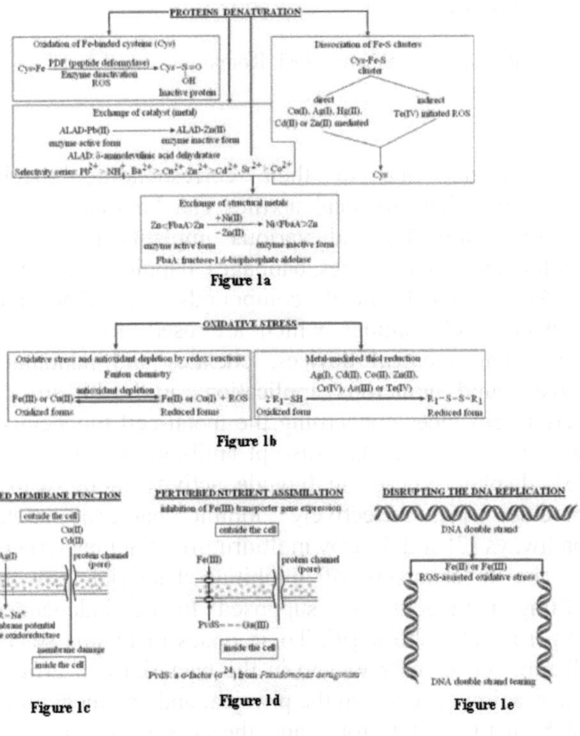

Fig. 1 Mechanisms of metal toxicity

A. Copper

Copper is known to be one of the most commonly used metal for antibacterial purposes, aside from silver. Model systems containing ascorbate, hydrogen peroxide and divalent copper or cobalt have been shown to possess marked bactericidal activity towards *Staphylococcus aureus* and *Staphylococcus epidermidis*, and low effect on *Serratia marcescens*, *Salmonella pullorum* and *Streptococcus faecalis* [10]. Cu-containing systems were more bactericidal than the corresponding mixtures containing Co. The mechanisms of action of these systems are not yet clear, but it may be assumed they act by generating superoxide, hydroxyl radicals or other free radicals. It seems that the two active metal-mediated systems function *via* separate mechanisms. The copper system acts with dehydroascorbate, whereas the cobalt system appears to depend upon the generation of free radicals.

In other formulation, Cu (as acetate $(CH_3COO)_2Cu$) was embedded in sol-gel coatings on textiles in combination with tetraethoxysilane (TEOS) and 3-glycidyloxypropyl-triethoxy-silane (GLYEO). Figure 2 shows a schematic of the sol-gel process.

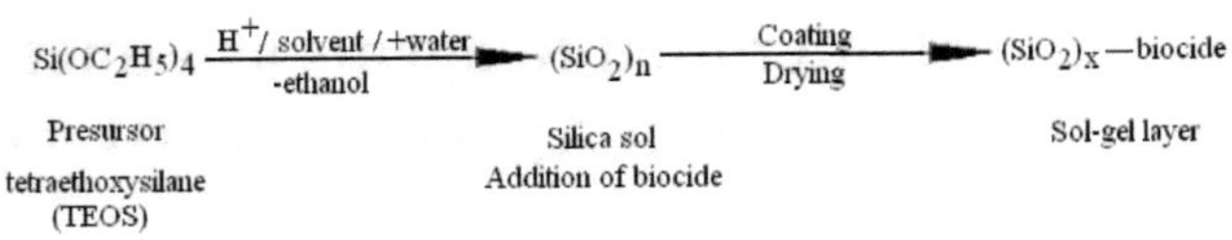

Fig. 2 The sol-gel process scheme

Its activity was tested on fungi (*Aspergillus niger*) and two types of bacteria (gram-positive *Bacillus subtilis* and gram-negative *Pseudomonas putida*) [11], and it was evaluated as function of concentration, thermal treatment and addition of epoxysilane compounds. A strong bactericide effect was observed against both bacteria, but a less fungicidal effect, either after a thermal treatment at 120 or 180°C. This may be explained by the strong release of copper from coatings.

Aqueous solutions of $CuSO_4 \cdot 5H_2O$ were employed to prepare copper-containing chitosan fibers [12], as well as zinc impreganted fibers, using $ZnCl_2$ solution. Significant weight gain was obtained as the zinc and copper ions were adsorbed onto the fibers through chelation with the primary amine groups from the chitosan structure. The zinc and copper-containing fibers were tested for biocide action against several species of bacteria (*Escherichia coli, Candida albicans, Staphylococcus aureus, Bacillus sabtilisin, Pseudomonas pyocyanea*) commonly found in wounds and on the skin. The zinc and copper ions can be released from fibers when exposed to aqueous solutions containing either NaCl or protein, indicating that these ions can be released from the fibers when they are used in wound and skin care applications. The protein component in the solution can greatly enhance the release of zinc and copper ions. Experimental results showed that these metal-containing chitosan fibers had much stronger antimicrobial properties than the original chitosan fibers.

A more recent approach was using Cu nanoparticles with a diameter of about 4 nm, obtained by the polyol method [13,14]. They were subsequently embedded into a polymer matrix (poly(vinyl chloride), PVC), and the resulting nanocomposite coating was tested on *Escherichia coli* with very good results, revealing a significant reduction of the adhered bacteria to the nanocomposite film after 4 days of incubation.

Concerning the mechanism of Cu toxicity, most studies indicated that it may be involved in metal-induced oxidative stress, by increasing the intracellular reactive oxygen species (ROS) [15]. In *Escherichia coli*, the exogenous addition of hydrogen peroxide (H_2O_2) or agents that catalyse the production of superoxide ($O_2^{\bullet -}$), such as paraquat, leads to DNA damage and the inhibition of particular enzymes activity that are vital for cell growth. *Pseudomonas aeruginosa* and *Saccharomices cerevisiae* mutants deficient in the production of ROS-scavenging enzymes, as well as other cellular antioxidants, frequently exhibit changes in sensitivity towards Cu [16]. In *Escherichia coli*, Cu(II) exposure abolishes enzymes activity that are, in some regards, similar to those lost during H_2O_2 and $O_2^{\bullet-}$ poisoning, although for Cu(II) it is now clear that this is ROS independent [17]. Moreover, many reports indicate that bacteria and yeast exposed to toxic doses of Cu(II) upregulate genes that are involved in the elimination of ROS [17]. Therefore, toxicity of Cu may be due to the ROS-mediated cellular destruction, as substantiated by various studies.

B. Iron

Another metal with bactericide properties is iron (Fe), known as a strong acid according to Pearson's acid-base theory, since it has a smaller ionic radius, a high oxidation state and weak polarizability. Due to its redox potential, it is present in *in vivo* Fenton chemistry [9] (metal-catalysed free radical chain reactions that are initiated by inadvertent by-products of aerobic respiration, such as hydrogen peroxide (H_2O_2) or superoxide ($O_2^{\bullet-}$), in which Fe(2+) is oxidized to Fe (3+). Thus, due to the high rate of Fe-catalysed oxidation, proteins, lipids, polysaccharides, and even DNA may be altered *in vivo*. A series of redox reactions illustrative for the *in vivo* Fenton chemistry is presented in Figure 3.

$$Fe^{2+} + H_2O_2 \rightarrow Fe^{3+} + OH^- + OH^\bullet \quad \text{Fenton reaction} \tag{1}$$

$$Fe^{3+} + O_2^{\bullet-} \rightarrow Fe^{2+} + O_2 \tag{2}$$

$$O_2^{\bullet-} + H_2O_2 \rightarrow O_2 + OH^- + OH^\bullet \quad \text{Haber–Weiss reaction} \tag{3}$$

$$Fe^{3+} + \underset{\text{antioxidant}}{\text{reduced}} \rightarrow Fe^{2+} + \underset{\text{antioxidant}}{\text{oxidized}} \tag{4}$$

$$OH^\bullet + RSH \rightarrow H_2O + RS^\bullet \quad \text{Thiyl radical} \tag{5}$$

$$OH^\bullet + (R)_3CH \rightarrow H_2O + (R)_3C^\bullet \quad \text{Carbon-centered radical} \tag{6}$$

$$(R)_3C^\bullet + O_2 \rightarrow (R)_3COO^\bullet \quad \text{Peroxyl radical} \tag{7}$$

Fig. 3 The in vivo Fenton metal-catalysed free radical chain reactions: radicals are hydrogen superoxide (O2•−), hydroxyl (HO•), thiyl (RS•), alkyl (R3C•) and peroxyl (R3COO•)

Some authors mentioned the use of iron oxides (Fe_2O_3, Fe_3O_4) as antibacterial compounds [18], but recently it was zero-valent Fe that suscited the interest in this field [19,20]. It is known that zero-valent iron (Fe^0) has been used in permeable reactive barriers for remediation of groundwater contaminated with halogenated solvents. Direct electron transfer (redox reaction) from metallic iron to contaminants has been acknowledged as the main pathway of contaminant interaction with Fe^0. In the presence of oxygen, the contaminants can also be oxidized by hydroxyl radical and other oxidants generated during the Fe^0 oxidation [21].

The zero-valent Fe was conditioned as nanoparticles and tested on *Escherichia coli* [20]. Compared to other Fe formulations (Fe II and III oxides – FeO, Fe_2O_3, Fe_3O_4, etc.), Fe^0 nanoparticles displayed an enhanced biocide effect, close to Ag nanoparticles. The mechanism of action assumes the oxidation of Fe^0 to Fe^{2+} and Fe^{3+}, subsequently. The penetration of the cell membrane is enabled by Fe^0 nanoparticles, although it is not yet clear in which step the oxidation of Fe^{2+} to Fe^{3+} occurred. The chemical interactions between the reactive surface of Fe^0 nanoparticles and *Escherichia coli* induce the disruption of the cell membrane by the reductive decomposition of functional groups of its proteins and lipopolysaccharides, given the highly reducing character of Fe. By TEM technique, it was possible to evidence the presence of iron oxides inside the cytoplasmatic membrane. The presence of Fe^{2+} substantiates the hypothesis of the oxidative stress by producing ROS inside the cell *via* Fenton reaction [22]. It is still not clear whether Fe(II) ions bind to DNA or specific proteins may directly prevent cell replication or hinder enzymatic function.

C. Noble Metals (Gold, Palladium)

Gold and palladium are known for their low solubility and biocompatibility, but are, however, used in medicine [23-25]. Still, some gold complexes show antimicrobial and antifungal activity [26,27] and gold or gold-palladium coated grafts were proven to prevent infection by *Staphylococcus epidermidis* [28]. Given that noble metal-coated silicone or polyurethane catheters have been shown to cause an *in vitro* reduction of *Escherichia coli*, *Pseudomonas aeruginosa*, *Enterococcus faecalis*, *Klebsiella*, and *Candida* [25], a further study was considered of interest. Therefore, Au and Pd-based coatings were applied through a wet dipping method that allowed different loads of noble metals. The study was performed in association with Ag. Low concentrations of Au and Pd decreased the anti-inflammatory properties of silver due, probably, to the electrochemical potential differences, which have been shown to have antibacterial effect, in particular by reducing/inhibiting the biofilm formation [29]. A further addition of Au and Pd enhanced the anti-inflammatory effect. Another possible explanation is that the anti-inflammatory properties of the coatings are related to the surface topography, which is known to be an important factor for protein adsorption and the subsequent cellular and bacterial adhesion. Thus, it may play a role in the modulation of the *in vivo* inflammatory response.

The anti-inflammatory activity of monovalent gold is well documented [30,31], but less data are available in the literature on the biological effects of palladium. Considering the current studies, it may be assumed that mainly Ag, but also Pd and Au, act as a catalytic hydrogen peroxide and superoxide sink. This effect may be efficient for antibacterial purposes *in vitro*, but can only moderately affect wound healing around implants. The mechanisms of reducing the inflammatory response during the first 24 h of implantation are still unclear, but they may be related, al least partly, to the catalytic degradation by ROS.

Coatings with a higher Au and Pd content were associated with the highest, but not significantly higher, cell death after 21 days of implantation and also displayed the thickest fibrous capsule, which is an indication of the enhanced ion release and toxicity, leading to the conclusion that using noble metals on implant surfaces enables modulation of specific events of inflammation and fibrosis during the integration in soft tissues [25].

Antifouling coatings based on noble metals, namely Ag and Au, with biocide effects were also developed, by the means of nanotechnology, with promising results in terms of applications in healthcare, as well as in nano-bio-technology [32].

D. Other Metals and Metallic Compounds

Along with other noble metals, platinum (Pt) is reported to have therapeutic applications, due to its ability to damage the DNA in tumoral cells and induce apoptosis [33], although it is necessary to convert Pt^{4+} (non-toxic) to Pt^{2+} in order to reach antitumoral activity.

Cobalt (Co) complexes with different ligands were reported in the literature [34], but Co (II) is the most studied probably because its complexes are stable in aqueous solutions, easy to prepare and available. Co(III) complexes were also reported and it was emphasized the increased effectiveness of Co ion coordination to a particular ligand when it comes to biocide properties.

Titanium dioxide (TiO_2) nanoparticles also have a bactericidal effect [35]. This is based on a photocatalytic reaction: under the influence of the UV radiation, and in the presence of water and oxygen, the nanoparticles generate extremely reactive radicals, such as hydroxyl and perhydroxyl, which have antimicrobial action by the ROS mechanism [36]. Titanium dioxide can be used to produce antiseptic surfaces that only operate in the presence of the UV radiation.

Another study reports on the preparation of TiO_2 and aluminum oxide (Al_2O_3) nanoparticles by laser pyrolysis [37]; they were tested in the presence of *Cupriavidus metallidurans* and *Escherichia coli*. Experimental data confirmed that the antimicrobial activity of TiO_2 and Al_2O_3 nanoparticles depends on their chemical composition, size, surface charge and shape, but not on their crystalline phase.

A special formulation of metallic compound and its antimicrobial activity against *Escherichia coli* was reported: CdTe quantum dots (QDs) [38]. QDs are nanoscale crystalline clusters synthesized from semiconducting materials. Currently, they are widely used as effectively alternatives or complementary tools to conventional fluorescent dyes because of their great photostability, bright photoluminescence, narrow emission, and broad UV excitation. Under UV irradiation, QDs yield in free radicals (depending on their core materials), which are known to be toxic to microorganisms, as well as the discharge of free heavy metal ions. It is unknown how CdTe QDs react with *Escherichia coli* to produce antibacterial effects. QDs can bind to the bacteria and transfer energy to nearby oxygen molecules to promote the formation of singlet molecular oxygen ($1O_2$) and hydroxyl free radicals (OH•) which can increase the level of ROS in bacterial culture medium.

Other QDs were also studied (CdSe and CdSe/ZnS) and showed bacteria growth inhibiting activity [39]. However, the use of QDs as antimicrobials should be limited at least until their toxic effects are completely understood and evaluated, mostly in correlation with their accumulation in human tissues and in the environment.

III. TEXTILE FINISHING APPROACHES

Antimicrobial agents can be incorporated into textile fibers, applied onto fibers surface or chemically bonded onto fibers. Natural and synthetic fibers can be loaded with metals and metallic compounds by conventional methods (such as exhaust, pad-dry-cure) or modern approaches (spraying, padding, coating, foam finishing). Recent developments include nanosized colloidal solutions, direct deposition of nanoparticles, impregnation by sol-gel technique etc. [40].

Thus, in the case of metal and metallic compounds nanoparticles it is necessary to stabilize the nanoparticles structure, to control the concentration of released nanoparticles or ions, to prolong the release time and to improve the durability and wash resistance of the finishing formulation. Various methods were used: incorporation in polymer matrices such as zeolites [41], a hydrogel network based on crosslinked poly(acrylamide) and carbohydrates [42], polyvinylalcohol in combination with a crosslinking agent [43], polyvinyl pyridine [44] or silica matrices [45,46] in which nanoparticles are embedded and held by physical forces.

In the case of a finish consisting of TiO_2 bound to aminopropyltriethoxy silane [47], a Ti-O-Si chemical bond is formed to stabilize the nanoparticles.

Another sol-gel approach was employed for multifilament yarn viscose fabrics finishing using silver nitrate ($AgNO_3$) and copper acetate ($Cu(C_2H_3O_2)_2$), in the presence of tetraethoxysilane (TEOS) and 3-glycidyloxypropyltriethoxy-silane (GLYEO) in water [11]. These viscose fabrics were treated with the modified silica sols by dipping at room temperature under atmospheric conditions. After removing from the liquid sols, the viscose fabrics were dried at room temperature for 2 h and then were submitted to a thermal treatment for 1 h at higher temperatures (80, 120 or 180°C).

Adsorption from salt solutions is applied for soluble metal salts, such as zinc chloride ($ZnCl_2$) or copper sulphate ($CuSO_4 \cdot 5H_2O$). Prior to their finishing, the chitosan fibers submitted to this treatment were dried under controlled conditions (at 105°C for 4 h) and then immersed into the aqueous solutions of metal salts [12]. The same procedure is used for Co (II) and manganese (Mn II) compounds.

Film casting was employed in the case of nanocomposites Cu-PVC [13], while QDs were suspended into a phosphate buffered solution (PBS) (pH=7.4) and diluted to desired concentrations prior to application and testing [38].

IV. CONCLUSIONS

Customers demand for healthcare, hygiene and wellness has created a new and rapidly developing market for antimicrobial textiles. Commercial antimicrobial textiles cover most of the major types of fibers, natural, as well as synthetic, in the textile industry. These products use broad-spectrum biocides, such as metals and metallic compounds as active agents. Specific treatments containing the biocide active component are applied to fabrics, either at the finishing stage or incorporated into synthetic fibers prior or during their extrusion. The stability of the finishing and the biocide activity depends on the type of fabric, the biocide and the finishing method used in the system. In some cases, the antimicrobial efficacy is lost or severely depleted after 10 or fewer washing cycles.

A relevant aspect of this subject is that many of the metals and metallic compounds are toxic to humans and hardly degradable through environmental mechanisms. Even more, most of them can promote bacterial resistance. The use of new antimicrobial substances can significantly minimize the side-effects of the metals used as biocides, especially when it comes to nanoparticles. It is common knowledge that nanoparticles have proved toxic to human tissues and cell cultures, due to the increased oxidative stress, inflammatory cytokine production and cell death. Studies demonstrated the potential for nanoparticles to cause even DNA mutations. Therefore, significant environmental, health, and safety issues must be considered along with the development in nanotechnology.

CONFLICT OF INTEREST

The authors declare that they have no conflict of interest.

REFERENCES

1. Gielen M., Tiekink ERT (Eds.) (2005) Metallotherapeutic drugs and metal-based diagnostic agents: The Use of Metals in Medicine. Wiley, Chichester
2. Thompson KH (2010) Medicinal Inorganic Chemistry: An Introduction. Encyclopedia of Inorganic Chemistry. Wiley.
3. Kozarich JW (2005) Medicinal Inorganic Chemistry: Promises and Challenges, Medicinal Inorganic Chemistry. ACS Symposium Series Chapter 2 193:4-14
4. Norman JF, Hambley TW (2011) Targeting strategies for metal-based therapeutics, in Bioinorganic Medicinal Chemistry. (ed. E. Alessio), Wiley-VCH Verlag GmbH & Co. KGaA, Weinheim.
5. Harrison JJ, Ceri H, Stremick C, Turner RJ, (2004) Biofilm susceptibility to metal toxicity. Environ Microbiol 6:1220-1227
6. Harrison JJ, Turner RJ, Joo DA, et al (2008) Copper and quaternary ammonium cations exert synergistic bactericidal and anti-biofilm activity against *Pseudomonas aeruginosa.* Antimicrob Agents Chemother 52:2870-2881
7. Macomber L, Elsey SP, Hausinger RP (2011) Fructose-1,6-bisphosphate aldolase (class II) is the primary site of nickel toxicity in *Escherichia coli.* Mol Microbiol 82:1291-1300
8. Kaneko Y, Theondel M, Olakanmi O et al (2007) The transition metal gallium disrupts *Pseudomonas aeruginosa* iron metabolism and has antimicrobial and antibiofilm activity. J Clin Invest 117:877-88
9. Lemire JA, Harrison JJ, Turner RJ (2013) Antimicrobial activity of metals: mechanisms, molecular targets and applications. Nat Rev Microbiol 11:371-384
10. Drath DB, Karnovsky ML (1974) Bactericidal activity of metal-mediated peroxide-ascorbate systems. Infect Immun 10:1077-1083
11. Mahltig B, Fiedler D, Fischer A, et al (2010) Antimicrobial coatings on textiles–modification of sol–gel layers with organic and inorganic biocides. J Sol-Gel Sci Technol 55:269-277
12. Qin Y, Zhu C, Chen J et al (2007) Absorption and release of zinc and copper ions by chitosan fibers. J Appl Polym Sci 105:527-532
13. Becerra A, Rodriguez-Llamazares S, Carrasco C et al (2013) Preparation of poly(vinyl chloride)/copper nanocomposite films with reduced bacterial adhesion. High Perform Polym 25:51-60
14. Park BK, Jeong S, Kim D et al (2007) Synthesis and size control of monodisperse copper nanoparticles by polyol method. J Colloid Interface Sci 311:417–424
15. Warnes SL, Caves V, Keevil CW (2012) Mechanism of copper surface toxicity in *Escherichia coli* O157:H7 and *Salmonella* involves immediate membrane depolarization followed by slower rate of DNA destruction which differs from that observed for Gram-positive bacteria. Environ Microbiol 14:730-1743
16. Teitzel GM, Geddie A, De Long SK et al (2006) Survival and growth in the presence of elevated copper: transcriptional profiling of copper-stressed *Pseudomonas aeruginosa.* J Bacteriol 188:7242-7256
17. Macomber L, Imlay JA (2009) The iron-sulfur clusters of dehydratases are primary intracellular targets of copper toxicity. Proc Natl Acad Sci USA 106:8344-8349,
18. Ryan JN, Harvey RW, Metge D et al (2002) Field and laboratory investigations of inactivation of viruses (PRD1 and MS2) attached to iron oxide-coated quartz sand. Environ Sci Technol 36:2403-2413
19. You Y, Han J, Chiu PC et al (2005) Removal and inactivation of waterborne viruses using zerovalent iron. Environ Sci Technol 39:9263-9269
20. Lee C, Kim JY, Lee WI et al (2008) Bactericidal effect of zero-valent iron nanoparticles on *Escherichia coli.* Environ Sci Technol 42:4927-4933
21. Joo SH, Feitz AJ, Sedlak DL et al (2005) Quantification of the oxidizing capacity of nanoparticulate zero-valent iron. Environ Sci Technol 39:1263-1268
22. Touati D (2000) Iron and oxidative stress in bacteria. Arch Biochem Biophys 373:1–6

23. Novelli F, Recine M, Sparatore F et al (1999) Gold(I) complexes as antimicrobial agents. Pharmacology 54:232-236
24. Thomsen P, Larsson C, Ericson LE et al (1997) Structure of the interface between rabbit cortical bone and implants of gold, zirconium and titanium. J Mater Sci 8:653-665,
25. Suska F, Svensson S, Johansson A et al (2010) *In Vivo* Evaluation of Noble Metal Coatings. J Biomed Mater Res Part B Appl Biomater 92B:86-94
26. Nomiya K, Noguchi R, Ohsawa K et al (2000) Synthesis, crystal structure and antimicrobial activities of two isomeric gold(I) complexes with nitrogen-containing heterocycle and triphenylphosphine ligands, [Au(L)(PPh3)] (HL 5 pyrazole and imidazole). J Inorg Biochem 78:363-370
27. Savvaidis I, Karamushka VI, Lee H et al (1998) Microorganism-gold interactions. Biometals 11:69-78
28. Saygun O, Agalar C, Aydinuraz K et al (2006) Gold and gold-palladium coated polypropylene grafts in a *S. epidermidis* wound infection model. J Surg Res 131:73-79
29. Roach P, Eglin D, Rohde K et al (2007) Modern biomaterials: A review - bulk properties and implications of surface modifications. J Mater Sci 18:1263-1277
30. Eisler R (2003) Chrysotherapy: A synoptic review. Inflamm Res 52:487-501
31. Bratt J, Belcher J, Vercellotti GM et al (2000) Effects of anti-rheumatic gold salts on NF-kappa B mobilization and tumour necrosis factor-alpha (TNF-alpha)-induced neutrophil dependent cytotoxicity for human endothelial cells. Clin Exp Dermatol 120:79-84
32. Senior K, Muller S, Schacht JV et al (2012) Antimicrobial Precious-Metal Nanoparticles and their Use in Novel Materials. Recent Pat Food Nutr Agric 4:200-209
33. Che CM, Siu FM (2010) Metal complexes in medicine with a focus on enzyme inhibition. Curr Opin Chem Biol 14:255-261
34. Chang EL, Simmers C, Knight DA (2010) Cobalt complexes as antiviral and antibacterial agents. Pharmaceuticals 3:1711-1728
35. Shrivastava S, Dash, D (2009) Applying nanotechnology tohuman health: revolution in biomedical sciences. J Nanotechnol, DOI:10.1155/2009/184702
36. Maness P-C, Smolinski S, Blake DM et al (1999) Bactericidal activity of photocatalytic TiO_2 reaction: toward an understanding of its killing mechanism. Appl Environ Microbiol 65:4094-4098
37. Simon-Deckers A, Loo S, Mayne-L`Hermite M et al (2009) Size-, composition- and shape-dependent toxicological impact of metal oxide nanoparticles and carbon nanotubes toward bacteria. Environ Sci Technol 43:8423-8429
38. Lu Z, Li CM, Bao H et al (2008) Mechanism of antimicrobial activity of CdTe quantum dots. Langmuir 24:5445-5452
39. Kloepfer JA, Mielke RE, Nadeau JL (2005) Uptake of CdSe and CdSe/ZnS quantum dots into bacteria via purine-dependent mechanisms. Appl Environ Microbiol 71:2548-2557
40. Ristić T, Zemljič LF, Novak M, et al (2011) Antimicrobial efficiency of functionalized cellulose fibers as potential medical textiles, *Science against microbial pathogens: communicating current research and technological advances,* vol. 1 FORMATEX "Microbiology Series No. 3", Badajoz, Spain, 2011, pp. 36-51
41. Matsumura Y, Yoshikata K, Kunisaki S et al (2003) Made of bactericidal action of silver zeolite and its comparison with that of silver nitrate. Appl. Environ Microb 69:4278-4281
42. Vimala K, Samba Sivudu K, Murali Mohan Y et al (2009) Controlled silver nanoparticles synthesis in semi-hydrogel networks of poly(acrylamide) and carbohydrates: A rational methodology for antibacterial application. Carbohydr Polym 75:463-471
43. Wasif AI, Laga SK (2009) Use of nano silver as an antimicrobial agent for cotton. AUTEX Res J 9:5-13
44. Kantouch A, Atel El-Sayed A (2008) Polyvinyl pyridine metal complex as permanent antimicrobial finishing for viscose fabric. Int J Biol *Macromol* 43:451-455,
45. Jeon HJ, Yi SC, Oh SG, (2003) Preparation and antibacterial effects of Ag-SiO_2 thin films by sol–gel method. Biomaterials 24:4921-4928
46. Wang JX, Wen LX, Wang ZH,et al (2006) Immobilization of silver on hollow silica nanospheres and nanotubes and their antimicrobial effects", Mater Chem Phys 96:90-97
47. Tomsič B, Orel B, Jovanovski V et al (2008) Bacterial growth reduction on cotton fabrics treated by titanium isopropoxide/aminopropyltriethoxy silane, *Magic World of Textiles: Book of Proceedings*, Dragcevic, Z. (Ed.), Faculty of Textile Technology, University of Zagreb, Zagreb, Croatia, 2008, pp. 465–470.

Author:	Madalina Zanoaga
Institute:	"Petru Poni" Institute of Macromolecular Chemistry
Street:	41A Grigore Ghica Voda Alley
City:	Iasi - 700487
Country:	ROMANIA
Email:	zanoaga@icmpp.ro

Biological Evaluation of Slip Casting Hydroxyapatite Intended for Cranioplasty

D. Talpeanu[1], Ch. Tardei[1], F. Grigore[1], M. Lucaci[1], G. Velciu[1], A. Dumitru[1], and D. Savu[2]

[1] National Institute of Research and Development for Electrical Engineering, INCDIE ICPE-CA, Bucharest, Romania
[2] National Institutte for C&D in Physics an Nuclear Engineering, IFIN–HH, Magurele, Ilfov, Romania

Abstract— **In this paper are presented the results obtained for biocompatibility tests on some cranial implant synthetic models destined for using in skull bone surgery applications. The samples were made of hydroxyapatite (HAp) powders and shaped by slip casting method. Specific tests were performed to evaluate the biological properties of the bioceramic samples. In this aim microbial load tests (sanitation/sterility) and cytotoxicity assays on cell cultures-agar diffusion tests performed on L929 cells, and respectively the, MTS viability tests performed on 2 types of cells (L929 and MG-63). The results obtained after the tests showed the biocompatible character of implants, the sterile environments from microbiological point of view and also observed the absence of pathogenic biological agents.**

Keywords— **hydroxyapatite, cranial implant, cytotoxicity.**

I. INTRODUCTION

As the world population aging, medicine is permanently facing new challenges; one of them is the increase of bone diseases. Plastic surgeons often are required to augment or repair bone structure to correct injury or birth defects. Cranioplasty is also necessary to repair bone defects in the cranium which are formed as a result of trauma and other surgery in response to disease [1]. The goal of surgery is a perfect fit of the implant without infection, and a good aesthetic result [2]. To resolve these complex problems at present are available for reconstruction different materials and different techniques including autogenous bone grafts and bone banks, bone matrix demineralised, pasty, 3D structures (scaffolds), and different materials [3, 4].

Synthetic materials can be bioactive or biologically inert. Some synthetic materials are easy to shape to desired configurations and can be made porous like biological bone. Different materials used for cranial implants are silicone, polytetrafluoroethylene (PTFE), metal (stainless steel, Ti and Ti alloys), ceramics and ceramic composites, materials of biological origin [4-5]. Generally, a porous material is preferred to allow tissue ingrowths and stabilize the implant in position [4].

For large and complex shapes of cranial bone defects are preferred ceramic materials on the basis of hydroxyapatite [$Ca_{10}(PO_4)_6(OH)_2$] (HAP) which is one of the most bioactive and biocompatible ceramics. In cranioplasty, especially for complex and large skull defect has recently increased applications because of the good cosmetic outcomes, biocompatibility, osteoconductive properties, strength, lack of risk of disease transmission, and last but not least avoid the problem of donor-site morbidity 6]. Porous bone synthetic implants for cranioplasty can be produced by different processes like manual-based processing techniques and computer-controlled fabrication processes, such as solid free-form (SFF) technology, known also as rapid prototyping (RP) [7]. The ideal bone substitute must form a strong bond with the tissues and allowing, new cells to grow and to penetrate [8,9].

Slip casting is a method by which the slurry (relatively thick) is cast into a porous mould, following by drying and finally sintering of green body Porous bone synthetic implants were produced by pore-forming agent addition.

The purpose of this study is to manufacture synthetic hydroxyapatite-based cranial implants through slip casting technique, and to investigate their biocompatibility by *in vitro* testing. For characterization and in vitro testing two types of these models, with dens and porous microstructures were prepared.

II. MATERIALS AND METHODS

A. Preparation of the Hydroxyapatite (HAP)

The starting materials $CaHPO_4\ 2H_2O$ and $CaCO_3$ with molar ratio of 1.5 as seen in equation (1) were used to produce hydroxyapatite stoichiometric powder via solid-state method.

$$4CaCO_3+6CaHPO_4H_2O\rightarrow Ca_{10}(PO_4)_6(OH)_2+8H_2O+4CO_2 \quad (1)$$

In brief, $CaCO_3$ (Fluka, Germany) and $CaHPO_4\ H_2O$ (Sigma-Aldrich, Germany) were milled using zirconia balls in deionized water medium for 6h and then dried in an oven at 90°C for 16h. The calcination step was achieved by heating the dried powder at 1100°C for 4h at the heating rate of 5°C/min and finally, the powder was milled for 4h to modify the properties of HAP powder suitable for the slurry preparation.

After calcinations the HAP powder has a specific surface area of 11.67m^2/g, average crystallite size d_{50} ~76nm, and crystallinity degree of about 92%.

Phase composition and crystallinity of the calcined HAP particles were determined by X-ray diffraction (XRD,

V. Sontea and I. Tiginyanu (eds.), *3rd International Conference on Nanotechnologies and Biomedical Engineering*,
IFMBE Proceedings 55,
DOI: 10.1007/978-981-287-736-9_76

Bruker D-8 Advance, Germany) with CuKα radiation, λ = 1·5406 Å, generated at 35 kV and 25 mA. Data were collected over 2θ range of 20–60° with a step size of 0.010° and a count time of 0.2s.

B. Preparation of Synthetic Cranial Implant Samples

Cranial implant models were made from hydroxyapatite (HAP) ceramic powders, shaped by molding plaster technique. Initially, ceramic slurries for molding were prepared from HAP powder, dispersed in distilled water, in the presence of specific additives (surfactant, Na-PAA, Fluka Analytical, and as binder, APV 7.5%sol.-Alfa Aesar). The mixture was stirred in a planetary mill to obtain homogeneous and stable ceramic slips, the optimal rheological characteristics for plaster molding. Casting products were finally dried and sintered at temperatures in the range 1250-1300˚C. For manufacturing of cranial implant models optimal hydroxyapatite ceramic slips with casting slip density ranging 1.73÷1.86g/L. Rheological behaviour of aqueous HAP suspensions is strongly dependent on dispersant concentration and phase volume (content of solid). To increase the porosity (micro and macroporosity) its used an pore forming agent (*"Ethocel", Fluka, AG, 7mPa.s viscosity-~5%wt.).*

Two types of cranial implant models have been developed, with low porosity (denser) and with high porosity (porous), as follows:

- high density cranial implant model ($P_T<10\%$), hereinafter named *"OS-D"*;
- low density cranial implant model (porous) (P_T = min. 26%), hereinafter named *"OS-P"*

Cranial implant models have been designed and manufacture as spherical calotte having a thickness of 3...6 mm (in accordance with the literature data).

Phase composition of the synthetic cranial implants prepared were determined by X-ray diffraction (XRD, Bruker D-8 Advance, Germany) and Fourier Transform Infra Red Spectroscopy (FTIR) with FTIR instrument, Tensor 27, of Bruker Optik GmbH. Germania).

C. Evaluation of Biocompatibility of Synthetic Cranial Implants Models

Materials aiming to be used in tissue engineering applications must adequately interact with cells and culture media *in vitro*, prior to their clinical implantation. The biocompatibility of the products was analyzed by the sanitation and sterility tests, and also the cytotoxic effect, using biochemical methods and cell cultures. *In vitro* cell culture studies allow to preliminary access the results of custom-made tissue engineering applications, and provide data regarding cell proliferation and differentiation events.

For bacteriological determinations (sanitation and sterility tests) were using germs such as molds, enterobachteria, coliforms bacteria, Escherichia coli, coagulase-positive staphylococc, bacillus Proteus (HG 1092/2006).

Two fundamental cell culture assays are used for evaluating cytocompatibility: direct contact and agar diffusion. In this aim cytotoxicity assays on cell cultures- agar diffusion tests performed on L929 cells, and respectively the, MTS viability tests performed on 2 types of cells (L929 and MG-63). The results obtained after the tests showed the biocompatible character of implants, the sterile environments from microbiological point of view and also observed the absence of pathogenic biological agents.

The cell viability was calculated by normalization of values from the treated cells compared to the same number of untreated cells.

III. Results and Discussions

A. Synthetic Implant Cranial Models Manufacture

Synthetic implant cranial models were performed by the casting technique in plaster mold of hydroxyapatite-based ceramic slips, with well-defined rheological characteristics. Achieving stable suspensions with optimal rheological characteristics of casting slips involve firstly, determining the optimal amount of dispersant.

Figure 1 presents difractograms for the two type of the cranial implant models. For comparasion is presented and the XRD difractogram for diffraction HAP compounds. These XRD patterns revealed the formation of the crystalline phase corresponding to hydroxyapatite compound, as the main component. Besides, there was evidence of the presence of a small trace of beta-tricalcium phosphate, usefully for the sintering and the densification process.

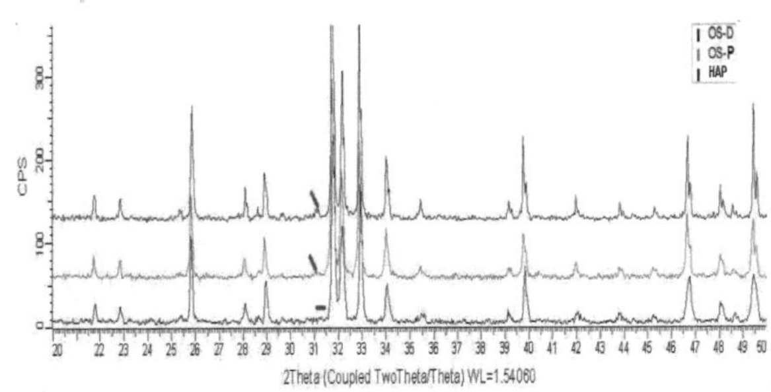

Fig. 1 XRD patterns of cranial implant synthetic models (OS-D; OS-P), developed by plaster molding and calcined at 1250 to 1300°C

IR spectra shows the characteristic peaks for HAP compound, slightly widened due to CO32$^-$ substitutions from the atmosphere, during the preparing process of the precursor, figure 2. Characteristic of the two versions of cranial implant models are the peaks at 1090; 965; 631cm^{-1} specific for PO43- group, peaks at 3573-3574cm^{-1}, characteristic for

OH- group, and peaks at 875 and 1469cm^{-1}, representative for CO32^{-}, respectively. Contamination with CO32^{-} are highlighted by the peaks group's presented by the wave number at 875.76, 1468.55 and 1412.27cm^{-1}.

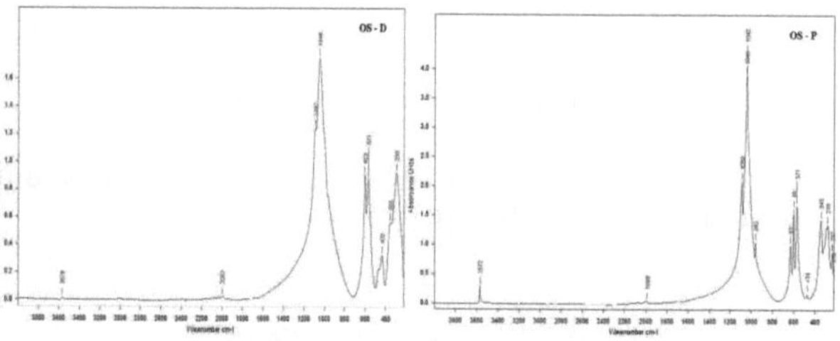

Fig. 2 IR spectra for the two models developed (OS-D / OS-P), by HAP slip casting in plaster molds, and calcined at 1250 to 1300°C

Cranial implant models have been designed and manufacture as spherical calotte having a thickness of 3÷6 mm, and are shown in figure 3.

Fig. 3 Cranial implant model manufactured by slip casting in plaster mold of hydroxyapatite ceramic powder, with average thickness of 3.5mm

Its thickness can be adjusted by altering the characteristics of ceramic slurries, and also by changing the main variables during the casting process, mainly the time of casting and the keeping period in the molds. The average thickness of the both models developed was ~ 3.5mm.

B. *Evaluation of Biocompatibility of Cranial Implants Sanitation and Sterility Tests*

The tested samples show an obvious biocompatibility character according with the obtained results, where the environments are sterile from microbiological point of view and they didn't present any pathogenic biological agents classified as HG 1092/2006.

C. *Cytotoxicity Assays*

A. *The Agar Diffusion Test*

The obtained results are presented in table 1 and figure 4 for the investigated materials.

Table 1 Agar diffusion test: results ceramic materials dense, OS-D and OS-P

Test	Tested parameters	Negative: control	Positive: control	Undiluted sample	½ undiluted sample	1/5 undiluted sample	1/10 undiluted sample
The agar diffusion test	Area index	0	2	0	0	0	0
	Lysis index	1	3	1	1	1	1
	Response index	0/1	2/3	0/1	0/1	0/1	0/1
Conclusion	Cytotoxicity	Non-cytotoxic	Moderate cytotoxic	Non-cytotoxic	Non-cytotoxic	Non-cytotoxic	Non-cytotoxic

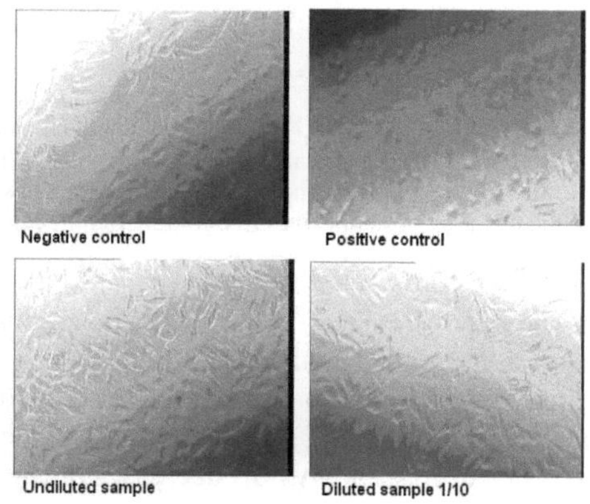

a)

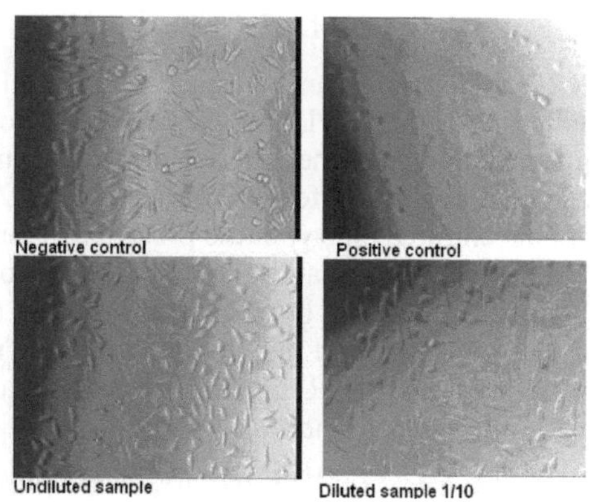

b)

Fig. 4 The fibroblast morphology L929 after application of the sample: a) *OS-D, b) OS-P* (microscope- magnification X10)

B. *MTS Assay*

The results obtained by using MTS assay (Figure 5) confirm the results obtained by the agar diffusion test. The tested ceramic materials (OS-D and OS-P) show a good biocompatibility for both cell lines (L929; MG-63).

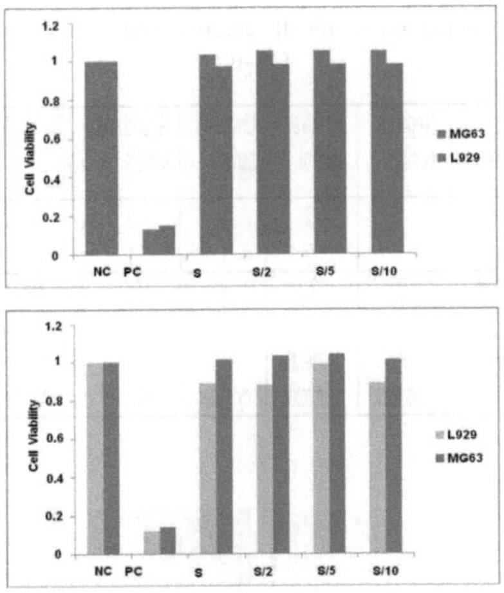

Fig. 5 The viability of L929 and MG-63 cells treated and untreated with the extract of the OS-D / OS-P materials (CN-negative control, CP-positive control; P-undiluted extract of the sample; P/2 diluted extract of sample by twice; P/5 diluted extract of sample was five-fold; P/10 diluted extract of sample by 10 times)

As it is shown in figure 6, both OS-D and the OS-P extracts that were in contact with both L929and MG-63 cells didn't induce any changes in their viability comparatively with the negative control (NC). Instead, as it was expected, the positive control (0.45% phenol) decreases drastically the viability of L929 and MG-63 cells.

IV. CONCLUSIONS

From XRD patterns, it can be observed that all of the peaks matched with the JCPDS pattern 09-0432 for HAP, which suggested that no other phases were present. HAP powders calcined at 1100°C for 4h were capable of achieving 67% of concentrated slurry, in the presence of surfactant (0.75÷1.5%). On the other hand, the combination of micro and macro pores, obtained by addition of pore forming agent, will provide a material that could be used as medical devices for bone ingrowths. Regarding the biological evaluation of the samples was found:

- There was no effect on cell toxicity of samples tested; cell monolayer remained intact without showing lysed cells, cytotoxic negative control has no effect, while positive control shows a moderate degree of cytotoxicity;

- Extracts from both samples in contact with L929 and MG-63 cells does not induce changes in the their viability compared to negative control. In return, as expected, the positive control (0.45% phenol) decreases drastically the viability of the cell L929 and MG-63.

In summary, dense and porous synthetic cranial implants with an porous structure were successfully fabricated using slip casting technique. Moreover, the attachment and proliferation of L929 and MG-63 cells on the both pieces indicated good cytocompatibility. A lot remains to be done to manufacture highly interconnected porous structures for bone tissue engineering with application in cranioplasty.

ACKNOWLEDGMENT

This work was financed by Nucleus ANCS Romanian Authority, Project no. 0935301/2009

CONFLICT OF INTEREST

The authors declare that they have no conflict of interest.

REFERENCES

1. Joël Brie, Thierry Chartier, Chapu Ch.t,, et.al., A new custom made bioceramic implant for the repair of large and complex craniofacial bone defects, Journal of Cranio-Maxillofacial Surgery, 2013, Volume 41, Issue 5, Pages 403–407South J, Blass B (2001) The future of modern genomics. Blackwell, London.
2. G. Staffa, A. Barbanera, A. Faiola et. al., Custom made bioceramic implants in complex and large cranial reconstruction: A two-year follow-up, Journal of Cranio-Maxillo-Facial Surgery 40 (2012) pp.65-70.
3. Lemée J.M., Petit D., Splingard M., Menei P., Autologous bone flap versus hydroxyapatite prosthesis in first intention in secondary cranioplasty after decompressive craniectomy: A French medico-economical study, Neurochirurgie, 59 (2013), 60-63.
4. A. Di Rienzo, M. Iacoangeli, Luca G.M. di Somma, L. Alvaro, et. al., Shape modification of porous hydroxyapatite prostheses to improve rigid implant fixation: Experience in 12 cases, Surgical Neorology International, 2012, 3:161.
5. Tardei, C; Moldovan, L; Craciunescu, O, Hybrid bioresorbable composites with high bioactivity properties, Romanian Journal of Materials, 2010, 40(1), 41-49.
6. Stefini R, Esposito G, Zanotti B, Laccarino C, Fontanella MM, Servadei F. Use of "custom made" porous hydroxyapatite implants for cranioplasty : postoperative analysis of complications in 1549 patients. Surg Neurol Int 2013; 4:12-12.
7. Muller A, Krishnan KG, Uhl E, Mast G (2003), The application of rapid prototyping techniques in cranial reconstruction and preoperative planning in neurosurgery. J Craniofac Surg 14:899–914.
8. H. Hardy, E. Tollard, S. Derrey, P. Delcampe, J.-M. Peron, P. Fréger, F. Proust. « Tolérance clinique et degré d'ossification des cranioplasties en hydroxyapatite de larges défects osseux ». Neurochir 58 (2012) 25-29.
9. Silva GA.,Marques AP., Gomes ME., Coutinho OP., Reis RL., Cytotoxicity screening of biodegradable polymeric systems. In: Reis RL., San Roman J. (eds) Biodegradable Systems in Tissue Engineering and Regenerative Medicine, 2005, pp. 339-349, CRC Press, Boca Raton Fl.

Author: Dorinel Talpeanu
Institute: INCDIE ICPE-CA Department of Advanced Materials
Street: Splaiul Unirii, 313
City: Bucharest, 030138
Country: Romania
Email: dorinel.talpeanu@icpe-ca.ro

Nanofibers for Tissue Engineering and Regenerative Medicine

N. Bölgen[1] and A. Vaseashta[2]

[1] Chemical Engineering Department, Mersin University, Mersin, Turkey
[2] Institute for Advanced Sciences Convergence, and Int'l Clean Water Institute, NUARI, Herndon, VA, USA

Abstract— **Electrospinning is a process in which polymer fibers are produced with diameters down to the nanometer range through the action of an electric field imposed on a polymer solution/melt. Distinct properties that make electrospun nanofibrous materials unique are their high surface area, porosity, tensile strength and high extensibility. Indeed the size of the fibers down to nanometer scale make the final structures unique. And since the structure of the nanofibers is very similar to the extracellular matrix, many applications of them are proposed to be used in biomedicine especially for tissue engineeri2ng applications. Many polymers were used to prepare electrospun nanofibers including natural and synthetic ones. Poly(ε-caprolactone) (PCL) was used extensively in the tissue engineering field as a synthetic polymer and alternative to natural polymers. The relatively inexpensive production routes, FDA approval, tailorable biodegradability, biocompatibility and easy manipulation make this polymer promising for electrospinning applications. Therefore PCL based electrospun matrices were used in several tissue engineering attempts including skin, bone, vascular and nerve regeneration. This review summarizes the use of electrospun PCL nanofibers in tissue engineering and regenerative medicine applications.**

Keywords— **PCL, nanofibers, electrospinning, tissue engineering, scaffold, biodegradable polymer.**

I. INTRODUCTION

Electrospinning is a unique method that produces polymer fibers with usually diameters in nanoscale and nonwoven fibrous structures composed of these fibers using electrically driven jet of polymer solution or melt [1]. Several parameters including polymer type, polymer melt properties, polymer solution properties, electrospinning conditions significantly affect fiber size, and morphology of the fibrous structure that are formed from these fibers, and in turn macroscopic properties [2,3].

The electrospinning method has been used for the nanofibrous fabrication from many nondegradable and biodegradable natural/synthetic polymers such as polyurethanes, polyamides, collagen, gelatin, chitosan, alginates, poly(α-hyroxy acids) [4]. Due to unique properties of the materials produced from these nanofibers, several novel applications in biomedicine including tissue engineering, controlled release systems, soft and hard tissue repair materials have been proposed and investigated [5-7].

Both soft and hard tissue can regenerate by natural pathways. During this regeneration period they need to be supported, preferentially by biodegradable repair materials, which should be degraded during or better after healing and disappear without need of elimination by a second surgery.

PCL is a semi-crystalline biodegradable polyester, and its co-polymers with lactides and glycolide are attracting increasing attention due to their controllable biodegradation rates and also more suitable and tailormade mechanical properties for tissue engineering applications [8].

In this review, properties of biodegradable and biocompatible PCL based electrospun matrices, and plain or drug loaded or cells seeded form of the nanofibrous scaffolds and their applications in tissue engineering and regenerative medicine will be presented.

II. POLY (ε-CAPROLACTONE)

PCL (Figure 1) is a hydrophobic, semi-crystalline polymer which can be synthesized by either ring opening polymerization of ε-caprolactone using a variety of anionic, cationic and co-ordination catalysts or via free radical ring-opening polymerization of 2-methylene-1-3-dioxepane [9]. The good solubility of PCL, and blend compatibility has stimulated extensive research into its potential processing with electrospinning method and application in tissue engineering and regenerative medicine.

PCL can be degraded both in vitro and in vivo by hdrolysis [8]. Enzymes, especially lipases, also play a role in the enzymatic degradation of PCL in vivo [4].

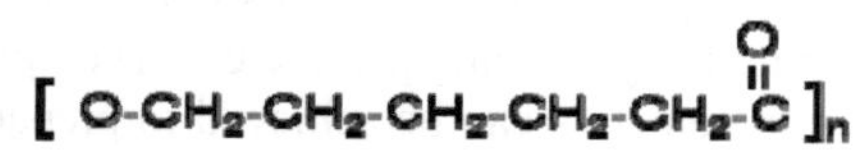

Fig. 1 Chemical structure of poly (ε-caprolactone) (PCL)

III. ELECTROSPINNING

Electrospinning is a process in which polymer fibers are produced with diameters down to the nanometer range through the action of an electric field imposed on a polymer solution/melt. There are basically three components in a simple electrospinning system: a high voltage supplier, a

V. Sontea and I. Tiginyanu (eds.), *3rd International Conference on Nanotechnologies and Biomedical Engineering*,
IFMBE Proceedings 55,
DOI: 10.1007/978-981-287-736-9_77

spinnerette with a small hole or a simple capillar and a collector screen (Figure 2). One electrode is placed into the spinning solution and the other one is attached to the collector. The voltage is from a variable high voltage source, which causes charging of the solution, which in turn creates repulsive forces. These repulsive forces overcome the surface tension of the polymer solution at a critical voltage which in turn creates an electrically charged jet of polymer solution out of the spinneret. The polymer jet undergoes an elongation process before reaching the collector screen, the solvent evaporates and the polymer is collected as a solidified interconnected web of nanofibers.

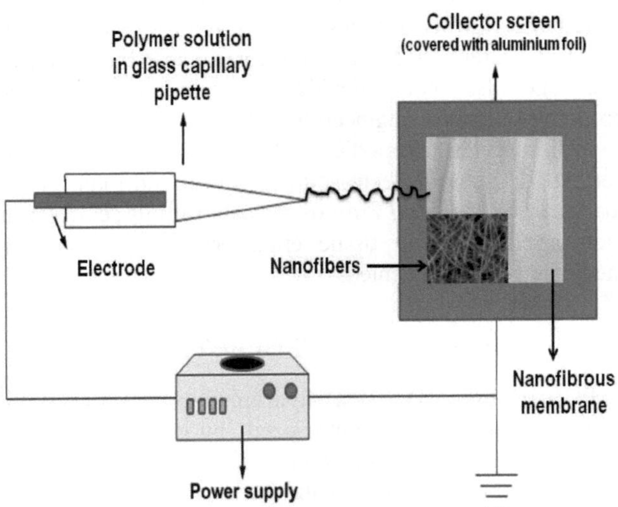

Fig. 2 Components of electrospinning system

There are process parameters that effect the morphology and size of the nanofibers such as viscosity conductivity and surface tension of the solution; or governing variables like electrical potential and tip to collector distance or ambient parameters such as temperature.

It is possible to prepare homogeniously distributed nanofibers by changing the governing parameters. The optimization of the process parameters can be achieved to prepare nanofibers without beads and with the same nanofiber size. Bead formation can occur during the process and creates nonhomogenity. To overcome bead formation, increasing the polymer concentration is one of the most preferred way [8].

IV. ELECTROSPUN PCL NANOFIBERS

Electrospinning technique enables to load drugs, growth factors or hormones inside the nanofibrous structures. Since the surface area of the electrospun membranes is very high compared to the conventionally prepared fibrous materials, they have a significantly increased embedding capability. Therefore, it either can be managed to embed an active substance in the nanofibers after electrospinning process is completed or the active substance can be added to the polymer solution before electrospinning and electrospun together with the polymer to form final loaded nanofibers [10].

Bölgen et al., polymerized ε-caprolactone to poly(ε-caprolactone) (PCL) by ring opening polymerization and prepared nonwoven films composed of nanofibers by electrospinning from PCL [8]. They loaded antibiotics, to nanofibers either after electrospinning process or by adding the drug into the polymer solution before electrospinning and the performance of antibiotic loaded electrospun membranes were evaluated in preventing intraabdominal adhesions [10]. The antibiotic loaded electrospun membranes significantly eliminated post surgery adhesions and the reorganization of the wound healing was quite favourable in the antibiotic embedded group compared to control.

Reconstruction of large bone defects is still a major problem. It is worth to mention that a limitation of the electrospun membranes is their two dimensional structure, and the bone is a 3-dimensional tissue. To overcome this incompatibility in dimension, Pişkin et al. prepared electrospun PCL membranes and rolled them to form disk shaped spiral-wounded scaffolds and sticked by using tissue glue [11]. The 3D scaffolds filled the bone defect and guided the new tissue formation.

There are many studies in the present literature demonstrating the interaction of electrospun biocompatible materials with cells such as stem cells, fibroblasts, chondrocytes, osteoblasts, muscle cells etc., to evaluate the *in vitro* biocompatibility of the nanofibers [12]. Electrospun nanofibers mimic very well the structure of original extracellular matrix, therefore, they create a favourable geometrical environment for the cells to attach, migrate and proliferate. However, the electrospun nanofibrous structures have a limitation for 3D applications due to their pore size which is smaller than a cellular diameter and can not allow cell migration within the structure. To overcome this disadvantage, Tuzlakoğlu et al., prepared starch/PCL micro/nano-fiber combination [13]. They prepared micro fibers by fiber bonding and then electrospun the polymer solution to form nanofibers within the micro fibers. Ko et al., combined melt electrospinning technique with particulate leaching to produce highly porous PCL fibers [14].

Potas et al., covalently bounded interleukin 10 (IL-10) to electrospun PCL nanofibers, to extend the in vivo bioavailability and bioactivity of IL-10 by attaching the protein onto nanofibrous scaffolds. The IL-10 conjugated scaffolds were wrapped around the sciatic nerves of rats. PCL biofunctionalised nanofibers were found to manipulate the cellular microenvironment and could potentially lead to new therapeutic strategies for nervous tissue regeneration [15].

Kim et al., prepared PCL fibers by electrospinning, modified the surface with heparin-dopamine and immobilized bone morphogenic protein-2 to induce osteogenic differentiation of periodontal ligament cells. Alkaline phosphatase activity and calcium deposition on the fibers demonstrated induced osteogenic differentiation, and no cytotoxicity against periodontal ligament cells were observed [16].

Ahn et al., fabricated a vascular scaffold by electrospinning PCL and type I collagen and combined this vascular scaffold with a smooth muscle cell sheet, instead of using a conventional cell seeding method. The smooth muscle cell sheet was wrapped around the vascular scaffold and this cell sheet combined electrospun structure maintained high cell viability (95.9 ± 2.7%) and phenotypes and improved cellular infiltration and mechanical properties [17].

Current developments related to the potential use of electrospun nanofibers in tissue engineering include the combination of functional nanoparticles within the scaffolds. Electrospraying technique can be combined with electrospinning method to achive nanoparticle and nanofiber including matrices. Zhu et al., produced aligned PCL fibrous framework with adjustable bioactive factor embedded poly (D, L-lactide-co-glycolide) core-shell nanospheres by combining electrospinning and electrospraying. A sustained bioactive factor release was achieved by the core shell nanospheres. Cells were seeded on the scaffolds for a possible neural tissue growth and regeneration application, cell proliferation was promoted on the nanocomposite scaffold [18].

Gautam et al., fabricated a PCL/gelatin nanofibrous scaffold by electrospinning and then modified the scaffold by collagen type I grafting for skin tissue engineering applications. L929 mouse fibroblasts were seeded on the scaffolds, and collagen type I-modified PCL/gelatin electrospun matrice was found to be favourable for maintaining characteristic shape of fibroblasts, and adequate cell proliferation [19].

Kobahashi et al., demonstrated the quantitative analysis of the cellular alignment inside implanted scaffolds with oriented fibers. They fabricated two layers of orthogonally aligned electrospun PCL scaffolds, and seeded intestinal smooth muscle cells on the scaffolds, and implanted the scaffolds with or without cells subcutaneously in mice. It was demonstrated that two layers of orthogonally aligned scaffolds could generate the histological organization of cells similar to that of intestinal circular and longitudinal smooth muscle [20].

V. CONCLUSION

Electrospinning is one of the simplest and most versatile methods to fabricate nanofibers with high surface area that mimic the extracellular matrix. Many polymers including synthetic, natural, biodegradable and non-degradable ones (gelatin, chitosan, collagen, alginate, silk, PVA, PCL, PLLA, PGA, PU) were electrospun from their solutions or from their melts.

Biomedicine is an important area of application of the nanofibers, they can be used as tissue engineering scaffolds in bone, cartilage, muscle, nerve, skin regeneration since they mimic very well the main structure of the original tissues.

Development of new electrospun matrices for future tissue engineering applications include the integration tissue regeneration inducing factors such as drugs, growth factors and hormones with nanofibers. Electrospinning enables loading active substances to the nanofibers easily, either after electrospinning is completed or before, by adding the functional substance to the solution and electrospinning with polymer. It is also possible to functionalize the electrospun nanofibers by combining them with nanoparticles.

Limitations of the electrospun nanofibrous matrices are their small pore size for cell penetration and two dimensional structure. Those limitations can be overcome by combining electrospinning with additional pore forming techniques or producing micro/nano fibers. A practical method to prepare three dimensional scaffolds can be spiral rolling the membrane to fill the bone defect for instance.

Electrospun PCL material is a good alternative for tissue engineering applications, because of its excellent properties; it is biodegradable(there is no need for a second operation to explant it), it is very easy to handle it during surgical operation, it is very light like a butterfly wing but mechanically stable at the same time, have drug loading capability, PCL has FDA approval.

The results of previous studies about electrospun PCL; production and characterization of electrospun nanofibers from this polymer; PCL's loading capacity with drugs and the release of the drugs from the produced matrices; their seeding with cells and *in vivo* performance studies suggest the potential of electrospun PCL biodegradable scaffolds for clinical applications in tissue replacement.

CONFLICT OF INTEREST

The authors declare that they have no conflict of interest.

REFERENCES

1. Huang ZM, Zhang YZ, Kotaki M, Ramakrishna S. (2003) A review on polymer nanofibers by electrospinning and their applications in nanocomposites. Comp Sci Tech 63:2223–2253.
2. Frenot A, Cronakis IS. (2003) Polymer nanofibers assembled by electrospinning. Curr Opin Colloid Interface Sci 8:64–75.
3. Larrondo L, Manley RSJ. (1981) Electrostatic fiber spinning from polymer melts. I. Experimental observations on fiber formation and properties. J Polym Sci Polym Phys Ed 19:909–920.
4. Pişkin E, Bölgen N, Egri S, İşoğlu İA. (2007) Electrospun matrices made of poly(α-hydroxy acids) for medical use. Nanomedicine 2(4): 441-457.
5. Lanutti J, Reneker D, Ma T, Tomasko D, Farson, D. (2007) Electrospinning for tissue engineering scaffolds. Materials Science and Engineering 27: 504-509.
6. Smith LA, Ma PX. (2004) Nanofibrous scaffolds for tissue engineering. Colloids Surf Bm39:125–131.
7. Li W, Laurencin C, Caterson E, Tuan R, Ko F. (2002) Electrospun nanofibrous structure: A novel scaffold for tissue engineering. J Biomed Mater Res 60: 613–621.
8. Bölgen N, Menceloğlu YZ, Acatay K, Vargel İ, Pişkin E. (2005) *In vitro* and *in vivo* degradation of non-woven materials made of poly(ε-caprolactone) nanofibers prepared by electrospinning under different conditions. Journal of Biomaterials Science Polymer Edition,16(12): 1537-1555.
9. Woodruf MA, Hutmacher DW. (2010) The return of a forgotten polymer-Polycaprolactone in the 21st century. Progress in Polymer Science, 35: 1217-1256.
10. Bölgen N, Vargel İ, Korkusuz P, Menceloğlu YZ, Pişkin E. (2007) *In vivo* performance of antibiotic embedded electrospun PCL membranes for prevention of abdominal adhesions. Journal of Biomedical Materials Research Part B: Applied Biomaterials 81B(2): 530-543.
11. Pişkin E, İşoğlu İA, Bölgen N, Vargel İ, Griffiths S, Çavuşoğlu T, Korkusuz P, Güzel E, Cartmell S. (2009) *In vivo* performance of simvastatin-loaded electrospun spiral-wound polycaprolactone scaffolds in reconstruction of cranial bone defects in the rat model. Journal of Biomedical Materials Research Part A 90(4): 1137-1151.
12. Mo XM, Xu CY, Kotaki M, Ramakrishna S. (2004) Electrospun P(LLA-CL) nanofiber: A biomimetic extracellular matrix for smooth muscle cell and endothelial cell proliferation. Bioma terials 25:1883–1890.
13. Tuzlakoğlu K, Bölgen N, Salgado AJ, Gomes ME, Piskin E, Reis RL (2005) Nano- and micro-fiber combined scaffolds: A new architecture for bone tissue engineering. Journal of Materials Science: Materials in Medicine 16: 1099-1104.
14. Ko J, Kan D, MBG Jun (2015) Combining melt electrospinning and particulate leaching for fabrication of porous microfibers. Manufacturing Letters 3: 5-8.
15. Potas JR, Haque F, Maclean FL, Nisbet DR (2015) Interleuki-10 conjugated electrsopun polycaprolactone (PCL) nanofibre scaffolds for promoting alternatively activated (M2) macrophages around the peripheral nevre in vivo. Journal of Immunological Methods 420: 38-49.
16. Kim SE, Yun YP, Han YK, Lee DW, Ohe JY, Lee BS, Song HR, Park K, Choi BJ (2014) Osteogenesis induction of periodontal ligament cells onto bone morphogenic protein-2 immobilized PCL fibers. Carbohydraye Polymers 99: 700-709.
17. Ahn H, Ju YM, Takahashi H, Williams DF, Yoo JJ, Lee SJ, Okano T, Atala A (2015) Engineered small diameter vascular grafts by combining cell sheet engineering and electrospinning technology. Acta Biomaterialia 16: 14-22.
18. Zhu W, Masood F, O'Brien J, Zhang LG (2015) Highly aligned nanocomposite scaffolds by electrospinning and electrospraying for neural tissue regeneration. Nanomedicine: Nanotechnology, Biology, and Medicine 11: 693-704.
19. Gautam S, Chou CF, Dinda AK, Potdar PD, Mishra NC (2014) Surface modification of nanofibrous polycaprolactone/gelatin composite scaffold by collagen type I grafting for skin tissue engineering. Materials Science and Engineering C 34;: 402-409.
20. Kobahashi M, Lei NY, Wang Q, Wu BM, Dunn JCY (2015) Orthogonally oriented scaffolds with aligned fibers for engineering intestinal smooth muscle. Biomaterials 61: 75-84.

Author: Nimet Bölgen
Institute: Mersin University
Street: Ciftlikkoy Campus
City: Mersin
Country: Turkey
Email: nimetbolgen@yahoo.com

Biocompatible SPIONs with Superoxid Dismutase/Catalase Immobilized for Cardiovascular Applications

L. Lacramioara[1], A. Diaconu[2], M. Butnaru[1], and L. Verestiuc[1]

[1] Gr.T.Popa University of Medicine and Pharmacy, Faculty of Medical Bioengineering, 9-13 Kogalniceanu Street, Iasi, Romania
[2] P.Poni Institute of Macromolecular Chemistry, 41 Grigore Ghica Voda Alley, Iasi, Romania

Abstract— **Antioxidant enzymes are involved in the therapy of the cardiovascular diseases associated with increased production of reactive oxygen species (ROS). The enzyme efficiency is dependent on the bioactive molecule's ability to achieve therapeutically adequate levels at the target site. In this paper, superoxide dismutase (SOD) and catalase, enzymes involved in decomposition of reactive oxygen species, have been immobilized onto superparamagnetic iron oxide nanoparticles(SPIONs) and characterized. FT-IR data, particles mean diameter and zeta potential confirmed the enzymes immobilization onto SPIONs. The particles dimensions vary from 400nm for catalase immobilization to 1µm for SOD immobilization. The particles morphology was studied by scanning electron microscopy (SEM) and suggests the particles tendency for clustering during the drying process. The biofunctinalized SPIONs are degraded by esterases. The citocompatibility tests (MTT) indicated normal values for the cells viability in presence of SPIONs with immobilized enzymes. Enzymatic activity of magnetic particles has been determined indirect by NTB-methionine-riboflavin (in the case of superoxide dismutase) and direct, with H2O2 (for catalase). EnzySPIONs preserve an enzymatic activity which is dependent of the magnetic particles concentration, the enzyme nature and biological molecule conformation onto magnetic particle.**

Keywords— **magnetic nanoparticles, biofunctionalization, antioxidant enzymes, cardiovascular applications.**

I. INTRODUCTION

The conventional methods used for administration of active compounds in vascular therapy usually involve the parenteral pathway [1]. This method has some disadvantages, like as a short half-time of the drugs in plasma or systemic side effects. The using of anticlotting drugs, which are orally and intravenous administrated, can induce coagulopathies or haemorrhages. Other types of therapeutically agents, like as fibrinolytics, does not maintain an adequate concentration at the target tissue, due to the hemodynamic fluctuations. New strategies have been considered for avoiding these disadvantages. Oxygen is necessary for aerobic life processes, but, paradoxically, produce and reactive oxidative stress due to his species. Reactive oxygen species (ROS) are chemical species with a single unpaired electron in an outer orbit, which makes them very reactive, especially in contact with lipids, proteins, carbohydrates [2]. A promising therapy for diseases associated with increased production of reactive oxygen species (ROS), such as hypertension, ischemia, myocardial infarction and restenosis, involve the antioxidant enzymes. The efficiency of enzyme is dependent on the ability to achieve therapeutically adequate levels at the site of ROS-mediated injury [3,4].

Enzymes immobilization on superparamagnetic iron oxide nanoparticles (SPIONs) can increase efficiency and limit adverse effects of thermal and proteolysis degradation [5]. To be used as a platform for therapeutics, the nanoparticulate systems have to meet specific requirements:

- to be stable in the circulating system;
- to improve the half-time of the biological compounds;
- to target the wounded vessel or to be involved in the in cellular mechanisms responsible for disease;
- to release the bioactive in the controlled manner, eventually activated by external stimulus/stimuli;
- to be biodegradable, resorbable and to be eliminated from body without toxic effects [6].

Although current clinically used contrast agents do not have targeting characteristics, a lot of work is in progress on the functionalization of nanoparticles in order to widen their diagnostic and therapeutically ability. Magnetic nanoparticle-aided drug delivery is still in its firsts steps. However, with continuing advancement in the field, biofunctionalized nanodevices play a significant role in biomedical research and nanomedicine [7].

In this paper, with the aim to design enzyme-immobilized SPIONs (enzySPIONs), catalase and superoxide dismutase (SOD) have been used as targeted therapeutic agents. These enzymes contain active centers with coordinated metals, which decompose superoxide anion and hydrogen peroxide, respectively, the most important reactive oxygen species [8]. For modulation the enzySPIONs efficiency the following parameters of the immobilization process were studied: the enzyme type, the ratio enzyme/SPIONs and enzyme conformation.

II. MATERIALS AND METHODS

A. Materials

SPIONs based on magnetite, coated with functional synthetic polymer (copolymer of maleic anhydride) have

V. Sontea and I. Tiginyanu (eds.), *3rd International Conference on Nanotechnologies and Biomedical Engineering*,
IFMBE Proceedings 55,
DOI: 10.1007/978-981-287-736-9_78

been used for biofunctionalization. Catalase, SOD from bovine liver (lyophilized powder), 1-ethyl-3-(3-dimethyl aminopropyl) carbodiimide hydrochloride (EDAC), N-hydroxysuccinimide (NHS), esterase and dimethyl sulfoxide (DMSO) were purchased from Sigma-Aldrich. Phosphate buffer (0.05M, pH7) and H_2SO_4solution (6N) was used to measure the enzymatic activity of catalase-functionalized particles. Solution 3% hydrogen peroxide and 0.005 M $KMnO_4$ have been used as received.

B. Methods

1. Biofunctionalization of the SPIONs

Functionalized SPIONs with antioxidant enzymes have been obtained by Catalase/SOD immobilization through direct reaction onto SPIONs based on magnetite, coated with functional synthetic polymer (copolymer of maleic anhydride). Briefly, 5 mg enzyme was dissolved in 20 mL phosphate buffer saline (PBS), at pH=8.00 and then poured within a suspension of SPIONs (100 mg nanoparticles dispersed in 30 mL PBS, pH=8.00, 0.05 M) and slowly mechanically stirred for 24 h at room temperature. The particles have been collected from suspension through magnetic separation, purified by repeated washes with PBS (2x) and deionised water (3x) and finally freeze-dried. The following parameters of the immobilization were studied: the enzyme type and the ratio enzyme/ SPION.

2. EnzySPIONs Characterization

EnzySPIONs mean diameter and Zeta potential were determined in water, at 25°C, using a Malvern Zetasizer NanoS (Malvern Instruments, UK). FT-IR spectra were obtained from a Varian 640 FTIR spectrometer coupled to a horizontal attenuated total reflectance (ATR) cell. Scanning electron microscopy (SEM TESCAN-VEGA; 30 kV; gold-sputtering) was performed in the aim to evaluate dimensions and morphology. Magnetization of saturation and superparamagnetic characteristics were evaluated using a vibrating sample magnetometer (Lakeshore VSM 7400 System).

3. Enzymatic Activity and Degradation Studies

Enzymatic activity has been determined indirect by NTB-methionine-riboflavin (in the case of superoxide dismutase) and direct, with H_2O_2 (for catalase)[9].

To study the EnzySPIONs degradation the following protocol has been used: in 5 ml of suspension (10 mg particles dispersed in 5 ml phosphate buffer), was added the enzyme (esterase). The mixture was placed into dialysis membrane (12 kDa) and the whole system was immersed in 20 ml of PBS solution at 37 °C. Every 24 hours, 2 ml of the external solution were collected and spectroscopic analyzed, on the whole range of wavelengths (from 190nm to 800nm).

4. Cell Culture and Cytotoxicity Assay

SNPN-SOD cytotoxicity was evaluated *in vitro*, on MG63 cell line, using standard MTT assay.

The cells were plated 24 hours prior the test, on 24 well culture plate, using $1{,}5x10^4$ cells/well cell density and DMEM culture media supplemented with 10% BFS and 1% P/S/N. SNPN-SOD were sterilized in ethanol 70%, then suspensions were prepared at concentration of 0.015 mg/mL and 0.030 mg/mL in PBS), and 100 µl of suspensions were added to the culture media of the preplated cells. The MTT test was performed at 24 hours and 48 hours of cells incubation with magnetic suspensions. Briefly, at each test time point the media was replaced with 0.5 mL of MTT work solution (0.25 mg/mL) in DMEM without BFS and P/S/N. The culture plate was incubated with MTT solution for 2.5 hours at 37 ºC, humidity and dark condition. After incubation the MTT solution was replaced with 1 mL/well of isopropanol, to solubilize the insoluble formazan crystals formed in the cells, as a result of MTT metabolizing. The absorbance of the blue formazan solution was measured at 570 nm wave length, using Boeco Spectrofotometer. The cell viability was calculated by normalizing absorbance results for culture cells incubated with magnetic suspensions to the negative control (culture cells incubated with the same volume of PBS).

III. Results and Discussions

FT-IR data, zeta potential and hydrodynamic particle mean diameter (dynamic light scattering evaluation) confirmed the enzymes immobilization onto SPIONs. In the FT-IR spectra (Figure 1), the absorption in the regions 1164.9 cm^{-1} and 1744 cm^{-1} are atributed to the polymer, representing CO vibrations and COC bonds within it. The amide bond formed between the enzyme and the copolymer is evidenced by changes in the characteristic bands, the peaks at 1567 cm^{-1} and 1618 cm^{-1}.

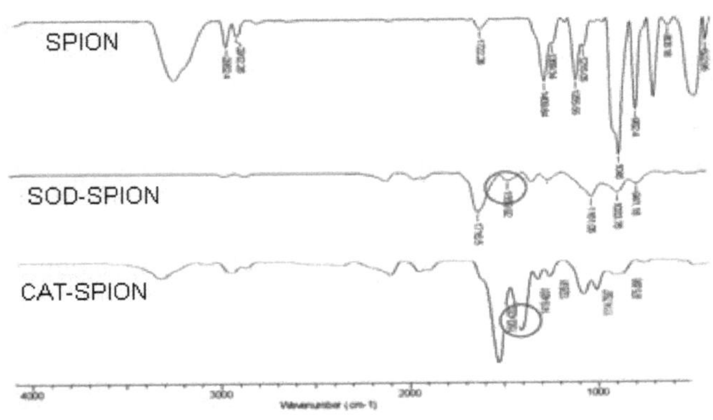

Fig. 1 FT-IR spetra for SPION, biofinctionalized SPION with SOD (SOD-SPION) and biofunctionalized SPION with catalase

The particles dimensions vary from 400nm for catalase immobilization to 1μm for SOD immobilization (Table 1). It is noticed, in general, an increase of the zeta potential, which confirms the polymer involvement in the binding of enzymes. Also, the values indicate the existence of a stable suspension, the values for Zeta potential being in the range [-30, + 30] mV. Positively charged nanoparticles are dangerous due to their potential to induce cytotoxic effects and disruption of the cell membrane.

Table 1 SPIONs dimension and Zeta Potential

SPIONs	Hidrodinamic diameter(nm)	Zeta Potential(mV)
Pure SPIONs	330±0.32	-20 ± 0.15
Catalase-SPIONs	432±3.12	-17 ± 0.39
SOD-SPIONs	970±0.27	-22 ± 0.32

data are presented as means ±standard deviation for n=3.

The particles morphology was studied by scanning electron microscopy and suggests that the obtained biofunctionalized SPIONs interact during the drying process and some clusters are formed (Figure 2). SEM image confirms dimensions obtained by DLS technique. It was found that the particles can be well customized and that few populations have been formed for the both enzymes. Also there is a low tendency to agglomerate in suspension, and the process of functionalization does not change the morphology of the particles.

Fig. 2 SEM morphology of SOD-SPIONs (A) and CAT-SPIONs (B)

Enzymatic activity of magnetic particles has been determined indirect by NTB-methionine-riboflavin (in the case of superoxide dismutase) and direct, with H_2O_2 (for catalase). Biofunctionalized SPIONs preserve an enzymatic activity which is dependent of the magnetic particles concentration, the enzyme nature and the biological molecule conformation onto magnetic particle. Example of the catalase and Cat-SPION concentration on the enzymatic activity is presented in Table 2.

Table 2 Enzymatic activity for Catalase-SPIONs

Sample	Concentration (mg/mL)	Activity (units)
Cat-SPION	0.030	10.22 ± 0.11
	0.100	27.25 ± 0.31
	0.600	32.14± 0.22
	1.000	45.41 ± 0.23
Catalase	0.010	36.04 ± 0.39
	0.600	71.87 ± 0.15
	1.000	164.28 ± 0.13

The degradation studies were performed on both SPIONs and Cat-SPIONs, in contact with esterase (enzyme which degrade the SPIONs in the human body [10]) and the degradation products were monitored spectrometrically (Table 3).

The results indicate the degradation of the polymeric matrix and enzyme too. In the unfunctionalized particles (SPION), peaks at 260 nm corresponding to the residue of maleic anhydride, and those around 200 nm, which increase during the firsts 72 h, correspond to polymeric matrix degradation.

Table 3 Degradation of SPION and EnzySPION

Time (h)	Cat-SPION		SPION	
	Wave number (nm)	Absorbance	Wave number (nm)	Absorbance
24	260	0,659	203	1,357
48	207,5	1,845	204,5	1,310
72	204	1,279	205,5	1,394
144	203	0,813	203,5	1,253
	585	0,022	-	-
168	574	0,024	205,5	1,001
240	203	0,780	203,5	1,284
312	548	0,011	220,5	0,879
	228	0,198	203	0,941
336	594	0,022	207	1,694
	203	0,960	-	-

In the case of catalase-functionalized particles, it was observed the presence of peptide bonds (180-230 nm) of the proteic residues and, after 312 hours, the appearance in the environment of novel compounds, indicated by the peaks at 548 nm and 594 nm, which corresponds to the heme groups in the catalase structure.

From the magnetization data was observed that the residual magnetization and the coercive force were zero, and there was no magnetic hysteresis loop. The lack of hysteresis is one of the criteria requirements for the identification of the particles as superparamagnetic.

The MTT assay revealed the citocompatibility of the enzyme-immobilized SPIONs (Figure 3).

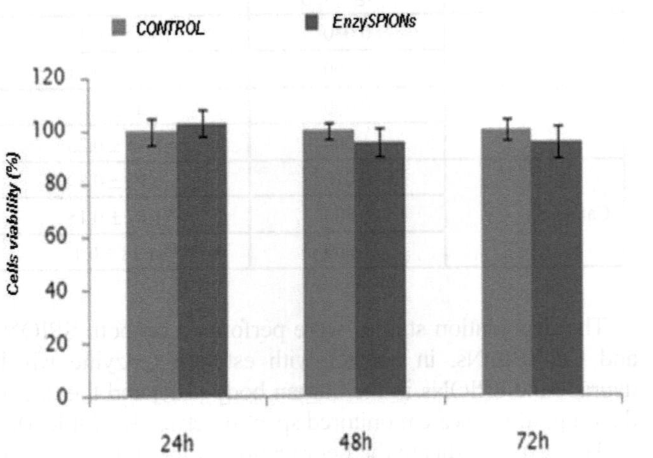

Fig. 3 Cells viability test results at 24 h, 48h and 72h

Cells viability in contact with functionalized magnetic particles was 90-97% after 72 h, and it was comparable with the control. The SPIONs not affect cell morphology and their proliferation, thus confirming their biocompatibility and the potential to be used in biomedical applications.

IV. CONCLUSIONS

A new type of functionalized superparamagnetic nanoparticles has been obtained by SOD/catalase immobilization onto SPIONs. The biofunctinalized SPIONs are degraded by esterases, enzymes responsible for polymer and enzymes degradation in the human body. Functionalized SPIONs are characterized by good magnetic properties, are biocompatible and exhibits enzymatic activity which recommend them in cardiovascular applications.

ACKNOWLEDGMENT

This work was financially supported by the Romanian Ministry of Education and Research; grant PN II- PT-PCCA-2011-3.2-0428 - INTERBIORES.

CONFLICT OF INTEREST

The authors declare that they have no conflict of interest.

REFERENCES

1. Wang B, Choudhry NK, Gagne JJ, Landon J, Kesselheim AS (2015) Availability and utilization of cardiovascular fixed-dose combination drugs in the United States, Am. Heart J. 169(3): 380-386.
2. Cai H, Harrison DG(2000) Endothelial Dysfunction in Cardiovascular Diseases: The Role of Oxidant Stress, Circ Res 87: 840-844
3. Wattanapitayakul SK, Bauer JA.(2001) Oxidative pathways in cardiovascular disease: roles, mechanisms, and therapeutic implications., Pharmachol Ther. 89:187-206.
4. Hola K, Markova Z, Zoppellaro G, Tucek J, Zboril R (2015)Tailored functionalization of iron oxide nanoparticles for MRI, drug delivery, magnetic separation and immobilization of biosubstances, Biotechnol Adv.,pii: S0734-9750(15)00028-2.
5. Netto CGCM. Toma HE, Andrade LH (2013) Superparamagnetic nanoparticles as versatile carriers and supporting materials for enzymes, J Mol Catal B: Enz., 85-86, 71.
6. Gupta AS (2011) Nanomedicine approaches in vascular disease a review, Nanomed Nanotechnol Biol Med 763-779
7. Moghimi SM, Hunter AC, Murray JC, Nanomedicine: current status and future prospects, The FASEB J, 19(3) 311-330.
8. Chorny M, Hood E, Levy RJ, Muzykantov VR (2010), Endothelial delivery of antioxidant enzymes loaded into non-polymeric magnetic nanoparticles, J Control Release. 146(1): 144–151.
9. Shangari N, O'Brien PJ (2006), Catalase Activity Assays, Current Protocols in Toxicology. Chapter 7:Unit 7.7.1-15
10. Mahapatro A, Singh DK (2011) Biodegradable nanoparticles are excellent vehicle for site directed in-vivo delivery of drugs and vaccines, J Nanobiotechnol 9:55, 1-11.

Author: Liliana Verestiuc
Institute: Gr.T.Popa University of Medicine and Pharmacy
Street: 9-13 Kogalniceanu
City: Iasi
Country: Romania
Email: liliana.verestiuc@bioinginerie.ro

Part III
Biomedical Instrumentation and Biosensors

Multilevel Signal Processing for Biomedical Nanodevices

S. Dadunashvili

Georgian Technical University, Tbilisi, Georgia

Abstract— **Ubiquitous healthcare has become one of the prominent areas of research in order to address the challenges encountered in healthcare environment. Ubiquitous healthcare requires networks of intelligent sensor nodes that could be deployed "by all anywhere and anytime" and is a permanent part of the human presence. Management of systemic diseases via technology based ubiquitous patient monitoring services has been widely proposed as a viable option for economizing healthcare resources, and providing efficient, quality healthcare.**

Ubiquitous sensors, devices and networks are paving the way towards a smart world in which computational intelligence is apply at all levels of the physical environment to provide reliable and relevant services to patients. This ubiquitous intelligence will change the computing landscape and environment around the patient, because it will enable new breeds of applications and systems to be developed and the realm of computing possibilities will be significantly extended. By enhancing everyday objects with intelligence, many tasks and processes could be simplified. The physical spaces where patients interact like the homes and hospitals could become more efficient, safer and more enjoyable. Ubiquitous computing, or pervasive computing, uses these many "smart things" to create smart environments, services and applications.

The Internet of Things is a new concept for telecommunication development. The Ubiquitous sensor Network is one of the general components of Internet of Things. A smart thing can be endowed with different levels of intelligence, and may be context-aware, active, interactive, reactive, proactive, assistive, adaptive, automated, sentient, perceptual, cognitive, autonomic and/or thinking. Research on ubiquitous intelligence is an emerging research field covering healthcare needs. A series of challenges exist to move from the current level of computing services in healthcare to the smart world of adaptive and intelligent services.

Keywords— **Ubiquitous healthcare, Ubiquitous sensors, Ubiquitous intelligence.**

I. INTRODUCTION

Management of diseases via technology based ubiquitous patient monitoring services has been widely proposed as a viable option for economizing healthcare resources, and providing efficient, quality healthcare. The process of ubiquitous patient monitoring and relevant new technologies are information intensive, the information generated is not only fragmented but also spans multiple processes, artifacts, parameters, and decision criteria. The key contribution is a framework that captures the complex processes, the parameters involved, and the decision criteria for ubiquitous patient monitoring. The decision protocols and enabling technologies supporting the processes and a conceptual model of ubiquitous patient monitoring is developed. Work began with the development of the system for recognizing activities in the home setting using a set of small and simple state-change sensors. The sensors are designed to be "tape on and forget" devices that can be quickly and ubiquitously installed in home environments. The proposed sensing system presents an alternative to sensors that are sometimes perceived as invasive, such as cameras and microphones. The system has been deployed in multiple residential environments with non-researcher occupants. Preliminary results on a small dataset show that it is possible to recognize activities of interest to medical professionals.

Another example is the fact that the number of senior citizens who are living alone and taking medicine is increasing due to population aging. Since these people have many risks of untimely and incorrect dosing, their caretakers living at a distance are concerned about their health. The system that is composed of intelligent medicine case and ubiquitous sensors can support timely correct dosing of recipient and dosing monitoring of their caretakers. The intelligent medicine case not only detects medicine in the medicine case by image processing but also notifies about dosing timing and dosage by touch display and speaker. The ubiquitous sensors assist to decide dosing timing.

II. MANAGEMENT OF DISEASES

Ubiquitous healthcare has become one of the prominent areas of research in order to address the challenges encountered in healthcare environment, based on modern scientific achievements. In contribution to this area, this study developed a system prototype that recommends medical services based on physiological, psychophysical and psycho-emotional data collected in real time from a distant patient. The system uses body sensors to be worn by the individual and a smart node as a personal server. Physiological, psychophysical and psycho-emotional data is collected and uploaded to a Medical Health Server via internet to be analyzed. Implemented prototype monitors the activity, location and physiological data of the elderly and

V. Sontea and I. Tiginyanu (eds.), *3rd International Conference on Nanotechnologies and Biomedical Engineering*,
IFMBE Proceedings 55,
DOI: 10.1007/978-981-287-736-9_79

patients in rehabilitation. The uploaded information can be accessed in real time by medical practitioners through a web application.

Sensor networks have the potential to greatly impact many aspects of medical care. By outfitting patients with wireless, wearable vital sign sensors, collecting detailed real-time data on psycho-physiological status can be greatly simplified. However, there is a significant gap between existing sensor network systems and the needs of medical care. In particular, medical sensor networks must support multicast routing topologies, node mobility, a wide range of data rates and high degrees of reliability, and security. Developing a combined hardware and software platform for medical sensor networks provides protocols for device discovery and publish/subscribe multi-hop routing, as well as a simple query interface that is tailored for medical monitoring.

An emerging application for wireless sensor networks involves their use in medical care. In a hospital or clinic, outfitting every patient with tiny, wearable wireless vital sign sensors would allow doctors, nurses and other caregivers to continuously monitor the status of their patients. In an emergency or disaster scenario, the same technology would enable medics to more effectively care for large numbers of casualties. First responders could receive immediate notifications on any changes in patient status, such as respiratory failure or cardiac arrest. Wireless sensors could augment or replace existing wired telemetry systems for many specific clinical applications, such as physical rehabilitation or long-term ambulatory monitoring.

Despite the increased interest in this area, a significant gap remains between existing sensor network designs and the requirements of medical monitoring. Most sensor networks are intended for deployments of stationary nodes that transmit data at relatively low data rates, with a focus on best-effort data collection at a central base station. By contrast, medical monitoring requires relatively high data rates, reliable communication, and multiple receivers. Moreover, unlike many sensor network applications, medical monitoring cannot make use of traditional in-network aggregation since it is not generally meaningful to combine data from multiple patients.

III. UBIQUITOUS MEDICAL SENSOR NETWORKS

The analysis showed, that "Ubiquitous Medical Sensor Networks" (UMSN) is used to describe networks of intelligent sensor nodes that could be deployed "by all anywhere and anytime". The results of research show that it is highly suitable for monitoring purposes in biomedical applications. However, there isn't any flexible and robust communication infrastructure to integrate these devices into an emergency care setting. An efficient wireless communication substrate for medical devices that addresses ad hoc or fixed network formation, naming and discovery, security and authentication, as well as filtration and aggregation of vital sign data need to be studied. The following example is aimed at developing a system that is able to provide continuous monitoring of patients outside the hospital environment.

As illustrated by Figure 1 different layers are used by a Ubiquitous Sensor Network (USN) to provide different services to different types of applications in a multi-technology, multi-devices and multi-protocol platform. These include an application layer (I), at which are embodied different applications to perform tasks related to diagnostic, therapy, hygiene, old-aged and many others. The First Generation Network (FGN) gateways (II) are USN middleware, residing inside or outside the gateway and used as an interface between the FGN and the application layer. A USN access networking layer (III) present USN-bridges which with sink nodes are used as an access network for the first-mile connectivity of a FGN gateways. A sensor networking layer (IV) represent sensor and RFID devices which placed into the environment to sense what is happening and report to sink nodes via USN-bridges.

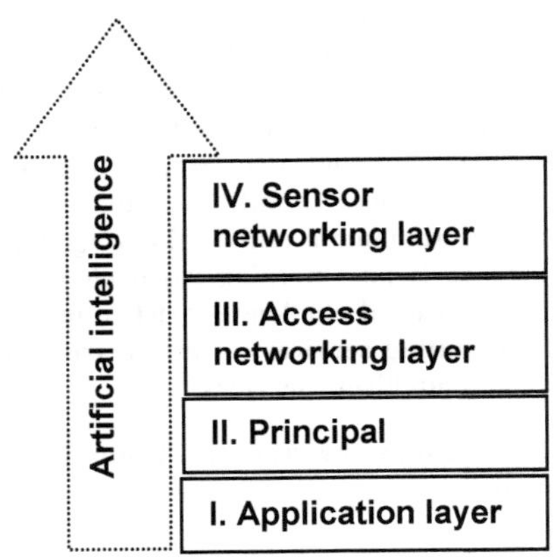

Fig. 1 Intelligence properties of USN

Building upon the layered approach depicted by Figure 1, adopted a "Ubiquitous Medical Sensor Networks" (UMSN) depicted by Figure 2 where software and hardware components are used to deliver different services to the different layers of the system. The resulting UMSN reveals sensors and actuators which are deployed to sense what is happening into the environment. Sensor readings stored into a set of databases to abstract the USN. A situation awareness component that uses the sensor readings which are stored into the databases and a user applications interface that receives as input deterministic and probabilistic data from the General Intelligence node respectively and outputs this data to different applications such as diagnostic, therapy, hygiene and old-aged monitoring. While sensor and actuators are located into the sensor networks layer, the databases and the situation awareness are software components which are embedded into a gateway located in the USN middleware layer. The databases provide an interface to the application layer in terms of deterministic data and input to the situation awareness component. The situation awareness is another important component of the middleware layer that provides an interface to the application layer in terms of probabilistic data and the sensor networking layer in terms of reactive actions to control the environment. In this way was adopted a model that combines the USN access networking and USN middleware into a General middleware layer where the USN is abstracted by the databases running into the gateways.

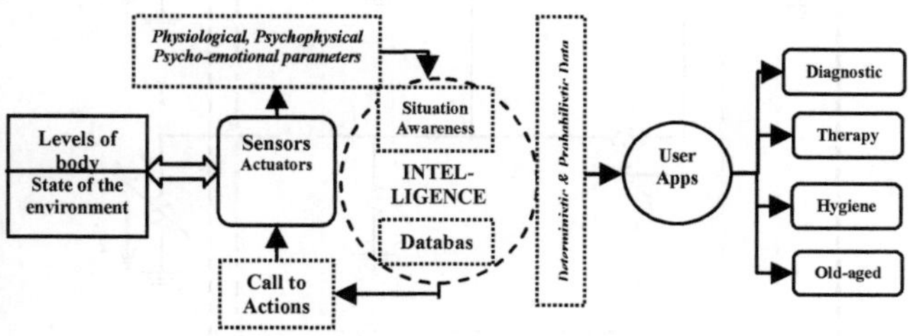

Fig. 2 Ubiquitous Medical Sensor Networks

IV. CONCLUSIONS

As we enter the era of ubiquitous sensing, we have the opportunity to monitor the world around us with unprecedented resolution and to leverage this vast wealth of data to make our environment smarter. Distributed remote sensors and computers in the new environment can sense surroundings and objects and subjects conditions and use this information in strategy of life. Smart meters enable power usage monitoring and can assist in demand-response in the smart grid, and cheap wireless motes that measure noise, light, and air pollution can be used as input into public health decisions.

The new fact is that currently available monitors the internal environment of the human body [1]. In addition, appear new methods of recording Psychophysical and Psycho-emotional parameters [2].

There is an opportunity to wear electronic communication, medical, geo position devices not in the human body, but inside it. In conclusion, note several new technologies, which will soon become familiar to us, and our body.

We do not consider here implantation technology, which all have become accustomed and which comes into our lives - all sorts of pacemakers, artificial joints and prostheses. We focus only on technologies related to miniature electronics and wireless communications.

- Implantable Smartphone;
- Implantable user interface;
- Attending chips, such as neurostimulator device;
- Robots in the blood vessels such as the fight against cancer;
- Smart tattoos, such as "tattoo with built-in sensors”;
- Electronic tablets with feedback. These miniature devices transmit a doctor "inside information";
- Built-in birth control, a very handy tool for those who are planning the composition of family;
- Brain-computer interface;
- Bionic vision and eye as a sensor;
- Implantable "smart dust";
- Implants for paralyzed;
- Soluble batteries.

The future is already here.

CONFLICT OF INTEREST

The authors declare that they have no conflict of interest.

REFERENCES

1. Dadunashvili S. Biosignals processing in the structural levels of body IFMBE Proc. vol. 39, World Congress on Med. Phys. & Biomed. Eng., Beijing, China, 2012, pp 581–584.
2. Dadunashvili S. System Pattern of Consciousness Functioning IFMBE Proc. vol. 38, International Symposium on Med. Phys. & Biomed. Eng., Riga, Latvia, 2013, pp 5–10.

Author:	Dadunashvili Sergo
Institute:	Georgian Technical University
Street:	Kostava 77
City:	Tbilisi
Country:	Georgia
Email:	dadu@gtu.ge

Transmission of Resistance Sensor Signals over Multi-wire Line with Losses

A. Penin and A. Sidorenko

"D. Ghitu" Institute of Electronic Engineering and Nanotechnologies, Academy of sciences of Moldova, Chisinau, Moldova Republic

Abstract— **The invariant relationship between the sets of load conductivity values at the output of the four wire DC communication line and the corresponding current values at the input of this line is shown. This relationship does not depend on accuracy of measuring devices. It allows transmitting signals over these wire communication line.**

Keywords— **multi-port, projective coordinates, load, invariant, transmission of signal.**

I. INTRODUCTION

Different sensors of physical values are used for monitoring technical or natural objects. For these usually remote devices, it is necessary to provide the transmission of measured signals [1, 2], for example, by multi- wire line [3].

The researches show that invariant relationship takes place between the sets of load conductivity (resistance) values at the output and the corresponding current values at the input of this line with losses [4]. The cross ratio of four points, known in projective geometry, is this invariant relationship. Therefore, the cross ratio value does not depend on parameters of communication line, accuracy of measuring devices and influence of noises. These invariants are used for transmission of electrical signals over the two and three wire lines [5, 6].

In this connection, it is naturally to consider the four wire line for transmission or, more precisely, for restoration of three signals.

II. STATEMENT OF THE PROBLEM

Let us consider the equivalent circuit or model of a four wire line with losses in Fig.1. Resistance sensors are the loads with conductivities Y_{L1}, Y_{L2}, and Y_{L3}, which connect to the output of this line. A mutual influence of the loads on the input currents I_4, I_5, I_6 takes place because of lateral loss resistances. For known values of the loss resistances or transmission parameters, we may calculate the load values by the measured input currents.

To determine the transmission parameters, we must test this line by complex manipulations at the input and output that is often inconvenience.

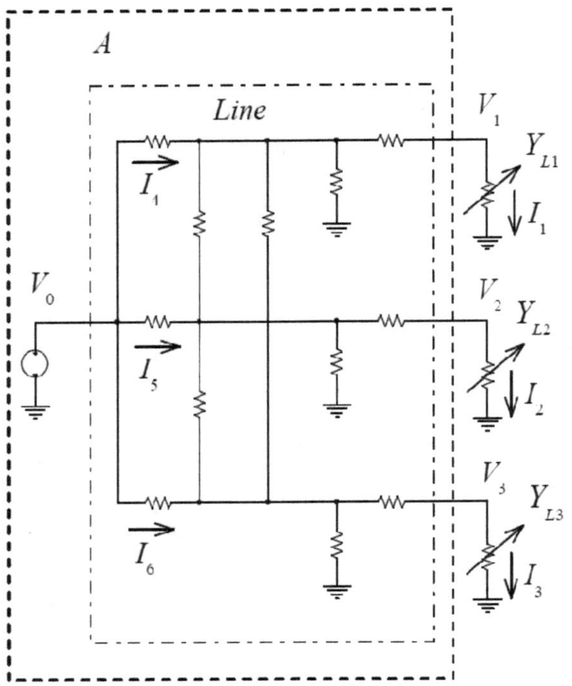

Fig. 1 Four wire line with longitudinal and lateral loss resistances

Further, we will show how to calculate the load values by the measured input currents by testing the line only by manipulations at the output terminals.

III. BALANCED MULTI-PORT CIRCUITS

We may consider our line as an active multiport A relatively to the load conductivities. The circuit is described by the following system of Y -parameters equation

$$\begin{pmatrix} I_1 \\ I_2 \\ I_3 \end{pmatrix} = \begin{pmatrix} -Y_{11} & Y_{12} & Y_{13} \\ Y_{12} & -Y_{22} & Y_{23} \\ Y_{13} & Y_{23} & -Y_{33} \end{pmatrix} \cdot \begin{pmatrix} V_1 \\ V_2 \\ V_3 \end{pmatrix} + \begin{pmatrix} I_1^{SC} \\ I_2^{SC} \\ I_3^{SC} \end{pmatrix}, \quad (1)$$

I_1^{SC}, I_2^{SC}, and I_3^{SC} are the short circuit SC currents. Taking into account the voltages $V_1 = I_1 / Y_{L1}$,

V. Sontea and I. Tiginyanu (eds.), *3rd International Conference on Nanotechnologies and Biomedical Engineering*,
IFMBE Proceedings 55,
DOI: 10.1007/978-981-287-736-9_80

$V_2 = I_2 / Y_{L2}$, and $V_3 = I_2 / Y_{L3}$, the equations of three bunches of planes are obtained in the form $(I_1, I_2, I_3, Y_{L1}) = 0$, $(I_1, I_2, I_3, Y_{L2}) = 0$, $(I_1, I_2, I_3, Y_{L3}) = 0$.

Crossing of the planes of one bunch among themselves defines a bunch axis. The equation of the axis of the bunch Y_{L1} corresponds to the condition $I_1 = 0$, $V_1 = 0$ and to equation $(I_2, I_3) = 0$. Therefore, this axis is located in the plane I_2, I_3 in Fig.2a.

(a)

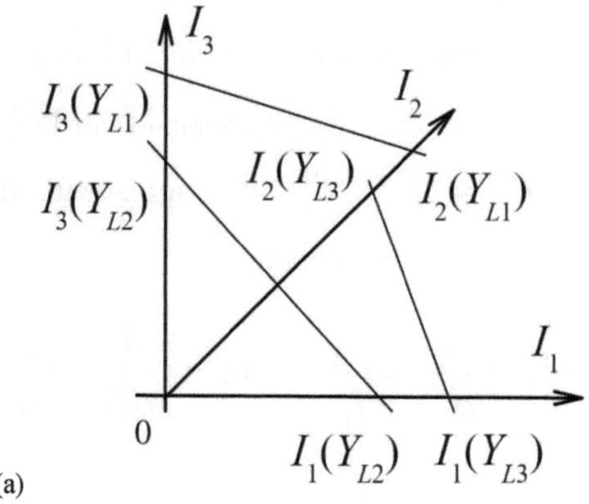

(b)

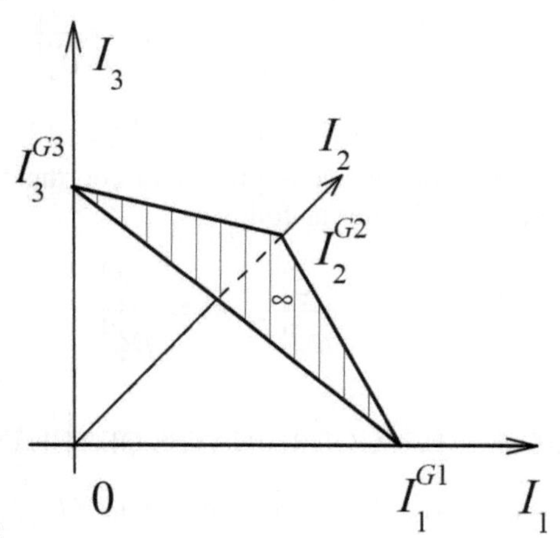

Fig. 2 a Points of intersection for each axis do not coincide among themselves. b Points of intersection coincide among themselves and form the plane of infinity ∞

The points $I_2(Y_{L1})$, $I_3(Y_{L1})$ are the points of intersection with corresponding axis. Similarly, we obtain the points $I_1(Y_{L2})$, $I_3(Y_{L2})$ of intersection of the bunch axis Y_{L2} and the points $I_1(Y_{L3})$, $I_2(Y_{L3})$ of intersection of the bunch axis Y_{L3}.

On the other hand, the projective system of coordinates has to represent a tetrahedron $0 I_1^{G1} I_2^{G2} I_3^{G3}$ in Fig.2b. In this case, we accept the plane, which passes through three base points I_1^{G1}, I_2^{G2}, and I_3^{G3} on the axes of coordinates, as the plane of infinity ∞. Therefore, the next conditions have to be satisfied

$$I_1(Y_{L2}) = I_1(Y_{L3}) = I_1^{G1},\ I_2(Y_{L1}) = I_2(Y_{L3}) = I_2^{G2},$$
$$I_3(Y_{L1}) = I_3(Y_{L2}) = I_3^{G3}.$$

We must determine requirements for Y-parameters.

Let us consider the base point or base values

$$I_1 = I_1^{G1},\ V_1 = V_1^{G1}.$$

Then, $I_2 = 0, V_2 = 0$, $I_3 = 0, V_3 = 0$. Using (1), we get

$$-V_1^{G1} = \frac{I_2^{SC}}{Y_{12}} = \frac{I_3^{SC}}{Y_{13}}.$$

Similarly, we consider

$$I_2 = I_2^{G2},\ V_2 = V_2^{G2},\quad -V_2^{G2} = \frac{I_1^{SC}}{Y_{12}} = \frac{I_3^{SC}}{Y_{23}}.$$

$$I_3 = I_3^{G3},\ V_3 = V_3^{G3},\quad -V_3^{G3} = \frac{I_1^{SC}}{Y_{13}} = \frac{I_2^{SC}}{Y_{23}}.$$

From here, we must obtain or complete the given line's conductivities.

We will identify this network as the balanced network for output terminals. The obtained conductivity values do not limit especially functional possibilities of such circuits, but allow simplifying essentially the calculation of loads, as it is shown further.

IV. Projective Coordinates of the Output

We accept that coordinate axes I_1, I_2, I_3 determine Cartesian coordinate system (I_1, I_2, I_3) in three-dimensional space in Fig.3.

Let an running regime correspond to a point M^1, which is set by load conductivities $Y_{L1}^1, Y_{L2}^1, Y_{L3}^1$ and currents I_1^1, I_2^1, I_3^1. The short circuit regime of both loads is presented by the point SC. As it was showed above, the plane $G_1 G_2 G_3$ passes through the base currents $I_1^{G1}, I_2^{G2}, I_3^{G3}$.

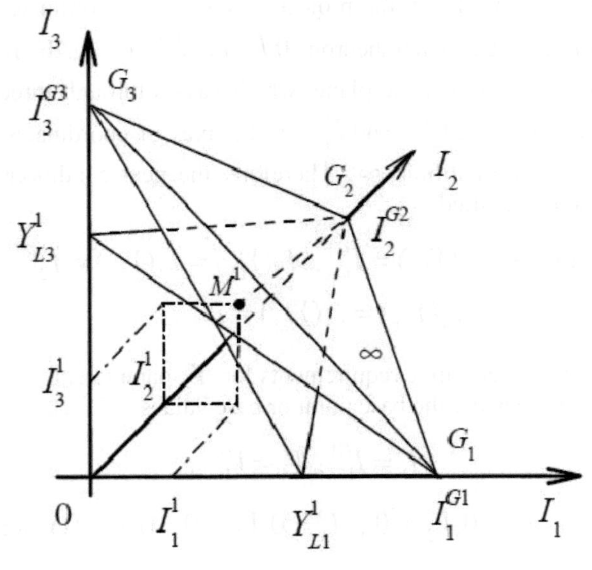

Fig. 3 Cartesian coordinate system (I_1, I_2, I_3) and projective coordinate $0G_1G_2G_3$

Then, the running value Y_{L3}^1 corresponds to the plane which passes through the point M^1 and straight line G_1G_2. This line corresponds to the intersection of the planes $G_1G_2G_3$ and $0G_1G_2$. Similarly, Y_{L2}^1 agrees to the plane, which passes through the point M^1 and straight line G_1G_3. Also, Y_{L1}^1 matches to the point M^1 and line G_2G_3. We remind that this line is the axis of the bunch plane Y_{L1}. In turn, the values $Y_{L1}=0$, $Y_{L1}=\infty$, Y_{L1}^{G1} are the characteristic values of Y_{L1}. Therefore, the running value Y_{L1}^1 corresponds to the non-uniform coordinate m_1^1 in the form of the cross ratio of the four points

$$m_1^1 = (0\, Y_{L1}^1\, \infty\, Y_{L1}^{G1}) = \frac{Y_{L1}^1}{Y_{L1}^1 - Y_{L1}^{G1}}. \tag{2}$$

There, $Y_{L1}=0$, $Y_{L1}=Y_{L1}^{G1}$ are the base points and $Y_{L1}=\infty$ is a unit point. The cross ratio for m_2^1, m_3^1 is expressed similarly

$$m_2^1 = \frac{Y_{L2}^1}{Y_{L2}^1 - Y_{L2}^{G2}},\ m_3^1 = \frac{Y_{L3}^1}{Y_{L3}^1 - Y_{L3}^{G3}}. \tag{3}$$

Besides the non-uniform coordinates m_1^1, m_2^1, m_3^1 of a point M^1, there are homogeneous projective coordinates $\xi_1^1, \xi_2^1, \xi_3^1, \xi_4^1$, which are set by a coordinate tetrahedron and a unit point SC. These homogeneous coordinates are defined as the ratio of the distances $\delta_1^1, \delta_2^1, \delta_3^1, \delta_4$ for the point M^1 and distances $\delta_1^{SC}, \delta_2^{SC}, \delta_3^{SC}, \delta_4^{SC}$ for a unit point SC to the planes of the coordinate tetrahedron $0G_1G_2G_3$. Then, the distances δ_1^1, δ_1^{SC} correspond to the plane $0G_2G_3$; therefore, $\delta_1^1 = I_1^1$, $\delta_1^{SC} = I_1^{SC}$. Similarly, δ_2^1, δ_2^{SC} match to the plane $0G_1G_3$ and $\delta_2^1 = I_2^1$, $\delta_2^{SC} = I_2^{SC}$. Also, δ_3^1, δ_3^{SC} correspond to $0G_1G_2$ and $\delta_3^1 = I_3^1$, $\delta_3^{SC} = I_3^{SC}$; δ_4^1, δ_4^{SC} agree with the plane $G_1G_2G_3$. Therefore, we obtain

$$\rho\xi_1^1 = \frac{\delta_1^1}{\delta_1^{SC}} = \frac{I_1^1}{I_1^{SC}},\ \rho\xi_2^1 = \frac{\delta_2^1}{\delta_2^{SC}} = \frac{I_2^1}{I_2^{SC}},$$

$$\rho\xi_3^1 = \frac{\delta_3^1}{\delta_3^{SC}} = \frac{I_3^1}{I_3^{SC}},\ \rho\xi_4^1 = \frac{\delta_4^1}{\delta_4^{SC}}, \tag{4}$$

where ρ is a coefficient of proportionality.

In turn, the homogeneous projective coordinates set the non-uniform coordinates as follows

$$m_1^1 = \frac{\rho\xi_1^1}{\rho\xi_4^1},\ m_2^1 = \frac{\rho\xi_2^1}{\rho\xi_4^1},\ m_3^1 = \frac{\rho\xi_3^1}{\rho\xi_4^1}.$$

V. Projective Coordinates of the Input

Let us superpose the coordinate system (I_4, I_5, I_6) with (I_1, I_2, I_3) in Fig.4. Then, any point (I_1, I_2, I_3) corresponds to a point (I_4, I_5, I_6). In particular, the running regime point M^1, a unit point SC and coordinate tetrahedron $0G_1G_2G_3$ correspond to the points $\overline{M}^1$, $\overline{SC}$ and $\overline{0G_1}\,\overline{G_2}\,\overline{G_3}$.

For convenience, the conformity of the output and input characteristic regime points is shown in Table 1.

It is known that projective coordinates of points M^1 and $\overline{M}^1$, concerning the own coordinate systems, are equal among themselves. Therefore, we must introduce the known coordinates $\xi_1^1, \xi_2^1, \xi_3^1, \xi_4^1$ of the point $\overline{M}^1$.

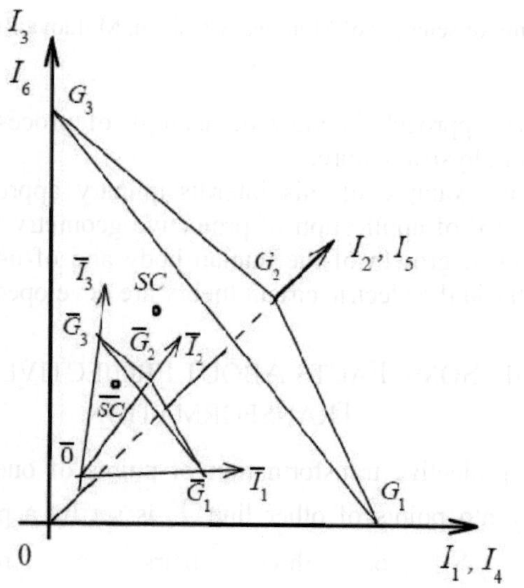

Fig. 4 Superposition of coordinate systems of output (I_1, I_2, I_3) and input (I_4, I_5, I_6) currents

Table 1 Output - input conformity of the multi-port currents

Regime points of		Input currents		
output	input	I_4	I_5	I_6
0	$\overline{0}$	I_4^{OC}	I_5^{OC}	I_6^{OC}
G_1	$\overline{G}_1$	I_4^{G1}	I_5^{G1}	I_6^{G1}
G_2	$\overline{G}_2$	I_4^{G2}	I_5^{G2}	I_6^{G2}
G_3	$\overline{G}_3$	I_4^{G3}	I_5^{G3}	I_6^{G3}
SC	$\overline{SC}$	I_4^{SC}	I_5^{SC}	I_6^{SC}

These homogeneous coordinates are defined as the ratio of the distances $\overline{\delta}_1^1, \overline{\delta}_2^1, \overline{\delta}_3^1, \overline{\delta}_4$ for the point $\overline{M}^1$ and distances $\overline{\delta}_1^{SC}, \overline{\delta}_2^{SC}, \overline{\delta}_3^{SC}, \overline{\delta}_4^{SC}$ for a unit point $\overline{SC}$ to the planes of the coordinate tetrahedron $\overline{0}\overline{G}_1\,\overline{G}_2\,\overline{G}_3$.

$$\rho\,\xi_1^1 = \frac{\overline{\delta}_1^1}{\overline{\delta}_1^{SC}}, \; \dots, \; \rho\,\xi_4^1 = \frac{\overline{\delta}_4^1}{\overline{\delta}_4^{SC}}.$$

In turn, in accordance to (4), (5), we obtain or restore the load conductivities by (2), (3)

$$Y_{L1}^1 = \frac{Y_{L1}^{G1} m_1^1}{m_1^1 - 1}, \; Y_{L2}^1 = \frac{Y_{L2}^{G2} m_2^1}{m_2^1 - 1}, \; Y_{L3}^1 = \frac{Y_{L3}^{G3} m_3^1}{m_3^1 - 1}.$$

VI. CONCLUSIONS

The offered choice of line's parameters permits to introduce projective coordinates. Input projective coordinates allow finding load conductivities only by the measured input currents without determination of the transmission parameters of a multi-port. It allows transmitting signals over a communication line.

CONFLICT OF INTEREST

The authors declare that they have no conflict of interest.

REFERENCES

1. Walker K. Ramplin M (2002) Two wire communication system. Patent US 6,459,363B1
2. Ovchinnikov V (2013) Method for transmitting discrete electric signals. Patent US 8,446,977B2
3. Aldereguia A. Richter G. Williams J (2009) Redundant 3-wire communication system. Patent US 7.502.991 B2
4. Penin A (2015) Analysis of Electrical Circuits with Variable Load Regime Parameters: Projective Geometry Method. Springer International Publishing
5. Penin A, Sidorenko A (2014) Transmission of measuring signals and power supply of remote sensor, Nanotech. Security Syst. NATO Sci. Peace and Security Series C: Environment Security. Springer Science+ Business Media Dordrecht, 20014, pp. 267–281
6. Penin A (2012) Invariant properties of cascaded six-pole networks. Int. J. Circ. Syst. Signal Proc. 6(5):305-312

Author: A. Penin,
Institute: "D. Ghitu" Institute of Electronic Engineering and Nanotechnologies, Academy of sciences of Moldova
Street: 3/3 Academiei st,
City: Chisinau
Country: Republic of Moldova
Email: aapenin@mail.ru

Projective Geometry Invariants of Human Body and Multi-port Electrical Circuits

A. Penin

"D. Ghitu" Institute of Electronic Engineering and Nanotechnologies, Academy of sciences of Moldova, Chisinau, Moldova Republic

Abstract— **The review of application of projective geometry for interpretation of growth of the human body is submitted and features of use of this geometry in the electric circuit theory are shown. Growth of the human body, the change of regime parameters of a network is interpreted as projective transformations. These transformations possess invariants, there are cross ratios of four points. The common mathematical apparatus represents interdisciplinary approach in view of analogy of processes of a different physical nature. The obtained results develop methodology of application of non-Euclidean geometries.**

Keywords— **human body, multi-port, projective coordinates, load, invariant.**

I. Introduction

The growth of the human body is essentially nonlinear for ordinary or Euclidean geometry. Analysis of the structure and growth reveals that separate parts or all the three-component kinematics blocks (fingers, extremities and body) change according to projective geometry transformations. Such transformations possess an invariant or invariable value. Therefore, all the called three-component blocks are characterized by a constant value throughout life [1], [2] by the cross- ratio of four points on a straight line. This value of wurfs, 1.31, in the body blocks is the function of the biological phyllotaxis laws of shaping; it coincides with the so-called golden wurf,

$$P = \frac{3+\sqrt{5}}{2} = \frac{\Phi^2}{2} = 1.309...,$$

conjugated with the Fibonacci numbers and the gold section,

$$\Phi = \frac{1+\sqrt{5}}{2} = \frac{\Phi^2}{2} = 1.618...,$$

which is so important in the esthetics of proportions and the mathematics of harmony [3].

The researches show that invariant relationship takes place between the load conductivity values at the output of circuit and the corresponding values at the input of this circuit [4, 5].

It is natural, when the common mathematical apparatus is applied in various areas of science; it represents interdisciplinary approach in view of analogy of processes of a different physical nature.

As an example of this interdisciplinary approach, the results [6] of application of projective geometry for interpretation of growth of the human body and of use of this geometry in the electric circuit theory are developed.

II. Some Facts About Projective Transformation

The projective transformation of points of one straight line V_L into points of other line Y_L is set by a projection centre S or three pairs of respective points $V_L^1,\, V_L^2,\, V_L^4 \to Y_L^1,\, Y_L^2,\, Y_L^4$ in Fig.1. Let points $V_L^3 \to Y_L^3$ be given as running points. Then, the projective transformations preserve the cross ratio of these four points

$$m_L^{32} = (Y_L^1\ Y_L^2\ Y_L^3\ Y_L^4\ ) = \frac{Y_L^2 - Y_L^1}{Y_L^2 - Y_L^4} \div \frac{Y_L^3 - Y_L^1}{Y_L^3 - Y_L^4}, \qquad (1)$$

$$m_L^{32} = (V_L^1\ V_L^2\ V_L^3\ V_L^4\ ).$$

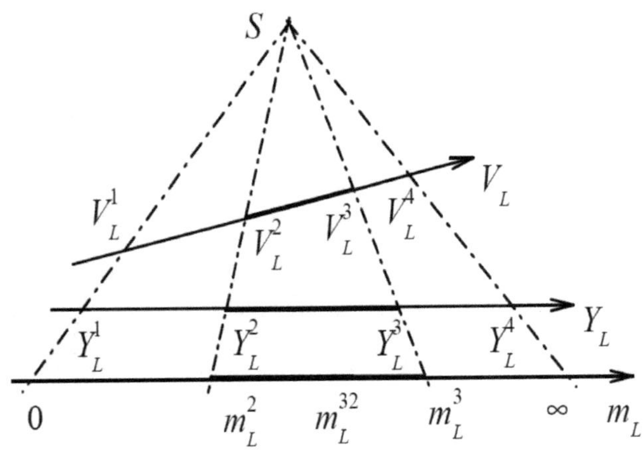

Fig. 1 Projective transformation of points of straight lines

The growth of the human body is essentially nonlinear, as it is shown in Fig.2[1]. For a human body, growth

V. Sontea and I. Tiginyanu (eds.), *3rd International Conference on Nanotechnologies and Biomedical Engineering*,
IFMBE Proceedings 55,
DOI: 10.1007/978-981-287-736-9_81

changes of three-component kinematic blocks (the parts of fingers, extremities, and body) are characterized by the constant value of cross-ratio (1)

$$W = \frac{B-A}{B-D} \div \frac{C-A}{C-D} = const\,.$$

The values of the cross-ratio of all the blocks, at least during individual development, are grouping around the benchmark 1.31. Thus, all these blocks of the human kinematics are projective equivalent and invariable during the lifetime.

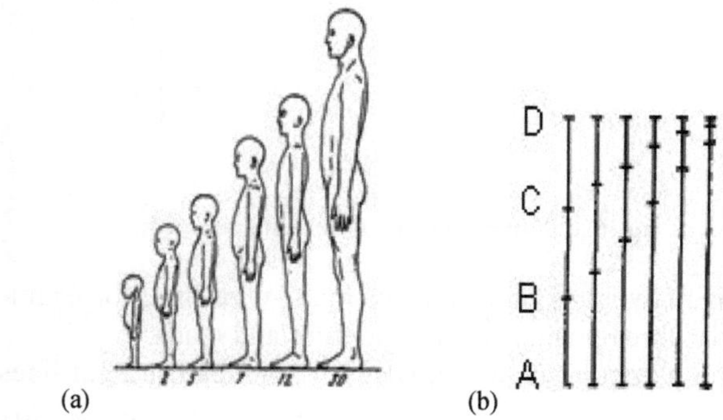

Fig. 2 a Changes of the human body with age. b The cross ratio of three-component bodies are equal to1,31

III. Projective Transformations of an Electrical Network for Lines and Planes

A. Projective Lines

Let us consider a two-port $TP1$ in Fig.3. The load change from the value Y^1_{L1} to the value Y^2_{L1} defines the correspond changes $Y^1_{IN1} \to Y^2_{IN1}$. The length of segments $Y^1_{L1}Y^2_{L1}$, $Y^1_{IN1}Y^2_{IN1}$ onto the corresponding load straight lines is different for the usually used Euclidean geometry.

If the mapping is viewed as the projective transformation, the cross-ratio of four points is performed and defines the same length of segments (1)

$$m^{21}_{L1} = (\infty \;\; Y^2_{L1}\; Y^1_{L1}\; 0\;) = Y^1_{L1} / Y^2_{L1}, \qquad (2)$$

$$m^{21}_{L1} = (\,Y^{SC}_{IN1}\;\; Y^2_{IN1}\; Y^1_{IN1}\; Y^{OC}_{IN1}\,),$$

$$m^{21}_{L1} = (\,I^{SC}_1\;\; I^2_1\;\; I^1_1\;\; 0\,),\; m^{21}_{L1} = (\,I^{SC}_0\;\; I^2_0\;\; I^1_0\;\; I^{OC}_0\,).$$

Thus, the parts of the networks of this kind are projectively – similar. Therefore, there is some the electro - biological analogy; disproportionate change of segments of load lines for different parts of a network corresponds to growth changes of parts of the human body.

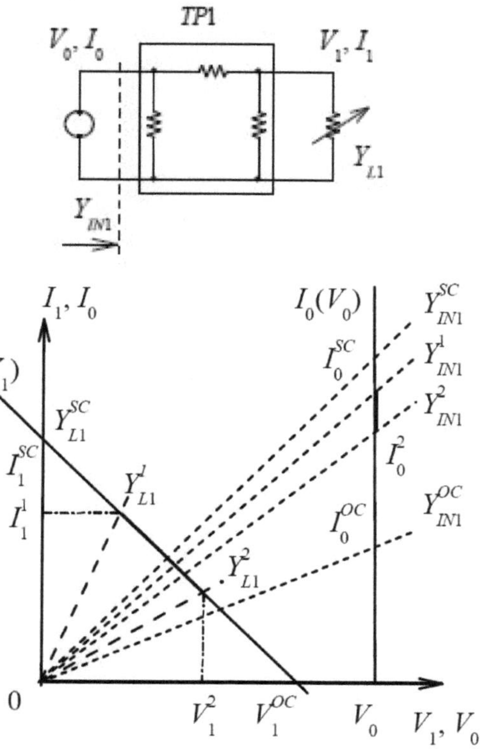

Fig. 3 Example of a two-port and its input - output load straight lines

B. Projective Planes

We consider a network with two changeable loads Y_{L1}, Y_{L2} in Fig.4. We may superpose the coordinates $(I_3\, 0\, I_4)$ with the coordinates $(I_1\, 0\, I_2)$ in Fig.5. In this case, the load straight lines form the coordinate or reference triangles $G_1\, 0\, G_2$, $\overline{G}_1\, \overline{0}\, \overline{G}_2$. The property of projective transformations shows that the point $\overline{M}^1$ coordinates are equal to the point M^1 coordinates, as these points $M^1, \overline{M}^1$ are set by the same loads Y^1_{L1}, Y^1_{L2} [5]. In addition, this point M^1 is defined by projective non-uniform coordinates m^1_1, m^1_2 and homogeneous coordinates $\xi^1_1, \xi^1_2, \xi^1_3$, which are set by the reference triangle $G_1\;\, 0\, G_2$ and a unit point SC.

Non-uniform projective coordinates have form (1)

$$m^1_1 = \frac{Y^1_{L1}}{Y^1_{L1} - Y^{G1}_{L1}}, \qquad m^1_2 = \frac{Y^1_{L2}}{Y^1_{L2} - Y^{G2}_{L2}}. \qquad (3)$$

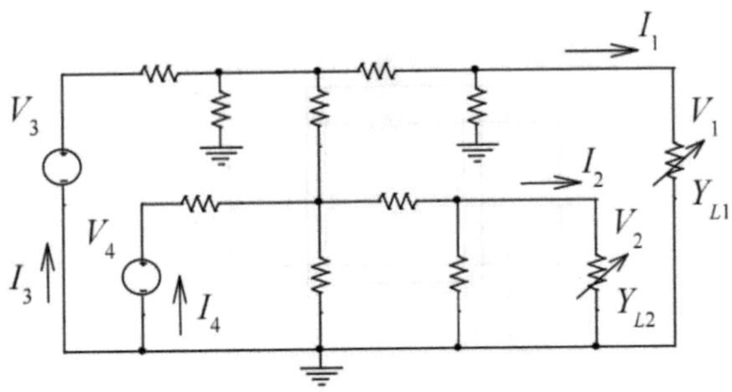

Fig. 4 Network with interference of two loads

In turn, homogeneous projective coordinates ξ_1, ξ_2, ξ_3 set the non-uniform coordinates

$$m_1 = \xi_1 / \xi_3,\ m_2 = \xi_2 / \xi_3 .$$

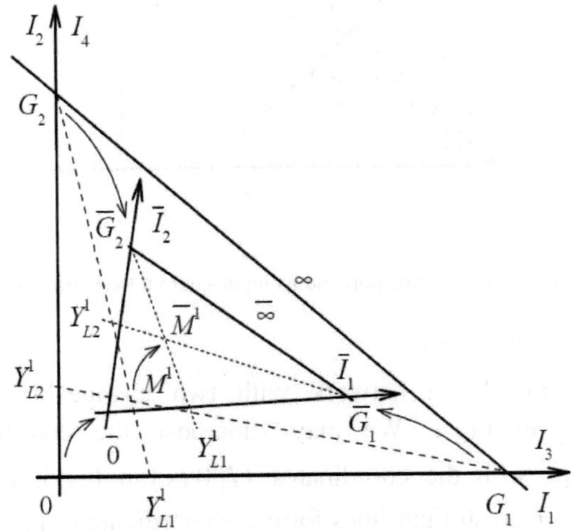

Fig. 5 Transformation of the plane (I_1, I_2) onto the plane (I_3, I_4)

The homogeneous coordinates are defined by the ratio of the distances for the points M^1, SC to the sides of the reference triangle $G_1\, 0\, G_2$. Similarly, these homogeneous coordinates are defined as the ratio of the distances for the points $\overline{M}^1$, $\overline{SC}$ to the sides of the reference triangle $\overline{G}_1\, \overline{0}\, \overline{G}_2$.

Therefore, this property gives required invariant relations between input – output of a network.

IV. Projective Space of a Multi-port Network

The circuit, important for practice, is presented in Fig. 6.

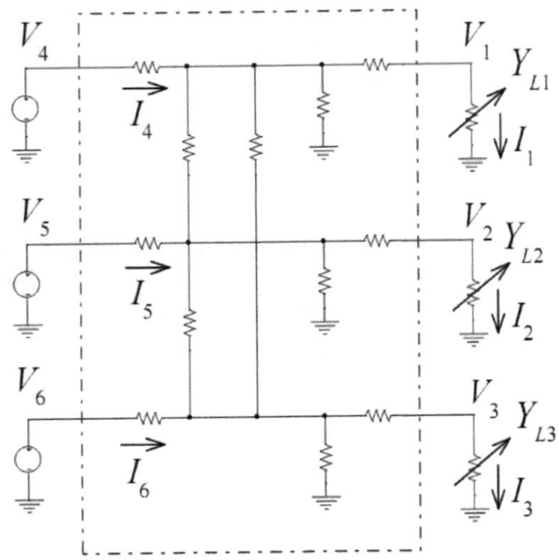

Fig. 6 Multi-port with interference of three loads

This network is multi-port or multi-wire line, for example, with three pairs of voltage sources and loads.

For a particular case of this circuit, the load straight lines form the coordinate tetrahedron $0 G_1 G_2 G_3$ in Fig.7. We accept that the coordinate axes I_1, I_2, I_3 determine the Cartesian coordinate system (I_1, I_2, I_3) in three-dimensional space.

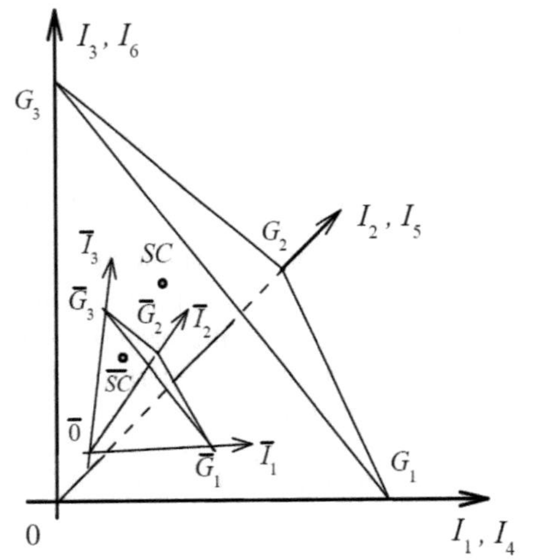

Fig. 7 Superposition of coordinate systems of output (I_1, I_2, I_3) and input (I_4, I_5, I_6) currents

Let an initial or running regime correspond to a point M^1 which is set by load conductivities $Y_{L1}^1, Y_{L2}^1, Y_{L3}^1$ and currents I_1^1, I_2^1, I_3^1. The short circuit regime of both loads is presented by the point SC. Then, the running value Y^1_{L3} corresponds to the plane which passes through the point M^1 and straight line $G_1 G_2$. This line corresponds to the intersection of the planes $G_1 G_2 G_3$ and $0G_1 G_2$ in Fig.8.

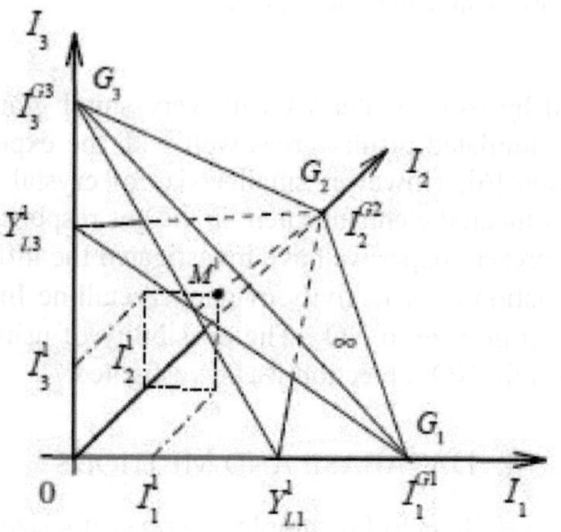

Fig. 8 Demonstration of the projective coordinate $0G_1 G_2 G_3$ relatively to the Cartesian coordinate system (I_1, I_2, I_3)

Similarly, the value Y^1_{L2} agrees to the plane which passes through the point M^1 and straight line $G_1 G_3$. Also, the value Y^1_{L1} matches to point M^1 and straight line $G_2 G_3$. Therefore, the running value Y^1_{L1} corresponds to the non-uniform coordinate m^1_1 in the form of cross ratio of four points (3). The cross ratio for m^1_2, m^1_3 is expressed similarly. There are also the homogeneous projective coordinates $\xi^1_1, \xi^1_2, \xi^1_3, \xi^1_4$, which are set by the coordinate tetrahedron $0G_1 G_2 G_3$ and a unit point SC.

Next, the running regime point M^1, a unit point SC and coordinate tetrahedron $0G_1 G_2 G_3$ correspond to the points $\overline{M}^1$, $\overline{SC}$ and coordinate tetrahedron $\overline{0G_1}\,\overline{G_2}\,\overline{G_3}$. Therefore, we must introduce the known coordinates $\xi^1_1, \xi^1_2, \xi^1_3, \xi^1_4$ of the point $\overline{M}^1$, which are set by the tetrahedron $\overline{0G_1}\,\overline{G_2}\,\overline{G_3}$ and a unit point $\overline{SC}$.

V. CONCLUSIONS

The relationship of regime parameters at different parts of the network or "movement" on these parts corresponds to projective transformations. There is some the electro - biological analogy; disproportionate change of segments of load lines for different parts of the network corresponds to growth changes of parts of the human body. The analysis of human growth and operating regimes of electric networks shows invariants of projective transformations. Established the electro - biological analogy develops a methodological basis of application of non-Euclidean geometries for these areas.

CONFLICT OF INTEREST

The author declares that he has no conflict of interest.

REFERENCES

1. Petukhov S (1989) Non-Euclidean geometries and algorithms of living bodies. Comp. Math. Applic. **17**(4-6): 505-534
2. Lundh T. Udagawa J. Hanel S. Otani H (2011) Cross- and triple-ratios of human body parts during development. The Anatomical Rec. **294**(8):1360–1369
3. Stakhov A (2009). The mathematics of harmony: From Euclid to contemporary mathematics and computer science. World Scientific Publishing Co. Pte. Ltd..
4. Penin A (2015) Analysis of Electrical Circuits with Variable Load Regime Parameters: Projective Geometry Method. Springer International Publishing
5. Penin A (2012) Invariant properties of cascaded six-pole networks. Int. J. Circ. Syst. Signal Proc. **6**(5):305-312
6. Penin A, Sidorenko A (2014) Projective geometry in electronics, techniques and living nature. Russia and Germany. **2**(6):267–281

Author: A. Penin,
Institute: "D. Ghitu" Institute of Electronic Engineering and Nanotechnologies, Academy of sciences of Moldova,
Street: 3/3 Academiei str.
City: Chisinau
Country: Republic of Moldova
Email: apenin@mail.ru

UV Effect on NO_2 Sensing Properties of Nanocrystalline In_2O_3

A. Ilin[1], N. Fantina[1], M. Martyshov[1], E. Forsh, P. Forsh[1,2], and P. Kashkarov[1,2,3,4]

[1] Physics Department, Lomonosov Moscow State University, Moscow, Russia
[2] National Research Centre "Kurchatov Institute", Moscow, Russia
[3] Department of Nano-, Bio-, Info- and Cognitive Technologies, Moscow Institute of Physics and Technology, Dolgoprudny, Russia
[4] Physics Department, Saint Petersburg State University, Saint Petersburg, Russia

Abstract— **Nanocrystalline indium oxide films with extremely small grains in range of 7 to 40 nm were prepared by sol-gel method. The influence of grain size on the sensitivity of indium oxide to nitrogen dioxide in low concentration at room temperature was investigated under the UV illumination and without illumination. The sensitivity increases with the decrease of grain sizes when In2O3 is illuminated while in the dark In2O3 with intermediate grain size exhibits the highest response. An explanation of the different behavior of the In2O3 with different grain size sensitivity to NO2 under illumination and in the dark is proposed. The pulsed illumination is demonstrated may be used for NO2 detection at room temperature that significantly reduces the power consumption of sensor.**

Keywords— **indium oxide, semiconductor gas sensor, photo-enhanced sensitivity, electric properties.**

I. INTRODUCTION

Indium oxide (In_2O_3) is attractive as solid gas sensor material in a view of excellent sensitivity and selectivity to combustion gases and typical air pollutants (CO, CH_4, H_2, NO_x) [1-3] As a rule, resistive gas sensors including indium oxide based sensors operate under high temperature condition (higher than 200˚C). High operating temperature is generally required to maximize the sensor response to the target gas either to activate the semiconductor surface toward chemisorption or to ensure the complete desorption of gaseous specious. Maintaining the sensor at high temperature requires a stable and consistent power source leading to high energy consumption. Moreover using the heated sensors is impossible in explosive environment and it can't be incorporated in electric boards of mobile devices. Therefore, obtaining highly sensitive resistive sensors that operate at low temperatures remains a challenging issue. In the some studies the decreasing of temperature was achieved by UV irradiation during the experiments. Some scientists studied the gas sensing properties of metal-oxide films subjected to UV radiation and found that films are capable to detect the gaseous compounds even at room temperature [4, 5].

Structural properties of semiconductor material utilized in gas sensors strongly affect the sensitivity to gas of these devices [5]. A sensor's sensitivity can be significantly increased by using materials with very small grains sizes, and this simulated result agrees well with the experimental observation [6]. However, smaller size of crystal does not necessary mean the enhancement in the gas response [7, 8].

In the present paper we have investigated the influence of UV irradiation on sensitivity of nanocryctalline In_2O_3 with different grain size to NO_2. The possibility of using pulsed irradiation for NO_2 detection was investigated.

II. DATABASE AND METHODS

Nanocrystalline In_2O_3 samples were synthesized by the sol–gel method, as described elsewhere [9]. After synthesis indium oxide powder annealed in the oven at 300°C, 500°C and 700°C for 24 hour, respectively.

Nanocrystalline indium oxide powder was mixed with a vehicle (α-terpineol in ethanol), deposited in the form of thick films (~1 μm) over glass substrates and dried in air. Au contacts were deposited by thermal evaporation over the films. The designations of the samples, annealing temperatures, average nanocrystals size and specific surface area are given in Table 1.

The specific surface area was determined by nitrogen adsorption using the BET model with the Chemisorb 2750 instrument (Micromeritics). The microstructure of the samples was characterized by high angle annular dark field scanning transmission electron microscopy (HAADF-STEM) with the Titan 80-300 "cubed" microscope operated at 200 kV.

Photoconductivity of the films was investigated in a cell shielded from the background light. Light irradiation of the nanocrystalline In_2O_3 samples was performed with UV LED with a maximum intensity at 380 nm. The intensity of light incident on the sample was 5 mW/cm^2.

Gas sensing tests were carried out by flow through technique under a controlled constant flux of 100 ml/min. The atmosphere composition was pre-assigned by means of electronic mass-flow controllers (Bronkhorst), mixing flows coming from certified bottles containing a given amount of the target gas diluted in synthetic air with the background flow. The measurements were effectuated during exposure to dry air or NO_2/air gas mixtures (8 ppm NO_2 in dry air)

V. Sontea and I. Tiginyanu (eds.), *3rd International Conference on Nanotechnologies and Biomedical Engineering*,
IFMBE Proceedings 55,
DOI: 10.1007/978-981-287-736-9_82

under continuous or periodic UV light illumination. The electrical resistivity was recorded in DC mode by means of Keithley 6487 picoamperemeter. Before each experiment the samples were annealed in dry air at 200 °C during 2 hours to return in initial state.

The gas response was calculated by formula

$$S = \frac{\sigma_{air}}{\sigma_{gas}} \quad (1)$$

where σ_{air} is conductivity of In_2O_3 in the air and σ_{gas} is conductivity after NO_2 adsorption.

III. RESULTS

TEM images of In_2O_3 samples are shown in Fig. 1(a)-(c). All samples consist of slightly agglomerated cubic particles. The histograms of the particle size distribution are presented in Fig. 1(d). As expected, the increase of annealing temperature leads to the particles growth and broadening of particle size distribution. The average size of In_2O_3 crystallites (*d*) varies from 8 to 41 nm (Table 1). The In_2O_3 powder treated at 300 °C is characterized by highest specific surface area S_{surf} of 100 m^2g^{-1}. This value decreases monotonously as annealing temperature increases that correlates with the crystallite size variation.

Table 1 Samples designation, annealing temperature and microstructure parameters

Sample	Annealing temperature, °C	Average crystallite size d, nm	Specific surface area S_{surf}, m^2/g
In_2O_3-300	300	8	100
In_2O_3-500	500	17	35
In_2O_3-700	700	41	10

At the nitrogen dioxide injection into the chamber where the sample was embedded the conductivity of sample decreased. When the experiment was carrying out without illumination the conductivity didn't return to its initial state even in several hours after nitrogen dioxide was deleted from the chamber (Figure 2). On the contrary when the sample was illuminated by UV light the conductivity returns to its initial state in the absence of nitrogen dioxide in the atmosphere. It should be noted that the magnitude of conduction decrease was large in experiment without illumination. The large magnitude of conduction decrease and the low conductance after the nitrogen dioxide exposure corroborate that the nitrogen dioxide doesn't desorb from sample's surface at the room temperature without illumination. The following gas sensor experiments were carried out under UV illumination.

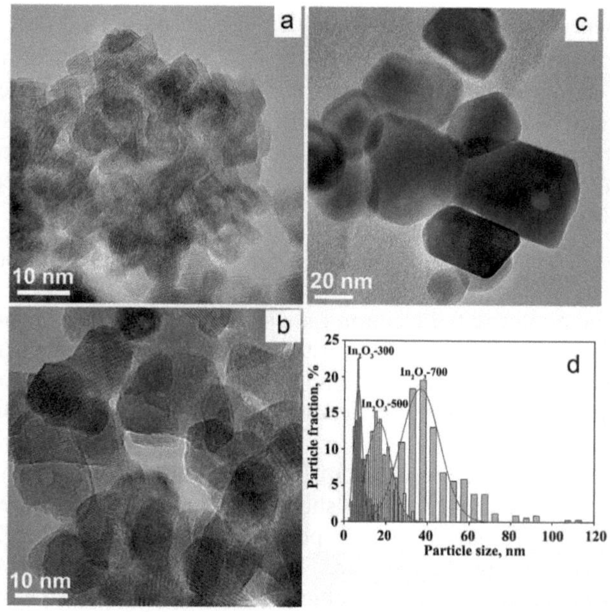

Fig. 1 TEM images of In2O3 samples: (a) In2O3-300, (b) In2O3-500, (c) In2O3-700, (d) histograms of particle size distribution.

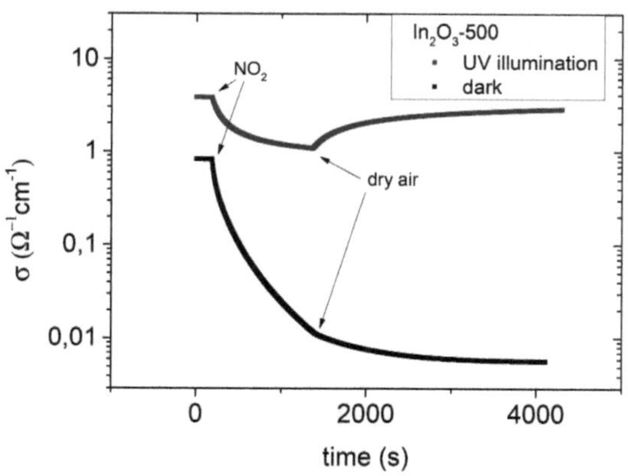

Fig. 2 Dynamic response of In2O3-500 sample to NO2. The moments when NO2 was injected and deleted from the chamber with sample are marked by arrows.

It is known that the dark conductivity of the In_2O_3 decreases when NO_2 appears in the gas phase due to adsorption of acceptor gas. $NO_{2(gas)}$ molecule captures the electron and become $NO^-_{2(ads)}$ that result in the decrease of In_2O_3 conductivity [8]:

$$NO_{2(gas)} + e^- \leftrightarrow NO^-_{2(ads)} \quad (2)$$

It can be assumed that analogous mechanism takes place when the sample is illuminated. Further we believe that when NO_2 is removed from the gas phase the adsorbed species $NO_{2(ads)}^-$ desorb by acting of UV light and the value of conductivity returns to its initial value before exposure. Indeed, under UV illumination the electron-hole pair is generated in the wide gap semiconductor oxide. The photoexcited hole can recombine with the electrons trapped by the chemisorbed $NO_{2(ads)}^-$:

$$NO_{2(ads)}^- + h^+ \leftrightarrow NO_{2(gas)} \quad (3)$$

while photoexcited electron remains in the conduction band to increase the conductivity. The dynamic equilibrium between the processes (2) and (3) determines the values of sample's conductivity depending on the NO_2 concentration in the gas phase.

The structure properties of In_2O_3 are known to determine the NO_2 sensitivity. Fig. 3 shows the conductivity changes of In_2O_3 samples upon exposure to 8 ppm NO_2 for 2 minutes at room temperature. The samples were illuminated during all time. It is noticeable that the sample with smaller nanocrystalls (and the higher specific surface area, respectively) exhibits the larger change of conductance under NO_2 adsorption. This dependence reveals the strong correlation between response to gas and the specific surface area. So it can be seen that the increase of the specific surface area leads to the increase of In_2O_3 response to NO_2 at illumination.

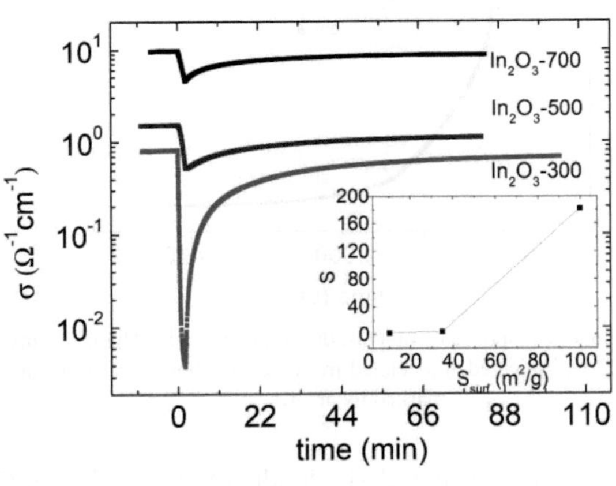

Fig. 3 Dynamic responses of samples to 8 ppm NO2. In the inset the response dependence on the specific surface ares Ssurf is shown.

Without illumination the sample with average nanocrystal size exhibit the highest response among the samples (Fig. 4). When the samples weren't illuminated responses were much higher because firstly, the duration of exposure to nitrogen dioxide was longer than in case of illumination and secondly, desorption of nitrogen dioxide didn't occur. For the explanation why the response exhibited nonmonotonically dependence on the nanocrystal size without illumination and monotomically dependence with illumination the following explanation may be proposed.

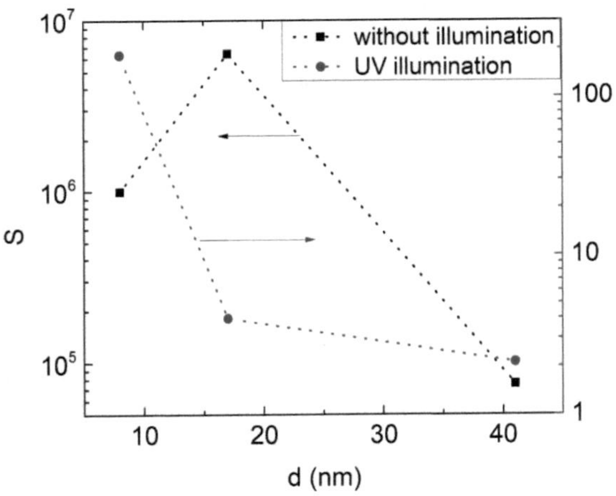

Fig. 4 Response to 8ppm NO2 versus average size of In2O3 nanocrystals with and without UV illumination.

The boundaries of nanocrystals generally contain fairly high densities of interface states which trap free carriers from the bulk of the grains and scatter free carriers by virtue of the inherent disorder and the presence of trapped charge. Interface charge gives rise to band-bending in the bulk of the grain and is therefore the origin of the interface potential barriers. The layer which is covered by barrier is depleted of electrons. The depleted layer width is related to carrier density n and determined by Debye length L_D ($L_D \sim 1/\sqrt{n}$).

The two different situations should be distinguished: when grain size succeeds the Debye length $d > 2L_D$ and when grain size is less the Debye length $d < 2L_D$. If $d > 2L_D$ the nanocrystals aren't fully depleted and the potential barriers are highest (Fig. 5a). Contrary, if $d < 2L_D$ the nanocrystals are fully depleted and the height of potential barriers decreases with grain size. In the limiting case when $d \ll Ld$, potential barriers disappear completely (Fig. 5c) and electron transport is no longer determined by the boundary area of nanocrystals. When the potential barriers exist, the NO_2 adsorption results in increase of potential barrier height due to the filling of surface states. However, when the height of potential barriers approaches to zero, the gas adsorption can't increase the potential barriers due to the large Debye length (in compare to grain size). So the sample with small nanocrystals can exhibit the smaller response due to the small barrier's height. In our case decreasing of the nanocrystal size In_2O_3 from 20 to 7 nm results in sensitivity decreasing and one can suppose that Debye length for In_2O_3-300 succeeds 7 nm.

Under the UV illumination the carrier density alters significantly and the Debye length decreases (because L_D is proportional to $1/\sqrt{n}$). So the small grains, which without illumination were fully depleted, under illumination become partly depleted and potential barriers appear. The density of surface states increase with decrease of grain size and under the illumination the response of In_2O_3-300 with smallest nanocrystals become greater and succeeds the response of samples with larger nanocrystals.

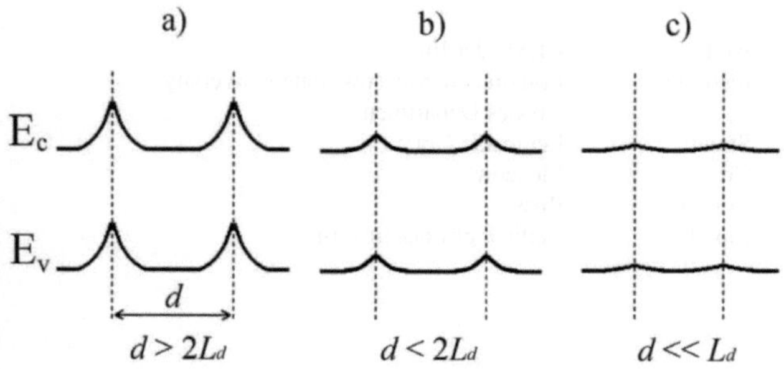

Fig. 5 Band diagram of In2O3 with different nanocrystal size.

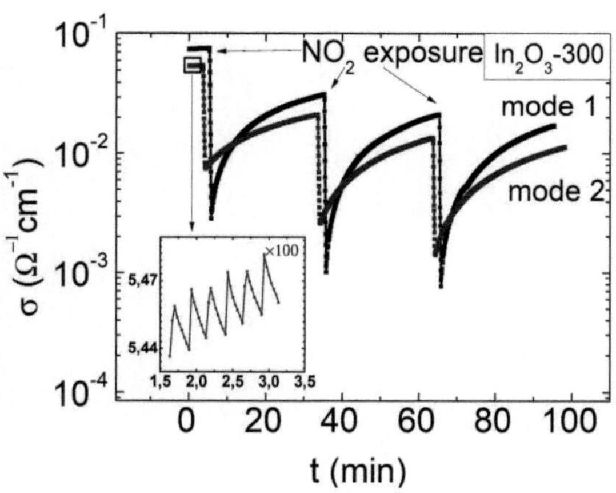

Fig. 6 Dynamic response of In2O3-300 to 8 ppm NO2 under blinking LED illumination. The moments of exposures are designed by arrows. In the inset dynamic conductivity affected by illumination before exposures is shown

The possibility of using nanocrystalline In_2O_3 as a sensor illuminated by blinking UV light was examined. The diminution of LED working time decreases the power consumption of sensor efficiently. In the experiments with blinking UV light the LED operated in a modulation mode with period of 15 seconds. The 2 modes of operation were observed: when LED is on for 4 seconds ("mode 1") and 1 second ("mode 2") during the period. The changes of conductivity become negligible and kinetic of conductance reaches stationary state in 2 hours after the blinking illumination was on. Since this point of time an increase in conductivity during illumination is equal to its reduction while LED is off (Figure 6, insert). It is noticeable that for both modes of illumination the photoconductivity is set almost at the same value. To determine the sensor signal the In_2O_3-300 sample was exposed to 8 ppm NO_2 atmosphere for 30 seconds and after the sample was exposed to dry air for 30 minutes at each illumination condition. To estimate the stability and reproducibility this procedure was repeated several times. As it can be seen in Figure 6 the conductance of In_2O_3-300 reacts to the presence NO_2 in atmosphere and it almost returns to initial value in dry air. The results indicate the possibility of twinkling illumination applying for detection of nitrogen dioxide at room temperature.

IV. CONCLUSION

Nanocrystalline In_2O_3 thin films were fabricated using sol-gel method and examined as UV light activated NO_2 gas sensors operating at room temperature. The results reveal that NO_2-sensitivity can be enhanced by using In_2O_3 with small grains. Moreover the change of In_2O_3 conductivity as a result of NO_2 adsorption depends nonmonotonically on the size of nanocrystals in the dark. But under the UV illumination the gas response increased with grain size reduction and monotonic size dependence of response to NO_2 was observed. We proposed that existence of potential barriers on the nanocrystal boundaries allows to effectively detect the NO_2 molecules. Without illumination the nanocrystalline In_2O_3 with grain size of 7 nm are fully depleted and can't be applied for effectively detection of NO_2. But under the UV illumination the same nanocrystalline In_2O_3 which were fully depleted become partly depleted and can be applied for NO_2 detection. So NO_2 gas of a concentration down to 1 ppm can be detected by nanocrystalline In_2O_3 at room temperature under UV illumination.

Moreover, we showed that pulsed UV radiation can be used to reduce the power consumption of the sensor

ACKNOWLEDGEMENTS

This work was supported by the Russian Foundation for Basic Research (project 15-29-01185) and the Russian Federation President Grant (NS-3461.2014.2).

CONFLICT OF INTEREST

The authors declare that they have no conflict of interest.

REFERENCES

1. G. Korotcenkov, V. Brinzari, A. Cornet, and J. Morante, "Experimental and theoretical studies of indium oxide gas sensors fabricated by spray pyrolysis," Sensors Actuators B, **106**(2), pp. 563–571, 2005.
2. T. Waitz, T. Wagner, T. Sauerwald, C.-D. Kohl, and M. Tiemann, "Ordered Mesoporous In_2O_3: Synthesis by Structure Replication and Application as a Methane Gas Sensor," Adv. Funct. Mater., **19**(4), pp. 653–661, 2009.
3. C. Cantalini, W. Wlodarski, H. . Sun, M. . Atashbar, M. Passacantando, and S. Santucci, "NO_2 response of In_2O_3 thin film gas sensors prepared by sol–gel and vacuum thermal evaporation techniques," Sensors Actuators B Chem., **65**(1–3), pp. 101–104, 2000.
4. T. Wagner, C.-D. Kohl, C. Malagù, N. Donato, M. Latino, G. Neri, and M. Tiemann, "UV light-enhanced NO_2 sensing by mesoporous {In_2O_3: Interpretation of results by a new sensing model," Sensors Actuators B, **187**(2), pp. 488–494, 2013.
5. S. Mishra, C. Ghanshyam, N. Ram, R. P. Bajpai, and R. K. Bedi, "Detection mechanism of metal oxide gas sensor under UV radiation," Sensors Actuators B, **97**, pp. 387–390, 2004.
6. G. Korotcenkov, "The role of morphology and crystallographic structure of metal oxides in response of conductometric-type gas sensors", Mater. Sci. Eng. R Reports, **61**, pp. 1–39, 2008
7. H. Liu, S.P. Gong, Y.X. Hu, J.Q. Liu, D.X. Zhou, "Properties and mechanism study of SnO_2 nanocrystals for H_2S thick-film sensors", Sensors Actuators B, **140**, pp. 190-195, 2009.
8. E. Comini, G. Faglia, G. Sberveglieri, "Solid state gas sensing", Springer, New York, 2009.
9. E. A. Forsh, A.V. Marikutsa, M.N. Martyshov, P. A. Forsh, M.N. Rumyantseva, A.M. Gaskov, et al., "Charge carrier transport mechanisms in nanocrystalline indium oxide", Thin Solid Films. **558,** pp.320–325, 2014.

Author:	Alexander Ilin
Institute:	Lomonosov Moscow State University, Physics Department
Street:	Leninskie Gory
City:	Moscow
Country:	Russia
Email:	as.ilin@physics.msu.ru

Sensitivity Evaluation of the Nanostructure-Enhanced BAW Mass Sensor

A. Zazerin[1], V. Ulianova[1], O. Bogdan[2], and A. Orlov[1]

[1] Microelectronics Department, National Technical University of Ukraine, Kyiv Polytechnic Institute, Kyiv, Ukraine
[2] Scientific and Research Institute, National Technical University of Ukraine, Kyiv Polytechnic Institute, Kyiv, Ukraine

Abstract— **The general implementation concept of the mass sensor based on membrane bulk acoustic wave resonator with nanostructured sensing layer was presented. The 3-dimensional finite element model of the thin film bulk acoustic wave resonator was built to estimate its sensitivity. It was shown that application of the nanostructured ZnO with large effective surface area provided the 20 times sensitivity increase over the polycrystalline sensing layer. The performance characteristics of the proposed sensor at the appointed conditions were described in details.**

***Keywords*— FEM modeling, BAW resonator, ZnO nanostructures, mass sensor.**

I. INTRODUCTION

Miniaturized mass sensors based on acoustic resonators [1, 2] have enabled the high-sensitivity, high-resolution detection of chemical particles, biological particles [3] such as antigens, DNA, proteins, gases [4], liquids [5], and a long and growing list of convergent applications. Mass sensors detect the amount of added mass of a film or an analyte deposited on the surface, which brings about down shifting of its resonance frequency and Q factor of the resonator proportionally to the amount of the deposited mass [2].

Commercially available mass sensors are implemented with acoustic quartz crystal microbalance, surface acoustic waves (SAW), and bulk acoustic waves (BAW) resonators, but sensitivity enhancement, recovery time reducing, size of the device, cost-effectiveness and performance efficiency, compatibility with conventional production techniques and integrability with active circuits on a single chip, e.g. for lab-on-chip application [6], present the challenge for developers and are of interest for consumers. A lot of modeling, testing, and optimization work is required to consolidate the sensor applications.

Based on the mass loading effect, the uniform thin-film or analyte deposition is a way of tuning the thin film bulk acoustic resonators (FBAR) resonance frequency. FBAR employed in silicon-integrated miniaturized gas sensors, have demonstrated high mass sensitivity [7]. The advantages of FBAR sensors over SAW sensors consist in manufacturability (photomask simplicity), low cost, small size, high quality factor, power handling capability, as well as integration with CMOS technology [8]. As has been fully proven, CMOS integration improves the impact and versatility of the sensors, boosting them to full systems-on-chip implementations.

Existing of nanostructured materials and opportunity to apply them for solving of different urgent problems adduce to necessity of performance and reliability prediction, which can be implemented by variety of simulation methods. Nanostructured materials application is promising for FBAR sensor performance improving. Such materials can be synthesized via inexpensive chemical methods [9] and provide high efficient surface area, and this way enhance sensitivity of the device.

Versatile flexible multiparameter model of nanostructure-enhanced SAW sensor is obtained in our previous research by finite element simulation method and allows the evaluation of the device performance parameters subject to morphometric parameters of nanostructures [10].

In this work the general implementation concept of the mass sensor based on membrane BAW resonator with nanostructured sensing layer was shown and sensor sensitivity was preliminary estimated.

II. SENSING LAYER PARAMETERS EXTRACTION

The most important property of nanostructured materials for application in modern sensors is a high surface-to-volume ration which ensures extension of sensing surface area at the same device size. Developed surface allows deposition of increasing amount of sensitive material, e. g. antigen, and thus increasing the sensitivity of sensor, e.g. probability of antibody-antigen interaction. Due to the design of FBAR resonator, where metal electrode is the contact interaction surface, the biocompatibility of nanostructured materials is desirable feature.

The nanostructured ZnO in a shape of rods provide a high efficient surface area and is especially suitable for sensitivity enhancement; moreover it is biocompatible and nontoxic material and appropriate for biomedical application [1]. The problem of evaluating the sensitivity and other characteristics of the devices with nanostructured materials is the complexity of modeling, because the parameters of nanostructures may significantly differ from the macroscale object parameters. Therefore the current trend and necessity is to develop new methods of nanostructure parameters

V. Sontea and I. Tiginyanu (eds.), *3rd International Conference on Nanotechnologies and Biomedical Engineering*,
IFMBE Proceedings 55,
DOI: 10.1007/978-981-287-736-9_83

modeling. One of the applicable methods is the creation of a thin layer equivalent model, which can be presented in the modeling environment as a macroscale object but characterized by the behavior and properties of nanostructures in each point of the volume. Similar model is developed and presented in [10] for SAW sensor based on ZnO nanorods sensitivity determination.

In this work the nanostructured layer was synthesized via hydrothermal method which provided the formation of nanostructures with different morphologies and structures by self-organization at the low-temperature chemical processes. Initially seed-layer was formed by sol-gel method to obtain the growing centers and bonding interface with substrates. Spin coating of the sol-gel based on dissolved in isopropanol zinc acetate dihydrate with the concentration of 0.3 mol/l on cleaned substrate at 3000 rpm for 30 s was carried out. After deposition of the fifth layer, the resulting thin films were annealed at 400°C in air for 1 h to obtain the homogeneous and stable seed-layer. The synthesis of the nanorods was carried out in zinc nitrate based water solution with the concentration of 0.03 mol/l at 95°C during 90 min. The top-view and cross-sectional scanning electron microscope SEM (Hitachi S4800) SEM images of the synthesized ZnO nanorods is obtained (Fig. 1).

a b

Fig. 1 Top-view (a) and cross-sectional (b) SEM images of the synthesized ZnO nanorods.

To sum up the model evaluation procedure, the approximate number of nanorods in the applied model was 3.19·105, the effective surface area was 1.04·10-7 m2 (more than 21 times exceed the area of the corresponding monolayer). Thus, with analyte concentration of 50 ng/mm2 the expected mass load value was around 5.2·10-12 kg.

Table 1 Equivalent Sensing Layer Model Parameters

Parameter type	Value
Sensing layer area	70x70 μm^2
Average nanorod radius	45 nm
Average nanorod height	1.1 μm
Nanorods array density	65 μm^{-2}
Effective surface area	$1.04 \cdot 10^{-7}$ m^2

III. Model Consideration for BAW Sensor

The membrane-type BAW resonator was used in this work. The conventional piezoelectric transformation theory that offers an equivalent circuit of the resonator (BVD, Mason etc.) is able to evaluate the thickness extensional vibration modes in simple one-dimensional configuration but is insufficiently adequate to predict the presence of spurious modes with lateral wave vector and other parasitic effects. That is why 3-dimensional finite element model of the FBAR was chosen for the modeling task. It is a versatile approach capable of modeling complicated device geometries, nonuniform material properties and various boundary conditions. FBAR is very similar to quartz crystal scaled down in size, which consists of piezoelectric thin film material sandwiched between two metal electrodes (Fig. 2).

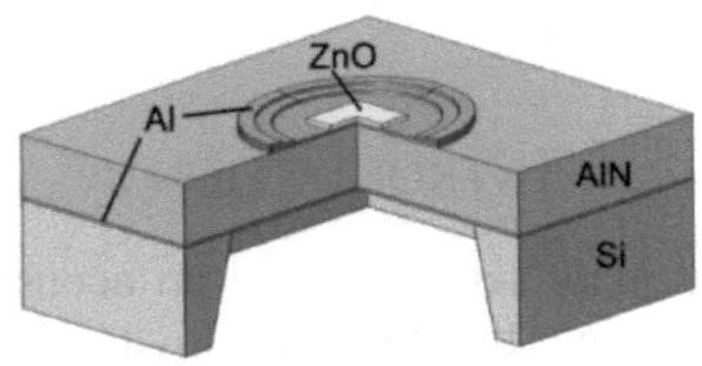

Fig. 2 The arbitrary scaled geometry of thin film bulk acoustic resonator used for simulation.

This is the most straightforward way to extend the working principle of thickness extensional vibrations up to GHz range and yields a resonator that needs a minimum of layers to be deposited. In this case air interfaces at the top and bottom electrodes represent almost ideal acoustic isolation and then potentially higher quality factors can be achieved.

It should be noted that 3-dimensional behavior of a real resonator is the cause of lateral waves propagation. These waves produce spurious resonances which may degrade the performance of FBAR that also affects the output performance of the sensor. For this purpose the resonator structure was upgraded using thickened-edge mass load method [11] to suppress the generation of the unwanted resonance modes.

The lowest layer of the resonator was silicon designed in the shape of supporting bridge: a large part of the bottom layer was etched away from the lower end of the central region to make an air gap acoustic isolation under electrode. On the top of silicon there was an aluminum layer that operated as the ground electrode. Above the aluminum layer was the active piezoelectric material made of aluminum nitride. The topmost layer of the resonator was a step-like circle-shaped aluminum electrode which led to more attenuated spurious lateral modes. Resonator geometry sizes were specified in Table 2.

Table 2 FBAR Materials And Layer Dimension

Structure element	Thickness	Width
Piezoelectric layer, AlN	11.4 μm	500 μm
Substrate bridge layer, Si	8 μm	500 μm
Bottom electrode, Al	200 nm	500 μm
Top electrode, Al	200 nm	Ø250 μm
Frame load ring 1, Al	200 nm	Ø250(35) μm
Frame load ring 2, Al	200 nm	Ø250(17) μm

The mechanical boundary conditions at the edges of attached perfectly matched layers represented that the horizontal and vertical displacement were set to zero. This approach increased the length of the resonator and simulated the effect of propagation and absorption of elastic waves in the adjoining regions which were not resolved in the true geometric scale. The acoustic losses of the piezoelectric material were accounted by damping factor η_c^E equal to 0.0007, and dielectric loss factor η_ε^S equal to 0.01. The proposed model was computed using finite element method in MEMS module of COMSOL Multiphysics software platform. The instrument utilizes complex numerical techniques to enable computationally effective and economic FEM simulations of piezoelectric devices including BAW resonators.

As could be noted from analysis, within a definite range of the electrode width, the resonator exploits the main resonant mode where the mechanical displacement is almost concentrated under the middle part of the top electrode surface. Due to optimized top electrode frame loading, only a small amount of energy contained in the fundamental mode leaks to the surrounding non-electroded region. Such a distribution of the displacement density prompts that the most crucial location of the sensing layer is in the middle of the top electrode.

In general the performance of the FBAR may be analyzed using the input impedance characteristics of the resonator. The input impedance electric and phase response of simulated model are shown in Figure 3. The input impedance has two distinctive peaks defining the resonant properties of the structure: the series resonance where the impedance is minimum and parallel resonant frequency where the impedance is maximum. The absence of spurious peaks in interresonant frequency range and phase response smoothness became the results of properly designed geometry and correctly specified model definitions.

The presented characteristics demonstrated the proposed resonator configuration is well suited for sensor parameters determination. To evaluate the possible improvement of the proposed sensor the simulation was performed in two steps. At first the output characteristics were calculated for the structure with polycrystalline ZnO sensing element.

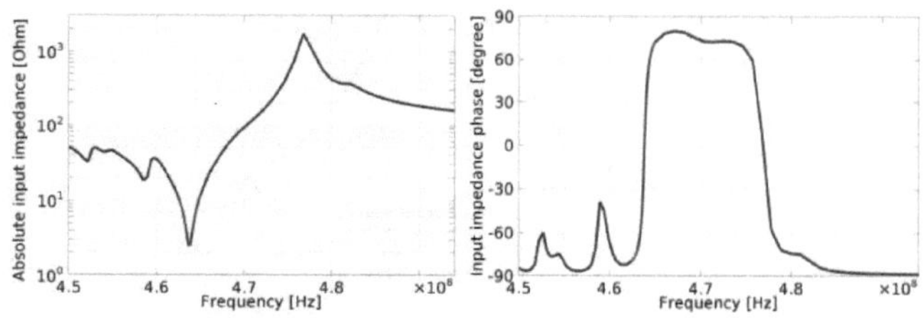

a b

Fig. 3 Input impedance (a) and phase (b) versus frequency of the simulated FBAR device.

Its effective surface can be determined by simple two-dimensional distribution. Then, the sensing element was replaced by the equivalent macromodel of the nanostructured film of highly developed surface. In case of this work the forward transfer admittance characteristics was used for measuring the frequency shift due to mass loading effect.

Figures 4 and 5 represent the processed results of the simulation. As expected, the analyte introduction (50 ng/mm^2) led to the change of resonance conditions in BAW sensor structure and thus affects the eigenmode frequency and resonator's quality factor. To elaborate the results, the forward transfer admittance peaks shifted to the lower frequencies in both cases. The polycrystalline sensing element provided the frequency detune of 13.4 kHz (0.003% fractional shift). The nanostructured sensing element provided frequency detune of 274.6 kHz (0.0592% fractional shift) which was 20 times greater for the same analyte concentration.

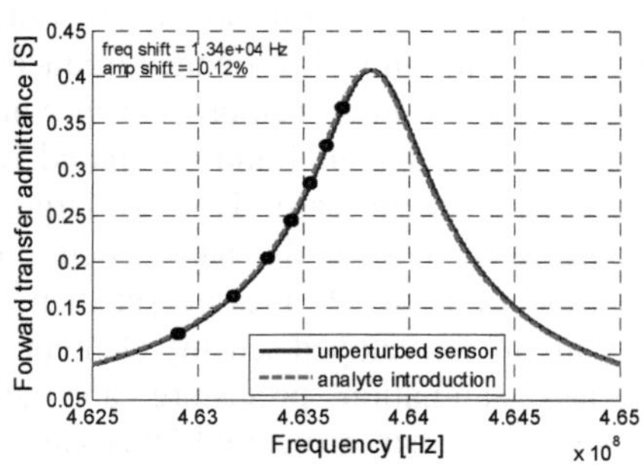

Fig. 4 Forward transfer admittance versus frequency of BAW sensor with polycrystalline ZnO sensing element.

In terms of integral device sensitivity evaluation the mass responsivity and mass responsivity per area were calculated for both cases and presented in Table 3. The results demonstrate that sensitivity of the simulated device is $2.59 \cdot 10^9$ Hz·cm^2/g for the resonant frequency of 463.7 MHz.

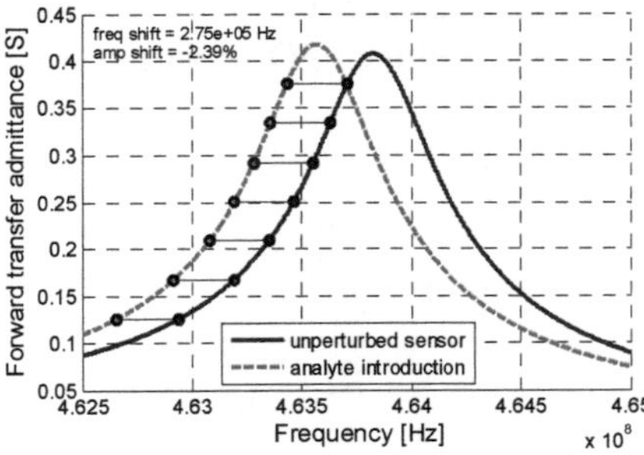

Fig. 5 Forward transfer admittance versus frequency of BAW sensor with nanostructure-enhanced ZnO sensing element.

Table 3 BAW Sensor Output Parameters

Parameter type	Polycrystalline	Nanostructured
Frequency shift, kHz	13.4	274.6
Frequency shift, %	0.0029	0.0592
|Y21| amplitude shift, %	0.1162	2.39
Mass responsivity, Hz/g	$2.584 \cdot 10^{12}$	$5.291 \cdot 10^{13}$
Mass responsivity per area, $Hz \cdot cm^2/g$	$1.2659 \cdot 10^{8}$	$2.588 \cdot 10^{9}$

This value can be further improved by increasing the operation frequency of the FBAR device due to higher power density.

IV. CONCLUSIONS

To determine the sensitivity of the nanostructure-enhanced BAW sensor the equivalent sensing element model was proposed as well as the design of FBAR resonator. The input parameters of the equivalent model were taken from the results of ZnO nanorods low-temperature chemical synthesis. FEM simulation of the proposed device proved the possibility of its application to obtain the values of deposited analyte mass in terms of resonance frequency shift.

Mass responsivity and mass responsivity per area of the proposed device were calculated and compared with polycrystalline sensing element case. Due to large effective surface area the sensitivity of the nanostructure-enhanced sensor reached the value of 20 times bigger than sensitivity of the sensor with non-developed sensing element.

The presented simulation procedure and appropriate equivalent model of the nanostructured layer can be applied also for optimization of the device performance and sensitivity. The parameters of synthesized nanostructures including average nanorod diameter, height and array density can be easily integrated to obtain the output sensor characteristics.

CONFLICT OF INTEREST

The authors declare that they have no conflict of interest.

REFERENCES

1. Ballantine D. S. Jr., Martin S. J., Ricco A. J. et al. (1996) Acoustic Wave Sensors: Theory, Design, & Physico-Chemical Applications, Academic Press
2. Campanella H., (2010) Acoustic Wave and Electromechanical Resonators: Concept to Key Applications. Norwood, MA: Artech House
3. Nirschl M., Blüher A., Erler C., et al. (2009) Film bulk acoustic resonators for DNA and protein detection and investigation of in vitro bacterial S-layer formation. Sen. Actuators A Phys. 156(1):180–184 DOI:10.1016/j.sna.2009.02.021
4. Chen D., Wang J., Li D. et. al (2011) A poly(vinylidene fluoride)-coated ZnO film bulk acoustic resonator for nerve gas detection. J. Micromechanics Microengineering 21:085017 DOI:10.1088/0960-1317/21/8/085017
5. Chen Y.-C., Chang W.-T., Kao K.-S. et al. (2013) The Liquid Sensor Using Thin Film Bulk Acoustic Resonator with C-Axis Tilted AlN Films. Journal of Nanomaterials 2013, e245095. DOI:10.1155/2013/245095
6. Zhang M., Cui W., Chen X. et al. (2015) Monolithic integrated system with an electrowetting-on-dielectric actuator and a film-bulk-acoustic-resonator sensor. Journal of Micromechanics and Microengineering 25:025002 DOI: 10.1088/0960-1317/25/2/025002
7. Gabl R., Green E., Schreiter M. et al (2003) Novel integrated FBAR sensors: a universal technology platform for bio- and gas-detection, in: Proceedings of IEEE Sensors, 2003. Presented at the Proceedings of IEEE Sensors, 2003, pp. 1184–1188 Vol.2. DOI:10.1109/ICSENS.2003.1279132
8. Johnston M.L., Kymissis I., Shepard K.L. (2010) FBAR-CMOS Oscillator Array for Mass-Sensing Applications. IEEE Sensors Journal 10, 1042–1047. DOI:10.1109/JSEN.2010.2042711
9. Orlov A., Ulianova V., Yakimenko Y. et al. (2014) Influence of process temperature on ZnO nanostructures formation, 2014 IEEE 34th International Conference on Electronics and Nanotechnology (ELNANO), 2014, pp. 51–53. DOI:10.1109/ELNANO.2014.6873960
10. Bogdan O., Orlov A., Pashkevich G. et al. (2014) Optimal parameters determination for nanostructure-enhanced surface acoustic waves sensor, Proceedings of the 2014 37th International Spring Seminar on Electronics Technology (ISSE), 2014, pp. 427–431 DOI: 10.1109/ISSE.2014.6887638
11. Yakimenko Y., Zazerin A., Orlov A. et al. Film bulk acoustic resonator finite element model in active filter design, Proceedings of the 37th International Spring Seminar on Electronics Technology (ISSE), Dresden, 2014, pp. 486–490 DOI: 10.1109/ISSE.2014.6887649

Author: Andrii Zazerin
Institute: National technical University of Ukraine «Kyiv Polytechnic Institute»
Street: Peremogy ave, 37
City: Kyiv
Country: Ukraine
Email: exeterza@gmail.com

Effect of Dopant on Selectivity of CuO Nanostructured Films – Based Sensors

V. Cretu[1], V. Postica[1], N. Ababii[1], N. Magariu[1], V. Sontea[1], F. Schütt[2], R. Adelung[2], and O. Lupan[1,2]

[1] Technical University of Moldova, Department of Microelectronics and Biomedical Engineering, 168 Stefan cel Mare Blvd., Chisinau, Republic of Moldova

[2] University of Kiel, Institute for Materials Science/Functional Nanomaterials, Kaiser Str. 2, D-24143, Kiel, Germany

Abstract— **In this work, the effect of different dopants on selectivity performance of gas sensor structures based on CuO nano-structured films that were synthesized from aqueous chemical solutions SCS is reported. The gas sensing studies were performed to investigate response and selectivity of the sensor structures. The un-doped CuO samples showed good selectivity to ethanol vapors with a response of about 149%, while Al-doped CuO samples demonstrated a response of about 175% to hydrogen gas and a better selectivity too. The gas sensing mechanism was proposed and discussed in details.**

Keywords— **CuO, films, p-type, cupric oxide, gas sensor.**

I. INTRODUCTION

In recent years, in order to replace expensive analytical instruments in gas sensing, investigation of different nano-materials for new types of electronic nose (e-nose) systems was carried out [1]. The semiconducting oxide based sensors are the most widely used for such type of applications due to their low-cost, smaller size, faster response and resistive response signal. To detect exact target gases from a mixture, a necessity of the highly sensitive and selective sensor appear. It is known that many of used sensing materials, like ZnO, SnO_2, In_2O_3, etc., are usually suffering by cross-sensitivity to several types of gases [2].

The control of charge carriers in the semiconducting oxide is known to be an efficient method to tune gas sensing of the nanomaterial, especially response and selectivity [3]. The change of the selectivity can be obtained by different approaches. The popular ways are the use of catalyst, change in operating temperature (OPT), doping of semiconducting oxides or a combination of these methods. Previous works [4] demonstrated a change in selectivity of the single MoO_3 nano-ribbon by variation of the OPT value and its sizes.

In this work, we report on the effect of donor or acceptor dopants, like Fe, Al, Ag, Zn, on the gas sensing properties, especially selectivity, of the sensor structures based on the cupric oxide (CuO) nano-layers, deposited on glass substrate by SCS method. The gas sensing mechanism was proposed to explain obtained results and the former is discussed in details.

II. EXPERIMENTAL

A. Growth of Metal-Doped CuO Nanostructured Films

Growth of undoped and metal-doped CuO nanostructured films were performed on the glass substrates (sizes of 76 mm × 26 mm × 1 mm). The substrate cleaning procedure and synthesis process by SCS were reported in our previous work [5]. For doping with Ag, Al, Zn and Fe, we added solutions of $AgNO_3$, $Al_2(SO_4)_3 \cdot 18H_2O$, $ZnSO_4 \cdot 7H_2O$ and $Fe_2(SO_4)_3 \cdot 7H_2O$, in complex solution, respectively.

B. Measurements of Samples

The thickness of all samples was in range of 0.8 - 0.9 μm, as measured in cross-section by SEM instrument.

The samples of un-doped and metal-doped CuO nano-structured films were analyzed using scanning electron microscopy (SEM) instrument Carl Zeiss (7 kV, 10 μA). The compositional analysis of the specimens was carried out by EDX analysis in combination with SEM.

III. RESULTS AND DISCUSSION

A. Morphological and Chemical Characterizations

Fig. 1(a)-(b) presents SEM images at different magnifications for pure CuO samples after thermal annealing (TA) at 650 °C for 30 min. In Fig. 1(c)-(f) are images recorded at higher magnification as in case (b), just for doped samples as follows: (c) CuO:Ag, (d) CuO:Al, (e) CuO:Zn, (f) CuO:Fe. It demonstrates a relative uniformity of the nano-structured films, as can be clearly seen from Fig. 1. According to our experimental observations, the films demonstrate good adhesion to the glass substrate and are composed of interconnected networks of nanocrystallites. The non-uniform grain size distribution is easily identified in micrographs, see Fig. 1. For un-doped samples the former is varied in the range of 50 - 400 nm. A slight variation in the size of the nanocrystallites of doped CuO is expected (Fig. 1), the doped films are more compact, probably due to the difference in ionic radii between Cu^{2+} (0.72 Å) and one of the dopants (Ag^{+} , Al^{3+}, Zn^{2+} and Fe^{3+}) [6] and other effects, that lead to agglomeration and coalescence of nano-crystallites during SCS deposition [6]. The Ag-doped and

V. Sontea and I. Tiginyanu (eds.), *3rd International Conference on Nanotechnologies and Biomedical Engineering*,
IFMBE Proceedings 55,
DOI: 10.1007/978-981-287-736-9_84

Zn-doped CuO nanostructured films demonstrate smaller crystallite sizes in the range of 20 - 100 nm, while for Al- and Fe- doped CuO nanostructured films their size is larger and is about 50 – 200 nm. The crystallite size is an important parameter for gas sensing properties of the oxide semiconductor films. The smaller size of crystallite increases the surface-to-volume ratio, that lead to a higher gas response, due to the higher amount of adsorbed oxygen species on the surface of oxide semiconductors and more potential barriers between them [6]. The EDX analysis revealed dopant impurities of 1.8 wt.%Ag, 0.2 wt.%Al, 1.9 wt.%Zn and 0.8 wt.%Fe for CuO:Ag, CuO:Al, CuO:Zn and CuO:Fe nanostructured films, respectively.

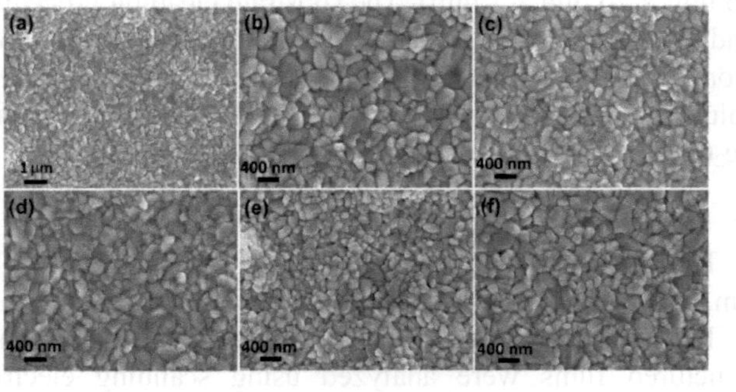

Fig. 1 SEM images of pure and doped cupric oxide nanostructured layers with × 10 K magnification of pure CuO films (a); and with × 20 K magnification of: (b) pure CuO; (c) CuO:Ag; (d) CuO:Al; (e) CuO:Zn; and (f) CuO:Fe films

B. Gas sensing Properties

The top metal contacts for the sensor structures were deposited in vacuum by evaporation of Al through a mask with meander configuration. The width of the meander free space is 1 mm. Gas sensing measurements were performed using the test system reported in our work [7-8].

Gas responses were calculated by the ratio $S = ((R_{air} - R_{gas})/R_{air}) \cdot 100\ \%$, where R_{air} is the resistance in air and R_{gas} is the resistance under gas exposure. All sensor structures exhibited responses common for *p*-type oxide semiconductors. Results on the gas response of doped CuO based sensors at the optimal operating temperature (OPT) are presented in Fig. 2a. Sensors were exposed to the 100 ppm of hydrogen H_2, ethanol (EtOH-C_2H_5OH) and methane CH_4. According to our experimental data, it was found that the un-doped CuO films have optimal OPT of 350 °C and a higher gas response to ethanol vapors (S ~ 149 %), while hydrogen response is three times lower (S ~ 51 %), while for CH_4 gas response is even lower (~34 %). Selectivity of the CuO films to ethanol vapors has been reported by other groups too [9]. However, the gas sensing properties of the doped CuO are less studied in the literature. For example, Kim and co-workers significantly increased response of the sensors to NO_2 gas by doping CuO nanorods with Cr [10].

In our case, to investigate gas sensing properties, CuO nanostructured films were doped with Ag, Al, Zn and Fe. By doping with Ag, the selectivity to ethanol vapors was not changed, but optimal OPT temperature was increased to 400 °C and ethanol response was decreased to 110 % (Fig. 2a). In case of doping CuO with Al and Zn, changes in selectivity from ethanol vapors to hydrogen gas were observed (Fig. 2a).

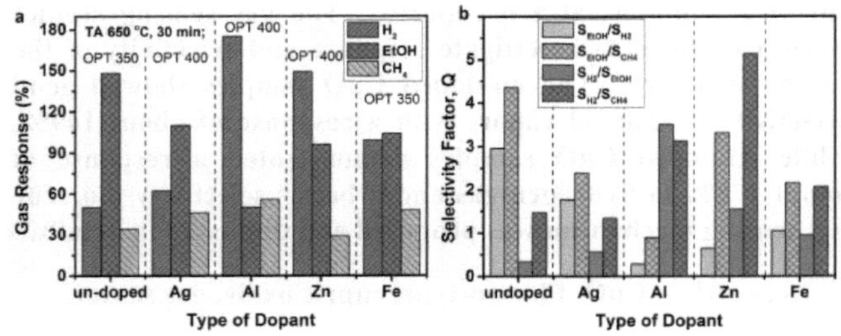

Fig. 2 (a) Gas response of the sensor structures based on the un-doped and doped CuO at their optimal operating temperature OPT versus type of dopant and tested gas (H2, EtOH, CH4); (b) Selectivity factor (Q) of the sensors versus type of dopant

Sensors based on the Al-doped CuO nanostructured films demonstrated a higher response to hydrogen gas (S ~ 175 %) in comparison to Zn-doped CuO (S ~ 143 %) and to un-doped CuO layers. Also, we found that Al-doped CuO possess a better selectivity to hydrogen gas. The optimal operating temperatures OPT for Al-doped and Zn-doped CuO are 400 °C, similarly to Ag-doped CuO. In case of Fe-doping in CuO, responses to H_2 gas (S ~ 99 %) and ethanol vapors (S ~ 104 %) are almost of the same value, without significant selectivity to any of the tested gases. The optimal OPT of 350 °C is the same, like for the un-doped CuO, thus by doping with Fe, the optimal OPT was not changed. For selectivity analysis of the investigated sensor structures, we will represent selectivity factor to ethanol vapors Q_{EtOH} by ratio of S_{EtOH}/S_{gas} (S_{H2} or S_{CH4}) and selectivity factor to hydrogen gas Q_{H2} by ratio of S_{H2}/S_{gas}(S_{EtOH} or S_{CH4}), where S_{EtOH}, S_{H2} and S_{CH4} are gas response values to ethanol vapors, hydrogen and methane gases, respectively. Results are plotted in Fig. 2b. As can be seen, the un-doped samples are more selective to ethanol vapors than Ag-doped samples with $S_{EtOH}/S_{H2} \sim 3$ and $S_{EtOH}/S_{CH4} \sim 4.3$.

For Ag-doped the ratios are $S_{EtOH}/S_{H2} \sim 1.7$ and $S_{EtOH}/S_{CH4} \sim 2.4$, respectively. In the case of Al- and Zn-doped, the higher selectivity to hydrogen is demonstrated by Al-doped CuO based sensor, with $S_{H2}/S_{EtOH} \sim 3.5$ and $S_{H2}/S_{CH4} \sim 3.1$, in comparison with $S_{H2}/S_{EtOH} \sim 1.5$ and

$S_{H2}/S_{CH4} \sim 5$ for Zn-doped CuO. Thus, we can conclude that sensors based on the un-doped CuO nanostructured films show a higher response to ethanol vapor and sensor structures based on the Al-doped CuO nanocrystallite films demonstrate optimal properties as hydrogen gas sensors.

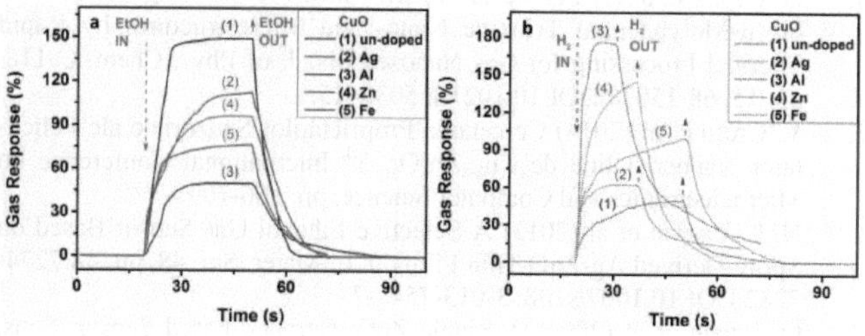

Fig. 3 Transient response to: (a) ethanol vapors; and (b) hydrogen gas of the un-doped and metal-doped CuO based sensor structures with a meander top contact.

Fig. 3a contains plots of transient responses of the un-doped and doped CuO based sensors to ethanol vapors at their optimal OPT. As can be seen, response time (τ_R) and recovery time (τ_F), ($\tau = |t_{90\%} - t_{10\%}|$), were faster for un-doped CuO based sensor, taking 6.5 s and 8.9 s, respectively. In the case of doped samples, time to reach 90% of the full response and recovery values are lower. Data for delay times for response τ_R and recovery-fall τ_F are generalized in Table 1. For Al- and Fe-doped CuO samples the τ_R values are comparable with un-doped samples, while for Ag- and Zn- doped samples the values are about twice lower. The lower τ_F value was observed for Al-doped CuO samples. In Fig. 3b transient response of the un-doped and doped CuO based sensors to hydrogen gas at their optimal OPT are plotted (Fig. 2a). Evidently, in this case results on the τ_R and τ_F are different from the case of ethanol vapor sensing, and results are summarized in Table 1. In case of un-doped samples the τ_R and τ_F are lower, demonstrating good performance for ethanol sensor. For Ag- and Al-doped samples, the τ_R is faster than for ethanol response, demonstrating good performance of the Al-doped sample for hydrogen gas sensor. However, recovery times τ_F for these samples are relatively low, taking about 50 s and 37.8 s for Ag- and Al-doped CuO, respectively.

Table 1 The response τ_R times and recovery τ_F times of sensor structure

	Ethanol (100 ppm)				
	CuO	:Ag	:Al	:Zn	:Fe
τ_R (s)	6.5	15.6	8.8	12.1	8.2
τ_F (s)	8.9	13.8	20.6	11.7	13.6
	Hydrogen gas (100 ppm)				
τ_R (s)	17.6	4.26	3.36	9.46	19.6
τ_F (s)	14.6	~50	37.8	14.3	18.4

C. Gas Sensing Mechanism Proposed for Our Samples

The gas sensing mechanism of the *p*-type semiconductors differs from *n*-type semiconductors due to significantly different conduction behavior [11]. At high operating temperatures (> 150 °C) there are adsorbed atomic oxygen species on the surface of CuO: $1/2O_{2(g)} \rightarrow O_{(ad)} \rightarrow O^-_{(ad)} + h^+$ [12]. In result, at the surface of the CuO, after adsorption and ionization of the oxygen, eq. (1), a hole accumulation layer (HAL) is formed with lower resistance than insulating core. In the formed core-shell structure of the nanocrystallites, electrical conduction will occur through the shell (HAL). At the exposure of the sensor structure to ethanol vapors and H_2 gas, that demonstrated a higher gas response depending on the dopant type, following reactions may occur [13]:

$$C_2H_5OH_{(g)} + 6O^-_{(ad)} + 6h^+ \rightarrow CO_{2(g)} + 3H_2O_{(g)} \quad (1)$$

$$H_{2(g)} + O^-_{(ad)} + h^+ \rightarrow H_2O_{(g)} \quad (2)$$

As can be seen from eq. (2) and (3), both gases will lead to increase of sensor resistance by trapping holes from the HAL layer. This gas sensing mechanism is known as ionosorption model [11], and involves reaction of target gas molecules with adsorbed oxygen species at the surface of oxide semiconductor. Considering that the nanostructured films are formed from CuO nanocrystallites, there is a need to take in consideration the interaction between nanocrystallites, i.e. variation of the surface band bending qV_S at the gas exposure [14]. For *p*-type semiconductors, the dependence of the conductance on the qV_S is expressed by the following equation [14]:

$$G_p \approx \exp\left(\frac{qV_S}{2kT}\right) \quad (3)$$

If we will note the variation of the qV_S in the air and under the gas exposure by $q\Delta V_S$, then the sensibility of the gas sensor can be defined as [14]:

$$S_p \approx \exp\left(-\frac{q\Delta V_S}{2kT}\right) \quad (4)$$

The change in selectivity by doping of CuO nanostructured films can be explained by electronic sensitization effect [11]. Substitution of the Ag^+ ions at a Cu^{2+} site can generate holes to compensate the electronic charge [15], which lead to resistance decrease of the sample. Taking into account that sensitivity of the gas sensor is $S = R_{air}/R_{gas}$, decrease in R_{air} will results in a lower gas response. Another factor influenced by Ag^+ doping is decrease of the Debye length $L_D = \sqrt{(kT\varepsilon\varepsilon_0/q^2p_b)}$ [16], due to increase of the bulk hole concentration p_b. Thus, Ag-doping of the CuO nanostructured films is reducing the gas sensing properties (Fig. 2a). On the other hand, aluminum- and iron- doping will lead to the electrons generation, which will increase sample resistance and Debye length. However, from Fig. 2a it can be observed that ethanol response is decreased for Al-doped and Fe-doped CuO, and this can be explained by dependence of the ethanol response on the acid-base properties of the sensing material, reported by Jinkawa et al. [17]. By addition of the acidic type of dopant, the gas response to ethanol is decreasing [17]. The Fe^{3+} ions are more acidic than Cu^{2+} [18], thus ethanol response of the Fe-doped CuO is decreasing in comparison with un-doped CuO, due to dehydrogenation of the ethanol [17]. In the case of doping with Zn and Al (Fig. 2a-b), the nature of higher selectivity to H_2 is still unknown and needs more detailed investigation like variation in the doping concentration, that will be reported in further works, which are in progress.

IV. CONCLUSIONS

In this work, un-doped and metal-doped CuO nanostructured films for gas sensing investigations were synthesized by SCS method on a glass substrate. A controlled change in selectivity to reducing gases of *p*-type CuO based sensors by acceptor and donor doping was demonstrated. The un-doped CuO sample shows a good selectivity to ethanol vapors with S ~ 149 %. By doping of films with silver a decrease in sensing was observed and by doping with aluminum, the selectivity was changed to hydrogen gas with a response S ~ 175 %. Change in selectivity to hydrogen gas was observed in case of doping with Zn^{2+}. The gas sensing mechanism was proposed for such sensor structures.

ACKNOWLEDGMENT

This work was financially supported by the STCU under the Grant no. 5989. Professor Dr. Lee Chow from University of Central Florida (USA) is acknowledged for support in sample characterization. Partial support of the research by Project 45inst-15.817.02.29A is acknowleged.

CONFLICT OF INTEREST

The authors declare that they have no conflict of interest.

REFERENCES

1. N. El Barbri et al (2007) Building of a metal oxide gas sensor-based electronic nose to assess the freshness of sardines under cold storage. Sens. and Actuat. B 128, pp.235-244 DOI 10.1016/j.snb.2007.06.007.
2. J. Liu et al (2005) Vanadium Pentoxide Nanobelts: Highly Selective and Stable Ethanol Sensor Materials. Adv. Mater. 17, pp. 764-767 DOI 10.1002/adma.200400993.
3. L. Chow et al (2013) Synthesis and characterization of Cu-doped ZnO one-dimensional structures for miniaturized sensor applications with faster response. Sens. and Actuat. A 189, pp. 399-408 DOI 10.1016/j.sna.2012.09.006.
4. O. Lupan et al (2014) Versatile Growth of Freestanding Orthorhombic α-Molybdenum Trioxide Nano- and Microstructures by Rapid Thermal Processing for Gas Nanosensors. J. of Phys. Chem. C 118, pp. 15068-15078 DOI 10.1021/jp5038415.
5. V. Crețu et al (2014) Cercetarea Proprietăților Senzoriale ale Peliculelor Nanocristaline de $Cu_{1-x}Zn_xO_y$. 8th International Conference on Microelectronics and Computer Science, pp. 106-109.
6. N. L. Tarwal et al (2013) A Selective Ethanol Gas Sensor Based on Spray-Derived Ag-ZnO Thin Films, J. of Mater. Sci. 48, pp. 48:7274-7282 DOI 10.1007/s10853-013-7547-7.
7. O. Lupan et al (2009) A Single ZnO Tetrapod-based Sensor. Sens. and Actuat. B 141 (2), pp. 511-517 DOI 10.1016/j.snb.2009.07.011.
8. L. Chow et al (2010) FIB fabrication of ZnO nanotetrapod and cross sensor. Phys. Status Solidi B 247:1628-1632 DOI 10.1002/pssb.201090013.
9. D. Barreca et al (2009) Chemical vapor deposition of copper oxide films and entangled quasi-1D nanoarchitectures as innovative gas sensors, Sens. and Actuat. B 141, pp. 270-275 DOI 10.1016/j.snb.2009.05.038.
10. K.-M. Kim et al (2012) Selective Detection of NO_2 Using Cr-Doped CuO Nanorods, Sensors, 12, pp. 8013-8025.
11. H.-J. Kim et al (2014) Highly sensitive and selective gas sensors using p-type oxide semiconductors: Overview. Sens. and Actuat. B, 192, pp. 607-627 DOI 10.1016/j.snb.2013.11.005.
12. A. Cruccolini et al (2004) Gas adsorption effects on surface conductivity of nonstoichiometric CuO. Sens. and Actuat. B, 98, pp. 227-232 DOI 10.1016/j.snb.2003.10.012.
13. Y.-H. Choi et al (2004) H_2 and C_2H_5OH sensing characteristics of mesoporous p-type CuO films prepared via a novel precursor-based ink solution route. Sens. and Actuat. B, 178, pp. 395-403 DOI 10.1016/j.snb.2012.12.096.
14. M. Hübner et al (2011) Influence of humidity on CO sensing with p-type CuO thick film gas sensors. Sens. and Actuat. B, 154, pp. 347-353 DOI 10.1016/j.snb.2010.10.046.
15. J.-W. Yoon et al (2006) Electronic sensitization of the response to C_2H_5OH of p-type NiO nanofibers by Fe doping. Nanotechnology, 24, 444005(8pp) DOI 10.1088/0957-4484/24/44/444005.
16. Z.-P. Sun et al (2006) Rapid synthesis of ZnO nano-rods by one-step, room-temperature, solid-state reaction and their gas-sensing properties. Nanotechnology, 17, pp. 2266-2270 DOI 10.1088/0957-4484/17/9/032.
17. T. Jinkawa et al (2000) Relationship between ethanol gas sensitivity and surface catalytic property of tin oxide sensors modified with acidic or basic oxides. J. of Molec. Catal. A: Chemical, 155, pp. 193-200 DOI 10.1016/S1381-1169(99)00334-9.
18. Y. Zhang (1982) Electronegativities of Elements in Valence States and Their Applications. 1. Electronegativities of Elements in Valence States. Inorganic Chemistry, 21, pp. 3886-3889 DOI 10.1021/ic00141a005.

Author: Lupan Oleg
Institute: University of Kiel, Technical University of Moldova
Street: Kaiser Str. 2
City: Kiel
Country: Germany
Email: ollu@tf.uni-kiel.de

Photocatalytic Applications of Doped Zinc Oxide Porous Films Grown by Magnetron Sputtering

L. Ghimpu[1], T. Reimer[2,3,4], D. Smazna[2], M. Hoppe[2], W. Benecke[4], A. Bejenari[1], A. Cojocaru[2,6,*], O. Lupan[2,5], R. Adelung[2], and I. Tiginyanu[1,5]

[1] Institute of Electronic Engineering and Nanotechnologies, Academy of Sciences of Moldova, Chisinau, Republic of Moldova
[2] Institute for Materials Science, Christian Albrechts University of Kiel, Kiel, Germany
[3] Technology for Silicon Based Micro- and Nanosystems, Institute for Electrical Engineering University of Kiel, Kiel, Germany
[4] Fraunhofer Institute for Silicon Technologies (ISIT), Itzehoe, Germany
[5] Department of Microelectronics and Biomedical Engineering, Technical University of Moldova, Chisinau, Republic of Moldova
[6] FUMT R&D Functional Materials GmbH, Kiel, Germany

Abstract— **Undoped, Al-doped and Sn-doped ZnO was sputter-deposited on glass substrates and on trenched Si substrates and some properties of these samples were evaluated. It was shown that the photocatalytic performance of the thin layers is improved by doping with Sn and Al. The samples doped with Al-doping provided a more pronounced effect compared to the control sample, as well as compared to the greater volume of undoped ZnO deposited on Si substrates. Morphology of produced particles was studied using transmission electron microscopy. Electron diffraction patterns were taken to confirm lattice parameters of the materials. Size and shape variations were monitored for un-doped, Al-doped and Sn-doped ZnO. EDX spectroscopy was utilized to assay chemical composition for doped and un-doped samples.**

Keywords— **ZnO, Sn-ZnO, Al-ZnO, Nanostructure, magnetron sputtering, photocatalys, TEM.**

I. INTRODUCTION

The materials suitable for photocatalytic applications have received interest for some time due to their potential for improvement in the industrial processes. Since one major component for their function, sunlight, is virtually free and inexhaustible, photocatalytic materials could help reducing energy expenditure and assist in the shift to "greener" economies based on renewable energies. The pro-environmental potential of photocatalysts is especially pronounced in applications such as waste water treatment [1].

Semiconductor metal oxide materials with band gaps ranging from 2.0-3.5 eV are widely studied materials for photocatalysis, especially TiO_2, which is one of the most widely studied photocatalyst to date due to its low cost and excellent chemical stability [1,2,3]. However, ZnO has also received increasing attention due to superior performance at sunlight wavelengths (even in the native material) and its possibly greater potential for adjusting the materials' properties, compared to TiO_2 [4,5,6,7].

ZnO can be easily grown in a wider range of morphologies than TiO_2, for example as 3D networks with structures in the micro- and nanoscale [8], and it is self-evident that a higher surface area provided by such structures helps to improve the performance of photocatalyst materials. Additionally, the material properties of ZnO can be tuned through doping with suitable dopant elements [4,5]. The most desired property would be the use of visible-range sunlight for effective photocatalysis, as opposed to the less available UV part of the spectrum, but doping could also enhance other properties, such as increasing the number of surface defects and reducing electron-hole recombination [9].

In this study, Al- and Sn-doped ZnO was magnetron sputter-deposited on glass substrates to assess photocatalytic performance, and un-doped ZnO was sputter-deposited in high aspect ratio trenches on Si substrates. In a later step, doped ZnO could instead be deposited in the trenches, for applications such as in a lab-on-a-chip. The photocatalytic performances of these materials were determined and compared.

II. EXPERIMENTAL

The films were deposited on different substrates by RF magnetron sputtering method in a vacuum chamber. The glass substrates were cleaned in mixed solution of chromium (7g K2Cr2O7; 10 ml H2O; 100 ml H2SO4) at room temperature. Afterwards, glass substrate cleaning was followed by rinsing with distilled water flow abundantly (18.2 MOhm cm). Substrates with silicon trenches were cleaned as reported before [10]. Next the latter were mounted on a rotatable substrate holder equipped with a resistive heater, all technical details can be found in our previous reports [11, 12, 13]

For the photocatalytic studies, ZnO samples were placed in a solution of 0.5 μmol/L (or 1 μmol/L, in case of the un-doped ZnO in Si-trenches) methylene blue and irradiated with UV light (370 nm, 170 mW per diode, 4 diodes; distance ~135 mm to sample) for 90 min. The fluid was stirred

V. Sontea and I. Tiginyanu (eds.), *3rd International Conference on Nanotechnologies and Biomedical Engineering*,
IFMBE Proceedings 55,
DOI: 10.1007/978-981-287-736-9_85

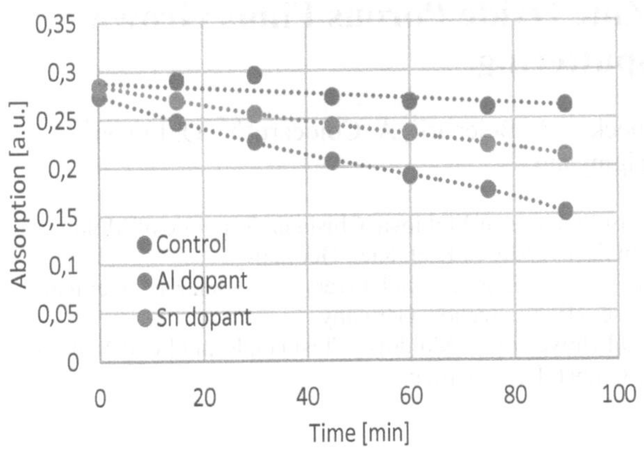

Fig. 1 Absorption of methylene blue at 664 nm wavelength over time. Blue: Control sample, sample holder without sample. Grey: 1,5 µm ZnO with Sn dopant. Orange: 1,5 µm ZnO with Al dopant. It is clearly evident that the samples are photocatalytically active and reduce the methylene blue concentration over time, if irradiated with UV light. In this case,ZnO with Al dopant proved superior to ZnO withSn dopant. Al-dopedt sample reduced meth. blue concentration by ~43% after 90 minutes, and Sn-doped one reduced it by ~28%. Respective activities per area are 0.00102 mmol/mm2*90min and 0.00090 mmol/mm2/90min.

with a magnetic stirrer. Aliquots of ~1ml were taken every 15 min and their absorption was measured in a Perkin Elmer UV/Vis/NIR Spectrometer Lambda 900. The signal at λ=664 nm was used to compare absorption to the value at t = 0 min and to the control sample (MB solution without ZnO).

For TEM studies a small amount of ZnO, Al-doped ZnO and Sn-doped ZnO was grinded and diluted with ~ 1mL Butanol solvent each. A droplet of ~100µL containing each material respectively was placed on a TEM copper grid with the carbon mesh and the solvent had been evaporated. The samples produced this way were studied using JEOL JEM-2100 LaB6 operating at 200 kV. EDX detector and the Software Aztec were utilized to acquire EDX spectra for three ZnO samples with different doping.

III. RESULTS

In the presented work, the glass slides with Al- and Sn-doped ZnO were prepared by sputter deposition and then exposed to the described protocol for the photocatalytic experiment.

Figure 1 shows the measured photocatalytic activity of Al- and Sn-doped samples compared to a non-active control. It is evident that doping with Al doping provides the stronger improvement in photocatalytic activity compared to Sn doping in ZnO.

Figure 2 offers a comparison between the doped samples and un-doped ZnO in trenches on Si – the un-doped ZnO has a slightly lower average activity, and it should be noted that due to a different production process, the amount of un-doped ZnO on Si samples greatly exceeded the amount of doped ZnO deposited on glass slides. This highlights that the doped samples have a superior specific activity compared to pure ZnO in our experimental evidences.

Since the material is bound to a solid substrate (glass or Si), it can employ all the advantages of heterogeneous catalysis, and the catalyst material can be easily retrieved from the reaction for reuse, while at the same time preventing the harmfull impact on the environment resulting from the catalyst material being disposed into the environment.

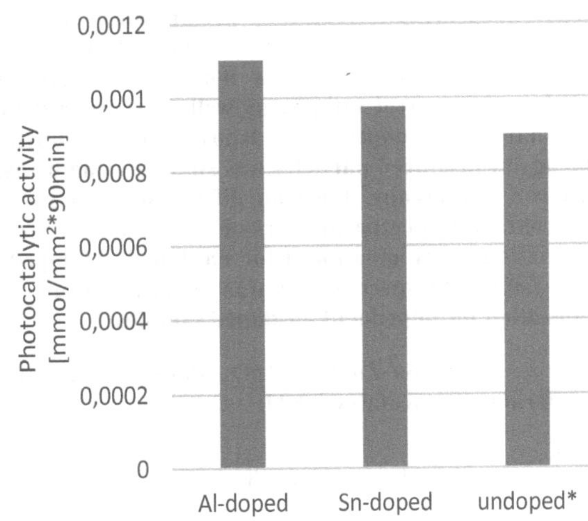

Fig. 2 Comparison of the average photocatalytic activity per area and time for Al-doped, Sn-doped samples and un-doped ZnO in trenches on Si. The activity of un-doped ZnO appears only slightly lower than the activites of the doped samples, however it must be noted that the layer thickness of un-doped ZnO was greater by at least a factor of two, and the material was porous (see Fig. 3). This highlights the improved performance of the doped ZnO layers, which achieve comparable or even better activity despite a much smaller amount of ZnO material and therefore reduced available surface area.

Morphology of the particles that was clearly observed using TEM was similar. TEM images for 3 samples (ZnO, ZnO:Al-doped, ZnO:Sn-doped) are presented in the Figure 4(a,b,c). Measured mean particle size was of ~105.26 nm for un-doped ZnO sample (Fig.4a), ~77.74 nm for Al-doped sample (Fig.4b) and ~59.30 nm for the Sn-doped sample (Fig.4c).

To compare lattice parameters of the samples selected area electron diffraction SEAD patterns were measured and results are depicted in Figure 5. Calculating the interplanar spacings for each sample out of their SAED was carried out and the results were matching the *d* values for ZnO.

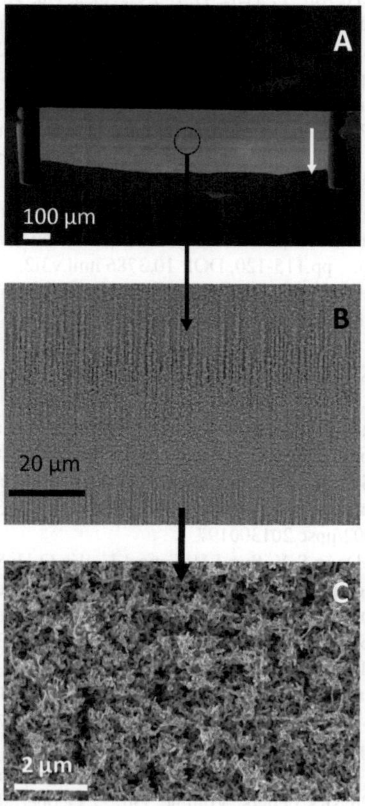

Fig. 3 High resolution SEM study of the vertical area of a ~350 µm deep trench in a silicon substrate with ZnO deposited on its surface. It is evident from the magnification that the vertical surface is covered with porous ZnO structures ; vertical structures in the center image are artifacts from the trench etching, which were also evenly coated with ZnO. The trenches, when viewed from above, are of a square shape. In this side-view image, one exposed vertical wall of a trench is visible in the center. To the left and right (top image) are perpendicular parts of the same trench, reaching into the substrate and also showing the true depth of the structures. Apparent irregularities in structure depth occurred because the Si piece was broken, not sawed, and a thin Si protrusion blocks the view on the lowest part of the exposed trench wall (white arrow).

In order to confirm the composition of the samples EDX spectra were recorded in TEM. Corresponding data is presented in the Figure 6. As expected, the spectra affirmed purity of the undoped ZnO (Fig.6.a), and presence of the dopant elements Al and Sn (Fig.6.b,c) amount of which was ~0.1at.% and ~0.7 at.%, those numbers may slightly vary along the samples. EDX peaks for Cu and C contained in the Fig.6.a,b clearly belong to the TEM grid.

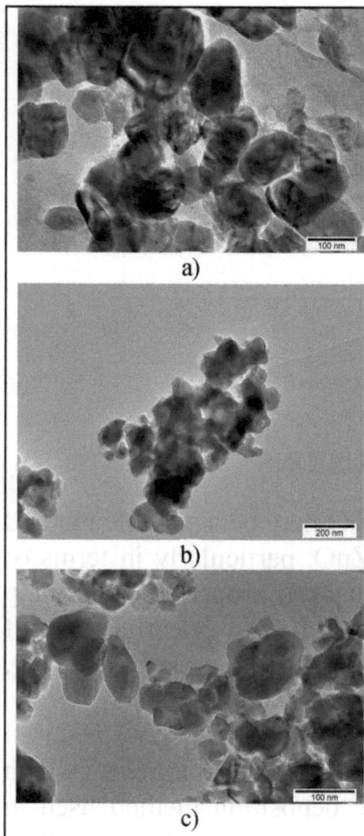

Fig. 4 TEM images for the ZnO (a), ZnO Al-doped (b) and ZnO-Sn doped (c) samples. Shapes of the particles point out the crystalline nature of the samples.

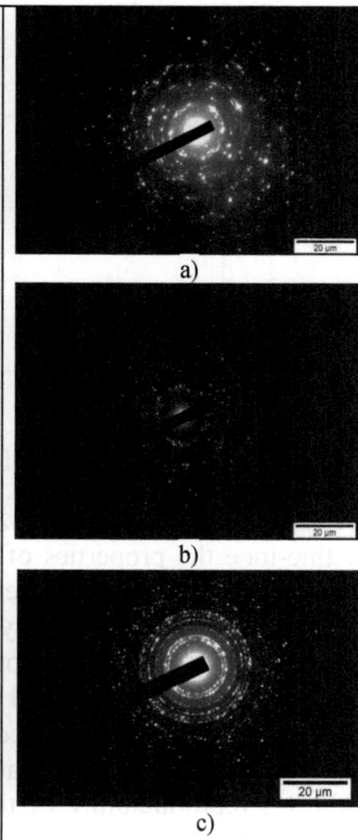

Fig. 5 Diffraction Patterns for the ZnO (a), ZnO Al-doped (b) and ZnO Sn-doped (c) samples.

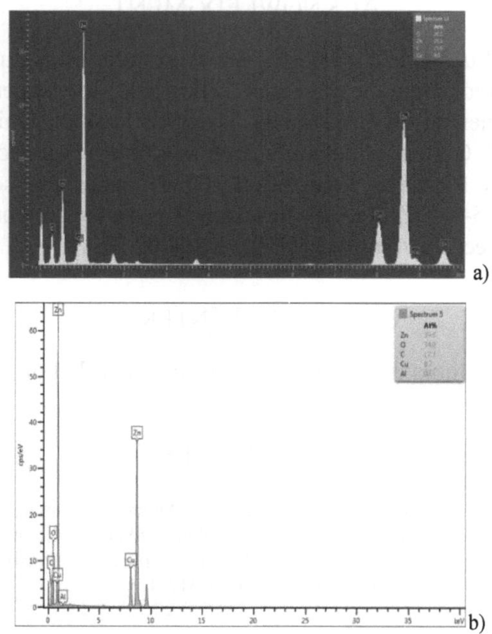

Fig. 6 EDX spectra for ZnO samples (a), ZnO-Al doped (b), ZnO-Sn doped (c)

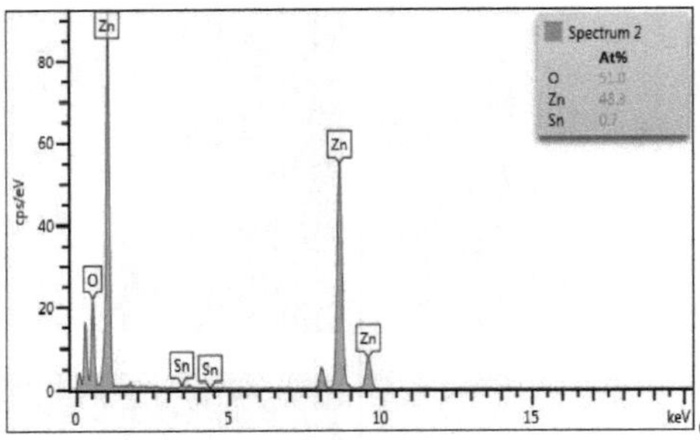

c)

Fig. 6 (*Continued*)

IV. Conclusions

Doping with Sn and Al in ZnO may be a suitable method to fine-tune the properties of ZnO, particularly in terms of photocatalytic activity. The enhanced properties could help improving the applicability and cost-effectiveness of photocatalytic materials for industrial processes, e.g. environmental protection and waste water treatment.

TEM studies on the doped and un-doped ZnO material revealed similar size variation of the particles which indicates reproducibility of the deposition method used in this study. Diffraction pictures confirmed the lattice parameters of the investigatedmaterials to match the ZnO crystal having a P63mc (186) hexagonal space group. EDX spectra showed the presence of the doping elements (Al and Sn respectively).

Acknowledgment

Dr. Lupan acknowledges the Alexander von Humboldt Foundation for the research fellowship for experienced researchers at the Institute for Materials Science, University of Kiel, Germany. This research was partly sponsored by German Research Foundation (DFG) under the schemes SFB 855-A5, Z-10. This research was partly supported by the project Institutional 45inst-15.818.02.29A.

Conflict of Interest

The authors declare that they have no conflict of interest.

References

1. Carlos A. K. Gouvêa, F. Wypych, S. G. Moraes, N. Durán, P. Peralta-Zamora, Semiconductor-assisted photodegradation of lignin, dye, and kraft effluent by Ag-doped ZnO, Chemosphere 40 (2000) pp. 427-432, DOI:10.1016/S0045-6535(99)00312-4
2. Reenamole Georgekutty, Michael K. Seery, and Suresh C. Pillai: A Highly Efficient Ag-ZnO Photocatalyst: Synthesis, Properties, and Mechanism, J. Phys. Chem. C, (2008), 112, pp. 13563–13570, DOI: 10.1021/jp802729a
3. A. Fujishima, X. Zhang, Titanium dioxide photocatalysis: present situation and future approaches, Comptes Rendus Chimie, 9 (2006) 750-760, DOI:10.1016/j.crci.2005.02.055
4. R. Wang, J.H. Xin, Y. Yang, H. Liu, L. Xu, J. Hu, The characteristics andphotocatalytic activities of silver-doped ZnO nanocrystallites, Appl. Surf.Sci. 227 (2004) pp. 312–317, DOI: 10.1016/j.apsusc.2003.12.012.
5. K. Vanhesuden, W.L. Warren, J.A. Voigt, C.H. Seager, D.R. Tallant, Impact of Pb doping on the optical and electronic properties of ZnO powders, Appl. Phys. Lett. 67 (1995) pp. 1280–1282, DOI:10.1063/1.114397.
6. Qu J, Luo C, Cong Q: Synthesis of multi-walled carbon nanotubes/ZnO nanocomposites using absorbent cotton. Nano-Micro Lett., (2011), 3(2), pp.115-120, DOI: 10.3786/nml.v3i2.
7. S. Sakthivel, B. Neppolian, M.V. Shankar, B. Arabindoo, M. Palanichamy, V. Murugesan, Solar Energy Solar Cells 77 (2003), pp. 65–82.
8. Yogendra K. Mishra, S. Kaps, A. Schuchardt, I. Paulowicz, X. Jin, D. Gedamu, S. Freitag, M. Claus, S. Wille, A. Kovalev, S. N. Gorb and R. Adelung, Fabrication of Macroscopically Flexible and Highly Porous 3D Semiconductor Networks from Interpenetrating Nanostructures by a Simple Flame Transport Approach, Particle & Particle Systems Characterization, Volume 30, Issue 9, 2013 pp 775–783, DOI: 10.1002/ppsc.201300197
9. Yang L-Y, Dong S-Y, Sun J-H, Feng J-L, Wu Q-H, Sun S-P: Microwave assisted preparation, characterization and photocatalytic properties of a dumbbell-shaped ZnO photocatalyst. J Hazard Mater 2010, 179: pp.438-443, DOI: 10.1016/j.jhazmat.2010.03.023.
10. O. Lupan, Th. Pauporté, I.M. Tiginyanu, V.V. Ursaki, H. Heinrich, L. Chow, Materials Science and Engineering: B, Volume 176, Issue 16, 25 September 2011, pp. 1277-1284, DOI: 10.1016/j.apsusc.2012.07.058
11. O. Lupan, V.M. Guérin, L. Ghimpu, I.M. Tiginyanu, T. Pauporté, Chemical Physics Letters, Volume 550, 22 Ocober 2012, Pages 125-129, DOI: 10.1016/j.cplett.2012.08.071
12. O. Lupan, L. Ghimpu, T. Reimer, M. Hoppe, I. Paulowicz, D. Gedamu, Y. Mishra, D. Hammerich, S. Chemnitz, V. Cretu, Journal of Nanoelectronics and Optoelectronics 9, 2014, pp. 257-264, DOI: 10.1109/SMICND.2013.6688111
13. O. Lupan, Y. Kumar Mishra, I. Paulowicz, D. Gedamu, A. Cojocaru, R. Adelung, Effect of Al, Sn - Doping on Properties of Zinc Oxide-Nanostructured Films Grown by Magnetron Sputtering. Semiconductor Conference (CAS), 2013 International, Sinaia, Romania, v. 1, pp.133-136 (2013), DOI: 10.1109/SMICND.2013.6688111

Authors to whom correspondence should be addressed:

Author:	A. Cojocaru
Institute:	FUMT R&D Functional Materials GmbH
Street:	Kaiserstr. 2
City:	Kiel
Country:	Germany
Email:	ac@fumt-rd.de

A DVG003 Medical Device for Millimeter Wave Therapy

Iu. Sainsus, A. Conev, Iu. Russev, I. Bejan, N. Tomşa, V. Babac, S. Piatighin, and A. Sidorenko

Institute of Electronic Engineering and Nanotechnology "D. Ghitsu", Chisinau, Republic of Moldova

Abstract— **We report on the patented technical solutions used in a new DVG003 millimeter-wave therapy device based on the experience accumulated on this subject at the Gitsu Institute of Electronic Engineering and Nanotechnolgies of the Academy of Sciences of Moldova for more than 8 years. Both the hardware and the generator head are described.**

Keywords— **Millimeter waves, the use of waves in medicine, EHF therapy, microstrip technology, nonthermal radiation.**

I. INTRODUCTION

The concept of the DVG003 device has been developed owing to the joint work with the Topaz plant (Chisinau), where the production of more than 500 devices for clinics and family doctors in Moldova was anticipated. The main difference of this instrument from other devices lied in the fact that it was originally conceived for large-scale production in a newly developed case, which implied a new design and a layout in the form of two boards instead of one. In addition, all units have been redesigned taking into account the frequency of occurrence and the possibility of purchasing the components by Topaz in the future. The working group of our Institute has worked with the design engineers of the Topaz plant for more than 2 weeks. The result was the development of a full set of specification documents.

II. OVERVIEW

Laboratory tests of the previously developed devices (DVG001 and DVG002) have shown that they have some disadvantages both in electronics and software, which lead to failures in the operation of the devices. The main disadvantages are as follows: low frequency stability; poor thermal stability, and short time of battery operation.

The following components have been subjected to modifications:

- the supply circuit of the ATMEGA8 microprocessor;
- the supply circuit of the "12 V" frequency setting;
- the inverter circuits to increase the efficiency to 88%;
- the EHF radiation detection circuit.

The entire structure was transferred from one main board to two main boards: a power supply board and a control board.

In addition, the software has been modified; new options for amplitude modulation rate and EHF radiation frequency deviation have been introduced in the new version.

All the modifications taken together have made it possible to introduce the following new options for both amplitude modulation and deviation from center frequency:

- an increase in the amplitude modulation bandwidth from 10 to 500 Hz;
- an increase in the percentage of EHF radiation frequency deviation from 10 to 15%.

Laboratory tests have shown that the device is stable in operation, the temperature effect is minimal, and the battery capacity is sufficient for operation for 8 h, provided that the setting of the operating parameters takes 30% of time (2.4 h).

III. DESCRIPTION OF PATENTED SOLUTIONS

The schematic diagram of the device has been constructed according to the principles described in [2] Brevet no. 466 and is shown in Fig. 1.

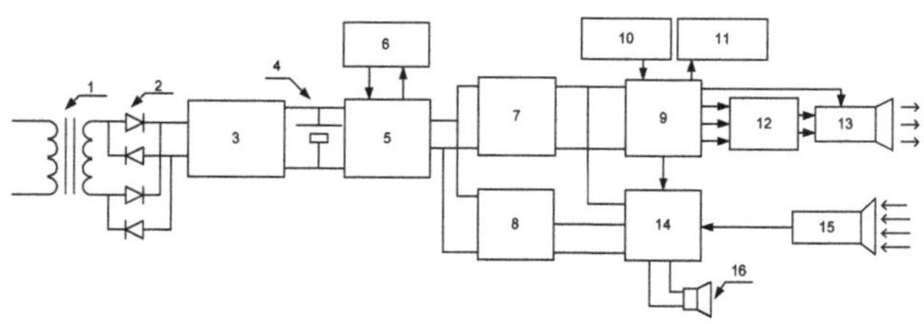

Fig. 1 Schematic diagram of the DVG003 device.

The device operates as follows: an alternating current of commercial line is applied to the input of voltaic transformer *1*, which reduces the alternating voltage to a desired value and supplies it to rectifier *2*, at the output of which a DC voltage is generated to feed DC inverter *3*, which in turn decreases the DC voltage to a level required for DC power supply of battery *4* and sequential cascades of the control circuit.

The output of battery *4* is connected to switching and protection unit *5* (described below); it connects and/or disconnects battery *4* to the sequential cascades of the control circuit by pressing the buttons on control board *6*. In addition, switching and protection unit *5* automatically disconnects battery *4* if it is discharged; the battery voltage

V. Sontea and I. Tiginyanu (eds.), *3rd International Conference on Nanotechnologies and Biomedical Engineering*,
IFMBE Proceedings 55,
DOI: 10.1007/978-981-287-736-9_86

is analyzed in unit *5* with respect to the minimum permissible value to protect the battery from deep discharge and subsequent failure. From the output of switching and protection unit *5*, the current flows to the input of negative DC inverter *8* and positive DC inverter *7*, which also simultaneously generates unipolar voltage applied to microprocessor *9* and bipolar voltage supplied to EHF radiation detector *14* to the input of which EHF radiation *15* is supplied.

The output of EHF detection cascade *14* is connected to sound unit *16*, which produces audio signals to indicate the presence of EHF radiation. Sound unit *16* is connected to command signals from processor *9*, which display the executing and/or issuing of commands.

Processor *9* is controlled by second control board *10*; display *11* shows which command is executed. The output of processor *9* is connected to switch *12* through three conductors, two of which constitute the power circuit, and the third is a signal conductor. Thus, the amplitude modulation of generator *13* is performed. The fourth conductor from the output on processor *12* is connected to generator *13* and provides digital information about the operating frequency of the generator and thereby varies the frequency.

The **switching and protection unit** (denoted by *5* in the schematic diagram) has been constructed according to the principles described in [3] **Brevet no. 474** and is shown in Fig. 2.

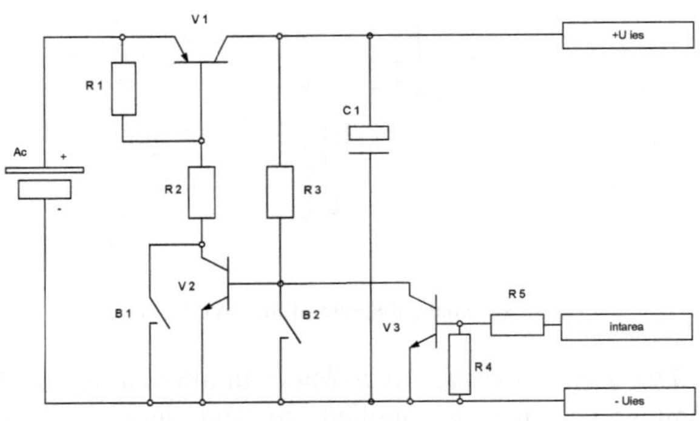

Fig. 2 Diagram of the switching and protection unit.

The unit operates as follows: terminals "+Uies" and "-Uies" are connected to the load (is not shown in the circuit), which includes a microprocessor, which in turn analyzes the voltage across the battery *Ac* and, if the battery voltage is less than the minimum permissible value, generates a high-level signal which is applied to the "intrarea" terminal. The initial position of buttons *B1* and *B2* is "off." To supply DC power to the load, it is necessary to press button *B1* for 1–3 s.

The voltage across the battery *Ac* is applied to the circuit composed of elements *R1, R2* and the base–emitter junction of transistor *V1*; as a result, this transistor is switched on and the battery current flows to the load. The same voltage at the output of transistor *V1* is applied to the base–emitter junction of transistor *V2* through resistor *R3*, which is switched on and thus grounds the current to the case through resistors *R1, R2* and thereby keeps field-effect transistor *V1* in the on state upon the release of button *B1*.

Electrical capacitor *C1* serves for storing the load voltage and reducing interferences during the transition of transistors *V1* and *V2* from the off to on state.

To manually turn off the supply of the load, it is necessary to press button *B2* for 1–3 s in order to switch off transistor *V2* and, accordingly, transistor *V1*. The capacity of electrical capacitor *C1* is selected so as to maintain a high voltage across the load for no longer than 0.5 s.

Another— automatic—algorithm of load shedding consists in supplying a high-level signal from the output of the microprocessor to the input of transistor *V3* through base divider *R5, R4* through terminal "intrarea." Switch-on of transistor *V3* leads to processes similar to the pressing of button *B2*, i.e., to load blackout.

The **EHF radiation detector** (denoted by *14* in the schematic diagram) has been constructed according to the principles described in [4] **Brevet no. 472** and is shown in Fig. 3.

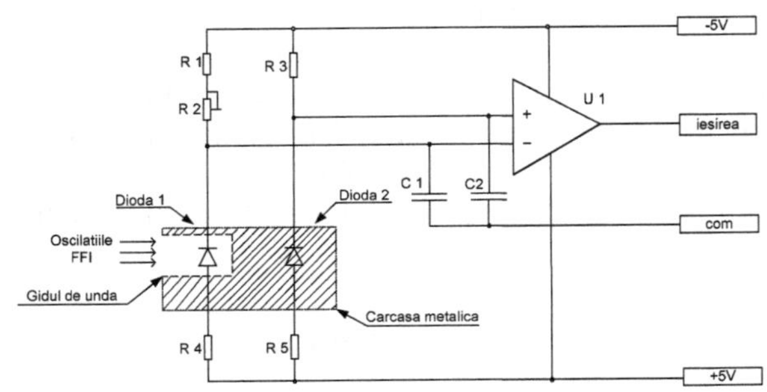

Fig. 3 Diagram of the EHF radiation detector

The unit operates as follows: the inverting input of comparator *U1* is connected to the diagonal of the measuring bridge comprising rectifying diode *Dioda 1* mounted in the waveguide of the metal case. A similar-type diode *Dioda 2* is mounted in the same case; this diode is connected to the other diagonal of the measuring bridge which is connected to the noninverting input of comparator *U1*. Using trimming resistor *R2*, at the inverting input of comparator *U1*, initial voltage is set at a level that is not much higher than the voltage across the non inverting input of comparator *U1*.

In this event, a relatively low level relative to bus "*com*" is set at the output of comparator *U1*. Resistors *R1, R3, R4*, and *R4* are the elements of the measuring bridge and should be of the same type to have the same temperature drift. Once EHF oscillations occur at the input of the waveguide, they are detected on diode *Dioda 1*, which in turn decreases the voltage across the inverting input of comparator *U1*; this in turn leads to a change in the voltage level at the output of comparator *U1* from low to high.

Electrolytic capacitors *C1* and *C2* are introduced to increase the noise immunity of the detector.

The introduction of two similar-type diodes in one case leads to a proportional change in the voltage drop on them under the effect of temperature because the diodes have the same temperature; this feature, along with the similar-type resistors *R1, R3, R4*, and *R4* mounted in one measuring bridge, contributes to the formation of an extremely precise temperature-independent measuring bridge.

The waveguide in the case where both diodes are mounted selectively passes the input oscillations. This design of the *EHF radiation detector* is characterized by high thermal stability and an extremely low detection level; the detection threshold is set by resistor *R2*.

IV. Design Description

A. Hardware

A version of the case that best meets the requirements of Topaz was proposed at the plant. According to this version of the case, we have developed the design and the drawings, which are shown in Fig. 4. After selecting the case and the components, a schematic diagram composed of two boards was developed.

For simplicity in the production and specialization of workplaces, the control board contains all the elements required for controlling and displaying:

- a microprocessor;
- connectors for the display board and the keyboard;
- voltage inverters "+5V"; "−5V" and "+12V";
- an EHF radiation detector;
- a battery protection circuit.

The power supply board includes the following components:

- a battery;
- a mains supply transformer;
- DC voltage circuits;
- a step-down inverter to supply and charge the battery;
- a connector for the control board.

The front panel also contains two boards:

- a display board (LCD display);
- a keyboard (buttons).

A communication protocol for the LCD display and the ATMEGA8 microprocessor and a program for setting the operating modes using the keyboard have been developed.

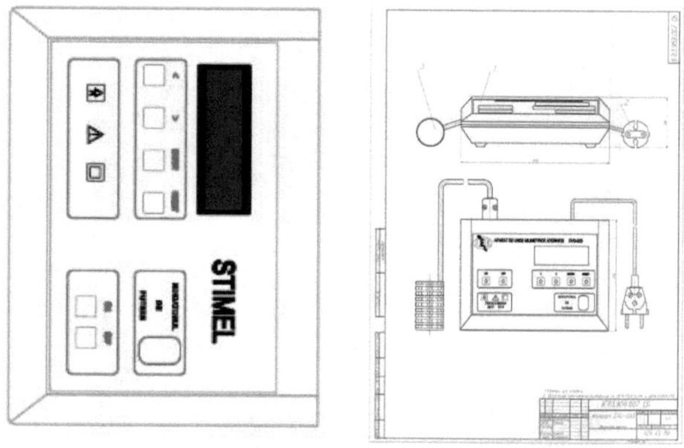

Fig. 4 Appearance and drawing of the case.

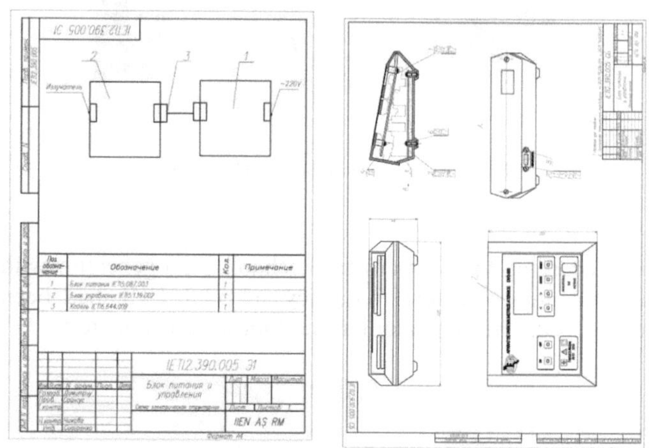

Fig. 5 Drawing of the case and schematic diagram of the hardware.

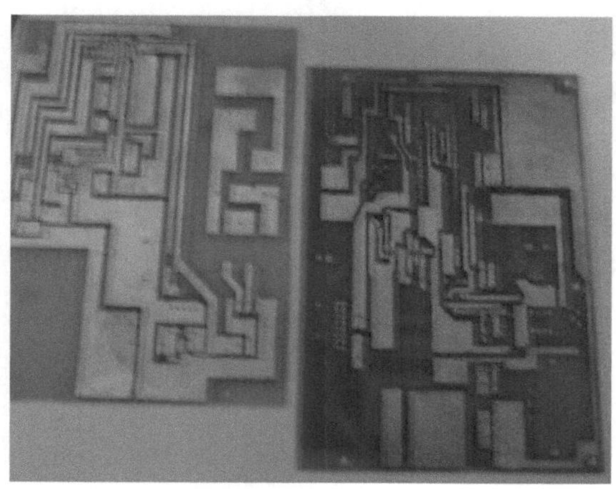

Fig. 6 Appearance of the power supply and control boards.

Boards have been developed and constructed. The development of printed circuit boards was conducted using the PCAD program. Figure 6 shows the control board and the power supply board of the hardware.

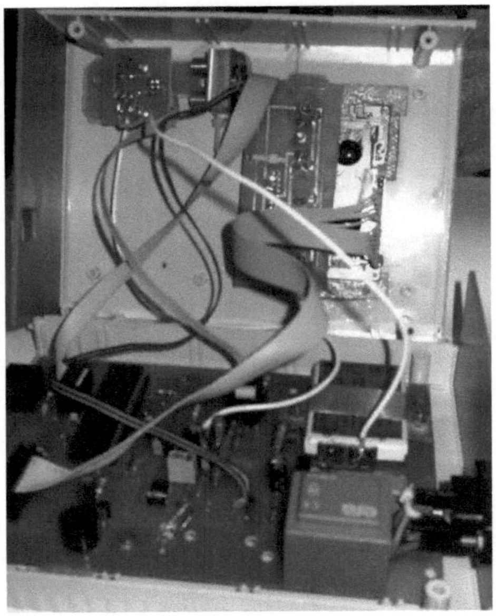

Fig. 7 Appearance of the assembled device.

B. EHF Generator

According to the microstrip technology, a generator with an operating (center) frequency of 53.3 GHz has been developed. The generator has the following parameters:

Fig. 8 Appearance of the generator with an operating frequency of 53.3 GHz.

- operating frequency of 52–55 GHz;
- minimum power of 2 μW;
- frequency deviation of ±10%;
- amplitude modulation of minus 20 Db.

An HMC739LP4E integrated circuit (United States) is used as a source of high-frequency oscillations with subsequent multiplication of the frequency by 2 [1].

This integrated circuit is a voltage-controlled oscillator with a relatively high intensity of output oscillations.

A generator, a multiplier by 2, and a matching circuit are arranged on a microstrip board, which also contains a probe that removes EHF oscillations in a circular waveguide. The microstripe board is made of a fluoroplastic-containing material that allows operation at frequencies of up to 100 GHz.

CONFLICT OF INTEREST

The authors declare that they have no conflict of interest.

REFERENCES

1. http://www.hittite.com/content/documents/data_sheet/hmc739lp4.pdf
2. **REVET MD 466.** ***Dispozitiv cu microunde,*** (11) 466(13)Z; A61N 5/02 (2006.01); G06F 9/00 (2006.01); G06F 9/22(2006.01); Sainsus Iu., Babac V., MD, Piatighin S., MD Rotaru A MD, Railean S., MD, Conev A., MD , Russev Iu., MD., , Postoronca S., MD, Scerbii D., MD , Cijov G., MD
3. **BREVET MD 474.** ***Dispozitiv de comutatie si protectie a acumulatorului,*** (11) 474(13); H02H 7/18 (2006.01); H02H 3/00 (2006.01); H02H 3/24 (2006.01); H02H 3/26 (2006.01); H020 3/28 (2006.01); H02J 7/00 (2006.01); H02J 7/36 (2006.01); Sainsus Iu.MD, Conev A., MD , Russev Iu., MD., Postoronca S., MD, Scerbii D., MD , Babac V, MD
4. **BREVET MD 472.** ***Dispozitiv de detectare a campului electromagnetic de frecventa foarte inalta si intensitate joasa,*** (11) 472(13)Z; G01R 1/12 (2006.01); G01R 1/24 (2006.01); G01R 1/44(2006.01); G01R 29/08 (2006.01); Sainsus Iu., Babac V., MD, Piatighin S., MD Rotaru A MD, Railean S., MD, Conev A., MD , Russev Iu., MD., , Postoronca S., MD, Scerbii D., MD

Author: Iurii Sainsus
Institute: Institute of Electronic Engineering and Nanotechnology "D. Ghitsu"
Street: 3/3 Academiei str.
City: Chisinau
Country: Moldova
Email: yenergy@mail.ru

Hydrogen Gas Sensor Based on Nanograined Pd/α-MoO_3 Belts

V. Cretu[1], V. Postica[1], D. Stoianov[1], V. Trofim[1], V. Sontea[1], and O. Lupan[1,2]

[1] Technical University of Moldova, Department of Microelectronics and Biomedical Engineering, Chisinau, Republic of Moldova
[2] University of Kiel, Institute for Materials Science/Functional Nanomaterials, Kiel, Germany

Abstract— **In this work, a simple and cost-effective method for nanostructurization of α-MoO3 micro-belts by using an aqueous solution of PdCl2 is reported. Obtained nanograined Pd/α-MoO3 belts, with 70-80 nm average diameters of nanocrystallites were integrated in sensor structures for application in hydrogen gas sensing. In dependence of method for nanostructurization, a gas response of about 2 – 3 at 200 °C operating temperature was obtained. In case of K+ doped samples, the hydrogen response was enhanced to about 7.5. To explain the obtained results, the gas sensing mechanism based on catalyst properties of Pd nanoparticles and generation of oxygen vacancies is discussed in details.**

Keywords— **MoO3, hydrogen, gas sensor, nanostructured, mechanism.**

I. INTRODUCTION

Molybdenum trioxide (MoO_3) has been recently attracted a major interest due to unique properties such as high chemical stability, specific gas sensing mechanism due to layered structure, which allow possibility of integration in a variety of applications, such as gas sensors, energy storage devices and catalysts [1-6]. Yang et al. [7] demonstrated that α-MoO_3 nanoribbons with higher content of Mo^{5+} exhibits high response to hydrogen gas at room temperature, due to high affinity for oxygen molecules of pentavalent Mo^{5+} ions. Imawan et al. [8] demonstrated improvement in hydrogen gas response and selectivity of MoO_3 thin films by addition of a V_2O_5-overlayer. Illyaskutty et al. [5] demonstrated excellent ethanol sensing performance of MoO_3:ZnO nanostructures with good selectivity. MoO_3 is also known to demonstrate profound piezoresisitive and photochromism effects [9-10].

In previous works our group demonstrated a new method for rapid synthesis of large amount of free-standing highly crystalline α-MoO_3 nano- and microbelts and ribbons in a single-step growth process from commercial bulk molybdenum metal [2]. Morphological, structural and optical properties for such type of structures were investigated in details in previous works [2,11]. Also quasi-electron-transparency in MoO_3 belt structures has been observed and investigated using SEM-ZEISS [2]. The formation mechanism by rapid thermal processing (RTP) of α-MoO_3 sheets of nanoscale thickness up to some extent allow growth of crystal structures with length up to several millimeters. The X-ray diffractograms and micro-Raman spectra of the synthesized specimens at 800 °C measured at room temperature shown the main α-MoO_3 phase with layered structure packed in the direction of its crystallographic axis [2].

For real applications in different technologies the free-standing nanostructures of metal oxides have the significant advantage because can be freely used in the *bottom-up* process for assembly into various nano- and microsystems. Another important advantage is possibility to use different type of substrate for further processing in device applications. A single nano- and microbelts of α-MoO_3 was integrated in a sensor structure, and has confirmed good gas response to ethanol and hydrogen gas in dependence of operating temperature with quite stable and moderately reversible response [2]. At 200 °C sensor structure demonstrated good selectivity to ethanol vapors, while at 350 °C was obtained an increase response to hydrogen gas with relatively fast response and recovery times. The main advantage of the synthesized α-MoO_3 belts and ribbons with much higher dimensions (with average length of up to several millimeters) is simplicity of the integration in the gas sensor structure, that don't need such expensive equipment like FIB-SEM [2]. However, due to very low surface-to-volume ratio, such types of sensors don't demonstrate any noticeable gas response. Thus, appear a necessity for development of cost-effective and simple methods to considerably increase surface-to-volume ratios of the α-MoO_3 belts and ribbons for enhancement of gas response. For example Park et al. [12] has demonstrated superior gas sensing properties of the nanograined ZnO NWs under single crystal ZnO NWs, due to the higher surface-to-volume ratio and specific gas sensing mechanism.

Here we report a simple method to obtain nanograined Pd/α-MoO_3 belts using aqueous solution of $PdCl_2$. Fabricated sensors demonstrated considerable increase in gas response and high selectivity to hydrogen gas, in comparison with pristine α-MoO_3 belts at relatively low operating temperature (OPT) of 150 °C – 200 °C. Also, by doping with K^+ was possible to further increase gas sensing performances of the nanograined Pd/α-MoO_3:K belts. Gas sensing mechanism based on catalytic properties of Pd nanoparticles and oxygen vacancies generation in α-MoO_3 is proposed to explain the obtained results.

V. Sontea and I. Tiginyanu (eds.), *3rd International Conference on Nanotechnologies and Biomedical Engineering*,
IFMBE Proceedings 55,
DOI: 10.1007/978-981-287-736-9_87

II. EXPERIMENTAL

A. Nanostructurization of the α-MoO_3 Belts

Figure 1 shows schematic illustration of the overall process for preparation of the nanograined Pd/α-MoO_3 belt based sensor. As-grown α-MoO_3 belt, synthesized by previously reported method at 800 °C [2], first was placed on the glass substrate. In the next step, both ends of belts were mechanicaly fixed on glass substrate by using silver paste (Fig 1 (a)). The obtained sensor structure is nanostructured by using $PdCl_2$ aqueous solution (0.4 mM) by two methods: (1) a drop of 10 μM $PdCl_2$ aqueous solution (100 μL) is placed on the α-MoO_3 belt surface and is left to dry for a one day in air (Fig 1 (b)); (2) sensor structure is dipped in a 10 μM $PdCl_2$ aqueous solution for a 10 s and then is dried in air flux at 400 °C for 5 min. In final step, sensor structure was exposed to Al metal deposition of external electrical electrodes through a mask on glass substrate (Fig 1 (c)). The resulted width between Al contacts is 1 mm. Gas sensing measurements of the devices at concentration of H_2 of 100 ppm were performed using the test system reported in previous work [2,13].

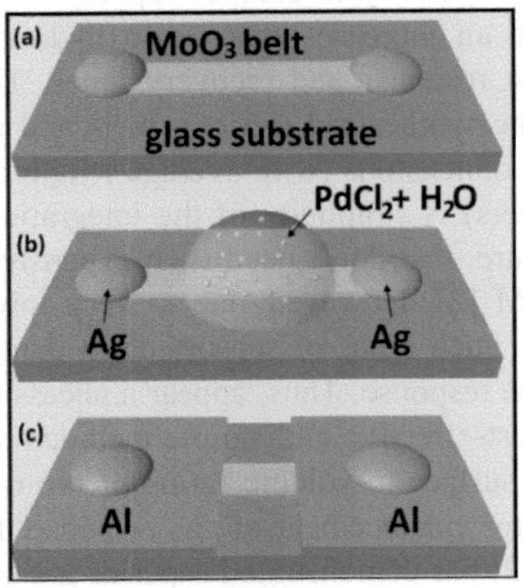

Fig. 1 Schematic illustration of the overall process used in this study for preparing of nanograined Pd/α-MoO3 belt based sensor: (a) mechanical fixation of the α-MoO3 belt on the glass substrate by using silver paste; (b) nanostructurization of α-MoO3 belt by using PdCl2 aqueous solution; (c) Al metal deposition of external electrical electrodes through a mask on glass substrate

III. RESULTS

A. Morphological Characterization

The morphology of samples was examined using a scanning electron microscopy (SEM) instrument (Carl Zeiss; 10 kV, 10 μA). After chemical reaction with $PdCl_2$ aqueous solution, surface of α-MoO_3 belts is nanostructured by formation of nanocrystallites (Fig. 2). Fig. 2 (a) and (c) shows overall view of the nanostructured belt by method (2), and as can be observed the layered morphology of belts is not modified after reaction with $PdCl_2$ and no evidence changes in morphology of the pristine α-MoO_3 belts is observed. There were not observed significant differences between morphology of α-MoO_3 belts nanostructured by method (1) and (2), showing only a slight variation in average diameter of nanocrystallites. Fig. 2 (e) shows a magnified area of the belt and demonstrates that entire surface is covered with nanocrystallites, which indicate that all surface of belt has reacted with $PdCl_2$ aqueous solution. Fig. 2 (b), (d) and (f) shows the interconnected networks of α-MoO_3 nanocrystallites that demonstrate different types of shapes with average diameter estimated in region of 70 - 80 nm. Fig. 2 (d) shows crossed by numerous convolutions surface of nanocrystallites, which also has big impact on increase of surface-to-volume ratio of the α-MoO_3 belt.

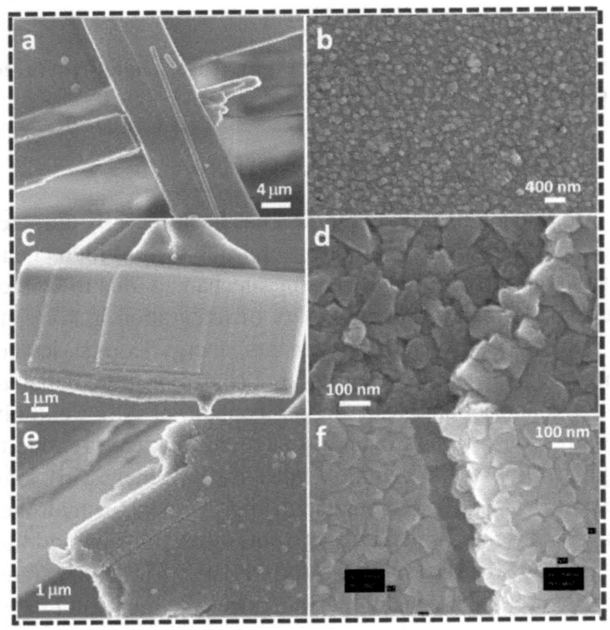

Fig. 2 SEM images of the nanograined Pd/α-MoO3 belts from different regions of the sensor structure and different magnifications: (a), (c) overall view of the ribbon showing layered morphology; (b) magnified area showing uniformity of the nanocrystallites; (e) magnified area of a belt showing nanograined morphology; (d), (f) high magnified area showing surface morphology of the nanograins

B. Hydrogen gas Sensing Properties

Sensors based on the Pd/α-MoO_3 belt nanostructured by method (1), (2) and (2+) doped with K^+ will be noted by sample #1, #2 and #3, respectively. Gas response versus time of all three samples are presented in Fig. 3, and is calculated by the ratio of I_{gas}/I_{air}, where I_{gas} is the current of the sensor structure under exposure to hydrogen gas and I_{air} is the current of the sensor structure exposed to the air.

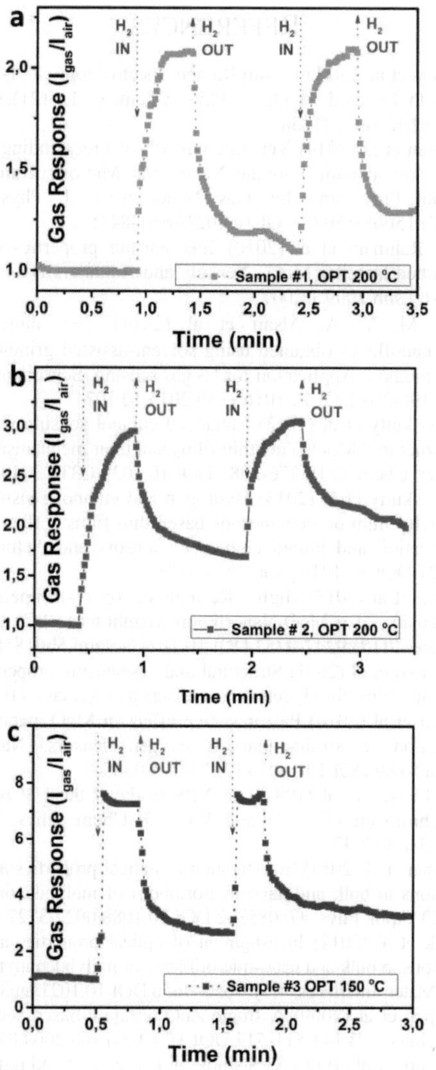

Fig. 3 Gas response to hydrogen gas versus time of the sample set: (a) #1; (b) #2; and (c) #3

As can be observed, samples prepared by method (2) shows higher gas response ($I_{gas}/I_{air} \sim 3$) in comparison with samples prepared by method (1) ($I_{gas}/I_{air} \sim 2.2$) (Fig. 3). In case of K^+–doped samples the gas response to hydrogen was enhanced by ~ 2.5 times ($I_{gas}/I_{air} \sim 7.6$). Due to fact that samples don't show complete recovery to the initial baseline at operating temperatures of 150 °C – 200 °C, was evaluated only response time (τ_r) of the sensors, defined as the time required to obtain 90 % of the full response. From Fig. 3 is evident that sample #3 demonstrate faster response (τ_r = 2.4 s) in comparison with sample #1 (τ_r = 10 s) and #2 (τ_r = 18 s).

C. Hydrogen Gas Sensing Mechanism

According to previous reports on the surface of α-MoO_3 are not chemisorbed oxygen species, due to peculiar layered structure, and gas sensing mechanism is occurred by reaction of gas molecules with lattice oxygen atoms [4-5]. However, this type of structure play important role in the gas sensing mechanism that facilitate interaction with hydrogen gas molecules, thus surface-to-volume ratio can be considered one of the most important factors [1]. It was observed that after reduction by H_2, the surface concentration of the Mo^{6+} is drastically decreased, by reduction to Mo^{5+} and Mo^{4+} [14]. Raman spectroscopy studies revealed that H^+ ions mainly interact with the doubly coordinated oxygen atoms, that lead to formation of hydrogen molybdenum bronze (H_xMoO_3) and substoichiometric MoO_3 (MoO_{3-x}) [1]. Fig. 4 show illustration of hydrogen gas sensing mechanism based on generation of oxygen vacancies in MoO_3. Overall process of hydrogen gas molecules interaction with α-MoO_3 can be described by the following equations [1,4,6]:

$$H_{2(g)} \rightarrow H_{2(ad)} \rightarrow H_{(ad)} + H_{(ad)} \tag{1}$$

$$H_{(ad)} \rightarrow H^+_{(ad)} + e^- \tag{2}$$

$$MoO_3 + xH^+ + xe^- \leftrightarrow H_xMoO_3 \tag{3}$$

$$2H_xMoO_3 \leftrightarrow xH_2O + 2MoO_{3-x/2} \tag{4}$$

where eq. (1) and (2) represent adsorption and dissociation of hydrogen gas molecules on surface of α-MoO_3 due to presence of Pd nanoparticles, and resulted H^+ ions interact with oxygen atoms from the lattice by eq. (3), that lead to theoretical formation of OH_2 groups and MoO_{3-x} structures (eq. (4)) [1]. High surface-to-volume ratio of the α-MoO_3 belt leads to more Pd catalytic surface sites for H_2 molecules dissociation [1]. As can be seen, oxygen vacancies (V_O) are created by formation of H_2O molecules, which generate free electrons, which lead to change in electrical conductivity of the sensor structure (Fig. 4 (a)). When nanograined α-MoO_3 belt is exposed to air, the oxygen molecules (O_2) are dissociated by Pd nanoparticles, leading to oxygen atoms (O) generation, that are transferred to MoO_3 surface for recombination with V_O (Fig. 4 (b)) [4]:

$$(x/4)O_2 + MoO_{3-x/2} \leftrightarrow MoO_3 \tag{5}$$

Incomplete recovery of the sensor signal was observed by many groups [4], and can be attributed to accumulation of H_2O molecules on surface of α-MoO_3 that decrease concentration of available surface sites for catalytic dissociation of O_2 [1,4]. Slow recovery time can be attributed to slower diffusion rate of O atoms in comparison with H^+ ions [1,4].

More comprehensive investigations on the influence of doping with K^+ and the methods of nanostructuring by $PdCl_2$ on morphology of α-MoO_3 belt will be reported in further works. However, we believe that increase of gas response in case of doped α-MoO_3 belt is due to decrease of electrons concentrations by generating of holes. As a result, the response value of the sensor structure in air is decreasing. According to the formula for gas response ($S = I_{gas}/I_{air}$), due to decrease of I_{air}, the gas response (S) can be enhanced [15].

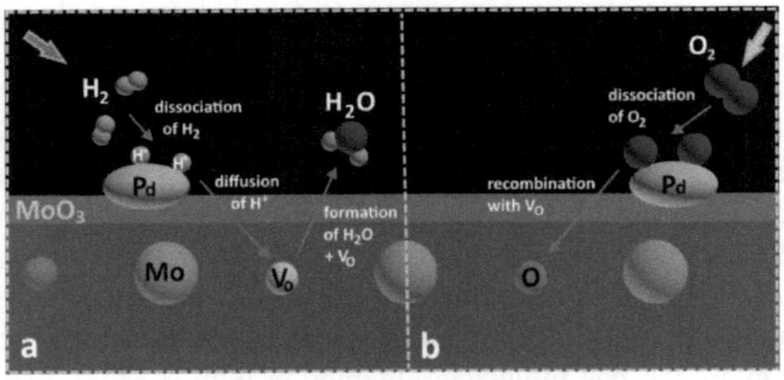

Fig. 4 Illustration of hydrogen sensing mechanism based on generation of oxygen vacancies: (a) under exposure to hydrogen gas; (b) underexposure to air that contain oxygen molecules

IV. CONCLUSIONS

In this work, nanograined Pd/α-MoO_3 micro-belts were integrated in the sensor structures for investigations of its hydrogen gas response performances. Significant increase in the surface-to-volume ratio of the α-MoO_3 belts with length up to several millimeters was performed by nano-structurization with $PdCl_2$ aqueous solution. At operating temperature of 150 °C – 200 °C, the prepared sensor structures demonstrated a gas response of about 3. Further increase in the gas sensor performances was performed by doping α-MoO_3 belts with K^+. Thus, gas response was increased to value of about 7.6 and response time was reduced to 2.4 s.

ACKNOWLEDGMENT

This work was financially partly supported by the STCU under the Grant no. 5989. Professor Adelung is acknowledged for his collaborative support of the Project 5989.

CONFLICT OF INTEREST

The authors declare that they have no conflict of interest.

REFERENCES

1. Z. H. Ou et al (2011) In Situ Raman Spectroscopy of H_2 Gas Interaction with Layered MoO_3. J. Phys. Chem. C 115(21):10757-10763 DOI 10.1021/jp202123a
2. O. Lupan et al (2014) Versatile Growth of Freestanding Orthorhombic α-Molybdenum Trioxide Nano- and Microstructures by Rapid Thermal Processing for Gas Nanosensors. J. Phys. Chem. C 118(27):15068-15078 DOI 10.1021/jp5038415
3. M. B. Rahmani et al (2010) Gas sensing properties of thermally evaporated lamellar MoO_3. Sensors and Actuators B 145:13-19 DOI 10.1016/j.snb.2009.11.007
4. Manal M. Y. A. Alsaif et al (2014) Two dimensional α-MoO_3 nanoflakes obtained using solvent-assisted grinding and sonication method: Application for H_2 gas sensing. Sensors and Actuators B 192:196-204 DOI 10.1016/j.snb.2013.10.107
5. N. Illyaskutty et al (2013) Enhanced ethanol sensing response from nanostructured MoO_3:ZnO thin films and their mechanism of sensing. J. Mater. Chem. C 1:3976-3984 DOI 10.1039/C3TC30408F
6. N. Illyaskutty et al (2013) Hydrogen and ethanol sensing properties of molybdenum oxide nanorods based thin films: Effect of electrode metallization and humid ambience. Sensors and Actuators B 187: 611-621 DOI 10.1016/j.snb.2013.05.092
7. S. Yang et al (2015) Highly Responsive Room-Temperature Hydrogen Sensing of α-MoO_3Nanoribbon Membranes. ACS Appl. Mater. Interfaces 7(17):9247-9253 DOI 10.1021/acsami.5b01858
8. C. Imawan et al (2001) Structural and gas-sensing properties of V_2O_5-MoO_3 thin films for H_2 detection. Sensors and Actuators B 77:346-351
9. X. Wen et al (2014) Piezoresistive effect in MoO_3 nanobelts and its application in strain-enhanced oxygen sensors. Nano research 7(2):180-189 DOI 10.1007/s12274-013-0385-8
10. T. H. Fleisch et al (1982) An XPS study of the UV reduction and photochromism of MoO_3 and WO_3. J. Chem. Phys. 76:780 DOI 10.1063/1.443047
11. O. Lupan et al (2014) Investigation of optical properties and electronic transitions in bulk and nano-microribbons of molybdenum trioxide. J. Phys. D: Appl. Phys. 47: 085302 DOI 10.1088/0022-3727/47/8/085302
12. S. Park et al (2012) Investigation of optical properties and electronic transitions in bulk and nano-microribbons of molybdenum trioxide. ACS Appl. Mater. Interfaces. 4(7): 3650–3656 DOI 10.1021/am300741r
13. O. Lupan et al (2009) A single ZnO tetrapod-based sensor. Sensors and Actuators B 141:511-517 DOI 10.1016/j.snb.2009.07.011
14. J.-G. Choi et al (1996) XPS study of as-prepared and reduced molybdenum oxides. Appl. Surf. Sci. 93:143-149DOI 10.1016/0169-4332(95)00317-7
15. C. W. Na et al (2011) Selective detection of NO_2 and C_2H_5OH using a Co_3O_4-decorated ZnO nanowire network sensor. Chem. Commun. 47:5148-5150 DOI 10.1039/c0cc05256f

Author: Lupan Oleg
Institute: University of Kiel, Technical University of Moldova
Street: Kaiser Str. 2
City: Kiel
Country: Germany
Email: ollu@tf.uni-kiel.de

Hypothermia Device Used in Medicine

V.P. Cojocaru, P.S. Tugui, T. Fedorisin, I.V. Postica, and R. Galus

D. Ghitu Institute of the Electronic Engineering and Nanotechnologies, Academy of Sciences of Moldova, Chisinau, Moldova

Abstract— **In this work we describe the construction and operation of a device intended to create the effect of hypothermia in the skull. This device is controlled through fuzzy algorithms for lowering or maintaining present temperature. The device is intended for use in emergency conditions, easily transportable for patients who have suffered ischemic attack.**

Keywords— **Peltier element, fuzzy, hypotermia, cooling system thermoelectrical effect.**

I. INTRODUCTION

Normal human body temperature varies within 36.1 - 37.2 °C. Values greater than 37.2 °C lead to hyperthermia, lower temperatures produce state of hypothermia. The hypothermic process began to be studied by empirical method, after cases of survival of people caught in avalanches of snow that could be brought to life despite extreme conditions. Deepening research and improving methods of hypothermia allowed it to become an effective method in modern treatments of anesthesia and intensive care.

Following research have established four categories of hypothermia.

- moderate (32-35 °C),
- medium (25-31 °C),
- deep (8-24 °C)
- freezing (0-8 °C)

After time length during hypothermic therapy distinguish: acute treatment (several hours), long (several hours), chronic (days or weeks). [1]

Methods for the production of hypothermia are invasive and non invasive .The treatment is the introduction into the bloodstream of a crystalloid fluid having a temperature of 4 °C for 30 minutes to form a closed circuit to lower the overall temperature of the body. Noninvasive method uses external sources of cooling, which are applied on the body where cooling aims. Cooling source can be: refrigerants, ice packs, cooling beds, but effective methods involve some inconveniences: Special conditions of use difficult to transport or intractable, environmentally harmful or difficult directing temperature. [2]

Peltier elements are used also as a small time noninvasive method that enable cooling while controlling temperature, until that can be used as heating source to restore after hypothermic treatment. [3]

The advantages of non-invasive methods (is that does not involve medication to which the patient may have adverse consequences, allergic states or immunity to drugs.

Also hypothermic treatments that are not applied correctly can occasionally cause thermal shock or tissue damage.

Therapeutic hypothermia indications are varied but ultimately addresses the same purpose, namely to protect the brain under conditions of neuronal aggression from various causes:

- Stroke Ischemic Burden
- Severe head trauma
- Spinal cord injury
- Status post cardiac resuscitation
- Facilitation of liver transplantation in patients with intracranial hypertension syndrome

For cases mentioned above the patient seeks treatment in the first hour after sustaining an accident. Note, however, that even after a few hours hypothermia started, it is still useful because it prevents reperfusion injury and counteract fever, detrimental factor head pathology.

The importance of creating and operating this device can be proved by statistics conducted by physicians showing that the use of hypothermia therapy reduces the mortality rate from 46.51% to 25.58%. Much of hypothermic treatment is applied to strokes, 70% of survivors restored and become infirm 15-30%. [4-6] It was noted that approximately 20% of patients with acute liver failure who are awaiting a liver transplant die of uncontrollable intracranial hypertension. [5-8] Further study author notes that the use of moderate hypothermia (32 °C) reduces intracranial pressure and allows patients to bear liver transplant.

II. EXPERIMENT

The Peltier effect is a thermoelectric phenomenon in which the heat is generated or absorbed at intersection of two conductive materials with different properties when there is a flowing electrical current. The elements can be connected in two modes: cooling and heating, difference is in connection, change wire between them. Cooling power is proportional to applied current, and becomes all the more effective depending how much heat is evacuated from the module's surface otherwise the accumulated heat returns to the cold side and performance of the cooling process significantly decreases [9-11].

V. Sontea and I. Tiginyanu (eds.), *3rd International Conference on Nanotechnologies and Biomedical Engineering*,
IFMBE Proceedings 55,
DOI: 10.1007/978-981-287-736-9_88

The advantage of Peltier modules consist of getting low values of temperature in a short time. Studying this criterion we conducted a dependence of time till temperature reaches 10 °C as a function of current. The measured temperature of the elements was equal to room temperature of 23 °C and it was not influenced by any heat source. It was observed that the current of 1000mA lowers cooling rate and the value of 10 °C could not be reached. With the current increase, cooling time decreases to 8 s as shown in the figure below[2].

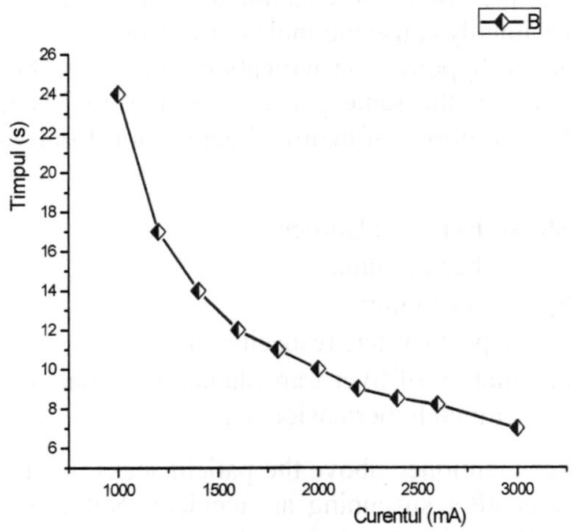

Fig.1 The time taken to reach 10 °C

In our device we use 4 items placed on top of the head to be individually controlled by the electronic system that will function as fuzzy algorithms developed in what follows.

III. Fuzzy Algorithm

A. *Fuzzifier*

Fuzzification at the introduction of fuzziness is the process of finding a function of fuzzy sets based on the stand and input data.

Defuzzification is the process of producing a quantifiable result in fuzzy logic given fuzzy sets and corresponding membership degrees. It is typically needed in fuzzy control systems. These will have a number of rules that transform a number of variables into a fuzzy result, that is, the result is described in terms of membership in fuzzy set. Defuzzification is interpreting the membership degrees of the fuzzy sets into a specific decision or real value.

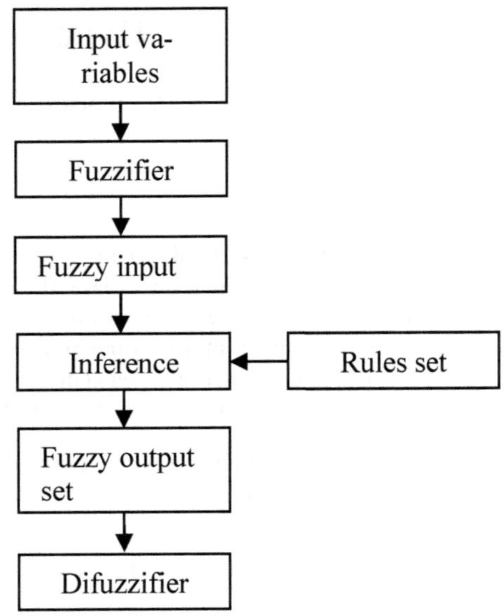

Fig. 2 Fuzzy algorithm

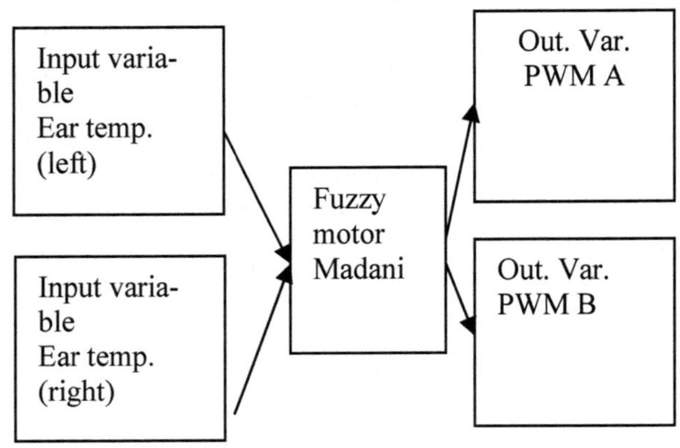

Fig. 3 Input and output variables

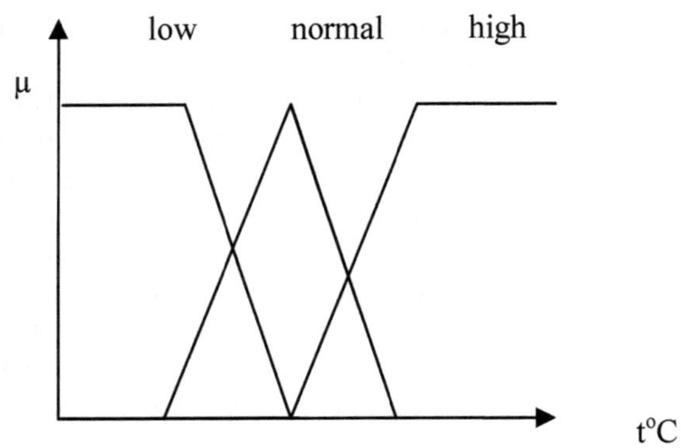

Fig. 4 Input variables Ear temp. (right

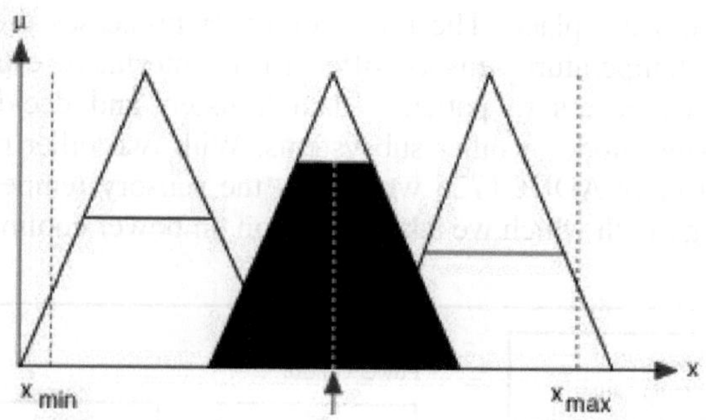

Fig. 5 Difuzzifier

Control signals should have a low frequency because the cooling effect fails to occur at fast switching, so we watch as their frequency to be low.

B. Rules Set

Set of rules to conduct Peltier elements is:

1. *If "Ear temp. (left)" is "low" and "Ear temp. (right)" is "low" then "PWM A" is "small" and PWM B" is "small".*
2. *"If "Ear temp. (left)" is "low" and Ear temp. (right)" is "normal" then "PWM A" is "small" and PWM B" is "normal*
3. *"If "Ear temp. (left)" is "low" and Ear temp. (right)" is "high" then "PWM A" is "small" and PWM B" is ""high".*
4. *If "Ear temp. (left)" is "normal" and "Ear temp. (right)" is "low" then "PWM A" is "normal" and PWM B" is "small".*
5. *If "Ear temp. (left)" is "normal" and "Ear temp. (right)" is "normal" then "PWM A" is "normal" and PWM B" is "normal".*
6. *If "Ear temp. (left)" is "normal" and "Ear temp. (right)" is "high" then "PWM A" is "normal" and PWM B" is "high".*
7. *If "Ear temp. (left)" is "high" and "Ear temp. (right)" is "low" then "PWM A" is "high" and PWM B" is "small".*
8. *If "Ear temp. (left)" is "high" and "Ear temp. (right)" is "normal" then "PWM A" is "high" and PWM B" is "normal".*
9. *If "Ear temp. (left)" is "high" and "Ear temp. (right)" is "high" then "PWM A" is "high" and PWM B" is "high".*

Fuzzy Logic algorithms, fuzzy systems are considered a special case of expert systems (which is why expert systems are also called fuzzy) systems that provide a flexible method for the treatment of uncertainty.

Programming fuzzy control method is very practical in that they are used linguistic variables, which are part of the real language, not technical. It is the same as communication between people. They have their own significance eg low temperature, high temperature, average temperature. By this method it can be easily programmed algorithms, more easily understood and modified [12-14].

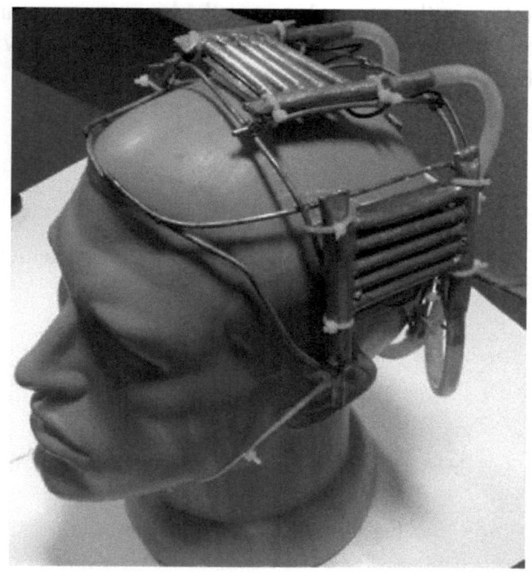

Fig. 6 Experimental setup

IV. CONSTRUCTION

Peltier modules operates as a heat pump, this heat accumulates on the external side. Heat elimination is done by copper heat sink, through which circulate water. The flow of water through the radiators form a closed circuit powered by an electric pump. Cooling efficiency is directly influenced by water temperature in circuit, at a lower water temperature, heat absorption is made with a high performance.

Because the helmet is designed for different patients, it will be adjusted depending on the size of the head. Arrange items in housing depends heavily on areas that will transfer low temperature without great losses to the brain and do not meet highly vascular areas that acts as a thermal insulator (see figure 3). Temporal and occipital areas are heavily vascularized so we will not succeed cooling these parts, but parietal and frontal area are available for cooling. It is known that the rate of about 50% of the excess heat generated in the brain is removed due to cooler blood flow in the carotid arteries. Another 50% of the heat is removed due to cooling of the blood flow to the scalp.

Because we want a cooling as quickly as possible, we will drop temperature value till 5 °C and we will consider it a minimum value. Temperatures lower than the amount specified above can damage tissue. The process of lowering temperature will be gradual and with duration of 15 minutes to not cause thermal shock. To measure brain temperature more precisely we need to install resistive thermo sensors in an area as little affected by the external environment. The temperature in the ear is closest to the brain temperature.

In the picture below is a 3D projection model previously described.

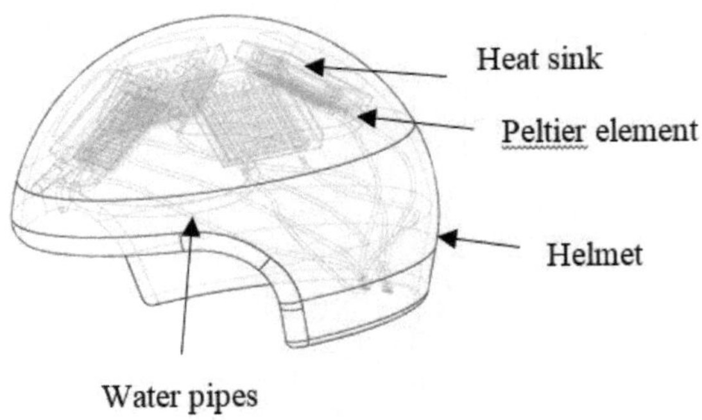

Fig. 7 Block diagram of system

V. CIRCUITS

Control of Peltier elements is achieved with a power block. It is made of semiconductor components that handle high currents. According to catalog TEC12706 maximum consumption is 6.4 A. The usage of elements at maximum power increases degradation. The output transistor of the power module holds a current of 20 A. The controlling signal transmitted from the controller is low frequency <100 Hz. We will not use PWM mode since switching speed is high and cooling phenomenon does not occur at the expected level. Peltier elements are individually controlled which allows localized cooling.

The system is built hierarchically, based on the LPC1768 microcontroller. The advantage of this choice is the reason it is an ARM architecture with advanced capabilities such as 512 kB of flash memory, up to 64 kB of on memory, Ethernet MAC, USB Device / Host / OTG interface, 8-channel DMA controller general purpose 4 UART, I2C-bus interfaces 3, 2-input 2-output addition I2S bus interface, 8-channel 12-bit ADC, 10-bit DAC, PWM motor control, Quadrature Encoder interface, 4 general purpose timers, 6-general purpose PWM output, ultra-low power Real-Time Clock (RTC) with separate battery supply, and up to 70 general purpose I / O pins, which would allow us the expansion of the project on a larger number of modules controlled in the same place. The microcontroller processes the data from temperature sensors, offer a user interfaces, establish internet access to patient's data transfer, and decide the working mode of other subsystems. With two other micro-controllers ADUC1726 we follow the sensory temperature change with which we take a decision on power cooling.

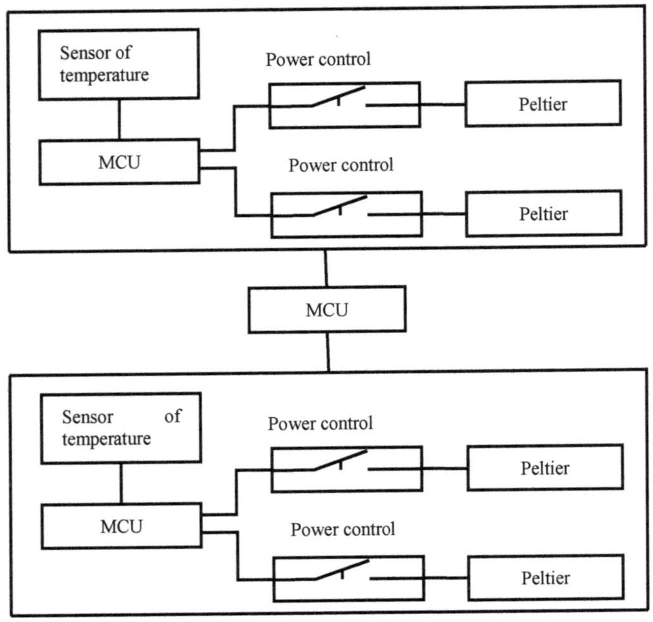

Fig.8 Block diagram of system

We chose this model given the number of ADC channels which could allow us to expand the number of sensors. The figure below provides one of the two symmetrical parts of the circuit with sensory interface and serial interface for connection to the computer.

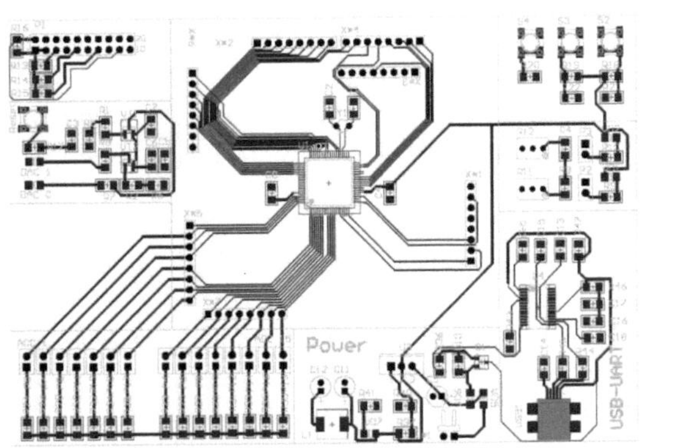

Fig.9 Sensory interface PCB

The device has a simple user interface for avoidance of unnecessary special instructing and risk of inducing the device in an inactivity state, which could be fatal for some patients. Also you can create a database in which to enroll patients list, medical conditions, treatment methods, recommendations however with the possibility of remote access via computer and the Internet. By this means the doctor can monitor the patient's condition and behavior of therapy remotely through access to appropriate web page.

For local communication between the helmet and the user it is used USB serial interface and a graphical interface through which items can be controlled individually for local cooling [14-16].

The operation will be applied in treatment in several modes such as lowering the temperature, temperature increase, or maintain at a constant level.

VI. CONCLUSION

The authors have built a sample device for therapeutic hypothermia used in medical needs for a controlled cooling of some tissues using Peltier elements

This helps to prevent ischemic attacks or if in case it already happened, helps stop bleeding and mobilization of the patient's condition during transportation to the hospital or as a treatment method. It can be used in mobile crews or in hospital wards providing remote data receiving from the patient. It's a small cooling device with no moving parts and selectively control the temperature.

CONFLICT OF INTEREST

The authors declare that they have no conflict of interest.

REFERENCES

1. Shigeo Yamaguchi , Hideyuki Homma, „ Fabrication of a unipolar Peltier device using a pair of N-type thermoelectric materials" Department of Electrical, Electronic, and Information Engineering, Kanagawa University, 3-27-1 Rokkakubashi, Kanagawa-ku, 221-8686 Yokohama, Japan Volume 129, 5 November 2014, Pages 77–80.
2. Victor COJOCARU,Vladimir MARDARI, *„Fuzzy controlled system for hypothermic brain therapy"* Proceedings of the Romanian Academy-series A : Mathematics, Physics, Technical Sciences, Volume 15, No. 4, pp. 396-402, Bucharest, 2014.
3. Victor Cojocaru, Anatolii Sidorenko, *„Device for hypothermic therapy based on fuzzy logic rules"* D. Ghitu Institute of the Electronic Engineering and Nanotechnologies of the Academy of Sciences of Moldova, Chisinau 2014.
4. J.L. Pérez–Aparicioa, R. Palmaa, R.L. Taylorb, *"Finite element analysis and material sensitivity of Peltier thermoelectric cells coolers"*, Volume 55, Issue 4, 31 January 2012, Pages 1363–1374.
5. R.P Huebenera, C.C Tsueib, *"Prospects for Peltier cooling of superconducting electronics"*, Volume 38, Issue 3, March 1998, Pages 325–328.
6. Dwain J. Reed,Alan D. Miller, *"Thermoelectric peltier device for local cortical cooling"*, Volume 20, Issue 2, February 1978, Pages 209–211.
7. Kei Morizanea, Tadanori Ogataa, Tadao Morinoa, Hideki Horiuchia, Gotaro Yamaokaa, Masayuki Hinob, Hiromasa Miurab, *A novel thermoelectric cooling device using Peltier modules for inducing local hypothermia of the spinal cord: The effect of local electrically controlled cooling for the treatment of spinal cord injuries in conscious rats*, Volume 72, Issue 3, March 2012, Pages 279–282.
8. Gianluca A. Mannella, Vincenzo La Carrubba,Valerio Brucato, Peltier cells as temperature control elements: *"Experimental characterization and modeling"* , Volume 63, Issue 1, 5 February 2014, Pages 234–245.
9. José Meseguer, Isabel Pérez-Grande, Angel Sanz-Andrés, *"Thermoelectric cooling, Spacecraft Thermal Control"*, 2012, Pages 263–273.
10. G. Casano, S. Piva, *"Experimental investigation of the performance of a thermoelectric generator based on Peltier cells"*, Experimental Thermal and Fluid Science, Volume 35, Issue 4, May 2011, Pages 660–669.
11. A. Martínez, D. Astrain, A. Rodríguez, *"Dynamic model for simulation of thermoelectric self cooling applications"*, Volume 55, 15 June 2013, Pages 1114–1126.
12. Horia-Nicolai L. Teodorescu, Abraham Kandel, Lawrence O. Hall, *"Report of research activities in fuzzy AI and medicine at USF CSE."*Artificial Intelligence in Medicine, Volume 21, Issues 1–3, January–March 2001, Pages 177–183.
13. Teodorescu H.N., *On fuzzy sequences, fixed points and periodicity in iterated fuzzy maps.* International Journal of Computers, Communications & Control, Vol. 6, Issue 4, pp. 752-763, Dec 2011.
14. TEODORESCU, H.N., On the characteristic functions of fuzzy systems. International Journal of Computers, Communications & Control, Vol. 8 Issue 3, pp. 469-476, Jun 2013.
15. Teodorescu, H.N., Cojocaru, V., *Three-transistor modulator-amplifier circuit works with swept-control frequencies*, Electronic Design (EDN), Vol. 55, Issue: 8, pp.: 72-+, Apr 22, 2010.
16. Teodorescu, H.N., Cojocaru, V. Experimental investigation of the reliability of reception of ultrasound signals in fire conditions. Fire Safety Journal, 2014, Vol. 66, p. 25-34

Author: Cojocaru Victor
Institute: Academy of Science of Moldova,
D. Ghitu Institute of the
Electronic Engineering and Nanotechnologies
Street: Academiei str., 3/3
City: Chisinau
Country: Moldova
Email: vcojocaru@nano.asm.ms

Accuracy Analysis of Measurements in Electrochemical Biosensing

P. Makiewicz[1], D. Matias[1], M. Jaskuła[1], M. Biegun[1], K. Penkala[1], E. Mijowska[2], M. El Fray[3], and J. Podolski[4]

[1] West Pomeranian University of Technology, Department of Systems, Signals and Electronics Engineering, Szczecin, Poland
[2] West Pomeranian University of Technology, Institute of Chemical and Environment Engineering, Department of Nanotechnology, Szczecin, Poland
[3] West Pomeranian University of Technology, Department of Biomaterials and Microbiological Technologies,Szczecin, Poland
[4] NZOZ „Meditest. Diagnostyka Medyczna" Jacek Podolski, Szczecin, Poland

Abstract— **In the paper the results of research on accuracy analysis of measurements in electrochemical biosensing are presented. The scope of performed tests includes EIS (Electrochemical Impedance Spectroscopy) and chronoamperometry, widely used for determination of glucose concentration. Measurement results obtained with two microcontroller systems, designed within a research project, and one reference platform (Autolab system, Metrohm, Netherlands) are compared and discussed. As a result, optimal design solutions may be chosen for microcontroller-based biomeasurement system. In glucosensing, promising results were obtained with the EIS technique in experiments performed using the Autolab system.**

Keywords— **Biosensing, electrochemical biosensing methods, accuracy analysis, impedance scanner, glucometry.**

I. INTRODUCTION

Accuracy analysis is an essential part of research on development of electronic control and measurement systems designed for biosensing. Unfortunately, in current literature this issue is not enough represented. Thus, selection of optimal design solutions is intuitive, and based on the designer's experience. The aim of the presented study was to compare measurements obtained with two currently being developed microcontroller systems, dedicated to different biosensing techniques using a novel graphene-based electrochemical sensor. This analysis should allow choosing design solutions, providing the best accuracy, quality to cost ratio, multifunctionality, and also optimizing measurement system performance.

II. MATERIAL AND METHODS

Performed analysis concerns two electrochemical biosensing methods: EIS (Electrochemical Impedance Spectroscopy) and chronoamperometry [1,2,4,6,7].

Two currently being developed microcontroller systems were compared. First of the two tested platforms is based on the AD5933 System-on-a-Chip device, which is a complete impedance scanner manufactured by Analog Devices. The analog front-end is driven by the board's controller, ATXmega128A1U which is a fast, compact 8 bit microcontroller [3,5].

Second system is based on fast microprocessor STM32F407, which is able to calculate impedance and send data to PC in real time. Data acquisition level includes DDS generator as a source of sinusoidal signal based on AD9850 integrated circuit.

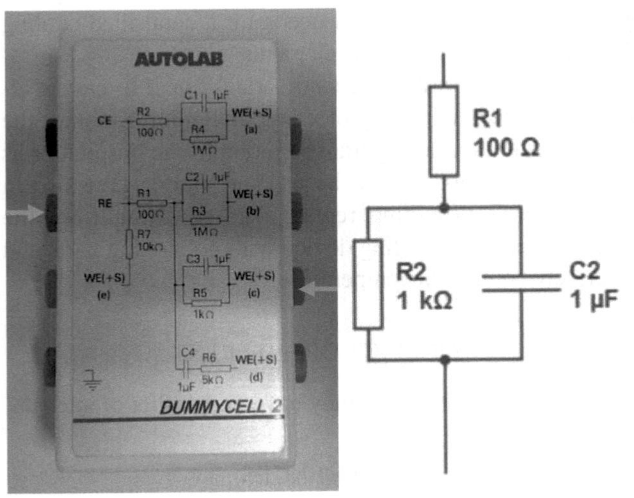

Fig. 1 Dummy Cell used for measurements (outside view, scheme)

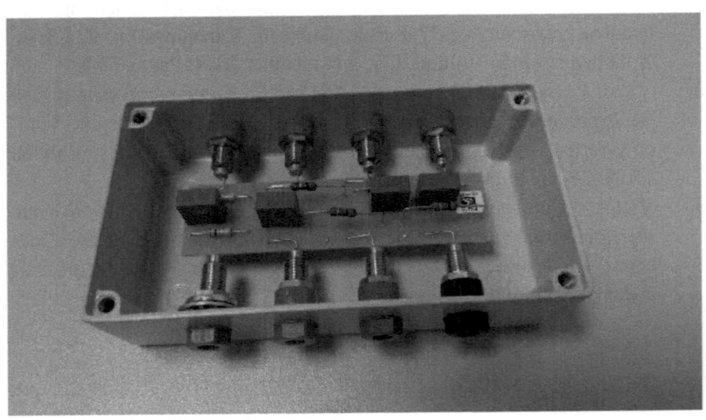

Fig. 2 Dummy Cell used for measurements (inside view)

In the EIS technique, Autolab system (Metrohm, Netherlands) was used as a reference platform. Original Autolab

V. Sontea and I. Tiginyanu (eds.), *3rd International Conference on Nanotechnologies and Biomedical Engineering*,
IFMBE Proceedings 55,
DOI: 10.1007/978-981-287-736-9_89

Dummy Cell was used as a model for electrical properties of fluids appearing in biosensing measurements. A scheme (Figure 1) and inside construction (Figure 2) of the Dummy Cell are shown on the pictures.

Two-electrode setup was used: CE – counter electrode closed with RE – reference electrode, and WE – working electrode.

Chronoamperometry was performed with the use of Autolab system and test strips from a commercially available glucometer (Gold Diagnostic). As the test samples, solutions of glucose in Phosphate-buffered saline - PBS were used (PBS is a buffer solution, which imitates osmolarity and ion concentrations of the human body).

III. Results, Discussion

A. Electrochemical Impedance Spectroscopy

Firstly, impedance measurements were performed using a model (Dummy Cell) with both tested platforms and the reference system. The results are shown on Figures 3-4.

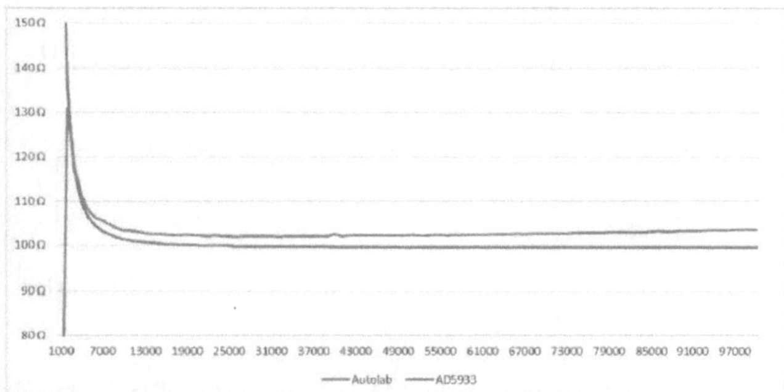

Fig. 3 Measurements performed with AD5933 system (impedance module as a function of frequency)

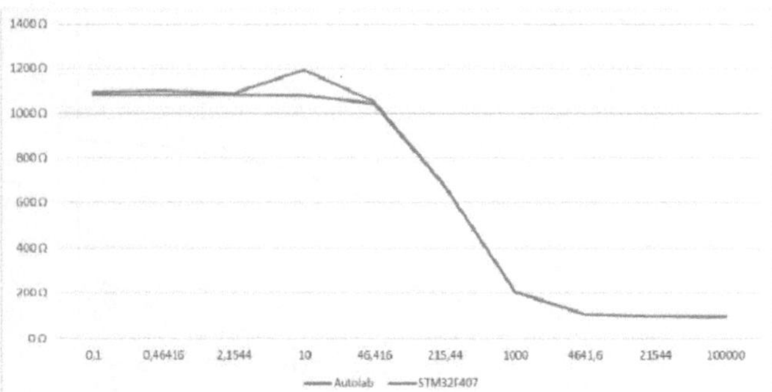

Fig. 4 Measurements performed with STM32F407 system (impedance module as a function of frequency)

Next Figures (5-6) represent impedance Nyquist plots.

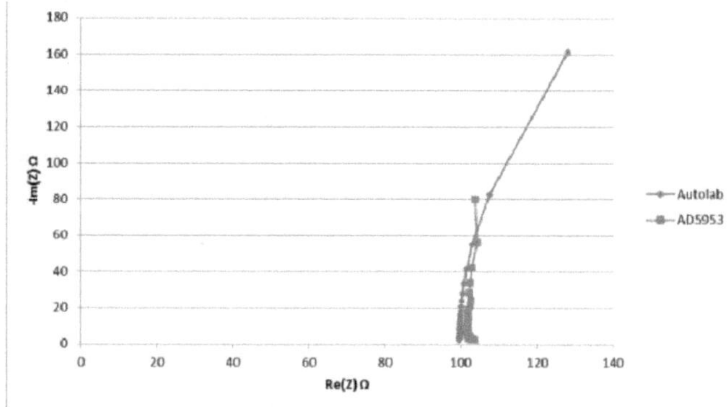

Fig. 5 Measurements performed with AD5933 system (Nyquist plot, frequency raises towards left, range: 1 kHz – 100 kHz)

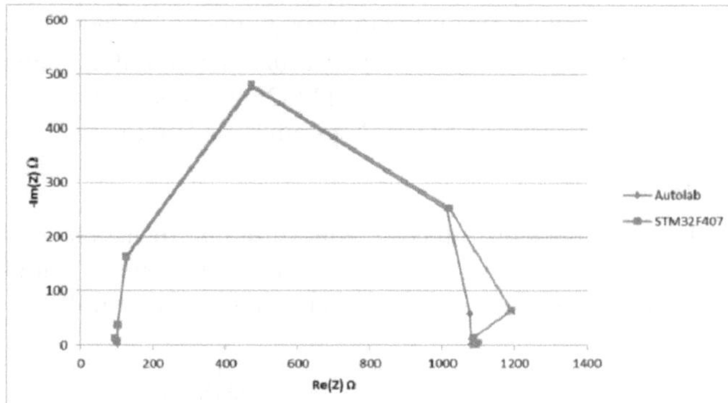

Fig. 6 Measurements performed with STM32F407 system (Nyquist plot, frequency raises towards left, range: 0,1 Hz – 100 kHz)

Then, relative errors of the measurement results were calculated. Frequency range presented on Figure 7 is equal to 2 kHz – 100 kHz. The result for 1 kHz frequency was excluded from the plot due to very high error. On both Figures 7-8, error raises for frequencies around 100 kHz.

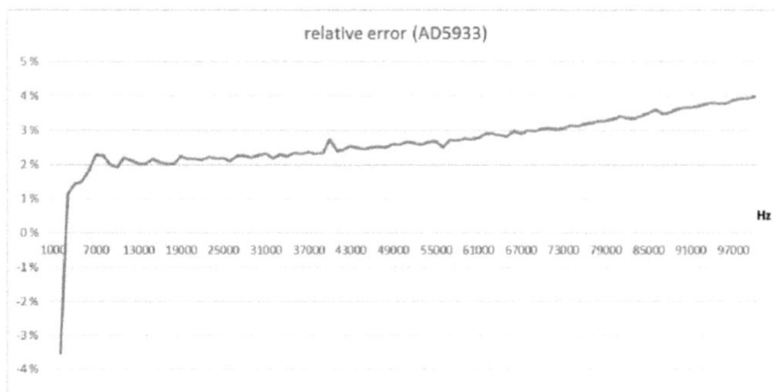

Fig. 7 Measurements performed with AD5933 system (relative error as a function of frequency)

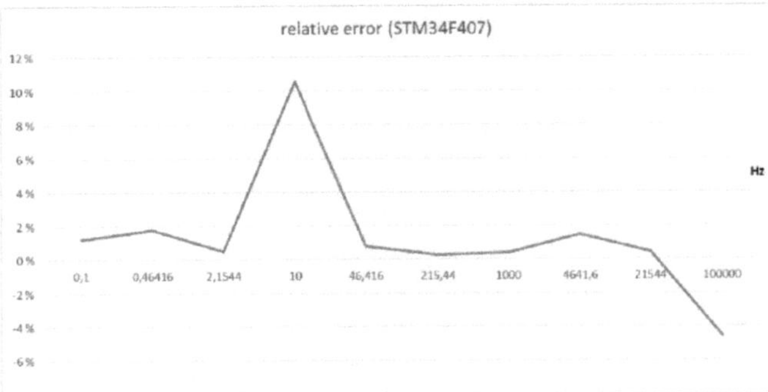

Fig. 8 Measurements performed with STM32F407 system (relative error as a function of frequency)

As seen on Figure 8, error of measurement performed with the STM32F407 system for 10 Hz frequency is disproportionately high (above 10%). It can be explained with some random mistake, which disturbed measurement for this single frequency. Generally, error of STM32F407 system was below 2%, while error of AD5933 system varied up to 4%.

B. Glucometry

Table 1 presents glucose concentration measurements performed with commercially available glucometer (Gold Diagnostic) for different concentrations of glucose in PBS.

Table 1 Glucose concentration measurements performed with commercially available glucometer (Gold Diagnostic)

glucose in PBS solution			Gold Diagnostic glucometer								
No.	glucose concetration [mmol/L]	glucose concetration [mg/dL]	measurement [mg/dL]				average value [mg/dL]	average value [mmol/L]	error [mg/dL]	error [mmol/L]	error [%]
1	2	36	12	16	16	21	16,25	0,90	-19,75	-1,10	-55%
2	4	72	69	80	79	78	76,5	4,25	4,50	0,25	6%
3	6	108	110	124	118	133	121,25	6,74	13,25	0,74	12%
4	8	144	151	165	162	165	160,75	8,93	16,75	0,93	12%
5	10	180	178	200	198	216	198	11,00	18,00	1,00	10%
6	12	216	215	221	235	248	229,75	12,76	13,75	0,76	6%
7	14	252	230	253	257	267	251,75	13,99	-0,25	-0,01	0%

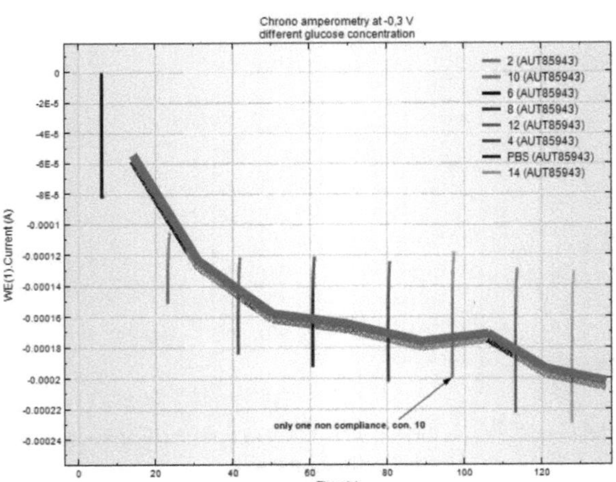

Fig. 9 Chronoamperometry performed with Autolab

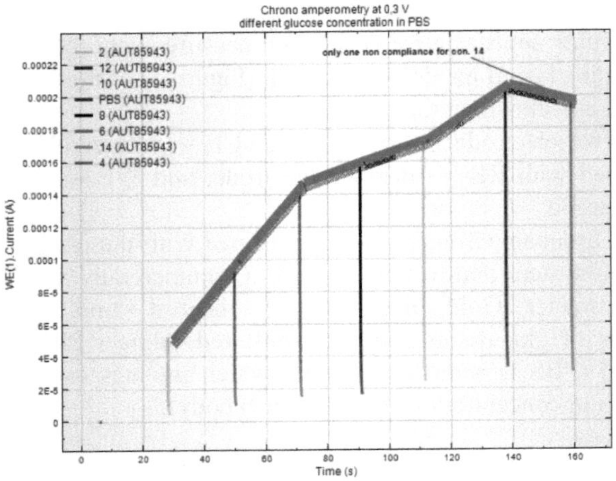

Fig. 10 Chronoamperometry performed with Autolab; reversed electrodes polarity

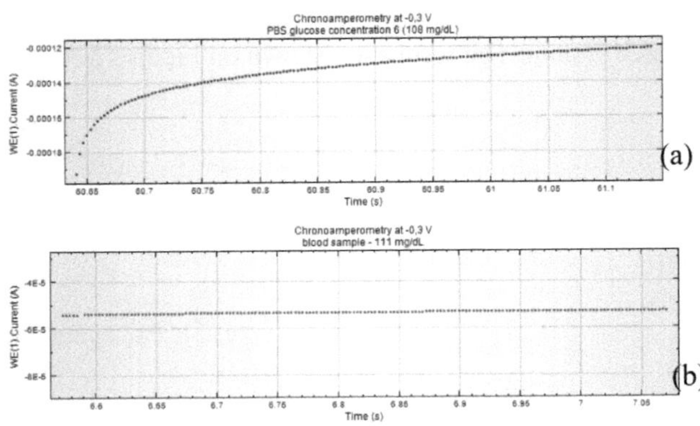

Fig. 11 Chronoamperometry performed with Autolab (a – PBS, b – blood)

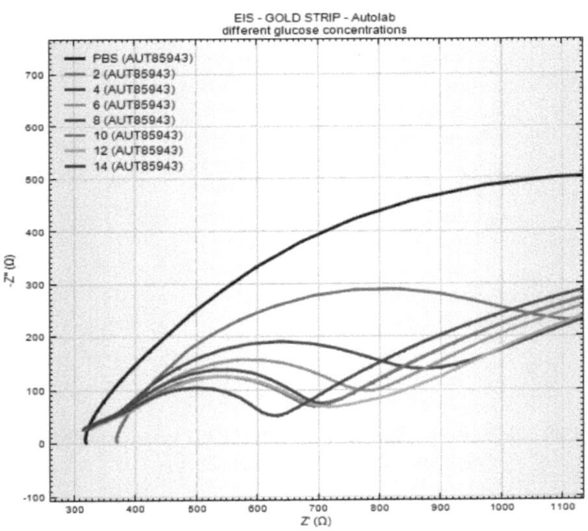

Fig. 12 EIS performed with Autolab – glucose in PBS (Nyquist plot, frequency raises towards left, range: 1 kHz – 100 kHz)

This same set of glucose solutions was measured with the Autolab system (Figure 9).

Additionally, the polarity of electrodes was reversed. This allowed getting higher current response (Figure 10).

Figure 11 shows chronoamperometry measurement for one sample solution (108 mg/dL glucose in PBS - a) with Autolab system. Electrochemical reaction reveals much higher dynamics in PBS than in blood (111 mg/dL - b).

Figure 12 presents EIS measurement for different glucose concentrations performed with the Autolab system. This technique is not widely used in glucometry. Clearly visible trend indicates that impedance analysis can be applied for *determination of glucose concentration in PBS.*

IV. CONCLUSIONS

The results of comparative impedance measurements justify the use of both platforms (based on AD5933 and STM32F407) in EIS research. However, the latter one provides clearly more accurate results as well as wider frequency range. Therefore, further development of multifunctional biosensing system using EIS technique should be based on STM32F407.

Chronoamperometric measurements performed with Autolab (reference platform) are consistent with the results obtained with commercially available Gold Diagnostic glucometer. However, PBS solution does not imitate blood well enough for the purposes of chronoamperometric measurement of the glucose concentration.

It appears also, that EIS technique may be useful in glucometry. For simple and fast glucosensing [3], the optimal frequency will be chosen in a separate study.

ACKNOWLEDGMENT

This project is financially supported by the National Centre for Research and Development within the framework of the GRAF-TECH Programme, grant No. GRAF-TECH/NCBR /08/06/2013.

CONFLICT OF INTEREST

The authors declare that they have no conflict of interest.

REFERENCES

1. Kyle U.G. et al.: Bioelectrical Impedance Analysis, part I: Review of Principles and Methods. Elsevier Ltd, 2004
2. Lewandowski D., Biedka A., Borkowska K., Penkala K.: Testy funkcjonalne dwóch wybranych urządzeń do analizy impedancji bioelektrycznej (BIA). Inżynieria Biomedyczna - Acta Bio-Optica et Informatica Medica, vol. 16, No. 2, 2010, pp. 83-87
3. Matias D., Przeniosło Ł., Miłosławski T., Jaskuła M., Makiewicz P., Biegun M., Penkala K., Mijowska E., El Fray M., Podolski J.: Development of the Mobile PoC Graphene-Based Biosensing Device Using Electrochemical Impedance Spectroscopy (EIS). Á. Jobbágy (ed.), First European Biomedical Engineering Conference for Young Investigators, IFMBE Proceedings vol. 50, DOI: 10.1007/978-981-287-573-0_5
4. Odenthal, Katherine J, Gooding, J Justin.: An introduction to electrochemical DNA biosensors., The Analyst, vol. 132, Issue 7, 2007
5. Przeniosło Ł., Walków M., Krzeszewska S., Pisarek S., Biedka A., Jaskuła M., Matias D., Penkala K.: Integrated impedance scanner in selected biomeasurement applications - control circuit. Krajowa Konferencja Studentów i Młodych Pracowników Nauki, Unieście, Poland, May 23, 2013, Zeszyty Naukowe Wydziału Elektroniki i Informatyki Politechniki Koszalińskiej, vol. V, pp. 63-69, Koszalin 2013, ISSN 1897-7421
6. Pumera M.: Graphene in biosensing, Materialstoday, vol. 14, no. 7-8, 2011
7. Yongchao Z., Yanique T., Eunkyoung Kim., Gregory F. Payne.: pH- and voltage-responsive chitosan hydrogel through covalent cross-linking with catechol. The journal of physical chemistry. B, vol. 116, Issue 5, 2012

Author: Przemysław Makiewicz
Institute: Department of Systems, Signals and Electronics Engineering, Faculty of Electrical Engineering, West Pomeranian University of Technology, Szczecin, Poland
Street: Sikorskiego 37
City: Szczecin, 70-313
Country: Poland
Email: makiewicz@zut.edu.pl

Electronic Circuits for Graphene-Based Biosensor

M. Raczyński[1,2], Ł. Przeniosło[1,2], M. Jaguszewski[1,2], E. Martinez Miguez[1,3], M. Jaskuła[1], D. Matias[1], A. Biedka[1], P. Makiewicz[1], M. Biegun[1], E. Mijowska[4], M. El Fray[5], J. Podolski[6], and K. Penkala[1]

[1] West Pomeranian University of Technology, Department of Systems, Signals and Electronics Engineering, Szczecin, Poland
[2] West Pomeranian University of Technology, Faculty of Electrical Engineering, Department of Systems, Signals and Electronics Engineering, Biomedical Engineering Student Research Group "AKSON",Szczecin, Poland
[3] Universidad Politécnica de Madrid, Madrid, Spain
[4] West Pomeranian University of Technology, Institute of Chemical and Environment Engineering, Department of Nanotechnology, Szczecin, Poland
[5] West Pomeranian University of Technology, Department of Biomaterials and Microbiological Technologies, Szczecin, Poland
[6] NZOZ "Meditest. Diagnostyka Medyczna" Jacek Podolski, Szczecin, Poland

Abstract— **In the paper the results of research on electronic measurement and control systems designed for a multifunctional graphene-based biosensor are presented. Design of different types of biosensing electrodes based on graphene technology is discussed. Development of electronic circuits for read-out of the biosensor signals based on two microcontroller platforms (ATXmega128A1U and STM32F407) is described, including hardware and software design solutions. Comparison of measurement results using both prototype systems in electrochemical biosensing techniques (EIS - Electrochemical Impedance Spectroscopy) is also presented.**

Keywords— **Biosensing, electrochemical biosensing methods, *Electrochemical Impedance Spectroscopy, EIS,* Bioimpedance, impedance scanner, biosensing electrode, microprocessor system, SoC, grapheme.**

I. INTRODUCTION

Various methods for electrochemical detection of specific particles in biological material (targets, including DNA of bacteria and viruses) are used [1,3,5,6]. Most common are Cyclic Voltammetry - CV, Differential Pulse Voltammetry - DPV, Square-Wave Voltammetry – SWV, and Electrochemical Impedance Spectroscopy – EIS. Field-Effect Transistor technology applied to biosensors (Bio-FET) may result in detecting even thousands of DNA targets at the same time on a small chip [5]. Several manufacturers offer biosensing equipment based on electrochemical techniques. However, these systems are extremely expensive and dedicated to operation in specialized labs, as stationary equipment, and cannot be used as portable Point-of-Care devices.

In the paper [2] published recently by our team running a research project on graphene-based biosensor for medical diagnostics, we described our work on development of a simple and relatively cheap mobile Point-of-Care version of the biosensor device based on the EIS technique. Bioimpedance (BI) measurements were performed using the integrated impedance scanner, and the 8-bit microprocessor as a control unit [4]. In the present study, current state is presented of our work on biosensor electrodes as well as electronic systems for reading the signal and for controlling measurement procedures - mainly based on more flexible, universal microcontroller platform.

II. DEVELOPEMENT OF BIOSENSING ELECTRODE

Construction of biosensor depends largely on the measurement technique that is used for detection. For accurate EIS, CV, DPV, SWV measurements, three electrode setup (WE – working electrode, CE – *counter electrode*, RE – reference electrode) is now a standard.

During the development of biosensors that will be sensitive and selective for effective electrochemical detection of DNA fragments of microorganisms (bacteria, viruses) and other biomolecules in biological material, four types of electrode were designed that can be divided into two design concepts.

The 1st design concept contained three different electrodes. One of them was especially designed for EIS (Fig. 1 a).

Second solution (Fig. 1 b) was based on Bio-FET technology. The idea was to use the thinnest available laminate (0,20 mm) to build a double-sided PCB electrode. Source (S) and drain (D) were manufactured in a special way to obtain good immobilization of DNA probes using the thiol linkages.

The third design solution (Fig. 1 c) includes the ability to operate in all electrochemical biosensing techniques. On the top side of the electrode two working electrodes are placed. One can be covered (functionalized) with biosensitive material - probes complementary with the target. The second can be used as a control sample (this working electrode is without any probes or enzymes).

The 2nd design concept was based on Bio-FET setup. Electrode from the 1st designed concept has been miniaturized. The base of the sensor is 300 nm SiO_2 layer on Si wafer. Silver drain and source were plotted using a microplotter (Fig. 2). Sensor size is 5x5 mm. Thanks to

V. Sontea and I. Tiginyanu (eds.), *3rd International Conference on Nanotechnologies and Biomedical Engineering*, IFMBE Proceedings 55,
DOI: 10.1007/978-981-287-736-9_90

miniaturization, less testing material will be needed for future experiments.

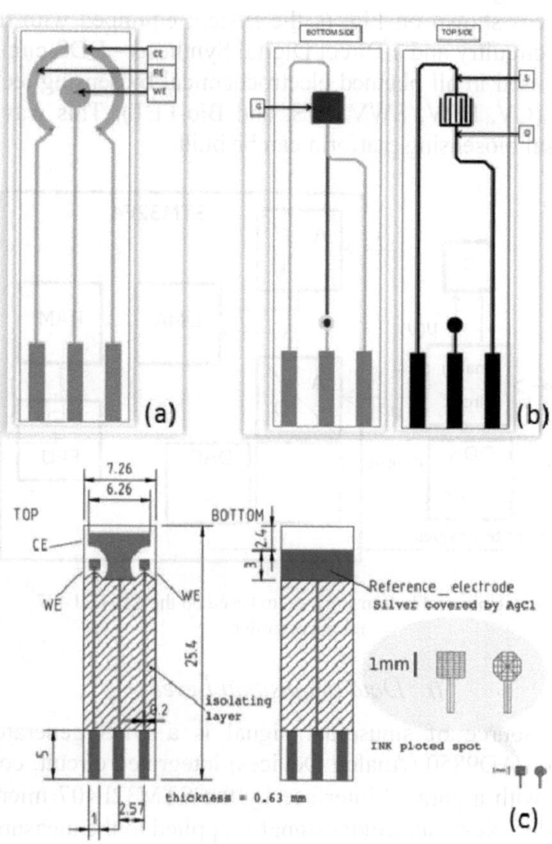

Fig. 1 Development of biosensing electrode – the 1st design concept (explanations in the text)

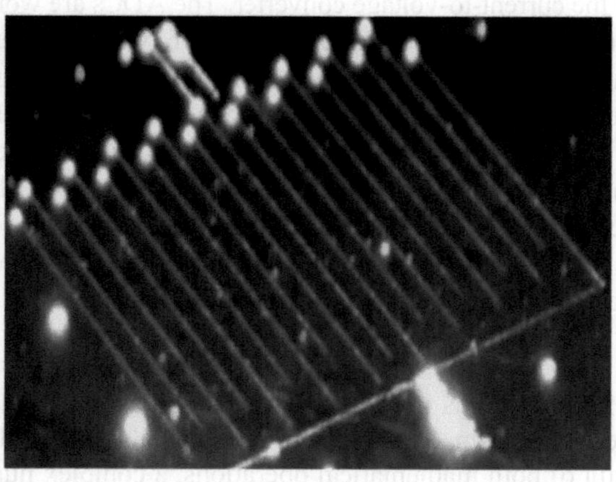

Fig. 2 The 2nd design concept of biosensing element (Bio-FET)

III. TWO CONCEPTS

A. *Using AD5933 Integrated Impedance Scanner*

Our first approach was to design a universal evaluation platform capable of measuring impedance in different electrode configurations, and with rich user interface. We built an instrument based on the AD5933 System-on-a-Chip (Fig. 3) which is a complete impedance scanner originally designed by the manufacturer (Analog Devices) to work in two-electrode configuration. The board contains a revised two-electrode circuitry, as well as extended four-electrode one. They can both work simultaneously as each one has its own AD5933. The analog front-end is driven by the ATXmega128A1U (Atmel) 8-bit microcontroller [2,5].

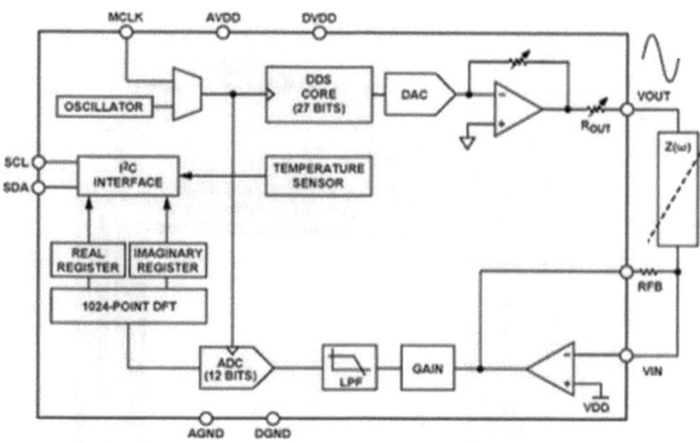

Fig. 3 AD5933 block schematic diagram

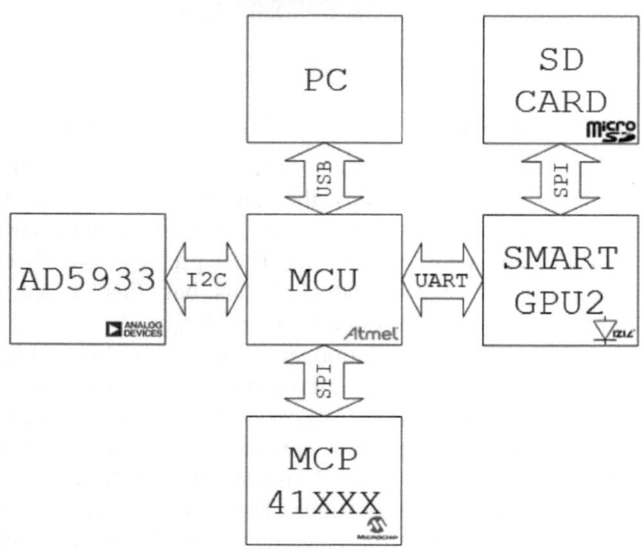

Fig. 4 Evaluation board - brief block schematic diagram

Digital potentiometers MCP41010 and MCP4100 (Fig. 4) are used to calibrate the analog circuitry using software. User can communicate with the device in two different ways. The first one, is to use Smart Gpu2, an integrated graphics processor located on a separate PCB board with LCD touch screen (Fig. 5). It is connected to the board directly and can be used to perform all measurements and display the results. The second way, is to connect the board to a PC using USB cable. The board will then appear on the PC as a portable storage device. The user can copy measurement data from the SD card and analyze it using proprietary software like MS Excel or similar. Firmware for the instrument is still under development, because we've been using it in various impedance measurements.

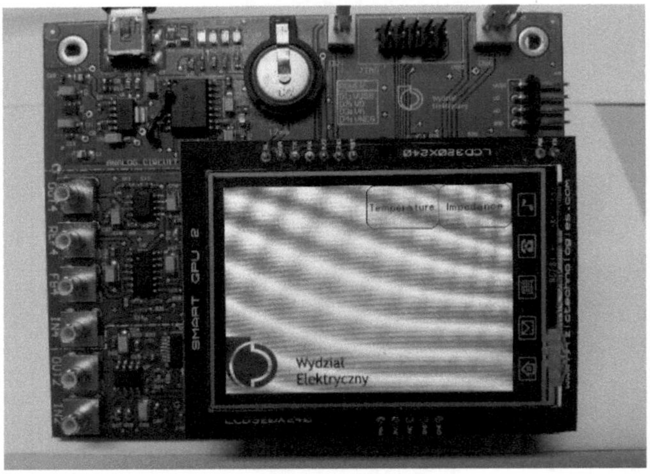

Fig. 5 Evaluation board

B. System based on 32-bit STM32F407 Microcontroller

Unfortunately, the system based on the AD5933 has some disadvantages (mainly, the range of frequency used in impedance scan is too narrow, the low limit for impedance measurement is not enough low). These problems have been solved using special additional circuits, but this has led to unwanted complications in electronics design.

As an alternative solution, we decided to choose high performance STM32F407 microcontroller (Fig. 6) which is based on a 32-bit ARM Cortex®-M4 with Floating Point Unit FPU core, 1 MB Flash memory, 192 KB RAM in an LQFP100 package. The DSP instructions and the FPU enlarge the range of applications. There are also two 12-bit DACs, three 12-bit ADCs reaching 2,4 MSPS or 7,2 MSPS in interleaved mode. Furthermore, STM32F407 has USB OTG, up to 15 communication interfaces (including 6x USARTs running at up to 11,25 Mbit/s, 3x SPI running at up to 45 Mbit/s, 3x I²C, 2x CAN, SDIO) which allows communicating the system with a PC in many ways; up to 17 timers of STM32F407 (16- and 32-bit running at up to 168 MHz) give enormous flexibility.

As it is shown on Fig. 6, the system equipped with the analog circuitry and a Direct Digital Synthesis - DDS circuit can be used in all planned electrochemical biosensing techniques (CV, DPV, SWV, EIS, and Bio-FET). This way a universal biosensing platform can be built.

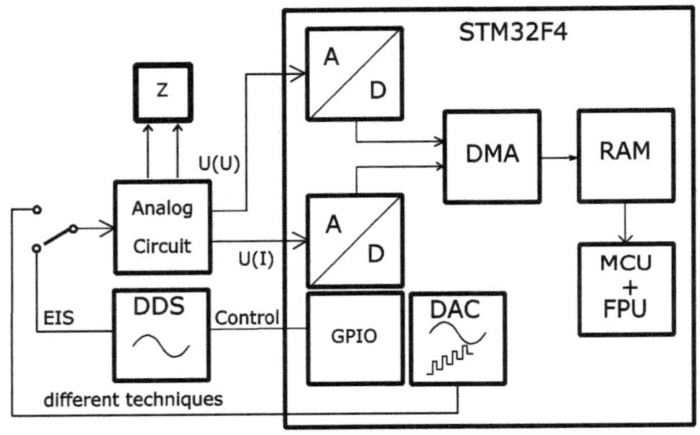

Fig. 6 Universal biosensing system based on the STM32F407 microcontroller

i) Data acquisition Level

The source of sinusoidal signal is a DDS generator, based on AD9850 (Analog Devices) integrated circuit, controlled with a parallel interface by the STM32F407 microcontroller. Next, sinusoidal signal is applied to the measured impedance Z, via analog circuit. It converts the current flowing through the impedance to voltage. The STM32F ADCs measure the voltage on the impedance, as well as voltage from the current-to-voltage converter. The ADCs are working in simultaneous mode – the samples of two measurement voltages are obtained at the same time. This way, the information about phase shift between voltage and current is not lost. The gain of current-to-voltage converter is variable – it is ensured by low noise, high precision resistors which are switched by the STM32F. Converted samples are transferred from ADC's registers by DMA to two separate tables (voltage and current, each containing 100 samples) in the RAM.

ii) Data Processing Level

DSP algorithms used at this level are based on discrete Hilbert transform and the method of analytic signal. As a result of both mathematical operations, a complex number including real and imaginary part of the impedance is obtained. This algebraic form of impedance can be easily converted to another form, represented by the module and the angle.

The described DSP algorithms, were implemented in the STM32F407VET6 microcontroller. This device includes a Floating Point Unit, which accelerates all FP calculations.

Communication protocols for wireless data transmission between biomeasurement system based on the STM32F and a PC were prepared, as well as visualization software modules.

IV. COMPARISON OF BOTH SYSTEMS IN EIS MEASUREMENTS

The results of impedance measurements for an electrical model of biological material performed with the use of both platforms (Fig. 7-8), were compared with the results obtained with the reference platform – the Autolab system. The detailed description of comparative accuracy analysis are presented in another paper from our team: "Accuracy analysis of measurements in electrochemical biosensing".

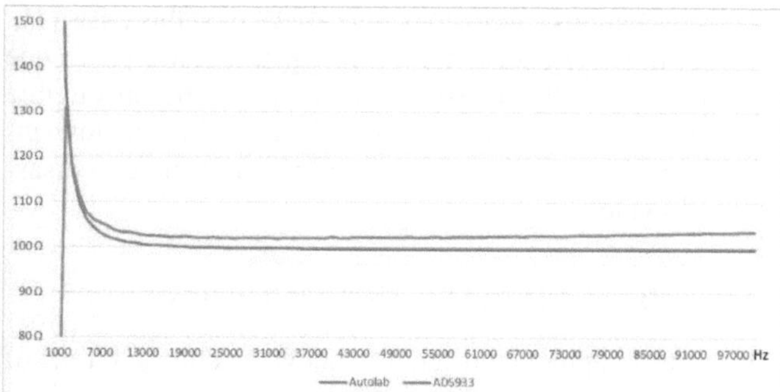

Fig. 7 Measurements performed with the system based on AD5933 (impedance module as a function of frequency)

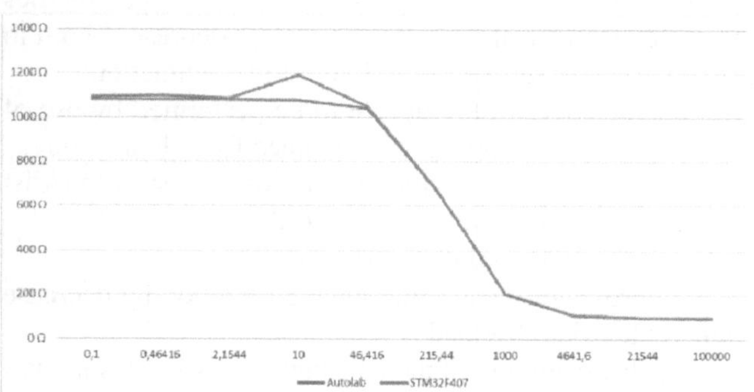

Fig. 8 Measurements performed with the system based on STM32F407 (impedance module as a function of frequency)

V. CONCLUSIONS

The results of presented research confirm the decision on choosing the STM32F407 platform for further development of universal multifunctional graphene-based biosensing system using different electrochemical techniques.

ACKNOWLEDGMENT

This project is financially supported by the National Centre for Research and Development within the framework of the GRAF-TECH Programme, grant No. GRAF-TECH/NCBR /08/06/2013.

CONFLICT OF INTEREST

The authors declare that they have no conflict of interest.

REFERENCES

1. Kyle U.G. et al.: Bioelectrical Impedance Analysis, part I: Review of Principles and Methods. Elsevier Ltd, 2004
2. Matias D., Przeniosło Ł., Miłosławski T., Jaskuła M., Makiewicz P., Biegun M., Penkala K., Mijowska E., El Fray M., Podolski J.: Development of the Mobile PoC Graphene-Based Biosensing Device Using Electrochemical Impedance Spectroscopy (EIS). Á. Jobbágy (ed.), First European Biomedical Engineering Conference for Young Investigators, IFMBE Proceedings vol. 50, DOI: 10.1007/978-981-287-573-0_5
3. Odenthal, Katherine J, Gooding, J Justin.: An introduction to electrochemical DNA biosensors. The Analyst, vol. 132, Issue 7, 2007
4. Przeniosło Ł., Walków M., Krzeszewska S., Pisarek S., Biedka A., Jaskuła M., Matias D., Penkala K.: Integrated impedance scanner in selected biomeasurement applications - control circuit. Krajowa Konferencja Studentów i Młodych Pracowników Nauki, Unieście, Poland, May 23, 2013, Zeszyty Naukowe Wydziału Elektroniki i Informatyki Politechniki Koszalińskiej, vol. V, pp. 63-69, Koszalin 2013, ISSN 1897-7421
5. Pumera M.: Graphene in biosensing, Materialstoday, vol. 14, no. 7-8, 2011
6. Yongchao Z., Yanique T., Eunkyoung Kim., Gregory F. Payne.: pH- and voltage-responsive chitosan hydrogel through covalent cross-linking with catechol. The journal of physical chemistry. B, vol. 116, Issue 5, 2012

Author:	Przemysław Makiewicz
Institute:	Department of Systems, Signals and Electronics Engineering, Faculty of Electrical Engineering, West Pomeranian University of Technology, Szczecin, Poland
Street:	Sikorskiego 37
City:	Szczecin, 70-313
Country:	Poland
Email:	makiewicz@zut.edu.pl

Pulsatile Mechanical Heart Assist Device

F.A. Pleşoianu[1], C.E. Pleşoianu[2], C. Corciovă[1], and G. Tinică[2,3]

[1] University of Medicine and Pharmacy "Gr. T. Popa"/Faculty of Biomedical Engineering, Iasi, Romania
[2] University of Medicine and Pharmacy "Gr. T. Popa"/Faculty of Medicine, Iasi, Romania
[3] Institute of Cardiovascular Disease "Prof. Dr. George I. M. Georgescu", Iasi, Romania

Abstract— **In this paper is described an innovative, easy to reuse heart assist device implying low production and maintenance costs, which needs minimally invasive techniques for the connection to the arterial system and that assists the failing heart according to the counterpulsation principle as an eligible solution for developing countries but also in case of emergency situations such as wars or natural disasters worldwide.**

Keywords— **assist device, heart failure, minimal invasive, low cost.**

I. INTRODUCTION

Cardiovascular disease (CVD) is the leading cause of death worldwide. Tremendous progress in the field of pharmaceutics and technology has contributed to the increase of healthcare performance translated into an increase of life expectancy with heart failure emerging as a challenge of modern cardiology in the context of population aging.

Due to the progress in optimal medical therapy, the mortality of patients with advanced heart failure from the waiting list of UNOS (the United Network for Organ Sharing) in the USA decreased from 43% in 1990 to 17% in 1999 [1].

Left ventricular assist device (LVAD) is used either as a bridge to transplant or as a final resource in case of comorbidities that contraindicate transplantation (Fig.1).

People in low and middle-income countries are more exposed to CVD risk factors and have less access to preventive efforts than those in high-income countries. As a result, over 80% of CVD deaths occur in low- and middle-income countries [2].

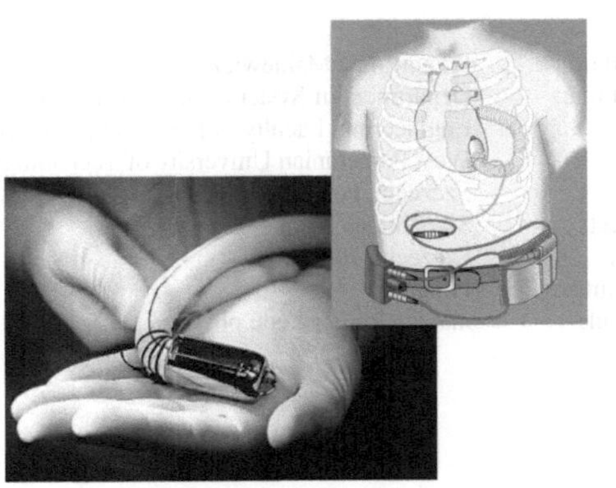

Fig. 1 Second-generation axial flow pump and intraventricular position of the device

Furthermore, in underdeveloped countries, the lack of financial resources deprives the health system of the latest technologies.

Even in developed countries, natural disasters and wars challenge the medical infrastructure with need of low cost, low technology devices. World Health Organization (WHO) has trained medical staff in Ukraine to deliver quality health services for internally displaced people and communities in conflict areas. Nowadays, 16 Mobile Emergency Primary Health-care Units are operating in Eastern Ukraine, giving over 1,600 consultations a week to people affected by this crisis [3].

Under these circumstances, a low price, low tech device requiring minimally invasive surgery could be an eligible alternative for the management of heart failure in low income countries but also in situations of war or natural disasters worldwide.

II. MATERIALS AND METHODS

In order to achieve our goal we have established a set of requirements grouped in three categories regarding medical, technical and economic aspects.

A. The Medical Aspects Include

- to assist the left ventricle in cardiovascular pathologies where the mechanical function is severely depressed both in case of temporary as well as in long-term dysfunction;
- the attachment of the device to be performed by use of minimally invasive surgery with no need for fluoroscopy;
- to ensure a stroke volume that can vary in order to assist patients with different body masses;
- to increase the coronary perfusion;
- to improve left ventricular ejection fraction by decrease of left ventricular afterload;
- to enable changing of the consumables without surgery.

B. The Technical Aspects Include

- power consumption < 1 W;
- total volume (including batteries) < 300 cm^3;
- total mass < 250 g;
- autonomy > 10 hours;
- rapid reuse by easy replacement of the single use assembly;
- to be manufactured with accessible technology;

V. Sontea and I. Tiginyanu (eds.), *3rd International Conference on Nanotechnologies and Biomedical Engineering*,
IFMBE Proceedings 55,
DOI: 10.1007/978-981-287-736-9_91

C. The Economic Aspects Include

- the production cost of the device < 100 euro;
- the production cost of the disposable components < 10 euro;
- the costs for maintenance <100 euro / month.

III. RESULTS

We have designed a reusable mechanical heart assist device operating on the principle of counterpulsation which is placed outside the body and is directly driven by an electrical motor. The device aspirates blood from the arterial system during the ventricular systole and returns it during the diastole. The connection with the arterial system is performed by attaching one extremity of a cannula to the subclavian artery (Fig. 4) while the other end is connected to the device.

A. The Disposable Assembly

The disposable assembly is made of a balloon provided with a cannula and 2 rigid ferromagnetic stainless domes (Fig 3). The balloon has an inner surface made of biocompatible materials (as it enters in direct contact with the blood of the patient). It is provided with a short cannula (that penetrates the upper dome) and is placed with its two poles between the two domes. The upper dome is convex in shape and has a central orifice through which the cannula of the balloon is introduced (Fig. 2).

The inferior dome is concave. Before use, the balloon is compressed (by vacuum) between the 2 structures.

A pressure sensor is placed inside the cannula and is connected with the main board of the device.

Fig. 2 Disposable assembly

B. The Reusable Device

Two domes made of permanent magnets are placed inside the device: the upper one with a central opening where the cannula of the single use assembly is introduced (Fig. 3) and the bottom one acting as a piston engaged by a rod-crank mechanism. The bottom dome performs a go forth translational movement compressing and decompressing the disposable assembly.

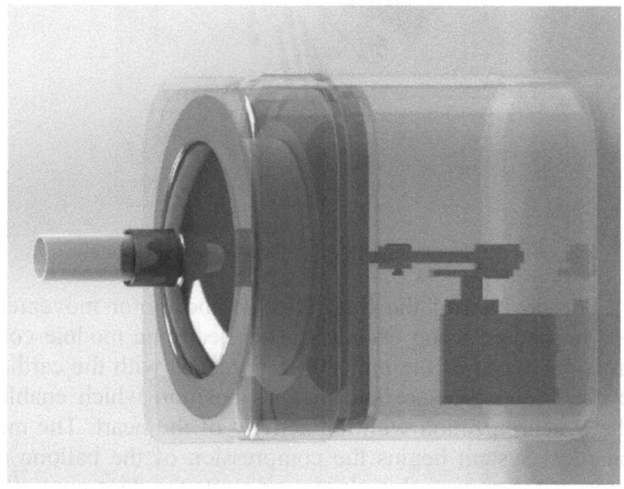

Fig. 3 Simulation of the proposed design

Once in use, the blood enters and leaves the interior of the balloon.

The translational movement is performed by use of a connecting rod-crank mechanism engaged to an electric motor through a speed reducer (Fig. 5).

The reducer lowers the rotation speed of the motor to the optimal level required by the application (20-100 rev/min) and increases the final transmission torque.

The gear ratio is 250:1. The electric motor converts electrical energy to mechanical work. Brushed DC motors are preferred for short-term and emergency applications while three-phase brushless DC motors are preferred for long term application.

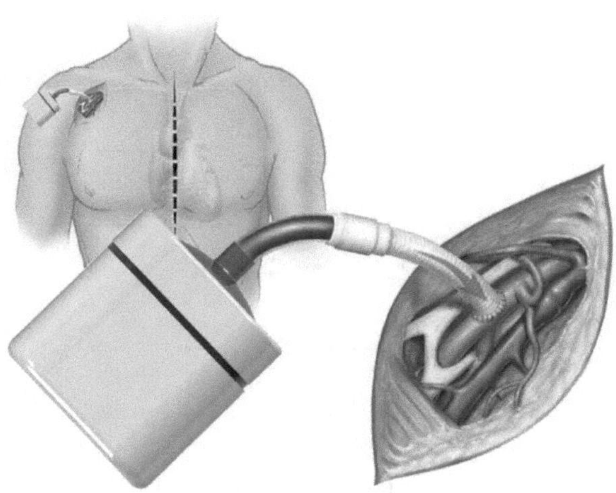

Fig. 4 Attachment of the device through the graft at the level of the subclavian artery

Fig. 5 Experimental model of the mechanical assembly

The position of the final effector and motor movement are monitored using encoders. The electronic module controls the speed of the motor in accordance with the cardiac cycle, blood pressure and piston's position which enables the synchronization with the activity of the heart. The mechanical system begins the compression of the balloon as soon as the aortic valve closes and starts the decompression when isovolumetric contraction of left ventricle occurs.

The synchronization of the pump with the heart is made by constant monitoring and assessment. The rotation speed of the motor is matched to the cardiac cycle timing, with higher rates in systole.

The motor driver provides power to the motor, according to the microcontroller (MCU) command.

The MCU estimates synchronization, evaluates ECG information, monitors encoders and pressure sensors, rectifies commands, takes safety decisions and communicates information through user interface.

The integrated power source is a 7.4 V, 1 Ah lithium-ion rechargeable battery, which is able to sustain the device up to 24 hours.

Different size balloons will provide different stroke volumes according to different body masses.

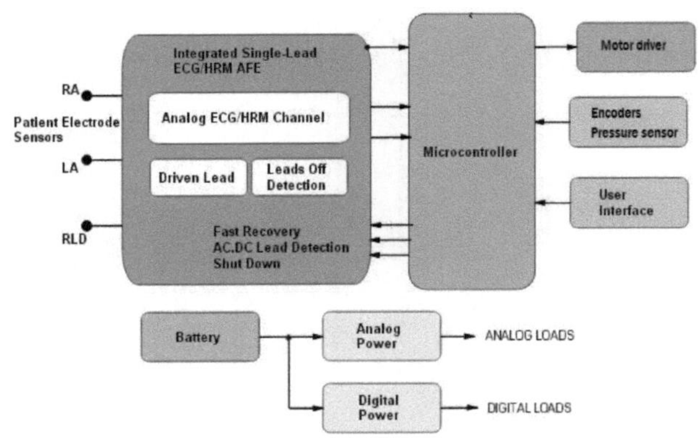

Fig. 6 Diagram of the electronic controller block

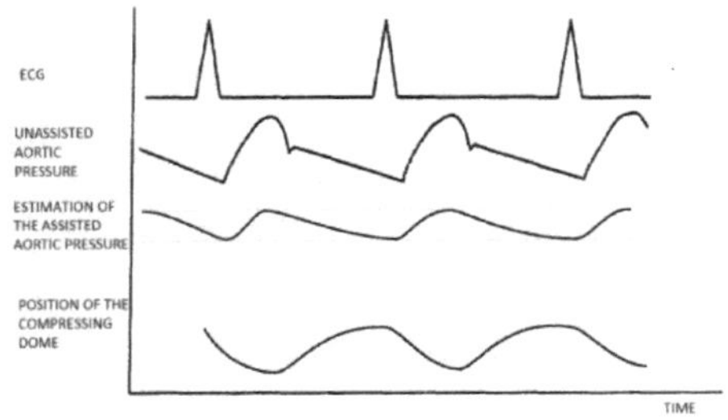

Fig. 7 Graphic estimation of pressure attenuation

In case of long term support, the balloon will be replaced at certain intervals in order to avoid the failure of the pump. The subclavian end of the cannula is provided with a tap (open/close system) which can assure closing position to enable the replacement of the single use assembly. Moreover, in case of pump failure, the replacement will require no surgical intervention as the subclavian cannula is left in place with the tap on close position.

IV. DISCUSSIONS

Intra-aortic balloon pump (IABP) counterpulsation reduces left ventricular afterload and myocardial work, improves diastolic coronary and systemic blood flow and increases cardiac output. Consequently, IABP use has grown significantly since its introduction in 1968 and is now the most commonly used form of mechanical circulatory assist support, with 70,000 devices annually inserted in the United States [4].

Complications related to IABPs occur in about 29% of cases, which is why routine radiographic surveillance is performed. IABP malposition is perhaps the most frequent complication. The superior migration of the IABP can occlude the left subclavian artery, resulting in limb ischemia. Conversely, inferior IABP malposition can lead to visceral complications such as mesenteric ischemia and, less commonly, hepatic failure. Predisposing factors may be anatomic and may include balloon length mismatch and spinal deformities. Aortic dissection occurs in about 1% of cases of IABP insertion [5].

As IABP can only be used for short term support, other devices that use the same principle of counterpulsation, have been developed for long term assistance in heart failure.

Sunshine Heart C-Pulse system is a non-blood contacting, non-obligatory assist device which is implanted externally on the ascending aorta and works on the principle of counterpulsation.

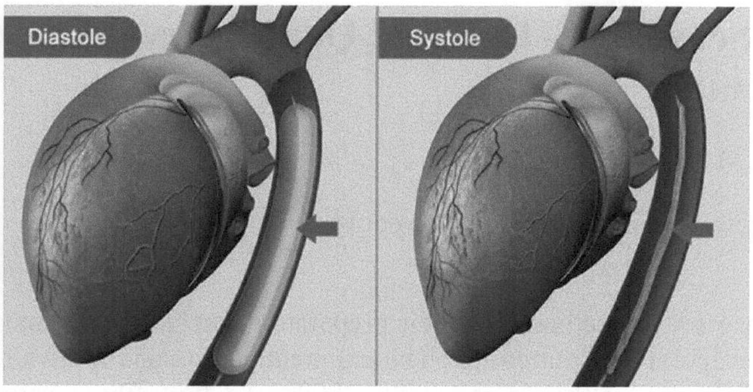

Fig. 8 Intra-aortic balloon pump (nmc heartcare)

Algorithms based on R-wave detection increase coronary bloodflow during diastole and reduce arterial impedance during systole with potential benefits on diastolic function and improve coupling of the ventricle and arterial system [6].

The pulsatile systems have proved to be prone to mechanical failure but the new features of C-Pulse that has no contact with blood reduce the risk of mortality due to device failure and other risks associated with the classical devices like thrombosis and blood mechanical stress. However, the short life of the balloon pump that requires surgical intervention for replacement, weak left ventricular afterload, and insignificantly increase in cardiac output, limit the benefits of this device to less severe cases.

Abiomed Symphony is another device that uses counterpulsation methods for assisting the heart. It coordinates the mechanisms of a pulsatile pump to suck blood from an artery through a conduit while the heart is in systole and expels the blood into the blood flow conduit and the artery while the heart is in diastole [7]. It is a minimal invasive, implantable pump for patients with chronic heart failure which immediately improves the cardiac output and renal function [8], but the low reliability of the system makes it useful only in short time support.

Our design uses the same principle of counterpulsation that was proven to reduce the mechanical strain over the ventricle (Fig. 9) [8] but with certain important differences:

- The device is not implantable;
- The device is mechanical driven providing continuous assistance for the heart.

The advantages of the mechanical system are translated in: low manufacturing cost, significantly higher efficiency with lower power need, better autonomy and high accuracy.

Despite the special attention to trauma, the position of the device outside the body assures great advantages such as the possibility of balloon and/or pump replacement without surgical intervention.

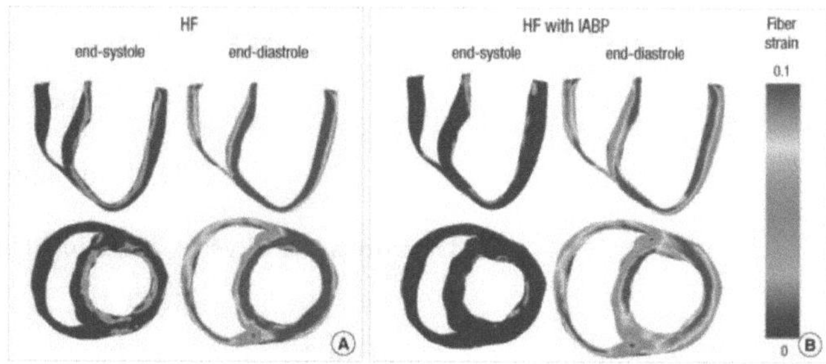

Fig. 9 Mechanical strain in unassisted and assisted heart failure

V. CONCLUSIONS

A new approach in design of a mechanical heart assist device, driven by the need of minimal invasiveness, low cost and low technology may offer an eligible solution not only to low-income countries but also in worldwide emergencies, with a tremendous life-saving potential.

CONFLICT OF INTEREST

The authors declare that they have no conflict of interest.

REFERENCES

1. Birks E (2011), Left ventricular assist devices, Heart 2010; 96:63–71
2. WHO at http://www.euro.who.int/en/health-topics/ noncommunicable-diseases/cardiovascular-diseases/cardiovascular-diseases2
3. WHO at http://www.euro.who.int/en/countries/ukraine/news/news/2015/06/ mobile-health-care-for-ukraine
4. Vales L, Kanei Y, Ephrem G, Misra D.(2011) IABP use and outcomes with current therapies. J Invasive Cardiology. 23:116-119
5. Ginat D, Massey HT, Bhatt S, Dogra VS.(2011) Imaging of Mechanical Cardiac Assist Devices. J Clin Imaging Sci;1:21
6. Slaughter M, Cecere R, Sun B, Huprikar S, Hotz H (2015) C-Pulse System Extra-Aortic Counterpulsation for Heart Failure: Driveline Infections and Management. International Society for Heart & Lung Transplantation (ISHLT) 35th Annual Meeting and Scientific Sessions
7. Cecere R, Ventricular assist devices today and tomorrow
8. Lim KM, Lee JS, Gyeong M-S, Choi J-S, Choi SW, Shim EB (2013), Computational Quantification of the Cardiac Energy Consumption during Intra-Aortic Balloon Pumping Using a Cardiac Electromechanics Model. Journal of Korean Medical Science 28(1):93-99.

Author: Pleşoianu Carmen Elena
Institute: University of Medicine and Pharmacy "Gr. T. Popa"
Street: G. Ibraileanu
City: Iasi
Country: Romania
Email: carmenplesoianu@gmail.com

Impedance Characterization of Gas Sensitive S-Te Based Quaternary Chalcogenides

D. Tsiulyanu and M. Ciobanu

Department of Physics, Technical University, Chisinau, Republic of Moldova

Abstract— **Impedance spectra of quaternary S -Te based alloys were investigated in both dry synthetic air and mixture with nitrogen dioxide in order to assess their use in future gas sensors working at room temperature. To elucidate the effect of tellurium, the quaternary compositions As2Te13Ge8S3 and As2Te130Ge8S3, with increasing concentration of Te have been considered along with pure tellurium films. Shown by AFM, SEM and X – ray analysis the nature of the films was predominantly amorphous but the morphology of the films grown on sintered alumina substrates appears to consist of interconnected islands and dots, which facilitate the solid-gas interaction. Analyses in Cole – Cole interpretation allowed evaluating the characteristic frequency, time constant, resistance and capacity of the films in both dry air and its mixture with NO2. It is shown that impedance spectra being strongly influenced by gaseous environment are also strongly influenced by material composition and phase - structural state of the film. Sensitivity spectra to nitrogen dioxide denote the competitive influence of carrier transport via states of allowed bands, hopping between localized states in the extended band tails or tunneling (variable range hopping) between localized states close to Fermi level. Impedance sensitivity being maximal for amorphous As2Te13Ge8S3 is assumed to be controlled by this competition of transport mechanisms.**

Keywords— **Chalcogenides, AC conductivity, Gas sensors, NO2.**

I. INTRODUCTION

The possibilities of detection of harmful gases using chalcogenide semiconductors are being intensively investigated since the effect of propylamine adsorption on dimorphite (As_4S_3) conductance was discovered [1]. Both glassy and crystalline chalcogenide semiconductors become attractive for gas sensing applications because of their room temperature operation abilities and fabrication simplicity. Although the gas sensitive films can be grown using a large number of binary, ternary and quaternary chalcogenide combinations, so far only elemental tellurium [2-9], its compounds [10,11] and Ag doped glasses [12,13] have been more or less investigated.

Pure polycrystalline tellurium based gas sensitive thin films were the subject of many investigations during the last decade due to number of advantages over compound semiconductors, one of them being the exemption from stoihiometry problems. This advantage allows the use of a large variety of methods for their preparation, including thermal evaporation, sputtering, chemical precipitation and hot wall epitaxy. Firstly, in the early 2000s, Te thin films were showed to exhibit a high sensitivity to nitrogen dioxide [2], but much smaller sensitivity to accompanying gases in the environment, such as oxygen, nitrogen and water vapors [3]. Detailed characterization of Te based gas sensors for nitrogen dioxide have been reported early, both for films fabricated by thermal vacuum evaporation [4, 5] and sputtering [6]. Then, the interaction of Te films with different gas species such as propylamine and carbon oxide [7], ammonia [8] or hydrogen sulphide [9] have been reported.

In contrast with elemental tellurium, the chalcogenide gas sensitive compounds, which work at room temperatures, are much smaller studied. To date only a few studies have attempt to clarify their practical viability. The binary $Te_{13}Ge_3$ glassy alloy was reported as a most stable material which exhibits the high sensitivity to nitrogen oxide [10], ternary Ge - As -Te films was shown to interact with different gas species but exhibit a high sensitivity to low concentration of nitrogen dioxide [11]. Amalgamation of a few elements including Ge and Te is considered as a way to stabilize the gas sensing properties of Te based sensor layer due to localization of electrons within a glassy network and reducing the metallic character at such compositions. On the other hand, the reduced metallic character of such compositions increases the electric layer resistance resulting in decreasing of signal / noise ratio. Some quaternary chalcogenide based DC sensors was reported [10] to exhibit so small signal / noise ratio that even sensing experiments failure. In the present work, to overcome the sensing disadvantage due to small signal / noise ratio, the a.c. sensing techniques was applied. We report the results of gas sensing measurements performed by investigation of impedance spectra in a large frequency range 5 *Hz* – 13 *MHz*, at room temperature. In order to get evidence on effect of tellurium, we have considered $As_2Te_{13}Ge_8S_3$ and $As_2Te_{130}Ge_8S_3$, as well as pure Te for comparison.

II. MATERIALS AND METHODS

The glassy alloys were prepared by melt-quenching method of pure (99,99%) As, S, Te and Ge in quarts ampoules evacuated up to $5 \cdot 10^{-5}$ Torr. The melting was

V. Sontea and I. Tiginyanu (eds.), *3rd International Conference on Nanotechnologies and Biomedical Engineering*,
IFMBE Proceedings 55,
DOI: 10.1007/978-981-287-736-9_92

performed at 900°C. The ampoule was rotated around the longitudinal axis at velocity of 7-8 rotations / min and was agitated for homogenization during the synthesis time (24 h) and then was quenched on a copper refrigerator with running water. Two quaternary compositions $As_2Te_{13}Ge_8S_3$ and $As_2Te_{130}Ge_8S_3$ with increasing concentration of Te were synthesized as well as the pure polycrystalline tellurium has been considered.

The chalcogenide thin films have been prepared by thermal "flash" evaporation of priory-synthesized materials from tantalum boats onto ceramic Al_2O_3 substrates containing the previously deposited platinum interdigital electrodes with an electrode width of 15 µm and interelectrode distances of 45 µm (Fig.1a).

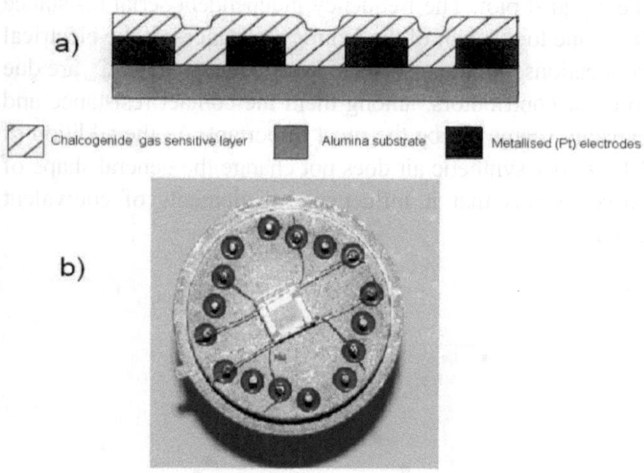

Fig. 1 Schematic representation of chalcogenide based thin film device (a) and its view encapsulated in a standard socket.

The evaporation of chalcogenide materials was performed at a working pressure of 10-4 Pa. The growth velocity of the film was in the order of 30 nm/s, and the area of deposition around 5 *mm*^{2}. The thickness of the films was about 60 *nm,* being determined using microinterferometer MII - 4. Structural investigations of grown films were carried out by the X –ray analyses and scanning electron microscopy (SEM). The films were encapsulated in standard TO – 16 sockets and their contacts were thermally bonded to socket pins by means of copper wires (Fig. 1b).

Impedance measurements were carried out in frequency range of 5 *Hz* to 13 *MHz* using a HP 4192A impedance analyzer. For these experiments the thin film sensing devices were put into a test cell (of 10 ml volume) in which the gases were injected parallel to the film surface. Constant flow (100 ml/min) was maintained by mass flow controllers (MFC, Wigha, Germany). The NO_2 vapor with a concentration of 1,5 *ppm* was obtained by using a calibrated permeation cylinder (Vici Metronics, USA), which was incorporated into the experimental set – up described everywhere. In this work, the a.c. gas sensitivity was defined as relative variation of impedance for a selected frequency in mixture of carrier gas with NO_2 and pure carrier gas, divided by gas concentration (C),in percent / ppm :

$$\eta = \frac{|Z_a - Z_g|}{Z_a C} \times 100\% \qquad (1)$$

III. RESULTS

A. X-ray Diffraction and SEM Measurements

The X-ray diffraction measurements were made using the DRON –YM1 diffractometer by FeK α radiation. Rotation velocity of the scintillation counter was 2 (or / and 4) angle degrees /min.

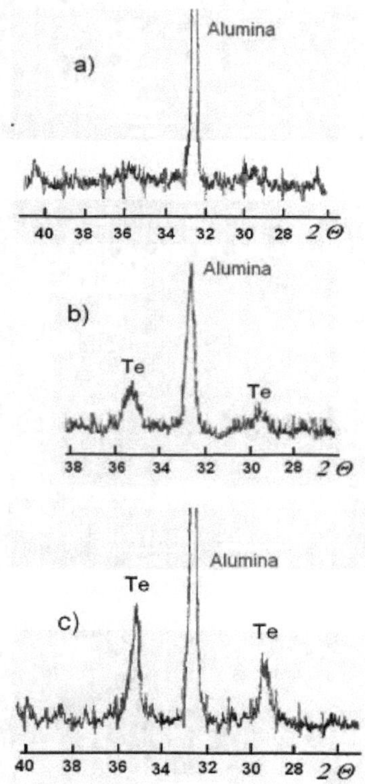

Fig. 2 XRD diffraction pattern of As2Te13Ge8S3 (a), As2Te130Ge8S3 (b) and pure Te (c) films.

Fig. 2 shows the XRD patterns of quaternary $As_2Te_{13}Ge_8S_3$, $As_2Te_{130}Ge_8S_3$ and pure Te films deposited on alumina substrates. It is evident that the $As_2Te_{13}Ge_8S_3$ films (Fig.2a) are in the amorphous state. The quaternary $As_2Te_{130}Ge_8S_3$ films with higher concentration of tellurium , as well as pure Te films show very small peaks corresponding to crystalline phases of Te [Figs. 2(b, c)] indicating their nearly amorphous nature. No peaks corresponding to oxides of tellurium are visible in the XRD patterns. These results are in agreement with SEM observations.

The surface morphology of the films was made visible with a VEGA TESCAN TS 5130 MM scanning electron microscope at an acceleration voltage of 30 kV. In Fig. 3 are shown the SEM images of chalcogenide films obtained in this work. It is seen that independently on composition, the films consists of agglomerated islands, resulting in the film with great surface roughness. No crystallite tracks are observed.

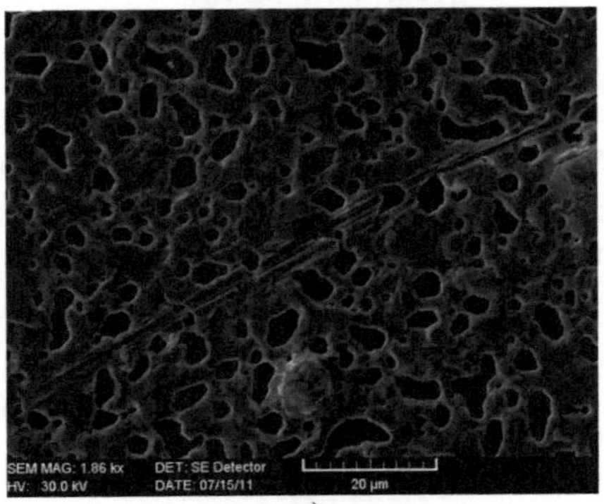

a)

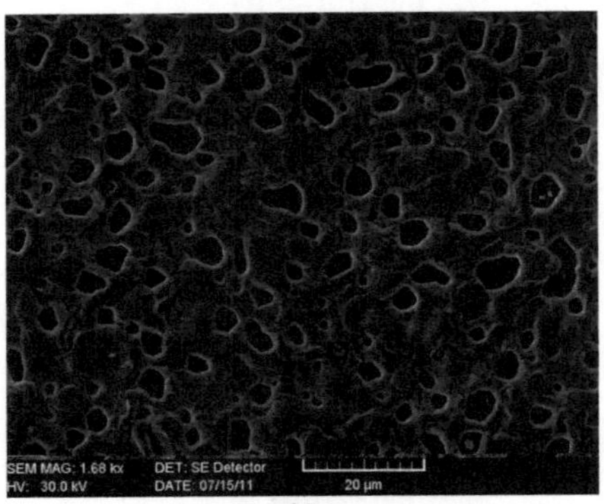

b)

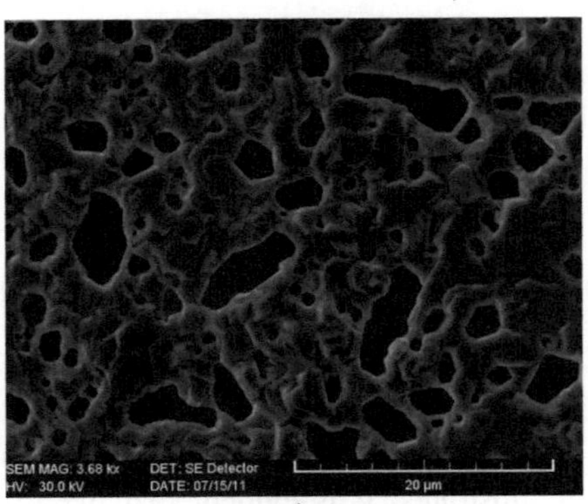

c)

Fig. 3 SEM micrographs of As2Te13Ge8S3 (a), As2Te130Ge8S3 (b) and pure Te (c) films.

B. A.C. Measurement

Tellurium Thin Films

Figure 4a shows the typical complex impedance diagrams in the Cole-Cole presentation (Nyquist diagram) obtained in pure synthetic dry air and upon exposure to NO_2, from a thin film device at room temperature (22°C). The diagrams shows the slightly depressed semi-circular arcs with the center placed below the real axis, owing to the presence of different circuit elements in the thin film device [14].

The simplified equivalent circuit in Fig. 4b can interpret the Nyquist plot. The frequency independent serial resistance R_0 is due to the sum of the Ohmic resistances of the electrical connections, but the resistance R_ω and the capacity C_ω are due to other contributors, among them the contact resistance and capacity seems to be the most important. As the addition of NO_2 to dry synthetic air does not change the general shape of curve, means that it influences all elements of equivalent circuit.

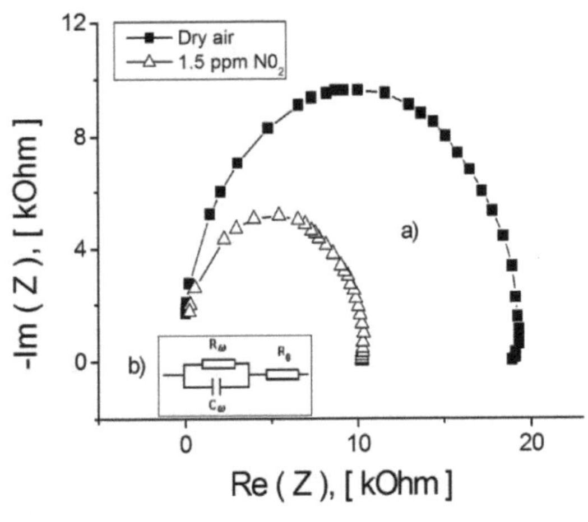

Fig. 4a) Nyquist diagrams of tellurium thin films in different environmental conditions; b) Suggested equivalent circuit.

The circles of the Nyquist diagrams shown in Fig. 4a are depressed owing to the dependence of both R_ω and C_ω on the frequency. For a simple parallel $R_\omega C_\omega$ circuit, the relaxation time (time constant) is given by:

$$\tau_m = \omega_m^{-1} = \frac{1}{2\pi f_m} = R_m C_m \, , \qquad (2)$$

where f_m is the characteristic frequency, at which the imaginary part Im(Z) reaches its maximum value, while R_m and C_m are the resistance and capacity of the film at the characteristic frequency f_m. The impedance Z_m and the time constant τ_m at the characteristic frequency of the sample in dry synthetic air and gaseous media with 1,5 ppm NO_2 are listed in Table 1.

The values of R_0 and $R_\gamma = R_\omega + R_0$ can be estimated using the left and right interception points of the semi – circles with the *Re(Z)* axis. Thus, R_0 was found to be very small, only about 5 *Ohm*. That is, the left intercept practically passes through the point of origin, and the right intercept gives the value of R_γ about 20.5 kOhm.

From simple analysis of the equivalent circuit the values of R_ω and C_ω of the film can be evaluated as:

$$R_\omega = \frac{\mathrm{Im}^2(Z) + \mathrm{Re}^2(Z)}{\mathrm{Re}(Z)} \tag{3}$$

and

$$C_\omega = \frac{\mathrm{Im}(Z)}{\omega\left[\mathrm{Im}^2(Z) + \mathrm{Re}^2(Z)\right]} \tag{4}$$

Estimated from equations (3) and (4) the resistance R_m and capacitance C_m of the film at a characteristic frequency f_m are also listed in Table 1.

Table 1 Impedance and R-C values at characteristic frequency for different environments

Composition	Te		$As_2Te_{13}Ge_8S$		$As_2Te_{130}Ge_8S$	
Environment	dry air	1,5ppm NO_2	dry air	1,5ppm NO_2	dry air	1,5ppm NO_2
f_m,[kHz]	450	950	-	140	1000	800
Z_m,[kΩ]	13,3	6,6	-	120	0,48	0,42
$\tau_m \times 10^{-7}$s	3,5	1,7	-	11,4	1,6	2,0
R_m, [kΩ]	18,5	9,3	-	167	0,49	0,43
C_m, [pF]	18,3	17,8	-	6,8	8,7	8,1

Addition of nitrogen dioxide to dry synthetic air does not change the general shape of impedance spectrum that is, it influences all elements of the equivalent circuit. The values of characteristic frequency, impedance and time constant τ_m of the film at this frequency, by indicated concentrations of NO_2 at room temperature, are summarized in Table 1.

From this table it is seen that as the environment is changed from dry air to its mixture with gas in question, the resistance R_m is mainly influenced while the capacitance C_m does not very essentially.

$As_2Te_{13}Ge_8S_3$

Figure 5 shows the typical complex impedance diagram in Nyquist plot obtained in pure synthetic dry air and its mixture (inserted in Fig.5) with 1,5 ppm NO_2 from quaternary $As_2Te_{13}Ge_8S_3$ (50 at.% Te) at room (22^0 C) temperature. It seems that the complex impedance diagrams, calculated from measured values of real and imaginary parts show the semi-circular arcs as well. Notwithstanding the diameter of the semicircle observed in dry air is so large, that it could not be plot completely, because of lack of data, due to device limitation. Calculated from equations (3) and (4) the resistance R_ω and capacity C_ω of the film at characteristic frequency, i.e. R_m and C_m as well as the film`s impedance for a NO_2 gaseous media are listed in Table 1. The respective values for dry air could not be calculated due to above-mentioned reason.

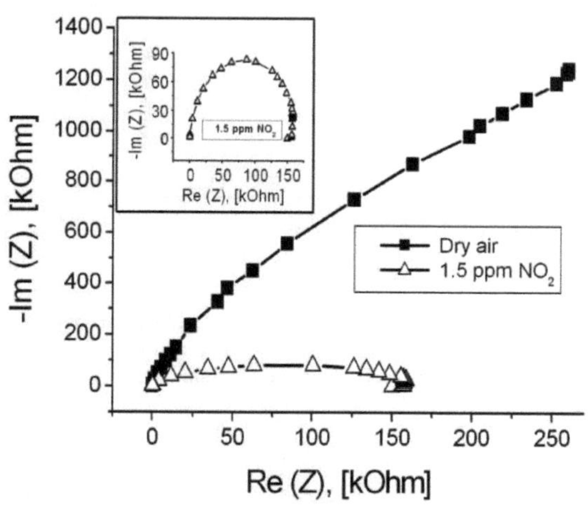

Fig. 5 Nyquist diagrams of quaternary As2Te13Ge8S3 thin films in both dry air and its mixture with NO2 at room temperature. Inserted picture shows separately only the diagram obtained under application of NO2.

It is seen, that under application of 1,5 ppm NO_2, the impedance and electrical resistance of $As_2Te_{13}Ge_8S_3$ films are about 20 times higher than their respective values of pure tellurium films , but the capacity is twice smaller.

$As_2Te_{130}Ge_8S_3$

The complex impedance diagram from a thin film $As_2Te_{130}Ge_8S_3$ (90.9 at.% Te) obtained at room temperature in pure synthetic dry air and upon exposure to NO_2, is depicted in Fig. 6. These diagrams not resemble with semi-circular arcs. That means, the equivalent circuit is much more complicated than a simple RC parallel one. Nevertheless, we have assessed the characteristic frequency f_m as frequency where the imaginary part of the impedance reaches the maximum value. Further, we have applied f_m to calculate the relaxation time τ_m , impedance Z_m , resistance R_m and capacity C_m of the film at this frequency. The obtained values are listed in Table 1 for both dry air and its mixture with nitrogen dioxide. Firstly, it should be mentioned that in spite of low concentration of other elements introduced in this alloy (2.1 at.% S; 1.4 at% As and 5.6 at % Ge), both the impedance and electrical resistance decrease dramatically in comparison with pure Te films: by two and three orders of magnitude respectively.

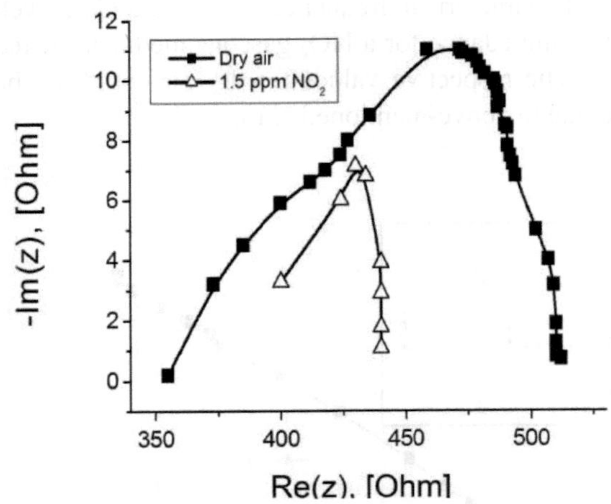

Fig. 6 Cole-Cole presentation of imaginary part v.s. real part of impedance for quaternary $As_2Te_{130}Ge_8S_3$ thin films in both dry synthetic air and its mixture with NO_2 at room temperature.

B. Gas Sensitivity trough the Impedance Change

The change of frequency - dependent impedance upon exposure to low concentration of NO_2 of chalcogenides in question, make evidence of their abilities to be used in sensors development. In the first approximation, the data given in Table 1 allow assessing the effect of nitrogen dioxide on device impedance and its sensitivity. However this assessing is valid only by characteristic frequency, which in its turn becomes essentially influenced by gas adsorption. That is why; more complete information concerning the gas sensing properties of the films can be obtained investigating their full impedance spectra both in dry air and its mixture with a target gas. Further, using relation (1) the spectra of gas sensitivity can be carried out and the frequency, at which the maximum sensitivity occurs, can be determined. Fig. 7 shows the frequency dependent sensitivity of the films in question towards NO_2 , at room temperature.

It is observed that the electric field modulation most strongly influences the gas sensitivity of Te and $As_2Te_{13}Ge_8S_3$ films, while the sensitivity of $As_2Te_{130}Ge_8S_3$ remains nearly independent on applied frequency.

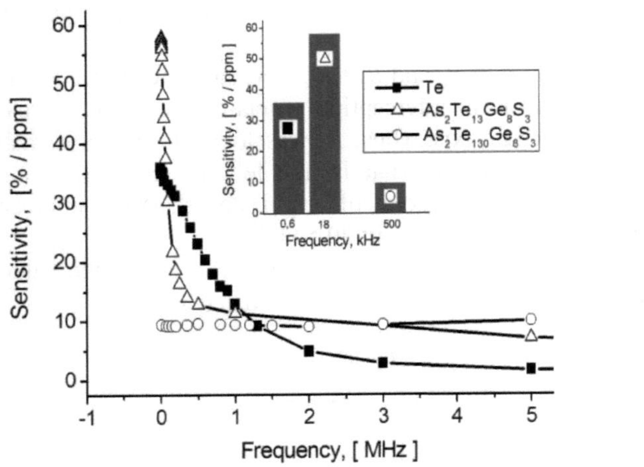

Fig. 7 Sensitivity spectra of quaternary S - Te based thin films to 1,5 ppm NO2 in dry air at room temperature. Inset show the comparison diagram of the maximum sensitivities at respective frequencies for the chalcogenides in question.

It is also noticeable the essential difference in velocity of decrease of sensitivity for $As_2Te_{13}Ge_8S_3$ and pure tellurium film. The sensitivity of $As_2Te_{13}Ge_8S_3$ film decreases sharply with frequency increase but tends to saturation already at approximately 500 kHz. On contrary, the sensitivity of Te film diminishes quite slowly and tends to saturation only at approximately 3 MHz. The inset in Fig.7 report for each composition the maximal value of sensitivity but abscissa on this diagram indicates the frequency, at which this maximum sensitivity occurs.

IV. DISCUSSION

Chalcogenide materials considered in this work belong to disordered, so called lone –pair semiconductors. The states of unshared p - electrons of chalcogen atoms in these materials display in a band, but the gap between this band and conduction one is the main forbidden gap, which control the semiconducting properties of the solid. Besides, the interaction between lone – pair electrons, as well as their interaction with "dangling" bonds, create the specific charge defects that operates as localization states in the forbidden gap, pinning the Fermi level closely to valence band, i.e. determining the p – type of conductivity. Disordering creates a random potential, which results in formation of both tails of extended bands and a finite value of localized states close to the Fermi level $N(E_f)$.

Thus, the a.c. conductivity can be realized by three mechanisms: band - to - band conduction with an activation energy E_g (forbidden gap of considered material), hopping between localized states in the extended band tails or tunneling (variable range hopping) between localized states situated close to Fermi level. The contribution of each mechanism depends on material composition, frequency range and temperature regime. As all experiments were performed at room temperature, only the effect of frequency on impedance can somewhat elucidate the predominant mechanism of current flow. A negligible variation of impedance with frequency increase in $As_2Te_{130}Ge_8S_3$ indicates that conductivity via extended states, i.e. band – to – band mechanism is the main. On the other hand, the strong fall of the impedance in pure Te film and $As_2Te_{13}Ge_8S_3$ by frequency increase, shows the hopping mechanism of conductivity through localized states to be dominant. In order to distinguish between hopping via localized states of the extended bands tails or variable range hopping over localized states near Fermi level, the additional investigation of effect of temperature on a.c. conductivity is required, that constitutes a subject of a future study.

The sensing properties of microcrystalline Te films toward toxic gases, including NO_2, has been reported in our previous publications, both for d.c. [4,7] and a.c. [15,16] current. It was suggested [3] that the response of tellurium films to NO_2 is due to a “strong” chemisorbtion , that is being adsorbed on the surface of the chalcogenide semiconductor the molecule of NO_2 acts as a dangling bond, that can accept a lone - pair electron to form an electron pair. Capture of a lone - pair electron means the transition of an electron from the upper part of the valence band to a NO_2 acceptor level, accompanied with releasing of an additional hole. Thus, the adsorption of nitrogen dioxide leads to increase in the majority carrier density on the surface, which results in increasing the film conductivity, including the a.c. one.

We assume that the basic interaction mechanism discussed for pure Te layers also occurs in Te alloys studied in this work, but addition of As, S and Ge atoms can result in structural - phase transformations, appearance of new mechanisms of current flow and obviously, in modification of gas sensing peculiarities. Indeed, as shown by X ray diffraction measurements, addition of only 5,6 at% Ge, 2,1 at% S and 1,4 at% As (alloy $As_2Te_{130}Ge_8S_3$) diminishes the fraction of crystalline tellurium, but leads to a dramatic decrease of impedance by two orders of magnitude, followed by decreasing of gas sensitivity by four times. On contrary, addition of higher concentrations of these elements: 30,8 at% Ge, 11,6 at% S and 7,6 at% As (alloy $As_2Te_{13}Ge_8S_3$) transforms the film into an amorphous state with a dramatic increase of impedance by at least two orders of magnitude, followed by an increasing of gas sensitivity more than twice. As have been mentioned above, these results shows how the composition and phase-structural state of the film control not only the mechanism of current flow but also their gas sensing properties and finally, their abilities to be used in development of gas sensing devices operate at room temperature.

V. CONCLUSIONS

Quaternary S – Te based chalcogenides exhibit high impedance sensitivity to NO_2 gas molecules in air, at room temperature. The phase –structural state of the films based on these materials and grown by thermal evaporation in vacuum is predominantly amorphous. Material composition and phase - structural state of the films, as well as gaseous environment strongly influences the shape of impedance spectra and their sensitivity to nitrogen dioxide Impedance sensitivity, being maximal for amorphous $As_2Te_{13}Ge_8S_3$ is assumed to be controlled by competition of several transport mechanisms.

ACKNOWLEDGMENT

This work was supported by SCSD AS of Moldova, grant 15.817.02.29A. The authors express their gratitude to Dr. M. Enache at NCMST of TUM for SEM analysis and Dr. G. F. Volodina at IAF ASM for XRD analysis.

CONFLICT OF INTEREST

The authors declare that they have no conflict of interest.

REFERENCES

1. Marian S, Potje – Kamloth K, Tsiulyanu D et al. (2000) Dimorphyte based gas sensitive thin films.Thin Solid Films 359: 108-112
2. Tsiulyanu D, Marian S, Miron V et al. (2001) High sensitive tellurium based NO_2 gas sensor. Sens. Actuators B 73:35 – 39
3. Tsiulyanu D, Stratan I, Tsiulyanu A et al.(2007) Investigation of the oxygen, nitrogen and water vapour cross-sensitivity to NO_2 of tellurium-based thin films. Sens. Actuators B 121: 406-413
4. Tsiulyanu D, Tsiulyanu A, Liess H.- D et al. (2005) Characterisation of tellurium based films for NO_2 detection. Thin Solid Films 485:252 -256
5. Bhandarkar V, Sen S, Muthe K.P et al.(2006) Effect of deposition conditions on the microstructure and gas-sensing characteristics of Te thin films. Materials Science and Engineering B131:156–161
6. Siciliano T, Giulio M. Di, Tepore Met al.(2008) Tellurium sputtered thin films as NO_2 gas sensors. Sens. Actuators B 135: 250-256
7. Tsiulyanu D, Marian S, Liess H. – D (2002) Sensing properties of tellurium based thin films to propylamine and carbon oxide. Sens. Actuators B 85: 232 – 238

8. Sen S, Muthe K.P, Joshi N et al. (2004) Room temperature operating ammonia sensor based on tellurium thin films. Sens. Actuators B 98: 154 -159
9. Sen S, Bhandarkar V, Muthe K.P et al.(2006) Highly sensitive hydrogen sulphide sensors operable at room temperature. Sens. Actuators B 115: 270 –275
10. Wüsten J. Potje-Kamloth K (2010) Chalcogenides for thin film NO sensors. Sens. Actuators B 145: 216–224.
11. Marian S, Tsiulyanu D, Liess H. – D (2001) Ge - As - Te based sensor selective to low NO_2 concentrations. Sens. Actuators B 78: 191 - 194.
12. Koleva K, Popov C, Petkova T et al.(2009) Complex $(As_2S_3)_{(100-x)}(AgI)_x$ chalcogenide glasses for gas sensors. Sens. Actuators B 143: 395–399
13. Popescu M, Velea A, Sava F et al. (2010) Structure and properties of silver doped $SnSe_2$ and $Ge_2Sb_2Te_5$ thin films prepared by pulsed laser deposition. Phys. Status Solidi A 207: 516–520
14. Macdonald J.R (1987) Impedance Spectroscopy. Willey, New York
15. Tsiulyanu D. Mocreac O (2011) Impedance spectroscopy of sensitive to harmful gases tellurium thin films. J. Non-Oxide Glass. 3:37-44
16. Tsiulyanu D, Mocreac O (2013) Hydrogen sensing behavior of tellurium thin films studied by A.C. measurements. Material protection 2: 107 -111

Author: Dumitru Tsiulyanu
Institute: Technical University
Street: Dacia 41
City: Chisinau
Country: Republic of Moldova
Email tsiudima@gmail.com

Part IV
Biomedical Signal and Image Processing

Automated Morphometry of Neutrophilic Granulocytes – A Simple and Reliable Method of Assessment of the Wound Process Activity

R.M. Chornopyshchuk[1], S.A. Sydorenko[2], and M.I. Burkovskyi[1]

[1] Vinnytsia National Pirogov Memorial Medical University, Vinnytsia, Ukraine
[2] State Medical and Pharmaceutical University "N. Testemitanu", Chisinau, Moldova

Abstract— **Absence of single algorithms of diagnostics and treatment of wounds points out the necessity of an individual approach to such patients with clear determination of the phase, nature and peculiarities of the wound process. One of the objective methods for assessment of the wound process is the determination of the functional activity of neutrophilic granulocytes by their morphometric analysis. Earlier studies required difficult calculations and special equipment. But development of computer technologies allowed to make this process easier after automation of the calculation of the morphometric indices. The use of the integral index of the neutrophilic granulocyte form factor (FFNG), providing determination of the area-perimeter ratio of these cells, is of great interest. To confirm the efficacy of the determination of the index of neutrophilic granulocyte form factor in case of a wound process we compared it with the results of the existing methods of control over the wound course. After the conducted statistical analysis, clear correlation relationships of the FFNG index with the results of the other studies were established that allowed to confirm the objectivity and reliability of this index for assessment of the wound process activity and to recommend it for the use in clinical practice.**

Keywords— **purulent wound, wound process, form factor of neutrophilic granulocytes.**

I. INTRODUCTION

Studying of the wound process as one of the oldest problems of the medical science remains urgent [1]. Despite a lot of diagnostic methods known to modern scientists, a number of issues concerning pathogenesis of this process remain unclarified [2]. Absence of single algorithms for diagnostics and treatment of wounds points out the necessity of an individual approach to such patients with clear determination of the phase, nature and peculiarities of the wound process [3]. As practice shows, exactly these factors become determinative for efficacy of the chosen treatment and prevention of possible complications. At present clinical characteristics of the wound process (patient's general condition, signs of perifocal inflammation, necrolysis, presence of granulations, wound contraction, marginal epithelization) supplemented by different instrumental (capillary measurements, tensiometry, echography, etc.) and laboratory methods of study (general blood analysis, biochemical, microbiological and cytological examination) remains the main criterion for diagnostics of the course of wound healing in the medical practice. But the majority of them are of little use for dynamic observation over the course of the wound process. They don't allow to determine the stage of the wound process clearly and are often of a subjective nature.

For complex assessment of the patient's condition a lot of integral systems of objective assessment of the patient's condition severity have been developed and used (TISS, APACHE I, II, III, SAPS, MPM) [4]. But the use of such methods is often complicated and does not take into account a number of important criteria typical of a surgical infection that, in fact, determine the patient's condition severity [5]. All this requires search of simple and reliable methods for assessment of the severity of patients with purulent pathology of soft tissues.

One of the methods for assessment of the wound process nature is assessment of the functional activity of phagocytes by their morphometric analysis at the light-optical level with determination of the form factor index [6]. This is about neutrophilic granulocytes – the most important cells of the nonspecific body protection that "struggle" first against foreign human agents [7]. To confirm the efficacy of the determination of the form factor index of neutrophilic granulocytes (FFNG) in case of a wound process we compared it with the results of the existing methods of control over the wound course.

The objective of the study was to determine the informativeness and expediency of determination of the morphometric criteria of neutrophilic granulocytes on diagnostics of the wound process based on the use of the index of the form factor of neutrophilic granulocytes in the patients with purulent wounds.

ii. MATERIALS AND METHODS

53 patients with purulent wounds of soft tissues were examined. They were treated at hospital for acute purulent pathology of soft tissues and underwent an operative intervention with further antibiotic treatment and local treatment of post-operative wounds considering the wound process phase. The main criteria for the patients' selection included age, localization of the suppurative focus and

V. Sontea and I. Tiginyanu (eds.), *3rd International Conference on Nanotechnologies and Biomedical Engineering*,
IFMBE Proceedings 55,
DOI: 10.1007/978-981-287-736-9_93

absence of a concomitant pathology. The species composition of causative agents identified in the wound content was also taken into account.

For dynamic control over the course of the post-operative purulent wounds the blood samples were taken for the study of laboratory (white blood count, ESR, leukogram [8], leukocytal intoxication index (LII) according to Ya.Ya. Kalf-Kalif [9]), immunologic (phagocytic parameter (PhP), phagocytic index (PhI), spontaneous NBT, stimulated NBT) indices [10] and wound content – for cytological and bacteriologial tests determined by traditional methods [11]. To determine FFNG we used blood smears stained according to Romanovsky-Giemsa, microscope "MIKMED-2" with MEDICAL IMAGE VIEW STATION, computer image analyzer UNHSCSA ImageTool v.3.0 and special program developed by the company "Eva" (Ukraine), "Factor M") [12]. Using this program the neutrophilic granulocyte was outlined in the picture using a cursive type. The perimeter (P) and area (S) were measured and the form factor index was calculated automatically ($NGFF = 4 \times S / P^2$). In each case 10 neutrophilic granulocytes in different areas of the object glass were studied.

In case of an ideal form of a circle the form factor should equal to 1. Abnormality of the circle form will lead to reduction of this index. Inactive neutrophilic granulocytes are bullet-like cells but if stimulated they change their form as a result of creation of pseudo-events.

The statistical processing of the obtained results was performed using the program package "STATISTICA 6.1." and calculating the average value M and Wilcoxon criterion (for comparison of the dependent indices in the studied groups). While comparing the indices in the studied groups the differences with $p \leq 0,05$ were considered significant.

In 30 healthy volunteers the average NGFF index was 0,96 ± 0,01.

III. RESULTS

During physical examination all the patients had: high body temperature (37,5 – 38,7 °C); increase of the size of the affected area with solid and painful infiltration on palpation; dermahemia, in some cases with a softening area (positive symptom of fluctuation), necroticly changed skin defects with discharge of purulent exudate; dysfunction of the affected extremity.

After the performed treatment the patients' condition improved, the pain syndrome disappeared on the 5,34 ± 0,19 day, the body temperature normalized on the 5,96 ± 0,37 day, necrolysis – on the 6,85 ± 0,18 day, discharges from the wound stopped – on the 8,24 ± 1,12 day, hyperemia disappeared on the 7,18 ± 0,26 day, infiltration and edema disappeared on the 7,59 ± 0,63 day, granulations appeared– on the 7,26 ± 0,48 day, epithelization signs appeared on the 8,92 ± 0,66 day.

The results of the routine laboratory blood examination allowed to establish changes only on admission of the patients to the department before the curative measures that included moderate leucocytosis (9,67 ± 0,33 G/l), ESR increase (34,6 ± 2,0 mm/h) and minor increase of the relative number of stab neutrophils (7,83±0,44 %), with further decrease of these indices on the 2nd day already and their dynamic normalization. There was no significant difference of changes in the dynamics of other blood cell populations.

The results of calculation of the integral haematological index LII that represents the degree of the endogenous intoxication showed that before the complex treatment it reached 2,43 ± 0,17. Starting from the second day of treatment a clear tendency to its rapid decrease with normalization on the third day already (1,37 ± 0,13) could be observed. During more remote periods of the observation there was no significant difference between the indices.

The stated hematologic indices point only at the degree of endogenous intoxication and does not reflect the real state of the processes in purulent wounds.

The performed bacteriological examination of the wound content allowed to establish that the main causative agent on the qualitative assessment of the wound content was S. aureus. Quantitatively the wound content showed 7,48 ± 0,49 log CFU/ml on the first day. Later the microflora amount in the wound decreased and on the 10th day already it was not revealed in the majority of patients.

The analysis of the obtained results of the cytological examination proved that reduction of the number of neutrophilic granulocytes and their destructive forms, increase of the number of fibroblasts and macrophages, reduction of the number of microorganisms, frequency of detection of uncompleted and increase of completed phagocytosis, increase of the phagocytic activity, displacement of the cytogram to the inflammatory type indicated a favorable course of the purulent wound leading to healing.

The study of the congenital nonspecific cell-bound immunity of patients with purulent pathology of soft tissues allowed to reveal decrease in the activity of this component of the immune system on admission: PhP– 38,33 ± 1,07 % and PhI – 4,17 ± 0,21 units. Later the increase of PhP and PhI indices could be observed with their approaching to the norm on the 10th day (55,17 ± 2,68 % and 6,83±0,41 %, respectively). The spontaneous index of the oxygen-dependent metabolism of neutrophils was higher – 14,17 ± 0,59 % with a tendency for gradual reduction during remote periods. The level of the spontaneous NBT remained within physiological norms on the 1st and 3rd day with further rapid increase on the 7th day (42,50 ± 1,49 %).

The results of the NGFF index determination in the patients' blood were the following: it was 0,8809 ± 0,0009 on admission. On the 2nd day after operative intervention performed in the patients with purulent wounds this index increased up to 0,8973 ± 0,0009 and it tended to increase in future. On the 10th day of the observation the blood FFNG index was 0,9239 ± 0,0010.

The statistical analysis of the obtained results allowed to determine the following correlation relationships with the FFNG index determined in the blood: inverse relationship with the quantity of leucocytes in blood (r = – 0,58; (p ≤ 0,05)), strong inverse correlation relationship with the quantity of microorganisms ((r = – 0,91; (p ≤ 0,05)), direct correlation relationship with PhP (r = 0,56; (p ≤ 0,05)), PhI (r = 0,62; (p ≤ 0,05)) and inverse relationship with spontaneous NBT (r = – 0,53; (p ≤ 0,05)).

The studied results of the cytological and morphological changes over time confirmed the regularity and staging of the processes in the wound and had a clear correlation relationship with the indices of FFNG (p ≤ 0,05).

IV. Conclusion

In patients with purulent wounds the index of the neutrophilic granulocyte form factor, determined in the peripheral blood, reflects the course of the wound process objectively and can be proposed for the use in clinical practice.

Conflict of Interest

The authors declare that they have no conflict of interest.

References

1. V.D. Fedorov, A.M. Svetukhin, S.P. Glyantsev, "The doctrine about a wound: from A.V. Vishnevsky to our days," Khirurgiia (Mosk), No. 8, 2004, pp. 56-61.
2. N.V. Zhadynsky, A.N. Zhadynsky, "Patho- and sanogenetic aspects of wound process," Ukrainian Journal of Surgery, vol. 21, 2013, pp. 158-162.
3. B.M. Datsenko, "Wound process as a fundamental problem of modern clinical surgery," Clinical Surgery, vol. 837, 2012, p. 14.
4. W.A. Knaus, E.A. Draper, D.P. Wagner & J.E. Zimmerman, "APACHE II: a severity of disease classification system," Critical Care Medicine, No. 13, 1985, pp. 818-829.
5. A.A. Zvyagin, A.M. Svetukhin, S.Y. Slepnyov, "Development of a system of objective evaluation of the severity of the patients with surgical infection," Intensive Care Herald, No. 2, 2003, pp. 34-38.
6. I.G. Gerasimov, T.M. Galburt, "Human blood neutrophil morphology during their phagocytosis," Reports of Donetsk National University. Series A: natural science, No. 1, 2009, pp. 377-382.
7. B.V. Pinegin, A.N. Mayansky, "Neutrophils: structure and function," Immunology, No. 6, 2007, pp. 374-382.
8. G.I. Nazarenko, A.A. Kishkun, Clinical evaluation of laboratory results. Moscow: Meditsina, 2000.
9. Ya.Ya. Kalf–Kalif, "About leukocyte index of intoxication and its practical significance," Vrachebnoe Delo, No. 1, 1941, pp. 31–33.
10. K.A. Lebedev, I.D. Ponyakina, Immunogram in clinical practice. Moscow: Nauka, 1990, pp. 75-79.
11. U.K. Abaev, Wound infection in surgery. Minsk: Belarus, 2003 pp. 140-150.
12. A.I. Godlevsky, S.I. Savolyuk, Diagnostics and monitoring of endotoxicosis in surgical patients. Vinnytsia: Nova Knyha, 2015, pp. 47-49.

Author: Roman Chornopyshchuk.
Institute: Vinnytsia National Pirogov Memorial Medical University.
Street: Pirogov str., 56.
City: Vinnytsia.
Country: Ukraine.
Email: ro_man@mail.ru.

Restoringn Spatial Phase Distribution of Complex Optical Fields for Biomedicine Application

C. Yu. Zenkova, P.A. Riabyi, and D.I. Ivanskiy

Optics, Printing&Publishing Department, Chernivtsy National University, Chernivtsy, Ukraine

Abstract— **The paper presents principal approaches to diagnosing the structure forming skeleton of the complex optical field. It is shown that intensity distribution smoothing and bicubic spline simulation allow to bring much closer the solution of the phase problem of localization speckle-field special points.**

Keywords— **phase skeleton, saddle points, gradient lines.**

I. Introduction

Great attention has been paid to solving the phase problem in optics, particularly to tackling the problems of diagnosing the structure of objects in microscopy, astrophysics and biomedicine [1,2]. As a rule, reference wave can be used for obtaining the phase information. The phase problem involves revealing spatial (coordinate) phase distribution in complex fields of the speckle-field type [3,4]. Phase restoring becomes actual by the development of non-invasive methods of medicine diagnostic. Optical beams as proofing tools are used in this case therely eliminating the introduction of any mechanical perturbation in the test medium. In the given situation the use of the reference wave becomes impossible. A separate new and promising line of tacking the phase task, i.e. finding the spatial (coordinate) phase distribution in complex fields of the speckle type is offered by singular optics. The approaches [5-8] in the framework of singular optics are based on the concepts of:

1. The so called field “skeleton” with “reference” structure-forming points of the field amplitude zeroes;
2. The visualization of the sign principal in the coordinate (spatial) distribution of the wave dislocation, i.e. of the field amplitude zeros;
3. The interconnection of intensity spatial distribution and field phase distribution.

This paper deals with the development of these concepts and presents different methods for tackling the problem of diagnosing spatial (coordinate) phase distribution with the help of registered intensity distribution in complex optical fields. The results of the computer simulation of the experiment fulfilled within the framework of scalar approximation form the basis of the investigation. The algorithm of solving the phase problem in the approximation of scalar coherent optical fields can be divided into several stages. The first one is to obtain complex amplitude-phase transformations in fields, which are generally called developed speckle fields with statistically distributed zeros of the field amplitude (phase dislocations). The second one is a search for optimal algorithms for diagnosing the zeroes of speckle fields (of complex spatial intensity distribution), registered by the help of CCD camera. The third one is searching for approaches and methods for restoring the phase coordinate distribution of the field by defining the principles and algorithm which can make possible to restore the phase grid (the coordinate phase distribution grid) in a complex speckle field in a justified optimal way.

II. Algorithms of the Phase Distribution Reconstruction

As a rule, phase problems of the above described type are solved rather successfully in the concrete case when using a reference field [9]. But in our case, for remotely diagnosing, the possibility of using a concrete reference field is excluded. Measurements with the help of the CCD camera give a spatial map of intensity distribution. The map presents a set of pixels. In the general case each of the pixels under consideration is surrounded by eight other neighbouring pixels (Fig.1).

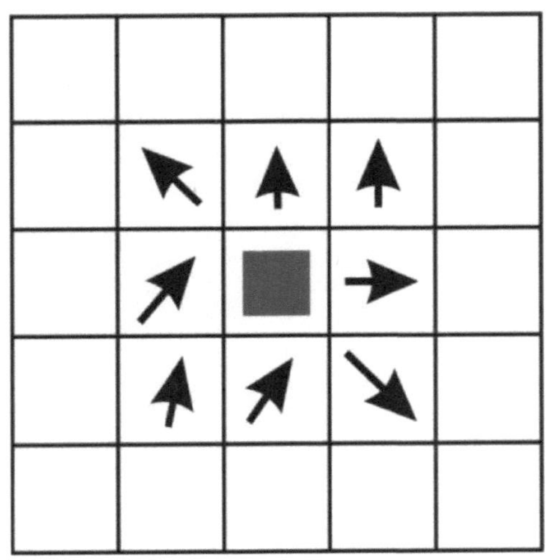

Fig. 1 Schematic sketch of a set of pixels.

V. Sontea and I. Tiginyanu (eds.), *3rd International Conference on Nanotechnologies and Biomedical Engineering*, IFMBE Proceedings 55,
DOI: 10.1007/978-981-287-736-9_94

The arrows in Fig. 1 point the direction of the gradient vectors of intensity. They are defined at each point by the equation

$$grad(I) = \frac{\partial I}{\partial x}\vec{i} + \frac{\partial I}{\partial y}\vec{j} \qquad (1)$$

where $\vec{i}$, $\vec{j}$ - are the unit vectors along the *x*- and the *y*-axes correspondingly, I – is the intensity.

If a standard camera is used the intensity of each pixel is varied from 0 to 255, i.e. it can take 256 different values. Three possible algorithms of defining the localization places of the field amplitude zeros were investigated.

A. Algorithm 1

The point under review was marked as a potential point of field phase singularity, if there were two intensity gradient vectors in the vicinity of the point and the angle between them was 180^0, i.e. two gradient vectors were oppositely directed. The effect of changing the image resolution (of the pixel size) on the potentialities of diagnosing the points of amplitude zeros (singularities) was considered. The size of the pictures was 200x200, 600x600 pixels. The influence of intensity discretizations (the number of possible intensity values for each pixel) was also investigated. The number of values was 256 (a standard camera), 500 and 1000.

The offered algorithm can be useful in searching for amplitude zero points in the field, but only if intensity discretization is absent. In this case many "excessive" points are diagnosed. At discretization the posibility to find singularity grows worse, particularly with the increase of the size of pictures.

B. Algorithm 2

It sprung from of the attempt to perfect algorithm 1. The sum of angles of the intensity gradient vectors was calculated for each point of potential singularity (the intensity minimum). Further the lines of sign change of this "total" angle were plotted. The given lines extend through the points of singularity, maxima and saddle points. They extend through all points, defined by algorithm 1. The influence of the image resolution variations was investigated (the values are in pixels). The sizes of the pictures are 200x200, 600x600. The influence of the intensity discretization (the number of possible intensity values for each pixel) was also investigated. As in the previous case the number of intensity values was 256 (a standard camera), 500 and 1000.

The proposed algorithm makes possible to search for singularity only when intensity discretization is absent. But even in that case a lot of "excessive" points are found. At discritization the possibilities for finding singularity grow worse, particularly with the increase of the size of the pictures. At the same time the proposed algorithm can be improved, and therefore it is worth to consider the feasibility of its future elaboration.

C. Algorithm 3

The standard algorithm of searching for points with a zero intensity is based on searching for pixels informing that the intensity in the given point is equal to zero. The effect of the change of the image resolution (the values are in pixels) was considered. The size of the pictures was 200x200, 600x600. The influence of intensity discretization was also considered (the number of possible intensity values for each pixel). The number of values was 256 (a standard camera), 500, 1000.

This algorithm makes it possible to search for fields containing singularity. But in this case regions without singularity are found as well. The algorithm is insensitive to the size of the imagine, but dependence on the discretization. It is quite possible to use it in combination with other algorithms.

D. Algorithm 4

The algorithm considered here is a logical continuation of algorithm №1, and is the most promising one of all algorithms discussed earlier. Two alternative procedures of smoothing – "cross" smoothing by two points and "circle" smoothing by two points were proposed (Fig.2 and Fig.3).

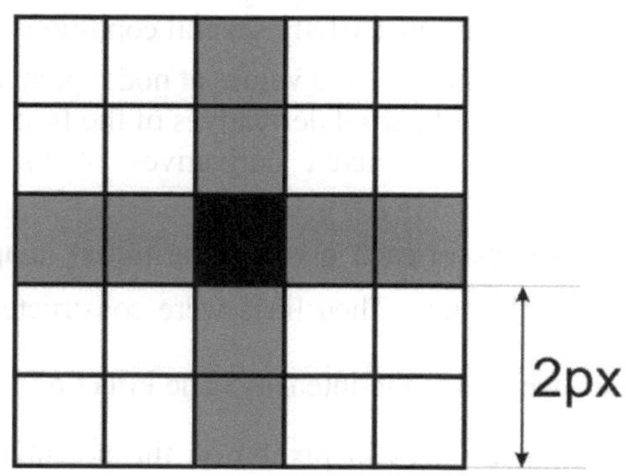

Fig. 2 "Cross" smoothing by two points

The point under consideration was marked off if there were two gradient vectors in the vicinity with an angle of 180^0 between them and the point intensity was zero. The procedure of "smoothing" was performed in addition. The effect of changing of the image resolution was studied (the sizes are given in pixels). The sizes of the pictures are 200x200, 600x600. The effect of the intensity discretion (the number of possible intensity values for each pixel) was studied as well. The number of values was 256 (standard camera) and 1000.

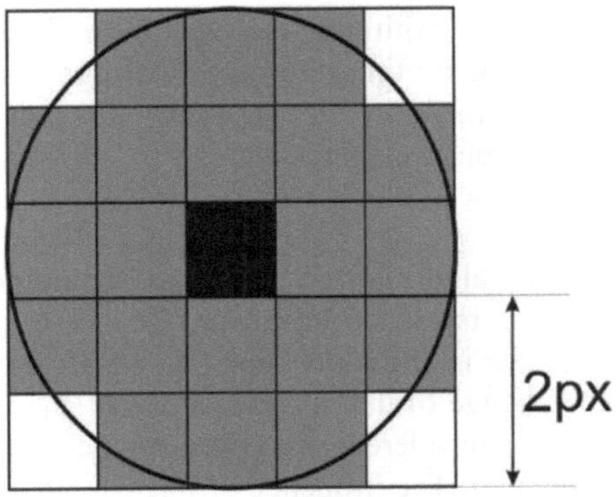

Fig. 3 "Circle" smoothing by two points

Smoothing provides greater scope for finding "zero points" for discrete image especially in the case of "circle" smoothing by 10 points. Optimum smoothing parameters depend upon the half-width of the image correlation function (in pixels) and the intensity discreteness of the image pixels. In the given situation the probability of faultless diagnostics of the image pixel intensities can be 95%. However the question of defining the sign of the field amplitude zero remain open and, correspondingly, the question of determining the trajectory of phase lines. In connection with this the following algorithm was proposed.

E. Algorithm 5

The intensity distribution image was smoothed and interpolated by bicubic splines [10].

The given splines must satisfy several conditions:

- equality of calculated values at nodal points,
- continuity of partial derivatives of the first order,
- continuity of mixed derivatives of the second order.

Intensity values in each pixel of the initial image were chosen as nodal points. Then lines were constructed dI/dx and dI/dy, where I is the intensity. The effect of the image resolution change (sizes in pixels) on the potentiality and exactness of diagnosing the localization places of intensity extrema and field phase singularity. The sizes of pictures were 200x200, 400x400, 600x600. The influence of the intensity discretization (the number of possible intensity values for each pixel) upon the feasibility of diagnosing the same field parameters were investigated as well. The number of intensity values was 256 (a standard camera), 1000 and 10000 (the maximum possible, i.e. the practically continuous one). Three different intensity distributions were considered (Fig. 4).

The given approach is only slightly sensitive to the size of the image and it can be well used with a low intensity discretization. It may allow to define the signs of singularity. A distinguish of feature of the given lines is their connection with the field "skeleton". If we investigate the physical meaning of the lines $dI/dx = 0$ and $dI/dy = 0$ it is seen that the given lines pass the intensity gradient in such a way that the gradient component "x" is equal to 0, analogously "the lines $dI/dy = 0$" connect the points in which the gradient component "y" is equal to 0. A schematic sketch of it is offered in Fig.5.

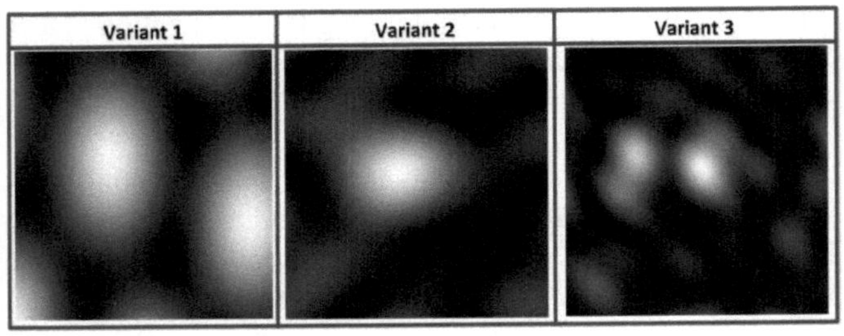

Fig. 4 Three different intensity distributions

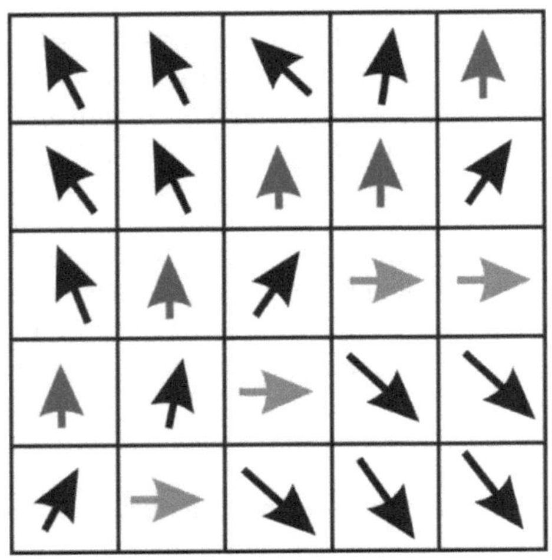

Fig. 5 Schematic representation of the lines of zero derivatives

The figure shows that the given lines are not invariant with respect to the system of coordinates. In other words, the position of the given lines will change with the rotation of the figure about the system of coordinates. At the same time the points formed by the intersection of such lines, i.e. the ones which meet the following conditions

$$\begin{cases} \dfrac{dI}{dx} = 0 \\ \dfrac{dI}{dy} = 0 \end{cases} \quad (2)$$

create a system of points invariant to the change of the system of coordinates. The given invariance has a quite definite physical meaning. The thing is that the special points in the intensity distribution, i.e. local minima (including "singular" points) and saddle points are reference points of the structure forming skeleton of the field under review. The position of these points doesn't depend upon the position of the figure relative the coordinate system, and condition (2) is a criterion for defining similar "special" points in mathematics.

III. RESULTS

The considered behavior of structure formation in the field of optical irradiation, as a result of simulating real experimental situations, has a quite definite real "life" analogy if a coaxial plane reference wave with a smoothly changing phase is used for diagnosing complex speckle fields. At the same time we have to state that the diagnostics of the zero signs of the field amplitude and the signs of vortices still present difficulties. Hence, when solving phase problems it is necessary to take into consideration different alternatives. Originally it was assumed that while moving along the connecting line from one singularity to the other, the second singularity "changes" its sign after meeting with the "saddle point". If an even number of saddle points is found along the line, the singularity sign will be identical to that of the point from which we began. If the number of the saddle points is not even the sign will be opposite. The figures below illustrate lines, singularities (red and blue points, the colour corresponds to the sign) and saddle points (marked off in yellow) (Fig.6).

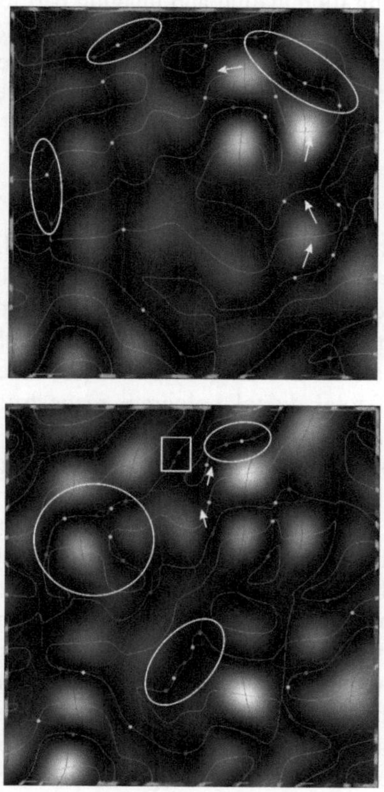

Fig. 6 Examples of distribution of intensity and localization «reference» points algorithm №5

The analysis showed that the primary assumption was not quite right. It can be observed when moving along the line marked off by white arrows. At the same time it is absolutely clear that the availability of one saddle point between singularities assures the difference of their signs. The cases are marked off by white ovals. But there are situations when it is impossible to diagnose the saddle point because it doesn't exist at all, as in the case marked off by the white square.

CONFLICT OF INTEREST

The authors declare that they have no conflict of interest.

IV. CONCLUSION

The analysis of different approaches to restoring phase distribution of the complex optical field allows to single out the best method for speckle-field skeleton reconstruction. The suggested method allows not only to diagnose singularity and saddle points, but to define the signs of singularity as well.

REFERENCES

[1] R.H.T. Bates, M.J. McDonnell, "Image Restoration and Reconstruction", Calerdon Press: Oxford; 1986.

[2] T. Acharya and K. Ray Ajoy, "Image Processing – Principles and Applications", Wiley: InterScience; 2006.

[3] J.W. Goodman, "Some fundamental properties of speckle", JOSA 1976; 66(11): 1145-1150.

[4] I. Mokhun, "Amplitude zeroes and structure of statistical optical fields. Correlation between the field's intensity and phase", Proc. SPIE 1998; 3573: 567-571.

[5] O.V. Angelsky, M.P. Gorsky, S.G. Hanson, V.P. Lukin, I.I. Mokhun, P.V. Polyanskii, P.A. Ryabiy, "Optical correlation algorithm for reconstructing phase skeleton of complex optical fields for solving the phase problem" Optics Express, Vol. 22 Issue 5, pp.6186-6193, 2014.

[6] O.V. Angelsky, C.Yu. Zenkova, M.P. Gorsky, P.A. Ryabyi, "Search for methods of restoring spatial phase distribution of complex optical fields", The Open Optics Journal, 1874-3285, 2014.

[7] C.Yu. Zenkova, M.P. Gorsky, P.A. Ryabyi, "The phase problem solving by the use of optical correlation algorithm for reconstructing phase skeleton of complex optical fields", Proc. of SPIE Vol. 9258 92582B-1, 2015.

[8] C.Yu. Zenkova, M.P. Gorsky, P.A. Ryabyi, "Different approaches to phase restoration of distant complex optical fields", Optica Applicata, Vol. 45, Issue 2, 2015 (in publishing).

[9] M.V. Berry, "Much ado about nothing: optical dislocation lines (phase singularities, zeroes, vortices…)", Proc. SPIE 1998; 3487: 1-5.

[10] R. Keys, "Cubic convolution interpolation for digital image processing", IEEE Transactions on Signal Processing, Acoustics, Speech, and Signal Processing 1981; 29 (6): 1153–1160.

Author: Riabyi Pavlo A.
Institute: Chernivtsy National University
Street: 2, Kotsyubinsky Str.
City: Chernivtsy
Country: Ukraine
Email: ryabyi.pavlo@gmail.com

Development of Digital Holographic Microscope for 3D Sensing of Biological Surface Morphology

E. Achimova

Institute of Applied Physics, Academy of Sciences of Moldova, Chisinau, Moldova

Abstract— **Sunflower broomrape Orobanche cumana Wallr.(O. cumana) is a parasitic angiosperm that infects the roots of sunflower plants, causing significant damages to this strategic oilseed crop. Combining microscopy and digital holography (DH) offers the unique advantage of simultaneously capturing complete three-dimensional (3D) information about the biological specimen, and under coherent illumination it has been investigated for computer acquisition and identification of biological specimens. Configuration of the digital holographic microscope optical setup provided with the microscope epi-objective for darkfield illumination by laser beam is described. The aim of this research is to develop microscopy 3D system to enhance the resolution and contrast in imaging of O. cumana seeds for they systemizing. Successful application of the modern digital holographic microscope allowed studying the broomrape seed in various layers, like surface and under surface of seed.**

Keywords— **digital holographic microscope, 3D imaging.**

I. INTRODUCTION

Optical imaging system using Digital Holography (DH) [1, 2] is attractive technique for the acquisition of 3D information by reconstructing both the amplitude and phase of the wave front that passes through or is reflected/scattered by the object. In order to establish DH in microscopy, the combination with common microscopy techniques is of particular advantage. In this case flexible and compact Digital Holographic Microscopy (DHM) modules for the integration into modern microscopy systems are required. Furthermore, for an automated evaluation of the measurement data, it is necessary to implement robust algorithms for the numerical reconstruction of digital holograms. Combining microscopy and DH offers the unique advantage of simultaneously capturing complete 3D information about the specimen, and under coherent illumination it has been investigated for continuous PC acquisition and identification of biological specimens. In contrast to conventional microscopy, no depth scanning focusing at various depths of the 3D specimen are necessary in range of some interference fringes. Thus it is possible to overcome the main limitation of confocal microscope to studying processes that occur on time scales slower than the acquisition time, which is on the order of one second. If a height of an object overcomes distance of several wavelengths of illumination it is possible to put object on stage with 100nm steps directed by PC with synchronous acquisition of the images. But the interferometric accuracy of altitude scanning is retained.

DHM is a promising interferometric technique allowing obtaining quantitative phase images of living samples dynamics with interferometric accuracy. There are many interferometry methods which have been developed for microscopy application. Among them, Digital Speckle Pattern Interferometry (DSPI) is widely used for deformation studies, and it is the most practical and powerful one [3]. DSPI combines the holographic method of image recording with digital processing of acquired image, and it can be used as for qualitative as for quantitative applications. One of the important features of DSPI is measuring object deformations in-plane and out-of plane.

In this paper, the DHM was developed based on DSPI with modifying illuminating and imaging part to do suitable for hundred nanometers sizes of the biological samples measurements. As applications of the proposed measurement technique, some experimental results are presented, which are measured with DHM developed.

II. ELABORATION OF OPTICAL SET-UP

For digital holographic measurement, DSPI, one of the temporal measurement methods, is preferred because this technique induces an object wave by plural interferograms with the relative phase-shift with respective to the reference wave which allows avoiding the power and time consuming procedure of total digital hologram reconstruction. Phase information recorded interferometrically allows studying of biological specimens that would otherwise die from staining or be invisible to conventional imaging means.

It is well known that when an optically rough surface is illuminated by coherent light, speckle can be observed due to phase difference from surface points imaged on a detector. Thus speckled image of object is medium about the object and the coherent laser beam. DSPI relies on the correlation between two speckle patterns; each one is created by the interference between a reference beam and a beam reflected from the object illuminated by a laser. Typically the two patterns are speckled images of an object before and after some deformation. Although there are several variations of the basic design, the interferometer of interest here

V. Sontea and I. Tiginyanu (eds.), *3rd International Conference on Nanotechnologies and Biomedical Engineering*,
IFMBE Proceedings 55,
DOI: 10.1007/978-981-287-736-9_95

uses image subtraction to correlate the two images. The process of measuring the correlation between the two images does not need to be accomplished electronically and was originally accomplished using film; however, the electronic subtraction process makes the interferogram easy to form, view, and record in real time. Due to relative low resolution of digital camera chip (about 200lp/mm) the beams angle must be as small as possible in order to the fringes period should be larger.

a.

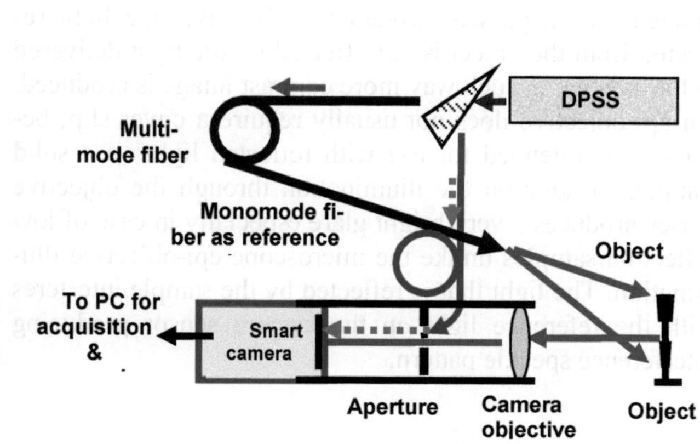

b.

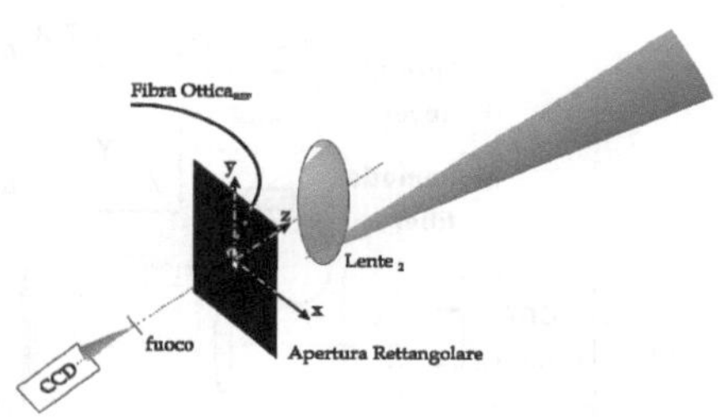

Fig. 1 Optical scheme of DSPI set-up (a), and camera objective part (b)

In our previous work [4, 5] the portable DSPI set-up was developed with optical fibers for laser beam delivering. Delivering laser energy in reference and object arm through the optical fibers gives, on the one hand, to attain hard-to-reach object under investigation, and, on the other hand, to get good vibration and environment immunity of setup. In addition two mechanically separated parts (illumination and monitoring parts) gives practical versatility of the set-up. The DSPI set-up with coherent source monomode DPSS CW laser (λ= 0. 53μm, power 100 mW) developed from off-the-shelf components is shown on Fig.1.

The object beam is obtained by illuminating the object under investigation with multimode optical fiber and then focusing the light scattered by the object onto a digital Smart Camera NI 1722 (monochrome 640 x 480 pixels CCD image sensor, pixel size 7.4 x 7.4μm, and Ethernet-based communication, on board processor 400 MHz Freescale PowerPC) using camera objective. A part of laser light was coupled into a polarization maintaining single mode "Bow- Tie" fiber and coupled directly with camera head module to serve as the reference beam arm. No beamsplitter was used in front of camera to avoid additional noises in form of Newton's rings which influence on interference pattern formed on camera sensor. The spot of reference beam is matched with the size of camera sensor whose the outlet monomode fiber position coincides with plane of aperture diaphragm, and they are at the same distance from the camera. So image plane hologram is formed on the camera sensor as a result of the interference between the slightly off-axis position (~200μm) of reference beam and object beam. Small angle between beams results wide interference fringes. Real time optical processor formed by objective lens and diaphragm yields Fourier filter processing of images (Fig.1b). Simplicity of the optical processor gives the possibility:

- do measurements with interferometric accuracy
- design compact device;
- use low power laser source.
- cut-off high frequencies carry no information about object

But the rate and uniformity of overall magnification are determined by the camera objective, and their values are not sufficient for microscopy application. So it is needed to fit optical system of DSPI with the respect to object dimensions.

Entocentric optics is rather unsuitable for measuring tasks and the inspection of microobjects because it introduces several adverse factors reducing measurement accuracy and repeatability: 1) a variation in magnification leads to coma; 2) the distortion of image due to aberration for parts of object lined away from the optical axis; 3) the perspective errors; 4) the poor resolution.

The telecentric objective helps to reduce most of these problems, and for this reason it has become a key optical component for high accuracy microscopic application. Our reasons for choosing telecentric objective in DSPI set-up design emerge from the requirements to cover all field of view (FOV) with the same magnification and permit an object move slightly within the depth of field. Telecentricity defines how the amount of magnification of an object within the FOV changes with object distance. Due to the telecentric objective acts like lens with infinite focal length, its magnification is independent of object distance. In our case the telecentric objective is supplied as object-space. Hence the requirement of precise repeatable measurements of microscopic object by the developing DSPI set-up is met.

The ZEMAX Optical Design Program is a comprehensive software tool [6] which integrates all the features required to conceptualize, design, optimize, analyze, tolerance, and document virtually any optical system.

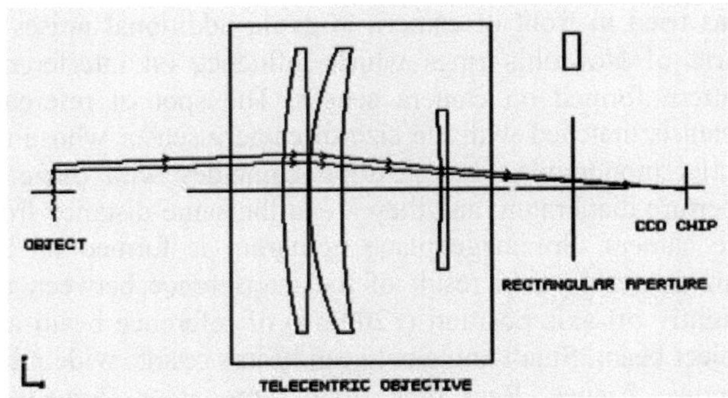

Fig. 2 Wireframe model of three lenses telecentric objective of DSPI set-up with F =27.7mm.

The ZEMAX's Image Analysis, Wavefront Map and Interferograms features have been applied for verifying and predicting the performance of optical part of fiber-optic DSPI set-up. First of all geometrical model of three lenses telecentric objective with F =27.7mm of the DSPI set-up was designed by optical code ZEMAX, which is presented on Fig.2. Object position is on the left side, and vertical line points out the image position. Wave front of objective was calculated for λ=0.53 μm.

An illumination system for microscopy must satisfy the following conditions:

1. The object field should be fully and evenly illuminated.
2. The contrast of the images should be high which is weighty for unstained biological samples when these are nearly transparent and for spoiled by speckle images.
3. The NA of the objective should be matching the variable range, from 0.1 to 1.25.

The developed object-space telecentric objective meets well first condition which proves by the wavefront and relative illumination. To satisfy 2) and partly 3) requirements we choose dark-field epi-illumination what renders the object as bright against a dark background, considerably enhancing the contrast and visibility of small objects. It would seem that epi-illumination can be used for most microscopy techniques including bright field, dark field, phase contrast etc., but careful setting up is required, and some adjustments have to be made in any case.

Our dark-field illumination scheme utilizes the microscope epi-objective in object arm of DSPI. The laser light is delivered through a 2000 μm diameter bundle of optical fibers and illuminates the sample under investigation (Fig. 3) through the condenser parabolic mirror of microscope epi-objective. The object beam formed by a hollow cone of illumination falls to the object, and only light backscattered by the object gets to camera. This backscattered light passes through the lenses of the epi-objective to be imaged. When passed through the objective, the light reflected from the object is not affected by the light delivered to the sample. In such way more contrast image is produced. An epi-objective does not usually require a cover slip, because it is intended for use with reflected light on a solid sample. In addition the illumination through the objective lenses produces a very bright glare especially in case of low reflection samples unlike the microscope epi-objective illumination. The light that is reflected by the sample interferes with the reference light on the camera sensor producing interference speckle pattern.

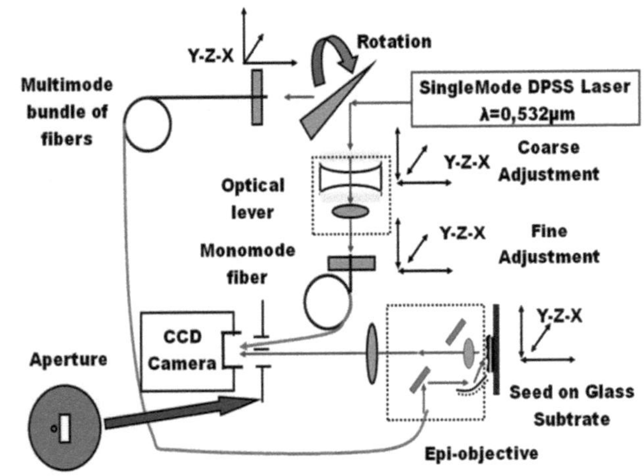

Fig. 3 Configuration of the DHM optical setup provided with the microscope epi-objective for darkfield illumination by laser beam

On the fig.3 the configuration of the developed DHM is presented. There are 100 mW DPSS laser, wedge prism as beamsplitter, multimode bundle of optical fibers as object arm, seed glued on dark glass substrate, optical "lever" composed from negative lens and microobjective (90^x and NA=1.25) as positive lens, monomode optical fiber as reference arm, rectangular aperture, CCD Smart Camera. Aligning laser light into the core of a monomode fiber requires submicron resolution. For this demanding task, with a simple concept that uses the optical "lever" effect, to increase the resolution of the system was used.

The optical "lever" provides accurate micron translation. A microobjective was used to maximize monomode coupling efficiency focusing to micrometrical size spot. The ratio of the focal lengths negative/ positive lenses is 23:1, and one rotation of the adjustment screw translates the focused spot by 2 micron. So an optical "lever" is a convenient device makes possibility an accurate coupling as a spot diameter and minimum translation step.

III. RESULTS AND DISCUSSION

In order to demonstrate the capabilities of our epi-illumination system for studying biological sample, we present a preliminary measuring of phase map, which are derived from interference fringes imaging of two state of the sample. The main procedure of phase map calculation described by many authors [1, 9, 11, 12] is follow.

For deformation measuring, an initial (reference) speckle pattern image from the surface target is taken. This reference image is subtracted from the loaded speckle pattern image when the object under study is deformed by any way. The two speckle patterns acquired by a digital imaging system have the phase difference φ. The intensities of the two speckle patterns can be represented as follows:

$$I_1 = I_r + I_o + 2\sqrt{I_r I_o} \bullet \cos\varphi .$$

$$I_2 = I_r + I_o + 2\sqrt{I_r I_o} \bullet \cos(\varphi + \Delta\varphi\ ).$$

where $I_o(x;y)$ and $I_r(x;y)$ are the intensities of the object's scattered light and reference beam, respectively. $\varphi_o(x;y)$ and $\varphi_r(x;y)$ are the phases of random speckles, and $\varphi(x;y)$ is the phase difference between the two speckle patterns resulting from loading. If the output camera signals I_1 and I_2 are proportional to the input image intensities, then the subtracted signal is given by

$$V_s = I_1 - I_2 = 2\sqrt{I_r I_o}\,[\cos\varphi - \cos(\varphi + \Delta\varphi)] = 4\sqrt{I_1 I_2}\,\sin(\varphi + \tfrac{1}{2}\Delta\varphi)\sin \tfrac{1}{2}\Delta\varphi_o.$$

This signal has negative and positive values. The monitor will, however, display negative signals as black areas; to avoid this loss of signal, V_s is rectified before being displayed on the monitor. The brightness on the monitor is proportional to $|V_s|$, so we have the brightness B at a given point in the monitor image given by

$$B = 4K\,[I_r I_o\ \sin^2(\varphi + \tfrac{1}{2}\Delta\varphi)\,\sin^2(\tfrac{1}{2}\Delta\varphi)]^{1/2}.$$

where K is a constant.

We apply the Fast Fourier Transformation (FFT) for spatial carrier phase measurement method which based on processing is performed in frequency domain. After the filtering procedure and inverse FFT the complex amplitude of the wave front is obtained. The operation described corresponds to the reconstruction called quasi Fourier hologram [11]. A deformation field in the shape of correlation fringes is obtained by subtracting the reference and deformed patterns.

Our microscope keep phase shift and intensity information separated in two speckled images. This enables biological samples to be observed and quantified. Phase shift is proportional to differences in reflection optical path length. In reflected light phase differences arise from relief on the spacemen surface. The height of an image points represent the optical path lengths and can be displayed in 3D.

In our experiments initial (reference) speckle pattern image of the seed samples glued on black glass substrate has been acquired first. Then the substrate has been heated by quartz Halogen Lamp with reflector, so the thermal excitation of glass substrate has been done. If a short heat pulse of lamp (5-10s) local hits at the substrate surface. The sample surface out-of-plane displaces the seed position by thermal expansion of glass. The temperature loaded speckle pattern image was done. It displaces sample depend on time of thermal excitation. So observing the phase map changing we detect the existing size and shape of seed surface. We acquire two consequent interference patterns before and after thermal excitation. Then the two speckled interferograms are subtracted to control interference quality. The image acquisition was accomplished using a Vision Builder Automatic Inspection and consequence speckle patterns processing based on LabVIEW 2011 program. The phase obtained from the evaluation of the interferograms is in the interval $-\pi$ to π, and is indefinite to an additive integer. Unwrapped phase map extracted by program gives 3D surface morphology.

The seed photo in white light on the Smart Camera trough epi-objective is presented on Fig.4.

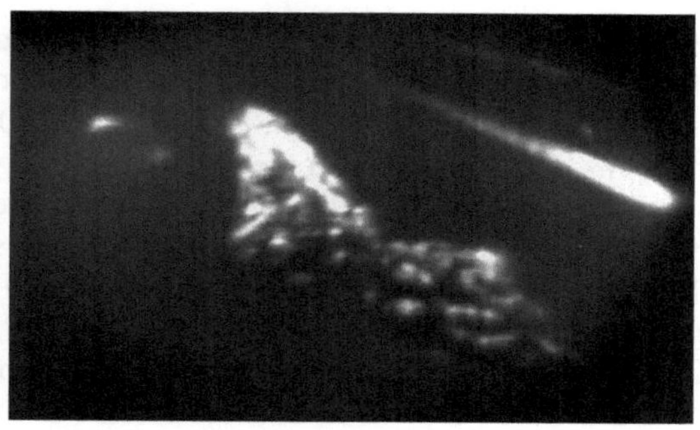

Fig. 4 The photo of the seed (~300μm long) glued on black glass substrate

To compare the results obtained with and without microscope epi-objective we have measured the phase map of the samples obtained by DSPI with one objective lens and DSPI with microscope epi-objective. Fig. 5(a), (b) shows succession of images which reflects the steps of acquiring and processing of data: (a) intensity interferogram obtained by subtracting images of heated and unheated (named reference) sample; (b) wrapped phase map obtained from the intensity interferogram. The images presented on fig.5 were acquired by DSPI.

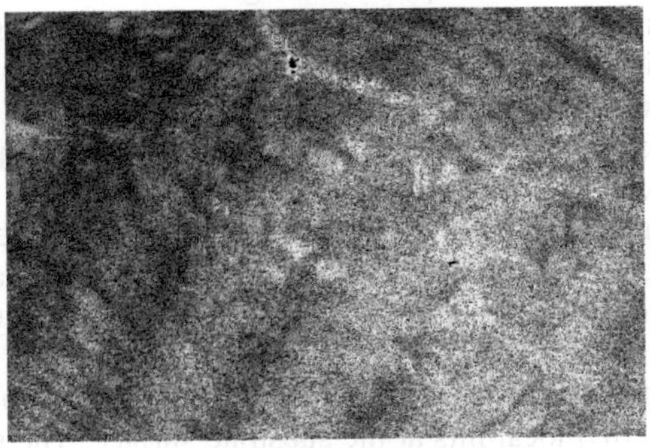

a

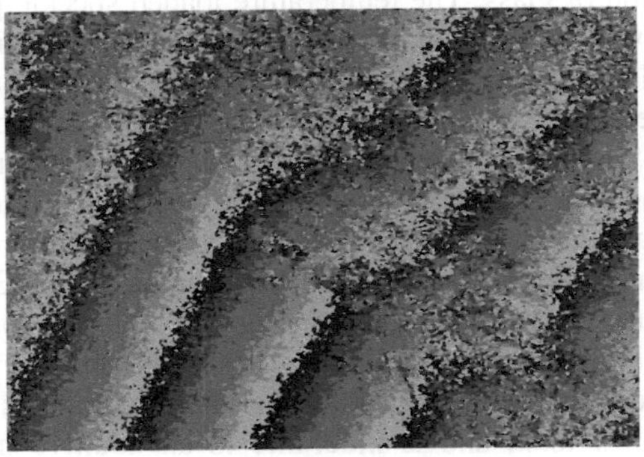

b

Fig. 5 The intensity subtraction interferogram (a), the wrapped phase map (b) recorded by DSPI without epi-microobjective. The seed position is marked by oval.

It is seen that the magnification of the DSPI (about 7^x) does not allow seeing the seed surface which introduces some distortion on the correlation fringes. The FOV is 10^6µm of this optical setup. On the fig.6 (b) intensity varieties from the black to the white spots in the oval region pointed out the surface phase shift. The surface phase shifts in reflection indicate the differences in height of the seed surface. The magnification of the DHM is about 70^x, and the FOV is 10^4µm. Unwrapped phase shift map will be the base for extract the 3D form of the seed surface in microscopic mode.

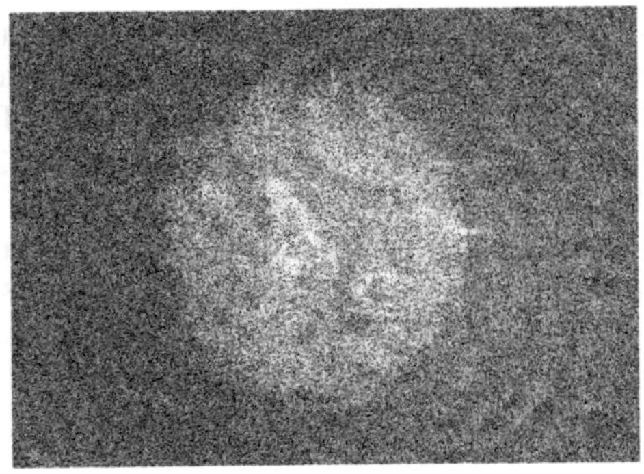

a

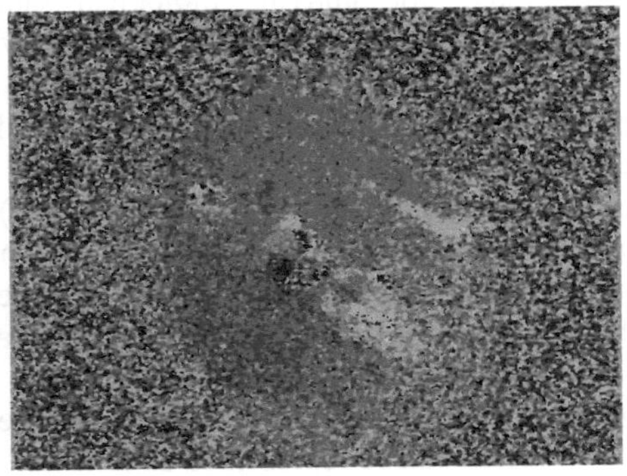

b

Fig. 6 The intensity subtraction interferogram (a), the wrapped phase map (b) recorded by DHM with epi-microobjective. The oval points the seed position.

IV. CONCLUSIONS

The technique performance with phase analysis of interferograms of the seed on a glass substrate by its thermal excitation was validated. The results obtained show to match our previously detailed telecentric illumination consideration [9]. The distribution of darkfield illumination was investigated, finding the regions of highest reflected intensity. The images acquisitions by smart camera using a vision builder automatic inspection were accomplished. Speckle patterns processing program based on labview 2011 was elaborated for phase map analysis. Finally, we have shown that the dhm technique is particularly useful for phase map analysis of scattering surface of sample in vivo, producing interferometric speckle images which are characterized the 3d form of surface.

ACKNOWLEDGMENT

This work is supported by the Bilateral Moldova-Germany Project (grant 13.823.15.10/GA).

CONFLICT OF INTEREST

The authors declare that they have no conflict of interest.

REFERENCES

1. B. Javidi, F. Okano, J.-Y. Son. Three-Dimensional Imaging, Visualization, and Display (Signals and Communication Technology); Springer-Verlag: New York, NY, USA, 2008.
2. S.A. Benton, V.M. Bove. Holographic Imaging; Wiley-Interscience: New York, USA, 2008.
3. K.M. Abedin, S.A. Jesmin, A.F.M.Y. Haider. Opt. Laser Technol. 32 (2000) 323.
4. V. Abaskin, E. Achimova. Non-destructive testing of wood defected samples by ESPI, Proc. "7th International Conference on Vibration Measurements by Laser Techniques: Advances and Applications", SPIE 5504–2006, pp. 322-329, Ancona, Italy.
5. V. Abaskin, E. Achimova. Portable electronic speckle interferometry device for the defect measurements in wood artworks, Journal of Cultural Heritage, V.IX, N3, pp.225-233, 2008.
6. ZEMAX Development Corporation, www.radiantzemax.com, ZEMAXDemo.exe.
7. V. Abaskin, E. Achimova. Rating of electronic speckle pattern interferometer by optical software. Proc. "8th Int. Conf. on Vibration Measurements by Laser Techniques: Advanced and Applications", 2008, Ancona, Italy, SPIE V.70980, pp. 2 - 8.
8. V. Abaskin, E. Achimova. Application of optical program for design of holographic speckle-interferometer, Proc. Sixth Internetional Conference "Holography. Science and Practice", 2009, pp. 165-169, Kiev, Ukraine.
9. V. Abaskin, E. Achimova. Using the Optical Code for Phase Analyses of the Fiber-Optic ESPI Set-Up. Proc. AIP –Int. Conf. on Advanced Phase Measurement Methods in Optics and Imaging, Locarno, Switzerland, 17-21 May 2010, pp 320-325.
10. V. Abaskin, E. Achimova. Design of the holographic fiber-optic electronic speckle pattern interferometer for measurements of glass optical constants. Proc. "Int. Conf. on Nanotechnologies and Biomedical Engineering". ICNBME-2011, 7-8th of July 2011, Chisinau, Moldova.
11. R.J. Collier, C.B. Burchard, and L.H. Lin, "Optical Holography," Academic, New York , 1971.
12. P. Almoro, G. Pedrini, and W. Osten, "Complete wavefront reconstruction using sequential intensity measurements of a volume speckle field," Appl. Opt. 45, pp.8596–8605 (2006).

Author:	E. Achimova
Institute:	Institute of Applied Physics
Street:	Academy str.5, MD-2028
City:	Chisinau
Country:	Moldova
Email:	achimova@phys.asm.md

Modeling of IMS Spectra in Medical Diagnostic Purposes

D.Y. Lipatov[1], Y.R. Shaltaeva[1], V.V. Belyakov[1], A.V. Golovin[1], V.S. Pershenkov[1], V.V. Shurenkov[1], and D.Y. Yakovlev[2]

[1] Department of Micro- and nanoelectronics, National Research Nuclear University MEPhI (Moscow Engineering Physics Institute), Moscow, Russian Federation

[2] Management of Business Projects department, National Research Nuclear University MEPhI (Moscow Engineering Physics Institute), Moscow, Russian Federation

Abstract— **This paper is about embedding of the mathematical methods and development of the software for automatic modeling of IMS spectra. This goal is urgent for IMS analysis of marker substances in human breath for disease diagnosis. The presented method based on the well-known physical model of IMS spectra that is well described IMS spectra using a set of Gauss functions. This paper present next points : 1) Development of the methods for automatic search and identification of Gauss peaks in IMS spectrum. 2) Development of the methods for determining of parameters for Gauss peaks model.**

***Keywords*— IMS spectrum, modeling, Gauss model, human breath, medical diagnosis.**

I. INTRODUCTION

Due of the perspective ways of diagnostics in medicine is the implementation of new methods of analysis of different chemical substances. The Ion Mobility Spectrometry (IMS) is one of such noninvasive methods of diagnostics based on a human inhaled breath analysis [1-4]. This method helps to recognize such dangerous diseases like lung cancer [5], diabetes and airborne infections [6]. Most examined exhalant substances are acetone and ethanol the concentration of which in an expiration of patients with diabetes and cardiovascular pathology significantly differs from those inherent in healthy people.

In brief, the essence of IMS method is that a vapor of analyzed substances are ionized according to atmospheric pressure chemical ionization processes and then the ions are separated on the basis of their mobility in an electric field. At IMS device output, we obtain the array of signal samples (several thousand samples) that represents a spectrum of human expiration substances which can be recognized by character peaks of the spectrum curve. The spectrogram of complex sample consists of a a few thousands of points, and is a superposition of the spectra of individual components of the mixture. Incorrect definition of structure of the sample can lead to errors in diagnosis. In the analysis of complex samples, such as air exhaled the person, it is especially important to improve the resolution of the equipment as a hardware and software methods. Simulation (modeling) of the spectrum is necessary for automation of spectrum processing and accurate identification of a peak position on the spectrum curve.

This paper is about embedding of the mathematical methods and development of the software for automatic modeling of IMS spectra. This goal is urgent for mobile IMS analysis systems where need the rapid analyze of the spectrum without operator or with its minimal participation. The presented program based on the well-known physical model of IMS spectra that is well described IMS spectra using a set of Gauss functions:

$$\hat{y}(t) = \sum_{i=1}^{N} B_i e^{-\left(\frac{t-E_i}{G_i}\right)^2} \quad (1)$$

The process of developing the IMS spectra simulation program can be divided into two stages:

1) Development of the methods for automatic search and identification of Gauss peaks in IMS spectrum.

2) Development of the methods for determining of parameters for Gauss peaks model.

To solve the first problem we have investigated properties of derivatives of Gauss functions. On the basis of this analysis was designed Gauss peaks identification method based on the behavior of the second and third derivatives of the analyzed spectrum. We also solved the problem of the spectrum derivatives calculation and spectrum derivatives noise filtering. The second stage was solved the problem of search the minimum of the quadratic merit function χ^2, that presented the quality of model in multi-dimensional space of Gauss model parameters {B, E, G}. To measure how well the model agrees with the data, we use the next chi-square merit function which in this case is

$$\chi^2(B,E,G) = \sum_{i=1}^{M} (y(t_i) - \hat{y}(t_i))^2 \quad (2)$$

where y(t_i) – is input data spectrum, y^(t_i) – Gaussian approximation of spectrum. The spectrum model can further be used to identify the test substance using the peaks parameters of the simulated spectrum.

II. MAIN PRINCIPLES AND GENERAL SETUP OF IMS

To start with IMS (Ion Mobility Spectrometry) spectrum simulation, it is important to observe general principles of IMS as well as a physical model of IMS spectra.

V. Sontea and I. Tiginyanu (eds.), *3rd International Conference on Nanotechnologies and Biomedical Engineering*, IFMBE Proceedings 55,
DOI: 10.1007/978-981-287-736-9_96

Ion Mobility Spectrometry is an analytic method and respective technical solutions which are used to identify ionized molecules of different chemical species in gas phase based on their mobility (drift speed) in applied electrical field [7]. IMS principles are very practical and convenient in implementation providing quick analysis along with compact realization and high reliability at relatively low cost.

III. PHYSICAL MODEL OF IMS SPECTRUM

Peak separation of IMS spectrum in time domain depends, in general, on a mass of ion, its shape, ambient temperature and electric field gradient. The form of peaks depends on an ion drift-diffusion process, and IMS spectrum can with sufficient accuracy be represented as a sum of Gauss functions (Gaussian) [8,9]:

$$y(t) = \sum_{i=1}^{N} B_i e^{-\left(\frac{t-E_i}{G_i}\right)^2} \tag{3}$$

where: Bi, Ei, Gi – amplitude, offset and width of i-th Gaussian, respectively. The width Gi of Gauss peaks directly proportional to dispersion σ of IMS spectrum and can be determined by the following physical processes [10]:

1) Initial width of ionizing pulse (time of shutter opening).
2) Ion cloud diffusion over time.
3) Mutual repulsion of ions with the same polarity.
4) The ion-molecule reaction is broadening.

Peak offsets Ei of IMS spectrum in time domain depends, in general, on a mass of ion, its shape, ambient temperature and electric field gradient. Identification of chemicals in IMS method is based on the ionized gas clouds offset Ei. This time is determined by position of corresponding peaks on the drift time scale. Hence, one of the data processing purposes of the IMS spectra is to build its Gauss simulation (3).

IV. IMS SPECTRUM SIMULATION

The whole process of IMS spectrum simulation consists of the following stages:

1) Preliminary identification of peaks in original IMS spectrum.
2) Spectrum simulation.

V. IDENTIFICATION OF PEAKS

Identification of peaks is a process of recognition and time determination of each Gauss peak in IMS spectrum. The qualitative solution of this task pre-determines the success of IMS simulation on the whole. The most obvious solution to this problem is to use IMS spectrum derivatives [11].

The 2nd derivative is most interesting see for us because it has more "acute" peaks than original Gauss sum function. Thus, it is more productive to identify peaks not by maximums of original function but determining minimums of the 2nd derivative of original IMS spectrum.

IMS spectrum simulation method suggests using the 3rd derivative of IMS spectrum for peak identification. A preliminary identification a peak occurs when zero of the 3rd derivative function is crossed in the negative value domain of the 2nd derivative function. All derivatives are formed from a raw IMS spectrum passed through differentiating Lanczos Filter [12]. Since such differentiation adds some noise, 2nd and 3rd derivatives are then filtered with the aid of a low-frequency filter. How to work presented peak identification method on real IMS spectrum can see in Fig. 1, 2, 3. Original spectrum and preliminary data about approx peaks position are further used in next step of IMS data processing for more accurate (optimal) peak identification.

VI. RESULTS OF IMS SPECTRUM SIMULATION

Optimization of peak parameters (Bi, Ei, Gi) is carried out by method of steepest descent to find a minimum of the sum of squares of difference between spectrum function y(t_i) and its Gaussian approximation y^(t_i) in multidimensional space of parameters (Bi, Ei, Gi):

$$\chi^2(B,E,G) = \sum_{i=1}^{M} (y(t_i) - \hat{y}(t_i))^2 \tag{4}$$

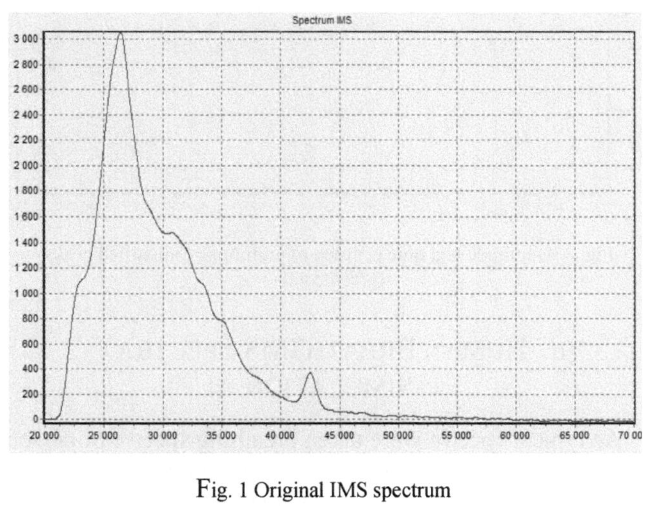

Fig. 1 Original IMS spectrum

The number of found peaks in the IMS spectrum is 25 (see Table 1). After cutting off insignificant peaks in the amplitude level (Bi > 50) the spectrum can be adequately represented by 19 peaks (see. Table 2)

Simulation of IMS spectrum is performed by optimization of the vector $\{B_{(i)},E_{(i)}\ \ G_i\ i=1..N\}$ for spectrum model (2.3) obtained during (preliminary) peak identification process. Since optimization χ^2 is a non-linear process due to non-linearity of Gauss function, Newton's method is used by least-squares method of modified Newton algorithm - Levenberg-Marquardt algorithm [13].

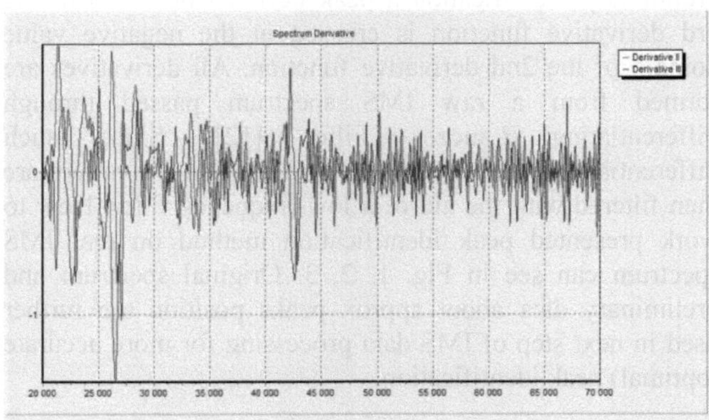

Fig. 2 2nd and 3rd derivatives of IMS spectrum

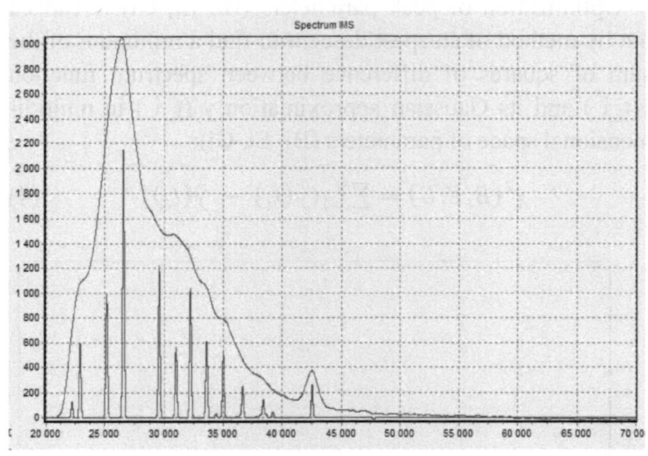

Fig. 3 Amplitude and time position of preliminary identified peaks (green color)

VII. HUMAN BREATH IMS SPECTRA SIMULATION

As typical spectra were taken breathing spectra of person working in the laboratory. It was a simulation of these spectra. The resulting model of the typical spectrum is represented in Figure 5.

According to the parameters of the peaks, detected the presence of substances such as to characterize the disease. As an example, cardiovascular diseases are recognized by the acetylene (C2H2) peaks in the model of IMS spectrum.

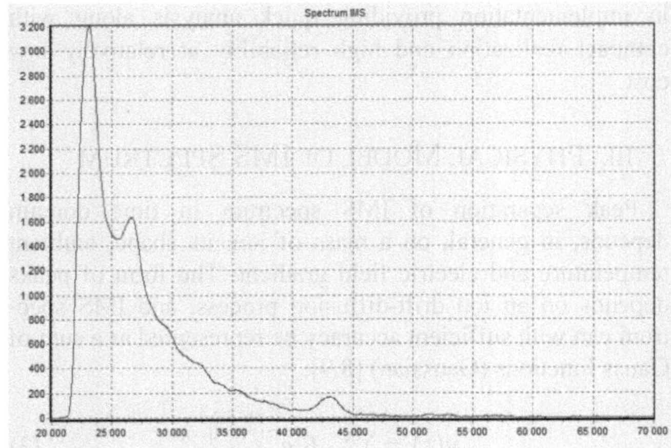

Fig. 4 A typical IMS breath spectrum of person working in the laboratory

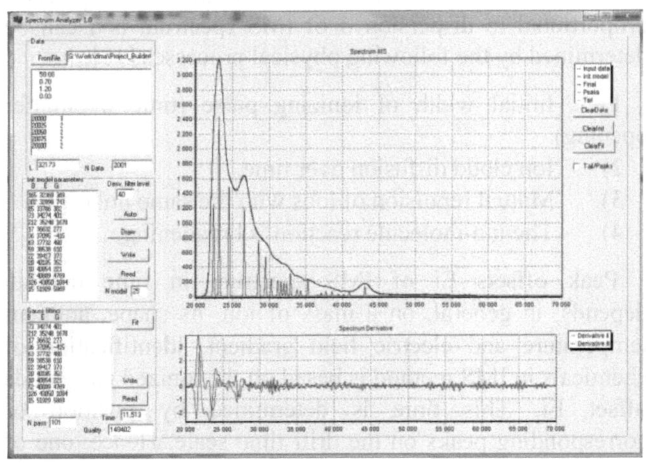

Fig. 5 Model of the IMS spectrum

Table 1 The found peaks in the IMS spectrum

I	Bi	Ei	Abs(Gi)
1	1 255	22346	549
2	2 432	23127	774
3	1 204	26579	1139
4	1 485	24474	1417
5	770	28548	1929

Table 1 (*Continued*)

6	548	29508	0,5
7	236	30264	1005
8	214	31145	547
9	192	31612	383
10	181	31989	324
11	165	32369	369
12	302	32898	743
13	85	33786	382
14	73	34274	401
15	212	35248	1678
16	17	36632	277
17	36	37095	415
18	63	37732	480
19	59	38538	610
20	22	39417	373
21	20	40185	352
22	10	40854	221
23	52	40889	4769
24	126	43050	1084
25	15	51929	5869

Table 2 The spectrum after cutting off insignificant peaks in the amplitude level

i	Bi	Ei	Abs(Gi)
1	1 255	22346	549
2	2 432	23127	774
3	1 204	26579	1139
4	1 485	24474	1417
5	770	28548	1929
6	548	29508	0,5
7	236	30264	1005
8	214	31145	547
9	192	31612	383
10	181	31989	324
11	165	32369	369
12	302	32898	743
13	85	33786	382
14	73	34274	401
15	212	35248	1678
16	63	37732	480
17	59	38538	610
18	52	40889	4769
19	126	43050	1084

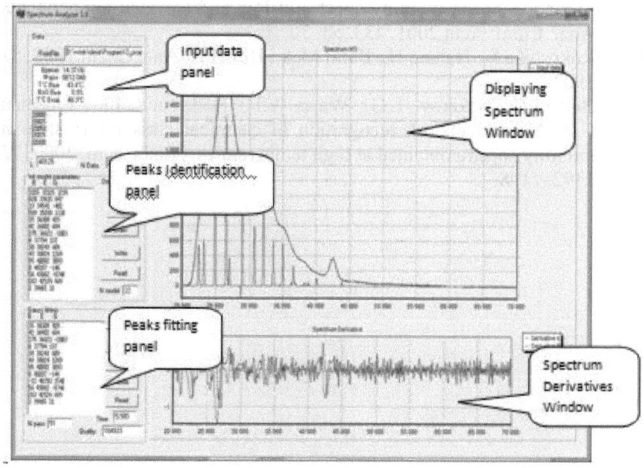

Fig. 6 Overall figure of IMS spectrum simulation program

VIII. Conclusions

It was developed method of simulation IMS spectra, which based on the representation of the spectrum by Gaussian functions. It was demonstrated simulation of human breath IMS spectrum. Parameters of these peaks enable to detect substances and diagnose human diseases.

It was developed the program for the PC, which enables complete cycle simulation spectra IMS (see Fig.6).

The questions of recognition and diagnosis diseases of are not considered in this article.

Conflict of Interest

The authors declare that they have no conflict of interest.

References

1. Vautz W., Breath analysis—performance and potential of ion mobility spectrometry // J. Breath Res. – 2009. - Issue 3. - P.25-33.
2. Ulanowska A., Ligor M., Amann A., Buszewski B., Determination of Volatile Organic Compounds in Exhaled Breath by Ion Mobility Spectrometry // Chem. Anal. (Warsaw). – 2008. - № 53. - P.953-965.

3. Madison N.J., Ion Mobility Diagnostic Test from Quest Diagnostics is First to Provide Direct Physical Measurement of Lipoprotein Particles, Cardiovascular Disease Indicators, Study Finds // DIABETES CARE. – 2008. - Volume 31. - № 4.
4. O'Donnell R.M., Sun Xiaobo, Harrington Peter de B., Pharmaceutical applications of ion mobility spectrometry // Trends in Analytical Chemistry.- 2008. – Volume 27. - № 1. - P.44-53.
5. Davies A.N., Baumbach J.I., Early lung cancer diagnostics by ion mobility spectrometry data handling // Spectroscopy Europe. – 2008. – Volume 20. - № 5. - P.18-21.
6. Baumbach J.I., Westhoff M., Ion mobility spectrometer to detect lung cancer and airway infections // Spectroscopy Europe. – 2006. – Volume 18. - № 6. - P.22-27.
7. Eiceman, G.A.; Karpas, Z., Ion Mobility Spectrometry, CRC Press, Boca Raton, FL, 2005.
8. Eiceman, G.A.; Nazarov, E.G.; Rodriguez, J.E. Chemical class information in ion mobility spectra at low and elevated temperatures, Anal. Chim. Acta 2001, 433, 53–70.
9. Rokushika S., Hatano H., Baim M.A., Hill H.H. Anal.Chem.1985, 57, 1902.
10. Bell, S.E.; Nazarov, E.G.; Wang, Y.F.; Rodriguez, J.E.; Eiceman, G.A. Neural network recognition of chemical class information in mobility spectra obtained at high temperatures, Anal Chem. 2000, 72, 1192–1198.
11. Tom O'Haver. A Pragmatic Introduction to Signal Processing with applications in Chemical Analysis. Cambridge University Press. 2002, pp.682-688.
12. Sysoev, A.A.; Ketola, R.A.; Mattila, I.; Tarkiainen, V., Kotiaho, T. Application of the Numerical Model Describing Analyte Permeation through Hollow Fiber Membranes into Vacuum for Determination of Permeation Parameters of Organic Compounds in a Silicone Membrane, Int. J. Mass Spectrom. 2001, 212 (1-3), 205-217 (doi:10.1016/S1387-3806(01)00440-7).
13. Press W.H., Teukolsky S.A., Vetterling W/T., Flannery B.P. Numerical Recipies in C. Cambridge University Press. 2002, pp.682-688.

Author: Y.R. Shaltaeva
Institute: Department of Micro- and nanoelectronics, National Research Nuclear University MEPhI (Moscow Engineering Physics Institute)
Street: 115409, Kashirskoe highway, 31
City: Moscow
Country: Russia
Email: shaltaeva@yandex.ru

Features in Infrared Image Processing of Biotissue with Internal Heat Source

A.P. Ivanov and V.V. Barun

Institute of Physics, Belarus National Academy of Sciences, Minsk, Belarus

Abstract— **On the base of the known solutions to the heat transfer equation for biotissue, the data of a thermal imager are related with thermophysical, optical and geometrical characteristics of the tissue as well as with power, depth, and temperature of an internal heat source. The main idea of the paper is to get insight into tissue depth by using observations of tissue surface. Observed quantities are discussed to provide a basis to include the features in infrared imaging into retrieval schemes for solving various inverse problems of biomedical optics.**

Keywords— **infrared biomedical imaging, biological tissue, temperature, heat source, heat transfer equation.**

I. INTRODUCTION

Currently thermo-graphic methods on the base of infrared (IR) images are actively incorporated into medical practice for dispensary investigations, early disease diagnostics, etc. The essence of these methods is non-invasive recording of thermal images of open biotissue surfaces to make a conclusion on inflammatory pathology of internal regions having an enhanced temperature or on dystrophic pathology with a reduced temperature. Opportunities are now demonstrated to expose a number of diseases with using IR thermographs at very early stages, when it is impossible to do so by other means. In spite of obvious benefits of these methods, however, they are not widely applied in medicine. One of the main limiting factors is the thermal image recording from a tissue surface only to indirectly bear information on the temperature regime of an internal region. At the same time, medical investigation results are highly influenced by many conditions, such as a psychophysical state of a person, physical loads, etc. The external conditions can, in principle, be avoided by a special procedure of patient training before the investigation. The effects of biotissues on the results cannot, naturally, be excluded. In other words, the thermal action of a tissue slab between the investigated tissue region and the thermal imager creates noisy images. Because of this, the techniques to correct the IR images for the extraction of information on temperature of an interested tissue area are very urgent. Such techniques would enable one to make conclusions on parameters of the thermal regime of internal organs and to derive their temperatures by a non-invasive way. The design of the correction method will enable one to improve the information content and the validity of diagnostic techniques for practical physicians.

This paper discusses the formation of IR images of tissue surfaces, gives some relations between tissue parameters and characteristics of an internal heat source, and outlines means to solve the inverse problem on retrieving the source characteristics from image data of a thermal imager.

II. SPATIAL DISTRIBUTIONS OF TEMPERATURE IN BIOTISSUE WITH AN INTERNAL HEAT SOURCE

We will consider a spherically symmetric stationary heat source. Let its thermal power be Q, W. Introduce the cylindrical coordinate set $\{r, \varphi, z\}$, where r is the radius-vector in plane z = const and φ is the azimuth. Let axis z be directed from the tissue surface ($z = 0$) inside the medium. Denote the depth of the heat source center by a. The source can be a pathological region of tissue having initially temperature T higher than that T_b of its environment. Then the stationary heat transfer equation in a homogeneous medium is [1]

$$\frac{\partial^2 \Delta T}{\partial z^2}+\frac{1}{r}\frac{\partial \Delta T}{\partial r}+\frac{\partial^2 \Delta T}{\partial r^2}=0 \quad (1)$$

and the boundary condition for Newtonian heat exchange with the ambient air at the surface is

$$\kappa\frac{\partial T(r,z)}{\partial z}=H[T(r,z)-T_a] \text{ at } z=0, \quad (2)$$

where $\Delta T(r,z) = T(r,z) - T^*(z)$, K is the temperature rise due to the source; T and T^* are the temperatures of the tissue with and without the source, respectively; $T^*(z) = T_s + hz(T_s - T_a)$; T_s and T_a are the tissue surface and air temperatures, respectively; κ, W/(cm·K) is the thermal conductivity; H, W/(cm^2·K) is the heat flux from the unit surface area at ΔT = 1 K; $h = H/\kappa$, 1/cm. Flux H includes energy exchange due to radiation, convection, and vaporization. Besides, additional conditions are $\Delta T \to 0$ at $z \to \infty$ and the similar relation for $r \to \infty$. The typical values of h are 0.05 to 0.5 1/cm [2, 3]

Equations (1) and (2) have the following solution [1, 4] that is obviously to be azimuthally independent

$$\Delta T(r,z)=\frac{Q}{4\pi\kappa}\left[\frac{1}{\sqrt{(z-a)^2+r^2}}-\frac{1}{\sqrt{(z+a)^2+r^2}}\right]+ \frac{Q}{2\pi\kappa}\int_0^\infty\frac{\exp[-p(z+a)]pJ_0(pr)}{p+h}dp \quad (3)$$

V. Sontea and I. Tiginyanu (eds.), *3rd International Conference on Nanotechnologies and Biomedical Engineering*, IFMBE Proceedings 55,
DOI: 10.1007/978-981-287-736-9_97

for a point source located at $\{r = 0, z = a\}$. Here J_0 is the Bessel function of zeroth order. In particular, at the surface

$$\Delta T(r, z=0) = \frac{Q}{2\pi\kappa}\int_0^\infty \frac{\exp(-pa)pJ_0(pr)}{p+h}dp \quad (4)$$

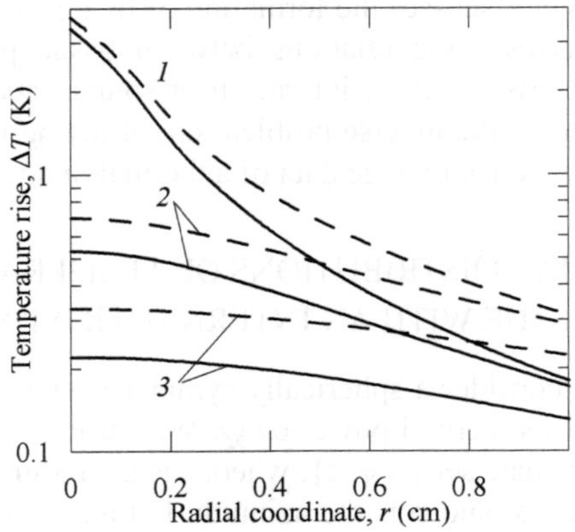

Fig. 1 Temperature distributions over tissue surface at a = 0.1 (1), 0.5 (2), 1 cm (3), h = 0.05 cм-1 (dashed), 0.4 cm-1 (solid), Q = 10 mW

Equation (4) gives the relation between the surface temperature and source parameters Q and a under known thermophysical quantities h and κ. If one measures the value of ΔT at two radial points (Fig. 1), there are two equations with two unknowns that can be, in principle, solved by ordinary means. Note that a point source for rather large distance a can be approximately treated as a spherical source of finite diameter d and finite surface temperature rise ΔT_s. Note that the ΔT_s quantity is the excessive temperature of the finite heat source as compared with the tissue temperature at the point of the source location. Then one has

$$Q = 2\pi\kappa\, \Delta T_s d. \quad (5)$$

In other words, measuring the tissue surface temperature rise ΔT can give parameters d and ΔT_s of a finite heat source by using Eqs. (4) and (5). Fig. 1 shows an example of the radial dependences of the temperature rise.

III. VOLUMETRIC GLOWING OF BIOTISSUE

While solving an inverse problem on retrieving parameters of a pathological tissue region, the input data are the relations between surface temperature and thermal, physical and geometrical characteristics of the system under investigation. Experimentally, the spectral luminosity E_0 (light flux per unit area per unit wavelength range) of skin surface in the IR wavelength range gives the information on the surface temperature by using the Plank formula:

$$E_0(\lambda, T) \equiv f_P(\lambda, T) = \frac{2\pi hc^2}{\lambda^5}\{\exp[hc/(k\lambda T)] - 1\}^{-1} = = C\lambda^{-5}\{\exp[D/(\lambda T)] - 1\}^{-1}, \quad (6)$$

where $C \approx 37500$ W/(cm^{2}μm), λ is the wavelength, μm, and $D \approx 14400$ μm·K. The fundamental constants in Eq. (6) have usual notations and are not decoded here. If one considers tissue as a black body with uniform temperature $T_0 = \Delta T(r,0) + T_s$ over its body volume, then its spectral luminosity is calculated according to Eq. (6) with $T = T_0$. But really the luminosity is formed by radiation coming from different tissue depths, where temperature $T(r,z)$ varies due to the heat source, see Eq. (3). So the spectral luminosity is determined by the volumetric glowing of all tissue layers as well as of the source itself. So one should use the integration

$$E(r,\lambda) = \int_0^\infty f_P(\lambda, T(r,z))\mu_a(\lambda)\exp[-\mu_a(\lambda)z]dz \quad (7)$$

where temperature $T(r,z)$ is determined by Eq. (3). Here $\mu_a(\lambda)$ is the spectral absorption coefficient of tissue. Since the main tissue component is water (75 to 90 % by volume) and other components have negligible absorption as compared with water, $\mu_a(\lambda)$ is assumed to be the water absorption coefficient [5]. Tissue scattering is by several orders of magnitude lower than its absorption. So the scattering is neglected in Eq. (7). Thus an imaging detector (e.g. a thermal imager) records a temperature other than the surface one to introduce some error δT_0 in the readings.

Before the derivation of δT_0, let additional remarks on the spatial temperature distribution be made. As one can easily see from Eq. (3), temperature rise ΔT becomes infinitely large near the source at $\{r = 0, z = a\}$ to lead to $f_P(\lambda, T) \to \infty$ in the vicinity of this point. However, the singularity is integrable. Really, it can be shown that spectral radiance values B_s and B_v of light coming to the tissue surface from the source itself and from its small vicinity, where $T >> D/\lambda$, are, respectively,

$$B_s = QC\lambda\mu_a(\lambda)\exp[-\mu_a(\lambda)a]/(3\pi^2 D\kappa) \quad (8)$$

and

$$B_v = QC\lambda\mu_a(\lambda)\exp[-\mu_a(\lambda)a]/(4\pi^2 D\kappa). \quad (9)$$

Note that radiance B is related with luminosity E by $B = E/\pi$. One can see that the radiance values are independent of the dimensions of the heat source and of its small vicinity. Due to Eqs. (8) and (9), there are no problems to compute the integral of Eq. (7).

Thermal imagers use often Schottky diodes on the base of PtSi/$Si_{1-x}Ge_x$ (detector 1) and PtSi/Si (2) with spectral sensitivities S [6]. Under imager calibration with respect to a black body, one usually relates electric signal α with T by

$$\alpha(T) = \sigma \int_{\lambda_1}^{\lambda_2} f_P(\lambda, T) S(\lambda) d\lambda, \qquad (10)$$

where σ is the surface area observed by the imager, λ_1 to λ_2 is the operating range of the imager. However, due to volumetric glowing of tissue, electric signal would be

$$\beta(T) = \int_{\sigma} dr \int_{\lambda_1}^{\lambda_2} E(r, \lambda) S(\lambda) d\lambda \qquad (11)$$

where $E(r,\lambda)$ is determined by Eq. (7). Introduce quantity $V = \beta/\alpha \approx 1 + \delta\alpha/\alpha$ characterizing the distortion of the surface temperature by the volumetric glowing. Then

$$V = 1 + \frac{1}{\alpha}\frac{\partial \alpha}{\partial T_0}\delta T_0 \qquad (12)$$

where δT_0 is the absolute error in the measured tissue surface temperature. Note that $\delta T_0 > 0$ always. In other words, the observations of the tissue volumetric glowing would always give temperature higher than T_0. Rewrite Eq. (12) in the following form with accounting for Eqs. (6) and (10)

$$\delta T_0 = (V - 1) T_0^2 \gamma(T_0)/D, \qquad (13)$$

where

$$\gamma(T) = \frac{\int_{\lambda_1}^{\lambda_2} f_P(\lambda, T) S(\lambda) d\lambda}{\int_{\lambda_1}^{\lambda_2} f_P(\lambda, T) S(\lambda) d\lambda / \lambda} \qquad (14)$$

One can see from Table that the typical values of V are very close to unity for the two detectors [6]. However, it will be showed below that such a small distinction of V from 1 gives an appreciable error in measured temperature.

Table 1 Quantity V for varying source locations and heat exchange conditions, $Q = 10$ mW

h, cm^{-1}	Detector 1, a, cm: 0.3	Detector 1, a, cm: 2	Detector 2, a, cm: 0.3	Detector 2, a, cm: 2
0.05	1	1	1.001	1
0.1	1	1	1.001	1
0.2	1.001	1	1.001	1.001
0.4	1.001	1.001	1.003	1.002
0.8	1.002	1.001	1.005	1.004

IV. EXAMPLES OF TEMPERATURE MEASUREMENT ERRORS

It follows from the above that a thermal imager observes not the tissue surface temperature, but some higher value due to the volumetric glowing. Fig. 2 illustrates the effect of systematic error δT_0 in measured surface temperature on the estimate of source temperature excess ΔT_s. Here are the plots of temperature rise ΔT at the central point $r = 0$ of the tissue surface versus excessive source temperature ΔT_s. The ΔT values were calculated for two cases, namely by using Eq. (4) (solid lines) or Eq. (7) (dashed ones). The data of Fig. 2 are given for several situations, when the values of $(V - 1)$ differ from zero in the third (or more) significant digit after decimal point. All the shown dependences are approximately linear ones. The slopes of the solid and dashed lines are about the same to tell that δT_0 is independent of ΔT_s. These features are due to the smallness of the temperature rise values as compared with the person and ambient temperatures.

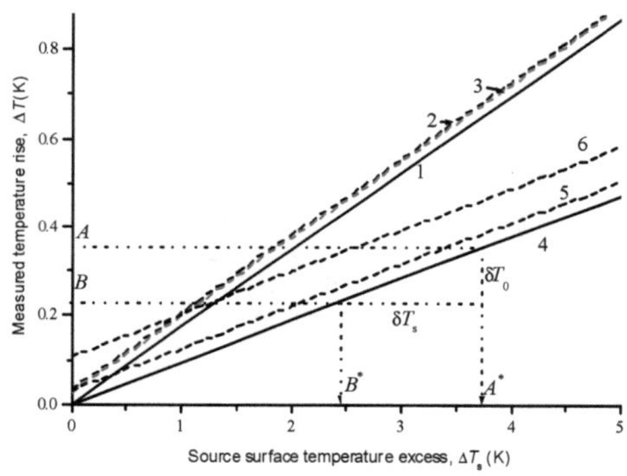

Fig. 2 Dependence of ΔT on ΔTs. Solid and dashed lines ΔT(0,0) are calculated by Eqs. (4) and (7), respectively. Curves 1, 2, and 3 – h = 0.2 cm-1; 4, 5, and 6 - h = 0.8 cm-1; 2 and 5 – detector 1; 3 and 6 – detector 2

Consider in detail the effect of volumetric tissue glowing due to a heat source on the error in retrieving the temperature excess ΔT_s of the latter. Let the experimentally measured value of ΔT correspond to point A of Fig. 2 (detector 1, $h = 0.8$ cm^{-1}). Then, according to line *4*, temperature ΔT_s corresponding to point A* will be determined. The ΔT value should be really decreased by δT_0, i.e. to go to point B, which corresponds to ΔT_s of point B*. In this case, the error δT_s in ΔT_s will be about 1 K. In other words, the volumetric tissue glowing provides enhanced source temperature, which can be important for solving inverse problems.

V. ON THE SOLUTIONS TO INVERSE PROBLEMS

Discuss here some opportunities of retrieving various characteristics of tissue and of internal heat source by using thermal images. Let us start with depth a. Consider the ratio of temperature rises at two different radial coordinates

$$t = \Delta T(r,0)/\Delta T(0,0). \quad (15)$$

As follows from Eq. (4), this value at the surface depends, in a general case, on depth a and heat exchange parameter h to be independent of Q and κ. Fig. 3 shows relative temperature t as a function of a for two values of h = 0.05 (curves *1*) and 0.4 cm^{-1} (*2*). These values cover approximately the whole possible range of parameter h. Note that dependence on h is rather weak. This enables one to directly determine the source depth by measuring the temperature rise at two radial coordinates. On the other hand, one can observe the temperature rise at more radial coordinates and use theoretical nomograms similar to those presented in Fig. 4. The nomogram opens an opportunity to retrieve simultaneously a and h values. Since the two techniques use the ratios of surface temperatures, error δT_0 has practically no effect on the retrieval of values a and h.

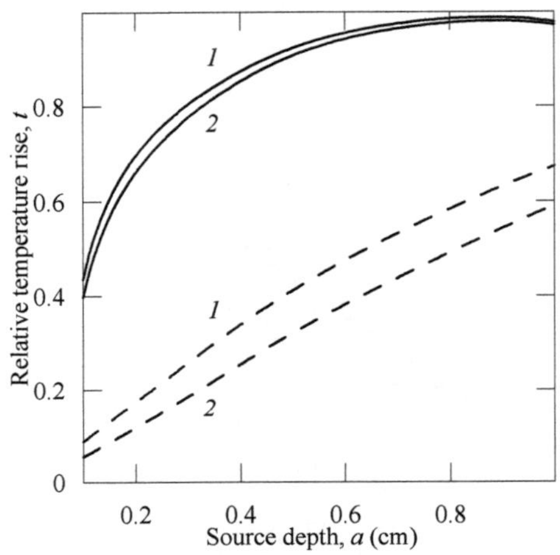

Fig. 3 Temperature rise t as a function of source depth for r = 0.2 (solid curves) and 1 cm (dashed), h = 0.05 (curves 1) and 0.4 cm-1 (2)

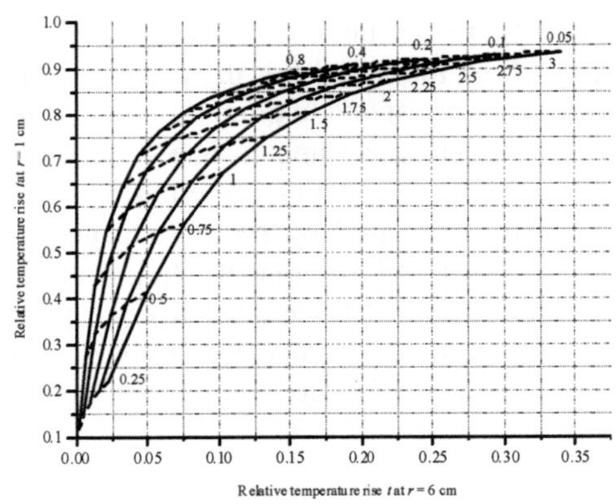

Fig. 4 Nomograms for retrieving source depth a and parameter h. Numbers near thin and bold curves give a and h values, respectively

When thermophysical characteristics of tissue and source depth are known, one can determine Q/κ value by using Eq. (4) for measured temperature rise dependence $\Delta T(r,0)$, see Fig. 1. Note the appreciable effect of parameter h on the temperature. It is interesting to point out that, at large r and h, the surface temperature increases with depth a decreasing. However one could assume the opposite effect. The explanation is physically transparent. Really, when the source is close to the tissue surface, the heat transferring to the radial direction decreases quickly with growing r due to intense heat exchange with the ambient air. If the source is deep in tissue, there is no so high heat leakage, and the temperature is rather uniformly spread over the surface. Generally, quantity Q/κ can be derived from the observations at any radial coordinate. But from practical viewpoint, it is more convenient to do so at r = 0, because the signal of a thermal imager and its sensitivity to changes in this quantity are maximal here.

VI. CONCLUSION

Recording a thermal image of a person body finds currently wide applications in medical practice, e.g. for diagnosing organism conditions, exposing pathological body regions, etc. Usually, the gathered image data are transformed into surface temperature maps. But there obviously arise a number of questions. For example, what a value we have measured, if temperature distribution over the body depth is non-uniform? This paper gives a quantitative answer to it. Namely, the estimates are provided here to relate data of a thermal imager with thermophysical, optical and geometrical parameters of a system under consideration. An internal heat source simulates a pathological tissue region. The second question occurs from the following physically transparent fact. Really, one observes a tissue surface, but desires to retrieve a temperature of an internal heated region. How are the observations related with a quantity of interest? Here are provided analytical tools to solve the inverse biooptical problem on deriving not only the source temperature, but also the heat source depth, its thermal power and dimensions, thermophysical parameters of tissue itself, etc. The presented data can be used for designing dedicated algorithms and program codes of IR biomedical image processing for various medical and biophysical applications.

CONFLICT OF INTEREST

The authors declare that they have no conflict of interest.

REFERENCES

1. Carslaw HS and Jaeger JC (1959) Conduction of Heat in Solids, 2nd ed., Clarendon Press, Oxford
2. Barun V.V. and Ivanov A.P. (2003) Thermal action of a short light pulse on biological tissues. Int J Heat Mass Transfer 46: 3243–3254
3. Gustafsson SE, Nilsson SK, and Torell LM (1975) Analytical calculation of the skin temperature distribution due to subcutaneous heat production in a spherical heat source. Phys Med Biol 20: 219–224
4. Draper JW and Boag JW (1971) The calculation of skin temperature distribution in thermography. Phys Med Biol 16: 201–211
5. Hale GM and Querry MR (1973) Optical constants of water in the 200-nm to 200-µm wavelength region. Appl Opt 12: 555–563
6. Maiti CK and Armstrong GA (2001) Applications of Silicon-Germanium Heterostructure Devices. Taylor & Francis, Bristol

Author:	Vladimir Barun
Institute:	Belarus National Academy of Sciences
Street:	66 Independence Avenue
City:	Minsk
Country:	Belarus
Email:	barun@dragon.bas-net.by

Fluorescent Nanoscale Structures for Selective Medical Diagnostics

V.S. Osipovich, K.D. Yashin, S.K. Dzik, and A.A. Bykov

Belarusian State University of Informatics and Radioelectronics (BSUIR),
Information Systems and Technologies for Industrial Safety, Minsk, Republic of Belarus

Abstract— **The development of nano-bio-information technologies for solving urgent problems of biomedical diagnostics: in particular, for imaging pathologically changed cells and tissues. Big Data technologies for this purpose is used.**

Keywords— **Nanoscale structures, fluorescence, visualization, nanobiomarkers.**

I. Introduction

The purpose of this research is the development of nano-bio-information technology for solving urgent tasks of improving the methods of disease diagnosis and biomedical imaging of biological objects. To achieve this aim we had to solve several problems: 1) to develop the construction of fluorescent nano-biomarkers applied for imaging tumor cells in vitro; 2) to work out the technology of nano-bio-markers production 3) to approbate the nano-bio-markers developed for imaging biological objects (cells, viruses, antigens); 4) to develop the scheme of reading, processing and storing visualized information; 5) to prepare the initial experimental results to process large statistical data of cell biological objects using the technology BIG DATA.

II. Writing the Paper

The principles of nanobioinformation system construction based on semiconductor nanocrystals (quantum dots) CdSe / ZnS. To visualize function processes, different cell-type interconnection, to find out the way of cell migration and other biological and medical research the labeling (tagging) of certain cell types is used. One of the most common methods of labeling a living cell is held by the introduction of fluorescent probes. Earlier organic fluorescent dyes were used for this purpose; bust last decade is more commonly known for quantum dots - fluorescent semiconductor nanoparticles which have significant advantages over traditional organic dyes, and in particular by fluorescence intensity and duration [1 – 4]. These days an intensive research on the development of methods of nanoparticles synthesis with given optical properties and design is held, including the study of their interaction with living cells and opportunities for practical use.

The concept of construction and operation nanoscale system in action for indicating and diagnostics of biological cells and cellular components is described in [3]. Information component of nanosized biological systems is based on fluorescent nanocrystals CdSe / ZnS. Penetrating inside diagnosed biological cells, nanocrystals "paint" it, making it light while irradiating the drug with ultraviolet (UV) light. This is how the informing based on concerning the presence of tumor cells (required biological objects). Biological and medical component of nanoscale semiconductor system is based on specific "recognition" of a certain type of cells (biological) in the analyzed samples. Such kind of diagnostics is widely used in immunluminescent and enzymatic analyses. Unfortunately this method has a number of limits and nanochrystals are not widely used. A semiconductor component of the biosystem is based on the application of nano-sized crystals of A_2B_6 semiconductors. Figure 1 is a schematic representation of immunluminescent analysis using nanocrystal CdSe / ZnS fixed with stabilizing shell which allows working in physiological solutions.

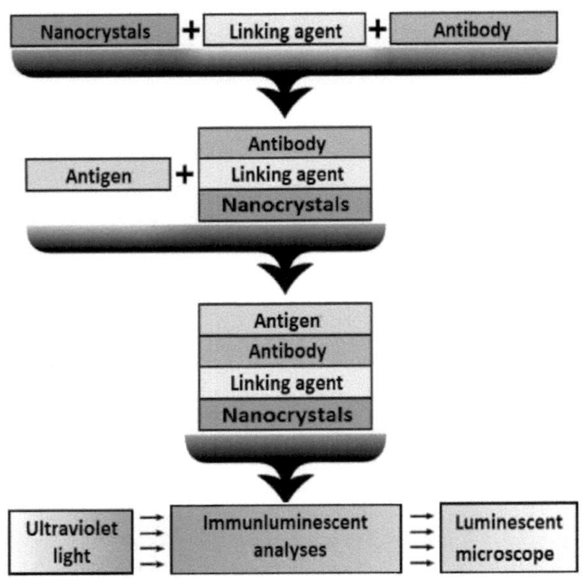

Fig. 1 The scheme of immunluminescent analysis using semiconductor nanobiomarkers (antigen / antibody system)

During the experiments the absorption of nanoparticles of various modifications was investigated (samples №1 – passivation film of closely packed molecules of

V. Sontea and I. Tiginyanu (eds.), *3rd International Conference on Nanotechnologies and Biomedical Engineering*,
IFMBE Proceedings 55,
DOI: 10.1007/978-981-287-736-9_98

mercaptoundecanoic acid; samples №2 – thioglycolic acid; samples №3 – cysteine) by the cells of tumor ascites and bone marrow, the dependence of the colouring quality from nanoparticle concentration in the incubation medium was revealed. It was discovered that the cells absorb Ehrlich ascites carcinoma (EAC) nanoparticles of all three types of samples (Figure 2), colouring was held during 1 hour under the temperature 37°C.

It was stated that the particles of samples № 1 and № 3 in the majority of the cells were being distributed in a membranous area; comparing to a relatively uniform fluorescence it was observed a bright glow of fine granules. At the same time after the cell incubation with the particles from samples №2 we managed to see larger brightly fluorescent granules. Various statistical distribution of the nanocrystals in the cytoplasm can be the evidence of the

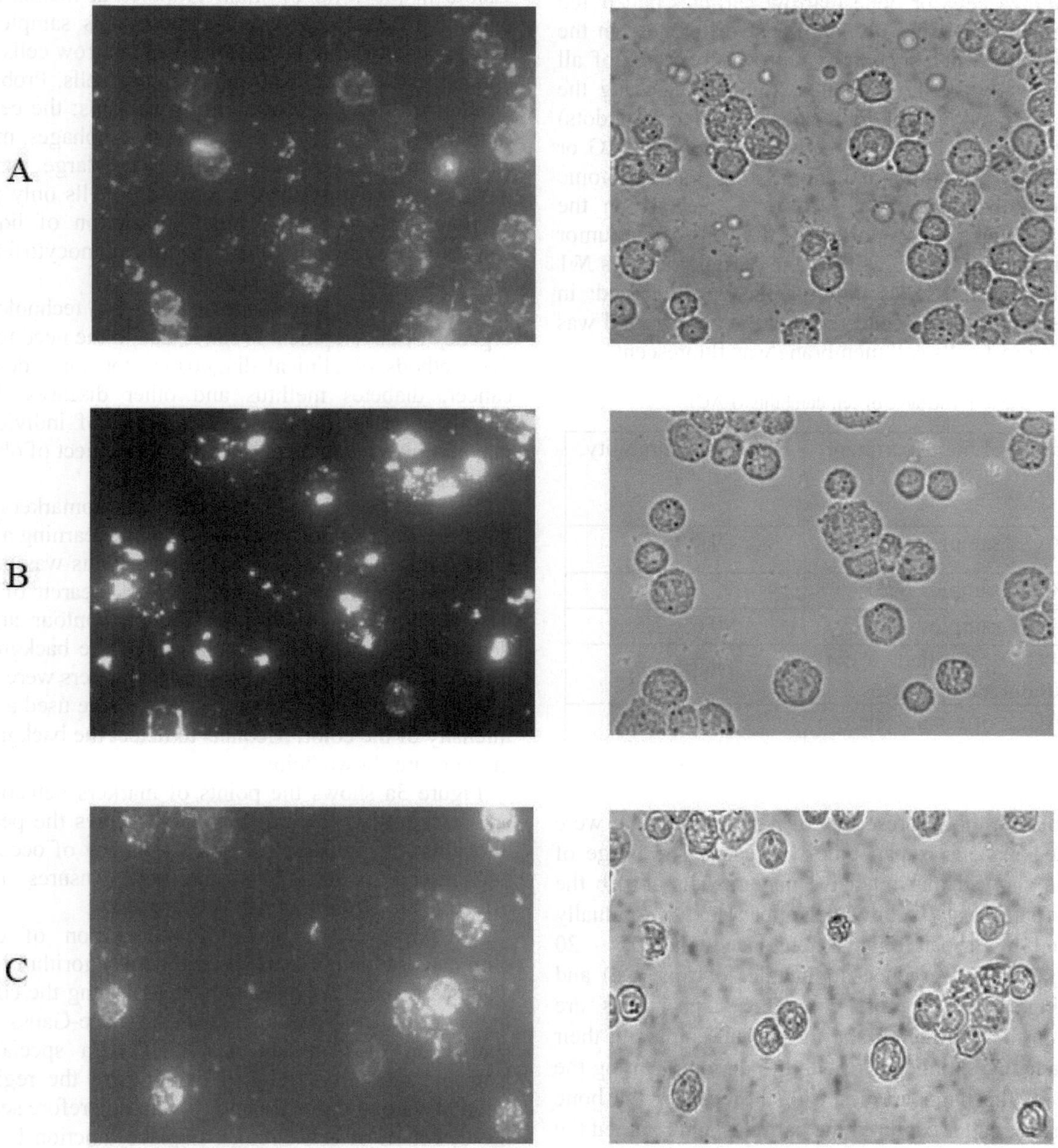

Fig. 2 The micropictures of EAC ascites cells stained by semiconductor nanocrystals in luminescent (left) and normal (right) lens: samples №1 (A); samples №2 (B); samples №3 (C)

differences in the mechanisms of absorption of nanoparticles by cells arising from the peculiarities of physical and chemical properties of the particles.

After a 5-hour incubation of Ehrlich ascites tumor cells with nanoparticles the cytotoxicity of the latter wasn't recorded. Statistically, cell viability decreased by only 5 - 9%, but differences between the experimental series (with nanoparticles) and control series (instead of nanoparticles was used distilled water) weren't observed (Table 1). While painting AG22a cells or bone marrow particles penetrated into the cells. There were no significant differences in the painting of cells when incubated with the particles of all three types for 60 or 120 minutes (Figure 3) During the incubation of nanobiomarkers suspension (quantum dots) for 3 hours at 37 ° C with the cell suspension 22a AG or bone marrow aggregation of nanoparticles in an isotonic solution was observed, which led to a decrease in the quality of painting, only a portion of the cells of the tumor ascites were stained. Thus, while using particle samples №1 fluorescence of the ascites tumor cells was observed; in single cells bright fluorescence around the background was observed, in most cells cell membrane was fluorescent

Table 1 The table of cell durability (EAC)

№ п/п	Series of the experiment	EAC cells durability, %
1	EAC + samples №1	89,4 ± 2,1
2	EAC + samples №2	92,9 ± 0,8
3	EAC + samples №3	90,6 ± 2,5
4	EAC + distilled water, without nanoparticles	90,5 ± 1,8
5	EAC, original, before painting	98,1 ± 1,6

To extinguish the fluorescence of nanocrystals that were not included in cells trypan blue was used. The usage of trypan blue showed that intensive nanocrystals absorb the dead cells (nanocrystals glow in the dead cells is gradually extinguished by trypan blue and replaced after 15 - 20 minutes by red fluorescence applicable for this dye) and single living cells. In most cells, the nanocrystals are deposited on the cell membrane to the outer side and their glow is gradually extinguished with trypan blue. During the incubation of the nanoparticles with the cells of the bone marrow and the latter captured the particles and retained the fluorescence while adding trypan blue which indicates the penetration of quantum dots inside cells.

While painting the samples №3 by the particles a similar result was observed, but the penetration of the quantum dots in the tumor cells was slightly better and while adding trypan blue the fluorescence in a membranous area in most cells remained.

While the incubation of ascites cells with the sample particles №2 bright luminescence of single living cells, and the fluorescence of quantum dots in the space inside the cell-membrane was observed; wherein the nanoparticles are visible in the form of small granules at membrane; bone marrow cells also absorbed nanocrystals samples №2. It should be noted that in all cases bone marrow cells absorbed nanoparticles better than ascites tumor cells. Probably, this is due to different absorption mechanisms: the cells of the reticuloendothelial system such as macrophages, monocytes and neutrophils, can absorb relatively large particles by phagocytosis, while for the remaining cells only processes of pinocytosis are applicable- absorption of liquids and solutes of up to 5 microns with macropinocytosis and 120 nm - at mikropinotsitoze [1].

Prospects for the development of technology. The urgency of development is stated due to the need to improve the methods of clinical diagnostics for early detection of cancer, diabetes mellitus and other diseases. However, visualization, monitoring and analysis of individual cells may not give a complete picture of the object of observation and research.

To analyze the visualized nanobiomarkers data a machine vision algorithms and machine learning algorithms were used. Initially, recognition algorithms was performed, then the pipelining including a quick search of markers, correction of the local color model, contour analysis to detect cell shape, adaptive analysis of the background and global optimization of the system parameters were used.

To recognize nanobiomarkers the image used a threshold intensity of the color. Medians to detect the background and markers are shown below

Figure 3a shows the points of markers detection (blue) and background (green). Figure 3b shows the peak of the allocation to the corresponding frequency of occurrence of background pixels. This threshold ensures the most effective recognition of nanobiomarkers.

To improve the accuracy recognition of cells and eliminate the points found by error the algorithm of contour analysis is used. As a basis for determining the contours of the algorithm the measure of the Laplace-Gauss was used (Laplacian of Gaussian, LoG). It is a specialized 2D isotropic measure, designed to compare the regions with rapid change in the color intensity and therefore selected for the detection of cell circuits. Laplace function L (x, y) for the image with intensity values of pixels (x, y) is detected by the formula:

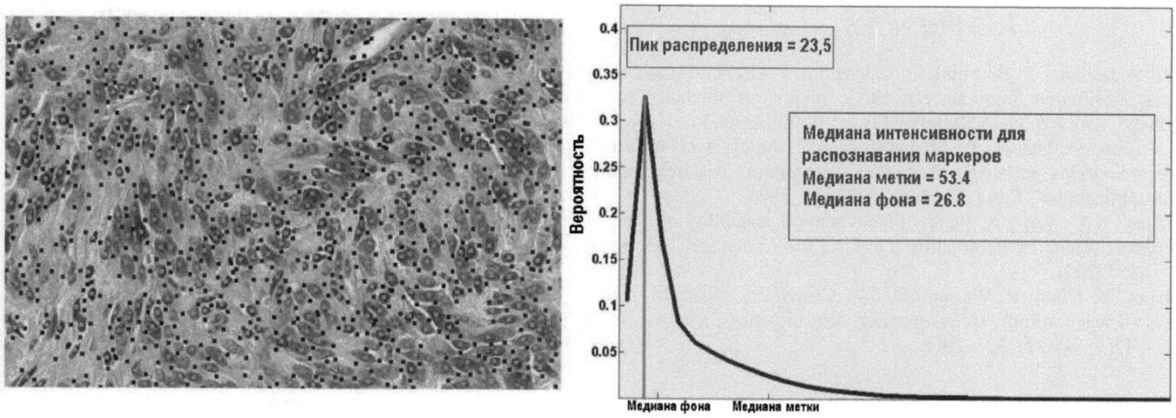

Fig. 3 (a) – dots of nanobiomarkers detection; (b) – median intensity of the cjlor for recognition

$$L(x,y)=\frac{\partial^2 I}{\partial x^2}+\frac{\partial^2 I}{\partial y^2}$$

Hence the input image is represented as a set of discrete points, the function finds a discrete convolution kernel, which can describe the second derivatives in the determination of Laplace. Based on the proposed measures a combined algorithm was used to search for the boundaries provided below. Figure 3 (a) shows the circuit cell space in discrete 2D area of points in the view of a grid, where each point has eight neighbors.

Figure 3 (b) shows obtaining the coefficients for detecting the direction of the contour vector in the point.

The direction of the cell circuit in a discrete 2D space of points in a view of the grid, where each point has eight neighbors, can be approximated as a solution to the problem of maximum flow of minimum length (max-flow / min-cut) with the following weights:

$$\omega_k(p)=\frac{\delta^2\cdot|e_k|^2\cdot\Delta\phi_k\cdot\det D(p)}{2\cdot(e_k^T\cdot D(p)\cdot e_k)^{3/2}},$$

where δ – mesh size pixels. $\Delta\phi_k$ – change in angle coefficient between k и $(k+1)$ grid line, $\Delta\phi_k = \phi_{k+1} - \phi_k$

$$D(p)=g(|\nabla I|)\cdot I+(1-g(|\nabla I|))\cdot u\cdot u^T,$$

where $u=\frac{\nabla I}{|\nabla I|}$ – unit vector in the direction of line gradients at p, I – the identity matrix and $g(x)=\exp(-\frac{x^2}{2\sigma^2})$.

Although contour analysis significantly improved to increase the accuracy recognition, the most important factor for streaming visual information processing is the use of nanobiomarkers. Let's observe the principles of markers formation based on nanoparticles.

III. CONCLUSION

Conclusion. A number of modifications of the semiconductor nanoparticles CdSe / ZnS with the passivation film of various compositions (molecules mercaptoundecanoic acid, tyoglinol acid and cysteine), suitable for imaging of live cells and vital staining technique was worked out. The results are shown in the visualization models of bone marrow cells of mice and ascites tumor of Ehrlich ascites tumor and ascites hepatoma 22a. There were no cytotoxic effect of nanoparticles during the 5-hour incubation. There were observed differences in morphological picture of absorbtion by tumor cells of nanocrystals with different passivation coating that is likely to be a reflection of differences in the mechanisms of absorption of these particles. The experimental results indicate the possibility of application of semiconductor nanoparticles for intravital imaging and labeling of cells, as well as for the study of patterns of cellular absorbtion of nano-objects of various size and construction. Large accumulated material was received in the form of cellular imaging of biological objects. The experimental material embodies large statistically structured and poorly structured data cell samples. The processing considerable diversity and vast amounts of images of cells is possible with the help of BIG DATA technology.

CONFLICT OF INTEREST

The authors declare that they have no conflict of interest.

REFERENCES

1 S. Poli, V. Barbaro, P. Bartolini, G. Calcagnini, F. Censi, "Prediction of atrial fibrillation from surface ECG: review of methods and algorithms", *Ann I-st Super Sanità,* 39(2), pp.195-203, 2003.

2 A. Bollmann, D. Husser, L. Mainardi, *et al.*, "Analysis of surface electrocardiograms in atrial fibrillation: techniques, research, and clinical applications", *Europace*, 8, pp. 911–926, 2006.

3 J. Aldana, Y.A. Wang, X. Peng, "Photochemical instability of CdSe nanocrystals coated by hydrophilic thiols", *J Am Chem Soc.*,123, pp. 8844 – 8850, 2001.

4 Y. Zhang, Y. Chen, P. Westerhoff, J.C. Crittenden, "Stability and removal of water soluble CdSe quantum dots in water", *Environ Sci Technol*, 42(1), pp. 321-325, 2008.

5 E. Etxeberrial, P. Gonzalezl, E. Baroja-Fernandez, J.P. Romero, "Fluid Phase Endocytic Uptake of Artifical Nano-Spheres and Fluorescent Quantum Dots by Sycamore Cultured Cells", *Plant Signaling & Behavior*, 1(4), pp. 196-200, 2006.

6 C.G. Hansen, B.J. Nichols, "Molecular mechanisms of clathrin-independent endocytosis", *Journal of Cell Science*, **122**(11), pp. 1713-1721, 2009.

Author: Vital Osipovich
Institute: Belarusian State University of Informatics and Radioelectronics
Street: P.Brovki
City: Minsk
Country: Republic of Belarus
Email: seth22@yandex.ru

Hepatoprotective Activity of Leaf Extract of *Laurus Nobilis* L. against CCL4 Induced Hepatotoxicity in Rats

H. Vardapetyan, S. Tiratsuyan, and A. Hovhannisyan

Department of Medical Biochemistry and biotechnology, Russian-Armenian (Slavonic) University, Yerevan, Armenia

Abstract— **The present study was conducted to evaluate the hepatoprotective activity of ethanolic extract of Laurus nobilis against CCL4 induced liver toxic damage in rats. To evaluate the degree of hepatocyte damage in experimental modeling hepatitis the lipid profile and histochemical assay were determined. Introduction of CCl4 in animals was always carried out in fatal to 36 hours. A single injection of the L. nobilis extract simultaneously with CCl4 leads to 100% survivability. L. nobilis leaves extract acts on the liver as a potent scavenger of free radicals to prevent the toxic effects of CCl4 induced hepatotoxicity in rats. Under the influence of CCl4 microscopically normal structure of the liver are violated, the signs of diffuse liver disease with extensive necrosis of hepatocytes, perivascular and pericellular edema are determined. Extract has angioprotective action on the capillary bed of the rat liver, vascular protective effect and prevented the progression of necroinflammation, which can be explained by the presence of antioxidants and antimicrobials both of flavonoids origin, such as quercetin and morin as well as terpens and terpenoids.**

Keywords— **L. nobilis extract, CCl4, hystochemistry, hepatotoxicity, cholesterol, triglycerids.**

I. INTRODUCTION

The problem associated with the liver disorder is still a worldwide health problem. Drugs used in the treatment of liver diseases are inadequate and sometimes can have serious side effects. Despite the large arsenal used hepatoprotectors, modern medicines are not always helpful to achieve an increase in reparative-regenerative activity, normalization of liver homeostasis and functional activity and prevent the development of fibrosis and cirrhosis. In view of severe undesirable side effects of synthetic agents, there is growing attention to the traditional herbal medicines that are claimed to possess hepatoprotective activity [1]. As one of the candidates for the role of hepatoprotector considered bay laurel, whose medicinal properties have been known for thousands of years and have been widely used in herbal medicine for the various diseases. The leaf of *L. nobilis* used as herbal medicine to treat rheumatism, earaches, indigestion, sprains, migraines, gastrointestinal diseases [2, 3]. Ethanolic extract of *L. nobilis* has antioxidant, antimicrobial and wound-healing effect [4-6]. *L. nobilis* is also used for preventing and treating of Type II-of diabetes, helps in reducing the level of total cholesterol and LDL-cholesterol simultaneously enhancing HDL-cholesterol [7]. There is a lack of scientific reports on the hepatoprotective role of ethanolic extract of *L. nobilis* leaves. One of the classic models of toxic liver damage is the intraperitoneal administration of CCl_4 [8]. The aim of the study was to evaluate the hepatoprotective activities of *L. nobilis* extracts in CCl_4 induced rats.

II. MATERIALS AND METHODS

The leaf of *L. nobilis*, were collected from the Zugdidi region of Georgia. The powdered plant materials were successively extracted with ethanol [5].

The GC-MS analysis of the ethanolic extracts using Hewlett Packard GCD (model 6890) and Hewlett Packard MS (model 5972) with mass selective detector (MSD) was performed by [5]. High-Performance Liquid Chromatography (HPLC) analisys of the samples was performed on an Agilent 1100 chromatograph (Agilent Technologies, USA) with UV - spectrophotometric detector and data processing software ChemStation [6].

Radical scavenging capacity (RSC) of tested extracts, quercetin and rutin solutions was assessed in a chemical model, i.e. DPPH system [5, 9].

The antibacterial activities of plant extracts were investigated by the disc-diffusion method on agar. As an indicator of antibacterial activity the inhibition zone formed after 24 – 48 hours of bacteria *E. coli* K-12 wild type strain development was taken [9]. The square ($pixel^2$) was calculated by special program "Image Repair" [10].

Male Wistar rats (170-200g) were used. Animals experiment was approved by the institutional Animal Ethics Committee. All the animals were divided into 3 groups. Animals of I^{st} group were served as a untreated control. Toxic liver injury was modeled by CCl_4 intraperitoneal injection in a dose of 0.2 ml per 100 g rats mass. IIIrd group were injected intraperitoneally with 0,2ml on CCl_4 and ethanolic extract of laurel leaves. The animals of the I^{st} and IInd groups were decapitated after 36 hours, while group IIIrd - were decapitated after 36, 240 and 480 hours, respectively [11].

Cholesterol was determined using a reagents of "kit for cholesterol," the company "Biostart" by [12]. Ttriglyceride was determined by a colorimetric enzymatic method at 546 nm [13].

V. Sontea and I. Tiginyanu (eds.), *3rd International Conference on Nanotechnologies and Biomedical Engineering*,
IFMBE Proceedings 55,
DOI: 10.1007/978-981-287-736-9_99

Hemorrhagic status of experimental animals liver was assessed by histochemical method using a digital microscope (Intel Play QX3). Liver fresh frozen sections were prepared 150 micrometers thick, which were incubated in the mixture prepared according to method [14].

Experimental results were expressed as means ± SD. All measurements were conducted three to five times. The data were analyzed by a one-way analysis of variance (ANOVA) and the value of $p < 0.05$ was considered as significant.

III. Results and Discussion

Phytochemical composition: The phytochemical composition of *L. nobilis* leaves were also determined by GC-MS analysis (Figure 1, left) has shown that the major compound of extract is an oxygenated monoterpene1.8-cineole (eucalyptol). Other predominant compounds α-thujene, α-pinene, β-pinene, D-limonene, *o*-cymene are. There also are smaller amounts of acyclic monoterpene – myrcene, and cyclic α- and β-phellandrene, camphene, α -terpinene and a terpineol.

HPLC analyses of ethanolic extracts are presented in figure 1 (right). Among the selected phenolic compounds in L. *nobilis* extract were identified: rutin, quercetrin, quercetin, morin, apigenin, kaempferol. As envisioned from the data in the table two major compounds of extract quercetin and morin.

Antiradical activity: There are many reports suggesting that the DPPH scavenging activity of the extracts is directly related with the total phenolic content. The isolated compounds were screened for their *in vitro* antioxidant activity by a DPPH free radical scavenging assay. Results indicate that three isolated compounds contribute importantly to the antioxidant activity of *L.nobilis* leaves, providing a scientific basis for the use of this plant in traditional medicine. The most active compound was found to be quercetin along with two flavonoid compounds, rutin and morin, showed significant scavenging activity with IC_{50} of 10±0.02, 30±0.05 and 54±0.20 μg/ml, respectively.

Antibacterial activity: Flavonoids are efficient antioxidants and plays an important role in augmenting the wound -healing process. Flavonoids and flavons also bind with adhesins that have been termed as the most important determinant of pathogenicity [15, 16]. Terpens and terpenoids are also known to promote the wound -healing process mainly due to their astringent and antimicrobial property[17, 18]. Significant antimicrobial activity of bay extract against *E. coli* strains virulence was observed connected with the adhesion activity [19]. The extracts of *L. nobilis* promote healing activities on induced by *E. coli* purulence wound mode [4]. The results of the antibacterial activity (via the disc diffusion method) showed that square of inhibition under influence of ethanolic extract, quercetin, rutin and morin were 2910 ± 197; 9500 ± 870, 3200 ± 299 and 142000 ± 970 pixel^2, respectively. Since antibacterial activity of morin was more than 15 times greater than the other two components, it can be assumed that the main input in antibacterial activity of extract belongs to morin. The antibacterial activity of bay can be also due compounds non-flavonoid origin, such as *p*-cymene – a precursor of carvacrol [20]. The high contents of eugenol, methyl eugenol and fatty acid methyl esters together with other active components [21] could contribute to its overall antibacterial activity [22].

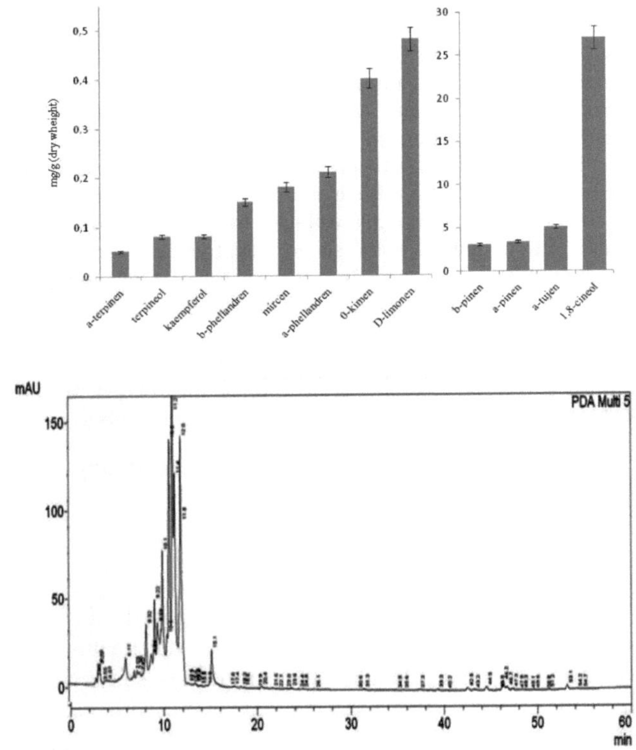

Fig. 1 GC-MS (left) and HPLC (right) analysis of L.nobilis leaf ethanolic extracts

Cholesterol and triglyceride determination: Triglyceride and cholesterol levels in blood are one of the most important determinants of lipid metabolism [23,24]. At liver pathogenesis hepatocytes esterification processes are also inhibited and cholesterol synthesis so accumulated acetic acid is the substrate for its formation. In large quantities acetic acid manifests a cytotoxic effect. Cholesterol is required for the synthesis of the lipid membrane structures, biliary acids [23].

CCL4 intoxicated groups showed increased levels of cholesterol in plasma and tissues attributed to increased β–hydroxyl methyl glutaryl CoA [HMG CoA] reductase activity which is the rate limiting step in cholesterol biosynthesis

[25].The administration of *L. nobilis* ethanolic extract carried out to lowering of total cholesterol and triacylgycerides levels (Table 1).

Table 1 Effect of ethanolic extract of *L. nobilis*on lipid profile in blood plasma of control and CCl_4 administered rats

	control	CCl_4	CCl_4 + *L. nobilis* extract		
			36 h	240 h	480 h
total cholesterol (mM/L)	1,4±0,10	1,8±0,20	1,1±0,05	0,5±0,03	0,7±0,01
triglycerides (mM/L)	1,6±0,06	1,7±0,10	1,2±0,04	0,7±0,03	0,6±0,04

Histochemical assay: Representative images of hematoxylin staining to visualize architecture of the liver and hepatocytes on sections of the liver (40-50 microns) are in the Fig.2.Under the influence of CCl_4 microscopically normal structure of the liver are violated, the signs of diffuse liver disease with extensive necrosis of hepatocytes, perivascular and pericellular edema are determined. It is also noted that diffuse lesion of hepatocytes with more severe changes in pericapillary areas. Often noticeable discomplexation of hepatic beams. Hepatocyte necrosis developed , in the foci of necrosis observed inflammatory infiltration, clearly visible in the central part of the hepatic beams. The single administration of CCl_4 and extract showed improvement in the preservation of hepatocytes both in the 10 and 20 day (Fig. 2). In primary liver tissue the processes of neutralization and reduction in the density of necrotic foci was observed. Much of the tissue did not differ from the normal liver tissue, signs of edema and tissue infiltration are absent. Thus extract has angioprotective action on the capillary bed of the rat liver, vascular protective effect and prevented the progression of inflammation and necrosis, which can be explained by the presence of antioxidants and antimicrobials flavonoid origin, such as quercetin and morin as well as terpens and terpenoids.

However, by the 20th day, there are marked signs lipidation of liver, or steatohepatitis. Steatosis initially causes oxidative stress could be the introduction of CCl_4, which leads to increased formation of free fatty acids. Its can lead to toxicity by stimulation of cellular stresses such as oxidative stress, apoptosis and gut-derived lipopolysaccharide (LPS) trigger an inflammatory response and progressive liver damage [26]. Extract of *L. nobilis* prevented ROS that are generated during free fatty acid metabolism and are responsible for initiating necroinflammation. This lead to increase of triglycerides level and accumulation in liver as a protective mechanism of liver detoxication.

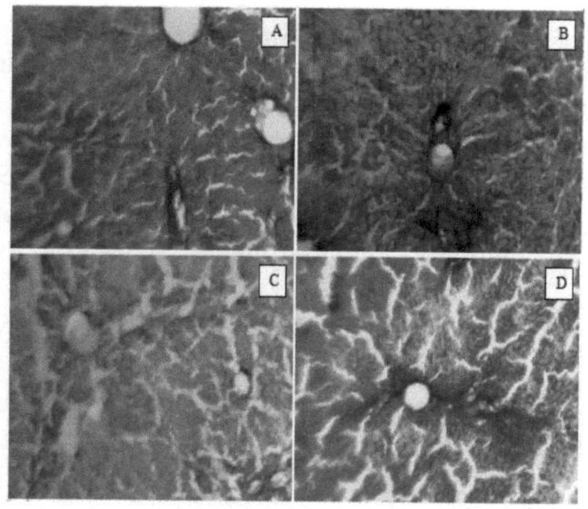

Fig. 2 Photomicrographs of rat liver similar structures sections of the: A - control,- B - CCl4, C -CCl4+ L. nobilis extract (240h), D - (480h) All photomicrographs were taken at a magnification of 200x. The scale bar represents 50μM.

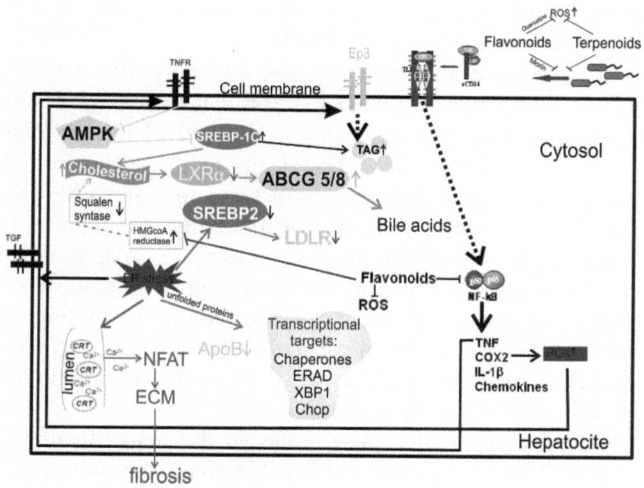

Fig. 3 Hypotetic scheme of multiple pathophysiological ways of hepatosteatosis induced by CCl4 and L.nobilis extract compounds action

A central role in cell metabolism by controlling synthesis of fatty acids, triglycerides, and cholesterol play transcription factors called sterol regulatory element-binding proteins (SREBPs) [23], that integrate multiple cell signals to control lipogenesis. SREBP-1a activates fatty acid and cholesterol synthesis, SREBP-1c fatty acid synthesis and SREBP-2 cholesterol synthesis and uptake. Under normal conditions, these pathways interact with each other and act as regulators of hepatic cholesterol levels and activate genes involved in the synthesis of cholesterol and free fatty acids carried out on the basis of feedback system regulation [27]. However during liver pathogenesis these systems are highly disorganized and in the context of liver damage the regulatory loop of SREBP is disturbed, regardless of the intracellular levels of cholesterol and/or fatty acids [28] and despite excess cholesterol accumulation in hepatocytes, de novo cholesterol synthesis remained greatly enhanced even though SREBP-2 expression while it uptake suppressed because of markedly downregulated expression of LDL receptor [27].

Toxic lipids such as free fatty acids, diacylglyceride, phospholipids and free cholesterol activate several cellular stress pathways [29. One epicenter for these stress responses is the endoplasmic reticulum (ER) stress as an emerging factor in fibrotic disease, although precise mechanisms are not clear [30]. Free cholesterol accumulation triggers ER stress by altering the critical free cholesterol-to-phospholipid ratio of the ER membrane. An important element of the fibrogenesis is endotoxin exposure, leading to activation of cytokines, including transforming growth factor (TGFβ), under which influence the transformation of fat accumulating cells in fibroblasts.

ER stress and fibrous tissues activated chaperone calreticulin (CRT), which regulates extracellular matrix by stimulation of TGFβ, with controlling the level of intracellular Ca^2 and NFAT signaling. This leads to the activation of calcineurin, which activates NFAT nuclear translocation and promotes that can directly or indirectly stimulate transcription of extracellular matrix proteins (ECM), such as type I collagen involved in the formation of fibrosis. Prolonged / or strong ER stress can also lead to misfolding of many secretory proteins in particular, ApoB100 due to failure ER chaperones, such as GRP 78, GRP94, protein-72 ER, calreticulin and calnexin.

Gut microbiota play many roles in the hepatosteatosis, hepatocellular carcinoma, cardiac functions, vascular atherosclerosis, diabetes and other conditions. Endotoxin or LPS produced by gut microbiota could be delivered to the liver via the portal vein through the intestinal barrier. LPS – compound of outer membrane of Gram- bacteria binds with LBP and transports to liver, where interact with $CD1_4$ and TLRs with following activation of nF-κB and its transcriptional targets –proinflammatory TNF, IL-6,- 1β, NOS, COX1/2, ROS, free fatty acids, etc. COX-1 is constitutively expressed in many tissues and is responsible for a variety of physiological functions, and does not change with the liver lipidation. COX-2 in the normal liver is not detected but induced *de novo* in fatty liver [31]. This increases prostaglandin synthesis that interacting with receptors Ep3, help to increase the synthesis of triglyceride and their accumulation in the liver by preventing lipolysis.

We have proposed a scheme (Fig.3), taking into account the multiple parallel pathophysiological ways of hepatic steatosis including microbiota LPS, oxidative stress and pro-inflammatory cytokines. A single administration of the extract L.nobilis led to a reduction of necroinflammation of the liver tissue possibly due lowering of ROS level, and the binding of endotoxin to the touch receptors, reducing induction of HMG-CoA reductase and can be one of the possible causes of lowering blood cholesterol levels.

ACKNOWLEDGMENT

We are grateful to thank for the financial support to the Ministry of Education and Science of the Republic of Armenia (grants № 10-2/I-1; 10-2/I-4).

CONFLICT OF INTEREST

The authors declare that they have no conflict of interest.

REFERENCES

1. Chatterjee T. Medicinal plants with hepatoprotective properties. In: Herbal Options. 3rd Edn. Books and Allied (P) Ltd. Calcutta. 2000, 135.
2. Buto S. Bay leaf impaction in the esophagus and hypopharynx. Ann Intern Med. 1990, 113(1), 82-3.
3. Santos F., Rao V. Antiinflammatory and antinociceptive effects of 1.8-cineole a terpenoid oxide present in many plant essential oils. Phytother. Res. 2000, 14, 240-244.
4. Rukhkyan M., et al. Wound healing activity of *Laurus nobilis* leaves extracts. Issues in theoretical and clinical medicine, 2013, 16, 20-24.
5. Vardapetyan H., et al. Phytochemical composition and biological activity of *L. nobilis*l. leaves collected from two regions of south Caucasus. JEBAS, 2013, 1(2), 45-52.
6. Vardapetyan et al., Quercetin content and antioxidant activity of *Armenian crataegus Laevigata, Plantago major* and *Artemisia absinthium* plants extracts. JEBAS, 2014, 2(2S), 220-225.
7. Alam K., et al.,Bay leaves improves glucose and lipid profile of people with type 2 diabetes. J. Clin.Biochem. Nutri. 2009, 44, 52–56.
8. Fischer-Nielsen A., et al. CCl_4 cirrhosis in rats: irreversible histological changes and differentiated functional impairment.1999, 12, 110-117.
9. Koleva I. et al. Screening of plant extracts for antioxidant activity: a comparative study on three testing methods. Phytochemical Analysis, 2002, 13, 8-17.
10. Asatryan D. et al. Technique for Coherent Segmentation of Image and Applications. Mathematical Problems of Computer Science, IIAP, 2007, 28, 88-93.
11. Gasparyan G., et al. Effect of Laurus Nobilis extract on the functioning of liver against CCl4 induced toxicity.JEBAS2015, 3(2), 174-183.
12. Maggi C., et al. Serum cholesterol and chronic hepatitis. J Gastroenterol. 1996, 28(8), 436-40.
13. McGowan M., Artiss J. Clin. Chem. 1983, 29 (3), 538–542.
14. Chilingaryan G. New Ca-adenosinetriphosphate method for determining intraorganic microvasculature in cats. 1986, 66-71.
15. Silva N., Fernandes J. Biological properties of medicinal plants: a review of their antimicrobial activity. The J. of Venomous Animals and Toxins including Tropical Diseases. 2010, 16, 402-413.
16. Croxen M., Finlay B. Molecular mechanisms of *E.coli* pathogenicity Nature. reviews microbiology. 2010, 8, 26-38.
17. Nayak B. et al. Experimental evaluation of Pentas lanceolata for ound healing activity in rats. Fitotherapia. 2005, 76, 671-675.

18. Hendra R, et al. Flavonoid Analyses and Antimicrobial Activity of Various Parts of *Phaleria macrocarpa* (Scheff.) Boerl Fruit. International Journal Molecular Sciences. 2011, 12, 3422-3431.
19. Sun D., et al. Influence of berberine sulfate on synthesis and expression of Pap fimbrial adhesin in uropathogenic *Escherichia coli*. Antimicrobial Agents Chemotherapy, 1988, 32, 1274–1277.
20. Braga P. et al. Thymol inhibits *Candida albicans* biofilm formation and mature biofilm. International Journal Antimicrobial Agents. 2008, 31, 472-477.
21. Marzouki H., et al. Extraction and Separation of Volatile and Fixed Oils from Berries of *Laurus nobilis* L. by Supercritical CO_2 Dessi 3. Molecules. 2008, 13, 1702-1711.
22. Ivanović J, et al. Supercritical CO_2 extract and essential oil of bay (*Laurus nobilis* L.) – chemical composition and antibacterial activity. Journal of the Serbian Chemical Society 2010, 1-10.
23. Goldstein J. et al. Protein sensors for membrane sterols. Cell. 2006, 124, 35–46.
24. Hashimoto E., et al. Hepatocellular carcinoma in patients with nonalcoholic steatohepatitis. Journal of Gastroenterology. 2009, 44, 89–95.
25. Shao W., Espenshade P. Expending roles for SREBPs in metabolism. Cell Metabolism. 2012, 414-419.
26. Csak, T., et al. Fatty acids and endotoxin activate inflammasome in hepatocytes which release danger signals to activate immune cells in steatohepatitis. Hepatology. 2011, 54, 133–144.
27. Enjoji M, et al. Nutrition and Nonalcoholic Fatty Liver Disease: The Significance of Cholesterol. International Journal of Hepatology. 2012 ID 925807, 6.
28. Donohue T. Alcohol-induced steatosis in liver cells. World Journal of Gastroen-terology. 2007, 13, 4974–4978.
29. Musso, G., et al. Cholesterol metabolism and the pathogenesis of non-alcoholic steatohepatitis. Prog. Lipid Res. 2013, 52, 175–191.
30. Zimmerman, K. et al. Calreticulin Regulates Transforming Growth Factor-_-stimulated Extracellular Matrix Production. The journal of biological chemistry. 2013, 288 (20), 14584–14598.
31. Takaki A et al, Multiple Hits, Including Oxidative Stress, as Pathogenesis and Treatment Target in Non-Alcoholic Steatohepatitis (NASH) Int. J. Mol. Sci. 2013, 14, 20704-20728.

Author:	Hrachik Vardapetyan
Institute:	Russian-Armenian (Slavonic) University
Street:	123 Hovsep Emin str.
City:	Yerevan
Country:	Armenia
Email:	hvardapetyan@mail.ru

An Automated Inertial Indoor Positioning and Fall Detection System for Elder

I.R. Edu[1], F.C. Adochiei[2], L. Grigorie[3], A. Pasarica[4], and N. Jula[1]

[1] Military Technical Academy, Bucharest, Romania, University Politehnica of Bucharest, Bucharest, Romania
[2] University of Craiova, Craiova, Romania
[3] "Gheorghe Asachi" Technical University of Iasi, Iasi, Romania

Abstract— **Statistic data suggest that in 2040 a third of Europeans will be more than 70 years old and that falling is responsible of 75% of accidental deaths in people over 75 years old. This problem leads to a continuous concern for developing an advanced fall detection systems. We are proposing a module for human fall detection based on a miniaturized inertial measurement unit, developed to be integrated in a complex tele-monitoring system for elder. For a better indoor estimation of position, we used a wavelet filtering mechanism, in order to calibrate our signal processing algorithm. By using this method we have succeeded to process and analyze signals received from the navigation unit and to provide a better estimation of the patients' indoor activities.**

Keywords— **Inertial indoor positioning, fall detections system.**

I. INTRODUCTION

Traumas caused by falls are dangerous especially for elderly people diagnosed with various ailments. Currently, researchers are working on developing integrated platforms for monitoring vital signs and the location of elderly people. It is important to monitor patients fall because it has been shown that, in general, the acceleration of the fall is approximately 0.9*g*, and the falling event is easy to be detected by using an accelerometer sensor. Falls are a common issue, but they are difficult to define rigorously.

Automated systems, accurate, fast and easy to use have been implemented in hospitals and homes of those who need medical supervision. In terms of recording vital parameters, these systems have registered real performances in this type of information collection, in processing and interpretation of information from specialized indoor environments, such as hospitals. In terms of precise localization of patients in other enclosed areas, not only hospitals, and in terms of rapid detection of the fall event, researchers are still working on obtaining fast and accurate equipment.

Another difficult aspect is to identify and isolate falls from daily activities because many activities can be interpreted as falls, such as sitting down or going down stairs fast. Information about body position can provide further insight in these cases but are not very useful when the final position of body is not horizontal, for example in the case of falling down from stairs. We introduced a new type of sensing device for the remote monitoring and for the prediction of falls in elderly population.

The main objective of our project was the achievement of an easy to wear integrated system, mainly composed of the following components: a detection cluster with tri-axial accelerometers, a tri-axial gyro, a magnetometer, a microcontroller, a wireless transmission module (Bluetooth) and an accumulator, mounted on the patient's shirt (Figure 1).

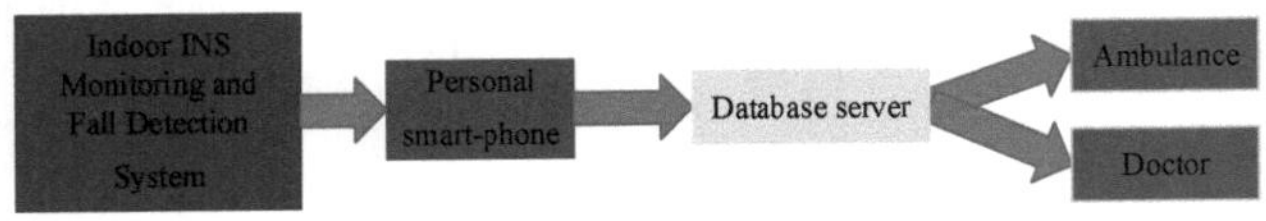

Fig. 1 The structure of inertial indoor fall risk detection system

To reduce costs, energy consumption and data processing time, the proposed architecture assumes linear positioning of two channels in the horizontal plane and angular positioning in relation to a reference directive also considered horizontally. In this respect two acceleration detectors were used in the horizontal plane, an acceleration detector and an angular speed detector in the vertical plane and a magnetometer, also positioned in the horizontal plane. In order to monitor the fall, all the three-dimensional accelerometer's outputs were used and threshold conditions were set up to trigger an alarm in the case of fall detection. As a certification of fall occurrence, the patient's angular attitude is also used. Data is obtained by mathematically combining the three accelerometer outputs which are sensitive to gravitational acceleration.

For the accurate localization of the patient's position, data achieved from the navigation system will be calibrated in relation to the information provided by the GPS. In other words, the position offered by the inertial navigation system will always be assessed in relation to the position provided by the smart-phone GPS module. Information from the device is sent to a smart-phone, through an interface module, which is used as a personal server where data packages, achieved from the remote device, are processed.

The smartphone will run processing algorithms and send all the results to a hospital database server and to the patient's personal doctor by using internet, GPRS or a telephone line. When the system will detect a fall, the patient's smart-phone will send an emergency message to the ambul-

V. Sontea and I. Tiginyanu (eds.), *3rd International Conference on Nanotechnologies and Biomedical Engineering*,
IFMBE Proceedings 55,
DOI: 10.1007/978-981-287-736-9_100

ance service and the patient's doctor. Having in mind that home coordinates are known by the monitoring station, when the patient will leave his home, the INS positioning will be switched on the GPS positioning while the fall detection unit will be permanently used.

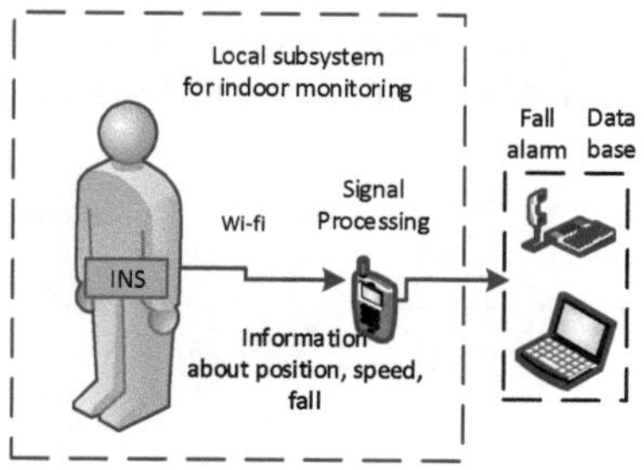

Fig. 2 Local subsystem for patient indoor monitoring

Until recently, falls detection solutions were divided into two classes: applications which were used to detect falls by using only accelerometers [1-4] and applications using both acceleration and body orientation information [4-6].

Linear acceleration and angular velocity are measured to determine whether motion transitions are intentional. If falling is defined as an unintended transition to the laying posture, if the system detects that the transition before a laying posture is not intentional, a fall event is detected.

To determine if a transition is intentional, by processing and analyzing information achieved from accelerometers, gyro and magnetometer, our fall detection module provides more accurate data than existing devices.

II. MATERIALS AND METHODS

A person's position and speed of displacement are obtainable by direct integration of the general equation of inertial navigation, with respect to the navigation referential. In inertial navigation systems with platform, acceleration sensors' axes are oriented respecting the navigation referential axes and, therefore, specific force components f are obtained directly in this referential system.

For indoor monitoring of patients, respectively indoor navigation, the local horizontal trihedral (SOL) was chosen as a navigation referential system, which can be considered quasi-inertial when taking into account the duration and the specific activity of the patient. Therefore, for solving this problem of navigation, two different reference systems were considered: the local horizontal trihedral (SOL) and patient device trihedral (PDT) (fig. 3).

Our basic navigation system is a bi-dimensional system in the horizontal plane, because we monitor patients in the horizontal plane, on x and y axes, for estimating their position and speed, which involves only three inertial sensors in the navigator system: accelerometers mounted on x_v and y_v axes of the *PTD* and a gyro mounted on the z_v axis. We added an accelerometer for detecting patient's fall on z_v axis. This sensor will be used for fall detection in emergency situations, not for monitoring the patient's activities, therefore the data from this sensor don't need to be integrated in the bi-dimensional navigation system for positioning.

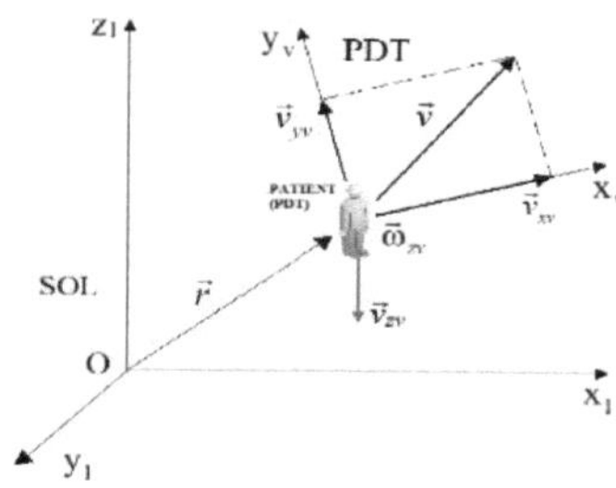

Fig. 3 Reference system and the evolution of PDT related to SOL system

A magnetometer will be used for the initialization and the continuous correction of the angular position. The components presented in Fig. 3 are: $\vec{r}$ - the position vector of patient in *SOL* system, $\vec{v}$ the patient speed reported to *SOL* system, $\vec{v}_{yv}$, $\vec{v}_{xv}$ - the speed components of patient related to the *PDT* axes, $\vec{\omega}_{zv}$ the angular speed component of the patient related to the z_v axis of *PDT* and $\vec{v}_{zv}$ the speed component of the patient registered when he falls.

The relationship which expresses the relationship between the specific force vector $\vec{f}$ ([5], [6], [10]) and $\vec{a}$, the kinematic acceleration of the patient, is:

$$\vec{f}=\vec{a} \tag{1.1}$$

Acceleration may be express using the relationship

$$\vec{a}=\frac{d\vec{v}}{dt}+\vec{\omega}\times\vec{v}, \tag{1.2}$$

where $\vec{\omega}$ is the angular speed of the patient compared to *SOL* system. Equation (1.1) is written, when taking into consideration equation (1.2), under the form

$$\vec{f}=\frac{d\vec{v}}{dt}+\vec{\omega}\times\vec{v}-\vec{g}. \tag{1.3}$$

Projecting equation (1.3) in the *PDT* system, on x and y axes we obtained the following equations:

$$f_{xv}=\frac{dv_{xv}}{dt}+\omega_{zv}v_{yv}, \qquad f_{yv}=\frac{dv_{yv}}{dt}-\omega_{zv}v_{xv}. \tag{1.4}$$

f_{xv}, f_{yv} are the specific force components $\vec{f}$ in *PDT* (accelerometer readings). Numerical integration of (1.4) provides the patient's speed component values in *PDT* trihedral. The conversion of these components into the *SOL* trihedral is made using (1.5), considering $v_{zv} = 0$:

$$\begin{bmatrix} v_{xl} \\ v_{yl} \\ v_{zl} \end{bmatrix} = R_v^l \begin{bmatrix} v_{xv} \\ v_{yv} \\ v_{zv} \end{bmatrix}, \tag{1.5}$$

To determine the R_l^v attitude matrix, prior calculations of ψ yaw angle are required, which is achieved by the numerical integration of the equation (1.6):

$$\dot{\psi} = \omega_{zv}, \tag{1.6}$$

where ω_{zv} is the gyro reading. For R_l^v matrix calculation roll and the pitch angles are set at $\varphi = 0$ and $\theta = 0$ values.

For determining the patient's position, the numerical integration of *SOL* trihedral speed components is performed,

$$x_l = x_{l0} + \int_{t_{n-1}}^{t_n} v_{xl}\, dt, \qquad y_l = y_{l0} + \int_{t_{n-1}}^{t_n} v_{yl}\, dt, \tag{1.7}$$

where x_{l0} are y_{l0} are the initial coordinates of the patient related to the *SOL* trihedral.

Also, in developing the indoor monitoring device we implemented a previously proposed tuning method for data achieved from accelerometers [NCAS], method based on the Joint Symbolic Dynamics (*JSD*) method; this method is used for the calibration of the level of decomposition of the wavelet filter implemented for filtering the signals received from accelerometers. The main idea of the method is to process and analyze signals received from the sensors by using the Wavelet transform until optimal levels of decomposition are established and the useful signals are achieved.

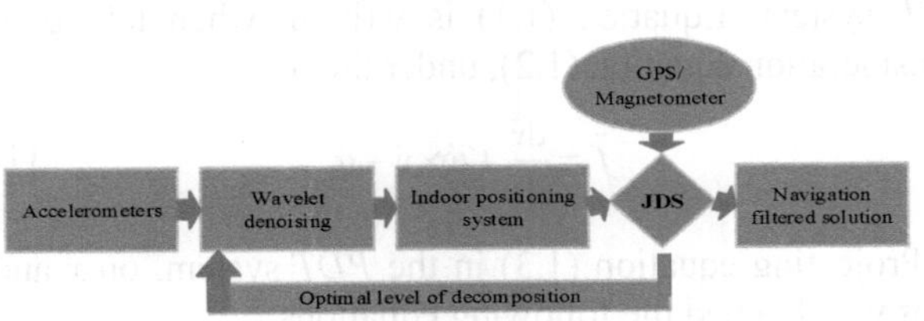

Fig. 4 Architecture of the tuning method for sensors denoising

The architecture of the tuning method for the inertial sensors denoising is presented in Fig. 4. For calibrating in real time, the reference signals are provided by a magnetometer and a *GPS*, while the disrupted input signals in *JSD* are the outputs of the inertial navigation system.

III. RESULTS

Starting from the functioning equations of the navigator and the previously presented theoretical algorithm, we developed the MATLAB/Simulink model presented in Fig.5.

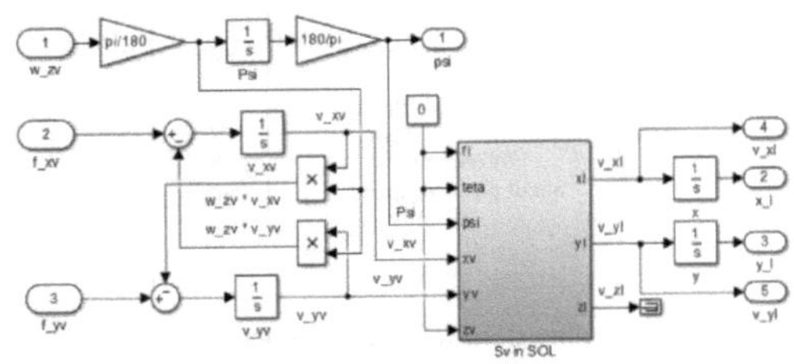

Fig. 5 Block scheme of the bi-dimensional navigator

The components of the bi-dimensional navigator presented are the components behind the navigator block represented in Fig. 6.This block has as inputs the angular velocity component read from the gyro, on the *z* axis, and the acceleration components, read by the accelerometers, on the *x* and *y* axes of the *PDT*. The outputs are ψ, the patient's attitude, given through the yaw angle, the position in *SOL* trihedral given through x_l and y_l components and the speed component, given through v_{xl} and v_{yl}.

The model is used in the numerical simulation of the strap-down inertial navigation system to reconstruct the patient's evolution. The Matlab/Simulink model represented in Fig.6, which uses the „Bi-dimensional navigator in the horizontal plane" subsystem, was implemented with the purpose of monitoring the patient's position, whose inputs are linear coordinates of the patient and attitude angles.

The simulation of the system represented in Fig. 6 is computed and the patient's movement is observed using the „Animation" block. The patient's evolution (speed, position) is monitored using the "Scope" blocks. Fig.7 represents the graphical features obtained for the patient's movement and fig. 8 and 9 are illustrating the fall detection, and respectively how are visualized the positioning signals on scopes.

For our indoor monitoring device we implemented a previously proposed tuning method for processing data achieved from accelerometers. We performed a number of numerical simulations for different noisy signals correlated with reference signals. This was done in order to test the *JSD* method capacity to set up a quality parameter (*QP*) and determine the optimal level of decomposition of the wavelet filter for indoor applications. Table 1 show the estimated

QP values achieved using *JSD* method for a *8* [m/s^2] acceleration signal and its wavelet filter level of decomposition (*D-Level*), for levels of decomposition between *1* and *7*.

In Fig. 10 is illustrated this signal evolution in time and also the time evolution of the filtered signal for *D-Level_opt* = *5*, for *D-Level_opt-1= 4* and for the filtered signal at *D-Level_opt+1*= 6, of the filtered signal.

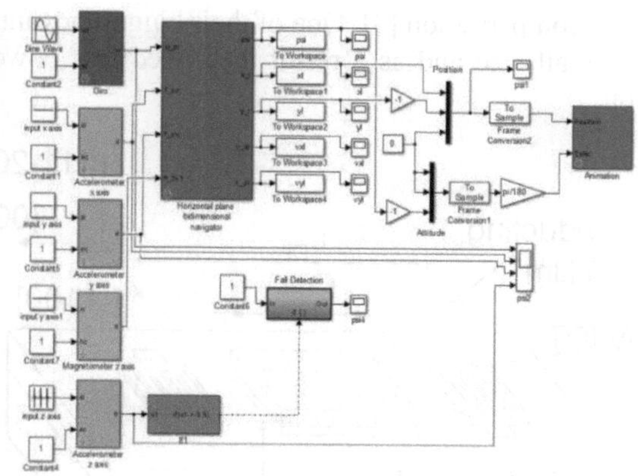

Fig. 6 Developed navigator simulation model

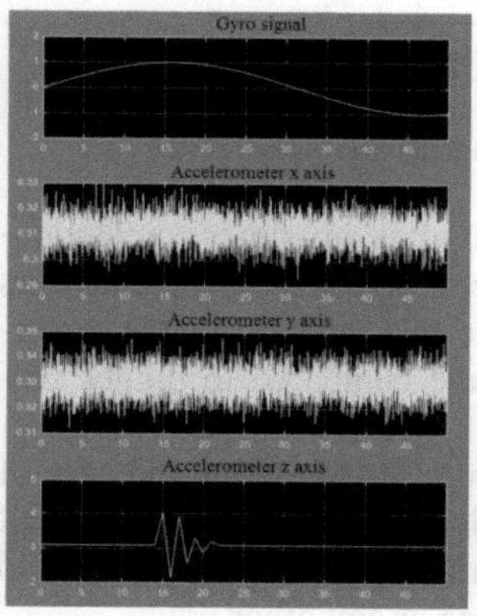

Fig. 7 Signals from sensors

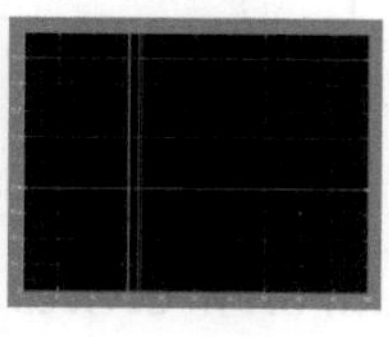

Fig. 8 Falling detection

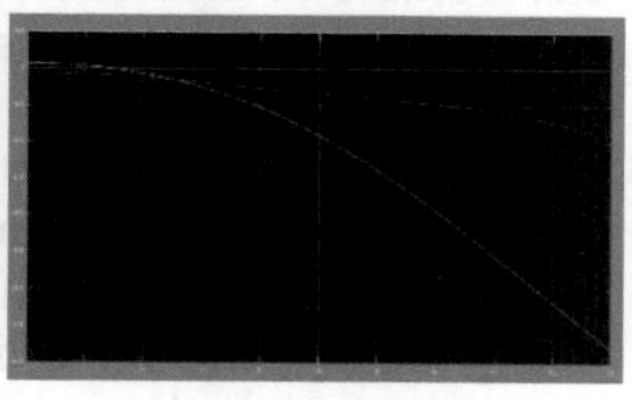

Fig. 9 Positioning signals

Table 1 - QP values estimated for various values of decomposition

D-Level	QP
1	0.1631
2	0.3583
3	0.5882
4	0.6725
5	0.7901
6	0.7527
7	0.7674

The best reproduction of the reference signal following wavelet filtering was achieved for the optimal decomposition level equal with 5.

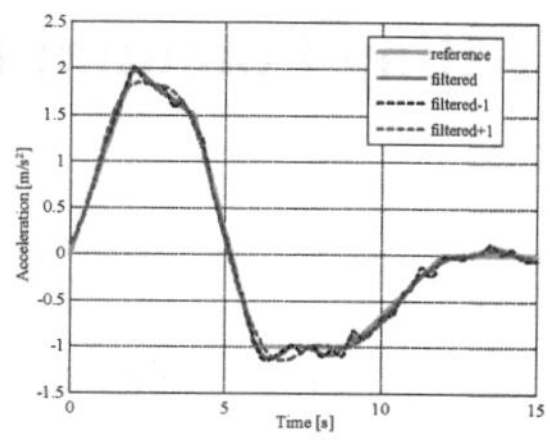

Fig. 10 The time evolutions of the reference and filtered signals

IV. CONCLUSIONS

We introduced a new type of sensing device for the remote monitoring and for the prediction of falls in elderly population. We are proposing a module for human fall detection based on an inertial measurement unit, containing accelerometers, gyros and magnetometers, integrated in a complex tele-monitoring system for identifying the activity of older patients. For a better indoor position monitoring, we used a wavelet filtering mechanism. By using this method we have succeeded to process and analyze signals received from the navigation unit and to provide a better estimation of the patients' indoor activities.

CONFLICT OF INTEREST

The authors declare that they have no conflict of interest.

REFERENCES

1. A. Dıaz, M. Prado, L. M. Roa, J. Reina-Tosina, and G. Sanchez, "Preliminary evaluation of a full-time falling monitor for the elderly.," *In Proceedings of the 26th Annual International Conference of the IEEE-EMBS, pp. 2180-2183, USA, Sept,* 2004.
2. M. Kangas, A. Konttila, I. Winblad, and T. Jamsa, "Determination of simple thresholds for accelerometry-based parameters for fall detection.," *In Proceedings of the 29th Annual International Conference of the IEEE EMBS, pages 1367–1370, Lyon, France, Aug* 2007.
3. U. Lindemann, A. Hock, M. Stuber, W. Keck, and C. Becker, "Evaluation of a fall detector based on accelerometers: A pilot study.," *Medical and Biological Engineering and Computing, 43(5):548–551, Oct* 2005.
4. M. J. Mathie, J. Basilakis, and B. G. Celler, "A system for monitoring posture and physical activity using accelerometers.," *Proc. of the 23rd Annual International Conference of the IEEE-EMBS, pp. 3654–3657, Istanbul, Turkey, Oct* 2001.
5. A. K. Bourke and G. M. Lyons, "A threshold-based fall-detection algorithm using a bi-axial gyroscope sensor.," *Medical Engineering and Physics, 30:84–90, Jan* 2008.
6. N. Noury, P. Barralon, G. Virone, P. Boissy, M. Hamel, and P. Rumeau, "A smart sensor based on rules and its evaluation in daily routines.," *In Proceedings of the 25th Annual International Conference of the IEEE EMBS, pp. 3286–3289, Mexico, Sept,* 2003.
7. J. Chen, K. Kwong, D. Chang, J. Luk, and R. Bajcsy, "Wearable sensors for reliable fall detection.," *In Proceedings of the 27th Annual International Conference of the IEEE EMBS, pages 3551–3554, Shanghai, China, Sept* 2005.

Author: Ioana – Raluca Edu
Institute: Military Technical Academy
Street: 39-49 George Coşbuc Avenue
City: Bucharest
Country: Romania
Email: edu_ioana_raluca@yahoo.com

Modelling Potential Distribution in ZnO with Different Thicknesses at GHz Frequencies

V.P. Cojocaru, D. Vrabii, E. Rusu, and N. Curmei

D. Ghitu Institute of the Electronic Engineering and Nanotechnologies, Academy of Sciences of Moldova, Chisinau, Moldova

Abstract— **A surface acoustic wave (SAW) device and it's behavior at high frequencies is presented. The experiment and the simulations were done on ZnO film and with the software Comsol Multiphysics. The simulations helped us understand the potential distribution in ZnO with three different thicknesses: 150µm, 200µm and 300µm. Based on this results we plotted resonance frequencies for every thickness on a range from 1 Ghz up to 10 Ghz.**

Keywords— **surface acoustic wave (SAW), piezoelectric material, resonance frequency, Comsol Multiphysics.**

I. INTRODUCTION

ZnO is an attractive material candidate for ultraviolet (UV) optoelectronics, as it has a direct energy band gap of approximately 3.32 eV at room temperature, and also a free exciton binding energy of 60 meV. ZnO is more resistant to radiation damage than GaN [1]. Piezoelectric ZnO thin films have been used for fabricating surface acoustic wave (SAW) devices on nonpiezoelectric or weakly piezoelectric substrates.

Over the past five decades, there has been an increasing demand for inexpensive, accurate, portable and reliable sensors. SAW sensors consist of two electrode pairs, which are deposited with a certain distance from each other on a substrate.

Figure 1 shows the approximate model used for simulation in this paper. Semiconducting ZnO film and Al_2O_3 were excluded to simplify the problem.

The region between the electrodes can be coated with a specific semiconductor thin film for different kinds of detection. The input transducer converts input voltage into an acoustic wave. This wave is less affected by other electromagnetic influences. The delay line is the region where the wave travels along the substrate until the output transducer where it is converted back into an electric wave.

Such kind of sensors offer some advantages compared to other conventional sensors. Their first big plus is their low cost and an easy fabrication. Data acquisition can be done passively or remotely via an RF antenna connected to input interdigital transducer and a reflector which replaces the output transducer [2]. They also show good performance at a wide range of temperature, but there are some applications which require some additional electric schemes for temperature error compensation [3]. One of their biggest advantages is their small size and as a result they need less power to function.

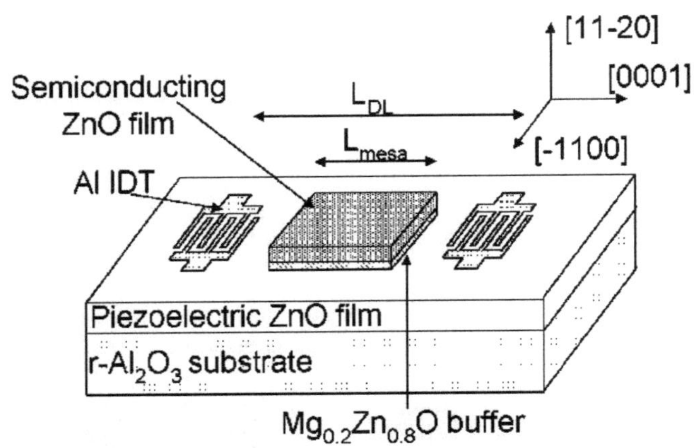

Fig. 1 Schematic reprezentation ZnO based UV

Such SAW devices can be used as any other conventional sensors such as accelerometers, gyroscopes, force sensors etc. Because of their small size they show a better performance in terms of response time compared to bigger sensors.

SAW devices show great promises for future industry of sensors manufacturing being low cost and with excellent performance with almost unlimited applications in any field of engineering, science, medicine or any other.

In this paper we have studied the potential distribution in ZnO, a piezoelectric substrate, which can lead to a more optimal construction of an SAW sensor based on the studied material. We have applied 10V at the input port and simulated the potential and frequency at the output port. Such modelling is rarely found despite its clear important use in designing a SAW device, sensor or not.

II. EXPERIMENT

ZnO films are deposited on p-Si substrates by spin-coating method. For this goal the solution is formed by dissolving zinc acetate dihydrate [Z $(CH3COO)_2$-H_2O] in 2-Methoxyethanol at room temperature. Diethanolamine (DEA) was added to the initial mixture as a stabilizer. The

V. Sontea and I. Tiginyanu (eds.), *3rd International Conference on Nanotechnologies and Biomedical Engineering*,
IFMBE Proceedings 55,
DOI: 10.1007/978-981-287-736-9_101

solution consisting of $InCl_3$ in 2-Methoxyethanol was used for doping. The concentration of indium in the solution was varied in the range 1-4% at. The final concentration of the solution was maintained at 0.35 - 0.50 m and the molar ratio of DEA to the zinc acetate is 1. The final mixture was subjected to ultrasonic treatment at T =50° C for 1 hour until complete homogenization.

In order to achieve the design of MSM type optical detector (metal-semiconductor-metal), it is widely used optical photolithography technique. In order to cover the surface of the substrate was used a positive photoresist AZ5214, deposited with spin-coating method with rotation speed of 3500 rev / min., for 40 sec. Exposure the photoresist through the mask is carried out by ЭМ-526 system. As ultraviolet light source serves ДРШ-250 mercury lamp. Process for 60 seconds exposure occurs. Etching regions designated for the metal contacts is performed in the concentration of 1% NaOH solution for 4 seconds.

Figure 2 shows the sequency process of obtaining structures type MSM, ZnO film deposited on the Si substrate (Figure 2), photoresist layer (fig.2b) exposure photoresist mask with the requested design (1c) and metallization layer for electrical contact in the interdigital design of the device (fig.2d). The ohmic contacts are deposited on the surface of the ZnO layer by means of thermal evaporation under vacuum. To increase adhesion between Al and ZnO, the sample was heated during metal deposition process at about 100 °C. After the removal of the remaining photoresist, aluminum contacts are treated in vacuum at 250 °C for 30 minutes.

The geometrical values of this model are as follows:

- The width of the interdigitated transducers is 10 μm and the space between them is 50 μm.
- The width of ZnO film is 1970 μm.
- A point of interest was the influence of the ZnO's thickness on voltage and frequency characteristics. To study it's effect, the proper simulation was done in Comsol Multyphysics for different size of thickness (150 μm, 200 μm, 300μm). Figures 2, 3, 4 shows the geometry of ZnO with different size of thickness.

The input port is located on left side of the device and the output port on the right side respectively. Here in the figures 3, 4, 5 we see the formation of the surface acoustic wave and its propagation along the X direction. We can observe how the potential is distributed not without losses across the material. Maximum potential can be found near the input port while only a small fraction of it reaches the output port.

The main cause of this losses resides in device's own geometry.

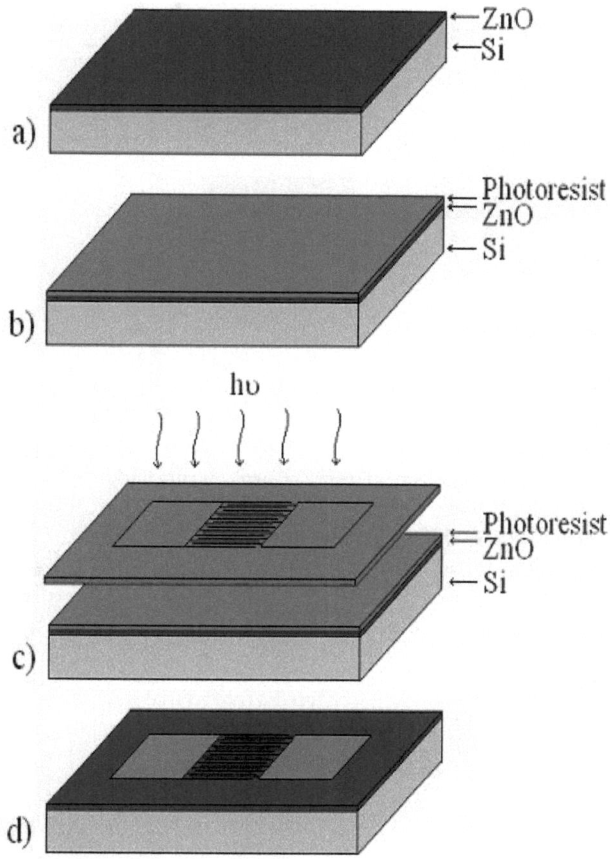

Fig. 2 The sequence of processes in the formation of the interdigital structures of type MSM

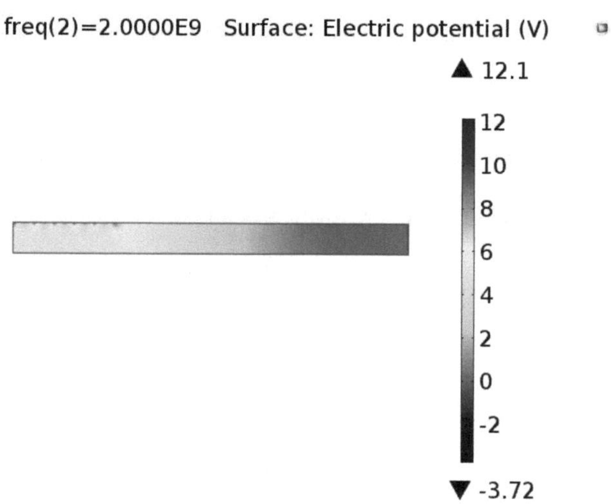

Fig. 3 Voltage distribution in ZnO with 150 μm thickness

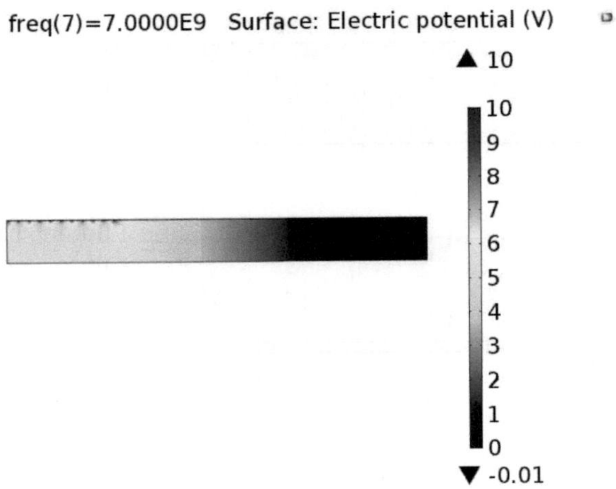

Fig.4. Voltage distribution in ZnO with 200 μm thickness

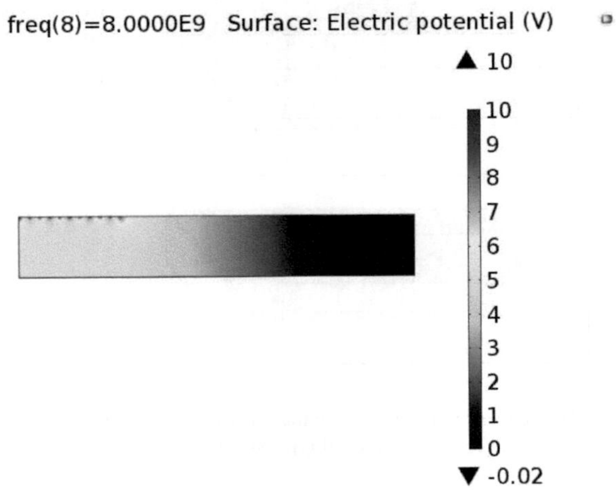

Fig. 5 Voltage distribution in ZnO with 300 μm thickness

III. Results

As a result of the technological processes in the formation of the interdigital structures of MSM type, we have obtained the following MEMS structures:

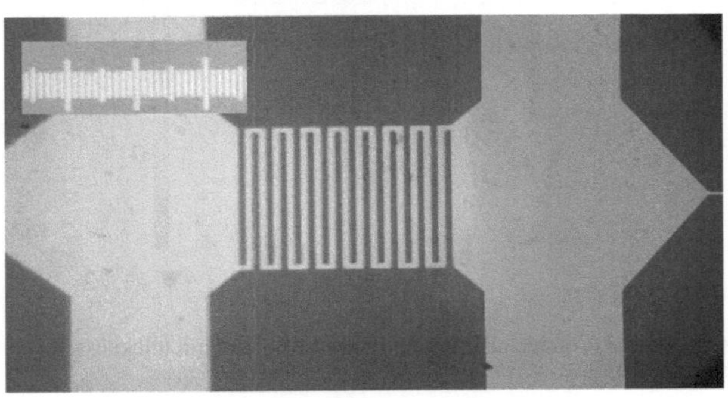

Fig. 6 The one port SAW device based on ZnO piezoelectric

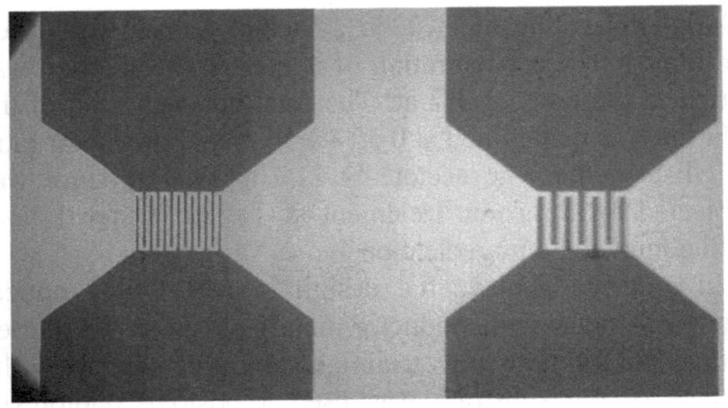

Fig. 7 The two port SAW device based on ZnO piezoelectric

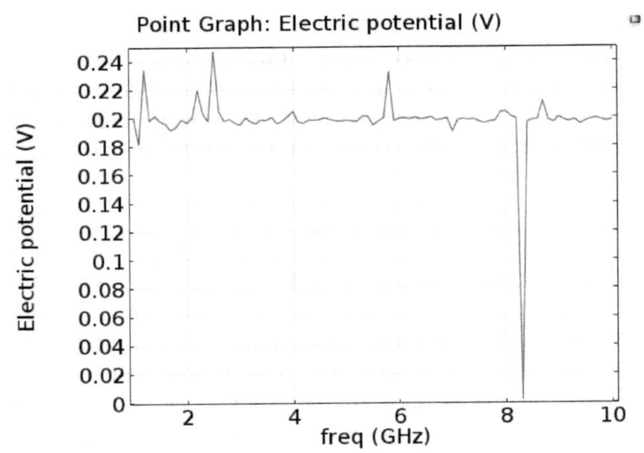

Fig. 8 The output voltage of the ZnO film with 150 μm thickness

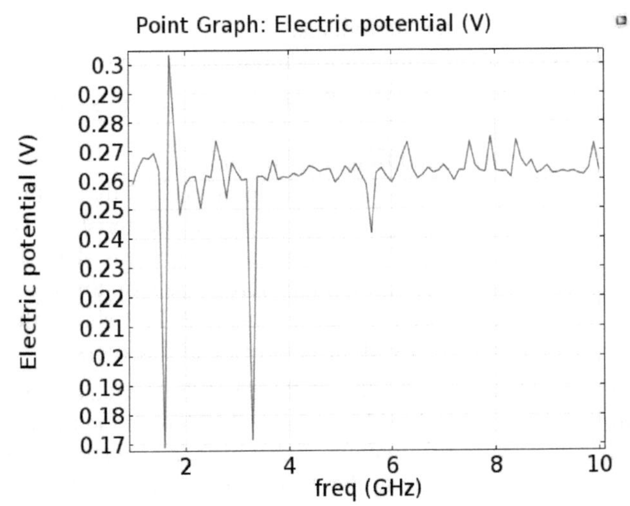

Fig. 9 The output voltage of the ZnO film with 200 μm depth

The port is made from alluminium mainly because it is a cheap, widespread material used as a conductor. Figure 6 shows a non symetrical dual port SAW device. The output port has a wider space between the digits than the input.

The simulations have shown that with the increase in thickness size, the output voltage also increases. The following figures show this characteristics.

As shown in the figure 8, there are three resonance frequency. It is also present an antiresonance frequency near 8 GHz. The maximum value of output voltage is 0.24 V.

In this figure there is one main resonance frequency at 1.8 GHz and and two antiresonance frequencies at 1.6 GHz and 2.6 GHz. The maximum output voltage is 0.3 V.

In figure 10 there is one major resonant frequency at 1.4 GHz and one antiresonance at 6.8 GHz. The maximum output voltage is 0.41 V.

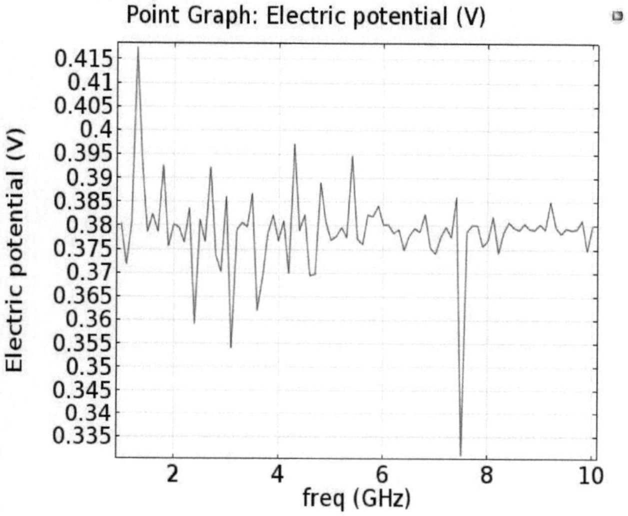

Fig. 10 The output voltage of the ZnO film with 300 µm depth

IV. Conclusion

In this study we have observed the behavior of output voltage for different size of material thickness. The input voltage was set to 10 V. As we can observe from figures above, such geometrical construction implies great losses for the SAW as only a small portion of the input voltage reaches the output transducer. On the other hand, it offers more than one resonance frequency in a wide range of frequency (1GHz to 10 GHz). As we observed from the simulations, the resonance frequency mainly is found in the range from 1 GHz to 2 GHz.

This kind of data is surely expectable because this is the range of operating frequencies of the majority SAW devices. Such simulation on various size of thicknesses can provide useful information upon potential distribution, resonance frequencies and on geometry of the device itself.

Conflict of Interest

The authors declare that they have no conflict of interest

V. References

1. D. C. Look, Mater. Sci. Eng., B **80**, 383 (2001).
2. J.Kirschner, "Surface Acoustic Wave Sensors, Design for Application", 2010
3. B.Morrisville, "Temperture compensated piezoelectric substrate", North Carolina
4. P. Sharma, A. Mansingh, and K. Sreenivas, Appl. Phys. Lett. **80**, 553, 2002
5. S. Muthukumar, J. Zhong, Y. Chen, Y. Lu, and T. Siegrist, Appl. Phys. Lett. **82**, 742 (2003)

Author:	Cojocaru Victor
Institute:	Academy of Science of Moldova, D. Ghitu Institute of the Electronic Engineering and Nanotechnologies
Street:	Academiei str., 3/3
City:	Chisinau
Country:	Moldova
Email:	vikcojocaru@gmail.com

Part V

Clinical Engineering, Health Informatics and Cellular and Tissue Engineering

Prophylaxis Monitoring of the State of Human Respiratory Organs

H.N. Rozorinov and N.I. Chichikalo

State University of Telecommunications, Kiev, Ukraine

Abstract— **The developed methodology of system modeling for prophylaxis monitoring of the state of human respiratory organs allows optimization of human listening means and means of spectral-time characteristics as a method for objectification of traditional auscultatory signs. The number of controlled points and their location have been explained. The registration and processing of human breathing signals have been held with the help of Adobe Audition application program package.**

Keywords— **Auscultatory sign, locus of control, prophylaxis, respiratory organ, respiration sensor.**

I. INTRODUCTION

The usage of physical research of auscultation is being generally accepted for almost two centuries. In practice each doctor uses the method of reduction of some sounds and amplification of other sounds for valuation of biological objects performance. The auscultation method can be applied in combination with percussion during determination of limits between neighbouring organs especially between hollow and solid, during determination of lower part of liver, neoplasms etc. The auscultation is also widely used for determination of arterial pressure, recognition of arteriovenous and intracranial aneurysms, during joints examination. Meanwhile, the major problem lays in differentiation and correct interpretation of sounds perceived by doctor caused by their poor audibility.

There are two types of auscultation:

– direct, which is lead by putting ear to patient's body;

– moderate – with the help of stethoscope, phonendoscope or stethophonendo-scope.

The most widely used methods of auscultation are those that use closed acoustic system which have air as major conductor [1]. They consider listening with the help of acoustic stethoscope. This is the tube with hard acoustic line and two-way expanded ends in the form of cone. One end is put to the doctor's ear, and the other end is put to the examined organ of the patient. Meanwhile, the free doctor's ear (either the occupied ear caused by non-tight fitting of phonendoscope) stays non hermetic because of the outside connection. In general, auscultation appears to be impossible or leads to critical mistakes. The doctor has to lean down to the patient that makes examination more difficult. So the binaural acoustic auscultation appeared. To accomplish such method the different types of stethophonendoscope are used. Phonendoscope percieves sounds through one funnel (head) across two flexible acoustic lines – acoustic channels connected to the general head and transferring to coaxial hard tubes named headband. To make it possible to hear low and middle frequency auscutatory characteristics stethophonendoscope is equipped with variable sockets: stethoscope and phonendoscope. Such auscultation doesn't allow to differentiate and has low informativity and accuracy of achieved results.

There is a type of auscultation that allows to hear two areas of the organ or organs of the patient with the help of two separately held equal independent acoustic channels [2]. This allows to increase faithfulness of the results of the auscultaion and to improve diagnostics and treatment effeciency. The examination of biological organ by means of stereoauscultation – simultaneous hearing from two places – helps to expand the amount of achieved information and to make comparative evaluation of the state of organ or organs. The stereoascultation can be used as a facility for prophylaxis control (express-diagnostics) as it takes 0,5–1,5 minutes in any conditions and gives such faithful results so generally it is not necessary to make any additional x-ray or ultrasound investigations. As known, the respiratory noises are the major informative resource at respiratory organs condition during prophylaxis monitoring.

At present the methods for automation of diagnostics of respiratory organs with the help of computer software tools are being developed. Meanwhile there is a variety of ways of sensor allocation on human body [3]. However, there is no general opinion around sensors allocation during multichannel data registration. Nevertheless the main requirement during their fitting is the ability to hear either vesicular (clear) or bronchial human breathing.

The aim of the work is the development of computer methods for processing and visualization of registered signals; detection of spectral time characteristics as a method for objectification of common asculatory signs.

II. MAIN PART

It is necessary to understand the noises generation aspects to create effective computer systems for signal

V. Sontea and I. Tiginyanu (eds.), *3rd International Conference on Nanotechnologies and Biomedical Engineering*,
IFMBE Proceedings 55,
DOI: 10.1007/978-981-287-736-9_102

registration and processing for the purpose of perfection of methods for diagnostics of respiratory diseases. The dependence of breathing noises characteristics from many parameters determines the necessity to record these parameters during the measuring process.

The analysis of the process of sound transfer from human tissue to sensitive elements of the pressure sensor of microphone type used for auscultation has shown that the presence of the air layer leads to decrease of the sensor sensitivity. The efficiency of the sensor can be raised by exception of the air sphere between human body and sensitive elements [4].

During multichannel data registration such places on human body in front and on back are usually used (Fig. 1, a,b).

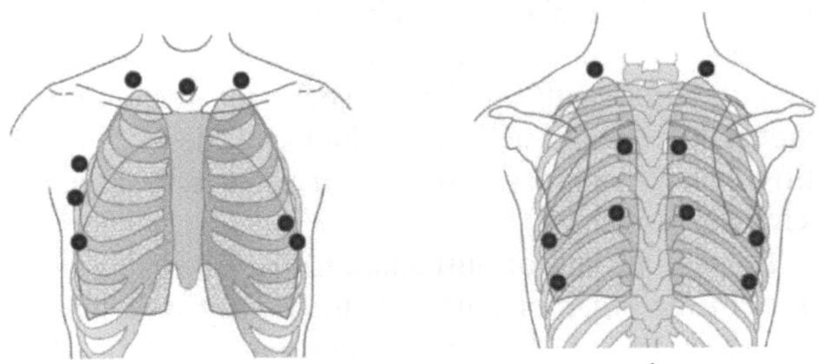

Fig. 1 Sensors allocation on human body: a – front, b – back

As a result the problem of development of efficient algorithm for respiratory noises classification obtained particular importance. Thus respiratory noises appear as a random process it is common to apply a statistic approach. The classification algorithm was used in this work. It considers present information about respiratory noises. At the same time it requires data bank, which includes verified references of recorded respiratory noises specific for one or another illness type. At that it is necessary to guarantee identity of conditions wherein reference and operating records of respiratory noises can be achieved. Particularly it subjects to scale coefficients which depend on sensitivity of sensors and coefficient of amplification of data registration channel. Moreover, the results of classification depend on jamming environment. In addition it is necessary to take into account the possible dispersion of physiological characteristics of patients, respiratory noises of which are used as reference data. Nonobservance of these requirements leads to classification mistakes. The most interesting are those classification algorithms which would have been efficient even while limited information about respiratory noises.

Practically, respiratory noises are divided into two classes:

– respiratory noises of healthy patient;

– respiratory noises of patients with respiratory organs pathologies.

So classification can be made this way. Let us specify through $u(t_i)$, $i = 1, M$ random realization, which includes M noises of breathing and interference descretized with interval $\Delta t = t_{i+1} - t_i$. On the ground of analysis of this realization it is available to apply one of two hypothesis:

– $H_0: u(t_i) = s_0(t_i) + N(t_i)$ - realization contains respiratory noises of healthy patient;

– $H_1: u(t_i) = s_1(t_i) + N(t_i)$ - realization contains respiratory noises of ill patient.

Herein $s_0(t_i) \neq s_1(t_i)$ – signals of respiratory noises, which match to examined hypothesis. These signals can be referred to nonstationary Gaussian processes of the type $s(t_i) = m(t_i)S(t_i)$, where $m(t_i)$ – modulatory function with period T_m equal to one respiratory cycle, and $S(t_i)$ – stationary Gaussian process with zero expectation function and correlation function $K_s(l\Delta t) = \langle S(t_i)S(t_{i+l})\rangle$. The interference $N(t_i)$ in the form of impact from surrounding noises during registration of one realization of the respiratory signal can be mentioned as quasi-steady Gaussian process with zero expectation function and correlation function $K_N(l\Delta t)$. As it is known the solution of this problem by criterion of classification mistake minimum is determined from the likelihood ratio (LR):

$$\varepsilon(u) = [W(u|H_1)]/[W(u|H_0)], \tag{1}$$

where $W(u|H_1)$ and $W(u|H_0)$ – relative probability densities of monitored processes in due hypothesis.

For practical usage of LR it is rational to execute linear transformation of vectors of input signals in such way that output could be considered as independent random processes.

Let us consider that there is a standard of healthy patient breathing. Then LR will be:

$$\hat{\varepsilon}(u, y_0) = [\widehat{W}(u, y_0|H_1)]/[\widehat{W}(u, y_0|H_0)], \tag{2}$$

where simultaneous conditional probability densities of operative $u(t_i)$ and standard $y_0(t_i)$ realizations are used and estimates of maximal likelihood are used instead of unknown modulation functions and spectral density power.

Graphs of spectral power density of breathing noises of healthy and ill patients are shown in Fig. 2.

Along the absciss axis there are numbers of spectral samples, and along the axis of ordinates – the levels of spectral power density in decibels.

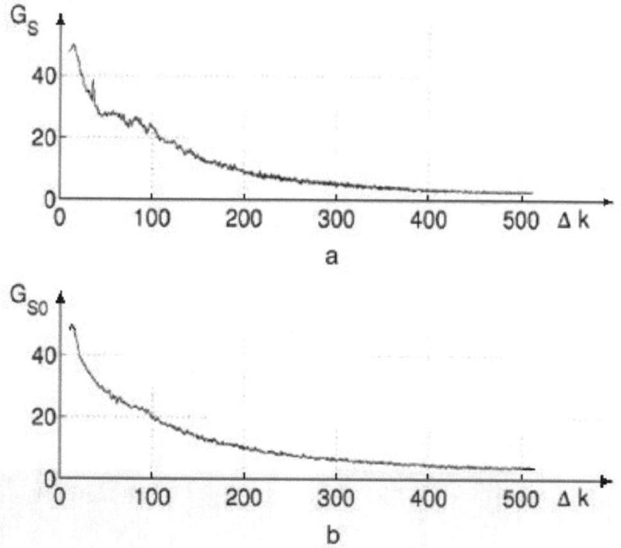

Fig. 2 Spectral power density of respiratory noises: a – healthy patient, b – ill patient

The curves of spectral power density of respiratory noises registered on the surface of the chest of healthy people are well described by monotonously recessive function of frequency measure. In case of respiratory organs illnesses the additional noises appear in the form of groans with different spectral coloring. This leads to the appearance of distortion in structure of spectral power density of breathing noises, which can be used as classified characteristic of illness presence.

For objectification of variants of sensors settings on human body during prophylaxis monitoring the experiment was held (the places for auscultation were chosen due to well known methods, eg [3]). Beyond that, the places nearby to the well known places have been analyzed. The examples of sensors allocation schemes on one and the other side of the human body are shown on Fig. 3–6.

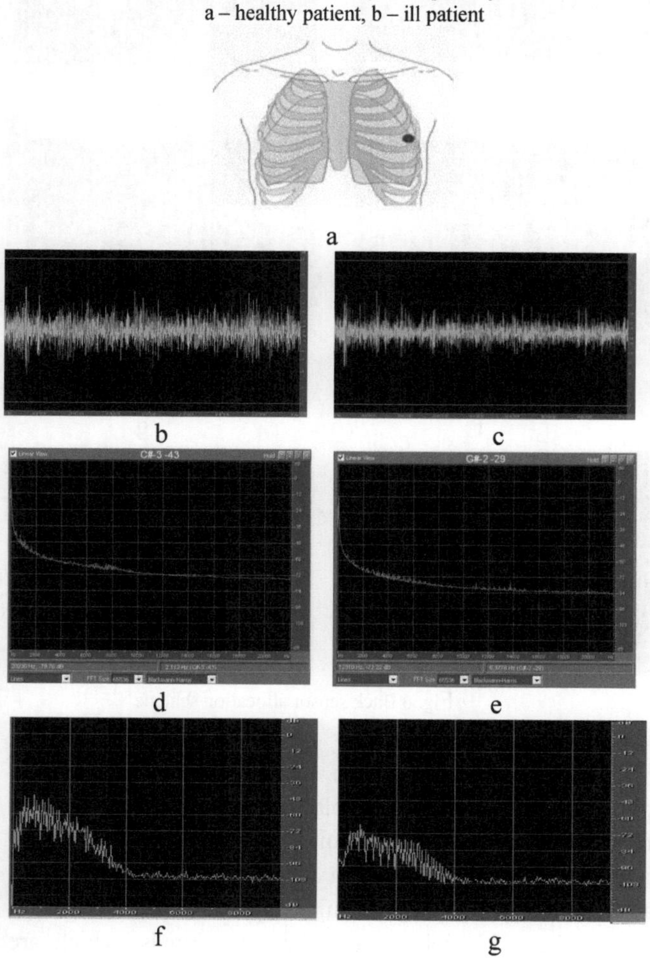

Fig. 3 Front sensor allocation scheme

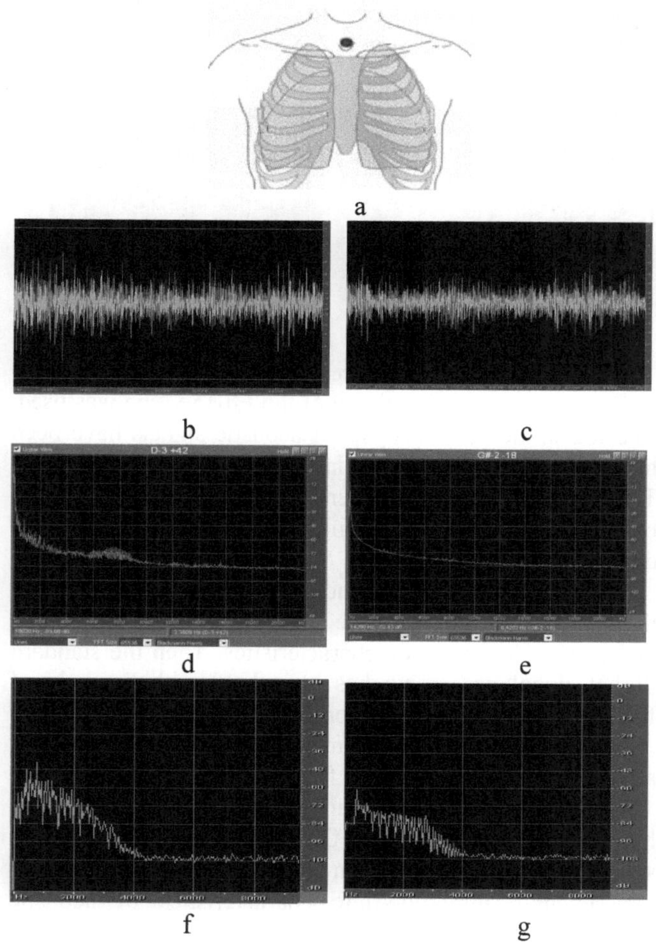

Fig. 4 Front sensor allocation scheme

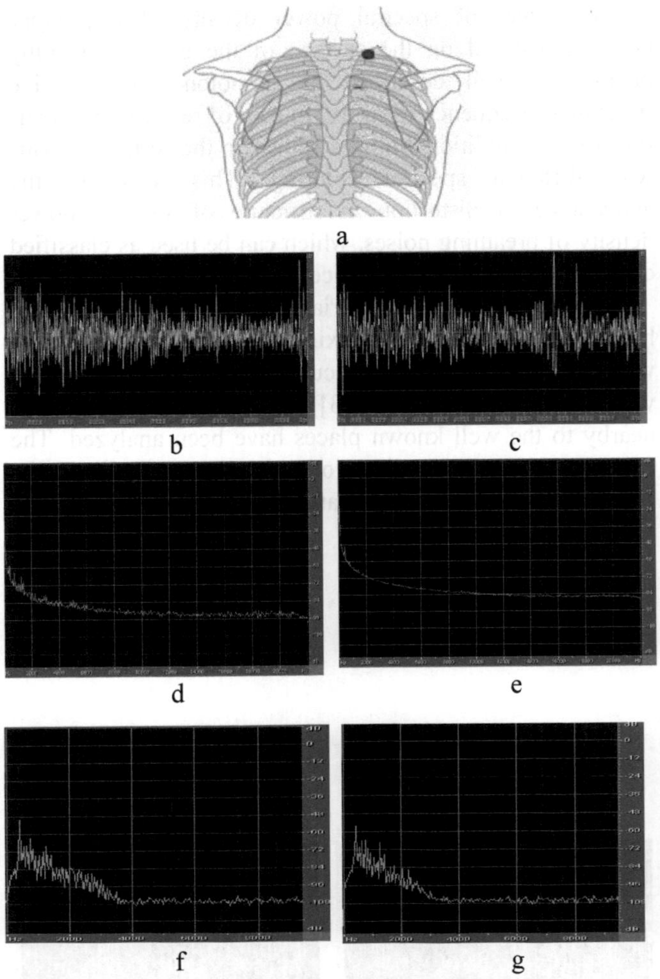

Fig. 5 Back sensor allocation scheme

As a sensor the microphone of the Fh-15e type was used. The registration ans analysis of acoustic signals have been held with the help of special software Adobe Audition. This software allows to either record and save big volume of audio files or make a multifunctional analysis of received data.

Registered and recorded audio signals were performed in the software mentioned above in the form of amplitude-time and amplitude-frequency characteristics. Then the standard digital signal processing has been held with the help of facilities set in Adobe Audition program [5].

This allowed to find informative parts of signals, which allocate in frequency interval from 300 to 4000 Hz.

On Fig. 3 – 5, a shown variants of sensor setting on the human body in front, and on fig. 6 shown variants of sensor setting on the human body on back. On Fig. 3 – 6, b shown experimental amplitude-time characteristics of acoustic signals taken "on bone", on Fig.3 – 6, c shown experimental amplitude-time characteristics on soft tissue. On Fig.3 – 6, d shown experimental amplitude-frequency characteristics taken "on bone", and on Fig. 3 – 6 shown experimental amplitude-time characteristics on soft tissue. On Fig. 3 – 6, f, g shown acoustic signals after digital low frequency filtration.

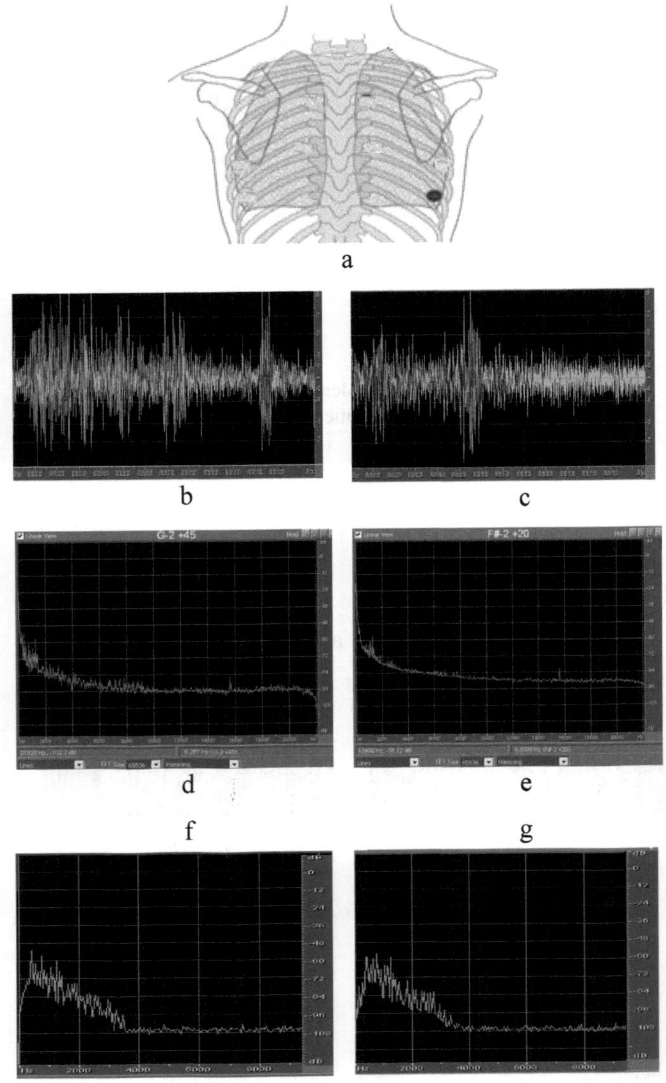

Fig. 6 Back sensor allocation scheme

It is easy to notice that the most informative signals are those, which are taken from places on bones. Signals, taken from soft tissues have more noisy appearance.

Moreover, signals, which are taken from different auscultation places have similar appearance and comply same rules. This relates to either the signals, which are registered in front of the chest or the signal taken from the back.

The analysis of signals taken from different places of auscultation has shown that from the point of view of simplicity and convenience of holding express-diagnostics of state of human respiratory organs without adequacy loss of results it is enough to use one acoustic channel with sound sensor. At that it is rational to put sensor as on Fig. 5, a, i.e. on back to the right on upper feather of the chest.

III. CONCLUSION

It is shown that prophylaxis monitoring of state of human respiratory organs can be held without loss of accuracy and adequacy of the results by means of one acoustic sensor. It is rational to put sensor on back to the right on upper feather of the chest.

CONFLICT OF INTEREST

The authors declare that they have no conflict of interest.

REFERENCES

1. Губергриц А.Я. Непосредственное исследование больного: М., 1972 г. –212 с.
2. Патент *РФ* №2229842, кл. А61В 7/00, 2004 г.
3. Gray H., Drake R., Vogl W., Mitchell A., Eds. Gray's Anatomy for Students. Churchill Livingstone, 2007, 1150 p.
4. Макаренкова А.А., Олийнык В.Н. Помехи сенсоров-виброакселерометров, используемых для аускультации дыхательных шумов // Акуст. вісник.– 2006.– 9, № 1.– С. 45–54.
5. Вовк И В , Гринченко В Т , Дахнов С Л , Крижановский В В , Олийнык В Н Шумы дыхания человека, объективизация аускультативных признаков // Акуст. Вісник. – 1999 – 2, № 3 – С 11-32.

Author: Heorhiy Rozorinov
Institute: State University of Telecommunications
Street: Solomenskaya, 7
City: Kiev
Country: Ukraine
Email: grozoryn@gmail.com

Author: Nina Chichikalo
Institute: State University of Telecommunications
Street: Solomenskaya, 7
City: Kiev
Country: Ukraine
Email: chichikalo@rambler.ru

BioR Medication in the Combined Treatment of Chronic Tonsillitis in Children

L. Danilov[1], I. Ababii[1], S. Ghinda[2], M. Maniuc[1], P. Ababii[1], S. Diacova[1], A. Bocan[1], and I. Cotelea[1]

[1] Department of Otolaryngology, State University of Medicine and Pharmacy "Nicolae Testemiţanu", Chişinău, Republic of Moldova
[2] Laboratory of Immunology and Allergology "Chiril Draganiuc", Chişinău, Republic of Moldova

Abstract— **Chronic tonsillitis therapy is directed towards normalizing the immune status of the organism, however conservative treatment does not provide conclusive results in this direction. 29 children with compensated chronic tonsillitis received standard conservative therapy and BioR medication. Obtained results coincide with those obtained by other authors, which established the BioR effect in the treatment of herpetic stomatitis and resistant forms of pulmonary tuberculosis. This study comes to demonstrate the high clinico-immunological efficiency of BioR medication in complex treatment of chronic tonsillitis in children.**

Keywords— **tonsillitis, immunity, therapy, children.**

I. Introduction

Conservative treatment methods of chronic tonsillitis, currently applied, do not always provide the desired therapeutic effects and do not lead to long-term remissions.

Contemporary complex therapy of chronic tonsillitis should be directed towards normalizing the general immune status of patient's body, restoring the drainage function of tonsil gaps, hypo-sensitization and reduction of local inflammatory events expressiveness [1].

One of the immunomodulatory preparations, considered adaptogen, is a medication obtained from the biomass of *Spirulina platensis* cyanobacteria – BioR, with antioxidant, cytoprotective, regenerative, anti-inflammatory, antiviral and especially immunomodulatory properties [2].

The positive effects of the application of this immunomodulator in the complex treatment of the pathologies, such as tuberculosis of the lungs, liver cirrhosis, diabetes mellitus, acute herpetic stomatitis and recurrent acute pneumonia, have been demonstrated by several researchers [3-7].

The purpose of our study was to analyze and demonstrate clinically and immunologically the efficiency of BioR medication application in children with chronic tonsillitis.

II. Materials and Methods

The study was performed at PMI Institute of Mother and Child, "E. Coţaga, Pediatric Otorhinolaryngology Clinic, Chişinău, and included: lot 1 – 29 children with compensated chronic tonsillitis (classification by I. B. Soldatov - 1975) [8], who received standard conservative therapy and BioR medication (orally, in the morning, 5 mg tablets, 10 days), and lot 2 – 29 children treated by standard method (removal of the infection focus off the palatine tonsils by washing the tonsillar gaps with Nitrofural solution 1:5000 in order to restore the role of drainage of the palatine tonsillar crypts and to stimulate their functional status, embrocation of the palatine tonsil surface with Lugol solution, oral administration of diphenhydramine desensitizer and vitamin complex).

All children benefited of complex immunological investigations before and after 3 months of treatment. To determine the functional status of T lymphocytes,it was usedthe blast transformation reaction (Ghinda S., 1982) [9]. Flow cytometry method (Partec PAS I) was used to determine the levels of T and B lymphocytesubpopulations (CD3, CD4, CD8, CD16, CD19). To determine phagocytic cells, the phagocytic index and phagocytic number were used (Pavlovici S., 1998) [10]. The phagocytic activity of neutrophils was assessed in NBT test (Nitro-Blue-Tetrazolium) (B. H. Park et al., 1968) [11].The content of circulating immune complexes was determined according to the procedure described by I. A. Grinevici and L.I. Kameneţ (1986) [12] in version adapted by S. Ghinda et al. (2008) [13]. Hemolytic complement activity (CH_{55}) was determined using the method proposed by L. S. Reznicova (1967) [14]. The content of immunoglobulins A, G, M was determined using enzyme immunoassay on solid support, using OOO "Vector-Best" reagents (Russia), according to the enclosed instructions. Total IgE was assessed using enzyme immunoassay on solid support, using UBI kits of reagents (Magiwel) (USA) according to the enclosed instructions. IL-8 titer was determined by enzyme immunoassay method on solid support, using IMMUNOTECH SAS reagent kits (France). The levels of IL-4, TNF-α, and IL-1β were determined using enzyme immunoassay on solid support, using OOO "Vector-BEST" reagents (Russia). The content of antinuclear antibodies (ANAcombi) was determined by enzyme immunoassay on solid support using DRG International Inc. reagents (USA).

III. Discussion

In last decades were accumulated a series of data according to which cytokines have an antiviral and antimicrobial effect.

V. Sontea and I. Tiginyanu (eds.), *3rd International Conference on Nanotechnologies and Biomedical Engineering*,
IFMBE Proceedings 55,
DOI: 10.1007/978-981-287-736-9_103

According to our data, BioR medication administered in case of chronic tonsillitis, contributes to CIC concentration reduction, reducing the levels of pro-inflammatory cytokines (TNF-α, IL-8, IL-1β) and increasing concentrations of serum inflammatory cytokines (IL-4).

Obtained results coincide with those obtained by Uncuţa D. (2014), which established the BioR effect in the treatment of herpetic stomatitis, where also decreased the levels of proinflammatory cytokines (IL-lα and IL-10).

Ghinda S. demonstrated a positive action of BIOR on CIC concentration dynamics in patients with resistant forms of pulmonary tuberculosis.

IV. RESULTS

The analysis of changes occuring in tonsilogenic intoxication expression demonstrated that these state indicators were similar before treatment.

The concentration of CIC (PEG – 8.0%) of low molecular weight and strong toxicity, in follow-up treatment was veridically reduced in children from lot 1 (before treatment – 300±22,0 u.d.o., after treatment – 213±15,8, $p<0.01$), compared with lot 2 (before treatment – 291±30,1 u.d.o., after – 246±18,9).

After performing the therapy, the phagocytic capacity of the engulfed bacteria (NBT test) increased more significantly in lot 1 in follow-up (before treatment – 0,115±0,003 c.u., after – 0,127±0,002, $p<0.001$) compared with lot 2 (before treatment – 0,116±0,003 c.u., after – 0,127±0,002, $p<0.01$). FN and FI, also, statistically significantly increased more pronounced in children from lot 1 compared with lot 2 (FN – lot 1 before treatment – 75,8±1,26 %, after – 81,9±1,18 %, $p<0.001$; FN – lot 2 before treatment – 75,0±1,45%, after – 78,7±1,00, $p<0.05$); (FI – lot 1 before treatment – 4,3±0,20 c.u., after – 5,2±0,16, $p<0.01$; FI – lot 2 before treatment – 4,6±0,22 c.u., after – 5,3±0,15, $p<0.05$).

CD16 lymphocytes titres (natural killers) increased truthfully after treatment only in the first lot of children (before treatment – 14.7±0.33%, after – 16.1±0.22%, $p <0.01$).

Total hemolytic complement activity (AHTC) rose more conclusively in the first lot of children (before treatment – 52.3±1.21 CH_{50}, after – 59,3±1,17, $p<0.001$) subjected to investigation with the children from the second lot (before treatment – 49,9±0,79; after – 53,5±0,97, $p<0.01$).

There was a true increase of the amount of natural antibodies after treatment, it being more expressive in children of the first lot (before treatment – 2,0±0,07, after – 2,3±0,06, $p<0.01$) compared with the second lot (before treatment – 2,1±0,11, after – 2,3±0,06, $p<0.05$).

The content of eosinophils, IgE, autoimmune reactivity test "ANAcombi" which was high before treatment, after therapy truthfully lessened in children in the first lot (eosinophils before treatment – 1,8±0,21%, after – 1,1±0,12%, $p<0.01$; IgE before treatment – 28,4±4,62IU/ml, after – 13,2±2,77, $p<0.01$; „ANAcombi" before treatment – 0,87±0,058 c.u., after – 0,47±0,041 c.u., $p<0.001$), while in children of the second lot, there were recorded decreasing trends of these indices (before treatment – 2,1±0,32%, after – 1,7±0,22%; before treatment – 49,4±17,4IU/ml, after – 24,0±4,79; before treatment – 0,99±0,056 c.u., after – 0,80±0,054 c.u. respectively).

Before treatment, the specific cell sensitivity to streptococcal antigen (TTBL-streptococcus) had almost identical-levels in both lots after treatment, their significant decrease was found only in the first lot (before treatment – 3,6±0,30 %, after – 2,7±0,23, $p<0.05$) compared with the second lot (before treatment – 3,5±0,32, after – 2,7±0,27).

During the treatment, the total lymphocyte contents increased by a higher level of authenticity in children in the first lot (before treatment – 34.3±0.63; after – 39.8±0.59, $p <0.001$) compared to second lot (before treatment 34.0 ±0.56, after – 37.2±0.73, $p<0.01$). So the functional activity of T lymphocytes (RTBL+PHA) increased in children in the first lot with the highest level of veracity compared to the second lot (before treatment – 65.6 ±0.58%, after – 69,0±0.54, $p <0.001$; before treatment – 66.7±1.04, after – 69.5±0.72, $p <0.05$ respectively).

In the first lot of children, the content decreased significantly CD20 (before treatment – 9.6±0.44%, after – 7.9±0.27, $p<0.01$), having favorable dynamics and in the second lot of children, this index had only a tendency of decrease (before treatment – 9.6±0.52%, after – 9.1±0.41). IgG, IgM and IgA titers after treatment were with nomajor changes.

Pro-inflammatory cytokine TNF-α in the course of their treatment decreased the titers significantly in the first lot (before treatment – 1.5±0.33 pg/ml, after – 0.8±0.07, $p<0.001$) in relation to the second investigational lot (before treatment – 1.9±0.22 pg/ml, after – 1.4 ±0.14, $p<0.001$).

After the treatment, the levels of the pro-inflammatory cytokine IL-8 decreased the titres in the first lot (before treatment – 133±12.2 pg/ml, after – 53±7.02, $p<0.001$), and in the second investigational lot just observing a trend of diminishing (before treatment – 124±13.1 pg/ml, after – 88±20.1). The samething was observed with IL-1β: in the first lot (before treatment – 133±12,2 pg/ml, after – 55±8.1, $p<0.001$); in the second lot (before treatment – 136±31.4 pg/ml, after – 80±21.3).

Anti-inflammatory cytokine IL-4 has increased true titers in both lots, the degree of increase being, however, higher in the first lot of children (before treatment – 7.3±0.92 pg/ml, after – 18, 1±1.85, $p<0.001$) than in the second lot (before treatment – 8.9±1.24 pg/ml, after – 14.5±1.69, $p<0.01$).

A positive clinical dynamics in catamnesis was appreciated in 93% of cases in patients in the first lot, who were applied general immunostimulation with BioR in the complex treatment. It was characterized by lack of acute upper respiratory tract infections, decrease of thepalatine tonsils in volume and lack of caseous masses in the tonsillar holes.

The treatment had a positive clinical effect in 85% of cases in children in the second lot treated by traditional method.

V. CONCLUSIONS

The increasing of indices, reflecting phagocytic ability test (NBT, phagocyte number and phagocyte index), accounts for the cause of a faster elimination of circulating immune complexes, and, therefore a better reduction of intoxication tonsilogenic levels in children treated by immunostimulation with standard therapy. Contribution of BioR immunocorrection ensures more effectively the normalization (decrease) of changed preimmune indices of the body's resistance, which effectively contributes to decreasing high indicators of allergic reactions and has an effective autoimmune desensitizing action by decreasing the high levels of streptococcal antigen specific cellular sensitivity, influences beneficially the process of normalization of T lymphocyte and B lymphocyte index change, efficiently contributing to normalizing the cytokine profile in children with compensated chronic tonsillitis, reducing the levels of pro-inflammatory cytokines (TNF-α, IL-8, IL-1β) and increasing concentrations of serum inflammatory cytokines (IL-4).

Therefore, therapeutic treatment of chronic compensated tonsillitis in children along with BioR immunomodulator have a more efficient clinical and immunological action compared to the standardt herapy.

CONFLICT OF INTEREST

The authors declare that they have no conflict of interest.

REFERENCES

1. Danilov L (2013) Problema amigdalitei cronice în otorinolaringologie şi pediatrie (sinteză de literatură). Buletin de perinatologie, 2 (58) - 3 (59): 143- 148
2. Rudic V, BulimagaV, Ghinda S, Chiriac T, Ghelbet V, Darii V, Gulea A, Ciapurina L, Melnic S (2004) Tehnologii de obţinere a noi bioremedii imunomodulatoare de origine algală. Buletinul academiei de ştiinţe a Moldovei (ştiinţe biologice, chimice şi agricole) 3 (294): 95-100
3. Ghinda S, Cula E, Lesnic T, Chiriac T, Chiroşca V, Guila A (2014) Dynamics of circulating immune complexes in patients with drugresistant tuberculosis under the influence of immunomodulatory drugs. 2nd International Conference on Microbial Biotehnology, Chisinau, Republica Moldova, 2014, pp.53-57
4. Gudumac V, Gulea A, Tagadiuc O, Bulimaga V, Chiriac T, Andronache L, Sardari V, Ştirba O (2011) Influenţa unor compuşi biologici activi autohtoni asupra restabilirii parenchimului hepatic în cirora hepatică experimental. Materialele Conferinţei Internaţionale „Biotehnologia microbiologică – domeniu scientointensiv al ştiinţei contemporane, Chisinau, Republica Moldova, pp.63-64
5. Lozovschi Z, Pelin Iu, Schibinschi C, Ghinda S, Rudic V, Ouatu V, Moscalenco D, Tropin O (2011) Caracteristica imunităţii umorale la pacienţii cu diabet zaharat pînă şi după imunoreabilitarea cu preparatul BioR. Materialele Conferinţei Internaţională „Biotehnologia microbiologică – domeniu scientointensiv al ştiinţei contemporane, Chisinau, Republica Moldova, pp.67-68
6. Uncuţa D (2014) Optimizarea diagnosticului şi tratamentului în stomatita herpetică. Autoreferatul tezei de doctor habilitat în medicină, Chişinău
7. Uglitskikh AK, Tsokova NB, Gmoshinskiĭ IV, Mazo VK, Kon' IIa, Ostreĭkov IF (2006) Experience with a selenium-containing biological active supplement used in children with pneumonias in an intensive care unit. Anesteziol Reanimatol, Jan-Feb(1), pp. 45-8
8. Danilov L (2014) Amigdalita cronică la copii [Chrnic tonsillitis in children]. Monografie. Chişinău, 2014, pp. 39-40
9. Ginda S S (1982) Modifikacija mikrometoda reakcii blasttransformacii limfocitov. Laboratornoe delo, (8), pp. 23-25
10. Pavlovich S A (1998) Osnovy immunologii. Vyshejshaja shkola", pp. 114
11. Park B.H. et al (1968) Infection and Nitroblue-tetrazolium Reduction by Neutrophils. The Lancet, 11, N 7567, pp. 532-534
12. Grinevich JU A, Kamenec L JA (1986) Osnovy klinicheskoj immunologii opuholej. Zdorovja, pp. 158
13. Ghinda S, Sofronie S, Chiroşca V, Privalova E, Calenda O, Smeşnoi V, Lesnic E, Barbuţa A, Rotaru N, Iaconi L (2008) Metoda concomitentă de determinare a CIC cu masa moleculară joase, medie şi înaltă. Innovation Certificate N. 46, was registered at the Institute of Phtiziopneumology at 12.05.2008.
14. Reznikova L S (1967) Komplement i ego znachenie v immunologicheskih reakcijah. Medicina. pp. 272

Author: L. Danilov
Institute: Department of Otolaryngology, State University of Medicine and Pharmacy "Nicolae Testemiţanu"
Street: St.V. Alecsandri, 2
City: Chişinău
Country: Republic of Moldova
Email: danilovlucian@yahoo.com

Water as a Receiver of Information from Digital Representations of Plant Objects Subjected to Thermal Stress Action: 2. Instrumental Testing

S. Kernbach[1], S. Maslobrod[2], O. Kernbach[1], and E. Maslobrod[2]

[1] Cybertronica Research, Research Center of Advanced Robotics and Environmental Science, Stuttgart, Germany
[2] Institute of Genetics, Physiology and Plant Protection of Academy of Sciences of Moldova, Chisinau, Moldova

Abstract— **It has been adduced the experimental data about the change in the condition of water in the plant – plant representation – water system under the stress action on seeds of this plant. Transmission of information from bio-objects via their representations to water, in our opinion, suggests the existence of long-distance communication between bio-objects and their representations, and thus presents experimental evidence of quantum-communication theory. The evaluation of the condition of water was implemented basing on biological indicator testing (by the number of left sprouts at seed germination in this water) and instrumental testing (by the change of the *pH* of water and electric current in the electrical double layer).**

Keywords—**water, pH meter, information, image, reciever.**

I. INTRODUCTION

In this work, at instrumental testing of water as receiver of information from digital images of seeds we studied changes in the physical and physico-chemical properties of water in the plant – plant image – water system under the stress action on seeds. evaluation of the condition of water was measured using processing equipment, designed at Cybertronica Research, Research Center of Advanced Robotics and Environmental Science. Precision differential *pH* meter with thermostat for long-term monitoring of liquids is intended for single-ended or differential *pH* measurement with high accuracy and high resolution [1,2]. Due to ultra-low-noise electronic components and precision measuring capabilities, the instrument is able to measure differential acid-base changes in the test fluids at 10^{-5}-10^{-7} *pH* level that is not possible for most of other devices. Main applications are long-term laboratory and field measurements of small and ultra-small pH changes. For precise pH measurements the device has a built-in two-channel thermostat with the PID controller, which allows keeping the temperature setpoint with the accuracy of < 0.01°C. Fig. 1 shows the structure of experiments.

II. DATABASE AND METHODS

Stuttgart team performed a detection of nonlocal impact based on the ENC effect that appears between a physical object and its digital representation. We used three *dpH* instruments [1] (6 parallel pH-measurement-channels, combined in three pairs of differential gauges). The researchers obtained three digital representations of grains from Chisinau, which were used in only one channel of the *dpH* device. The second channel acted as a control channel (i.e., under the same conditions, but without impact). In the first experiment we also used three additional conductometric devices with 9 parallel channels. As shown in [2] [3], such parallel measurements allow obtaining statistically significant results for low-repeatable-processes.

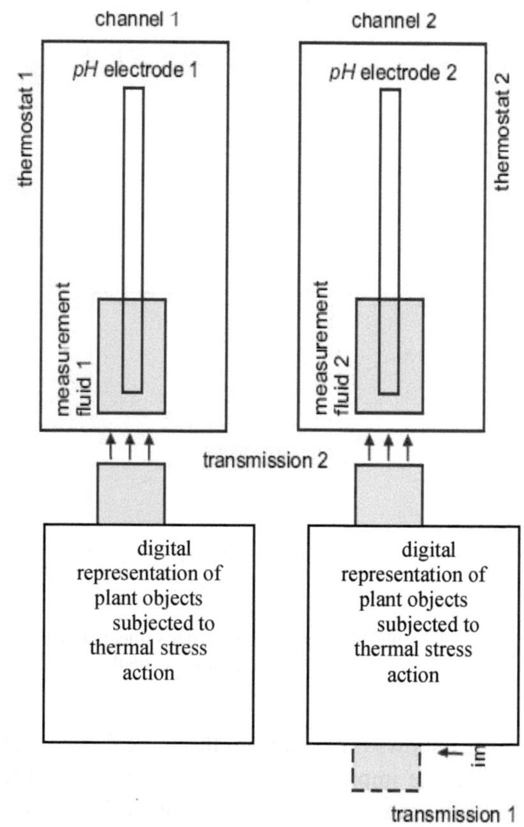

Fig. 1 Structure of the experiments

Following the experimental methodology, all devices started data recording 24 hours before the experiment and

V. Sontea and I. Tiginyanu (eds.), *3rd International Conference on Nanotechnologies and Biomedical Engineering*,
IFMBE Proceedings 55,
DOI: 10.1007/978-981-287-736-9_104

finished 24 hours after. The experiment starts from the moment where digital representations were inserted into the device. The interval between this point and the beginning of impact is about 24 hours. This is necessary in order to distinguish between *dpH-changes caused by the representation and the changes transmitted through the image.* The impact time was calculated in Stuttgart on the basis of data without a priori knowledge of the true impact time. Based on the measured data we elaborated hypotheses about the most probable impact time in Chisinau. Two experiments were performed on 05-06.03.15 and 16-17.03.15, two further experiments are scheduled for a later time.

III. RESULTS

Figure 2 displays *pH* and *dpH* data of the device N2 on 5.03.15 at the moment when images are inserted into the device (at 11.00 marked by a gray bar). As well visible, the channel 1 with the image responded by deviating the trend on 200μV for ~6 hours (points A, B, C).

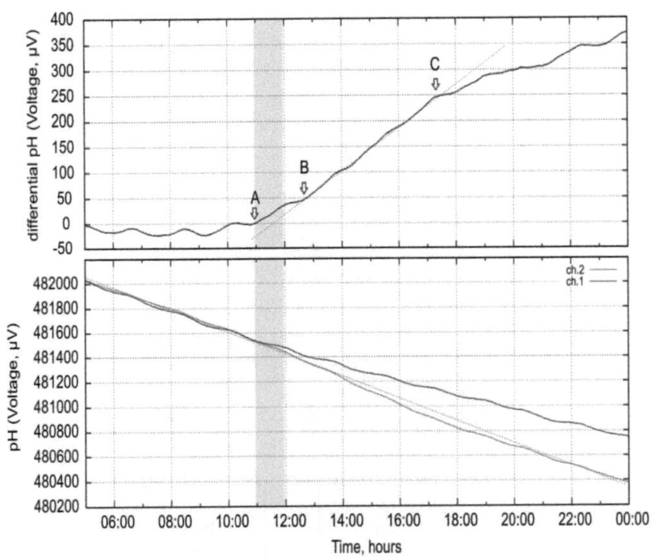

Fig. 2 Measurements of *pH* and *dpH* of water subjected to emission from digital representations done on 05.03.15

Figure 3 shows the data from the same device on the next day during the impact in Chisinau. Changes of differential trend are observed around 9.00, 12. 00 and 16.00 (points A, B, C), and only at 9.00 in the channel 1. The true impact time was 8.30. The level of trend deviation is about 100μV for 3 hours, which roughly corresponds to measurements from previous day.

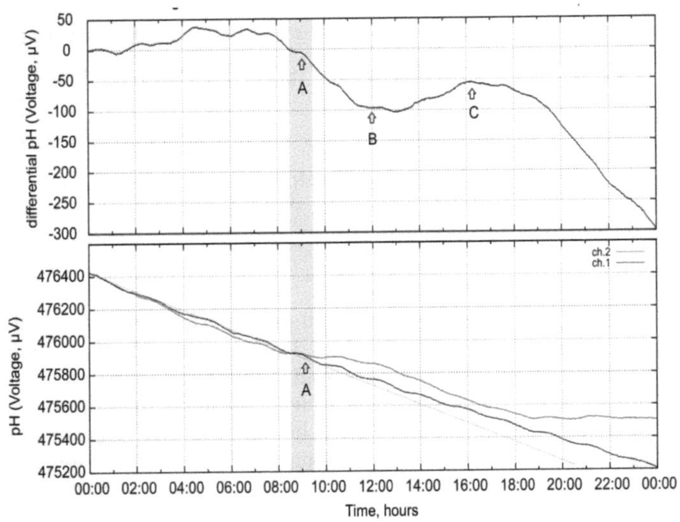

Fig. 3 Measurements of *pH* and *dpH* of water subjected to emission from digital representations done on 06.03.15

Figure 4 shows the dynamics of two *pH* channels from the device N1 in the experiment on 17.03.15. In order to demonstrate that this is not a temperature artifact, the graphs show the temperature dynamics inside the thermostats (*pH* electrodes and measuring liquids are thermostatic regulated), and in the laboratory.

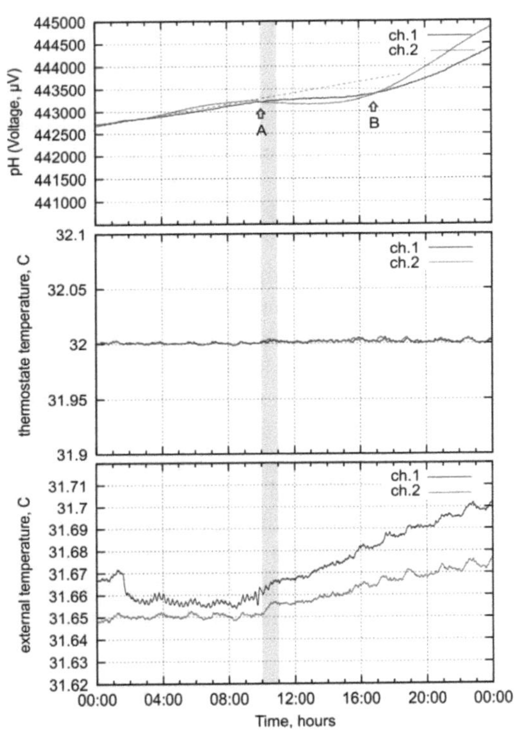

Fig. 4 Dynamics of two pH channels from the device in the experiment on 17.03.15.

Fluctuations in the thermostat are ~0.01°C and in a lab ~0.2°C with no abrupt changes, i.e. the *pH* dynamics cannot be explained by temperature fluctuations (such data are common to all *dpH* measurements). Based on *pH* curves, a hypothesis about potential impact on 17.03.15 was expressed towards 10.00 and 17.00. The true exposure time was 10.00.

IV. CONCLUSION

1. Using instrumental testing of the condition of water it was established experimentally that water has the ability to receive information signals from remote objects when the signals are transmitted via the following communication line: seeds subjected to thermal stress → digital images of these seeds → tap water.

2. Existence of long-distance communication between bio-objects and their images has been supported by the experimental evidence.

CONFLICT OF INTEREST

The authors declare that they have no conflict of interest.

REFERENCES

1. S.Kernbach, O.Kernbach, On precise pH and dpH measurements, International Journal of Unconventional Science, 83-103,5(2), 2014
2. S. Kernbach, V. Zamsha, Y. Kravchenko, " Long and Super-Long Range device-device and operator-device Interactions", International Journal of Unconventional Science, 1(1), pp. 24–42, 2013.
3. S. Kernbach, "Replication attempt: Measuring water conductivity with polarized electrodes", Journal of Scientific Exploration, 27(1), pp. 69-105, 2013.

Author: Serge Kernbach
Institute: Research Centre of Advanced Robotics and Environmental Science
Street: Universitätstr. 38
City: Stuttgart
Country: Germany
Email: serge.kernbach@yahoo.com

Water as Receiver of Information from Digital Representations of Plant Objects Subjected to Thermal Stress Action: 1. Biological Indicator Testing

S. Maslobrod[1] and S. Kernbach[2]

[1] Institute of Genetics, Physiology and Plant Protection of Academy of Sciences of Moldova, Chisinau, Moldova
[2] Cybertronica Research, Research Center of Advanced Robotics and Environmental Science, Stuttgart, Germany

Abstract— **It has been adduced the experimental data about the change in the condition of water in the *plant – plant image – water* system under the stress action on seeds and sprouts. Transmission of information from bio-objects via their representations to water, in our opinion, suggests the existence of long-distance communication between bio-objects and their images, and thus presents experimental evidence of quantum-communication theory. The evaluation of the condition of water was implemented basing on biological indicator testing (by the number of left sprouts at seed germination in this water) and physico-chemical testing (by the change of the pH of water and electric current in the electrical double layer). Ability of water to receive genetic information from bio-objects was demonstrated. Change in the condition of water is more pronounced at the implementation of top down view representations of sprouts. Tap water better receives information from the bio-object image than distilled water.**

Keywords— **sprouts, water, image, information, receiver.**

I. INTRODUCTION

Nowadays, researchers who explore the role of quantum mechanics in physical and biological systems are directing their attention to the effects of quantum entanglement and long-distance communication [1-4]. According to Vlatko Vedral, professor of Quantum Information Science at the University of Oxford, “quantum mechanics is not just about teeny particles. It applies to things of all sizes: birds, plants, maybe even people” [2]. There is a growing scientific literature describing the effects of quantum entanglement in living organisms [1-4]. As our studies have shown, long-distance communication exists either between separate organisms or between groups of organisms (seeds of plants) [5]. Evidence for this comes from the experimental data on long-distance communication between seeds, which in the course of contact interaction during imbibition become part of the same system (phytocenosis).

Communication between the system components (seeds) continues even if they are spaced from each other (positioned at a distance approximately equal to 7 km). Subsequently, it was realized that there is a distinct connection between plant objects (seeds) and their digital representations [6,7]. Stress (chemical, thermal, radiative, etc.) action impact on seeds representations led to changing of the morpho-physiological parameters of these seeds at germination (germinative energy and viability of seeds, bio-isomerism and growth of sprouts). Maximum distance between the objects of research (seeds located in Germany and images in Moldova) was 1500 km. We also indicated presence of long-distance communication in the opposite direction in the *seed – its image* system, at stress action on seeds [8]. In this case, image informative characteristics changed, which was indicated by change in the condition of water placed over the digital representation of seeds, about 1 mm from it, in Petri dishes. The seeds grown in this water essentially differed by morphological characteristics (sprout length and number of left sprouts) from the control seeds germinated in non-activated water.

In this work, we decided to continue and expand our studies in this field. Our objectives are: 1) use not only seeds, but also sprouts as emission sources; 2) find out whether a digital representation of sprouts transmits to water information about the morphological traits of the plant; and 3) define whether this information depends on the viewing directions of the representation (side or top).

II. DATABASE AND METHODS

In experiment 1 data transmission was studied, we have considered two communication lines for comparing sensitivity of different kinds of water and seeds to the information signal from digital representations. The communication lines under study were: 1) wheat seeds (Nikoniya) subjected to thermal stress → digital representations of these seeds → seeds (Nikoniya and Accent) data receivers; 2) wheat seeds (Nikoniya) subjected to thermal stress → digital representa-tions of these seeds → tap water → wheat seeds (Nikoniya and Accent) data receivers. Three identical digital images stacked on one another were placed under Petri dishes (fig. 1) during 48 hours of seeds imbibition (300 seeds in a single image). Wheat seeds - data receivers were grown in Petri dishes with tap water (10 Petri dishes with 50 seeds in each) until the first leaf sprouting from each germinating seed. Using this leaf, we studied bio-isomerism of every plant

V. Sontea and I. Tiginyanu (eds.), *3rd International Conference on Nanotechnologies and Biomedical Engineering*,
IFMBE Proceedings 55,
DOI: 10.1007/978-981-287-736-9_105

determining leftness or rightness of its sprout. Right sprouts are those with the first leaf wrapped clockwise and left sprouts are with the first leaf wrapped counter-clockwise (fig. 2).

Fig. 1 Digital representation of swelling seeds

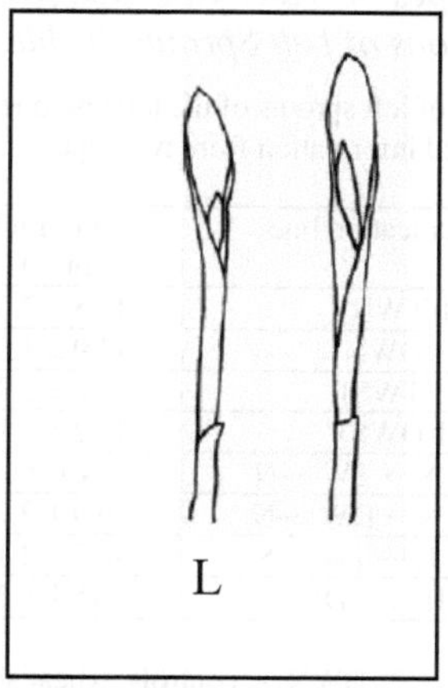

Fig. 2 Left (L) and right (D) sprouts of wheat

The aim of experiment 2 was to assess the effect of plant morphological traits on the information received by water from the digital representations of this plant seeds (Accent and Select). Other steps of the test were the same that in experiment 1.

In experiment 3, seven-day sprouts acted as sources of signal from the bio-object to its image and further to water. Sprouts were previously divided into two groups: one containing left sprouts, other – right ones (fig. 2). Plants of each group were placed in three glasses (50 pieces each) and photographed showing them with a top view and a side view (fig. 3). Then we put three identical digital images stacked on one another under each Petri dish with tap water. The signal was transmitted via the following communication line: wheat sprouts (Nikoniya) subjected to thermal stress → digital representations of these sprouts (top and side view) → tap water → wheat seeds (Nikoniya) - data receivers. In each embodiment, we used 12 dishes each with 50 seeds. Combined effect was estimated by number of left and right sprouts grown from the seeds data receivers.

In experience 4, considering the communication line from experience 3, we used tap and distilled water to germinating test seeds (the last component of system of information transfer).

TW

Fig. 3 Digital representation of wheat sprouts: top view (TW), and side view (SW)

III. Results

A. Studying of the Response of the Test Seeds to the Signal from Digital Representations of Seeds and From the Water Which Received a Signal from the Seeds Representations

Table 1 Part of left sprouts of the total number of plants grown from the seeds that had received a signal from seeds representations directly and indirectly through water, %

№	Communication line	Part of left sprouts
1	C1 (N)	44.9 ± 1.77
2	C2(A)	47.9 ± 1.27
3	N1→IN1→N2	46.5 ± 2.03
4	N1→IN1→TW→N2	52.7 ± 2.21*
5	N1→IN1→A1	47.8 ± 1.55
6	N1→IN1→TW→A1	45.6 ± 1.80

Notes: C1 (N), C2 (A) – control, wheat seed varieties Nikoniya and Accent, respectively; N1 – Nikoniya seeds subjected to thermal stress action; IN1– digital representation of Nikoniya seeds; TW - tap water; N2, A1 – test seeds varieties, Nikoniya and Accent, respectively

1) According to the data of Table 1, the main receiver of information from the seeds representations under the thermal stress action on seeds becomes tap water, but not directly seeds. With this, the condition of water changes considerably: it becomes capable to increase the number of left sprouts grown from the seeds that swelled in this water.

2) Water is capable to receive genetic information of a bio-object (morpho-physiological parameters of the plant) from its digital representation. The use of water that had received information from a bio-object to grow a different variety gave no effect (number of left sprouts did not increase).

B. Studying of the Response of the Test Seeds to the Signal from Water as Receiver of Genetic Information from Digital Representations of Seeds (Table 2)

Table 2 Part of left sprouts of the total number of plants grown from seeds of different varieties, %

№	Communication line	Part of left sprouts
1	C1(A)	45.4 ±2,49
2	C2(S)	44.0 ± 2.82
3	A1→IA1→TW →A2	50.3 ± 2.52
4	A1→IA1→TW →S	45.3 ± 2.13
5	S1→IS1→TW →S2	53.6 ± 2.13*
6	S1→IS1→TW →A	42.7 ± 2.02

Notes: C1 (A), C2 (S) – control, wheat seed varieties Accent and Select; N1 – Nikoniya seeds subjected to thermal stress action; IA1, IS1– digital representations of Accent and Select seeds, respectively; TW - tap water; A, A2, S, S2 – test seeds varieties, Accent and Select, respectively

1) Ability of water to receive genetic information from a bio-object has been also confined by comparative tests.

2) Communication intensity depends on variety. The effect is more pronounced with Nikoniya (the number of left sprouts is significantly higher than in the control) and less pronounced with Accent.

C. Studying of the Response of Test Seeds to the Signal from Water Received from the Digital Representations of Left and Right Sprouts (Table 3)

Table 3 Part of left sprouts of the total number of plants grown from different varieties that had received information from digital representations of left and right sprouts, %

№	Communication line	Part of left sprouts
1	C(N)	46.8 ± 1.99
2	LN→ ILN top veiw→ TW →N	54.1 ± 2.14*
3	RN→ IRN top veiw→ TW →N	52.0 ± 2.30
4	LN→ ILN side view→ TW →N	47.3 ± 2.10
5	RN→ IRN side view→ TW →N	47.5 ± 2.26

Notes: C(N) – control, wheat seed varieties Nikoniya; LN, RN – left and right Nikoniya sprouts subjected to thermal stress action; ILN, IRN– digital representations of left and right Nikoniya sprouts, respectively; TW - tap water; N – test Nikoniya seeds

1) Digital representations of left sprouts change the condition of water. When immersed in this water the same variety seeds give significantly greater number of left sprouts in comparison with the control. Thus, the ability of water to receive genetic information from a bio-object representation was confined.

2) The essential role plays direction from which the image was taken, was it top view or side view. The effect was observed only with the top view. Therefore, the signal from the digital representation is emitted through the tops of sprouts shown in this image, but not from their sides.

3) The effect significantly depends on the bio-isomerism of sprouts, when using the right sprouts the effect was not observed. If the experiment was carried out without division of sprouts into two groups (left and right) the expected effects of data transmission from images to water did not show up.

D. Studying of the Response of Test Seeds to the Signal from Two Types of Water Received from the Digital Representations of Left Sprouts (Table 4)

Table 4 Part of left sprouts of the total number of plants that had received information from two types of water, %

№	Communication line	Part of left sprouts	mean value
1	C1(TWN)	40.5 ± 2.07	44.2 ± 1.42
2	C2(DWN)	47.9 ± 1.09	
3	C3(TWS)	42.0 ± 2.90	44.1 ± 2.00
4	C4(DWS)	46,2 ± 2.87	
5	LN→ ILN→ TW →N	51.8 ± 2.67*	51.4 ± 1.67*
6	LN→ ILN→ DW →N	50.9 ± 2.04	
7	LN→ ILN→ TW →S	48.5 ± 2.12	46.8 ± 1.70
8	LN→ ILN→ DW →S	45.1 ± 2.61	

Notes: C1 (TWN), C2 (DWN) – controls, wheat seed variety Nikoniya grown in tap and distilled water, respectivily; C3 (TWS), C4 (DWS) – controls, wheat seed varieties Select grown in tap and distilled water, respectively; LN – left sprouts of Nikoniya subjected to thermal stress action; ILN – digital representations of left Nikoniya sprouts; N, S – test seeds varieties, Nikoniya and Select, respectively

1) Communication intensity depends on the type of water. The effect is more pronounced with the tap water received the information from the images of left sprouts and less pronounced with the distilled water.

2) Positive results reconfirm the ability of water to receive genetic information from a bio-object representation. The information received from wheat sprouts is better transferred to the same variety.

IV. CONCLUSION

The provided data confirm the evidence of existence of long-distance communication between bio-objects and their images, the results have enhanced our expertise in this field. Water is an effective indicator, which reflects the information received from a bio-object subjected to stress. In its turn, change of the condition of water as receiver of information from digital representations was tested in two ways: 1) estimating the effect basing on the known relationship between received information and biological indicators; 2) measuring the physico-chemical parameters of water by means of supersensitive devices.

We emphasize the fact that the biological indicator testing of water, which had received the information from the digital representation of the bio-object, has revealed a number of the new interesting facts concerning the nature of the transmitted data from the bio-object to its image:

First, we have found that the digital representation of bio-objects (seeds, sprouts) carries genetic information of these particular bio-objects (their morpho-physiological parameters). Water is capable to receive this information and transfer it to the same variety.

Second, the information received by the digital representation from the bio-object (sprout) significantly depends on the bio-object isomerism, water better receives information from the images of left sprouts than from the images of right sprouts. When we carried out the experiment without preliminary division of the bio-objects (sprouts) into two groups (left and right sprouts), the expected effects of data transmission from images to water did not show up. In view of the above, the following problem arises: it is necessary to determine how the information on spatial organization of the object at the organismal level (plant bioisomerism) can be received at the molecular level (structure of water) and retransmitted to the live object thus turning back to the organismal level.

Third, communication intensity essentially depends on the type of water, the effect is more pronounced with the tap water compared with the distilled water.

The obtained results, besides their significant scientific value as experimental evidence of quantum-communication theory, might also have potential for use in a large number of diverse application areas. In principle, not merely plant objects, but objects of any origin are capable of being implemented as information carriers in the *object – its digital representation* system. Such experiments were also carried out using water as a means for transmitting information from one object to another [9,10]. Measuring the conductivity of water with deeply polarized electrodes the researchers have revealed change in the condition of water exposed to a high-penetrating emission generated by objects of different origin.

In the long term, *object – its representation* systems based on long-distance communication may serve as prototypes for developing effective, real-time communication systems between Earth and space objects.

CONFLICT OF INTEREST

The authors declare that they have no conflict of interest.

REFERENCES

1. V. Vedral, "Quantifying entanglement in macroscopic systems", Nature, 453(7198), pp. 1004–1007, 2008.
2. V. Vedral, "Living in a Quantum World", Scientific American, (6),
3. pp. 38–43, 2011.
4. R. Graham, K. Fleming, et.al., "Quantum entanglement in photo-synthetic light-harvesting complexes", Nature Physics, (6), pp. 462-467, 2010.
5. K. C. Lee, M. R. Sprague, et. al, "Entangling Macroscopic Diamonds at Room Temperature", Science, 334(6060). pp. 1253–1256, 2011
6. S. N. Maslobrod, "Effect of non local communication between germinating seeds occurring in the course of their contact during seed imbibitions", Surface Engineering and Applied Electrochemistry, 48(6), pp. 99-113, 2012.
7. S. N. Maslobrod, S. Kernbach, E. S. Maslobrod, " Non local communication in plant object – its digital representation system" , part 1, International Journal of Unconventional Science, 2(4), pp. 26-46,2014.
8. 7. S. N. Maslobrod, S. Kernbach, E. S. Maslobrod, " Non local communication plant object – its digital representation system" , part 2, International Journal of Unconventional Science, 2(5), pp. 56-78, 2014.
9. S. N. Maslobrod, S. Kernbach, "Experimental evidence of direct and indirect connection in poto representation of seeds – seeds system" , Nonconventional plant growing, Simferopol, pp.743-747, 2014.

Author:	Serghei Maslobrod
Institute:	Institute Of Genetics Physiology and Plant Protection of ASM
Street:	20, Padurii St.
City:	Chisinau
Country:	Moldova
Email:	maslobrod37@mail.ru

SonaRes Platform for Development of Medical Informatics Applications

S. Cojocaru, C. Gaindric, O. Popcova, and I. Secrieru

Institute of Mathematics and Computer Science, Academy of Sciences of Moldova, Chisinau, Moldova

Abstract— **SonaRes platform is a complex of tools. On its basis a number of informational systems for medical assistance were elaborated. These systems are designed for ultrasound diagnostics in ordinary conditions, in emergency situations and learning. The developed systems and those under development are briefly described.**

Keywords— **decision support systems (DSS), ultrasound diagnostics, medical knowledge, formalization methodology, assistance in medical diagnostics.**

I. INTRODUCTION

The prediction "Ultrasound [sonography] is the stethoscope of the future", made by Roy A. Filly [1] 25 years ago, today becomes a reality. SonaRes technological platform is designed to support diagnostic process using ultrasound examination methods. Diagnostics is one of the most difficult tasks that clinicians face every day. Determination of the correct diagnostics involves a multitude of factors, that characterize the patient's state of health. Omission or misinterpretation of them may lead to mistakes with unpredictable consequences.

No less important is to reveal the relationship between these factors, as well as interaction between organs and its manifestation mode.

In difficult cases, and not only, a second opinion, an advice, is useful to any physician, even to an experienced one. It is natural to ask for a help, to make sure that we proceed in the right way or to find a solution, when we do not know how to proceed.

II. SONARES PLATFORM

The problem, associated with the medical diagnostics activities, acquires a special relevance in modern circumstances. First of all, it is connected with the fact that physicians have to work with poorly structured and weakly formalized information. Besides, the volume of information is in a continuous growth thanks to the appearance of new methods of medical examination.

The ultrasound investigation domain is not an exception. The appearance of new ultrasound devices or the improvement of old scanners doesn't simplify, but even complicates the physician's diagnostic thinking, because one has to analyze a number of diagnostic data, that increases the time of diagnostics determination and sometimes may reduce the accuracy. The SonaRes platform consists of four main components (Fig. 1): expert shell, knowledge base, collection of annotated images, image processing tools. Its scope is limited to hepato-pancreato-biliary region.

Expert shell is designed to support effective communication with experts in the development of knowledge base. The main stages of the development are: problem identification, knowledge acquisition, structure definition, formalization and implementation. The experience of experts and the medical specialty literature served as the main sources of knowledge.

The knowledge base contains information related to organ's localization, its pathological states (for example: chronic cholecystitis, compressed gallbladder, etc.), principal characteristics for organ description (for example: dimensions/volume, shape, tonicity, gallbladder contour, etc.), description of organ's anomalies and pathologies.

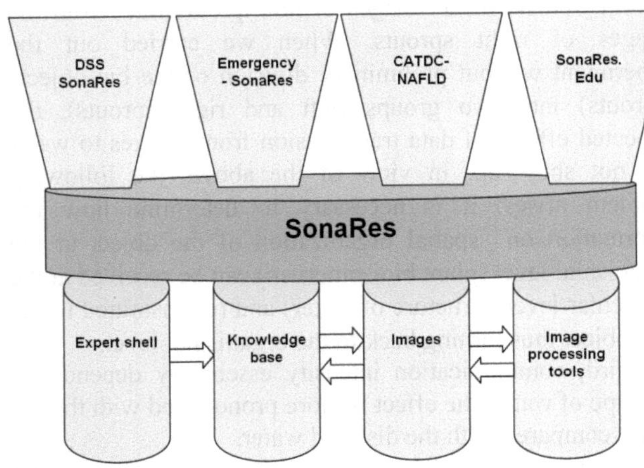

Fig. 1 SonaRes platform and its derivatives

SonaRes platform combines two basic approaches of ultrasound diagnostic systems development, assisting the decision making process on the base of both rules and images. It contains a collection of model images associated to the corresponding rules, with ROIs (region of interest for a particular pathology) defined. The role of annotated images in decision making is to provide help by illustration, in cases when the medical specialist is not sure how to interpret correctly the images [4,9].

V. Sontea and I. Tiginyanu (eds.), *3rd International Conference on Nanotechnologies and Biomedical Engineering*,
IFMBE Proceedings 55,
DOI: 10.1007/978-981-287-736-9_106

Basing on this platform a number of applications were developed including the following: SonaRes – a decision support system for ultrasound examination [2], Emergency-SonaRes – a system for ultrasound examination in the case of mass casualty and disaster setting [3], SonaRes.Edu – a system oriented to learning process.

III. DSS SONARES

The SonaRes decision support system [5,8] is designed to assist the doctor-diagnostician in the examination process of the abdominal region – a particularly difficult both because of the large number of organs, and the need to take into account the interaction between them.

The system operates the knowledge represented by rules and images [6,7]. It has an integrated database, inhomogeneous elements of which are decision rules, original and processed images with annotations, etc.

In case of ultrasound images the most fruitful method seems to be the application of both medical (obtained from knowledge base) and visual features for retrieving. Medical similarity allows confirmation of the diagnosis assumption by obtaining the gallery of images marked by experts as containing supposed pathology or fact (for example: gallstone). Visual similarity allows identifying the supposed fact by presenting the list of those, which correspond in the knowledge base to the found graphical features.

The system is designed for different categories of physicians: from the novices to the experienced ones. Obviously, its usefulness can be appraised only after an assessment made by the users-physicians who give their views on relevance of the conclusions offered by the system, the impact on their conclusions, standardization of protocols, etc. For this purpose the system was tested by two groups of physicians: the experienced ones and interns, being conducted on a test sample of 132 patients with cholecystopathy.

Users' attitude was examined according to the following criteria: quality of conclusion, its argumentation, obtaining a detailed report (Fig. 2) on detected organic ultrasound changes [10]. Experienced physicians mentioned the more evident impact in standardization of medical documents (descriptions, reports) and the role of the second opinion in difficult cases, including the role of theoretical source and images collection value. After using the system the physicians-beginners mentioned both an increase of diagnosis accuracy and positive influence for training.

It is worth mentioning that the physicians-beginners emphasized an increased influence of the system over the working mode, in fact due to the abilities of young specialists in the uptake of new technologies and proper use of knowledge provided by the system.

Both groups evidenced high appreciation of conclusions argumentation. The more experienced physicians showed less interest for conclusions generation.

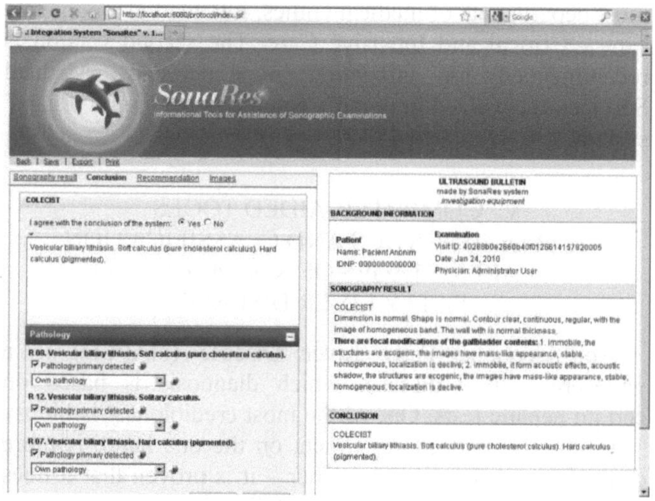

Fig. 2 The reports generator

Generally the system was well accepted, but it remains debatable if there will be accepted all the solutions proposed by the system, which don't coincide with those of physicians.

IV. EMERGENCY-SONARES

Human life is the most precious and the most fragile thing that we must take care in disaster situations. Each emergency case, when people suffer, requires much effort from the rescue teams that can be facilitated and accelerated by applying technologies capable for saving lives and health of victims.

The Emergency-SonaRes system, will be useful both in on-the-fly diagnostics and training in the critical care domain. It allows qualitative and fast primary sorting of victims in order to provide the necessary, but differentiated assistance.

Statistics show that in cases of natural disasters, catastrophes and accidents about 70% of affected persons need specific healthcare approach limited in time. Ultrasound diagnostics at the site of disaster is aimed at determining the level of urgency to save lives and to prevent any complications for people at risk. The Emergency-SonaRes system provides help to physicians from emergency crews in determination of adequate diagnosis in proper time for saving lives of patients. It will offer on-the-fly instructions to obtain qualitative images of affected organs and recommendations for victims sorting: to provide medical assistance in the site or after quick evaluation of the patient state to send him immediately to clinic Service, Emergency Department, Imaging Department.

There are planed two versions of the system: the first one, which can be used at the accident site, as well as in most departments of medical clinics, involved in emergency care – Critical and Intensive Care. The second version is recommended for utilization in Emergency Medical Services specialized in organ's pathology or injured organs' system.

V. COMPUTER-AIDED TOOLS FOR DIAGNOSTICS AND CLASSIFICATION OF EARLY STAGES OF NON-ALCOHOLIC FATTY LIVER DISEASE

Non-Alcoholic Fatty Liver Disease (NAFLD) is a widespread liver disease, which diagnosis is based on certain hepatic tests. One of the most credible test is biopsy – an invasive procedure, which, on the one hand, is more difficult to perform, on the other – it is proven that it does not provide a complete differential diagnosis.

NAFLD in its early stage is asymptomatic, often with normal liver function tests. It is difficult to diagnose, and prevalence studies are difficult to carry out. It is estimated that NAFLD is the most common liver disease both in countries with high income per capita from Western Europe and USA and in poor countries.

This pathology has different stages of evolution, the main milestones being the following: simple steatosis (fatty liver), non-alcoholic steatohepatitis and cirrhosis.

Table 1 CATDC-NAFLD knowledge base evolution

	CATDC-NAFLD KB with only NAFLD rules	CATDC-NAFLD KB adjusted to NAFLD ultrasound diagnostics	CATDC-NAFLD KB adjusted to NAFLD diagnostics ultrasound & hepatologic findings
Total facts	207	120	246 120 ultrasound findings 126 hepatologic findings
Total rules	8	12	12
Total model US images	24	24	24
Total US images with ROIs marked	31	31	31

Our idea is to join liver tests with ultrasound examination in order to elaborate a methodology of formalization and analysis of vast volume of corresponding data, to find and explore the processes of onset and progress of NAFLD. It is used as a base for the development of Computer-Aided Tools for Diagnostics and Classification of Early Stages of NAFLD with Predictive Models for Risk Assessment of Complications (CATDC-NAFLD).

Using the SonaRes platform, a classification of early NAFLD stages was made. A number of scenarios, describing the NAFLD progression, was created (Table 1). An effective algorithm for examination of patients with suspected NAFLD and corresponding predictive models were proposed.

VI. SONARES.EDU

Sonares.Edu is a knowledge-based system aimed to support learning process in ultrasound diagnostics domain. The learning process in Sonares.Edu system is based on two main components: (i) knowledge base and rules obtained during the preliminary domain formalization; (ii) vast collection of model images annotated by the leading experts.

The preliminary formalization of the application domain [6] allowed to organize various type of learning for students with different levels of being grounded.

This system offers adaptive learning mode for beginner users and consulting assistance mode for experienced physicians-sonographists.

VII. CONCLUSIONS

We conclude that general concepts can help, accelerating the development of some concrete systems and, on the other hand, particular solutions of the specific systems are useful for the platform as a whole.

In our development we started with the solutions for DSS SonaRes and, creating other systems, we formulated general ideas, which were included in other described systems.

All these derivatives of the SonaRes platform were evaluated both by experienced clinicians and novices. The applications demonstrated that diagnosis accuracy is at least similar to those formulated by specialists as free text and can be qualified as satisfactory for adequate diagnosis support.

ACKNOWLEDGMENT

This research is supported by ASM-BMBF Project Nr. 13.820.18.02/GA "Computer-Aided Tools for Diagnostics and Classification of Early Stages of Non-Alcoholic Fatty Liver Disease with Predictive Models for Risk Assessment of Complications (CATDC-NAFLD)".

CONFLICT OF INTEREST

The authors declare that they have no conflict of interest.

REFERENCES

1. Roy A. Filly (1988) Ultrasound: The Stethoscope of the Future. Editorial. May, 167–400
2. Burtseva L, Cojocaru S, Gaindric C, Jantuan E, Popcova O, Secrieru I, Sologub D (2007) SonaRes – A decision support system in ultrasound investigations. Computer Science Journal of Moldova, vol. 15, N2 (44):153–177
3. Gaindric C, Cojocaru S, Popcova O, Puiu S, Secrieru I (2014) Emergency-SonaRes: A system for Ultrasound Diagnostics Support in Extreme Cases. In: Teodorescu H-N, Kirschenbaum A, Cojocaru S, Bruderlein C (Eds.) Improving Disaster Resilience and Mitigation – IT Means and Tools, NATO Science for Peace and Security Series C: Environmental Security, Part I, Fundamentals and Modeling, Chapter 18: 283-292 DOI 10.1007/978-94-017-9136-6__18
4. Rîbac E, Cojocaru S, Gaindric C, Puiu S, Ţurcanu V (2005) The examination support in sonographic investigations. Proceedings of the 4th International Conference on Microelectronics and Computer Science, Chisinau, Moldova, 2005, vol. II, pp 271–274
5. Gaindric C, Cojocaru S, Puiu S, Turcanu V (2008) Development peculiarities of the system for physician support in the process of ultrasound examination. Izvestia Orel STU, Series "Fundamental and applied problems of engineering and technology: information systems and technologies", Orel, Russia, N.1-3/269(544), V. 2, pp 51–55 (in Russian)
6. Popcova O, Puiu S, Secrieru Iu (2008) Knowledge structurization and formalization for gallbladder ultrasound investigation domain. Proceedings of the II International Conference "Advanced Information and Telemedicine Technologies for Health", Minsk, Belarus, 2008, pp 98–102
7. Secrieru Iu, Popcova O, Puiu S, Sologub D (2008) Knowledge Structure Modeling in Ultrasound Investigation Domain. Proceedings of the 5th European Conference on Intelligent Systems and Technologies, Iasi, Romania, 2008, CD
8. Gaindric C (2009) Decision support systems in ultrasound diagnostics. Computer Science Journal of Moldova, vol. 17, N3 (51):255–277
9. Burtseva L, Cojocaru S, Gaindric C, Popcova O, Secrieru I (2011) SonaRes – Diagnostic decision support system for ultrasound examination. Proceedings of the 3rd International Conference on e-Health and Bioengineering, Iasi, Romania, November 24-26, 2011, pp 239–242
10. Cojocaru S, Gaindric C, Puiu S (2013) The evaluation of DSS SonaRes – the initial results. Proceedings of the 4th IEEE International Conference on e-Health and Bioengineering, "Improving Quality of Life through Research and Innovation", Iasi, Romania, November 21-23, 2013, ISBN 978-1-4799-2372-4, CD

Authors: S. Cojocaru, C. Gaindric, O. Popcova, I. Secrieru
Institute: Institute of Mathematics and Computer Science
Street: Academiei, 5
City: Chisinau
Country: Republic of Moldova
Email: gaindric@math.md

The Modality of the Regeneration of the Intervertebral Lombar Disc in Osteochondrosis

A. Cociug[1],V. Nacu[2], and O. Macagonova[2]

[1] Morphopathology, State Medical and Pharmaceutical University of Moldova "NicolaeTestemiţanu", Chişinău, Moldova
[2] Laboratory of Tissue Engineering and Cells Cultures, State Medical and Pharmaceutical University of Moldova "NicolaeTestemiţanu", Chişinău, Moldova

Abstract— **The repeated studies of the intervertebral disc degeneration pathogenesis shows that there is a high degree of the neovascularization in the terminal lamellas or the periphery of the herniated content. We studied 87 postoperative intervertebral discs obtained from the patients operated in the section of Vertebrology in Republican Clinical Hospital of Orthopedics and Traumatology in January 2008 - December 2009 and after the sampling of the corpse material (lumbar vertebral segments) of 28 cases with various somatic diseases, age and sex in the Municipal Clinical Hospital "Holy Trinity". The immunohistochemical staining of the lumbar spinal artery wall with CD34 and MCT shows that at different stages of the atherosclerosis development, in the atherosclerotic plaques the mast cells are found in their degranulation phase with the release of the proinflammatory enzymes that enhance the neovascularization process in the vascular wall. Histologically the moderate degenerative changes are better visualized: a high proliferation of fibroblasts, the tissue reorganization trend of the nucleus pulposus as well annulus fibrosus, mainly in the terminal blades where there is an abundance of capillaries with the neovascularization at this level. This explains the tendency to the disk tissue regeneration.**

Keywords— **osteochondorosis, neovascularisation, regeneration, degeneration.**

I. INTRODUCTION

Rätsep T. et al. (2013) [1], in their studies at the degenerated intervertebral disc, they have proved that there is a high level of neovascularization in terminal lamellae or on the periphery of the hernial content. The level of neovascularization was significantly associated with score of the histological degeneration of the herniated disc.

The severe degeneration of the lumbar disc with the herniation is accompanied by neovascularization which is a favorable factor in the regression of the hernia inducing the regeneration.

Although there have been numerous experimental studies at the disk -the more samples contained of human cadaver tissues, animal models and cohort investigations of the postoperative material in the department vertebrology. The several studies have demonstrated the histological tissue granulation with abundant neovascularization, which serves as the hallmark expressed by prolapsing of the disc material [2, 3, 4, 5] .Weidner N. and colab. [5] stated that neovascularization has a marginal location. Other authors reject this assumption considering it as insignificant for research [6, 7, 8]. Some histological studies focused on histological criteria of differentiation, focused on the composition of the hernia contents (nucleus pulposus, annulus and terminal lamellae) [9, 10, 11, 12, 13, 14].

Nerlich A.G. et al. [15] in 2007 using CD31 immunohistochemical staining of the lumbar disc changed degenerative-dystrophic, there have proved to neovascularization. The most abundant is neovascularized the posterior region of the annulus fibrosus. They also made discs examining to the newborns and noted that terminal blades are perforated by the capillaries directed toward the annulus fibrosus bypassing nucleus pulposus.

T. Hasegawa et al. [16] in experiments on dogs showed that the nucleus pulposus fragments degenerated there is an inflammatory reaction which attract lymphocytes, macrophages, fibroblasts and mast cells. Others have examined human samples and found the presence of macrophages in the majority of acute and chronic inflammation of hernias [17, 18, 19].

Mast cells present in inflammation following degranulation eliminated the pro-inflammatory enzymes - interleukins, tumor necrosis factor (TNF), histamine, prostoglandine, phospholipase A2, PGE 46 - which contributes to the stimulation of neovascularization [20, 21, 22, 23].

In the study we used immunohistochemical staining by MCT of the atherosclerotic vertebral artery wall and noticed that in arteries with an advanced degree of impairment the degranulate mast cells are present in different stages and perifocaly by the plaques and vasavasorum- the neovascular primordia which demonstrates that the vessels tend to the pathological regeneration.

V. Sontea and I. Tiginyanu (eds.), *3rd International Conference on Nanotechnologies and Biomedical Engineering*,
IFMBE Proceedings 55,
DOI: 10.1007/978-981-287-736-9_107

From the above it is clear that the CD31 immunohistochemical staining of material taken postoperatively from the degenerated disc elements, in it we find neovascular formations. We have not used this staining, but by the histological study we found many fibroblast in the proliferation phase which indicates a favorable sign with the regenerative character and the formation of the neovascular granulations.

II. DATABASE AND METHODS

We studied 87 postoperative intervertebral discs obtained from the patients operated in the section of Vertebrology in Republican Clinical Hospital of Orthopedics and Traumatology in January 2008 - December 2009 and after the sampling of the corpse material (lumbar vertebral segments) of 8 cases with various somatic diseases, age and sex in the Municipal Clinical Hospital "Holy Trinity". The hematoxein-eosin and picrofuxin histological staining were performed in Laboratory of Pathological Anatomy State Medical and Pharmaceutical University of Moldova „ NicolaeTestemiţanu". The age of operated pacients in the study group (figura 1): by till 20 years – 7 (8,04±1,07%); 21–30 years– 14 (16,09±3,94%), 31–40 years–23 (26,43±4,73%), 41–50 years – 24 (27,58±4,79%), 51–60 years – 14 (16,09±3,94%), after 61 years – 5 (5,73±3,55%).

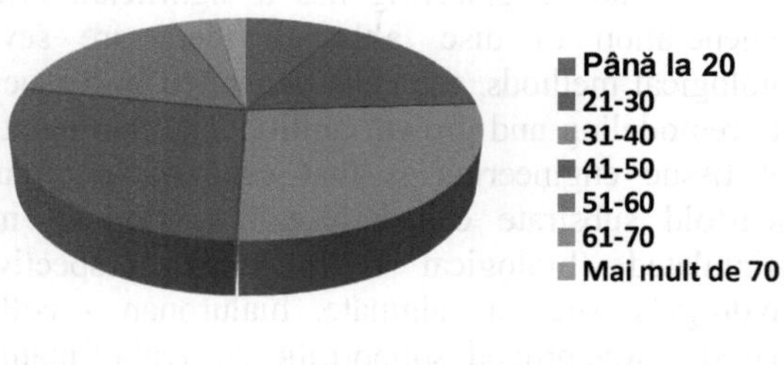

Fig. 1 Structura of group L II in age depending (age) (%).

Average age for group LII is 40,1±1,41 years

Histological study revelead analysis of the morphological evaluation parametres of intervertebral segments studied – of **macroscopicalevaluation** (the dimension of the hernial sac contents and macroscopic aspect it) and **microscopical evaluation** (histopatological type, morfologal changes at level of the pulposus nucleus şi annulys fibrosus, morfological changes at the level of the terminal blades, microscopical aspect of the edges of the vertebral bodies and fibrillar state of the ligaments).

III. RESULTS

Histologically, at the level of the intervertebral disc stained by hematoxilin–eozina there are observed severe forms of degeneration of the intervertebral disc (figura 2, A) with hyaline disc cartilage (a), calcifications inside of the disc or ossification areas (b), lymphocytic infiltrate (c).

Media forms of the degeneration a discului intervertebral (figura2, B) reveals: fibrosed discuss tissue (**a**), with interfibrillar edema areas (**b**); cartilage tissue of disc with condrocytes in chondroblasts (**c**), papilar aspect (**d**) and hemorrhagic areas (**e**), hyaline zone (**f**), and fibroblast proliferation (**g**).

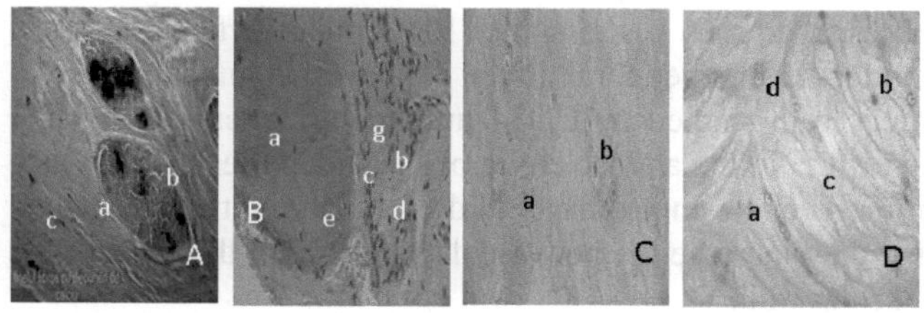

Fig. 2 A. The annulus fibrosus from the intervertebral disc 5) The hyalinisation of intervertebral disc cartilage, B) the intradiscal calcification or ossification areas, c) the lymphocytic infiltrate, HxE staining, x 140. (Pacient K., 82 years).

B. The pulposus nucleus of lumbar intervertebral disc L4, L5 a) the disc tissue fibrosis, with b) the areas of edema between fibers, c) the intervertebral disc cartilage with chondrocytes into chondroblasts, d) and the papillary aspect e) the haemorrhagic areas, f) the hyalinisation areas, g) the fibroblast proliferation, H-Ex staining, x 140. (Pacient A., 62 years) .

C. a) themyxoid degeneration, b) the chondrocytes and chondroblasts are in the center, H-Ex staining, x 140. (Pacient D., 51 years).

D. Nucleus pulposus from the intervertebral lombar disc 2: a) condrocytes and b) chondroblasts conturated, c) organized fibrillar structure, with the persistence of the the fundamental chondrogenic substance (d); H–E x 140.(Pacient C., 32 years) .

Histologically (figure 2, C), stained by hematoxylin–eosin, there are revelead initial changes de degenerescenţă of the intervertebral disc: the aspect of the mixoid degeneration (a), central – chondroblasts with condrocytes (b).

The disc unchanged hematoxylin-eosin stained (Figure 2, D), there are well viewed the chondrocytes (a) and chondroblasts (b) unchanged, with structural elements of fibrillar tissue well defined (c) chondrogenic persistence of the fundamental substance (d).

From here revealed that: histologically, 10 pacients (11,49±3,42%) had incipient degeneration forms of the intervertebral disc, with myxoid degeneration aspects, central with chondroblasts to chondrocytes, 32 pacients (36.78 ± 4.32 %) with the moderated forms of degeneration of the intervertebral disc, where it was found: disc tissue fibrosis, with areas of interfibrilar edema; disc cartilage tissue with chondrocytes in chondroblasts, papillary aspest and hemorrhagic, and hyaline areas. 45 pacients (51.72 ± 5.36 %) in age more 60 years had the advanced forms of intervertebral disc degeneration, where it was determined: hyaline cartilage disc, intradiscal calcification or areas of ossification, lymphocyte infiltrat.

There are well visualized the moderate degenerative - the intense fibroblast proliferation, with tissue reorganization trend of nucleus pulposus and fibrosus annulus, mainly in the terminal lamellae where it is an abundance of neovascular capillaries at this level. This explains the tendency of disk tissue regeneration.

A study performed in the Laboratory of Immunohistochemistry in the Angiogenesis Research Center, Timisoara, Romania, on a sample of 20 cases of the lumbar vertebral arteries L1-L5 spine (100 lumbar vertebral arteries) and L1-L5 intervertebral disc at this level, in lumbar vertebral artery wall studied by CD34 immunohistochemical staining and MCT in the aterosclerotic placues identifies the presence of mast cells with its phases of degranulation, it is determined that they increase the neovascular process, in this case in the vascular wall.

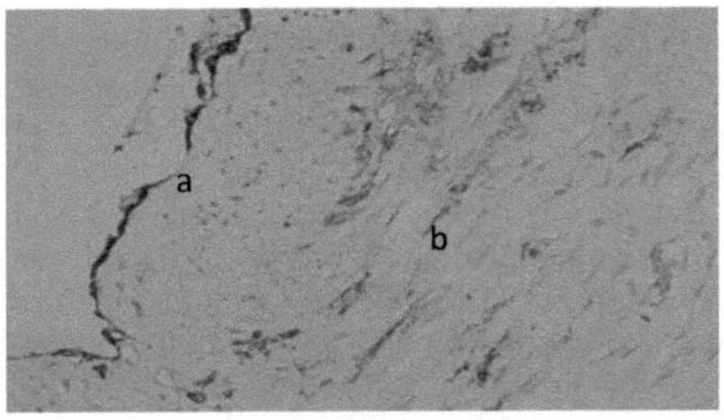

Fig. 3 A 42-year-old patient B. The 5 lumbar vertebral artery a) the intima is sometimes interrupted, thinned with the irregular outline of the fibrillar structures b), CD34 IHC staining, x140.

Immunohistochemical staining with antibody CD34 and MCT (mastocyte cell tryptase) performed in five veretbral arteries (5 ± 1.12%) identifies the presence of mast cells with its phases of activity. The sample of investigation shows the well advanced atherosclerotic changes (Figure 4), the presence of mast cell in step of the degranulation (a), particularly in the L5 and L4 vertebral artery wall at the level plates with the fibroblast proliferation (b), surrounded by an mucoid intumescent (c).

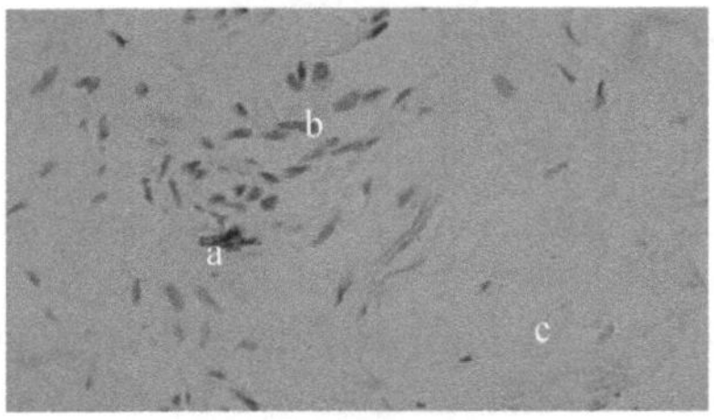

Fig. 4 Vertebral lombar artery 4: a) degranulated, mastocyte b) proliferation of fibroblastes, c) mucoid intumescent, IHC MCT x 90; (Pacient G., 64 years).

In similar cases, the multiple inflammatory cellular elements are present in the intervertebral disc with tend to the sclerosis of the tissue affected with substitution of the connective tissue and at the periphery of the lumbar intervertebral disc, in the terminal blades, the intense proliferation of the fibroblasts apart fibrous protein synthesis and secretion of the extracellular matrix , produce the collagenase wich cleaves the collagen fibers, regulating the anabolism and catabolism of the connective tissue and the tissue pathological neovascularization, and performing the prolongation of viability of the intervertebral disc.

The cell therapy with mesenchymal stem cells plays a significant role in regeneration, being introduced locally in the degenerated disc, aimed, these cells proliferate in specific cells of disc with the same phenotype and as a result improve the degenerative effects, and lead to the production of extracellular matrix designed to restore the function of healthy disc [17, 25]. The repairing treatment is appropriate for more advanced stages of degeneration, which are characterized by the structural degradation of pulpous nucleus and annulus fibrous [16, 22, 23, 24]

The injection of the growth factors proteins as (BMP-7), β (TGFβ) leads to the transformation of tissue having anmetaplasial effect, the cell differentiation with 5 (GDF-5) and others, all of which are widely studied, stimulating an extracellular production of matrix and a cell proliferation [18]. The injected directly into the disc of the anabolic factors effect are limited because of the short half-life and its rapid diffusion.

The tissue engineering has a significant effect in the regeneration of disc, although there are several non-biological methods, the cells generated by tissues preserve its remodeling and growth ability. The dominant paradigm of tissue engineering is that cells on a biomaterial or scaffold substrate can induce formation of new tissue stimulated (biological or physical) respectively. The hydrogels, such as alginate, hialuronan - collagen gels based, it was proved, support the survival of mature nucleus pulpous and promote extracellular matrix formation [23, 24]. Although, the engineering of pulpous nucleus and fibers of the annulus tissues was a special focus throughout the years, by increasing of discs (*in vitro*), including the translation of these technologies in large animal models (bovines), for the clinical trials, respecting the nutritional requirements (*in vivo*)[15, 19, 22].

IV. CONCLUSION

1. The histopathological changes shows in the early forms the degeneration of the intervertebral disc, identified to 10 (11.49 ± 2.42%) operated patients, were as follows: myxoid degeneration in the cartilage disc, with chondroblasts and chondrocytes in the center of the image; - Hyalinisation and areas of discal tissue edema in the disc fibrosed tissue.

2. The histopathological changes in the second phase, the media forms of degeneration of the intervertebral disc, were detected in 32 (36.78 ± 3.67%) patients: discal cartilage tissue with the chondrocytes in chondroblasts with the papillary aspect and the hemorrhagic areas with the chondroblasts and chondrocytes associated in area of the myxoid degeneration.

3. The histopathological changes of third type, in the advanced forms of degeneration of the intervertebral disc were found in 45 patients (51.72 ± 5.36%): myxoid degeneration, hyalina discal cartilage with the areas of the intradiscal calcification, the presence of young cells (fibroblasts and chondroblasts), cartilage tissue with areas of fibrosis and chronic inflammatory infiltrate.

4. An important element revealed by us was the identification of new young cells (fibroblasts) in the moderate degenerative lesions, offering the opportunity of the disc repair in the injuries.

5. Immunohistochemistry with anicorpi MCT (mastocyte cell tryptase) identifies in the arteries with an advanced degree of damage are present mast cells in different stages of degranulation, but perifocal by the aterosclerotic placues and vasa vasorum – the neovascular primordia which demonstrates that the vessel pretend to regeneration.

CONFLICT OF INTEREST

The authors declare that they have no conflict of interest.

REFERENCES

1. T.Rätsep, A.Minajeva, T. Asser,"Relationship between neovascularization and degenerative changes in herniated lumbar intervertebral discs. Source Department of Neurology and Neurosurgery, University of Tartu", *J. Eur Spine*, **80**, pp. 510–14, 2013.
2. L.Y.Carreon, T.Ito, M.Yamada, S.Uchiyama, H.E Takahashi,"Neovascularization induced by anulus and its inhibition by cartilage endplate. Its role in disc absorption",*Spine*, **22**, pp. 1429–1434, 1997.
3. T.Ito, M.Yamada, F.Ikuta, T.Fukuda, S.I.Hoshi, Y.Kawaji, et al.,"Histologic evidence of absorption of sequestration–type herniated disc",*Spine*,**21**, pp. 230–234.262, 1996.
4. M.Repanti, P.G.Korovessis, M.V.Stamatakis, P.Spastris, P.Kosti,"Evolution of disc degeneration in lumbar spine: a comparative histological study between herniated and postmortem retrieved disc specimens",*J Spinal Disord*,**11**, pp. 41-45, 1998.
5. T.Videman, M.C. Battie,"The influence of occupation on lumbar degeneration", *Spine*,**24**, pp. 1164–1168, 1999.
6. T.Yasuma, K.Arai, Y.Yamauchi,"The histology of lumbar intervertebral disc herniation. The significance of small blood vessels in the extruded tissue", *Spine*, ,**18**, pp. 1761–1765, 1993.
7. D.S.Choy,"Familial incidence of intervertebral disc herniation: an hypothesis suggesting that laminectomy and discectomy may be counterproductive", *J Clin Laser Med Surg*, **1**,pp.29-32, 2000.
8. R.W. Moskowitz, I.Ziv, C.W.Denko,"Degeneration lumbar intervertebral disc tissue",*Spine*,**21**, pp. 2149-2155, 1996.
9. C.Eckert, A. Decker,"Pathological studies of intervertebral discs",*J Bone Joint Surg Am.*, **29**, pp. 447—454, 1947.
10. 10 R.E.Willburger, U.K.Ehiosun, C.Kuhnen, J.Kramer, G. Schmid,"Clinical symptoms in lumbar disc herniations and their correlation to the histological composition of the extruded disc material",*Spine*,**29**, pp. 1655–1661, 2004.
11. T.Yasuma, E.Makino, S.Saito, M.Inui,"Histological development of intervertebral disc herniation", *J Bone Joint Surg Am*, **68**, pp.1066–1072, 1986.
12. Yasuma T., Arai K., Yamauchi Y. The histology of lumbar intervertebral disc herniation. The significance of small blood vessels in the extruded tissue. Spine. 1993, 18:1761–1765.
13. A.Zortam, M.A.Vettori, C.P.Trevisan, S.Bellini, G.Vazza, Armani M., Simon A., Mostacciuolo M.L,"Genetic mapping of a susceptibility locus for disc herniation, 6q23.3–q24.1 spastic paraplegia on",*J Med Genet*, **39**, pp. 387–390,2002.
14. Y.Harada, S. Nakahara,"A pathologic study of lumbar disc herniation in the elderly", *Spine*, **14**, pp. 1020-1024, 1989.
15. A.G.Nerlich, R.Schaaf, B.Walchli, N.Boos,"Temporo-spatial distribution of blood vessels in human lumbar intervertebral discs."*J. Eur. Spine.*;**16**, pp. 547–555, 2007.
16. T.Hasegawa, H.S An., A.Inufusa, et al.,"The effect of age on inflammatory responses and nerve root injuries after lumbar disc herniation: an experimental study in a canine model."*Spine*,**25**, pp. 937–940, 2000.
17. M.Gronblad, J.Virri, J.Tolonen, et al.,"A controlled immunohistochemical study of inflammatory cells in disc herniation tissue", *Spine*,;**19**, pp. 2744–2751, 1994.
18. A.Habtemariam, M.Gronblad, J.Virri, et al.,"A comparative immunohistochemical study of inflammatory cells in acute-stageand chronic-stage disc herniations."*Spine,***23**, pp. 2159–2166, 1998.
19. T.Ikeda, T.Nakamura, T.Kikuchi, et al.,"Pathomechanism of spontaneous regression of the herniated lumbar disc: histologic and immunohistochemical study",*J Spinal Disord*, **9**, pp. 136–140, 1996.
20. M.Doita, T.Kanatani, T.Harada, et al.,"Immunohistologic study of the ruptured intervertebral disc of the lumbar spine",*Spine*, **21**, pp.235–241, 1996.
21. M.Gronblad, J.Virri, S.Ronkko, et al.,"A controlled biochemical and immunohistochemical study of human synovial–type (Group II) phospholipase A2 and inflammatory cells in macroscopically normal, degenerated, and herniated human lumbar disc tissues."*Spine.*;**21**, pp.2531–2538, 1996.
22. J.S Saal,"The role of inflammation in lumbar pain."*Spine*;**20**, pp. 1821–1827, 1995.
23. P.Boutin, H. Hogshead,"Surgical pathology of the intervertebral disc. Is routine examination necessary? ",*Spine*, **17** pp.1236-1238, 1992.
24. R.Paul, R.C.Haydon,H. Cheng, A.Ishikawa, N.Nenadovich, W.Jiang, L.Zhou, B.Breyer, T.Feng, P.Gupta, et al.,"Potential use of Sox9 gene therapy for intervertebral degenerative discdisease."*Spine***28**, 755–763, 2003.
25. D.Sakai,"Future perspectives of cell-based therapy for intervertebral disc disease." *Eur. Spine J.***17**, pp. 452–458. [PMC free article] [PubMed],2008.
26. J.S. Saal, R.C. Franson, R.Dobrow, et al,"High levels of inflammatory phospholipase A2 activity in lumbar disc herniations."*Spine.*;**15**, pp. 674–678, 1990.

Author:	Cociug Adrian
Institute:	State Medical and Pharmaceutical University of Moldova „ NicolaeTestemiţanu"
Street:	bd. Ştefancel Mare şiSfânt, 165
City:	Chisinău
Country:	Republic of Moldova
E-mail:	adriancociug77@gmail.com

Middle Ear Monitoring in Children

S. Diacova[1], I. Ababii[1], M. Maniuc[1], L. Danilov[1], A. Chiaburu[1], P. Ababii[1], O. Sosnowska[1], and V. Desvignes[2]

[1] Department of Otorhinolaryngology, Pediatric Clinic, SMPhU "Nicolae Testemitanu", Chisinau, Moldova
[2] Society "Médecines du Monde", Clermont-Ferrand, France

Abstract— **The aim of our study was to analyze the possibility of middle ear monitoring in children to predict chronicity and recurrence of otitis media. Material and Methods: Children aged 1 to 7 years of life were monitored by screening tympanometry and screening otoscopy during 2 years. Workup also included complete audiological examination: Impedance Audiometry, Conventional Audiometry, Oto-Acoustical Emissions and Brainstem Electrical Response Audiometry and X-ray examination of paranasal sinuses, bacteriologic examination, spiral CT with 3D reconstruction of tympanic cavity and mastoid process. Oto-microsurgical intervention by modified method was done in children with recurrent and chronic otitis media. Results. Pre-chronic and recurrent forms were diagnosed in 36 % and 18 % of children with respiratory pathology, in difference from healthy group, 3% and 4 % (p<0,001). In children with gastrointestinal problems chronic middle ear pathology was diagnosed rare, but recurrence-chronicity rate in both groups were more than 60 %. Conclusions: Effectiveness of middle ear monitoring in children with chronic recurrent somatic pathology is confirmed by high recurrence / chronicity rate of otitis media. Middle ear monitoring in practically healthy children is not justified. Information from complex audiological examination is sufficient for diagnostics of conductive hearing loss but not sufficient for differentiation of otitis media forms and indication of surgery. Complex audiological, otomicroscopical, radiological and microbiological examination is necessary for children with chronic otitis media with effusion and recurrent otitis media for indication of correct treatment.**

Keywords— **Middle ear monitoring, Impedance Audiometry, BERA, children with somatic pathology.**

I. INTRODUCTION

Global middle ear (ME) monitoring is introduced in preschool and school institutions in USA, Australia, Finland, etc. for early otitis media (OM) diagnostics. Its positive and negative aspects are discussed in literature. [1]. Positive one consists of early detection of children with ME pathology which is characterized by conductive hearing loss and leads to language and learning retard and in some cases to chronic suppurative OM and even to intracranial complications. But practically all children in small age are at risk of OM, because of well-known anatomical and functional features of ME and other systems of child's organism. [2,3]. Monitoring for OM diagnostics confirms that almost every child has at least one episode of OM during his first 7 years of life. Moreover majority of OM cases resolves spontaneously after treatment of concomitant pathology or even without any treatment. Therefore global OM screening may lead to hyperdiagnostics and unnecessary treatment, including antibiotics and surgery. The effectiveness of global childhood screening for OM is questionable.

Existing surgical treatment largely used in Europe and USA is indicated for early OM forms. OM is the most common illness in children, and tympanostomy (surgery for OM treatment) is the most common surgical procedure performed (in USA and Europe). Effectiveness of this classical intervention is not high because of very large indications in early OM forms and limited possibilities in late forms. [3,4,6]. The range of risk factors for OM development is very large and unspecific. The same group of risk factors is on the base of more evident pathological conditions of health than OM, such as recurrent and chronic respiratory tract pathology. Recurrent and chronic gastrointestinal pathology connected to another risk factors group of chronicity in childhood because of the nutritional misbalance and gastro-esophageal reflux which they provoke. Additionally, evolution of OM accompanied by the recurrent and chronic somatic pathology in children is silent what predetermines its late diagnostics. Absence of timely diagnostics and adequate treatment of OM early forms predisposes to chronic OM formation with progressive hearing loss and high risk of intracranial complications. [1,7].

The most important aim of OM monitoring is the identification of children with high risk of recurrent and chronic OM, which need intensive treatment. [1,2,4,5].

In Moldova the system of management of OM in childhood is in an ongoing development. Together with pediatricians from Society "Medicines due Monde", Clermont-Ferrand, France, we conduct ME monitoring in different groups of children with recurrent and chronic somatic pathology in order to determine groups of children who really need surgical intervention. [1,6].

Surgical treatment - myringotomy with insertion of tympanostomy tube by modified version (BOPI MD, 2013), was elaborated on the basis of researches which we conducted during Joint bilateral Project at Mayo Clinic, Rochester, MN, USA [7]. Our surgery modification is indicated for late OM forms, such as chronic otitis media with effusion (COME) and recurrent otitis media (ROM) because it

V. Sontea and I. Tiginyanu (eds.), *3rd International Conference on Nanotechnologies and Biomedical Engineering*,
IFMBE Proceedings 55,
DOI: 10.1007/978-981-287-736-9_108

improves possibility to eliminate all the pathological content from the tympanic cavity and mastoid cells. Moreover, our method enlarges visualization of tympanic cavity and possibility to precise morphology changes of tympanic cavity in children with late OM forms. [7].

The ***aim*** of our study was to analyze the possibility of ME monitoring in children.

II. MATERIALS AND METHODS

A. Materials

Patients: Children from 1 to 7 years of life with chronic somatic pathology: group R – respiratory tract pathology, group G – gastro-intestinal pathology, and group H – healthy children.

B. Methods

Monitoring: Every child was monitored by screening methods such as screening tympanometry and screening otoscopy 4 times during 1 year (once in 3 months).

Complete work up: Presence of pathologic changes for at least 3 months or repeatedly during 6 months was indication for complete audiological examination which consisted of Impedance Audiometry, Conventional Audiometry, Oto-Acoustical Emissions and Brainstem Electrical Response Audiometry. Workup also included X-ray examination of paranasal sinuses, bacteriologic examination of palatine tonsils, rheumatic tests, spiral CT with 3D reconstruction of tympanic cavity and mastoid process.

Confirmation: Surgical exploration of ME is the gold standard middle in ME diagnostics. In case of OM myringotomy in our modification with tynpanostomy tube insertion was performed as a treatment intervention. Detailed description of otomicroscopical findings from surgery by modified method were compared with preoperative noninvasive data.

III. RESULTS

A. Monitoring

Patients: Totally 2138 children at age 1 to 7 years of life with somatic pathology and healthy children were recruited to a prospective ME monitoring. The mean age of children was 4.1 years (SD 1.5).

ME pathology: Majority of children had ME pathology during these 2 years of observation (Fig. 1.). The cumulative data of ME pathology, clinically typed as acute otitis media (AOM), otitis media with effusion (OME), chronic otitis media with effusion (COME) recurrent otitis media (ROM), adhesive otitis media (AdOM) and chronic suppurative otitis media (CSOM) demonstrated that pathological ME processes were characteristic for children till 7 years with the prevalence of short duration forms, such as AOM and OME during first 3-4 years of life. Formation of subchronic forms such as COME, ROM began from 2nd year of life with the maximum at the age 4-5 years of life. Chronic forms AdOM and CSOM became evident from 3-4 years of life, but their background was built earlier.

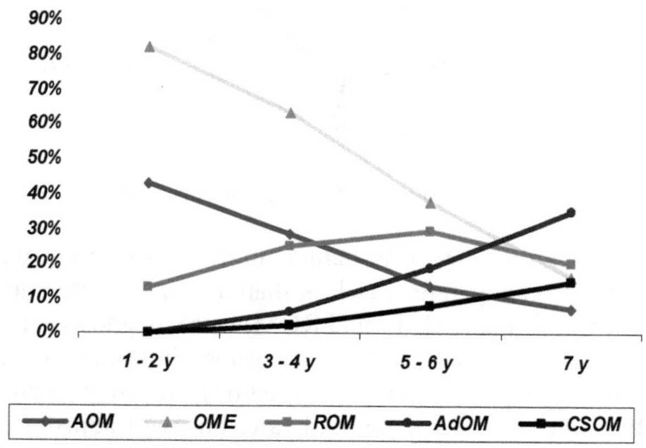

Fig. 1 ME pathology during 2 years

Influence of chronic somatic pathology: According to our research, recurrent and chronic somatic pathology is an important predisposing factor. Majority of children with OM were from the group with recurrent and chronic respiratory tract pathology (group R). Children with gastrointestinal (group G) pathology and healthy children (group H) in majority cases develop OM in 28 – 29 % of cases.

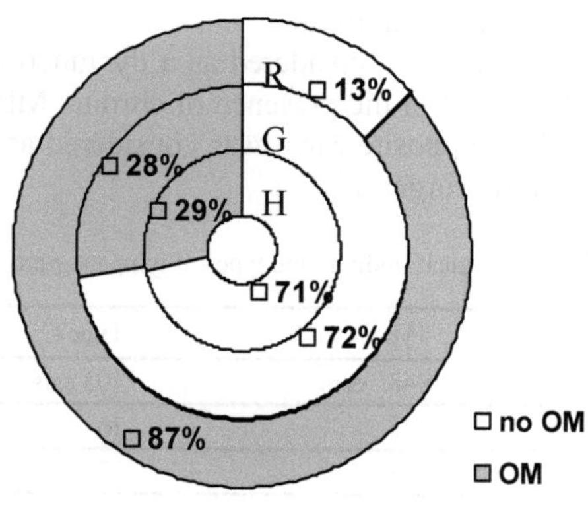

Fig. 2 ME pathology in 3 cohorts of children

Types of OM: Analysis of OM types demonstrated (Fig 3) the tendency to recurrence and chronicity in R and G groups.

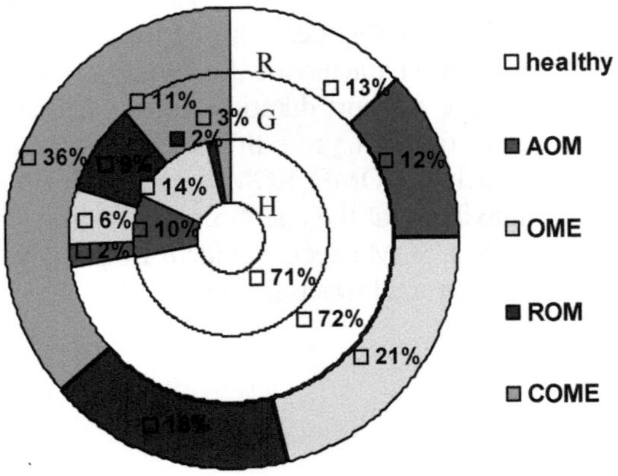

Fig. 3 Forms of OM in 3 cohorts of children

In H group OM was benign in 75 % of cases. Recurrence-chronicity rate was less than 15 %. Pre-chronic and recurrent forms (COME and ROM) were diagnosed in 36 % and 18 % of children from R group, in difference from healthy group, 3% and 4 % (p<0,001). In group G chronic ME pathology was diagnosed rare, but with adjustment of the total ME pathology, recurrence-chronicity rate in both groups was more than 60 %.

B. Agreement between Methods

Impedance Audiometry: Pathological tympanogram during more than 6 months with the confirmation of ME pathology by others noninvasive methods and absence of it normalization after medical treatment was the indication for surgery. All cases of OM and duration at least 6 months were included in surgical group.

Type C classically is considered as a dysfunction of the Eustachian tube. But in the presence of chronic ME pathology as Type B in opposite ear it was considered as an indication for bilateral surgery.

Table 1 Surgical findings and types of tympanogram

Types of tympanogram	Type B	Type C
Pathological changes	448 ears	103 ears
Effusion	426	36
Granulations	123	22
Retraction pockets	87	35
Cholesteatoma	2	6

Analyzing surgical data in children with type C of tympanogram we found in majority cases signs of AdOM and CSOM (Fig. 4.), even more often than in Type B subgroup.

Audiology: Conductive hearing loss is the most characteristic sign of OM. It level in majority of cases is relatively low therefore hypoacusis is inappreciable by small children and their parents. Conventional audiometry revealed presence of mild to moderate unspecific conductive hearing loss, 25 – 45 dB, which was practically identical for all forms and duration of OM.

Oto-acoustical emissions as a screening audiological test were registered in case of impossibility of conventional audiometry in small children or in patients with psycho-neurological disorders. This method was sensitive to all cases of OM without dependence of OM forms.

Brainstem Electrical Response Audiometry was used for examination of very small children. Peak Latencies (I, III, V) and BERA threshold levels were registered; Function Latency-Intensity (V) was analyzed. The registration of potential electrical activity on 20 dB or more was considered as significant for conductive hearing loss, characteristic for OM. Additionally, wave PL I prolongation in serial recordings and the shortness of I - V inter-peak latency were checked as an indicator of OM. [10, 12]

Audiology results correlated to presence of ME pathology independently of the OM form.

Radiologic examinations: X-ray examination of paranasal sinuses was done for all children with ROM or COME.

Majority of patients with OM - 78 %, had inflammatory process in paranasal sinuses. Presence of sinusitis was an indication for intensive prolonged treatment, which included antibiotics, antihistamines, mucolytics, local procedures such as nasal irrigation and therapeutic droops.

Spiral CT with 3D reconstruction of tympanic cavity and mastoid process was performed for children with ROM and COME. The most evident pathological changes were found in ROM group, such as effusion in tympanic cavity and mastoid process, soft tissue similar to granulations.

Bacteriologic examination of palatine tonsils was done for all children with ROM and COME. The most common organisms isolated were in COME group *Streptococcus pneumoniae* - 43 %, *Haemophilus influenzae* 29%. In ROM group predominant microbes were *Staphylococcus aureus* 53 % and *Streptococcus b-hemolytic gr A* - 32 %. In healthy children microbiological spectrum was large, predominated nonpathogenic types such as *alfa-hemolytic Streptococci* in 37%. Recurrence and chronicity were related to *Staphylococcus aureus* and *Streptococcus b-hemolytic gr A*.

Rheumatic tests were positive in 18 % of children with ROM, 3 % of children with COME and 7 % of healthy children.

IV. DISCUSSIONS

Our data confirmed that children with recurrent and chronic somatic pathology had otitis media more often than healthy children. Chronic respiratory tract pathology had relatively stronger predisposition to OM than chronic gastro-intestinal pathology. But recurrence / chronicity rate is approximately equal. Acute otitis media was diagnosed approximately with the same frequency in all groups of patients. Otitis media in children with chronic somatic pathology in majority of cases had an oligosymptomatic course such as otitis media with effusion. Adhesive OM and CSOM may develop from early childhood and progress with minimal functional changes, such as type C of tympanogram.

Medical treatment which was indicated to children with RAOM and COME and sinusitis was effective for patients without somatic pathology. Treatment of paranasal sinuses pathology is important for differential diagnostics and presurgical preparation. Majority of patients with chronic OM with effusion or recurrent acute OM and somatic pathology needed surgical treatment.

Morphologic changes in tympanic cavities which are characteristic for chronic suppurative and adhesive processes were found in both forms of OM – chronic OM with effusion and recurrent acute OM but relatively more often in the last one.

V. CONCLUSIONS

1. Effectiveness of ME monitoring (or repeated screening) in children with chronic recurrent somatic pathology is confirmed by high recurrence / chronicity rate of otitis media.
2. ME monitoring in practically healthy children is not justified because otitis media in this patients' category is rare condition with benign course.
3. Information from complex audiological examination is sufficient for diagnostics of conductive hearing loss but not sufficient for differentiation between otitis media forms and indication of surgery.
4. Type C of tympanogram is pathological and corresponds as to Eustachian tube dysfunction thas to adhesive otitis media and together with some other changes may be an indication for surgical treatment.
5. Complex audiological, otomicroscopical, radiological and microbiological examination is necessary for children with chronic otitis media with effusion and recurrent otitis media for indication of medical and surgical treatment.

VI. COMPLIANCE WITH ETHICAL REQUIREMENTS

CONFLICT OF INTEREST

The authors declare that they have no conflict of interest.

STATEMENT OF INFORMED CONSENT

The informed consent was obtained for every participant from his or her parents.

STATEMENT OF HUMAN AND ANIMAL RIGHTS

The study was approved by the Ethics Committees at the SMPhU "Nicolae Testemitanu" and Republican Hospital for Children "Em. Cotaga".

REFERENCES

1. Lieberthal A.,Carroll A., Chonmaitree T. et al. The diagnosis and management of acute otitis media. *Pediatrics Vol. 131, No.3, March 1, 2013 pp. e964-e999 (doi: 10.1542/peds.2012-3488)*
2. Grevers G. Challenges in reducing the burden of otitis media disease: an ENT perspective on improving management and prospects for prevention. Int J Pediatr Otorhinolaryngol. 2010 Jun;74(6):572-7. doi: 10.1016/j.ijporl.2010.03.049. Epub 2010 Apr 20.
3. Shaikh N., Hoberman A., Rockette H., Kurs-Lasky M., Development of an Algorithm for the Diagnosis of Otitis Media. Academic Pediatrics, 2012;12:214–218
4. E.Lambert, S.Roy Otitis media and ear tubes. Ped Clin North Amer v. 60 (2013), p. 809 – 826.
5. Marchica C., Pitaro J., Daniel S. Recurrent tube insertion for chronic otitis media with effusion in children over 6 years. Int J Ped ORL, v. 77 (2013), p. 252 – 255.
6. Cho Y-S., Hong S. D.,, Chung K. W., Hong S. H., Chung W-H., Park S. H. Revision surgery for chronic otitis media: characteristics and outcomes in comparison with primary surgery. Auris Nasus Larynx, v. 37 (2010), p. 18 – 22.
7. Diacova S., Ababii I. Method for treating exudative otitis media in children. BOPI N9 (2013), p. 24.

Author: Diacova Svetlana
Institute: State University of Medicine and Pharmacy "Nicolae Testemitanu"
Street: bd. Stefan cel Mare, 165
City: Chisinau
Country: Moldova
Email: svetdiac@yahoo.com

An Evaluation of the Accuracy and Reproductibility of Cephalometric Measurements Using Two Different Versions of Romexis Software

O. Ladunca (Rusu)[1], A. Petcu[2], D. Haba[1], I. Zetu[2], L.V. Boiculese[3], C. Corciova[4], and M. Moscalu[3]

[1] University of Medicine and Pharmacy "Grigore T. Popa", Department of Dento-Maxillo-Facial Radiology, Iasi, Romania
[2] University of Medicine and Pharmacy "Grigore T. Popa", Department of Orthodontics, Iasi, Romania
[3] University of Medicine and Pharmacy "Grigore T. Popa", Preventive Medicine and Interdisciplinarity Department, Iasi, Romania
[4] University of Medicine and Pharmacy "Grigore T. Popa", Biomedical Sciences Department, Iasi, Romania

Abstract— **The present study was designed to determine the reliability and reproductibility of angular and linear cephalometric measurements obtained with two different versions of Romexis Planmeca imaging software. 48 pre-treatment orthodontic X-rays were selected for the present study. All the radiographs were measured by two observers, and traced under Steiner and Tweed analyses. The digital tracing was performed using Romexis imaging software version 3.2.0 and the updated year 2014 – compatible version 3.6.0 R. When compare the results found for Romexis 3.2.0 and the other one found for Romexis 3.6.0 R, our reserche sustain, based on the statistical results, that Romexis version 3.2.0 is more accurate and reproductible.**

Keywords— **cephalometry, image resolution, statistical analysis, Steiner analysis, Tweed analyses.**

I. INTRODUCTION

Cephalometry is mostly used in orthodontics, to describe the morphology and the growth of the facial skeleton, predict growth, plan treatment and evaluate treatment results [1, 2]. To performe a cephalometric analysis, the orthodontists mark specific and standard anatomical landmarks on the radiographs, and measure various angular and/or linear parameters. Image thickness and resolution, anatomical complexity and superimposition of hard and soft tissue, and the experience of the observers when looking for a particular landmark are important factors that can influence the identification of the landmark [3]. Currently, radiology field has developing towards digitization and computerization.

Planmeca RomexisTM Software is one of multiple digital software present on the medical market in these days. The software was invented and designed by Planmeca Oy, Finland and can very easily to capture, view, manipulate and process the 2D or 3D images, depend what is nedeed. Romexis software functionalities are concepted to solve many directions which are included in dental area. Moreover, Planmeca Romexis developed different module for 2D and 3D imaging, 2D and 3D diagnosis, cephalometric assessment, orthodontists and dental labs, dental model analysis, treatment planning in 3D, export of digital dental models in STL format, cephalometric analyses and superimpositions, implant planning, dental imaging needs, and the last but not the least, special modules for dental education and unique concept for clinic management and maintenance. A comparison between digital to conventional radiography demonstrates several advantages, particullary in implant and endodontic treatment [4]. Several studies have examined the accuracy and reproducibility of measurements using different cephalometric methods, and then reported the fact that direct method is considerate more reproducible and more accurate; moreover, the conclusion was that the differences between the methods are insignificant for majority of them.

II. MATERIALS AND METHODS

Digital cephalometric exams of 48 pre-treatment orthodontic radiographs were selected for the present study from the archive of the private radiological center. The patients distribution was 21 males and 27 females, and their ages ranged between 7 - 33 years. All the radiographs were measured individually by two observers, in which linear and angular measurements were made using the same 2D computerized cephalometric X-ray, and traced with 2 different version of Romexis imaging software, both under Steiner and Tweed analyses for all the sample.

For the acquisition of the lateral cephalometric images, we use in the study the same device, a Planmeca ProMax 3D Mid (Planmeca, Helsinki, Finland) and the tehnician use the same protocol: the head of the subjects were positioned in the cephalostat with the sagittal plane perpendicular to the path of the ground, Frankfort Horizontal plane was parallel to the floor, the teeth were in their centric occlusion and the lips were in repose. For the next step the digital images were transfered into Romexis Planmeca software database.

The digital tracing was performed using Romexis imaging software version 3.2.0 and the updated year 2014 – compatible version 3.6.0 R (Planmeca, Helsinki, Finland); No more 10 radiography were traced on a single day. Digital Imaging and Communications in Medicine

V. Sontea and I. Tiginyanu (eds.), *3rd International Conference on Nanotechnologies and Biomedical Engineering*,
IFMBE Proceedings 55,
DOI: 10.1007/978-981-287-736-9_109

(DICOM) files obtained with our device were subsequently visualized, separatelly by each observers, with the Planmeca Romexis program and digitized on a 16 : 9, 23 - inch, Light Emitting Diodes (LED) backlighting monitor display (Dell, USA) at a screed resolusion of 1920 × 1080 pixels. Both version of software were instaled on the same computers, first one was version 3.2.0 and after the updated process was used the second one, version 3.6.0 R. Observer 1 and observer 2 traced on the same computers. The software, regardless of the version used for cephalometric tracing, permitted to the resercher to adjust the images using functions for brightness, zoom and contrast; in plus, version 3.6.0 R have one more function than 3.2.0 version, filter function, which help the observer to offer better clarity to the radiographs. Before starting to identify the anatomical landmarks, both observers made individually, the calibration of the images, by selected two points on the X-rays ruler from up-right side. 13 angular and 11 linear (Fig. 1 and Fig. 2), commonly used to assess the dentofacial relationships, were selected to sustain the present reserche.

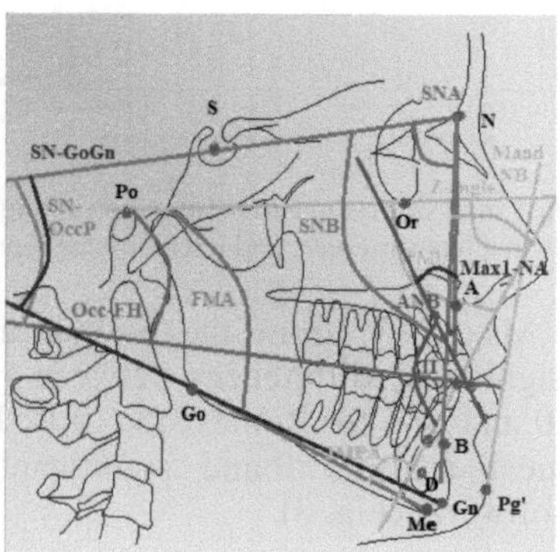

Fig. 1 Angular measurements developed under Steiner and Tweed analysis

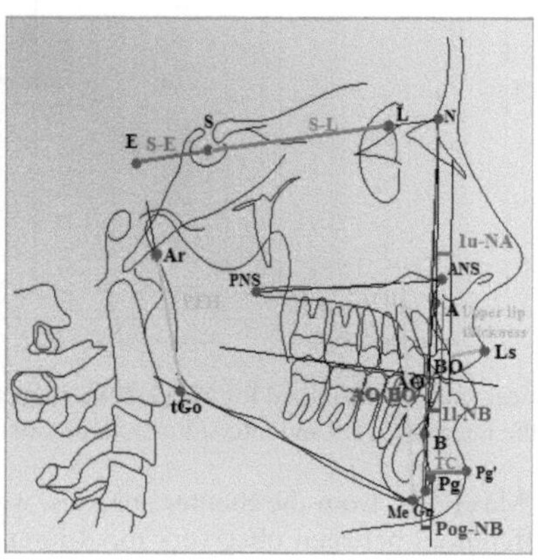

Fig. 2 Linear measurements developed under Steiner and Tweed analysis

Next step, which were performed by the observers, was landmark identification which was carried out manually on the digital cephalometric images using a mouse cursor. Following digitalization of landmarks, measurements were automatically generated by the imaging software and a report sheet model was issued for each radiograph.

III. RESULTS AND DISCUSSIONS

Statistical analysis for the present research was based on the Scheffe test. The parameter p was our reference and the meaning of this parameter is the significance level of the test (p = 0.05), for a confidence interval of 95%. The results were presented as mean ± SD. The results from our study show significant differences in both type of analysis, Steiner and Tweed analyses. Moreover, in Tweed analysis, interobservers reproducibility for Romexis 3.2.0 were found significant for parameters AO-BO, Z-Angle and Facial height index, and for Romexis 3.6.0 R the interboservers reproducibility were found significant for SNA and Facial height index. A comparison between Romexis 3.2.0 and Romexis 3.6.0 R measurements showed significant differences in AO-BO and Z-Angle.

For parameter SNA, from Tweed analysis, there were no significant differences between the mean values of Observer 1 and Observer 2 for Romexis 3.2.0; In Romexis 3.6.0 R SNA shows significant differences between Observer 1 and Observer 2 (p=0.0074); and the statistical analysis revealed no significant differences between the measurements of the two versions of software (Fig. 3).

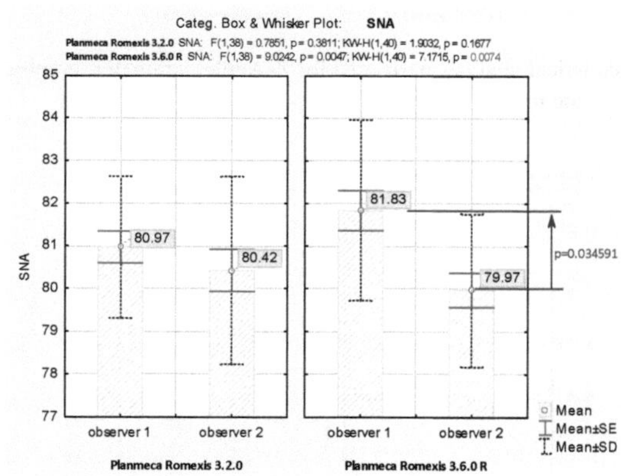

Fig. 3 Statistical analysis perfomed for **SNA** parameter; the graphics show the interobservers and intersoftware reproducibility.

For parameter AO-BO, from Tweed analysis, there were significant differences between the mean values of Observer 1 and Observer 2 for Romexis 3.2.0 and for Romexis 3.6.0 R. A comparison between Romexis 3.2.0 and Romexis 3.6.0 R for Observer 2 shows significant differences for AO-BO measurements (p=0.03) (Fig. 4).

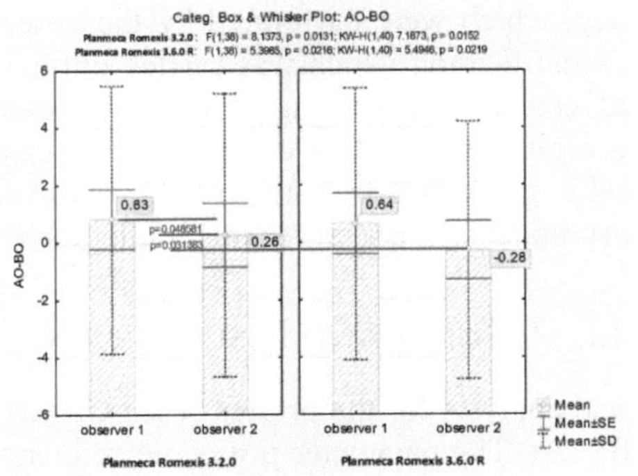

Fig. 4 Statistical analysis perfomed for **AO-BO** parameter; graphics show the interobservers and intersoftware reproducibility.

For parameter Z-Angle, from Tweed analysis, were significant differences between mean values of Observer 1 and Observer 2 for Romexis 3.2.0. A comparison between Romexis 3.2.0 and Romexis 3.6.0 R show significant differences between Observer 2 and Observer 1 (Fig. 5).

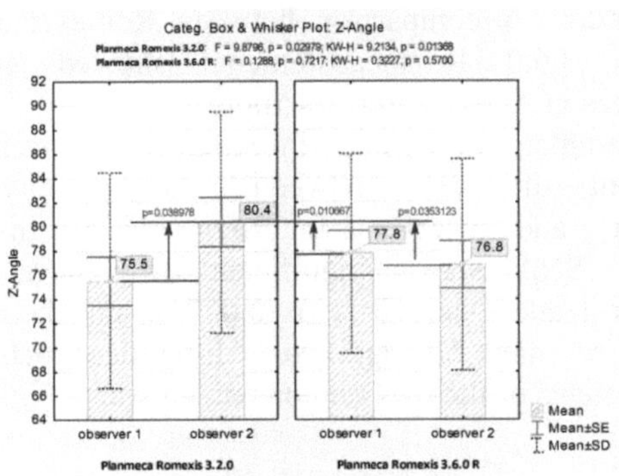

Fig. 5 Statistical analysis perfomed for Z-Angle parameter; graphics show the interobservers and intersoftware reproducibility.

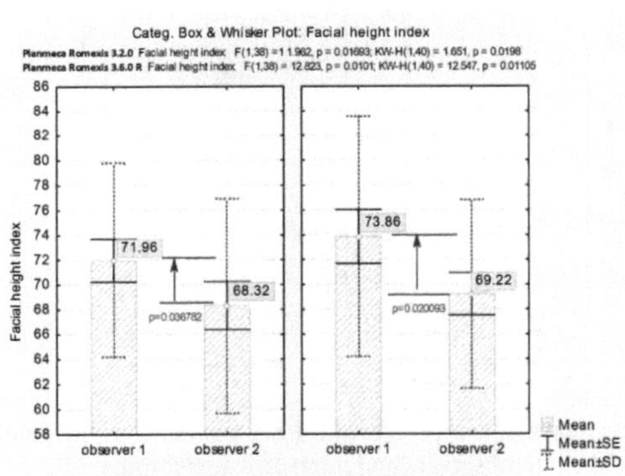

Fig. 6 Statistical analysis perfomed for **Facial height index** parameter; graphics show the interobservers and intersoftware reproducibility.

For parameter Facial height index, from Tweed analysis, were significant differences between mean values of Observer 1 and Observer 2 for Romexis 3.2.0 and Romexis 3.6.0 R; No significant differences were observed between the measurements of the two versions of software (Fig. 6).

On the other hand, for for the second analysis used in our study, Steiner analysis, a comparison between observers, for Romexis 3.6.0 R, showed significant differences for Interincisal angle, Max1-NA and Max1-SN measurements. A comparison between Romexis 3.2.0 and Romexis 3.6.0 R showed significant differences for Observer 1 for Interincisal angle, Max1-NA and Max1-SN measurements.

For the parameter Interincisal angle, from Steiner analysis, there were significant differences between Romexis 3.2.0 and Romexis 3.6.0 R for Observer 1. A comparison between Observer 1 and Observer 2 for Romexis 3.6.0 R shows significant differences for the Interincisal angle (Fig. 7).

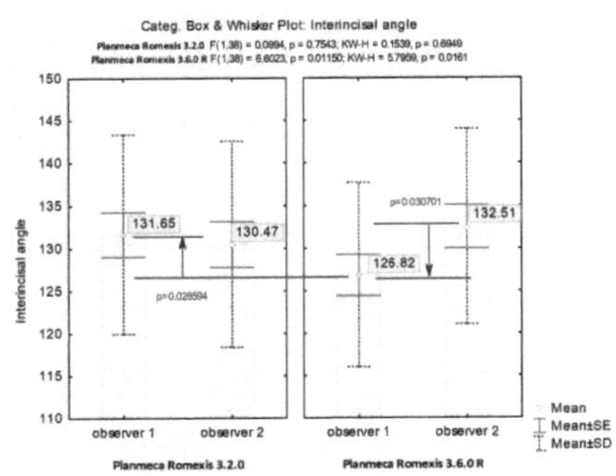

Fig. 7 Statistical analysis perfomed for **Interincisal angle** parameter; graphics show the interobservers and intersoftware reproducibility.

Parameter Max1-NA, from the Steiner analysis, was found with significant differences between observers for Romexis 3.2.0 and also for Romexis 3.6.0 R; For Observer 1 the statistical analysis found significant differences between the softwares (Fig. 8).

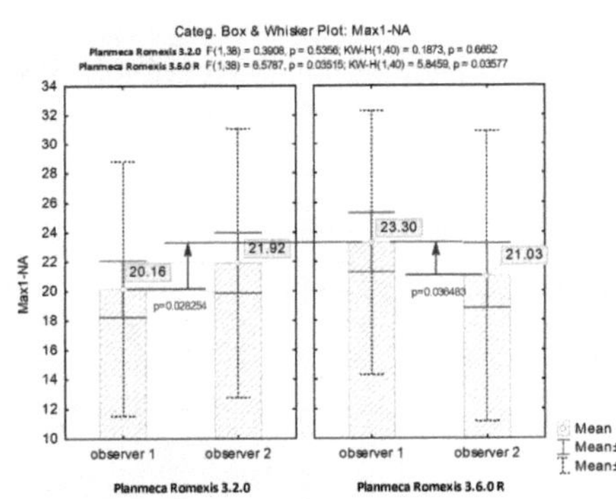

Fig. 8 Statistical analysis perfomed for **Max1-NA** parameter; graphics show the interobservers and intersoftware reproducibility.

Parameter Max1-SN, from the Steiner analysis, was found with significant differences between observers for Romexis 3.6.0 R; A comparison between softwares found significant differences for Observer 1 (Figure 9).

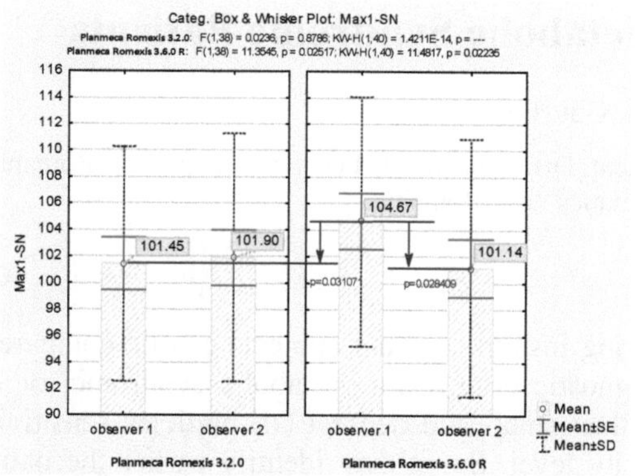

Fig. 9 Statistical analysis perfomed for **Max1-SN** parameter; graphics show the interobservers and intersoftware reproducibility.

However, for Romexis 3.6.0 R, Tweed analyses, there were significant differences in SNA and Facial height index. SNA values between the Observer 1 (81.83 ± 2.12) and Observer 2 (79.97 ± 1.79) measurements with p-value Scheffe test=0.034591; Facial height index between Observer 1 (73.86 ± 9.71) and Observer 2 (69.22 ± 7.61) measurements with p-value Scheffe test = 0.020093

For Steiner analysis there were significant differences in Interincisal angle, Max1-NA and Max1-SN. Interincisal angle values between the Observer 1 (126.82 ± 10.83) and Observer 2 (132.51 ± 11.46) measurements with p-value Scheffe test=0.030701; values between the Observer 1 (23.3 ± 9.0) and Observer 2 (21.0 ± 9.8) measurements with p-value Scheffe test=0.036483; values between the Observer 1 (104.67 ± 9.41) and Observer 2 (101.14 ± 9.75) measurements with p-value Scheffe test=0.028409;

For Romexis 3.2.0, Tweed analyses, there were significant differences in AO-BO, Z-Angle and Facial height index. AO-BO values between the Observer 1 (0.83 ± 4.67) and Observer 2 (0.26 ± 4.94) measurements with p-value Scheffe test=0.048581; Z-Angle values between Observer 1 (75.53 ± 8.93) and Observer 2 (80.38 ± 9.13) measurements with p-value Scheffe test=0.038978; Facial height index values between Observer 1 (71.96 ± 7.80) and Observer 2 (68.32 ± 8.61) measurements with p-value Scheffe test=0.036782. For Steiner analysis there were found no statistically significant differences between observers.

In this reserche, the comparison of Romexis 3.2.0 and 3.6.0 R imaging program for cephalometric analyses demonstrated that only the variales SNA, Interincisal angle, Max1-NA, Max1-SN, AO-BO, Z-Angle and Facial height index presented significant differences; one of them presented a small differences between versions but was considerated statistical significant. The other variables did not present any siginificant differences between the two versions of Romexis software. Owing to these results it can be assumed that both version of Romexis imaging program could be fairly relible methods, noting that Romexis 3.6.0 R holds more functions that may be helpful than 3.2.0 version.

A possible explication for these significant differences between these two versions of imaging program can be caused by the clarity of the radiographs, which was better for Romexis 3.6.0 R after apply the filter function; another possible explanation for the existence of these errors might be caused by the difficulty identification of points A, N, Pg' and occlusal plane. Similar results were evidenced also by other researchers [5, 6].

IV. CONCLUSIONS

In our computerized digital cephalometric analysis, the differences of measurements between observers and the two versions of Romexis imaging program were authentic. The reliability of variables was comparable except for the SNA, Interincisal angle, Max1-NA, Max1-SN, AO-BO, Z-Angle and Facial height index measurements. When compare the results found for Romexis 3.2.0 and the other one found for Romexis 3.6.0 R, our reserche sustain, based on the statistical results, that Romexis version 3.2.0 is more accurate and reproductible.

CONFLICT OF INTEREST

The authors declare that they have no conflict of interest.

REFERENCES

1. Baumrind S, Frantz RC. The reliability of head film measurements. 1. Landmark identification. Am J Orthod 1971; 60: 111–127;
2. Adams GL, Gansky SA, Miller A, Harrell WE, Hatcher DC. Comparison between traditional 2-dimensional cephalometry and a 3-dimensional approach on human dry skulls. Am J Orthod Dentofacial Orthop 2004; 126: 397–409;
3. Trpkova B, Major P, Prasad N, Nebbe B. Cephalometric landmarks identification and reproducibility: a meta analysis. Am J Orthod Dentofacial Orthop 1997; 112(2):165-70;
4. Fidanoski. Digital radiography. Retrieved 1st February 2007 from the World Wide Web;
5. Graber T, Vanarsdall LR, Vig Katherine. Orthodontics: Current Principles and Techniques. Elsevier Mosby, 2005; Ricketts RM. Orthodontic diagnosis and planning. vol.1, Philadelphia: Saunderes 1982; 107-125;
6. Buschang PH, Stroud J, Alexander RG. Differences in dental arch morphology among adult females with untreated Class I and Class II malocclusion. Eur J Orthod 1994; 16: 47-52;
7. Versteeg, Sanderink, Stelt. Efficacy of digital intra-oral radiography in clinical dentistry. J Dent 1997; 25, 215-224;

Author:	Calin Corciova
Institute:	University of Medicine and Pharmacy "Grigore T. Popa"
Street:	Universitatii, No 16
City:	Iasi
Country:	Romania
Email:	calin.corciova@bioinginerie.ro

Using CHAID Algorithm in Low-Risk Metabolic Syndrome Patients

M.G. Felea[1], V. Felea[2], and C.M. Gavrilescu[1]

[1]'Gr.T. Popa' University of Medicine and Pharmacy, Ist Medical Department, Discipline of Medical Semiology, Iasi, Romania
[2]'Al. I. Cuza' University, Faculty of Informatics, Iasi, Romania

Abstract— **Metabolic Syndrome (MetS), a cluster of more than three cardio-metabolic risk factors, is becoming a major worldwide health problem. CHAID analysis can bring to front the role of pathogenesis – represented by insulin resistance and measured by Reaven index, and the importance of two-cumulated criteria – as the association of hypertension and obesity, in revealing the likelihood of developing MetS in patients that do not yet account for all condition criteria. The aim of our research was to stress the order of influence of a specific two-cumulated criterion comparative to other MetS criteria comprised into a pathogenic index – the Reaven index (comprising TG and HDL that are two independent MetS diagnosis criterion). CHAID Decision Tree Algorithm can become an intelligent system to support wise decision-making and to predict the likelihood of developing MetS in patients at low-risk for this condition.**

Keywords— **CHAID Decision Tree Algorithm, metabolic syndrome, insulin resistance, elderly, two-cumulated criterion.**

I. INTRODUCTION

Metabolic Syndrome (MetS) is a cluster of more than three cardio-metabolic risk factors [1], becoming a major worldwide health problem considering its clinical and economical burden [2].

Although insulin resistance is not a MetS diagnostic criterion, it plays a major pathogenic role in the development of metabolic syndrome. Metabolic syndrome may include one or all of the following conditions: type 2 diabetes or impaired glucose tolerance, central obesity [3], hypertension, dyslipidemia (include increased levels of triglycerides), reduced levels of HDL-cholesterol and high level of small dense LDL particles.

Insulin resistance indicators include waist circumference, abdominal circumference/hip ratio and triglyceride levels. Adiposity and insulin resistance are related, but not synonymous although obesity is the most frequent cause of insulin resistance; each one is independently contributing to an increased cardiovascular risk [4]. Among the acquired causes of insulin resistance there are: sedentary lifestyle, diet, medication, hyperglycemia (the sugar toxicity), increased free fatty acids and aging [5].

Reaven index (the ratio triglycerides/HDL-Cholesterol) is a reliable indicator for insulin resistance in case of values above 3.5 [6]. The sensitivity and specificity of this index in identifying insulin resistant subjects can be compared with the diagnostic criteria for metabolic syndrome, or can be higher than small and dense LDL particles sensitivity and specificity level. So, it may identify earlier the patients at high cardiometabolic risk, cumulating dyslipidemia and insulin resistance [7].

II. STUDY GROUP

The initial study group was formed from 399 patients that attended some general physician's offices in Iasi, in the North-East part of Romania. Patients under the age of 40 and/or with a BMI < 18.5 kg/m² were excluded from our study. Thus, data from 259 patients remained to be analyzed, the study group including patients over 40 years old with none or at least one component of the metabolic syndrome.

The mean age in the study group was 68.47, the age ranging from 44 to 95 years old. The majority were living in the urban area.

We collected information on gender, waist circumference (cm), triglycerides (TG) values (mg/dl), high density lipoprotein cholesterol (HDL) values (mg/dl), low-density lipoprotein cholesterol (LDL) values (mg/dl), and hypertension (HTN) degree (Table 1).

Table 1 Laboratory values in the study group

Parameters	Mean	Min	Max
Age (yr)	68.47	44.00	95.00
Triglyceride (mg/dl)	131.26	37.00	770.00
HDL (mg/dl)	49.00	23.20	102.00
LDL (mg/dl)	135.44	39.20	318.20

Table 2 Categorical variables: modal classes

Categorical variable	Modal class	Percentage
Age categories (years)	65-69 yrs	19.30%
Chronological age classification	60-74 yrs (young old)	42.50%
Gender	Female	52.10%
HTN severity	HTN 2nd degree	36.30%
Obesity WC	pathological	73.40%

V. Sontea and I. Tiginyanu (eds.), *3rd International Conference on Nanotechnologies and Biomedical Engineering*,
IFMBE Proceedings 55,
DOI: 10.1007/978-981-287-736-9_110

Other features of the study group are described in Table 2. In the sample, young old patients were predominant, with a slight prevalence of women, and a very high frequency of abdominal obesity.

III. IMPLEMENTATION OF CHAID ALGORITHM

The CHAID algorithm designs a decision tree using the analysis of variance. The tree explains the interactions between the dependent variable and predictors [8].

In young adults (20 and 39 years old), waist circumference was proved to be the most important predictor of MetS occurrence by using CHAID Decision Tree Algorithm [9].

The aim of our research was to stress the order of influence of a specific two-cumulated criterion comparative to other MetS criteria comprised into a pathogenic index – the Reaven index (comprising TG and HDL that are two independent MetS diagnosis criterion) in a predominantly old population.

The dependent variable was defined as 'Metabolic syndrome diagnosis by criteria'. As for the explanatory variables, we considered the following independent variables: gender, a two-cumulated criterion for MetS comprising hypertension and obesity, the Reaven index and LDL predefined ranges.

Hypertension (HTN) was defined as values above 140 for systolic blood pressure and above 90 for diastolic blood pressure. Abdominal obesity was identified by waist circumference (WC) value, with specific gender pathologic cut-offs – more than 88 cm for women and more than 102 cm for men. The pathological waist circumference is linked to visceral obesity and cardiovascular risk [10]. In our study group, the most frequent associations of two-cumulated MetS criterion were in the following decreasing order: HTN-WC (156 cases), HTN-hyperglycemia (145) [11], hyperglycemia-WC (135). The high frequency of HTN-WC in our sample was the reason for choosing a two-cumulated criterion in the CHAID Decision Tree Algorithm.

The Reaven index encompassing two metabolic syndrome criteria, as TG and HDL values, can show by the level of their ratio the presence of one of the underlying mechanism of MetS, respectively the insulin resistance. HDL-cholesterol was set up for normal ranges accordingly to gender.

LDL pre-defined ranges comprised the cut-off limits by intervals, depending on the absence or presence of other cardiovascular diseases or risk factors.

IV. USING CHAID ALGORITHM IN LOW-RISK METABOLIC SYNDROME PATIENTS

We implemented CHAID classification tree algorithm in order to reveal which of the risk factors or two-cumulated criteria for developing MetS are more prone to explain the likelihood to have the metabolic syndrome condition (Fig. 1). The algorithm built 21 nodes, among which 12 formed terminal nodes. Risk factors for developing MetS or two-cumulated criteria of MetS diagnosis were merged and than split into 8 parent nodes by a sequence that highlights their influence ranking. The resulting decision tree had 4 levels in depth. The overall classification accuracy of the designed model was 70.3%, but the ability to detect MetS grew to 85.3%.

We chose "gender" as the user-specified variable.

The CHAID decision tree algorithm displayed a synchronous split by the same influence factor on every level, independent of gender. After splitting by gender, the significant predictor on the second level was the association between hypertension and obesity – named as "HTN+WC". For the third level, "Reaven index" was the next most important predictor for the previous "HTN+WC" positive splitting nodes. So, the tree stopped here for male gender, but continued with another splitting factor – as LDL, for female gender.

Given the gender, the prevalence of MetS was homogenous, slightly higher in men (67.7%) than in women (63.7%).

With respect to the split by the two-cumulated criteria – Hypertension and Obesity measured by waist circumference (WC), nodes 4 and 7 (corresponding to male and female respectively) comprised patients with a low likelihood to develop MetS.

In the presence of one risk factor, either Hypertension or Obesity, the likelihood of MetS reached 50% in men group, and 42.9% in women sample.

The coexistence of Hypertension and Obesity induced a risk of MetS above 80% (86.8% for men, and 83.8% for women).

For parent nodes with both (3, 6) or one of the two-cumulated criteria (5, 9), the next important predictor was represented by the Reaven index. For each parent node mentioned above (3, 6, 5, and 9), every child node with a Reaven index more than 3.5 explained the MetS occurrence in 100% of the cases. So, the coexistence of an abnormal TG on HDL ratio with two-cumulated criteria Hypertension-Obesity is equal to the presence of the metabolic syndrome condition.

The normal Reaven index values below 3.5 with only one criterion of MetS (HTN or Obesity) were associated with a 33.3% and 26.4% (nodes 11 and 15) probability for MetS.

In patients with two-cumulated criteria, the normal values for Reaven index below 3.5 (nodes 9 and 13) were explaining the occurrence of MetS in 81.1% and 78.7% of cases in men and women, respectively.

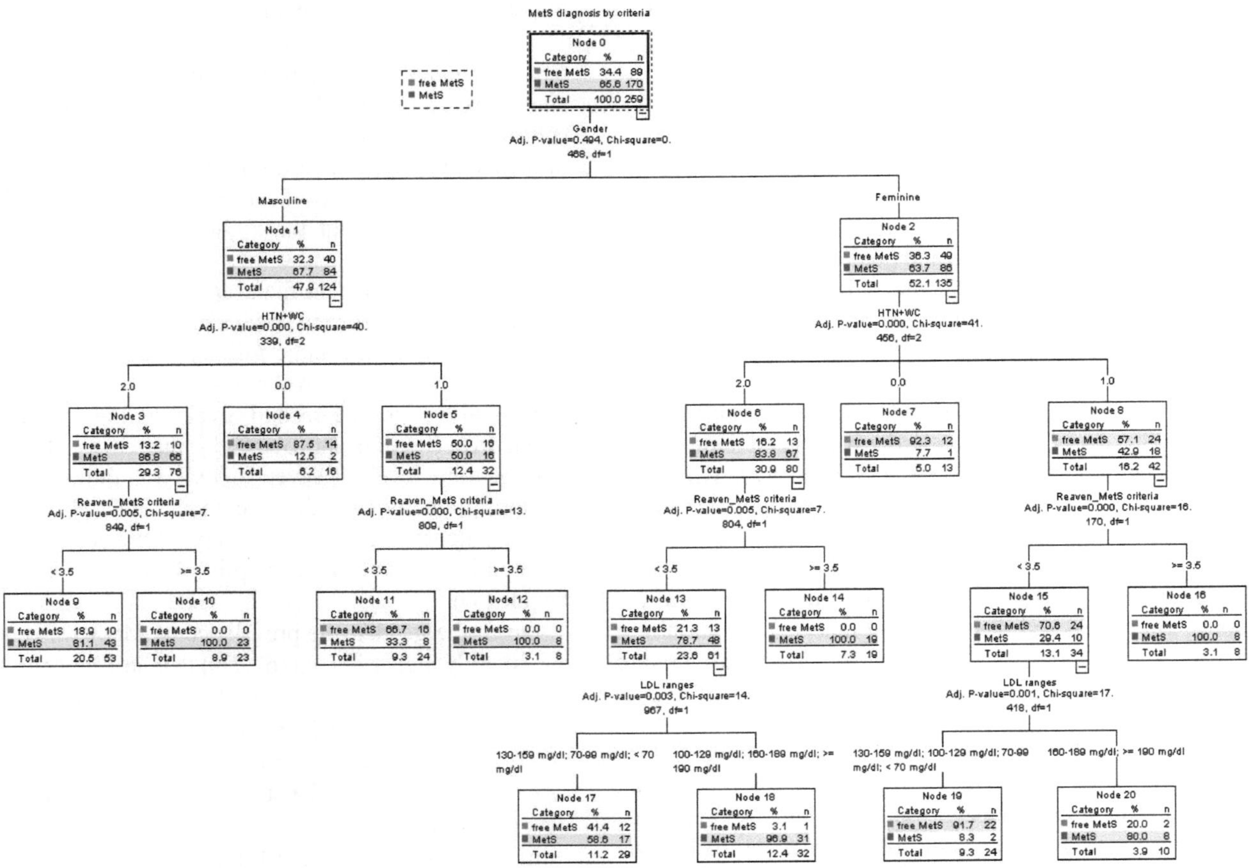

Fig. 1 CHAID decision tree model for stratifying risk factors for the likelihood of developing MetS

V. CONCLUSIONS

The presence of only two criteria for the diagnosis of Metabolic Syndrome can raise the likelihood of developing MetS towards 80%. The two-combined criteria comprise Hypertension and Obesity. Blood pressure measurement is part of every clinical examination, and it is frequently performed when the patient addresses to primary care office or by the patient himself in home settings. Obesity is a clinical feature that can be easily and immediately noticed.

To evaluate obesity by measuring the waist circumference doesn't cost time or money. Reaven index as it is does not represent a MetS diagnosis criterion, but it encompasses two other MetS diagnosis criteria, TG and HDL. As in many patients, the TG and HDL values might be very slightly above, and respectively below the normal ranges, physicians do not consider further investigations. But in these conditions, the ratio between TG and HDL could exceed the normal limit of 3.5, thus showing the presence of the insulin resistance – an important underlying mechanism for hypertension, as well as for diabetes and metabolic syndrome.

In our sample, the specific clinical feature was the high prevalence of hypertension together with obesity (HTN-WC), thus supporting the choice of a two-cumulated criterion in the CHAID Decision Tree Algorithm.

CHAID analysis can bring to front the role of pathogenesis – represented by insulin resistance and measured by Reaven index, and the importance of two-cumulated criteria – as the association of hypertension and obesity, in revealing the likelihood of developing MetS in patients that do not yet account for all criteria of MetS.

CHAID Decision Tree Algorithm can become an intelligent system to support wise decision-making and to predict the likelihood of developing MetS in adults as in old patients at low-risk for this condition.

VI. COMPLIANCE WITH ETHICAL REQUIREMENTS

A. Conflict of Interest

There was no relationship, financial or otherwise between the authors or the institutions of the authors with individuals or organizations that could influence the author's work inappropriately.

B. Statement of Informed Consent

Patients have a right to privacy that should not be infringed without informed consent. Identifying information, including patients' names, initials, or hospital numbers have not been published in written descriptions. The patients have given written informed consent for any medical data inserted in the patient file during hospitalization and their use in the research papers.

Identifying details were omitted because they are not essential. Anonymity of each patient was protected.

C. Statement of Human and Animal Rights

The procedures followed were in accordance with the ethical standards of the responsible committee on human experimentation (institutional and national) and with the Helsinki Declaration of 1975, as revised in 2000 and 2008.

CONFLICT OF INTEREST

The authors declare that they have no conflict of interest.

REFERENCES

1. Reaven GM (1988) Banting lecture 1988. Role of insulin resistance in human disease. Diabetes 37:1595–1607
2. O'Neill S, O'Driscoll L (2015) Metabolic syndrome: a closer look at the growing epidemic and its associated pathologies. Obes Rev 16: 1–12
3. Reaven G, Abbazi F, McLaughlin T (2004) Obesity, insulin resistance, and cardiovascular disease. Recent Prog Horm Res 59:207–223
4. Lee SH, Park SA, Ko SH et al (2010) Insulin resistance and inflammation may have an additional role in the link between cystatin C and cardiovascular disease in type 2 diabetes mellitus patients. Metabolism 59(2):241–246
5. Lutsey PL, Steffen LM, Stevens J (2008) Dietary intake and the development of the metabolic syndrome: the Atherosclerosis Risk in Communities study. Circulation 117(6):754–761
6. Reaven GM (1995) Pathophysiology of insulin resistance in human disease. Physiol Rev 75(3):473–486
7. McLaughlin T, Reaven G, Abbasi F et al (2005) Is There a Simple Way to Identify Insulin-Resistant Individuals at Increased Risk of Cardiovascular Disease? Am J Cardiol 96(3):399–404
8. Kass GV (1980) An Exploratory Technique for Investigating Large Quantities of Categorical Data. Applied Statistics 20(2):119–127
9. Miller B, Fridline M, Liu PY et al (2014) Use of CHAID Decision Trees to Formulate Pathways for the Early Detection of Metabolic Syndrome in Young Adults. Comput Math Methods Med 1–7 DOI 10.1155/2014/242717.
10. Despres JP (2012) Obesity: Body Fat Distribution and Risk of Cardiovascular Disease: An Update. Circulation 126:1301–1313 DOI 10.1161/CIRCULATIONAHA.111.067264
11. Reaven PD, Barrett-Connor EL, Browner DK (1990) Abnormal glucose tolerance and hypertension. Diabetes Care 13:119.

Author: Victor Felea
Institute: 'Al. I. Cuza' University, Faculty of Informatics
Street: 6, Aleea Copou Street
City: Iasi
Country: Romania
Email: felea@info.uaic.ro

Study of Interoceptive Signals Perception in Patients with Panic Disorder and Eminent Respiratory Symptoms

A. Ganenco, T. Besleaga, S. Lozovanu, and V. Vovc

State University of Medicine and Pharmacy "Nicolae.Testemiţianu", Chisinau, Republic of Moldova

Abstract— **Strong sensations of suffocation, chest pain, tingling, and fear of death are characteristic in a group of patients with panic disorder (PD) and they occur during attacks. This group of patients is designated as the respiratory subtype. In our study, we evaluated symptoms of autonomic disorders that occur during panic attacks and outside in a group of 10 patients with PD respiratory subtype and 10 patients with PD that did not manifest symptoms mentioned above. The results allow us to state that people with PD respiratory typeshow more symptoms of autonomic disorders outside attacks. Delimitation of respiratory subtype of PD can help us to apply special tactics of medical and non-drug treatment of these patients.**

Keywords— **respiratory subtype of panic disorder, symptoms of autonomic disorders**

I. INTRODUCTION

The basic theory that explains the development of panic disorder is the increased sensitivity of suffocation alarm system [1]. The anxiety of patients with panic disorder is focused on their own symptoms occurring during attacks, the patient is afraid of making a new panic attack. Incorrect interpretation of the body sensations in patients with panic disorder [2] seems to completethe theory ofsuffocation alarm system activation, proposed by Klein.

Some researchers noted the pathogenetic and clinical heterogeneity of panic disorder taking as a basis the prevailing symptoms [3]. Patients with panic disorder with eminent respiratory symptoms areespecially relevant; itis evenproposed therespiratory subtype [4, 5]. Although the etiologic role of hyperventilation in the development of panic disorder is denied by most researchers, however, hyperventilation and hypocapnia appear to play an important role in the pathogenesis of panic disorder [6]. Dyspnea is a common symptom of panic disorder [7]. The presence of eminent respiratory symptoms, chest pain or discomfort, numbness or tingling and fear of dying during a panic attack - these are criteria that distinguish the respiratory subtype [8]. Furthermore, high sensitivity to respiratory challenge test can be added - inhalation of mixtures with increased content of CO2 and voluntary hypocapnic hyperventilation, high familial comorbidity, prevalenceof agoraphobia [9.10].

Activation of anxiogenic structures can be a rich source of symptoms of autonomic disorders that occur outside the panic attacks. Our hypothesis is that activation of the central suffocation alarm system and emphasized interpretationof suffocation sensations, with hypocapnia on the background, is the cause of autonomic dysfunction that create vicious loops of maintaining the anxiety and genesis of symptoms even outside ofattacks.

Generationof symptoms of autonomic disorders without any organic changes in patients with elevated levels of anxiety has an impact on all life aspects - personal, social and professional. Autonomic motor profile (AMP) is a subjective clinical test for assessment of autonomic disturbances expressed through bodily sensations [11]. Hyperactivation of anxiogenic structures and centers of autonomic control appears to be the pathogenetic nucleus in such nosology as panic disorder.

The objectives of the study.The assessment of autonomic instability and its perception in subjects with panic disorder with eminent respiratory symptoms (respiratory group of panic disorder) and in subjects without eminent respiratory symptoms.The assessment of other features –level of anxiety, depression, stress factors in patients with panic disorder.

The aim of the study.To highlight differences in scores of symptomsof autonomic functional disorders in daily life in patients with panic disorder with and without eminent respiratory symptoms.

II. SUBJECTS AND METHODS

The research was conducted on a sample of 20 patients with panic disorder (9 male) aged 20 to 35 years,patients of Republican Institute of Neurology in Moldova. Selection of patients was carried out according to certain criteria.

The criteria for the formation of the groups were:

1. The diagnosis of panic disorder and lack of related respiratory or cardiovascular disease.
2. The presence of eminent sensations of suffocation during panic attacks, chest pain or discomfort, numbness or tingling, and fear of dying during a panic attack. 10 patients that met these requirements were included in PDr (respiratory) group.

V. Sontea and I. Tiginyanu (eds.), *3rd International Conference on Nanotechnologies and Biomedical Engineering*,
IFMBE Proceedings 55,
DOI: 10.1007/978-981-287-736-9_111

3. Lack of eminent sensations of suffocation during panic attacks and other symptoms listed above. 10 patients that met these requirements were included in PD group.

The psychophysiological assessment included evaluation of personal and situational anxiety scores (Spielberger State and Trait Anxiety Inventory) and functional symptoms of autonomic disorders (AMP Moldovanu). Also the level of depression (Beck questionnaire) and stressful events during the last year (Holmes questionnaire) were assessed.

The tests (questionnaires) used in the study:

Autonomic motor profile includes items – affirmations and questions expressing individual peculiarities of perception and appreciation of autonomic disorders symptoms, bodily sensations, inclusively the sensations during panic attacks. Items are grouped into scales. For each item the examinee must a corresponding answer: almost never (0), rarely (1), sometimes (2), often (3), almost permanent (4). We used 11 scales of autonomic motor profile.

Scales of autonomic motor profile,the number of items is indicated in brackets:

1. Anxiety and panic attacks (7)
2. Vertigo and syncope states (9)
3. Fatigue and hypersensitivity (9)
4. Dyspnea (12)
5. Respiratory behavior (8)
6. Sensomotor disorders (20)
7. Tetany–neuromuscular hyperexcitability(14)
8. Cardiovascular dysfunction (15)
9. Gastrointestinal dysfunction(17)
10. Disorders of thermoregulation and sweating (18)
11. Pain disorders (11)

Sensations scores were estimated for each group of items.

Spielberger test is based on 40 questions for self-evaluation. It assesses two types of anxiety: a) momentary anxiety level (STATE) includes 20 items for each item; examinee chooses the answer - all (1), slightly (2), pretty much (3), very much (4);

b) trait anxiety level or level of anxiety as a characteristic personal parameter (TRAIT) – 20 items. Examinee chooses the answer for each item -never (1), sometimes (2), often (3), almost always (4). For each scale, the difference between the amount of items that confirms increased anxiety and amount of items that confirms the absence of anxiety plus 35 is calculated.The score lower than 31 indicates a low level of anxiety, the 31-45 is average anxiety, more than 45 - elevated anxiety.

Holmes questionnaire includes items that allow assessment of stress factors that subject has suffered in the last year.

Beck questionnaire includes items to assess level of depression in the last week.

Subjects completed all questionnaires at once, without the break, in a given order. There was no time limit for subject

Statistical analysis of data obtained in the present study was performed using parametric and nonparametric tests (impaired t-testand Mann-Whitney test).

III. RESULTS

The comparisons between groups showed no significant differences for the values of state and trait anxiety,for the levels of depression and stress.

All results are included in Table 1.

Table 1 Anxiety, stress, depression and APM scores PDr and PD groups.

	PDr	PD
Trait anxiety	54,2 ± 7,5	51,7 ± 7,0
State anxiety	41,2 ± 9,8	38,5 ± 8,9
Holmes' score	278 ± 118	200 ± 81
Beck's score	10.4 ± 6,9	7,9 ± 4,4
APM		
Scale	PDr	PD
Anxiety and panic attacks	26,0 ± 6,7	21,5 ± 6,5
Vertigo and syncope states	13,5 ± 6,4	8,4 ± 3,2
Fatigue and hypersensitivity	12,7 ± 3,5	9,6 ± 3,9
Dyspnea	24,8 ± 3,8	13,8 ± 5,9
Respiratory behavior	19.9 ± 5,7	11.6 ± 5,6
Sensomotor disorders	19,8 ± 5,3	11,9 ± 4,7
Tetany–neuromuscular hyperexcitability	15,9 ± 5,3	8,4 ± 4,1
Cardiovascular dysfunction	11,6 ± 4,1	9,7 ± 5,4
Gastrointestinal dysfunction	17,3±7,9	11,4 ± 8,6
Disorders of thermoregulation and sweating	18,9 ± 6,5	15,5 ± 7,5
Pain disorders	14,0 ± 5,5	11,3 ± 6,8

The assessment of APM scores showed significantly increased scores for scales dyspnea, respiratory behavior, vertigo and syncope states, tetany, senso-motor disorders in PDr group (Fig. 1).

IV. DISCUSSION

Higher scores of symptoms that occurduring panic attacks in patients in PDr group are consistent with data obtained by some researchers about higher severity of panic attacks inpatients with PD respiratory subtype [12].

Strong sensations of suffocation that develop in some patients can be a powerful source of anxiety perpetuation and genesis of new symptoms that occur outside the attack.

Some authors prove the existence of hereditary peculiarity of respiratory control in patients with panic disorder that causes instabilityof respiratory volumes with frequent inspirations (sighs) [13]. Hereditary factors may explain the high family incidence in the respiratory subgroup of patients with panic disorder [4].

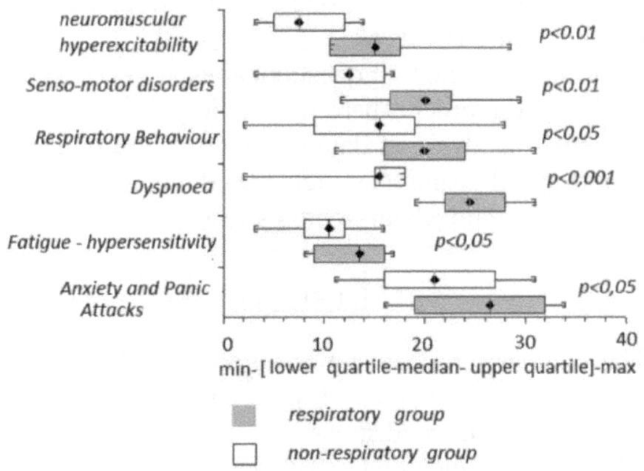

Fig. 1 APM score for dyspnea, respiratory behavior, vertigo and syncope states, tetany and senso-motor disorders.

Focus on the unpleasant respiratory sensations of some patients, with erroneous interpretation, can also be ananxiogenicsource generating symptoms outside the attacks and disorder of ventilation mode. Any hyperventilation may also be a link in the perpetuation of anxiety and genesis of new symptoms. Higher scores in some scalesin our study support this assumption.

Drug treatment reduces some symptoms such as anxiety, depression and could affect our results regarding the estimation of anxiety, depression and symptoms of autonomic disorders. Another factor that could affect our results islow number of patients included in our study.

Treatment with selective serotonin reuptake inhibitors (SSRI) was found to be more efficient in PDr group compared toPD group [8]. Using special drug treatment strategies in respiratory subgroup seems to be an important direction in clinical studies. Another possibility is using non-pharmacological methods of respiratory reeducation with biofeedback techniques for correction of ventilation mode in these patients.

V. CONCLUSION

1. Patients with eminent respiratory symptoms present more symptoms that prove suprasegmentary autonomic dysfunction in daily life outside panic attacks.

2. Autonomic motor profile proves to be a useful subjective test for estimation of symptoms of autonomic disorders.

3. Tactics of medical treatment and respiratory reeducation could be beneficial in these patients.

CONFLICT OF INTEREST

The authors declare that they have no conflict of interest.

REFERENCES

1. Klein DF. False suffocation alarms, spontaneous panics, and related conditions. An integrative hypothesis. Arch Gen Psychiatry. 1993 Apr; 50(4):306-17. Review
2. Ley R. Dyspneic-fear and catastrophic cognitions in hyperventilatory panic attacks.Behav Res Ther. 1989;27(5):549-54. Review
3. Roberson-Nay R, Kendler KS. Panic disorder and its subtypes: a comprehensive analysis of panic symptom heterogeneity using epidemiological and treatment seeking samples.Psychol Med. 2011 Nov;41(11):2411-21
4. Freire RC, Perna G, Nardi AE. Panic disorder respiratory subtype: psychopathology, laboratory challenge tests, and response to treatment.Harv Rev Psychiatry. 2010 Jul-Aug; 18(4):220-9
5. Abrams K, Rassovsky Y, Kushner MG. Evidence for respiratory and nonrespiratory subtypes in panic disorder.Depress Anxiety. 2006; 23(8):474-81
6. Sikter A, Frecska E, Braun IM, Gonda X, Rihmer Z. The role of hyperventilation: hypocapnia in the pathomechanism of panic disorder.Rev Bras Psiquiatr. 2007 Dec;29(4):375-9. Epub 2008 Jan 8. Review
7. American Psychiatric Association (2000). Diagnostic and Statistical Manual of Mental Disorders, Fourth Edition: DSM-IV-TR®. American Psychiatric Pub. ISBN 978-0-89042-025-6
8. Song HM, Kim JH, Heo JY, Yu BH. Clinical characteristics of the respiratory subtype in panic disorder patients.PsychiatryInvestig. 2014 Oct; 11(4):412-8. doi: 10.4306/pi.2014.11.4.412. Epub 2014 Oct 20
9. Nardi AE, ValençaAM, Nascimento I, Lopes FL, Mezzasalma MA, Freire RC, et al. A three-year follow-up study of patients with the respiratory subtype of panic disorder after treatment with clonazepam. Psychiatry Res 2005;137:61-70
10. Freire RC, Nardi AE. Panic disorder and the respiratory system: clinical subtype and challenge tests.Rev Bras Psiquiatr. 2012 Jun;34Suppl 1:S32-41. Review
11. Moldovanu I. Profilulvagetativ motor (ghidpractic). Chişinău 2010
12. Onur E, Alkin T, Tural U. Panic disorder subtypes: further clinical differ¬ences. Depress Anxiety 2007;24:479-486
13. Abelson JL, Weg JG, Nesse RM, Curtis GC.Persistent respiratory irregularity in patients with panic disorder.Biol Psychiatry. 2001 Apr 1;49 (7):588-95

Author: Svetlana Lozovanu
Institute: State University of Medicine and Pharmacy „Nicolae.Testemiţianu",
Street: 165, Stefan cel Mare si Sfant bd.
City: Chisinau
Country: Republic of Moldova
Email: slozovanu@yahoo.com

Collective Behavior of Water Molecules in Microtubules

A. Nistreanu

Institute of Applied Physics, Academy of Sciences of Moldova, Academiei Street 5, Chisinau MD 2028, Republic of Moldova

Abstract— **A theoretical model for the description of a collective behavior of water molecules as an assembly of two-level quantum biological system is proposed. In this model, MicroTubules are considered as quantum cavities. Their role is to provide a single mode of biophoton field, in such a way that water molecules to be considered not as independent individuals, but rather as whole, in this manner water molecules are embedded in and interacting with a common radiation field. In the model proposed, collective behavior of water molecules is characterized by coherent water states analogous to Bloch states, whose main feature is to trap biophotons in a collective fashion. Finally some applications to electroencephalography are considered.**

Keywords— **MTs, coherent H_2O states, biophoton, collective inversion function.**

I. INTRODUCTION

We mention that in the last decade many theories and papers have been published concerning the biophysical properties of MTs including the hypothesis of MTs implication in coherent quantum states in the brain evolving in some form of energy and information transfer e.g. Hameroff[1], Mavromatos[2]. In this context applying quantum mechanics gives us the possibility to apply some well known models such as cooperative quantum phenomena [3,4], to water (H_2O) molecules inside MicroTubules(MTs) which show rich, ordered and systematic dynamics. Thus we mention that Dicke[5] was one of the first authors to realize the importance of treating the assembly of radiating molecules as a single quantum system on the grounds that the molecules, which are interacting with a common radiation field, should not be treated as independent individuals, but rather as a totality or assembly.

Next we need to take into account that the formation of "electromagnetic cavities" in living matter appears as an essential task for elucidating the biomolecular interaction dynamics [6,7]. The appearance of a coherent field occurs in the presence of an external supply of energy, a pump in the physical jargon. Moreover an externally supplied cavity, whose size allows us to select a specific wavelength of the coherent field, is needed. Coherence appears when suitable boundary conditions are met and only when the system is pumped. For living systems a good candidate is MTs whose surface consist of dimers arranged in spiral shape which ensure pumping, and at the same time the MTs can be considered as cavity as well.

II. QUNTUM MECHANICAL TREATMENT OF BIOLOGICAL SYSTEM

A. Biophoton Emission from Free Radical Reactions

Biophotons are weak emissions of light radiated from the cells of all living things. During natural metabolic processes, in all types of living systems, lasting spontaneous photon emission has been detected without any external excitation [8,9,10]. Bioluminescent photon emission ranges from a few up to hundreds of photons per second per cm^2 within the spectral range of radiation from ultraviolet to near infrared. This ultraweak bioluminescent photon emission is referred to using various terms such as ultraweak photon, dark luminescence, low intensity chemiluminescence, ultraweak electromagnetic light, spontaneous autoluminescence, ultraweak bioluminescence and biophotons. Biophotons originate from bioluminescent reactions of reactive oxygen species (ROS) and reactive nitrogen species (RNS) and their derivatives, and also from simple cessation of electronically excited states. As examples we may list the mitochondrial respiration chain, lipid peroxidation, peroxisomal reactions, oxidation of catecholamines, oxidation of tyrosine and tryptophan residues in proteins etc. [11,12].

B. MTs as Quantum Cavities in Biological Systems

Microtubules (MTs) are cylindrical protein polymers and are key constituents of all cells of the eukaryotic cytoskeleton. These are formed by reversible polymerization of the tubulin protein see Fig.1.

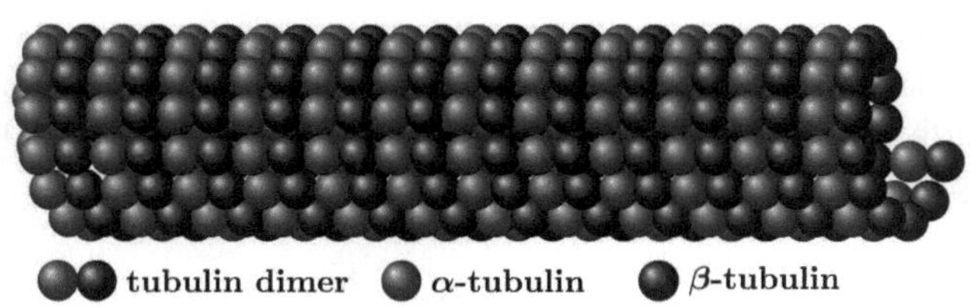

Fig. 1 MTs are hollow cylinders composed of protein units called tubulin. Up to 30,000 - 40,000 dimers can assemble in to MTs [13,14].

Tubulin protein has two parts, α and β, both appear similar, connected face-to-face Nogales[13],(see Fig.1 and 2). They assemble in hexagonal close packing in to a 2D sheet which folds in to a hollow cylinder wrapped around a water channel (see Fig. 2 and 3).

V. Sontea and I. Tiginyanu (eds.), *3rd International Conference on Nanotechnologies and Biomedical Engineering*,
IFMBE Proceedings 55,
DOI: 10.1007/978-981-287-736-9_112

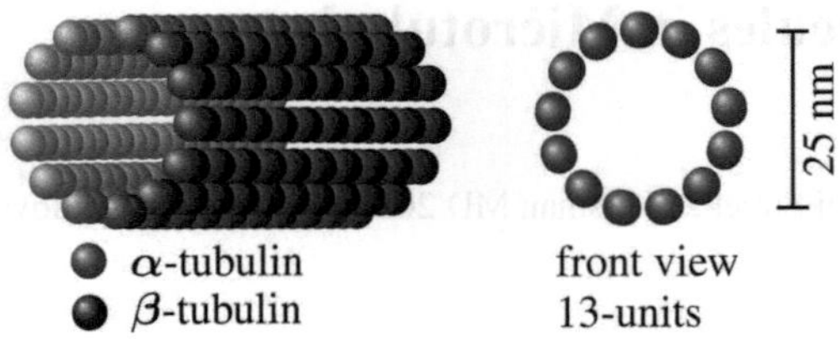

Fig. 2 Visualization of MTs from different angles

Taking into account that a tubulin dimer is an intrinsically fluorescent molecule, mainly due to eight tryptophan residues it contains, as can be seen, e.g., in the α, β-tubulin dimer [15], we may assume that the main source of biophotons is coming by means of oxidation of tryptophan residues in proteins, it follows that the biophotons created in MT cavity allow creation of field cavities modes. Thus we have a good background for our model to state that MTs are indeed cylindrical quantum cavities, which provide a formation of coherent electromagnetic fields.

C. Collective Dynamical Variables for H_2O Molecules

Jibu et al. suggested that the dynamics of ordered H_2O molecules and the quantizated electromagnetic field confined inside the hollow MTs core can give rise to the collective quantum optical modes responsible for the phenomenon of superradiance. Here we reconsider their model in terms of coherent H_2O states.

We initially represent the H_2O molecules as a collection of N identical, but distinguishable, two-level systems each with level splitting ω_0. The spin-1/2 operators describe the ith H_2O molecule. $\{s_k^{(i)}, k = z, \pm i\}$ obeying the commutation rules $[s_z, s_\pm] = \pm s_\pm$; $[s_+, s_-] = 2s_z$. These two-level atoms interact with one biophoton mode, which has a frequencies ω, and a coupling strengths λ, and are described by the bosonic creation and annihilation operators $a^\dagger$ and a.

We now specialize the Hamiltonian to consider a single-mode bosonic field. The analysis of this Hamiltonian is further simplified by the introduction of collective H_2O operator

$$J_z = \sum_{i=1}^{N} s_z^{(i)}; \quad J_\pm \equiv \sum_{i=1}^{N} s_\pm^{(i)} \tag{1}$$

D. Holstein-Primakoff Transformation and the Coherent States of MTs

Next we consider the Dicke Hamiltonian(DH) in the thermodynamic limit, in which the number of H_2O molecules becomes infinite, $N \to \infty$. In this analysis, we shall make an extensive use of the Holstein-Primakoff representation Holstein[16] of the angular momentum operators, which represents the operators in terms of a single bosonic mode in the following way

$$\bar{J}_\pm = (2j)^{-1/2} J_\pm \tag{2}$$

where j=N/2, is the total spin of the H_2O molecules. In this limit the Hamiltonian in Rotating Wave Approximation (RWA) may be written as

$$H_{RWA} = \omega_0 \bar{J}_z + \omega a^\dagger a + \hbar\lambda\sqrt{N}(a^\dagger \bar{J}_- + a\bar{J}_+) \tag{3}$$

where fluctuation operator or $\bar{J}_z = |z\rangle\langle z| - |0\rangle\langle 0|$ causes transitions between excited and ground states of collective water molecules, $a^\dagger, a$ denote the creation and annihilation boson operators associated to the field of biophotons and $\bar{J}_\pm$ respectively denote the creation and annihilation boson operators.

By analogy to the coherent field states we introduce the coherent H_2O states, where these states are obtained from Bloch states, Arecchi[17]. They obey the following rules Glauber[18]

$$\bar{J}_-|z\rangle = z|z\rangle, \quad \langle z|\bar{J}_+ = z^*\langle z| \tag{4}$$

We may infer from above results, that in the large N limit the collection of two-level H_2O molecules appears as a collective bosonic system (see Fig.3). The original coupling of the individual atoms to the field gets enhanced by the factor $\sqrt{N}$ and appears as the coupling of the collective modes (the system as a whole to the field).

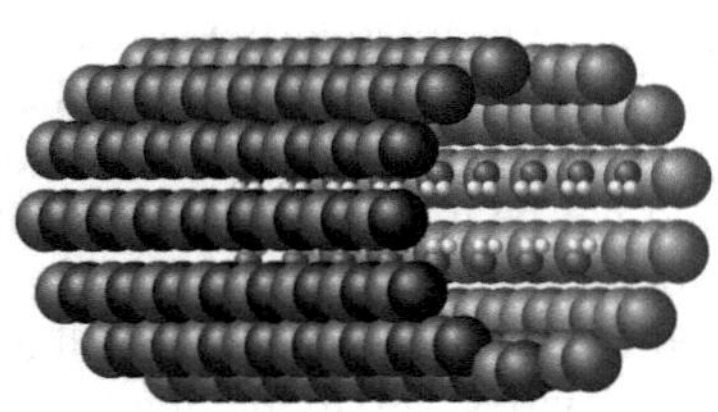

Fig. 3 Collective behavior of water molecules inside MTs. Green particles represents the oxygen ions and white represent hydrogen ions. The ordered water molecules (high polarization) inside MTs manifest themselves as single dipole oscillator.

We conclude that in HP approximation at high polarization(see Fig.3), a collection of two level H_2O molecules and a collection of oscillators behave identically, moreover collective behavior of water molecules inside MTs can be thought as a single dipole oscillator.

III. Quantum Mechanical Formalism of Intercation of Biophotons with Collective H_2O Molecules

A. Collective Inversion Function

Let us consider an ensemble of H_2O molecules interacting with a single mode quantized field of biophotons, the total Hamiltonian is given by Eq.(3). In this case the state of a collective water molecules can be written as a superposition of the ground state and a coherent state as

$$|\psi(0)\rangle_{\mathrm{H_2O}} = c_g|0\rangle + c_e|z\rangle \qquad (5)$$

where $|z\rangle$ is our representation for the coherent sate of collective H_2O molecules. We now investigate its interaction with N biophotons. The state of biophotons is considered as

$$|\psi(0)\rangle_{\text{biophotons}} = \sum_{n=0}^{\infty} c_n|n\rangle \qquad (6)$$

Inserting into the Schrödinger equation we get the solution as the time dependent wave function Gerry[19]

$$|\psi(t)\rangle = \sum_{n=0}^{\infty} \left\{ \begin{array}{l} \{[c_{z,n}\cos(\lambda t\sqrt{n+1}) - ic_{0,n+1}\sin(\lambda t\sqrt{n+1})]|z\rangle \\ +[-ic_{z,n-1}\sin(\lambda t\sqrt{n}) + c_{0,n}\cos(\lambda t\sqrt{n})]|0\rangle\} \end{array} \right\} |n\rangle$$

We consider the case when collective state of water is in the exited state, it means $c_e = 1$ and $c_t = 0$.

To determine when the MT is in the coherent state and when in the ground state, we use the collective inversion function. We define the collective inversion function as

$$F(t,z) = \langle \psi | \bar{J}_z | \psi \rangle \qquad (7)$$

This function determines the rate of transitions between the coherent state and the ground state. Eventually, the collective inversion function for this state is

$$F(t,z) = \sum_{n=0}^{\infty} |c_n|^2 \left(1 - e^{|z|^2}\right) \cos\left[2\lambda t|z|^2 \sqrt{n+1} e^{\frac{|z|^2}{2}}\right] \qquad (8)$$

B. Biophoton Statistics and Simulation

The coefficient describing the biophoton $|c_n|^2$ is the probability of detection of biophotons. We may infer information about coherent light from photocount statistics (PCS) Gerry[19]. One calculates the probability p(n,Δt) of registering n photons in preset time interval Δt.

Finally, the collective inversion function becomes

$$F(t,z) = \sum_{n=0}^{\infty} e^{-\bar{n}} \frac{\bar{n}^n}{n!} \left(1 - e^{|z|^2}\right) \cos\left[2\lambda t|z|^2 \sqrt{n+1} e^{\frac{|z|^2}{2}}\right] \qquad (9)$$

where $\bar{n}$ is the average number of detected photons per time interval Δt.

Next we simulate the collective inversion function, in Fig. 4, we present a three-dimensional plot which shows that in the vicinity of zero has strong fluctuations. The presence of the coherent biophoton field has a constructive effect on the trapping phenomenon. It means that by increasing the coherent H_2O states the exited $|z\rangle$ level become supersaturated and biophothons which come from eyes can pass trough MTs cavities without loses, in this way decreasing the amplitude of fluctuation when eyes are opened and vice versa when eyes are closed.

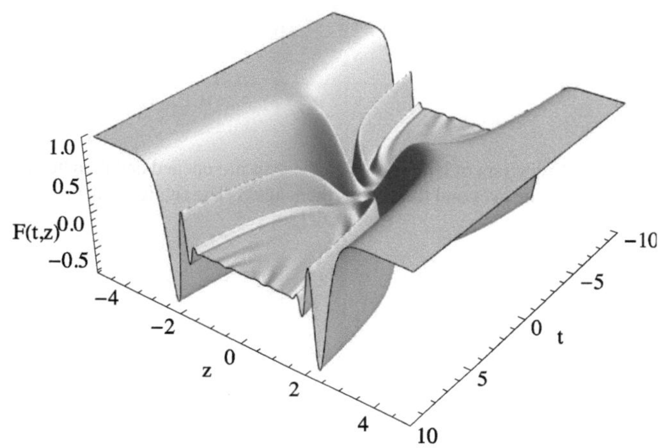

Fig. 4 Collective inversion function shows strong fluctuations when coherent H_2O states approach zero, this situation infer a new phenomenon for biological species i.e. collective coherent population trapping.

A conceptually simple approach to quantum memory is to "store" quantum states of single photons in individual H_2O molecules. Thus our model predicting a collective trapping of biophotons, may play a significant role in formation and storage of quantum memory.

In our support we add the Popp arguments who postulates that destructive and constructive interference between radiating sources of biological systems (biomolecules, organelles, cells, organs, organisms, and populations) may play a central role in biological organization Popp[8]. Within the coherence volume between the emitting systems, the radiation may sustain a coherent interference field for considerable lengths of time. By the mechanism of destructive interference, which leads to the cancellation of force vectors, the system can trap biophotons and thereby store energy. On the other hand, by constructive interference, in the coherence volume between the radiators, stored biophotons are released.

C. Electroencephalography

Electroencephalography(EEG) is one of the most widely used noninvasive techniques for recording electrical brain activity. Electroencephalographs, measure membrane potential variations occurring in neurons. The polarity of the signal changes according to the location of the synapses, being excitatory and inhibitory synapses inversely correlated.

We attempt to describe EEG by a collective inversion function thus for comparison between experiment and theory we present some analogies see Fig.5 and 6.

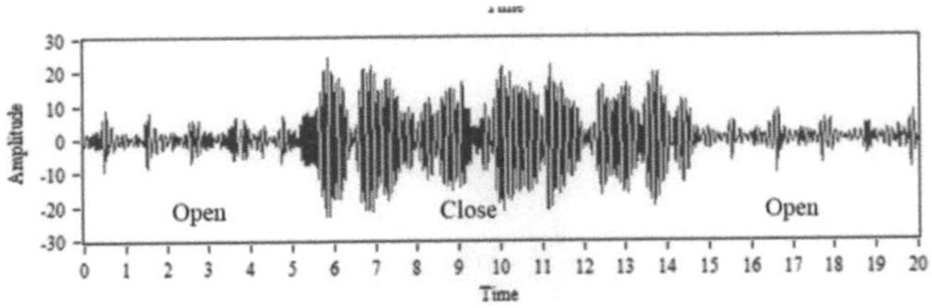

Fig. 5 EEG diagram of a person when the eyes are opened and then he/she closes the eyes and then opens them again. The figure is taken from Aziz[20]

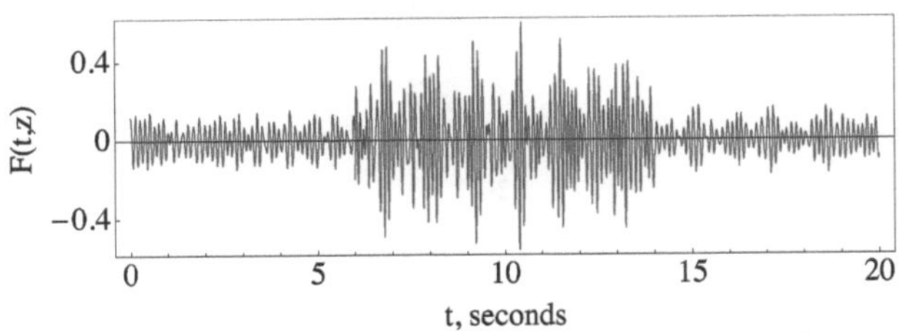

Fig. 6 Collective inversion function at constant z, with strong coupling between biophoton field and H_2O molecules. This is a theoretical prediction of EEG diagram. Compare with Fig.5

Neuroscientists or brain specialists record the EEG diagrams of their patients when their eyes are closed, because when their eyes are open, the amplitudes of diagrams would be dramatically reduced and they cannot diagnose changes in the amplitudes. In Fig.5 and 6 we represent the EEG diagrams of a person with different situations as explained in the figure caption. When the eyes are opened, the number of photons interacting with the biomolecules becomes high and the amplitudes of EEG become low. In contrast to Rahnama et al.[21] our model explain fairly well the mechanism which is in essential a collective biophoton trapping. Thus we strongly believe that the reducing the amplitudes when the eyes are opened is a consequence of the biophoton trapping as explained above.

In our theoretical model with the decreasing of coupling strength between biophotons and coherent H_2O states the fluctuations disappear, e.g. a sleeping person.

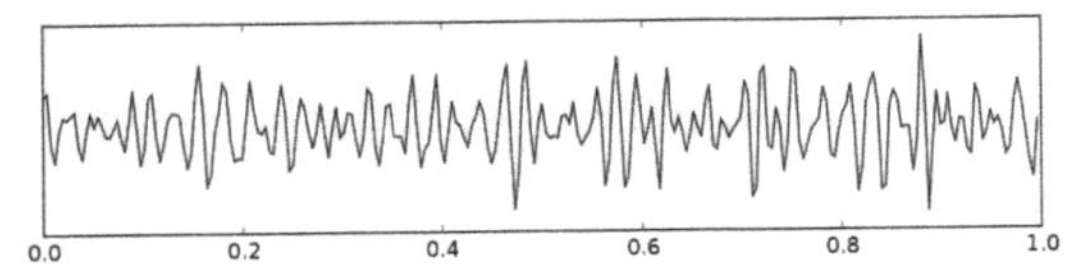

Fig. 7 EEG gamma waves describes motor functions, higher mental activity. Neurosky[22]

Another example we take gamma waves wich describes motor functions or higher mental activity Fig.7 experiment and his theoretical prediction Fig.8.

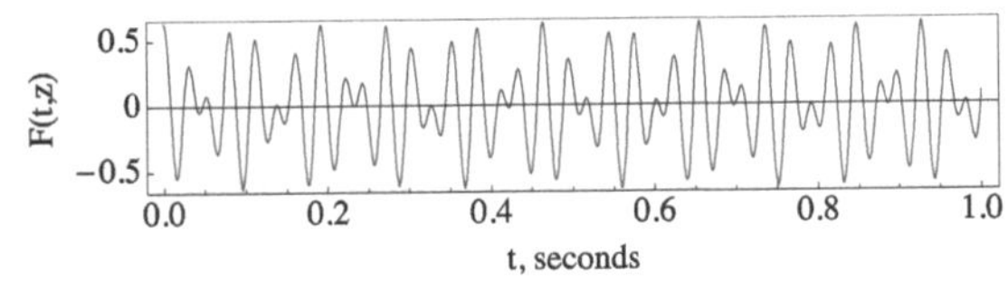

Fig. 8 Collective inversion function for gamma waves, at constant z, with weak coupling between biophoton field and H_2O molecules

In conclusion to this section we mention that a new mechanism of collective biophoton trapping explaining EEG diagrams is proposed, which is based on the interaction of biophotons and coherent H_2O molecules inside the neurons, thus when the eyes start to be opened, the number of photons incident onto the eyes increases and according to the collective biophoton trapping diagrams in Figs. 4 and 6, the amplitudes are reduced.

IV. CONCLUSIONS

We have proposed a theoretical model for the description of a collective behavior of H_2O molecules as an assembly of two-level quantum biological system. In HP approximation at high polarization (see Fig.3), a collection of two level H_2O molecules and a collection of oscillators behave identically, thus HP transformation ensures the assignment of coherent Bloch states to the assembly of H_2O molecules. Further, taking MTs as quantum cavities with a single mode of biophoton field, enables the coherent H_2O molecules to be considered not as independent individuals, but as a totality or in other words the coherent H_2O molecules act like single dipole oscillator. Moreover it is worthy to note that our model is strongly supported by experimental evidences Sahu et al. [14] confirming that the role of water molecules in MTs is to increase conductivity 1000 times. In the end we have introduced a new mechanism of collective biophoton trapping explaining EEG diagrams, from which we see a good experimental and theoretical agreement.

Conflict of Interest

The author declares that he has no conflict of interest.

References

1. Hameroff, S., Penrose, R., 1996. Orchestrated reduction of quantum coherence in brain microtubules: A model for consciousness. Mathematics and Computers in Simulation 40, 453–480. doi:10.1016/0378-4754(96)80476-9
2. Mavromatos, N.E., Nanopoulos, D.V., 1998. On Quantum Mechanical Aspects of Microtubules. Int. J. Mod. Phys. B 12, 517–542. doi:10.1142/S0217979298000326
3. Jibu, M., Hagan, S., Hameroff, S.R., Pribram, K.H., Yasue, K., 1994. Quantum optical coherence in cytoskeletal microtubules: implications for brain function. Biosystems 32, 195–209. doi:10.1016/0303-2647(94)90043-4
4. Jibu, M., Pribram, K.H., Yasue, K., 1996. From conscious experience to memory storage and retrieval: the role of quantum brain dynamics and boson condensation of evanescent photons. Int. J. Mod. Phys. B 10, 1735–1754. doi:10.1142/S0217979296000805
5. Dicke, R.H., 1954. Coherence in Spontaneous Radiation Processes. Phys. Rev. 93, 99–110. doi:10.1103/PhysRev.93.99
6. Bischof, M., Del Giudice, E., 2013. Communication and the Emergence of Collective Behavior in Living Organisms: A Quantum Approach. Mol. Biol. Int. 2013, e987549. doi:10.1155/2013/987549
7. Mavromatos, N.E., Mershin, A., Nanopoulos, D.V., 2002. Qed-cavity model of microtubules implies dissipationless energy transfer and biological quantum teleportation. Int. J. Mod. Phys. B 16, 3623–3642. doi:10.1142/S0217979202011512
8. Popp, F.A., Nagl, W., Li, K.H., Scholz, W., Weingärtner, O., Wolf, R., 1984. Biophoton emission. Cell Biophys. 6, 33–52. doi:10.1007/BF02788
9. Popp, F.A., Gu, Q., 1992. Recent Advances in Biophoton Research and Its Applications. World Scientific
10. Bokkon, I., Salari, V., Tuszynski, J.A., Antal, I., 2010. Estimation of the number of biophotons involved in the visual perception of a single-object image. J. Photochem. and Photobiol. B: Biol. 100, 160–166. doi:10.1016/j.jphotobiol.2010.06.001
11. Pollet, E., Martínez, J.A., Metha, B., Watts Jr., B.P., Turrens, J.F., 1998. Role of Tryptophan Oxidation in Peroxynitrite-Dependent Protein Chemiluminescence. Arch. Biochem. Biophys. 349, 74–80. doi:10.1006/abbi.1997.0436
12. Kruk, I., Lichszteld, K., Michalska, T., Wronska, J., Bounias, M., 1989. The Formation of Singlet Oxygen during Oxidation of Catechol Amines as Detected by Infrared Chemiluminescence and Spectrophotometric Method. Zeitschrift für Naturforschung C 44, 895–900
13. Nogales, E., Wolf, S.G., Downing, K.H., 1997. Structure of the alpha beta tubulin dimer by electron crystallography. Nature 391, 199–203. doi:10.1038/34465
14. Sahu, S., Ghosh, S., Ghosh, B., Aswani, K., Hirata, K., Fujita, D., Bandyopadhyay, A., 2013. Atomic water channel controlling remarkable properties of a single brain microtubule: Correlating single protein to its supramolecular assembly. Biosens. Bioelectron. 47, 141–148. doi:10.1016/j.bios.2013.02.050
15. Craddock, T.J.A., Friesen, D., Mane, J., Hameroff, S., Tuszynski, J.A., 2014. The feasibility of coherent energy transfer in microtubules. J. Roy. Soc. Inter. 11, 20140677. doi:10.1098/rsif.2014.0677
16. Holstein, T., Primakoff, H., 1940. Field Dependence of the Intrinsic Domain Magnetization of a Ferromagnet. Phys. Rev. 58, 1098–1113. doi:10.1103/PhysRev.58.1098
17. Arecchi, F.T., Courtens, E., Gilmore, R., Thomas, H., 1972. Atomic Coherent States in Quantum Optics. Phys. Rev. A 6, 2211–2237. doi:10.1103/PhysRevA.6.2211
18. Glauber, R.J., 1963. The Quantum Theory of Optical Coherence. Phys. Rev. 130, 2529–2539. doi:10.1103/PhysRev.130.2529
19. Gerry, C.C., Knight, P.L., 2005. Introductory Quantum Optics. Cambridge University Press.
20. Aziz, F., Arof, H., Mokhtar, N., Mubin, M., 2014. HMM based automated wheelchair navigation using EOG traces in EEG. J. Neural Eng. 11, 056018. doi:10.1088/1741-2560/11/5/056018
21. Rahnama, M., Tuszynski, J.A., Bókkon, I., Cifra, M., Sardar, P., Salari, V., 2011. Emission of mitochondrial biophotons and their effect on electrical activity of membrane via microtubules. J. Integr. Neurosci. 10, 65–88. doi:10.1142/S0219635211002622
22. Brain Wave Signal (EEG) of NeuroSky, Inc. at http://sleephat.com/wp-content/uploads/2014/10/neurosky-vs-medical-eeg.pdf

Author:	Nistreanu Andrei
Institute:	Institute of Applied Physics
Street:	Academiei Street 5
City:	Chisinau
Country:	Moldova Republic of
Email:	nistreanu.andrei@gmail.com

Medical Devices Management Strategy in the Republic of Moldova

V.Sontea[1], S.Morgoci[2], Gh. Turcanu[3], and C.Pislaru[4]

[1] Technical University of Moldova, Chisinau, Moldova
[2] Center for Health Policies and Studies, Chisinau, Moldova
[3] Ministry of Health, Chisinau, Moldova
[4] Mother and Child Institute, Chisinau, Moldova

Abstract— **Advanced medical devices are an indispensable part of medical care in the processes of prevention, diagnosis and treatment of diseases. Using them effectively requires, as a priority, increasing the number of investigations for cost-effective and quality treatment. "Management of Medical Devices for a 2015-2020 period" Strategy aim is to enhance quality of care through improving of medical device management, providing a high level of safety and performance of medical devices in medical institutions, to facilitate the development of a functional and efficient management of medical device in Republic Moldova. The expected result from implementation of the Strategy is providing quality of medical devices and optimizing the use of public funds by way proposed by a rational and efficient management.**

Keywords— **medical devices, management, maintenance, strategy, biomedical engineering.**

I. INTRODUCTION

Guaranteeing the rights of citizens to quality health care remains a priority of contemporary medical institutions. The quality of services depends on the material and technical basis, including provision with medical devices / apparatus, and the professional competence of involved personnel (human factor).

Evaluation of global experience in this field has demonstrated that advanced medical devices are indispensable part of medical care in the correct prevention, diagnosis and treatment of diseases. Using them effectively requires, as a priority, increasing the number of investigations and cost effective and qualitative treatment

The global medical devices industry with a turnover of over 226 billion dollars in 2011 will reach in 2017 a business volume of over 434 billion dollars with a growth rate about 7.1% annually. Global medical device industry includes over 13 600 registered producers, about 10 000 generic devices and over 500 000 products and brands. 10 years ago there were about 50% of all diagnostic and treatment methods currently used. It should be mentioned that period of medical device elaboration is estimated to average 1-3 years, and the period of aging medical devices, starting with the production release is on average 5-7 years. [1]

WHO estimates showed that despite the billions of dollars allocated annually for procurement and management of medical devices, most countries consider Medical Device Management (MDM) is a part of public health policy. According to WHO estimates, in medical institutions in developing countries, about 50% of medical equipment is not functioning or is not used and maintained correctly - a situation with serious consequences for services and care. [2] For these reasons, WHO recommends and is essential to have a national policy on MDM, including the provision of MD, ensuring the maintenance, inspection and proper use of medical technologies, training specialists and creating a training system, etc...[3]

In most EU countries this area is strictly regulated by the European Directives through adjusting national legislation and through the approving of standards recognized at European level. The level of medical equipment provisioning in the countries of Europe is on average 5-15 medical devices to a bed and responsible for medical devices management is Clinical Engineering Department, which is a part of medical institution structure, identical with clinical departments and auxiliary ones[4].

Management of medical devices has become a priority in health policy in many countries, many studies proved that consistent policies in this area improve the cost/effectiveness of advanced medical technologies use, increase patient safety and not least, improve the quality of medical act. The level of medical institutions provision with medical devices and level of ensuring of an appropriate level of medical staff professionalism are key tools in ensuring the proper functioning of the health system; and indirect impacts on the functional efficacy of the system, quality of service and the degree of satisfaction of the beneficiary.

Strategy "Management of Medical Devices for period 2015-2020" in the Republic of Moldova (hereinafter - Strategy) is part of the socio-economic state policy, directed towards health development and builds on the experience of European Union member countries, strategic documents of the World Health (hereinafter WHO), other international organizations active in health care.

V. Sontea and I. Tiginyanu (eds.), *3rd International Conference on Nanotechnologies and Biomedical Engineering*,
IFMBE Proceedings 55,
DOI: 10.1007/978-981-287-736-9_113

II. CURRENT SITUATION DESCRIPTION

Currently the health system in Moldova includes 912 hospitals, of which 278 public health facilities (59 hospitals with 19840 beds), where it operates 12 794 27 407 doctors and medical staff with secondary education. In the current economic crisis, the health system makes it difficult to ensure adequate access to quality healthcare and services. [5]

Limited financial resources on the one hand, and increasing demands from the professionals in order to use advanced medical devices, as well as those of the population to have access to quality service, on the other hand, make existing needs are only partly covered. The share of national public budget for health expenditure is steadily increasing, while the cost of medical devices is increasing, the requirements and possibilities for assessment of their quality and safety are lacking.

Currently the technological level of medical devices within the public medical institutions is outdated morally and physically, with major deficiencies, and it's one of the root causes that determine quality in the curative and diagnostic process. The low quality of medical services obtained as a result of such a situation therefore creates problems of disease evolution, extension of optimal treatment initiation, chronization of diseases and population disability.

According to evaluations carried out through electronic Registry, the public health system has over 30 thousand active medical devices taken from the accounts as fixed assets. [6] The degree of wear-out of medical devices is high both: in district institutions and in the republican; over 37% of devices were over the age of 10 years, are not functional or requires major repairs 8.7% of all devices provided.

43.2% of the medical devices included in the registry are equipping medical institutions in the Chisinau municipality. Medical devices taken to balance value is only 16% of their procurement costs.

In the country as a new direction develops scientific-practical elaboration and production of new biomedical equipment (Academy of Sciences, Technical University of Moldova, research institutes and production branch companies, including private).

Since 2005, at the insistence of the Ministry of Health and the Technical University of Moldova Nomenclature of areas and specialties in higher education was included the specialty "Biomedical Systems Engineering" which provides basic and specialized training of biomedical engineering specialists strictly necessary for Republic Moldova; and since 2010 master studies in the field of "Biomedical engineering». Despite inadequate funding have been trained 102 medical bioengineers and 46 specialists with master studies, most of whom are employed in the health system. Recently, biomedical engineering and clinical engineer professions were included in clinical Classification of Occupations of the Republic of Moldova, but not covered by medical personnel standards of medical institutions.

With the support of the Swiss Agency for Development and Cooperation through projects REPEMOL PERINAT were analyzed all existing procedures related to MDM at hospital level, has been defined and developed set of procedures for managing and administering of Medical Devices and Systems openMEDIS (SIMDM openMEDIS), which were initially implemented in five pilot medical institutions. MDM evaluation results at medical school have demonstrated the positive impact on the efficient use of medical devices, reducing costs of maintaining medical devices through internal appropriate services in a timely manner, increasing the cost - effectiveness and safety of care.

In the same context, the was set up pilot center - "Management of Medical Technologies," on the basis of Mother and Child Institute jointly with the Technical University of Moldova, which aims to design and develop Medical Devices Management procedure models according to international standards, development of maintenance procedures, diagnosis and repair of medical devices.

Moldova adopted the Law on medical devices (no.92 / 04.26.12), which aims to adjust the legal framework of the Republic of Moldova to the acquis communautaire for the implementation of European standards in the field of medical devices and medical services consumer protection offered by the application of medical devices. The law provides definitions correspond with European Directives on medical devices, states that describe how medical devices may be placed on the market, put into service or used only if they are certified and registered, not to affect the safety and health of patients, users, and others, and the environment.

The Ministry of Health has the power to develop and promote policies on medical devices.

There is created Medicines and Medical Devices Agency with the status of public institution with autonomous budget under the Ministry of Health, which aims to ensure the presence of pharmaceutical products and qualitative medical devices on Moldovan market, efficient and safe. Agency is the competent and decisional authority in the field of medicines and medical devices and ensures the promotion of state policies and achieving objectives.

III. PROBLEMS DEFINITION

The health system lacks a strategic document on the management of medical devices, technical regulations, and minimum standards of endowment of medical institutions at different levels. As a result, there are not determined real needs of medical devices for healthcare institutions and non-optimal allocation of purchased devices, inefficient use. Devices in the endowment of medical institutions have a life span up to 40 years (international rules are up to 10 years), produced by over 300 manufacturers, provided by over 300 operators in Moldova. Nowadays are missing the criteria of institutions selection, which require modernization, based on the priorities of the health sector.

Procurement planning of medical devices is not based on calculations and objective analysis. As a result, medical devices are purchased mainly of the same type, which doubles the services, which are used insufficiently.

Expensive medical devices are not used to optimum efficiency, with real opportunities to improve their operation. Lack of proper monitoring of medical devices makes impossible to exclude negative trends of use.

Lack of medical devices management, which include all actions necessary to ensure effective use of medical devices, including: needs assessment, procurement planning, procurement, maintenance, and replacement of medical devices, reduce the quality of health services.

Medical institutions have the necessary capabilities and means of medical devices maintenance and repair, as well as, personal capacity and suitable devices for testing and calibration. The services provided by private companies, in the absence of fair competition, are largely unqualified, preventive maintenance is shallow, the duration of work is high and cost is extremely high.

Healthcare institutions do not have sufficient competent staff in medical devices, which contributed to the purchase of medical devices with technical parameters, which do not fully correspond to contracts, but also contributed to the technical maintenance, repair, verification, diagnosis and insufficient metrological assistance.

There is an acute lack of specialists in biomedical engineering concerning to the maintenance, inspection and use of medical devices.

Not all medical devices are included in the electronic register, which creates great difficulties in implementing rules in Medical Devices Management. No Medical Device Registry created under Law 92/2012.

Lack of continuous compulsory regulation for users and professionals working in the service, marketing and commissioning of medical devices, generates deficiencies in the use and maintenance of ones, in ensuring patient safety, in decreasing cost effectiveness of care.

The health system lacks sufficient institutional and operational capacities to fully implement the Law on Medical Devices (Law no. 92 / 04.26.12). Lack of normative acts regulating the area, and an effective monitoring and evaluation of MDM in IMS and MS generates deficiencies in procurement, planning and improper distribution of medical devices nationwide.

IV. General and Specific Objectives of the Strategy

The overall objective of the Strategy is to improve health by providing health services accessible and at high quality level.

The aim of the Strategy is to enhance quality of care through improved management of medical devices providing a high level of safety and performance of medical devices in medical institutions.

This requires achieving of three specific objectives until 2020, namely:

1. Adjusting of medical device regulatory framework to EU requirements and WHO recommendations;
2. Improving the effectiveness of medical devices based on management skills, so that by the end of 2020 planning the replacement of medical devices will be based on cost-effectiveness and safety; the average downtime for most damaged MD reduced to 5 hours; During the use of MD rate reaches 90% of its potential; 80% of PHI will be equipped according to new standards until 2020.
3. Strengthening capacities in MDM, managers capacities, medical and technical staff, medical institutions by modernizing initial and continuing professional training, so that by the end of 2020 90% of staff involved in MDM will have suitable training.

Action Plan for the implementation of the Strategy represents the basic tool for achieving systemic and continuing the Strategy goals, including all components, specific objectives, planned actions, performance indicators, sources of financing, execution schedules and responsible for the implementation.

To reach the goals are provided the following:

1. Developing of policies in accordance with EU rules and recommendations of the World Health Organization, of the management of medical devices by adjusting the legal and regulatory framework concerning:

a) placing the medical devices on market;

b) harmonizing the technical regulations of medical devices with European directives for medical devices;

c) endowment uniforming, planning and maintenance of medical devices within the IMS;

d) procurement procedures of medical devices;

e) implementing of surveillance and vigilance system for medical devices;

f) finding and sanctioning contraventions in medical devices.

2. Analysis of current endowment with medical devices of different level of medical institutions, the development and implementation of standards for medical devices

equipped within medical institutions in conformity with the new conditions of health reform and health services;

3. Creating Clinical Engineering departments / sections within PHI with advanced technology platforms, development and implementation of maintenance procedures, diagnostics and repair of medical devices;
4. Developing and implementing a plan to equip medical institutions with devices;
5. Developing and implementing a new financing mechanisms based on medical device management and optimization of financing sources;
6. Developing of medical devices information system.
7. Developing and implementing accountability mechanisms in relation between IMS managers and DM managers;
8. Developing continuous training programs mandatory for those who use MD, as well as specialists, who put MD into service;
9. Promote the profession of medical bioengineer in healthcare, with the requirements and quality standards of care;
10. creation and development, of the National Center of biomedical engineering, based on Technical University of Moldova, responsible for undergraduate and postgraduate training and expertise of medical devices;

The plan includes medium-term actions and is flexible, allowing operation changes according to changing circumstances and the emergence of possible problems in the implementation of the Strategy.

V. CONCLUSIONS

Developing and implementing a strategy based on systematic approach to health care reform, coupled with the vector of socioeconomic development and European integration of Moldova. This approach determines the strategy advantages, thanks to which we will produce a significant positive impact on the quality and efficiency of health care services.

These are:

1) improving health by maintaining a high level of safety and performance of medical devices in medical institutions and health services accessible and high quality.

2) adduction the medical device regulatory framework to EU requirements and WHO recommendations, needs and requirements of the national health system;

3) formation of the institutional management of medical devices by creating templates and procedures for the management of medical device manufacturers according to international standards and requirements;

4) the improvement of the education system in biomedical engineering adjusted to Medical technology trends.

CONFLICT OF INTEREST

The authors declare that they have no conflict of interest.

REFERENCES

1. J. Sanders, "Where is Technology Taking Healthcare?Where is Technology Taking Healthcare? "CADTH Symposium 2007.
2. www.who .Technologies1 WHA60.29Health http://apps.who.int/gb/ebwha/pdf_files/WHASSA_WHA60-Rec1/E/reso-60-en.pdf
3. www.who .Technologies1 WHA60.29Health http://apps.who.int/gb/ebwha/pdf_files/WHASSA_WHA60-Rec1/E/reso-60-en.pdf
4. Development of Medical Device Policies. (WHO Medical device series) 2011 http://whqlibdoc.who.int/publications/2011/9789241501637_eng.pdf ?ua=1
5. www.cnms.md/rapoarte

Author:	Victor Sontea
Institute:	Technical University of Moldova
Street:	168, Stefan cel Mare bd.
City:	Chisinau
Country:	Moldova
Email:	sontea@mail.utm.md

Method of Treatment of Immune Cell Disorder Caused by Ionizing Radiation

L. Coretchi, I. Bahnarel, and C. Spinu

National Centre of Public Health, Chisinau, Republic of Moldova

Abstract— **New method to optimize the medical treatment algorithm of the patients exposed to ionizing radiation have been elaborated. Inclusion of the „Pacovirina-plus" in the medical treatment algorithm of the radiation exposed patients, demonstrates the improvement of the health indices, manifested through the significantly decreasing of asthenia, dyspeptic and depressive syndromes; restoration of the CD4/CD8 ratio and antioxidant activity through the return to norm limits of ALAT and ASAT parameters with the diminishing of the activity of lipid's peroxide oxidation. For investigation, was selected 69 persons, participating at the diminution of consequence of nuclear accident from Chernobyl (PDCNAC), with place of residence in the department Orhei (town Orhei – 18 (26, 1%)), 51 villages of department Orhei – (79,9%), 68 men and one women, with 48,6±8,0 average age. The group of patients was divided in 2 groups by the randomized method: experimental 33 patients and control 36 patients. The data shows that "Pacovirina-plus" administrated together with pyridoxine and essential have positive influence on the immune system and lead to normalization of CD4/CD8 ratio. The results of the influence of "Pacovirina-plus" preparation on some biochemical parameters of participants in diminishing of consequences of the Chernobyl nuclear accident (PDCCNA) also are presented in this paper. A tendency of normalization of parameters concerned with their comeback in the normal range, have been elucidated. The administration of the „Pacovirin-plus" preparation had a beneficial effect on the evolution of the main clinical signs after three months of treatment. According to the results recorded in the experimental group a clear trend towards increasing the total number of lymphocytes in peripheral blood, have been demonstrated, which shows the immunoregulatory action of the preparation. "Pacovirina-plus" is a biological product of plant origin, ecologically pure, obtained by simple technology. This remedy is easily tolerated by patients, without adverse effects.**

Keywords— **ionizing radiation, immune system, medical treatment.**

I. INTRODUCTION

An essential issue is the development of new micro/nano technologies to reduce the onset of reactions resulting from the influence of ionizing radiation to patients. The aim of the paper is to study the influence of new secondary metabolites on the immune system response to the action of ionizing radiation.

About three decades has gone after the Chernobyl disaster, but it still remains in the human memory. It is a tragic fact that this accident was not a natural catastrophe, but a technological emergency. Chernobyl nuclear accident represents the most catastrophic nuclear accident in the history of mankind, characteristic not only by a large number of emergency workers but an important part of population affected in nearby regions. Radioactive substances derived from Chernobyl nuclear accident also affected extensive areas in Europe, the Republic of Moldova inclusively, with subsequent deterioration of the environment. There are 2 groups of people that suffered from CE-derived ionizing radiation exposure consequences in the Republic of Moldova. The group of PDCNAC seems to be more representative, conditioning 3500 persons. The second group comprises 306 evacuated persons [1, 4, and 7].

Somatic and neurologic modifications developed in this group of PDCNAC were influenced by a number of factors besides psycho-emotional stress, for example social and economic problems, traditional risk factors, preexisted somatic pathologies. These pathologies diagnosis and treatment is of primordial importance in current conditions [2, 5, and 6].

II. MATERIALS AND METHODS

For investigation, was selected 69 persons, participating at the diminution of consequence of nuclear accident from Chernobyl, with place of residence in the department Orhei (town Orhei – 18 (26, 1%)), 51 villages of department Orhei – (79, 9%), 68 men and one women, with 48, 6 ± 8, 0 average age. The group of patients was divided in 2 groups by the randomized method: experimental 33 patients and control 36 patients.

At the control group traditional treatment was made with vitamin therapy of group B, essential strengths, and in some cases, as required, spasmolithics, sedatives and analgesics.

Studying the effectiveness of "Pacovirina-plus" used in the recovery process of people participating in mitigating of the consequences of the nuclear accident at Chernobyl was based on clinical and laboratory surveillance of this patient. Clinical investigations included: state of severity, neurological status (conscience, accuse, skin and mucous membranes, lymph nodes, condition of the respiratory, cardiovascular, digestive, urinary at the initiation and

V. Sontea and I. Tiginyanu (eds.), *3rd International Conference on Nanotechnologies and Biomedical Engineering*,
IFMBE Proceedings 55,
DOI: 10.1007/978-981-287-736-9_114

completion of the study. Laboratory investigations included: investigations of ALT, AST, bilirubin, total protein, acid-base balance, glucose, thymol, ECG, EUS, hemoleucograms, urine analysis, viral exams (HSV, CMV, EBV), of immunological investigation: CD4+ (T-helper), CD8+ (T-suppressor), CD3+ (T-cells). It was performed basic supplementation therapy for persons participating in the liquidation of the Chernobyl accident (patients in the experimental group) with local phytoremediation with antiviral, immunomodulatory, interferonogen and antioxidant action: 5α-furostan-3β.22.26–triol–3-[O–β–D-glucopiranozil(1→2)–β–D-glucopiranozil(1→4)-β–D-galactopiranozil]–26–O–β–D-glucopiranozid ("Pacovirina-plus") [8], 50 mg (one tablet) 1 time per day for 90 days while pyridoxine and essential medicinal products as standard treatment, while the control group only pyridoxine and essential as standard treatment. For approval of method of treatment of persons participating in the liquidation of the Chernobyl accident, clinical and laboratory investigations were conducted randomized for assessing the effectiveness of the new variant.

Blood samples (5 ml) for immunological investigations were harvested individualy with 5 days prior to treatment initiation and after 5 days from the end of it, where were examined the following markers of T-lymphocytes: CD4+, CD8+, CD3+, using the lymphocyte phenotyping by flowcitometry, FACS count system [2, 3, 9, 10].

III. Results and Discussion

At the people exposed to radiogenic stress under the action of ionizing radiation in the radioliza process of the cell, free radical formation occurs, which makes the modification of the indices of links T-cell immunity, which testifies about the immunological cellular disorders and weakening of the body. The deregulation of the lymphocytes component population which contribute to expression of the significantly T-immunological deficience, is characterized by decreased blood content of T-lymphocytes, especially of the helpers (CD4 +). So, it was necessary to develop the methods of regulating immune responses to prevent the cancer diseases, which usually is starting on the immunity disorders background.

The new elaborated method of treatment of immune cell disorder caused by ionizing radiation, consisted in administration of oral antioxidant preparations as: Vitamin B6 with Essential, is very practice. For example: Vitamin B6 (Pyridoxinum) 5% 1 ml intramuscularly once daily for 3 weeks, 2 capsules Essential strength 3 times a day course of treatment - 2 months [11]. But in the case of the traditional treatment, the disorders of cellular immunity is not fully normalized and may be due to frequent relapses in people exposed to ionizing radiation of different diseases and may contribute to starting the cancer diseases.

The problem which solved the elaboration of new method consists in the treatment of disorders cellular immunity at the persons exposed to ionizing radiation using the preparation "Pacovirina-plus" co-administered with pyridoxine and essential. "Pacovirina-plus" is a biological product of plant origin, ecologically pure, obtained by simple technology. This remedy is easily tolerated by patients, without adverse effects.

The above mentioned disadvantages are eliminated in that it proposes the use of preparation "Pacovirina-plus" co-administered with pyridoxine and essential as treatment of people exposed to ionizing radiation, to stimulate the immunocorrection.

"Pacovirina-plus" is a product of plant adaptive origin, obtained by an original method of extraction, where biologically active substance from the technological process of production of the stock, reached a value of 92.8%.

For investigation, was selected 69 persons, participating at the diminution of consequence of nuclear accident from Chernobyl, with place of residence in the department Orhei (town Orhei – 18 (26, 1%)), 51 villages of department Orhei – (79, 9%), 68 men and one women, with 48, 6 ± 8, 0 average age. The group of patients was divided in 2 groups by the randomized method: experimental 33 patients and control 36 patients. The results are presented in tables 1-4.

The study of the initial data of the immune status at the patients exposed to ionizing radiation has been established that practically, all patients had signs of chronic immune process.

Table 1 The influence of the „Pacovirina-Plus" preparation on some biochemical parameters of participants in diminishing of consequences of the Chernobyl nuclear accident

Indices	$X \pm m_x$	Norm
ALAT[1]	47,9±21,6	0 - 41,0 un/l
ALAT[2]	41,1±17,4	
ASAT[1]	58,3±28,9	0 - 37,0 un/l
ASAT[2]	36,8±19,1	
Bilirubin[1]	10,6±3,0	8,0 - 23,0
Bilirubin[2]	12,3±3,3	
Glucose[1]	4,9±0,6	3,3 - 6,5 mmol/l
Glucose[2]	4,7±1,2	
Tymol[1]	1,7±0,7	1,5 - 5,0 mmol/l
Tymol[2]	1,7±0,7	

Note: 1 – The results to the administration of „Pacovirina-plus" preparation. 2 – The results after administration of „Pacovirina-plus" preparation.

The data presented in the table 2 shows that "Pacovirina-plus" administrated together with pyridoxine and essential have positive influence on the immune system and lead to normalization of CD4/CD8 ratio.

Table 2 The characteristic of the immune status indices CD4/CD8 of participants in diminishing of consequences of the Chernobyl nuclear accident

In–dice		Experimental group, n=33					Control group, n=36				
		norm		<;> norm		p	norm		<;> norm		p
		abs.	$M\pm m_x$	abs.	$M\pm m_x$		abs.	$M\pm m_x$	abs.	$M\pm m_x$	
CD 4 400-1100 cell/µL	1	12(650)	34,6±8,4	21 (360)	63,6±8,4	<0,01	11(530)	30,5±7,7	25 (340)	69,5±7,7	> 0,05
	2	30 (800)	90,1±5,2	3 (390)	9,9±5,2		13 (580)	36,1±8,0	23(360)	63,9±8,0	
CD8 300-700 cell/µL	1	12 (400)	36,4±8,4	21 (750)	63,6±8,4	<0,01	11 (350)	30,5±7,7	25(370)	69,5±7,7	> 0,05
	2	30 (350)	90,1±5,2	3 (740)	9,9±5,2		13 (370)	36,1±8,0	23(400)	63,9±8,0	
CD3 1100-1700 cell/µL	1	31 (1200)	93,9±4,2	2 (900)	6,1±4,2	<0,01	33(1300)	91,6±4,6	3(850)	8,4±4,6	> 0,05
	2	33 (1500)	100	-	-		33(1350)	91,6±4,6	3(900)	8,4±4,6	
CD 4/ CD8 1,5-2,5	1	12 (1,6)	36,4±8,4	21 (0,5)	63,6±8,4	<0,01	11 (1,5)	30,5±7,7	25(0,9)	69,5±7,7	>0,05
	2	30 (2,3)	90,1±5,2	3 (0,5)	9,9±5,2		13 (1,6)	36,1±8,0	23(0,9)	63,9±8,0	

Note: 1– before the treatment, 2 – after the treatment

Table 3 The characteristic of the clinic indices of participants in diminishing of consequences of the Chernobyl nuclear accident (Control and experimental group) before and after the treatment

Indice	Experimental group (n=33)				Control group (n=36)				p
	Up to treatment		after treatment		Up to treatment		after treatment		
	Abs.	$M\pm m_x$	Abs.	$M\pm m_x$	Abs.	$M\pm m_x$	Abs.	$M\pm m_x$	
Asthenia syndrome, increased irritability, irritability, emotional liability, mental and physical fatigue, exhaustion, vegetative disorders	31	93,9±4,5	26	78,8±7,1	33	91,7±4,6	29	80,6±6,6	>0.05
Depressive syndrome, sleep and appetite disorders, lack of joy and pleasure	32	96,9±3,0	18	54,5±8,6	33	91,7±4,6	33	91,7±4,6	<0.05
Digestive disorders - sensation of heaviness, pain below the right costal margin	24	72,7±7,7	12	36,4±8,4	25	69,4±7,9	21	58,3±8,2	<0.05
Hepatomegaly	16	48,5±8,7	7	21,2±7,1	17	47,2±8,3	15	41,7±8,2	<0.05

The dates presented in table 1, presents the results of the influence of "Pacovirina-plus" preparation on some biochemical parameters of participants in diminishing of consequences of the Chernobyl nuclear accident. A tendency of normalization of parameters concerned with their comeback in the normal range, have been elucidated.

Analyzing the data presented from table 3 shows, that the administration of the “Pacovirin-plus” preparation had a beneficial effect on the evolution of the main clinical signs after three months of treatment.

According to the results recorded in the experimental group a clear trend towards increasing the total number of lymphocytes in peripheral blood, have been demonstrated, which shows the immunoregulatory action of the preparation (table 4).

Table 4 The comparative evolution of laboratory main indices of participants in diminishing of consequences of the Chernobyl nuclear accident (experimental and control group) before and after the treatment

Indice	Experimental group, n=33					Control group, n=36				
	norm		<;> norm		p	norm		<;> norm		p
	abs.	$M \pm m_x$	abs.	$M \pm m_x$		abs.	$M \pm m_x$	abs.	$M \pm m_x$	
Hemoglobin 130-175 g/l	16(145)	48,4±8,7	17 (120)	51,6±8,7	<0,05	18 (140)	50,0±8,3	18 (123)	50,0±8,3	>0,05
	26(150)	78,7±7,1	7 (125)	21,3±7,1		19(139)	52,8±8,3	17 (121)	47,2±8,3	
Erythrocytes $4\text{-}5{,}6\times10^{12}$/l	20(4,5)	60,6±8,5	13(3,0)	39,4±8,5	<0,05	19 (4,3)	53,0±8,3	17 (3,2)	47,0±8,3	>0,05
	27 (5,0)	81,8±6,7	6 (3,2)	18,2±6,7		17 (4,0)	47,2±8,3	19 (3,1)	52,8±8,3	
Leucocytes $4\text{-}5{,}6\times10^{12}$/l	32(5,0)	96,9±3,0	1 (3,5)	3,1±2,9	<0,05	35 (4,5)	97,3±2,7	1 (3,0)	2,7±2,7	<0,05
	33(5,2)	100	-	-		36 (5,0)	100	-	-	
Lymphocytes 19-37%	33 (30)	100	-	-		36 (33)	100	-	-	
	33 (35)	100	-	-		36 (37)	100	-	-	
VSH 1-14 mm/o	33 (6)	100	-	-		36 (7)	100	-	-	
	33 (5)	100	-	-		36 (6)	100	-	-	

Note: 1– before the treatment, 2 – after the treatment.

IV. Conclusions

"Pacovirina-plus" administrated together with pyridoxine and essential have positive influence on the immune system and lead to normalization of CD4/CD8 ratio.

A tendency of normalization of biochemical parameters concerned with their comeback in the normal range after the influence of the „Pacovirina-Plus" have been elucidated.

The results shows, that the administration of the "Pacovirin-plus" preparation had a beneficial effect on the evolution of the main clinical signs after three months of treatment.

According to the results recorded in the experimental group a clear trend towards increasing the total number of lymphocytes in peripheral blood, have been demonstrated, which shows the immunoregulatory action of the "Pacovirin-plus".

"Pacovirina-plus" is a biological product of plant origin, ecologically pure, obtained by simple technology. This remedy is easily tolerated by patients, without adverse effects.

Conflict of Interest

The authors declare that they have no conflict of interest.

References

1. Bahnarel I., Coreţchi L., Moldovanu M. (2006) Medical and biological aspects of the Chernobyl nuclear accident influence on the population of the Republic of Moldova, Chisinau, 150 p.
2. Bahnarel I., Samotîia E., Coreţchi L., Moldovan M. Patent MD. C2, MD, A 61 B 5/145. The process of evaluation of immune status, BOPI No. 1/2005. (In Romanian).
3. Coreţchi L.Ş. et all. Evaluation of the immunological effects of medical exposure staff. In: *International Conference Radiation Protection in Medicine*. Varna, 1-3.09.2010. p. 55.
4. Coreţchi L.Ş., Bahnarel I.N., Apostol I., Samotiia E.E., Sofroni V.G. (2015) Health effects of children of the Chernobyl NPP accident consequences liquidation participants. In: NATO Science for Peace and Security. Series E: Human and Societal Dynamics, vol. 120, p. 275-282. (Proceedings of the NATO Advanced Research Workshop on Engaging the Public to fight the Consequences of Terrorism and Disasters, Tbilisi, Georgia, 2-4 June, 2014). ISBN 978-1-61499-492-3 (print), ISBN 978-1-61499-493-0 (online). IOS press.
5. Coreţchi L., Bahnarel I., Samotîia E., Chirca L. (2010) The characteristic of stochastic effects on participants to mitigate the consequences of the Chernobyl nuclear accident. In: Bulletin of ASM, Medical Sciences, 2010, nr. 5(28), 74-82. ISSN1857-0011. (In Romanian).
6. Coreţchi L.Ş. DNA damage-related gene expression as biomarkers to assess low dose of radiation exposure. In: *International Conference Radiation Protection in Medicine*. Varna, 1-3.09.2010. p. 55.
7. Samotîia E., Chirca L., Coreţchi L., Gabunea M. (2010) The structure of malignant tumors in Chernobyl NPP accident consequences liquidation participants. In: Proceeding of III National Cancer Congress, Chişinău, p.32-33. (In Romanian).
8. Spînu C., Chintea P. A process for preparing a 5α-furostan-3β,22,26–triol–3-[O–β–D-glucopiranozil(1→2)–β–D-lucopiranozil (1→4)-β–D-galactopiranozil]–26–O–β–D-glucopiranozid (Pacovirin). Patent nr 2700 MD (In Romanian).
9. Samotîia E., Coreţchi L. et all. Immunophenotype aberrations surface of T-lymphocytes and their clinical significance. In: Proceeding of III National Cancer Congress, Chisinau, 2010, p.320-321 (In Russian).
10. Schmidt R.E. Monoclonal antibodies for diagnosis of immunodeficiency. Blut. (1989) 59:200-206.
11. Moldovanu M. Diseases clinical polymorphism and immune and genetic peculiarities at participants overcome the consequences of the Chernobyl accident. PhD thesis, Chisinau, 2001, 20 p. (In Romanian).

Author: Coretchi Liuba
Institute: National Centre of Public Health, Academy of Sciences of Moldova
Street: Gh. Asachi str., nr. 67 A
City: Chisinau
Country: Republic of Moldova
Email: lcoretschi@mail.ru

Part VI
Biomedical Engineering Education

Management and Implementation of the TEMPUS IV BME-ENA Project in the Field of Biomedical Engineering Education

Z. Bliznakov, A. Rodina-Theocharaki, and N. Pallikarakis

Biomedical Technology Unit, Department of Medical Physics, Faculty of Medicine, University of Patras, Rio - Patras, Greece

Abstract— **The future of Biomedical Engineering (BME) profession has been shown to be in high demand and countries, which do not invest in BME education will be unable to use efficiently the forthcoming healthcare technologies.**

The TEMPUS IV BME-ENA: "Biomedical Engineering Education Initiative in Eastern Neighbouring Area" project aims to promote BME Education in four partner countries, Armenia, Georgia, Moldova and Ukraine, through the creation of four joint postgraduate programs in line with the European policy and the most recent developments.

Effective coordination and management has been performed in order to provide smooth project implementation. Management structures have been established since the very project beginning and regular meetings have taken place in order to ensure successful project progress. The quality assurance has been regarded an important issue for every task performed within the project. The joint BME programs curricula have been developed taking into account the needs, the existing know-how, and the available expertise in the participating institutions.

The newly created joint MSc BME programs are expected to play an important role in strengthening the BME education in Armenia, Georgia, Moldova and Ukraine, and extending the bilateral collaboration with EU universities in the BME field.

Keywords— **Biomedical Engineering Education, Curricula Reformation, Postgraduate Joint study programs, TEMPUS program.**

I. INTRODUCTION

Health care delivery today is technology driven and Biomedical Engineering (BME), as one of the fastest growing technological areas in the developed world, is the driving force of this development. As a consequence, the role of biomedical engineers is becoming essential in both the industrial and the health care delivery sectors [1]. The future of the BME profession has been shown to be amongst the "hottest" jobs by the end of this decade [2]. It is clear that countries which will not invest in the BME education will be unable not only to develop but even to use efficiently existing and forthcoming health care technologies. In the Eastern Neighbouring Area (ENA) countries, medical technology manufacturing is extremely limited and this is partly due to the inexistence of specialized educational programs in BME. Lack of specific educational programs in BME results in the absence of research and development in this field, a fact that has a negative effect on the local production of even simple medical devices.

The wider objective of the TEMPUS IV BME-ENA: "Biomedical Engineering Education Initiative in Eastern Neighbouring Area" project, is to promote BME Education in the ENA countries through the creation of new postgraduate programs in line with the European policy and the most recent development and demands in the field. The BME-ENA project aims to create four comprehensive joint MSc programs in the BME field in one of each four ENA countries, Armenia, Georgia, Moldova and Ukraine, that will provide the possibility to link local educational institutions with high experience EU universities.

The BME-ENA consortium, established to meet the needs and achieve the project's objectives and goals, consists of 7 EU and 10 ENA Partner Institutions, from 11 countries (Fig. 1).

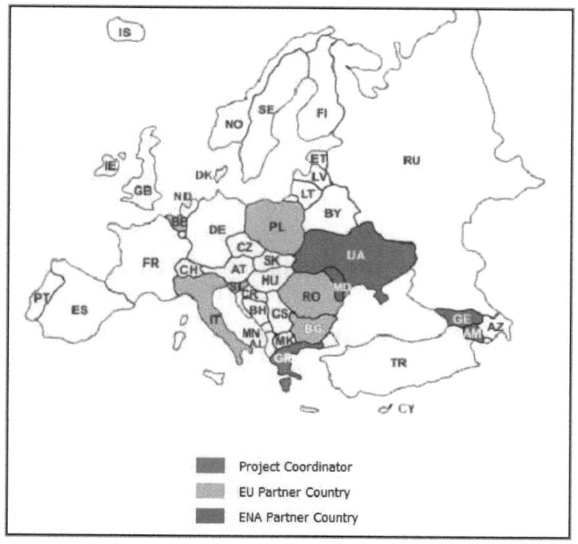

Fig. 1 The BME-ENA project consortium

All consortium members have high priority to promote education in BME and an extensive experience in international projects and collaboration. The core group of the consortium consists of universities collaborating since the 90's in the framework of the ERASMUS and TEMPUS Programs: Vrije Universiteit Brussel (VUB) – Belgium,

V. Sontea and I. Tiginyanu (eds.), *3rd International Conference on Nanotechnologies and Biomedical Engineering*,
IFMBE Proceedings 55,
DOI: 10.1007/978-981-287-736-9_115

University of Bologna (UNIBO) – Italy, and University of Patras (UPAT) – Greece that started their collaboration over 20 years ago, have well developed graduate and/or postgraduate programs in BME. This initial core group has been joined by Technical University of Varna (TUV) – Bulgaria, "Grigore T. Popa" University of Medicine and Pharmacy (POPA) – Iasi, Romania, West Pomeranian University of Technology (ZUT) – Szczecin, Poland, and University of Ljubljana (UL) – Slovenia. These universities also offer BME study programs and have recently implemented the TEMPUS IV CRH-BME "Curricula Reformation and Harmonisation in the field of Biomedical Engineering" project recommendations [3,4]. All EU universities have already well established links with ENA universities and are looking forward to promoting their collaboration. The 10 participating institutions from ENA countries are: National Polytechnic University of Armenia (NPUA) (former State Engineering University of Armenia (SEUA)), Russian-Armenian State University (RAU) and Armenian Association of Telemedicine (AATM) in Armenia; Georgian Technical University (GTU), Tbilisi State Medical University (TSMU), and Akaki Tsereteli State University (ATSU) in Georgia; Technical University of Moldova (TUM) and State University of Medicine and Pharmacy (SUMPh) in Moldova, as well as National Technical University of Ukraine "Kyiv Polytechnic Institute" (NTUU_KPI) and Sumy State University (SSU) in Ukraine.

The BME-ENA project has overall duration of three years (36 months) and currently the project has been under implementation for 18 months. The purpose of this work is to report on the current project progress implementation, present the most significant project achievements up-to-date, while focusing on several important aspects concerning the project coordination management and quality assurance.

II. BME-ENA Project Coordination and Management

This project activity deals with the working methodology adopted for the effective coordination and management, proven to provide smooth project implementation.

During the 1st General Assembly and Kick-off Meeting, the project management responsibilities have been assigned to the following project bodies: General Assembly (GA), project Steering Committee (SC) and the Project Coordinator (PC). Local coordination, day-to-day management and administrative tasks at country level has been assigned and performed by 4 Local Working Groups (LWGs), one in each ENA country. The Project Coordinator (UPAT) has undertaken general project management, has an overall responsibility of the project and all partners, manages project resources, monitors overall project performance, reports to the EU Commission, promotes project visibility and is the primary contact point for all formal communications. The GA is the highest decision-making body and consists of 17 representatives – one from each partner. The responsibility of the project implementation in the partner countries is under the respective LWG. Each of the LWG is composed of 2 representatives from each partner institution from the respective ENA country and one expert from EU country has been assigned to each LWG for assistance in Joint MSc BME program's creation. Each LWG has elected a country coordinator who is also a member of the project SC. The SC is composed of 7 members: 4 from the ENA countries, 2 from EU Partners appointed by the GA and the PC. The SC overlooks and continuously evaluates the project progress, decides on high level management issues including planning and control matters, and has in general the role of a leadership group. The GA has the task of the strategic planning of the project and also decides by consensus on the main actions and other project issues that may arise, on the basis of proposals introduced by the PC or any other project body or partner.

To assure the smooth project implementation and effective overall day-to-day management, the BME-ENA working groups/teams (WG) have been established in each institution. Regular LWG and each institutions' WG meetings have been arranged in order to tackle all relevant issues for ensuring the successful implementation of the project. All the partners at the institutional and LWG levels have duly conducted daily administrative and financial management of the project: planning and distribution of the tasks from the different work packages between the team/working group members, organizing of the working group meetings, coordinating of project activities, providing communication and collecting the information from project staff, regular communication with partner institutions within the BME-ENA project, preparing, signing, archiving and mailing the administrative and financial documents and reports.

During the first few months of the project implementation the PC has performed bilateral visits to ENA countries' partner institutions in order to explore the situation, discuss with local partners on particular points that need attention and better prepare the collaboration and action plans in order to facilitate the progress of the joint MSc programs under creation. The 1st GA and Kick-off meeting was planned to take place in Ukraine in February 2014, but it had to be cancelled due to the political situation in the country. Therefore, it was held in Patras, Greece, on 28-29 April 2014, and representatives from all partners participated. With this occasion, a Workshop on "Biomedical Engineering Education in Europe", endorsed by the International Federation for Medical and Biological Engineering

(IFMBE), the International Academy of Medical and Biological Engineering (IAMBE) and the Hellenic Society for Biomedical Technology (ELEVIT), was held on 28 April 2014 at the Conference and Cultural Centre of the University of Patras, Greece. Following the 1st GA meeting, the 1st SC meeting was held in Dubrovnik, Croatia, with the occasion of the 6th European Conference of the International Federation for Medical and Biological Engineering (MBEC 2014), 7-11 September 2014. Within the framework of the Conference, a Special Session on Biomedical Engineering Education (BME-ENA) has taken place on 9 September 2014. In the first 18 month, more than 20 Local Coordination and/or Project Working Group meetings with the participation of ENA and EU representatives (including PC) have taken place. On the local level, partners have held regular meetings, whenever was necessary in order to ensure the smooth project implementation progress.

III. BME-ENA Project Implementation

The aim of the project is to develop tailor made curricula for four MSc programs in BME involving 10 partner institutions in Armenia, Georgia, Moldova and Ukraine, in order to support a systematic approach to education and research in this very important area.

As a first step, requirement analysis on the particular partner country demands has been performed. Detailed analysis of the state of education in BME and related areas of training and higher education, as well as the state of BME sector in all four countries have been made. It has been also analyzed the current legislation in the field of higher education and regulatory documents on the creation and implementation of new training programs for MSc studies. The healthcare domain in all four countries is represented by a number of public and private institutions, and increasing penetration of modern technologies, while the medical industry is underdeveloped and mostly represented by local dealers of the international brands.

The bases for curricula development have been placed on the results of the previous TEMPUS IV CRH-BME project [3]. The development of the curricula has been based on knowledge transfer from the EU partners and has been adapted according to the needs, the existing know-how and infrastructure in the participating institutions. Upon the results of the analysis a final list of BME programs' courses has been elaborated and approved for the four MSc BME curricula. Particular attention has been paid to ensure a balanced subdivision between basic biomedical topics, core topics and elective topics, and an appropriate distribution of the topics across the semesters for smoothly schedule running, from basic medical, to core engineering, and highly specialized elective topics, towards final thesis elaboration.

Each of the Partner Countries university has been assigned to develop a number of BME program topics based on their available expertise, including the course syllabi and teaching materials, and subsequently to deliver them during the implementation of the study program. As a result, the complete syllabi for the BME programs including both core courses and elective topics have been developed. New teaching materials have been prepared where necessary, while existing and available teaching materials have been adapted to satisfy the particular needs.

Special attention has been given to the procedures for appropriate selection of teaching staff as well as preparation of training seminars for qualification improvement. Staff selection of ENA partner countries' universities has been performed during visits of the representatives of Partner' Countries universities to EU partners and vice-versa. These activities have contributed and highly facilitated the knowledge transfer from EU partners to the Partner Countries institutions. Furthermore, majority of them have also been performed on a local level at each university site. Teaching staff has been selected from a number of qualified professionals with a high experience in the corresponding BME programs courses. Emphasis has also been given to the implementation and preparation of educational materials. Each teacher has also been responsible for the development and preparation of the educational and teaching materials in the corresponding topics, in most cases with the cooperation of EU partners. Each EU partner university has examined and identified possibility for active involvement of their teaching staff in every aspect concerning the implementation of four BME programs in ENA countries. Specifically, for the preparation of educational materials, EU partners have adapted available teaching materials and will prepare ad-hoc new educational materials for the use in the joint MSc BME programs.

Next steps include the completion of the accreditation procedures for the BME programs and the official establishment according to the respective national legislation, the programs announcement, the procedures for students registration, the running of the BME programs according to the schedule and education of young engineers. Invited lecturers from EU partners have been involved to teach specific topics for which local expertise is missing or insufficient. All necessary activities have been coordinated by the LWG in each ENA country, together with the PC (UPAT).

All EU partners have actively participated in the discussions on the design and implementation of joint MSc BME programs, from the perspectives of their expertise. Specific compliance problems have been discussed and possible solutions have been proposed in an attempt to assist in the successful implementation of the designed programs.

IV. BME-ENA PROJECT QUALITY ASSURANCE

This project activity addresses the need for the projects' continuous quality plan, the preparation of the project's internal guidelines for reporting and reviewing procedures and the implementation of the project's Quality Assurance (QA) system. A contingency plan aiming to create effective methods of maintaining project operations against unforeseen eventualities, sustain the project's results, and ensure that project activities are performed on time, has been developed since early beginning of the project.

The project quality and its pursuit is regarded an important issue for every single individual task within the project. The quality of the project has been continuously monitored and assessed through the establishment of an appropriate QA system that has been designed to address both the quality of the project documentation and the project activities. All partners have agreed by consensus and follow the quality control and monitoring procedures that have been adopted. The produced outcomes and outputs have been monitored and assessed through reviewing procedures and regular reporting from all partners submitted to the project coordinator. The use of telematics communication means automate many of the internal monitoring procedures and facilitate management tasks. Review procedures for project outcomes, have been implemented. The project deliverables are being reviewed and approved prior to their finalization and publication. During the planned GA and SC meetings, milestone review takes place, the progress of the project is being examined and the project results thoroughly reviewed.

Furthermore, the majority of the project partners have implemented and use QA and control systems at local university level. All the activities within the framework of the project are also performed according to the requirements of the home institution quality systems that further on increases the quality of the project outputs. In the Partner Countries, National Erasmus+ Offices are also involved in the quality control and monitoring process of the BME-ENA project.

Concerning the QA of the MSc BME programs to be implemented, support has been provided at all stages of the development of the programs, in compliance with the QA plan. During the second half of the project duration, continuous improvement of the educational process is planned to be achieved through evaluation and QA procedures. Aggregated feedback data will be discussed at the meetings of the project consortium representatives. Peer-reviewing measures are foreseen to be implemented, including an external assessment of the project activities and results.

V. CONCLUSIONS

The biomedical engineer applies Engineering knowledge and skills to solve problems in biology and medicine. The BME-ENA project is expected to play an important role in strengthening the education in BME in the ENA countries through knowledge transfer from the EU universities. The joint MSc BME programs are expected to become a model in applying the European Higher Education Area policy, state-of-the-art educational approaches, use of ECTS and extensive bilateral collaboration agreements with EU countries. The BME education in ENA countries will be offered the opportunity to enlarge its horizons and extend the activities beyond the national borders through the creation of new international collaborations in this field.

ACKNOWLEDGMENT

This project has been funded with support from the European Commission. This publication reflects the views only of the author, and the Commission cannot be held responsible for any use which may be made of the information contained therein.

CONFLICT OF INTEREST

The authors declare that they have no conflict of interest.

REFERENCES

1. Bliznakov Z, Pallikarakis N, Overview of Biomedical Engineering education programs in Europe: The results of the CRH-BME project survey, 5th European Conference of International Federation on Medical and Biological Engineering, IFMBE Proceedings, Vol. 37, 2012, pp 1414-1417
2. The Wall Street Journal "What Will Be the Hot Jobs of 2018?", May 26, 2010, at http://online.wsj.com/news/articles/ SB100014240527487040262045752663429354l8962
3. TEMPUS IV project CRH-BME: Curricula Reformation and Harmonisation in the field of Biomedical Engineering, at http://www.crhbme.upatras.gr/
4. Jarm T, Miklavcic D, Pallikarakis N, Bliznakov Z, Magjarevic R, Lackovic I, Pecchia L, Stagni R, Jobaggy A, Barbenel J, Proposal for Generic Biomedical Engineering Programs Based on European Experience, 5th European Conference of International Federation on Medical and Biological Engineering, IFMBE Proceedings, Vol. 37, 2012, pp 1418-1421

Corresponding author:

Author: Zhivko Bliznakov
Institute: University of Patras
Street: University Campus
City: Rio - Patras
Country: Greece
Email: bliznakov.zhivko@gmail.com

Application of Computational Phantoms and their 3D Print-outs for Educational Purposes

A. Marinov[1], D. Ivanov[1], Z. Bliznakov[1], H. Bosman[2], I. Buliev[1], and K. Bliznakova[1]

[1] Department of electronics and microelectronics, Technical University of Varna, Varna, Bulgaria
[2] Department of Radiology, Katholieke Universiteit Leuven, Leuven, Belgium

Abstract— **The paper presents an approach to be used in the face-to-face and the on-line phases of one of the chapters in module from the EUTEMPE-RX project. The module and the project are aimed at the training of medical physics experts in radiology. The approach consists of: (a) generation of computational anthropomorphic phantoms, (b) program scripts, specifically developed to convert computational phantoms to suitable formats used with 3D printing devices, and (c) 3D printing of software models. In this paper, the design and printing of a 3D breast glandular system has been demonstrated. Two small size breast phantoms differing in the complexity of their duct tree system were generated with a dedicated program. A set of image processing techniques were applied on the obtained breast phantoms to obtain an STereo-Lithography (STL) file format of the breast's duct tree that was imported within the software of the 3D printer used to print the sample breast tree.**

Keywords— **3D modelling, 3D printing, computational phantoms, e-based learning.**

I. INTRODUCTION

EUTEMPE-RX is an EU FP7 project within the Euratom in 'Nuclear Fission, Safety and Radiation Protection' financed by the Commission of the European Communities [1]. The main reason for this challenging initiative was the need for a platform devoted to the education of Medical Physics Experts in Radiology. In December 2014, the on-line phase of the first course module was launched and the face-to-face phase took place in February 2015. The second module started in March 2015.

Ten more modules with radiation safety related topics in diagnostic and interventional radiology are under development (http://www.eutempe-rx.eu/). Individual modules will take typically one month of online preparation with a web-based teaching platform and one week of face-to-face teaching for a limited number of participants or students. The leaders of the modules need to present the module content during on-line and face-to-face phases as attractively as possible to the participants who are both medical physicists and biomedical engineers.

The aim of this paper is to present the approach used in "Anthropomorphic phantoms" module not only to teach attractively the content of chapter "Computational Anthropomorphic Phantoms as Prototypes of Physical Objects" but also to demonstrate the feasibility to produce real physical phantoms for use in the clinical practice from their computational models. This approach includes the creation of software anthropomorphic phantoms and the implementation of a set of image processing operations used to prepare the phantom file in a suitable format for 3D printing devices.

II. THE APPROACH

The proposed approach for teaching "*Computational Anthropomorphic Phantoms as Prototypes of Physical Objects*" is outlined in figure 1 and can be summarized in the following eight steps:

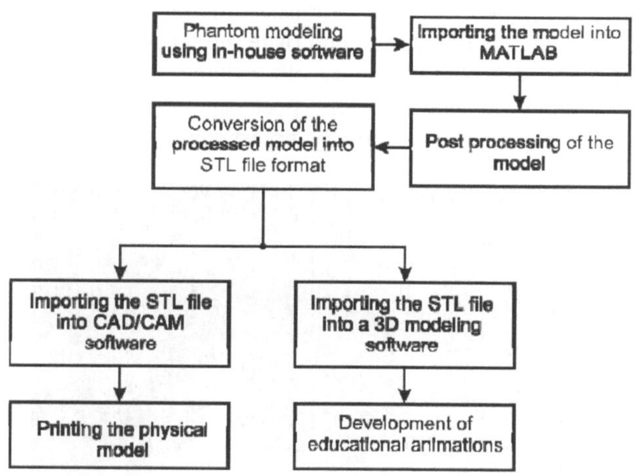

Fig. 1 Flow-chart diagram of the presented approach

(a) Generation of a computational model by using the in-house developed BreastSimulator software, described in details in [2, 3]. The software application for this step will be available for student training and education. Demonstrations, organised in the form of screen capture videos, will accompany the teaching process and will focus on the interaction of the students with the software interface in order to generate successful computational models of the breast. The output file from this software application is a binary file containing a voxelised version of the computational model.

V. Sontea and I. Tiginyanu (eds.), *3rd International Conference on Nanotechnologies and Biomedical Engineering*,
IFMBE Proceedings 55,
DOI: 10.1007/978-981-287-736-9_116

(b) Importing the model in MATLAB environment.

(c) Performing of relevant post processing, e.g. image enhancement for smoothing the voxel-based software model and decreasing the possible artefacts from voxelisation in the final printed version.

(d) Conversion of the processed model to STereoLithography (STL) file format. The obtained STL file can be used for either printing the physical model or for development of educational material. Specifically:

(e) Importing the STL file into dedicated Computer Aided Design (CAD) or Computer Aided Manufacturing (CAM) software applications, processing of the model and appending additional supportive structures and finally printing the physical model during the face-to-face training.

(f) Importing the STL file into a dedicated 3D modelling software for producing computer animations for the purposes of model visualization to be used during the on-line phase.

The approach was experimentally implemented and subjectively validated taking into account the opinion of colleagues from the University.

III. IMPLEMENTATION

The first step in the implementation of the proposed approach is the generation of breast phantoms using the in-house developed software - BreastSimulator. The computational breast phantom is a composite model of the woman's breast and includes: the breast outer shape, its duct system, the Cooper's ligaments, the pectoral muscle, mammographic texture and eventually breast abnormalities [2, 3].

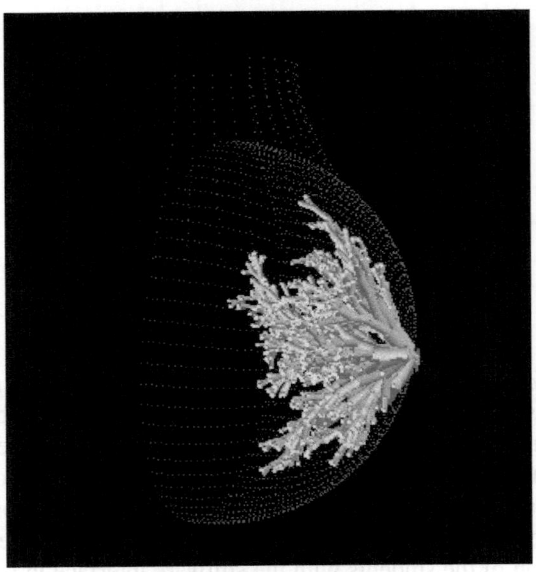

Fig. 2 Computational model of a duct system in an uncompressed breast shape

For the purposes of this work, the computational phantom included only the breast duct system and breast shape as shown in Figure 2. The duct system is modelled using a network of cylinders, probabilistically arranged in the breast as branches, in a tree like arrangement. Extensive explanation on the mathematical development of the duct model can be found in [2].

So far, two such breast phantoms were generated that differ in their duct tree complexity: (1) a breast phantom consisting of a simpler breast system configuration composed of two main duct trees as shown in figure 3 and (2) a breast phantom composed of a more complex breast duct tree model as shown in figure 4. For the first 3D printings, the external breast shape was skipped, as the reproducing of the duct tree in details was considered as much more challenging.

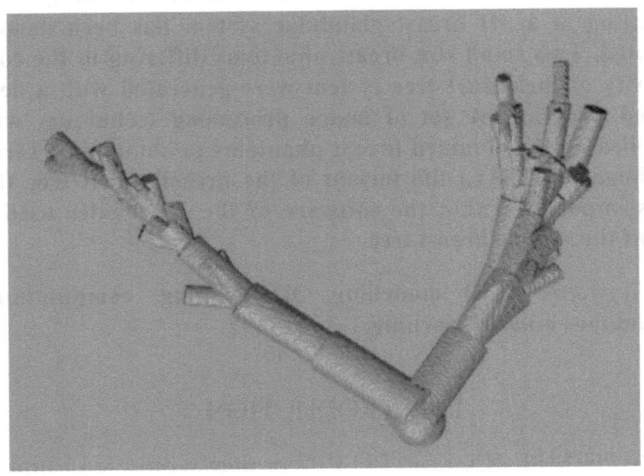

Fig. 3 Software model of breast duct system, generated with the BreastSimulator software- configuration of two main ducts

Fig. 4 Software model of breast duct system, generated with the BreastSimulator software - configuration of four main ducts

The 3D presentations of the generated breast ducts were additionally processed. The detailed reproducing of the walls of the individual voxels (although precise) is actually unrealistic and a drawback in this case. Therefore, a 3D surface smoothing was applied. The processed computational models are shown in figures 5 and 6.

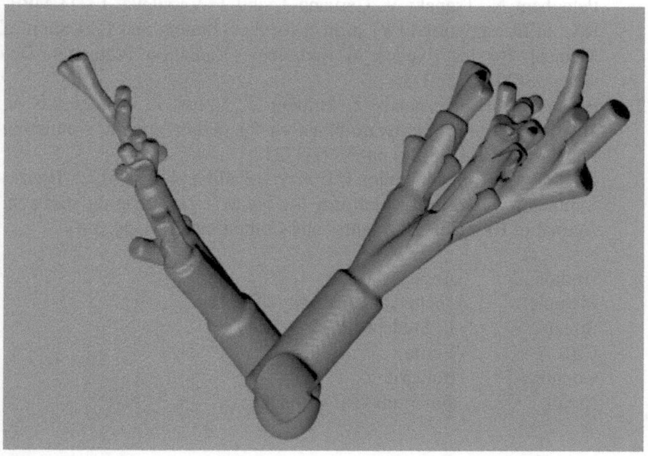

Fig. 5 Processed computational model of breast ducts - simple duct system

Fig. 6 Processed computational model of breast ducts - sophisticated duct system

The processed computational models were next used for developing of computer animation videos and for manufacture of the 3D physical models. Stereolithography printing technology was used. The two printed samples are shown in figures 7 and 8.

The selected technology allowed the use of minimum number of supporting structures during the 3D printing.

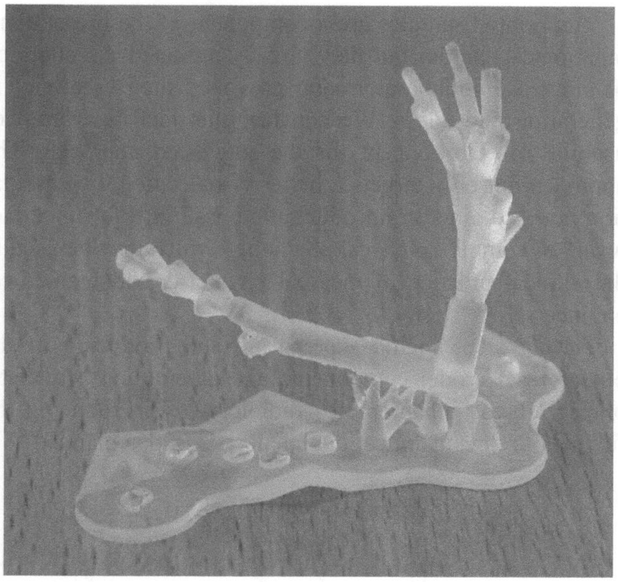

Fig. 7 Printed breast duct model - simple duct system

Fig. 8 Printed breast duct model - sophisticated duct systems

IV. Conclusions and Future Work

This paper presented a completed approach for education relevant to computational phantoms, where several new and modern technologies are used – computer animation videos and 3D printing. In the paper the approach was limited to the duct system breast phantoms, but it can also be successfully implemented for other anatomical parts as well.

The printed samples are a component of the breast computational phantom but they already attracted the attention of our close collaborators and provoked their creativity in generating new ideas. We consider this fact as a proof of usability of the concept for the suggested approach. The approach not only targets a better visualisation or presentation but also aims to introduce the possibility for manufacturing of complex physical phantoms, which may be used in the clinical practice of medical physicists and biomedical engineers.

Further steps includes preparing and producing of the other breast model components and assembling a realistic physical breast phantom for radiology use. Producing the full set of relevant educational materials is also planned.

ACKNOWLEDGMENT

This work is supported by EUTEMPE-RX Project of the European Union Seventh Framework Program (FP7/2007-2013), under Grant Agreement No: 605298

CONFLICT OF INTEREST

"The authors declare that there are no known conflict of interest".

REFERENCES

1. Bosmans H, Bliznakova K, Padovani R, Christofides S, Van Peteghem N, Tsapaki V, Caruana J, and J. Vassileva, EUTEMPE-RX, an EC supported FP7 project for the Training and Education of Medical Physics Experts in Radiology, Radiation Protection Dosimetry, (2015), pp 1-5.
2. Bliznakova K, Bliznakov Z, Bravou V, Kolitsi Z, Pallikarakis N, 2003, A 3D breast software phantom for mammography simulation Phys. Med. Biol. 48(22) pp. 3699-3721
3. Bliznakova K, Sechopoulos I, Buliev I, Pallikarakis N, 2012, Breast-Simulator: A software platform for breast x-ray imaging research. Journal of Biomedical Graphics and Computing, 2(1), pp. 1-14

Author: Angel Marinov
Institute: Technical University of Varna
Street: 1 "Studentska"
City: Varna
Country: Bulgaria
Email: a.marinov@tu-varna.bg

Development of the BME MSc Study Program in Georgia within the BME-ENA TEMPUS IV Project

T. Sanikidze[1], I. Gotsiridze[2], G. Gigilashvili[2], S. Dadunashvili[2], D. Gegechkori[3], I. Pkhakadze[3], D. Nadareishvili[1], S. Kiparoidze[1], E. Shekiladze[1], E. Gogilidze[2], T. Pertaia[3], E. Nyssen[4], W. Chlewicki[5], J. Górecka[5], P. Makiewicz[5], and K. Penkala[5]

[1] Tbilisi State Medical University (TSMU), Tbilisi, Georgia
[2] Georgian Technical University (GTU), Tbilisi, Georgia
[3] Akaki Tsereteli State University (ATSU), Kutaisi, Georgia
[4] Vrije Universiteit Brussel (VUB), Brussels, Belgium
[5] West Pomeranian University of Technology (ZUT-WPUT), Szczecin, Poland

Abstract— **In the paper current state of the progress on curricula and syllabi development for joint multidisciplinary MSc BME Study Program in Georgia is described. The work is carried out by the Georgian Working Group within the TEMPUS IV BME-ENA Project "Biomedical Engineering Education Tempus Initiative in Eastern Neighbouring Area", realized in years 2013-2016. The MSc BME program has been prepared jointly by three Georgian universities: Akaki Tsereteli State University (ATSU), Tbilisi State Medical University (TSMU) and Georgian Technical University (GTU), with support of academic teachers from partner EU universities, mainly the West Pomeranian University of Technology (ZUT-WPUT), Szczecin, Poland. In the paper problems related to the project design and implementation are discussed (structure of the program, accreditation issues, preparation of teaching material). Results of three meetings of the GE WG which were held in Szczecin, Brussels and Tbilisi are presented. Sustainability activities carried out by the GE WG are also described: application for the ERASMUS+ credit mobility project coordinated by the West Pomeranian University of Technology, Szczecin.**

Keywords— **BME-ENA, CRH-BME, MSc Study Program, curriculum, syllabus, Higher Education**

I. INTRODUCTION

Project BME-ENA - Biomedical Engineering Education Tempus Initiative in Eastern Neighbouring Area (Project number: 543904-TEMPUS-1-2013-1-GR-TEMPUS-JPCR) aims to expand the training of practitioners specializing in biomedical engineering from Europe to its eastern neighbouring countries. In the project implementation of educational program is planned, which will ultimately deliver successful and competitive specialists in medical and technical areas. Therefore, the joint educational program seems to be an optimal model of such expediency. Curricula and syllabi of the MSc Joint Program in Biomedical Engineering (BME) are being designed according to the recommendations elaborated within the TEMPUS IV CRH-BME Project run in 2009-2012, and within previous EU educational projects [1-6]. The content of the study program is adjusted to the national priorities and needs, as well as to the expertise available at the Universities involved.

Presently, continual modifications have been made to the education system of traditional degree programs. That has been done to make the process of attaining a degree more personally and professionally valuable for students. One such modification includes the implementation of the joint programs.

Towards the European Higher Education Area: responding to challenges in a globalised world, Conference of European Ministers Responsible for Higher Education, London, 17-18 May 2007:

At national level, we will work to implement fully the agreed recognition tools and procedures and consider ways of further mobility for both staff and students. This includes encouraging a significant increase in the number of joint programs and the creation of flexible curricula, as well as urging our institutions to take greater responsibility for staff and student mobility, more equitably balanced between countries.

There are several alternative definitions/models of joint study programs. Under the Project realization we developed the model of joint program which refers to situations where three institutions collaborate on a joint study programs, but where each institution is responsible for admission module implemented in particular University. This program leads to a degree at all of the partner institutions. Each institution has academic responsibility for the contents of the study program. The course descriptions must be approved by the institution that issues the degree, and this institution is responsible for assessing the basis for the diploma in relation to the descriptions of learning outcome in the program description. The institution issuing the diploma is responsible for academic approval of the course and for the quality of the program.

V. Sontea and I. Tiginyanu (eds.), *3rd International Conference on Nanotechnologies and Biomedical Engineering*,
IFMBE Proceedings 55,
DOI: 10.1007/978-981-287-736-9_117

While collaboration on a joint study program is less of a commitment than a joint degree, many of the same academic, administrative and practical challenges must be discussed and solved by the partner institutions.

At the same time we had consultation with National Centre for Educational Quality Enhancement for compatibility of our educational program with the accreditation standards in Georgia. Standards of accreditation are as follows: Aim of educational program, results of learning and compliance of a program with those; Methodology and organizing of teaching, adequacy of evaluation of learning of a program; Students' achievements, individual work with them; Provision of teaching resources; Capabilities of development of quality of teaching. For this time, procedures are finished, and we are going to submit an accreditation application.

While working on the program, simultaneously we carried out a research among the potential employers. The survey was held with managements of medical corporation "Evex" clinics and managements of different pharmaceutical firms. The aim of the survey was to find out what is the profession of persons, who design the apparatus and deliver medical equipment services.

The research demonstrated the following picture: in Georgia, service of medical equipment in various hospitals is carried out by general specialization engineers in most cases. In accordance, the demand on Biomedical engineers is very strong, which further confirms the need and urgency of our joint program.

Because the structure and function of a joint degree program will differ based on the rules and regulations of each learning institution, it is important to understand the different types that exist.

Name of the program: Biomedical Engineering

Qualification: Master of Biomedical Engineering/Master Sc in Biomedical Engineering

Educational program will be given at least 120 credits.

The volume of the program by the ECTS: 120 ECTS credits include educational program. The Master's program is carried out jointly by three institutions of higher education: Akaki Tsereteli State University (ATSU), Tbilisi State Medical University (TSMU) and the Georgian Technical University (GTU). At ATSU one credit equals 25 academic hours, at TSMU - one credit equals 30 academic hours of students, at GTU one credit equals 27 academic hours. The MSc Joint Program in Biomedical Engineering resides equally within the Faculties of Health Sciences and the Faculties of Engineering at the collaborating Universities.

Educational program duration is 2 academic years and it includes 4 semesters. Its general structure is shown on Figure 1.

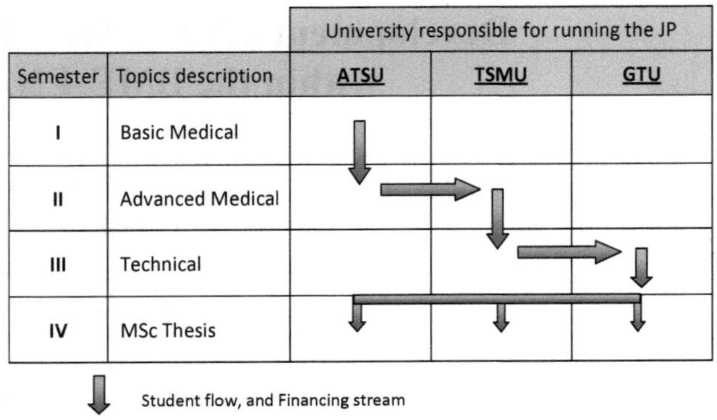

Fig. 1 General structure of the MSc Joint Program in Biomedical Engineering in Georgia

- The first semester (30 ECTS) is carried out at the Akaki Tsereteli State University and is devoted to human organ systems in order to study their structure and normal operation. This should give theoretical basics for medical training courses, public health and health care technology as well as create the basics for medical legislation.

- The second semester (30 ECTS) is carried out at the Tbilisi State Medical University and is devoted to Biomedical Physics and High Mathematics, Physical Principles of diagnostic and therapeutic methods in Medicine, Modern Methods of Medical Visualization and Health Risk, Biomedical Statistics, Research Methodology and the basics of project planning. After completing these courses, graduate students will be able to make determination, interpretation and evaluation of physical characteristics of the structures and processes in living organisms. They will gain biomedical knowledge and skills necessary for realization in practice professional tasks in the field of Biomedical Engineering.

- The third semester (30 ECTS) is carried out at the Georgian Technical University and is devoted to Medical Electronics, Biomedical Signal Processing, Medical Imaging Systems and modelling studies.

- The fourth semester (30 ECTS) is devoted to the work on Master's thesis, and to carrying out studies on the topic chosen by the student.

Educational program includes elective courses to choose transparency 8 credits.

Master's program in Biomedical Engineering will be given to graduates of academic degree, a higher education qualification, which will be assigned to each of the three higher education institutions: ATSU, TSMU and GTU jointly. A graduate is awarded a diploma certifying the joint qualification. If the program is part of the Diploma Supplement, accurately determine the partner university. Students will be enrolled at Master's degree program under the rules set by the Georgian legislation, as well as on the basis of the Unified National Entrance Examinations and a particular program.

The study program developed within the BME-ENA project should attract more candidates, leading to significant increase of the number of employees well prepared to the demands of today job market [7]. We anticipate that graduating students will be in a position to make enormous contributions to the field, developing innovative solutions to medical and health-related problems in the Georgian Healthcare System, as a result of their interdisciplinary exposure and training.

BME-ENA Project activities of the Georgian Working Group are supported by the West Pomeranian University of Technology (ZUT-WPUT), Szczecin, Poland. This is the result of long lasting co-operation between the Faculty of Electrical Engineering, ZUT-WPUT and the GTU as well as TSMU, including collaboration of student research groups from these three Universities. Additional support is offered by another BME-ENA Partner – Vrije Universiteit Brussel (VUB), Belgium.

Three meetings of the Georgian Working Group were organized up to now:

- The 1st BME-ENA Georgian Working Group Meeting & Workshop was held on 26th – 31st October 2014 in Szczecin, Poland, under the auspices of Rector of the ZUT-WPUT, Polish Federation of Engineering Associations (FSNT NOT), and was technically co-sponsored by the Association of Polish Electrical Engineers (SEP). Main activities of the meeting participants were focused on preparing final curricula for the Georgian BME Joint Program, discussions on the content of courses, investigating the possibility for creation of Bilateral Agreements between ENA and EU partners, and agreements regarding creation of common educational and teaching materials. Within the accompanying workshop on education in the field of BME eight papers were presented on links between education, research and industry. Two of them demonstrated activities of the BME Student Research Group Akson (FEE ZUT-WPUT) in this field. In the workshop numerous students participated, along with representatives of Pomeranian Medical University in Szczecin (Department of Medical Physics), professional and scientific societies as well as hospitals. During the meeting, one-day visit to Poznań was organized. Labs of the Poznań University of Technology, Adam Mickiewicz University, and Karol Marcinkowski University of Medical Sciences were visited as well as Nanobiomedical Centre (Figure 2). It was also the opportunity to present the BME-ENA project to the academic staff of Poznań HEIs.

Fig. 2 Visiting labs of the Nanobiomedical Centre in Poznań

- The 2nd BME-ENA Georgian Working Group Meeting took place on 1st – 4th February 2015 in Brussels, Belgium, and was organized by the academic staff from Vrije Universiteit Brussel (VUB). The meeting, supported by the BME-ENA Project Coordinator, was devoted mainly to the problems concerning accreditation of the Georgian BME Joint Program. Particular steps of the accreditation procedure were discussed as well as technical and financial aspects of the program. During the meeting, participants delivered three presentations (on the joint Biomedical Engineering teaching program of VUB and Ugent, on scientific activities in the field of BME run at VUB, and on Erasmus+ mobility schemes). Also, a visit to the TeraHz research lab was organized (Figure 3).

Fig. 3 Visiting the TeraHz research lab of prof. Johan Stiens

- The 3rd BME-ENA Georgian Working Group Meeting was organized on 20th – 23rd April 2015 in Tbilisi and Kutaisi, Georgia, with participation of the BME-ENA Project coordinators. Specific problems associated with the implementation and accreditation of the BME Joint Program in Georgia were discussed during the meetings at all three Partner Georgian Universities, and at the NEO+ of Georgia in Tbilisi. Workshops "Biomedical Engineering Perspectives" were organized at these HEIs (Figure 4), and meetings with the teaching staff in the new joint BME program. Also, the discussion on technical and financial problems related to implementation of the program was run (the structure of the Georgian BME Joint Program was re-defined – as it is shown on Figure 1), and on elaboration of common educational material in English language. The last two days of the meeting were dominated by this issue – we agreed the most important topics that should be covered within teaching materials which are being prepared for the BME-ENA Project by the ZUT-WPUT team (an academic textbook on BME, and several presentations in English language for lectures run at the Partner Universities). We also discussed the possibilities for organizing staff training visits at the FEE ZUT-WPUT, in co-operation with the Pomeranian Medical University in Szczecin.

Fig. 4 Poster announcement of the Workshop on Biomedical Engineering Perspectives at TSMU, Tbilisi, Georgia.

All three Partner Universities from Georgia, and the National Technical University of Ukraine "Kyiv Polytechnic Institute", also a Partner in the BME-ENA Project, expressed their intention to participate in the ERASMUS+ credit mobility project coordinated by the West Pomeranian University of Technology, Szczecin. All these HEIs from the Eastern Partnership Countries are interested in cooperation for the exchange of students and staff with the West Pomeranian University of Technology, Szczecin, in the context of the Erasmus+ Programme, Key Action 1, Learning Mobility of Individuals, Higher education student and staff Mobility between Programme and Partner Countries. Inter-Institutional Agreements for academic years 2015/16 – 2017/18 were signed between the ZUT-WPUT and Partner Universities. They also provided input necessary to develop Application Form – call 2015, with regard to Quality Questions, and established a structure and procedures supporting the project implementation (selection, preparation, recognition and dissemination). Application for the ERASMUS+ credit mobility project was submitted by the ZUT-WPUT to the Foundation for the Development of the Education System (FRSE), which has been appointed the Polish National Agency of the Erasmus+ Programme implemented in the years 2014-2020. We believe that success of our common application will create a good sustainability platform for the results of our BME-ENA Project. Erasmus+ credit mobility framework should provide effective schemes for student and staff exchange in the field of BME.

ACKNOWLEDGMENT

BME-ENA – Biomedical Engineering Education Tempus Initiative in Eastern Neighbouring Area, Project Number: 543904-TEMPUS-1-2013-1-GR-TEMPUS-JPCR, is a Joint Project within the TEMPUS IV program and is 90% financed by the Commission of the European Communities.

This project has been funded with support from the European Commission. This publication reflects the views only of the author, and the Commission cannot be held responsible for any use which may be made of the information contained therein.

CONFLICT OF INTEREST

The authors declare that they have no conflict of interest.

REFERENCES

1. Bliznakov Z., Pallikarakis N.: Review of the Biomedical Engineering education programs in Europe within the framework of TEMPUS IV, CRH-BME project. Medicon 2010 Conference, Porto Carras, Chalkidiki, Greece, 27-30 May 2010, www.springerlink.com

2. Dadunashvili S., Sanikidze T., Pkhakadze I., Pertaia T., Gegechkori D., Nyssen E., Makiewicz P., Chlewicki W. and Penkala K.: BME-ENA TEMPUS IV Project - Curriculum for the BME MSc Study Program in Georgia. 6th European Conference of the International Federation for Medical and Biological Engineering (MBEC2014), Dubrovnik, Croatia, 7-11 September 2014
3. Griva V., Pallikarakis N.: A European Biomedical Engineering Postgraduate Program - From Evaluation to Continuous Improvement. Medicon 2010 Conference, Porto Carras, Chalkidiki, Greece, 27-30 May 2010, www.springerlink.com
4. Kolitsi Z. (Ed.): Towards a European Framework for Education and Training in Medical Physics and Biomedical Engineering. Studies in Health Technology and Informatics, Vol. 82. IOS Press, Amsterdam, 2001
5. Penkala K.: TEMPUS IV Project „Curricula Reformation and Harmonisation in the field of Biomedical Engineering CRH-BME". Bio-Algorithms and Med-Systems (BAMS), UJ, 2011, 47-50
6. Penkala K.: Tempus Project CRH-BME "Curricula Reformation and Harmonisation in the field of Biomedical Engineering as example of structural cooperation enhancing mobility". Eastern Dimension of Mobility Conference, Eastern Partnership Conference, Warsaw, 6-7 July 2011, www.eap-mobility.pl

Author: Przemysław Makiewicz
Institute: Department of Systems, Signals and Electronics Engineering, Faculty of Electrical Engineering, West Pomeranian University of Technology, Szczecin, Poland
Street: Sikorskiego 37
City: Szczecin, 70-313
Country: Poland
Email: makiewicz@zut.edu.pl

The Implementation of the BME-ENA Tempus Project in Ukraine

A. Orlov[1], T. Volkhova[1], V. Maksymenko[2], and T. Jarm[3]

[1] National Technical University of Ukraine "Kyiv Polytechnic Institute", Department of Microelectronics, Kyiv, Ukraine
[2] National Technical University of Ukraine "Kyiv Polytechnic Institute", Department of Biomedical Engineering, Kyiv, Ukraine
[3] University of Ljubljana, Department of Biomedical Engineering, Ljubljana, Slovenia

Abstract— **Some specific characteristics of the Biomedical Engineering Education Tempus Initiative in Eastern Neighbouring Area (BME-ENA) project in Ukraine are discussed. Some major problems encountered by the Ukrainian working groups for the BME-ENA Project are disclosed. The work on new Bachelor's and Master's programs for the BME-ENA project in Ukraine is detailed.**

Keywords— **Tempus project, higher education, biomedical engineering, BME Masters.**

I. INTRODUCTION

Currently in Ukraine there is a growing need for expertise in biomedical engineering. BME-ENA (Biomedical Engineering Education Tempus Initiative in Eastern Neighbouring Area) Tempus Project implementation in Ukraine will undoubtedly serve the successful solution of this problem, as well as improve the academic and scientific mobility of students and academic staff. One of the four special Programs for Ukraine will be developed as a part of the Project [1]. BME-ENA project consortium includes seven EU and ten institutions from four Eastern Neighbouring Area Countries.

In the project there are two working groups involved from the two Ukrainian Partner universities: National Technical University of Ukraine "Kyiv Polytechnic Institute" (NTUU "KPI") and Sumy State University (SSU).

II. SOME SPECIFIC CHARACTERISTICS OF THE BME-ENA PROJECT IMPLEMENTATION IN UKRAINE

Ukraine has much greater total number of high level educational institutions (325 including 198 universities), greater number of universities related to BME (medical and technical) universities, and finally a correspondingly higher number of graduates. In addition, more than 15 large and middle-size enterprises-producers of medical equipment have remained in Ukraine. Significantly, there are more than 2900 medical institutions and hospitals with only a small number of private clinics (about 1% share) in Ukraine [2].

A new law on higher education is currently under consideration, which will significantly change the specialties accreditation system. Furthermore, in Ukraine has already been some progress in the field of BME training. Since the academic year 2013-14, five universities have officially launched a BME Bachelor's program and will expect about 250 graduates. By Orders of Ukrainian government in 2006 and 2010 the direction of bachelors training "Biomedical Engineering" and masters training in the specialties "Biomedical Engineering", "Biotechnique and Biocompatible materials" and "IT in Medicine" have been confirmed.

The work on the BME-ENA project began and took place in a period of political and economic instability in Ukraine and the subsequent political and economic reforms, a reform of higher education, the development and introduction of a new law on higher education, and the regulatory acts of the Ministry of Education and Science and the Academic Council of the University (NTUU "KPI"). Because of the political situation in Ukraine the BME-ENA project kick-off meeting originally planned for February 2014 in Kiev had to be relocated to Patras.

As a result of the reform of higher education the number of higher education institutions in Ukraine was reduced by more than of 2.5 times. At the same time, the fact that the Commission developing a new Higher Education Law, worked on the basis of NTUU "KPI" and it was headed by the Rector and Vice-Rector, NTUU "KPI" allowed quick decisions on the Project management. Recently, the Ukrainian Government approved a new list of specialties, where thanks to the efforts of the project team with the support of university administration we were able to incorporate BME as a new specialty though the total number of majors was reduced several times. At the same time before the full entry into force of the Law on Higher Education the transitional provisions are still acting. SSU obviously will not be able to start preparing for the Masters BME joint program at the time of the project, but is going to do it with a delay of 1-2 years.

Changed distribution of hours and credits and instead of the three topics types there was introduced two types of disciplines (compulsory and elective) and changed the ratio between them, a new list of mandatory humanities and general-technical (not special) disciplines (was expected to decline, but was added one more extra discipline) and the relationship between them and special. In this connection, a

V. Sontea and I. Tiginyanu (eds.), *3rd International Conference on Nanotechnologies and Biomedical Engineering*,
IFMBE Proceedings 55,
DOI: 10.1007/978-981-287-736-9_118

corrected curriculum for BME MSc has been developed till June 2015.

However, there is no reason to doubt the successful and sustainable implementation of the project and the improvement of its results upon completion.

In January 2015, thanks to the efforts of the working group, submitted a package of documents to the Ministry of Education and Science (amended in March 2015 at the request of the Ministry) on the licensing of the master's program at the BME Department and accreditation for the Masters' preparation. Also, great efforts of the working group and NTUU "KPI" First Vice-Rector were needed to promote BME specialty in a new list of specialties approved by the Cabinet of Ministers on April 29, 2015, despite the fact that the specialty has been included for the first time, and the overall list was reduced by more than 2 times compared to the previous situation.

III. Development of Programs and Courses

The curricula design of a joint BME masters program with SSU followed the recommendations resulting from the successful completion of the previous Tempus Project on harmonization of BME programs (CRH-BME - "Curricula Reformation and Harmonization in the field of Biomedical Engineering"). But we also had to consider the following features associated with the specific requirements and the current situation in Ukraine:

- Preparation of BME Master can be conducted sequentially step by step and based on BME bachelor training, while it was necessary to take into account that the significant part of the recommended by CRH-BME project documents disciplines have got into the curriculum of bachelor previously;
- Except NTUU "KPI" four universities of Ukraine four years ago began preparing BME bachelors based on the general Curricula (developed in NTUU "KPI" as a head high school in this specialty). These BME bachelors were graduated in June 2015;
- Even after the entry into force of the new Ukrainian Law on Higher Education, which will make join into BME Master Program after BSc of other specialties available, the first years of the vast majority of applicants will be "native" specialty bachelors graduated from BME. But at the same time due to the new Law the SSU students will be able to enter the NTUU "KPI" Masters' preparation;
- By virtue of the BME-ENA Project working plan paragraph WP4.4, SSU will begin preparations for the joint masters SSU Curricula no earlier than after 2 years, so the training will take place on the basis of NTUU "KPI" with SSU participation in the teaching of some subjects, practical training and the provision of laboratory base for the preparation of master's theses;
- Participation of SSU teachers in the training in Kiev will be possible using distance learning or solutions requiring a number of organizational and financial issues of internal mobility;
- Consideration should be given to specific rules for Ukraine Master training programs in accordance with applicable at the time of drawing up requirements such as 3 groups of topics (obligatory, at the option of the University, at the student's choice), and the availability of the required humanitarian and general engineering disciplines.

In accordance to the Universities' differences at first stage two joint training programs with 90 and 120 credits have been developed. After taking into account a number of changes in the legislation and regulatory guidelines of the Ministry and of Ukrainian partner Universities it was stayed one final modernized joint training Curriculum based on 120 credits.

There are a number of new requirements, in particular, it became clear that the amount of the humanities is not reduced (one extra discipline plus- Innovations and Startups Management), there will not be three but two groups of subjects (compulsory and elective), the changes will be in the amount of credits and types of examinations etc. We consider that after an adjustment related to new modifications to the requirements of the joint curriculum will meet the special requirements of Ukraine and at the same time follow almost completely CRH-BME Project recommendations.

In accordance with the above the final Joint Curricula shown below in Table 1 have been developed with 120 ECTS credits.

A universal educational laboratory was also designed and updated with the list of equipment. Required equipment Due to objective reasons (the situation in Ukraine) the required equipment has not been acquired yet and will be supplied in the near future.

A characteristic feature is the versatility of the BME laboratory design for a variety of training disciplines. Changing of labs topics will be reached by changing the sensor and the object of study. The data acquisition kit and universal part of software will not be changed. In some cases the object of study would be the physical model, obtained as a result of the digital-to-analog conversion of the biomedical object's digital model. Such digital-to-analog converters with included objects imaging on special monitor are manufactured by Ukrainian HOLIT Data Systems Company. The first topic for labs will be "Electronic sensors and biochips" discipline. The room number 147-12 has been prepared for the laboratory.

Table 1 BME MSc 120 ECTS Credits Joint Curricula

Semester	Lecture (**Core topics**, institution's self-selection topics, *specialization topics)*	ECTS
	Occupational safety in industry	1
1	**Basics of sustainable development**	2
	Design of biotechnological systems	5,5
	Medical Physics	5
	Innovation Management	3
	Foreign language professional direction	1,5
	Telemedicine and computer networks	3
	Cell and tissue biotechnology	4
	Biomedical nanoelectronics	5
	Subtotal	30
2	**Intellectual Property**	1
	Civil protection	1
	Research on the topic of master's thesis	1,5
	Foreign language professional direction	1,5
	Marketing of the field	1,5
	Use of professional software in biomedical engineering	4
	Engineering support of healthcare technologies	5
	Digital signal processing and analysis in biomedical systems	4
	Diagnostic and therapeutic methods in arrhythmology and electrophysiology	3,5
	Biocompatible materials with desired properties	4,5
	Medical microprocessor systems	4
	Subtotal	31,5
3	**Philosophical problems of scientific knowledge**	1,5
	Biomedical Information Imaging Systems	5,5
	Research on the topic of master's thesis	4,5
	Foreign language professional direction	1,5
	Pedagogy of Higher School	1,5
	Factors for successful employment in the specialty	1
	Diagnostic and therapeutic methods in arrhythmology and electrophysiology	3
	Bionics, biocybernetics and robotics	4
	Artificial blood circulation and artificial organs	4
	Electronic sensors and biochips	3,5
	Subtotal	30
4	**Research - practice**	6
	Elaboration of the master's thesis	22,5
	Subtotal	28,5
	TOTAL	120

The list of includes the following groups of equipment: computer hardware, software, presentation equipment, complete sets of digital laboratory, instruments, equipment for printing. Complete digital electronic laboratories (sets of physical, chemical and biomedical sensors and USB-Data Logger) allow creating universal workstations for several laboratory courses.

IV. CONCLUSIONS

Information on the progress of the project was added to the general BME-ENA project website and sites of BME departments, faculties of biomedical engineering and electronics at NTUU "KPI". Presentation of the project and the results of its implementation in Ukraine were discussed at the Workshop «Biomedical Engineering Education in Europe» (April 2014, Patras); 6th European Conference of the International Federation for Medical and Biological Engineering, Dubrovnik, September, 2014; 5th and 6th Anniversary International Medical Forums "Innovations in Medicine - Health of the Nation" (October 2014 and April 2015, Kyiv, Ukraine); Workshops on Biomedical Engineering Perspectives (April 2015, Sumy, Kyiv); IEEE 35th International Conference on Electronics and Nanotechnology (ELNANO-2015, April 2015, Kyiv).

We believe the work to implement the project during the reporting period was successful in spite of the political situation in Ukraine and the continuous reform of the higher education system. The main results is the creation of conditions for the start of training BME Masters since September 1, 2015, the inclusion of BME as a new specialty in the new List of professions approved by the Government, the preparation and launching of the procedure for licensing BME specialty and accreditation of BME Department by the Ministry of Education and Science, preparation of curricula, programs, teaching materials and teaching staff, providing visibility and wide dissemination of Project progress and results in Ukraine.

Particular attention is paid to the selection and training of teachers who will start teaching on September 1st of the academic year 2015/2016. Personnel is coming from a number of qualified professionals (mainly teachers) who have experience in giving lectures and classes, as well as textbooks or manuals on the relevant or similar subjects. Knowledge of English was an advantage. Each trainer has prepared a training program on the appropriate discipline, which has been included in the package of documents submitted to the Ministry of Education for accreditation. On the basis of these programs a brief description of the credit modules has been prepared. Special meetings (seminars) for teachers are planned in the aim of matching of Master's training courses.

ACKNOWLEDGMENT

The authors are very thankful to European Commission's Education, Audiovisual and Culture Executive Agency for their financial and logistical support of the Project "Biomedical Engineering Education Initiative in Eastern Neighboring Initiative" (543904-TEMPUS-1-2013-1-GR-TEMPUS-JPGR) and University of Patras, Greece, as a Project Coordinator from EU for providing necessary guidance concerning Project implementation.

CONFLICT OF INTEREST

The authors declare that they have no conflict of interest.

REFERENCES

1. Bliznakov Z, Rodina-Theocharaki A, Pallikarakis N (2014), Fostering the Biomedical Engineering Education in Eastern Neighbouring Countries: The TEMPUS IV BME-ENA Project, Program & Abstract Book, 6th European Conference of the International Federation for Medical and Biological Engineering, Dubrovnik, Croatia, 2014, p 161
2. Orlov A, Maksimenko V, Sukhodub L et al (2014) BME-ENA Tempus IV Project Progress in Ukraine, Program & Abstract Book, 6th European Conference of the International Federation for Medical and Biological Engineering, Dubrovnik, Croatia, 2014, p 163

Author: Anatolii Orlov
Institute: National Technical University of Ukraine "Kyiv Polytechnic Institute", Department of Microelectronics
Street: Peremogy ave. 37
City: Kyiv, 03056
Country: Ukraine
Email: atorlov@rambler.ru

Medical Bioengineering Education in Iasi, Romania

H. Costin[1,2], L. Verestiuc[1], D. Zaharia[1], R. Ciorap[1], C. Corciova[1], and G. Andruseac[1]

[1] Grigore T. Popa University of Medicine and Pharmacy, Faculty of Medical Bioengineering, Iasi, Romania
[2] Institute of Computer Science of Romanian Academy, Iasi Branch

Abstract— **Besides medical and pharmacy specialists, the engineering branch has a main role in modern health care, characterized by high medical technology, powerful software programs for data analysis, and top management tools to increase efficiency of the health care system. Among many engineering specialties, bioengineers and biomedical engineers, even if somehow different, have a main common feature: the strong inter-, multi- and transdisciplinarity of their work. The paper analyses some characteristics of the specific education process in both fields, shows their sub-domains of activity, presents opportunities related to designing a career, presents specific aspects of 21 years of medical bioengineering education in Iasi city, Romania, and concludes that medical bioengineering and biomedical engineering belong to the top jobs, with a constant growths in demand.**

Keywords— **education, medical bioengineering, biomedical engineering, career development.**

I. INTRODUCTION

The existence of so many terrible and incurable diseases, among which cardiovascular, neurodegenerative and psychic diseases, cancers, even a newcomer – obesity – emphasizes the permanent need of very skilled staff in health care.

Medical bioengineering (*MBE*) and *Biomedical engineering* (*BME*) are two of the prominent scientific and economic domains nowadays. This statement is supported including by one of the most increasing of the job market in the world. For instance, between 2010 and 2020, the number of biomedical engineers is projected to rise by about 62 percent in the U.S. [1].

To fix notions, "*Medical bioengineering* integrates physical, chemical, mathematical, and computational sciences and engineering principles to study biology, medicine, behavior, and health. It advances fundamental concepts, creates knowledge from the molecular to the organ systems levels, and develops innovative biologics, materials, processes, implants, devices, and informatics approaches for the prevention, diagnosis, and treatment of disease, for patient rehabilitation, and for improving health." (NIH, USA, Working Definition of Bioengineering - July, 1997, [1]). MBE includes the study of cellular engineering, tissue engineering, cell signaling, cell imaging, bioinformatics, molecular engineering, bioacoustics, biomaterials, genetic engineering etc.

Biomedical engineering (BME) is the application of engineering principles and design concepts to medicine and biology for healthcare purposes (e.g. diagnostic, therapeutic or preventive). This domain seeks to close the gap between engineering and medicine, i.e. combines the design and problem solving skills of engineering with medical and biological sciences to advance healthcare treatment, including diagnosis, monitoring, and therapy. It is highly inter- and multidisciplinary, i.e. brings together concepts from different branches of engineering like electrical, mechanical, chemical and computer science. Prominent examples are biomedical devices and instrumentation, biomechanics and biotransport, biomaterials, medical biotechnologies, biomedical optics, neural engineering, etc.

II. MOTIVATION OF EDUCATION IN MBE AND BME

The aging of the population and the focus on health issues increase the demand for better medical devices and equipment designed by bioengineers. For example, rehabilitation engineering is growing rapidly to meet the increasing needs of aged patients and patients with disabilities.

For instance, the U.S. Department of Labor's Occupational Outlook Handbook states that employment opportunities for bioengineers were better and are also expected to grow faster than the average for all occupations in the coming years. In fact, *BME is expected to be the fastest-growing job market in the United States during the next seven years*, according to the U.S. Bureau of Labor Statistics (BLS) [1].

Combined with a continuous growing job market and financial reward, bioengineers and biomedical engineers have the gratification coming from working to meet the needs of society. They choose the bioengineering field to be of service to people, to be a part of the excitement of working with living systems, and to apply advanced technology to the complex problems of biology and medicine.

A. General Educational Aspects

The core action lines of the Bologna process was the adoption of a system of easily readable and comparable degrees for promoting European citizens employability and the international competitiveness of the European higher education system. Qualifications frameworks came into the

V. Sontea and I. Tiginyanu (eds.), *3rd International Conference on Nanotechnologies and Biomedical Engineering*,
IFMBE Proceedings 55,
DOI: 10.1007/978-981-287-736-9_119

Bologna agenda between 2001 and 2003. Several Bologna policy seminars were organized on qualifications frameworks, describing qualifications in terms of level, workload, and competences, and should be useful both at national level and at the level of the European Higher Education Area. Qualifications frameworks have the potential to make higher education systems more transparent, providing common reference points for levels of qualifications, and also strengthening links between qualifications and learning outcomes. Romania and other few countries (Belgium, Denmark, Germany, Ireland, Malta, the Netherlands, Portugal, and United Kingdom) have fulfilled all steps in implementation of qualifications frameworks. MBE and BME programs in Romania are based on national recommendations and correlated with similar programs in EU.

Fields of specialization within undergraduate MBE and BME programs usually include: Bioinformatics, Bioinstrumentation, Biomaterials, Biomechanics, Biomedical devices, Bio-Micro/Nano-Electro-Mechanical Systems, Biosignal and biomedical image processing, Biotechnologies, Cellular, tissue and molecular engineering, Clinical engineering, Medical imaging, Medical electronics, Molecular imaging, Rehabilitation engineering, Healthcare management. This is also the situation at Grigore T. Popa University of Medicine and Pharmacy in Iasi, Romania, where the first and single Faculty of Medical Bioengineering in Romania was founded 21 years ago. Now this faculty offers undergraduate and master programs in MBE.

B. MBE and BME Education Programs in Romania

In Romania MBE and BME directions are developed first as undergraduate programs: bioengineering, in Iasi, at University of Medicine and Pharmacy and medical engineering that are implemented in Cluj at Technical University, Bucharest at Politehnica University, in Timisoara, Aurel Vlaicu University of Arad, Transilvania University of Brasov, Oradea, Constanta. The MBE/BME qualification and competencies in MBE/BME is included in RNCIS (National Higher Education Qualifications Registry): qualification title, graduation title, identification elements for the qualification, qualification summary (professional and transversal competencies, possible occupations for the owner of the diploma. The programs are included in Fundamental domain Engineering sciences, and the study domain is Applied engineering sciences).The Master Degree in MBE/BME is conferred in two years of study after the BEng or BSc were obtained; at present the curricula and syllabi are varied and diverse. During the years of study, elective courses are offered in packages (modules) for more specialization in the domain of medical engineering.

III. MBE Education Programs at Grigore T. Popa University of Iasi

The Faculty of Medical Bioengineering at Grigore T. Popa University of Medicine and Pharmacy in Ia□i offers academic programs and research in health and applied engineering fields. The undergraduate program in medical bioengineering (MBE), initiated in 1994, is among the pioneering programs in East-Europe. The faculty contributes to the raising of the standards of quality in the health care system in Romania. To achieve this goal, the Faculty of Medical Bioengineering aims to:

- promote quality and efficiency in education;
- shape the personality and development of student's creative skills;
- advance continuous development of scientific research, international standards and assert scientific performance of the academic community;
- integrate itself in national and international academic networks based on performance criteria and complementarity;
- develop partnerships with businesses, local government or national institutions and international organizations.

With top resources and facilities, our programs provide an ideal, stimulating environment to develop the necessary foundation for a career in Medical Bioengineering.

A. Undergraduate Program in Medical Bioengineering

Based on broad bioengineering competencies, that must be profound in selected topics, the BSc program teaches and develops the ability for basic and application oriented research in bioengineering and promotes analytical, creative and constructive skills for the development and improvement of complex biomedical systems and methods.

The specialization of Bioengineering, accredited by ARACIS, aims to train specialists in the following areas:

- equipment for diagnosis and treatment – training is done by accumulating knowledge in the fields of electronics, computers, instrumentation, biomedical signal processing, physiological measurements;
- prosthetics – devices that aim to replace damaged structures of the body – the study program consists of the following disciplines: chemistry, biomaterials, tissue engineering, biocomposites, biomechanics, prosthetics and orthotics;
- medical biotechnology – combines and applies knowledge from biology, biochemistry, biological active substances, chemical engineering and biochemistry, in order to produce drugs, natural products, cosmetics, food, and stimulators.

Medical Bioengineering meets the real need to increase medical performance by approaching some of the activities that, at present, are carried out by physicians, e.g. good knowledge and maintenance of medical equipment, innovation of prosthetics and development of products derived from biotechnology.

B. Master Programs in MBE/BME

The Faculty of Medical Bioengineering offers the following Master Degree Programs (2 years, 120 ECTS):

- Clinical Bioengineering;
- Medical Biotechnologies and Advanced Biomaterials;
- Medical Rehabilitation;
- Prosthetic Bioengineering;
- Advanced Medical Biotechnologies;
- Health Care Management.

Master degree studies assure new skills in bioengineering and rehabilitation, practical and theoretical knowledge, that allow graduates to reach specialized positions in hospitals and laboratories, to continue their professional growth with the doctoral studies or to obtain research positions.

Graduates of the program *Clinical Bioengineering* are prepared to: provide technical and operational support using clinical equipment and medical technology; offer consultancy regarding the assessment, purchase and installation of medical equipment; offer know-how to perform the periodic planned maintenance of medical equipment; solve issues concerning the management of medical equipment in hospitals and to ensure the achievement of high quality results in their operation; learn how to identify, formulate and solve problems that could occur in the operation of the medical equipment as part of the routine functioning of a hospital; design and coordinate experiments in the biomedical domain and to analyze and interpret the obtained data.

The program *Medical Biotechnologies and Advanced Biomaterials* prepare graduates to: monitor and control process parameters in obtaining of biologically active compounds and biomaterials; use methods and techniques of analysis and ensure the operation of equipment in preclinical and clinical laboratories, research centers and production units; designs complex biomaterials for tissue engineering and advanced techniques of investigation and therapy and apply these techniques; apply specific techniques used in the development of culture; execute models of biological systems and implement the models in biomedical investigations; identify the functional characteristics and properties of implants and prosthesis in accordance with the applicable standards; make activities in control and quality assurance according to national and international ethical standards.

Additional to the programs in MBE, the Master program *Rehabilitation* is addressed to the BSc in Health, Sport Sciences but also to MBE/BME graduates which are interested to work in rehabilitation teams. The graduates of the program: apply evaluation methods, measure various parameters and interpret the data describing the status of disabled patients; devise and conduct scientific experiments; work in interdisciplinary teams alongside medical staff and engineers; understand the bioethical aspects involved in the use of technologies addressing medical issues specific to disabled patients; apply knowledge regarding normal function of different human biologic systems at rest and during exercise and physiological limitations of the systems; understand the impact of various pathological conditions on the functioning of the apparatus and systems; apply basic knowledge regarding the technical and construction specifications of medical devices with the purpose of correcting or compensating a reduced or missing bodily function.

Graduates of the program *Prosthetic Bioengineering*: execute models of biological systems and models of biomechanical structures, in order to implement models in biomedical investigations; offer know-how to design prosthetic devices, create advanced investigation and therapy techniques and apply them in the health care environment; design and use medical devices that can substitute biological functions or assist persons with disabilities; can analyze, assess and process biomaterials and prosthetic devices; identify and use equipment and procedures belonging to investigation and therapy techniques.

Graduates of the program *Advanced Medical Biotechnologies*: are able to intervene in the production process in order to correct technological parameters, to optimize efficiency and to evaluate risks; offer know-how to identify the bacteria and the viruses involved in human infectious pathology by spotting their particular microscopic characteristics, metabolic cultivation characteristics, antigenic structure and/or genetic configuration; use the methods for the analysis of medical active principles; participate in quality control and assurance, and apply appropriate norms.

The program *Health Management* graduates specialists who: plan, organize and run managerial structures active in health care systems; offer know-how to identify and solve complex issues regarding clinical management and performance management for health care; participate in the implementation of health care strategies and health care programs; participate in control and managerial auditing actions; develop and implement financial plans in the health insurance systems; use information technology solutions in order to improve the efficiency of professional and administrative activities in the health care system.

One of the most important international educational projects in which our faculty takes part is the TEMPUS IV project "BME-ENA Biomedical Engineering Education Tempus Initiative in Eastern Neighbouring Area", with 17 partners (from which 7 are from EU). The wider objective of this project is to promote BME Education in East Neighbouring Area Countries (Armenia, Georgia, Moldova and Ukraine) through the creation of new postgraduate programs in line with the European policy and the most recent development in the field. We consider that this project will have major positive effects in both the health technology industrial sectors and the health delivery systems.

C. Research Activity

The optimization of scientific research results and the technological transfer towards the corresponding industries represent major objectives for the programs financed by the EU. The Faculty of Medical Bioengineering assumed scientific research as a priority that shows the profile of the faculty at national and international levels.

Research directions include: Biomedical instrumentation and physiological measurements; Biomedical signals and imaging processing; e-Health and telemedicine; Bioelectromagnetism; Tissue engineering (scaffolds for tissue engineering - bone, cartilage, skin, blood vessels); Biotechnology (special bioreactors and biomass processing); Nanotechnologies for diagnosis and therapy; Rehabilitation (functional stimulation; assistive technologies; biomechanical analysis of the posture and locomotion).

With modern research centers and laboratories, which are fitted with modern equipment and located in a new building, the Faculty of Medical Bioengineering offers an excellent environment for research.

D. Working in MBE/BME in Romania

As a result of the developments in industry and the health care systems, there is a rapidly increasing need for the educational systems to provide the necessary human resources, which satisfies the needs of the employers, also implying the need for patient safety in health care delivery.

The Faculty of Medical Bioengineering database includes 91% of BE graduates from 2000 to 2013 who are working in various areas such as: hospitals, research, higher education, prosthetic departments, public health department, home health insurance, medical representatives for companies of medical equipment, drugs or pharmaceutical plant and other categories of companies with related activity profiles. The Bioengineering graduate specialization insertion in the labor market includes: Health system and medical companies – 63%, Non-medical companies-5%, Master and PhD students – 25%, Research and education – 7%.

Medical Bioengineers who work in hospitals are involved in technical departments for medical equipment manipulation and maintenance, in medical technology management. They are also involved in selection of the devices for patient care. Bioengineers also work carefully to ensure that all of the hospital's equipment is safe and reliable. Some bioengineers work in design and research laboratories in companies for designing monitoring equipment, medical dedicated software, or processes to deliver therapeutic drugs safely and effectively. Rehabilitation bioengineers are medical bioengineers who work with technology and computers to help individuals with disabilities reach toward their maximum potential for an enjoyable and productive life. Helping the mentally disabled to learn, providing a voice for those who cannot speak, and transportation of the physically disabled are some of the activities of the rehabilitation bioengineer. Other bioengineers work in research laboratories located in medical schools, universities and government facilities. Still others make mathematical models of how parts of the body work together. Some bioengineers make man-made materials to repair and replace damaged organs. These materials may be used e.g. for contact lenses, dental implants, replacement hip joints and replacements for limbs.

IV. CONCLUSIONS

Medical Bioengineers work spans all of engineering and medicine. The great potential, challenge, and promise in this endeavor offer not only significant technological benefits but humanitarian benefits as well. The Faculty of Medical Bioengineering of Iasi, Romania, reached its maturity and proved its usefulness, professionalism and framing into European curricula in MBE/BME. Nowadays our alumni are deeply engaged in various international research programs and work in prestigious universities, research centers and companies all over the world.

CONFLICT OF INTEREST

The authors declare that they have no conflict of interest.

REFERENCES

1. IEEE, http://theinstitute.ieee.org/career-and-education/careerguidance/what-it-takes-to-be-a-bioengineer
2. BLS, http://www.bls.gov/oes/current/oes172031.htm#(2)
3. Report-Bioengineering and Biomedical Engineering in Europe -
4. Overview, Education, Standards and Professional Competences – ELearning in Biomedical Engineering, Project 2011-1-RO1-LEO05-15321, Contract LLP-LdV/ToI/2011/RO/008, 201

Author: Hariton Costin
Institute: Grigore T. Popa University of Medicine and Pharmacy, Faculty of Medical Bioengineering
Street: 9-13, M. Kogalniceanu street
City: Iasi
Country: Romania
Email: hcostin@gmail.com

Design and Content of Biomedical Curriculum for Biomedical Engineering Master's Program in the Republic of Moldova

V. Vovc[1], S. Lozovanu[1], A. Ganenco[1], and N. Ciobanu[1]

[1] Department of Human Physiology and Biophysics, State University of Medicine and Pharmacy "Nicolae Testemitanu", Chisinau, Republic of Moldova

Abstract— **This article presents the design and content of biomedical disciplines included in biomedical engineering master's program for universities of the Republic of Moldova. The biomedical curriculum of the biomedical education program contains two fundamental disciplines: 1.Functional anatomy and physiology; 2.Biophysics. To make more familiar future biomedical engineers with technical part of medical practice one applied discipline is proposed: Diagnostics and treatment methods in medicine.**

Keywords— **biomedical curriculum, biomedical engineering, master's program.**

I. INTRODUCTION

Biomedical Engineering is concerned with applying the engineering science to medical innovation and bridges the gap between engineering and medicine with the purpose of improving the healthcare area [1]. Amazing progress and development has been made in the fields of diagnosis and therapy thanks to the newly established discipline of Biomedical Engineering through the production of advanced medical equipment and devices that incorporate the biological dimension. Bioprostheses, microimplants, imaging machinery like the MRI and tissue growth are some of the notable achievements of this study field. Bringing engineering to a new level, Biomedical Engineering offers unlimited research, development and growth perspectives [2].

Training of specialists in biomedical engineering requires a special curriculum because it must contain necessary engineering disciplines and disciplines of biology and medicine. For several years the department of human physiology and biophysics of the State University of Medicine and Pharmacy collaborates closely with the Department of microelectronics, nanotechnology and biomedical engineering of the Technical University of Moldova in the development and implementation of the interdisciplinary educational program in biomedical engineering [3]. The curriculum in biomedical engineering was developed and approved for Cycle 1 license and Cycle 2 master degree. With the experience gained in the training of specialists in biomedical engineering the curriculum in this specialty was developed referred to the International Project "Tempus Biomedical Engineering Education Initiative in Eastern Neighbouring Area" (BME-ENA). We propose design and content of biomedical curriculum for biomedical engineering master's qualification in the project BME-ENA.

II. DESIGN AND CONTENT OF BIOMEDICAL CURRICULUM

The biomedical curriculum of the biomedical education program contains two fundamental disciplines: 1.Functional anatomy and physiology; 2.Biophysics. To make more familiar future biomedical engineers with technical part of medical practice one applied discipline is proposed: Diagnostics and treatment methods in medicine.

Disciplin: *Functional anatomy and physiology;* ***Total hours*** *40/150;* ***Lectures*** *30;* ***Practical lessons*** *10;* ***Credits*** *5;* ***Destination:*** *Cycle II. Master in Biomedical Engineering.* **Educational objectives in the framework of disciplin:** ***At the level of knowledge and understanding:*** to explain physical phenomena involved in the functioning of biological systems; to understand the importance of mechanisms regulating physiological functions in coordination bodies and separate systems for normal activity of the whole organism; to gather information on the normal functioning of the human body; to know the physiological constants in conjunction with new developments in the biomedical sciences; ***At the level of application and integration***: develop skills in recording, measurement and interpretation of physiological parameters; understand how to use physico-mathematical techniques in laboratory and clinical investigations; use modern methods of research physical phenomena and processes in the human body; use theoretical and practical knowledge obtained from studying the physiology course by correlating them with the professional activity. **Syllabus:** The structure of biological membranes. Electrogenesis, excitation process. Properties of peripheral nerves. Classification and function of nerve fibers. Reflexes and nerve center. Structure and function of neuromuscular synapses. The physiological properties of skeletal and smooth muscles. Inhibition in the central nervous system. Types of inhibition. Endocrine glands. The

V. Sontea and I. Tiginyanu (eds.), *3rd International Conference on Nanotechnologies and Biomedical Engineering*,
IFMBE Proceedings 55,
DOI: 10.1007/978-981-287-736-9_120

mechanism of hormones action. Hormones effectors. Hypo- and hypersecretion. The hormones of the adrenal medulla. The autonomic nervous system. Physiology of the heart. The peculiarities of the heart muscle. Hemodynamics. Blood pressure, blood flow and resistance. Systemic and local blood pressure control. Clinical and physiological methods of cardiac activity assessment. Heart auscultation. Electrocardiogram. Digestion functions motility, secretion and absorption in the gastro-intestinal tract. Metabolism and thermoregulation. External breathing. Biomechanics of breath. Gas exchange in lungs and tissues. Pulmonary function tests - volume, lung capacity and flow rates. Gas transport in the blood. Nervous and humoral control of breathing. The formation of urine. Glomerular filtration, renal blood flow. Final urine. Hydro-saline metabolism. Volemia. EAB. Erythrocytes. Anemias. Blood groups and transfusion. Rh-factor. Leukocytes. Immunity and allergy. Hemostasis and blood clotting. Somatosensory system. Tactile sensitivity and position. Pain. Thermal sensitivity. Visual analyzer. Auditory analyzer. Chemical senses - taste sensitivity and olfactory sensitivity. Motor functions of the spinal cord. The cortex and brainstem levels of motor functions control. The role of the cerebellum and basal ganglia in controlling motor function. Cerebrospinal fluid. The cerebral cortex. Intellectual functions of the brain. Learning and memory. The limbic system and the hypothalamus in controlling behavior and motivation.

Disciplin: Medical Biophysics; **Total hours** 40/150; **Lectures** 30; **Practical lessons** 10; **Credits** 5;
Destination: Cycle II. Master in Biomedical Engineering.
Educational objectives in the framework of disciplin: At the level of knowledge and understanding: to familiarise students with fundamental biophysical methods and techniques. ***At the level of application and integration***: to develop the skills of practical application of the theoretical concepts and experimental techniques on the interaction of various types of radiations with biomedical systems. **Syllabus:** Introduction to Medical Biophysics. Definition and classification of solutions. The electrical properties of solutions. The degree of dissociation, ionic strength, activity of concentration and the number of transport. The electrical conductivity of solutions. Methods of determination of ions concentration in solutions or biological cells. The basic properties of ions. Molecular diffusion and transport phenomena: simple diffusion. The diffusion coefficient. Definition of chemical and electrochemical potential. Diffusion through artificial membranes and cell. Facilitated diffusion and peculiarities. Membrane permeability and susceptibility. Physical and biological properties of cell membranes. Active and passive transport of ions through molecules and cells. The structure of vascular system and blood circulation. Blood pressure and pulse. Determination of rate of blood flow through the vascular system. Auditory analyzer. Physical properties of sound wave (propagation velocity, sound intensity, sound level). Characteristics of sound: sound timbre, sound strength, auditory field. Visual analyzer. The main parameters of the visual system in normal conditions, reduced eye, visual acuity, field of vision. Electromagnetic waves: mechanism of action of electromagnetic waves on the bio-medical systems. Classification of electromagnetic waves. Production and emission of X-ray tube Coolidge. The diffraction and scattering radiation radiography and fluoroscopy X-ray Computed Tomography X. Fundamental characteristics of thermal radiation. Black body and the laws of thermal radiation. Ultraviolet radiation properties and their action on living systems. Infrared radiation. The mechanism of infrared radiation absorption by living bodies. Basic principles of stimulated emission (Laser). The mechanism of operation of the various classes of lasers. The types of lasers used in medicine. Electrophoresis. Ultrasound. Radiation. Lasers. Magnetic Therapy. Artificial cardiac stimulation. Defibrillator. UHF currents.

Disciplin: Medical diagnostic and treatment methods; **Total hours** 40/150; **Lectures** 40; **Practical lessons** -; **Laboratories** -; **Credits** 5; **Destination** Cycle II. Master in Biomedical Engineering.
Educational objectives in the framework of disciplin: *At the level of knowledge and understanding:* to explain the theoretical basis of medical diagnostic and treatment methods; to know the morphological diagnostic methods; to know the functional diagnostic methods; to know the methods of laboratory diagnosis; to know the basis of therapeutic and surgical treatment methods with using of special technical devices. ***At the level of application and integration***: to select a special technical device appropriate for method of medical investigation; to select a special technical device suitable for method of medical treatment; to select appropriate accessories and consumables for good function of devices for diagnosis and medical treatment; to apply the correct conditions and technical protocols of medical diagnosis and treatment devices using. **Syllabus: Introduction.** Objectives of physical methods of investigation used in medicine. A brief history of the emergence and development of modern methods. Invasive and noninvasive method concept. Classification of physical methods used in biology and medicine – according to the physical phenomenon used; - according to the research object; - according to the objectives (research, diagnosis, treatment). **Investigation of the structure of organs and tissues - the notion of morphology**. Optical methods. Electron microscopy and its role in molecular biology and medicine. Ultrasound using. Features of propagation of ultrasound in living tissues. Diagnostic methods using ultrasound. Radiography. X-ray tomography and its shortcomings. Computed tomography. Nuclear mag-

netic resonance (NMR). Imaging MRI. Radionuclide investigation: technique, equipment, field use, advantages. **Investigation of organ and tissue function**. Biomechanics. Principles of experiments in the study of biological mobility. The correlation between the method chosen and the level of organization of the object. Methods of molecular biology and organ level. Experimental modeling methods of isolated objects (cells, tissues, organs) and animals. Ultrasound as a method for studying biological objects' movement function. Doppler effect and measuring the speed of blood and organs. Rate of speed measurement accuracy. Bioelectrical processes. Origin of electrical processes in various biological tissues. Propagation of excitation. Method of electrocardiography. Types of recorded signals and their parameters. **Using of physical methods for treatment.** Electroplating and electrophoresis. Ultrasound. Radiation. Lasers. Magnetic Therapy. Artificial cardiac stimulation. Defibrillator. UHF currents.

III. CONCLUSIONS

We hope this design and content of biomedical curricula will be optimal for master program in Bioengineering. The current phase of training in biomedical engineering in the Republic of Moldova is predominantly characterized by preparing them for management, efficient use and repair of medical technologies for diagnosis and treatment.

An important part of the qualification of engineers specialized in biomedical apparatus are basic knowledge in biology and medicine.

ACKNOWLEDGMENT

The design and content of biomedical curriculum for biomedical engineering master's program in the Republic of was elaborated in the framework of the International Project "Tempus Biomedical Engineering Education Initiative in Eastern Neighbouring Area" (TEMPUS BME-ENA).

CONFLICT OF INTEREST

The authors declare that they have no conflict of interest.

REFERENCES

1. LaPlaca M. C., Newstetter W. C., Yoganathan A. P. (2001) Problem-based learning in biomedical engineering curricula. 31st ASEE/IEEE Frontiers in Education Conference, October 10 - 13, 2001 Reno, NV. P.16-21.
2. World Bank An Overview of Medical Device Policy and Regulation, February 2007.
3. Sontea V., Buzdugan T., Railean S., et al. (2013) Continuous training in biomedical engineering: necessities, effects and possibilities. 2nd International Conference on Nanotechnologies and Biomedical Engineering, Chisinau, Republic of Moldova, April 18-20, 2013. P.679-680.

Author:	Victor Vovc
Institute:	State University of Medicine and Pharmacy "N.Testemitanu"
Street:	Stefan cel Mare 165
City:	Chisinau, MD2004,
Country:	Republic of Moldova
Email:	victor.vovc@usmf.md

Part VII

Nuclear and Radiation Safety and Security

Ion Mobility Spectrometer for Rapid Simultaneous Detection of Positive and Negative Ions

V. Vasilyev, V. Pershenkov, V. Belyakov, N. Samotaev, A. Golovin, E. Malkin, E. Gromov, I. Ivanov, M. Matusko, A. Ivanova, and D. Lipatov

Micro- and Nanoelectronics Department, National Research Nuclear University MEPhI (Moscow Engineering Physics Institute), Moscow, Russian Federation

Abstract— **Ion Mobility Spectrometry (IMS) is a common technology used for rapid detection of trace amounts of explosives (mostly forming negative ions) and drugs (forming positive ions). Emergency chemically hazardous substances widely used in modern industry. Some emergency chemically hazardous substances form positive ions, others form negative ions. Therefore, for the detection of different substances in real time is required to apply the structure of two IMS-detectors. The article describes a device for the simultaneous detection of ions of both polarities with the ionization source based on a pulsed corona discharge. Authors investigated the advantages and disadvantages of such construction compared to unipolar cell in terms of sensitivity. Experimental results have demonstrated.**

Keywords— **hazardous industrial chemicals, ion mobility spectrometry, on-site detection, spectrometric drift tube.**

I. INTRODUCTION

During the ionization process analyte may form ions of the positive and negative polarity. The processes taking place in the ionization chamber of the IMS-device depend on many factors[1-3]. However, we know that many explosives, e.g. relating to the nitro compounds, are characterized by negative ions, drugs form in the ionization chamber mostly positive ions, toxic substances and hazardous chemical emergency can form both positive and negative ions [4-5]. Taking into account that a priori information about the analyzed sample is usually absent, to improve the efficiency of analysis is necessary to ensure the simultaneous analysis of both negative and positive ions, that means, to ensure operation of the device, both in negative and in a positive polarity modes. Most easily it can be achieved by applying the device in two operating simultaneously modes of the drift tube.

II. DEVELOPMENT

As the main unit for the implementation of this design was taken ion mobility spectrometer, developed in National Research Nuclear University MEPhI [6]. The design of the spectrometer is a monoblock shown in Figure 1. The basis of design is thermally insulated drift tube. On the back side of the drift tube container with the molecular sieves is placed. It is necessary to drain and clean the air entering into the drift region. Storage battery, which is one of the most massive of the detector units is disposed under the drift tube and lowers the center of gravity, providing mechanical stability of the device.

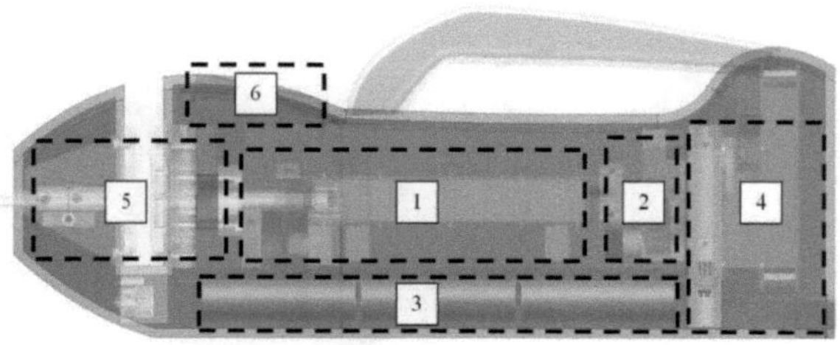

Fig. 1 Accembly of monoblock detector (1 - thermally insulated drift tube, 2 - container with the molecular sieve, 3 - battery, 4 - computer, 5 - sampling device, 6 – display and control unit).

Built-in computer located at the back of the device, is a complete embedded computer format PC104 processor with a frequency of 800 MHz, has 256 MB RAM, hard drive, integrated graphics, USB, Ethernet, Com and LPT controllers. Before the drift tube located a sampling device, over which the control and display unit is mounted. Control and display unit is used for displaying information and control device in the scanning substances mode and consists of LCD graphical display, LED indicator and control buttons. For setting on the substances and the changes in the control program on the hard disk embedded computer can connect the USB keyboard and the external display with VGA interface through connectors located on the rear panel of the unit.

Internal structure of the compact monoblock spectrometer is shown in Fig. 2. Constructive basis of design is formed by two large printed circuit boards, going along the entire length of the sides of the device. On the one of these boards microcontrollers managing the gas flow, the sampling device operation, temperature measurement and control and processing signals from pressure, temperature and humidity sensors of the

V. Sontea and I. Tiginyanu (eds.), *3rd International Conference on Nanotechnologies and Biomedical Engineering*,
IFMBE Proceedings 55,
DOI: 10.1007/978-981-287-736-9_121

external air as well as inlet and outlet channels of the container with the molecular sieve are located. Microcontrollers on the second board manage the collection and processing of spectrometric data from the collector current amplifier, high voltage level control in the resistive divider circuit, the puller electrode and ion gate, and source control based on the ionization of the corona discharge. The two printed circuit boards are connected to the control computer by the appropriate connectors without the use of cables.

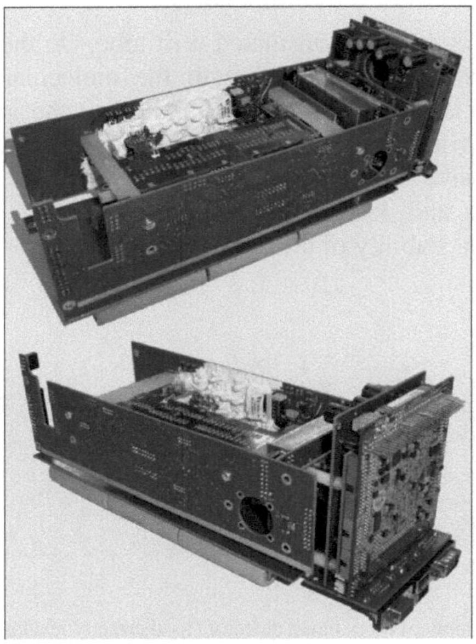

Fig. 2 The internal structure of compact hand-held ion mobility spectrometer with a built-in computer and the battery.

Bipolar detector is based on technical solutions embodied at the monoblock detector described above, and with its basic elements: sampling device, two spectrometric tubes with molecular sieve to drain and clean the air entering the drift region, the modified electronic unit and software. For combining drift tubes into a single structure and allow separation of the gas flow coming from the sampling device into the spectrometric drift tubes the thermostated connecting unit was developed. Flow separation unit allows mechanical and gas dynamic connection of the sampling device with two spectrometric drift tubes in order to implement simultaneous detection of ions of both positive and negative polarities. Connecting unit represents a structure consisting of a heated flow distributor, two PTFE insulating gaskets, heating and pressing PTFE tubes forming the gas channels. General view of the unit in the composition of ion mobility spectrometry shown in Figure 3.

Designed switching circuit interfaces and unified power management allows the transmission of data from two spectrometric detector cells to control computer. The data transmission system is built on switching lines of asynchronous serial interface UART, providing connection up to 256 independent external devices to one COM port control computer. Communication between devices implemented in full-duplex mode and is set solely by the computer request.

III. RESULTS AND DISCUSSION

Verifying the developed device was carried out on organic substances, which can be roughly considered imitators explosives and toxic (by the presence of certain atoms and functional groups) – nitrobenzene, dichloromethane, diethylamine. The choice of materials was also due to the possibility of their registration and subsequent identification as a positive or in a negative mode at the same time [12, 14-17]. During the test analytes (in liquid phase) were deposited on the surface of tissues made of aluminum foil using microsyringe. Then the tissues have been introduced into the sampling device or supplied directly in vapor form from the vessel with the substance (solution). The results of the work are shown in Figures 4-7 (RIP is reactant ion peak).

Fig. 3 Connection unit in the composition of ion mobility spectrometry.

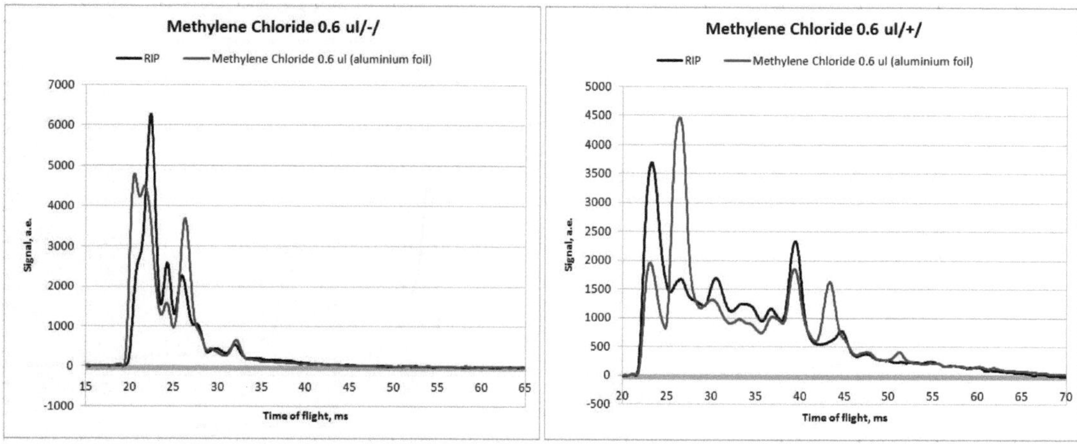

Fig. 4 Methylene chloride CH_2Cl_2 ion mobility spectra.

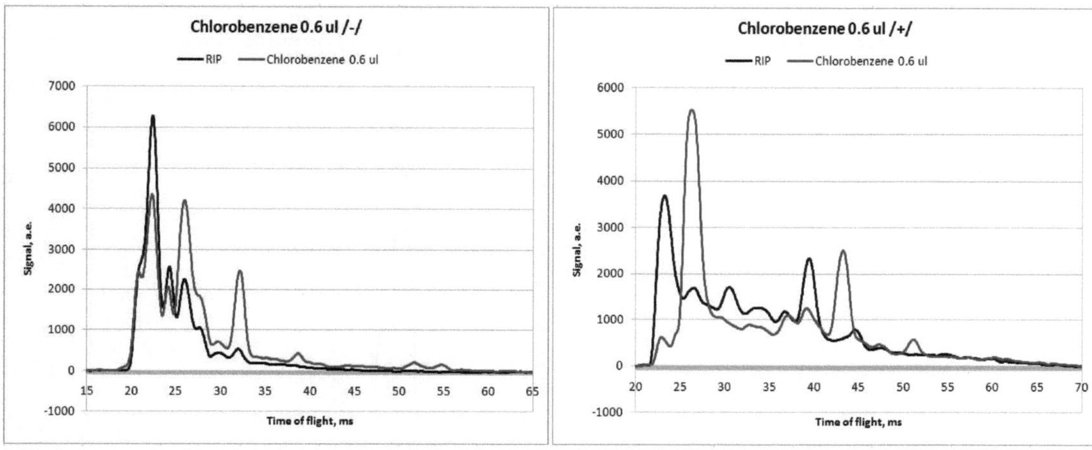

Fig. 5 Chlorobenzene C_6H_5Cl ion mobility spectra.

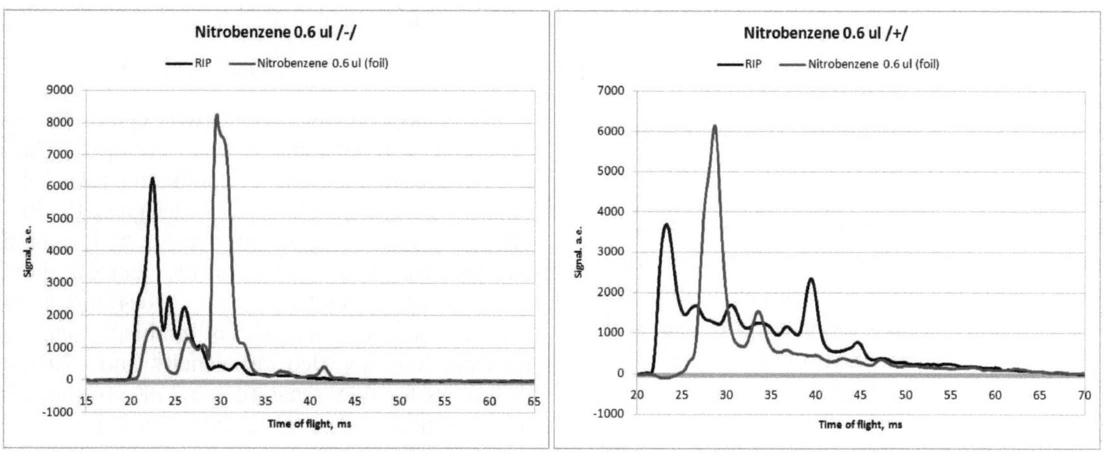

Fig. 6 Nitrobenzene $C_6H_5NO_2$ ion mobility spectra.

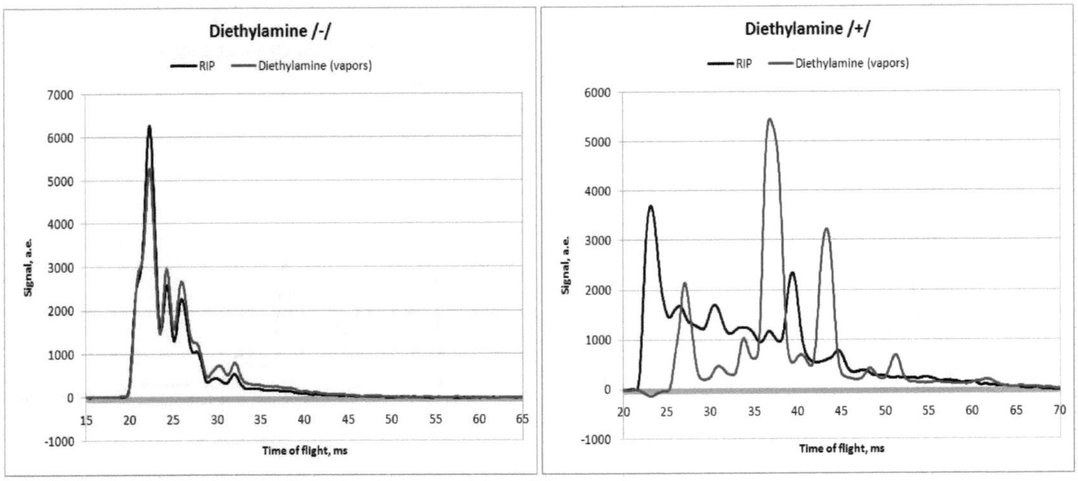

Fig. 7 Diethylamine $C_4H_{11}N$ ion mobility spectra.

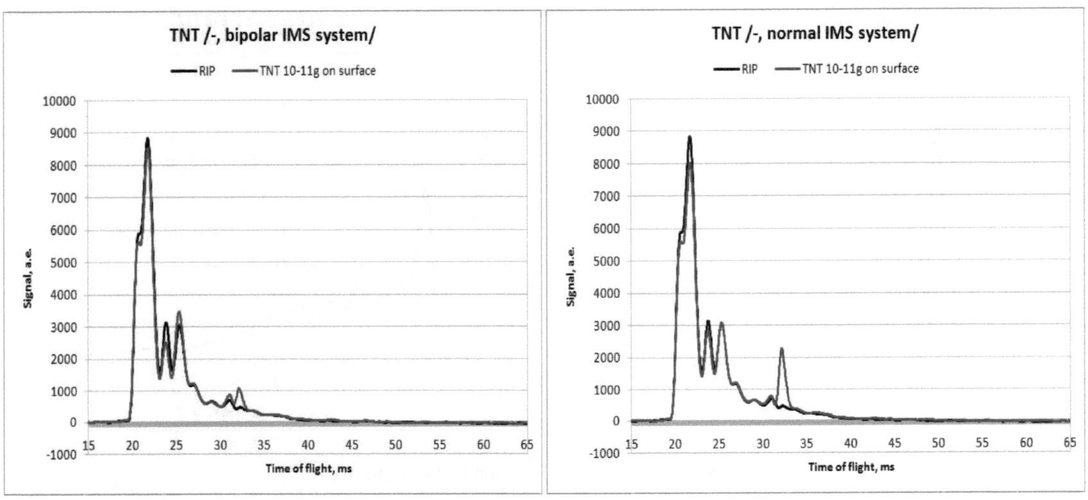

Fig. 8 TNT ion mobility spectra.

Also compared the sensitivity of the developed device for TNT in comparison with the classical design of the spectrometer (one cell). A solution of TNT in acetonitrile dispensed from a micropipette onto the foil, the foil is then maintained at a temperature of 50°C for 30 seconds to evaporate the solvent. TNT spectra are shown in Figure 8, we can estimate the sensitivity threshold of the order of 10-11 g at the surface. For comparison, the same concentrations were introduced into the spectrometer with a single cell, do not differ from those, but it works in the mode of detection of ions of negative polarity without a system of dividing the stream. The obtained peak amplitude (and hence sensitivity) differ by almost three times. This may be explained not only by reducing the amount of the substance by dividing the flow, but also because of the decreased flow rate through the ionization chamber.

IV. CONCLUSION

The spectrometer with two cells for simultaneous operation in bipolar mode was developed. The mechanical and electronic components combining two drift tubes have described. As a result of tests designed device showed its efficiency with simultaneous sample analysis in the negative and positive mode both when the device is in the mode of selection of the particles with tissues and in the mode of selection vapors (analysis of ambient air). Experimental investigations revealed deficiencies and structure with two drift tubes for flow separation and of sensitivity, in particular trinitrotoluene and certain toxic substances.

CONFLICT OF INTEREST

The authors declare that they have no conflict of interest.

REFERENCES

1. Borsdorf H., Eiceman G.A., Ion Mobility Spectrometry: Principles and Applications, Taylor & Francis, – 2006.
2. H. Borsdorf, T. Mayer, M. Zarejousheghani, Recent developments in ion mobility spectrometry, Applied Spectroscopy Reviews, 46, 2011, pp. 472-521.
3. Matsaev V., Gumerov M., Krasnobaev L., Pershenkov V., Belyakov V., Chistyakov A., Boudovitch V., IMS Spectrometers with Radioactive, X-ray, UV and Laser Ionization, International Journal for Ion Mobility Spectrometry, 5 (3), 2002, pp. 112-114.
4. P. Rodacy, The minimum detection limits of RDX and TNT, Explosive Subsystems & Materials, Sandia National Laboratories, 1993.
5. Wu C., Steiner W.E., Tornatore P.S., Matz L.M., Siems W.F., Atkinson D.A., Hill H.H., Construction and characterization of a high-flow, high-resolution ion mobility spectrometer for detection of explosives after personnel portal sampling, Talanta, vol. 57, 2002, pp.123–134.
6. Samotaev, N., Golovin, A., Vasilyev, V., Malkin, E., Gromov, E., Shaltaeva, Y., Mironov, A., Lipatov, D., IMS development at NRNU MEPhI, Lecture Notes in Electrical Engineering, 268, 2014, pp. 447-451.
7. Lapthorn, C., Pullen, F., Chowdhry, B., Ion mobility spectrometry-mass spectrometry (IMS-MS) of small molecules: Separating and assigning structures to ions, Mass Spectrometry Reviews 32 (1), 2013, pp. 43-71.
8. A.A. Nanji, A.H. Lawrence, N.Z. Mikhael, Use of skin sampling and ion mobility spectrometry as a preliminary screening method for drug detection in an emergency room, Clin. Toxicol., 25, 1987, pp. 501-515.
9. Phillips M., Herrera J., Krishnan S., Zain M., Greenberg J., Cataneo R.N., Variations in volatile organic compounds in the breath of normal humans, Journal Chromatograph. B. Biomed. Sci. Appl., vol. 729(1–2), 1999.
10. A.N. Davies, J.I. Baumbach., Early lung cancer diagnostics by ion mobility spectrometry data handling, Spectroscopy Europe, 5, 2008, pp.18-21.
11. Ungethüm B., Walte A., Münchmeyer W., Matz G., Comparative measurements of toxic industrial compounds with a differential mobility spectrometer and a time of flight ion mobility spectrometer, International Journal for Ion Mobility Spectrometry, vol. 12, no. 4, 2009, pp. 131-137.
12. Eiceman G.A., Ion mobility spectrometry as detector and sensor for chemical warfare agents and toxic industrial chemicals, Abstracts of Papers of the American Chemical Society 2002, 224, U145-U145.
13. S. Sielemann, F. Li, H. Schmidt, J.I. Baumbach, Ion Mobility Spectrometer with UV-Ionization Source for the Determination of Chemical Warfare Agent Simulants, IJIMS 4 (2001), 81-84, p. 44.
14. B.M. Kolakowski, P.A. D'Agostino, C.Chenier, Z.Mester, Analysis of Chemical Warfare Agents in Food Products by Atmospheric Pressure Ionization-High Field Asymmetric Waveform Ion Mobility Spectrometry-Mass Spectrometry, Analytical Chemistry, Vol. 79, No. 21, November 1, 2007.
15. Oberhuettinger C., Langmeier A., Oberpriller H., Kessler M., Goebel J., Mueller G., Fast detection of toxic industrial compounds by laser ion mobility spectrometry, Proceedings of SPIE - The International Society for Optical Engineering, 2009.
16. Zimmermann S., Barth S., Baether W.K.M., Ringer, J., Miniaturized low-cost ion mobility spectrometer for fast detection of chemical warfare agents, Analytical Chemistry, vol. 80, no. 17, 2008, pp. 6671-6676.
17. M.A. Makinen, O.A. Anttailainen, M. E. T. Sillanpaa, Ion Mobility Spectrometry and Its Application in Detection of Chemical Warfare Agents, Analytical Chemistry vol. 82, No. 23, December 1, 2010.
18. Rodi Steropoulos, A review of CWA Detector Technologies and Commercial-Off-The-Shelf Items, Australian Government, Department of Defence - AR-014-414 March 2009.
19. K.M. Roscioli, M.R. Lamabadusuriya, C.S. Harden, A.J. Midey, Ching Wu, W.F. Siems, H.H. Hill Jr, Structure selective ion molecule interactions (SSIMI) in ion mobility spectrometry, International Journal for Ion Mobility Spectrometry, Volume 17, Issue 1, 2014, pp 43-53.

Author:	Igor Ivanov
Institute:	National Research Nuclear University MEPhI
Street:	Kashirskoe shossee 31
City:	Moscow
Country:	Russian Federation
Email:	ivanovnew@yandex.ru

Experimental Equipment for Extraction of ELDRS Conversion Model Parameters and its Application for Estimation of Radiation Effects in Bipolar Devices

A.S. Bakerenkov

National Research Nuclear University MEPhI (Moscow Engineering Physics Institute), Department of Micro- and nanoelectronics, Moscow, Russian Federation

Abstract— **Experimental equipment for extraction of ELDRS conversion model parameters was described. The equipment involves measure device with three voltage sources, three current sensors, one voltmeter and two relay switchboards. All the components of the device are controlled by computer program, which enable to provide measurements during irradiation in automatic mode. The program is able to perform math processing of experimental results to obtain conversion model parameters. The availability of the equipment for experimental extraction of the parameters was demonstrated for voltage comparator LM111. The estimation of the dependence of input current of the comparator on total dose at low dose rate and elevated temperature was presented and experimentally verified.**

Keywords— **measuring systems, voltage-current characteristic, nanoelectronic devices, semiconductor, integrated circuits.**

I. INTRODUCTION

At present a large number of different models for description of enhanced low dose rate sensitivity (ELDRS) effect in bipolar integrated circuits are developed [1]. The conversion model of ELDRS effect [2] differs from other models because it enables to provide a procedure for extraction of fitting parameters, which can be used for prediction of radiation hardness of bipolar integrated circuits at defined dose rates and temperatures during irradiation. It is important, that the procedure for extraction of the fitting parameters includes: irradiation at high dose rate (5-300 rad (SiO_2)/s), post-irradiation annealing at two different temperatures and elevated temperature irradiation at high dose rate. The extraction procedure can be performed during one day because low dose rate irradiation is not requiring for the extraction of the fitting parameters. The experimental extraction of the fitting parameters can be used for numerical estimation of radiation degradation of electrical parameters of bipolar integrated circuits at any dose rate from 10^{-1}rad(SiO_2)/s to 10^{-5} rad(SiO_2)/s and at any irradiation temperature up to 125^0C. This advantage of the conversion model enables to create numerical prediction methods for ELDRS effect in bipolar integrated circuits.

The experimental equipment for extraction of ELDRS conversion model parameters in automatic mode is developed in this work. The equipment includes a measure device for automatization of measure the process during irradiation and control software, which is able to provide math processing of experimental results to obtain the conversion model parameters.

II. THE FITTING PARAMETER EXTRACTION TECHNIQUE FIGURES AND TABLES

The conversion model is based on the assumption that the positive charge, which is accumulated by ionizing radiation, converts to interface traps. The shallow and deep positive charge traps are supposed to exist. During short period of high dose rate irradiation only shallow traps are converted. An additional conversion of deep traps occurs during long period of low dose rate irradiation. It leads to increase of base current degradation. Conversion process operates like a pump, which pumps radiation induced trapped holes into interface traps continuously. Mathematically in a bipolar transistor base current degradation dependence versus dose rate is described by the following expression [2]:

$$\Delta I_B=(K_D+K_S)\cdot D+\gamma\cdot K_D\cdot \tau_D(Exp(-D/\gamma\tau_D)-1), \qquad (1)$$

where ΔI_B is excess current; D is total dose; γ is a dose rate; fitting constants: K_D is excess base current per unit of total dose at low dose rate, K_S is excess base current per unit of total dose at high dose rate, τ_D is deep traps conversion time.

The Arrhenius law describes temperature dependence of deep trap conversion time

$$\tau_D = \tau_{D0}\cdot Exp(E_A/kT), \qquad (2)$$

where τ_{D0} is a pre-exponential constant, E_A is an activation energy of deep trap anneal.

The conversion model of ELDRS effect has four fitting constants: K_D, K_S, E_A and τ_{D0}. Experimental extraction of these parameters enables to estimate radiation response of bipolar integrated circuits for any dose rate, temperature and total dose level.

Fitting constants extraction was performed by the following way:

V. Sontea and I. Tiginyanu (eds.), *3rd International Conference on Nanotechnologies and Biomedical Engineering*, IFMBE Proceedings 55,
DOI: 10.1007/978-981-287-736-9_122

1. Constant K_S, which determines the contribution of shallow trapped charge conversion to the base current degradation, shall be estimated as a ratio of the base current degradation after irradiation to the specified total dose at 10 rad(SiO_2)/s .
2. The deep traps conversion time τ_D shall be estimated from post-irradiation anneal, which follows after high dose rate irradiation to the specified total dose. Pre-exponential constant τ_{D0} and activation energy E_A in (2) shall be obtained from the data for two different temperatures of elevated temperature post-irradiation anneal.
3. Constant K_D, which determines the contribution of deep trapped charge conversion to base current degradation at low dose rate, shall be estimated from elevated temperature irradiation data. Constant K_D shall be derived from (1), where the constant τ_D for corresponding elevated temperature is calculated from (2) (values of τ_{D0} and activation energy E_A are determined in stage (2).

The application of described technique for extraction of conversion model parameters using designed experimental equipment is presented in follow for LM111 comparator.

III. Structure Diagram of the Measuring Equipment

Structure diagram of the measuring equipment is presented in fig. 1. To automate the measurement process software control measuring system with a computer (PC) is used. The control program, which is installed on the computer, includes math functions to provide numerical processing of experimental results for extraction of the conversion model fitting parameters. COM-port of the computer is connected to board 1 (fig. 1) of the device. Power supply is provided by standard power module (12V, 25W) produced by MW. A commutation device was designed for comfort connection of different types of nanoelectronic devices during the measurements. It allows a user to connect devises in different packages.

Measuring functions are provided by board 1 and board 2 (Fig. 1). Board 1 includes three digital to analog conversion modules E1, E2, E3 and three current sensors A1, A2, A3. Therefore there are three measure channels on the board 1. Together with analog ground and digital voltmeter V the channels are connected to inputs of switch II, which provides connection of any input to any output, which is selected by user. If it is necessary, it is possible to control an external device. Digital input-output block is designed on board 1 of measurement system to provide this possibility. It can transmit and receive standard logic signals (0..5V) through terminals DIO1 ...7. According to a user-defined configuration, each terminal is able to perform both input and output operations. Integrated circuits often have a large number of lids. Board 1 of the measuring system has 5 output lines, which can be connected to analog lids of IC. In majority cases it is not enough and it is necessary to have a greater number of output lines. Board 2 of measuring system is connected to board 1 to solve this problem. Board 2 includes switch III, which is more powerful than switch II. Switch III has 16 output lines.

Each board of measuring device includes a power circuit for stabilization of the temperature of the IC during measurements. Detailed description of the temperature

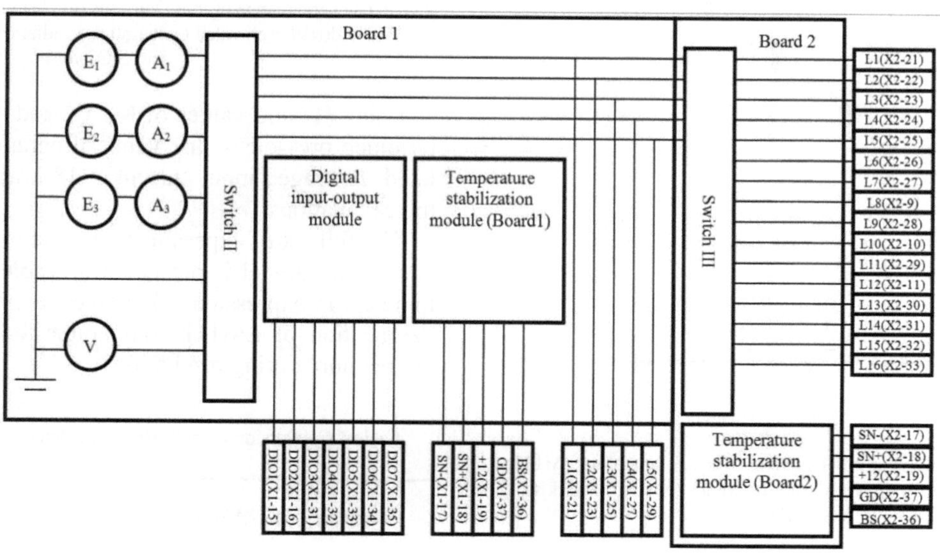

Fig.1 Structure diagram of measure equipment

stabilization technique is presented in [3]. The technique uses LM335 temperature sensor and BCX51 transistors for heat up and enables to set temperature of the IC with ± 0,1°C accuracy. The terminals SN-, SN+, +12V, GD, BS (Fig.1) are used to provide this possibility.

IV. EXPERIMENTAL EXTRACTION OF ELDRS CONVERSION MODEL PARAMETERS FOR VOLTAGE COMPARATOR LM111

Send Experimental extraction of the conversion model fitting parameters was performed with experimental irradiation data, which was obtained in RISI (Lytkarino, Moscow Region). All the irradiations were performed by Co-60 isotope radiation sources.

Two sample lots of LM111 voltage comparators were irradiated in passive electrical mode (all lids grounded) to total dose level 72 krad(SiO_2) at high dose rate 10 krad(SiO_2)/s. Each sample lot consists of two samples of LM111 devices. The dependences of input currents of irradiated comparators on total dose are presented in fig. 2.

From presented dependences it is clear, that variation of input current degradation between samples is less, than 10% in all the total dose range. The value of parameter K_S was determined from this data. The calculations were performed in linear region of the dependence for total dose level 24 krad(SiO_2). The averaged value of excess input current was used in the calculations.

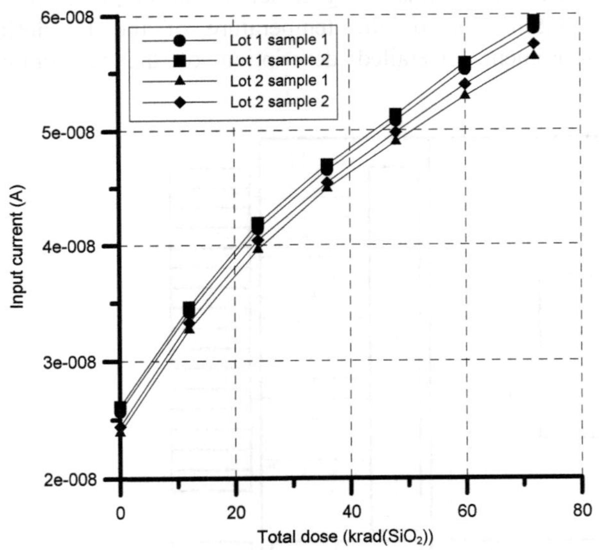

Fig. 2 The dependences of input currents of the comparators LM111, which were irradiated at high dose rate 10 krad(SiO2)/s to total dose 72 krad(SiO2)

For extraction of parameter K_D, elevated temperature irradiation of the third sample lot for LM111 comparators was performed at 100^{0}C. The dose rate during irradiation was 7,8 rad(SiO_2)/s. The dependences of input currents of both samples of the lot on total dose are presented in fig. 3.

For determination of E_Aи τ_{D0} parameters post-irradiation annealing of both sample lots was performed. The first and the second sample lots ware annealed at 90^{0}C and at75 ^{0}C respectively. During annealing process input currents of the comparators continued to increase. According to the conversion model it is connected with a conversion of positive charged oxide traps to interface traps. Using this data the time constants of conversion processes were determined for both temperatures of post-irradiation annealing. The values of E_A and τ_{D0} were obtained from expression (2).

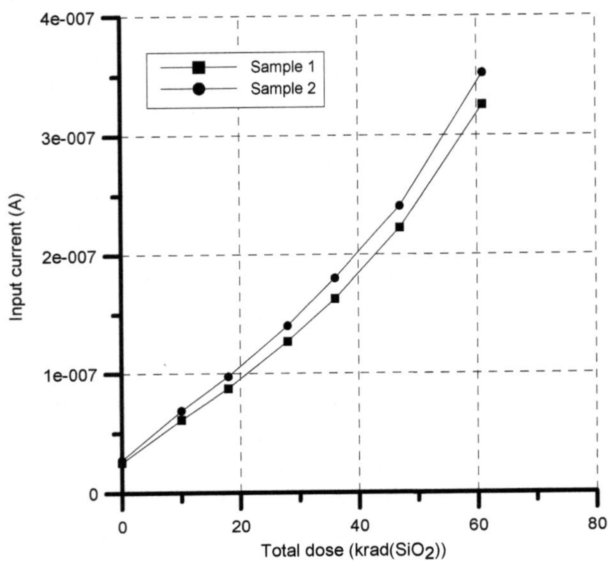

Fig. 3 The dependence of input currents of LM111 voltage comparators on total dose for elevated temperature irradiation at 1000C and 7,8 rad(SiO2)/c

Using (1) and values of K_S, E_A and τ_{D0}, which were determined previously, the value of parameter K_D was calculated. Averaged input current at 18 krad(SiO_2) was used in the calculations.

The full set of experimentally obtained parameters of the conversion model is presented in Table 1. Using these parameters in expression (1) enables to estimate input current degradation of LM111 comparator for any dose rate and temperature during irradiation.

Table 1 Conversion model fitting parameters for LM111 comparator

Parameter	K_S, A/rad(SiO_2)	E_A, eV	τ_{D0}, s	K_D, A/rad(SiO_2)
Value	$1{,}40 \cdot 10^{-11}$	0,83	$3{,}74 \cdot 10^{-8}$	$6{,}6 \cdot 10^{-13}$

For experimental verification of the technique for estimation of radiation hardness of bipolar devices at low dose rate irradiation, which is based on conversion model fitting parameter extraction procedure, the dependence of input current of LM111 comparator on total dose was obtained from low dose rate irradiation at 3 mrad(SiO_2). In the experiment the fourth sample lot of the LM111 devices, which consists of two samples, was used. Experimental results for both samples of the sample lot are presented in fig. 4 with corresponding simulation data, which was obtained from expression (1) using parameters from Table (1). From fig. 4 it is clear, that simulation data and experimental results are in good correlation.

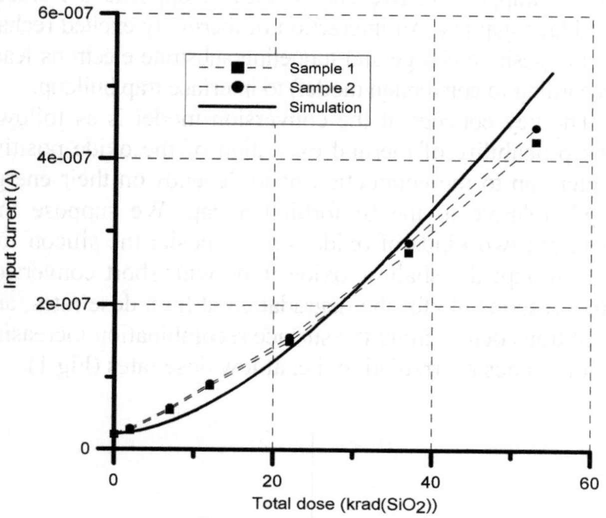

Fig. 4 Experimental and calculated dependence of comparator input current on total dose for low dose rate irradiation at 3 mrad(SiO2)

The conversion model fitting parameters from Table 1 with expression (1) can be used to estimate radiation degradation of input current for LM111 comparator for given irradiation temperature as well as for given dose rate. To experimentally verify this statement an irradiation of the fifth sample lot of LM111 comparators was performed at elevated temperature 85°C. As well as other lots, this lot consists of two comparators. The experimental and simulation results are presented in fig. 5. It is clear, that conversion model describes the dependence of input current of the comparator during elevated temperature irradiation correctly.

V. CONCLUSIONS

An automatic equipment for measurement of electrical characteristic of semiconductor devices was designed. The technique for experimental extraction of the conversion model parameters was presented. Electrical, temperature and dose rate conditions, which were used in the experiments, were described. The developed equipment was used for experimental extraction of the conversion model fitting parameters for bipolar voltage comparator LM111. The dependence of input current of the comparator was calculated for low dose rate irradiation by the use of the fitting parameters, which were obtained experimentally. The comparison of the calculation results with measuring data, which was obtained in low dose rate irradiation experiment for the comparators, was performed. From the comparison it is clear, that experimental and estimation results are in good correlation. It was experimentally verified, that conversion model fitting parameters are available for estimation of radiation hardness of bipolar integrated circuits for any defined irradiation temperature as well as for any defined dose rate.

CONFLICT OF INTEREST

The authors declare that they have no conflict of interest.

REFERENCES

1. R.P. Pease, R.D. Schrimpf, D.M. Fleetwood, ELDRS in bipolar linear circuits: a review, IEEE transactions on Nuclear Science, vol. 56, № 4, pp. 1894-1908, 2009, DOI: 10.1109/TNS.2008.2011485
2. V.S. Pershenkov, D.V. Savchenkov, A.S. Bakerenkov, V.N. Ulimov, A.Y. Nikiforov, A.I. Chumakov, A.A. Romanenko, The conversion model of low dose rate effect in bipolar transistors, RADECS 2009, Conference Proceedings, pp. 286-393, DOI: 10.1109/RADECS.2009.5994661
3. Bakerenkov A.S., Belyakov V.V., Kozyukov A.E., Pershenkov V.S., Solomatin A.V., Shurenkov V.V., Temperature control system for the study of single event effects in integrated circuits using a cyclotron accelerator, Nuclear Instruments and Methods in Physics Research, Section A: Accelerators, Spectrometers, Detectors and Associated Equipment Volume 773, 11 February 2015, Pages 33-38, DOI: 10.1016/j.nima.2014.11.034

Author: Alexander Bakerenkov
Institute: National Research Nuclear University MEPhI (Moscow Engineering Physics Institute)
Street: Kashirskoe shosse 31
City: Moscow
Country: Russian Federation
Email: AS_Bakerenkov@list.ru

Numerical Estimation of the Radiation Hardness of Bipolar Integrated Circuits in Various Irradiation Conditions of Space Environment

A.S. Bakerenkov, V.S. Pershenkov, A.S. Rodin, V.A. Felitsyn, and A.G. Miroshnichenko

National Research Nuclear University MEPhI (Moscow Engineering Physics Institute)/Department of Micro- and nanoelectronics, Moscow, Russian Federation

Abstract— **The conversion model of low dose rate effect in bipolar devices was used for numerical simulation of total dose effects in bipolar devices for various radiation conditions of space environment. The numerical simulation was performed for cyclic temperature irradiation, which is typical for space applications, and for solar flare impact.**

Keywords— **measuring systems, voltage-current characteristic, nanoelectronic devices, semiconductor, integrated circuits.**

I. INTRODUCTION

The estimation of long time operation bipolar devices in the space environment is based on the dose rate dependence of bipolar transistor electrical characteristic radiation degradation (low-dose rate effect or ELDRS [1]). In real operating conditions on-aboard space apparatus is exposed to a cyclical ionizing radiation during crossing the radiation belts of Earth, and it is possible a cyclic variation of the device operating temperature, not comparable to random short high-intensity radiation impact from a solar flare. Consideration of those factors combined with low-dose rate effect (ELDRS) does not fulfill in known literature data but can be important in practice. In this work, on example of the bipolar LM111 comparator, which conversion model fitting parameters were extracted, a numerical estimation of this circuit behavior during space mission is demonstrated. The quantitative numerical analysis of the LM111 comparator input current radiation degradation is considered at dose rate variation corresponding to 12-hour orbit, cyclic device temperature variation and impact of solar flare. The basic conception of the ELDRS physical model (so-called conversion model [2,3]) is described for better understanding.

II. ELDRS CONVERSION MODEL

The model assumes that ELDRS is governed by the relation of interface trap buildup to dose rate. The generation of interface traps attributes with neutralization of positive oxide charge due to "conversion" mechanism, i.e. due to conversion of radiation-induced positive charges to interface traps [4]. The name of "ELDRS conversion model" connects with "conversion" mechanism of interface-trap buildup.

The interaction of captured positive charge (may be hydrogen related species) and substrate electron has crucial importance for interface trap generation. We suppose that an interface-trap buildup connects with a conversion of rechargeable part of trapped positive charge located opposite the silicon forbidden gap [5]. An interaction of thermally excited rechargeable positive charge and tunneling substrate electrons leads (according to conversion model) to interface trap buildup.

The key concept of the conversion model is as follows. The probability of thermal excitation of the oxide positive centers up to the conduction band depends on their energy depth relative to the Si forbidden gap. We suppose that there are two kinds of oxide traps opposite the silicon forbidden gap: the shallow oxide traps with short conversion time, responsible for the degradation at high dose rates, and deep traps determining the surface recombination increasing at long times of irradiation, i.e. at low dose rates (Fig.1).

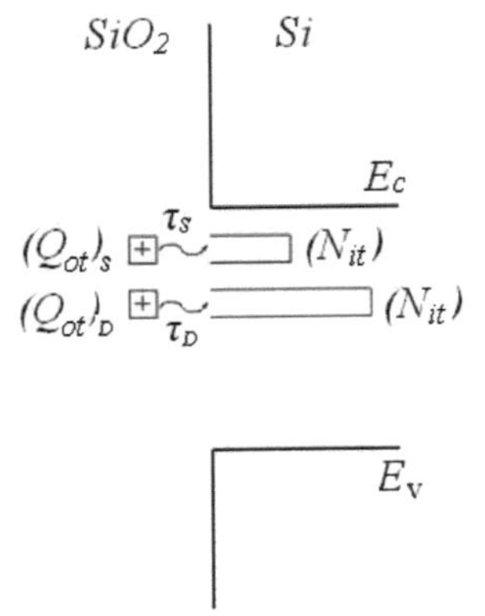

Fig. 1 The shallow (Qot)S and deep (Qot)D oxide trapped charges with conversion time τ_S and τ_D

Based on assumption that the radiation-induced interface trap concentration is proportional to excess base current, in [2,3] the following relationship for input current degradation I_{in} as function of dose rate γ and total dose D was obtained:

$$I_{in} = (K_D + K_S)D + \gamma\tau_D K_D\left(\exp\left(-\frac{D}{\gamma\tau_D}\right) - 1\right), \quad (1)$$

where $D = \gamma t$; K_S, K_D and τ_D are fitting parameters.

V. Sontea and I. Tiginyanu (eds.), *3rd International Conference on Nanotechnologies and Biomedical Engineering*,
IFMBE Proceedings 55,
DOI: 10.1007/978-981-287-736-9_123

The time constant τ_D is function of temperature and is described by Arrhenius equation:

$$\tau_D(T) = \tau_{D0} \exp\left(\frac{E_A}{kT}\right), \quad (2)$$

where T is temperature; E_A is the activation energy of deep oxide trap thermal excitation; k is Boltzmann's constant; τ_{D0} is pre-exponential coefficient, q is the electron charge.

Fitting parameters (K_S, K_D, E_A and τ_{D0}) extraction was performed by the following steps presented in [6]:

1. Constant K_S determining the contribution of shallow trapped charge conversion to input current degradation is estimated as a ratio of input current degradation to the specified total dose at 10 rad(SiO_2)/s irradiation.
2. The deep traps conversion time or constant τ_D is estimated from data of post-irradiation anneal following high dose rate irradiation to the specified total dose. Pre-exponential constant τ_{D0} and activation energy E_A in (2) are derived from the data for two different temperatures of elevated temperature post-irradiation anneal.
3. Constant K_D determining the contribution of deep trapped charge conversion to input current degradation at low dose rate is estimated from elevated temperature irradiation data. Constant K_D is derived from (1), where the constant τ_D for using elevated temperature is calculated from (2) (values of τ_{D0} and activation energy E_A are determined on step 2).

The fitting parameters for LM111experimentally extracted using above described technique are presented in Table 1.

Table 1 Conversion model fitting parameters for LM111 comparator

Parameter	K_S, A/rad(SiO_2)	E_A, eV	τ_{D0}, s	K_D, A/rad(SiO_2)
Value	$1{,}40 \cdot 10^{-11}$	0,83	$3{,}74 \cdot 10^{-8}$	$6{,}6 \cdot 10^{-13}$

The relationship (1) is valid when γ and T are constants. A comparison of calculations made within the model [2,3] and previously reported experimental data has shown the model to adequately represent the behavior of bipolar devices by different processes [7,8,9]. The validation of the conversion model allowed spreading it application for other practical interesting purposes when the dose rate and temperature are not constant. It takes place during real operating conditions on-aboard space apparatus, when the dose rate cyclically varies during crossing the radiation belts of Earth and temperature changes according to sun apparatus position. It is probable short high-intensity radiation impact from solar flare. Below using the conversion model for the quantitative numerical analysis of radiation degradation LM111 comparator (as example) for possible operation on-aboard space apparatus condition on is described.

The relationship (1) in [2,3] was received from convolution integral. Consider the common case. A kinetic of oxide trapped charge and interface trap buildup can be described by following differential equations:

$$\frac{d(Q_{ot})_D}{dt} = K'_D\gamma - \frac{(Q_{ot})_D}{\tau_D}, \quad (3,a)$$

$$\frac{dN_{it}}{dt} = K'_S\gamma + \frac{1}{q}\frac{(Q_{ot})_D}{\tau_D}, \quad (3,b)$$

where $(Q_{ot})_D$ is the deep trapped oxide charge; γ is the dose rate; τ_D is the conversion time of the deep oxide traps; N_{it} is the interface trap density; K'_S and K'_D are constants.

In (3,a) $K'_D\gamma$ and $\frac{(Q_{ot})_D}{\tau_D}$ are radiation induced oxide charge accumulation and annealing rates respectively. Interface trap buildup rate $\frac{dN_{it}}{dt}$ in (3,b) includes two components. The first component is $(K'_S\gamma)$ and it attributed to conversion of shallow oxide traps. The second component $(\frac{1}{q}\frac{(Q_{ot})_D}{\tau_D})$ in (3,b) is a degradation rate produced by annealing of deep oxide traps and their conversion to interface traps. It provides a worst-case approximation because corresponds to 100% conversion of oxide trapped charge to interface traps. The equations (3,a) and (3,b) are correct for arbitrary values of the dose rate and temperature.

As well as [2,3] this paper based on assumption, that excess base current should be a linear function of excess interface trap density only. This assumption is based on numerical calculation of dependence of surface recombination current on interface trap density, which presented in [10]. Calculation results of [10] enable to conclude, that for total dose level, which are typical for space applications (less, than 100 krad(SiO_2)), the dependence of input current on N_{it} is linear and it is correct with sufficient accuracy level. Mathematical representation of this assumption is:

$$I_{in} = AN_{it}, \quad (4)$$

where A is a constant.

The solution of equations (3) and (4) for I_{in}(t) is:

$$I_{in} = \left(AK'_s + \frac{1}{q}AK'_D\right)\gamma t + \frac{1}{q}AK'_D\gamma\tau_D(\exp\left(-\frac{t}{\tau_D}\right) - 1). \quad (5)$$

It is possible to rewrite (5) without t, taking into account, that total dose is $D = \gamma t$:

$$I_{in} = \left(AK'_s + \frac{1}{q}AK'_D\right)D + \frac{1}{q}AK'_D\gamma\tau_D(\exp\left(-\frac{D}{\gamma\tau_D}\right) - 1), \quad (6)$$

Comparing (6) and (1) leads to:

$$K_s = AK'_s, \quad (7,a)$$

$$K_D = \frac{1}{q}AK'_D. \quad (7,b)$$

Substitution of (4) and (7) to (3) enables to rewrite (3) without K'_s and K'_D, using K_s and K_D only, which values were extracted for LM111. The substitution of (4) and (7) to (3) leads to following expressions:

$$\frac{dQ'_{ot}}{dt} = qK_D\gamma - \frac{Q'_{ot}}{\tau_D}, \quad (8,a)$$

$$\frac{dI_{in}}{dt} = K_s\gamma + \frac{1}{q}\frac{Q'_{ot}}{\tau_D}, \quad (8,b)$$

where $Q'_{ot} = AQ_{ot}$.

Equations (8) are available for simulation of time dependence of input current degradation I_{in} during variable dose rate and temperature. Numerical solution of (8) is the base of our calculation. The relationship (1) follows from (8) for constant dose rate γ and constant temperature T, under zero initial conditions ($I_{in} = 0$ and $Q'_{ot} = 0$).

During short time Δt the dose rate γ and temperature T can't change significantly. Using (8) for this interval Δt it is possible to obtain following expression for time-step simulation:

$$Q'_{ot}(t+\Delta t) = \left(qK_D\gamma(t) - \frac{Q'_{ot}(t)}{\tau_D(T)}\right)\cdot \Delta t + Q'_{ot}(t), \quad (9,a)$$

$$I_{in}(t+\Delta t) = \left(K_s\gamma(t) + \frac{1}{q}\frac{Q'_{ot}(t)}{\tau_D(T)}\right)\cdot \Delta t + I_{in}(t), \quad (9,b)$$

where values $Q'_{ot}(t)$, $I_{in}(t)$ are taken for time t, and values $Q'_{ot}(t+\Delta t)$, $I_{in}(t+\Delta t)$ are taken for time $t+\Delta t$.

The simulation technique includes following steps:

1. Use initial values of $Q'_{ot}(0)$, $I_{in}(0)$ and time-step Δt to calculate $Q'_{ot}(0+\Delta t)$ and $I_{in}(0+\Delta t)$ by expressions (9).
2. Perform step 1 with new initial values of Q'_{ot} and I_{in}, which were obtained as the result of the previous step.
3. Repeat this procedure until reaching defined time or total dose level.

Described technique was used to calculate radiation degradation of input current of LM111 comparator in variable dose rate and temperature condition during irradiation, which are typical for space environment.

III. CYCLIC TEMPERATURE VARIATION

During electronic device operation on the board of space apparatus it is possible a cyclic variation of environmental temperature. The simulation was performed for sinusoidal temperature variation from 268 K (-5 ^{0}C) to 328 K (+55 ^{0}C), the period of the variation is 12-hours. Average temperature during the period is 298 K (25 ^{0}C).

The results of the simulation with using ELDRS conversion model are presented in Fig.5 for variable and constant averaged temperatures. The obtained results mean that using of the averaged temperature can lead to large error. The correct estimation of radiation induced degradation of bipolar device parameters at low dose irradiation needs in accurate considering of the temperature time dependence during device operation. For total dose level 1.6 krad(SiO_2) (see Fig.2) estimation error may be more than 300% when averaged temperature level is used instead of real dependence temperature on time. If parametric failure level for LM111 input current is defined as 100nA, failure total dose level could be overestimated more than three times when using averaged temperature value.

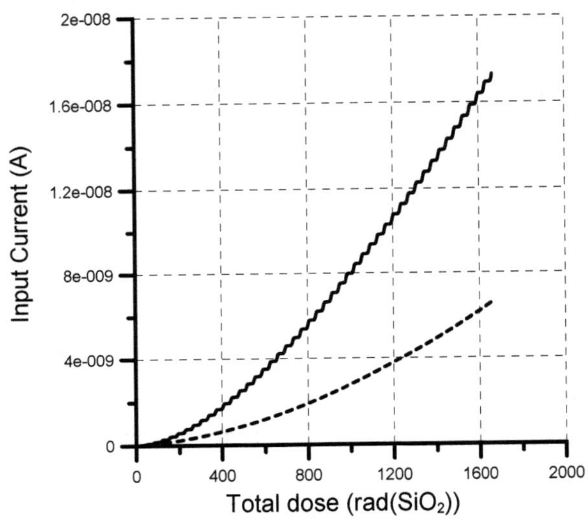

Fig. 2 Radiation degradation of LM111 comparator input current with a total dose for variable (solid) and constant averaged (dashed) device temperature operation.

IV. SOLAR FLARE

Proposed conversion model can be used for the estimation of solar flare impact on characteristics of bipolar devices during their long time mission in space. If emission efficiency of solar flare is known it is possible to estimate the margin for operation life-time of devices. In Fig.3 the simulation results are presented for the October 1989 solar flare (data from [12]). It follows from simulation data that the increasing of input current after the flare can be very large and operation time of devices in space significantly reduces. Input current increasing is the result of following the conversion of oxide trapped charge, which was accumulated during short-time and high dose rate solar flare.

According to [12] total dose, which was accumulated during October 1989 solar flare, is 3.5 krad(SiO_2), duration of the flare is approximately 70 hours. Averaged dose rate was 14 mrad(SiO_2)/s. Comparison of simulation results for degradation LM111 input current without solar flare (curve 1 in Fig.3) and with impact of solar flare leads to the conclusion, that solar flare impact increases degradation of input current more over 30%. It is equivalent reducing of total dose radiation hardness at 30%. Next solar flare (curve 3 in Fig.3) leads to the same reducing of radiation hardness. During ten years space mission it is a necessity to allow the effect of solar flare on radiation degradation device characteristics.

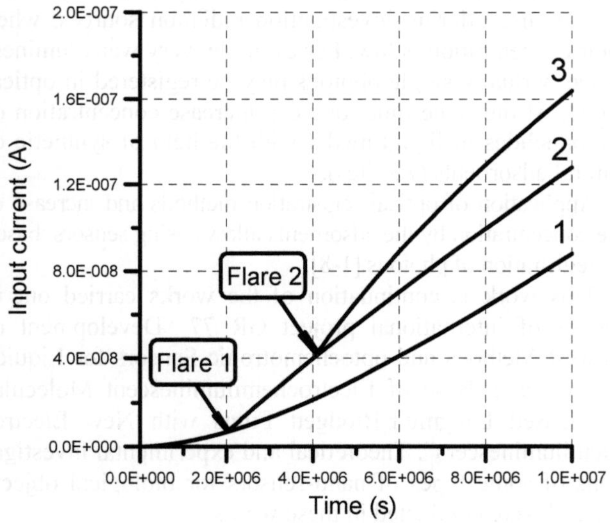

Fig. 3 Time dependence of LM111 comparator input current without solar flare (1), with one solar flare (2) and two solar flares (3).

V. CONCLUSIONS

The applicability of the proposed ELDRS conversion model for estimation of radiation degradation of bipolar devices on impact of the low dose rate irradiation is demonstrated. On the example of one type of microcircuit (LM111 comparator), which fitting parameters of ELDRS conversion model were extracted, we try to demonstrate a numerical estimation of this circuit behavior during space mission. The numerical analysis of comparator input current radiation degradation during long time operation in space is simulated during cyclic device temperature variation and impact of solar flare. All these quantitative numerical results may have practical application.

CONFLICT OF INTEREST

The authors declare that they have no conflict of interest.

REFERENCES

1. R. L. Pease, R.D. Schrimpf, D.M. Fleetwood, Recent advances in understanding total-dose effects in bipolar transistors, IEEE Trans. Nucl. Sci., 2009, NS-56, no.4, pp.1894-1908.
2. V.S. Pershenkov, D.V. Savchenkov, A.S. Bakerenkov, V.N. Ulimov, A.Y. Nikiforov, A.I. Chumakov, The conversion model of low dose rate effect in bipolar transistors, RADECS 2009, Conference Proceedings, pp. 286-393.
3. V.S. Pershenkov, D.V. Savchenkov, A.S. Bakerenkov, V.N. Ulimov, Conversion model of enhanced low dose rate sensitivity in bipolar ICs, Russian Microelectronics, v.39, № 2, 2010, pp. 91-99.
4. S.K. Lai, Interface trap generation in silicon dioxide when electrons are captured by trapped holes. J Appl. Phys. Vol.54, pp. 2540-2546, May 1983.
5. V.V. Emelianov, A.V. Sogoyan, O.V. Meshurov, V.N. Ulimov, V.S. Pershenkov, Modeling the field and thermal dependence of radiation-induced charge annealing in MOS devices. IEEE Trans. Nucl. Sci., 1996, vol. NS-43, no.6, pp.2572-2578.
6. A.S. Bakerenkov, V.V. Belyakov, V.S. Pershenkov, A.A. Romanenko, D.V. Savchenkov, V.V. Shurenkov, Extracting the fitting parameters for the conversion model of enhanced low dose rate sensitivity in bipolar devices, Russian Microelectronics, v.42, Issue 1, January, 2013, pp. 48-52.
7. S.C. Witczak, R.D. Schrimpf, K.F. Galloway, D. M. Fleetwood, R.L. Pease, J.M. Puhl, D.M. Schmidt, W.E. Combs, J.S. Suehle, Accelerated tests for simulating low dose rate gain degradation of lateral and substrate pnp bipolar junction transistors. IEEE Trans. Nucl. Sci., 1996, vol. NS-43, no.6, pp.3151-3160.
8. R.K. Freitag, D.B. Brown, Study of low-dose-rateeffets on commercical linear bipolar IC's, IEEE Trans. Nucl. Sci., 1998, vol. NS-45, no.6, pp.2649-2658.
9. J. Boch, F. Saigne, R.D. Schrimpf, J.-R. Vaille, L. Dusseau, E. Lorfevre, Physical Model for the Low-Dose-Rate Effect in bipolar Devices. IEEE Trans. Nucl. Sci., 2006, vol. NS-54, no.6, pp.3655-3660.
10. V.S. Pershenkov, D.V. Savchenkov, A.S. Bakerenkov, A.S. Egorov, Calculation of surface recombination current in bipolar microelectronic structures subjected to ionizing radiation, Russian Microelectronics, v.38, № 1, 2009, pp. 21-33.
11. T.L. Turflinger, W.M. Schemichel, J.F. Krieg, J.L. Titus, A.B. Campbell, M. Reeves, R.J. Walters, P.W. Marshall, and R.L. Pease, ELDRS in Space: An Updated and Expanded Analysis of the Bipolar ELDRS Experiment on MPTB, IEEE Trans. Nucl. Sci., 2003, vol. NS-50, no.6, pp.2328-2334.
12. R.D. Harris, S.S. McClure, B.G. Rax, R.W. Evans, and I. Jun, Comparison of TID effects in space-like variable dose rates and constant dose rates, IEEE Trans. Nucl. Sci., 2008, vol.55, no.6, pp.3088-3095.

Author: Bakerenkov Alexander
Institute: National Research Nuclear University MEPhI (Moscow Engineering Physics Institute)
Street: Kashirskoe shosse 31
City: Moscow
Country: Russian Federation
Email: AS_Bakerenkov@list.ru

Theoretical Investigations of Nano-sensors for Radiation Processes

T. Marsagishvili, M. Machavariani, G. Tatishvili, R. Khositashvili, E. Tskhakaia, N. Ananiashvili, J. Metreveli, and M. Kikabidze-Gachechiladze

R.Agladze Institute of Inorganic Chemistry and Electrochemistry of I.Javakhishvili Tbilisi State University, Tbilisi, Georgia

Abstract— **Problem of fixing radiation sources occurs in case of environmental contamination by radionuclides. Because of low concentrations their detection is connected with difficulties. At the same time, long-term exposure of radiation sources, even at their low concentration, brings to essential changes in biological processes.**

Possibilities of modern technical registration facilities may be used during investigation radiation sources, when their concentration is low. For example, very weak luminescence, virtually single photons may be registered in optical range. Concentration of radionuclides in liquid media may be increased with the help of synthetic or natural adsorbents.

Application of optical methods of registration allows using sensors, based on registration of the photons.

The processes passing in investigated system can be divided in two groups. First group - charge transfer and phototransfer processes in condensed medium between the particles and between electrode and particle. Second group - processes of electron excitation of separated particles, electron excitation energy transfer between the particles in condensed medium.

Analytical expressions for rate constant and kinetic parameters of charge transfer processes and electron excitation transfer processes between impurity particles are obtained. A dipole moment of transfer may be calculated when known distribution function of reagents and transfer interval being rather substantial during investigation of complicated systems can be evaluated.

This work is continuation of the works carried out in frames of international project GR 77. Theoretical and experimental investigations of some types of nano-sensors for biological objects were conducted in these works.

Keywords— **Nano-sensor, Radiation, Photon, Charge transfer, Energy transfer.**

I. INTRODUCTION

Different processes occur in substances under the influence of high energy radiation (heavy charged particles, neutrons, electrons, X-ray beams, etc.). As a result of such influence different short-lived formations occur in the substance, such as exited molecules, free radicals, solvated electrons, etc.

Dosimetry, based on registration of the products created as a result of radiation influence has essential value when investigation radiation processes. For example, changes which take place in solid (glass, crystal, polymer materials) as a result of radiation influence may be fixed by physical methods.

Other method for fixing of radiation influence is based on chemical changes in substance, which is under radiation influence, for example photoplate or photographic film.

In connection with environmental contamination by radionuclides a problem of fixing of radiation sources occurs. Because of very low concentrations of the lasts their detection is connected with some difficulties. At the same time, long-term exposure of radiation sources, even at their low concentration, brings to essential changes in biological processes.

Possibilities of modern technical registration facilities may be used during investigation radiation sources, when their concentration is low. For example, very weak luminescence, virtually single photons may be registered in optical range. At the same time, one can increase concentration of radionuclides in liquid media with the help of synthetic or natural adsorbents (zeolites).

Application of optical registration methods and increase of the concentration by the adsorbents allows using sensors, based on registration of photons [1-8].

This work is continuation of the works carried out in frames of international project GR 77 "Development of Newest Methods and optochemotronic Sensors for Liquids Assays on a Base of Electrochemiluminescent Molecular Condensed Langmuir-Blodgett Films with New Electrochemiluminescers". Theoretical and experimental investigations of some types of nano-sensors for biological objects [17-21] were conducted in these works.

Principle of operation of nano-sensors may be the following. Under the influence of radionuclide, electron or heavy particle excitation of molecules of the electrode coating takes place. When molecules return to basic state photon irradiation takes place, which is registered by receiving device of the sensor.

This simplifies analysis and makes it fast and cheap. Substances harmful for environment are not used for creation of electrodes for sensors (materials with repeated processing and regeneration possibilities will be used). Optical analytical signal of these sensors may be transmitted to large distances by optical guides without any previous processing. That allows placement of such devices in dan-

V. Sontea and I. Tiginyanu (eds.), *3rd International Conference on Nanotechnologies and Biomedical Engineering*,
IFMBE Proceedings 55,
DOI: 10.1007/978-981-287-736-9_124

gerous or inaccessible zones for service personnel. Another advantage is flow analysis mode for systems under consideration. This will considerably decrease measurement time and make analysis quicker. Working electrode with mono and multimolecular layers is one of the most challenging, among the modern technologies, for use in ECL (electrochemiluminescence) analysis technologies and devices.

Light emission in such systems is due to ECL radiated from this film layer. ECL excitation is realized by electrochemical oxidation and/or reduction and following recombination of molecules between ionic forms of the luminescer fixed in the film layer and appropriate nucleophilic or electrophilic particles of the other reactants (*i.e.*analytes) placed in solution. The specified thin-layer systems in solutions having similar mechanisms of radiation compared with mentioned EL samples and traditional "volumetric" ECL systems possess a number of new positive properties. This allows using them, first of all, for analysis of liquids, liquids analysis ECL methods and creation of optochemotronic electrochemiluminescer sensors using Langmuir-Blodgett deposition technology on the plane surfaces of metal or optically-transparent semiconductor electrodes. Usage of given sensors and ECL methods for definition of some substances existing in biological liquids (with radioactive substances) and harmful radioactive impurities existent in water ecosystems is suggested.

The processes passing in investigated system [9-16] may be divided into several groups. First group - charge transfer and phototransfer processes in condensed medium between the particles and between electrode and particle. Second group - processes of electron excitation of separated particles, electron excitation energy transfer between the particles in condensed medium.

Analytical expressions for kinetic parameters of charge transfer processes and electron excitation transfer processes between impurity particles are obtained.

II. Rate Constant of Charge Transfer Process

Probability of electron transfer process with participation of photon has the form:

$$dW_a = \sum_{n(N_{\vec{k}\sigma})} \sum_{n'(N_{\vec{k}'\sigma'})} e^{\beta(F_i - F_{in})} \phi_i(N_{\vec{k}\sigma}) \cdot$$
$$\cdot 2\pi \left| \left\langle (N_{\vec{k}'\sigma}) n' \right| V_{fi} \left| (N_{\vec{k}\sigma}), n \right\rangle \right|^2 \cdot \qquad (1)$$
$$\cdot \delta(E_{in} - E_{fn'} + \sum_{\vec{k}\sigma} \omega_k (N_{\vec{k}\sigma} - N_{\vec{k}'\sigma}))$$

$(N_{\vec{k}\sigma})$ is number filling of photons; $\phi_i(N_{\vec{k}\sigma})$ is function of photon distribution in initial state; F_i is free energy of system in initial state; $E_{in}, E_{fn'}$ are energy spectra of the system in initial and final states; V_{fi} is matrix element of interaction Hamiltonian.

$\vec{E}$ is operator of intensity of photon field, which can be represented as:

$$\vec{E}(\vec{r}) = i \sum_{\vec{k}\sigma} \left(\frac{2\pi\omega_k}{V} \right)^{1/2} \vec{e}_{\vec{k}\sigma} (a_{\vec{k}\sigma} e^{ik\vec{r}} - a^+_{\vec{k}\sigma} e^{-ik\vec{r}}) \qquad (2)$$

Here $a^+_{\vec{k}\sigma}$, a are operators of birth and annihilation of photons, $\vec{e}_{\vec{k}\sigma}$ - is vector of photon's polarization, V – volume of the system.

After simple transformation we'll obtain:

$$dW_a = \frac{\beta}{i} \int_{C_\theta} d\theta \int d\vec{r} d\vec{r}' \langle E_\beta(\vec{r}',\theta) E_\alpha(\vec{r},0) \rangle_i \cdot \qquad (3)$$
$$\cdot \left\langle e^{\beta\theta \bar{H}_i} P^*_\beta(\vec{r}')_{fi} e^{-\beta\theta \bar{H}_f} P_\alpha(\vec{r})_{fi} \right\rangle_i$$

where H_i is system Hamiltonian in the initial state, H_f is system Hamiltonian in the final state. It is assumed, that there are sufficiently many photons in the system and as processes with absorption, so processes with radiation of photon may be realized.

$$E_\beta(\vec{r}';\theta) = e^{\beta\theta H_f} E(\vec{r};\theta = 0) e^{-\beta\theta Hi} \qquad (4)$$

$$\langle E_\beta(\vec{r}';\theta) E_\alpha(\vec{r},0) \rangle = \sum_{\vec{k}\sigma} \frac{2\pi\omega_k}{V} e^\alpha_{\vec{k}\sigma} e^\beta_{\vec{k}\sigma} \cdot$$
$$\cdot \left\{ \left(1 + \langle N_{\vec{k}\sigma} \rangle_i \right) e^{-\beta\theta\omega_k} e^{ik(\vec{r}'-\vec{r})} + \langle N_{\vec{k}\sigma} \rangle e^{\beta\theta\omega_k} e^{-ik(\vec{r}'-\vec{r})} \right\}$$

Processes with radiation *Processes with absorption*

(5)

Calculation of probability of charge transfer process with radiation and absorption of photons may be conducted in general form, in the same model approximations, which are for thermal processes of electron transfer.

As a result for rate constant of electron transfer process with photon radiation we'll obtain the following expression:

$$K = \sum_\sigma^2 \frac{\left| L^\sigma_{fi} \right|^2}{\sqrt{\left| \psi''_\sigma \right|}} \phi^V_\sigma A^V_\sigma \cdot \exp\{ -\beta\theta^*_\sigma \Delta F_\sigma - \qquad (6)$$
$$- \beta\theta^*_\sigma \hbar\omega_k - \Psi^m_\sigma(\theta^*_\sigma) - \Psi^V_\sigma(\theta^*_\sigma) - \Psi^{mV}_\sigma(\theta^*_\sigma) \}$$

This expression describes the processes with photon radiation as during electron transfer between particles, so during transfer of electron-excited particle in ground state with photon radiation. Parameters in the formula have the same meaning as above. Naturally, reorganization of the both subsystems must be implied by medium reorganization function. L_{fi}^{σ} is dipole moment of transfer (matrix element of interaction of impurity molecules and medium molecules with photon field, calculated with the help of wave functions of final and initial states of the system).

Expressions for simplified models are obtained similarly to above-mentioned.

More extensive information may be obtained from absorption curve form of photon radiation by the system as function of photon frequency. In case of implementation of extinction coefficient notion for photon absorption of radiation processes during electron transfer process between particles or on particle, for photon radiation process during electron transfer from impurity particle in the film on impurity particle in solution will be obtained the following expression:

$$\varepsilon_k = \frac{2\pi\omega_k}{3}\frac{\left|L_{fi}^{\sigma}\right|^2}{\sqrt{|\psi''|}}\phi^v A^v \cdot \exp\{-\beta\theta^*\Delta F - \quad (7)$$
$$-\beta\theta^*\hbar\omega_k - \Psi^m(\theta^*) - \Psi^v(\theta^*) - \Psi^{mV}(\theta^*)\}$$

Application of this correlation allows realization of analysis for wide range of processes. It is necessary to select concrete model to simplify analytically general expressions, in framework of this model. From comparison with experimental data obtaining of numerical values of reorganization energies, transfer dipole moment, reorganization of quantum subsystem etc. is essential.

Rules of sum for determination of transition dipole moment may be used first of all. For example, if during electron transfer process motion on translational coordinates may be considered as classic and transfer takes place at contact of reagents, for bimolecular process we'll obtain:

$$\int_0^\infty d\omega_k \frac{\varepsilon_k}{\omega_k} = \frac{4\pi^2}{3} 4\pi R^2 \delta R^* \Phi(R) d_{fi}^2(R) \quad (8)$$

here $\Phi(R)$ is distribution function of reagents, δR^* is spread of characteristic distances, on which electron transfer takes place.

At known distribution function of reagents the last correlation allows to calculate dipole moment of transfer and correspondingly to evaluate transfer interval being rather substantial during investigation of complicated systems.

III. Conclusions

Analytical expressions for rate constant and kinetic parameters of charge transfer processes and electron excitation transfer processes between impurity particles are obtained for the first time. At known distribution function of reagents the last correlation allows to calculate dipole moment of transfer and correspondingly to evaluate transfer interval being rather substantial during investigation of complicated systems.

Acknowledgment

This work is continuation of the works carried out in frames of STCU international project GR 77.

Conflict of Interest

Authors declare that they have no conflict of interest.

References

1. N.N. Rozhitski, A.I. Bykh, M.A. Krasnogolovets. Electrochemical luminescence. (in Russian). Kharkov, Kharkov Tech. Univ. of Radio Electronics. 2000, 320p.
2. L.M. Blinov. UspekhiFizicheskikhNauk (in Russian). 1988, v. 155, ed. I, p.443.
3. M.J. Melnikov, V.L. Ivanov. Experimental methods of chemical kinetic. Fotokhimia (in Russian). Moscow, 2004, Ed. Moscow State Univ.125p.
4. Y. Fu, M. Forman, C.C. Leznoff, A.B.p. Lever. J.Phys. Chem., 1994, v.98, p.8985.
5. L.M. Goldenberg. UspekhiKhimii (in Russian). 1997, v.66, #12, p.1141.
6. V.V. Arslanov. Uspekhi Khimii (in Russian). 1994, v.63, #1, p.3.
7. Y. Fu, K. Jayaraj, A.B.P. L Lever. Langmuir, 1994, v.10, p. 3836.
8. A.A. Abrikosov, L.P. Gorkov, & I.E. Dzyaloshinski. Methods of Quantum Field Theory in Statistical Physics. Dover Publ. INC. New York. 1963, 341p.
9. F. Plattsman, P. Volf. Waves and interactions in plasma of solid. (in Russian). Moscow: Mir, 1975, 436 p.
10. R.R. Dogonadze, T.A. Marsagishvili. The Chemical Physics of Solvation. Part A. Amsterdam. Elsevier. 1985. p. 39-76.
11. V.M. Agranovich, M.D. Galanin. Electron excitation energy transfer in condenced medium. (in Russian). Moscow: Nayka, 1978, 383 p.
12. V.L. Ermolaev, E.N. Bodunov, E.B. Sveshnikov, T.A. Shakhverdov. Nonradiative electron excitation energy transfer. (in Russian). Leningrad: Nayka, 1977, 311p.
13. R.R. Dogonadze, R.R. Kuznetsov, T.A. Marsagishvili. The present state of the theory of charge transfer processes in condensed phase. Electrochimica Acta, 1980, v.25, #1, p. 1-28.
14. Z. Flügge. Practical Quantum Mechanics II. V.2, Springer-Verlag, Berlin-Heidelberg-New York, 1971.
15. R. Kubo. Statistical Mechanics. North-Holland Publishing Company – Amsterdam, 1965.
16. Marsagishvili T., Machavariani M., Tatishvili G. Conversion of substances by the application of laser radiation. Eur. Chem. Bull., 2013, 3(2), 127-132.

17. Marsagishvili T.A., Chikovani R., Mkheidze T., Berakchjan V., Tatishvili G., Machavariani M.N., Rozhitski N. Investigation of optical properties of optochemotronic sensors at low intensity of radiation. Bull. of National Academy of Sci. of Georgia, Chemical series, 2007, v. 33, №3., p. 310-313.
18. Chankashvili M., Marsagishvili T.A., Machavariani M.N., Japaridze J., Rozhitski N. Investigation of electrochemical properties of optochemotronic sensors. Bull. of National Academy of Sci. of Georgia, Chemical series, 2007, v. 33, №3., p. 314-317.
19. Marsagishvili T., Machavariani M., Tatishvili G. Processes passing in optochemotronic sensors. Nanochemistry –Nanotchnologies. Tbilisi. 2010, ed. "Universali", p 95-99.
20. Marsagishvili T., Machavariani M., Tatishvili G. Nanotechnological sensors for ecological investigations. Georgian Chemical Journal. 2010, vol.10, #4, p. 80.
21. Marsagishvili T., Machavariani M., Tatishvili G. To quantum theory of the work of nanosensor. Bull. of National Acad. of Sci. Georgia, Chemical series, 2011, v. 37, №3.

Author: Grigor Tatishvili
Institute: TSU R.Agladze Institute of Inorganic chemistry and Electrochemistry
Street: 11 Mindeli
City: Tbilisi
Country: Georgia
Email: tati-iice@caucasus.net

Portal Monitor for Human Body Alpha-Radioactive Contamination Control

N. Samotaev, B. Gurkovskiy, V. Miroshnichenko, E. Onischenko, and A. Simakov

Micro- and Nanoelectronics Department, National Research Nuclear University MEPhI (Moscow Engineering Physics Institute), Moscow, Russia

Abstract— **Pedestrian portal monitor was designed for operational remote control of an alpha radioactive surface contamination of clothing, head and hands of the nuclear industry personnel. The method and device described in this report provides the reliable fast detection of alpha-particles at a distance from the radiation source significantly longer than the alpha-particle run in air by measuring the ions clusters, produced on the tracks of the heave charged particles in the ambient air. The alpha-radioactive pollutants are detected by gas-discharge detectors, opened to the air. It allows quickly controlling the alpha-radioactive contamination by artificial air flux, produced in closed portal, on the irregular surfaces of any indoor objects. This portal monitor may be used for safety inspection as the nuclear plants employers, so at radiation terroristic danger. The experiments shown that portal provides stable remote detecting of low alpha particles radioactivity contamination - about 50 Bq (decays per second) in 5 minutes time.**

Keywords— **alpha radiation, air ion, remote detection, portal monitor, nuclear contamination.**

I. INTRODUCTION

Currently, in many areas of practical human activity (especially in the energy, marine transport, science and medicine) radioactive materials are widely used, which are sources of ionizing radiation. Man-made accidents and catastrophes, such as, for example, the Chernobyl accident in 1986 and Fukushima in 2011 have a significant impact on the level of radioactive contamination of the environment.

Today many existing pedestrian portal monitors (usually portal monitors are made for gamma contamination detection: Canberra Industries Inc., Ludlum Measurements Inc. [1, 2]) do not allow to remote detect very small concentrations of alpha-particles on the human's clothes, hair, hands and etc, especially it argent for passing gates of nuclear objects and in public places with existence of nuclear terroristic danger. In the most critical cases usually monitoring systems allow to measure an alpha radiation on hands. The most similar commercial portal monitor TwoStep™-Gas II of company Mirion [3] is also a whole body contamination monitor, but on the basis of gas flow proportional detector technology will have a good efficiency for beta particles, but for alpha particles the subject should be near the detector due to attenuation of alpha particles in air. So, using direct methods of alpha-particles detection it is possible to find it when the source of alpha contamination is situated near the detector (for 10 MeV particles it is about 10 cm in air, for 2 MeV it's about 1 cm [4]).

II. METHOD OF THE REMOTE ALPHA-PARTICLES DETECTION

Most existing methods of alpha-active isotopes like isotopes of Plutonium (Pu-238, Pu-239, Pu-240) or Americium (Am-241) monitoring involve either laboratory analysis of field samples or employment of devices for direct detection of alpha-particles which implies direct contact with contaminated surfaces [5]. These methods fail to ensure complete and rapid examination of a territory or an object due to much labor-consuming measurement. Besides, they have certain limitations associated with type of surfaces examined. Radiation detection may use ionization effect of radioactive particles in air. Atmospheric air contains both neutral atoms and molecules and their ions generated as a result of air ionization. In lower atmospheric layers the ionization is mainly due to natural radiation of radioactive substances and, to a lower degree, to outer space background. Air gas molecules combine with the ions to make up molecule complexes or air ions. Equilibrium ion concentration is determined by the rates of ion generation in the air and decrease in ion number due to recombination. Lifetime of light negative air-ions is determined by recombination with light positive ions as well as by atmospheric aerosol content [6-7]. Under normal conditions the values of lifetime light negative air-ions are within 10-1000 seconds [8-9], which is sufficient for ion collection and detection.

Activity of natural ionization sources such as soil, rocks, outer space nuclides and radiation, is a determinant factor for air ion concentration in a particular place. Deviations from local standards are suggestive of an increased ionization source, e.g. an emergency radioactive discharge. Measurement of the air ion concentration is a method of radiation control. A more accurate approach involves registration of ionization clusters arising on charged particle tracks. Distribution of ionization density is generally determined by ionization source and may be indicative of the ion origin. This principle underlies the alpha-radiation distant detection method to be embodied in developed device.

V. Sontea and I. Tiginyanu (eds.), *3rd International Conference on Nanotechnologies and Biomedical Engineering*,
IFMBE Proceedings 55,
DOI: 10.1007/978-981-287-736-9_125

Typical 5 MeV alpha particle produced by Plutonium (Pu-238, Pu-239, Pu-240) or Americium (Am-241) isotopes during traveling in air creates 140000 pair of electrical charges. Some of them combine with electronegative molecules and create the negative charged ions. In results, so called, light ions complexes have aroused. Life time of the light ions ranges from 10 to 1000 seconds, it is enough for their transportation by artificial air flow from alpha-particle tracks to the working volume of the discharge detector. The light ions are accelerated in the high strength field of the wire chamber and free electrons arising at impact collisions multiplying the avalanche pulse current. As the ionization density on the alpha tracks substantially exceeds density on the tracks of electrons, it is possible to reject background radiation by choosing the proper counter voltage. These method and open on air gas-discharge detector were suggested by the authors of this report [10, 11, 13]. The diagram of all scheme and detector chamber is shown in Fig.1.

III. INSTRUMENTS FOR REMOTE ALPHA-PARTICLES DETECTION

The system for directing airflow is needed to forward ions and clusters of ions from a man with a source of alpha radiation to the detector chamber. It is based on a large amount of fans with adjustable flow rate. The detector works in the pulse counting mode in ambient air [10, 12]. That allows achieving greater selectivity of ions occurring in the presence of alpha particles. The simplified schematic model of the portal is shown in Fig. 1. Wire netting and filter allow to get rid of dust particles and pass negatively charged ions to the anode which is connected to the charge sensitive amplifier and pulse shaper.

The internal view of portal monitor gate is demonstrated in Fig. 4. The fore ventilators on the right side of portal gate create the air flow and the detector unit placed on the left side catch up by own air fan flux the air ions clusters at presence of alpha contamination. The person is coming in portal for about fifteen seconds. The doors are closed and measurement starts.

Block-diagram of the control system of the portal is presented in Fig. 5. Control system of the portal is built on Atmel ARM9 microcontroller. For data display LCD panel of the 4D SYSTEMS company was used (markered "4" in Fig. 5). This allows to use LCD panel for keyboard. For this device we wrote special core module on OS Linux that allows to operatively trigger events from the keyboard (markered "3" and "5" in Fig. 5), detector (markered "8" in Fig. 5) and counter (markered "9" in Fig. 5). Using in control system of the portal conbination of external computer and web-server interfaces (markered "7" and "15" on Fig. 5) is easy for getting measurements dates, watching continuously current count and changing settings. When you need to make a backup for the data and setting you just need to plug in the USB-flash drive (markered "6" in Fig. 5). All this is options controlled by standard linux core modules (markered "10" in Fig. 5). The core of the control system program is the main control sub-program (markered "1" in Fig. 5) responsible for qality of mesurement. That sub-program allows to control for the execution of main modes such as "Calibration" (in this mode device gets coefficients and sets appropriate high voltage according to environmental conditions), "Background", "Search" (posible continuously search for radioactive alpha sources, usually in experiments using 1s period for measurement, but user can change this value from the browser's menu "Settings") and "Measure" (using when needs more accurate assessment of radioactive decay activity). This modules is markered "11-15" in Fig. 5 respectively.

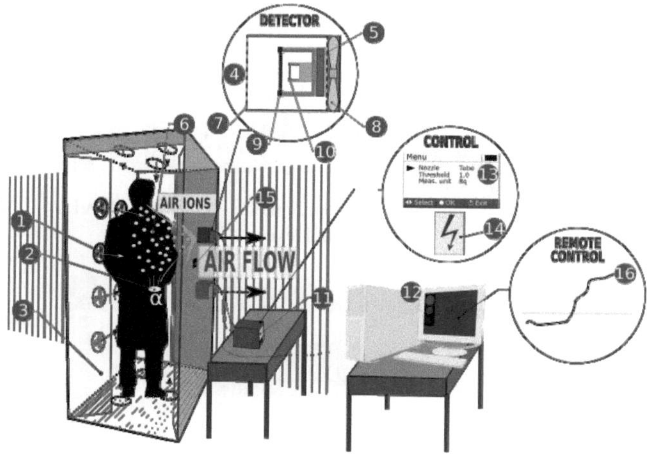

Fig. 1 Scheme of the portal. 1 – system for directing airflow, 2 – source of an ionizing radiation (α), 3 – doors, 4 – enlarged view of the detector, 5 – charge-sensitive amplifier and pulse shaper, 6 – air ions, 7 – wire netting and filter, 8 – fan, 9 – cathode, 10 – anode, 11 – data processing system, 12 – remote computer, 13 – LCD display with keyboard, 14 – stable and precise high voltage source, 15 – temperature, humidity and atmospheric pressure sensors, 16 – remote control system (possible to use from anywhere according to the permissions).

From figure 2 it can be seen that when somebody with an alpha particle source (in this case, with a Pu-239 source with the activity of 760 Bq) enters into portal the count rate value changes from background value (for this picture it is about 4 counts) to approximately 670 counts that much more than the threshold level (for this picture is 13). The threshold value allows for the program more likely to distinguish the presence of an ionization source, such as, for example, the alpha source from the possible ionization fluctuations. The use of moving average with a period of 10 (for the measuring time of 1 second) also allows to avoid fluctuations. Calibration process is depicted in figure 3.

Detector cover is shown in figure 4. Such construction allowed to increase sensitivity of the device. Now air ions don't enter directly to the detector. The panel, to which a detection unit is attached, from the outside is covered with a thin foiled laminate. This is necessary because during work the insulator can accumulate charge that creates an electric field that prevents the negative ions from flowing to the window.

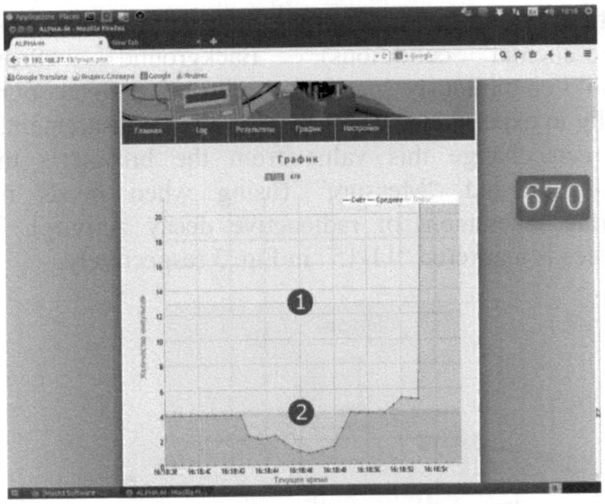

Fig. 2 Continuous monitoring in search mode. 1- threshold level, 2 – background level.

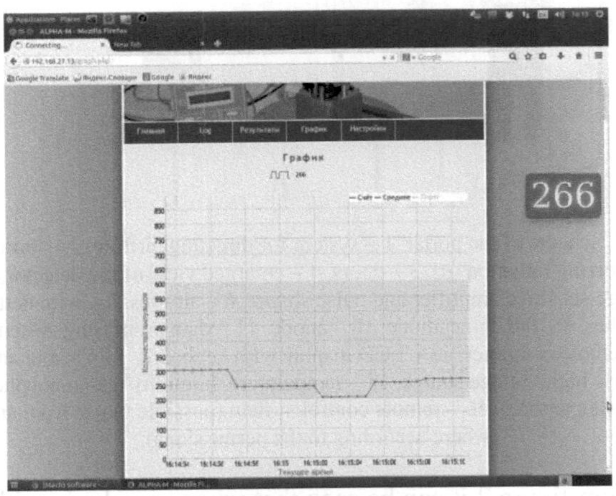

Fig. 3 Calibration mode. An example of a successful calibration (calibration curve lies in the allowable range).

From Settings panel user can easily control the device by changing the parameters of main modes such as "Calibration", "Background", "Search" and "Measure". For example program gives possibility to change operating range, number of measurements, average parameters, device information, date and time etc. Also the software gives possibility to test the modules of the portal. As example implementation of user software is presented in Fig. 3, there are the detector of portal is going to be successfully calibrated, because the measurements get into the corridor.

Fig. 4 Internal view of the portal monitor gate and detector cover.

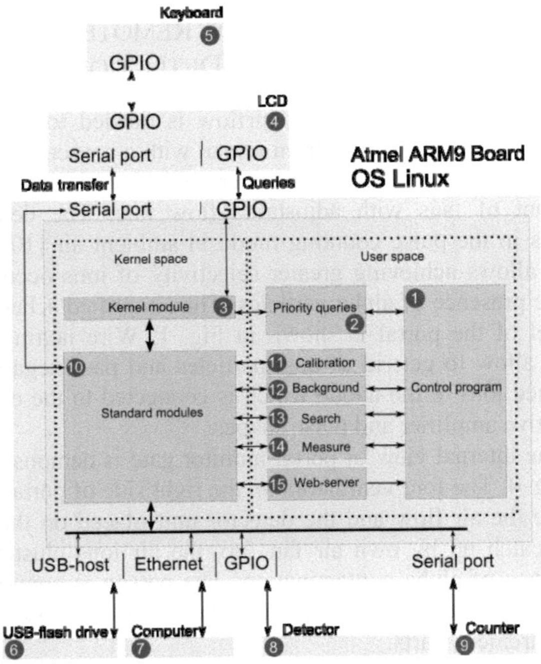

Fig. 5 Block-diagram of the device control system. Description of the numbered nodes is presented in the main text.

IV. EXPERIMENTAL RESULTS

The automatically controlled block provides preliminary calibration and setting up the proper working voltages depended on ambient temperature, pressure, humidity and so on. The count date after processing is displayed on special portal screen and on computer display that is shown in Fig. 2. This screen picture demonstrates the case where the person with an alpha-radioactive source enters into the portal camera and the amount of pulses increased from the background level to the high activity. The portal has also the sound and light alarm signal at detection of alpha-radioactive contamination on clothes, hair, hands of the person. The experimental portal gate has the following characteristics:

- portal gate sizes – H210xL90xW80 cm;
- single detection time – 15 s;
- low level of detectable alpha radioactivity - 50 Bq (decays per second).

In an experiment to determine the detection area (one of the zones with a minimum sensitivity) man with an attached source of alpha radioactivity activity (760 Bq) enters in the portal (figure 6). The high voltage on the anode during the measurements was the same. For such arrangement of fans the efficiency was decreased when placing the calibration source of alpha radiation from the shoulder level to waist level (figure 7).

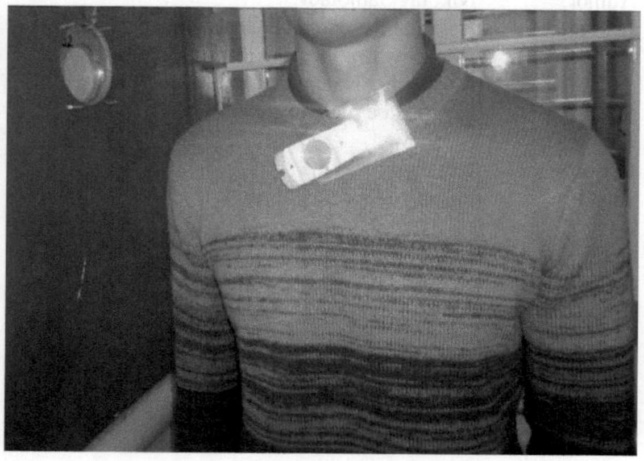

Fig. 6 Measurement process.

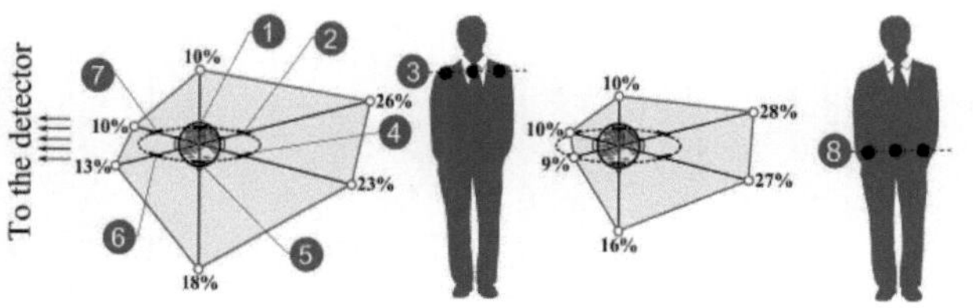

Fig. 7 Decrease in the efficiency of the registration of count rate on pulse ion detector when placing the calibration source of alpha radiation from the shoulder level (3) to waist level (8) in the next places: back (1), back side of the left shoulder (2), front side of the left shoulder (4), chest (5), front side of the right shoulder (6), back side of the right shoulder (7).

Response time is defined (the time of determining the presence of alpha-radioactive contamination). The measurements were repeated 6 times and the average response time is determined by the system. Table 1 shows data to determine the response time for one of the zones with the lowest sensitivity. The same table shows the measurements and the data for the zone with the highest sensitivity (see figure 8).

Table 1 Measurement of the pedestrian portal monitor response time

Number of measurement	Response time for a zone with the lowest sensitivity, sec	Response time for a zone with the highest sensitivity, sec
1	13	4
2	16	3
3	16	5
4	14	5
5	12	4
6	17	5

Fig. 8 Defining areas with the highest and lowest sensitivity.

V. CONCLUSIONS

Searching of alpha-radioactive nuclides by registering ions arising on the tracks of alpha particles is a promising way to create a pedestrian portal monitors for the personnel of alpha-radioactive nuclear plants and research laboratories contamination detection. The pedestrian type portal monitor for human body alpha-radioactive contamination control has been developed. The measurement system of portal allows detecting of alpha sources with very small activities (50 Bq). The main purpose of the portal measurement system is detection of ions occurring as a result of interaction alpha particles with an air medium. That allows finding alpha particles from relatively large distances for such class of measurement system – around half a meter. Also measuring system of portal have following features and advantages: detection of air ions in an event count model; high selectivity against different types of penetrating radiation; high accuracy of measurement under conditions of varying atmospheric parameters; operation in a real time mode; portability; absence of limitations associated with irregular configuration of surfaces; fast examination of any objects, materials in addition to the human body.

Conflict of Interest

The authors declare that they have no conflict of interest.

References

1. Personnel Portal Monitor - Model 52-1-1 http://www.ludlums.com/component/virtuemart/market-1/environmental-113/personnel-portal-monitor-202-detail?Itemid=0.
2. MiniSentry - Transportable Gamma Portal Monitor http://www.canberra.com/products/emergency_response/minisentry.asp.
3. Whole Body Personnel Contamination Monitors https://mirion.app.box.com/s/ribrpov0grfq3tur6sis.
4. Jesse, W. P., Forstat, H., & Sadauskis, J. (1950). The ionization in argon and in air by single alpha-particles as a function of their energy. *Physical Review*,*77*(6), 782.
5. Z. Varga, G. Suranyi, N. Vajda, Z. Stefanka. Determination of plutonium and americium in environmental samples by inductively coupled plasma sector field mass spectrometry and alpha spectrometry. Microchemical Journal, 85, 2007, pp. 39–45.
6. Massey, H.S.W., Negative Ions, Cambridge: Cambridge University Press, 1976.
7. McDaniel, E.W.,Collision Phenomena in Ionized Gases, New York: Wiley, 1964.
8. Tammet, Kh.F., Mobile Light Air Ion Spectra Dependence on Air Trace Contaminant, Uch. Zap. TGU, Tartu, 1975, no. 348, p. 3.
9. Miroshnichenko V.P, Rodionov B.U, Chepel V.Yu. Letters to Journ. of Experimental and Theoretical Physics, 12, 1989 pp.15-16.
10. Simakov, A. (2005). A Device for the Remote Detection of Alpha Emitters under Field Conditions, *Instruments and Experimental Techniques*: 387-391, Vol.48, No 3.
11. Gurkovskiy, B., Miroshnichenko, V., Onishchenko, E., Simakov, A., Streil, Th. (2014). The Remote Detection Of Alpha-Radioactive Nucleus Decay, Advances in Radioactive Isotope Science, Tokyo, 1-6 June 2014.
12. Charpak, G., Benaben, P., Breuil, P., & Peskov, V. (2008). Detectors for alpha particles and X-rays operating in am-bient air in pulse counting mode or/and with gas amplifica-tion. Journal of Instrumentation, 3(02), P02006.
13. Gurkovskiy B, Miroshnichenko V, Onishchenko E, Simakov A, Streil Th (2015) Long-distance detection of alpha-radioactivity: method and device. Journal of Environmental Radioactivity DOI 10.1016/j.jenvrad.2015.07.019.

Author:	Nikolay Samotaev
Institute:	National Research Nuclear University MEPhI (Moscow Engineering Physics Institute), Micro- and Nanoelectronics Department
Street:	Kashirskoe highway, 31
City:	Moscow
Country:	Russia
Email:	nnsamotaev@mephi.ru

The Circuit Method for Decreasing of Sensitivity to ASET Effect for Bipolar Operational Amplifiers

A.A. Lebedev, V.A. Felitsyn, V.A. Komleva, and A.A. Komlev

National Research Nuclear University MEPhI (Moscow Engineering Physics Institute), Moscow, Russian Federation

Abstract— **The schematic method for increasing of radiation hardness of bipolar operational amplifiers for ASET effects is presented. The method is based on the use of improved current mirror and modified input stage circuit.**

Keywords— **single-event transient, operational amplifier, common mode rejection ratio, differential stage.**

I. INTRODUCTION

Since operational amplifiers have good precision characteristics, low supply current and are able to work in single supply (for example from 3 to 32 V) and in dual supply modes (from ±1.5 V to ±16 V), the industrial integrated circuits of LM124 operational amplifiers (Op Amp) [1] are used widely in electronic systems in space environment. However, these circuits have some disadvantages: low phase margin and common mode rejection ratio (CMRR). Moreover they are very susceptible to SET effects [2]. The purpose of majority of works [2-5] is to estimate the sensitivity of the operational amplifiers to ASET exposure at given total collected charge in defined critical transistor nodes of the circuit, which impulse with maximal amplitude and duration occurs in.

The schematic method for reduction of ASET susceptibility of bipolar operational amplifiers, which are based on the use of two differential stages in parallel connection with common feedback in the input stage of the circuit, is presented in this paper. The radiation hardened current mirror, which output current increases, while current gain of bipolar transistors reduces, is used to increase total dose and ASET radiation hardness of the operational amplifier circuit.

II. THE REDUCTION OF ASET SENSITIVITY OF THE LM124 OP AMP BY USING THE RADIATION HARDNED CURRENT MIRROR

The full circuit modeling of LM124 OpAmp for identification of critical node, which is most sensitive to ASET, was performed in a PSPICE simulator. The results of the simulations are matched with the results in works [2, 3]. It means the second stage, which is formed by transistors Q6 and Q9, which are mentioned in [2], makes the main contribution to ASET sensitivity.

The modified Widlar current mirror is used in this paper for reduction of sensitivity to ASET. The modified schematic of the current mirror is presented in Fig.1. The dependence of current transfer ratio K_I of the current mirror on current gain of bipolar transistors β is described by following expression:

$$K_I = Iout/I_0 = 1+ (\beta - 2)/ (\beta+2), \qquad (1)$$

The current transfer ratio K_I of the modified current mirror is increased by 11% in comparison with original Widlar current mirror's one, especially for small values of β, as it follows from expression (1). The dependence of the K_I on β is presented in Fig.2. Since the second stage is the most sensitive to HCP influence [3], the greatest effect of reduction of ASET duration is reached by inclusion modified current mirror between the out of the first stage and the input of the second stage. The ASET time is reduced approximately by 3 times in comparison with similar time in LM124 Op Amp.

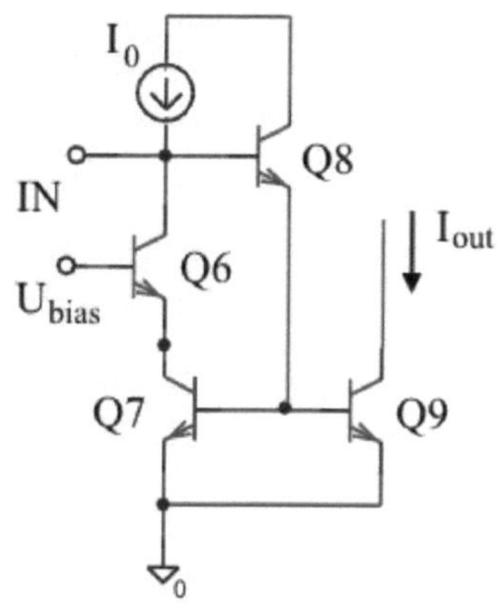

Fig. 1 Modified circuit of the current mirror

This approach requires three additional transistors in comparison with traditional LM124 circuit. The reduction of the ASET duration occurs due to reduction of accumulation of a radiation-generated charge in the Q9 transistor in comparison with the same one in traditional circuit of LM124 Op Amp.

V. Sontea and I. Tiginyanu (eds.), *3rd International Conference on Nanotechnologies and Biomedical Engineering*,
IFMBE Proceedings 55,
DOI: 10.1007/978-981-287-736-9_126

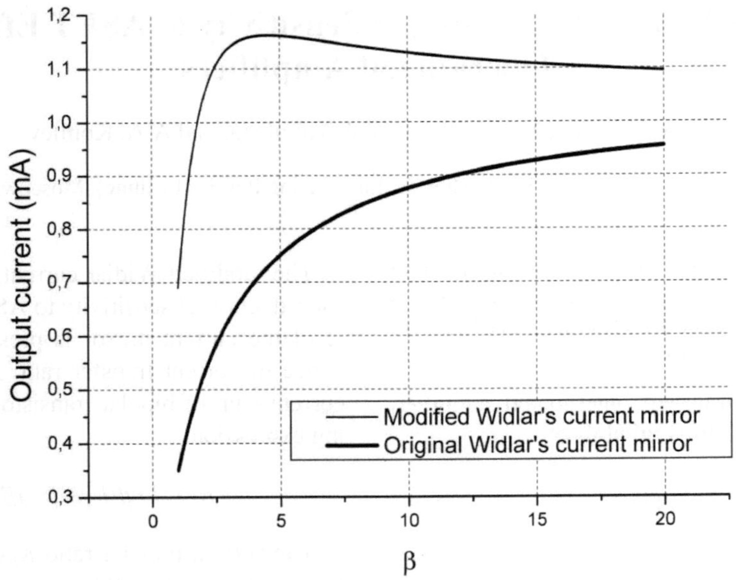

Fig. 2 The dependence of current transfer ratio KI of the current mirror on current gain of bipolar transistors β

III. Modified Circuit of LM124 Op Amp

The original and modified circuits of LM124 Op Amp are presented in Fig.3a and Fig.3b respectively. The circuit of the parallel connection of two conventional differential stages with a negative voltage feedback is presented in Fig.3. The only different between proposed Op Amp and traditional LM124 Op Amp is in input differential stages. The second stage, the output stage and the current source are kept the same ones, as in the prototype. It allows to improve the precision characteristics and to decrease the ASET duration and amplitude. The comparison of ASET characteristics of original and developed LM124 Op Amps are presented in Fig.4a and Fig.4b respectively. The proposed circuit allows to increase the bandwidth of the OP Amp up to 1.9 MHz, to improve the phase margin by 20-30° and to increase the CMRR. The simulated dependences of CMRR of the original and developed circuits on frequency are presented in Fig.5.

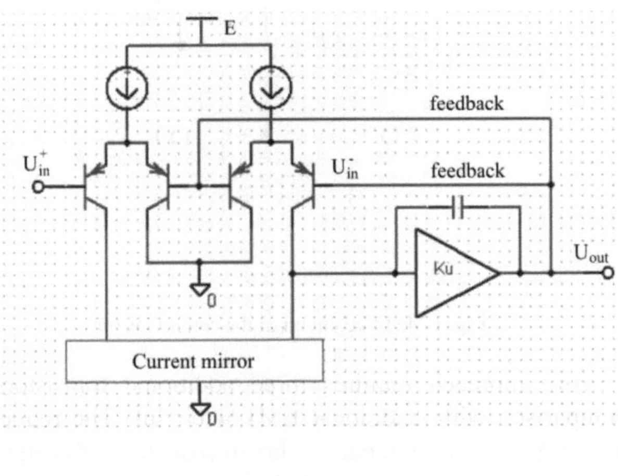

Fig. 3a Original circuit of the input stage of the Op Amp

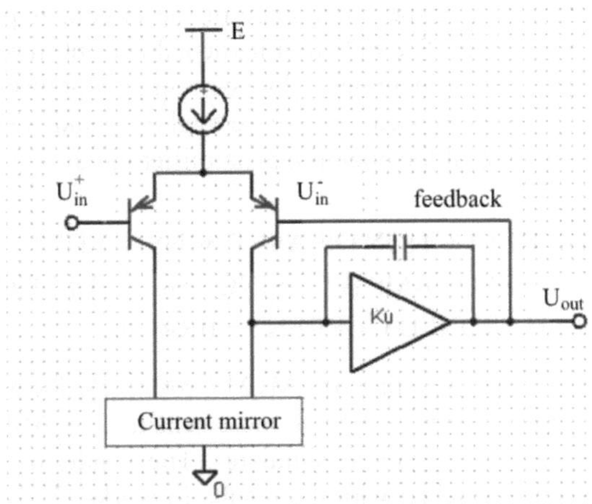

Fig. 3b Modified circuit of the input stage of the Op Amp

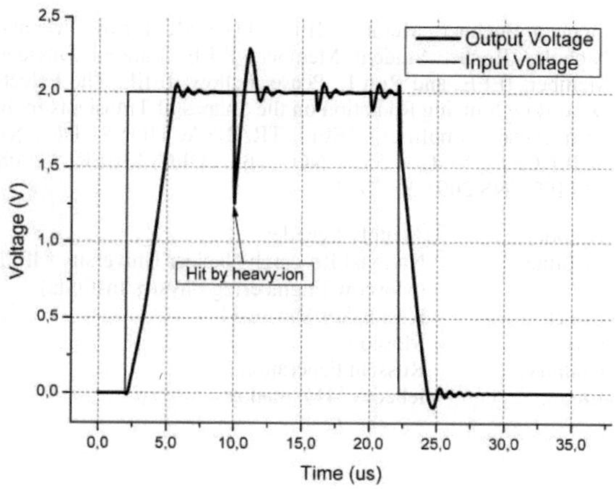

Fig. 4a ASET effect in conventional circuit of LM124 Op Amp

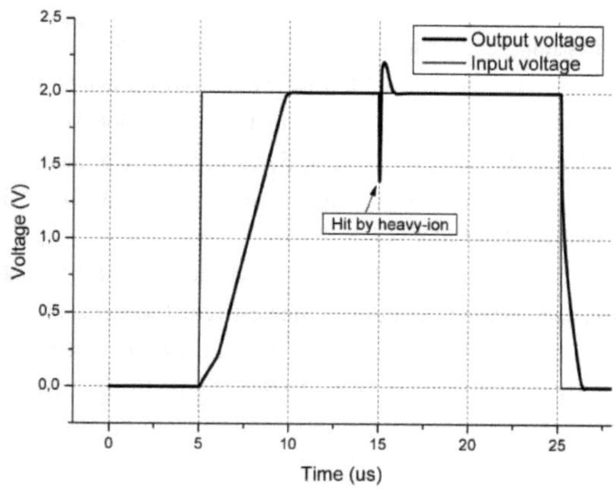

Fig. 4b ASET effect in modified circuit of LM124 OP Amp

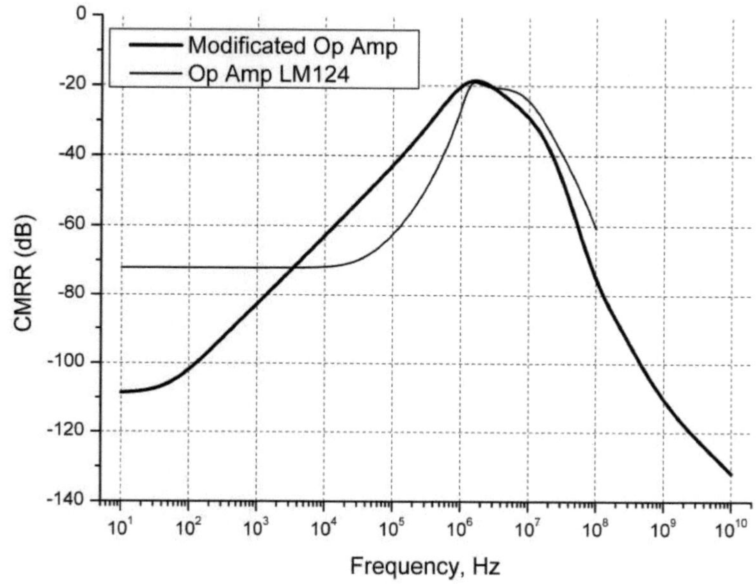

Fig. 5 The simulated dependences of CMRR of the conventional and modified circuits of the LM124 Op Amp on frequency

IV. CONCLUSIONS

The schematic methods for reducing of the ASET sensitivity, which are based the use of radiation-hardened current mirror is proposed in this work. The current mirror was connected between the first and second stages to reduce the duration and amplitude of ASET effect. However this schematic method doesn't improve the precision characteristics of the developed circuit. The use of two conventional differential stages in parallel connection with negative voltage feedback in the input stage of the developed circuit allows to reduce the duration of ASET by 3 times, to reduce the amplitude of ASET by 1,5 times, to increase the CMRR for 10 20 dB, to extend the bandwidth and to improve the frequency stability of the developed circuit.

REFERENCES

1. "LMx24-N, LM2902-N Low-Power, Quad-Operational Amplifiers", Texas Instruments, January 2015, http://www.ti.com/lit/ds/symlink/lm124-n.pdf
2. F. Roig, Student Member, IEEE, L. Dusseau, Senior Member, IEEE, P. Ribeiro, G. Auriel, Member, IEEE, N. J-H. Roche, Member, IEEE, A. Privat, Student Member, IEEE, J.-R. Vaillé, Member, IEEE, J. Boch, Member, IEEE, F. Saigné, Senior Member, IEEE, R. Marec, P. Calvel, F. Bezerra, R. Ecoffet, Member, IEEE and B. Azais, Study of Manufacturing Variations Impact on TID-ATREEs Synergistic Effect in LM124 Operational Amplifier, Radiation and Its Effects on Components and Systems (RADECS), 2013 14th European Conference on, Pages 1-4, DOI: 10.1109/RADECS.2013.6937386 .

3. Y. Boulghassoul, L. W. Massengill, A. L. Sternberg, R. L. Pease, S. Buchner, J. W. Howard, D. McMorrow, M. W. Savage and C. Poivey, Circuit Modeling of the LM124 Operational Amplifier for Analog Single-Event Transient Analysis, Nuclear Science, IEEE Transactions on, Volume 49, Issue 6, Pages 3090 – 3096, DOI: 10.1109/TNS.2002.805400.
4. P. Adell, R. D. Schrimpf, H. J. Barnaby, R. Marec, C. Chatry, P. Calvel, C. Barillot, and O. Mion, Analysis of Single-Event Transients in Analog Circuits, IEEE TRANSACTIONS ON NUCLEAR SCIENCE, VOL. 47, NO. 6, DECEMBER 2000, DOI:10.1109/23.903817.
5. Stephen Buchner, Member, IEEE, Dale McMorrow, Member, IEEE, Nicholas Roche, Student Member, IEEE, Laurent Dusseau, Senior Member, IEEE, and Ron L. Pease, Fellow, IEEE, The Effects of Low Dose-Rate Ionizing Radiation on the Shapes of Transients in the LM124 Operational Amplifier, IEEE TRANSACTIONS ON NUCLEAR SCIENCE, VOL. 55, NO. 6, DECEMBER 2008, DOI: 10.1109/TNS.2008.2007952.

Author: Anatoly Lebedev
Institute: National Research Nuclear University MEPhI (Moscow Engineering Physics Institute)
Street: Kashirskoe shosse 31
City: Moscow
Country: Russian Federation
Email: lebedev741@mail.ru

Cyber Security in the Nuclear and Radiological Domain: Case Study of Republic of Moldova

A.A. Buzdugan[1] and A.I. Buzdugan[1,2]

[1] National Nuclear Security Support Center, Technical University, Moldova
[2] National Agency for Regulation of Nuclear and Radiological Activity, Moldova

Abstract— **In this paper we will reflect upon the importance of cyber security in the functioning of physical security systems in nuclear/radiological entities, as well as the corresponding changes that are foreseen in the process of authorization and monitoring by the National Agency for Regulation of Nuclear and Radiological Activity (Regulatory Body). The recent developments in information technology (IT) have led to the fact that each system that contains IT components, is subject to cyber security risks, including industrial control systems which are often used in nuclear/radiological facilities. As a case study we will present the current situation in Moldova, and we will present briefly various methods to enforce cyber security control implementation in order to reduce cybersecurity risks in entities that deal with nuclear/radiological data.**

Keywords— **cyber security development, nuclear security, Republic of Moldova.**

I. Introduction

Information technology has a tremendous growth and implementation in all sectors of the industry in the last decade. The new technologies offer the possibility to automate the functioning of a component via software program, so that it could receive data, analyze and take decisions independently based on the input provided. However, this also opens new risks for the correct functioning of a device, or in the storage or transport of data. Information security refers to all this aspects. Taking into account that any device or entity that uses nuclear or radiological materials contains as well elements of IT, makes the entity subject to cyber security risks as well.

In this paper we will present the international context and motivation upon recognizing cyber security as intrinsic part of the nuclear security [1], as well as the actions that the Republic of Moldova has taken so far in this aspect. We will reflect upon current legislation, and will propose future steps in order to support the practical application of the existing regulations. This can be done via cooperation with national organizations, such as the NSSC, TSOs as well as with international assistance from donor countries.

II. Context and Motivation

Moreover, according to [2,3] the nuclear domain can be a driver or a motivation for cyber-attacks, in combination with political and economic reasons. Report [4] contain brief link between cyber threat groups and the nuclear motivation, therefore this data can be used during risk assessments for NR facilities. On the other site, the cyber-attacks targeting the energy sector, or more precisely the targeting for industrial control systems [5] [6] can also be linked to nuclear/radiological facilities, since these tend to use the same kind of equipment in order to control and monitor the facility functionality. This is one of the main reasons why cyber security is nowadays taking into consideration upon designing and maintaining various industrial systems, including those used in the NR facilities, or such as physical security systems used by operators.

On another note, according to cyber-attack report of the Red October campaign [7], various domains such as research, government, military as well as nuclear agencies were targeted. This information can be analyzed in the context of other reports [8] where it is mentioned that Republic of Moldova was also in the list of affected countries by this cyber campaign, it can easily be deducted that cyber-attacks nowadays have an international character and targets can be from any domain. Therefore, from our perspective it is necessary to take adequate measure first on the legal and after on the practical side, in order to mitigate cybersecurity risks for NR operators in Republic of Moldova. This refers to physical security systems as well as any other system that has IT components and can be susceptible to a cyber-attack.

The multitude of efforts in cyber security in the physical security systems in NR context, is another motivation to consider the implications of IT security in the nuclear domain. For example, according to the NRC Regulation 10.CFR.73.1 cyber-attacks can be a tool for radiological sabotage or theft or diversion of formula quantities of strategic special nuclear material [9]. Taking into account that Republic of Moldova does not have sufficient resource to conduct more in-depth studies upon the implications of cyber security in nuclear/radiological entities, as well as the specific of the country, the external publications and guides, for example IAEA Nuclear Security [10] papers, can serve as a baseline of best practices that could be implemented in the national legislation and practical application in Republic of Moldova.

V. Sontea and I. Tiginyanu (eds.), *3rd International Conference on Nanotechnologies and Biomedical Engineering*, IFMBE Proceedings 55,
DOI: 10.1007/978-981-287-736-9_127

III. National Legislation in Cyber/ Nuclear Domain

The Moldovan legislation concerning IT security is currently being reviewed and refined, due to the significant progress the country has in this domain. However, the technological development took place much faster compared to the institutional and human resources development, which can be translated into higher cyber security risks of being compromised. Among these, we can mention the low level of security culture, lack of specialized courses/master programs for enhancing IT security at a national level, as well as the lack of a responsible body in case of a cyber-attack and coordination in such events.

On the other side, in the nuclear and radiological security domain, the cyber security component was recognized as part of the physical protection systems in nuclear/radiological operators [1]. This new aspect will be taken into account during the authorization by the Regulatory Body, and will consist the base for future legislation development according to international community recommendations. As an example, the Regulatory Body has developed a draft on Government Regulation [12] for on describing the actions and responsibilities of all actors. However, the lack of resources on the state and operator level, as well as cyber security being a foreign domain of activity the Regulatory Body, sets new impediments in implementing and accomplishing the regulatory duties. Therefore, the possible solutions are multilateral, and require actions from various stakeholders, some of which might be time-consuming. Moreover, usually on the legislative framework are set minimum requirements an operator has to meet in order to fulfill authorization requirements, however in practice these are the basic or only requirements that the operator has controls or tools for. Moreover, the nature of information technology development sets new challenges for the legislative framework, as this can become obsolete in a short period of time. Therefore, it is necessary to conduct future studies on how to enforce minimum requirements not just as a baseline, but to allow and encourage the operators to continuously adapt and implement controls and/or tools in order to mitigate IT security risks.

In the case study of Moldova must be taken into account the fact that it is a non-nuclear country and resources are limited, but fortunately there is a Regulatory Body enforced to authorize NR activities. Also, relevant to this study is that cyber security has been acknowledged on the legal side as part of the physical security systems in NR operating entities, as well as there is a law project towards approval for describing in more details how the authorization process will be met, as well as the actions required from the operators in case of a cyber-attack. The main issue that comes into attention is the capacity of the Regulatory Body to conduct this activity related to IT, as well as the lack of resources on the operator side in order to fulfill these requirements. We would like to present the future possible solutions in three parts: external assistance, staff development and cyber nuclear education curricula development.

The Regulatory Body can be supported either externally, from specialized organizations or technical support programs, or internally, via promoting and creating a special service and staff position of experts in IT security, as from the US NRC model. This new position would also help maintain a security level within the Regulatory Body, as well as help in the authorization process in auditing and coordination the cybersecurity plans and infrastructure of the operators. Also, this would help the Regulatory Body in improving the law on physical security systems with more details and requirements from the cyber security aspect, as it would have in-house subject matter expert in this domain. It is also mandatory during the authorization process conducted by the Regulatory Body to check whether the cyber security plan the operator has developed, as well as to conduct technical audit of random IT components of the operator systems that are in charge or physical security systems. As the country is non-nuclear and many of the nuclear/radiological activities are in the health system, and only few are in nuclear waste and management, this could be a viable proposal.

On the side, it is necessary to have a specialized team that has the required knowledge in both cyber and nuclear/radiological domain, in order to respond and mitigate a cyber-attack in such type of infrastructures. As improving cyber security in industrial control systems in Moldova is a priority of the action plan of the national cyber security program, it is necessary to consolidate the legislative initiative into practical word. At the moment we consider there are not enough specialized human resources for monitoring and detecting cyber-attacks in this type of infrastructures. However, the establishment of the NNSSC [13] opens new opportunities for cooperation between cyber security and nuclear security experts that would lead to staff development in this niche of the IT domain.

Another short to medium term plan is to support NNSSC efforts in NR curricula development, as well as to promote cyber security education in this programs. This would allow splitting the courses according to the functions and responsibilities of the staff, for example for students, practitioners or decision makers.

IV. Conclusions

The last smuggling attempts, as well as the regional security could be other reasons to enforce the implementation of cyber security control in NR operators, as well as to enhance the authorization process of the Regulatory Body in this domain. The approval of the Law 132 [1] is a step ahead towards improving physical security systems by taking into account cyber security issues. However, it is necessary to be adopted by Government Decree [12] which defines the actions and responsibilities from the Regulatory Body and the operators. The implementation of the new university curricula, with subject from cyber security will allow to raise the user awareness level among the nuclear security relevant personnel, as well as to minimize the human factor risk upon using IT in NR domain. The NNSSC capacity development to conduct trainings as well as to support the Regulatory Body as well as operators in awareness raising would help to minimize the gap in this domain in public authorities dealing with NR. Also, international cooperation in the domain and data exchange will assure a high level of cyber security. There are a number of channels to cooperate at international level with other organizations in the cyber security area. These can be via national organizations that are specialized in IT security and have established information exchange agreements, such as regarding threat data or joint trainings, as well as via the Regulatory Body which has very good cooperation with IAEA and other similar bodies from abroad. Overall, the integration of computer systems in technologies used in the NR field, requires a comprehensive action plan in IT security, starting from awareness training for basic users, up to specialized trainings for practitioners and NR operators. At this moment this can achieved by the Regulatory Body only with external support, such as from the NNSSC as well as from external partners that have already established an educational plan in this area.

REFERENCES

1. Law No. 132 from 08 June 2012 - On safe deployment of nuclear and radiological activity (English version): http://anranr.gov.md/wp-content/uploads/2010/04/legea-132-En-11.pdf
2. CrowdStrike Global Threat Report 2013 YEAR IN REVIEW http://www.crowdstrike.com/wp-content/uploads/cs_downloads/CrowdStrike_2013_Global_ Threat_Intel_Report .pdf
3. CrowdStrike Global Threat Intel Report http://go.crowdstrike.com/rs/crowdstrike/images/GlobalThreatIntelReport.pdf
4. MANDIANT M-Trends Report 2014 http://www2.fireeye.com/rs/fireye/images/fireeye-m-trends-report-2014.pdf
5. ICS-CERT, ICS-ALERT-14-281-01B https://ics-cert.us-cert.gov/alerts/ICS-ALERT-14-281-01B
6. Trend Micro - The SCADA That Didn't Cry Wolf http://www.trendmicro.com/cloud-content/us/pdfs/security-intelligence/white-papers/wp-the-scada-that-didnt-cry-wolf.pdf
7. Fireeye WORLD WAR C: Understanding Nation-State Motives Behind Today's Advanced Cyber Attacks https://www.fireeye.com/resources/pdfs/fireeye-wwc-report.pdf
8. Kaspersky "Red October" Diplomatic Cyber Attacks Investigation, 2013https://securelist.com/analysis/publications/36740/red-october-diplomatic-cyber-attacks-investigation/
9. NRC Regulation 10 CFR http://www.nrc.gov/reading-rm/doc-collections/cfr/part073/part073-0001.html
10. IAEA Nuclear Security Series No.17, Computer Security at Nuclear Facilities - Technical Guidance, 2011 http://www-pub.iaea.org/MTCD/Publications/PDF/Pub1527_web.pdf
11. Fireeye Report - WORLD WAR C: Understanding Nation-State Motives Behind Today's Advanced Cyber Attacks http://www-pub.iaea.org/MTCD/Meetings/PDFplus/2013/cn203/cn203Ministerial Declaration.pdf
12. Regulation regarding physical protection in nuclear and radiological activities - Draft, visited 05.05.2015 http://anranr.gov.md/wp-content/uploads/2010/04/Regulament-securitatea-fizica.doc -in Romanian
13. NNSSC - National Nuclear Security Support Center was established in 2014 in the Technical University from Moldova by joint actions of the Regulatory Body and Technical University. Now, it is the INSEN member from 2014 and its activity is sustained by technical support from the IAEA Office of Nuclear Security and Swedish Radiation Safety Authority (SSM). The Board of Governors of the NNSSC consists of 5 nominated persons from Regulatory Body (Chairman), Technical University from Moldova (Executive Director & member), State University of Medicine and Pharmacy (member), Association of Biomedical Engineers from Moldova (member). Additional there are invited counsellors on cybersecurity, radiation protection, nuclear security field.

Author: Artur Buzdugan
Institute: National Agency for Regulation of Nuclear and Radiological Activity
Street: 1 A Russo, of. 67 C.
City: Chisinau
Country: Republic of Moldova
Email: artur_buzdugan@yahoo.com

Nuclear Security as an Ongoing International Process

M. Gerlini

Machiavelli Center, Department of Political and Social Sciences, University of Florence, Florence, Italy

Abstract— **The paper focuses on the Nuclear Security Summit process and its implication toward an expected world Nuclear Security regime, which shall involve medical facilities as well. It is based on communiqués and on the joint statements of participants and references from the Nuclear Security Summit web site. It aims to specify the features of Nuclear Security as prevention of malicious acts, compared to Nuclear Safeguards (counter proliferation) or Nuclear Safety (avoiding nuclear accidents). It adopts a historical analysis methodology, based on empirical research on official sources.**

Keywords— **Nuclear security, international agreements, IAEA.**

I. INTRODUCTION

Nuclear Security is the third concept in order of appearance compared to the earlier Nuclear Safety and Nuclear Safeguards. Because of some overlaps among the three concepts, it is important to put Nuclear Security in the right place of this threefold nuclear matter. In order to clarify the specific field of Nuclear Security and in particular the need to implement a Nuclear Security regime, this paper will analyze the historical and political perspective of such emerging challenge.

II. NUCLEAR SECURITY, AN EMERGING ISSUE

A. The Risk of Nuclear Terrorism

The risk of nuclear terrorism stems from major changes occurred in the international system and in the development of nuclear technology. Nuclear technology and materials are no longer in the exclusive capabilities of the States only. The wider space for criminal organizations active in the nuclear black market followed the world disorder. The US government sensitiveness toward this actual possibility of a nuclear terrorism act forced Americans decision-makers to go further compared to the International Convention for the Suppression of Acts of Nuclear Terrorism (ICSANT) of April 13, 2005 [1]. The first Nuclear Security Summit recognized that the threat of a "nuclear 9/11" was stressed in the speech Barack Obama, President of the United States, gave in Prague. It was clear that it referred to the long-standing program of the network of Abdul Qadeer Khan, the Pakistani nuclear scientist. It was evident that the nuclear black markets were not only businesses created to meet proliferation goals but were something larger and also referred to a non-state level of proliferation. For this reason, the illicit trafficking of nuclear materials and equipment could be a tool for a nuclear terrorist action. The safeguards of the Non-Proliferation Treaty didn't tackle the risk of non-State actors' malicious or terrorist acts. This is certainly significant because the Nonproliferation Treaty is the main architect of the proliferation regime that may prevent a state from proliferating. Also, the safety measures of a nuclear facility didn't avoid all the risks of such terrorist or malicious acts.

B. Radiological and Nuclear Malicious Acts

Radiological dispersal devices are **based** on the use of conventional explosives aimed to spread radioactive material in the environment. If detonated in a city, they would cause mass hysteria and force the evacuation of the affected area, followed by a lengthy and costly clean up. Another malevolent use of radioactive sources involves radiation emission devices, which consist of radioactive sources planted in a location where they can irradiate a target, while remaining undetected for a long period of time.

The use of both radiological dispersal devices and radiation emission devices is considered to be a high nuclear security risk, since radionuclides are extensively used in medicine, industry, and science, and therefore they are potentially accessible to criminals and terrorists on a large scale. There are more than 100,000 radioactive sources in the world. They are mostly used in hospitals and industry, and are considered as very dangerous by the IAEA. As an example, more than 10,000 cobalt-60 and caesium-137 sources are used for teletherapy in hospitals, mainly in developing countries, with intensities that can reach tens of thousands GBq (one gigabecquerel equals one billion becquerel). In addition, there are more than one million weaker radioactive sources, which are also eligible for malicious use.

In the past, before the terrorist attack to the World Trade Centre that took place on September 11, 2011, in New York, security measures applied to radioactive sources were mainly focused on preventing accidental access or petty theft. Before that day, only very few took the threat of radiologic terrorism seriously. As a result, most systems were poorly regulated. Many states have been addressing the problem ever since, but there are still thousands of unaccounted radioactive sources worldwide to look after. These 'orphaned radioactive sources' are either operating outside

V. Sontea and I. Tiginyanu (eds.), *3rd International Conference on Nanotechnologies and Biomedical Engineering*,
IFMBE Proceedings 55,
DOI: 10.1007/978-981-287-736-9_128

official regulatory control or hidden altogether. They may be lost, discarded, or stolen. In any case they all constitute potential weapons for terrorists [2].

C. *The International Legal Framework Background*

For the international legal framework background we had at first the United Nations agreements and resolutions. We should recall the Convention on the Physical Protection of Nuclear Material of March 1980 [3] but obviously also ICSANT. We should recall also the United Nations Security Council Resolutions 1540 of 2004 [4]. In addition to these United Nations agreements and resolutions, there was an international background of partnerships and initiatives. It is worth recalling the Global Initiative to Combat Nuclear Terrorism. This was an initiative promoted by the governments of Russia and the United States by President Bush and President Putin, before the Ukranian crisis exploded. The Global Partnership against the Spread of Weapons and Materials of Mass Destruction of 2002 is a G8 promoted partnership. It has been very interesting to follow. Mr. Obama, President of the United States, called for all vulnerable nuclear material to be secured in four years from 2010, the year of the first Nuclear Security Summit.

III. The Nuclear Security Summits

A. *Washington 2010*

The first Nuclear Security Summit was held in Washington DC. The Nuclear Security Summit is the high-level political process to secure vulnerable nuclear material in all the participant states. There was a commitment for participant states to this vision. Four sessions were planned: Washington in 2010, Seoul in 2012, The Hague in 2014 and Washington in 2016. One of main focuses of this process shall be the contingency plans relevant to highly enriched uranium and plutonium and reactors converted to low enriched uranium and MOX.

In the First Nuclear Security Summit work plan, the final statement reaffirmed the value of IAEA actions, as the nuclear security plan promoted by the IAEA. It welcomed international cooperation among participant states, in the fields of capacity building, sharing of information, education and training, and national law enhancement. Most importantly, the nuclear industry was involved in this process. [5].

B. *Seoul 2012*

In the Second Nuclear Security Summit in Seoul 2012, we had to consider that it was the Fukushima accident, which marked the need for a coherent improvement of both nuclear safety and nuclear security [6]. All the international instruments (ICSANT, GICNT, CPPNM, and so on) were understood in the proposed Global Nuclear Security Architecture. The IAEA actively promoted nuclear security culture. The Nuclear Security Symposium - with a large academic and think tank participation, and the Nuclear Industry summit were events on the edge of the Nuclear Security Summit that marked the commitment of the nuclear community to this process.

During the Second Nuclear Security Summit, there were several agreed statements, and I selected a few to share with you. Groups of participant states agreed to cooperate in specific topics as:

- •IAEA nuclear security fund contribution
- •Elimination of HEU and separated Plutonium
- •Transportation security
- •Nuclear forensics
- •Information security
- •Nuclear security training and support centers (NSSCs)

This was called "Gift basket" diplomacy, a smart frame to favor regional or thematically aggregation between participant states to achieve some outcomes that are supposed to meet the designs of NSS. In the diplomacy it was a new phenomenon, yet to be thoroughly studied.

C. *The Hague 2014*

The Third Nuclear Security Summit at The Hague in 2014 marked an advancement compared to the state of the works [7]. In the Third Communiqué, every participant state presented a progress report of national advancement toward the Nuclear Security Summit aims as a collective document. A new cluster of agreements among participant States was issued. In general, the leading role of the IAEA in the coordination of such efforts was recognized as well as the importance of the International Conference on Nuclear Security held in Vienna in July 2013 promoted by the IAEA Nuclear Security department. During this Third Nuclear Security Summit, there were a lot of "gift baskets." For instance, there were Integrated Nuclear Security Support Plans (INSSP) in which each of the participant states jointly with the IAEA would assist the states in the making up of national Nuclear Security plans. The making and the implementation of Nuclear security plans are supposed involve all the nuclear material, equipment and radiological sources as a whole. For this reason not just nuclear power or research reactors, but also medical facilities and food processing lines using radiological sources shall be taken into account in a nuclear security plan. Furthermore, a group of states stated that a considerable amount of nuclear material has been secured. This was a joint statement by countries free of high enriched uranium. Also, a statement was made to undertake activities and cooperation to counter nuclear smuggling.

The 2014 Communiqué in The Hague reinforced the concept of the nexus between nuclear safety and security and emphasized the importance of addressing both when considering emergency preparedness, response and mitigation.

IV. CONCLUSIONS

It is worth to conclude by presenting some highlights of the Fourth Nuclear Summit planned to be held in Washington in 2016. Towards the end of the process, participant states shall secure all the nuclear materials. The education and training of experts already has scored some results: the International Nuclear Security Education Network and the IAEA-ICTP NS School held yearly in Trieste in Italy, as well as Nuclear Security Support Centers. Participant states will provide the IAEA with relevant information for the Incident and Trafficking Database of the Agency. This issue was raised in the latest Nuclear Security Summit, because Nuclear Security culture will be strengthened in information and cybersecurity.

ACKNOWLEDGMENT

The author acknowledges Swedish SSM for its support.

CONFLICT OF INTEREST

The author declares that he have no conflict of interest.

REFERENCES

1. http://www.un-documents.net/icsant.htm
2. Tuniz C., "Security of Radiactive Sources in Use, Storage and Transport" in Chetaine A., Gerlini M. (2015), Non Proliferation from an International Perspective, IOS Press, Amsterdam, The Netherland.
3. https://www.iaea.org/publications/documents/conventions/convention-physical-protection-nuclear-material http://www.nss2014.com/en
4. http://www.un.org/en/sc/1540/IFMBE at http://www.ifmbe.org
5. http://fpc.state.gov/c35775.htm
6. http://www.mofa.go.kr/ENG/policy/disarmament/nss/references/idex.jsp?menu=m_20_70_30&tabmenu=t_3
7. http://www.nss2014.com/en

Author: Matteo Gerlini
Institute: Department of Political and Social Sciences
Street: Via delle Pandette, 21 - 50127
City: Firenze
Country: Italy
Email: matteo.gerlini@unifi.it

Effects of Electromagnetic Field on Human's Health – A Short Review

N. Gubceac[1,2,3], V. Vovc[2], and G. Lazar[1]

[1] "Vasile Alecsandri" University of Bacau, Bacau, Romania
[2] State University of Medicine and Pharmacy "Nicolae Testemițanu", Chisinau, Moldova
[3] Moldova State University, Chisinau, Moldova

Abstract— **Nowadays, a lot of scientific publications are focused on the effects of electromagnetic field on environment. Especially during the last years has been intensively studied the influence of radiofrequency electromagnetic fields, mostly used in telecommunications. However, published information is often contradictory or incomplete. Thereby, we aimed to perform a fundamental study of both, literature published in internationally recognized journals as well as results published in national scientific journals and available reports of investigations conducted by research institutions in the former USSR. It is known that starting with `70s this research direction has developed very quickly and an impressive number of papers have been published in this field.**

Keywords— **electromagnetic field, non-ionizing radiation, human health.**

I. INTRODUCTION

The effects of radiofrequency electromagnetic field (RF-EMF) on biological systems have been reported in many works. Evidence suggests that there is a large lacuna in research regarding the ecological effects of RF-EMF and little is known about the possible mechanism of interaction between EMF and living organism. In this paper we are looking for determination of EMF frequencies with application in treatment and, along with some positive effects – the most dangerous nonionizing radiation for human health. We analyzed a wide number of reports and the main conclusions of influence of different wavelength EMF on humans were highlighted. Was tacked into account papers published in reviewed international journals and reports of investigations conducted within research institutions in the former USSR. At initial stages of research of low frequency EMF influence on human health was discovered a wide application in treatment of muscular traumas, bones and skin injuries. However, in studied papers were selected some specific frequency for EMF under the action of which the influence on measured quantities can be observed without any argumentation of choosing that particular values for EMF parameters. Specific applications of EMFs in medicine are fairly well reflected in the scientific literature and will not be deeply analyzed in this paper.

II. EMF INTERACTION WITH HUMAN'S ORGANISM

A. EMF of Extremely Low Frequencies

Based on scientific researches published in '90, many researchers are convinced that the humans are extremely sensitive on EMF of extremely low frequencies (ELF EMF) in 2003 was published the report of Malchevsky et al. [1]. Here were synthetized results what present a positive effect of ELF EMF on skin and the musculoskeletal system during a post traumatic period as well as the electroacupuncture application studies. The fields in range of 0.1– 30 Hz and the amplitude of 100 mV seems to have a strong influence on bioactive points and therefore contribute to the functions of specific organs or systems in the human body.

In a wide number of papers were reported the laboratory observation of ELF EMF influence on human bone marrow stromal cells (hBMSCs, also known as bone marrow-derived mesenchymal stem cells). The hBMSCs are a population of progenitor cells that contain a subset of skeletal stem cells, able to recreate cartilage, bone, stroma that supports hematopoiesis and marrow adipocytes. As such, they have become an important resource in developing strategies for regenerative medicine and tissue engineering due to their self-renewal and differentiation capabilities [2]. In the *Table 1* were presented few frequencies for ELF EMF with effect on human reflected in a large number of papers based on laboratory observations.

Table 1 Frequency specific effects of EMF on hBMSCs [2]

Frequency, Hz	Outcome
7.5	Significantly increased ALP beginning at day 7 and reaching the highest level at day 28; increased early expression of osteogenic marker, RUNX2;
15	Increased expression of osteogenic markers ALP, SMAD1, RUNX2, OSTEOPONTIN, OSTEOCALCIN, BMP2, TGF-β, MMP1, MMP3, and BONE SIALOPROTEIN.
5, 25, 50, 75, 100, 150	50 Hz was most effective at differentiation of hBMSCs to osteoblasts via significant increases in ALP, OSTEOCALCIN, COLLAGEN I, and Ca_2^+;

V. Sontea and I. Tiginyanu (eds.), *3rd International Conference on Nanotechnologies and Biomedical Engineering*, IFMBE Proceedings 55,
DOI: 10.1007/978-981-287-736-9_129

Obviously it is not proposed the replacement of classical treatments by applying of EMF, which actually can promote cellular responses synergistically and be used as some auxiliary treatment applied to stimulate tissue regeneration under favorable physical conditions. A great issue in using EMF concerns the fact that investigators rarely discuss why they selected specific parameters in their studies, leaving the reader to assume that their selections were random [2].

B. EMF Effect on Brain Electrical Activity

On the other hand, an impressive series of papers are focused on the effect of EMF on the brain electrical activity. In the report [3] dealing with the stimulus-response relationship between low-level, low-frequency electromagnetic fields (EMFs) and changes in brain electrical activity permit assessment of the hypothesis that EMFs are detected by the body via the process of sensory transduction. These reports, as well as those involving effects on brain activity observed after a fixed time of exposure, are critically reviewed here. A consistent stimulus-response relationship between EMFs and changes in brain activity has been demonstrated in animal and human subjects. The effects, which consisted of onset and offset evoked potentials, were observed under conditions permitting the inference that the fields were transduced like ordinary stimuli such as light and sound. However, unlike the changes in brain activity induced by these stimuli, the changes induced by EMFs were governed by nonlinear laws. The studies involving attempts to determine whether a period of EMF exposure caused a metabolic effect reflected in pre-exposure/post-exposure differences in brain activity were generally inconclusive.

If brain electrical activity is altered in the presence of an EMF, it is reasonable to suspect that the effect would occur in subjects with epilepsy because their brain electrical activity is labile and vulnerable to changes caused by imperceptibly subtle internal and external factors. This possibility was investigated in a series of studies involving exposure of subjects with mesial temporal lobe epilepsy (MTLE) who were stimulated with 1–40 G [4-7]. The authors [7] favored the opinion that the effect on epileptiform activity arose not from the presence of the field but from the fact that it was switched on and off. However, it is equally possible that the effect could have been due (wholly or partly) to the presence of the field. In the study in which both individual and group analyses were performed [4,5], they found that the individual effects were averaged away when the subjects were analyzed as a group. This result is consistent with an inference that the effects were nonlinear, but it could also be explained by assuming that the effects were linear but weak, and hence were not averaged away but simply diluted by the 50% of the subjects who did not show statistically significant results. At influence on brain (neural) tissue with a EMF of frequency close to the natural frequency of brain biorhythms, the biological effects of is increased [8]. Previously were studied [9] the influence of a EMF in range of 0.2 – 1Gauss at low flequencies (<60 Hz) by short and long term action. Some known relations between EMF frequency and human reactions on exposure are presented in the *Table 2*. These studies highlight the ability of organisms to adapt even to EMF what can influence the brains electrical activity proved by electroencephalogram (EEC) records. Therefore, the influence of weak EMF and the effect on electrical activity of the brain remains an open subject.

Table 2 Dangerous and harmful frequencies (by D. Kontorov et al., 1993)

Frequency, Hz	The negative effect
0.02	Increase the response time to excitation;
0.06	Persistent mental inhibition;
1-3 (delta rhythm of brain)	Stress;
5-7 (theta rhythm of the brain)	Mental fatigue; Stress; The negative emotional arousal;
8-12 (alpha rhythm of the brain)	Affects the reactivity and excitement; Convulsive activity;
12-31 (beta rhythm of the brain)	Mental fatigue; Increased stress;
40-70	At high field strength - deterioration of metabolic processes; Individual physiological changes, anxiety;

C. Radiofrequency EMF

McDovall reported on a case-control study involving 537 deaths in England and Wales (males greater than 15 years of age) that died from acute myeloid leukemia in 1973 [10]. The control group included all causes of death except leukemia. They found a consistently increased relative risk for the occupations that involved exposure to electromagnetic fields: occupationally-exposed men had a relative risk of acute myeloid leukemia of 2.3 ($P < 0.05$). In a similar case-control study Pearce et al., also reported that electrical workers in New Zealand were at increased risk of leukemia [11]. Milham studied the leukemia deaths that occurred in male members of the American Radio Relay League, a group of amateur radio operators [12]. In 1971-83, 1,691 deaths among League members occurred in Washington and California. Twenty-four deaths from leukemia were observed, compared to the expected 12.6. Robinette et al., studied the effects of exposure to electromagnetic fields among Navy servicemen who graduated from training schools and served aboard ships during 1950-1954 [13]. No differences in a variety of mortality and morbidity indices were found by the investigators that could be attributed to the difference in exposure (1000 pW/cm^2 and occasionally 100,000 pW/cm^2). A typical computed penetration depth at

60 Hz is 9 mm, and at 100 MHz it is 0.01 mm. Thus, with the possible exception of magnetic fields, the body's detection system for external electromagnetic fields probably exists within the first few millimeters of skin and underlying tissue. Only magnetic fields penetrate into the body organs. Therefore, the presence of electromagnetic fields in tissue cannot be measured directly [14]. In concerned work, a lot of publications are analyzed in the corresponding field of study, authors reveal numerous observations on growth (almost doubling) the number of cancer cases registered within workers exposed to the influence of the strong EMF compared to workers in areas that are not exposed. However, due to the influence of different complex factors, a certain conclusion that the cancer risk was increased due to exposure to radiofrequency electromagnetic field can't be made.

Table 3 Picture of clinical manifestations of the impact of microwaves on the human body at different radiation intensities [8, 17]

The intensity of microwaves mW/cm^2	The changes observed
600	Painful sensations during exposure.
200	Inhibition of redox processes of tissue.
100	Increasing of blood pressure followed by it's decrease. In the case of chronic exposure - sustainable hypotension. Bilateral cataracts.
40	Feeling warmth, vasodilation, irradiation for 0.5-1.0 h is increasing the pressure with 20-30 mmHg.
20	Stimulation of redox processes of tissue.
10	Asthenization after 15 minutes of irradiation.
8	Undefined changes in the blood with a total exposure time of 150 hours, a change in blood coagulation.
8	Electrocardiographic changes, changes in the receptor system.
4-5	The change in blood pressure during multiple exposures, leukopenia of short duration erythropenia.
3-4	Vagotonic reaction with symptoms of bradycardia, the slowing of heart's electrical conductivity.
2-3	Pronounced reduction in blood pressure, increased heart rate, variations in blood volume of the heart.
1	Reduction of blood pressure, a tendency to increased heart rate, minor variations in blood volume of the heart, reduction of ophthalmotonus at daily exposure for 3.5 months.
0.4	Auditory effects when exposed to pulsed EMI.
0.3	Some changes in the nervous system at chronic exposure for 5-10 years.
0.1	Electrocardiographic changes.
< 0.5	The trend to lower blood pressure under chronic exposure.

The influence of radiofrequency EMF (E_0 = 1.39 V/m) of the heart beat rate was studied in [15] and the authors argue that "obtained data suggest that the electromagnetic radiation has an adverse effect on the cardiovascular system and psychophysiological state of adolescents". Nevertheless, by analyzing the presented experimental data it is hard to considerer the conclusion as being reliable and the assumptions that radiofrequency EMF influence the cardiovascular activity can be considered as poorly-founded.

Dovgusha et al. (2009) performed a study on the influence of EMF on the human body as a whole. The authors assume that water structures in live organism are most sensitive to influence of EMF of bellow 4 Hz and magnetic strength ten times lower than the earth magnetic field magnitude [16]. They are worried about the fact that using a specially created and oriented Schumann frequency it is possible to insensitive and invisible (as radiation) effects on humanity, his mood, emotions, health, and brain activity. A useful systematic analysis of EMFs interaction with the human body in dependence of frequency and field strength is shown in *Table 2* and *Table 3*, respectively.

III. CONCLUSIONS

Without any doubts EMF exert a significant influence on all living organisms starting from microorganisms, plants and ending with human. It is well known that all living organisms exhibit adaptation properties regarding to changing environment conditions. But in case when the organism's adaptation period is much longer than the period of environmental conditions variation, then the respective condition can produce a negative influence on organisms and their development. From the large number of published investigations we can determine the influence of EMF with certain parameters on some living organisms. While the medical application of EMF is well described in papers, the most of researches are in the narrow domains and usually it seems that frequencies and other field parameters is basically randomly selected. Hence, this does not allow proposing of a complex picture of the influence of EMF on the humans, in particular and on the environment, in general. On other hand, in many papers published mostly in non-reviewed journals, the influence of EMF on living organisms (i.e. human organism) is presented in a popular way and usually the connection between the observations, experimental data and conclusions is not obviously. Therefore, the understanding of EMF influence is very important for both - safety operating measurements and calming the spirits of people panicked by modern technologies. In this context we consider that existing knowledge in the presented research field is still unaccomplished and it is necessary to conduct a comprehensive study of literature and developing a portrait of EMC influence on the environment for a wide range of frequencies.

CONFLICT OF INTEREST

The authors declare that they have no conflict of interest.

REFERENCES

1. Mal'chevskiy V.A., Sergeyev S.V., Semenov A.V., (2003) Peremennoye elektromagnitnoye pole i mekhanizmy yego vozdeystviya na organizm cheloveka (Literature Review). Vestnik Tyumenskogo Gosudarstvennogo Universiteta, no. 2, url: http://cyberleninka.ru/article/n/peremennoe-elektromagnitnoe-pole-i-mehanizmy-ego-vozdeystviya-na-organizm-cheloveka-obzor-literatury (accesed: 30.05.2015).
2. Ross C.L., et.al (2015) The effect of low-frequency electromagnetic field on human bone marrow stem/progenitor cell differentiation, Stem Cell Research, vol. 15, 2015, p. 96–108.
3. Carrubba S., Marino A.A., (2008) The Effects of Low-Frequency Environmental-Strength Electromagnetic Fields on Brain Electrical Activity: A Critical Review of the Literature. In: Electromagnetic Biology and Medicine, vol. 27, 2008, p. 83–101.
4. Fuller M., Dobson J., (1995) On the sensitivity of the human brain to magnetic fields: evocation of epileptiform activity. Brain Res. Bull, vol. 36, p.155–159.
5. Fuller M., Wilson C.L., (2003) On the confirmation of an effect of magnetic fields on the interictal firing rate of epileptic patients. Brain Res. Bull., vol. 60, 2003, p. 43–52.
6. Dobson J., Pierre T. St., (2000) Changes in paroxysmal brainwave patterns of epileptics by weak-field magnetic stimulation. Bioelectromagnetics, vol. 21, 2000, p. 94–99.
7. Dobson J., Pierre T.G. St., (2000) Analysis of EEG data from weak-field magnetic stimulation of mesial temporal lobe epilepsy patients. Brain Res., vol. 868, 2000, p. 386–391.
8. Tikhonov M.N., Dovgusha V.V., Dovgusha L.V., (2014) Influence Mechanism of the Natural and Technogenic Electromagnetic Fields on Life Safety, J. Health Risk Analysis, vol 4, 2014, p. 85-100.
9. Bell G., Marino A., Chesson A., (1994) Frequency-specific blocking in the human brain caused by electromagnetic fields, NeuroReport, vol. 5, no. 4, 1994, p. 510-512.
10. MoDowall M.E., (1983) Leukemia Mortality in Electrical Workers in England and Wales, Lancet, 1983, p. 246.
11. Pearce N.E., Sheppard R.A., Howard J.K., Fraser J., Lilley B.M., (1985) Leukemia in Electrical Workers in New Zealand, Lancet, 1985, p. 811.
12. Milham S., Keys Jr.S., (1985) Leukemia Mortality in Amateur Radio Operators, Lancet, 1985, p. 812.
13. Robinette C.D., Silverman C., Jablon S.,(1980) Effects upon Health of Occupational Exposure to Microwave Radiation (Radar), Amer. J. Epidemiol., vol. 112, 1980, p. 39-53.
14. Marino A.A. and Morris D.H., (1985) Chronic electromagnetic stressors in the environment: a risk factor in human cancer, J. Environ. Sci. Health, vol. 3, no. 2, 1985, p.189-219.
15. Dashieva D.A., (2012) The Inflence of Electromagnetic Radiation on the Cardiovascular System and Psychophysiological Indices of Children Aged 15–16. Scientific notes of ZabGGPU, Natural Sciences Series, no.1 (42), 2012, p. 115-121.
16. Dovgusha V.V., Tikhonov M.N., Dovgusha L.V., (2009) Vliyaniye yestestvennykh i tekhnogennykh elektromagnitnykh poley na bezopasnost' zhiznedeyatel'nosti. Ekologiya cheloveka, no.12, 2009, url: http://cyberleninka.ru/article/n/vliyanie-estestvennyh-i-tehnogennyh-elektromagnitnyh-poley-na-bezopasnost-zhiznedeyatelnosti (accesed: 01.06.2015)
17. Minin B.A., (1974) Sverkhvysokochastotnyye volny i bezopasnost' cheloveka, Mosk. Sovetskoye radio, 1974, 348 p.

Author: Natalia Gubceac
Institute: State University of Medicine and Pharmacy "Nicolae Testemiţanu"
Street: 165 Stefan cel Mare si Sfant Bd.
City: Chisinau
Country: Moldova
Email: natalia.gubceac@usmf.md

Methods for the Self Calibration of Ion Mobility Spectrometer

V. Vasilyev, V. Pershenkov, N. Samotaev, V. Belyakov, A. Golovin, E. Malkin, E. Gromov, I. Ivanov, Y. Shaltaeva, and D. Lipatov

Micro- and Nanoelectronics Department, National Research Nuclear University MEPhI
(Moscow Engineering Physics Institute) Moscow, Russian Federation

Abstract— **In transport systems like airports, subway or railway stations widely used ion mobility spectrometers for hazardous toxic gases or explosives traces detection. At the present time ion mobility spectrometers use ionization sources based on the corona discharge. These sources of ionization require frequent calibration, especially if the device is used in the portable embodiment where rapid changes in ambient atmosphere are possible. Error false positives in such systems are critical – the evacuation of passengers transport terminal and emergency calls. To improve situation with false errors we develop system for the self calibration of ion mobility spectrometer based on combination of explosives vapors of and products released during the corona discharge combustion. Designed calibration system is based on corona discharge for ion mobility spectrometry. The control system is integrated into the overall spectrometer structure.**

Keywords— **ion mobility spectrometry, explosives, calibration.**

I. Introduction

Currently, ion mobility spectrometry is a globally recognized standard for the detection of trace amounts of substances [1-3]. Devices based on the principles of ion mobility spectrometry, are used in public places (airports, train stations, stadiums) and during search and examination activities [4]. First of all, this type of devices designed to detect traces of banned transportation, storage and use of substances – it comes about explosives, narcotics, and chemical warfare agents. The operation principle of ion mobility spectrometry devices is the selection of the sample from the surface or from the air, the ionization of the sample and following drift of ions formed in a constant electric field. The result is formed by the combination of the sample output spectrum. The ion of each substance has a definite mobility, so the measurement time of flight characteristics of ion allows to determine their type. Stability of the output spectrum is determined by the efficiency of the spectrometer. As a result, changes in ambient temperature, humidity and pressure causes the shift of the spectrum. Therefore, a periodic calibration of the spectrometer, allows the influence of external factors reduction. Calibration is carried out by previously known substances which are periodically introduced into the ionization chamber. It ensures the accuracy of the measured time of flight instrument characteristics. This article describes the working principle, advantages and disadvantages of the calibration method using steam vapors of explosives or their imitators. Furthermore, in an alternative embodiment of the work calibration of ion mobility spectrometers operating in the detection of explosives by means of corona discharge of the reaction products (oxides of nitrogen) is considered. This method is the most suitable one for use in portable detectors, due to the small weight and size parameters and low power consumption, as well as the possibility of constructive integration of calibrants into the device.

II. Calibration with Explosive Vapors

Exact reproduction of minute quantities of explosive vapors is difficult for several reasons. First, the vapor pressure of these compounds is usually very low – making it difficult to create the desired concentration of molecules in the air. For example, trinitrotoluene (TNT) has a vapor pressure approximately equal to 1 ppb, and a vapor pressure of hexogen (RDX) by almost three orders of magnitude below [5]. To increase the number of the molecules near the surface, the heated material is applied. It should be taken into account that explosives are very active compounds, which tend to decompose during heating. Therefore, there is a possibility that the produced vapors contain much larger molecular decomposition products, than molecular explosive. Finally, a pair of output from the steam generator is to be delivered to the input of the detector without significant losses. This is difficult because of the molecules deposition on the cold surfaces of the connecting paths. This process is difficult to control.

Existing generators of standard concentrations developed by the Institute of Inorganic Chemistry (Siberian Branch of the Russian Academy of Sciences), based on a particular class of substances – fluorinated graphite intercalation compounds (ISFG). ISFG possess the ability to retain substantial amounts (up to 20 wt%) target substances and allocate them constant (for a given temperature) speed. These sources have a number of drawbacks: a bulky, complex work, inertia, and characterized by poor reproducibility concentrations.

Variant of steam, based on the explosive diffusion of molecules from the solid phase, was first proposed in 1976

V. Sontea and I. Tiginyanu (eds.), *3rd International Conference on Nanotechnologies and Biomedical Engineering*,
IFMBE Proceedings 55,
DOI: 10.1007/978-981-287-736-9_130

USA [6] and is widely used for calibration of this type until now. In Russia, generators based on this principle have not been developed. In this paper, a study of trace amounts of steam generator using ordinary explosives trinitrotoluene (TNT) or RDX.

The general scheme of the steam generator is shown on Fig. 1. The source is a thin metal tube, in one end of which a small amount (a few mg) of solid is placed. The tube is placed in a specially designed heater which temperature can vary from room temperature to 120°C with an accuracy of 0.5 degrees. Upon heating, the substance vapors diffuse to the top edge of the tube. The heated tube is joined to a channel through which the purified air moves toward the entrance of the spectrometer and an ion mobility spectrometer to fall through the sample collection device.

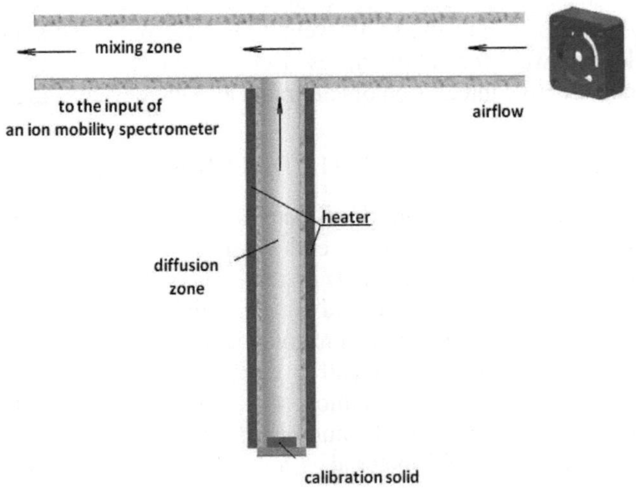

Fig. 1 The diffusion steam generator.

A study of the steam generator based on the evaporation of the substance from the solid phase, and subsequent diffusion of gas molecules in the volume. Produced by a steam generator flow of matter-calibrants is expressed in weight of substance transported per unit time. This parameter depends on the temperature of the heating source, temperature and size of the region in which the diffusion process occurs:

$$M_{output} = N_0 * D * H/L \qquad (1),$$

where M_{output} – flow agent at the source output, g/s; N_o – saturated vapor pressure at a given temperature material heat, g/cm^3; D – the diffusion coefficient, cm^2/s; H – diffusion section area, cm^2; L – size in the direction of the diffusion region, cm. H and L are determined by the geometry of the experimental setup [7].

According to the results of theoretical calculations using (1), the output of TNT at room temperature is about $1.2 \cdot 10^{-10}$ g/cm^3. A heating temperature of the source of the tube 80°C substance increases the amount of effluent to a value of $5 \cdot 10^{-8}$ g/cm^3. These values correspond to the order of trace amounts of explosives, remains on the surface of various items during transportation. Detection of trace amounts of substances should provide an ion mobility spectrometer.

At the exit from the heated tube molecule substances into the air, flow rate is regulated in the range from 4 to 30 cm^3/s. This allows you to vary the degree of dilution and create the desired volume concentration of the substance.

TNT spectrum at the output of the device is presented on Fig. 2.

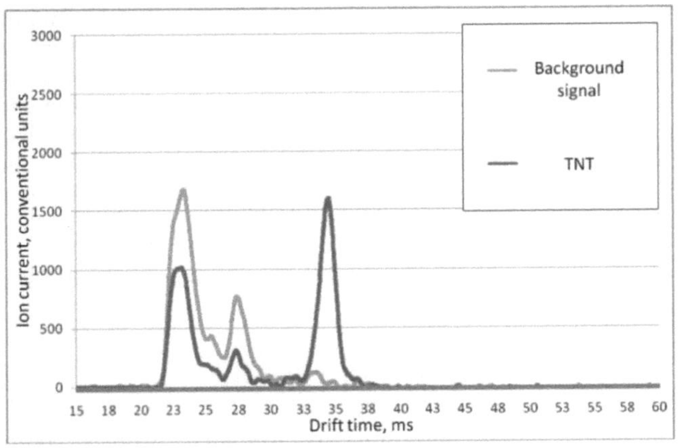

Fig. 2 Spectrogram of the trinitrotoluene (TNT).

During the experiments it was determined that in conventional laboratory without facilities further purified air developed ion mobility spectrometer with an ionization source based on corona discharge [8] is capable for detecting the concentrations of TNT up to 10^{-11} g/cm^3. A further increase in sensitivity is possible with improving of the sampling device.

Despite its relative simplicity, this calibration method is applicable only for laboratory conditions. In the field, it's hard to use because it does not meet certain requirements. Namely, the source has a relatively large mass and size parameters, high power consumption, time-to-order mode for 5-7 minutes and uses their work with explosives or their imitators.

To address these shortcomings, has been tasked to develop a method for calibrating ion mobility spectrometry in negative ion detection mode (mode detection of explosives) that satisfies the following requirements:

built-in portable spectrometer design, small size, low power consumption, the calibration less than 10 seconds, avoid the use of simulators explosives, fully automatic operation, long service life of the system. The following describes a calibration method using corona discharge.

III. CALIBRATION USING CORONA DISCHARGE

The possibility of chemical reactions under the influence of an electric discharge was discovered almost a century ago. Since then, detailed studies of the processes occurring during combustion of the corona discharge in air were realized [9-12]. By-products resulting from combustion of the corona discharge, such as nitrogen oxides, compete with the analyte in the electron capture reactions. The concentration of nitrogen oxides can be quite large and have to be taken into account when working with ion mobility spectrometry. These compounds have high oxidizing properties and complicate the detection of explosives. However, because of combustion stability nitrogen oxides can be used as a calibration substance.

The study of the reaction products of the corona discharge was carried out on a specially-designed layouts. The design consists of two electrodes made of stainless steel with thickness of 0.5 mm using the laser cutting technology. Tip electrodes are placed at a distance of 2 mm. Figure 3 shows a diagram of the corona discharge formation and combustion control.

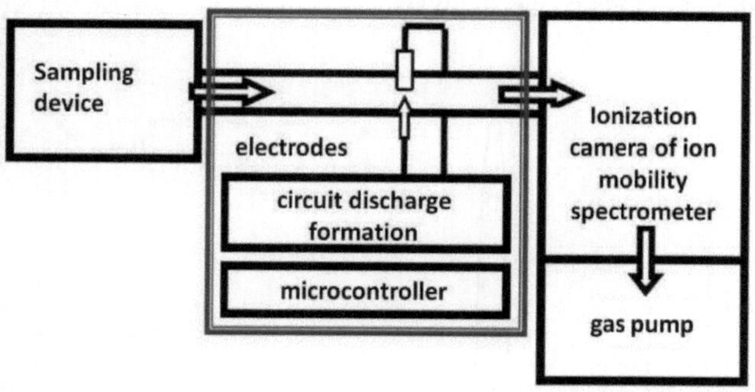

Fig. 3 A block diagram of calibration system based on corona discharge.

Scheme for generating high-voltage pulses is presented on Fig. 4.

Managing pulse-width signal COR_CTL is formed by microcontroller. After filtering and linear transformation it is applied as control signal for voltage regulator (integrated circuit U1) switching with an output voltage in the range from 0 to 12 V. The output switching regulator voltage is applied to the resonant self-oscillating sine wave generator. The oscillation frequency is determined by the capacitor C8

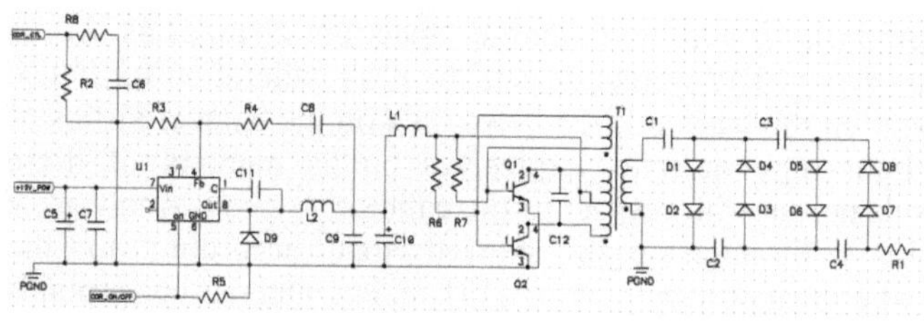

Fig. 4 Scheme of the corona discharge formation.

and the inductance of the coil is connected between the collectors of transistors Q1 and Q2. Working frequency is 40 kHz. Steady voltage across the primary winding is directly proportional to the input voltage and determined by the equation 2:

$$U2 = U1 * \pi/2 \quad (2).$$

Increase of the winding of the transformer T1 generator provides a signal amplitude of 800 V. The output of the transformer is supplied to the two-stage diode-capacitor voltage multiplier, at each stage of which is formed by a double amplitude of the sine wave. The output generated by the high voltage direct current, which has the maximum value of 3200 V.

A current limiting resistor voltage is applied to the needle corona source. As used in this paper based on the ionization source of the corona discharge has the geometry of the tip-arc. Corona discharge current is fed to the filtration element comparisons forming a feedback signal. The difference signals from the current sensor and is used to control COR_CTL pulsed voltage regulator.

Functionally, this unit provides stabilization of the corona discharge current. The parameters of the circuit are described by equation 3:

$$I_{CORONA}\,[mA] = DAC\ code * 0.014 \quad (3),$$

where DAC code is 10-bit microcontroller digital to analog converter code with a reference voltage of 5 V. A feature of this device is the possibility of forming a short pulse corona discharge. The system was tested during the generation of pulses with pulse widths ranging from tens of milliseconds to a continuous mode. This characteristic, together with the possibility of adjusting the corona discharge current, provides high flexibility of control signal calibrator.

Mechanical system arrester consists of two electrodes made of stainless steel thickness of 0.125 mm using laser cutting technology. Electrodes are placed in the Teflon tube through which air flow 150 cm^3/min is formed. One end of the tube is connected to the input of an ion mobility spectrometer. Tip electrodes are placed 2 mm apart.

In the transition from setup mode to the calibration mode, the device suspends detection of explosives. In the ionization chamber of the device entered the combustion products of corona discharge. Spectrum changes as shown on Fig. 5. The system detects the peak calibration substance and on the basis of the known data on the characteristics of flight time detected explosives builds markers of these substances with respect to the calibration peak. Analysis of the calibration peak showed that according to the experimental data given in [13], sizing material corresponds to oxides of nitrogen. Spectrograms display the presence of a vast number of samples generated during the discharge of matter.

By varying the voltage applied to the electrodes, possible to vary the amount of combustion products generated, as it illustrated on Fig. 6.

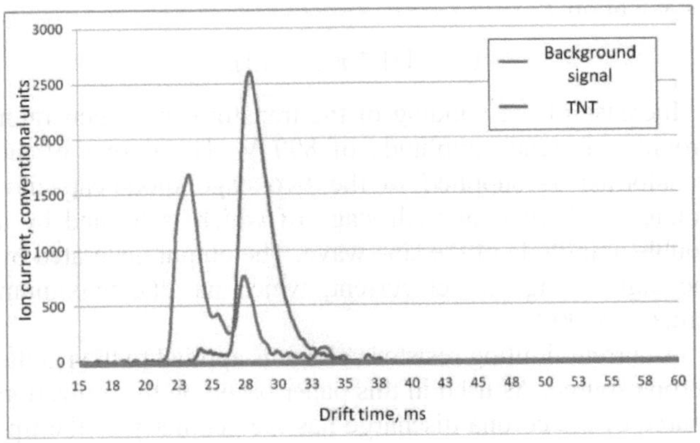

Fig. 5 Spectrogram of the spectrum changes in calibration mode.

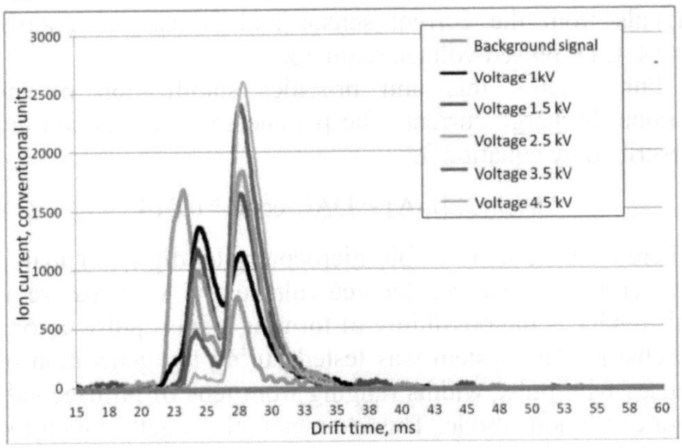

Fig. 6 Calibration spectra for different discharge voltages.

Figure 6 shows that the number of outgoing materials is directly proportional to the voltage of the discharge. However, since a certain level, the spectrum changes slightly. This means that changes little amount of ionized nitrogen oxides. With increasing of applied voltage current increases as well, but the electric field on the surface of the corona electrode is kept constant, or nearly equal to the intensity of occurrence of corona. At high voltages applied to the source of the corona discharge, sizing material pulls charge from the peaks of the background signal. This suggests that the calibration substance has strong oxidizing properties. In most cases, the electron affinity is a fraction of eV, but in the case of NO_2 value exceeds 3 eV. In turn, the group OH (intermediate negative reactant ions OH^- ion mobility spectrometer mode), this value is 2 eV and confirms that observed in the experiment charge transfer. By changing the voltage applied to the electrodes, the configuration of needles and the distance between them can achieve stable combustion of corona discharge, ensure the full combustion of the corona. The developed system allows to change the number of products of combustion generated by varying the voltage applied to the electrodes. In addition, the pulsed nature of the corona calibrants work ensures reliability and extends the service life of needles.

The proposed calibration method successfully used in a portable ion mobility spectrometry in a form of module weighing 2.9 kg (without battery). Designed calibration system based on corona discharge is built-in part of ion mobility spectrometer. To verify compliance with the mobility of the resulting peak nitrogen dioxide an experiment was carried out. Figure 7 shows a comparison of calibrants spectrogram peak and detection of nitrogen dioxide. The arrival time of the ions and thus, the mobility of the peak obtained when the work input calibrants and nitrogen dioxide in the ionization region are identical. From these results it can be assumed that the resulting corona discharge combustion agent is nitrogen dioxide. Confirmation of this assumption can be obtained by using other methods of gas analysis.

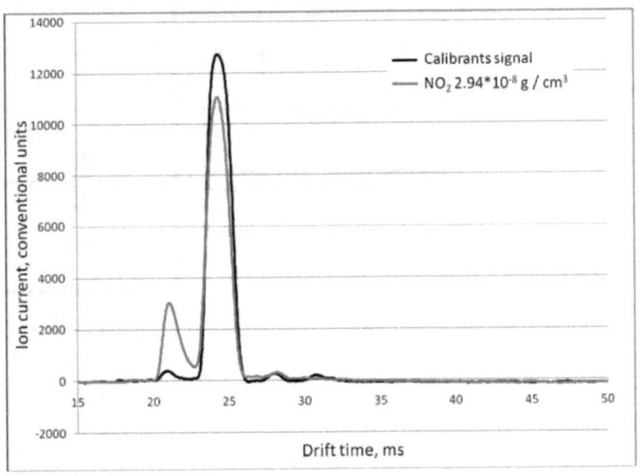

Fig. 7 The spectrum of NO2 and calibration in negative polarity.

IV. CONCLUSIONS

The study of calibration methods of ion mobility spectrometer mode detection of explosives found:

1. The calibration method using a steam generator based on vapor diffusion heated substance allows to create minute quantities of explosive vapor, leaving the masses to make calculations generated substances and quantify the sensitivity of ion mobility spectrometry.
2. The calibration method using corona discharge has a number of advantages, allows to build calibrants in portable ion mobility spectrometer and use in the field conditions. The developed control system based on the ionization source microcontroller implements the possibility of changing the parameters of the combustion of the corona discharge (duration of burning and discharge voltage).
3. In the future, it is necessary to carry out research in the field of embedded calibrants (ion mobility spectrometry detection of drugs).

CONFLICT OF INTEREST

The authors declare that they have no conflict of interest.

REFERENCES

1. Eiceman G.A., Karpas Z., Ion Mobility Spectrometry (Second edition), CRC Press, 2005.
2. Vautz, W., Şlodzynski, R., Hariharan, C., Seifert, L., Nolte, J., Fobbe, R., Sielemann, S., Lao, B.C., Huo, R., Thomas, C.L.P., Hildebrand, L., Detection of metabolites of trapped humans using ion mobility spectrometry coupled with gas chromatography, Analytical Chemistry, 85 (4), 2013, pp. 2135-2142.
3. J. Laakia, A. Adamov, M. Jussila, C.S. Pedersen, A.A. Sysoev, T. Kotiaho, Separation of Different Ion Structures in Atmospheric Pressure Photoionization – Ion Mobility Spectrometry – Mass Spectrometry (APPI-IMS-MS), Journal of the American Society for Mass Spectrometry, 21 (9), 2010, pp. 1565-1572.
4. Alexey A. Sysoev, S.S. Poteshin, D.M. Chernyshev, A.V. Karpov, Y.B. Tuzkov, V.V. Kyzmin, Alexander A. Sysoev, Analysis of new synthetic drugs by ion mobility time-of-flight mass spectrometry, European Journal of Mass Spectrometry, 20 (2), 2014, pp. 185-192.
5. P. Rodacy, The minimum detection limits of RDX and TNT, Explosive Subsystems & Materials, Sandia National Laboratories, 1993.
6. P.A. Pella, Generator for Producing Trace Vapor Concentrations of 2,4,6- Trinitrotoluene, 2,4-Dinitrotoluene, and Ethylene Glycol Dinitrate for Calibrating Explosives Vapor Detectors, Analytical Chemistry Division, National Bureau of Standards,1976.
7. J.P. Davies, L.G. Blackwood, S.G. Davies, L.D. Goodrich and R.A. Larson, Design and Calibration of Pulsed Vapor Generators, Analytical Chemistry, 65, 1993, pp. 3004-3009.
8. D.C. Leggett, T.F. Jenkins, R.P. Murrmann, Composition of vapors evolved from military TNT as influenced by temperature, solid composition, age and source, June 1977, Cold Regions Research and Engineering Laboratory, Hanover, New Hampshire.
9. N. Samotaev, A. Golovin, V. Vasilyev, E. Malkin, E. Gromov, U. Shaltaeva, A. Mironov, D. Lipatov, IMS development at NRNU MEPHI, Lecture Notes in Electrical Engineering, 268 LNEE, 2014, pp. 447-451.
10. Hickling A., Modern Aspects of Electrochemistry, Butterworths, 6, 1971, p. 329.
11. Matsaev V., Gumerov M., Krasnobaev L., Pershenkov V., Belyakov V., Chistyakov A., Boudovitch V., IMS Spectrometers with Radioactive, X-ray, UV and Laser Ionization, International Journal for Ion Mobility Spectrometry, 5 (3), 2002, pp. 112-114.
12. J. Laakia, C.S. Pederson, A. Adamov, J. Viidanoja, A. Sysoev, T. Kotiaho, Sterically hindered phenols in negative ion mobility spectrometry – mass spectrometry, Rapid Communications Mass Spectrometry, 23, 2009, pp. 3069-3076.
13. M. Stano, M. Sabo, M. Kučera, S. Matejčík, H.Y. Han, H.M. Wang, Y.N. Chu, IMS study of Negative Corona Discharge in N2/O2 Mixtures in N2 Drift Gas, Acta Physica Universitatis Comenianae, vol. L-LI, Number 1&2 (2009-2010), pp. 77-83.

Author: Igor Ivanov
Institute: National Research Nuclear University MEPhI
Street: Kashirskoe shossee 31
City: Moscow
Country: Russian Federation
Email: ivanovnew@yandex.ru

Author Index

Keyword Index

R

S

T

U

V

W

X

Z

MIX
Papier aus verantwortungsvollen Quellen
Paper from responsible sources
FSC® C105338

If you have any concerns about our products,
you can contact us on
ProductSafety@springernature.com

In case Publisher is established outside the EU,
the EU authorized representative is:
Springer Nature Customer Service Center GmbH
Europaplatz 3, 69115 Heidelberg, Germany

Printed by Libri Plureos GmbH
in Hamburg, Germany